coronal suture

temporal lines

parietal bone

squamous portion of temporal bone

occipital bone

external acustic meatus

external occipital protuberance

mastoid process

styloid process

frontal bone

sphenoid bone

nasal bone

ethmoid bone

lacrimal bone

zygomatic bone

nasal spine

zygomatic process of temporal bone

maxilla

condyle

coronoid process

ramus

mandible

body

B

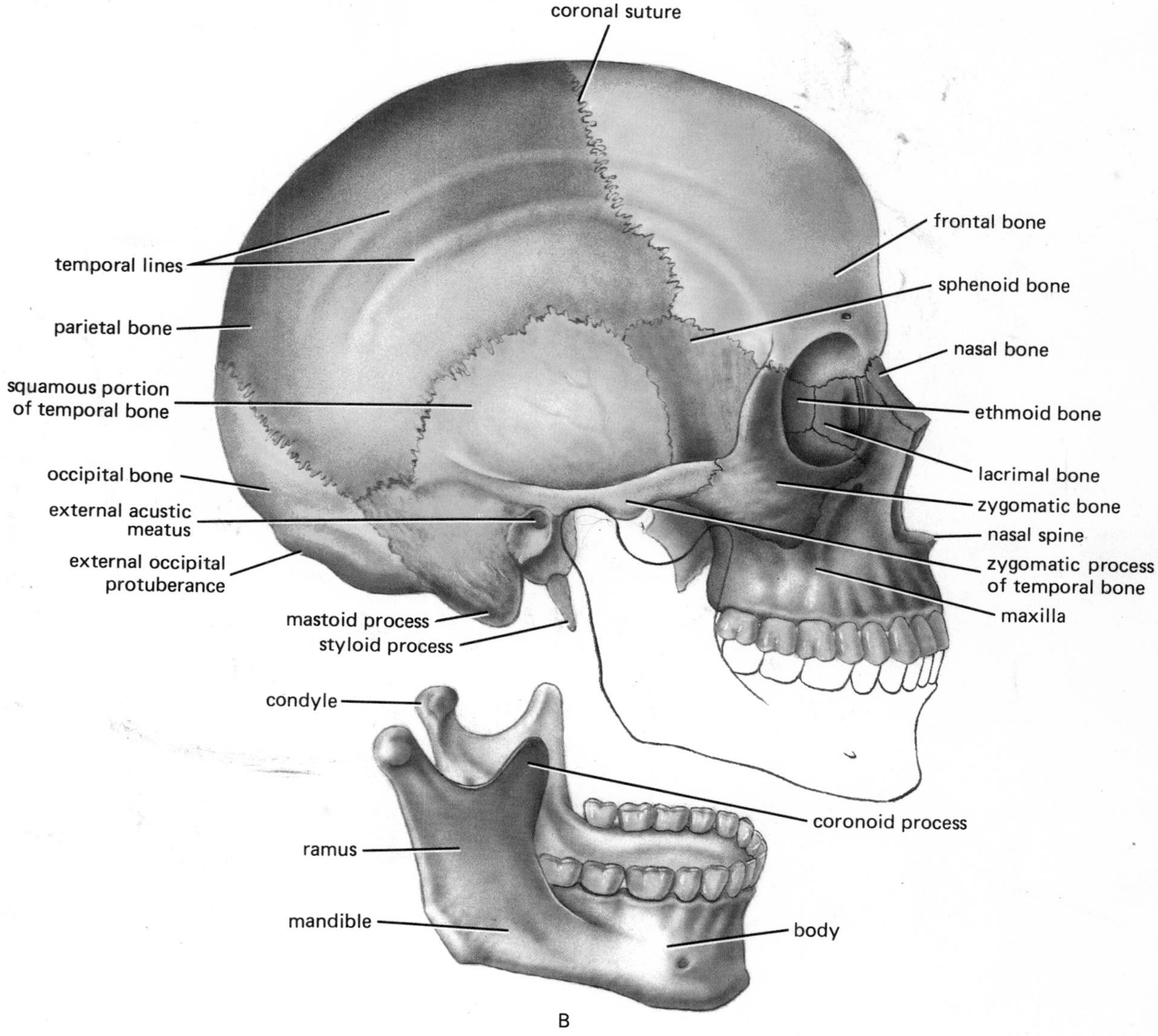

LEFT LATERAL VIEW OF THE SKULL. (See Fig. 7-15B on page 171 of the text.)

Human Function and Structure

Human Function and Structure

Dorothy S. Luciano, PhD
Formerly of the Department of Physiology, University of Michigan

Arthur J. Vander, MD
Professor of Physiology
University of Michigan

James H. Sherman, PhD
Associate Professor of Physiology
University of Michigan

McGraw-Hill Book Company
New York • St. Louis • San Francisco • Auckland • Bogotá • Düsseldorf • Johannesburg • London • Madrid • Mexico • Montreal • New Delhi • Panama • Paris • São Paulo • Singapore • Sydney • Tokyo • Toronto

Human Function and Structure

2 3 4 5 6 7 8 9 0 V H V H 7 8 3 2 1 0 9 8

This book was set in Times Roman by Ruttle, Shaw & Wetherill, Inc. The editors were James E. Vastyan, William J. Willey, Thomas A.P. Adams, and J. W. Maisel; the designer was Merrill Haber; the production supervisor was Charles Hess. Von Hoffmann Press, Inc., was printer and binder.

The Illustrations

The anatomical illustrations of bone and muscle in Chapter 7 were rendered by Lou Barlow and Douglas Cramer; those in all other chapters were rendered by Judy Glick, with the assistance of Jean Jaekel. The schematic illustrations were prepared by J & R Services, Inc.

The Cover

The stone sculpture is *Standing Figure,* by the Russian-born American sculptor Alexander Archipenko (1887–1964). The original stands in the Darmstadt Museum, which kindly supplied the photographs of both views.

Library of Congress Cataloging in Publication Data

Luciano, Dorothy S.
Human function and structure.

Includes index.
1. Human physiology. 2. Anatomy, Human.
I. Vander, Arthur J., date joint author.
II. Sherman, James H., date joint author.
III. Title. [DNLM: 1. Physiology. 2. Anatomy.
QT104 L938h]
QP34.5.L83 612 77-13472
ISBN 0-07-038942-X

**To Horace W. Davenport
with respect and affection**

Contents

Preface

The primary purpose of this book is to present the fundamental facts and concepts of human physiology and anatomy. It is intended for introductory courses in anatomy-physiology taken by students of biology, nursing, and allied health. It is suitable for them regardless of their science background: although a previous introductory course in either biology or chemistry would be useful, it is not essential, since relevant background is presented where necessary.

This book was written in response to a large number of requests from teachers of combined anatomy-physiology courses who have expressed a desire for a book with the approach to physiology presented in our *Human Physiology: The Mechanisms of Body Function* but with extensive anatomy as well. Accordingly, the book is our attempt at a balanced integrated presentation of the two disciplines. The physiology is more extensive than that in most other combined anatomy-physiology texts, emphasizing as it does mechanisms and control systems. It stresses that the body's various coordinated functions – circulation and respiration, for example – result from the precise control and integration of specialized cellular activities, serve to maintain relatively constant the internal composition of the body, and can be described in terms of control systems similar to those designed by engineers. Consequently, the book progresses from the structure and function of cells and tissues to the anatomy and physiology of the integrated organ systems of the body. Part I is devoted to an analysis of the discrete components of the body, beginning with basic cellular biology and concluding with the anatomy and physiology unique to individual tissues. Part II describes how the structures and activities of these discrete tissues are integrated to achieve overall form and function of the body.

The presentation of material is flexible enough that the book can be used in several ways. A course which has basic biology as a prerequisite could omit Part I completely. If gross anatomy is to be the major anatomical focus, the chapters of Part I could be skimmed; alternatively, time could be saved by eliminating from Part II whole topics such as Consciousness and Behavior or Defense Systems. A longer course could use the book in its complete form. The anatomy is presented at two levels of complexity: the illustrations and their accompanying descriptive material are at a general introductory level, and the extensive tables serve as more detailed sources of information.

Several anatomists have provided valuable assistance to us. We would like to thank Sara Winans and Ray Kahn who helped us to develop our overall approach to anatomy. There is no way that we can adequately express our gratitude to Thomas Oelrich, who was our major anatomy consultant. Dr. Oelrich graciously devoted an immense amount of time and effort to going over anatomical illustrations and tables; merely having him to work with was a source of great reassurance to us. Any errors in these materials almost certainly represent our failure to follow his advice. Thomas Adams, our editor at McGraw-Hill, continues to astonish us with his patience, attention to detail, and creative suggestions. Finally, our incredible typist, Helen Mysyk, has once more accomplished her mission impossible, and we only hope that she realizes how grateful we are to her.

We are also grateful to the following persons for their helpful comments upon reading the manuscript: Dr. John P. Harley, Eastern Kentucky University; Dr. William E. Dunscombe, Union College; Dr. Annabelle Cohen, College of Staten Island (CUNY); Professor Donald Bertucci, Shasta College; Professor Robert H. Catlett, University of Colorado at Colorado Springs; Professor Ray Kahn, University of Michigan; Professor Charles A. Meszoely, Northeastern University; Dr. David H. Noyes, Ohio State University; and Professor David S. Smith, San Antonio College. Dr. Patricia A. Lorenz, Penn. Valley Community College.

Arthur J. Vander
Dorothy S. Luciano
James H. Sherman

Human Function and Structure

PART ONE

COMPONENTS OF THE BODY: CELLS AND TISSUES

CHAPTER 1

The Cell: Structure and Function

The cell is the simplest unit of biological structure into which an organism can be divided and still retain the characteristics we associate with life, yet each is so small that a microscope is required to observe its structure. An analysis of the enormously complex activities of the human body must begin with an understanding of the general structure and function of these basic units.

One of the unifying generalizations of biology is that certain fundamental activities that represent the minimal requirements for maintaining the life of a cell are common to all cells. A human liver cell and an ameba are remarkably similar in their means of exchanging materials with their immediate environments, obtaining energy from organic nutrients, synthesizing complex molecules, and duplicating themselves. This is not to say that there are no significant differences between an ameba and a liver cell or between a muscle and a nerve cell. These differences in cell function, however, represent a specialized development of one or more of the general properties common to all the trillions of cells in the body. For example, the ability to generate the forces which produce movement is a common property of all cells; muscle cells represent a highly specialized development of this common force-generating mechanism. Nerve cells have made use of electric properties common to all cell surfaces to develop a specialized mechanism for conducting signals in the body. Liver cells represent a specialization of the general ability to chemically transform organic nutrients into various chemical products. These specializations in cell function occurred during the evolution of single cells into the coordinated society of cells which make up the human body. This chapter describes those elements of cell structure and function that most cells in the body have in common. Subsequent chapters will describe the properties of specialized cells and their organization into tissues and organs.

CELL STRUCTURE

As cells became specialized in function, their structures became adapted to perform these specialized functions. Thus, the sizes and shapes of the many kinds of cells in the body vary tremendously. Large cells, e.g., skeletal muscle cells, may have a volume that is a billion times larger than that of small cells, e.g., red blood cells. Whereas a few cells are essentially spherical in shape, and some are long cylinders, the majority of cells are box-shaped as a result of being surrounded by other cells (the term *cell* is derived from this roomlike appearance as seen through a microscope). Most cells, whatever their size or shape, are composed of a number of basic structural elements (Fig. 1-1), each structure having a particular role to play in maintaining the life of the cell.

Table 1-1 provides a summary of the various cell structures and their associated functions. A cell, seen at low magnification, appears to be filled with a number of granules which slowly oscillate back and forth, suggesting that they are suspended in a fluid medium. At much higher magnification, these granules are found to have characteristic structures of their own, and the entire cell is seen to be covered by a very thin barrier, the *plasma membrane*. Like a very thin sheet of plastic, the plasma membrane is flexible but nonextensible, i.e., it can be folded and bent into a number of shapes but it cannot be stretched without being torn.

The interior of a cell is composed of two large regions: (1) the *nucleus*, a spherical or oval body generally located in the center of the cell, and (2) the *cytoplasm*, the region of the cell located outside the nucleus. The cytoplasm contains many more structures than does the nucleus. Most of these structures, the *cell organelles*, are surrounded by membranes which act as barriers to the movements of molecules from one area of the cell to another. Thus, particular types of molecules and their associated functions are confined to specific types of cell organelles. Even at the cellular level of organization, there exists a subcellular specialization of structure and function.

The nucleus (Fig. 1-2) is the largest structure in the cell. It is surrounded by a barrier, the *nuclear envelope*, which consists of two membranes separated by a small space. At regular intervals along the surface of the nuclear envelope, the two nuclear membranes become joined, forming the rims of circular openings known as *nuclear pores*. These pores provide access to and from the nucleus for large molecules which otherwise could not cross the membranes of the nuclear envelope. The interior of the nucleus consists of granular and filamentous elements in various states of aggregation. The most prominent structure within the nucleus is the densely staining *nucleolus*, a highly coiled filamentous structure associated with numerous granules, but not surrounded by a membrane. Fibrous threads, known as *chromatin*, are distributed throughout the remainder of the nucleus. These chromatin threads, which carry genetic information from parent to offspring and from parent cell to daughter cell, are coiled to a greater or less degree, producing a variation in the granular density of the nucleus.

The most extensive cytoplasmic cell organelle is the system of membranes which forms the *endoplasmic reticulum* (Fig. 1-3). Like the nuclear envelope, the endoplasmic reticulum consists of two opposing membranes separated by a small space. These membranes form a series of relatively flat sheets that are distributed throughout the cytoplasm and interconnect with each other. The small space between the membranes of the reticulum appears to be continuous throughout this membranous network and with the space located between the two nuclear membranes. Two types of endoplasmic reticulum can be distinguished: *granular* (rough-surfaced) and *agranular* (smooth-surfaced) *endoplasmic reticulum*. The granular endoplasmic reticulum has small particles, *ribosomes*, embedded in its membrane surface (see Figs. 1-1 and 1-5). Other ribosomal particles, *free ribosomes*, occur in the cytoplasm as free particles not attached to membranes. The agranular endoplasmic

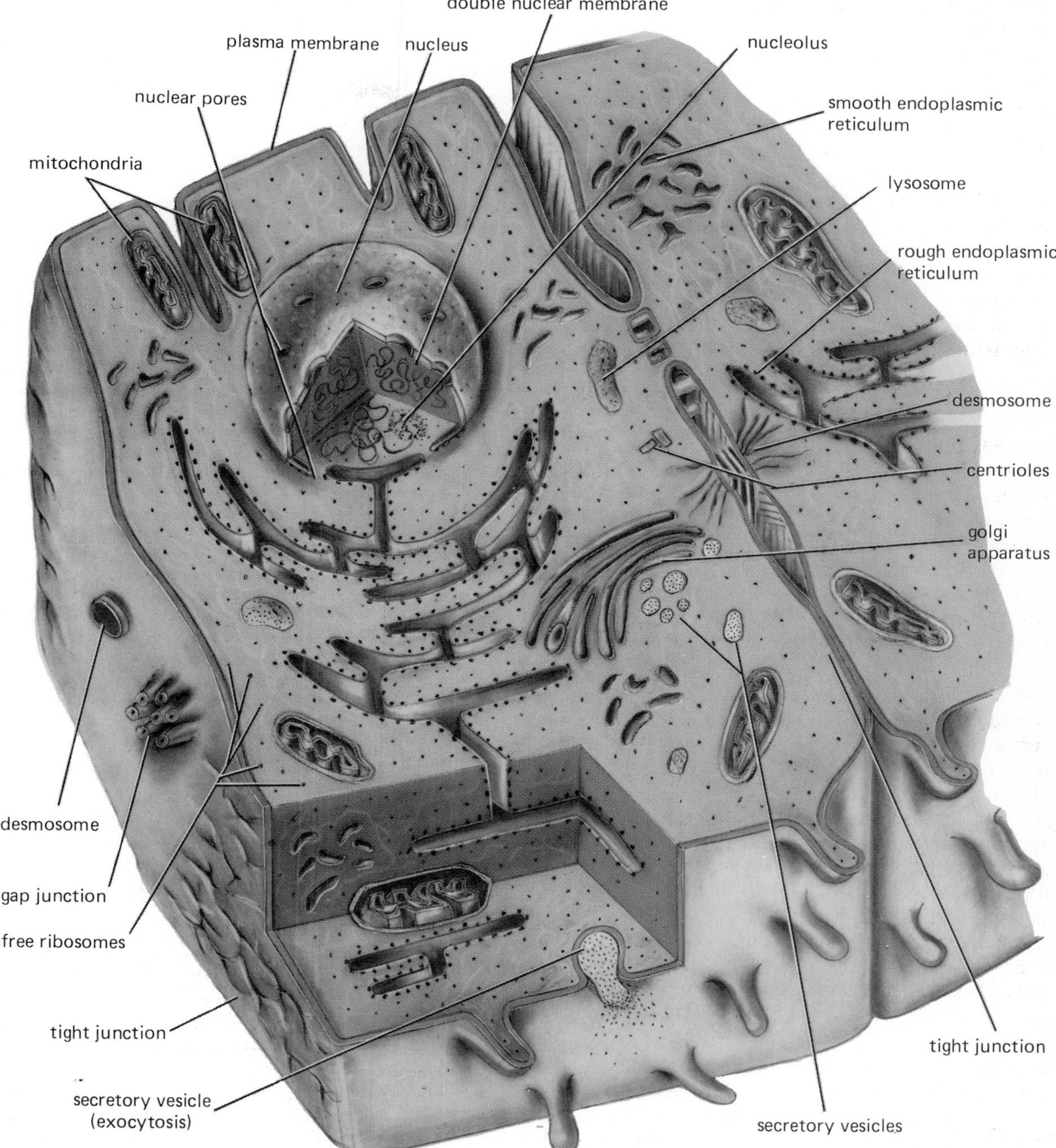

FIGURE 1-1 ***Diagram of the structures and organelles found in most cells of the body.***

TABLE 1-1
CELL STRUCTURES†

Structure	*Number per cell*	*Structural organization*	*Function*
Plasma membrane	1 Cell surface	100 Å thick. Composed of a phospholipid bilayer and protein.	Selective barrier to movement of ions and molecules into and out of cell.
Nucleus			
Nuclear envelope	1 Surrounds nucleus	Two opposed membranes separating small space. Nuclear pores, 500–700 Å diam.	Barrier to movement of most molecules. Messenger RNA passes to cytoplasm through pores.
Chromatin	46 Strands per nucleus	Coiled threads composed of DNA and protein.	DNA stores genetic information determining amino acid sequence of cell proteins. Chromatin condenses into *chromosomes* at time of cell division.
Nucleolus	Usually 1	Coiled filamentous structure associated with granules. Not surrounded by membrane.	Site of ribosomal RNA synthesis.
Cytoplasm			
A. Membrane-bound cell organelles			
Endoplasmic reticulum (ER)	1 Interconnected cell organelle	Two opposed membranes separating a space continuous throughout organelle and interconnecting with space of nuclear envelope.	
Granular ER		Ribosomal particles bound to ER membrane.	Synthesis of proteins to be secreted from cell.
Agranular ER		Smooth membrane; no ribosomes.	Fatty acid and steroid synthesis. Calcium storage and release in muscle cells.
Golgi apparatus	Usually 1 located near nucleus	Cup-shaped series of closely opposed membranous sacs and vesicles.	Concentration and modification of protein prior to secretion.
Secretory vesicles (granules)	Many	Membrane-bound sacs containing concentrated solution of proteins.	Protein secretion.
Mitochondria	Many	Rod- or oval-shaped bodies surrounded by two membranes. Inner membrane folds into inner *matrix* forming *cristae*.	Major site of ATP production, oxygen utilization, and CO_2 formation. Contains enzymes of *Krebs cycle* and *oxidative phosphorylation*.
Lysosomes	Several	Densely staining oval body surrounded by membrane, containing hydrolytic enzymes.	Digestive organelle, specialized for breakdown of engulfed bacteria and damaged cell organelles.

TABLE 1-1 (*Continued*)
CELL STRUCTURES

Structure	*Number per cell*	*Structural organization*	*Function*
B. Nonmembranous structures			
Ribosomes	Many	200-Å-diameter particles composed of RNA and protein.	Site at which amino acids are assembled into proteins.
Free ribosomes		Not bound to membranes.	Site at which proteins to be used intracellularly are assembled.
Bound ribosomes		Bound to membranes of granular endoplasmic reticulum.	Site at which proteins to be secreted from cell are assembled.
Filaments	Many	50–150-Å-diameter protein threads of variable length.	Cell movements, especially in muscle cells. Also used to provide structural support at cell junctions.
Microtubules	Many	250-Å-diameter protein tubules with 150-Å-diameter hollow core.	Maintenance of shell shape (*cytoskeleton*). Associated with movements of cilia, flagella, and mitotic spindle.
Centrioles	2 Located near nucleus	Two small cylindrical bodies composed of nine sets of three fused microtubules.	Formation of *spindle apparatus* at poles of cell during cell division. Also associated with formation and movement of *cilia*.
Granules	Few to many	Aggregates or crystals of chemical substances.	Storage of specialized end products of metabolism. *Glycogen* granules most common.
Fat droplets	Few	Spherical globule of triglyceride (sometimes surrounded by a membrane).	Storage of fat.

† Much of the information in this table will be developed in later sections of this and other chapters.

reticulum (Fig. 1-3) has no ribosomal particles on its surface and is more fragmented in appearance, being less likely to exist as extended sheets of membranes. Both granular and agranular endoplasmic reticulum can exist in the same cell and appear to be continuous with each other, but the relative amounts of the two types vary in different cells and even within the same cell with changes in cell activity.

The *Golgi apparatus* (Fig. 1-4), named in honor of Camillio Golgi who first described this cell organelle, consists of a series of closely opposed, flattened membranous sacs which are slightly curved, forming a cup-shaped organelle. Associated with this organelle, particularly near its concave surface, are a number of membrane-enclosed vesicles. These vesicles, which are produced by the Golgi apparatus, move to the periphery of the cell where they fuse with the plasma membrane and empty their contents to the outside of the cell during cell secretion. The vesicles are therefore referred to as either *secretory granules* (when they contain densely staining material) or *secretory vesicles* (when enclosing less dense material). Most cells have a single Golgi apparatus located near the nucleus, although some cells may have several.

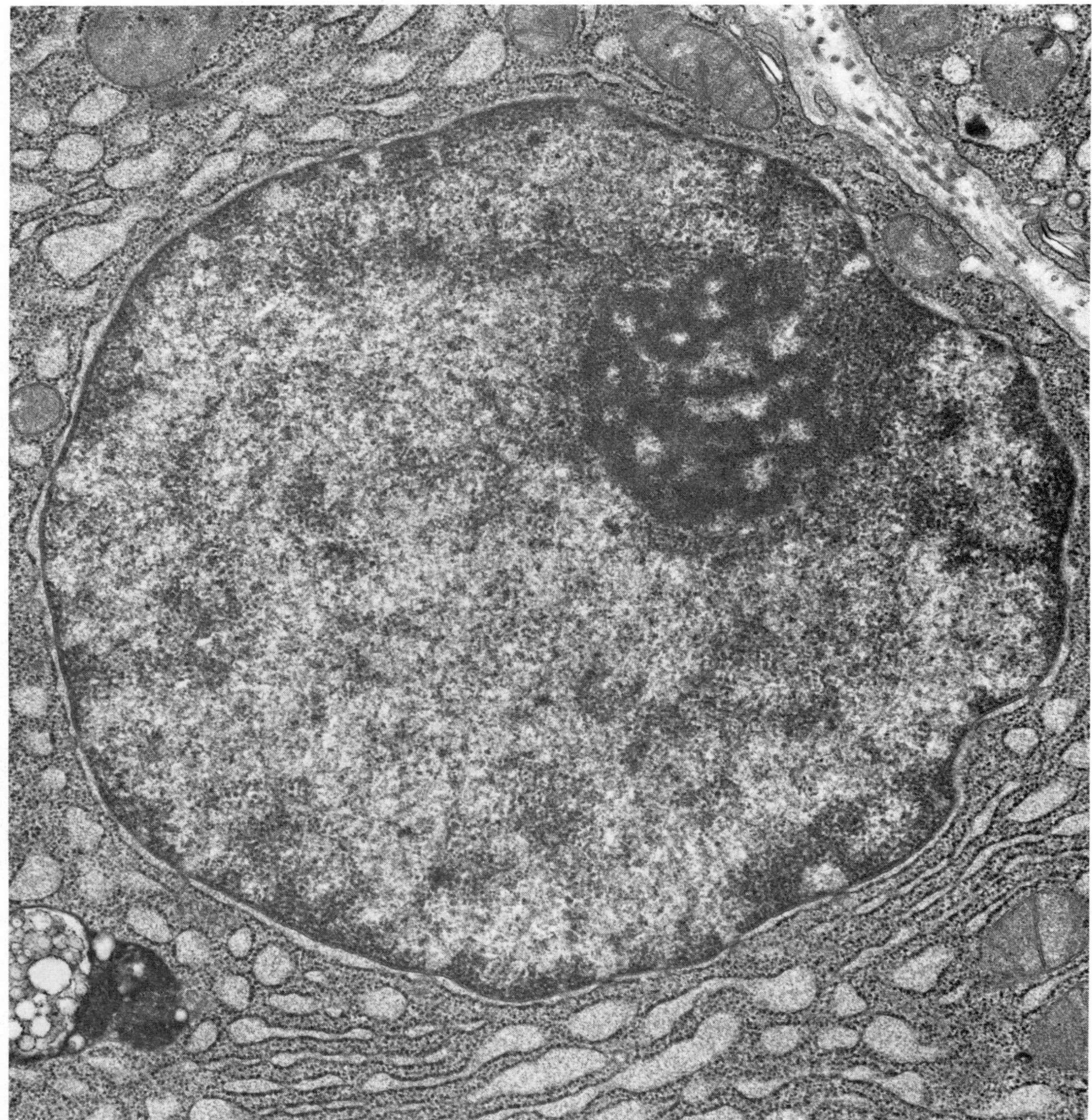

FIGURE 1-2 ***Electron micrograph of the nucleus of a cell.*** **(Courtesy of Keith R. Porter.)**

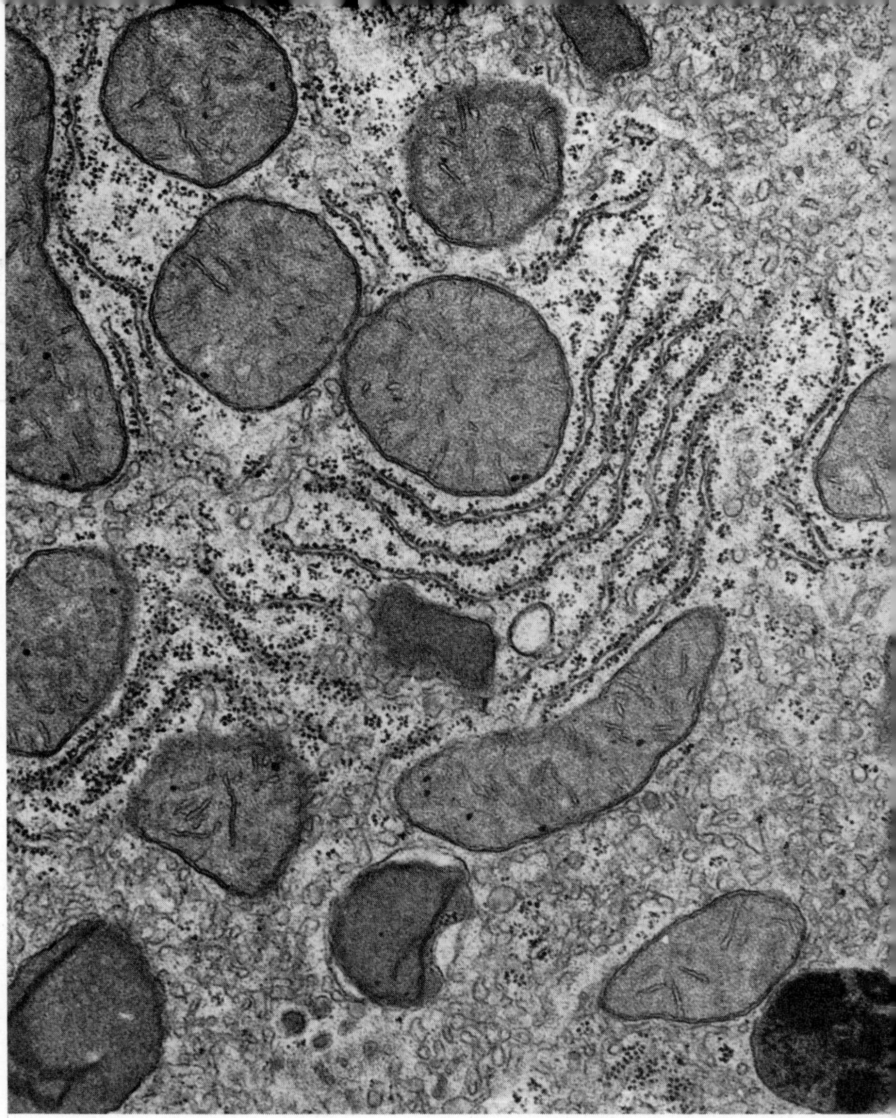

FIGURE 1-3 ***Electron micrograph of a portion of the cytoplasm of a liver cell showing both granular and agranular endoplasmic reticulum, as well as mitochondria and lysosomes.*** **(From Keith R. Porter, in T. W. Goodwin and O. Lindberg (eds.), "Biological Structure and Function," vol. 1, Academic, New York, 1961.)**

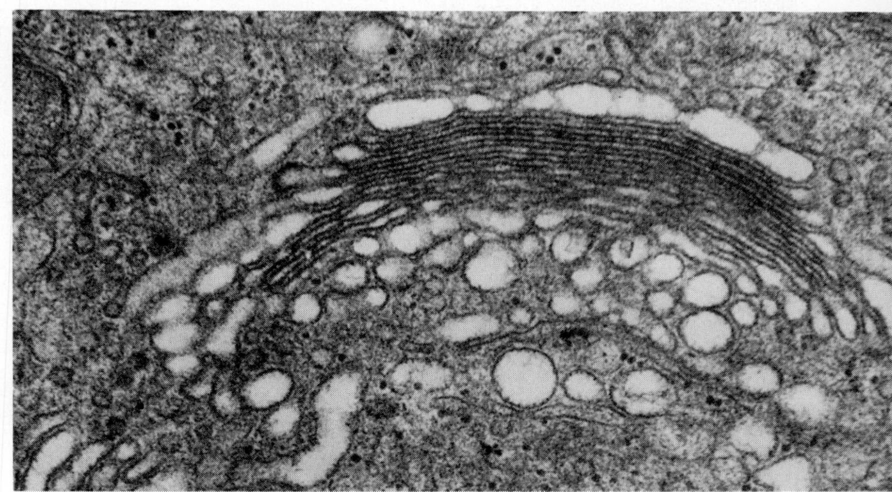

FIGURE 1-4 ***Electron micrograph of the Golgi apparatus.*** **(From W. Bloom and D. W. Fawcett, "Textbook of Histology," 9th ed., Saunders, Philadelphia, 1968.)**

The *mitochondria* (Greek *mitos,* thread; *chondros,* granule) are the cell organelles which provide the major source of chemical energy utilized by cells in performing their various functions. These organelles, which are usually rod- or oval-shaped (Fig. 1-5), are surrounded by two membranes, an inner membrane and an outer membrane. The outer membrane is smooth, whereas the inner membrane is folded into sheets or tubules, known as *cristae,* which extend into the inner space (*matrix*) of the mitochondrion. The mitochondria are found scattered throughout the cytoplasm. Large numbers of them are present in cells that utilize large amounts of energy; for example, a single liver cell may contain 1,000 mitochondria. Less active cells contain fewer mitochondria.

Lysosomes (Fig. 1-3) are small, spherical or oval bodies, surrounded by a single membrane which encloses a densely staining, granular matrix. They often cannot be distinguished structurally from the secretory vesicles derived from the Golgi apparatus and are in fact probably formed in the Golgi region of the cell. In contrast to the secretory vesicles which export material from the cell, the lysosomes function intracellularly to break down various complex structures, such as bacteria and cellular debris, that have been engulfed by the cell or, in some cases, to break down other intracellular organelles which have been damaged and are no longer functioning normally. The lysosomes are thus a highly specialized intracellular digestive system.

The cytoplasmic organelles we have described thus far—the endoplasmic reticulum, Golgi apparatus, secretory vesicles, mitochondria, and lysosomes—all have one structural element in common: They are surrounded by membranes. Membranes are the major structural elements in cells, and we will discuss their structure and function in later sections of this chapter.

In addition to these membranous organelles, several other structures are found in the cytoplasm. Most cells contain rodlike *filaments* and hollow *microtubules.* The microtubules appear to be more rigid than the filaments and may provide structural support (a *cy-*

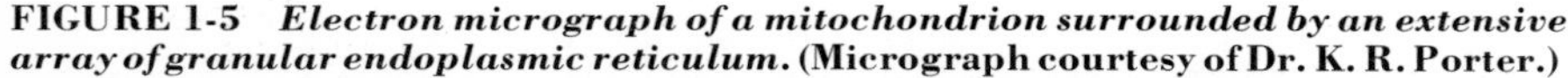

FIGURE 1-5 ***Electron micrograph of a mitochondrion surrounded by an extensive array of granular endoplasmic reticulum.*** **(Micrograph courtesy of Dr. K. R. Porter.)**

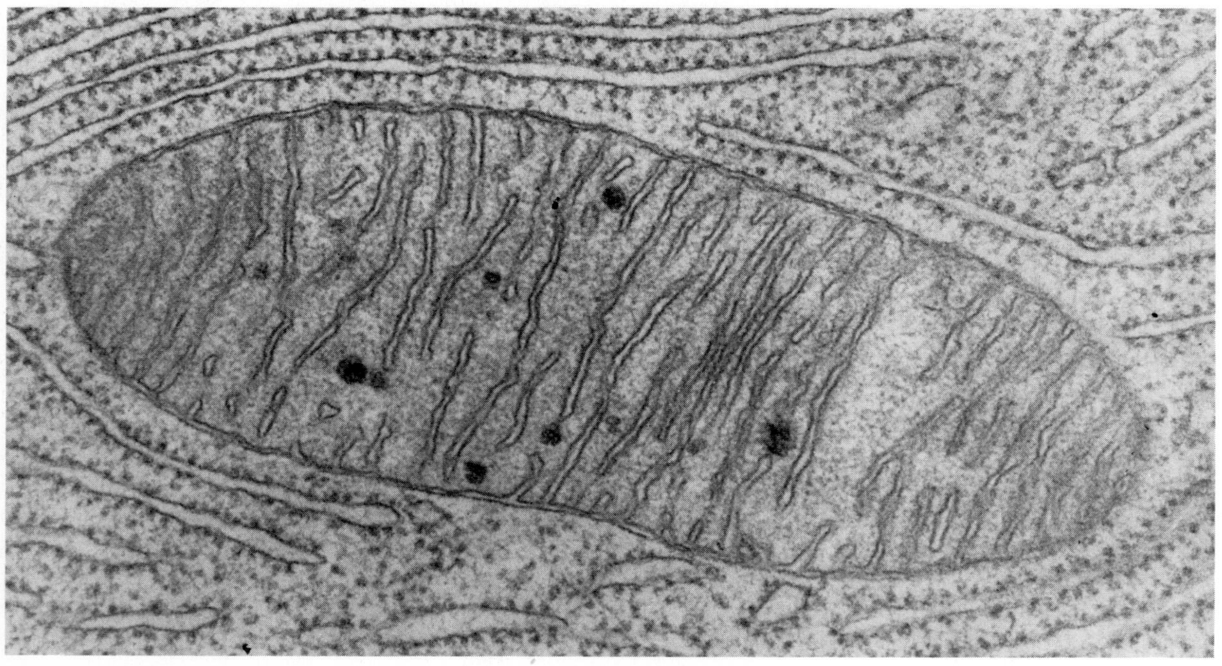

toskeleton) for maintaining various cell shapes. Both filaments and microtubules have been implicated in cell processes which involve movement, whether it be movement of whole cells (the myofilaments of muscle cells, for example), the movement of organelles within cells, or the specialized movements of cell division. The *centrioles* (Fig. 1-1), microtubular structures involved in the process of cell division, consist of two small cylindrical bodies generally located near the nucleus in the region that contains the Golgi apparatus. Finally, the least specialized of the cytoplasmic structures are the fat droplets and various granules which are aggregates or crystals of particular chemical compounds.

CHEMICAL COMPOSITION OF CELLS

Atoms and Molecules

Cells, like all matter, are composed of chemical substances formed by the interactions between the atoms of the various chemical elements. Although there are 105 different chemical elements, only 24 are known to play essential roles in the human body (Table 1-2). In fact, just four elements, hydrogen, carbon, nitrogen, and oxygen, account for 96 percent of the body weight and about 99 percent of the atoms in the body. These four major elements are combined to form water and organic molecules. Most of the seven mineral elements (Table 1-2) are found either in crystallized solid structures, such as teeth and bone, or as charged particles known collectively as *electrolytes* which are dissolved in the fluids of the body. The 13 trace elements are present in very small quantities but perform specific chemical functions essential for the normal growth and functions of the body.

Atoms are linked together by *chemical bonds* to form molecules. Two atoms are held together by a chemical bond created by the electric forces associated with the sharing of two negatively charged electrons, one provided by each atom. In some cases, rather than sharing electrons to form a neutral chemical bond, one atom will completely transfer an electron to another atom, forming two electrically charged atoms known as *ions*.[1] Ions have a net positive charge if the electrically neutral atom has lost an electron, as in the case of the sodium (Na^+) and potassium (K^+) ions, or a net negative charge if the atom has gained an electron, as in the case of the chloride (Cl^-) ion.

[1] *Ion* is the general term used for any electrically charged particle; *electrolyte* refers to mineral ions.

TABLE 1-2
ESSENTIAL ELEMENTS IN THE BODY

Symbol	*Element*
Major elements: 99.3% total atoms	
H	Hydrogen
O	Oxygen
C	Carbon
N	Nitrogen
Major minerals: 0.7% total atoms	
Ca	Calcium
P	Phosphorus
K (Latin, *kalium*)	Potassium
S	Sulfur
Na (Latin, *natrium*)	Sodium
Cl	Chlorine
Mg	Magnesium
Trace elements: less than 0.01%	total atoms
Fe (Latin, *ferrum*)	Iron
I	Iodine
Cu (Latin, *cuprum*)	Copper
Zn	Zinc
Mn	Manganese
Co	Cobalt
Cr	Chromium
Se	Selenium
Mo	Molybdenum
F	Fluorine
Sn (Latin, *stannum*)	Tin
Si	Silicon
V	Vanadium

The formation of ions as a result of gaining or losing electrons is known as *ionization*. The most commonly encountered groupings of atoms which undergo ionization in organic molecules are the *carboxyl group* (R—COOH) and the *amino group* ($R—NH_2$), where R stands for the remainder of the molecule. The hydrogen-oxygen bond in the carboxyl group ionizes to form a carboxyl ion and a hydrogen ion:

$$R-\overset{\overset{\displaystyle O}{\|}}{C}-OH \rightleftharpoons R-\overset{\overset{\displaystyle O}{\|}}{C}-O^- + H^+$$

The amino group is able to combine with a hydrogen ion to form an ionized amino group:

$$R-NH_2 + H^+ \rightleftharpoons R-NH_3^+$$

The ionization of carboxyl and amino groups provides some organic molecules with a net electric charge.

In addition to forming neutral chemical bonds and ions, atoms can interact to form *polarized chemical bonds*. Such polarized bonds are formed when the negative electrons forming the chemical bond reside closer to one atom (which becomes slightly negative) than to the other atom (which becomes slightly positive). The atoms remain bonded together but the chemical bond is electrically polarized. The most commonly encountered polarized chemical bond occurs between hydrogen and oxygen in the *hydroxyl group* (R—OH), in which the oxygen atom is slightly negative and the hydrogen atom slightly positive. Polarized chemical bonds play important roles in determining the solubility of molecules and their interactions with other molecules.

In summary, the atoms in the body chemically interact through the sharing or transfer of electrons to form chemical bonds or ions. The resulting molecules may be completely uncharged or may have regions which are electrically polarized or completely ionized. The hundreds of chemical substances found in the body can be categorized according to their chemical composition and function (Table 1-3), and we turn now to a brief description of these major categories.

TABLE 1-3
CHEMICAL COMPOSITION OF THE BODY

Category	*Percent of body weight*
Water	60
Proteins	17
Lipids	15
Minerals (Na^+, K^+, Cl^-, Ca^{2+}, Mg^{2+}, etc.)	5
Carbohydrates	2
Nucleic acids	1

Water

Water is the most abundant molecule in the body; 99 out of every 100 molecules is a water molecule, and 60 percent of the total body weight is water. As a first approximation, the cells of the body can be viewed as water-filled sacs in which various ions and molecules are dissolved to form the intracellular medium. These cells are surrounded by a second fluid medium, the *extracellular fluid,* which bathes all the cells of the body.

The primary role of water in the body is to provide a fluid medium in which chemical reactions can occur. The two chemical bonds between hydrogen and oxygen in the water molecule (H—O—H) are polarized bonds, with the oxygen atom being slightly negative and the hydrogens slightly positive. Water molecules are electrically attracted to each other through these polarized bonds (Fig. 1-6). A molecule that dissolves in a liquid is known as a *solute,* and the liquid in which it dissolves is known as the *solvent*. Water is the solvent for the hundreds of solutes in the body. The high solubility of many types of molecules in water is due to the interactions between the electrically polarized bonds of the water molecules and the charged regions of the dissolved molecules. For example, table salt (NaCl) dissolves in water because the polar water molecules are attracted to the charged sodium and chloride ions, forming clusters of water molecules around each ion. These water molecules decrease the electric attraction between the sodium and chloride ions which had held them together in the crystalline state (Fig. 1-6). Thus, in order for a molecule to dissolve in water, its molecular structure must contain a certain number of polar or ionized groups which can associate with the polar water molecules. Conversely, molecules which have few, if any, polar or ionized groups are insoluble in water. Most molecules found in the body are soluble in water, with the exception of one general class of compounds known as lipids. The chemical bonds of lipid molecules are minimally polarized or ionized and thus will not interact with water.

Organic Molecules

The chemistry of living organisms is centered around the chemistry of the carbon atom. The carbon atom is able to form four separate bonds with other atoms, such as oxygen, nitrogen, or hydrogen, but most importantly, with other atoms of carbon. The ability of the carbon atom to form neutral (nonpolarized) chemical bonds with other carbon atoms permits the formation of an unlimited variety of molecules by the linkage of more and more carbon atoms. When carbon atoms are linked together, the molecule grows in three dimensions. Although we draw diagrammatic structures of molecules on flat sheets of paper, these molecules have a three-dimensional shape. Within limits, these molecules are flexible and can change their shape without breaking their bonds, because the bonds linking the carbon atoms behave like an axle about which the atoms can rotate.

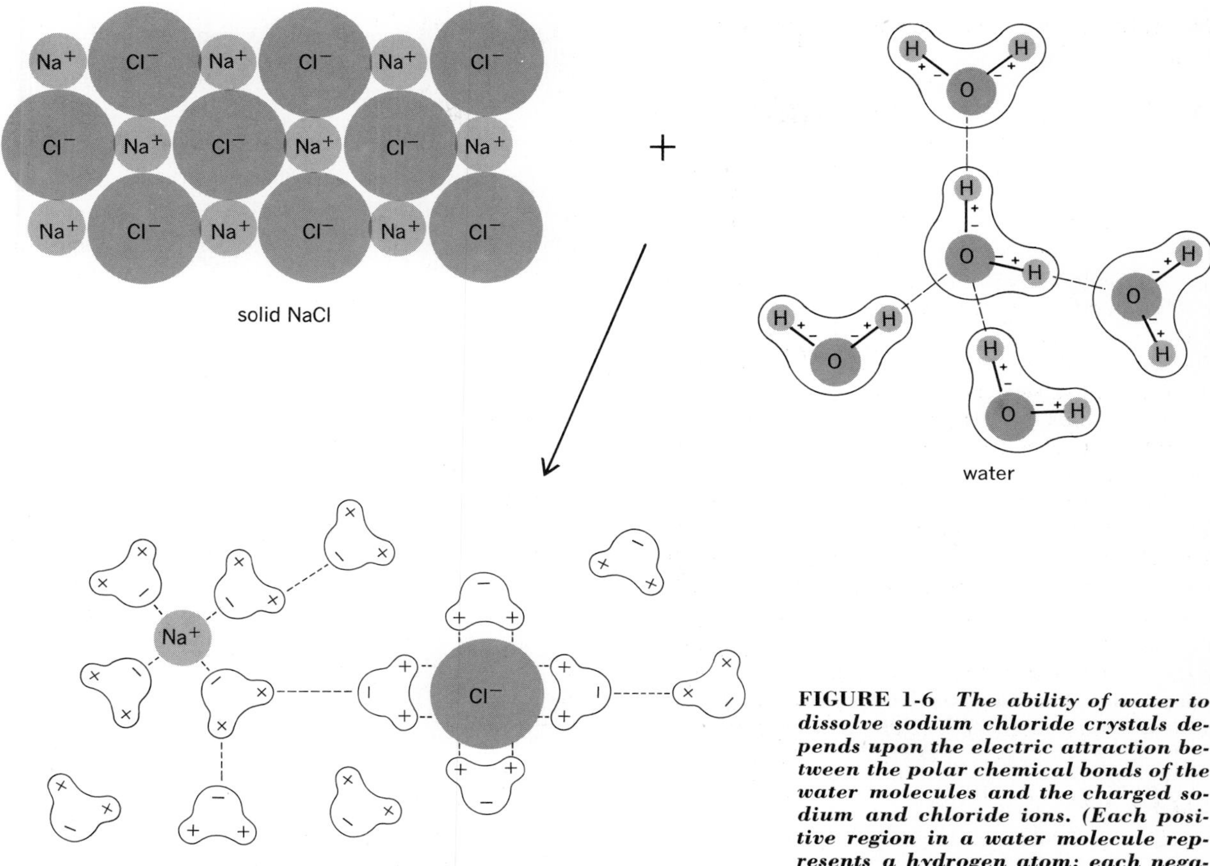

FIGURE 1-6 ***The ability of water to dissolve sodium chloride crystals depends upon the electric attraction between the polar chemical bonds of the water molecules and the charged sodium and chloride ions. (Each positive region in a water molecule represents a hydrogen atom; each negative region, an oxygen atom.)***

The major types of organic (carbon-containing) molecules in the body are the *proteins, lipids, carbohydrates,* and *nucleic acids* (Tables 1-3 and 1-4).

Proteins The term *protein* comes from the Greek *proteios* (of the first rank) which aptly describes their importance. Proteins, which account for about 50 percent of the organic material in the body, are components of most of the body structures and play critical roles in almost all the chemical transformations that occur within the body. Proteins are extremely large molecules, formed by linking together much simpler molecular subunits, the *amino acids*. There are 20 different amino acids in proteins, all of which have one structural feature in common: The terminal carbon atom is attached to both an amino group ($—NH_2$) and a carboxyl group (—COOH) (Fig. 1-7). The amino group of one amino acid reacts with the carboxyl group of the next amino acid, splitting off a molecule of water and forming a chemical bond between the two amino acids, known as a *peptide bond*. Thus, proteins are linear sequences of amino acids linked together by peptide bonds. The remainder of the amino acid, referred to as the amino acid side chain, has a different structure for each of the 20 amino acids. Some of these side chains contain ionized and polarized chemical bonds whereas others are electrically neutral.

The properties of a given protein depend upon the types of amino acid side chains that are present and their position in the amino acid sequence. Since there are 20 different amino acids, there are 20 different

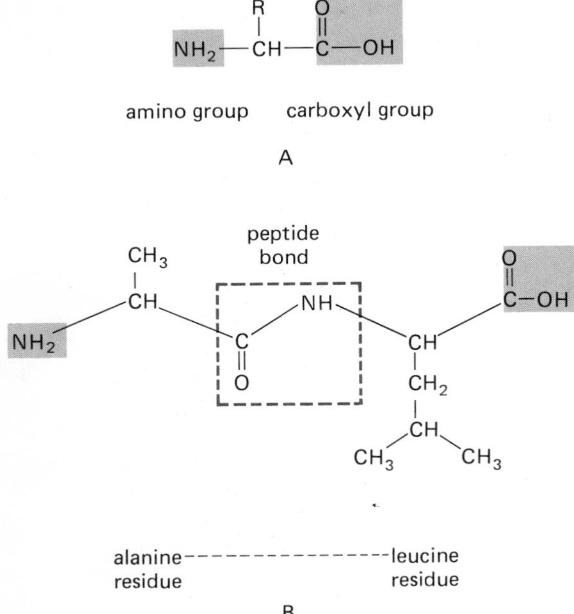

FIGURE 1-7 (A) ***General structure of amino acids, each of which consists of a carboxyl group, an amino group, and a side chain, symbolized by R, which is different for each of the 20 amino acids.*** (B) ***Carboxyl group of the amino acid alanine reacts with the amino group of the amino acid leucine to form a peptide bond linking the two amino acids together.***

types of side chains that can branch off the protein backbone, although all 20 amino acids need not be present in any given protein. We shall see examples later of how changing a single amino acid in a sequence of several hundred can completely alter the function of a protein molecule. Thus, starting with 20 different amino acids, an almost unlimited variety of protein molecules can be constructed by simply rearranging the sequence and altering the total number of amino acids in the sequence. If we consider a very small protein consisting of three amino acids and assume that only two different types of amino acids, A and B, occur in the protein, then $2 \times 2 \times 2 = 8$ different molecules can be formed, each having a different amino acid composition and sequence. If any of the 20 different amino acids normally found in proteins occupy any of the three different positions in our small protein, then the number of possible molecules that can be formed becomes $20 \times 20 \times 20 = 8{,}000$. If we consider proteins that are six amino acids long, we find that $20^6 = 64{,}000{,}000$ different proteins can be formed. But even this number is infinitesimal when we consider that, with few exceptions, the smallest proteins in the body contain a sequence of about 50 amino acids, and the larger proteins have sequences of several hundred.

Thus far we have described the primary structure of a protein molecule as a linear sequence of amino acids which would produce a molecular structure analogous to a long piece of rope. Since segments of the protein chain can rotate about their chemical bonds, a variety of three-dimensional shapes can be assumed just as a long piece of flexible rope can be twisted into many types of configurations (Fig. 1-8). The three-dimensional shape of any given protein is primarily determined by the types and locations of the polarized and ionized amino acid side chains, since these will electrically interact with each other, causing the protein chain to fold into various configurations. Thus, some proteins are rod-shaped, and others are coiled into almost spherical globules. The three-dimensional shape of a protein molecule plays an important role in its ability to interact with other molecules having a complementary shape.

Lipids The second major category of organic molecules is the *lipids* (Greek *lipos,* fat) which make up about 40 percent of the organic matter in the body. The lipids are defined primarily by a single property: their insolubility in water. The lipids are insoluble because their chemical structure, composed mostly of carbon and hydrogen atoms, contains few, if any, polar chemical bonds or ionized groups that can interact with polar water molecules. The lipids can be divided into three subclasses on the basis of their chemical structure: the *neutral fats,* the *phospholipids,* and the *steroids.*

The *neutral fats* constitute the largest subclass of lipids in the body and are generally referred to simply as fat. There are two chemical units which make up the structure of a neutral fat molecule: *fatty acids* and *glycerol.* Fatty acids consist of a sequence of carbon atoms, most commonly 16 or 18 carbons, combined with hydrogen and have an acidic carboxyl group at one end (Fig. 1-9). Although the carboxyl group can ionize and thus associate with water, the long hydrocarbon chain, which constitutes the bulk of the fatty acid, cannot because it contains no polar or ionized groups. Some fatty acids contain one or more double bonds (two chemical bonds linking the same two

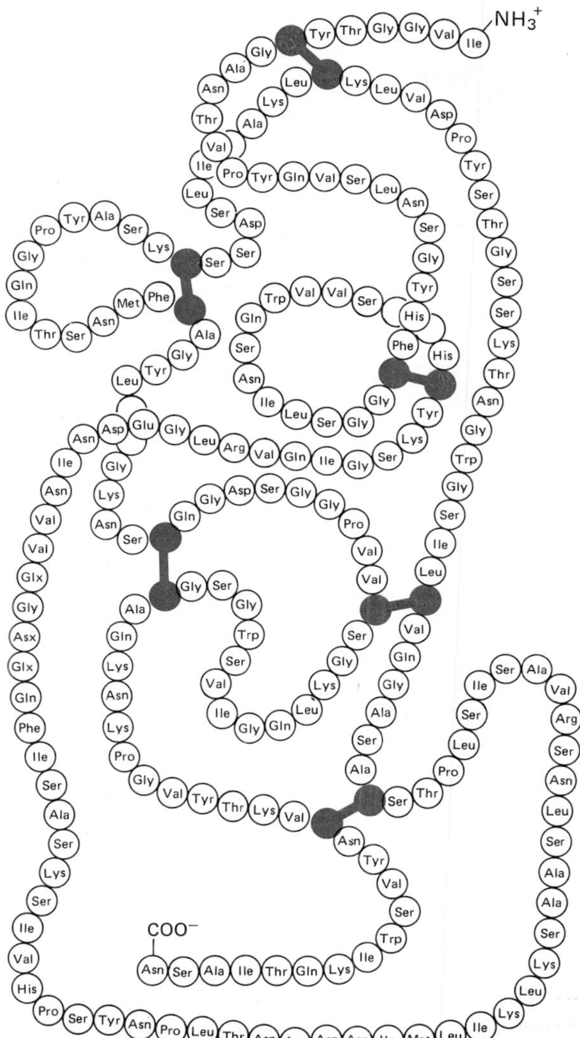

FIGURE 1-8 ***Protein molecule (trypsin) illustrating the complexity of its three-dimensional shape that is formed by the folding of the linear sequence of amino acids (circles). The bars cross-linking various segments of the protein chain represent chemical bonds between the side chains of particular amino acids.*** **(Redrawn from Sidney Bernhard, "The Structure and Function of Enzymes," W. A. Benjamin, Inc., New York, 1968.)**

atoms) at various positions in their hydrocarbon chain (Fig. 1-9) and are said to be *unsaturated* (not saturated with hydrogen atoms). If more than one double bond is present they are known as *polyunsaturated fatty acids*. There are many different fatty acids in the body, varying in number of carbon atoms (length of the fatty acid chain) and number and location of double bonds.

The second unit of a neutral fat molecule is glycerol (Fig. 1-9), a three-carbon molecule which is soluble in water because of the three polar hydroxyl groups it contains. Thus, glycerol, by itself, is not a lipid and, in fact, belongs to the category of organic molecules known as carbohydrates. A molecule of neutral fat is formed by linking the carboxyl group of three fatty acids to the three hydroxyl groups of glycerol, resulting in a molecule which no longer contains any chemical groups that can interact with water and is therefore insoluble (Fig. 1-9). Neutral fats are often referred to as *triglycerides* because of the three fatty acids attached to the glycerol molecule. The primary function of neutral fat is to provide a store of chemical fuel which can be used by cells to provide energy.

The second subclass of lipids, the *phospholipids,* are similar to the neutral fats (Fig. 1-9). However, these molecules contain only two fatty acids attached to glycerol, the third carbon of glycerol being linked to a phosphate group to which, in turn, is usually attached a small polar or ionized nitrogen-containing compound. Thus, unlike the totally nonpolar neutral fat molecule, a phospholipid contains several charged or polar groups. The combination of a polar region at one end of the molecule and nonpolar fatty acid chains at the other allows the molecule to associate with both polar and nonpolar molecules, a property best illustrated when describing the arrangement of these molecules within the structure of cell membranes. Most of the phospholipids in the body are incorporated into the structure of cell membranes and are responsible for many of the membranes' special properties.

The third class of lipids, the *steroids,* are quite different structurally from the neutral fats and phospholipids (Fig. 1-10). Four interconnected rings of carbon atoms form the basic structure of all steroids. Attached to this ring structure are small chemical groups or short hydrocarbon chains. Few, if any, polar groups are present, and thus the steroids are insoluble in water. Cholesterol is a steroid, as are many of the hormones in the body, including the male and female sex hormones, testosterone and estrogen, and the adrenal corticosteroids.

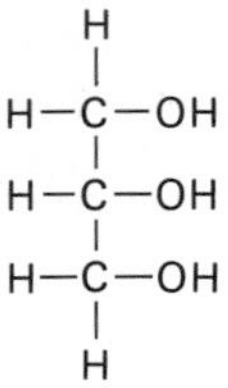

glycerol

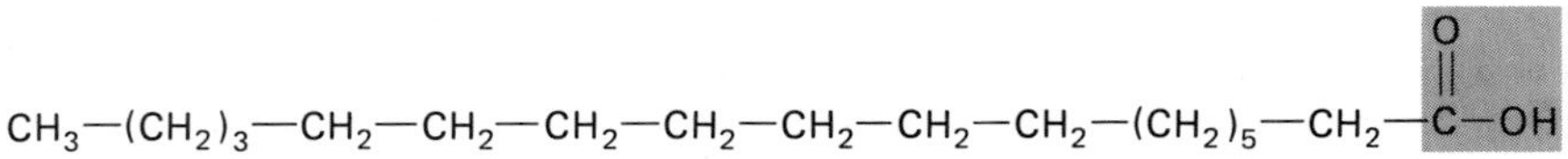

$CH_3—(CH_2)_3—CH_2—CH_2—CH_2—CH_2—CH_2—CH_2—CH_2—(CH_2)_5—CH_2—C(=O)—OH$

a saturated fatty acid

$CH_3—(CH_2)_3—CH_2—CH{=}CH—CH_2—CH{=}CH—CH_2—(CH_2)_5—CH_2—C(=O)—OH$

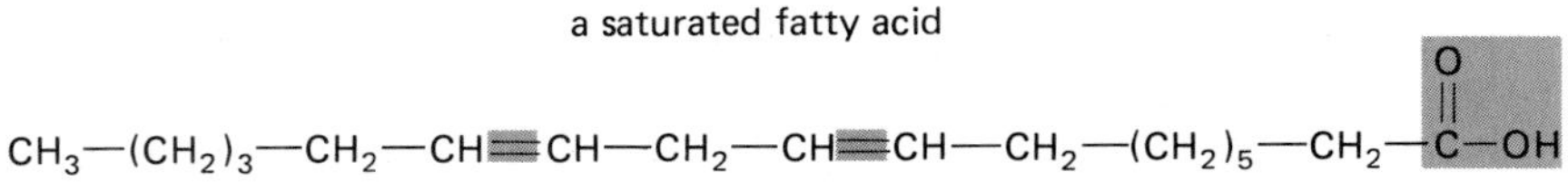

a polyunsaturated fatty acid

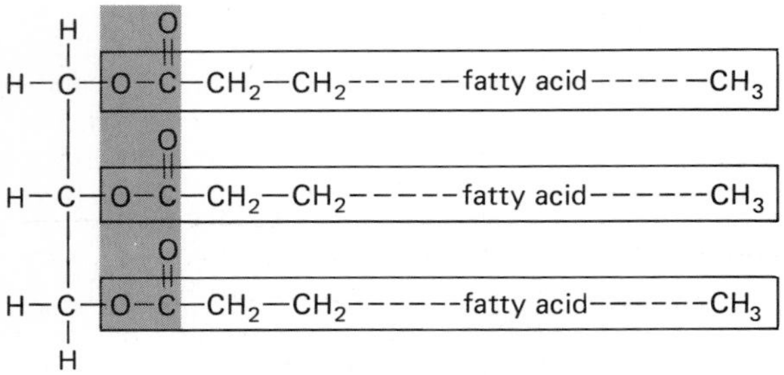

neutral fat (triglyceride)

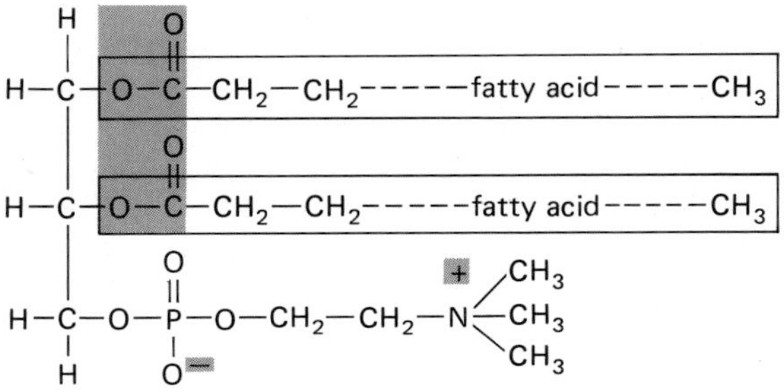

phospholipid

FIGURE 1-9 ***Glycerol and fatty acids are the major subunits that are combined to form neutral fats and phospholipids.***

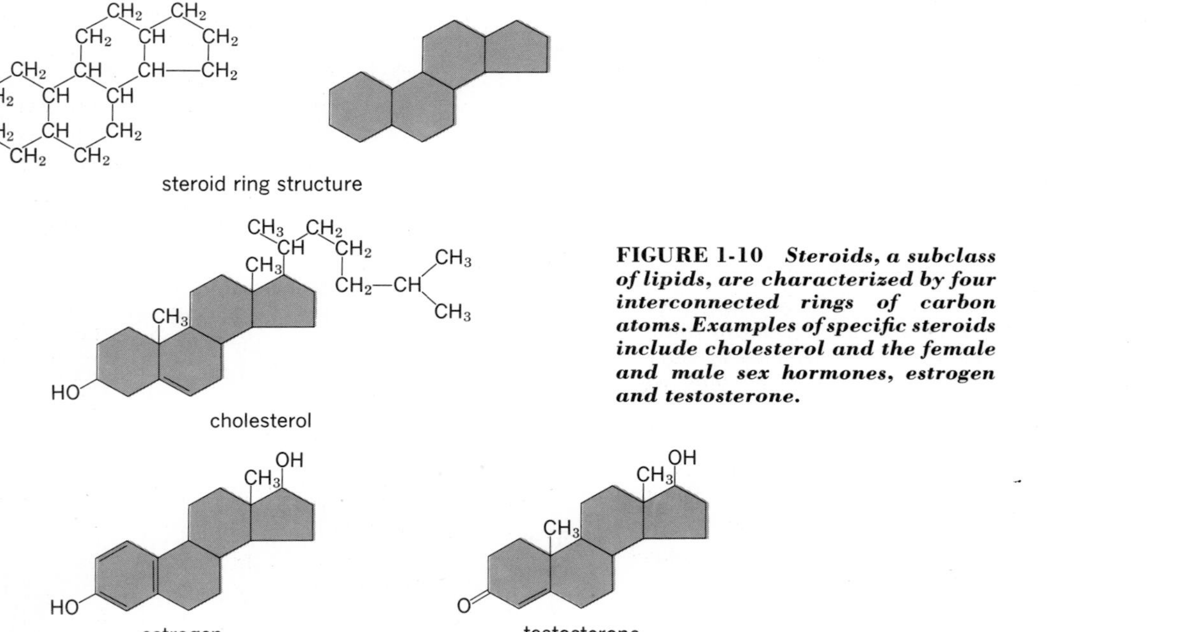

FIGURE 1-10 ***Steroids, a subclass of lipids, are characterized by four interconnected rings of carbon atoms. Examples of specific steroids include cholesterol and the female and male sex hormones, estrogen and testosterone.***

Carbohydrates The third category of organic molecules in the body is the carbohydrates. These molecules account for only a small percentage of the total body weight, but they play a central role in the chemical reactions which provide cells with energy. Carbohydrates are water-soluble molecules composed of carbon, hydrogen, and oxygen in the proportions represented by the general formula $C_n(H_2O)_n$, where n is any whole number. Thus, the carbohydrates contain as many oxygen atoms as carbon atoms, in contrast to the lipids which have few oxygen atoms. The oxygen in carbohydrates is found primarily in the form of hydroxyl groups attached to the carbon atoms. These polar hydroxyl groups account for the water solubility of the carbohydrates.

The most important group of carbohydrates in the body are the *sugars,* most of which consist of five or six carbon atoms formed into a ring, closed by an atom of oxygen (Fig. 1-11). Hydroxyl groups and hydrogen atoms attached to the carbon atoms extend above or below the plane of the ring. Different sugars may have the same number and type of atoms but differ in the orientation of their individual hydroxyl groups above or below the plane of the ring. The most important sugar in the body is the six-carbon sugar *glucose* (Fig. 1-11). The chemical breakdown of glucose into carbon dioxide and water provides much of the energy utilized by cells.

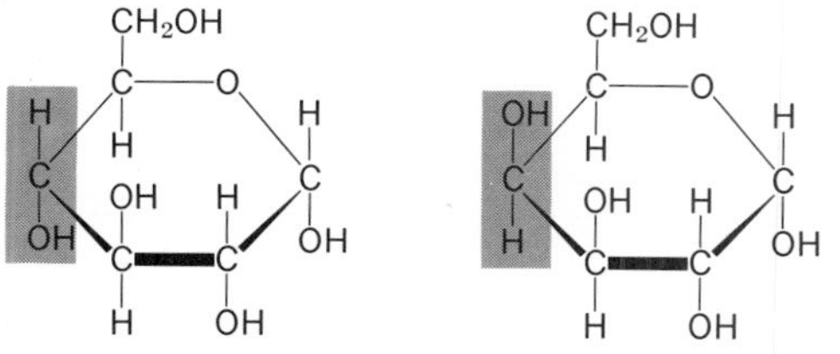

FIGURE 1-11 ***Difference between the two sugars, glucose and galactose, depends upon the position of the hydroxyl group on the fourth carbon atom. It is below the plane of the ring in glucose and above the ring in galactose.***

Two sugars, *monosaccharides* (Greek *sakcharin,* sugar), can be linked together to form a *disaccharide.* Common table sugar, sucrose, is a disaccharide composed of two sugars, glucose and fructose. When many sugars are linked together, the resulting molecule is known as a *polysaccharide.* Two similar polysaccharides, *starch* and *glycogen,* composed entirely of glucose, are found in the plant and animal kingdoms, respectively. The human body has only a limited ability to store carbohydrate, mainly in the form of glycogen, compared with its almost unlimited capacity for the storage of neutral fat. Plants, in contrast, store fuel mainly as starch whose structure is similar to glycogen, differing only in the number of branched chains and the length of these chains. Starch provides the high carbohydrate content of most foods of plant origin. Table 1-4 provides a summary of some of the general characteristics of proteins, lipids, and carbohydrates. The fourth class of organic molecules, the nucleic acids, will be considered in a later section of this chapter.

CELL METABOLISM

Metabolism (Greek: change) refers to all chemical reactions that occur within a cell or living organism. Virtually all the organic molecules in most cells undergo a continuous transformation as some molecules are broken down while others of the same type are being synthesized. These transformations occur within a relatively short time compared with the total life-span of the cell. Therefore, the human body is in a dynamic chemical state. Those chemical reactions which result in the fragmentation of a molecule into smaller and smaller parts are known as *degradative* or *catabolic reactions,* and those reactions which put molecular fragments together to form larger molecules are known as *synthetic* or *anabolic reactions.* Accompanying many of the catabolic reactions is a release of chemical energy; this in turn is used by the cell for the synthesis of new molecules and the performance of other energy-requiring functions such as the contraction of muscle.

Energy, Chemical Reactions, and ATP

Energy can most simply be defined as the ability of a physical or chemical system to undergo change. Accordingly, energy is measured in terms of the magnitude of change that occurs when various forces are applied to matter. The *law of the conservation of energy* states that energy can be transferred from one system to another but cannot be created or destroyed. All physical and chemical change is the result of the transfer of energy from one system to another.

The total energy content of any object consists of two components: *kinetic energy,* associated with the object because of its motion, and *potential energy,* as-

TABLE 1-4
MAJOR CATEGORIES OF ORGANIC MOLECULES IN THE BODY

Category	Majority of atoms	Subclass	Subunits	Characteristics
Proteins	C, H, O, N		Amino acids	Large polymers of amino acids. Major source of nitrogen. Many functions.
Lipids	C, H			Insoluble in water.
		Neutral fats	3 Fatty acids + glycerol	Major store of reserve fuel which can be used by most cells to provide energy.
		Phospholipids	2 Fatty acids + glycerol + phosphate + small charged nitrogen molecule	Molecule has a polar and nonpolar end. A major component of membrane structure.
		Steroids		Cholesterol and steroid hormones.
Carbohydrates	C, H, O	Monosaccharides (sugars)		Glucose is major fuel used by cells to provide energy.
		Polysaccharides	Sugars	Small amounts of glucose can be stored as glycogen.
Nucleic acids	C, H, O, N	DNA	Nucleotides containing the bases adenine, cytosine, guanine, *thymine*	Store genetic information.
		RNA	Nucleotides containing the bases adenine, cytosine, guanine, *uracil*	Translation of genetic information into protein synthesis.

sociated with the object because of its position or internal structure. The most elementary form of kinetic energy is the motion of individual molecules. The hotter an object, the faster its molecules move and the greater their kinetic energy. Thus, heat is a form of energy—the kinetic energy of molecular motion.

The potential energy of an object or molecule has the potential of becoming kinetic energy. A book resting on a table has only potential energy; as the book falls to the floor, the potential energy is converted into kinetic energy. Chemical energy is a form of potential energy locked within the structure of a molecule. It can be released during a chemical reaction which breaks the bonds between the atoms in the molecule. The released chemical energy appears mainly in the form of heat, i.e., as increased molecular motion. Heat energy is measured in units known as *calories,* one calorie being the amount of heat energy required to raise the temperature of one gram of water one degree Celsius. The energies associated with most chemical reactions are of the order of several thousand calories and are given as kilocalories (1 kcal = 1,000 calories).

The reaction between hydrogen and oxygen to form water releases a large quantity of heat energy.

$$\text{H—H} + \text{O} \rightleftharpoons \text{H—O—H} + 68 \text{ kcal/mol}$$

Just as 68 kcal/mol of heat energy is released during the formation of 1 mol of water, water can be broken down into hydrogen and oxygen if 68 kcal/mol of energy is added to the system. In other words, the sum of the potential-energy content of hydrogen and oxygen is higher than the potential-energy content of water; therefore, energy must be added to water to break it down into hydrogen and oxygen. Any chemical reaction is, at least in theory, a *reversible chemical reaction* (the double arrow in the above equation indicating the forward and reverse reaction), provided energy equal to the amount released when a reaction proceeds in one direction is put back into the reaction when it goes in the opposite direction.

When energy must be added to a chemical reaction, where does the energy come from? One source is the kinetic energy of molecular motion: heat energy. During the collisions that occur between moving mole-

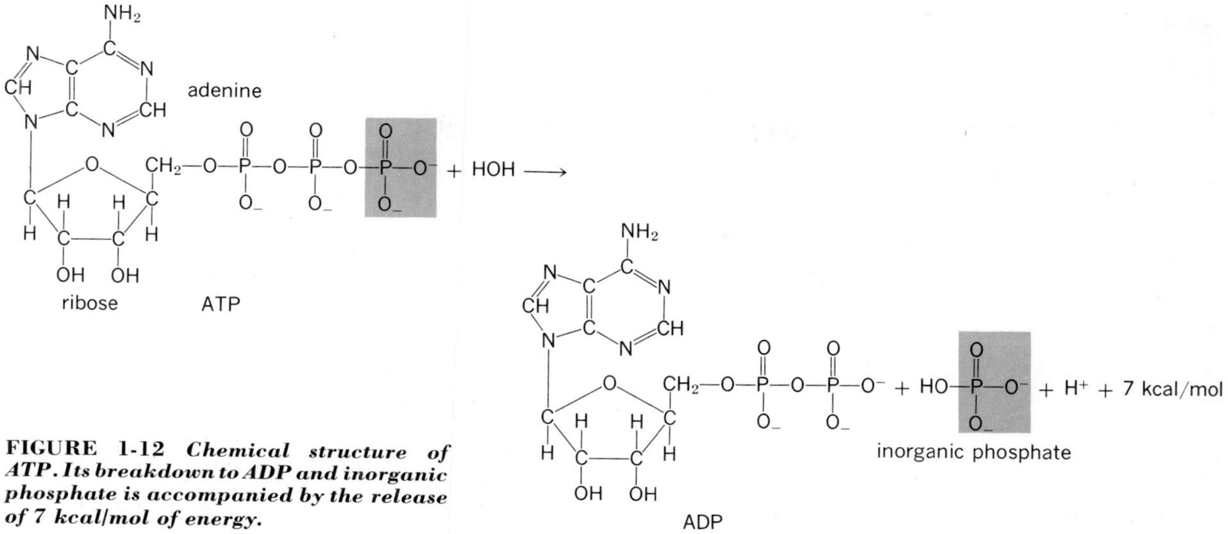

FIGURE 1-12 ***Chemical structure of ATP. Its breakdown to ADP and inorganic phosphate is accompanied by the release of 7 kcal/mol of energy.***

cules, kinetic energy can be transferred into the potential energy of a chemical bond. However, for those reactions requiring a large input of energy, this would require a much higher temperature than the body temperature of 37°C (98.6°F). In the body, the energy for high-energy-requiring reactions can be obtained by coupling the reaction to a second reaction which releases a large amount of energy. During the catabolism of carbohydrates, lipids, and proteins, chemical energy associated with these molecules is released. About half of this energy appears directly as heat, and the remainder is coupled through a special set of chemical reactions to the synthesis of *adenosine triphosphate* (ATP). ATP (Fig. 1-12) has three components: *adenine, ribose* (a sugar), and three *phosphate* groups. ATP is synthesized by the addition of inorganic phosphate (P_i) to the terminal phosphate of ADP (*adenosine diphosphate*); 7 kcal/mol is required to form this bond. Conversely, 7 kcal/mol of energy is released when ATP breaks down into ADP and inorganic phosphate.

$$ATP + H_2O \rightleftharpoons ADP + P_i + 7 \text{ kcal/mol}$$

This energy is universally used by living organisms to carry out energy-requiring cellular functions, whether it be the contraction of muscle or the synthesis of a new molecule. The function of ATP is to transfer the energy released from the breakdown of carbohydrates, lipids, and proteins to the many different energy-requiring processes in the cell (Fig. 1-13).

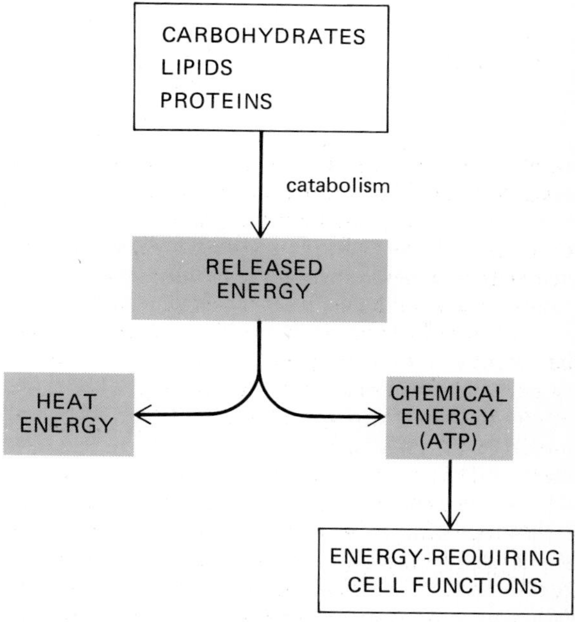

FIGURE 1-13 ***ATP is the chemical intermediate through which about half of the energy released during the catabolism of carbohydrates, fats, and proteins flows to the energy-requiring processes in the cell. The other half of the energy released during catabolism appears as heat energy.***

Enzymes

Chemical substances which accelerate the rate of a chemical reaction without themselves being chemically altered by the overall reaction are known as *catalysts*. The molecules in the body which act as chemical catalysts are the proteins known as *enzymes* (meaning "in yeast," since it was from yeast cells that the first protein catalysts were extracted and studied). Although all enzymes are proteins, not all the proteins in the body act as enzymes.

The molecules acted upon by an enzyme are known as *substrates,* and the chemically modified molecules resulting from the reaction are known as *products*.

$$\underset{\text{Substrates}}{A + B} \xrightarrow{\text{Enzyme}} \underset{\text{Products}}{C+D}$$

After an enzyme has reacted with its subtrates and released its products, it can combine with new substrate molecules and repeat the reaction. Thus, enzymes are not used up during the reaction and can be used over and over again. Therefore, only small quantities of a particular enzyme are required to transform large amounts of substrate into product. When a substrate (S) binds to an enzyme (E), a temporary enzyme-substrate complex (ES) is formed which then breaks down into the enzyme and product (P) molecules.

$$E + S \rightleftharpoons ES \rightleftharpoons E + P$$

The binding of the substrate molecule to the enzyme weakens particular chemical bonds in the substrate attached to the enzyme. New chemical bonds may be temporarily formed between the substrate molecule and the enzyme which are then broken as the products are released from the enzyme (Fig. 1-14). These interactions between the substrate and enzyme are responsible for the accelerated rate at which the reaction occurs in the presence of an enzyme.

Most of the chemical reactions in the body are catalyzed by enzymes which have a high *specificity* for the type of chemical reaction they catalyze. What property of enzymes is responsible for the specificity of their action, i.e., the ability to catalyze one type of reaction but not another? Every enzyme, being a protein, has a characteristic shape resulting from the twisting and folding of its primary amino acid sequence. As a result, its surface has a region, known as the *active site,* which has a geometrical shape complementary to the geometrical shape of the substrate molecule with which the enzyme interacts (Fig. 1-14). The substrate molecule binds to the active site much as a key fits into a lock. If a particular molecule does not fit the shape of the enzyme active site, the enzyme will not catalyze the chemical transformation of that molecule. Just as only one key will open a particular lock, only one type of substrate molecule will, in general, be able to interact with a particular enzyme molecule. Thus, the geometrical shapes of the enzyme and substrate molecules are the determining factors governing enzyme specificity.

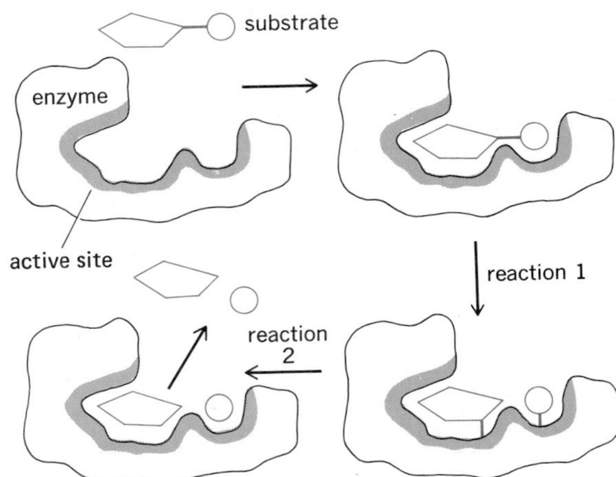

FIGURE 1-14 ***Reaction between a substrate molecule and the active site on an enzyme molecule to form an intermediate enzyme-substrate complex which then undergoes a second reaction to produce the product molecules and return the enzyme to its original state.***

There are almost as many different types of enzymes in a cell as there are chemical reactions. The specificity of enzymes accounts for the particular sequence of reactions that occur within each cell. Each enzyme acts on only one or two chemical bonds in a particular molecule. The products of one enzyme-mediated reaction become the substrates for the next enzyme reaction in the sequence. By a sequence of small changes catalyzed by enzymes, a molecule can be transformed into a totally different chemical structure. The sequence of enzyme-catalyzed reactions leading from an initial major substrate, such as glucose, to a final end product, such as carbon dioxide, is known as a *metabolic pathway*. Thus, the total metabolism of the cell can be analyzed in terms of a number of separate and interrelated metabolic pathways.

Particular types of enzymes are associated with particular types of cell organelles, either bound into the structure of the membrane surrounding the organelle or free within the internal compartment of the organelle. Therefore, the types of enzymes associated with different cell organelles and the substrate's ability to move across the organelle membrane serve to segregate different types of metabolic pathways to different regions of the cell.

Sources of Metabolic Energy

Carbohydrate Catabolism As mentioned above, the catabolism of carbohydrates, lipids, and proteins is accompanied by the release of chemical potential energy; slightly more than half of this energy appears directly as heat and the remainder is coupled to the synthesis of ATP. Glucose plays a central role in these processes because the pathway for its breakdown serves as a common pathway for the metabolism of chemical intermediates derived from proteins and lipids as well.

The catabolism of glucose to carbon dioxide and water in the presence of molecular oxygen releases 686 kcal of energy per mol[2] of glucose.

$$C_6H_{12}O_6 + 6O_2 \longrightarrow 6CO_2 + 6H_2O + 686 \text{ kcal/mol}$$

This overall reaction involves 19 separate enzyme-mediated steps which are grouped into two sequences based upon the location of the enzymes, the requirement for molecular oxygen, and the amount of ATP formed. The first ten reactions constitute the metabolic pathway known as *glycolysis* (Fig. 1-15), and the last nine are referred to as the *Krebs cycle* (see Fig. 1-17) in honor of the biochemist Hans Krebs. Both glycolysis and the Krebs cycle illustrate a general principle applicable to all metabolic pathways: Each step in the metabolic sequence results in only a small modification of chemical structure, generally the breakage and formation of only one or two chemical bonds.

Glycolysis is mediated by soluble enzymes present in the intracellular fluid of the cytoplasm and not associated with any particular cell organelle. A net synthesis of 2 mol of ATP per mol of glucose occurs during glycolysis. Since the synthesis of 1 mol of ATP "traps" 7 kcal of energy, a total of only 14 kcal (2 percent) out of the 686 kcal of potential energy available from glucose has been transferred to ATP. Despite this limited capacity for ATP generation, the glycolytic pathway provides a mechanism for ATP generation during brief periods of oxygen lack.

The reaction occurring at step six is essential for maintaining the flow of glucose through the glycolytic pathway. In this reaction two atoms of hydrogen are transferred from 3-phosphoglyceraldehyde to a molecule of NAD (nicotinamide adenine dinucleotide) to form $NADH_2$. NAD participates in many chemical reactions in the cell, its function being to transfer hydrogen from one chemical reaction to another (several other molecules in the cell perform a similar role and will be referred to simply as *carrier–H_2* molecules). If $NADH_2$ were unable to transfer its hydrogens to another molecule to re-form NAD, the supply of NAD in the cell would become depleted, and glucose breakdown, which requires NAD at step six, would come to a halt. As we shall see in the next section, the major substrate to which $NADH_2$ transfers its hydrogen is molecular oxygen. However, in the absence of molecular oxygen, $NADH_2$ can still transfer its hydrogens to pyruvic acid to form *lactic acid* (Fig. 1-16) and regenerate NAD which can now return to reaction six and maintain the breakdown of glucose and generation of ATP through the glycolytic pathway. Thus, in the absence of molecular oxygen (*anaerobic conditions*), lactic acid becomes the end product of glycolysis because it provides the mechanism for regenerating NAD. In contrast, in the presence of molecular oxygen (*aerobic conditions*), the end product of glycolysis is pyruvic acid because NAD can be regenerated by reacting with molecular oxygen. Although all the cells in the body contain the enzymes of the glycolytic pathway and can thus generate limited amounts of ATP in the absence of oxygen, the small amounts of ATP formed are not sufficient to meet the large energy demands of most cells for more than a short time.

The Krebs-cycle reactions (Fig. 1-17) convert the carbons in a molecule of pyruvic acid into three molecules of carbon dioxide and in the process form five molecules of carrier–H_2. The removal of the carboxyl group from pyruvic acid leaves a two-carbon molecule, *acetate,* which combines with a molecule of *coenzyme A* (CoA) to form *acetyl coenzyme A*. Acetyl coenzyme A is one of the main sources of carbon fragments used by the cell as a starting material to synthesize a variety of organic molecules. In the Krebs cycle, acetyl coenzyme A transfers its two-carbon acetate to the four-carbon molecule, oxaloacetic acid, to form the six-carbon molecule of citric acid. In the remainder of

[2] A mol is the number of grams of a substance equal to its molecular weight; a mol of any substance contains 6×10^{23} molecules.

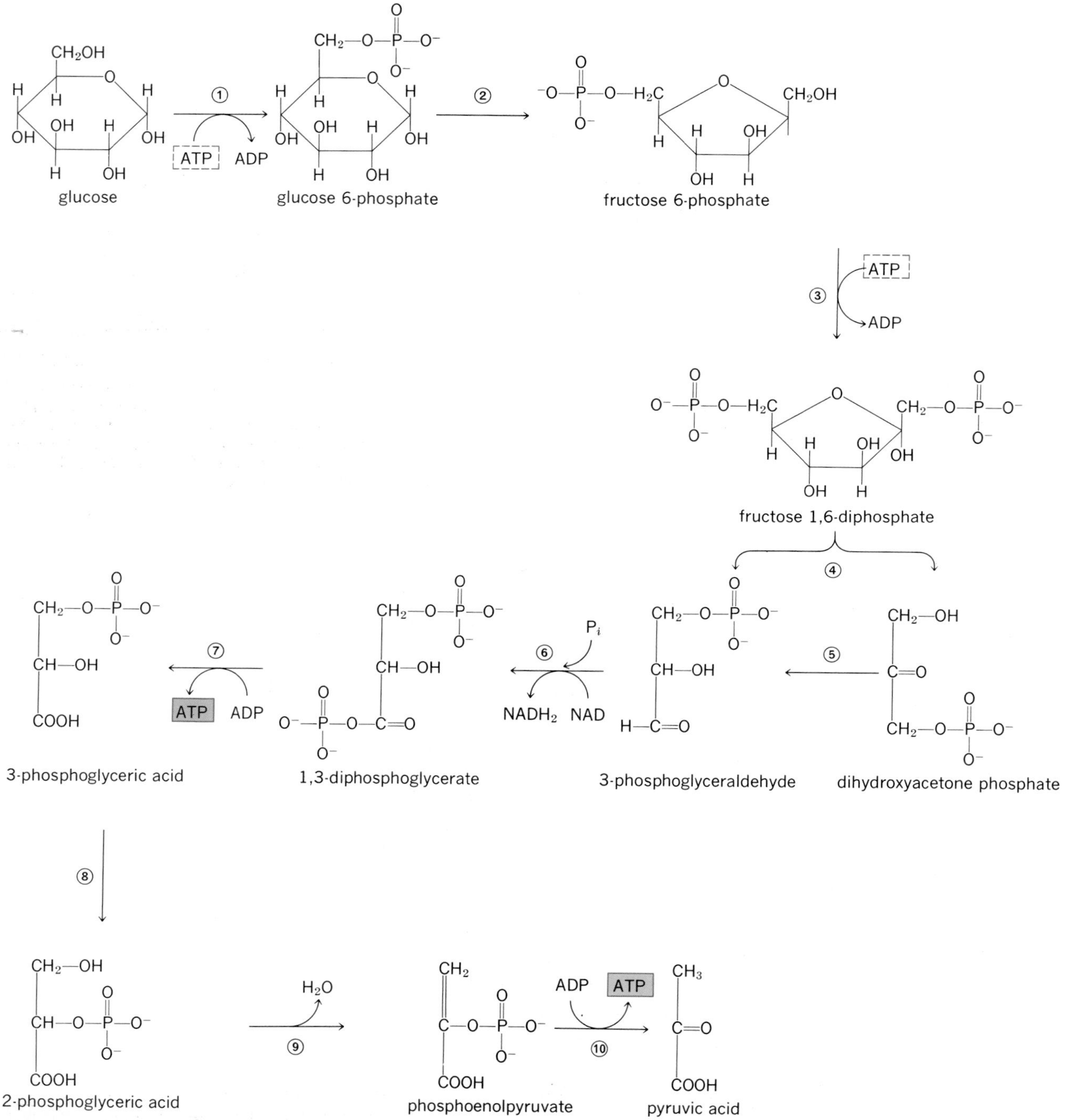

glucose → (1) ATP → ADP → glucose 6-phosphate → (2) → fructose 6-phosphate

(3) ATP → ADP → fructose 1,6-diphosphate

(4) → 3-phosphoglyceraldehyde + dihydroxyacetone phosphate

(5) dihydroxyacetone phosphate → 3-phosphoglyceraldehyde

(6) P_i, NAD → $NADH_2$: 3-phosphoglyceraldehyde → 1,3-diphosphoglycerate

(7) ADP → ATP: 1,3-diphosphoglycerate → 3-phosphoglyceric acid

(8) 3-phosphoglyceric acid → 2-phosphoglyceric acid

(9) H_2O: 2-phosphoglyceric acid → phosphoenolpyruvate

(10) ADP → ATP: phosphoenolpyruvate → pyruvic acid

FIGURE 1-15 ***Glycolytic pathway by which glucose is broken down into two molecules of pyruvic acid with the net formation of two molecules of ATP.***

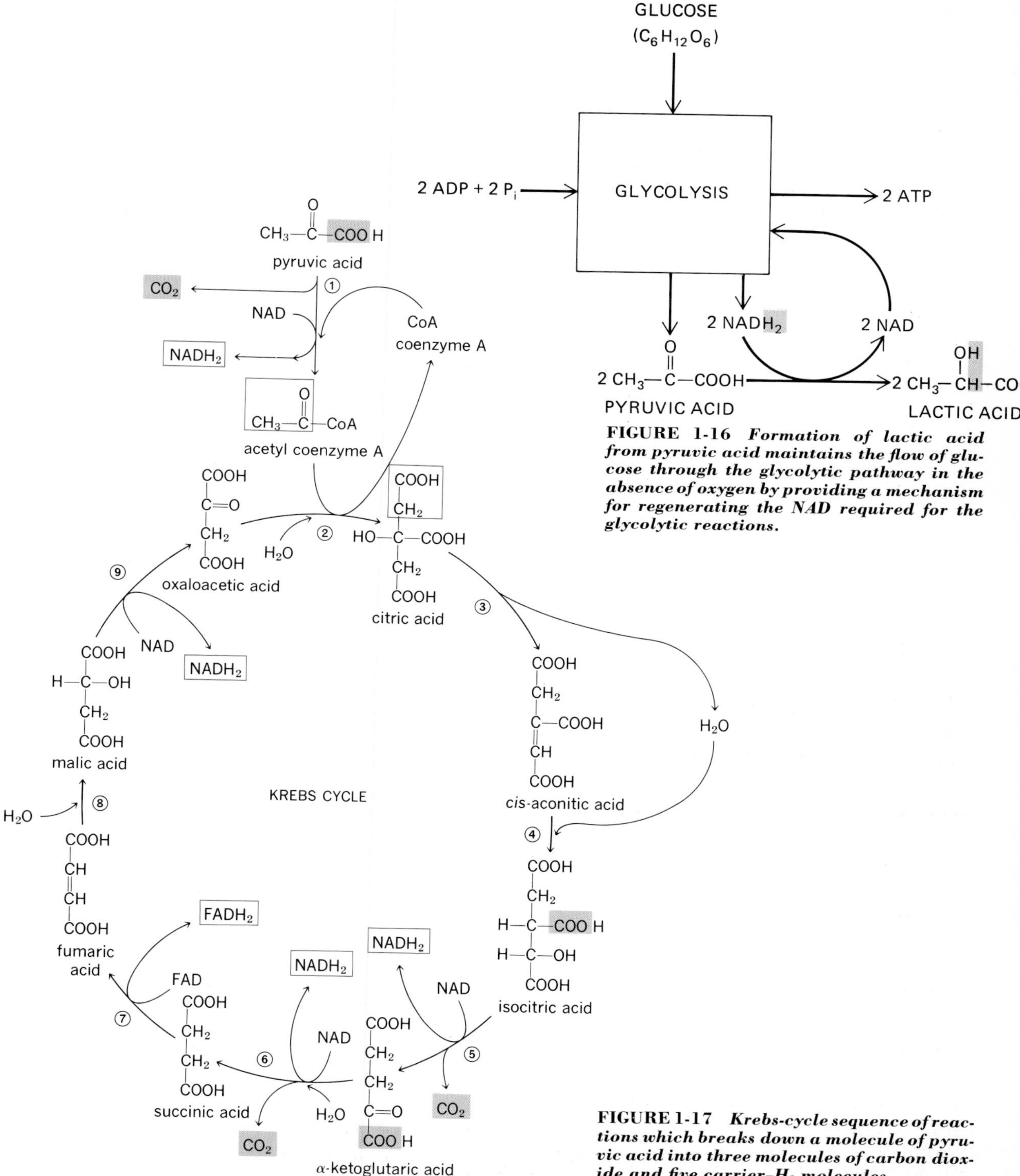

FIGURE 1-16 *Formation of lactic acid from pyruvic acid maintains the flow of glucose through the glycolytic pathway in the absence of oxygen by providing a mechanism for regenerating the NAD required for the glycolytic reactions.*

FIGURE 1-17 *Krebs-cycle sequence of reactions which breaks down a molecule of pyruvic acid into three molecules of carbon dioxide and five carrier-H_2 molecules.*

the Krebs cycle, two of the carboxyl groups of citric acid are eventually converted into carbon dioxide, leaving at the end of the cycle the starting material, oxaloacetic acid, which can begin the cycle again by combining with a new molecule of acetyl coenzyme A. In contrast to the enzymes for glycolysis, the enzymes of the Krebs cycle are localized within the mitochondria. Most of the enzymes are located in the central matrix of the mitochondria; the remaining enzymes of the cycle are bound to the inner membrane.

Oxidative Phosphorylation The inner membranes of the mitochondria also contain *cytochrome* enzymes which mediate the chemical process known as oxidative phosphorylation: the reaction between hydrogen and molecular oxygen which releases energy that is coupled to the synthesis of ATP. The oxygen for the reaction is molecular oxygen taken into the body during the process of breathing, and the hydrogen, derived from the breakdown of carbohydrates, fats, and proteins, enters the reaction in the form of carrier–H_2.

$$\text{Carrier–}H_2 + \tfrac{1}{2}O_2 \longrightarrow \text{carrier} + H_2O + 52 \text{ kcal/mol}$$

Almost 50 percent of the energy released in the reaction is coupled to ATP synthesis.

The overall reaction between carrier–H_2 and oxygen proceeds in a series of reactions, each of which releases a small amount of energy and is mediated by one of the cytochrome enzymes. The carrier–H_2 molecule donates two electrons to the first cytochrome in the enzyme chain. These electrons are then passed on to successive cytochromes until, at the end of the sequence, they are added to molecular oxygen which then combines with the two hydrogen ions to form a molecule of water. At three separate points in the the cytochrome chain, the energy released during the transfer of electrons from one cytochrome to the next is coupled to the phosphorylation of ADP to produce ATP (Fig. 1-18).

We can now calculate the total amount of ATP formed during the breakdown of one molecule of glucose into carbon dioxide and water in the presence of oxygen. Two molecules of carrier–H_2 ($NADH_2$) are formed during glycolysis of glucose to two molecules of pyruvic acid, and five molecules are formed during the Krebs-cycle reactions from each molecule of pyruvic acid, giving a total of 12 carrier–H_2; these lead to the formation of 36 molecules of ATP.[3] Since two molecules of ATP were also directly formed during glycolysis; a total of 38 ATP is produced during the breakdown of each molecule of glucose. This represents about 39 percent ($38 \times 7 = 266$ kcal) of the total energy (686 kcal) released during glucose breakdown. The remaining 61 percent of the energy appears as heat. Since 36 out of the total of 38 ATP were formed by the process of oxidative phosphorylation in the mitochondria, the majority of the cells' ATP production (95 percent) is coupled to molecular oxygen. The interactions among glycolysis, the Krebs cycle, and oxidative phosphorylation are summarized in Fig. 1-19.

Neutral Fat Catabolism The metabolic pathway for the breakdown of neutral fat into carbon dioxide and water feeds into the metabolic pathway for glucose breakdown. Neutral fat catabolism begins with an enzyme that splits off the three fatty acids linked to glycerol in the triglyceride molecule. Glycerol is a

[3] Only 34 molecules of ATP are actually generated by oxidative phosphorylation because two molecules of ATP are formed by a substrate-phosphorylation reaction within the Krebs cycle prior to the transfer of a special hydrogen carrier to the cytochrome chain.

FIGURE 1-18 ***Energy is coupled to the formation of ATP at three points in the cytochrome chain during oxidative phosphorylation.***

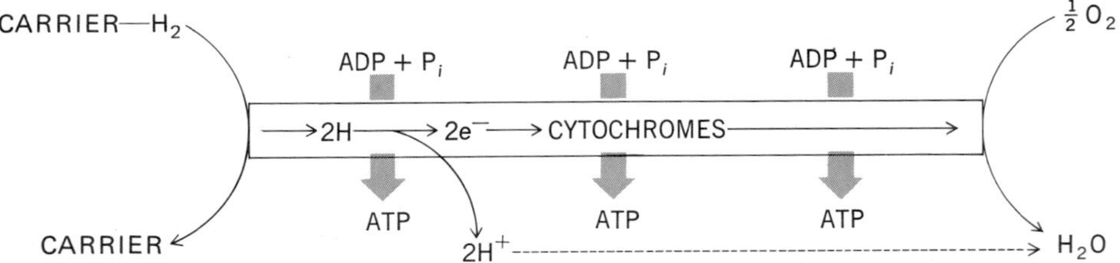

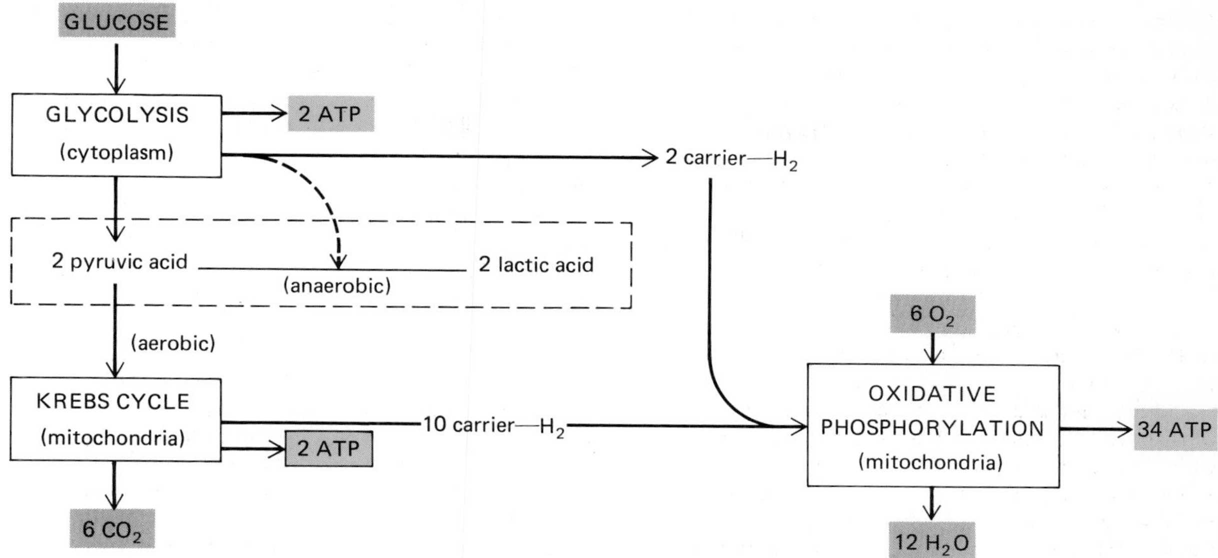

FIGURE 1-19 ***Summary diagram of the pathway of glucose catabolism to carbon dioxide and water, and the coupling of the released energy to ATP synthesis.***

three-carbon carbohydrate that can be converted into one of the three-carbon intermediates in the glycolytic pathway, from which point it can be metabolized through the rest of the glycolytic pathway and on into the Krebs cycle just like a molecule of glucose. The breakdown of the fatty acids proceeds through a pathway in which the two carbons at the carboxyl end of the fatty acids are split off (by enzymes located in the mitochondria) and transferred to a molecule of coenzyme A, and hydrogen is transferred from the fatty acid to carrier–H_2 molecules. This reaction is repeated over and over until the entire fatty acid molecule is converted, two carbons at a time, into molecules of acetyl coenzyme A which enter the Krebs cycle to undergo further breakdown to carbon dioxide. Likewise, the carrier–H_2 molecules formed during fatty acid breakdown can donate their hydrogens to the cytochrome system, giving rise to ATP synthesis by the oxidative phosphorylation pathway.

Since all the ATP formed from the breakdown of fatty acids results from the hydrogens donated to the cytochrome system, the production of ATP by fatty acid catabolism requires molecular oxygen. Each 18-carbon fatty acid forms a total of 147 molecules of ATP. The three fatty acids from one neutral fat molecule can thus form 441 molecules of ATP; adding the 22 molecules of ATP formed from glycerol gives a total of 463 molecules of ATP formed from each molecule of neutral fat catabolized. Taking the molecular weights into account (842 for a molecule of neutral fat composed of three 18-carbon fatty acids and 180 for glucose), one can calculate that, on a per gram basis, fat catabolism provides about three times as many molecules of ATP as does glucose. Therefore, fat is a much more efficient means of storing fuel than is carbohydrate.

Protein Catabolism The major difference in the composition of proteins, compared with carbohydrates and fats, is the presence of nitrogen atoms in the amino groups of the amino acids. Once this nitrogen has been removed, the remaining carbon, hydrogen, and oxygen portions of the molecule can be metabolized via the glycolytic and Krebs-cycle reactions. For example, removal of the amino group from the amino acid alanine yields pyruvic acid, the end product of glycolysis.

$$\underset{\text{Alanine}}{CH_3\text{—}\underset{\substack{|\\ NH_2}}{CH}\text{—}COOH} + H_2O + NAD \longrightarrow$$

$$\underset{\text{Pyruvic acid}}{CH_3\text{—}\overset{\overset{O}{\|}}{C}\text{—}COOH} + \underset{\text{Ammonia}}{NH_3} + NADH_2$$

The initial step in protein catabolism is the breakdown of the protein into its subunit amino acids, followed by the removal of the amino group as ammonia, NH_3. This ammonia is combined with carbon dioxide in a series of reactions which occur in the cells of the liver to form *urea*, which is then excreted in the urine.

$$\underset{\text{Ammonia}}{2NH_3} + CO_2 \longrightarrow \underset{\text{Urea}}{NH_2-\overset{\overset{\displaystyle O}{\|}}{C}-NH_2} + H_2O$$

Metabolic Regulation

The rates at which molecules pass through metabolic pathways vary and depend on two primary factors: the concentration of the substrates and the activity of the enzymes mediating the reactions, both of which are subject to physiologic control.

Consider a simple reversible reaction in which an enzyme catalyzes the splitting of a substrate molecule A into product molecules B and C.

$$A \overset{\text{enzyme}}{\rightleftharpoons} B + C$$

In order for the reaction to occur, a molecule of substrate must become bound to the active site of the enzyme as a result of random collisions between the substrate and the enzyme molecules. The probability of a collision is increased if there are more molecules of substrate in the solution. Therefore, the rate of the reaction will be increased by increasing the substrate concentration. This is an example of the principle of *mass action*, which states that increasing the concentration of any molecule in a chemical reaction will cause the reaction to proceed in a direction which tends to decrease the concentration of that molecule. Increasing the concentration of A causes the reaction to proceed toward the formation of B and C; increasing the concentrations of either B or C will cause the reaction to proceed toward the formation of A. This principle of mass action is often the determining factor governing the direction in which reactions proceed along a reversible metabolic pathway. However, there is an upper limit to the rate at which an enzyme-mediated reaction can be increased by increasing substrate concentration. This point is reached when all the enzyme molecules in a solution are bound to substrate; the enzyme is then *saturated* with substrate and the reaction proceeds at a maximal rate.

A number of molecules in a cell undergo a repeated cycle of reactions from one chemical state to another and back again. We have already encountered examples of such molecules in the form of NAD and coenzyme A, which function as carriers of hydrogen and acetate, respectively. Such molecules, known as *coenzymes*, participate in chemical reactions as a substrate, since hydrogen or acetate is added to or removed from them in the course of the reaction, but unlike other substrates, they are not degraded by the overall cycle of reactions. For example, glucose is degraded to carbon dioxide whereas NAD is transformed into $NADH_2$ and back again to NAD. The total concentration of the two forms of a coenzyme, for example, NAD + $NADH_2$, remains fairly constant in a cell, but the concentration of one form may increase at the expense of the other. Thus, under one set of conditions a cell may contain a high concentration of NAD with little $NADH_2$, and under other conditions the concentration of $NADH_2$ may be high and the concentration of NAD low. Accordingly, reactions in which coenzymes participate are often subject to mass-action control, depending on the concentration of the various forms of the coenzyme. We have already seen one example of this: the increased formation of lactic acid from pyruvate when cell concentration of $NADH_2$ is increased under anaerobic conditions.

The activity of the enzyme mediating a reaction provides the second major factor governing the rate of a reaction. By *enzyme activity* we mean the maximum rate at which the enzyme-mediated reaction proceeds when the enzyme is saturated with substrate. One way to vary the activity of an enzyme is to vary its actual concentration. Since all enzymes are proteins, the concentration of an enzyme depends on the rates of protein synthesis and degradation. For most enzymes, these rates remain essentially constant, and thus the concentration of the enzyme remains constant. In contrast, other enzymes undergo marked changes in their rates of synthesis (and therefore their concentrations) under various conditions. Some of these enzymes are present in a cell only when its substrate is also present. The substrate molecule, in this case, appears to interact with the protein synthetic system to turn on the synthesis of the enzyme; this process is called *enzyme induction*. In other cases (*enzyme repression*), product molecules inhibit the synthesis of the enzyme, and the concentration of the enzyme decreases when the products of the reaction it catalyzes increase. Enzyme repression allows a cell to turn off the synthesis of an end product when its concentration rises above a certain level.

Since it takes time to synthesize and degrade a protein, the processes of enzyme induction and repression cannot be used to regulate minute-to-minute changes in the flow of substrates through various metabolic pathways. However, it is possible to alter the activity of already synthesized enzymes. In order for an enzyme to be active, it must be able to bind substrate to its active site. Accordingly, the shape of the active site and its accessibility to substrate are the primary factors governing the activity of already synthesized enzymes. The shape of the active site of some enzymes can be altered by a nonsubstrate molecule known as a *modulator molecule,* which binds to the enzyme at a site other than the active site. The activity of the enzyme may be either increased or decreased, depending on the change in shape of the active site. Enzymes whose activities are altered by the binding of modulator molecules are known as *allosteric enzymes* (Greek *allos,* other; *steric,* shape). The end products of a metabolic pathway often act as modulator molecules which inhibit the activities of allosteric enzymes in the same pathway, a form of regulation known as *feedback inhibition.* In some cases the end products of one metabolic pathway are the modulator molecules which interact with the allosteric enzymes in a completely different pathway. Such interaction allows various metabolic pathways in a cell to be coordinated. Figure 1-20 summarizes the various factors we have discussed which influence the rates of enzyme-mediated reactions in a cell.

Interacting Metabolic Pathways Consider the following arbitrary metabolic pathway from A to F, consisting of five steps:

$$\underset{\text{I}}{A \overset{e_1}{\longleftrightarrow} B} \underset{\text{II}}{\overset{e_2}{\longleftrightarrow}} C \underset{\text{III}}{\underset{e_4}{\overset{e_3}{\rightleftharpoons}}} D \underset{\text{IV}}{\overset{e_5}{\longleftrightarrow}} E \underset{\text{V}}{\overset{e_6}{\longrightarrow}} F$$

Three of the reactions (I, II, and IV) in the pathway are reversible; only small changes in energy accompany these reactions, and a single enzyme mediates the flow of metabolites in either direction at a particular step. Reaction V is irreversible, because it is associated with the release of a large amount of energy. Without a large source of energy to put back into the reaction, F cannot be converted back into E. One of the most important examples of such an irreversible reaction is the formation of carbon dioxide and acetyl coenzyme A from pyruvic acid, the reaction which links the end of glycolysis to the beginning of the Krebs cycle. Once acetyl coenzyme A has been formed, it cannot be converted back into pyruvic acid.

Reaction III represents a reaction that is usually present at one or two steps in a metabolic pathway and is often the major control point regulating the flow of metabolites through the pathway. The conversion of C to D is an irreversible reaction similar to reaction V, and if e_3 were the only enzyme present, D could not be converted directly back into C. However, D can be

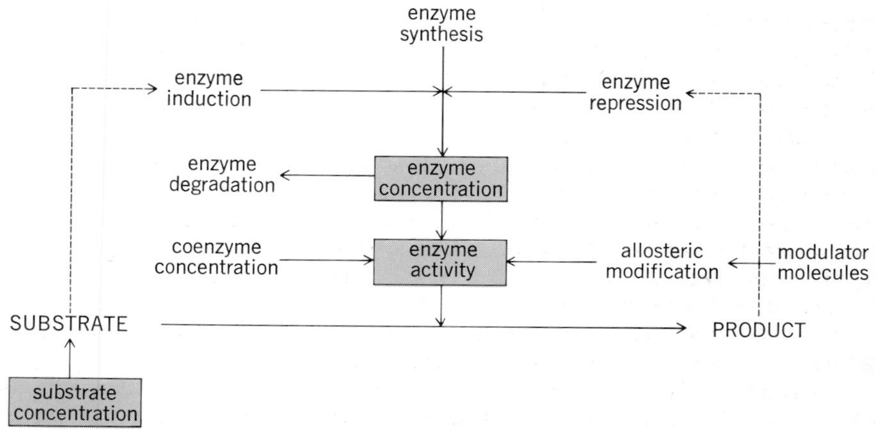

FIGURE 1-20 ***Factors which affect the rate of enzyme-mediated reactions.***

to C by a second enzyme e_4 which couples to a source of energy: the breakdown of ... ntrolling the activities of the two enzymes e_3 and e_4, the direction of flow through the pathway can be regulated so that it occurs from C to D under one set of conditions and in the reverse direction under other conditions.

When a series of reactions are linked together in a metabolic pathway, the rate at which A gets converted into F depends on the activity of the one enzyme in the pathway which has the lowest enzyme activity, since the overall rate cannot be faster than the reaction proceeding at the slowest rate, the so-called *rate-limiting reaction*. By regulating the activity of the enzyme mediating the rate-limiting step, the overall rate of flow of metabolites through the entire pathway can be controlled. The rate-limiting reaction in most pathways is mediated by an allosteric enzyme whose activity can be altered by modulator molecules.

Interconversion of Carbohydrate, Fat, and Protein Each of the three major classes of organic molecules—carbohydrates, fats, and proteins—can be broken down to provide the energy for ATP synthesis; conversely, cells are also able to synthesize these organic molecules. The sequences of chemical reactions used to synthesize carbohydrates, fats, and amino acids are in large part the reverse of those utilized in catabolism. However, a major difference between the anabolic and catabolic pathways occurs at certain steps in which large amounts of energy are released during catabolism; during anabolism, this energy must be restored to the molecule by coupling the reactions to the breakdown of ATP. Thus, the catabolic reactions of a cell provide the energy (ATP) that is utilized by the anabolic pathways. Two separate enzymes must mediate these high-energy reactions in the catabolic and anabolic directions. All the enzymes of an anabolic pathway may occur in separate cell

FIGURE 1-21 ***Interrelations between the metabolism of carbohydrates, fats, and proteins.***

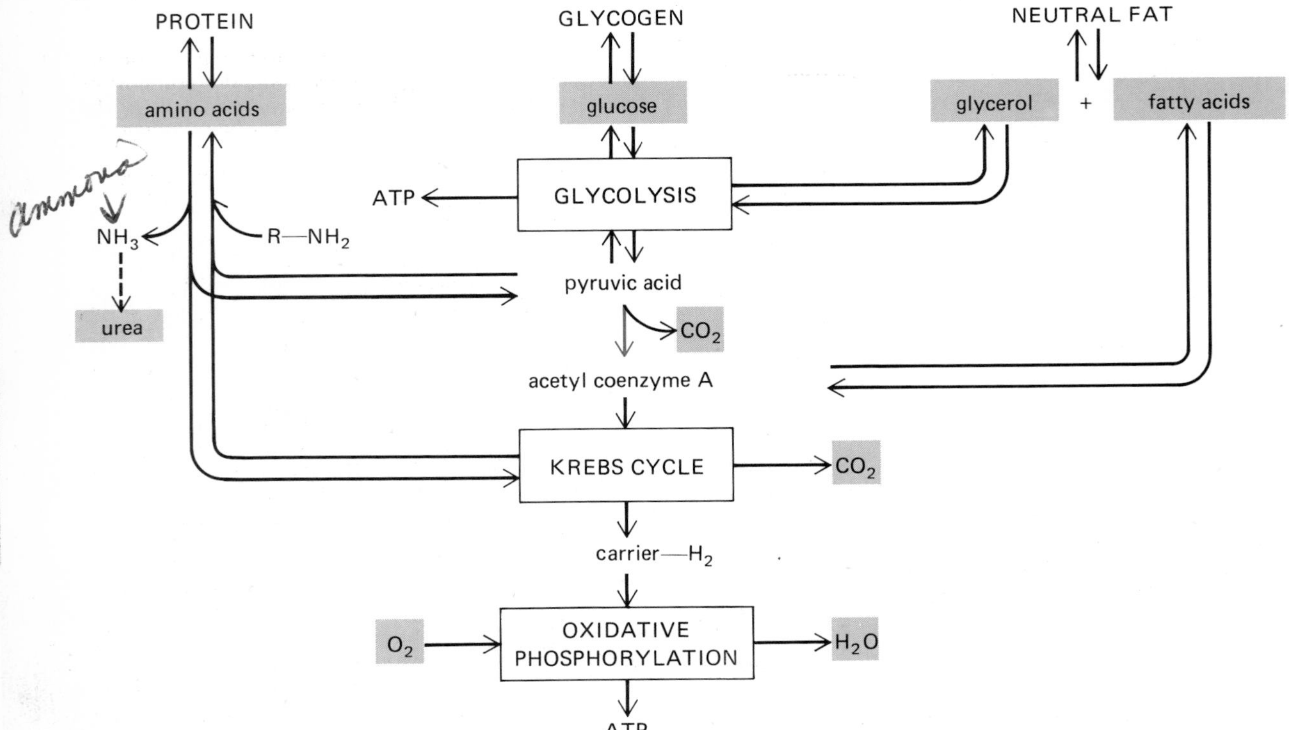

organelles from those mediating the catabolic pathway, or the two pathways may occur in the same organelle and utilize some of the same reversible enzymes at various points in their respective pathways.

Figure 1-21 illustrates the interrelations of the catabolic and anabolic pathways for carbohydrate, fat, and protein metabolism. Note that the starting point for the synthesis of a particular organic molecule may be a chemical intermediate derived from the catabolism of a different class of organic molecule. Thus, neutral fat can be synthesized from glycerol and acetyl coenzyme A, both derived from the catabolism of glucose; amino acids can be formed by adding an amino group to a metabolic intermediate of glucose catabolism; and a fatty acid can be synthesized from an intermediate formed by removing the amino group from certain amino acids. There is, however, one major restriction on the interconversion of carbohydrates, fats, and proteins: Fatty acids cannot be used to synthesize net amounts of glucose or amino acids. This restriction results from the fact that the reaction which converts pyruvic acid to acetyl coenzyme A and carbon dioxide is irreversible. Acetyl coenzyme A, derived from the breakdown of fatty acids, cannot be reconverted, in net amounts, into pyruvic acid and thus cannot proceed through the pathways from pyruvic acid to glucose and amino acids. On the other hand, glucose and amino acids can be broken down to acetyl coenzyme A and used to synthesize fatty acids.

The total metabolism of the body depends upon the supply and distribution of nutrients to the cells where they are metabolized. A cell's metabolic activity may be primarily anabolic following the ingestion of a meal and primarily catabolic during a period of fasting. Under one set of conditions carbohydrate catabolism may provide the major source of a cell's ATP whereas under other conditions fatty acid catabolism may predominate. Some of the organs of the body, notably the liver and fat tissue, have become specialized in their ability to store and release various nutrients into the blood. The various mechanisms regulating these aspects of total body metabolism will be the subject of a later chapter.

PROTEIN SYNTHESIS

Enzymes mediate the hundreds of chemical reactions responsible for cell structure and function. Therefore, differences in cell structure and function can be traced largely to differences in the enzyme composition of cells. Since all enzymes are proteins, the cellular processes associated with protein synthesis are the ultimate regulators of cell structure and function since they determine which enzymes and thus which metabolic pathways exist in a given cell type.

The synthesis of a protein molecule requires linking together a sequence of amino acids. A cell must therefore contain some form of information determining which particular sequence of amino acids must be linked together to form a particular protein. One of the outstanding accomplishments of twentieth-century biology has been the discovery of the mechanism by which cells store this information and translate it into the process of protein synthesis (Fig. 1-22). This process involves the fourth class of organic molecules found in the body, the *nucleic acids* deoxyribonucleic acid (DNA) and ribonucleic acid (RNA).

Nucleic Acids

Deoxyribonucleic acid molecules, the largest molecules in the cell, consist of a sequence of molecular subunits linked together to form long chains. These subunits, the *nucleotides,* each consists of three parts:

FIGURE 1-22 ***Relation between metabolism and the synthesis of protein enzymes which requires that information be transferred from DNA in the nucleus to the site of protein synthesis in the cytoplasm by way of RNA.***

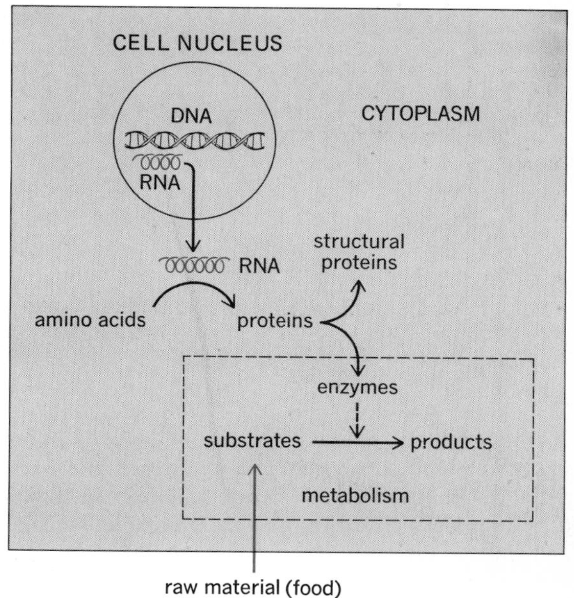

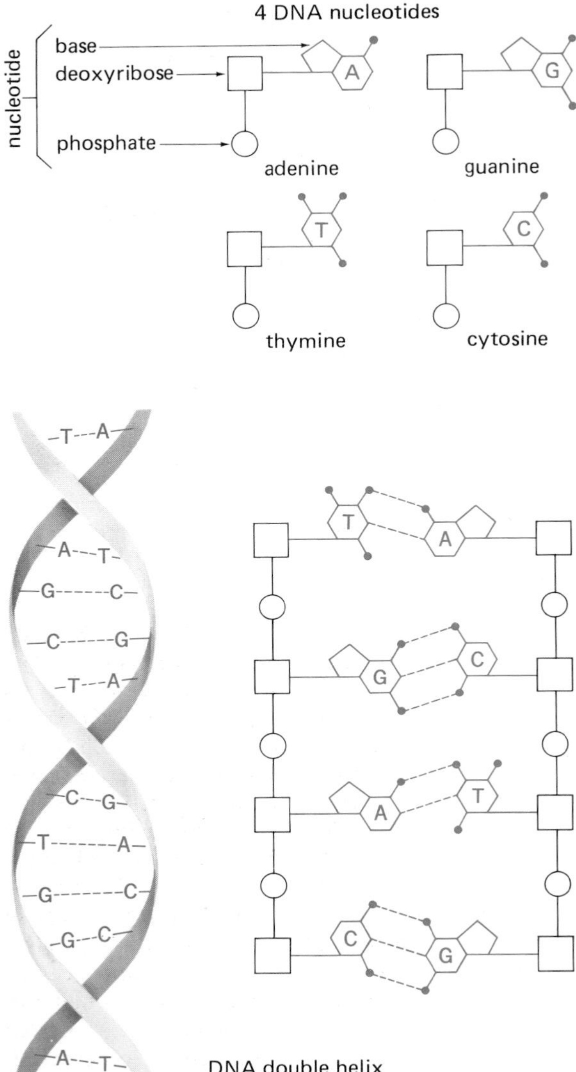

FIGURE 1-23 ***Base pairings between the two nucleotide chains account for the double-helical structure of DNA.***

phosphate, the sugar deoxyribose (hence the name deoxyribonucleic acid), and one of four ring-shaped organic molecules known as *bases*, adenine, guanine, cytosine, or thymine. Nucleotides are linked together to form the DNA chain by bonds from the phosphate groups of one nucleotide to the deoxyribose sugar of the next nucleotide (Fig. 1-23). A DNA molecule consists of two chains of nucleotides coiled about each other to form a double helix. The structures of the nucleotide bases are such that thymine (T) can pair only with adenine (A), whereas cytosine (C) pairs only with guanine (G). These two sets of base pairs, TA and CG, link the two nucleotide chains together to form the double helix of DNA. The specificity of the base pairing between A and T, and between G and C, has important implications for both the replication of DNA during cell division and for the mechanism by which information is transferred from DNA to RNA during protein synthesis.

The structure of *ribonucleic acid* is similar to DNA in that nucleotides composed of phosphate, sugar, and base form the repeating subunits of RNA. However, the sugar in the nucleotides is *ribose* rather than deoxyribose—thus the name ribonucleic acid. Three of the bases in the nucleotides of both DNA and RNA are the same: adenine, guanine, and cytosine. The fourth DNA base, thymine, is replaced in RNA by the base *uracil* (U). Another difference is that RNA consists of only a single chain of nucleotides in contrast to DNA's double chain.

The function of DNA is to store information which codes the sequences in which amino acids are linked together to form proteins. This information is coded in terms of the sequence of the four bases, A, T, G, and C. Since there are 20 different amino acids in proteins but only four different bases, a single base cannot be the code word for a single amino acid. Rather, a specific sequence of three bases, for example, CGG or ACT, serves as the code word, or *codon*, corresponding to a specific amino acid. Thus, each codon specifies a particular amino acid, and a sequence of codons along a single chain of the DNA double helix specifies the sequence of amino acids in the protein. The sequence of codons which code one protein is known as a *gene*. A single molecule of DNA contains the genetic information corresponding to many different proteins arranged as a sequence of genes. Some of the three-base codons, rather than specifying a particular amino acid, act as punctuation marks, indicating the points along the DNA chain where a gene begins and ends (Fig. 1-24).

Almost all of a cell's DNA is located within the cell nucleus where it exists as long, coiled chromatin threads, giving the nucleus a granular appearance. The DNA molecules are too large to pass through the

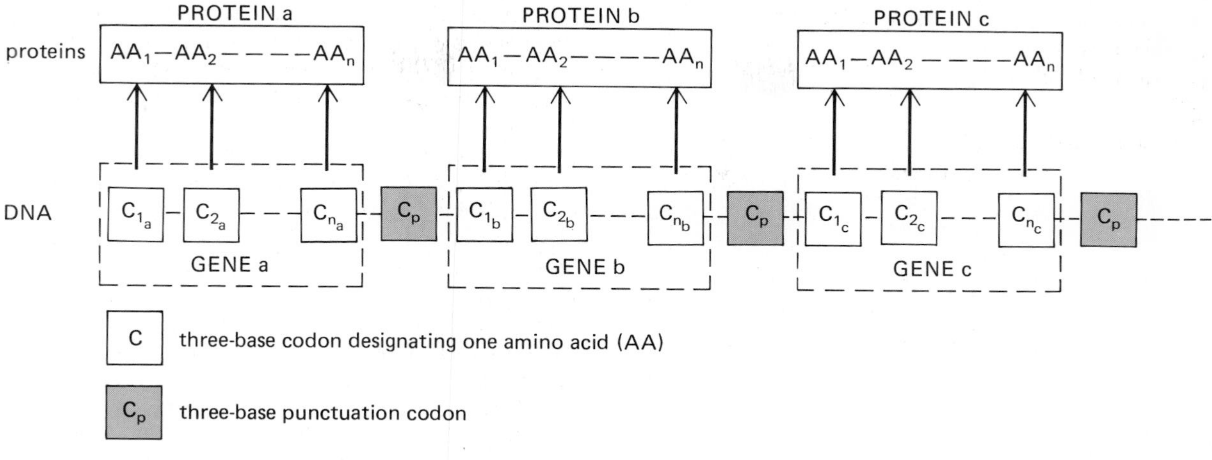

FIGURE 1-24 ***Relation between the linear sequence of three-base codons and punctuation marks along the DNA molecule and the sequence of amino acids in the corresponding proteins.***

nuclear pores into the cytoplasm. Since the actual assembly of a protein from individual amino acids occurs in the cytoplasm, a message containing the information specifying the sequence of amino acids must be carried from the nuclear DNA to the cytoplasm. This message is carried by molecules of RNA, known as *messenger RNA* (mRNA), which are able to pass through the nuclear pores into the cytoplasm.

Messenger RNA is synthesized in the nucleus on the surface of the DNA molecules by linking together the appropriate sequence of nucleotides. The nucleotide sequence in mRNA is determined by base pairings between free nucleotides and corresponding nucleotides present in one of the two chains of a DNA double helix. Thus, a free adenine nucleotide pairs with thymine in the DNA chain, C with G, U with A, and G with C, creating a sequence of bases in mRNA which is the mirror image of the base sequence in DNA. For example, when a codon sequence of bases in DNA is ATC, the corresponding codon sequence in mRNA is UAG.

Protein Assembly

Once a molecule of messenger RNA has been synthesized, it leaves the nucleus through the nuclear pores and enters the cytoplasm where it becomes attached to a *ribosome*, composed of both protein and RNA. Free amino acids cannot bind directly to their corresponding codon in messenger RNA. Each amino acid must first be attached to a second type of RNA molecule known as *transfer RNA* (tRNA), which contains within its nucleotide sequence a specific three-base sequence, known as an *anticodon*, which is able to base-pair to its mirror-image codon in a messenger RNA molecule. There are several bases in tRNA that are not found in other RNAs, such as the base inosine (I). There is a different tRNA molecule for each of the 20 different amino acids, as well as a different cytoplasmic enzyme specific for catalyzing the linkage of each type of amino acid with its corresponding tRNA. Once a given amino acid has been attached to a tRNA molecule containing the anticodon for that particular amino acid, the tRNA–amino acid complex base-pairs with the appropriate codon in messenger RNA on the surface of the ribosome (Fig. 1-25). This recognition system allows the amino acids to be linked onto the protein chain in a sequence determined by the codon sequence in messenger RNA.

We have now identified all the major ingredients required for protein systhesis. A messenger RNA molecule carrying the codon sequence for a particular protein becomes attached to a ribosomal particle. Each ribosome also has, in addition to a binding site for mRNA, two binding sites for tRNA; one holds the tRNA molecule already attached to a growing chain of amino acids, and a second site holds the tRNA con-

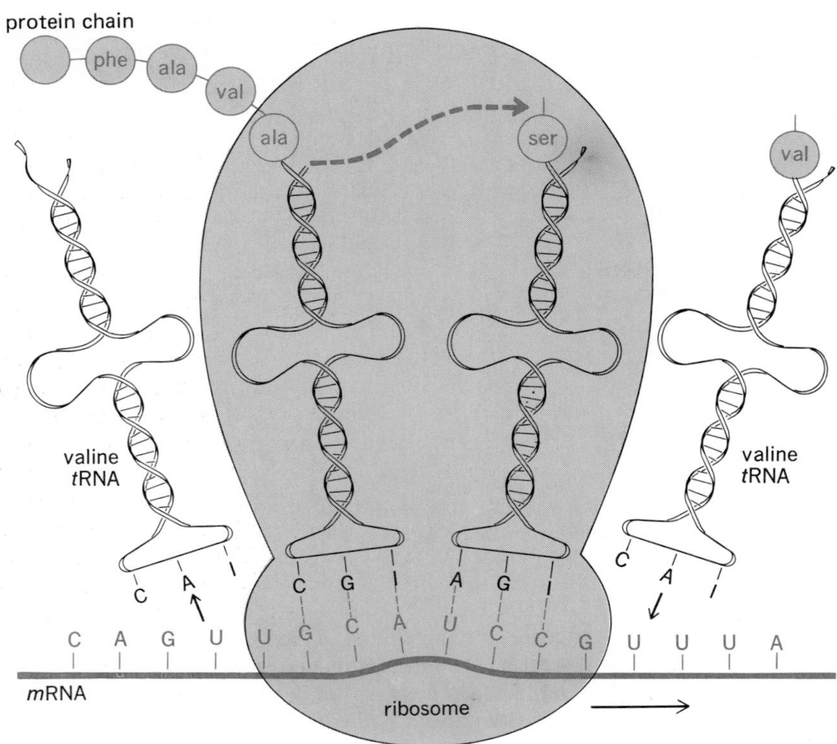

FIGURE 1-25 ***Sequence of events during the synthesis of a protein molecule on the surface of a ribosome as it moves along a strand of mRNA. The three-base anticodon of tRNA carrying one amino acid binds to the corresponding codon of mRNA and transfers its amino acid to the growing protein chain.***

taining the next amino acid to be added to the sequence (Fig. 1-25). Ribosomal enzymes catalyze the formation of a peptide bond between the last amino acid attached at one site and the amino acid linked to its tRNA at the adjacent site. The ribosome now moves a distance of one codon down the messenger RNA chain, releasing a molecule of tRNA and binding the next tRNA–amino acid complex. This process is repeated over and over as each amino acid is added in succession to the growing peptide chain. When the ribosome reaches the end of the mRNA or the end of a coded sequence for a single protein, the completed protein is released from the surface of the ribosome. It takes about 1 min to synthesize a protein containing 100 amino acids, that is, 1 or 2 s to add each amino acid to the growing peptide chain. Table 1-5 summarizes the sequence of events leading from DNA to the completed synthesis of a protein molecule.

Protein Secretion

Most proteins are unable to cross cell membranes and are therefore confined to cells or to various membrane-bound cell organelles. However, certain types of cells synthesize proteins to be released (secreted) from the cell into the extracellular fluid. Examples are the digestive enzymes secreted into the intestinal tract, protein hormones secreted into the blood, and proteins secreted by connective-tissue cells to form the extracellular fibrous matrix of tissues. Accordingly, special mechanisms are required to secrete these proteins into the extracellular fluid.

When cells which secrete large amounts of protein are compared with nonsecretory cells, one immediately notices a difference in the structural organization of their cytoplasm. Nonsecretory cells have a smaller number of ribosomes, and these ribosomes exist as free particles distributed throughout the cytoplasm of

TABLE 1-5
SEQUENCE OF EVENTS FROM DNA TO PROTEIN SYNTHESIS

Transcription of DNA to mRNA
1 The two strands of the DNA double helix separate in the region of the gene to be transcribed.
2 Free nucleotides base-pair with their complementary bases in DNA.
3 The base-paired, free nucleotides are linked together by an enzyme to form a strand of mRNA containing a base sequence complementary to the base sequence in DNA.
4 The newly synthesized molecule of mRNA passes from the nucleus, through a nuclear pore, to the cytoplasm where it binds to a ribosome.

Binding of amino acids to the ribosomal–mRNA complex
5 Free amino acids are linked to molecules of tRNA by enzymes in the cytoplasm specific for each amino acid and corresponding tRNA.
6 Two of the amino acid–tRNA complexes become bound to the surface of a ribosome. The base pairing between the three-base anticodon in each tRNA with the complementary three-base codon in mRNA determines the sequence in which amino acid–tRNA complexes bind to the ribosome.

Elongation of protein chain
7 A ribosomal enzyme mediates the linkage of the two adjacent amino acids on the ribosomal surface.
8 The linked amino acids remain attached to one of the tRNA molecules while the other is released into the cytoplasm.
9 The ribosome moves one codon step along the mRNA strand.
10 The next amino acid–tRNA complex in the sequence binds to the vacated site on the ribosome.
11 Steps 7 to 10 are repeated over and over, adding one amino acid at a time to the growing protein chain.

Termination of protein-chain assembly
12 The completed protein is released from the ribosome when the termination codon in mRNA is reached.

the cell. These free ribosomal particles synthesize the proteins and enzymes used internally by the cell. In contrast, the cytoplasm of protein-secreting cells contains numerous ribosomes, most of which are attached to the membranes of *granular endoplasmic reticulum.* The proteins to be secreted by a cell are synthesized on the ribosomes attached to the endoplasmic reticulum. As the protein is being synthesized, it crosses the reticulum through a channel formed by the ribosome embedded in the reticulum membrane. The protein then moves along the space enclosed by the reticulum membranes to the Golgi apparatus which concentrates the secretory proteins and attaches sugars to some of the amino acid side chains to form *glycoproteins*. Once these operations are completed, portions of the Golgi membranes break away, forming small secretory granules consisting of a concentrated solution of the secretory protein surrounded by a membrane. The secretory granule then migrates to the surface of the cell where its membrane fuses with the plasma membrane, and the contents of the granule are released into the extracellular fluid (Fig. 1-26). Thus, the secretory protein does not actually have to cross the plasma membrane, since the membrane of the secretory granule simply opens up on fusion with the plasma membrane. This process of membrane fusion associated with secretion is known as *exocytosis.* A similar process occurs in reverse when extracellular materials are engulfed by an infolding of the plasma membrane (*endocytosis*), which then breaks away from the plasma membrane, thereby carrying a membrane-bound vesicle of extracellular material into the cell.

CELL REPRODUCTION

The growth of the human body from a single fertilized egg cell to an adult person weighing 100 pounds or more involves three processes: (1) *cell division,* which increases the number of cells; (2) *cell growth,* which increases the size of individual cells; and (3) *cell differentiation,* the process by which cells develop specialized functional properties. Even after the adult body size has been reached, these three processes continue to operate in replacing cells that have become damaged or have a short life-span.

The growth of a multicellular organism and the propagation of the species from generation to generation depend on the fundamental property of cell division. However, all the cells in the human body are not able to divide. As cells become highly specialized through the process of cell differentiation, they often lose their capacity for division. Such specialized cells can be replaced only by the division of less specialized, undifferentiated cells, followed by differentiation into the specialized cell type, or, as is the case with nerve cells, they cannot be replaced at all.

Replication of DNA

The initial event leading to cell division is the replication of the DNA molecules in the cell nucleus, followed by the distribution of these replicates of the

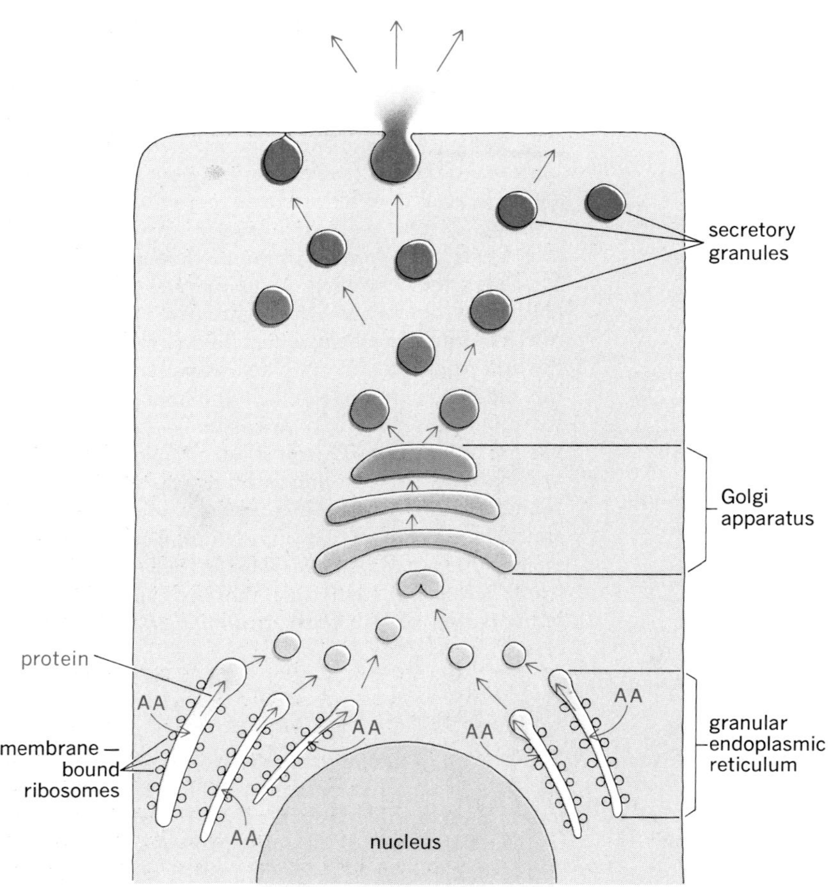

FIGURE 1-26 ***Protein secretion begins with the synthesis of the protein from free amino acids (AA) on the ribosomes of the granular endoplasmic reticulum. The synthesized protein enters the lumen of the reticulum and passes to the Golgi apparatus where it is concentrated and packaged into secretory granules which move to the cell surface, fuse with the plasma membrane, and release the protein into the extracellular fluid.***

cell's genetic information to each of the two new daughter cells. Replication of DNA (Fig. 1-27) begins with the separation of the two base-paired strands of the DNA double helix. The exposed bases of the two strands can now act as templates which bind the corresponding bases of free nucleotides. Enzymes catalyze the formation of bonds between successive nucleotides of the new strands, resulting in the formation of two identical molecules of DNA, each containing one strand of nucleotides that was formerly present in the original molecule of DNA and one new strand formed by the nucleotides that have base-paired with the old strand.

Mistakes, although rare, can occur during the process of DNA replication. Any alteration in the genetic message carried by DNA is known as a *mutation.* If, for example, a nucleotide containing the base guanine (G) is inserted at a point normally occupied by adenine (A) in the codon sequence CAT, the three-base codon at that site will be altered to CGT. Since the new codon may specify a different amino acid from that originally coded, the protein synthesized by instructions from the mutated gene may have a different chemical activity in the cell. A number of factors in the environment increase the rate at which genes mutate. These factors, known as *mutagens,* include certain chemical substances and various forms of ionizing radiation, such as x-rays and atomic radiation. Most of these agents cause the breakage of chemical bonds, allowing the substitution of new bases in the DNA sequence.

Mutation is the mechanism that underlies the evolution of living organisms. If an organism carrying a mutant gene is able to perform some function more

FIGURE 1-27 ***Replication of DNA begins with the separation of the two strands of the DNA double helix, followed by the pairing of the bases of free nucleotides with the exposed bases of the old DNA strands, giving rise to two identical double-helical molecules of DNA, each containing one old and one new nucleotide strand.***

efficiently than an organism lacking such a gene, it has a better chance of surviving and passing the mutant gene on to its descendants (this is the principle of *natural selection*). However, although some mutations may enable a cell to function more efficiently in a given environment, the majority of mutations lead to modifications in protein structure which result in less effective functioning or even the death of the cell carrying the mutant gene. Medical science is beginning to identify a number of human diseases that are the result of abnormal enzymes produced by mutant genes. Such inherited genetic diseases have been termed *inborn errors of metabolism*.

Cell Division

Rapidly growing cells divide about once every 24 h, whereas other cells may go for weeks or months before undergoing division, and highly specialized cells may not divide at all. Although the general morphologic changes in cell structure that occur during cell division have been studied for the past hundred years, little is known about the chemical events which initiate and regulate this process.

The life-span of a reproductively active cell passes through three periods: interphase, mitosis, and cytokinesis. The first period, *interphase,* is by far the longest and represents the time between cell divisions. Most of the cells in the body are in the interphase state, and the major functions of the body are carried out by cells in this state. Once the chemical events which initiate cell division have begun, the physical process of splitting the cell in two lasts only about 1 to 2 h. The first phase of this division process, *mitosis* (Greek *mitos,* thread), involves the division of the cell nucleus with the accompanying passage of the replicated DNA threads (chromosomes) to the two new daughter cells. Mitosis is then followed by *cytokinesis,* the division of the cytoplasm of the whole cell into two daughter cells, each containing a replica of the original DNA of the parent cell.

Mitosis The interphase nucleus contains 46 long chromatin threads composed of DNA and protein. Half of these chromatin threads were derived from the female parent and half from the male parent at the time of conception, and the genetic information they contain has been passed on from cell to cell at each subsequent cell division. Each of the 23 chromatin threads from one parent carries a unique set of genes. Thus, each cell contains two genes corresponding to every

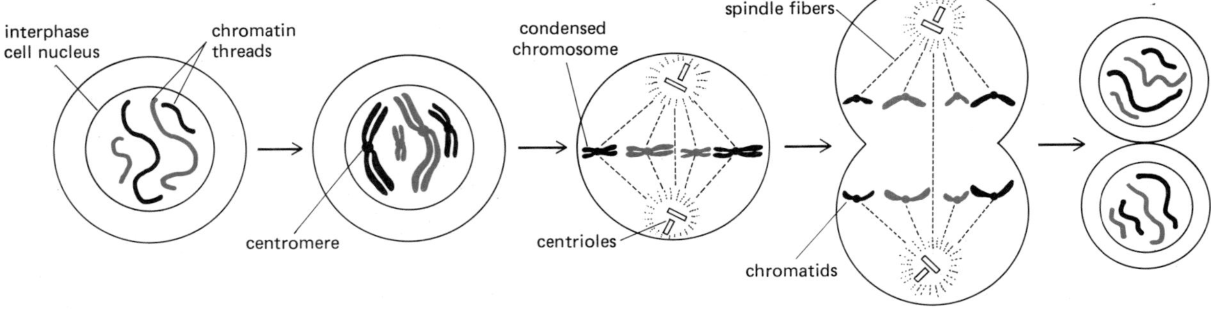

FIGURE 1-28 ***Sequence of events during cell division.***

protein in the cell, one gene from each parent.

Near the end of interphase, the DNA molecules in the 46 chromatin threads replicate, but the duplicate copies of each chromatin thread remain joined together at one point, the *centromere.* Mitosis begins when these duplicated chromatin threads start to coil up, forming 46 highly condensed, rod-shaped bodies known as *chromosomes* (colored bodies). At this stage (Fig. 1-28) each chromosome consists of two *chromatids* (the duplicated chromatin threads) joined together at the centromere.

While the chromatin threads are condensing into chromosomes in the nucleus, other events are taking place in the cytoplasm in preparation for cell division. The two microtubular *centrioles,* located just outside the nucleus, replicate to form four centrioles. The two sets of centrioles then begin migrating to opposite sides of the nucleus. During this migration, a number of new microtubules are formed, the *spindle apparatus,* which link the two sets of centrioles together.

By the time the centrioles have migrated to the poles of the nucleus, the chromosomes have condensed and the nuclear envelope begins to disintegrate. At this point, the centromere region of the chromosomes attaches to the fibers of the spindle apparatus, causing the chromosomes to become aligned in a plane midway between the two sets of centrioles. The centromeres, holding the two duplicate chromatids of each chromosome together, now separate, and one chromatid moves toward each of the opposing centrioles along the spindle fibers. This separation of the chromatids provides each daughter cell with an identical set of DNA molecules. A special type of nuclear division, known as *meiosis,* occurs in the germ cells (ova and spermatozoa) and will be described in Chap. 21 on reproduction.

Cytokinesis The movement of the chromatids to the opposite poles of the cell marks the end of mitosis, and the cell now begins to undergo cytokinesis. The surface of the cell begins to constrict along a plane perpendicular to the spindle apparatus. This constriction continues until the cell has been pinched in half, forming two separate daughter cells. The condensed chromatids in each of the daughter nuclei now uncoil, forming the extended interphase chromatin threads; the spindle apparatus disintegrates; and the nuclear envelope re-forms.

Most of the cytoplasmic cell organelles are distributed randomly between the two daughter cells during cytokinesis. As the daughter cells grow, new membrane material is synthesized, probably by the endoplasmic reticulum, to form the membranes of the nuclear envelope, the Golgi apparatus, and lysosomes. The mitochondria, on the other hand, appear to be able to duplicate themselves. Small amounts of DNA are found in these organelles, and morphologic evidence suggests that new mitochondria are formed by the growth of a membranous partition across an old mitochondrion, separating it into two new organelles.

MEMBRANES AND MOLECULAR MOVEMENT

Membrane Structure

The plasma membrane and the membranes surrounding the various cell organelles provide barriers which limit the movement of various molecules between intracellular compartments and between the extracellular fluid and the inside of the cell. Regulation of the movements of molcules across these membranes is a major mechanism by which cell activities can be controlled.

Membranes are composed of lipids and proteins in about equal proportions. The lipids provide the passive barrier to molecular movements through the membrane and form the basic structural framework of the membrane. The majority of these lipids are phospholipids. Because they have both a polar and a nonpolar end (see Fig. 1-9), they tend to organize themselves into a bimolecular layer; the nonpolar ends composed of the fatty acid chains associate with each other, and the polar ends associate with water molecules at the two surfaces of the bilayer (Fig. 1-29). This phospholipid bilayer is an effective barrier to the passive transmembrane movements of polar and charged molecules because these molecules are unable to associate with the nonpolar (fatty acid) layer of the membrane.

Some of the membrane proteins function as enzymes, and others selectively regulate the flow of various molecules across the membrane. The amino acid sequence and coiling of the peptide chains of membrane proteins are such that one portion of the protein contains charged and polar amino acids and the remainder contains nonpolar amino acids. This gives the membrane proteins a bipolar structure which is able to associate with the bipolar phospholipid bilayer, giving rise to the lipoprotein membrane structure illustrated in Fig. 1-30. Membranes are quite thin, varying from about 60 to 100 Å in thickness, depending on the amount and types of protein associated with each lipid surface. Each molecule of the membrane is free to move within the plane of the membrane independently of the other molecules; thus, the membrane proteins appear to float within the lipid bilayer. This dynamic membrane model is known as the *fluid mosaic model.* The fluid nature of the membrane gives it a flexibility not possessed by more rigid structures. Moreover, the membrane is an asymmetric structure, the types of proteins of the outer surface being quite different from those on the inner surface. However, some of the membrane proteins do extend all the way through the phospholipid bilayer. These proteins may form pores through the membrane, allowing passage of small charged molecules that would otherwise be unable to pass the phospholipid barrier. They may also be involved in transmitting signals from one surface of the membrane to the other.

FIGURE 1-29 ***Bilayer organization of phospholipid molecules in water.***

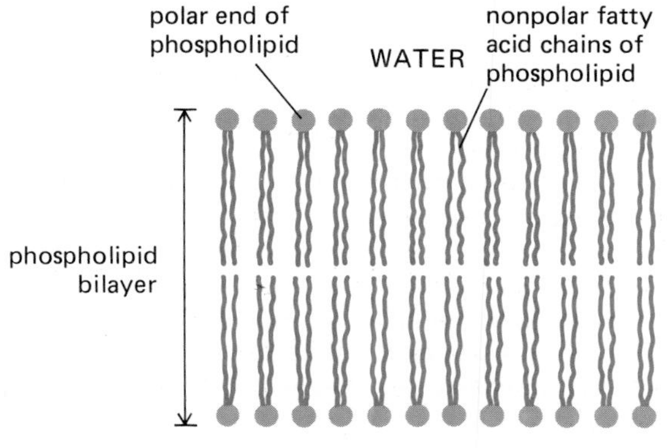

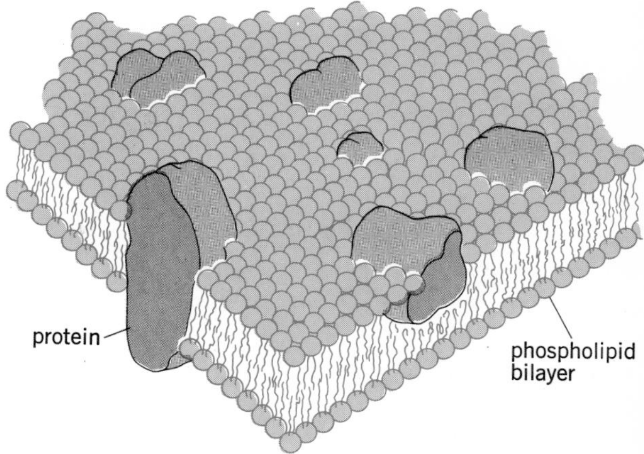

FIGURE 1-30 ***Fluid mosaic model of cell membrane structure.*** **(Redrawn from S. J. Singer and G. L. Nicholson, *Science*, 175:723. Copyright 1972 by the American Association for the Advancement of Science.)**

Diffusion

The movement of molecules within a cell, across cell membranes, or from one area of the body to another is essential for the normal functioning of the body. When molecules must be moved over long distances, the contraction of muscles provides the required forces. This type of movement, known as *bulk flow,* is the mechanism producing blood circulation and the flow of air into and out of the lungs.

In contrast to bulk flow is *diffusion.* The individual molecules in a solution are in a state of continuous random movement, bouncing off each other like rubber balls, each collision altering the direction of molecular movement. This frenzied molecular motion is the physical equivalent of heat; the hotter an object becomes, the faster its molecules move. The random

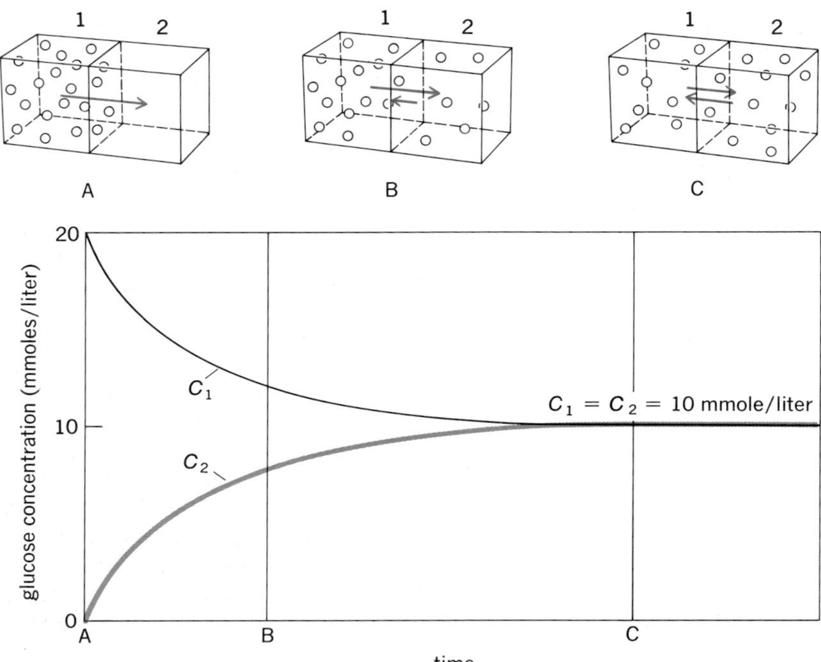

FIGURE 1-31 ***Diffusion of glucose between two compartments of equal volume. (A) Initial conditions: No glucose is present in compartment 2. (B) Some glucose molecules have moved into compartment 2, and some of these are moving at random back into compartment 1. (C) Diffusion equilibrium has been reached, the flux of glucose between the two compartments being equal in the two directions.***

molecular motion which carries individual molecules over short distances from one region to another is known as *diffusion*. Unlike bulk flow, in diffusion there is no external force pushing each individual molecule from one region to another.

Figure 1-31 illustrates the characteristics of the diffusion process. A solution containing glucose is separated from pure water by an imaginary boundary across which glucose can freely move. Initially, as a result of their random motion, some of the glucose molecules move across the boundary into compartment 2. The amount of material crossing a unit of surface area in a unit of time is known as a *flux* whose units are millimoles per square centimeter per second. The magnitude of this initial one-way flux of glucose depends on the glucose concentration in compartment 1; the higher the glucose concentration, the greater is the number of glucose molecules randomly moving in the direction of compartment 2 at any instant. As the concentration of glucose in compartment 2 increases, some of the glucose molecules will randomly move back into compartment 1. The magnitude of this reverse, one-way flux depends on the concentration of glucose in compartment 2. Therefore, the *net flux* F_D (the difference between the two one-way fluxes) of glucose from compartment 1 into compartment 2 depends on the difference in the glucose concentration between the two compartments, $\Delta C = C_1 - C_2$ (often referred to as the *concentration gradient*). Accordingly, the net flux of molecules across a boundary can be written in the form of the following equation:

$$F_D = k_D \, \Delta C$$

where k_D is the *diffusion coefficient,* a number which reflects the molecular weight of the diffusing molecule and the temperature of the solution.

To reiterate, the concentration gradient determines the magnitude of the net flux and its direction. The net flux always proceeds from a region of high concentration to a region of lower concentration. When the concentrations of the two compartments become equal, the system is said to be in *diffusion equilibrium;* there is no concentration gradient ($\Delta C = 0$) and thus no net flux.

Although individual molecules travel at very high speeds, the number of collisions they undergo prevent them from traveling very far in a straight line. Thus, diffusion can lead to the rapid movement of molecules over short distances but is a very slow process for transferring molecules over long distances. For example, the diffusion of glucose from a blood vessel to a point one cell diameter away would take only about 3.5 s to reach 90 percent of the glucose concentration in the blood, whereas it would take about 11 years for this glucose concentration to be reached 10 cm (about 4 in) away from the vessel. Fortunately, all the cells in the body are located within a very short distance of a blood vessel so that nutrients and waste products of cell metabolism can rapidly diffuse between the cells and the blood. Because of the slowness of diffusion over long distances, multicellular organisms had to evolve bulk-flow systems, such as the circulatory system, to rapidly move molecules over long distances throughout the body.

Membrane Permeability

The plasma membrane surrounding the cell separates the extracellular chemical environment from the intracellular fluid. All exchanges of material between the cell and its immediate environment must occur across this structure. The plasma membrane acts as a selective barrier preventing certain molecules from entering or leaving the cell, while allowing others to diffuse freely across its surface. The rates at which molecules diffuse through the plasma membrane are found to be a thousand to a million times slower than the rates at which these same molecules diffuse through water, indicating that the plasma membrane acts as a partial barrier to diffusion. Oxygen, carbon dioxide, water, lactic acid, and urea are examples of molecules which cross plasma membranes by simple diffusion. The net flux, F_p, of a solute crossing a membrane by simple diffusion can be described by the same type of equation that was used to describe the diffusion of a solute through water. The difference in concentration (ΔC) between the inside of the cell (C_i) and the outside (C_o) determines the magnitude of the net flux across the membrane.

$$F_p = k_P \, \Delta C$$

The proportionality constant k_P is known as the *permeability constant*. The numerical value of this constant depends not only on the molecular weight of the penetrating molecule and the temperature of the medium but also upon the thickness and chemical composition of the plasma membrane. The net diffusion of molecules across a cell membrane can be altered by changing either the concentration gradient across the membrane or the permeability properties of the cell membrane, i.e., changing the value of the permeability constant.

As we have seen, the plasma membrane is composed of a bimolecular layer of phospholipid molecules in which proteins are embedded. Most of the molecules which rapidly cross the membrane by diffusion have one chemical property in common: They are composed almost entirely of nonpolar, uncharged chemical groups. These nonpolar molecules can pass across the membrane by dissolving in the nonpolar fatty acid layers. As would be predicted, the permeability of the membrane to various molecules decreases in proportion to the number of polar and charged groups in their molecular structure. Highly charged molecules may be unable to diffuse through the membrane at all and are either excluded from entering the cell or trapped within it. Many of the phosphorylated and ionized intermediates of the various metabolic pathways fall into this category, as well as most proteins.

Although most of the molecules which are able to diffuse rapidly across the plasma membrane are nonpolar molecules, there are a few important exceptions to this general rule. Water, which is a highly polar molecule, diffuses across most membranes, as do the ions Na^+, K^+, and Cl^-. The distinguishing characteristic of this group is their small size, which has led to the hypothesis that the plasma membrane contains a number of small holes through which these small charged ions and polar molecules can pass. Because of their size, larger polar and charged molecules are excluded from these *membrane pores*. These pores may be channels between clusters of membrane protein molecules which extend all the way through the membrane.

Membrane pores are too small to be seen with the electron microscope, but their size has been estimated from the largest charged particle that can diffuse through the membrane. They appear to be about 8 Å in diameter (about three times the size of a single water molecule), and they occupy less than 1 percent of the membrane surface area. The membrane pores of certain cells provide a site where the permeability of the plasma membrane to ions can be regulated. By selectively blocking and unblocking pores, the flow of Na^+ and K^+ ions across the membrane can be altered. As we shall see in later chapters, this is the underlying basis of the electric activity of nerve and muscle cells.

Mediated-transport Systems

Some of the major cell metabolites, including glucose and the amino acids, are not in the class of molecules that can cross cell membranes by simple diffusion, since they are polar or charged and too large to pass through pores. Yet these essential molecules do enter cells but not by simple diffusion. They cross cell membranes by *mediated transport* which involves chemical interactions between the transported molecule and molecules in the cell membrane. The details of these molecular interactions are unknown, but the experimental evidence supports what is known as the *carrier hypothesis,* whereby the transported molecule binds to a specific molecule in the membrane, probably a membrane protein, called a *carrier*. The carrier molecule has a binding site on its surface, much like the active site on an enzyme molecule, that will bind a specific type of solute molecule. There are a number of different types of carrier molecules in the membrane, each specific for a different type of solute.

Once the solute is bound to the carrier on one surface of the membrane, the carrier moves within the membrane, possibly by rotation, so that it carries the bound solute molecule to the other surface of the membrane, where it is released. Thus, the carrier molecule acts like a ferryboat, carrying specific solute molecules across the lipid stream of the membrane.

Figure 1-32 summarizes the three major routes by which molecules may cross cell membranes: (1) by simple diffusion through the lipid regions of the membrane if the molecule is nonpolar; (2) by simple diffusion through membrane pores if the molecule, though polar, is small; and (3) by a carrier-mediated transport system if carrier molecules specific for the solute exist in the membrane.

The carrier model accounts for a number of the properties of mediated-transport systems. Since there are a finite number of carriers of any specific type in the membrane, there exists a maximal rate of solute transport through the membrane which is reached when all the carriers are combined with solute. The saturation of the carriers in a mediated-transport system places an upper limit on the rate of solute movement through the membrane. Increasing the solute concentration on one side of the membrane will increase the flux through the membrane until the carriers become saturated. A further increase in the concentration will produce no further increase in the flux (Fig. 1-33). Contrast this to simple diffusion in which the flux will continue to increase in proportion to the concentration. One way to alter the maximal rate of transport is to add or remove carriers from the membrane, and in fact some hormones influence cell activity through the regulation of the synthesis of carrier molecules. Two categories of carrier-mediated transport can be distinguished: *facilitated diffusion* and *active transport.*

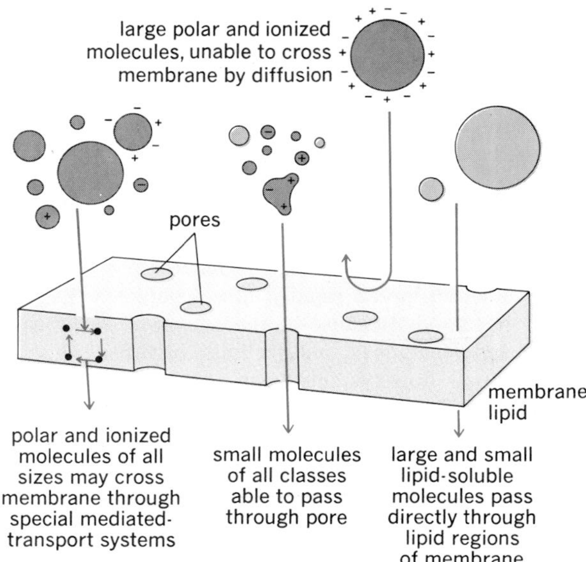
large polar and ionized molecules, unable to cross membrane by diffusion
pores
membrane lipid
polar and ionized molecules of all sizes may cross membrane through special mediated-transport systems
small molecules of all classes able to pass through pore
large and small lipid-soluble molecules pass directly through lipid regions of membrane

FIGURE 1-32 ***Summary of the three pathways by which molecules can cross cell membranes: (1) diffusion through the lipid matrix of the membrane, (2) diffusion through a pore, and (3) carrier-mediated transport.***

FIGURE 1-33 ***Flux of molecules diffusing across a membrane increases in proportion to the extracellular concentration, whereas the flux of molecules entering by a mediated-transport system reaches a maximal value which no longer increases with increasing concentration. This maximal transport flux corresponds to the saturation of all the available carriers in the membrane by the transported solute.***

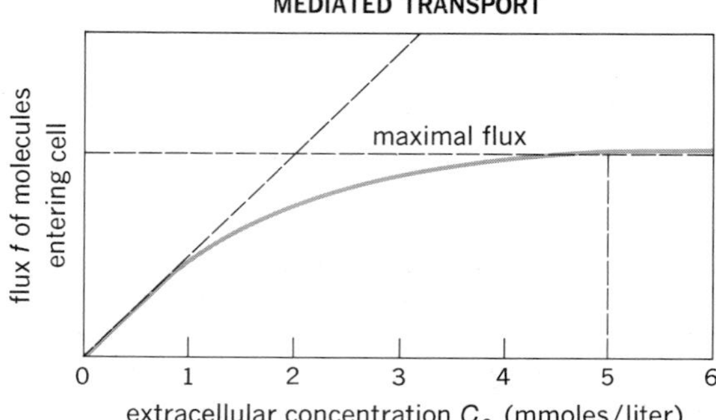
MEDIATED TRANSPORT
maximal flux
flux f of molecules entering cell
0
1
2
3
4
5
6
extracellular concentration C_o (mmoles/liter)

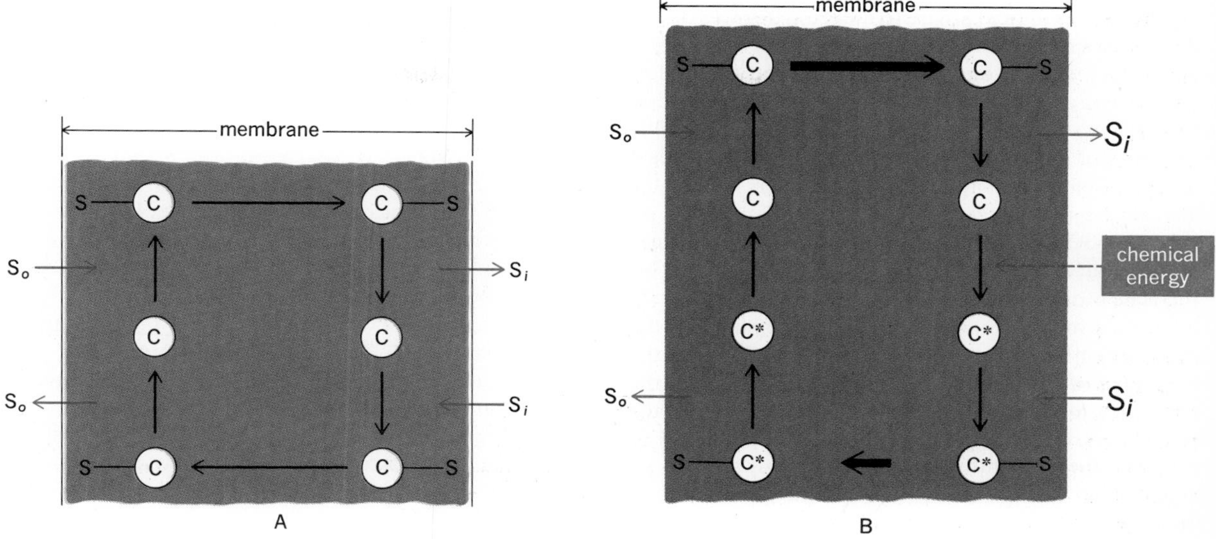

Figure 1-34 (A) ***Carrier model of facilitated diffusion. Identical reactions between the membrane carrier molecule (C) and the transported solute (S) occur at both surfaces of the membrane.*** (B) ***Carrier model of active transport. The reaction between the membrane carrier molecule and the transported solute differs at the two surfaces of the membrane. At one surface the carrier molecule is modified, by an energy-requiring reaction, to a form that will less readily bind solute (C*). At the opposite surface the modified carrier is returned to its high-affinity form (C).***

Facilitated Diffusion In a facilitated-diffusion system, the carrier undergoes identical binding reactions with the transported solute at both membrane surfaces (Fig. 1-34A). Therefore, a net flux of solute across the membrane occurs only when a concentration gradient exists for that solute. If the solute concentrations on the two sides of the membrane are equal, the carriers will move equal amounts of solute into and out of the cell. In other words, facilitated diffusion is similar to simple diffusion in two respects: It achieves net transport only when a concentration gradient already exists, and it continues until equal concentrations of the solute have been reached on the two sides of a membrane. Accordingly, the real function of a facilitated-diffusion system is to facilitate the movement through the membrane of molecules which are unable to penetrate by simple diffusion because they are polar and too large to pass through the membrane pores. For example, glucose crosses most cell membranes by facilitated diffusion. Without such a system, glucose would be unable to enter or leave cells.

Active Transport Whereas a facilitated-diffusion system proceeds only in the direction which decreases the concentration gradient across a membrane, an active-transport system leads to an increase in the concentration gradient because it is able to move solute molecules from a lower to a higher concentration. This is analogous to pumping water uphill, whereas simple and facilitated diffusion are analogous to the flow of water downhill. In order for net movement of a molecule to occur against a concentration gradient, energy must be expended, just as energy is required to pump water uphill. Accordingly, these transport systems are often referred to as "pumps." There are a number of different active-transport systems in cell membranes, some of which pump molecules into the cell whereas others pump molecules out.

Movement against a concentration gradient is possible because of the asymmetry of the reactions which occur between the carrier and solute on the inner and outer surfaces of the membrane (Fig. 1-34B). The structure of the carrier molecule on the inner and outer membrane surface is different, allowing the carrier to bind solute to different degrees on the two sides, and thus the carrier is less likely to carry a solute molecule in one direction than the other. Energy, usually in the form of ATP, is coupled to the active-transport system by a reaction which modifies the structure of the carrier molecule at one surface. When the modified carrier moves to the other surface of the membrane, it undergoes a second reaction which returns it to its initial state. As a result of these changes in the ability of the carrier to bind solute on the two surfaces of the membrane, more solute molecules will be moved in one direction by the carrier than in the other, even when the concentrations of solute on the two sides of the membrane are equal. Therefore, the concentration of solute on one side of the membrane will rise above the concentration on the other side. For example, the concentrations of amino acids are 2 to 20 times higher inside cells than in the blood, as a result of their active transport across the plasma membrane.

Since we have already seen that Na^+ and K^+ can diffuse through the small pores in the membrane, one might expect these ions to reach diffusion equilibrium across the membrane. However, the concentration of K^+ in the extracellular fluid is about 5 mmol/l, and the concentration of K^+ in cells is about 150 mmol/l, 30 times higher. Given this high concentration gradient and the presence of membrane pores, there is a net diffusion of potassium out of cells. However, simultaneous with this leak of potassium is the active transport of potassium into the cell, so that the rate of net diffusion out is equal to the rate of active transport into the cell, and the large concentration gradient is maintained. If energy, in the form of ATP, were not being supplied to the pump, it would stop and the large intracellular potassium concentration would decrease as potassium diffused out of the cell.

The situation is reversed for sodium. The concentration of sodium ions in the blood is about 145 mmol/l, whereas the intracellular concentration is quite low, about 15 mmol/l. There is a continuous net diffusion of sodium into the cell through the pores, but an equal active transport out of the cell by the sodium pump. Here again, energy must be continuously expended by the cell to operate the sodium pump and prevent the concentration of sodium within the cell from rising.

This unequal distribution of Na^+ and K^+ ions is found across all plasma membranes in the body, and a considerable portion of the cells' metabolic energy is used to maintain these ion pumps. As we shall see in later chapters, these ion concentration gradients lead to the electric activity associated with nerve and muscle cells. These ion pumps also help to maintain a stable cell volume, which leads us to a discussion of the phenomenon of osmosis.

Osmosis

The small water molecule is able to diffuse rapidly through the pores in the cell membrane. If a water concentration gradient occurs across a plasma membrane, a net diffusion of water will occur, causing the cell to either swell or shrink. This net diffusion of water, known as *osmosis*, is a direct result of the basic principles of diffusion. How can the concentration of water in a solution be altered? By the addition of solute. If a solute molecule, such as glucose, is dissolved in water, the concentration of water molecules in the resulting solution will be less than that of pure water. Each molecule of glucose that is added to the solution adds an element of volume that cannot be occupied by a water molecule. The more solute molecules that are added, the less water there will be in a given volume of solution and thus the water concentration will be lowered.

The degree to which the concentration of water is decreased by the addition of solute depends upon the *number* of solute molecules that are added to the solution and not upon the chemical nature of the individual solute molecules. The total solute concentration, irrespective of the type of solutes that may be present, is known as the *osmolarity* of the solution. The *higher* the osmolarity of a solution, the *lower* will be its water concentration. Normally, water is in diffusion equilibrium across the cell membrane; thus the concentration of water inside and outside of cells is the same. Note that this implies that the total solute concentration (osmolarity) of the intracellular and extracellular solutions is also the same. If the osmolarity of the extracellular solution were to be increased or decreased by adding or removing solute, this would alter the water concentration in the extracellular fluid and produce a water concentration gradient across the cell membrane, leading to the net diffusion of water into or out of the cell, or osmosis.

Note that in order to produce a water concentration gradient across a membrane there must also be a difference in solute concentration across the same membrane. If the solutes are able to diffuse across the membrane down their concentration gradient, the solute and water concentration gradients both decrease as the solute concentration on the two sides of the membrane approaches equality. If the solutes are unable to cross the membrane, then water will be the only molecule able to diffuse down its concentration gradient. Therefore, it is the concentration of *nonpenetrating solutes* on the two sides of a membrane that determines the net concentration gradient for water diffusion during osmosis.

The total solute concentration (osmolarity) of the extracellular fluid in the body is about 300 mmol/l. About 80 percent of this concentration is due to sodium and chloride ions. Thus, the extracellular fluid is essentially a saltwater solution. Inside the cell, the major solute is K^+, which accounts for about half of the intracellular osmolarity. If the membrane Na^+ and K^+ pumps were stopped because of a lack of ATP, more solute would enter the cell by diffusion than would leave. As a result, the intracellular osmolarity would increase, water would diffuse into the cell, and the cell would swell. Thus, the ion pumps are essential for maintaining a stable cell volume.

If a cell is placed in a solution of sodium chloride having a total osmolarity of 300 mmol/l (150 mmol of Na^+ plus 150 mmol of Cl^-), equivalent to the osmolarity of the extracellular fluids in the body, the cell will neither swell nor shrink. Such a solution is said to be *isotonic*. Any solution containing the same concentration of nonpenetrating solutes as the extracellular fluid is an isotonic solution. Isotonic saline (NaCl) is often used as the basic solution in which small amounts of drugs or other substances which are to be injected into the body are dissolved. Since this solution is isotonic, it will not cause osmotic swelling or shrinking of the cells at the site of injection.

A solution which contains a lower concentration of nonpenetrating solutes than the extracellular fluid is said to be *hypotonic*. The water concentration in a hypotonic solution is greater than the water concentration within cells and thus water will diffuse into the cell, causing the cell to swell. Injection of a drug dissolved in pure water would lead to osmotic swelling and possible rupture of the cells at the site of injection. Solutions which contain a higher concentration of nonpenetrating solutes than the extracellular fluid are known as *hypertonic* solutions and will cause cells to shrink as water diffuses out of the cell into the hypertonic solution. As we shall see, the body has evolved mechanisms for maintaining the osmolarity of the extracellular fluid nearly constant by regulating the rates at which the kidneys excrete salt and water into the urine. ▬

CHAPTER 2
Organization of the Body

The human organism begins as a single cell, the fertilized ovum, which gives rise to the entire body. The body is composed of various types of specialized cells. Cells which have a similar origin and structure and subserve the same general function are frequently found grouped together to form *tissues*. Sometimes a single cell or tissue may function fairly independently of all others, but more commonly a number of tissues are intimately associated to form larger units called *organs:* heart, liver, kidney, pancreas, etc. Finally, the last order in classification is that of the *organ system,* a collection of organs which subserve an overall function. For example, the kidneys, the bladder, and the tubes leading from the kidneys to the bladder and from the bladder to the exterior constitute the urinary system.

In essence, then, the human body can be viewed as a complex society of cells of many different types which are structurally and functionally combined and interrelated in a variety of ways to carry on the functions essential to the survival of the organism as a whole. Yet the fact remains that individual cells still constitute the basic units of this society and that almost all these cells individually exhibit the fundamental activities common to all forms of life.

There is a definite paradox in this analysis. If each individual cell performs the fundamental activities required for its own survival, what contributions do the different organ systems make? How can we refer to a system's functions as being "essential to the survival of the organism as a whole" when each individual cell of the organism seems to be capable of performing its own fundamental activities? The resolution of this paradox is found in the isolation of most of the cells of a multicellular organism from the environment surrounding the body (*external environment*). An ameba and a human liver cell both obtain most of their required energy by the breakdown of certain organic nutrients; the chemical reactions involved in this intracellular process are remarkably similar in the two types of cells and involve the utilization of oxygen and the production of carbon dioxide. The ameba picks up required oxygen directly from its environment and eliminates the carbon dioxide into it. But how can the liver cell obtain its oxygen and eliminate the carbon dioxide when, unlike the ameba, it is not in direct contact with the external environment? Supplying oxygen to the liver is the function both of the respiratory system (comprising the lungs and the airways leading to them), which takes up oxygen from the environment, and of the circulatory system, which distributes oxygen to all parts of the body. Conversely, the circulatory system carries the carbon dioxide generated by the liver cells and all the other cells of the body to the lungs, which eliminate it to the exterior.

Similarly, the digestive and circulatory systems, working together, make nutrients from the external environment available to all the body's cells. Wastes, other than carbon dioxide, are carried by the circulatory system from the cells which produced them to the kidneys, which excrete them from the body. Thus, the overall effect of the activities of organ systems is to create *within* the body the environment required for all cells to function.

Clearly, the society of cells which constitutes the human body bears many striking similarities to a society of people (although the analogy must not be pushed too far). Each person in a complex society must perform an individual set of fundamental activities (eating, excreting, sleeping, etc.), which is virtually the same for all persons. In addition, because the complex organization of a society makes it virtually impossible for individuals within the society to raise their own food, arrange for the disposal of their wastes, and so on, each individual participates in the performance of one of these supply-and-disposal operations required for the survival of all. A specialized activity, therefore, becomes an *additional* part of each person's daily routine, but it never allows him or her to cease or to reduce the performance of the fundamental activities required for survival.

THE INTERNAL ENVIRONMENT

A cell is a very fragile chemical machine. Large fluctuations in the physical and chemical properties of the fluid medium immediately surrounding it can disrupt the regulated flow of metabolism maintaining the life of the cell. Seawater, whose temperature and chemical composition do not change rapidly (because of its large volume), provided the stable environment for the first living cells which appeared on earth about 3 billion years ago. These single, free-living cells obtained nutrients from and excreted wastes directly into the external environment, seawater. Thus, life at this early stage depended upon chemical exchanges between two fluid environments separated by a plasma membrane.

A loose association of independent cells into small clusters was the first step in the evolution of multicellular organisms. Only the cells at the surfaces of such clusters were in immediate contact with the external environment, seawater. Within the cluster, cells were surrounded by other cells and by *extracellular fluid* which had been trapped between the cells. Thus, this extracellular fluid provided the immediate environment for the interior cells of the cluster.

Even though organisms have grown in size and complexity from these simple clusters, they all have within them a thin layer of extracellular fluid which bathes each of the cells in the body. In other words, the environment in which the trillions of cells in the body live is not the *external environment* surrounding the total organism but is the local fluid environment that immediately surrounds each individual cell. It is from this fluid, known as the *internal environment* (Fig. 2-1), that a cell receives nutrients and into which it excretes wastes. A multicellular organism can survive only as long as it is able to maintain the composition of its internal environment in a state compatible with the survival of its individual cells. The French physiologist Claude Bernard first clearly described, in 1857, the central importance of the extracellular fluid of the body: "It is the fixity of the internal environment which is the condition of free and independent life. . . . All the vital mechanisms, however varied they may be, have only one object, that of preserving constant the conditions of life in the internal environment." This concept of an internal environment and

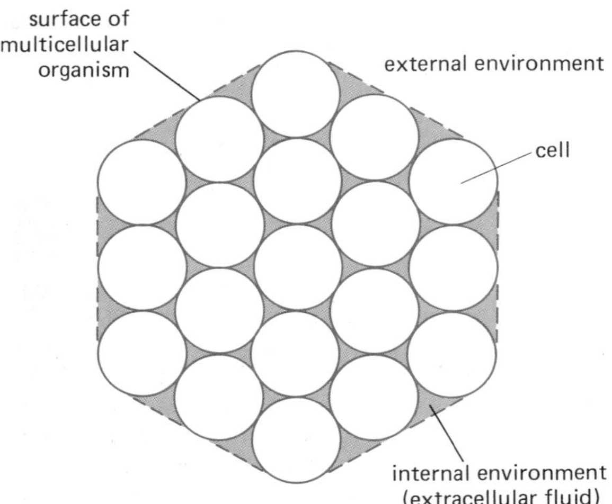

FIGURE 2-1 ***Extracellular fluid is the internal environment of the body.***

the necessity for maintaining its composition relatively constant is the single most important unifying concept to be kept in mind while attempting to unravel and understand the function and structure of the human body.

Body-fluid Compartments

Multicellular organisms obtain nutrients and oxygen from the external environment and deliver them to their cells by way of the extracellular fluids. Likewise, waste products of metabolism produced by cells must travel the reverse route through the extracellular fluids to the surface of the organism where they are released into the external environment. The extracellular fluids of the internal environment thus provide the medium by which substances are exchanged between cells and the external environment.

If the distance between a cell and the surface of an organism is small—a few cell diameters—the diffusion of nutrients and wastes will be sufficiently rapid to meet the requirements of the cell's metabolism. However, as seen in Chap. 1, the rate of net diffusion becomes progressively slower as the distance over which the substance must diffuse increases. Beyond a distance of a few cell diameters, diffusion is not sufficiently rapid to meet the metabolic requirements of cells. If multicellular organisms were to evolve to a size larger than a microscopic cluster of cells, some mechanism other than simple diffusion would have to provide the means of rapidly transporting molecules over the long distances between the cells and the body's surface. This problem was solved in the animal kingdom by the evolution of a *circulatory system.*

In human beings, the circulatory system consists of a fluid, *blood,* which is pumped by a muscular pump, the *heart,* through a branching network of progressively smaller- and smaller-diameter tubes, *blood vessels,* the smallest of which, the *capillaries,* are located within a few cell diameters of most of the cells in the body. From the capillaries, blood returns to the heart through a coalescing network of larger and larger blood vessels. The rapid bulk flow of blood through these blood vessels results from pressures created by the contraction of the heart. The blood which circulates through this closed network consists of two phases: (1) a cellular component made up of red and white blood cells, and (2) a fluid component, *plasma.*

Thus, two separate compartments contain extracellular fluid (internal environment) (Fig. 2-2): (1) the *blood plasma,* which is the fluid surrounding the cells in the circulatory system, and (2) the *interstitial fluid,* consisting of the fluid surrounding the cells outside the blood vessels. The blood plasma, which rapidly circulates throughout the body by bulk flow, is the dynamic component of the internal environment. Diffusion across the capillary walls and through the interstitial fluid links the plasma and the cells surrounding

FIGURE 2-2 ***Plasma and the interstitial fluid constitute the internal environment of the body.***

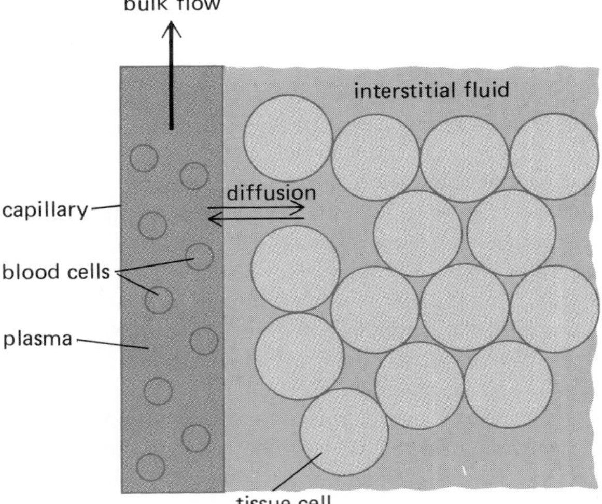

the capillary. This diffusion is sufficiently rapid to meet the metabolic requirements of the cells because the cells are located within a short distance of a capillary and molecules can be delivered to or removed from a capillary rapidly by the bulk flow of blood. Moreover, as a result of the rapid exchange between the plasma and the interstitial fluid, the composition of the interstitial fluid is almost identical to that of plasma.[1]

To generalize one step further, the human body can be viewed as containing three fluid compartments: (1) blood plasma, (2) interstitial fluid, and (3) intracellular fluid, the fluid in the cells of the body (Fig. 2-3). The major molecular component of all three compartments is water, which accounts for about 60 percent of the normal body weight, or 42 l, in an average man. Two-thirds of the total body water (28 l) is located inside the cells: the intracellular fluid. The remaining one-third of the total body water (14 l) comprises the two extracellular fluids: 80 percent as interstitial fluid (11 l) and 20 percent as plasma (3 l).

The above figures for the volumes of fluids in the various compartments of the body are those of a 70-kg man. Obviously, body measurements vary between different individuals. Some of these values may vary directly with body weight and can be expressed as a percentage of the total body weight, whereas others depend on age, sex, and state of health as well as body weight. The typical values for a healthy, 70-kg (154-lb), 21-year-old male have been used for years as representative of an average individual. This standard was chosen because a majority of the data collected over the years on normal healthy individuals has come from measurements made on students attending medical school. The 21-year-old, 70-kg male represents the average body size, sex, and age of the medical students from which the data were obtained. Measurements on females of the same age and corrected for differences in total body weight give values of a similar magnitude, although slight sex-specific differences do exist. Thus, a female has a slightly lower total-body-water content than males of the same weight because her body is composed of a higher percentage of fatty tissue containing less water than other tissues. Unless otherwise stated, quantitative values given in this text refer to a healthy 70-kg, 21-year-old male.

UNITS OF STRUCTURAL ORGANIZATION

The cells of the body are combined to form a hierarchy of structural organization. Individual *specialized cells*

[1] There is, however, much more protein in plasma than in interstitial fluid.

FIGURE 2-3 ***Fluid compartments of the body. Volumes are for the "average" 70 kg man.***

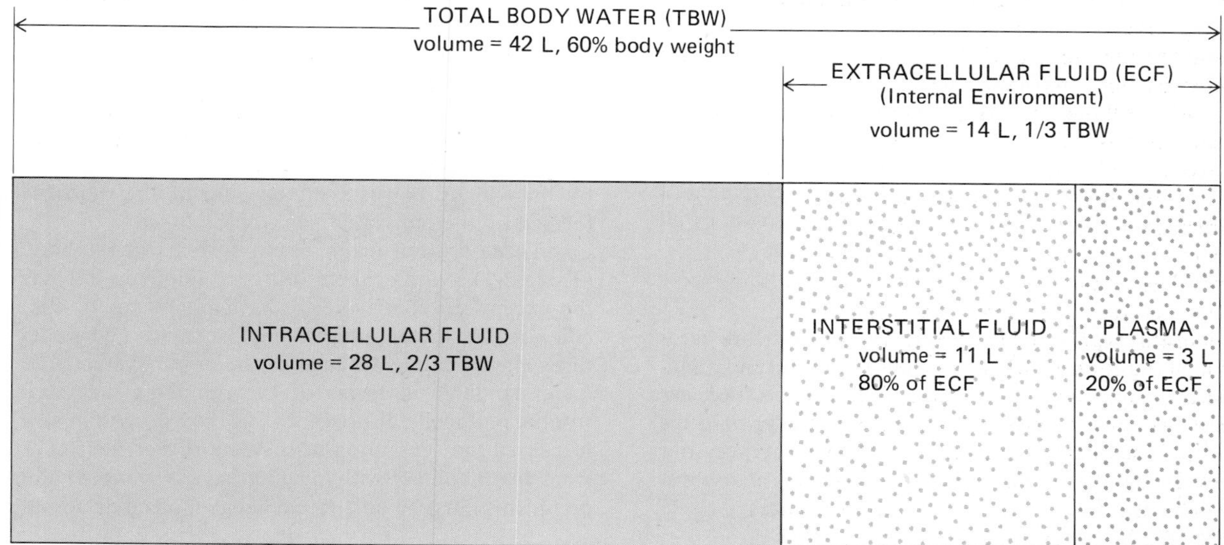

are arranged into *tissues,* which are combined to form *organs,* which are linked together to form *organ systems.*

Specialized Cell Types

Each cell in the body functions at two levels: (1) It performs those basic processes essential for its own survival—energy metabolism, synthesis of proteins and various structural end products, and exchange of nutrients, salt, and water across its plasma membrane; and (2) it performs a specialized function which, in cooperation with other specialized cells, serves to maintain the composition of the internal environment in which all the cells must live. Rather than being a jack-of-all-trades, as was the case for free-living, single-celled organisms, the cells of a multicellular organism became specialists in performing one or more particular types of jobs, leaving other jobs to other types of specialized cells.

Four categories of specialized cells have evolved: (1) *muscle cells,* specialized for the production of forces which produce movement; (2) *nerve cells,* specialized for initiation and conduction of electric signals over long distances; (3) *epithelial cells,* specialized for the selective secretion and absorption of organic molecules and ions; and (4) *connective-tissue cells,* specialized for the formation and secretion of various types of extracellular connective and supporting elements. In each of these functional categories is a variety of cell types which perform variations of the general type of specialized function. For example, there are three different types of muscle cells—skeletal, cardiac, and smooth muscle cells—all of which generate forces and produce movements but which differ from each other in shape, mechanisms controlling their contractile activity, and location in the various organs of the body.

Note that each of the functions of the specialized cell types is the result of an augmented development of one or more of the essential functions carried out by all cells.

Tissues

Most of the specialized cells in the body are associated with other cells of a similar kind, forming multicellular aggregates known as *tissues.* Just as there are four general categories of specialized cell types in the body, there are, corresponding to these cell types, four general categories of tissues: *muscle tissue, nerve tissue, connective tissue,* and *epithelial tissue.*

Histology is the branch of microscopic anatomy that deals with the structure of these tissues and their patterns of arrangement in the various organs of the body, whereas *cytology* is concerned with the structure of individual cells. Histologic structure and functional characteristics of these tissues will be developed in Chaps. 3 to 5.

It should be noted that the term "tissue" is frequently used in several different ways. It is formally defined as described above, i.e., as an aggregate of a single type of specialized cell; however, it is also commonly used to denote the general cellular fabric of any given organ or structure, for example, kidney tissue, lung tissue, etc., which are in fact usually composed of all four specialized cell types. In the latter case, the general term *parenchyma* is used to refer to the functional components of a tissue; *stroma* refers to the supportive components.

Organs

The organs of the body are composed of the four kinds of tissues arranged in various proportions and patterns: sheets, tubes, layers, bundles, strips, etc. For example, the kidney consists largely of (1) a series of small tubules, each composed of a single layer of epithelial cells; (2) blood vessels, whose walls consist of an epithelial lining and varying quantities of smooth muscle and connective tissue; (3) nerve fibers with endings near the muscle and epithelial cells; and (4) a loose network of connective-tissue elements which are interspersed throughout the kidney and which form an enclosing connective-tissue capsule. The structural components of many organs are organized into small, similar subunits, each performing the function of the organ. For example, the kidneys' 2 million functional units, the *nephrons,* are the tubules with their closely associated blood vessels. The total production of urine by the kidney consists of the sum of the amounts formed by the individual nephrons.

There are ten major organ systems in the body (Table 2-1). Nine of them (the reproductive system is the exception) contribute to the maintenance of specific aspects of the internal environment. The structures and functions of each of the organ systems are understandable in terms of the problems that confronted multicellular organisms as their growth in size increased the separation of the majority of their cells from direct contact with the external environment. All organisms must be able to exchange molecules across

TABLE 2-1
ORGAN SYSTEMS OF THE BODY

System	*Major organs or tissues*	*Primary function(s)*
Circulatory	Heart, blood vessels, blood	Rapid bulk flow of blood throughout the body's tissues.
Respiratory	Nose, throat, larynx, trachea, bronchi, bronchioles, lungs	Exchange of carbon dioxide and oxygen and regulation of hydrogen-ion concentration.
Digestive	Mouth, pharynx, esophagus, stomach, intestines, salivary glands, pancreas, liver, gallbladder	Digestion, absorption, and processing of nutrients.
Urinary	Kidneys, bladder, ureters and urethra	Regulation of plasma composition through excretion of organic wastes, salts, and water.
Musculoskeletal	Cartilage, bone, ligaments, tendons, joints, skeletal muscle	Support, protect, and move the body.
Immune	White blood cells, lymph vessels and nodes, spleen, thymus, bone marrow, reticuloendothelial cells	Defense against foreign invaders. Also functions in the return of extracellular fluid to blood. In addition to forming some white blood cells, the bone marrow forms red blood cells.
Nervous	Brain, spinal cord, peripheral nerves and ganglia, special sense organs	Control and integration of the body's many activities. Responsible for states of consciousness.
Endocrine	All glands secreting hormones into the blood: Pituitary, thyroid, parathyroid, adrenal, pancreas, testes, ovaries, intestinal glands, and kidneys	Control and integration of many activities in the body.
Reproductive	Male: Testes, penis, and associated ducts and glands Female: Ovaries, oviducts, uterus, vagina, and mammary glands	Production of egg and sperm cells, transfer of male sperm to female, and provision of a nutritive environment for the developing embryo.
Integumentary	Skin	Protects against injury and dehydration, defense against foreign invaders, and temperature regulation.

their body surface between the internal and external environments. As we have seen, the evolution of a rapid-transit circulatory system allowed an increase in body size by rapidly distributing molecules to the cells once the molecules had entered the body across the body's surface.

However, the increase in body size presented another problem related to the rate at which molecules can be moved across the body's surface. As an organism increases in size, the volume of its body increases more rapidly than the surface area; i.e., the ratio of surface area to volume decreases. Eventually a size would be reached at which the rate of exchange of molecules across a unit area of the body's surface would not be sufficient to supply the metabolic requirements of the large volume of underlying tissue, in spite of a rapid circulatory system. The solution to this problem was to increase the surface area in contact with the external environment without significantly increasing the volume of the organism. This was achieved either by an extensive folding of the body's surface or by forming highly coiled small-diameter tubes which, although "inside" the body, remain in direct contact with the external environment. Both of these solutions provide a large surface-to-volume ratio.

Such specialized regions of the body's surface evolved over millions of years into those organ systems involved with the exchange of molecules between the external and internal environments: the *respiratory system,* which exchanges oxygen and carbon dioxide between the air and blood; the *digestive system,* which is the site of entry for nutrients, salts, and water from the external environment; and the *urinary system,* which excretes organic waste products, salts, and water into the external environment. Both the respiratory and digestive systems achieve a large surface area by an extensive folding of the layers of cells lining the organs of these systems, whereas the large surface area of the urinary system results from numerous small tubules which make up the kidneys. As these organ systems have evolved, their surface areas have increased, become internalized, and become connected to the exterior of the body through tubes. Each provides an interface between the external environment and the blood plasma, which, in turn, is linked to the cells of the body through exchanges across the capillaries (or red-blood-cell membranes) (Fig. 2-4).

The *musculoskeletal system* supports and moves the body; in addition, it provides the delicate tissues of the brain and spinal cord, lungs, heart, and great vessels with bony protection. Such a system makes it possible for an organism to defend itself against large predatory organisms, but there are also microscopic invaders, the bacteria and viruses, which can injure the body from within, just as a larger organism will do so from without. Defense against the microscopic invaders is provided by the *immune system*.

As the complexity of organisms increased, it became necessary to coordinate and regulate the multiple activities of the body so that a stable internal environment could be maintained. Leon Fredricq, in 1885, amplifying the concepts of Claude Bernard, noted: "The living being is an agency of such a sort that each disturbing influence induces by itself the calling forth of compensatory activity to neutralize or repair the disturbance. The higher in the scale of living beings, the more numerous, the more perfect, and the more complicated do these regulatory agencies become. They tend to free the organism completely from the unfavorable influence and changes occurring in the environment." Two organ systems play the major role in this regulation and coordination: the *nervous system*

FIGURE 2-4 ***Exchanges of matter occur between the external environment and the circulatory system via the digestive system, respiratory system, and urinary system.***

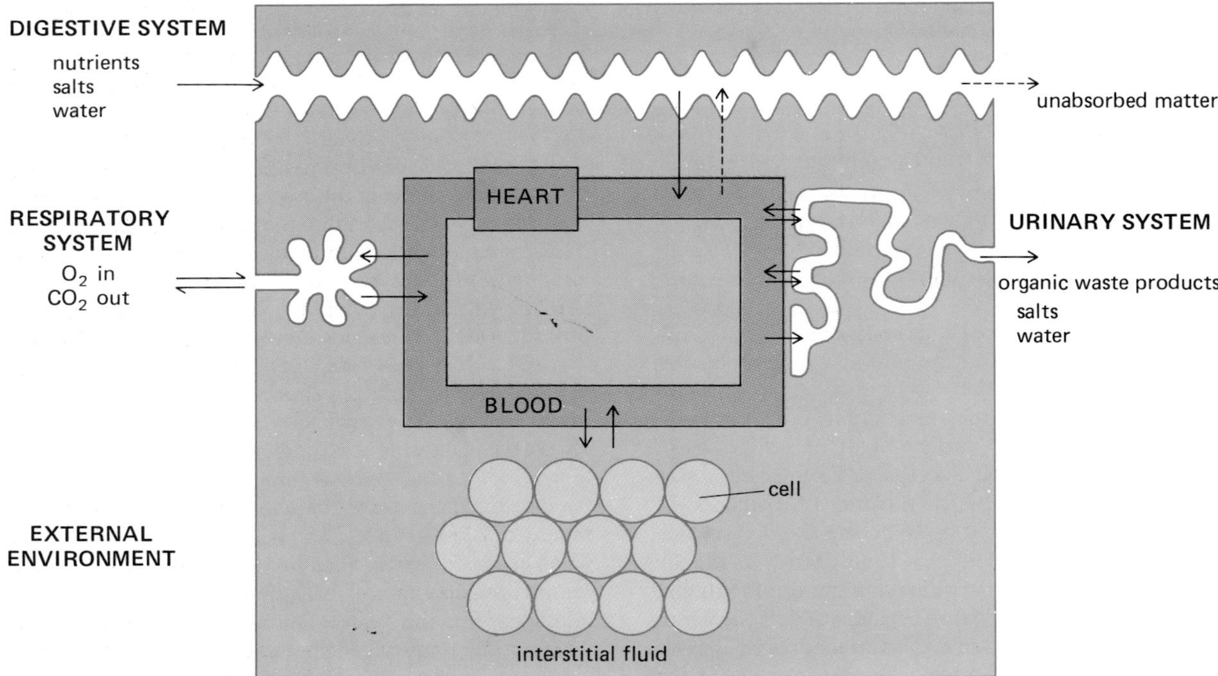

and the *endocrine system*. The nervous system makes use of electric signals conducted along the plasma membranes of nerve cells; the endocrine system makes use of chemicals, *hormones*, secreted into the blood. These two organ systems provide the major routes for linking the activities of cells in one part of the body with those in another. In addition, the nervous system transmits to the interior of the body information received at its surface by specialized receptor cells of the eyes, ears, nose, and skin and is responsible for the various states and contents of consciousness.

The failure of any of these organ systems can be fatal because adequate maintenance of the composition of the internal environment depends on the proper functioning of each of these organ systems. In the case of a few organs, such as those of the immune and endocrine systems, a major malfunction may not directly kill the organism but will severely limit the types of external environment in which it can survive. For example, in the absence of an immune system, an organism can survive only in an environment free of microscopic organisms.

The one remaining organ, the *reproductive system*, did not evolve to regulate the composition of the internal environment. It can be removed from the body without affecting the ability to survive. It is, however, obviously essential for the survival of the species. Multicellular organisms must, in essence, repeat the whole process of evolution by developing from a single cell, the fertilized egg cell. The reproductive organs produce the primary sex cells, the sperm and ova in the male and female, respectively, and provide a special environment in the female for the stages of embryonic and fetal development.

This completes our general survey of the patterns and levels of cellular organization seen in the body. Figure 2-5 summarizes this material in terms of the urinary system.

ORIENTATION TO THE BODY

Terminology

In order to describe the location of a particular structure, a number of terms are used which specify directions. In the *standard anatomic position* (Fig. 2-6), the body is erect with the feet together, the arms hanging at the sides, the palms of the hands facing forward, and the thumbs pointing away from the body. All directions, including the various movements of the limbs, are described relative to this standard anatomic position. Because of the erect posture of human beings, the terms *posterior*, toward the back, and *anterior*, toward the front, are used, as well as the terms *dorsal* and *ventral*, but the latter have slightly different meanings when applied to human beings as compared with an animal that stands on four legs, in which case the ventral surface is closest to the ground and the dorsal surface is the upper surface of the animal. In human beings, the *cranial* or *superior* direction is toward the head, and the *caudal* or *inferior* direction is toward the feet. In addition to the four primary directions—anterior (front), posterior (back), cranial (top), and caudal (bottom)—one other set of directions is necessary to locate a position across the width of the body. The *median sagittal plane* divides the body symmetrically into right and left halves; directions away from it are *lateral*, and directions toward it are *medial*. For example, the eyes are lateral to the nose, and the nose is medial to each eye.

Two other primary planes of the body are the *coronal plane* and the *transverse plane* (Fig. 2-6). Note that a transverse plane can be placed through the body at any point along the cranial-caudal axis, passing through the head, the chest, or the legs, depending upon its location. Numerous sagittal and coronal planes can also be passed through various segments of the body in their respective planar orientations.

The terms used thus far refer to directions relative to the standard anatomic position. The two terms *proximal* and *distal* have a more generalized meaning, referring to directions toward (proximal) or away from (distal) the origin of a particular structure. For example, the hands and feet are located at the distal ends of the limbs; the elbows and knees are located proximal to the hands and feet, respectively, but still remain distal to the shoulder and hip.

The Body Cavities

Most of the body's organs are located within two large cavities, the *dorsal* and *ventral cavities* (Fig. 2-7), each of which has smaller subdivisions. The dorsal cavity consists of the connected *cranial cavity* in the skull and smaller *vertebral canal* which runs through the vertebral bones. The bones surrounding these cavities protect the delicate tissues of the central nervous

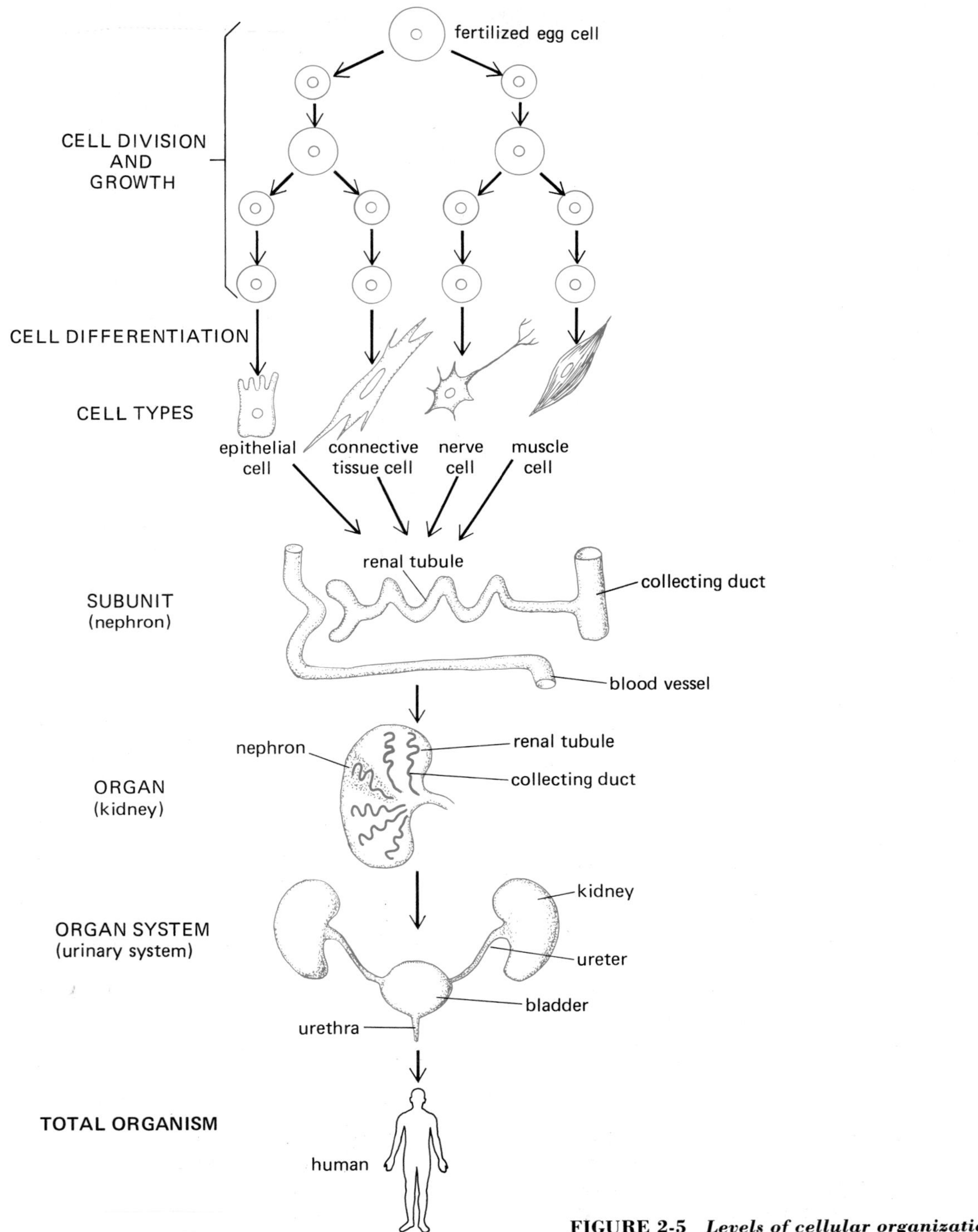

FIGURE 2-5 ***Levels of cellular organization.***

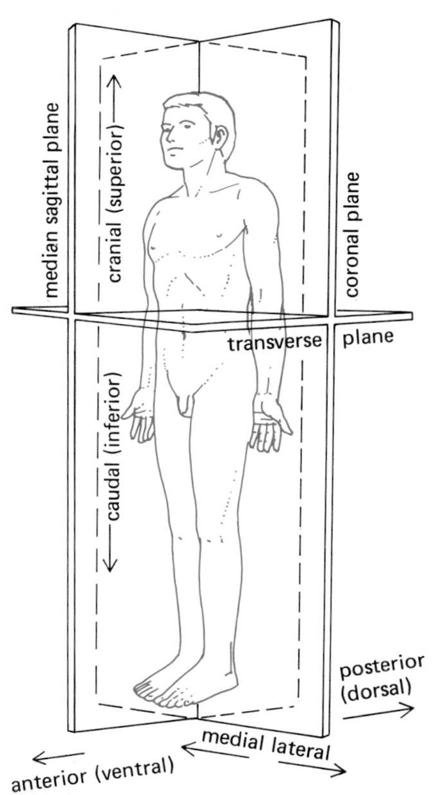

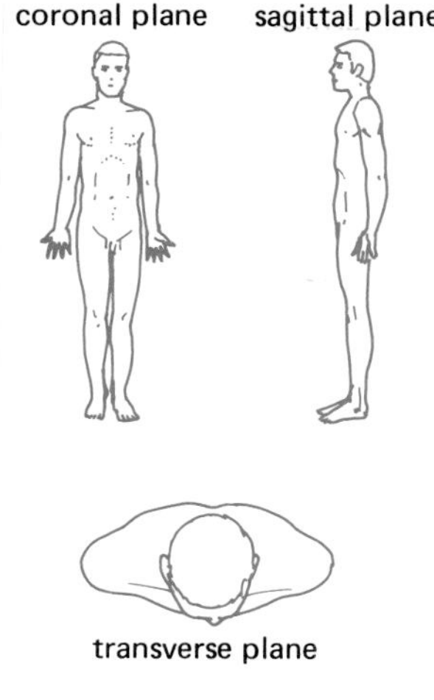

FIGURE 2-6 ***Anatomic directions and planes.***

system, the brain located in the cranial cavity, and the spinal cord extending from the base of the brain through the vertebral canal. The dorsal cavity, being surrounded on all sides by solid bone, has a fixed volume; any abnormal growth of tissue or accumulation of fluid within it exerts pressure on the brain or spinal cord which can affect their functioning.

In contrast to the dorsal cavity, the ventral cavity can vary in capacity and shape, depending on its contents and the muscular activity of the surrounding walls. The ventral cavity is divided into two distinct chambers, the *thoracic cavity* and the *abdominopelvic cavity*, by a sheet of muscle, the diaphragm.

The thoracic or chest cavity is surrounded by a protective rib cage, the diaphragm forming its floor. The major organs located in the thoracic cavity are the heart and lungs, each of which is enclosed in its own separate chamber surrounded by a membranous lining.

FIGURE 2-7 ***Body cavities.***

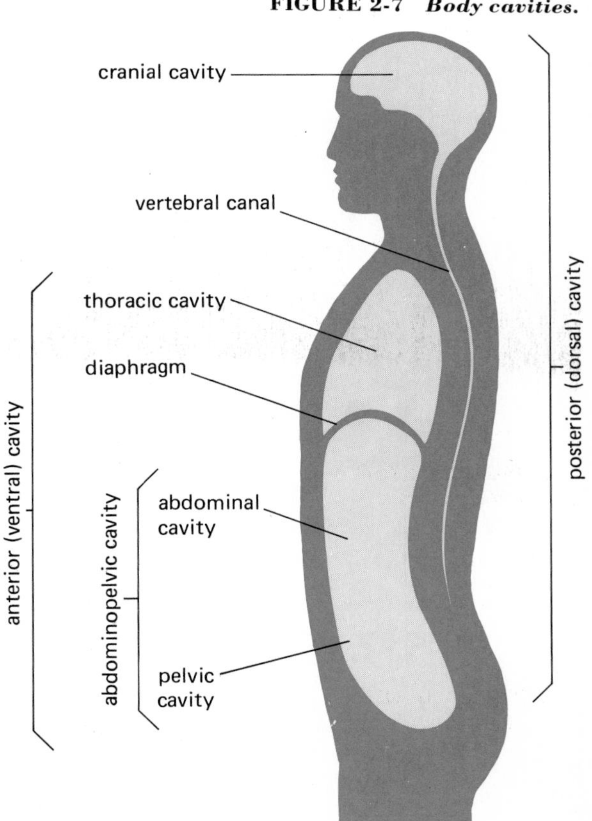

The heart is suspended in a fluid-filled sac, the *pericardial cavity*. On either side of the pericardial cavity are the two lungs, each of which is enclosed in a separate chamber, a *pleural cavity* (Fig. 2-8). The region of the thoracic cavity located between the pericardial cavity and the two pleural cavities is the *mediastinum*. The major blood vessels to the lungs and heart, the trachea, and the esophagus pass through the mediastinum; the thymus gland is also located there. The volume of the thoracic cavity and the pressures in this cavity vary with each respiratory cycle, as the muscles of the diaphragm and surrounding rib cage contract and relax.

FIGURE 2-8 ***Thoracic cavity.***

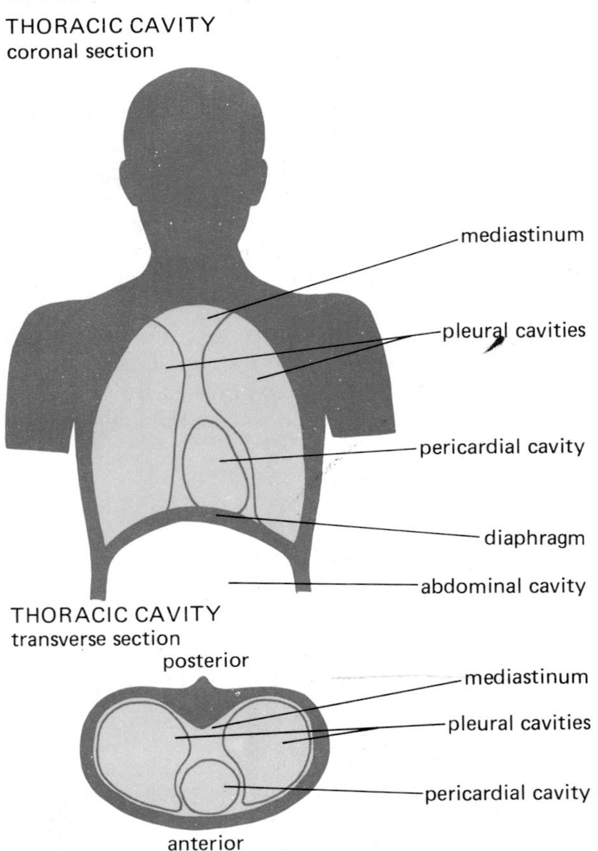

The largest single cavity in the body is the abdominopelvic cavity, which forms the second portion of the ventral cavity. It is within this cavity that the organs often referred to as the *viscera* are located. Although the abdominopelvic cavity forms one continuous, undivided chamber, it has arbitrarily been divided into an upper *abdominal cavity* and a lower region, the *pelvic cavity*, which is partially protected by the surrounding pelvic bones.

The abdominal cavity is the least protected by bony structures of any of the body's cavities. Located between the rib cage of the thoracic cavity and the pelvic bones of the pelvic cavity, its walls are composed of layers of muscle and connective tissue but without any bony elements except for the segments of the vertebral column at the back. The major organs in the abdominal cavity are the liver, stomach, pancreas, spleen, and intestines. The inner surface of the abdominal cavity is lined with a membranous tissue, known as the *parietal peritoneum*. The peritoneum also extends from the walls of the cavity to cover various abdominal organs (*visceral peritoneum*) and provides a loosely folded sheet of tissue to which the various organs and blood vessels are attached. The space between the parietal and visceral peritoneum is known as the *peritoneal cavity;* in a normal individual this space contains only a very small amount of serous fluid. The two kidneys lie behind the peritoneum on the dorsal abdominal wall and are thus *retroperitoneal*. On top of each kidney is located an adrenal gland. The stomach, spleen, and pancreas are located in the upper left region of the abdomen, and the liver occupies most of its upper right segment. The highly coiled small intestine and the shorter, straight segments of the large intestine occupy most of the remaining space of the abdominal cavity.

The pelvic cavity contains the bladder as well as portions of the intestines; the latter terminate in the short segment of the rectum leading to the anus. Also located in the pelvic region are the reproductive organs of the female—the vagina, uterus, oviducts, and ovaries—or, in the male, the prostate gland, seminal vesicles, and portions of the vas deferens. The gonads of the male, the testes, are located outside the pelvic cavity in a separate scrotal sac suspended from the lower abdomen.

The other major organs not located in one of the body cavities are the special sense organs associated with the eyes, ears, nose, and mouth, and the glands of the neck, the thyroid and parathyroid glands.

EMBRYONIC DEVELOPMENT

Thus far, the patterns and levels of organization seen in the body have been described. We now describe the chronologic development of this organization.

The major shaping of the body takes place during the first 2 months following fertilization. This early stage of development is divided into two periods: the first 3 weeks, the *period of germ disk formation,* followed by the *embryonic period,* weeks 4 through 8, during which the germ disk gives rise to the organs of the body. By the end of 2 months, the embryo has assumed a human form with a face, arms, legs, fingers, and toes, although it is only about 3 cm long, weighs about 10 g, and is incapable of independent survival outside the mother's womb. The embryo now becomes known as a *fetus*. During the remaining 7 months of pregnancy—the *fetal period*—the organ systems continue to mature, developing the detailed organized structures that will enable them to function at birth, independently of the maternal support provided during the 9 months of pregnancy.

Formation of the Trilaminar Germ Disk (Weeks 1 to 3)

The processes leading to the production of the sperm or ova—the *gametes*—and to fertilization will be described in Chap. 21. Fertilization brings together the 23 chromosomes of the ovum and the 23 chromosomes of the sperm cell to form the full complement of 46 chromosomes of the new individual. These 46 chromosomes will be passed on to each of the cells of the developing embryo by mitotic cell divisions. The fertilized egg cell, now known as a *zygote,* undergoes a sequence of four mitotic cell divisions, resulting in a spherical cluster of 16 cells known as the *morula*. During these initial cell divisions, known as the *cleavage divisions,* there is no growth in the resulting daughter cells.

On the third day following fertilization, the morula undergoes transformation into a *blastocyst* as fluid begins to accumulate in the central space between the cells of the morula. The outer cell layer of the blastocyst (Fig. 2-9) is the *trophoblast;* it will attach to the uterus and eventually form the placenta, the organ which links the embryo with the mother, providing the pathway for the exchange of nutrients and waste products (Chap. 21). The *inner cell mass,* or *embryoblast,* of the blastocyst forms the embryo itself.

One week after fertilization the embryo implants in the tissue of the uterus. By the eighth day the embryoblast has become organized into two distinct layers of cells: a single layer of large columnar-shaped cells, the *ectodermal layer,* and a single layer of smaller cuboidal cells, the *entodermal layer*. Together, these two layers of cells form the *bilaminar germ disk* (Fig. 2-10). The ectodermal layer is separated from contact with the trophoblast by a small fluid-filled cavity, the *amniotic cavity*. As the embryo develops, the amniotic cavity comes to completely surround it, encasing it in a fluid-filled, protective environment in which the embryo will live during its intrauterine development. (It is the breaking of this amniotic sac that produces the watery discharge of fluid at the time of birth.) On the entodermal side of the germ disk is the blastocele, which will develop into the *yolk sac* and eventually form the inner lumen of the gastrointestinal tract. By the end of the second week, a third fluid-filled

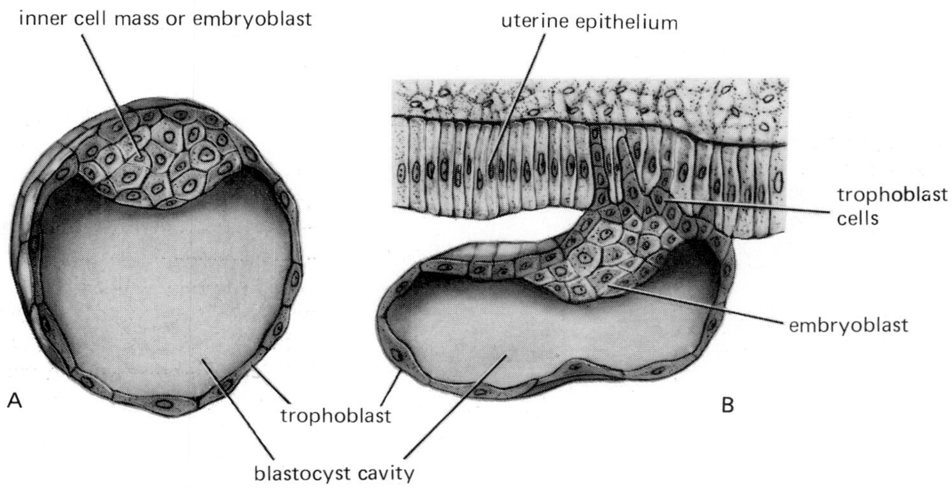

FIGURE 2-9 (A) *Blastocyst.* (B) *The attachment of the blastocyst to the uterine epithelium.* (Adapted from Langman.)

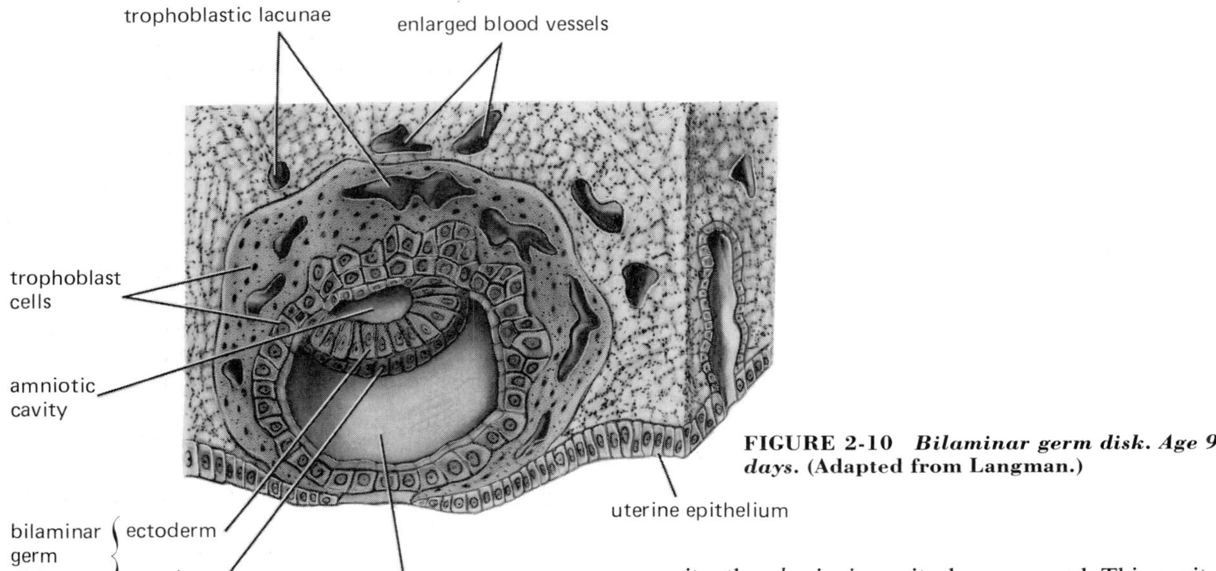

FIGURE 2-10 ***Bilaminar germ disk. Age 9 days.*** **(Adapted from Langman.)**

FIGURE 2-11 ***Bilaminar germ disk. Age 13 days.*** **(Adapted from Langman.)**

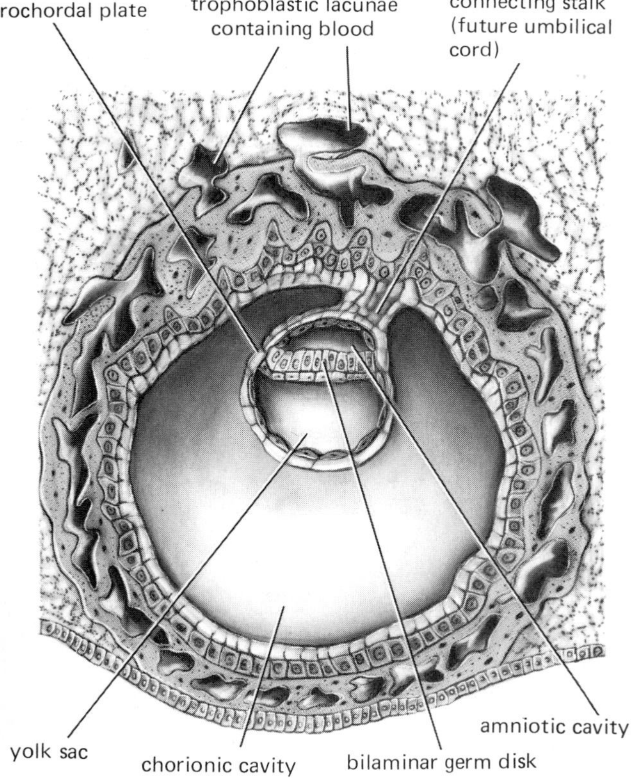

cavity, the *chorionic cavity,* has appeared. This cavity forms between the trophoblast layer and a thin layer of cells lining the yolk sac. As the chorionic cavity expands, it leaves the germ disk and its associated amniotic cavity and yolk sac suspended in the middle of the chorionic cavity, connected to the trophoblast layer by a thin connecting stalk that will eventually develop into the *umbilical cord,* containing the blood vessels that connect the fetus to the placenta (Fig. 2-11).

At the beginning of the third week, the germ disk is a two-layered oval disk situated between the amniotic cavity and the yolk sac. At this time, a narrow groove, the *primitive streak,* appears in the ectodermal layer facing the amniotic cavity and extends about half the length of the germ disk. Ectodermal cells migrate toward the primitive streak, move into the groove, and thence between the ectodermal and entodermal layers to form a third layer of cells, the *mesodermal layer* (Fig. 2-12). As cells continue to move between the ectoderm and entoderm, they migrate laterally and to the front of the primitive streak until the germ disk becomes a three-layered structure. By the end of the third week, having formed the third of the three layers of the germ disk, the primitive streak begins to regress.

The Embryonic Period (Weeks 4 to 8)

During the 5-week embryonic period the three germ layers give rise to the various tissues and organs

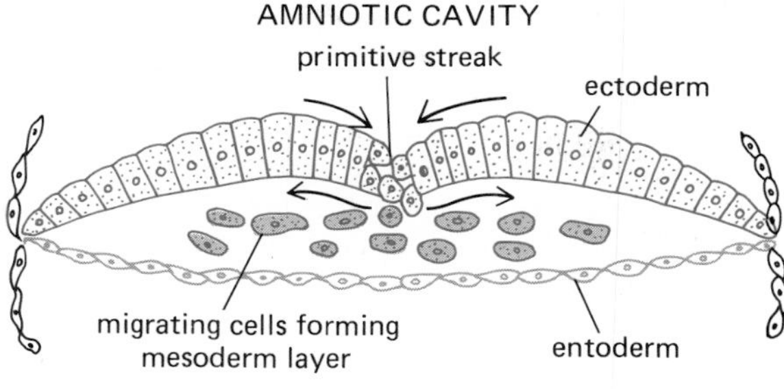

FIGURE 2-12 ***Migration of ectodermal cells through the primitive streak to form the mesodermal layer.***

of the body. Each germ layer makes a distinct contribution to the morphogenetic development of the embryo and is associated with the formation of specific organs or parts of organs. In the following sections we shall examine the contributions of each of the three germ layers separately.

Ectoderm The ectodermal layer forms a hollow tube of ectodermal cells, the *neural tube,* which will evolve into the brain, spinal cord and associated nerves, and the central mass of cells of the adrenal gland known as the *adrenal medulla.* The special sense organs of the eyes, ears, nose, and mouth are formed by infoldings of ectodermal cells at the appropriate locations during various stages of development. In addition to these structures associated with the nervous system, the ectodermal layer, which eventually covers the entire outer surface of the embryo, gives rise to the *epidermis* which forms the outer layer of the skin. Associated with this epidermal layer are additional structures of ectodermal origin: the *hair, nails, sweat* and *sebaceous* (oil) *glands,* and, in the female, the glandular portion of the breast. The ectoderm also covers the interior of the oral cavity where it gives rise to the salivary glands and the enamel layer of the teeth. One of the major endocrine glands of the body, the *pituitary,* located at the base of the brain is formed from ectodermal tissue by the invagination of a pouch from the roof of the mouth which meets an invagination from the overlying neural tube to form the *anterior* and *posterior* portions of the pituitary gland, respectively.

Mesoderm By the end of the third week the mesoderm layer along the neural tube begins to break up into distinct segments, forming pairs of *somites* (Fig. 2-13). By the middle of the fifth week there are 42 to 44 pairs of somites located along the neural tube. As development proceeds, cells migrate from the somites to the region underlying the neural tube where they form a loosely woven tissue known as *mesenchyme.* Mesenchyme differentiates into a number of cell types belonging to the general class of connective tissue. Additional cells from the somites migrate toward the ectodermal surface where they give rise to the *dermal layer* which underlies the epidermis of the skin. The remaining cells of the somites multiply, forming masses of tissue known as *myotomes,* which will evolve into the skeletal muscles of the body.

The paired, segmented organization of the somites in the early embryo has important implications for understanding the distribution of sensory and motor functions in the adult organism. In association with each somite there develops a pair of nerves which extend from each side of the neural tube and will become the spinal nerves of the spinal cord in the fully developed organism. Each of these nerves eventually innervates the muscles and skin developing from its corresponding somite. As the embryo develops, the simple geometric relation between the spinal nerves and corresponding somites becomes distorted because different regions of the embryo develop at different rates, but the link between the segmented nerves and the segmented somites remains. In addition to those

FIGURE 2-13 ***Organization of the mesodermal layer during the third to fifth weeks into somite segments along the neural tube.***

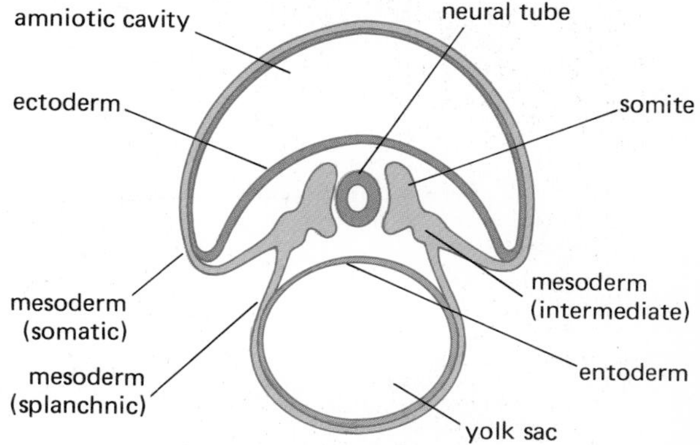

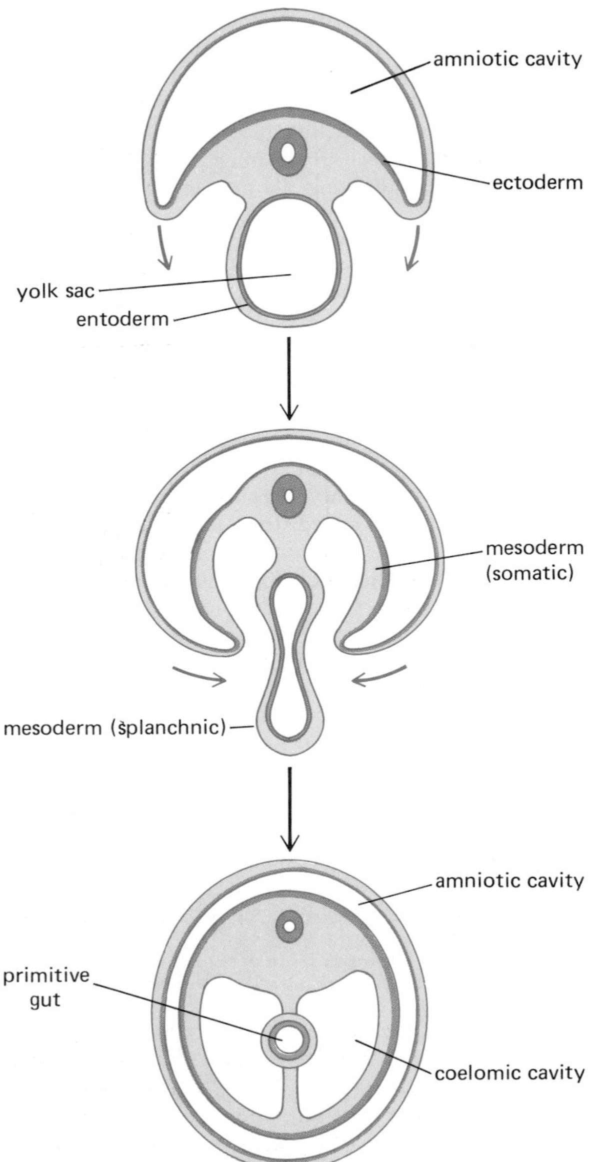

FIGURE 2-14 ***Folding of the embryonic layers during the fourth week to form the coelomic cavity and primitive gut.***

structures arising from the somites, the mesoderm gives rise to the large number of other important structures listed in Table 2-2.

Entoderm The third layer of the trilaminar germ disk, the entoderm, forming the roof of the blastocyst cavity, eventually grows to cover the inner surface of the yolk sac. During the fourth week, the embryo undergoes extensive folding, both in the longitudinal direction in which the head and the tail regions fold toward each other and laterally in which the edges of the germ disk fold downward until they meet underneath the embryo (Fig. 2-14), forming two intraembryonic spaces: the *coelomic cavity,* which is lined with mesoderm and will become the thoracic and abdominal cavities, and the *primitive gut,* formed by sealing off a portion of the yolk sac, which is lined with entoderm. A portion of the yolk sac remains outside the embryo, connected to it by a thin tube in the region of the midgut which also contains the blood vessels of the umbilical cord linking the embryo to the placenta. The gut, which will become the gastrointestinal tract, thus begins as a simple tube suspended in the center of the embryo. Initially the ends of this tube, in the region of the mouth and anus, respectively, are sealed. As development progresses, the ends of the tube become open and are continuous with the amniotic fluid in the cavity surrounding the embryo. Considerable amounts of amniotic fluid are swallowed by the embryo.

Arising from the entodermal lining of the gut wall are a number of organs, all of which begin as small pouches or ducts that extend out from the surface of the gut into the surrounding coelom. These include the organs directly associated with gastrointestinal function such as the liver, gallbladder, bile ducts, and pancreas, as well as organs with other functions related to the external environment: the lungs, urinary bladder, and the auditory tubes which link the inner ears to the back of the mouth cavity.

Some of the organs arising from the entoderm of the gut completely lose their association with the lumen of the gut and become separate glandular tissue; these include the thyroid and parathyroid endocrine glands which arise in the neck as outpouchings of the gut wall.

A summary of the organs and tissues arising from the three germ layers is provided in Table 2-2.

During the 5 weeks of the embryonic period many *congenital malformations* may arise. A congenital

TABLE 2-2
ORGANS AND TISSUES ARISING FROM THE THREE GERM LAYERS

Ectoderm	*Mesoderm*	*Entoderm*
Nervous system	*Muscle*	*Epithelium of:*
Brain	Cardiac	Larynx
Spinal cord	Smooth	Trachea
Peripheral nerves	Skeletal	Lungs
Ganglia		Esophagus
Special sensory receptors of eye, ear, nose, and mouth	*Connective tissue*	Stomach
General sensory receptors	Fibrous connective tissue	Intestines
Posterior pituitary (neurohypophysis)	Adipose tissue	Liver
Adrenal medulla	Bone and bone marrow	Gallbladder
	Blood cells	Pancreas
	Lymphatic tissue	Bladder
Skin (epithelium of the integument)	Reticuloendothelial system	Urethra
Epidermis		Vagina
Hair	*Skin* (fibrous connective tissue)	Inner ear cavity
Nails	Dermis	Auditory tubes
Sweat glands		Thyroid
Sebaceous glands	*Epithelium of:*	Parathyroid
Mammary glands	Thoracic and abdominal cavities (pleura, peritoneum, pericardium)	Thymus
Epithelium of:	Kidneys	
Salivary glands	Ureters	
Nasal cavity	Gonads and associated ducts	
Mouth	Adrenal cortex	
Anal canal	Lining of heart (endocardium) and vessels (endothelium)	
Enamel of teeth		
Anterior pituitary (adenohypophysis)		

malformation is a gross structural defect present at the time of birth, such as a cleft palate, absence of fingers or toes, blindness, deafness, defects in the structure of the heart and major blood vessels, etc. Approximately 5 percent of all infants are born with one or more congenital malformations. A number of factors, including certain viral and bacterial infections, radiation, a variety of drugs and chemical pollutants, the nutritional state of the mother, and genetic and chromosomal abnormalities have been implicated as possible causes of these congenital malformations. The type of malformation induced is critically dependent upon the stage of embryonic development at the time the embryo is exposed to the agent. If exposed early in the embryonic period, a large number of major structural defects may result, whereas at later stages the same agent may affect only the particular organs that are undergoing a critical step in the morphogenic development at the time of exposure.

The Fetal Period (Weeks 9 to 38)

By the end of the eighth week the embryo has acquired a human form and is now known as a *fetus* (Fig. 2-15). The remaining 30 weeks of pregnancy, the *fetal period,* is a period of major growth and maturation. Rarely will the fetus survive outside the mother until it has undergone about 28 weeks of development, at which time the lungs have become sufficiently developed to function in gas exchange, and the appropriate neural connections have been made to the respiratory muscles to regulate the rhythmic patterns of breathing. The lungs do not function in gas exchange during development and are, in fact, filled with amniotic fluid.

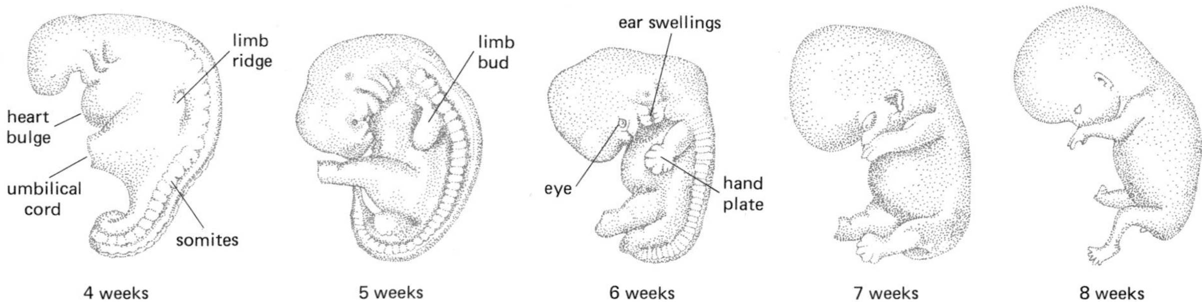

FIGURE 2-15 ***Changes in embryonic form during the last 4 weeks of the embryonic period.*** **(Redrawn from Langman.)**

Although most of the initial events associated with the formation of the various organs occur during the embryonic period, the tissues of the various organs continue to differentiate and shape the fine structures of the organs during the fetal period. The brain, in particular, has a very long period of differentiation and development, which continues, in part, even after birth. Exposure of the fetus to various environmental agents, which may cause significant anatomic malformations during the embryonic period, often produce much less obvious morphologic damage during the fetal period but still institute significant functional impairment, particularly in the developing nervous system, and can lead to various forms of mental retardation.

MECHANISMS OF DEVELOPMENT: CELL DIFFERENTIATION

Thus far we have described the process of development from a purely anatomic standpoint. Cells multiply, migrate, and specialize, forming layers which fold, fuse, separate, and elongate in the process of shaping the human body. What is the nature of the cellular processes that determine these patterns?

The process by which a cell is transformed into a specialized cell type is known as *cell differentiation.* The general sequence of development is thus one of cell multiplication by mitotic cell division, followed by cell differentiation into the various types of specialized cells. Cells which have not undergone cell differentiation are known as *undifferentiated* or *stem* cells. Although most cells become differentiated during the period of embryonic and fetal development prior to birth, some undifferentiated cells remain and multiply in the various organs of the adult body where, upon receiving an appropriate stimulus, they may differentiate into a specialized cell type.

Beginning with the division of the fertilized egg cell, identical copies of the genetic information obtained from the male and female parent at the time of fertilization are distributed to each of the daughter cells of the growing embryo. Thus each cell contains an identical set of genes. How then is it possible for one cell to differentiate into a muscle cell which synthesizes muscle proteins, while another cell containing the same set of genes differentiates into a nerve cell which synthesizes the proteins characteristic of nerve cells? We must conclude that not all the genes in a given cell transcribe their genetic information into protein synthesis; otherwise, all cells would synthesize the same proteins and presumably would have the same characteristics. Thus, in order to understand how cells differentiate, we must discover the mechanisms which allow some of the genes in one type of cell to be "turned on" (i.e., form messenger RNA and the corresponding proteins) whereas in another type of cell these particular genes are "turned off" and another set "turned on" (Fig. 2-16).

In addition to those genes which are turned on or off in a particular type of differentiated cell, there are other genes which are active in all cells, for example, the genes responsible for synthesizing the enzymes of glycolysis. Even these genes may be more active in one type of cell than another, in the sense that they may produce larger quantities of a particular type of enzyme in one cell than in another. Thus there is a quantitative as well as a qualitative control of gene expression associated with cell differentiation. In some cases the process of differentiation appears to be

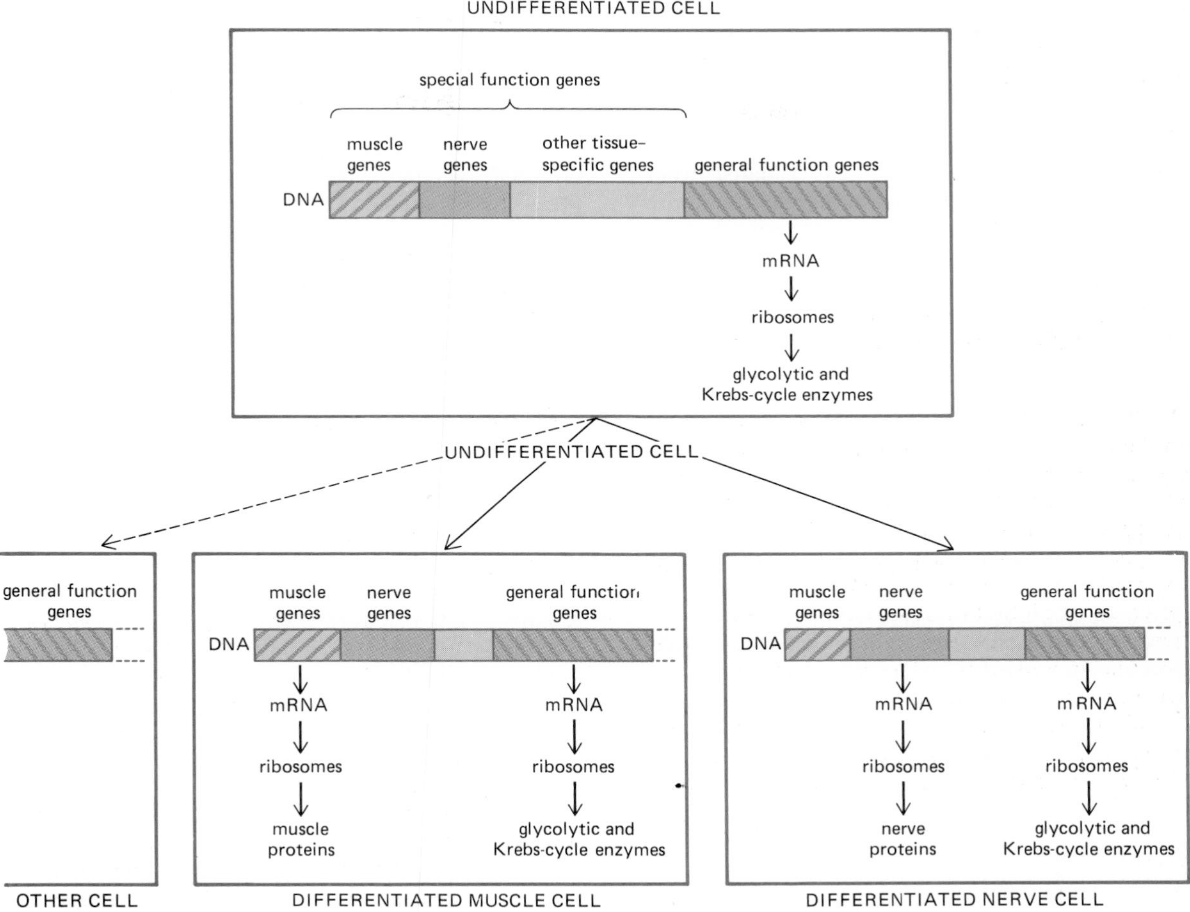

FIGURE 2-16 ***Cell differentiation depends upon the selective transcription of those genes associated with the specialized functions of the differentiated cell.***

programmed by chemical events occurring entirely within the cell and independent of influences from the extracellular environment, whereas in other cases cells may remain in an undifferentiated state until they receive a specific chemical signal from an extracellular source. It remains for future biologists to work out the detailed chemical mechanisms controlling these many aspects of genetic expression.

As a result of differential gene expression, cells acquire differences in their locomotion, adhesiveness, shape, and ability to divide; these differences account, in large part, for the patterns of cell organization we have described.

Cell Locomotion

Almost all cells contain contractile proteins capable of producing ameboidlike motion. Ameboid motion involves the controlled flow of cytoplasm from one region of the cell to another. The flowing cytoplasm produces a fingerlike extension of the cell surface, known as a *pseudopod*. Progressive motion is achieved by extending a pseudopod, which adheres to the surface over which the ameba is moving, followed by the flow of cytoplasm from the rear of the cell into the extended pseudopod and the detachment of the rear of the cell from the surface upon which it is moving. By repeatedly extending one end and pulling up its rear, the ameba moves along a surface. During development, many of the cells in the embryo exhibit ameboidlike movements as they migrate from one area to another; different cells become actively mobile at different stages of development.

Some aspects of motility appear to be regulated by contacts with other cells. Cells isolated from chicken or mouse embryos can survive and multiply when placed in a glass bottle containing a nutrient fluid with oxygen and warmed to body temperature. This technique of *cell* or *tissue culture* provides a convenient method of directly observing cell growth, since the cells adhere to the glass surface of the flask and can be observed under a microscope. When embryonic cells are added to a flask, they attach themselves at random locations on the glass surface, begin to multiply, and move along the surface in random directions by an ameboidlike motion. When a cell comes in contact with another cell, its movement in that direction ceases; it will not crawl over the contacted cell, although it may back off and begin moving in a new direction. Such inhibition of forward movement upon contact with another cell is called *contact inhibition;* it seems to result from a localized inhibition of the contractile mechanism in the region of the cytoplasm adjacent to the region of contact. As the cells continue to multiply and move about the surface, they come into contact with more and more cells, until the surface of the glass is covered with a monolayer of cells. At this point all further movement ceases, since the cells are in contact with other cells on all sides and movement in all directions is inhibited.

As cells differentiate into specialized types with specialized functions to perform, they generally lose their ability to undergo independent locomotion. In some cases the contractile machinery remains in the cell but is not activated. Such a cell remains immobile until it receives an appropriate activating sign to turn on its contractile machinery. One such example occurs during wound healing where the cells surrounding an open wound, which previously were immobile, begin moving into the wound area, filling up the opening and drawing the edges of the wound together. When the cells from opposing sides of the wound meet in the middle, contact inhibition halts all further motion. Some of the cells in the adult have retained the ability to undergo ameboid-type locomotion as a basic aspect of their primary function. These include various cell types which belong to the defense system of the body. For example, the white cells of the blood are able to move between the cells of the capillary walls and enter the surrounding tissue spaces.

Cell Adhesion

Most of the cells in the body adhere to neighboring cells or to an extracellular fibrous matrix produced by connective-tissue cells. If cells did not adhere to each other, the body would collapse like a bag of dry sand. At least three classes of adhesive mechanisms hold cells together (Fig. 2-17): (1) electrostatic attraction between charged surfaces of the cell, (2) specific proteins which cross-link the surfaces of cells, and (3) specialized cell junctions.

The weakest and least specific type of cell adhesion is the electrostatic interaction between charged cell surfaces. The phospholipids and proteins that form the surface structure of cell membranes, along with various sugar residues that are attached to these molecules, contain ionized chemical groups, primarily negatively charged carboxyl groups. Thus the surface of cells has a net negative, fixed charge built into the structure of the plasma membrane. (This fixed electric charge should not be confused with the separation of sodium and potassium ions across cell membranes that is responsible for membrane potentials (Chap. 4). Two negatively charged surfaces, when brought together, should repel each other, since like electric charges repel, and opposite charges attract. However, two negatively charged surfaces can be linked together if a positively charged material is placed between them. The divalent cation calcium Ca^{2+} appears to play this role in some cases, since the removal of calcium from the extracellular medium causes some cells to separate from each other.

Proteins also play an essential role in linking certain cells together. Several proteins, known as *aggregation factors,* that will cause a population of separated cells to aggregate together have been isolated from embryonic tissues. Furthermore, such proteins appear to be specific for the types of cells they will bind together. The binding together of cells by such proteins depends on the presence of not only the specific type of aggregation factor but also a specific binding site on the cell surface to which the aggregation factor can become attached. The strength of cell adhesion can be varied by varying the number of binding sites on the cell surface and thus the number of cross links that can be formed. Again it appears to be the differential expression of genetic elements during cell differentiation which determines the types of aggregation factors that are formed and the number and types of binding sites that are present on a given cell's surface.

When cells that adhere to each other are examined under the electron microscope, one finds that the

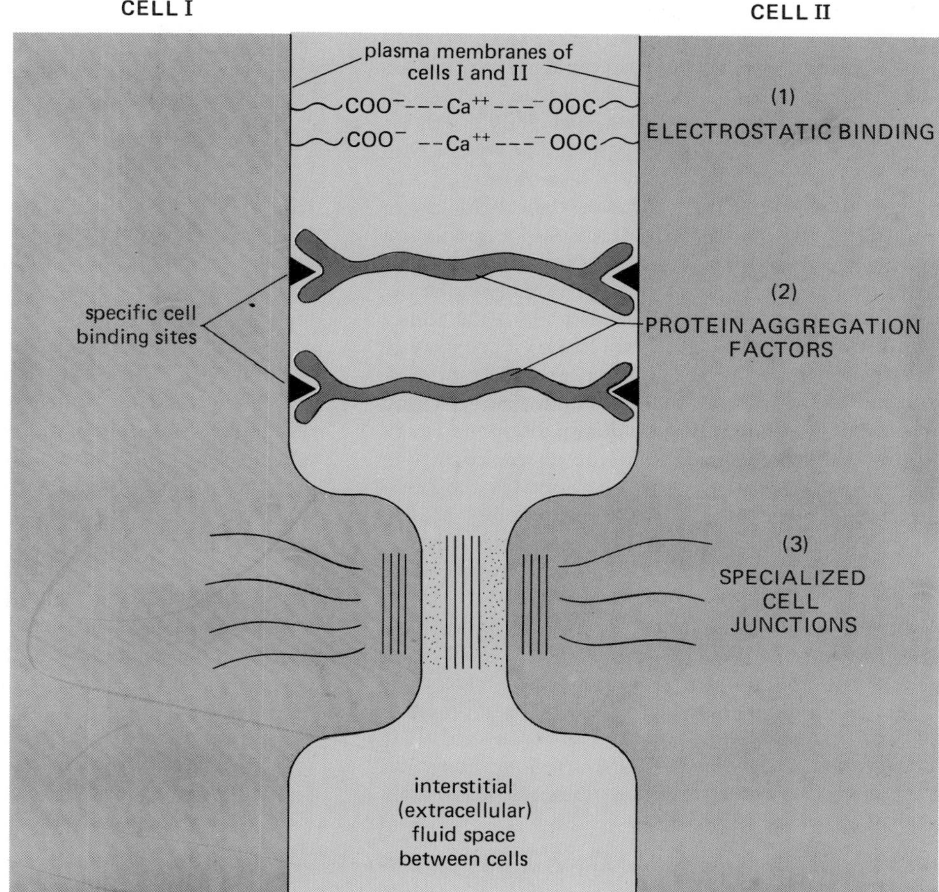

FIGURE 2-17 ***Three types of linkages responsible for cell adhesion.***

surfaces of the opposing plasma membranes are not in physical contact with each other but are separated by a small extracellular space that is several hundred angstroms wide. It is assumed that the proteins that bind the cells together span this small space, forming a loose network of fibers holding the cells together but at the same time allowing interstitial fluid to percolate between them.

The third mechanism of binding cells together involves specialized junctions which have a definite anatomic structure that can be observed by using an electron microscope. In some cases these junctions involve the fusion of opposing membranes. In other cases a space remains between the cells but is filled with a dense fibrous material. In addition to holding cells together, these junctions perform a number of other functions associated with the flow of molecules between cells and from one cell to another; their structures will be described in later chapters.

The combination of cell locomotion and the specificity of cell adhesion provides mechanisms that can explain a great deal of the patterned cellular organization that occurs during embryonic development. For example, when the cells of an embryo are dissociated and placed in tissue culture, the interactions of the cells can be followed over a period of days. Within 1 or 2 days the cells have arranged themselves in various specific patterns. Most of the cells of a similar cell type are bound together in clusters, whereas aggregates of different cell types are arranged relative to each other, so that a cluster of one cell type completely surrounds a cluster of another cell type. This sorting out of cell types in tissue culture is due to the specificity of cell adhesion as cells of various types randomly collide with each other as they move over the surface of the glass. Weakly adhering cells will break their contacts in favor of forming stronger adhesions until a population of strongly adhering cells comes to occupy the center of the cluster surrounded by more weakly adhering cells. Such experiments clearly indicate that the ability of cells to associate in various patterns is a property of individual cells and is not imposed on the system by some external force. In the developing embryo, the various cell types are not as randomly distributed as in the above experiment, and thus the degree of sorting out is not as extreme, although presumably the same types of selective adhesion and cell mobility are involved. In some cases, however, embryonic cells migrate over very large distances, and other forces must be active.

Cell Shape

Cell differentiation is usually accompanied by a change in the shape of the cell. Many types of differentiated cells can be readily identified because of their characteristic shapes; nerve cells have long branched processes, muscle cells are long cylinders, some cells are thin and flat, and others are cuboidal. Two general factors are responsible for the shape of a cell: (1) the attachment of the cell to other cells and extracellular materials, and (2) structural elements within the cell or associated with the cell membrane which provide a certain amount of rigidity.

Most cells, when suspended in a fluid medium, tend to become spherical; when they become attached to a surface, however, they become flattened, often with an irregular cell outline. When cells are surrounded by neighboring cells, the close association may give the cells a polyhedral shape. The more characteristic shapes of cells, however, are determined by structural elements in the cell itself. Microfilaments and microtubules appear to play a major role in maintaining specific cell shapes. The microtubules, which appear to be relatively stiff rods, are usually found in abundance in regions of the cell where processes extend from the cell surface, for example, the long cylindrical processes of nerve cells. Microfilaments, which appear to have a contractile function, are often found in bands underlying the plasma membrane and may even be attached to the membrane. Contraction of these filaments may alter the shape of the cell surface.

Figure 2-18 provides two examples of how changes in cell shape in turn contribute to the morphologic shaping of tissues and organs during development. Beginning with a layer of cuboidal cells, these

FIGURE 2-18 ***Morphologic shaping of an epithelial layer as a result of* (A) *increased adhesion between the epithelial layer and underlying cells, resulting in flattening of layer, and* (B) *contractile microfilaments at the base and top of the cell, producing conical-shaped cells causing invagination of the epithelial layer into the underlying tissue.***

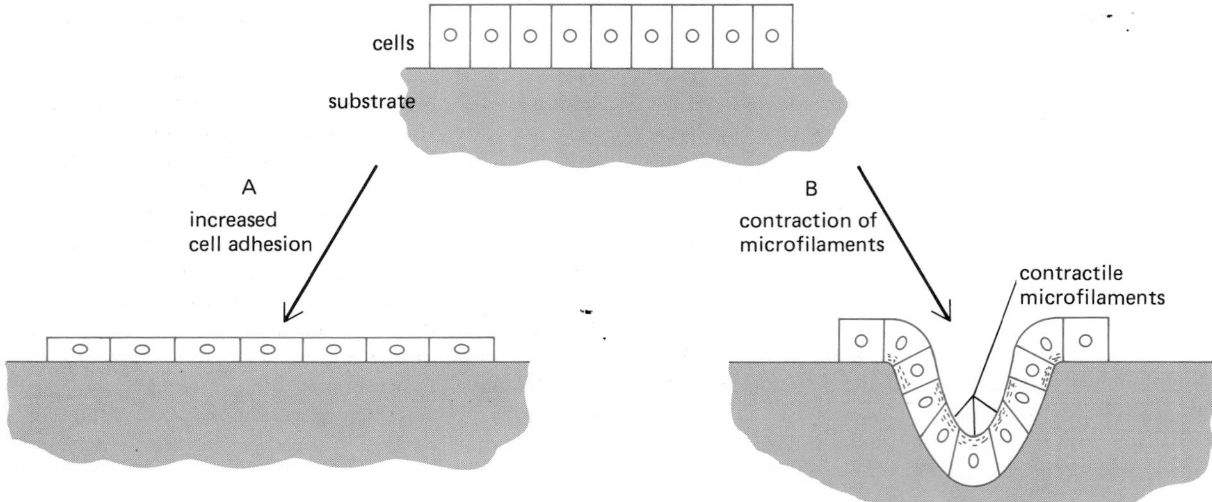

cells may become flattened, thus extending the layer of cells. Such flattening may arise because of a change in the adhesive properties of the cell surface for the underlying layer of cells, which leads to a greater area of surface contact between the two cell layers. The second example illustrates the role of microfilaments arranged in bands just below the plasma membrane at the two ends of the cell. Contraction of these microfilament bands in certain cells in an epithelial layer produces a purse-string type of action, converting the cuboidal cell into a more cone-shaped cell, with the result that the layer of cells invaginates into a pocket.

Cell Division

Obviously the whole process of development could not occur without the multiplication of cells to provide the units that become organized into tissues and organs. Like other cellular processes, cell division is regulated and does not occur at the same rate in all cells. Since the body does not continue to increase in size indefinitely, either cells stop dividing or cells are destroyed as rapidly as they are replaced by cell division. Actually, both processes occur in the adult. Many types of cells, such as nerve and muscle cells, lose the ability to divide following cell differentiation; other cell types, including most epithelial, glandular, and connective-tissue cells, retain this ability. In general, cells which develop highly specialized internal structures lose the ability to divide; an example is muscle cells with their highly organized array of contractile filaments. In contrast, cells specialized for the secretion or absorption of materials use structures normally present in all cells and retain the ability to divide.

Growth of the embryo is the result of two processes: (1) an increase in the number of cells resulting from mitotic cell division, *hyperplasia,* and (2) an increase in the size of individual cells, *hypertrophy*. By the time of birth many of the cells of the body have differentiated into cell types that are no longer capable of cell division. This means that at birth the body contains all the cells of certain types that it will ever have; yet from birth to the adult state there is a considerable increase in the size of these tissues. Thus, this increase in size is due to the hypertrophy of individual cells and not to an increase in their number.

Both hyperplasia and hypertrophy are regulated processes, as can be illustrated by a few examples. The cells of the liver are capable of cell division; yet the liver stops growing when it reaches the adult size. However, if a large section of the liver is removed from an adult, the remaining liver tissue will undergo hyperplasia until the lost tissue has been replaced and the liver has returned to its original size. Muscle provides an example of the control of tissue hypertrophy. If a muscle is repeatedly exercised, its size increases as a result of an increase in the diameter of its individual muscle cells, the latter due to the synthesis of new contractile filaments; there is no increase in cell number because the adult muscle cell is incapable of cell division. If the exercise which leads to muscle hypertrophy is discontinued, the size of the muscle cells will decrease as the contractile filaments within the cells are broken down but not replaced.

The general concept that overall size depends upon the number of cells as well as the size of individual cells can also be applied to the development of the functional units within the various organs of the body. In some organs, the number of functional units is fixed by the time of birth. Throughout the remainder of life, their number will not increase but the size of an individual unit may increase, both by cell division and hypertrophy of the cells composing the unit. If a portion of the organ is damaged, the remaining portion may be able to maintain the organ's function by increasing the size of the remaining functional units. In other organ systems, particularly glands, the number of functional units is not fixed at birth, and new units can differentiate and become organized in the adult body.

Cancer

Now that we have examined some of the factors involved in the normal growth and organization of tissues, it is instructive to examine briefly the abnormal growth of cells that occurs in *cancer.* Cancer is the uncontrolled growth of cells in a multicellular organism. Although the initiating causes of cancer at the molecular level are still unknown, the weight of evidence points to an alteration in the mechanisms controlling the genetic elements of the cell associated with cell motility, cell adhesion, cell shape, and cell division.

Cancer arises when a normal cell is transformed into a cell which no longer responds to the various regulating mechanisms controlling cell growth, with the result that the cancer cell undergoes unlimited growth. Viruses, chemical agents, and radiation have all been shown to be capable of inducing the transformation from a normal to a cancerous cell. All these agents have in common the ability to alter the genetic elements contained in the DNA molecules of the cell. Any cell in the body has the potential of being transformed into a cancer cell. Cancer cells multiply by cell division, producing a growing mass of cells known as a *tumor*. It is not the ability to divide or the rate of cell division that characterizes a cancer cell but its failure to respond to the processes which regulate growth.

Cancer cells are less adhesive than are corresponding normal cells; as a result they tend to break away from the initial site of the growing tumor and travel through the bloodstream to become lodged in other tissues and grow into secondary tumors. This process of the spreading of cancer cells throughout the body is known as *metastasis*. Because of this tendency to metastasize, it is important to detect the growth of cancerous cells as early as possible before they have spread to other tissues. At this early stage, the small nodule of growing cancer cells can often be completely removed from the body by surgery, a process which is practically impossible once the cancer has spread to multiple organs in the body.

Not only do cancer cells show alterations in their ability to adhere to other cells but also in their lack of contact inhibition. In tissue culture, cancer cells move over the tops of other cells, forming multiple layers of cells. This lack of contact inhibition in cancer cells increases the ability of these cells to invade surrounding tissues during the tumor's growth and metastasis.

Although cancer may arise in a differentiated cell, the degree to which the cancerous cell retains the differentiated properties varies considerably between different types of cancer cells. Some cancer cells retain many of the differentiated characteristics of their cell of origin, such as the synthesis of characteristic proteins or the response of the cell to those hormonal agents which influence the activity of the normal cell. In contrast, other cancer cells tend to lose many of the differentiated characteristics of their cell of origin and resemble more an undifferentiated embryonic cell. For example, the growth and differentiation of the normal cells which constitute the tissue of the female breast are controlled by specific sex hormones secreted by the female gonads; in some breast cancers, the cancer cells are still regulated by these sex hormones, and the removal of the hormone (by surgically removing the gonads) will inhibit the growth of the cancer. In other cases of breast cancer, however, in which the cells have lost their ability to respond to the sex hormones, the removal of these hormones has no effect upon the growth of the cancer cells. It is hoped that a better understanding of the processes that control normal cell growth and differentiation in the body will contribute to our understanding and treatment of the cancerous process.

CHAPTER 3

Epithelium, Connective Tissue, and Skin

The four categories of specialized tissues—epithelium, connective tissue, nerve, and muscle—which are arranged in various proportions to form the body's organs, arise during embryonic growth and development by the process of genetic differentiation, as described in Chap. 2. This chapter will examine the structural and functional characteristics of epithelial and connective tissues as well as skin, which is largely a composite of these two tissues; separate chapters will be devoted to nerve and muscle.

EPITHELIUM

The external and internal surfaces of the body are covered by layers of specialized cells arranged in flat or tubular sheets to form boundaries between various compartments; these constitute the *epithelia*. Just as the plasma membrane of a single cell provides a selective barrier to the flow of molecules between the cell and the surrounding extracellular fluid, the layers of epithelial cells covering or lining the body's surfaces are selective barriers regulating the exchange of molecules across these surfaces. For example, the epithelial cells of the skin form a barrier at the body's surface; the epithelial lining of the lungs, gastrointestinal tract, and kidneys regulates the exchange of molecules between the blood and the external environment; the epithelial cells lining the smallest blood vessels (capillaries) form the barrier between the blood plasma and the interstitial fluids. In addition to subserving these functions, epithelial tissue forms almost all the *glands* of the body.

Epithelia are associated with the processes of *ab-*

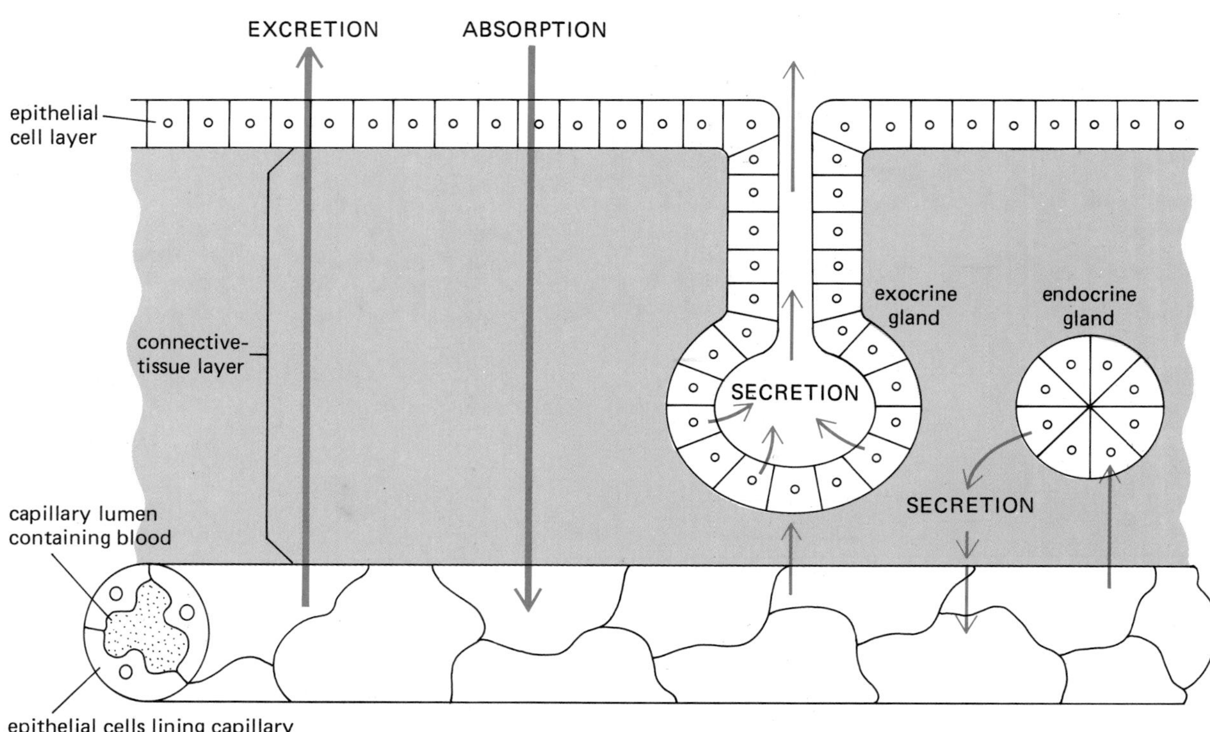

FIGURE 3-1 ***Epithelial surfaces and their relation to excretion, absorption, and secretion.***

sorption, the movement of molecules from the external environment into the internal environment (interstitial fluid and plasma); *excretion,* the movement of molecules from the internal environment to the external environment; and *secretion,* the release of specific molecules from a cell into the extracellular medium (Fig. 3-1). Each of these processes is associated with the movement of molecules from one compartment to another.

Types of Epithelia

Epithelial cells are organized into sheets of one or more layers of cells. The *apical* surface of the epithelial sheet is exposed to the aqueous or gaseous medium at the outer surface of the epithelium, and the *basal* region rests upon the underlying tissues (Fig. 3-2). At the boundary between the basal region of the epithelium and the underlying tissues is an extracellular structure known as the *basement membrane.* It consists of a matrix of polysaccharides (secreted by the overlying epithelial cells), in which are embedded fine filaments formed by the underlying connective-tissue cells. The basement membrane forms a continuous layer along the basal surface of the epithelium. In some types of epithelia the basement membrane may be quite thin, whereas in others it is thick and may offer considerable resistance to the diffusion of large molecules, such as proteins. All blood vessels are located below the basement membrane and do not penetrate into the epithelial layers. Nerve fibers, however, may penetrate the basement membrane and terminate near the plasma membranes of the epithelial cells.

Epithelial cells are classified according to the shape of the cells and the number of layers of cells constituting the epithelial surface. Epithelia consisting of a single layer of cells are *simple epithelia* (Fig. 3-3). The

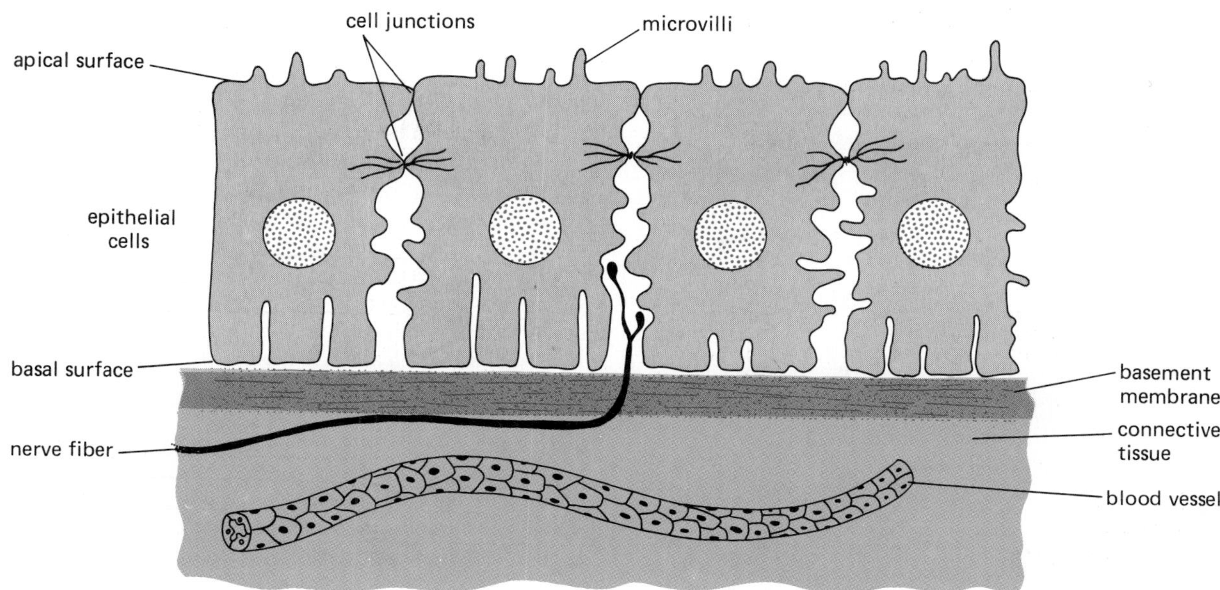

FIGURE 3-2 ***Typical arrangement of epithelial cells and underlying tissues at an epithelial boundary.***

cell shapes may be thin and flat, termed *squamous;* roughly symmetric in shape, *cuboidal;* or tall and thin, *columnar.* Simple epithelia are usually found at surfaces that move molecules across the epithelial layer; examples are the columnar epithelium lining inner portions of the kidney tubules and the gut. A single layer of squamous epithelium lines the inside of the heart and blood vessels and is known as *endothelium;* a similar layer of squamous epithelium (*mesothelium*) is found in the serous membranes which line the pleural, peritoneal, and pericardial cavities. One form of simple epithelium known as *pseudostratified* appears to be composed of several layers but is, in fact, a single layer of cells of unequal heights, some of which do not reach all the way to the apical surface. This gives the single layer of cells the appearance of being composed of more than one layer.

Epithelia composed of more than one layer of cells are known as *stratified* epithelia (Fig. 3-4). Such multilayered epithelia are found in regions of the body where the surface is subject to considerable mechanical stress; examples are the skin and various ducts that link parts of the body to the external environment. The shapes of the cells in the various layers of stratified

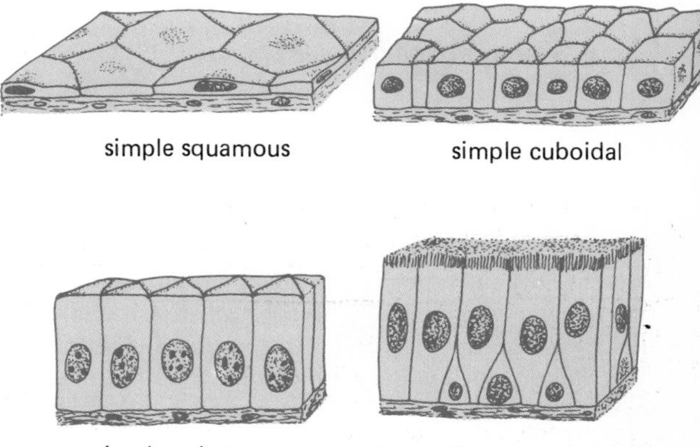

FIGURE 3-3 ***Types of simple epithelium (composed of a single layer of cells).***

epithelia may differ, but the type of stratified epithelium is designated as squamous, cuboidal, or columnar on the basis of the shape of the cells at the apical surface. Thus, *stratified squamous epithelia* have thin, flat cells

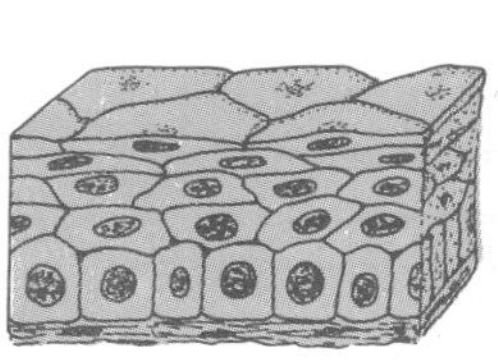
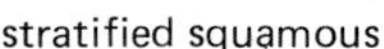

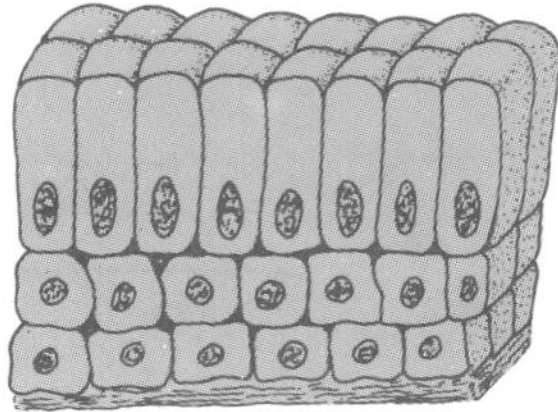

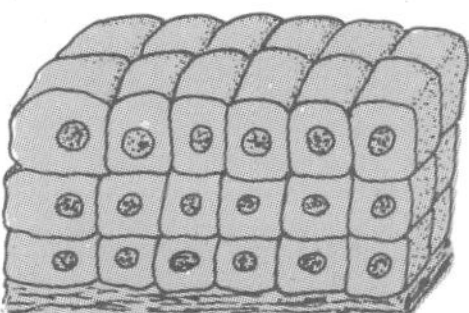

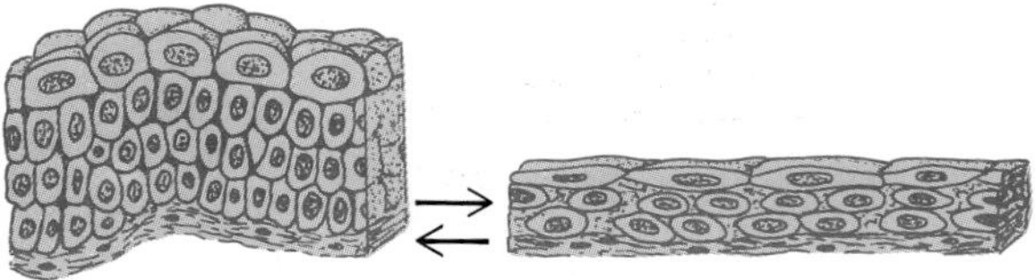

FIGURE 3-4 ***Types of stratified epithelium (composed of more than one layer of cells).***

at their apical surfaces although the cells in the underlying layers may be cuboidal or columnar.

In regions of the body, such as the urinary bladder, where the epithelium is subject to varying degrees of stretch, a special type of stratified epithelium, known as *transitional* epithelium, is found. As transitional epithelial membranes stretch, the cuboidal superficial cells flatten out and resemble squamous epithelium; thus, there is a considerable increase in surface area without rupturing the cells.

Epithelial cells divide at a fairly rapid rate, and some epithelial linings (for example, the luminal surface of the intestinal tract) are entirely replaced every few days. In those epithelia composed of several cell layers, the cells at the basal surface provide the stem cells that undergo division; the new cells then migrate toward the apical surface. Since most epithelial layers are subject to varying degrees of mechanical stress, the continual replacement of damaged cells is essential to maintaining the integrity of the epithelial boundary.

Epithelial Membranes

The term *epithelial membrane* is often used to denote the combination of an epithelial-cell surface and its underlying connective-tissue layer. Thus, it refers to a multicellular structure located at the surface of an organ, not to the thin lipoprotein plasma membrane at the surface of individual cells. There are two types of epithelial membranes in the body: (1) mucous membranes, and (2) serous membranes.

Mucous membranes line the internal surfaces which are connected to the outside of the body; examples are the lining of the mouth and gastrointestinal tract, the airways leading to the lungs, and the passages from the reproductive and urinary systems to the exterior. The epithelial cells at the surface of a mucous membrane secrete *mucus,* a mixture of proteins and polysaccharides having a thick, viscous consistency which moistens and lubricates the epithelial surface.

Serous membranes line the body cavities that are not connected to the exterior of the body; these include the abdominal, thoracic, and pericardial cavities. These membranes secrete a fluid containing salts and proteins, similar in composition to blood serum. These secretions moisten and lubricate the surfaces of the body cavities, allowing the organs within these cavities to move smoothly relative to each other and the surfaces of the cavity, as, for example, during inflation of the lungs.

Cell Junctions

Adhesions between epithelial cells play important roles in maintaining the mechanical integrity of the surface layers and forming a continuous barrier to the movement of molecules between the apical and basal surfaces. In addition to the electrostatic attractions between surfaces and specific binding proteins which assist in the aggregation of cells, as described in Chap. 2, a number of specialized junctions are found between the cells in an epithelial layer. Three types are illustrated in Fig. 3-5.

Desmosomes (Greek *desmos,* binding) are disk-shaped and have been compared to rivets or spot-welds firmly linking two adjoining cell surfaces at discrete points. In the region of a desmosome, the extracellular space is filled with a densely staining material; numerous fibers extend from the inner surfaces of the adjoining plasma membranes into the cytoplasm. The function of desmosomes is to provide regions of high mechanical strength between cells. Desmosomes are quite numerous between the cells in the epithelial layers

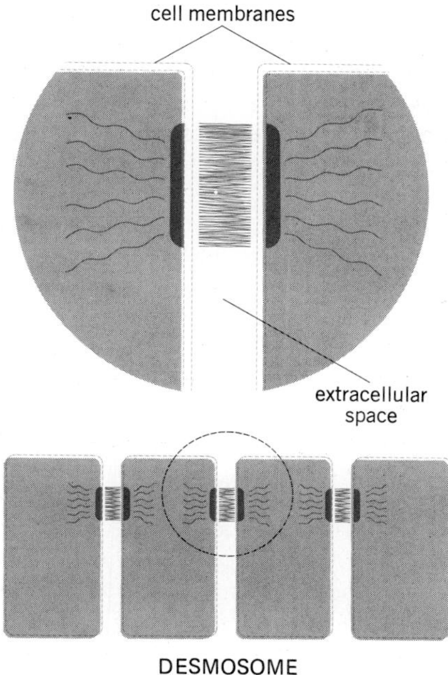

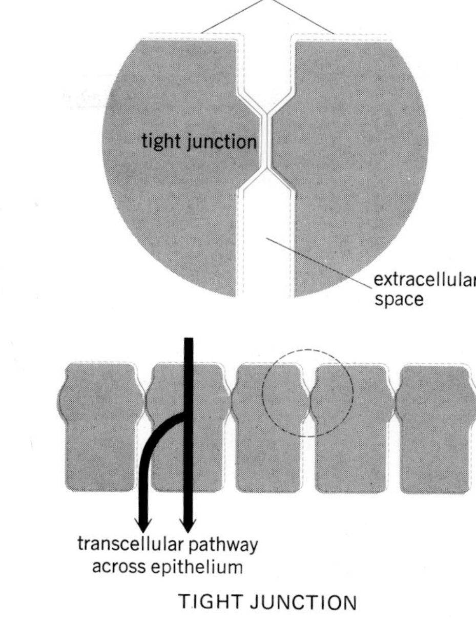

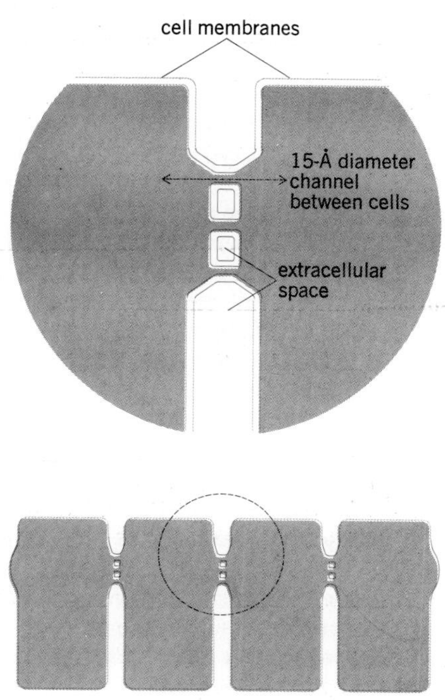

FIGURE 3-5 ***Schematic diagram of the three types of specialized cell junctions.***

of the skin, a tissue that is subject to considerable mechanical stress.

Tight junctions are formed by the fusion of the outer surfaces of two adjoining plasma membranes, leaving no extracellular space between the cells in the region of these junctions. This type of junction forms a bandlike collar completely around the circumference of the cell near its apical surface and seals off the extracellular space between the cells, eliminating it as a route for diffusion of molecules across the epithelium. In order to move across a layer of cells linked by tight junctions, a molecule must first cross the plasma membrane of an epithelial cell, pass through its cytoplasm, and exit through the plasma membrane on the opposite side. Wherever the extracellular passageways are blocked with tight junctions, epithelial cells are able to regulate the flow of molecules across the epithelial boundary by using the selective permeability and mediated-transport processes associated with their plasma membranes.

The third type of special junction is the *gap junction*. Like the desmosome, it is localized to a discrete region or spot between two adjoining cells and does not therefore block the flow of molecules around the junction through the extracellular space. At a gap junction, the two opposing plasma membranes come within 20 to 40 Å of each other; small cylindrical channels, like small pipes, extend across this space to link the cytoplasms of the two cells together. These small channels, about 15 Å in diameter, provide a pathway for the diffusion of small molecules and ions directly from the cytoplasm of one cell to the adjoining cell; this eliminates passing through the plasma membranes or entering the extracellular medium. Molecular exchanges between cells, by way of these gap junctions, may provide a way of chemically coordinating the activities of the cells within the epithelial layer. Certain types of muscle cells are also linked by gap junctions; as we shall see, the flow of ions through these junctions provides the pathway for the flow of electric currents between these cells.

Cell Surfaces

A characteristic of most epithelial cells, especially those associated with the transport of molecules across epithelial layers, is the extensive folding of the plasma membrane (Fig. 3-6). At the apical surface of the cell, numerous fingerlike projections, *microvilli*, extend into the surrounding medium. The microvilli may be short and irregularly spaced, or fairly long and densely packed, giving a brushlike appearance to the surface of the cell, hence the term brushborder. The foldings of the plasma membrane may also occur on the lateral and basal surfaces of the cell; in the basal region they often form extensive infoldings deep into the cytoplasm of the cell. By increasing the surface area of the epithelium, these infoldings increase the flux of various molecules across it since this flux is proportional to the total available surface area. Carrier-mediated transport systems are also located in these plasma membranes, and the types and concentrations of carrier molecules may differ at the apical and basal regions of the cell. Numerous mitochondria are often found in close association with the basal infoldings of the membrane where they supply ATP to the active-transport mechanisms concentrated in that region of the cell.

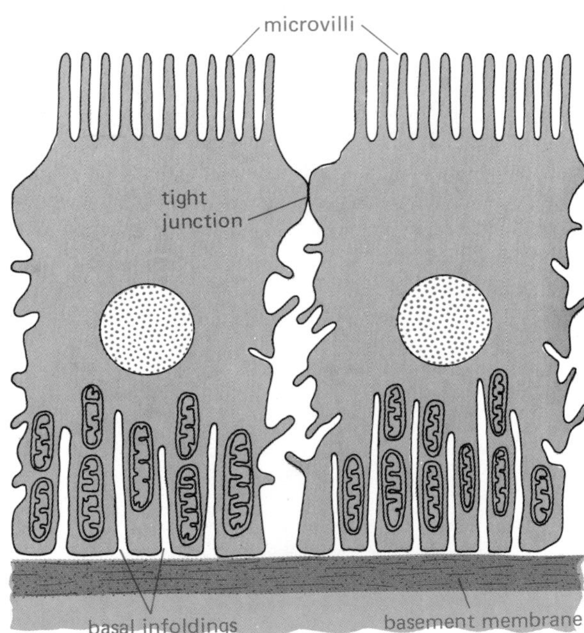

FIGURE 3-6 ***Extensive foldings of the plasma membrane are characteristic features of most epithelial cells.***

A further modification of the cell surface occurs in certain types of epithelia which have motile, hairlike *cilia* extending from their apical surface. In cross section a cilium consists of nine sets of paired microtubules surrounding a central pair and resembles the microtubular structure of the cell's centrioles from which the cilia are derived (Fig. 3-7). The beat of a cilium consists of a power stroke, during which the cilium behaves like

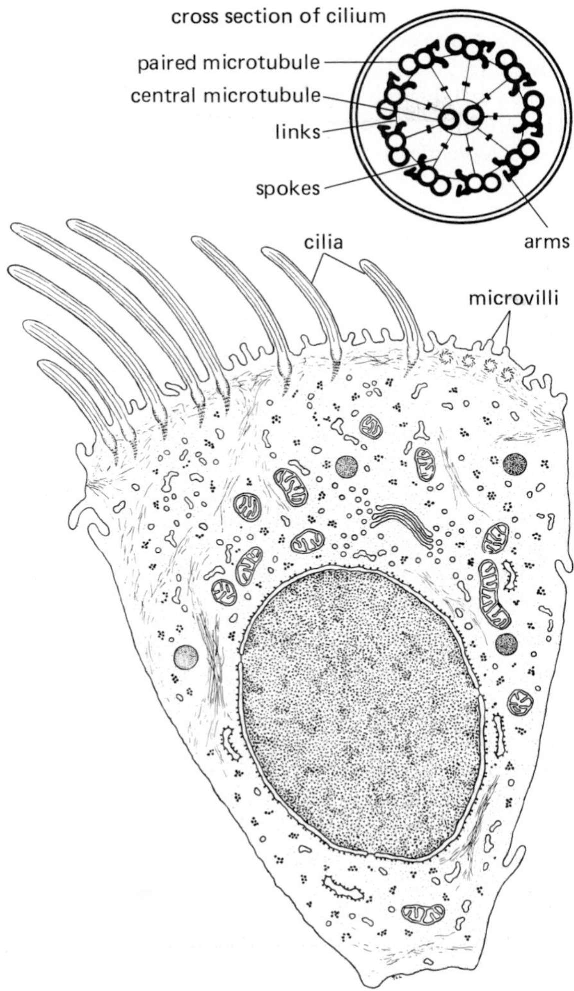

FIGURE 3-7 ***Cilia extending from the surface of an epithelial cell. The microtubular substructure of a single cilium is shown in cross section at top. The arms, links, and spokes hold the microtubules together in a functional unit. The arms are thought to be responsible for mobility. (Redrawn from Lentz.)***

a relatively stiff rod, and a return stroke, when it resembles more a piece of flexible rope (Fig. 3-8). Cycles of ciliary beating thus provide a unidirectional propulsive force at the cell surface. In single-cell organisms, such as a paramecium, the beating cilia propel the cell through the water. In the case of ciliated epithelia, the cell remains anchored to the epithelial surface and the beating cilia propel any overlying material at the surface along the epithelial layer in the direction corresponding to the power stroke of the ciliary beat.

Glands

The glands of the body are formed during embryonic development by the invagination of the epithelial surfaces of the ectoderm and entoderm. Many of the glands remain connected by ducts to the epithelial surface from which they were formed and are known as *exocrine glands*. The secretions of the exocrine gland cells pass along the ducts and are discharged into the lumen of the organ lined by that particular epithelium, or, in the case of skin glands, onto the surface of the skin. Thus, the functions of the secreted substances are carried out at the external surface of the epithelial layer onto which they are discharged.

A second group of glands, known as *endocrine glands*, are ductless in that they have lost, during embryonic development, the ducts connecting them to the epithelial surface from which they were formed by invagination. The molecules secreted by the endocrine gland cells are released into the interstitial fluid sur-

FIGURE 3-8 ***A beating cilium is rigid during the power stroke of the cycle and flexible during the return stroke.***

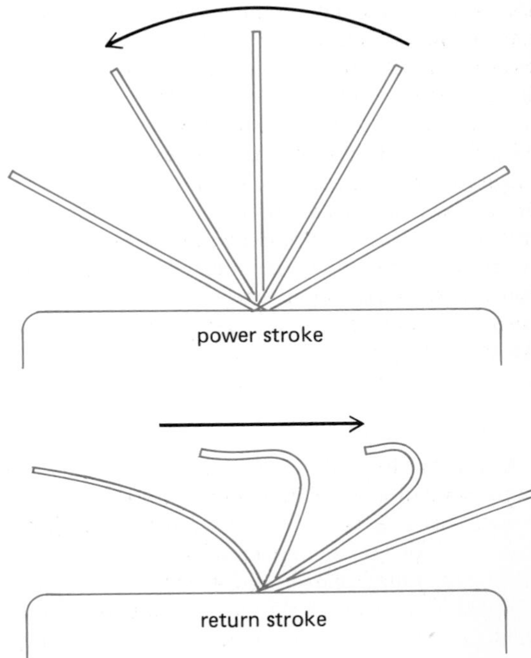

rounding the cells, diffuse into the blood, and are carried by the blood to other cells in the body. The special molecules synthesized and secreted by the endocrine glands are known as *hormones*. Hormones are blood-borne chemical messengers which, along with the nervous system, coordinate the activities of different cells in the body. The general properties of hormones and their mechanism of action will be described in Chap. 12.

In practical usage, the term "ductless gland" has come to be synonymous with endocrine gland. However, it should be noted that there are cells which secrete nonhormonal, nonwaste, organic substances into the blood; for example, fat cells secrete fatty acids and glycerol into the blood, and the liver secretes glucose, amino acids, fats, and proteins. These substances serve as nutrients for other cells or perform special functions in the blood but they do not act as hormones.

Structure of Glands *Exocrine Glands* Most exocrine glands are multicellular; however, one of the few types of unicellular gland is the *goblet cell*, many of which are interspersed throughout the epithelia of mucous membranes and secrete mucus. Most multicellular exocrine glands are invaginations of the epithelial sheet into the underlying connective tissue. The nature of the ducts and terminal portion of an exocrine gland determines its classification (Fig. 3-9). If the ducts pass directly to the surface, the glands are *simple;* if the ducts branch, the glands are *compound*. If the terminal portions are tubular in shape, the glands are *tubular;* if they are saclike, the glands are *acinar* (*alveolar*), the sacs themselves are *acini* (*alveoli*). If the terminal portions of a compound gland are both tubular and acinar, the gland is *tubuloacinar*. In a tubular gland, the secretory cells may be located throughout the tubule or restricted to its terminal portion. In an acinar gland, the primary secretion is produced by the cells of the acini, but the composition of the secreted material may be altered by additional secretion or reabsorption by the cells lining the duct during its passage from acinus to surface. In all cases, the epithelial cells of the gland are usually separated from the surrounding connective tissue by a basement membrane.

Endocrine Glands As described above, endocrine glands lose their connection with surface epithelium during embryonic development. In most cases, the separated clusters of cells lose their original lumen as well and form compact masses of cells penetrated by blood vessels and connective tissue. Certain endocrine cells (for example, those which produce the gastrointestinal hormones) remain embedded in the connective tissue underlying the epithelial lining from which they derived. Others constitute distinct organs which have migrated considerable distances from their epithelium of origin.

Mixed Glands Some organs consist of both exocrine and endocrine glands. In most cases (the pancreas, for example), the cells constituting the different gland types are quite distinct.

Gland Secretion The secretion of salts by exocrine glands is accomplished by ion-transport systems in the plasma membranes of these cells. Various ions are delivered to the cells by way of the blood and are actively transported into the lumen of the gland where they reach higher concentrations than are present in the blood or cytoplasm of the cell. Water diffuses into the lumen because the high osmolarity of the salt solution there, produced by the active ion transport, has lowered the water concentration in this region. The composition of this primary secretion may be modified as it flows along the excretory ducts as a result of the secretion and reabsorption of ions and water by the cells lining these ducts.

Some of the organic molecules secreted by exocrine cells are merely transported from the blood into the lumen of the gland in a manner similar to that described above for ions. However, many organic molecules secreted by exocrine cells and all the hormones secreted by endocrine cells are synthesized within the cells themselves.

Gland cells differ in the manner in which they release their secretory products. The type of secretion in which the gland cell remains intact during release is known as *merocrine* secretion (Fig. 3-10). The membrane of the secretory vesicles fuses with the plasma membrane, allowing the secretory product to diffuse into the extracellular medium (exocytosis). The secretion of specific proteins, as described in Chap. 1, occurs primarily by merocrine secretion. In *apocrine* secretion, in contrast, the secretory droplets accumulate at the apical end of the cell and, at the time of discharge, the entire top portion of the cell breaks away, releasing the secretory product and some of the cytoplasm. The remaining portion of the membrane of the ruptured cell reseals, and the cell repeats the process after synthesizing and accumulating more secretory product. The secretion of lipid droplets by the cells forming the mammary glands (breasts) is an example of apocrine secre-

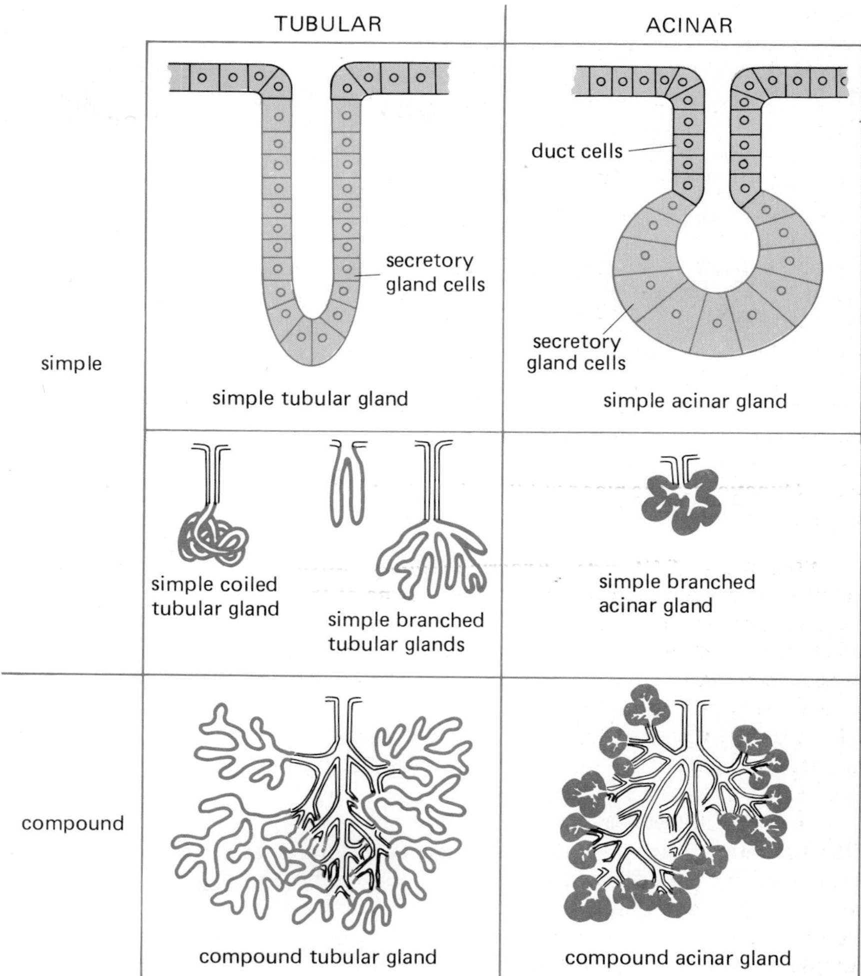

FIGURE 3-9 ***Classification of tubular and acinar glands.***

tion. A third way of releasing secretory products, *holocrine secretion,* involves the accumulation of secretory products within the cell followed by the complete disintegration of the cell, thereby releasing the entire cell contents and at the same time destroying the cell. The secretion of oils by the sebaceous glands in the skin occurs by this mechanism.

CONNECTIVE TISSUE

Of the four types of specialized tissues, connective tissue is the most abundant and diverse in its function and structure. In essence, it is all the cells and associated extracellular structures that remain after one has identified the cells and structures associated with the other three types of tissues. Unlike nerve, muscle, and epithelia, connective tissue has a considerable amount of extracellular material located between widely spaced connective-tissue cells. It ranges from the loose meshwork of cells and fibers underlying most epithelial layers to the solid structure of bone, and it includes cell types as diverse as fat-storing (adipose) cells and red and white blood cells. As its name implies, a major function of connective tissue is to connect, anchor, and sup-

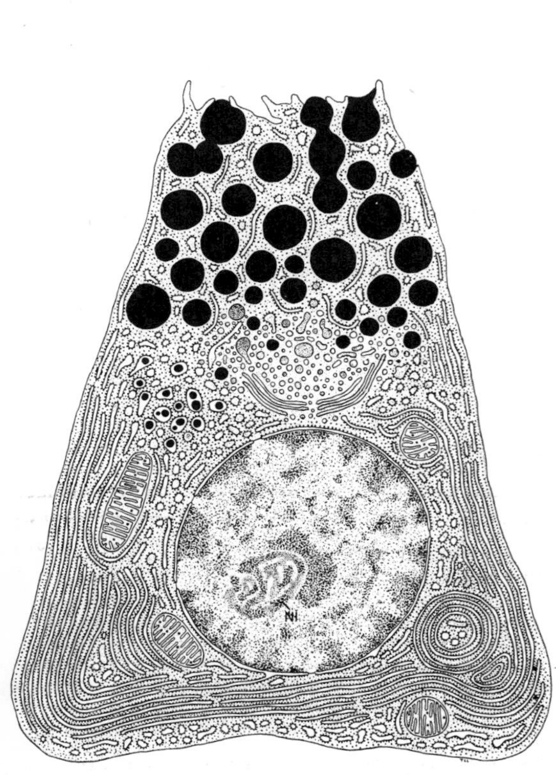

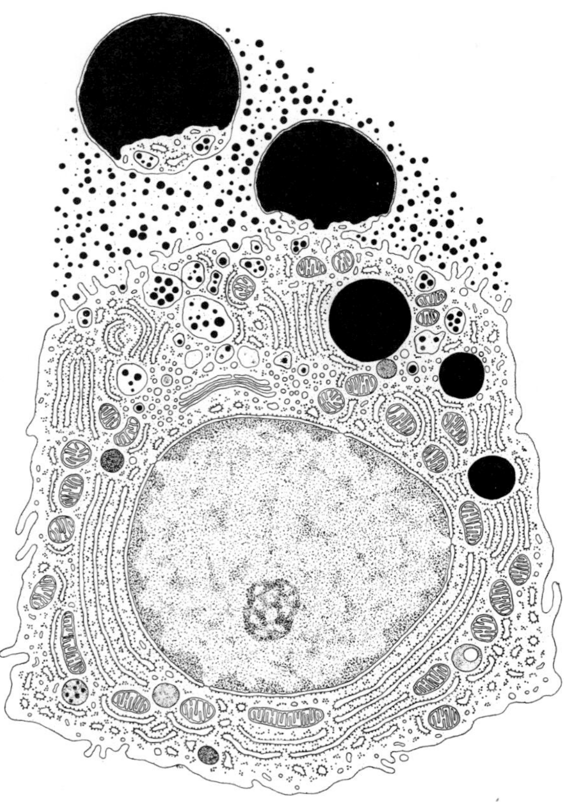

FIGURE 3-10 ***Examples of merocrine and apocrine discharge of secretory products. (Redrawn from Lentz.)***

port the other tissues of the body. The actual structures that provide this physical support are located in the extracellular spaces surrounding the connective-tissue cells. These supporting elements are secreted by the connective cells into the extracellular medium where they form a matrix consisting of various types of fibers embedded in a ground substance; this matrix may vary in consistency from a semifluid gel to the solid crystalline structure of bone. The three characteristic features of most connective tissues are thus (1) fibers, (2) ground substance, and (3) cells. The latter will be described in the discussion of the specific types of connective tissue.

Connective-tissue Fibers

The extracellular fibers of connective tissue are formed from protein molecules synthesized by the most abundant connective-tissue cells, known as *fibroblasts*. The proteins are synthesized by the ribosomes of the rough endoplasmic reticulum and released at the cell surface by exocytosis. Once outside the cell, they aggregate to form filaments (Fig. 3-11). The assembly of these filaments from individual protein molecules occurs spontaneously because of the inherent ability of the molecules to bind together. However, the rate of assembly and the thickness of the fibers formed are also influenced by the chemical environment surrounding them which is created by the fibroblasts. These fibers may slowly undergo further modifications in structure which are associated with an increase in the number of chemical bonds that cross-link the proteins in the filaments. These modifications contribute to the progressive changes in the flexibility and elasticity of many regions in the body associated with the process of aging.

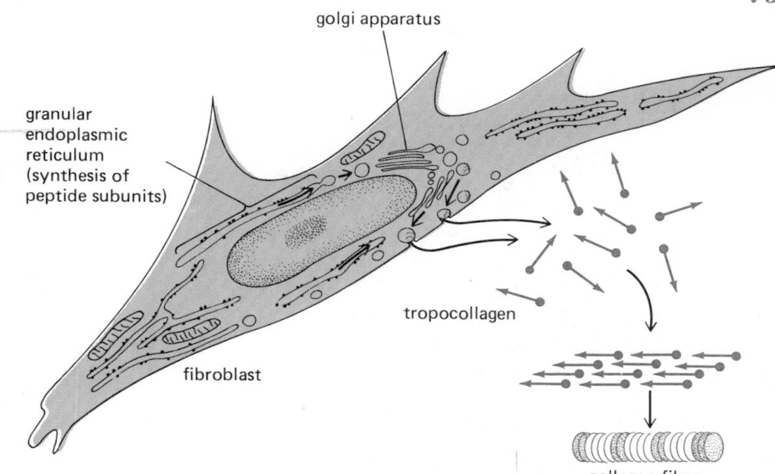

FIGURE 3-11 ***Formation of collagen fibers in the extracellular matrix from tropocollagen subunits synthesized and secreted by fibroblasts. (Redrawn from Bloom and Fawcett.)***

Connective-tissue fibers are classified as *collagenous, elastic,* or *reticular. Collagenous* fibers are composed of the protein collagen, which is secreted by fibroblasts as *tropocollagen,* a molecule composed of three rod-shaped polypeptide chains helically coiled about each other. The tropocollagen rods aggregate in a parallel organization in which each rod overlaps about one-quarter of the adjacent rods, giving rise to a staggered parallel arrangement which is manifested as a banded appearance in the fibers (Fig. 3-12). Collagen fibers range in diameter from 1 to 20 μm and may be many centimeters long; they do not branch. The physical properties of collagen fibers resemble those of a piece of rope; they are flexible and have a high tensile strength which resists stretching.

Elastic fibers are formed from the fibrous protein *elastin.* These fibers, which are thinner than the collagen fibers, do not have a periodic banding pattern.

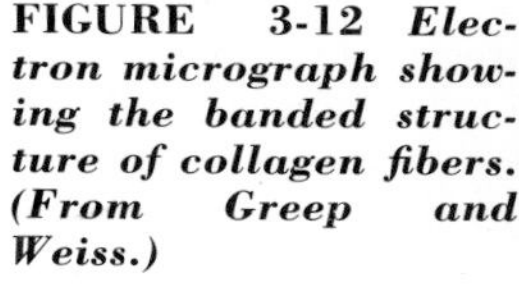

FIGURE 3-12 ***Electron micrograph showing the banded structure of collagen fibers. (From Greep and Weiss.)***

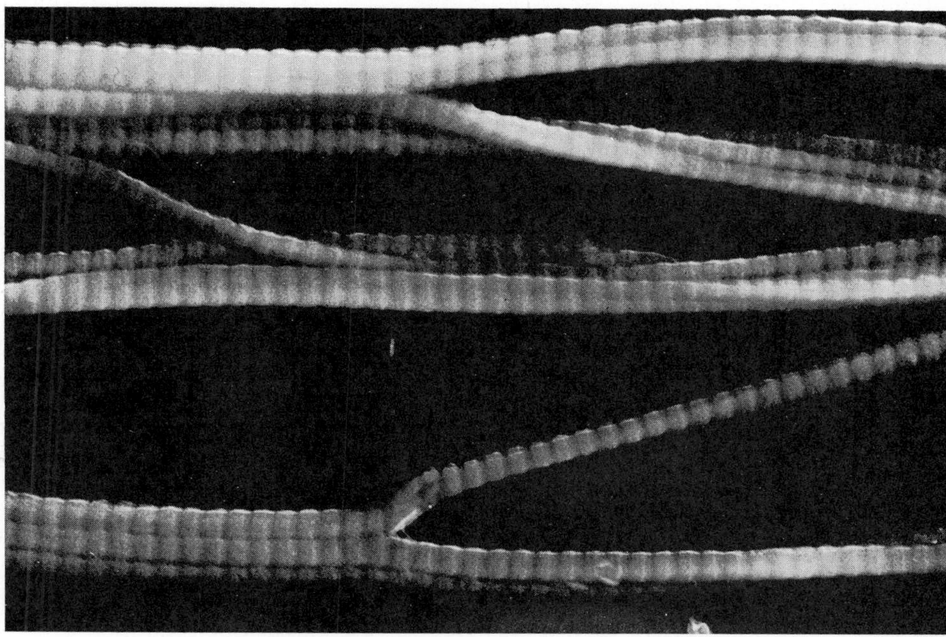

They have numerous branches which interconnect with other elastic fibers, producing a loose, fibrous network. Like rubber bands, they can be stretched and upon release return to their original length. It is the presence of these fibers in the connective-tissue layers of various structures, such as the walls of the blood vessels, which gives rise to their elastic behavior. Aging is accompanied by a progressive decrease in the number of elastic fibers and an increase in the number of collagen fibers in the connective tissues, leading to such changes as the hardening of the arteries.

Reticular fibers are very fine, highly branched fibers. They show the same periodic banding pattern as collagen and may actually represent an early stage in the formation of the larger mature collagen fibers. They can be identified in histologic preparations by their ability to bind silver stains; collagen and elastic fibers do not bind these stains.

Ground Substance

In addition to secreting fibrous proteins, connective-tissue cells release several types of polysaccharides and proteins which form the nonfibrous, homogeneous matrix, known as *ground substance*, surrounding them. These substances convert the extracellular matrix from a liquid into the semisolid state of a gel. The physical consistency of the extracellular medium varies considerably, depending on the amount of ground substance, the number and type of fibers embedded in it, and the possible deposition of various inorganic salts to form a more solid, crystallized extracellular material such as bone.

The structural polysaccharides which form the ground substance are known as *mucopolysaccharides;* they contain sugar residues which have both amino and carboxyl groups as part of their structure. The proteins of the ground substance contain similar sugar residues attached to some of their amino acid side chains, and these proteins are known as *mucoproteins*. Mucoproteins and mucopolysaccharides form most of the structure of basement membranes.

Mesenchyme

Embryonic connective tissue, or *mesenchyme*, first appears in the mesodermal germ layer, i.e., the layer between the entoderm and ectoderm. Mesenchyme consists of a loose arrangement of stellate-shaped cells surrounded by large extracellular spaces filled with interstitial fluid and a meshwork of extracellular fibers. Mesenchyme fills in the spaces between the developing organs of the embryo, and it is from these mesenchymal cells that all the connective cells of the adult body are ultimately derived (Table 3-1).

TABLE 3-1
CONNECTIVE-TISSUE TYPES

- I. Mesenchyme
- II. Adult connective tissue
  - A. Connective tissue proper
    - 1. Loose (areolar)
    - 2. Dense
    - 3. Reticular
    - 4. Adipose
  - B. Cartilage
    - 1. Hyaline
    - 2. Elastic
    - 3. Fibrous
  - C. Bone
    - 1. Spongy (cancellous)
    - 2. Compact
  - D. Dentin
  - E. Hemopoietic tissue, blood and lymphoid cells

Types of Connective Tissue Proper

Loose (Areolar) Connective Tissue Loose connective tissue contains large extracellular spaces composed of a loose network of collagenous, elastic, and reticular fibers embedded in a gel-like ground substance (Fig. 3-13). It serves as a "filler" throughout the body's structures and also forms the connective-tissue layer underlying most epithelia, including the skin. Reticular fibers are particularly prominent in the spleen and lymph nodes where numerous white blood cells lie in the interfiber spaces. A number of different cell types are distributed throughout this loose network, the majority of which are the *fibroblasts* which synthesize and secrete the fibers and ground substance of the extracellular matrix. Variable numbers of adipose-tissue cells may be present as well as a number of specialized cells which participate in defense mechanisms. These are the macrophages, plasma cells, mast cells, and some white blood cells. *Macrophages* (or *histiocytes*) are cells which are able to engulf and digest bacteria and other particulate matter that may invade the connective-tissue layers. The macrophages are known as "fixed" macrophages when they remain anchored to one position and as "free" or wandering macrophages if

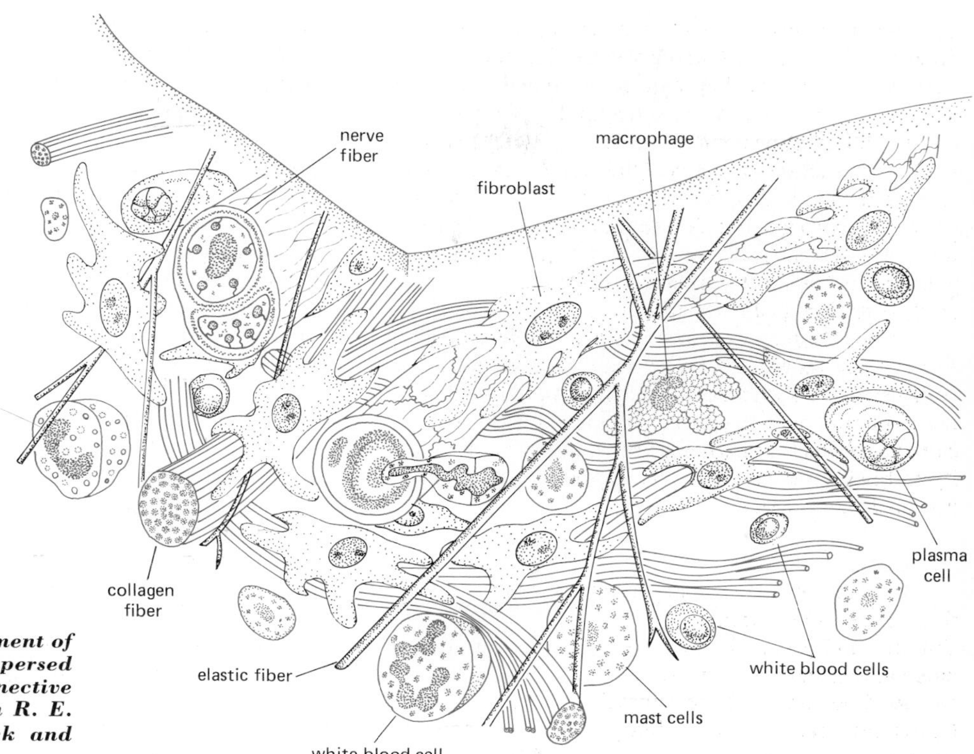

FIGURE 3-13 ***Assortment of cells and fibers dispersed throughout loose connective tissue. (Redrawn from R. E. M. Moore, in Warwick and Williams.)***

they migrate throughout the connective-tissue layers.

Plasma cells synthesize and secrete antibodies, proteins which play an important role in the defense mechanisms of the body and whose properties will be described in Chap. 22. These cells are found primarily in the lymph nodes but also occur in smaller numbers throughout the loose connective tissues of the body. *Mast cells* contain numerous dense secretory granules which they release into the surrounding medium in response to various local stimuli. The contents of the secretory granules include *heparin,* a substance which inhibits the coagulation of blood, and *histamine* and *serotonin,* chemicals which have powerful effects upon the smooth muscles and glands in the regions surrounding the connective tissues. These effects will be described in later chapters.

Dense Connective Tissue This type is distinguished from loose connective tissue by the greater abundance of fibers, particularly large collagen fibers, in its extracellular matrix. Fewer cells are found between the fibers than in loose connective tissue, and these are mostly fibroblasts. Dense connective tissue forms the capsules, which cover the outside of most internal organs, and the partitions or septa, which subdivide many organs into subunits. A particularly dense form of connective tissue occurs in the tendons and ligaments which link muscles to bones and bones to each other, respectively. These fibrous structures are composed almost entirely of compact bundles of parallel collagen fibers.

Adipose Tissue This tissue consists of large numbers of adipose cells embedded in a loose or reticular network of extracellular fibers. These specialized cells store large quantities of triglycerides which accumulate to form large droplets of fat surrounded by a thin layer of cytoplasm containing the cells' nuclei; these cells have the general appearance of a ring with the nuclear region corresponding to the setting in the ring (Fig. 3-14). Various hormones in the body regulate the synthesis and release of fat from these cells in accordance with the energy requirements of other cells in the body. Adipose-tissue cells are found throughout the

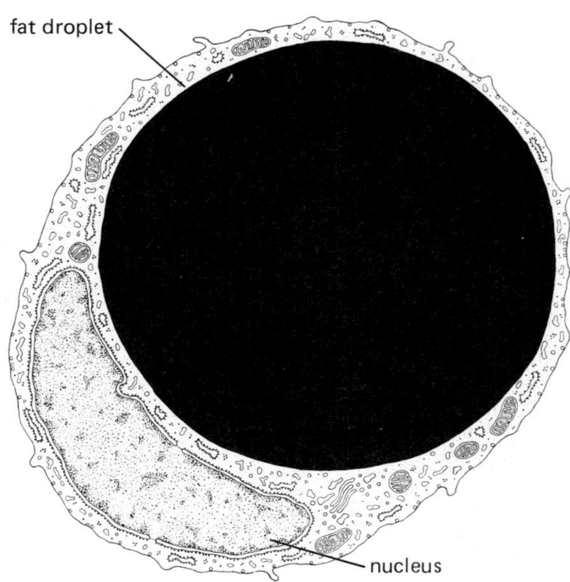

FIGURE 3-14 ***Adipose (fat) tissue cell. A thin rim of cytoplasm surrounds the large lipid droplet composed predominantly of triglyceride molecules. (Redrawn from Lentz.)***

loose connective tissues of the body, being especially numerous in the subcutaneous layers of the skin where they provide a protective cushion for the overlying skin as well as forming a layer of insulating material at the body's surface. Large amounts of fat are also found around the kidneys and heart and as padding around the joints.

Cartilage

Cartilage is a firm but flexible tissue. *Chondroblasts* arise from undifferentiated cells in the mesenchyme and form an extracellular matrix composed of bundles of collagen fibers embedded in a ground substance consisting of acid mucopolysaccharides. As this matrix is laid down around each cell, the chondroblasts become further and further separated from each other. Each cell comes to exist in a small cavity or *lacuna* enclosed by the surrounding matrix. These isolated cells scattered throughout the matrix of fully developed mature cartilage are now called *chondrocytes* (Fig. 3-15). Surrounding the outer surface of most cartilaginous tissue is a sheath of dense connective tissue known as the *perichondrium,* composed of fibroblasts and bundles of collagen fibers. The fibroblasts in this sheath can undergo differentiation to form chondroblasts which then deposit new layers of matrix along the surface of the cartilage.

Most cartilage is avascular; thus, nutrients can reach the chondrocytes only by diffusing through the matrix from blood vessels in the perichondrium. Since diffusion cannot deliver nutrients rapidly over long distances, the survival of the chondrocytes depends upon their proximity to the perichondral surface. As a result, cartilage occurs in relatively thin plates or sheets.

There are three types of cartilage in the body: (1) hyaline, (2) fibrous, and (3) elastic. *Hyaline cartilage* is the most abundant form; its matrix of collagen fibers and mucopolysaccharides is translucent, and the individual collagen fibers cannot be seen with the light microscope because of their small size and the density of the surrounding mucopolysaccharides. This type of cartilage is found in the nose, in supporting rings surrounding the large airways to the lungs, between the ends of the ribs and the sternum (breast bone), and on the surfaces of many bones where they articulate at joints. Hyaline cartilage is also found in regions of bone growth, as will be described in the next section.

Fibrous cartilage has a structure that is intermediate between hyaline cartilage and dense connective tissue, and its collagen fibers are large and quite visible. Fibrous cartilage forms the intervertebral disks of the spinal column and blends with the tendons and ligaments at their junctions with bone.

Elastic cartilage is similar to hyaline cartilage but contains an additional elastic material within its matrix, which consists of a loose meshwork of branched thin fibers and granules embedded in an amorphous material. The elasticity of the matrix material gives this type of cartilage greater flexibility than is found in other forms. It is found in the external ear and auditory tube.

Bone

Bone tissue is the special type of connective tissue which forms the solid matrix of bones, the latter being technically organs composed of bone tissue, other forms of connective tissues, nerves, and blood vessels. Bones form the internal supporting and protecting structures of the body's skeleton and provide the attachments for the muscles that move the skeleton. In addition to these mechanical functions, the central cavity of some bones contains the connective tissue, *hemopoietic tissue,* which forms a variety of blood cells. Bone also performs an important role in the regulation of the calcium and phosphate concentrations in

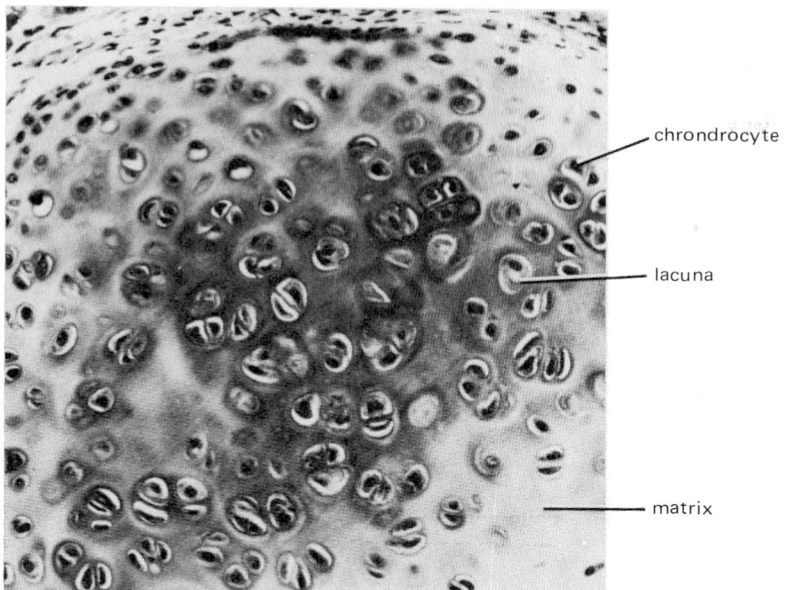

FIGURE 3-15 ***Cartilage. Each chondrocyte lies in its own lacuna encased by the surrounding matrix of collagen fibers and acid mucopolysaccharides. (From Greep and Weiss.)***

the internal environment through the controlled deposition of these salts in the bone matrix or their release. Living bone is a dynamic structure that is continually resorbed, re-formed, and remodeled.

Gross Structure of Bone The basic features of a typical long bone can be seen in Fig. 3-16. The long, central shaft is the *diaphysis,* and the two ends are the *epiphyses.* The portion of each epiphysis which contacts other bones is the *articular surface;* it is covered with hyaline cartilage. The rest of the bone is covered by a tightly adhering sheath of connective tissue, the *periosteum.*

A longitudinal section through a long bone reveals a central *medullary cavity,* which contains the *yellow marrow,* composed largely of fat. The dense, ivorylike surface layer of the bone is *compact bone.* Below this layer is the region of *cancellous* (or *spongy*) bone, which has a latticelike structure containing many small cavities which merge with the large central cavity. Spongy bone is arranged like struts to resist forces that tend to compress the bone or put tension on it; the spaces between the spongy bone are filled with *red marrow,* consisting mainly of hemopoietic tissue, some macrophages, and fat cells.

The long bones are subject to the greatest bending and twisting forces in the central region of the diaphysis, and here the compact bone is relatively thick and the spongy bone almost nonexistent. Spongy bone occurs mainly in the epiphyses and near the ends of the diaphysis where the bone is subject to the greatest weight-bearing stresses; here the shell of compact bone is relatively thin.

The tensile strength of bone, i.e., its resistance to being stretched under applied forces, is comparable to that of cast iron, yet it is only a third the weight of iron. In its flexibility, bone resembles steel more than cast iron; it has about half the strength of steel. The tubular shape characteristic of most long bones is the strongest, yet lightest, arrangement that can be designed, and in bone this shape is modified to withstand the types of stresses to which the particular bone is exposed.

As we shall see in Chap. 7, there are bone types other than long bones. Their structure tends to be similar to that of the epiphyses of long bones: primarily cancellous bone (with no medullary cavity) underlying a

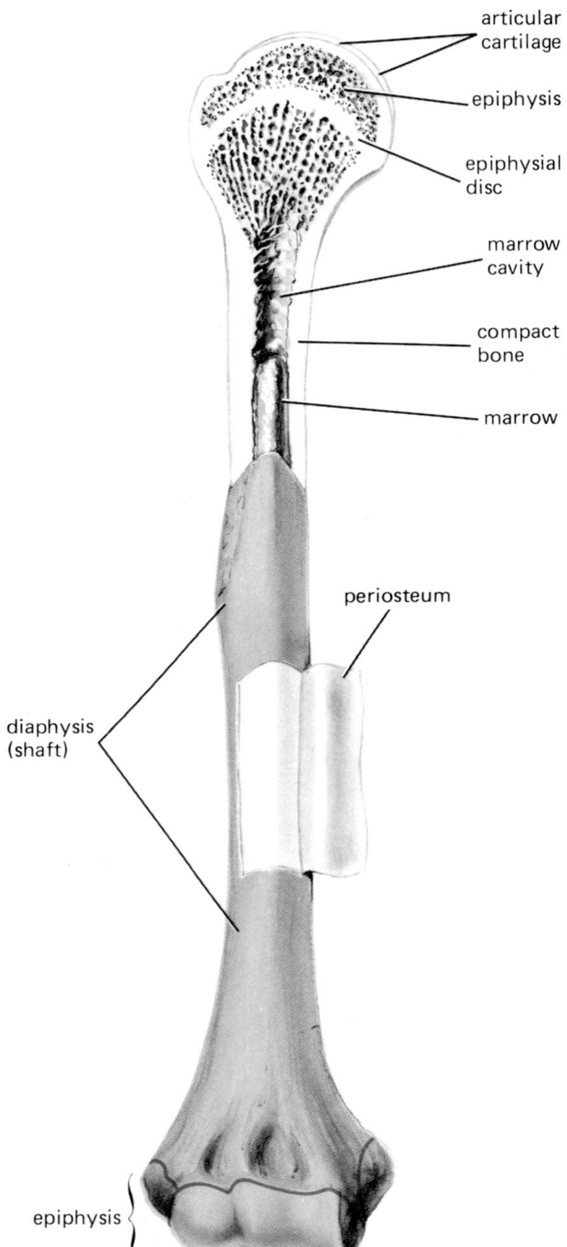

FIGURE 3-16 ***Cross structure of a typical long bone.***

thin layer of compact bone. As with long bones, the entire bone, except for the surfaces involved in joints, is encased in periosteum.

Structure of Bone Tissue The composition of the matrix surrounding the bone-forming cells is basically similar in both compact and cancellous bone. The organic components of the bones' matrix are collagen fibers embedded in an amorphous mucopolysaccharide ground substance. Bound to these fibers are needlelike crystals known as hydroxyapatite, which consists of inorganic minerals, mostly calcium phosphate, calcium carbonate, and sodium salts in proportions given by the formula $Ca_{10}(PO_4)_6(OH)_2$. These minerals, composing 65 percent of the bone's weight, give bone its hard consistency.

The microscopic unit of a transverse cross section of compact bone is a *haversian system.* Each system contains at its center a small channel, a *haversian canal,* which runs essentially parallel to the long axis of the bone. One or two small blood vessels, as well as nerve fibers, lie within each canal. Surrounding each haversian canal is a concentric series of rings, *lamellae,* composed of mineralized matrix (Fig. 3-17). Scattered along the boundaries of each lamella are small cavities known as *lacunae,* each of which contains a single bone cell, or *osteocyte.* Extending radially from the lacunae are numerous small channels, *canaliculi,* which link the lacunae of adjacent rings and ultimately extend into the haversian canal. Cell processes extend from the osteocytes into the canaliculi, and it is through these channels that nutrients from the blood vessels in the haversian canal reach the osteocytes. The relatively thin, interconnected columns of bone tissue which form the cancellous regions of bone do not contain haversian canals, although the tissue is still arranged in concentric lacunae-containing lamellae; in this region nutrients are delivered to the osteocytes through the canaliculi from blood vessels that pass through the bone marrow and hollow regions of the spongy bone.

Bone Development Calcified bone is formed by the process known as *ossification;* although it occurs in two different ways, intramembranous bone formation and endochondral bone formation, the resulting structures of the calcified bone are similar. *Intramembranous bone formation* involves direct calcification of the organic bone matrix; *endochondral bone formation* involves the replacement of a cartilaginous structure by calcified bone. The bones of the

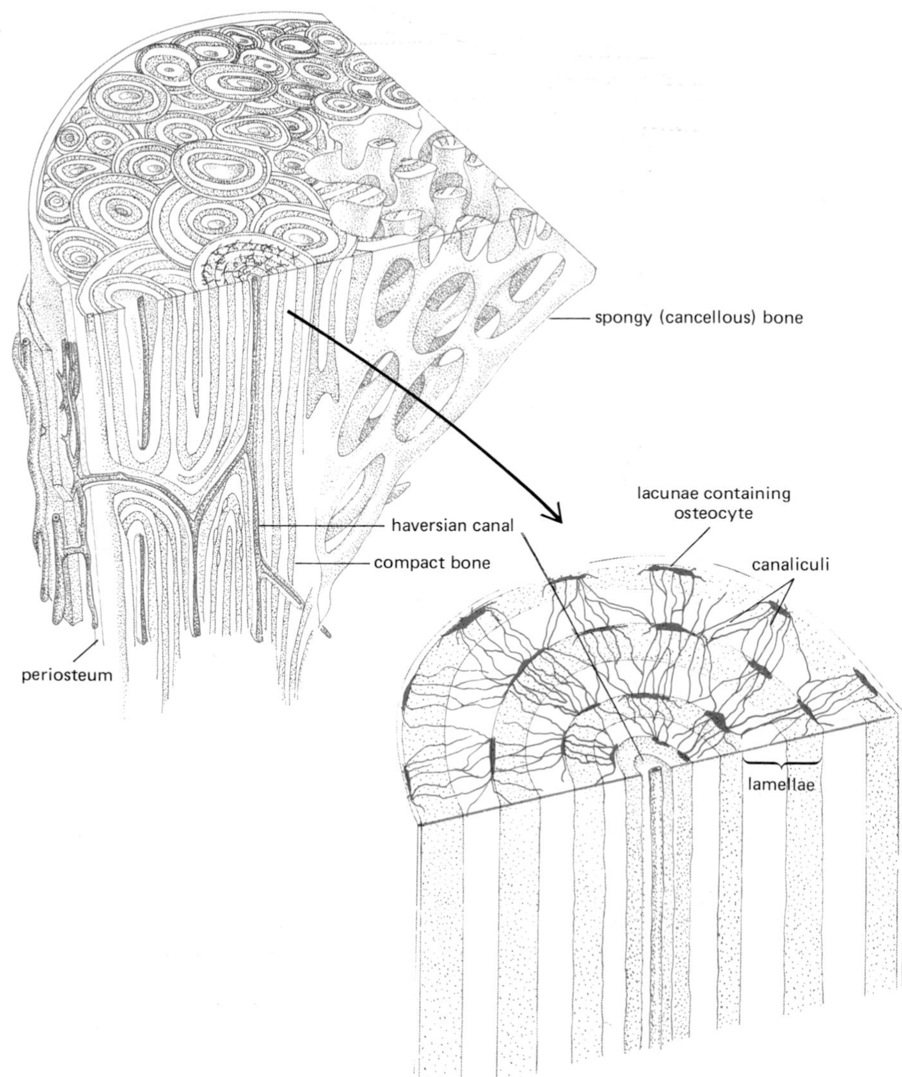

FIGURE 3-17 ***Arrangement of haversian systems in compact bone, showing an enlarged segment of one haversian canal and surrounding lamellae. (Redrawn from R. E. M. Moore, in Warwick and Williams.)***

skull are the only bones that are formed entirely by intramembranous bone formation; the diaphysis of a long bone is also formed in this way, but the epiphyses utilize endochondral bone formation.

Intramembranous bone formation begins in the embryo with the differentiation of mesenchymal connective-tissue cells into bone-forming cells, or *osteoblasts*. These osteoblasts synthesize collagen fibers and an organic ground substance and surround themselves with this nonmineralized matrix, known as *osteoid*. As the osteoid becomes mineralized, small islands (spicules) of bone are formed, each surrounded by a single layer of osteoblasts. Adjacent spicules then fuse at various points to form a spongelike structure. As the process continues, some of the osteoblasts are trapped within the mineralized matrix and become osteocytes. Mesenchymal tissue trapped within the spaces of the spongy bone eventually gives rise to the hemopoietic tissues of the bone marrow. The layer of mesenchymal tissue surrounding the developing bone becomes the *periosteum;* from this periosteal sheath further osteoblasts arise and add new concentric layers of ossified bone to the growing network of spicules. As the bone matures, the spaces between the spongy-bone tissue near the periosteum become filled with calcified-bone tissue to form compact bone.

Endochondral bone formation begins with the embryonic differentiation of mesenchymal cells into chondroblasts which deposit a matrix of cartilage in the shape of the developing bone. The next step is the transformation of the outer sheath of cartilage-forming cells, the *perichondrium,* into a sheath of bone-forming cells, the periosteum; this occurs initially around the mid-region of the bone shaft. At the inner surface of the periosteum, adjacent to the cartilaginous bone, osteoblasts begin to lay down a mineralized matrix of bone in a manner resembling the process of intramembranous bone formation. When this stage is completed, the developing bone consists of a mineralized shaft with two large masses of cartilage at the epiphyses.

The next stage involves the invasion of the epiphyses by osteoblasts to form secondary centers of ossification at the two ends of the bone. The formation of these secondary centers of ossification begins in different bones sometime between birth and the first 3 years of life. Between the primary ossification center along the shaft of the bone and the two secondary ossification centers at the epiphyses are layers of growing cartilage. In these regions, known as the *epiphyseal plates,* the growth of the cartilage approximately keeps pace with the rate at which it is being replaced by mineralized bone.

Four zones exist within the epiphyseal plate (Fig. 3-18). Resting cartilage constitutes the most distal zone and contains chondrocytes distributed at random throughout its matrix. Next is the zone of chondrocyte multiplication which is the major area for new cartilage deposition; in this zone the cells become arranged in rows parallel to the long axis of the bone. In the third zone, the zone of chondrocyte hypertrophy, the cells become enlarged and show signs of beginning disintegration. In the last zone, that of cartilage calcification, mineralized bone replaces the disintegrating cartilage; this zone, which is closest to the diaphysis, receives a rich blood supply from the underlying marrow. By this continuous process of cartilaginous growth and replacement by mineralized bone, the ends of the long bone are able to elongate during the years of growth. Eventually, the epiphyseal plates at the ends of the bone become completely replaced by bone tissue (epiphyseal closure) and further lengthening of the bone ceases. This occurs at various ages in the different bones but is usually complete in all the long bones by the age of 20; x-raying the skeleton to determine which epiphyses are still open permits the assessment of the "bone age" of an individual.

The formation of new bone tissue and bone mineralization do not actually cease with closure of the epiphyseal plates. The deposition of newly mineralized bone occurs throughout life although normally at a rate balanced by the removal of old bone; thus, there is a continual resorption and re-forming of bone. This removal and deposition of bone occur most prominently both along the outer surface of the bone underlying the periosteum and on the inner surface of the marrow cav-

FIGURE 3-18 ***Schematic representation of active cartilage growth and transformation into bone in the region of the epiphyseal plate. (Redrawn from R. E. M. Moore, in Warwick and Williams.)***

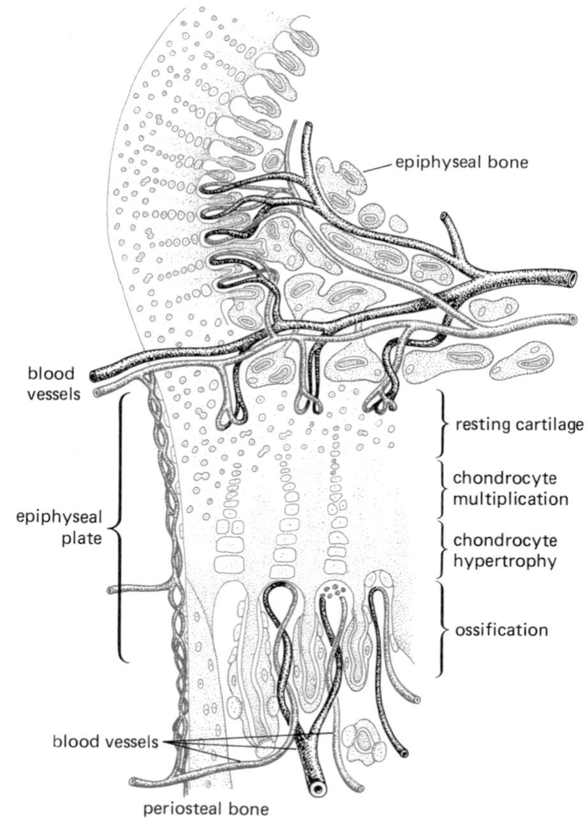

ity. Large, multinucleated cells, known as *osteoclasts*, have long been thought to be a distinct cell line mediating the demineralization of bone; however, it is more likely that osteoblasts, osteoclasts, and osteocytes are actually phases of a single cell type. A number of hormones are known to influence bone growth, epiphyseal closure, and the rates at which calcium and phosphate are exchanged between the bone matrix and the blood, by acting on these cell types; these hormones will be described in later chapters.

It should be noted that bone takes up lead and other heavy metals in a manner similar to that for calcium. This is often a protective mechanism in that storage in bone removes the element from circulation, thereby reducing its toxicity for other body sites. However, when the element is a radioactive one (radium, plutonium, or strontium, for example), the result is harmful since the local radioactivity may cause death or cancerous changes in the bone cells.

Bone Fractures When a bone is cracked or broken, a series of events occurs which leads to the repair of the broken bone over a period of several weeks through the formation of new bone tissue in the region of the break. The repair process is initiated by the migration of blood vessels and connective tissue from the periosteum into the region of the break. This dense fibrous tissue fills in the break and forms a temporary union, or *callus* (Fig. 3-19A). The cells near the broken edges differentiate into osteoblasts; those further away become chondroblasts and begin to deposit cartilage between the broken surfaces of the bone. This cartilage is slowly replaced by mineralized bone tissue, thereby completing the repair. Some common fractures are shown in Fig. 3-19B.

Metabolic Bone Disease There are a large number of diseases in which abnormal bone metabolism is due not to an inherent defect in the bone cells but

FIGURE 3-19 ***(A) Stages in the healing of a bone fracture. (B) Various bone fractures. A*** **fracture** ***is any break in the continuity of a bone. If the break crosses the entire width of the bone, it is a*** **complete** ***fracture; if it does not, it is*** **incomplete, or partial.** ***Thus, a*** **greenstick** ***fracture, in which the cortex opposite to the break is still intact, is a type of incomplete fracture. A*** **closed, or simple,** ***fracture is one in which the skin or mucous membrane covering the bone is not open. Both the bone and the overlying skin or mucous membrane are broken in an*** **open, or compound,** ***fracture. A*** **compression** ***fracture results in decreased length or width of a bone or portion of bone. A*** **comminuted** ***fracture contains more than two fragments or potential fragments of bone.***

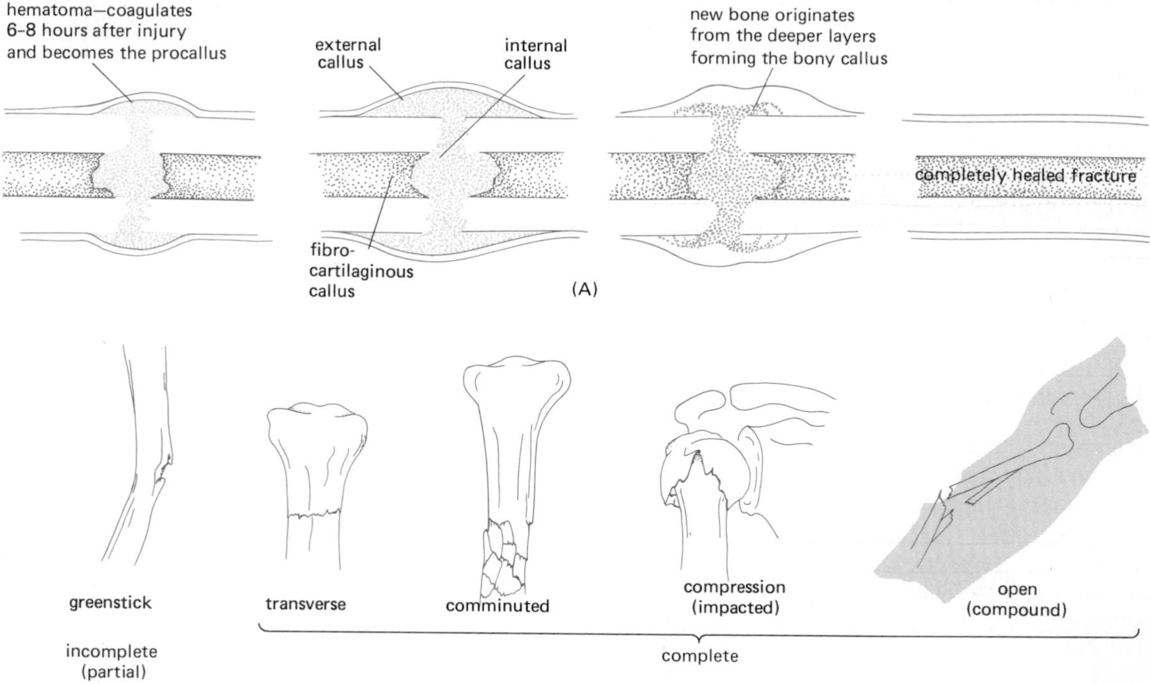

is secondary to a metabolic disturbance originating elsewhere in the body. For example, *osteosclerosis*, increased amounts of calcified bone, occurs in patients with certain tumors, lead poisoning, and deficient parathyroid gland function. *Osteomalacia*, inadequate mineralization per unit matrix, occurs in rickets (vitamin D deficiency) and a variety of other diseases. *Osteoporosis* is a decrease in bone mass with no change in the mineralization:matrix ratio; it may occur as a result of immobilization but is most common and important in postmenopausal women, in whom it accounts for serious bone malfunctions.

SKIN

The multiplicity of structure and function of epithelium and connective tissue can be illustrated by examining the largest organ in the body, the *skin*. The skin, or *integument*, is composed of two layers of tissues: an outer epithelial layer, the *epidermis*, and an underlying layer of connective tissue, the *dermis*, which also contains numerous nerve fibers and blood vessels.

Functions

In lower forms of animals, many of the exchanges of molecules between the external and internal environments occur across the skin. In human beings and other higher organisms, these exchanges occur mainly across epithelial surfaces that have invaginated from the surface of the body into the interior (forming the lungs and intestinal tract, etc.), and the skin forms a barrier that prevents most substances from entering or leaving the body. For example, potentially harmful bacteria are barred from entering, and loss of water by evaporation from the moist surfaces of the underlying tissues is greatly diminished. The skin also protects the tissues of the body from damage by the ultraviolet radiation present in sunlight. The toughness of this flexible outer covering also protects internal organs from physical damage that would result from frictional contacts between the body and objects in the external environment.

Much of our awareness of the external world results from stimulation of the nerve fibers in the skin by various environmental stimuli, leading ultimately to the sensations of touch, pressure, and temperature.

The skin also assists in the regulation of body temperature, since heat, released during the metabolism of cells, is carried by the blood to the skin, from which heat loss can be controlled. The secretion of sweat by glands in the dermis of the skin also assists in temperature regulation by carrying heat away from the body in association with the process of water evaporation. Although the skin is not a major excretory organ, small amounts of water, salts, and a variety of organic molecules are secreted onto the surface of the skin by the sweat and sebaceous (oil) glands. During periods of intense physical exercise, large quantities of salt and water may be lost from the body through sweating. If sustained for a long period of time, such losses may significantly alter the composition of the internal environment.

The adipose tissues underlying the dermis are one of the major sites in the body where fat is stored, and the skin performs a number of other metabolic functions, including the formation of vitamin D_3 (a precursor of vitamin D) in the presence of sunlight. (Further modifications of vitamin D_3 must occur in the cells of the liver and kidney before it is able to function in the body.)

Damage to large areas of the skin can be life-threatening. Excessive amounts of water and salt pass through the damaged area of skin and are lost from the body, and infection becomes a major problem, as bacteria readily penetrate the damaged surface.

Epidermis

The outer surface of the skin consists of a layer of stratified squamous epithelium which varies from 30 to 50 cell layers thick. Proceeding from the innermost to the outermost zone, four major zones (Fig. 3-20) can be distinguished within this epithelial layer: (1) stratum germinativum, (2) stratum granulosum, (3) stratum lucidum, and (4) stratum corneum.

The *stratum germinativum*, composed of 8 to 10 layers of cells, is separated from the underlying dermis by a basement membrane. The cuboidal cells resting upon this membrane are the stem cells whose mitotic activity continually replaces the cells in the upper epidermal layers. It takes about 14 days for a cell to move from the stratum germinativum to the stratum corneum, where it may reside for another 30 days before being sloughed. In the upper layers of the stratum germinativum the cells are irregularly shaped and interconnected by multiple desmosomes. These cells contain numerous ribosomes where the fibrous protein *keratin* is synthesized. As the cells continue to be pushed toward the surface of the epidermis, they become flattened, the amount of keratin in them progressively increases, and numerous granules of keratin begin to appear. These flattened cells containing the granules form the *stratum granulosum*.

Scattered throughout the stratum germinativum and the stratum granulosum are pigment-synthesizing

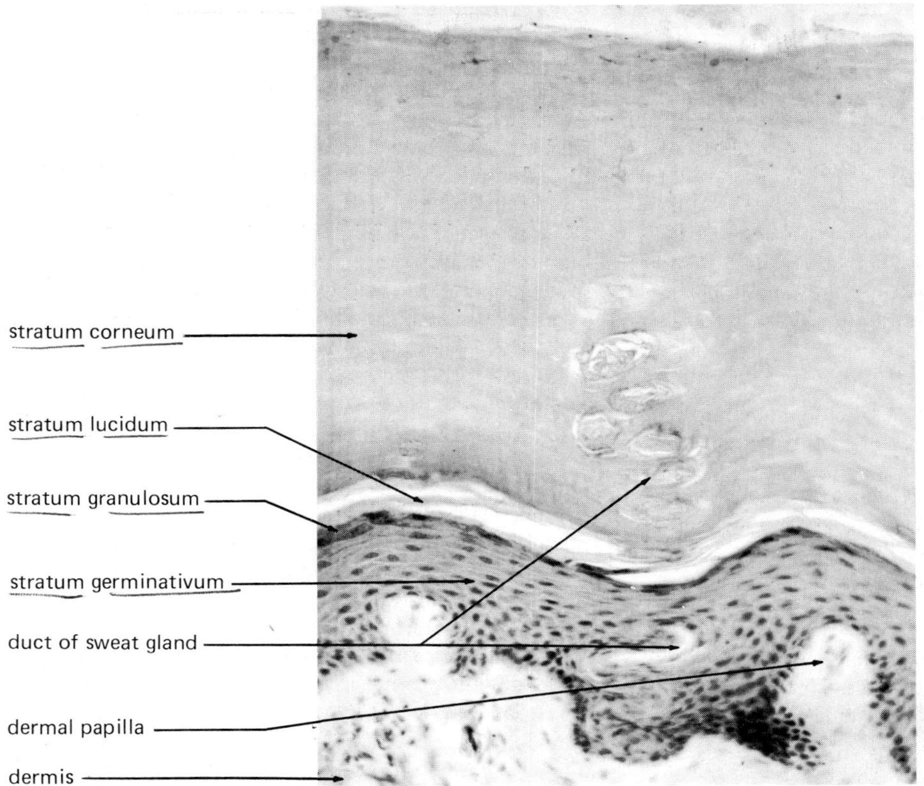

FIGURE 3-20 ***Outer surface of the skin, showing the four layers of the epidermis. (From Langley et al.)***

cells known as *melanocytes*. These cells synthesize a black pigment, *melanin,* which is derived from the amino acid tyrosine. Melanin accumulates in the dense, membrane-bound vesicles within the melanocytes. These vesicles are then secreted and taken up by the surrounding epithelial cells by phagocytosis (endocytosis). The number of melanocytes relative to the number of epithelial cells is approximately the same in all skin colors; the wide range of skin colors found among the human population is determined by the numbers, size, and distribution of the melanin granules. The pink color of Caucasian skin is due to the failure of the melanin pigment to completely mask the blood vessels in the underlying dermis. Ultraviolet radiation in sunlight stimulates the synthesis of melanin, producing an increase in the melanin content and thereby a darkening or tanning of the skin. As a result of genetic mutation, some individuals are unable to synthesize the enzyme that converts tyrosine into melanin. Such individuals, known as albinos, have no pigment in their skin, although melanocytes are present. This lack of pigment makes the albino's skin particularly sensitive to the damaging effects of ultraviolet light.

The *stratum lucidum* is a thin, clear area between the stratum granulosum and the stratum corneum that is most pronounced in the thick epidermal layer overlying the palms of the hands and soles of the feet. By the time the outwardly migrating cells enter the stratum lucidum, they show signs of disintegration and have lost their nuclei and organelles. The cells in this layer are filled with a transparent substance which appears to be a precursor of keratin.

The outermost layer of the epidermis is the *stratum corneum,* or horny layer. It is composed of non-nucleated disintegrated flat, squamous cells which are filled with keratin fibers and contain very little water. It is

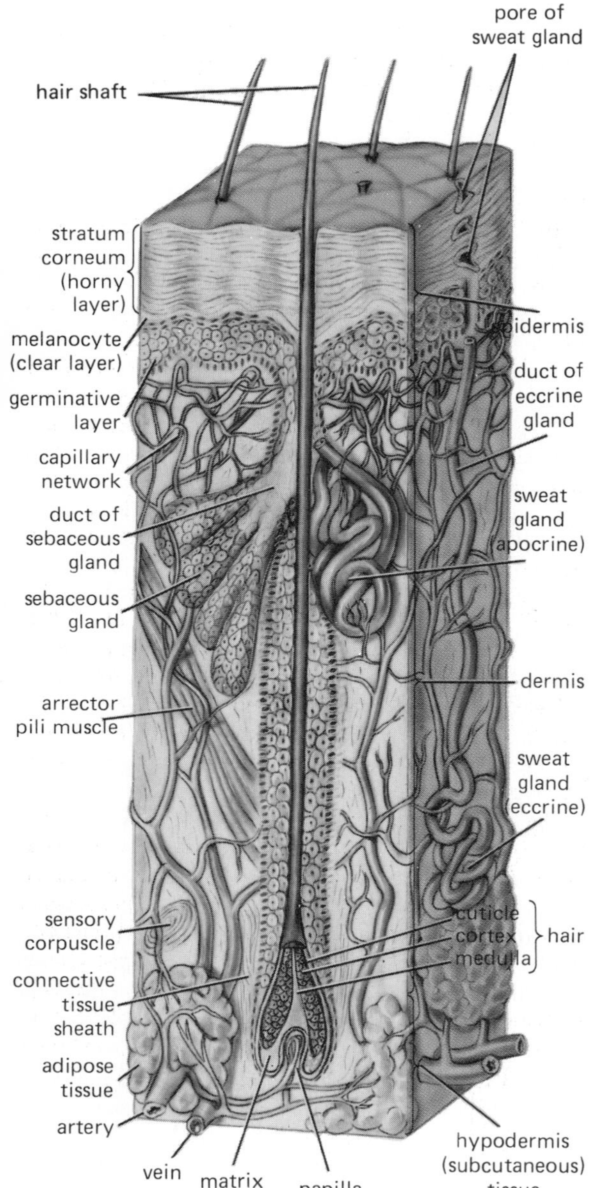

FIGURE 3-21 ***Cross section of skin, showing the organization of glands and hair.***

this compact, fibrous layer of dead cells that forms the protective barrier of the skin.

At the tips of the fingers and toes, the epidermis forms the horny plates of the *nails,* which, like the stratum corneum of the skin, are composed of a compact layer of dead, flat cells filled with keratin. The growing portion of the nail is located below a thin fold of skin, the *cuticle,* and extends below the *lunula,* the white, half-moon-shaped region at the base of the nail, a region containing proliferating cells of the stratum germinativum. As the cells accumulate keratin, they become flattened, die, and form the hard compact structure of the nail plate, which grows at a rate of about 0.5 mm per week. Nails appear pink because underlying blood vessels can be seen through the translucent, keratinized layer of the nail. Unlike superficial layers of the skin, keratinized surface of the nail is not shed; the nail will continue to grow until trimmed or broken.

Dermis

The epidermis lies below the connective-tissue layer of skin, the *dermis* (Fig. 3-21), which is from 1 to 3 mm thick. The outermost region of the dermis is the papillary layer; the remaining portion, which occupies about 80 percent of its thickness, is the reticular layer.

The *papillary layer* consists of loose connective tissue, containing many fine collagen and elastic fibers and numerous fibroblasts, but few fat cells. This layer is folded into ridges, or papillae, which extend into the epidermis and produce ridges on the surface of the skin. These are especially noticeable on the palms of the hands and fingers where they produce the patterns of the fingerprints and enable the hand to form a firmer grip on objects than would be possible if the surface of the skin were smooth. Numerous small blood vessels extend into the papillary folds, supplying nutrients to the overlying epidermis, which is avascular. The papillary layer also contains numerous nerve fibers and associated structures which respond to stimuli applied to the surface of the skin.

There is no distinct boundary between the papillary and reticular layers. The *reticular layer* contains denser connective tissue with fewer fibroblasts but many bundles of thick collagen fibers. It is this region that is responsible for the toughness of leather formed by processing animal skins.

The reticular layer borders on the underlying subcutaneous tissue known as the *hypodermis,* which is technically not part of the skin. This region is composed of loose connective tissue and contains an abundance of fat cells. In addition to providing a store of fat that can be metabolized by the body, the hypodermis provides a layer of insulation, preventing heat loss through the skin from the underlying organs of the body. The various amounts of adipose tissue underlying the skin smooth out the angular surfaces of the body.

Hair

One of the characteristics of mammals, including human beings, is the presence of hair on most of the body's surface except for the palms of the hands, soles of the feet, and lips. In furry mammals, this hairy surface provides an extra layer of thermal insulation, but human beings are not sufficiently hairy to benefit from this insulating property. The hair on the top of the head provides protection from the rays of the sun, and the eyelashes partially protect the eyes from various airborne particles. Likewise, the hairs in the nostrils and ears help to prevent particles from entering these passages.

Hair is formed by epidermal cells which invaginate into the underlying dermal layer, forming a *hair follicle* (Fig. 3-21). Hair consists of keratin that is synthesized by cells in the *papilla* at the base of the hair's shaft. Similar to the process of forming the stratum corneum and the nails, the cells in the papilla divide, synthesize keratin, flatten, and die, leaving behind compact layers of keratin which form the shaft of the hair. Hair does not continue to grow indefinitely; after it reaches a certain length, the cells in the papilla stop dividing, and the cells at the base of the hair become completely keratinized. Subsequently, the papillary cells begin to divide again, forming a new hair which forces the old hair out of the follicle.

Hair color is determined by the amount of melanin present or absent. Blond hair does not contain melanin, nor does gray hair which has many air spaces located between the fibrous layers of keratin, giving the hair a gray or white appearance.

Attached to the middle region of the hair follicle is a small smooth muscle, the *arrector pili,* which is anchored to the connective tissue of the dermis. The hair follicles invaginate into the dermis at an angle, so that when the arrector pili contracts, it moves the hair to a more vertical position, causing the hair to "stand on end." At the same time it causes an indentation of the skin around the hair, producing a "goose pimple." In furry animals, this erection of the body hair produces a thicker layer of insulation.

Skin Glands

The skin contains several types of exocrine glands which are formed by invagination of the epithelial cells into the dermis. The secretory portions of these glands are connected to the surface of the skin by ducts that pass through the epidermis or empty into a hair follicle.

The *sebaceous (oil) glands* have a branched acinar structure; their ducts empty into hair follicles, from which point the secretion reaches the surface of the skin along the hair shaft (Fig. 3-21). These glands secrete a mixture of various lipids known as *sebum* which lubricates the surface of the skin. The secretory process in these glands is holocrine, involving the complete disintegration of the cell at the time of release.

The most common type of sweat gland is the *eccrine* gland, which secretes water and salts. The secretory unit of the gland consists of a highly coiled tube which lies in the dermis; it sends a long, relatively straight duct to the surface of the skin (Fig. 3-21). The secretory activity of these glands is regulated by the nervous system via nerve fibers that end in the secretory region of the gland.

A second type of sweat gland is found in the armpits, nipples, anus, and pubic regions. These sweat glands are known as odoriferous, or *apocrine,* glands. They are considerably larger than the eccrine glands, and their ducts usually empty into hair follicles. In addition to water and salts, the secretions of these glands also contain a variety of organic molecules. Although these secretions are initially odorless, the action of bacteria on the skin's surface decomposes the organic molecules, producing the distinctive body odors.

CHAPTER 4
Neural Tissue

The nerve cell (*neuron*) is the basic unit of neural tissue. Neurons perform the specialized function of rapidly transmitting messages in the form of electric signals (*action potentials*). Action potentials can be initiated spontaneously, by stimuli from the external or internal environment, or by signals from other neurons. Once initiated, the action potential is transmitted along the length of the neuron to the specialized junctions between neurons and other cells. These junctions are called *synapses* when they are between two neurons, or *neuroeffector junctions* when they are between a neuron and a muscle or gland cell, these latter cell types being known as *effector cells*. At these junctions neurons are separated from each other and from effector cells by an extracellular space; action potentials reaching the end of a neuron release chemical transmitter agents which cross the space and influence the activity of the postjunctional cell.

Section A. The Neuron

Neurons can be divided structurally into three parts, each associated with a particular function (Fig. 4-1): (1) the cell body and dendrites, (2) the axon, and (3) the axon terminals. The cell body contains the nucleus,

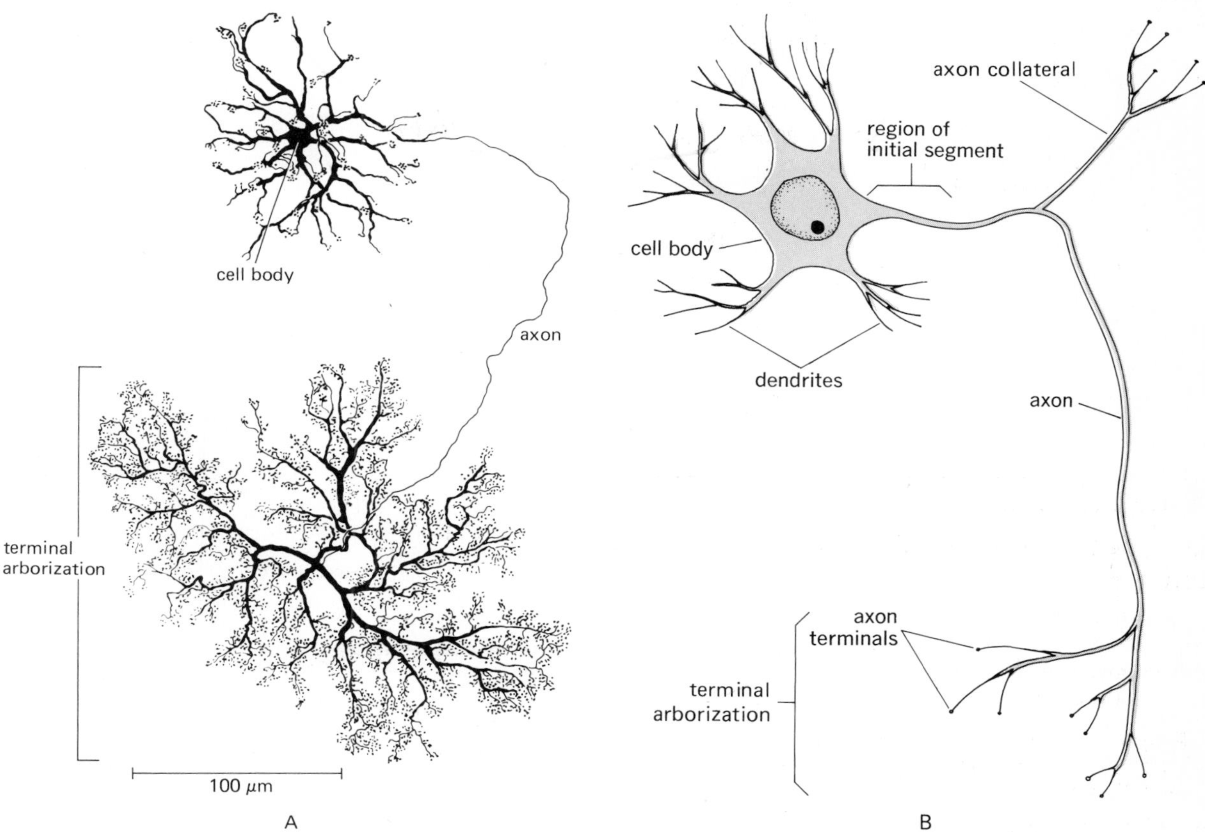

FIGURE 4-1 (A) ***Silver-stained neuron from a cat retina. (The retina is the neural layer at the back of the eye.) Although not typical, it shows clearly the cell body and the terminal arborization. The calibration line is 100 μm long.*** **[Part A, redrawn from R. Nelson et al., *Science,* 189:137 (1975).] (B) Highly diagrammatic representation of a neuron.**

Golgi apparatus, mitochondria, granular endoplasmic reticulum (Nissl substance), and occasionally fat droplets and melanin pigment granules. It also contains fine *neurofilaments* and *neurotubules* (microfilaments and microtubules). The dendrites are highly branched processes of the cell body and may be looked upon as extensions of the cell membrane of the neuron cell body. They, too, contain granular endoplasmic reticulum and microtubules. The dendrites and cell body are the site of most of the synapses with other neurons through which signals are passed to the cell.

The *axon,* or *nerve fiber,* is a single long process extending from the cell body and is usually considerably longer than the dendrites. The first portion of the axon plus the part of the cell body where the axon is joined is known as the *initial segment.* The axon may give off branches, called *collaterals,* along its course, and near the end it undergoes considerable branching, forming a *terminal arborization.* Each branch ends in an enlarged *axon terminal,* which is responsible for transmitting a signal from the neuron to the cell contacted by the axon terminal. The axons also contain mitochondria, neurofilaments, and neurotubules although they lack endoplasmic reticulum, a Golgi apparatus, and pigment. The axons of some neurons are incredibly long relative to the diameter of their cell bodies; for example, axons may extend approximately 1 m from the base of the spinal cord either to the tip of the big toe or to the brain.

Neurons assume many different shapes (Fig. 4-2). *Multipolar cells* are similar to the typical neuron described above; one long axon and multiple short, highly branched dendrites extend from the cell body. *Bipolar*

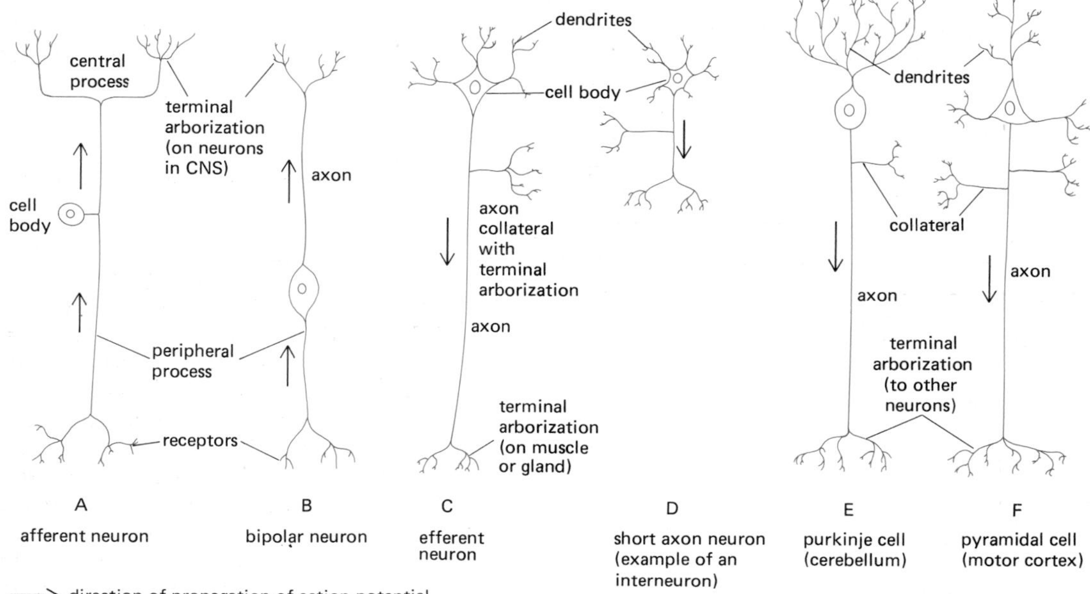

FIGURE 4-2 ***Examples of the different shapes of neurons. Cell A is a unipolar neuron, cell B a bipolar neuron, and cells C to F multipolar neurons.***

cells have, as their name suggests, only two processes which extend from opposite ends of the cell body, one process (the axon) transmitting away from the cell body and the other (nominally a dendrite) transmitting toward it. *Unipolar cells* have only a single short process connected to the cell body; this process divides into two processes which extend in opposite directions. In unipolar cells, it is hard to know what to call these two processes; for example, the long process of cell A in Fig. 4-2 carries action potentials from receptors in the periphery toward the cell body and is therefore functionally a dendrite, yet it looks like an axon and is, in fact, often called an axon, or an axon-dendrite. Note that in unipolar cells the action potential travels directly from one process to another without first passing into the cell body.

FIGURE 4-3 ***Three classes of neurons. Note that the interneurons are entirely within the central nervous system.***

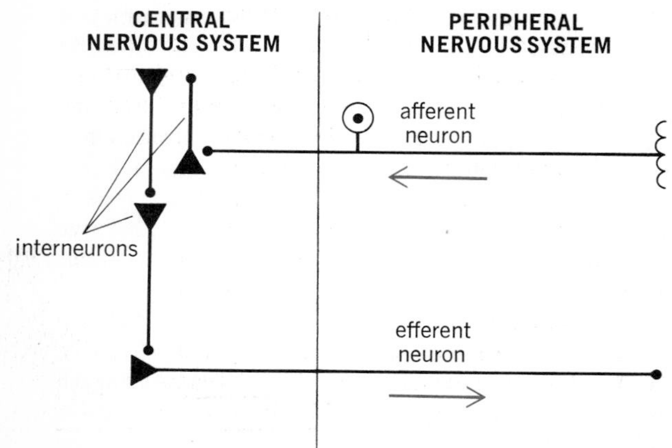

Neurons can be divided into three classes: *afferent neurons, efferent neurons,* and *interneurons* (Figs. 4-2 and 4-3). Afferent neurons carry information from all parts of the body *into* the central nervous system (the brain and spinal cord), where they transmit signals to other neurons. At their peripheral endings the afferent neurons have, or communicate with, *receptors* which, in response to various physical or chemical changes in their environment, cause action potentials to be generated in the afferent neuron; these action potentials are the messages transmitted to the central nervous system. In contrast, efferent neurons transmit information *from* the central nervous system out to the effector cells (muscles or glands). The third group of nerve cells, the *interneurons,* both originate and terminate within the central nervous system, and 99 percent of all nerve cells belong to this group. Their function is to transmit signals from one neuron to another within the central nervous system.

Neurons, like all cells, must maintain their integrity through energy-yielding metabolic reactions. Except for several situations, all this energy is supplied by the metabolism of glucose. Most of this energy is transferred to ATP (adenosine triphosphate) molecules during oxidative phosphorylation. Since the nervous system's glycogen stores are negligible, its function is completely dependent upon a continuous blood supply of glucose and oxygen. Within just a few minutes of deficient blood supply, deterioration of nervous system function can be detected. The energy used by neurons goes mainly for maintaining their electric activity and for synthesizing the chemical transmitter agents released from their endings.

Neural tissue develops from the ectoderm layer of the embryo. With differentiation, the cell's axon and dendrites begin to extend from the cell body. The growth of these processes is associated with formation of a network of neurotubules and neurofilaments which form a cytoplasmic skeleton for them. How these processes end up in their appropriate locations (some of which may be 1 m distant from the cell body) is still not clear.

Another problem faced by neurons is supplying proteins (for example, enzymes) to their processes. Only the nucleus, located in the cell body, contains the genetic information required for protein synthesis; therefore, either the message (as mRNA) or the proteins themselves must be transported from the cell body to all portions of the processes. The neurotubules and neurofilaments are involved in this transport process. Whatever the mechanism, this dependence on the cell body explains why the peripheral portion of a severed axon degenerates whereas the proximal segment (that attached to the cell body) survives and, over a period of weeks to months, may reextend to its original destination.[1] In contrast, should the cell body be destroyed, the axon and dendrites degenerate as well. Moreover, this cell will not be replaced because fully differentiated neurons do not divide, and neurons are not formed after birth from stem cells (neuroblasts).

All neurons are activated ultimately by electric currents; such currents are also utilized in the transmission of information from one region of the neuron to another. Before discussing the electric activity in nerve cells, we turn to a review of the basic principles of electricity.

[1] This regrowth process reestablishes functional connections only when the neurons are severed outside the central nervous system.

Section B. Membrane Electric Activity

BASIC PRINCIPLES OF ELECTRICITY

All chemical reactions are basically electric in nature since they involve exchanging or sharing negatively charged electrons between atoms to form ions or bonds. Most chemical reactions result in neutral molecules, containing equal numbers of electrons and protons, but in some cases ions, which have a net electric charge, are formed, such as the sodium, potassium, and chloride (Na^+, K^+, and Cl^-) ions. With the exception of water, the major chemical components of the extracellular fluid are the sodium and chloride ions, whereas the intracellular fluid contains high concentrations of potassium ions and organic molecules (particularly proteins and phosphate compounds) containing ionized groups. Since the environment of the cell contains many charged particles, it is not too surprising to discover that electric phenomena resulting from the interaction of these charged particles play a significant role in cell function.

According to the laws of physics, all physical and chemical phenomena can be described using only five fundamental units of measurement: length, time, mass temperature, and electric charge. Each of these units is independent in the sense that none can be defined in terms of a combination of the others. For example, velocity is not a fundamental unit since it can be defined in terms of a length moved in a given period of time. Force is not a fundamental unit since it is defined by Newton's law as $F = ma$, force equals mass times acceleration, and thus is defined in terms of mass, length, and time. In contrast, electric charge is a fundamental unit of measurement and cannot be defined in terms of length, time, mass, or temperature.

There are two types of charge, arbitrarily called *positive* and *negative*. This labeling occurred before it was known that atoms consist of protons and electrons. When electrons were discovered, they behaved like the charged particles which had been arbitrarily labeled negative. Protons, on the other hand, behave like the electric charges labeled positive. If the original labeling had been the opposite, we would call electrons positive and protons negative.

When positive and negative charges are separated,

an electric force draws the opposite charges together; yet positive charge repels positive charge and negative charge repels negative charge. Why there is such a force cannot be answered in terms of other physical properties of matter since electric charge is a fundamental property of matter. However, the force can be measured, and the relation between the amount of force, the quantity of charge, and the distance separating the charges can be studied. It is found that the amount of force acting between electric charges increases when the charged particles are moved closer together and with increasing quantity of charge.

Energy E is measured by its ability to do work, and work W is defined as the product of force F and distance X: $W = FX$. When oppositely charged particles come together as a result of the attracting force between them, work can be done by these moving particles; conversely, to separate oppositely charged particles, work must be done, i.e., energy added. The amount of work depends upon both the total number of charges involved and the distance between them. Thus, when electric charges are separated, they have the "potential" of doing work if they are allowed to come together again. *Voltage* is a measure of the potential of separated electric charge to do work and is defined as the amount of work done by an electric charge when moving from one point in a system to another. Voltage is always measured with respect to two points in a system; thus, one refers to the potential difference between two points, and the units of measurement are known as volts. The terms voltage and *potential difference* are synonymous. Since the total amount of charge that can be separated in most biological systems is very small, the potential differences are small. The voltage measured across a nerve cell membrane is approximately 70 millivolts (mV). A millivolt is $\frac{1}{1000}$ volt.

The movement of electric charge is known as *current*. If electric charge is separated between two points, there is a potential difference between these points, and the electric force of attraction between the opposite charges tends to make charges flow, producing a current. The amount of charge that does move, the current, depends upon the nature of the material lying between the separated charges. The space between the two points may be occupied by copper wire, a solution of water and ions, glass, or rubber, or it may be entirely empty, a vacuum. The ability of an electric charge to move through these different media varies, the flow depending upon the number of charged particles in the material that are able to move and thus carry current. The amount of current also depends upon the interactions between the moving charges and the material, such as frictional interactions if the charges collide with other molecules while moving through the material. The hindrance of the movement of electric charge through a particular material is known as *resistance*. Thus, for a given amount of charge separation (a given voltage), the amount of current flow depends upon the resistance of the material between the separated charges. The relationship between current (I), voltage (E), and resistance (R) is given by Ohm's law, $I=E/R$. The higher the resistance of the material, the lower is the amount of current flow for any given voltage. Some materials, like glass and rubber, have such high electric resistance that the amount of current flow through them, even when high voltages are applied, is very small. Such materials, known as insulators, are used to prevent the flow of current. Thus, the rubber insulation around electric wires prevents the flow of current from the wire to areas outside it. Materials having low resistance to current flow are known as conductors.

Pure water is a relatively poor conductor because it contains very few charged particles, but when sodium chloride is added, the solution becomes a relatively good conductor with a low resistance because the sodium and chloride ions provide charges that can carry the current. The water compartments inside and outside the cells in the body contain numerous charged particles (ions) which are able to move between areas of charge separation. Lipids contain very few charged groups and thus have a high electric resistance. The lipid components of the cell membrane provide a region of high electric resistance separating two water compartments of low resistance.

ELECTRIC AND CHEMICAL PROPERTIES OF A RESTING CELL

One can determine the presence of an electric-potential (voltage) difference across cell membranes by inserting a very fine electrode into the cell, another into the extracellular fluid surrounding the cell, and connecting the two to a voltmeter (Fig. 4-4). In this manner it has been found that all cells of the body exhibit a membrane potential oriented so that the inside of the cell is negatively charged with respect to the outside. (As we shall see, this is true for nerve and muscle cells only when they are not being stimulated.) This potential is called the *resting membrane potential;* its magnitude varies from 5 to 100 mV, depending upon the type of cell and its chemical environment.

The normal ionic composition of the fluid bathing

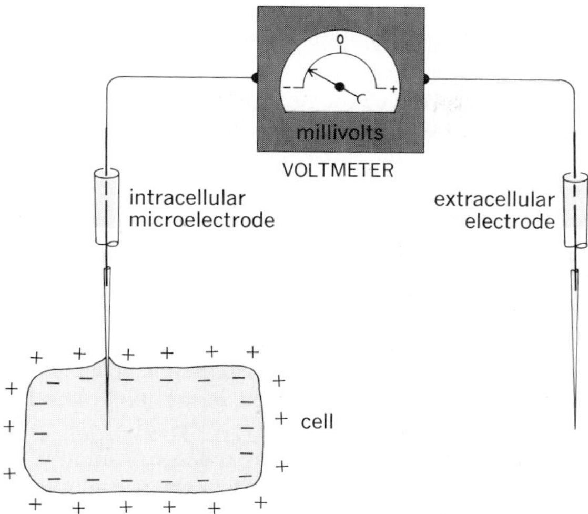

FIGURE 4-4 ***Intracellular microelectrode used to measure electric-potential difference across the cell membrane.***

the cells (the *extracellular fluid*) is approximately that listed in Table 4-1. (Note that the chemical composition of the intracellular fluid is entirely different.) There are many other substances in the extracellular fluid, such as Mg^{2+}, Ca^{2+}, HCO_3^-, PO_4^{2-}, SO_4^{2-}, glucose, urea, amino acids, and hormones, but sodium and potassium play the most important roles in the generation of the resting membrane potential.

TABLE 4-1

DISTRIBUTION OF IONS ACROSS THE CELL MEMBRANE OF A CAT NERVE CELL

Ion	*Extracellular concentration, mmol/l of water*	*Intracellular concentration, mmol/l of water*
Na^+	150	15
Cl^-	110	10
K^+	5	150

Diffusion Potentials

Given that we are dealing with two solutions of different ionic composition and, for the moment, ignoring the nature of the membrane which separates them, consider how the solutions interact. Figure 4-5 depicts two dilute solutions of sodium chloride, the solution on side 1 at a concentration of 0.1 *M* and that on side 2 at 0.01 *M*. The barrier separating the solutions is extremely

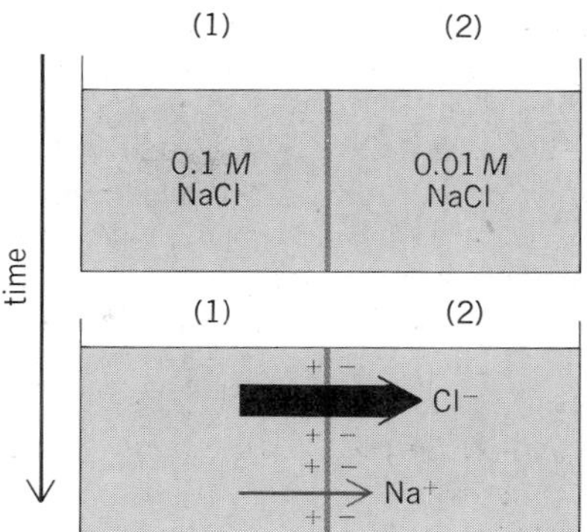

FIGURE 4-5 ***Generation of a diffusion potential by differential ion movement through a completely permeable membrane.***

permeable to all ion species. Both sodium and chloride are more concentrated on side 1, and they will diffuse down their concentration gradients, moving from side 1 to side 2. However, the mobility of chloride, i.e., the ease with which it can move through the solution, is about 50 percent greater than that of sodium; thus, chloride will move to side 2 more rapidly than sodium and side 2 will, at least transiently, become slightly negatively charged with respect to side 1. This electric gradient is due to the differential diffusion of charged particles in solution and is called a *diffusion potential.* The potential developed by a system such as that in Fig. 4-5 will disappear over time as the concentrations of sodium and chloride in sides 1 and 2 become equal.

What would happen at the junction between sides 1 and 2 if side 1 contained a solution of 0.15 *M* NaCl (similar to extracellular fluid) and side 2 contained 0.15 *M* KCl (similar to intracellular fluid) (Fig. 4-6)? Again, we ignore the membrane that separates the two compart-

FIGURE 4-6 ***Generation of a diffusion potential by potassium movement through a completely permeable membrane.***

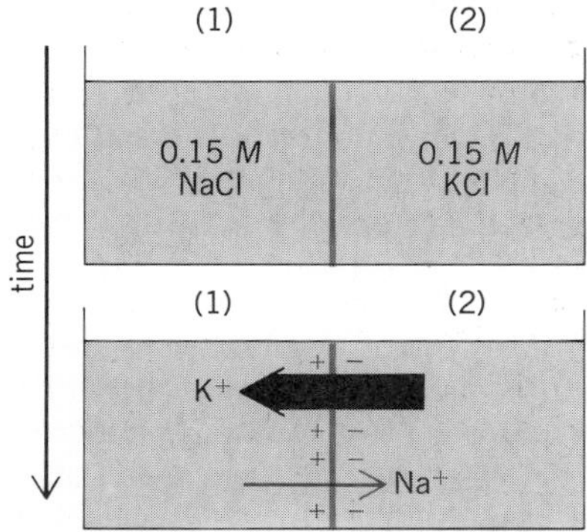

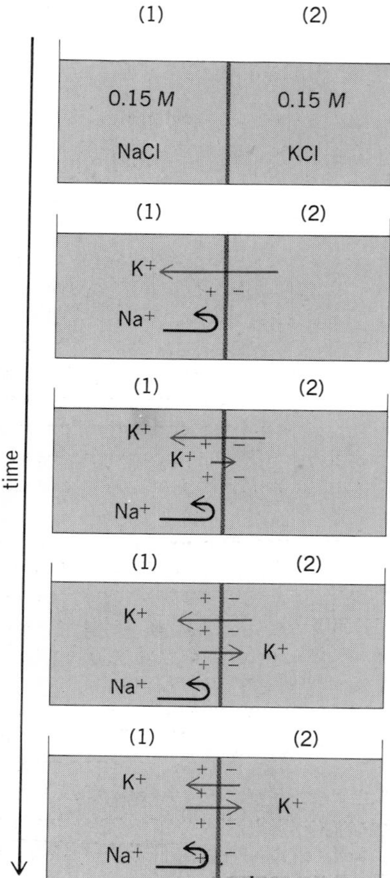

FIGURE 4-7 ***Generation of a diffusion potential across a membrane permeable only to potassium.***

ments. Chloride concentrations on the two sides are equal, but those of sodium and potassium are not. The mobility of potassium, like that of chloride, is about 50 percent greater than that of sodium; thus, potassium will diffuse down its concentration gradient faster than sodium, i.e., positive charge will initially leave side 2 faster than it will enter, and side 2 will become electronegative with respect to side 1. Again, with time, equilibrium will be reached and the potential difference will disappear.

Equilibrium Potentials Now consider the situation of Fig. 4-7 but assume that a selectively permeable membrane separates the two compartments, such that potassium can pass through but sodium and chloride cannot. In this case, all the sodium will remain on side 1, and some of the potassium, diffusing down its concentration gradient, will be added to it; side 1 will become relatively positive. In contrast to the previous examples, this diffusion potential will not disappear with time and will be of greater magnitude, the actual magnitude depending upon the concentration gradient for potassium.

The concentration gradient, which causes net diffusion of particles from a region of higher to a region of lower concentration, is called the *concentration force*. As this force moves potassium from side 2 to side 1 and side 1 becomes increasingly positive, the electric-potential difference itself begins to influence the movement of the positively charged potassium particles; they are attracted by the relatively negative charge of side 2 and repulsed by the positive charge of side 1. This attraction, because of a difference in electric charge (or repulsion because of a similarity of charge), is the *electric force*. As long as the concentration force driving potassium from side 2 to 1 is greater that the electric force driving in the opposite direction, there will be net movement of potassium from side 2 to 1 and the potential difference will increase. Side 1 will become more and more positive until the electric force opposing the entry of potassium equals the concentration force favoring entry. The membrane potential at which the electric force equals in magnitude and opposes in direction the concentration force is called the *equilibrium potential*. At the equilibrium potential there is no net movement of the ion because the forces acting upon it are exactly balanced.

It can be seen that the value of the equilibrium potential for any ion depends upon the concentration gradient for that ion across the membrane; if the concentrations on the two sides were equal, the concentration force would be zero and the electric potential required to oppose it would also be zero. The larger the concentration gradient, the larger is the equilibrium potential. Using potassium concentrations typical for neurons and extracellular fluid, the equilibrium potential for potassium is close to 90 mV, the inside of the cell being negative with respect to the outside.

If the membrane separating sides 1 and 2 is replaced with one permeable only to sodium, the initial net flow of the positively charged sodium will be from side 1 to 2 and side 2 will become positive (Fig. 4-8). This movement of sodium down its concentration gradient is opposed by the electric force generated by that movement. A sodium equilibrium potential will be established with side 2 positive with respect to side 1, at which point net movement will cease. For neurons, the

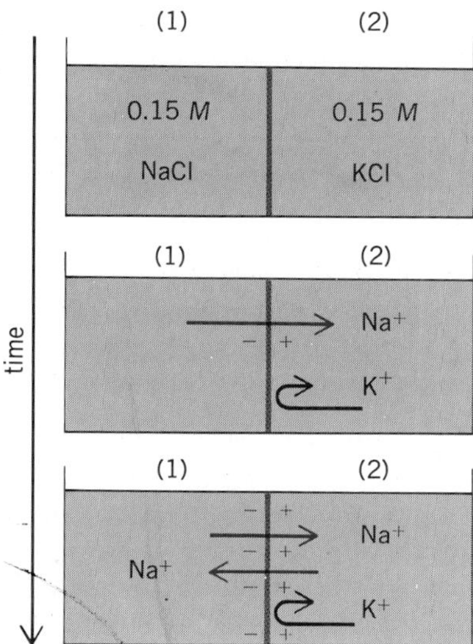

FIGURE 4-8 ***Generation of a diffusion potential across a membrane permeable only to sodium.***

sodium equilibrium potential is 60 mV, inside positive. *Thus, the direction of the diffusion potential across the membrane is determined both by the permeability properties of the membrane and the orientation of the concentration gradients.* The diffusion potential for each ion species is different from those for other ion species since the concentration gradients are different.

The Resting Cell Membrane Potential It is not difficult to move from these hypothetical experiments to a nerve cell at rest where (1) the potassium concentration is much greater inside the cell than out and the sodium concentration gradient is the opposite and (2) the cell membrane is some 50 to 75 times more permeable to potassium than to sodium. Given these characteristics, it should be evident that a diffusion potential will be generated across the membrane largely because of the movement of potassium down its concentration gradient so that the inside of the cell is negative with respect to the outside. The experimentally measured membrane potential is not, however, equal to the potassium equilibrium potential because the membrane is not perfectly impermeable to sodium and some sodium continually diffuses down its electric and concentration gradients, adding a small amount to the inside of the cell. Thus the measured membrane potential, of a neuron at least, is closer to −70 mV than to the potassium equilibrium potential of −90 mV. An important result of this fact is that, since the membrane is not at the potassium equilibrium potential, there is a continual net diffusion of potassium out of the cell.[2]

The Membrane Pump If there is net movement of sodium into and potassium out of the cell, why do the concentration gradients not run down? The reason is that active-transport mechanisms in the membrane utilize forces derived from cellular metabolism to pump the sodium back out of the cell and the potassium back in. For each sodium pumped from the cell, a potassium is moved in; one positive ion is exchanged for another. Under these circumstances the pump itself does not contribute directly to the membrane potential.[3] The pump does make an essential indirect contribution, however, because it maintains the concentration gradients down which the ions diffuse. The membrane potential is then due directly to the diffusion of these ions.

Summary Some potassium ions diffuse out of the cell down their concentration gradient and some move in down the electric gradient, but because the membrane potential is not as negative as the potassium equilibrium potential (−70 mV rather than −90 mV), less potassium enters passively than leaves. The difference is relatively small and is made up by active transport via the membrane pump; therefore the *total* potassium entering equals that leaving, and the resting cell neither gains nor loses potassium (Fig. 4-9). Sodium is driven passively into the cell by both electric and concentration forces but, because the membrane permeability of a resting cell to sodium is so low, the amount entering is small. There is no passive force to remove sodium from the cell, and that which enters must be actively transported out by the membrane pump. The amount of sodium pumped out equals, in most cases, the amount of potassium pumped in (Fig. 4-9).

The membrane permeability properties as well as the concentration gradient of each ion species must be considered when accounting for the membrane potential. For a given concentration gradient, the greater the membrane permeability to an ion species is, the greater the influence that ion species will have on the diffusion

[2] In contrast, chloride is at its equilibrium potential (−70 mV); accordingly, there is no net chloride flux in the resting neuron.

[3] Whenever a pump does not exchange the ions on a one-to-one basis, it directly produces charge separation; such a pump is called an electrogenic pump. In those cases in which an electrogenic pump has been identified in excitable tissues its overall contribution to the generation of the membrane potential is small relative to that contributed by the diffusion potentials.

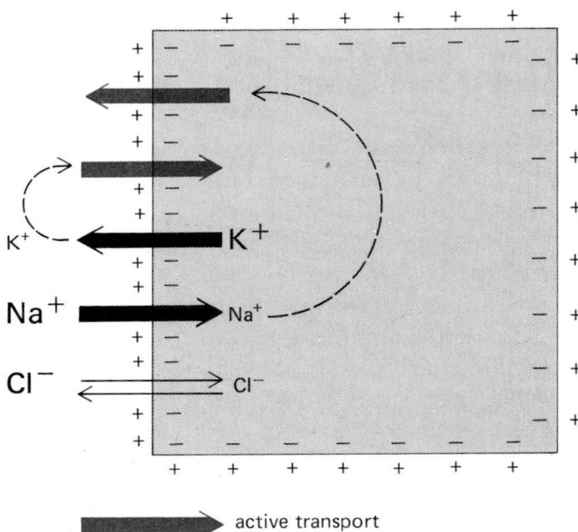

FIGURE 4-9 ***Steady-state total fluxes of sodium, potassium, and chloride ions across the cell membrane. The net flux of sodium and potassium ions by diffusion is balanced by the active transport of these ions in the opposite direction across the membrane (i.e., potassium in and sodium out). There is no net flux of chloride ions because they are in electrochemical equilibrium across the membrane.***

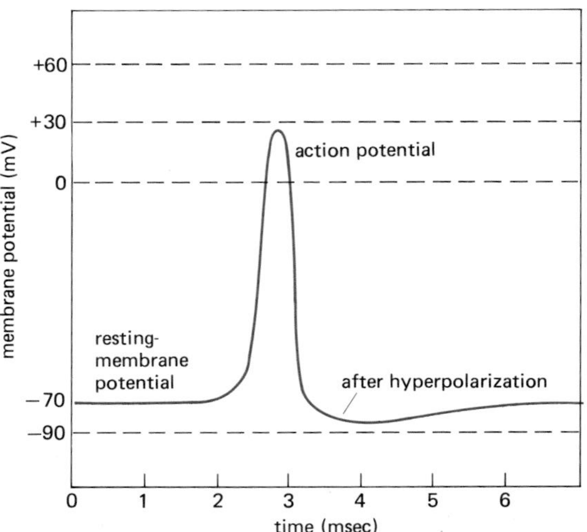

FIGURE 4-10 ***Changes in membrane potential during an action potential. (The sodium equilibrium potential is +60 mV; the potassium equilibrium potential is −90 mV.)***

potential. Since the resting membrane is much more permeable to potassium than to sodium (and the concentration differences are approximately similar), the resting membrane potential is much closer to the potassium equilibrium potential than to that of the sodium.

ACTION POTENTIALS

During periods when nerve and muscle cells appear to be physiologically active the membrane potential undergoes rapid alteration (Fig. 4-10), suddenly changing from −70 to +30 mV and then rapidly returning to its original value. This rapid change of membrane potential, which may last only $\frac{1}{1000}$ s, is called an *action potential.* Of all the types of cells in the body, only nerve and muscle cells are capable of producing action potentials; this property is known as *excitability.* Excitable membranes, besides generating action potentials, are able to transmit them along their surfaces. Thus, the action potential is the signal which is transmitted from one part of the nerve or muscle cell to another.

How is an excitable membrane able to make rapid changes in its membrane potential? How does a change in the environment (a stimulus) interact with an excitable membrane to bring about an action-potential response? How is an action potential propagated along the surface of an excitable membrane? These questions will be discussed in the following sections. The terms *depolarize, hyperpolarize,* and *repolarize* will be used frequently. The membrane is said to be *depolarized* when the membrane potential is less negative than the resting membrane potential, i.e., closer to zero, and *hyperpolarized* when it is more negative than the resting level. When the membrane potential is changing so that it moves toward or even above zero, it is *depolarizing,* and when it moves away from zero back toward its resting level, it is *repolarizing* (Fig. 4-11). When it moves beyond its resting level it is *hyperpolarizing.*

Ionic Basis of the Action Potential

Action potentials can be explained by the concepts already developed for the origins of resting membrane potentials. This explanation, known as the *ionic hypothesis,* was developed mainly by the English scientists A. L. Hodgkin and A. F. Huxley, who received the Nobel Prize in 1963. We have seen that the magnitude of the resting membrane potential depends upon the concentration gradients of and membrane permeabilities to ions, particularly sodium and potassium. This situation is true for the period of the action potential as well. Obviously, then, the action potential must result from a transient change in either the concentration gradients or the membrane permeabilities. The latter is the case. In the resting state the membrane is 50 to 75 times more permeable to potassium than to sodium ions. Thus, the magnitude and polarity of the resting potential are due almost entirely to the movement of potassium ions out of the cell. During an action potential,

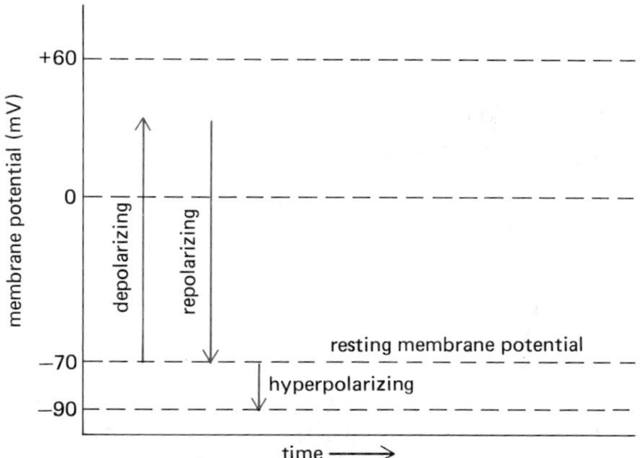

FIGURE 4-11 ***As potentials become more positive inside than the resting potential, they are said to be* depolarizing; *those returning toward the resting potential are said to be* repolarizing; *and those becoming more negative inside than the resting membrane potential are* hyperpolarizing.**

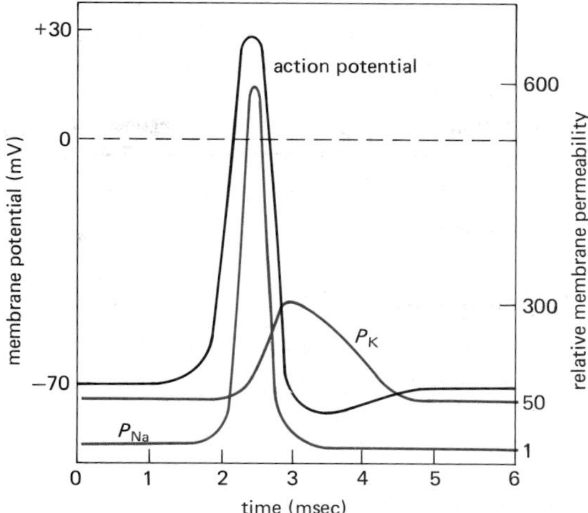

FIGURE 4-12 ***Changes in membrane permeability to sodium (P_{Na}) and potassium (P_K) ions during an action potential.***

however, the permeability of the membrane to sodium and potassium ions is markedly altered. In the rising phase of the action potential the membrane permeability to sodium ions undergoes a 600-fold increase and sodium ions rush into the cell, whereas there is little change in the potassium permeability of the membrane. During this period more positive charge is entering the cell in the form of sodium ions than is leaving in the form of potassium ions, and thus the membrane potential decreases and eventually reverses its polarity, becoming positive on the inside and negative on the outside of the membrane. In this phase the membrane potential approaches but does not quite reach the sodium equilibrium potential.

Action potentials in neurons last about 1 ms (0.001 s). What causes the membrane to return so rapidly to its resting level? The answer to this question is twofold: (1) The increased sodium permeability (*sodium activation*) is rapidly turned off (*sodium inactivation*), and (2) the membrane permeability to potassium increases over its resting level. The timing of these two events can be seen in Fig. 4-12. As the membrane becomes more positive inside and the drive for sodium entry is reduced, sodium permeability decreases toward its resting value, and sodium entry rapidly decreases. This alone would restore the potential to its resting level. However, the entire process is speeded up by a simultaneous increase in potassium permeability which causes more potassium ions to move out of the cell down their concentration gradient. These two events, sodium inactivation and increased potassium permeability, allow potassium diffusion to regain predominance over sodium diffusion, and the membrane potential rapidly returns to its resting level. In fact, while the potassium permeability is greater than normal, there is generally a small hyperpolarizing overshoot of the membrane potential (*after hyperpolarization,* Figs. 4-10 and 4-12).

In our description it may have seemed as though the sodium and potassium fluxes across the membrane involved large numbers of ions. Actually, only one of every 100,000 potassium ions within the cell need diffuse out to charge the membrane potential to its resting value, and very few sodium ions need enter the cell to cause the depolarization during an action potential. Thus, there is virtually no change in the concentration gradients during an action potential. Yet if the tiny number of additional ions crossing the membrane with each action potential were not eventually restored, the concentration gradients for sodium and potassium across the cell membrane would gradually disappear. As might be expected, an accumulation of sodium and loss of potassium are prevented by the continuous action of the membrane active-transport system for sodium and potassium. This restoration occurs mainly after the action potential is over. The number of ions that cross the membrane during an action potential is so small, however, that the pump need not keep up with the action-potential fluxes, and hundreds of action potentials can occur even if the pump is stopped experimentally. Action potentials would eventually disappear in such circumstances as the sodium and potassium concentration gradients ran down. Note, however, that the pump plays no direct role in the generation of the action potential itself.

Mechanism of Permeability Changes

The cause of the permeability changes which underlie action potentials has been elucidated by experiments in which membrane permeability and ion fluxes are monitored as action potentials are triggered by means of electric stimulation applied to the nerve cell membrane. The stimulation is accomplished through the use of two electrodes, one placed in the cell and the other in the extracellular fluid surrounding it. With appropriate settings, the electrodes can be made to add positive charge to the outside of the membrane while simultaneously removing positive charge from the inside of the cell, thereby causing the membrane potential to increase, i.e., become hyperpolarized. If the settings are reversed, the electrodes remove positive charge from the outside while adding it to the inside, thereby reducing the membrane potential, i.e., depolarizing the cell membrane.

Measurement of membrane permeability changes during these experiments revealed that the permeability to sodium was altered whenever the membrane potential was changed; specifically, hyperpolarization of the membrane caused a decrease in sodium permeability whereas depolarization caused an increase in sodium permeability. In light of our previous discussion of the ionic basis of membrane potentials, it is very easy to confuse the cause-and-effect relationships of the statement just made. Earlier we pointed out that an increase in sodium permeability *causes* membrane depolarization; now we are saying that depolarization *causes* an increase in sodium permeability. Combining these two distinct causal relationships yields the positive-feedback cycle (Fig. 4-13) responsible for the rising phase of the action-potential spike: Depolarization alters the cell membrane structure so that its permeability to sodium increases; because of increased sodium permeability, sodium diffuses into the cell; this addition of net positive charge to the cell further depolarizes the membrane, which, in turn, produces a still greater increase in sodium permeability, which, in turn, causes . . . We still have no physicochemical explanation of how the sodium permeability of the membrane is altered when the level of the membrane potential changes. Whatever mechanisms may be responsible, they appear to be present only in the cell membranes of nerve and muscle cells. In other types of cells, changing

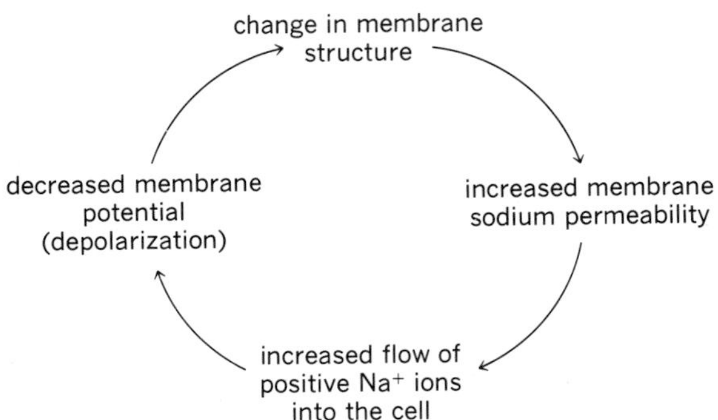

FIGURE 4-13 ***Positive-feedback relationship between membrane depolarization and increased sodium permeability which leads to the rapid rising phase of the action potential. This relationship is called* sodium activation.**

the membrane potential does not cause a change in membrane permeability.

Finally, as described above, the rising phase of the action potential is terminated and repolarization is brought about mainly by a rapid shutting off of the increased sodium permeability. The cause of this sodium inactivation is unknown. Similarly, the cause of the changes in potassium permeability which contribute to repolarization is unknown.

Threshold

Given the positive-feedback cycle just described, one might conclude (wrongly) that an action potential will be triggered whenever the cell membrane is even slightly depolarized, either by electric input, as in our experiment, or by the physiologic means to be described later. But such is not the case, as demonstrated in Fig. 4-14. In this experiment, a series of depolarizing stimuli is delivered electrically to the cell membrane. The first stimulus, which is fairly weak, causes a transient, small membrane depolarization *but no action potential.* The next stimulus is doubled in strength and causes twice the amount of depolarization but again the change is transient and does not trigger an action potential. Only when the stimulus strength has been increased even more does an action potential occur. Thus, partially depolarizing an excitable membrane initiates an action potential only when the strength of the stimulus is sufficient to depolarize the membrane potential to a critical level, known as the *threshold potential.* Such a stimulus is known as a *threshold stimulus.* Stimuli weaker than this are known as *subthreshold stimuli* and do not initiate an action potential. Stimuli of more

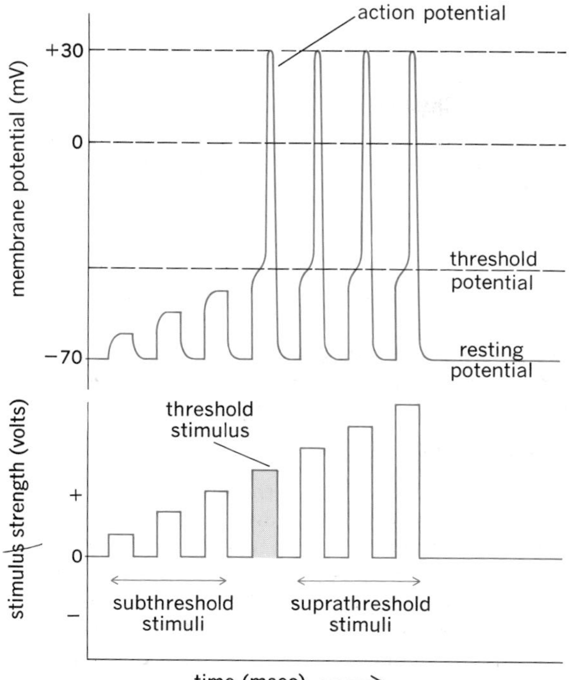

FIGURE 4-14 ***Decreasing membrane potential with increasing strength of depolarizing stimulus. When the membrane potential reaches the threshold potential, action potentials are generated. Increasing the stimulus strength above threshold level does not alter the action-potential response. (The afterhyperpolarization has been omitted from this figure.)***

than threshold magnitude (*suprathreshold stimuli*) elicit action potentials but, as can be seen in Fig. 4-14, the action-potential response is not different from that following a threshold stimulus. The threshold potential of most excitable membranes is 5 to 15 mV less polarized than the resting membrane potential. Thus, if the resting potential of a neuron is −70 mV, the threshold potential may be −60 mV; in order to initiate an action potential in such a membrane, the potential must be decreased by at least 10 mV.

What is the explanation for the phenomenon of threshold; i.e., why does not any slight amount of initial depolarization trigger an action potential? The answer lies in the fact that, at any potential between resting and threshold, sodium movement into the cell is less than potassium movement out, despite the increased sodium permeability. This prevents any further depolarization beyond that induced directly by the stimulus and drives the potential rapidly back toward the resting level as soon as the stimulus is removed. In contrast, at potentials just barely above threshold, the sodium permeability has been increased so much (as a result of the stimulus-induced depolarization) that its inflow exceeds potassium outflow. This net inflow of positively charged ions depolarizes the membrane still further, which increases sodium permeability even more. Thus the positive-feedback relationship can be effective in causing an action potential only after the membrane has been initially depolarized to the critical threshold value by a stimulus. The nature of the physiologically occurring stimuli will be described in the subsequent sections on receptors and synapses.

It should be reemphasized that, once threshold is reached, the process is no longer dependent upon stimulus strength. The depolarization continues on to become an action potential solely because the membrane permeability changes allow sodium ions to diffuse down the electric and concentration gradients which exist across the membrane. Thus, action potentials triggered either by stimuli just strong enough to depolarize the membrane a bit above threshold or by very strong stimuli are identical. Action potentials occur either maximally as determined by the electrochemical conditions across the membrane or they do not occur at all. Another way of saying this is that action potentials are *all or none*.

Because of the all-or-none nature of the action-potential response, a single action potential cannot convey any information about the magnitude of the stimulus which initiated it, since a threshold-strength stimulus and one of twice threshold strength give the same response. Since the function of nerve cells in the body is to transmit information by propagating action potentials, one may ask how a system operating according to an all-or-none principle can convey information about the strength of a stimulus. How can one distinguish between a loud noise and a whisper, a light touch and a pinch? The answer, as we shall see later, depends upon the number of action potentials transmitted per unit time, i.e., the frequency of action potentials, and not upon their size.

Refractory Periods

How soon after firing an action potential can an excitable membrane be stimulated to fire a second one? If we apply a threshold-strength stimulus to a membrane and then stimulate the membrane with a second threshold-strength stimulus at various time intervals following the first, the membrane does not always respond to the second stimulus. Even though identical stimuli are applied to the membrane, it appears unresponsive for a certain time. The membrane during this period is said to be *refractory* to a second stimulus.

Instead of applying the second stimulus at threshold strength, if we increase it to suprathreshold levels, we can distinguish two separate refractory periods associated with an action potential. During the 1-ms period of the action-potential spike a second stimulus will not produce a second action-potential response no matter how strong it is. The membrane is said to be in its *absolute refractory period.* Following the absolute refractory period there is an interval during which a second action-potential response can be produced but only if the stimulus strength is considerably greater than threshold level. This is known as the *relative refractory period* and can last some 10 to 15 ms or longer.

The mechanisms responsible for the refractory periods are related to the membrane mechanisms that alter the sodium and potassium permeability. The absolute refractory period corresponds with the period of sodium permeability changes, and the relative refractory period corresponds roughly with the period of increased potassium permeability. Following an action potential, time is required to return the membrane structure to its original resting state. The system may be likened to the cocking of a spring; until the spring (sodium permeability mechanism) has been reset, it cannot be released again.

The refractory periods limit the number of action potentials that can be produced by an excitable membrane in a given period of time. Recordings made from nerve cells in the intact organism that are responding to physiologic stimuli indicate that most nerve cells in the body respond at frequencies up to 100 action potentials per second although some nerve cells may produce much higher frequencies for brief periods of time.

Action-potential Propagation

Everything we have discussed so far concerns a signal, in the form of an electrochemical change across the membrane, brought about by a stimulus, but for this signal to serve as a means of communication there must be a way for it to travel from one part of the cell to another, and this brings us to the topic of action-potential propagation.

One particular action potential does not itself travel along the membrane; rather, each action potential triggers, by local current flow, a new one at an adjacent area of membrane. The old action potential provides the electric stimulus that depolarizes the new membrane site to just past its threshold potential. Once this has happened, the sodium activation cycle at the new membrane site takes over and an action potential occurs there. Once the new site is depolarized to threshold, the action potential generated there is solely dependent upon the electrochemical gradients and membrane permeability properties at the new site. Since these factors are identical to those involved in the generation of the old one, the new action potential is virtually identical to the old. This is an important point because it means that no distortion occurs as the signal passes along the membrane; the signal (action potential) arriving at the end of the membrane is precisely identical to the initial one.

Follow the steps in action-potential propagation, concentrating initially on the shaded area in Fig. 4-15. Ignore for the present the question of how the first action potential was initiated and start with the fact that sodium permeability has changed so that sodium rushes in across the membrane, making the inside of the cell relatively more positive and leaving the outside more negative (Fig. 4-15A and B). Looking at the phenomenon in tiny fragments of time as we are, we can assume that at this first instant the remainder of the cell membrane is at its normal resting potential; accordingly, the shaded area of membrane has an electric potential different from that of the adjacent areas. Like charges repel, and unlike charges attract; thus, current (defined as the flow of positive charge) flows *away* from the activated membrane region through the cytoplasm and *toward* the activated region through the extracellular fluid (Fig. 4-15C).

What is the effect of these local current flows on the as-yet-unexcited regions of the membrane adjacent to the shaded area? The addition of positive charge to the inside of the cell and the removal of positive charge from the outside decrease the potential difference across the membrane; this initial depolarization, due to local current flow, acts as the stimulus to trigger an action potential. Meanwhile, at the original membrane site, sodium inactivation is occurring and potassium permeability is increasing so that the membrane is repolarizing. These processes repeat themselves until the end of the membrane is reached (Fig. 4-15D to F).

The step illustrated in Fig. 4-15C is an example of local current flow. Let us look at this process in more detail. Note that, as in all biological systems, the current is carried by ions such as K^+, Na^+, Cl^-, and HCO_3^-. By convention, the direction of movement of

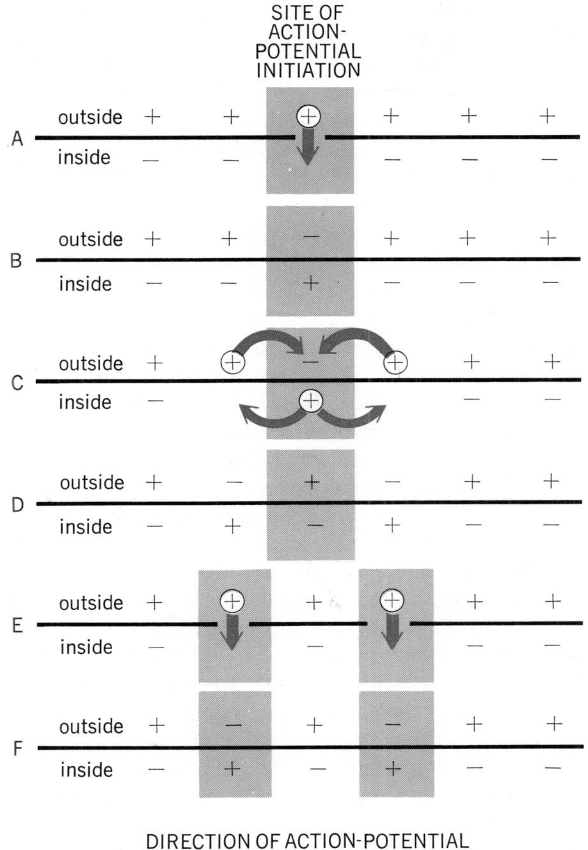

FIGURE 4-15 ***Mechanisms of action-potential propagation.***

the positive ions is designated the direction of current flow, but negatively charged particles can and do move in the opposite direction.

The change in electric charge or potential, however, passes from one point to another much faster than the ions themselves move between these same two points. This can best be understood by considering water movement through a water-filled pipe. If the pressure is changed at one end of the pipe, e.g., by putting in more water, the water flow at the opposite end is changed long before those molecules that were put in could have traveled to the outflow end. This occurs because the influence of the molecules (in this example the increase in water pressure) is transmitted much faster than the molecules themselves.

In much the same way, the influence of electric charge is transmitted from one area to another much faster than the ions could move between the same two points. But here the analogy between water flow in the fluid-filled pipe and current flow in the cytoplasm or extracellular fluid begins to break down. In the case of the pipe, the flow out one end equals the input. But current flow in the cytoplasm is more like water flow through a leaky hose; charge is lost across the membrane, with the result that the flow out the other end is less than the input. In fact, when conducting current, cells are so leaky that the current almost completely dies out within a few millimeters of its point of origin. For this reason, local current flow is *decremental;* i.e., its amplitude decreases with increasing distance.

But it does not matter that cytoplasm is such a poor conductor of electric current and that local current flow is decremental because membranes of nerves and muscles depend upon local current flow only over very short distances. Only that amount of local current flow which is necessary to depolarize adjacent membrane areas to threshold is required. Once threshold is reached, an action potential occurs at the new site. The action potentials are transmitted *without decrement,* because new action potentials are continually generated along the membrane.

Direction of Action-potential Propagation

In the example discussed above (Fig. 4-15) the membrane is stimulated in the middle; in this case the action potential spreads in both directions away from the site of stimulation. Local current flow occurs in all directions in which there is an electric gradient (meaning, of course, that it also flows toward the original site of stimulation); however, the membrane areas which have just undergone an action potential are refractory and cannot undergo another; thus, the only direction of action-potential propagation is away from the stimulation site (Fig. 4-15).

The action potentials in skeletal muscle membrane are initiated near the middle of the cell and propagate from this region toward the two ends, but in most nerve-cell membranes action potentials are initiated at one end of the cell and propagate in only one direction toward the other end of the cell. Realize, though, that this unidirectional propagation of action potentials is determined by the stimulus location rather than an intrinsic inability to conduct in the opposite direction.

Velocity of Action-potential Propagation

The velocity with which an action potential is transmitted down the membrane depends upon fiber diameter

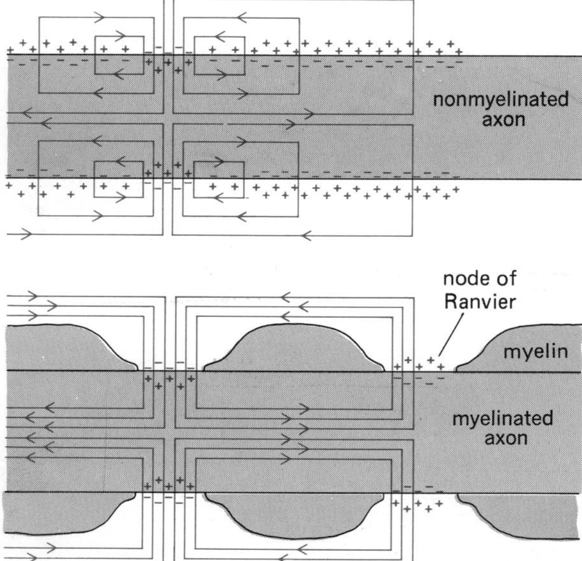

FIGURE 4-16 ***Ion current flows during an action potential in a nonmyelinated and a myelinated axon.***

and whether or not the fiber is myelinated. The larger the fiber diameter, the faster is the action-potential propagation, because a large fiber offers less resistance to local current flow; therefore adjacent regions of the membrane are brought to threshold faster.

Myelinization is the second factor influencing propagation velocity. *Myelin* is a fatty material covering the axons of many neurons. Myelin electrically insulates the membrane, making it more difficult for current to flow between intra- and extracellular-fluid compartments. In effect, it "reduces the leak in the hose"; less current passes out through the myelin-covered section of the membrane during local current flow so that there is a lesser change in the voltage gradient along the fiber (Fig. 4-16). Action potentials do not occur along the sections of membrane protected by myelin; they occur only where the myelin coating is interrupted at regular intervals along the axon (the *nodes of Ranvier*). At each node the membrane is exposed to the extracellular fluid (Figs. 4-16 and 4-17). Thus the action potential appears to jump from one node to the next as it propagates along a myelinated fiber, and for this reason this method of propagation is called *saltatory conduction,* from the Latin *saltare,* to leap. The membrane of nodes adjacent to the active node is brought to threshold faster and undergoes an action potential sooner than if myelin were not present. The velocity of action-potential propagation in large myelinated fibers can exceed 250 mi/h.

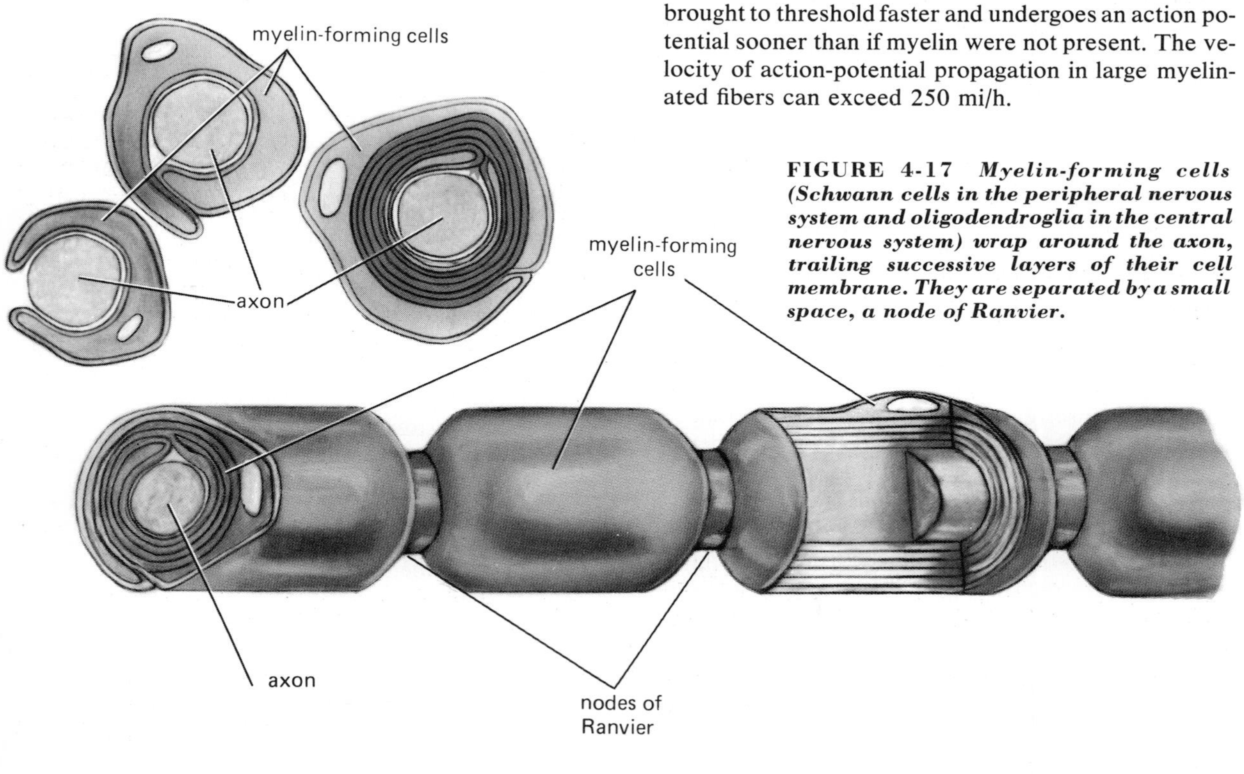

FIGURE 4-17 ***Myelin-forming cells (Schwann cells in the peripheral nervous system and oligodendroglia in the central nervous system) wrap around the axon, trailing successive layers of their cell membrane. They are separated by a small space, a node of Ranvier.***

Section C. Functional Organization of Neurons

Information is relayed along neurons in the form of action potentials. Action potentials are initiated physiologically in three ways: by the activation of receptors, by synaptic input from other neurons, or (in some neurons) by spontaneous activity. The frequency of action potentials can be altered to relay different types of information. The mechanisms of action-potential initiation at receptors and synapses and the determinants of action-potential frequency form the subject of this section. The spontaneous generation of action potentials in certain neurons occurs in the absence of any identifiable external stimulus and thus appears to be an inherent property of these neurons, manifested as oscillating changes in membrane permeability to sodium or potassium; such neurons are thought to play important roles in rhythmic events such as breathing and the menstrual cycles.

RECEPTORS

Information about the external world and internal environment exists in different energy forms—pressure, temperature gradients, light, sound waves, etc.—but only receptors can deal with these energy forms. The rest of the nervous system can extract meaning only from action potentials (or, over very short distances, from small, subthreshold changes in membrane potential). Thus, regardless of its original energy form, information must be translated into the language of action potentials.

The devices which do this are receptors. They are either specialized peripheral endings of neurons or separate cells intimately connected to them. Not all neurons have specialized receptor regions, only that small class of neurons called afferent neurons. There are several types of receptors, each of which is specific; i.e., it responds more readily to one form of energy than to others, although virtually all receptors can be activated by several different forms of energy. For example, the receptors of the eye normally respond to light, but they *can* be activated by intense mechanical stimuli like a poke in the eye. Usually much more energy is required to excite a receptor by energy forms to which it is not specific. On the other hand, most receptors are exquisitely sensitive to their specific energy form. Olfactory receptors can respond to as few as three or four odorous molecules in the inspired air, and visual receptors can respond to the smallest known quantity of light.

Receptor Activation: the Generator Potential

Here we describe only the general mechanisms for receptor activation and use the simple case in which the receptor is a bulb composed of concentric layers of cells and connective tissues specialized to transmit information about pressure or mechanical deformation to the peripheral ending of the afferent neuron (Fig. 4-18). (In subsequent chapters, we shall describe the many other kinds of receptors which respond to such stimuli as light, sound, and heat.) Mechanical stimuli bend, stretch, or press upon the receptor membrane, and somehow, perhaps by opening up pores in the membrane, increase membrane permeability. With the increased permeability, ions move across the membrane down their electric and concentration gradients. Although these ion movements have not been worked out as clearly as those associated with action potentials, the permeability increases appear to be nonselective and to apply to all small ions.

Remember that the intracellular fluid of all nerve cells has a higher concentration of potassium and a lower concentration of sodium than the extracellular fluid and that the inside of a resting neuron is about 70 mV negative with respect to the outside. The effect of a

FIGURE 4-18 **(A)** ***An afferent neuron with a mechanoreceptor (pacinian corpuscle) ending.*** **(B)** ***A pacinian corpuscle showing the nerve-ending modification by cellular structures.*** **(C)** ***The naked nerve ending of the same mechanoreceptor. The generator potential arises at the nerve ending, 1, and the action potential arises at the first node in the myelin sheath, 2. (Adapted from Lowenstein.)***

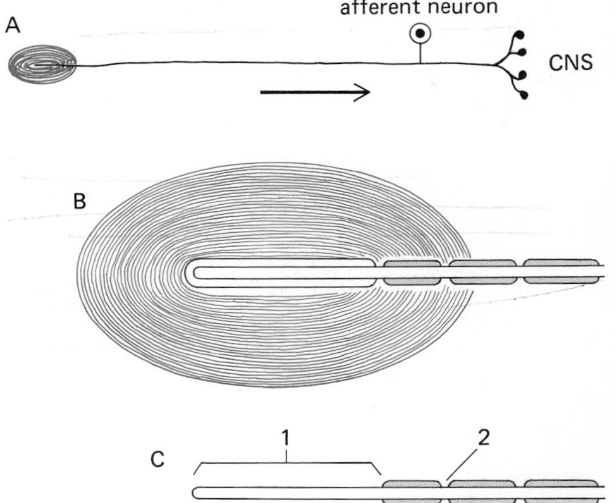

nonselective increase in membrane permeability at a stimulated receptor is a net outward diffusion of a small number of potassium ions and the simultaneous movement of a larger number of sodium ions in. The result is net movement of positive charge into the cell, leading to a decrease in membrane potential (depolarization). The movement of sodium into the nerve fiber thus plays the major role in depolarizing the nerve ending, potassium and other ions such as chloride being less important. This initial depolarization of the receptor is known as the *generator potential*. It occurs at the unmyelinated nerve ending where the cell membrane has a very high threshold; thus, the generator potential cannot, itself, depolarize the nerve ending enough to cause an action potential there. Rather, the depolarization, i.e., the generator potential, is conducted by local current flow a short distance from the nerve ending to the first node in the myelin sheath where the membrane's threshold is lower. The magnitude of the generator potential decreases with distance from its site of origin (this is true of all potentials due to local current flow) but, if the amount of depolarization which reaches the first node is large enough to bring the membrane there to threshold, an action potential is initiated. The action potential (not the generator potential) then propagates along the nerve fiber.

If, after one action potential has been fired, the depolarization at the first node is still above threshold, another action potential will occur; as long as the first node is depolarized to threshold, action potentials continue to fire and propagate along the membrane of the afferent neuron. In fact, it is much more common for stimuli to cause trains, or bursts, of action potentials than single ones.

Generator potentials are not all or none; their amplitude and duration vary with stimulus strength and other parameters to be discussed shortly. A change in the generator potential is reflected via local current flow in a similar change in the degree of depolarization at the first node. Because of this relationship, the amplitude and duration of the generator potential determine action-potential frequency in the afferent neuron.

Before discussing the factors that can alter the amplitude of the generator potential, we want to emphasize the fact that action potentials and generator potentials represent quite separate events. It is the action potential which travels along the nerve fiber to its terminations, not the generator potential. The latter is a local response whose only function is to trigger the action potential. The differences between them are summarized in Table 4-2.

TABLE 4-2
DIFFERENCES BETWEEN GENERATOR POTENTIALS AND ACTION POTENTIALS IN NEURONS

Generator potentials	*Action potentials*
1 Graded response; amplitude increases with increasing stimulus strength or increasing velocity of stimulus application.	1 All-or-none response; once stimulus strength is great enough to bring the membrane to threshold, further increases in stimulus strength do *not* cause an increase in amplitude.
2 Can be added together; if a second stimulus arrives before the generator potential of the first stimulus is over, the generator potential from the second stimulus is added to the depolarization from the first.	2 Cannot be added together.
3 Has no refractory period.	3 Has a refractory period of about 1 ms.
4 Is conducted passively and decreases in magnitude with increasing distance along the nerve fiber.	4 Is propagated without loss of amplitude along the length of the nerve fiber.
5 Duration is greater than 1 to 2 ms and varies.	5 Duration is 1 to 2 ms.

Amplitude of the Generator Potential

Since the amplitude and duration of the generator potential determine the number of action potentials initiated at the first node, the factors contolling these parameters are important. They vary with the stimulus intensity, rate of change of stimulus application, summation of successive generator potentials, and adaptation. Note that generator-potential amplitude determines action-potential *frequency*, i.e., number of action potentials fired per unit time; it does *not* determine action-potential *size*. The action potential is all or none; its amplitude is always the same regardless of the size of the stimulus.

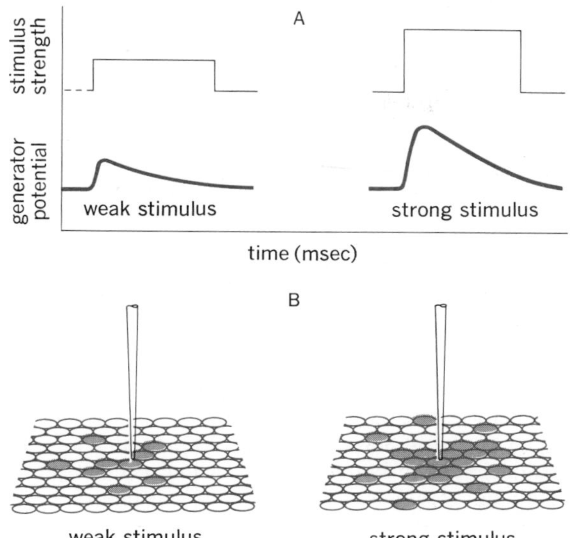

FIGURE 4-19 ***(A) A weak stimulus produces a small generator potential. The amplitude of the generator potential increases as stimulus intensity increases. (B) This increase in amplitude possibly occurs because more pores in the membrane open.***

Intensity and Velocity of Stimulus Application Generator potentials become larger with greater intensity of the stimulus (Fig. 4-19) because the permeability changes increase and the transmembrane ion movements are greater. The explanation for this phenomenon may be an increase in the number or size of pores which permit transmembrane ion flow. The amplitude of the generator potential also rises with a greater rate of change of stimulus application. This applies also to the rate of removal of the stimulus. Thus, although one would expect the amplitude of generator potentials and the rate of firing of action potentials in the afferent neuron to decrease as the stimulus is removed, some receptors give rise to an *off response,* which is a further depolarization of the receptor membrane leading to a burst of action potentials.

Summation of Generator Potentials Another way of varying the amplitude of the generator potential is by adding two or more together. This is possible because generator potentials are graded phenomena and because they last 5 to 10 ms; indeed, they may last as long as the stimulus is applied. If the nerve ending is stimulated again before the generator potential from a preceding stimulus has died away, the two potentials sum and make a larger single generator potential.

Adaptation *Adaptation* is a decrease in frequency of action potentials in the afferent neuron despite a constant stimulus energy. Adaptation is not due to nerve fatigue. Fatigue (in the sense of wearing down from overuse) does not occur in nerve membrane. Adaptation occurs for three reasons: (1) The stimulus energy can be dissipated in the tissues as it passes through them to reach the receptor. As the energy loss gradually increases, the amount of energy reaching the receptor decreases. Thus, the membrane permeability changes and generator-potential amplitudes decrease with time. (2) The responsiveness of the receptor membrane can decrease with time so that the generator-potential amplitude drops even though the energy reaching the receptor remains unchanged. These two reasons for adaptation are probably the most frequent. In both the lower action-potential frequency is due to lower amplitude of the generator potential. (3) Even if the generator-potential amplitude remains unchanged, there can be a decreased frequency of action potentials in response to a constant stimulus because of membrane changes at the first node. The actual reasons for adaptation vary in different receptor types. Adaptation of an afferent neuron in response to the constant stimulation of its mechanoreceptor ending can be seen in the first line of Fig. 4-20. Some receptors adapt completely so that in spite of a constantly maintained stimulus, the transmission of action potentials stops. In some extreme cases, the receptors fire only once at the stimulus application or release. In contrast to these rapidly

FIGURE 4-20 ***Action potentials in a single afferent nerve fiber in response to the application of various constant-pressure stimuli to the mechanosensitive receptor ending. The action potentials are unequal heights because of a recording artifact.*** **(Adapted from Hensel and Bowman.)**

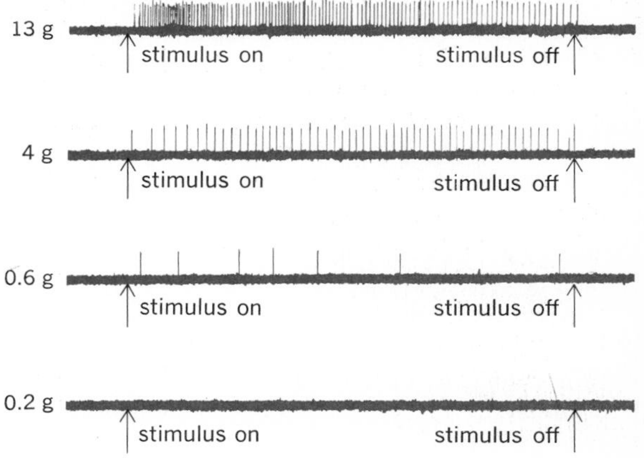

adapting receptors, slowly adapting types merely drop from an initial high action-potential frequency to a lower level, which is then maintained for the duration of the stimulus.

Summary The magnitude of a given generator potential can vary with stimulus intensity, rate of change of stimulus application, summation, adaptation, and cessation of the stimulus, but in spite of these, the amplitude of the generator potential in no way determines the amplitude of the action potentials in the afferent neuron. It does determine how many, if any, action potentials will occur.

Intensity Coding

We are certainly aware of different stimulus intensities. How is information about stimulus strength relayed by action potentials of constant amplitude? One way is related to the frequency of action potentials; increased stimulus strength means a larger generator potential and higher frequency of firing of action potentials. A record of an experiment in which increased stimulus intensity is reflected in increased action-potential frequency in a single afferent nerve fiber is shown in Fig. 4-20.

There is an upper limit to this positive correlation between stimulus intensity and action-potential frequency. When stimulus strength becomes very great, the generator potential reaches a maximum and a further increase in rate of firing action potentials by that receptor cannot occur. However, even though that particular receptor cannot generate a higher frequency of action potentials in the afferent neuron, receptors at other branches of the same neuron can be stimulated. Most afferent neurons have many branches, each with a receptor at its ending, but the receptors at different branches do not respond with equal ease to a given stimulus; some are less easily excited and respond only to stronger stimuli. Thus, as stimulus strength increases, more and more receptors begin to respond. Action potentials generated by these receptors propagate along the branch to the main afferent nerve fiber, and if the membrane is not refractory, they increase the frequency of action potentials there. The maximal firing frequency of a neuron is limited only by the membrane refractory period.

In addition to the increased frequency of firing in a single neuron, similar receptors on the nerve endings of other afferent neurons are also activated as stimulus strength increases, because stronger stimuli usually affect a larger area. For example, when one touches a surface lightly with a finger, the area of skin in contact with the surface is small and only receptors in that area of skin are stimulated. But pressing the finger down firmly upon the surface increases the area of skin stimulated. This "calling in" of receptors on additional nerve cells is known as *recruitment*. These generalizations are true of virtually all afferent systems: Increased stimulus intensity is signaled both by an increased firing rate of action potentials in a single nerve fiber and by recruitment of receptors on other afferent neurons in the surrounding area.

By the mechanisms discussed in this section different energy forms activate specific receptors to supply information about the kind of stimulus and its duration and intensity. This information is then transmitted in the form of action potentials along the afferent neuron. Next to be considered is the mechanism by which the activity is transferred from one neuron to another.

SYNAPSES

A synapse is an anatomically specialized junction between two neurons where the electric activity in one neuron influences the excitability of the second. Most synapses occur between the axon terminals of one neuron and the cell body or dendrites of a second. The neurons conducting information toward synapses are called *presynaptic neurons*, and those conducting information away are *postsynaptic neurons*. Figure 4-21 shows how, in a multineuronal pathway, a single neuron can be postsynaptic to one group of cells and, at the same time, presynaptic to another.

Every postsynaptic neuron has thousands of synaptic junctions on the surface of its dendrites or cell body so that information from hundreds of presynaptic nerve cells converges upon it. A single motor neuron in the spinal cord probably receives some 15,000 synaptic endings, and it has been calculated that certain neurons in the brain receive even more. Each activated synapse produces a small electric signal, either excitatory or inhibitory, in the postsynaptic cell for a brief time. The picture we are left with is one of thousands of synapses from many different presynaptic cells *converging* upon a single postsynaptic cell. The level of excitability of this cell at any moment, i.e., how close the membrane potential is to threshold, depends upon the number of synapses active at any one time and how many are excitatory or inhibitory. If the postsynaptic neuron reaches threshold and generates a response, action potentials are transmitted out along its axon to the terminal

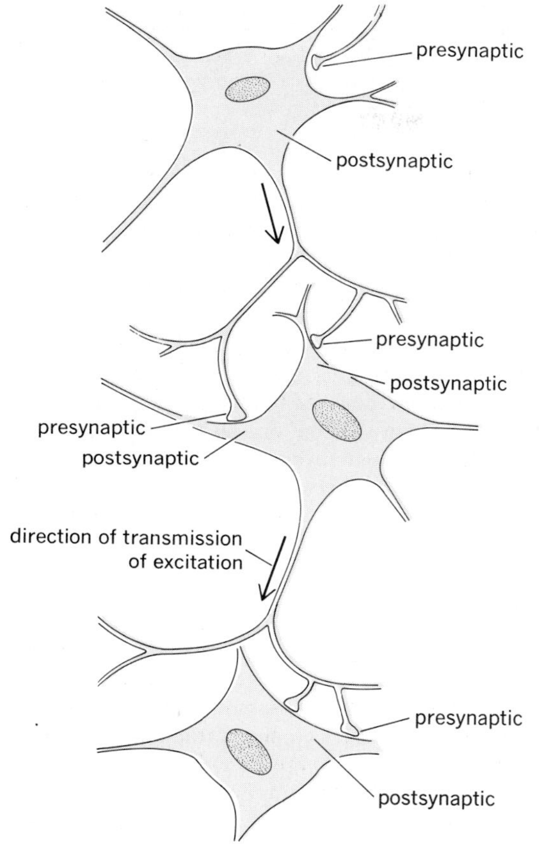

FIGURE 4-21 ***Single neuron postsynaptic to one group of cells and presynaptic to another.***

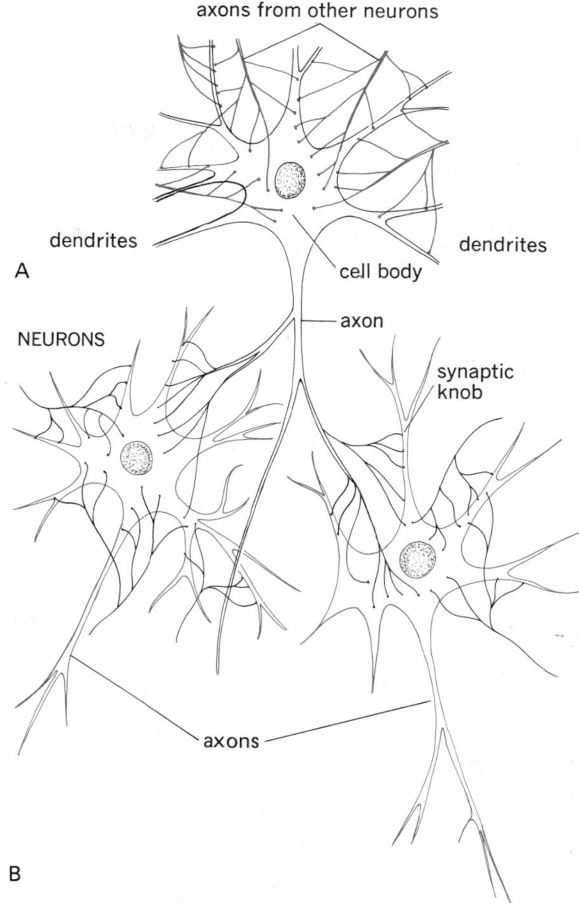

FIGURE 4-22 **(A)** ***Convergence of neural input.*** **(B)** ***Divergence of neural output.***

branches, which *diverge* to influence the excitability of many other cells. Figure 4-22 demonstrates the neuronal relationships of convergence and divergence.

In this manner, postsynaptic neurons function as neural *integrators;* i.e., their output reflects the sum of all the incoming bits of information arriving in the form of excitatory and inhibitory synaptic inputs.

Functional Anatomy

Figure 4-23 shows the anatomy of a single synaptic junction. The axon terminal of the presynaptic neuron ends in a slight swelling, the *synaptic knob.* A narrow extracellular space, the *synaptic cleft,* separating the pre- and postsynaptic neurons prevents direct propagation of the action potential from the presynaptic neuron to the postsynaptic cell. Information is transmitted across the synaptic cleft by means of a chemical agent (in most cases its exact chemical nature is not known) stored in small, membrane-enclosed *vesicles* in the synaptic knob. When an action potential in the presynaptic neuron reaches the axon terminal and depolarizes the synaptic knob, small quantities of the chemical transmitter are released from the synaptic knob into the synaptic cleft. Little is known about the events which couple action potentials of the presynaptic-cell membrane with the secretion of the transmitter substance. Once released from the vesicles, the transmitter diffuses across the synaptic cleft and combines with membrane sites[4] on the part of the postsynaptic cell

[4] As will be described in Chap. 6, these membrane sites are one type of the membrane molecules known as "receptor sites." We have avoided this term in the present chapter to prevent confusion with the totally different afferent receptors described in the previous section.

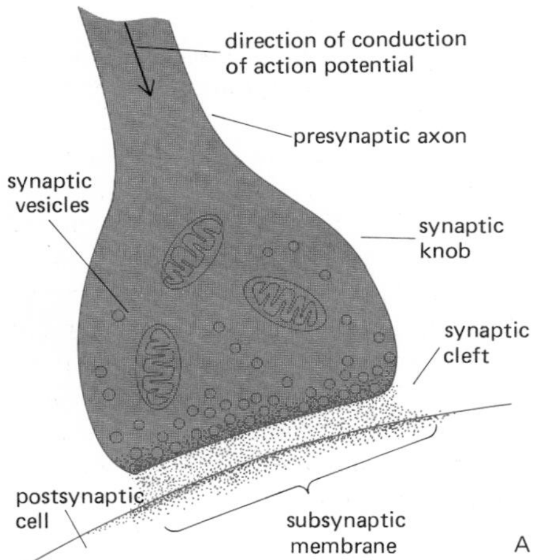

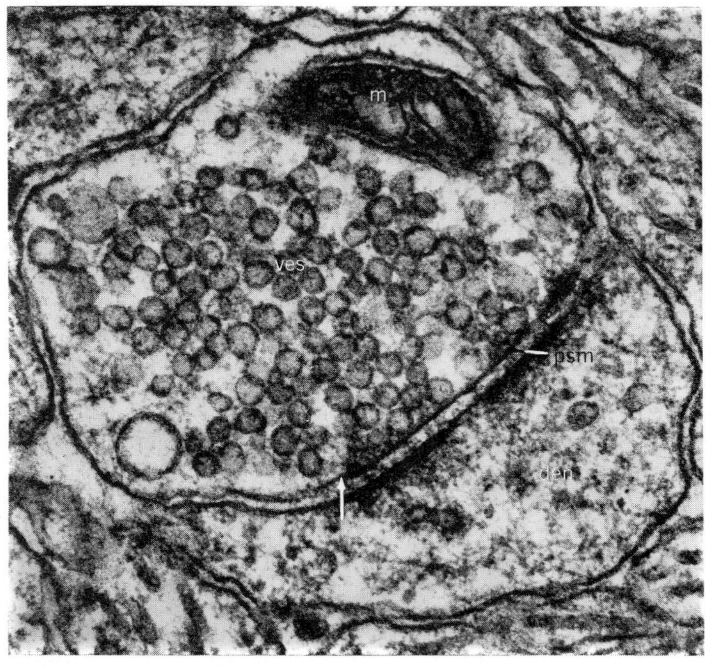

FIGURE 4-23 (A) ***Diagram of synapse.*** **(B)** ***Electron micrograph of synapse. Axon terminal contains vesicles (ves) and a mitochondrion (m), and dendrite (den) shows thickening of postsynaptic membrane (psm). Arrow marks beginning of synaptic cleft.*** **(From W. F. Windle, "Textbook of Histology," 5th ed., McGraw-Hill, New York, 1976.)**

lying right under the synaptic knob (*subsynaptic membrane*). The combination of the transmitter with the membrane sites causes changes in the permeability properties of the subsynaptic membrane and in the membrane potential of the postsynaptic cell. There is a delay between excitation of the nerve terminal of the presynaptic neuron and membrane-potential changes in the postsynaptic cell; it lasts less than $\frac{1}{1000}$ s and is called the *synaptic delay*. Its duration is a function of the release mechanism which frees transmitter substance from the synaptic knob, since the time required for the transmitter to diffuse across the synaptic cleft is negligible. Synaptic activity is terminated when the transmitter is chemically transformed into an ineffective substance, simply diffuses away from the membrane sites, or is taken back up by the synaptic knob.

Excitatory Synapse The two different kinds of synapses, excitatory and inhibitory, are classified by their effect on the postsynaptic cells. An excitatory synapse, when activated, increases the likelihood that the membrane potential of the postsynaptic cell will reach threshold and that the cell will undergo an action potential. Here the effect of the chemical-transmitter–membrane site combination is to increase the permeability of the subsynaptic membrane to positively charged ions so that they are free to move according to the electric and chemical forces acting upon them. The mechanisms responsible for the increased permeability are not known.

FIGURE 4-24 ***Excitatory postsynaptic potential (EPSP). Stimulation of the presynaptic neuron is marked by the arrow. Note the short synaptic delay before the postsynaptic cell responds.***

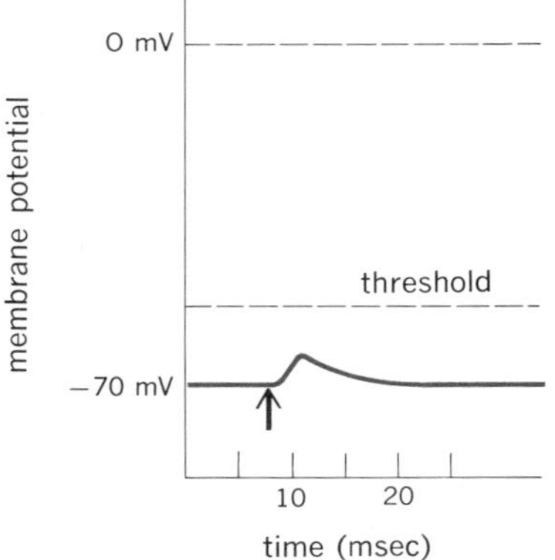

At the subsynaptic membrane of excitatory synapses there occurs the simultaneous movement of a relatively small number of potassium ions out of the cell and a larger number of sodium ions in. The *net* movement of positive ions is into the neuron, which slightly depolarizes the postsynaptic cell. This potential change, called the *excitatory postsynaptic potential* (EPSP), brings the membrane closer to threshold (Fig. 4-24). The EPSP, like the generator potential of the receptor, is a local, passively propagated potential; its only function is to help trigger an action potential.

Inhibitory Synapse Activation of an inhibitory synapse produces changes in the postsynaptic cell which lessen the likelihood that the cell will undergo an action potential. At inhibitory synapses the combination of the chemical transmitter with the membrane sites on the subsynaptic membrane also changes the permeability of the membrane, but only the permeabilities to potassium and chloride ions are increased; sodium permeability is not. The greater permeability to potassium is responsible for the changes in membrane potential associated with the activation of an inhibitory synapse. Earlier it was noted that if the membrane were permeable only to potassium ions the resting membrane potential would equal the potassium equilibrium potential; i.e., the resting membrane potential would be −90 mV instead of −70 mV. The increased potassium permeability at an activated inhibitory synapse makes the postsynaptic cell more like the hypothetical cell that is permeable only to potassium ions. Consequently, the membrane potential becomes closer to a true potassium equilibrium potential. This increased negativity (hyperpolarization) is an *inhibitory postsynaptic potential* (IPSP) (Fig. 4-25). Thus, when an inhibitory synapse on a neuron is activated, the neuron's membrane potential is moved farther away from the threshold level.

The rise in chloride-ion permeability lessens the likelihood that the cell will reach threshold, the reason being that it increases the tendency of the membrane to stay at the resting potential because the equilibrium potential of chloride is very close to the resting membrane potential. The greater chloride permeability is important when EPSPs and IPSPs arrive at the postsynaptic cell simultaneously because stabilization of the membrane at its resting potential makes it less likely that it will change toward threshold.

FIGURE 4-25 ***Inhibitory postsynaptic potential (IPSP). Stimulation of the presynaptic neuron is marked by the arrow. Note the short synaptic delay.***

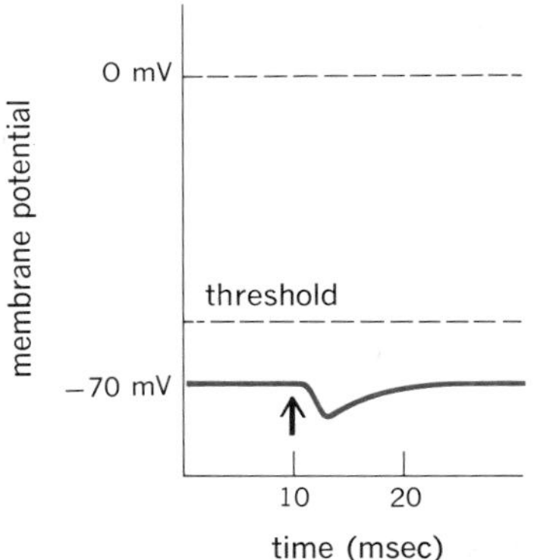

Activation of the Postsynaptic Cell

A feature that makes postsynaptic integration possible is that, in most neurons, one excitatory synaptic event is not enough by itself to change the membrane potential of the postsynaptic neuron from its resting level to threshold; e.g., a single EPSP in a motor neuron is estimated to be only 0.5 mV whereas changes of up to 25 mV are necessary to depolarize the membrane from its resting level to threshold. Since a single synaptic event does not bring the postsynaptic membrane to its threshold level, an action potential can be initiated only by the combined effects of many synapses. Of the thousands of synapses on any one neuron, probably hundreds are active simultaneously (or at least close

FIGURE 4-26 ***Intracellular recording from a postsynaptic cell during episodes when*** **(A)** ***excitatory synaptic activity predominates and the cell is facilitated, and*** **(B)** ***inhibitory synaptic activity dominates.***

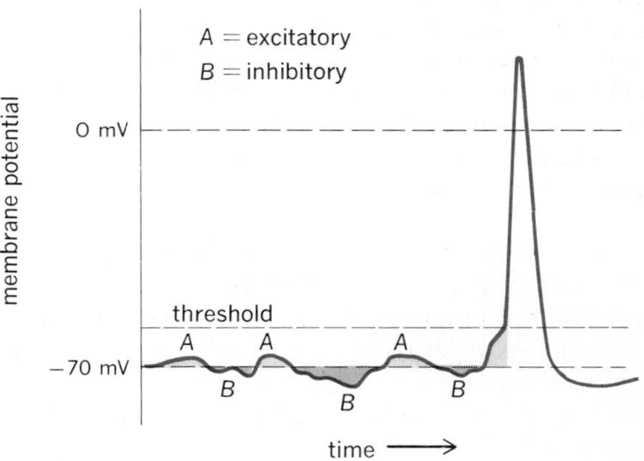

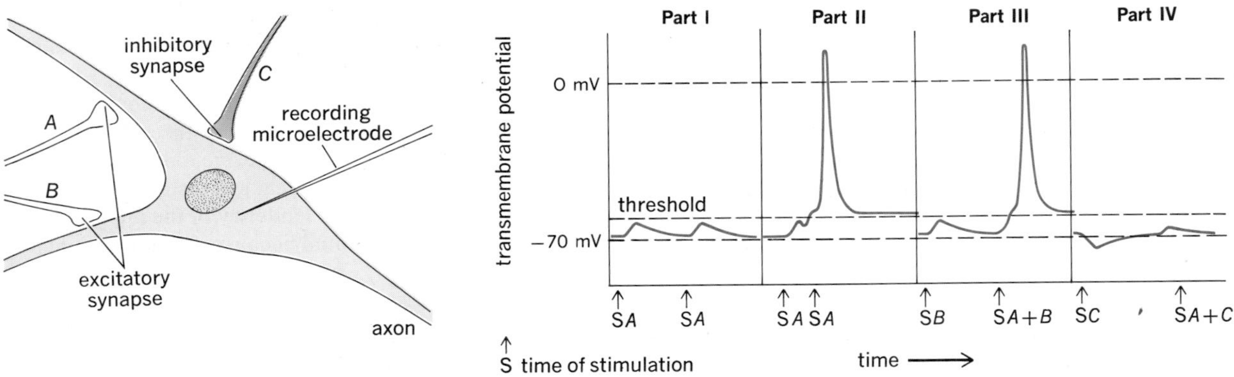

FIGURE 4-27 ***Interaction of EPSPs and IPSPs at the postsynaptic neuron.***

enough in time that the effects of later synaptic events occur before the potential changes caused by the first disappear), and the membrane potential of the postsynaptic neuron at any one moment is the resultant of all the synaptic activity affecting it at that time. There is a general depolarization of the membrane toward threshold when excitatory synaptic activity predominates (this is known as *facilitation*) and a hyperpolarization when inhibition predominates (Fig. 4-26).

Let us perform a simple experiment to see how two EPSPs, two IPSPs, or an EPSP plus an IPSP interact (Fig. 4-27). Let us assume that there are three synaptic inputs to the postsynaptic cell; *A* and *B* are excitatory and *C* is an inhibitory synapse. There are stimulators on the axons to *A*, *B*, and *C* so that each of the three inputs can be activated individually. A very fine electrode is placed in the postsynaptic neuron and wired to record the membrane potential. In part I of the experiment we shall test the interaction of two EPSPs by stimulating *A* and then, a short while later, stimulating *A* again. Part I of Fig. 4-27 shows that no interaction occurs between the two EPSPs. The reason is that the change in membrane potential associated with an EPSP is fairly short-lived. Within 100 ms the permeability properties of the subsynaptic membrane return to normal, and the excess positive charge in the cell moves back out into the extracellular fluid. By the time axon *A* is restimulated, the postsynaptic cell has returned to its resting condition. In part II of the experiment, axon *A* is stimulated again before the effect of the first stimulus has died away, and the potential change from the second stimulation of axon *A* can be added to the EPSP caused by the first. The two synaptic potentials summate, and because this summation is due to successive stimulation of the same presynaptic fiber, it is called *temporal summation*. In part III, axons *A* and *B* are stimulated simultaneously. The potential change attributable to the activity of synapse *B* can be added to the EPSP caused by the activity of synapse *A*. The two EPSPs summate, and, because they originate at different places on the postsynaptic neuron, this is called *spatial summation*. The summation of EPSPs can bring the membrane to its threshold so that an action potential is initiated. So far we have tested only the patterns of interaction of excitatory synapses. What happens if an excitatory and inhibitory synapse are activated so that their effects occur at the postsynaptic cell simultaneously? Since the EPSP and IPSP are due to the movement of different ions, they do not exactly cancel each other, and there is a slight depolarization (Fig. 4-27, part IV). Inhibitory potentials can also show temporal and spatial summation.

In the preceding examples we referred to the threshold of the postsynaptic neuron. However, the fact is that different parts of a neuron have different thresholds. The neuronal cell body and larger dendritic branches reach threshold when their membrane is depolarized about 25 mV from the resting level, but in many cells the *initial segment* [that part of the neuron cell body from which the axon leaves plus the first, unmyelinated portion of the axon itself (see Fig. 4-1B)] has a threshold which is less than half that.

As described before, the subsynaptic membrane is depolarized at an activated excitatory synapse and hyperpolarized at an activated inhibitory synapse. By the mechanisms of local current flow described earlier, current flows through the cytoplasm from an excitatory synapse and toward an inhibitory synapse (Fig. 4-28). Thus, the entire cell body, including the initial segment, becomes slightly depolarized during activation of an ex-

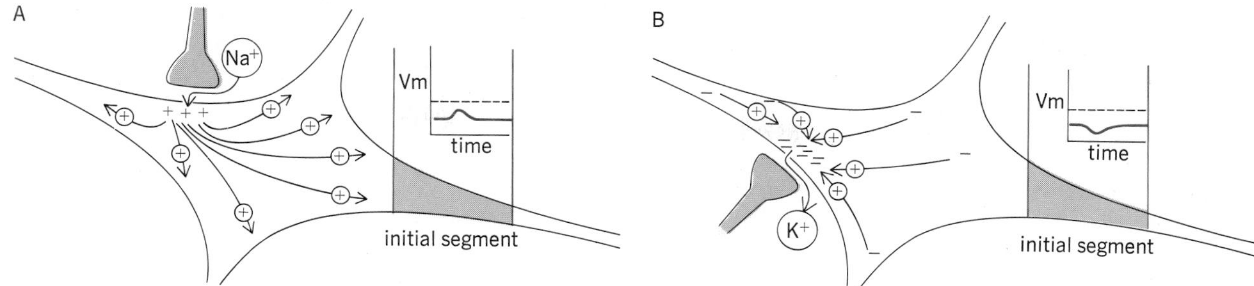

FIGURE 4-28 ***Comparison of excitatory (A) and inhibitory (B) synapses, showing direction of current flow through the postsynaptic cell following synaptic activation. (A) Current flows through the cytoplasm of the postsynaptic cell away from the excitatory synapse, depolarizing the cell. (B) Current flows through the cytoplasm of the postsynaptic cell toward an inhibitory synapse, hyperpolarizing the cell. Arrows show the direction of positive-ion flow.***

citatory synapse and slightly hyperpolarized during activation of an inhibitory synapse. In cells whose initial-segment threshold is lower than that of their dendrites and cell body, the initial segment is activated first whenever enough EPSPs summate, and the action potential originating there is propagated both down the axon and back over the cell body.

Synaptic events last more than 10 times as long as action potentials do. In the event that the initial segment is still depolarized above threshold after an action potential has been fired and the refractory period is over, a second action potential will occur. In fact, the greater the depolarization due to synaptic events, the greater is the number of action potentials fired (up to the limit imposed by the duration of the absolute refractory period). Neuronal responses are almost always in the form of so-called bursts or trains of action potentials.

The significance of the lower threshold of the initial segment can best be demonstrated if we suppose for a moment that the threshold were the same over the entire neuron so that an action potential could be initiated with equal ease at any point of the cell body or dendrites. In Fig. 4-29 the greatest input to the cell is clearly inhibitory, but in the upper left corner three active excitatory synapses are clustered together. At this one point there is sodium-ion movement into a relatively small portion of the cell. If this small region had a low threshold, an action potential could be initiated at this site and conducted over the entire cell membrane despite the fact that most of the input of the cell is inhibitory. This possible bias of the cell's activity by synapse grouping is greatly lessened by the fact that the initial segment acts to average all the synaptic input.

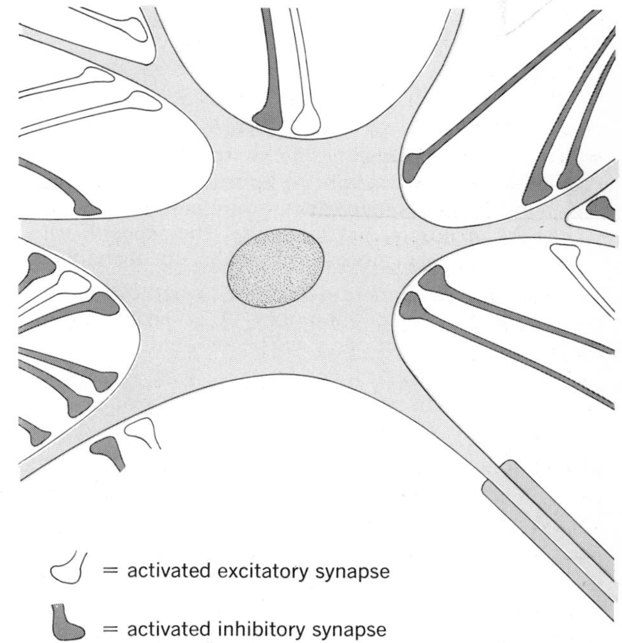

FIGURE 4-29 ***Synaptic input to a neuron.***

On the other side of the coin, these synapses right next to the initial segment have a greater influence upon cell activity than those at the ends of the dendrites, and thus synaptic placement provides a mechanism for giving different inputs a greater or lesser influence on the postsynaptic cell's output. Finally, it should be noted that, in some cells, action potentials can be initiated in regions other than the initial segment.

This discussion has no doubt left the impression that one neuron can influence another only at a synapse. However, this view requires qualification. We have emphasized how local currents influence the membrane in which they are generated (as, for example, in action-potential propagation and synaptic excitation). In addition, local currents can also affect the membrane potentials of other nearby neurons, producing small potential charges which may significantly influence the activity of these neurons in a variety of ways. This phenomenon is particularly important in areas of the central nervous system which contain a high density of unmyelinated neuronal processes.

Chemical Transmitters

As described above, presynaptic neurons influence postsynaptic neurons by means of chemical transmitters. This section lists some of the most important facts relating to them.

1 There appear to be many different synaptic transmitters. However, all synaptic endings from a single presynaptic cell probably liberate the same one. Some of the substances thought to be transmitters are acetylcholine, norepinephrine, γ-aminobutyric acid (GABA), serotonin, and dopamine. Their specific roles will be described in more detail in subsequent chapters.

2 The membrane site determines whether the synapse is excitatory or inhibitory. A given transmitter may be excitatory at one synapse and inhibitory at another.

3 Synapses can operate in only one direction because all the transmitter is stored on one side of the synaptic cleft and all the membrane sites are on the other side. This is in contrast to action potentials, which can travel along a nerve fiber in either direction. Because of the one-way conduction across synapses, action potentials pass along chains of neurons in only one direction.

4 The chemical transmitter continues to combine with the membrane sites on the subsynaptic membrane until it is inactivated by combining chemically with another substance, is taken up again by the presynaptic endings, or simply diffuses away from the synaptic region. In some cases, the transmitter substance is not removed immediately from the membrane site, and a single volley of impulses in the presynaptic fibers causes prolonged firing in the postsynaptic neuron.

5 Synapses are vulnerable to many drugs and toxins which can modify the synthesis, storage, release, inactivation, or uptake of the transmitter substance, or block the membrane sites on the subsynaptic membrane to prevent combination with the transmitter. For example, the toxin produced by the tetanus bacillus acts at the inhibitory synapses upon the motor neuron, presumably by blocking the membrane sites. This eliminates inhibitory input to the motor neuron and permits unchecked influence of the excitatory inputs, leading to muscle spasticity and seizures. The spasms of the jaw muscles appearing early in the disease are responsible for the common name of lockjaw.

NEUROEFFECTOR COMMUNICATION

Efferent neurons innervate muscle or gland cells. Information is transmitted from axons to these cells, the effector cells, by means of chemical transmitters. When an action potential reaches the terminal portions of an axon, it causes the release of transmitter which diffuses to the effector cell and alters its activity. The structures of these axon terminals and their anatomic relationships to the effector cells vary, depending upon the effector cell type; they will be described in detail in subsequent chapters. The neuroeffector transmitters are well characterized (in contrast to the synaptic transmitters); they are either acetylcholine or norepinephrine.

CHAPTER 5
Muscle Tissue

Three different types of muscle cells can be identified on the basis of structure and contractile properties: (1) *skeletal muscle,* (2) *smooth muscle,* and (3) *cardiac muscle.* Most skeletal muscle, as the name implies, is attached to the bones of the body, and its contraction is responsible for the movements of parts of the skeleton. Skeletal muscle movement is also involved in other activities of the body such as the voluntary release of urine and feces. Skeletal muscle is under the control of the somatic nervous system. The movements produced by skeletal muscle are primarily those involved with interactions between the body and the external environment.

Smooth muscle surrounds such hollow chambers in the body as the stomach and intestinal tract, the urinary bladder, blood vessels, and uterus. The contraction of smooth muscle is controlled in large part by the autonomic nervous system and thus is not normally under direct conscious control. The contraction of smooth muscle is associated with processes which regulate the internal environment of the body. The third type of muscle, cardiac muscle, is the muscle of the heart and, like smooth muscle, is primarily under the control of the autonomic nervous system. Although there are significant differences in the structure, contractile properties, and control of these three types of muscle, the physicochemical principles underlying their contractile activity are similar.

STRUCTURE OF SKELETAL MUSCLE

Skeletal muscle is the largest tissue in the body, accounting for 40 to 45 percent of the total body weight. An individual muscle, such as the biceps or gastrocnemius, is made up of single cells, called *muscle fibers* (Fig. 5-1). Each fiber is cylindrical, having a diameter of 10 to 100 μm and a length of up to 30 cm (about 1 ft). In contrast to most cell types, each fiber contains hundreds of nuclei, which are located just beneath the cell membrane. The fibers, each surrounded by a loose connective-tissue *endomysium*, are organized into various-sized bundles, or *fascicles* (Fig. 5-1). These in turn are surrounded by a dense connective-tissue sheath known as the *perimysium*; the entire muscle, which is composed of many fascicles, is surrounded by the *epimysium* (Fig. 5-1). The relation between a single muscle fiber (cell) and a whole muscle is similar to that between a single nerve fiber (axon) and a nerve composed of many axons. Finally, blood vessels and nerves follow these sheaths into the muscle interior.

FIGURE 5-1 ***Various levels of organization of skeletal muscle from whole muscle down to molecules. (Redrawn from Warwick and Williams.)***

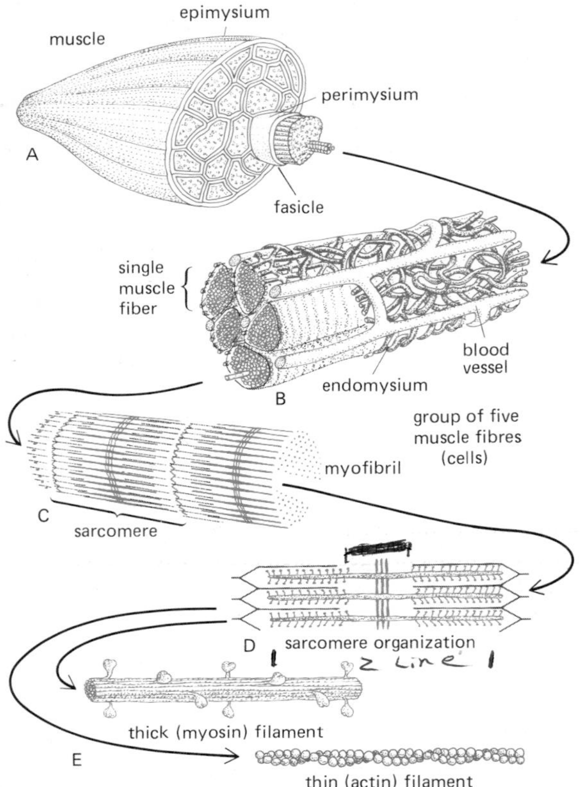

Over 600 muscles can be identified in the human body. Some of them are very small, consisting of only a few hundred fibers; larger muscles may contain several hundred thousand fibers. In some muscles the individual fibers are as long as the muscle itself; but most fibers are shorter than the total muscle, and their ends are attached to the connective-tissue network interlacing the muscle fibers.

Generally each end of the whole muscle is attached to a bone by bundles of collagen fibers known as *tendons*, which have great strength but no active contractile properties. The collagen fibers in the connective tissues of the perimysium and epimysium are continuous with those in the tendons, and together they act as a structural framework to which the muscle fibers and bone are attached. The forces generated by the contracting muscles are transmitted by the connective tissue and tendons to the bones. The transmission of force from muscle to bone is like a number of people pulling on a rope, each person corresponding to a single muscle fiber and the rope to the connective tissue and tendons.

Subcellular Organization of Skeletal Muscle Fibers

Figure 5-2 shows a section through a skeletal muscle as seen with a light microscope. In between the individual muscle fibers are capillary blood vessels containing red blood cells. The most striking feature of the muscle fibers is the series of transverse light and dark bands forming a regular pattern along the fiber. Both skeletal and cardiac muscle fibers have this characteristic banding and are known as *striated muscles*; smooth muscle cells show no banding pattern. Although the pattern appears to be continuous across a single fiber, the fiber is actually composed of a number of independent cylindrical elements, known as *myofibrils*, in the cytoplasm of the fiber (Fig. 5-1C). Each myofibril is about 1 to 2 μm in diameter and continues through the length of the muscle fiber. Myofibrils occupy about 80 percent of the fiber volume and vary in number from several hundred to several thousand, depending on the fiber diameter. The myofibrils form a longitudinal striation of the muscle; the banding patterns in the myofibrils form transverse striations.

Viewed with the electron microscope, the structures responsible for the banding patterns become evi-

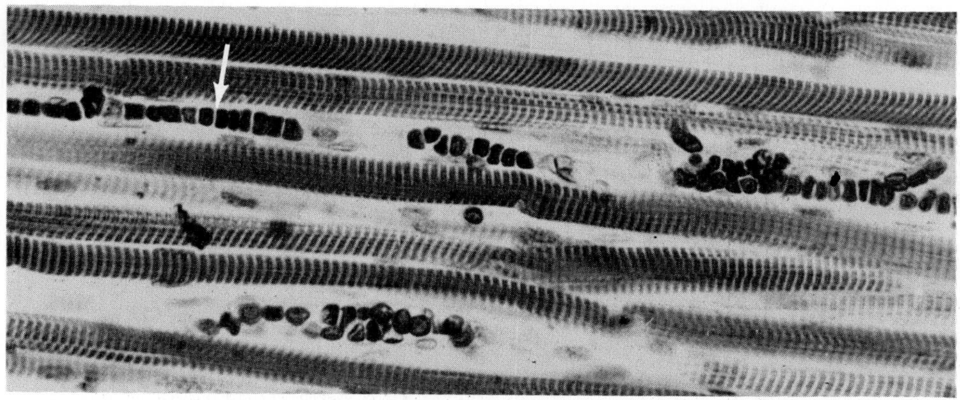

FIGURE 5-2 ***Photomicrograph of skeletal muscle fibers. This level of organization is comparable to Fig. 5-1B. Arrow indicates capillary blood vessel containing blood cells.*** **(From Edward K. Reith and Michael H. Ross, "Atlas of Descriptive Histology," Harper & Row, New York, 1968.)**

dent. The myofibrils consist of smaller *filaments* (Figs. 5-1D and E and 5-3A), which form a regular repeating pattern along the length of the fibril. One unit of this repeating pattern is known as a *sarcomere.* It is the functional unit of the contractile system in muscle, and the events occurring in a sarcomere are duplicated in the other sarcomeres along the myofibrils.

Each sarcomere contains two types of filaments: thick and thin. The thick filaments, 150 Å in diameter, are located in the central region of the sarcomere where their orderly parallel arrangement gives rise to the dark bands, known as *A bands,* that are seen in striated muscles. These thick filaments contain the protein known as *myosin.* The thin filaments, 50 Å in diameter, contain the protein *actin* and are attached at either end of the sarcomere to a structure known as the *Z line.* Two successive Z lines define the limits of one sarcomere. The Z lines consist of short elements which interconnect the thin filaments from two adjoining sarcomeres and thus provide an anchoring point for the thin filaments. The thin filaments extend from the Z lines toward the center of the sarcomere where they overlap with the thick filaments.

In addition to the A band and Z line already mentioned, two other bands will be identified so that changes in the banding patterns occurring during contraction can be related to the relative positions of the thick and thin filaments in the sarcomere. The *I band* (Fig. 5-3) represents the region between the ends of the A bands of two adjoining sarcomeres. This band contains that portion of the thin filaments which do not overlap with the thick filaments and is bisected by the Z line. Because it contains only thin filaments it usually appears as a light band separating the dark A bands. Finally, the *H zone* appears as a thin, lighter band in the center of the A band which corresponds to the space between the ends of the thin filaments. Only thick filaments are found in the H-zone region.[1]

A cross section through myofibrils in the region of the A band where both thick and thin filaments overlap (Fig. 5-3B) shows their hexagonal arrangement. Each thick filament is surrounded by six thin filaments, and each thin filament is surrounded by three thick. Thus there are twice as many thin as thick filaments in the region of overlap.

At higher magnification, in the region of overlap in the A band, the gap between thick and thin filaments appears to be bridged by projections at intervals along the filaments. Although it is not evident from this electron micrograph, these projections, or *cross bridges,* are part of the structure of the myosin molecules in the thick filaments. These cross bridges are arranged in a spiral around the thick filament. One turn of the spiral gives rise to six bridges which are able to interact with the actin in the six thin filaments surrounding each thick filament (Fig. 5-3B). Figure 5-1 summarizes the levels of structure within a muscle fiber.

[1] A thin dark band can be seen in the center of the H zone. This is known as the M line and is produced by processes which link the thick filaments together, maintaining their orderly, parallel arrangement.

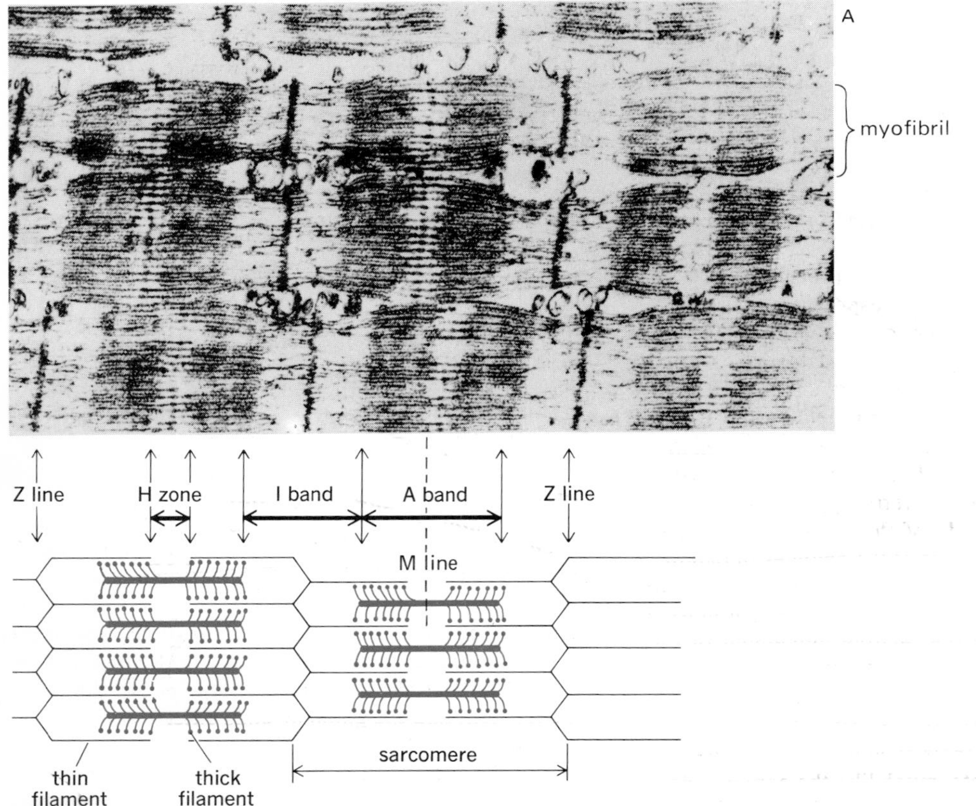

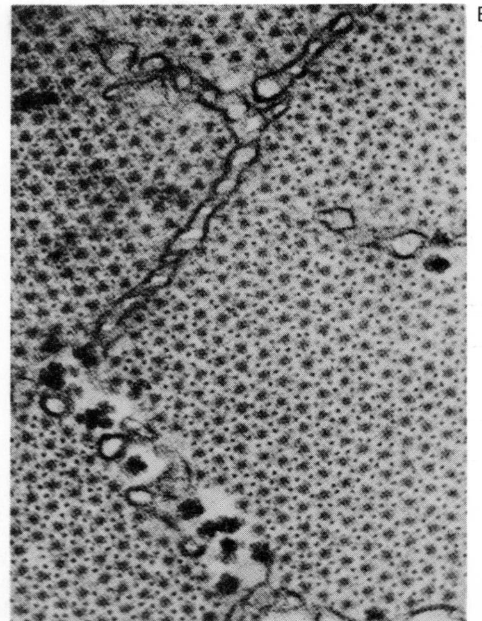

FIGURE 5-3 **(A)** ***Electron micrograph showing three myofibrils corresponding to Fig. 5-1C (although Fig. 5-1C shows only one myofibril). Diagramed below is the organization of thick and thin filaments.*** **(B)** ***Electron micrograph of a cross section through portions of five myofibrils. Diagramed to the right are the arrangements of thick and thin filaments seen in cross sections through different regions of the sarcomere. Part 1 would be seen in cross sections through the H zone; part 2, through the A band in regions of overlap of the thick and thin filaments (this corresponds to the section in the electron micrograph in B); and part 3, through the I band.*** **[Part A, from H. E. Huxley and J. Hanson, in G. H. Bourne (ed.), "The Structure and Function of Muscle," vol. 1, Academic, New York, 1960.; part B, from H. E. Huxley, *J. Mol. Biol.*, 37:507–520 (1968).]**

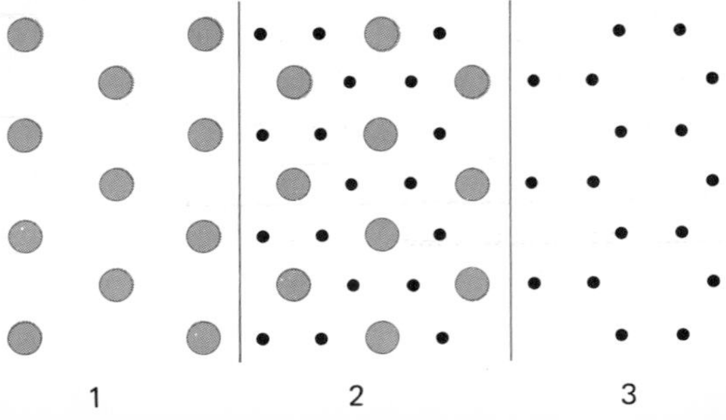

MOLECULAR BASIS OF CONTRACTION

Sliding-filament Theory

H. E. Huxley used the electron microscope to examine muscle in a resting, relaxed state and after different degrees of shortening. The changes in sarcomere structure found at the different muscle lengths is shown in Fig. 5-4. His crucial observation was that, as the muscle becomes shorter and shorter, the thick and thin filaments slide past each other, but the lengths of the individual filaments do not change. Thus, the width of the A band remains constant, corresponding to the constant length of the thick filaments. The I band narrows as the thick filaments approach the Z line. As the thin filaments move past the thick filaments, the width of the H zone between the ends of the thin filaments becomes smaller and may disappear altogether when the thin filaments meet at the center of the sarcomere. With further shortening, new banding patterns appear as thin filaments from opposite ends of the sarcomere begin to overlap. These observations of the changes in banding pattern during contraction led to the *sliding-filament theory of muscle contraction,* which states that muscle shortening results from the relative movement of the thick and thin filaments past each other.

The structures which actually produce the sliding of the filaments are the myosin cross bridges which swivel in an arc around their fixed positions on the surface of the thick filaments, much like the oars of a boat (Fig. 5-5). The movement of the cross bridges in contact with the actin thin filaments produces the sliding of the thick and thin filaments past each other. Since one movement of a cross bridge will produce only a small displacement of the thin filament relative to the thick, the cross bridges must undergo many repeated cycles of movement during a contraction. During contraction each cross bridge undergoes its own independent cycle of movement so that at any one instant during contraction only about 50 percent of the bridges are attached to the actin thin filaments; the others are at intermediate stages of the cycle.

Actin, Myosin, and ATP

What are the properties of actin and myosin which produce this cyclic activity of the cross bridges? Actin is a globular-shaped molecule about 55 Å in diameter, which has a receptor site on its surface that is able to combine with myosin. The actin molecules are arranged in two chains which are helically intertwined to form the thin filaments. Myosin is a much larger molecule that is shaped like a lollypop (Fig. 5-1E) with a

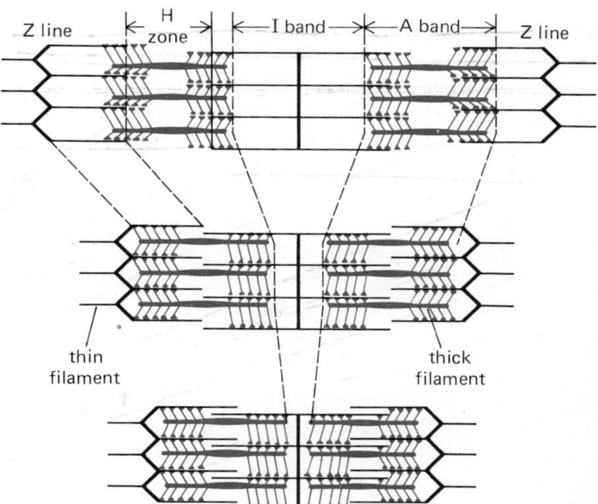

FIGURE 5-4 ***Changes in banding patterns resulting from the movements of thick and thin filaments past each other during contraction. This organizational level corresponds with that in Fig. 5-1D.***

large globular end attached to a long tail. The myosin molecules are arranged within a thick filament so that the molecules are oriented tail to tail in the two halves of the filament and the globular ends extend to the sides, forming the cross bridges.

The globular end of myosin contains a receptor site that is able to bind to the receptor site on the actin mole-

FIGURE 5-5 ***Movements of myosin cross bridges produce the relative movements of the thick and thin filaments. The level of tissue organization corresponds to Fig. 5-1E.***

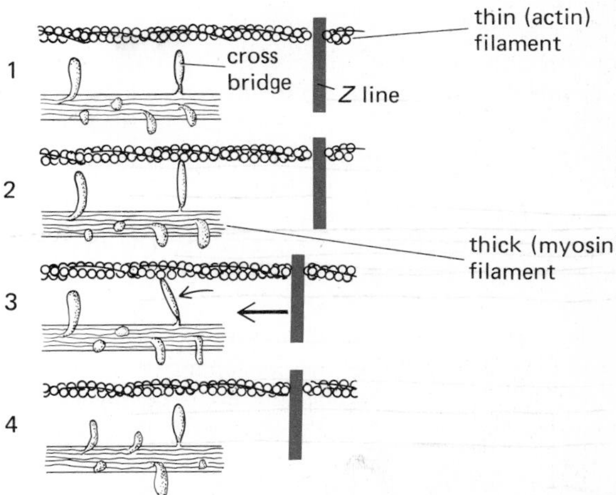

cule. In addition, the globular end contains a separate active site that is able to split ATP. Thus myosin is an enzyme (myosin ATPase) whose substrate is ATP. (Magnesium is a cofactor required to bind ATP to the active site of myosin ATPase.) However, myosin alone has a very low ATPase activity (slow rate of splitting ATP), but when a myosin cross bridge combines with actin in the thin filaments, the activity of myosin ATPase is increased considerably. The energy released from the splitting of ATP produces cross-bridge movement although we still do not understand, in molecular terms, how this occurs. Presumably this involves a change in the shape of the globular end of the myosin molecule while it is attached to the actin thin filament.

As just described, since a single movement of the myosin bridge produces only a small displacement of the thin filament with respect to the thick, these bridges must undergo repeated cycles of activity to produce the degree of shortening observed during muscle contraction. This means that the myosin bridge must be able to detach itself from actin, rebind to a new actin site, and repeat the cycle of bridge movement. What causes the dissociation of myosin bridges from actin at the end of a bridge movement?

The dissociation is achieved by the binding (not splitting) of a molecule of ATP to myosin. The process of binding ATP appears to break the linkage between actin and myosin.

$$A \cdot M + ATP \longrightarrow A + M \cdot ATP$$

The reaction returns the bridge to its initial state so that it can now undergo binding to a new actin site and repeat the cycle of cross-bridge movement. The three basic reactions in the cross-bridge cycle are summarized in Fig. 5-6.

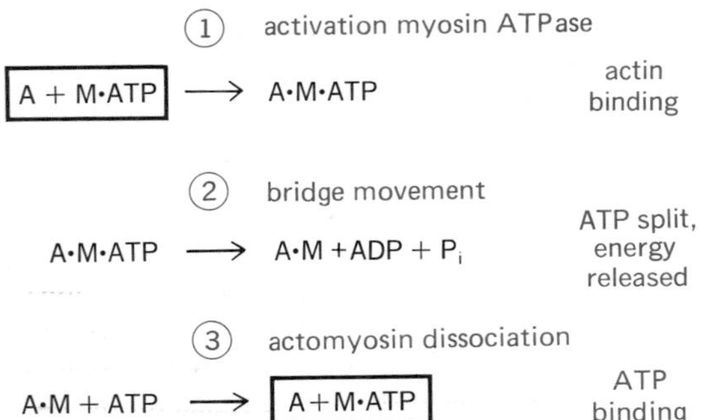

FIGURE 5-6 ***Sequential steps in the interaction of actin, myosin, and ATP leading to cross-bridge movement.***

The importance of ATP in dissociating actin and myosin at the end of a bridge cycle is illustrated by the phenomenon of *rigor mortis* (death rigor), in which the muscles of the body become very stiff and rigid shortly after death. This results directly from the loss of ATP in the dead muscle cells. In the absence of ATP the myosin cross bridges are able to combine with actin but the bond between them is not broken. The thick and thin filaments become cross-linked to each other and cannot be passively pulled apart by stretch, thus the rigid condition of the dead muscle. At the molecular level of actin and myosin, we can identify two very specific roles for ATP: (1) The splitting of bound ATP by myosin ATPase provides the energy for the movement of the cross bridges, and (2) the binding (not splitting) of ATP to myosin dissociates actin from the myosin cross bridges during the cycle of the bridges.

Regulator Proteins

Since a muscle cell contains all the ingredients necessary for cross-bridge activity—actin, myosin, ATP, and magnesium ions—the question arises: Why are muscles not in a continuous state of contractile activity? The reason is that the myosin bridges can be prevented from combining with actin by the two regulator proteins *troponin* and *tropomyosin*. These proteins are associated with the thin filaments in muscle. The interactions between these proteins and actin prevent actin from combining with myosin in a resting muscle, perhaps by blocking or changing the shape of the actin receptor site. These regulator proteins thus act as natural inhibitors of the contractile process. The problem now becomes, not what prevents contractile activity from occurring continuously, but whatever starts the process in the first place? The role of turning contractile activity on and off falls to the calcium ion.

Calcium inhibits the inhibitory effects of troponin and tropomyosin. For example, if calcium ions are injected into a muscle fiber, it immediately contracts. The site of action of calcium ions appears to be the troponin molecule. The binding of calcium to the troponin molecule produces a change which is transmitted through the tropomyosin molecule to the actin molecules, the overall effect of which is to restore the ability of actin to combine with myosin and initiate cross-bridge activity. Thus, muscle contraction is initiated when calcium is made available to troponin and ceases when calcium is removed. The mechanisms which regulate the availability of calcium ions to the contractile machinery are coupled to electric events that occur in the muscle membrane.

Excitation-Contraction Coupling

The cell membranes of muscle are excitable membranes capable of generating and propagating action potentials by mechanisms very similar to those discussed in nerve cells (Chap. 4). An action potential in the muscle cell membrane provides the signal for the initiation of contractile activity. The mechanism by which an electric signal in the membrane triggers off the chemical events of contraction is known as *excitation-contraction coupling* and is mediated by calcium ions.

Sarcoplasmic Reticulum In a resting muscle the concentration of free calcium ions is very low, and the regulatory proteins, in the presence of ATP, are able to maintain actomyosin in its dissociated state. An action potential in the cell membrane leads to an increase in the intracellular calcium-ion concentration. These calcium ions which react with the regulator proteins, blocking their inhibitory action, initiate the cycle of cross-bridge activity which leads to shortening of the muscle. More than sufficient calcium is released by a single action potential to inhibit all the troponin-tropomyosin in the muscle fiber. The source of these calcium ions is the *sarcoplasmic reticulum*.

The sarcoplasmic reticulum in muscle is homologous in its general structure to the sheets of intracellular membranes (endoplasmic reticulum) found in most cells. In muscle the sarcoplasmic reticulum forms a sleevelike structure which surrounds each of the myofibrils (Fig. 5-7). Its structure is closely associated with the repeating pattern of the sarcomeres in the myofibrils. At regular intervals, associated with each sarcomere, the reticulum enlarges to form what are known as *lateral sacs*. These lateral sacs contain the calcium ions that are released following membrane excitation. In between the lateral sacs is a small tubule that runs transversely around the myofibril, known as the *trans-*

FIGURE 5-7 ***Three-dimensional view of transverse tubules and sarcoplasmic reticulum surrounding six myofibrils in skeletal muscle fiber. Corresponds to Fig. 5-1C, although Fig. 5-1C shows only one myofibril.*** **(Adapted from Bloom and Fawcett.)**

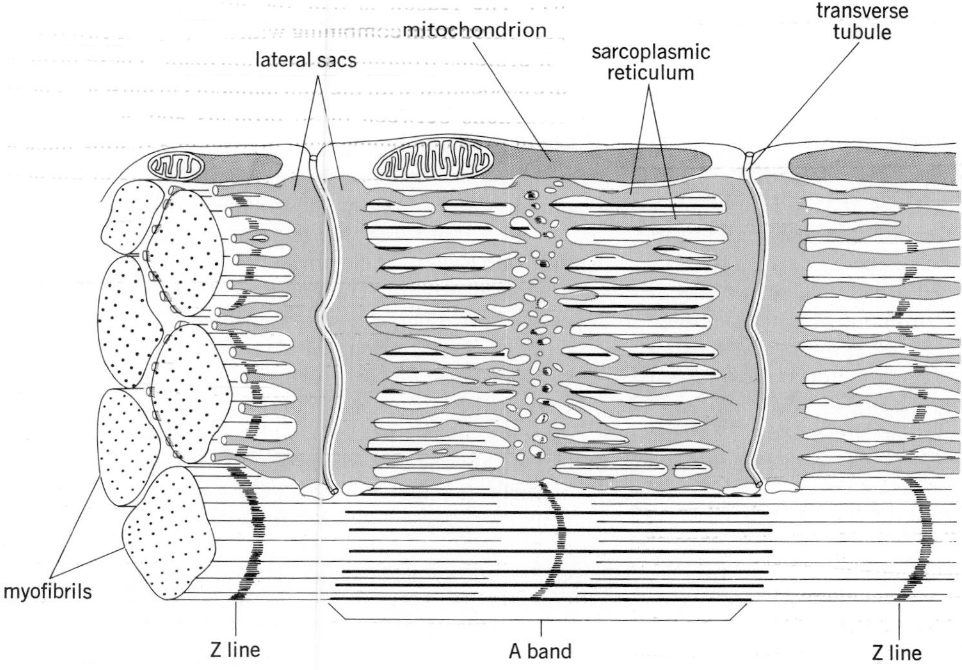

verse tubule (t tubule). This t tubule interconnects with other t tubules surrounding myofibrils and eventually joins the surface membrane of the muscle fiber. The lumen of the t tubule is thus continuous with the extracellular medium surrounding the muscle fiber. The repeating organization of the sarcoplasmic reticulum and t tubules with the sarcomere structure varies in different species of animals and in different types of muscle in the same animal. In human skeletal muscle the combination of the lateral sacs and t tubule (which is known as a *triad* because of its appearance in sections seen with the electron microscope) is located opposite the junctions of the A and I bands (Fig. 5-7), but in human cardiac muscle, which is very similar to skeletal muscle in its organization of thick and thin filaments into myofibrils, the t tubules are found associated with the Z lines rather than at the A-I junctions.

The t tubules are involved in transmitting the electric signal in the membrane into the muscle fiber where it can trigger contraction. The details of this process are still unclear, but it appears that the action potential in the membrane produces an electric signal that is passed along the t tubules to the lateral sacs of the sarcoplasmic reticulum, causing them to release calcium into the muscle cytoplasm in the immediate vicinity of the regulator proteins, thereby initiating a cycle of cross-bridge activity by the mechanisms described previously.

How is contraction, once initiated by the release of calcium from the sarcoplasmic reticulum, turned off? It is turned off by lowering the intracellular calcium concentration, thereby removing calcium from the regulator proteins. The membranes of the sarcoplasmic reticulum have the ability to concentrate calcium in the lumen of the lateral sacs, using energy derived from the splitting of ATP. This is the third important role of ATP in the process of muscle contraction. The action potential and its release of calcium from the lateral sacs last only a few milliseconds. Immediately following this electric activity, the calcium pumps in the membranes of the reticulum begin pumping the released calcium back into the lateral sacs. This process of reaccumulating the released calcium takes much longer than the initial release, and contractile activity of the cross bridges proceeds for several hundred milliseconds after the release of calcium until the concentration of free calcium becomes so low that the regulator proteins are no longer inhibited and they again begin to block actin and myosin interactions. Thus, if contractile activity is to last for more than a few hundred milliseconds, repeated action potentials must occur to maintain the free-calcium concentration surrounding the myofibrils at a high enough level to inhibit troponin and tropomyosin inhibition of actomyosin. Figure 5-8 summarizes the role of calcium in excitation-contraction coupling.

Drugs are known which will interfere with various stages in this process of excitation-contraction coupling. Thus some drugs can block the release of calcium from the reticulum and maintain the muscle in a relaxed state even in the presence of action potentials in the muscle membrane. Other drugs, such as caffeine, in higher concentrations than are found in coffee or tea, can cause the release of calcium and produce contractures of muscle in the absence of action potentials. Some of the drugs which are used to increase the force of cardiac muscle contraction in heart disease act by increasing the release of calcium from the sarcoplasmic reticulum.

Membrane Excitation The whole process of contraction begins at the cell surface where the muscle interacts with its environment. What are the mechanisms that lead to the initiation of action potentials in muscle cell membranes? There are three answers to this question, depending on the type of muscle that is being considered: (1) stimulation by a nerve fiber, (2) stimulation by hormones and chemical agents, and (3) spontaneous electric activity within the membrane itself. Stimulation by nerve fibers is the only one of the three mechanisms by which skeletal muscles are normally excited in the body. We shall find that the other two mechanisms as well as nerve fibers are involved in initiating excitation in smooth and cardiac muscle.

The axonal process of a nerve fiber forms a junction with a skeletal muscle membrane which resembles in general structure and function the synaptic junctions between two nerve fibers described in Chap. 4. These junctions between nerve and skeletal muscle are known as *neuromuscular* or *myoneural* junctions. The nerve cells which form myoneural junctions with skeletal muscles are known as *motor neurons* (somatic efferent), and the cell bodies of these neurons are located in the brain and spinal cord. The axons of these motor neurons are myelinated and are generally the largest-diameter axons in the body. They are thus able to propagate action potentials at high velocities, sending signals to the muscle that can rapidly initiate muscle activity.

As the motor axon approaches the muscle, it divides into many branches, each of which forms a

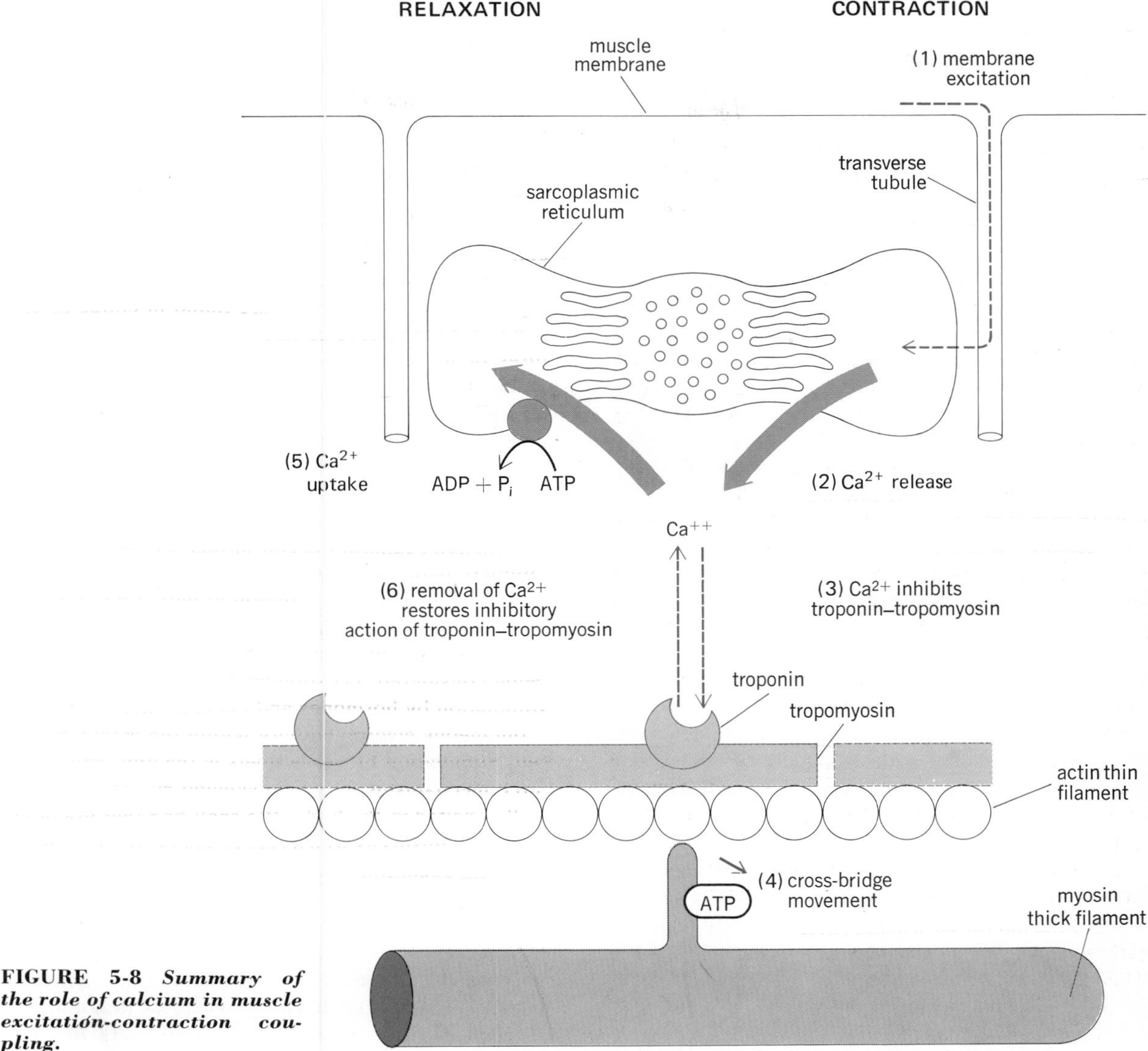

FIGURE 5-8 ***Summary of the role of calcium in muscle excitation-contraction coupling.***

single myoneural junction with a muscle fiber (Fig. 5-9). Thus, each motor neuron is connected through its branching axon to several muscle fibers. The combination of the motor neuron and the muscle fibers it innervates is known as a *motor unit*. Although each motor neuron innervates many muscle fibers, each muscle fiber is innervated by only a single motor neuron.

As a branch of the motor axon approaches the muscle surface, it loses its myelin sheath and further divides into a fine terminal arborization which lies in grooves on the muscle membrane. The region of the muscle membrane which lies directly under the terminal portion of the axon has special properties and is known as the *motor end plate*.

The terminal ends of the motor axon contain membrane-bound vesicles resembling the synaptic vesicles found at synaptic junctions (Fig. 5-10). These vesicles contain the chemical transmitter *acetylcholine*

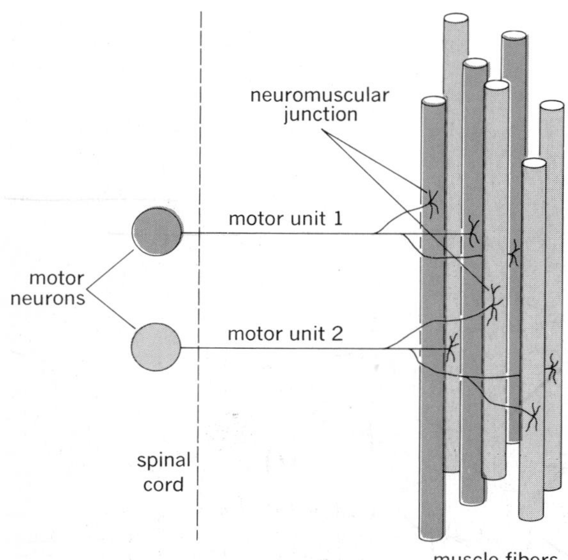

FIGURE 5-9 ***Muscle fibers associated with two motor neurons, forming two motor units.***

(abbreviated ACh). When an action potential in the motor axon arrives at the myoneural junction, it depolarizes the nerve membrane, initiating the reactions which lead to a fusion of the transmitter vesicles with the nerve membrane, allowing them to release their acetylcholine into the space separating the nerve and muscle membranes.

Once acetylcholine is released, it diffuses across the extracellular cleft and combines with receptor sites on the motor-end-plate membrane. This combination causes the membrane permeability to sodium and potassium ions to increase, leading to a depolarization of the muscle end plate known as the *end-plate potential* (EPP). The mechanism leading to the formation of the EPP in muscle is analogous to that of the EPSP (excitatory postsynaptic potential) produced at synaptic junctions. The magnitude of a muscle EPP, however, is much larger than a single EPSP because the surface area between the nerve and muscle is larger and much larger amounts of transmitter agent (ACh) are released. The magnitude of a single muscle EPP is sufficiently large to exceed the threshold potential of the muscle membrane and initiate an action potential which is propagated over the surface of the muscle membrane by the same mechanism described for the propagation of action potentials in nerves. Thus, there is a one-to-one transmission of nerve action potentials to muscle action potentials at the myoneural junction. In contrast, at synaptic junctions (between two neurons) a single EPSP is not sufficient to depolarize the postsynaptic membrane to threshold, and several EPSPs must occur (temporal and spatial summation) in order to activate the postsynaptic neuron.

A further difference between synaptic and myoneural junctions should be noted. At a synaptic junction it is possible to produce an inhibitory postsynaptic potential (IPSP) which hyperpolarizes the postsynaptic membrane and decreases the probability of firing an action potential. No such inhibitory potentials are found in human skeletal muscle; all myoneural junctions are excitatory. Thus the only way to inhibit the electric activity in the muscle membrane is to inhibit the initiation of action potentials in the motor neuron by synaptic activity (or lack of activity) at the level of the motor neurons in the spinal cord and brain.

In addition to the receptor sites for acetylcholine, the motor-end-plate membranes also contain the enzyme *acetylcholinesterase* which destroys ACh. The molecules of ACh released from the motor-neuron endings have a lifetime of only about 5 ms before they are destroyed by this enzyme. Once ACh is destroyed, the muscle membrane permeability to sodium and potassium ions returns to its initial state and the depolarized end plate returns to its resting potential.

There are many ways in which events at the neuromuscular junction can be modified by disease or drugs. For example, the deadly South American Indian arrowhead poison, curare, is strongly bound to the acetylcholine receptor site, but it does not change membrane permeability, nor is it destroyed by acetylcholinesterase. When a receptor site is occupied by curare, acetylcholine released from an axon cannot interact with the motor end plate. Thus, the motor nerves conduct normal action potentials and release acetylcholine, but there is no resulting muscle action potential or contraction. Since the skeletal muscles responsible for breathing movements depend upon neuromuscular transmission to initiate their contraction, death comes from asphyxiation.

Neuromuscular transmission can also be altered by inhibition of acetylcholinesterase, e.g., by some organophosphates, which are the main ingredients in many pesticides and some nerve gases developed for biological warfare. When the enzyme is inhibited, acetylcholine is not destroyed and its prolonged action

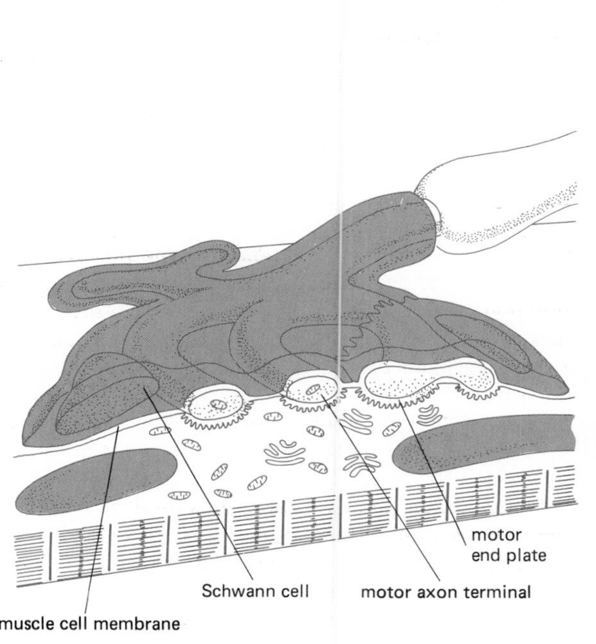

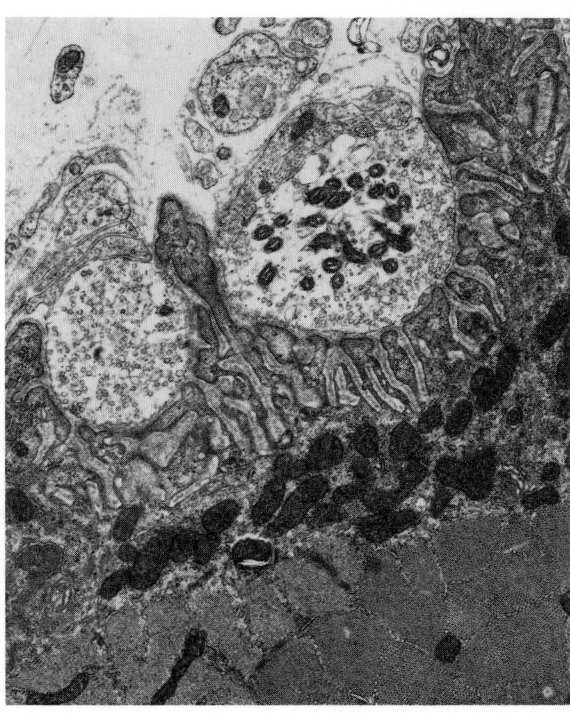

FIGURE 5-10 ***Diagram* (A) *and electron micrograph* (B) *of a neuromuscular junction.* (Part A redrawn from Warwick and Williams, part B courtesy of Dr. D. N. Landon, National Hospital for Nervous Diseases, London.)**

maintains depolarization of the muscle cell. The failure of repolarization prevents new action potentials from being initiated, so that the muscle does not contract in response to nerve stimulation. The result is paralysis of the muscle and death from asphyxiation.

A third group of substances affects the release of acetylcholine from the nerve terminals, thereby interfering with normal action at the neuromuscular junction. Botulinus toxin, produced by the bacterium *Clostridium botulinum,* blocks the release of acetylcholine in response to an action potential and thus prevents excitation of the muscle membrane. Botulinus toxin is responsible for a type of food poisoning and is one of the most deadly poisons known. Less than 0.0001 mg is sufficient to kill a person, and half a pound could kill the entire human population.

The molecular events leading to muscle contraction are summarized in Table 5-1.

MECHANICS OF MUSCLE CONTRACTION

Contraction refers to the active process of generating a force in a muscle. This force, generated by the contractile proteins, is exerted parallel to the muscle fiber. The force exerted by a contracting muscle on an object is known as the muscle *tension,* and the force exerted on a muscle by the weight of an object is known as the *load.* Thus, muscle tension and load are opposing forces. To lift a load the muscle tension must be greater than the load.

When a muscle shortens and lifts a load, the muscle contraction is said to be *isotonic* (constant tension) since the load remains constant throughout the period of shortening. When shortening is prevented by a load that is greater than muscle tension, or when a load is supported in a fixed position by the tension of the muscle, the development of tension occurs at constant

TABLE 5-1

SEQUENCE OF EVENTS BETWEEN NERVE ACTION POTENTIAL AND CONTRACTION AND RELAXATION OF A MUSCLE FIBER

1 An action potential is initiated and propagated in a motor axon as a result of synaptic events on the cell body and dendrites of the motor neuron in the central nervous system.
2 The action potential in the motor axon causes the release of acetylcholine from the axon terminals at the neuromuscular junction.
3 Acetylcholine is bound to receptor sites on the motor-end-plate membrane.
4 Acetylcholine increases the permeability of the motor end plate to sodium and potassium ions, producing an end-plate potential (EPP).
5 EPP depolarizes the muscle membrane to its threshold potential, generating a muscle action potential which is propagated over the surface of the muscle membrane.
6 Acetylcholine is rapidly destroyed by acetylcholinesterase on the end-plate membrane.
7 Muscle action potential depolarizes transverse tubules.
8 Depolarization of transverse tubules leads to the release of calcium ions from the lateral sacs of the sarcoplasmic reticulum surrounding the myofibrils.
9 Calcium ions bind to troponin-tropomyosin in the thin actin filaments, releasing the inhibition that prevented actin from combining with myosin.
10 Actin combines with myosin ATP:

$$A + M \cdot ATP \longrightarrow A \cdot M \cdot ATP$$

11 Actin activates the myosin ATPase, which splits ATP, releasing energy used to produce a movement of the myosin cross bridge:

$$A \cdot M \cdot ATP \longrightarrow A \cdot M + ADP + P_i$$

12 Movements of the cross bridges lead to relative movement of the thick and thin filaments past each other.
13 ATP binds to the myosin bridge, breaking the actin-myosin bond and allowing the cross bridge to dissociate from actin:

$$A \cdot M + ATP \longrightarrow A + M \cdot ATP$$

14 Cycles of cross-bridge contraction and relaxation continue as long as the concentration of calcium remains high enough to inhibit the action of the troponin-tropomyosin system.
15 Concentration of calcium ions falls as they are moved into the lateral sacs of the sarcoplasmic reticulum by an energy-requiring process which splits ATP.
16 Removal of calcium ions restores the inhibitory action of troponin-tropomyosin, and in the presence of ATP, actin and myosin remain in the disassociated relaxed state.

muscle length and is said to be an *isometric* contraction (constant length). The internal physicochemical events are the same in both isotonic and isometric contractions. Supporting a weight in a fixed position involves isometric contractions whereas body movements involve isotonic contractions.

Figure 5-11 illustrates the general method of recording isotonic and isometric contractions of isolated muscles. During an isotonic contraction the distance the muscle shortens and the time of the contraction are recorded. To measure an isometric contraction the muscle is attached at one end to a rigid support and at the other to a force transducer, which controls the movement of a recording pen in proportion to the force exerted. Thus, the two parameters recorded during an isotonic contraction are distance shortened and time, and the two parameters recorded during an isometric contraction are muscle tension and time.

Single Twitch

The mechanical response of a muscle to a single action potential is known as a *twitch*. Figure 5-12 shows the main features of an isometric and an isotonic twitch. Following excitation there is an interval of a few milliseconds known as the *latent period*, before the tension begins to increase in an isometric twitch. The time from the start of tension development to the peak of tension is the *contraction time*. Not all skeletal muscles contract at the same rate. Some fast fibers have contraction times as short as 10 ms, whereas slow fibers may take 100 ms or longer. The time from peak tension until the tension has decreased to zero is known as the *relaxation time*. In the example of a typical isometric twitch, shown in Fig. 5-12, the entire sequence of contraction and relaxation lasts about 150 ms.

Comparing an isometric twitch with an isotonic twitch in the same muscle, one can see from Fig. 5-12

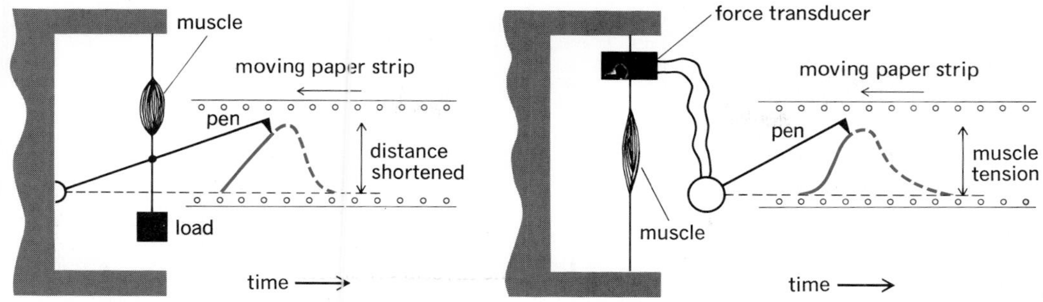

FIGURE 5-11 ***Methods of recording isotonic and isometric muscle contractions.***

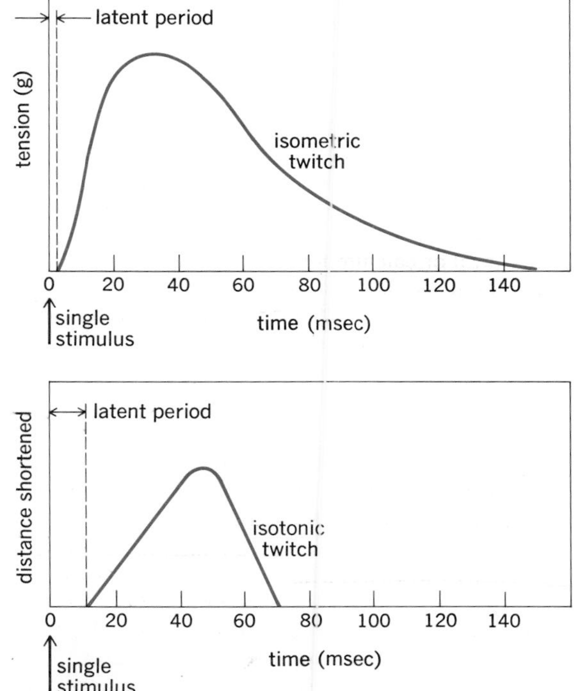

FIGURE 5-12 ***Isometric and isotonic skeletal muscle twitches as a function of time.***

that the duration of the isotonic twitch is considerably shorter. The latent period, on the other hand, is considerably longer than in an isometric twitch. Once shortening begins, it proceeds at a constant velocity over about 70 percent of the total distance shortened by the muscle, as can be seen from the straight-line relation between distance shortened and time. The slope of this line gives the velocity of shortening. Both the velocity of shortening and the duration of an isotonic twitch depend upon the magnitude of the load being lifted by the muscle (Fig. 5-13). At heavier loads the latent period lasts longer but the velocity of shortening, the duration of the isotonic twitch, and the distance shortened all decrease. Eventually a load is reached that the muscle is unable to lift. At this maximum load, the velocity of shortening is zero, and the contraction of the muscle becomes isometric. The maximum velocity of shortening occurs when there is zero load on the muscle. The relation between the load on a muscle and the velocity at which the muscle lifts the load is shown in Fig. 5-14.

FIGURE 5-13 ***Change in the isotonic-twitch response of a muscle with different loads.***

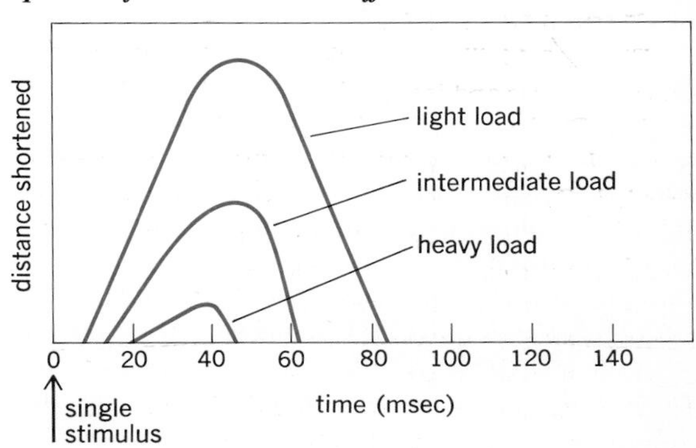

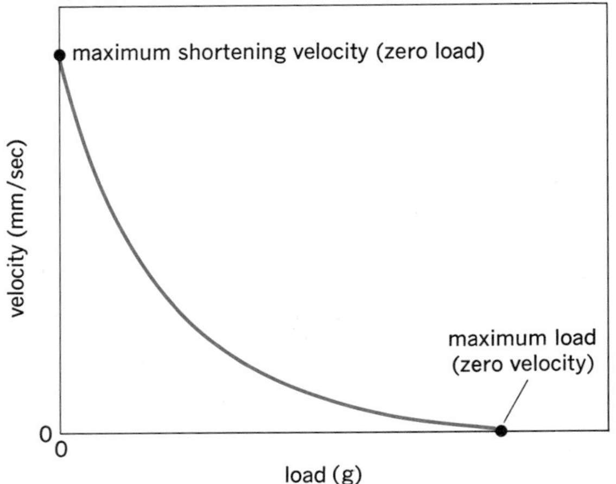

FIGURE 5-14 ***Shortening velocity as a function of load.***

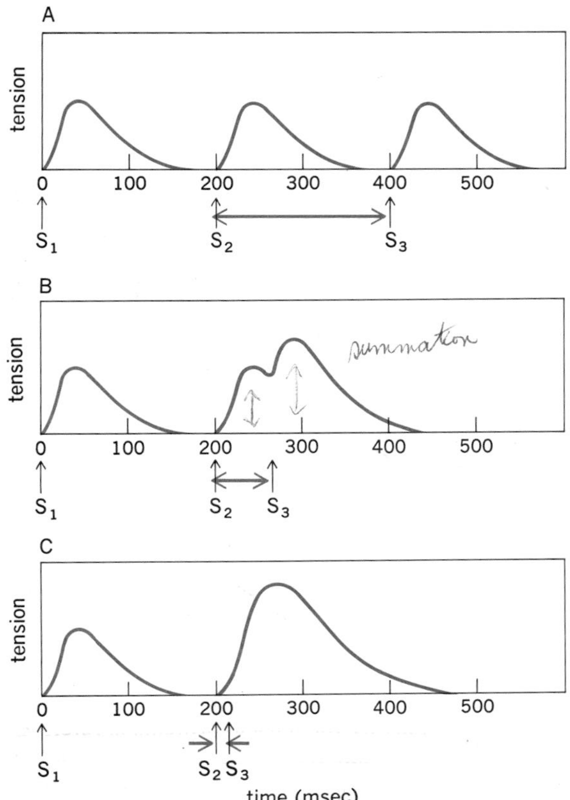

FIGURE 5-15 ***Summation of isometric contractions produced by shortening the time between stimuli*** **S_2** ***and*** **S_3.**

Summation of Contractions

The muscle action potential lasts about 1 to 2 ms and is over before muscle tension has even begun to increase; however, the mechanical response (twitch) which follows may last several hundred milliseconds. Thus it is possible for a second action potential to be initiated during this period of mechanical activity. Figure 5-15 illustrates the isometric contractions of a muscle in response to three successive stimuli. In Fig. 5-15A the isometric twitch following the first stimulus S_1 lasts 150 ms. The second stimulus S_2, applied to the muscle 200 ms after S_1 when the muscle has completely relaxed, causes a second identical isometric twitch. In Fig. 5-15B the interval between S_1 and S_2 remains 200 ms, but a third stimulus is applied 60 ms after S_2, when the mechanical response resulting from S_2 is beginning to decrease. Stimulus S_3 induces a contractile response and the resulting peak tension is greater than that of a single isometric twitch. In Fig. 5-15C the interval between S_2 and S_3 is further reduced to 10 ms. The resulting peak tension is even greater than in Fig. 5-15B, and the rise in tension forms a smooth curve. Here the mechanical response to S_3 appears as a continuation of the mechanical response already induced by S_2.

The property of skeletal muscle contraction in which the mechanical response to one or more successive stimuli is added on to the first is known as *summation*. The greater the frequency of stimulation, the greater is the tension produced until a maximal frequency is reached beyond which the tension no longer increases (Fig. 5-16). This is the greatest tension the muscle can develop and is generally about three to four times greater than the isometric-twitch tension produced by a single stimulus. A sustained maximal summation is known as *tetanus*.

Just as the contraction time of different muscle fibers varies considerably, so does the stimulus frequency that will give a maximal tetanus. The slower the contraction of the fiber, the lower is the frequency of stimulation needed. Frequencies of about 30 per second may produce a tetanus in slow fibers, but frequencies of 100 per second or more are necessary in very rapidly contracting fibers.

Summation and tetanus also occur when a muscle contracts isotonically, repetitive stimulation leading in this case to greater shortening. During a maximal isotonic tetanus, a lightly loaded muscle fiber shortens

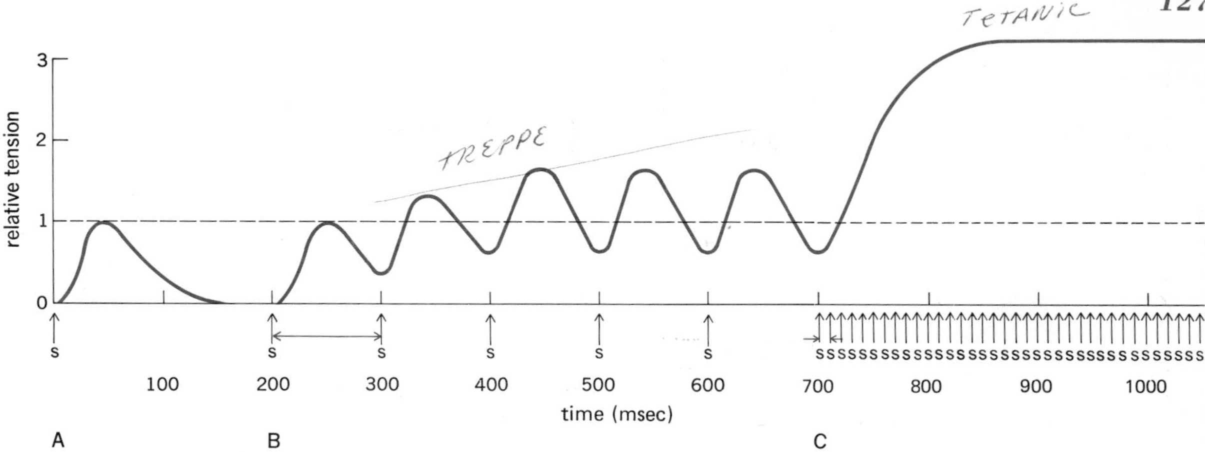

FIGURE 5-16 ***Isometric tetanic contractions produced by multiple stimuli of* (B) *10 stimuli per second and* (C) *100 stimuli per second as compared with a single isometric twitch* (A).**

about 40 percent of its resting length. At heavier loads the maximum degree of shortening during tetanic stimulation is less.

The explanation of summation involves the passive elastic properties of the muscle. The tension produced *internally* by the muscle proteins is referred to as the *active state* and follows the time course of the calcium release and reuptake. This tension following a single action potential is the maximal amount of tension the muscle proteins are capable of producing. The time course of the internal active state, as compared with the *external* tension exerted by a muscle on an external object, is illustrated in Fig. 5-17. Note the differences in magnitude and time course of these two tensions.

The discrepancy between the external tension and the internal tension of the active state is due to the structure of the muscle and the element of time. Tension is transmitted from the cross bridges through the thick and thin filaments, across the Z lines, and eventually through the extracellular connective tissue and tendons to the bone. All these structures have a certain amount of elasticity and are collectively known as the *series elastic element*. The series elastic element is equivalent to a spring that is placed between the contractile components of the muscle (the cross bridges) and the external object. When the cross bridges contract, the tension produced stretches the spring, which in turn transmits the tension to the external object. To illustrate the consequence of this linkage, consider what would happen if a strong man attempted to lift a brick that was attached to a very weak spring. Although a great deal of force may be exerted by the man on the spring, he would at first succeed only in stretching the spring, not in lifting the brick, until a point is reached at which the tension in the stretched spring becomes equal to the weight of the brick.

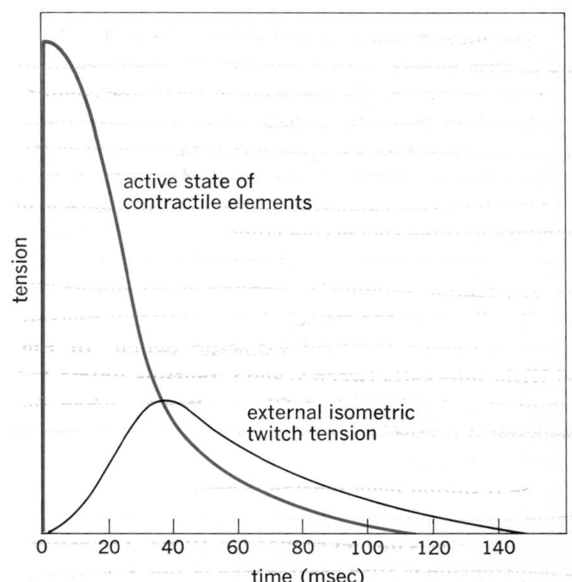

FIGURE 5-17 ***Active-state tension within the muscle in response to a single stimulus as a function of time.***

In the muscle the contractile components in their fully active state begin to stretch the series elastic element immediately following the release of calcium from the sarcoplasmic reticulum. While the contractile elements are stretching the series elastic element, the tension of the active state is decreasing as calcium is being pumped back into the sarcoplasmic reticulum. In a single twitch, the active state decreases before it is able to stretch the series elastic element to a tension equal to the maximal active state and thus less than the full internal tension is transmitted to the external object. When tetanus occurs, the active state is maintained long enough to completely stretch the series elastic element to a tension equal to that of the active state. Thus the increase in external tension that accompanies an increased frequency of stimulation is the result of the increased amount of time during which the contractile elements are maintained in their active state, enabling them to fully stretch the series elastic component. In an isotonic contraction the increased length of the latent period with increasing load is also explained by this same mechanism. The long latent period represents a period during which the series elastic element is being stretched to a tension equivalent to the load before the muscle will begin to lift the load. The heavier the load, the longer it takes to develop the amount of stretching of the series elastic element required to lift the load, and thus the longer the latent period.

What physiologic role does summation play? In general, when a motor neuron is stimulated by synaptic input, it discharges a burst of action potentials rather than a single action potential. Thus the contractions of the muscle fibers in the body are not usually single twitches but brief summations or more prolonged tetanic contractions.

Length-Tension Relationship

One of the classic observations in muscle physiology is the relationship between muscle length and the tension the muscle can develop at that length. A relaxed muscle has properties similar to a rubber band; when an outside force is applied to the muscle, it is stretched, and the greater the force stretching the muscle, the longer it becomes. If the maximal tetanic tension developed by a muscle is measured at various initial muscle lengths, the relationship shown in Fig. 5-18 is obtained. If the length at which the muscle develops maximum tension, l_0, is used as a reference point, the different muscle lengths at which isometric tension is measured can be expressed in terms of the percent of the l_0 muscle

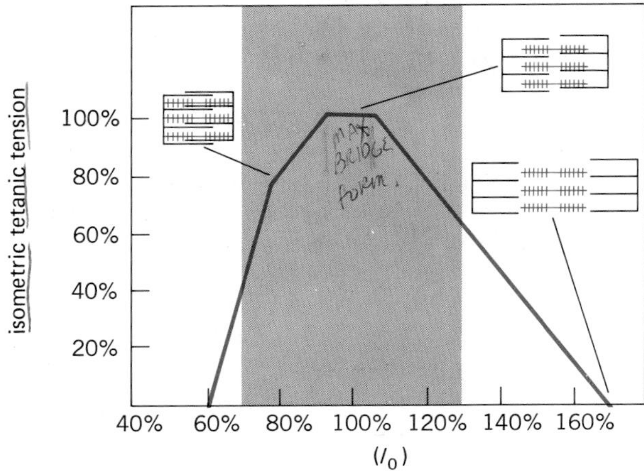

FIGURE 5-18 ***Variation in isometric tetanus tension with muscle length. The shaded band represents the range of length changes (from 70 to 130 percent) that can occur in the body while the muscles are still attached to bones.*** **(Adapted from Gordon, Huxley, and Julian.)**

length. If the muscle is set at a length equal to about 60 percent of l_0, it develops no tension when stimulated. As the length of the muscle is increased, the isometric tension rises to a maximum at l_0 and further lengthening of the muscle causes a drop in active tension. When the muscle is stretched to about 175 percent of l_0 or beyond, it no longer develops tension. Thus, the length of a muscle determines the amount of isometric tension it can develop.

This relationship is explained by the sliding-filament model. Passively stretching a muscle changes the amount of overlap between the thick and thin filaments in the myofibrils. Passively stretching a muscle 175 percent of l_0 pulls the thick and thin filaments so far apart that there is no overlap between the two. Since active tension is developed by the interaction of the myosin bridges with the actin molecules in the thin filaments, when there is no overlap there can be no bridge interaction and no tension is developed. At shorter muscle lengths more and more bridges overlap with thin filaments, increasing the active tension in proportion to the number of active cross bridges. At l_0 there is a maximal overlap of thick and thin filaments, and tension is maximal. At lengths less than l_0 two factors lead to decreasing tension: (1) The thin filaments in the two halves of the sarcomere begin to overlap, interfering with cross-bridge interaction, thus decreasing the total number of active cross bridges, and (2) the thick filaments become compressed against the two Z lines.

In the body, where muscles are attached to bones, the relaxed length of the muscle is very nearly l_0 and thus at the optimal length for force generation. The total range of length changes that a skeletal muscle can undergo while still attached to bone is limited to a maximum of about 30 percent of the resting length and is often much less. Thus, even at maximal extension or flexion of a limb, its muscles are still able to develop more than 50 percent of their maximum tension. However, in a muscular organ such as the heart, which is not attached to bones, the range over which muscle fiber length can be varied during the filling of the heart with blood is considerably greater.

Recruitment

Since a muscle is composed of many muscle fibers, the total tension a muscle can develop depends upon the number of muscle fibers in the muscle that are contracting at any given time, which depends, in turn, upon the number of motor neurons to the muscle that are being stimulated. Recall that each motor neuron innervates several muscle fibers, forming a motor unit (Fig. 5-9). Stimulation of a motor neuron produces a contraction in all the muscle fibers in the motor unit. The total tension a muscle develops can thus be varied by varying the number of motor units that are activated. This is determined by the balance between excitatory and inhibitory synaptic input to the motor neurons in the brain and spinal cord. The process of increasing the number of active motor neurons (and thus the number of active motor units) is known as *recruitment.*

The number of muscle fibers associated with a single motor axon varies considerably in different types of muscles. In muscles which are able to produce very delicate movements, such as those in the hand and eye, the size of the individual motor units is small but in the more coarsely controlled muscles of the back and legs, each motor unit contains hundreds of muscle fibers. For example, a motor neuron innervates only 13 fibers or so in a muscle of the eye whereas one motor unit in the large calf muscle of the leg contains about 1,730 muscle fibers. The smaller the size of the motor units, the more precisely the tension of the muscle can be controlled by the recruitment of additional motor units.

The motor neurons to a given muscle fire in an asynchronous pattern determined by their individual synaptic inputs. Thus, the muscle fibers in some motor units may be contracting while other motor units are relaxing. This asynchronous activity in the pool of motor units has several consequences for the development of muscle tension. In muscles which are active for long periods of time, such as the postural muscles which support the weight of the body, the asynchronous activity of their motor units tends to prevent fatigue that might otherwise result from prolonged continuous activity. Some units are active while others rest briefly, only to return to activity as others rest. Furthermore, this pattern of asychronous activity is able to maintain a nearly constant tension in the muscle, as illustrated in Fig. 5-19. If the motor units ever were to fire simultaneously, the resulting movement would be a jerky series of contractions and relaxations. The asynchronous activity of motor units is one of the factors responsible for the smooth movements produced by contracting muscles in the body.

In summary, the preceding three sections have described major factors determining the total tension developed by a whole muscle when it contracts. In addition, as we shall discuss in more detail later, individ-

FIGURE 5-19 ***Asynchronous motor-unit activity maintains a nearly constant tension in the total muscle.***

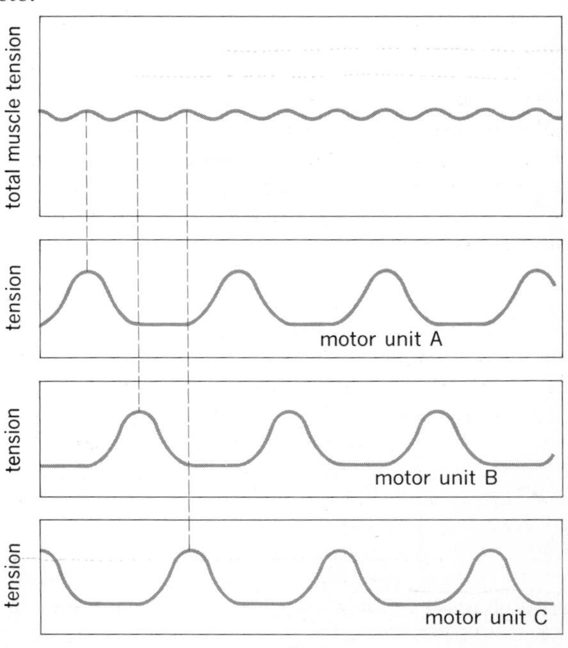

TABLE 5-2
FACTORS DETERMINING TOTAL MUSCLE TENSION

I Number of muscle fibers that are contracting
  A. Recruitment of motor units (synaptic input to motor neurons)
  B. Size of motor units (number of muscle fibers per axon)
  C. Asynchronous activity of motor units
II Tension produced by each contracting muscle fiber
  A. Frequency of action potentials in motor neuron (summation and tetanus)
  B. Muscle fiber length (degree of overlap of thick and thin filaments)
  C. Duration of activity (fatigue)
  D. Chemical composition and characteristics of individual fibers

ual fibers show distinct chemical differences which result in marked differences in their rates of contraction, tension development, and ease of fatigue. The multiple factors involved in the control of total muscle tension are summarized in Table 5-2.

ENERGY METABOLISM OF MUSCLE

If a muscle is to contract and relax, ATP must be available to perform three major functions: (1) The energy released from ATP splitting is directly coupled to the movement of the cross bridges; (2) binding ATP to myosin without splitting is necessary to break the actomyosin bond and allow the cross bridge to operate cyclically; and (3) energy released from ATP splitting is utilized by the sarcoplasmic reticulum to accumulate calcium ions, producing relaxation.

If an isolated muscle is given an adequate supply of oxygen and nutrients which can be broken down to provide ATP, it can continue to give a series of twitch responses to low-frequency stimulation for long periods of time. Under these conditions the muscle is able to synthesize ATP at a rate sufficient to keep up with the rate of ATP breakdown. If the rate of stimulation is increased, the twitch responses soon begin to grow weaker and eventually fall to zero (Fig. 5-20). This drop in tension following prolonged stimulation is *muscle fatigue*. If rates of stimulation produce tetanic contractions, fatigue occurs even sooner. If the stimulation is stopped and a period of rest is allowed before resuming stimulation, the muscle briefly recovers its ability to contract before again undergoing fatigue. When the muscle is completely fatigued and is unable to develop tension, the concentration of ATP in the muscle is very low. During recovery the ATP concentration rises as metabolism replaces the ATP broken down during contraction.

If a muscle had to rely on its supply of previously synthesized ATP for contraction, it would be completely fatigued within a few twitches. Therefore, if a muscle is to maintain its contractile activity, molecules of ATP must be synthesized as rapidly as they are broken down. There are three sources for supplying this ATP: (1) *creatine phosphate*, (2) substrate phosphorylation during glycolysis, and (3) oxidative phosphorylation in the mitochondria.

When contraction is initiated by the release of calcium, the myosin ATPase begins to break down ATP at a very rapid rate. The increase in ADP and P_i concentrations resulting from this breakdown of ATP leads, ultimately, to increased rates of oxidative phosphorylation and glycolysis by the mechanisms described in Chap. 1. However, a short period of time elapses before these multienzyme pathways begin to deliver newly formed ATP at a high rate. It is the role of creatine phosphate to provide the energy for ATP formation during this interval.

Creatine phosphate (CP) provides the most rapid means of forming ATP in the muscle cell. This molecule contains energy and phosphate, both of which can be transferred to a molecule of ADP to form ATP and creatine (C):

$$CP + ADP \rightleftharpoons C + ATP \qquad (5\text{-}1)$$

A single enzyme catalyzes this reversible reaction. Energy is stored in creatine phosphate in resting muscle by the reversal of reaction (5-1). The high levels of ATP in a resting muscle favor, by mass action, the formation of creatine phosphate, and during periods of rest the muscle builds up a concentration of creatine phosphate that is about five times that of ATP. When the ATP level rapidly falls at the beginning of contraction, mass action favors the rapid formation of ATP from creatine phosphate mediated by this single enzymatic reaction. The creatine phosphate system is so efficient that the actual concentration of ATP in the cell changes very little at the start of contraction but the concentration of creatine phosphate falls rapidly.

If contractile activity is to be continued for more than a few seconds, the muscle cell must be able to derive ATP from sources other than creatine phos-

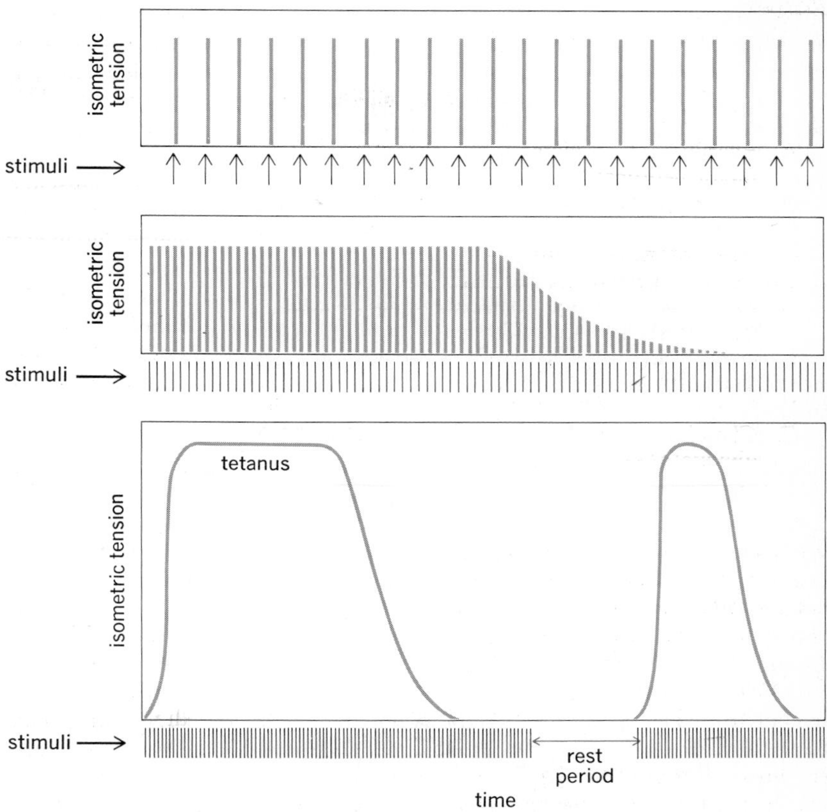

FIGURE 5-20 ***Muscle fatigue resulting from prolonged stimulation and recovery of ability to contract after a period of rest.***

phate. At moderate levels of muscle activity (moderate rates of ATP breakdown) most of this ATP can be formed by the process of oxidative phosphorylation. Carbohydrates, fats, and proteins can all provide sources of energy for this process.

During very intense exercise, when the breakdown of ATP is very rapid, a number of factors begin to limit the cell's ability to replace ATP by oxidative phosphorylation: (1) the delivery of oxygen to the muscle, (2) the availability of substrates such as glucose, and (3) the rates at which the enzymes in the metabolic pathways can process these substrates. Any of these may become rate-limiting under various conditions. Since oxidative phosphorylation depends upon the utilization of oxygen, the continued formation of ATP by this process depends upon an adequate delivery of oxygen to the muscle by the circulatory system. It is this delivery of oxygen to the cell which eventually becomes rate-limiting for most forms of prolonged intense exercise, such as running.

Even when adequate oxygen is delivered to the muscle, the rate at which oxidative phosphorylation can produce ATP may be inadequate during very intense exercise. When the level of exercise exceeds about 50 percent of maximum (50 percent of the maximal rate of ATP breakdown), anaerobic glycolysis begins to contribute an increasingly significant fraction of the total ATP produced by the muscle.

Although the aerobic process of oxidative phosphorylation produces large quantities of ATP (36 of the 38 ATP formed from each molecule of glucose) the enzymatic machinery of this pathway is relatively slow. The glycolytic pathway, although producing only small quantities of ATP from the breakdown of glucose, can operate at a much higher rate. Thus, in the same amount of time that oxidative phosphorylation can produce 36 molecules of ATP from 1 glucose molecule, about 64 molecules of ATP can be formed by glycolysis through the breakdown of 32 molecules of glucose to lactic acid. Not only is glycolysis faster than oxidative phosphorylation, but it can proceed in the absence of oxygen, leading to the formation of lactic acid as its end product. Thus, during intense exercise, even if adequate oxygen is available, anaerobic glycolysis becomes an additional source for rapidly supplying the muscle with ATP, and lactic acid, the end product of this process diffuses out of the muscle tissue and accumulates in the blood.

Although anaerobic glycolysis can produce ATP very rapidly, it has the disadvantage of requiring very large quantities of glucose to produce relatively small amounts of ATP. The ability of muscle to store glucose in the form of glycogen provides the muscle with a certain degree of independence from externally supplied glucose. During intense exercise, the glycogen content of the muscle falls progressively, the rate of fall depending upon the intensity of the exercise. The onset of fatigue from exercise lasting more than a few minutes correlates closely with the depletion of the muscle glycogen stores. Finally, in very intense exercise, myosin ATPase may break down ATP faster than even glycolysis can replace it from existing glycogen stores, and fatigue occurs rapidly as the cells' ATP is depleted. The pathways providing ATP for cross-bridge activity are illustrated in Fig. 5-21.

In contrast to true muscle fatigue, psychologic fatigue may cause an individual to stop exercising even though his muscles are not depleted of ATP and are still able to contract. An athlete's performance depends not only on the physical state of his or her muscles but also upon the "will to win."

Following an intense period of exercise a number of changes have occurred in the muscle cell; creatine phosphate levels have decreased, and much of the muscle glycogen may have been converted to lactic acid. To return the cell to its original state, the glycogen stores must be replaced and creatine phosphate resynthesized; both processes require energy. Thus, even though the muscle has stopped contracting, it continues to consume oxygen at a high rate to provide the energy necessary for these synthetic processes. Increased oxygen uptake may proceed for quite some time after the end of exercise, as seen by the fact that one continues to breathe deeply and rapidly after a period of intense exercise. The longer and more intense the exercise, the longer it takes to restore the muscle to its original state. The efficiency with which muscle converts chemical energy to work (movement) is about 40 percent. However, if the energy necessary to return glycogen and creatine phosphate to their original levels in the muscle is included, the overall efficiency is only about 20 percent, the remaining 80 percent of the energy released appearing as heat. The more intense the exercise, the greater is the amount of heat produced. This increased heat production may place a severe stress upon the body's ability to maintain a constant body temperature, especially on a hot day. On the other hand, the process of shivering reflects the body's use of this same source of heat energy to maintain body temperature in a cold environment. A more detailed discussion of temperature regulation will be found in Chap. 20.

MUSCLE GROWTH AND DIFFERENTIATION

All skeletal muscle fibers are not identical in such properties as their speed of contraction or their enzymatic capacity to produce ATP. Furthermore, it is well known that the capacity of an individual for exercise, both in terms of strength and endurance, can be increased by regular exercise (training). These variations in muscle's contractile properties, growth, and development are strongly dependent upon the activity of the motor neurons to the muscle.

Muscle fibers are formed during embryologic development by the fusion of a number of small, mononucleated myoblast cells to form long, cylindrical, multinucleated muscle fibers. It is only after the fusion of the myoblasts that the muscle fibers begin to form actin and myosin filaments and are capable of contraction. Once the myoblasts have fused, the resulting muscle fibers no longer have the capacity for cell division and any increase in the size of the muscle results from a growth in the length and diameter of these preformed fibers. This stage of muscle differentiation appears to be completed about the time of birth, and a newborn child already has all the skeletal muscle fibers it will ever have.

Following the fusion of the myoblasts to form muscle fibers, the motor neurons begin to send axon processes into the muscle, forming myoneural junctions and bringing the muscle under the control of the central nervous system. From this point on there is a very critical dependence of the muscle fiber on its motor neuron, not only to provide a means of initiating contraction but also for the continued survival and development of the muscle fiber. If the nerve fibers to a muscle are severed or the motor neurons destroyed, the

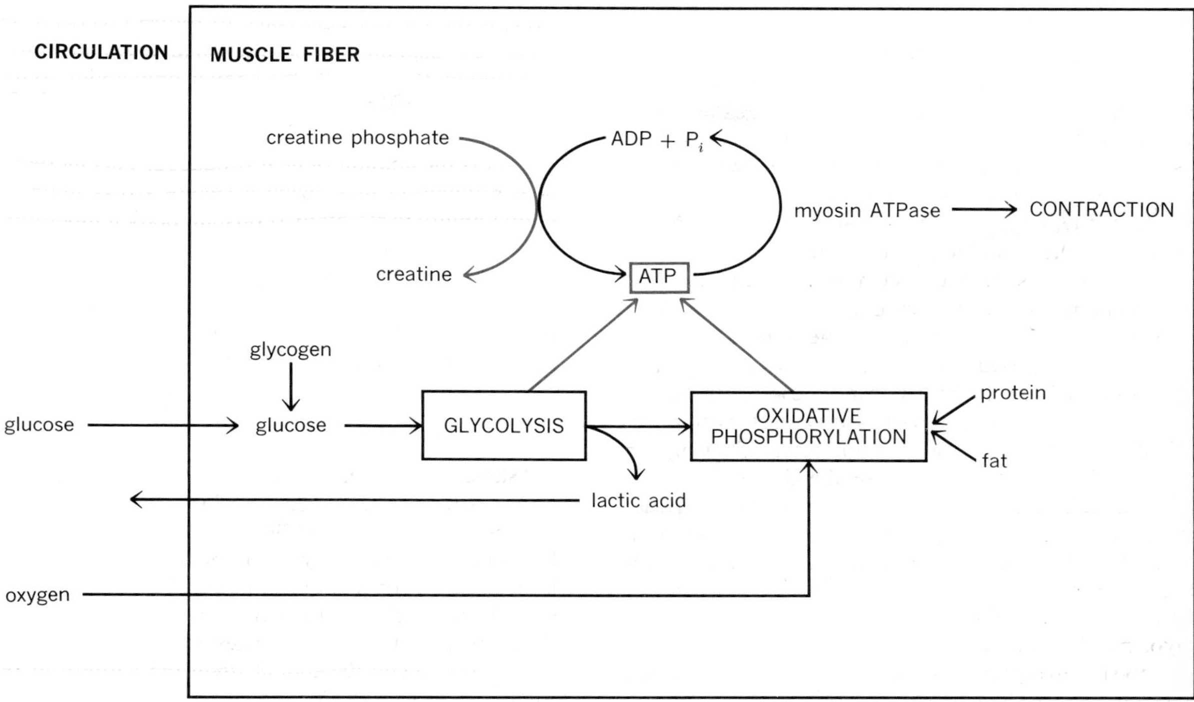

FIGURE 5-21 ***Biochemical pathways producing ATP utilized during muscle contraction.***

denervated muscle fibers become progressively smaller, their content of actin and myosin decreases, and the connective tissue around the muscle fibers proliferates. This decrease in muscle mass following denervation is known as *denervation atrophy*. A muscle can also atrophy with its nerve supply intact if it is not used for a long period of time, as when a broken arm or leg is immobilized in a cast. This is known as *disuse atrophy*. Since atrophy can be prevented by electrically initiating action potentials in the muscle membrane in the absence of nerve fibers, some degree of electric activity in the muscle membrane is necessary for the maintenance of the functional state of the muscle. In contrast to the decrease in muscle mass that results from a lack of regular neural stimulation, increased amounts of neural activity, which accompany repeated exercise, produce changes in the chemical composition of muscle fibers, leading in some cases to a considerable increase in the size (*hypertrophy*) of the individual muscle fibers. Action potentials in nerve fibers appear to release chemical substances which influence the biochemical activities of the muscle fiber. The chemical identity of these tropic agents and their site of action in the muscle are unknown.

Because of this strong influence of the motor neurons on the chemical properties of muscle fibers, it is not surprising to find distinct differences in the enzymatic composition of the muscle fibers in the body which result from long-term (weeks or months) differences in the amount of neural activity to the muscles. Three classes of skeletal muscle fibers can be identified on the basis of their rate of utilizing ATP (speed of contraction) and their capacity to generate ATP. These three classes are (1) high-oxidative slow-twitch fibers, (2) high-oxidative fast-twitch fibers, and (3) low-oxidative fast-twitch fibers.

The speed with which a muscle fiber contracts is determined primarily by the rate at which myosin ATPase splits ATP and thus the rate at which cross bridges can undergo repeated cycles of activity. Although dif-

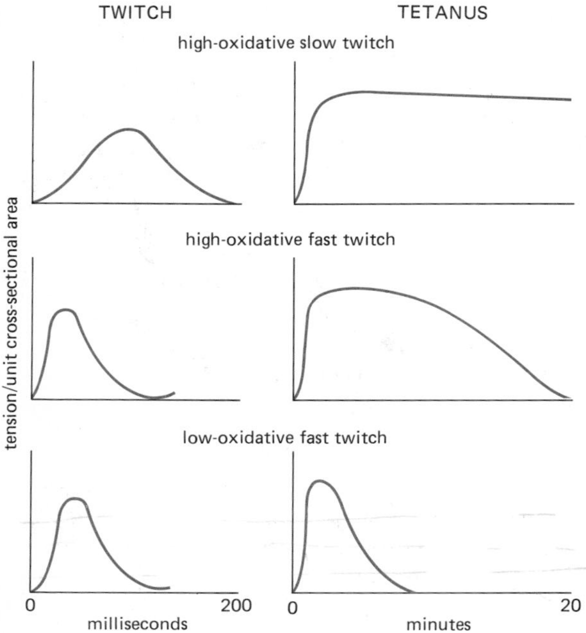

FIGURE 5-22 ***Twitch and fatigue characteristics of the three types of skeletal muscle fibers.***

ferences in the myosin ATPase activity produce different speeds of contraction in the different muscle types (Fig. 5-22), the total amount of tension generated depends on the number of actin and myosin filaments in the cross-sectional area of the muscle. Thus large-diameter fibers contain more actin and myosin and produce more tension, regardless of the myosin ATPase.

The second major difference between the various types of muscle fibers is the type of enzymatic machinery available for synthesizing ATP. The high-oxidative fibers have a high rate of oxidative phosphorylation and contain numerous mitochondria. The activity of the glycolytic enzymes in these cells is relatively low. Thus, most of the ATP produced by these cells must be derived from oxidative phosphorylation and is dependent upon an adequate supply of oxygen to these cells. These fibers are surrounded by numerous capillary blood vessels which deliver oxygen and nutrients to the muscle fiber (Fig. 5-23). These high-oxidative fibers also contain a protein known as *myoglobin* which is red in color. Myoglobin is very similar to the protein, hemoglobin, found in the red blood cells. Myoglobin binds oxygen and increases the rate of oxygen diffusion into the muscle cell as well as providing a mechanism for storing small amounts of oxygen within the muscle. Muscle fibers containing myoglobin are often referred to as red muscle.

Low-oxidative fibers hve an enzymic composition that is geared to the utilization of glucose and the glycolytic pathway for the production of most of its ATP. These fibers contain few mitochondria and have a high content of glycolytic enzymes. Because they contain little myoglobin they are often referred to as white muscle. Very few capillaries are found in the vicinity of these fibers (Fig. 5-23). These fibers also contain large quantities of glycogen to provide an immediately available source of glucose for the anaerobic glycolytic pathway. In contrast, the high-oxidative fibers contain relatively little glycogen.

Recalling the various factors associated with ATP production and utilization which may become rate-limiting for muscle contraction, we can predict some of the properties of a muscle fiber that we would expect to result from placing a given type of myosin ATPase in an environment which has a particular capacity for ATP production. The low-oxidative fast-twitch white fibers split ATP very rapidly and are able to produce ATP rapidly by anaerobic glycolysis. However, these fibers also fatigue very rapidly (Fig. 5-22), as their high rate of ATP splitting quickly depletes their glycogen stores. These fibers are generally found to have a large diameter (Fig. 5-23) and are thus able to produce large amounts of tension but only for short periods of time before they fatigue. At the other extreme there are the high-oxidative, slow-twitch red fibers which have a high rate of oxidative phosphorylation and are able to keep pace with the relatively slow rate of ATP breakdown. These fibers are very difficult to fatigue (Fig. 5-22) since the high rate of blood flow to these fibers delivers oxygen and nutrients at a sufficient rate to keep up with the relatively slow rate of ATP breakdown by myosin ATPase. These fibers are relatively small in diameter (Fig. 5-23) and thus do not produce the large amounts of tension developed by white fibers.

The third class of fibers, the high-oxidative fast-twitch, has properties intermediate between the other two types. These fibers can maintain their contractile activity for longer periods than the fast white fibers because of their capacity to utilize oxidative phosphorylation for some of their ATP requirements and their well-developed blood supply which can deliver oxygen and nutrients to the fibers. However, at high

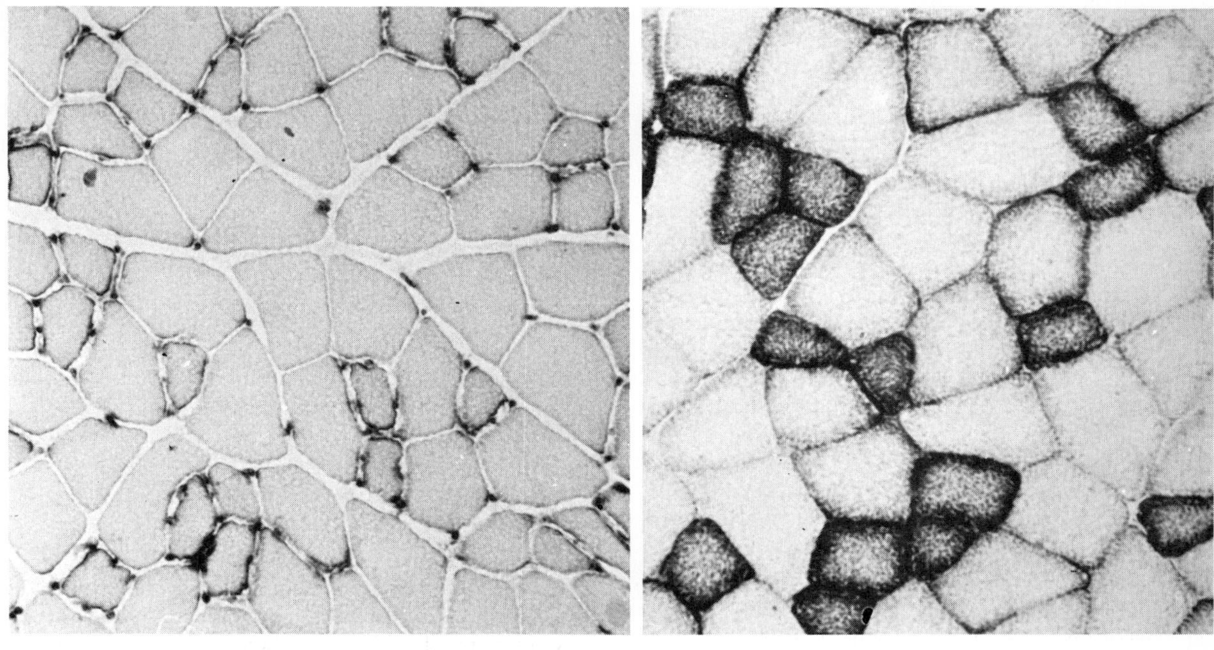

FIGURE 5-23 ***Cross sections of skeletal muscle, showing individual muscle fibers which have been stained according to their chemical composition. (A) The capillaries surrounding the muscle fibers have been stained. Note the large number of capillaries surrounding the small-diameter (high-oxidative) fibers. (B) Darkly stained fibers reveal the presence of high concentrations of oxidative enzymes in small-diameter, high-oxidative fibers.*** **(Courtesy of Dr. John Faulkner.)**

TABLE 5-3

PROPERTIES OF THREE TYPES OF SKELETAL MUSCLE FIBERS

	High-oxidative slow-twitch	*High-oxidative fast-twitch*	*Low-oxidative fast-twitch*
Speed of contraction	Slow	Fast	Fast
Myosin-ATPase activity	Low	High	High
Primary source of ATP production	Oxidative phosphorylation	Oxidative phosphorylation	Anaerobic glycolysis
Glycolytic enzyme activity	Low	Intermediate	High
Number of mitochondria	Many	Many	Few
Capillaries	Many	Many	Few
Myoglobin content – O_2 storage	High	High	Low
Muscle color	Red	Red	White
Glycogen content	Low	Intermediate	High
Fiber diameter	Small – don't need space for glycogen	Intermediate	Large
Rate of fatigue	Slow	Intermediate	Fast
TENSION DEVELOPED	LOW	INT.	HIGH

rates of activity, the high rate of ATP splitting by the fast myosin ATPase exceeds the capacity of oxidative phosphorylation to supply ATP and these fibers eventually fatigue (Fig. 5-22). Table 5-3 provides a summary of the characteristics of these three types of skeletal muscle fibers.

The contractile activity that skeletal muscles are called upon to perform varies in different locations in the body. The muscles which support the weight of the body, the postural muscles of the back and legs, must be able to maintain their activity for long periods of time without fatigue, whereas the muscles in the arms are intermittently called upon to rapidly produce large amounts of tension associated with the lifting of objects. Thus, there is a spectrum of activities that a skeletal muscle may be called upon to perform, ranging from long-duration, low-intensity endurance-type activity, to short-duration, high-intensity strength activities. Some muscles which perform predominantly one form of contractile activity are often composed of only one type of muscle fiber, usually of the high-oxidative type. More commonly, however, a muscle is required to perform endurance-type activity under some circumstances and high-intensity strength activity under others. These muscles generally contain a mixture of the three types of muscle fibers.

Different patterns of neural activity to a muscle occur during different types of exercise. During exercise of short duration and low intensity, just the high-oxidative slow-twitch fibers are activated. Stronger contractions result from the recruitment of the high-oxidative fast-twitch and eventually the low-oxidative fast-twitch fibers. Thus, a high-intensity short-duration exercise such as weight lifting is characterized by a pattern of high-frequency short-duration neural discharges to all types of motor units of the involved muscles. In contrast, so-called endurance exercises such as jogging, which are of long duration and, of necessity, low intensity, are characterized by a pattern of low-frequency long-duration neural discharges mainly to high-oxidative motor units. Changes in the type or amount of activity a muscle is called upon to perform alter the pattern of neural activity to the muscle and gradually produce changes in the chemical composition of the muscle.

A muscle can be altered in two ways: (1) by transformation of one biochemical type of fiber into another and (2) by the growth in size (hypertrophy) of the muscle fibers. Endurance types of exercise (running and swimming) are associated with a transformation of low-oxidative fast-twitch fibers into high-oxidative fast-twitch fibers. Endurance exercise also leads to an increase in the number of mitochondria in the high-oxidative fibers and an increase in the number of capillaries surrounding these fibers. These changes are accompanied by relatively small increases in the mass (strength) of the muscle and result in a muscle which has an increased capacity for long-duration, relatively low-intensity activity.

High-intensity, short-duration exercise, such as weight lifting, produces quite a different pattern of change in the muscle. The short-duration, high-frequency discharges to the low-oxidative fast-twitch fibers induces hypertrophy in these fibers, with increased synthesis of actin and myosin filaments and a large increase in the mass and strength of the muscle. The extreme result of this type of exercise is the bulging muscles of a professional weight lifter. Because different types of exercise produce quite different chemical changes in skeletal muscle, an individual performing regular exercises to improve his muscle performance must be careful to chose a type of exercise that is compatible with the type of activity he ultimately wishes to perform. Thus lifting weights will not improve the endurance of a long-distance runner and jogging will not produce bulging biceps in a weight lifter. As we shall see in later chapters, endurance exercise not only produces changes in the skeletal muscles of the body but also produces changes in the respiratory and circulatory systems which improve the delivery of oxygen and nutrients to the muscle fibers.

SMOOTH MUSCLE

Smooth muscle is found in the walls of hollow organs such as the intestinal tract, blood vessels, the air passages to the lungs, the urinary bladder, and the uterus. It may also be found as single cells distributed throughout an organ such as the spleen or in small groups of cells attached to the hairs in the skin. Because of the diversity of smooth muscle function we shall not attempt to describe the specific properties of any one type of smooth muscle but shall identify their general properties. The reader must keep in mind that any one specific smooth muscle may not exhibit all these properties. In later chapters, as we discuss each organ system, the specific factors affecting the activity of the smooth muscle in that specific organ will be described.

Smooth Muscle Structure

Smooth muscle fibers are considerably less broad than skeletal muscle fibers, being only 2 to 20 μm in di-

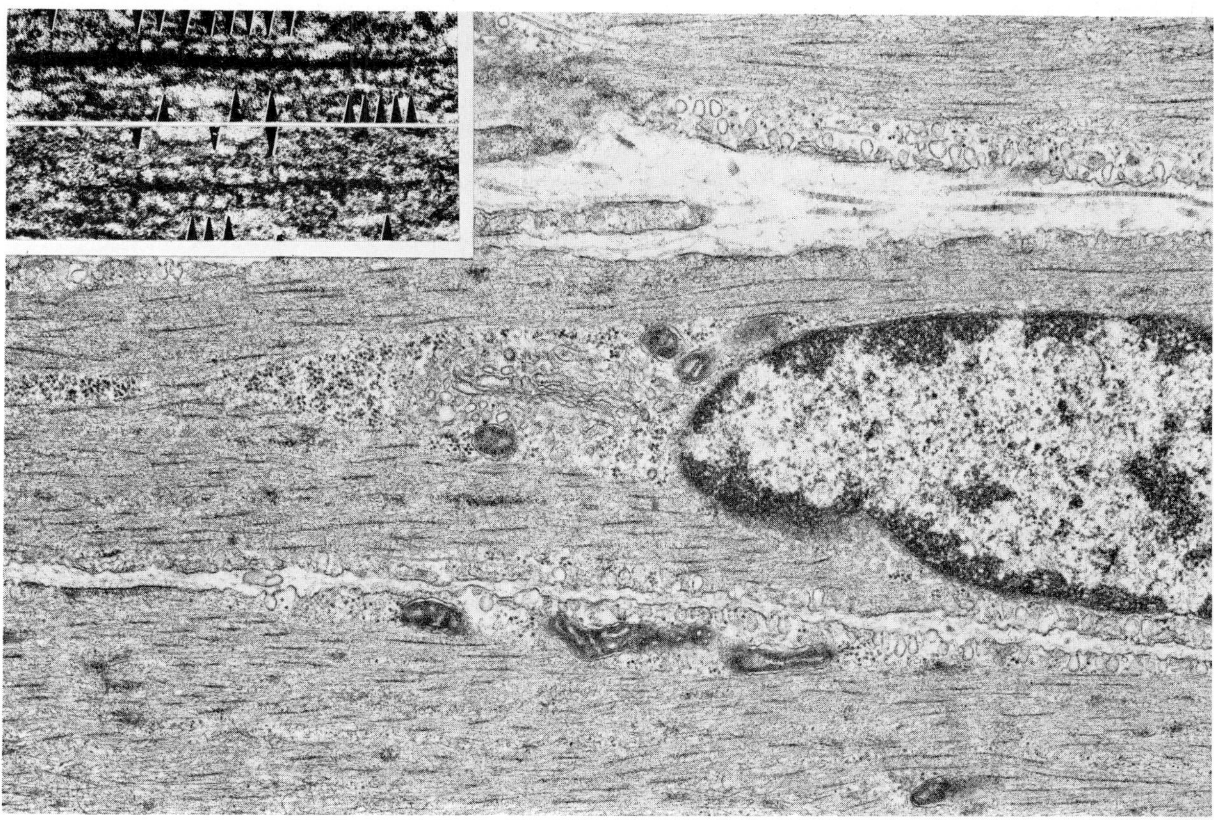

Figure 5-24 ***Electron micrograph of portions of three smooth muscle fibers.*** **(Insert)** ***Higher magnification of thick filaments with projections*** **(arrows)** ***suggestive of cross bridges connecting adjacent thin filaments.*** **[From A. P. Somlyo, C. E. Devin, A. V. Somlyo, and R. V. Rice, *Philos. Trans. R. Soc. London, Ser. B*, 265:23–229 (1973).]**

ameter, and are spindle-shaped rather than cylindrical. Each fiber has a single nucleus located in the central portion of the cell. The most noticeable morphologic factor distinguishing smooth from either skeletal or cardiac muscle is the absence of striated banding patterns in the cytoplasm (thus the name smooth muscle). The banding patterns of striated muscle arise from the regular arrangement of thick and thin filaments in the myofibrils of these muscles. As can be seen from Fig. 5-24, smooth muscle does not contain myofibrils. However, myosin thick filaments and actin thin filaments can be seen distributed throughout the cytoplasm, oriented parallel to the muscle fiber but not organized into regular units of filaments as in striated muscle. At high magnification the thick filaments can be seen to possess cross bridges, just as in skeletal muscle. Since troponin and tropomyosin can be isolated from smooth muscle, there is every reason to believe that the molecular events of force generation in these cells, following the release of calcium ions, are similar to those in skeletal muscle. It is still unclear exactly how the thick and thin filaments are anchored in smooth muscle so that the relative sliding of the filaments past each other leads to a shortening of the cell.

The random arrangement of thick and thin filaments in smooth muscle is such that there is almost always some overlap between the two sets of filaments; this may provide part of the explanation for the observation that smooth muscle can develop active tension over a wide range of muscle lengths (Fig. 5-25). This property allows the smooth muscles surrounding hollow organs, such as the stomach and bladder, whose walls become stretched as the organs are filled, to exert a force on the contents of their cavities. If such organs

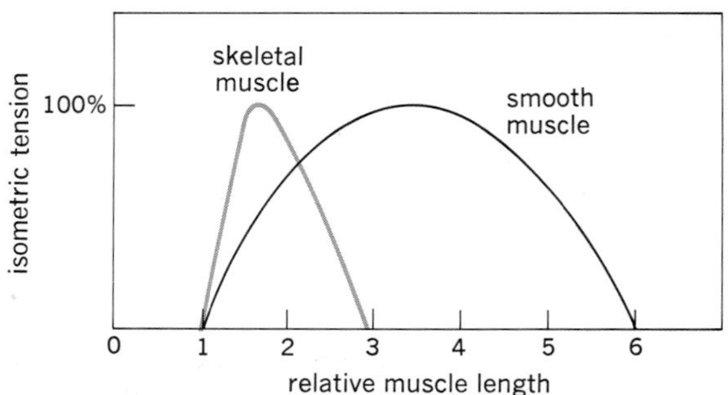

FIGURE 5-25 ***Variation in isometric tension with muscle length for skeletal and smooth muscle.***

were surrounded by skeletal muscle, a moderate amount of stretching would pull the thick and thin filaments beyond the range of overlap and no active tension could be developed.

A gram of smooth muscle contains only about 10 percent of the amount of actin and myosin found in a gram of striated muscle. The concentrations of ATP and creatine phosphate are also lower in smooth muscle. Furthermore, the actin-activated myosin-ATPase activity of most smooth muscle is relatively low. Thus, both the rate of contraction and the total tension developed by smooth muscles are generally much less than in striated muscle.

A further morphologic difference between smooth muscle and skeletal muscle is the lack of a well-developed sarcoplasmic reticulum and transverse tubule system. A series of vesicles can be seen just below the cell membrane (Fig. 5-24) and may be involved in the storage and release of calcium, but smooth muscles do not have the well-developed system of intracellular membranes that is associated with excitation-contraction coupling in striated muscle.

In many types of smooth muscle, the surface membrane appears to have properties similar to the membranes of the sarcoplasmic reticulum in striated muscles; they allow calcium to enter the cell when the membrane is stimulated and pump calcium out of the cell to produce relaxation. These cells will not contract when stimulated unless calcium is present in the extracellular medium. Skeletal muscle fibers have a large diameter; if they depended upon only their surface membrane to regulate the access of calcium to the cell, there would be a long delay between the electric activity in the membrane and the contraction (because of the slow rate of calcium diffusion over long distances). This does not present a problem for smooth muscle because of its small diameter.

Classification of Smooth Muscle

When the total range of smooth muscle properties is examined there emerge two general classes of smooth muscle, known as *single-unit* and *multiunit smooth muscle*. A given smooth muscle generally exhibits properties which are characteristic of one of these two classes. However, these classifications are not absolute, and some smooth muscles may show some characteristics common to both classes. Multiunit smooth muscle has functional properties which resemble those of skeletal muscle, whereas, as we shall see, single-unit smooth muscle more closely resembles cardiac muscle in its functional activity.

Multiunit Smooth Muscle Smooth muscles showing multiunit type of activity are found in the larger arteries and in some areas of the intestinal and reproductive systems and make up the pilomotor muscles that are attached to the hairs of the skin. A major characteristic of multiunit smooth muscles is that contractile activity is initiated in them by electric activity in the nerve fibers to the muscle, just as in the case of skeletal muscle. The innervation of the smooth muscle is by way of the sympathetic and parasympathetic branches of the autonomic nervous system rather than the somatic nerves which innervate skeletal muscle. However, the autonomic nerve fibers do not form discrete junctions with the smooth muscle membrane as the somatic neurons do on skeletal muscle.

Distributed along the terminal branches of the autonomic nerves are a series of regions where the axon appears swollen. These regions are filled with membrane-bound vesicles which are presumably the location of the chemical transmitters. Upon stimulation of the nerve, these vesicles fuse with the nerve membrane and release their transmitter into the extracellular space. Since a single axon releases transmitter from several such regions along its length, it affects numerous surrounding smooth muscle cells. Moreover, the released transmitter is able to diffuse to a number of smooth muscle cells in the vicinity. Thus, a single autonomic nerve fiber may affect the activity of a number of smooth muscle fibers even though it does not form an anatomic junction with any given cell.

Because of the relatively large distances between the nerve terminals and the smooth muscle membrane, there is a considerably longer period of time between nerve stimulation and contractile response in smooth muscle than is found in skeletal muscle where only a few hundred angstroms separate the nerve ending from the motor end plate.

The total surface of the smooth muscle membrane appears to contain receptor sites that can combine with the nerve transmitter agents; this contrasts with the localization of receptors in one region under the nerve terminal in the motor end plate of skeletal muscle. Binding of the transmitter to the membrane receptor alters the permeability of the smooth muscle membrane to ions, which leads to changes in the membrane potential resembling those seen at synaptic junctions between nerve fibers. Either excitatory or inhibitory potentials may be produced, depending on the nature of the chemical transmitter released. Just as with synaptic junctions, a single action potential in the nerve fiber produces only a small subthreshold change in the smooth muscle membrane potential. Multiple action potentials in the nerve are required to depolarize the smooth muscle membrane to threshold and initiate an action potential. Once an action potential is produced, it releases calcium into the cell which initiates contraction.

In addition to being stimulated by autonomic nerves, multiunit smooth muscle may also be induced to contract by hormones, such as epinephrine, which reach the cell by way of the circulatory system. These hormones apparently react with receptor sites on the membrane surface, producing changes in the membrane potential by mechanisms similar to those produced by neurotransmitter agents at synaptic junctions.

Single-unit Smooth Muscle Single-unit smooth muscle resembles cardiac muscle since both types of muscle can undergo spontaneous, rhythmic contractions in the absence of nerve or hormonal input. Spontaneous electric and mechanical activity occurs in single-unit smooth muscles that have been removed from the body and are no longer attached to nerves. Thus, this spontaneous activity appears to be an inherent property of the muscle cell itself. The molecular properties of the cells which produce this spontaneous activity are not understood. The membrane potentials in these cells show spontaneous fluctuations in potential which are responsible for initiating the cells' mechanical activity. Examples of smooth muscle showing single-unit-type activity are the vascular smooth muscle in small arteries and veins, intestinal smooth muscle, and the smooth muscle of the uterus.

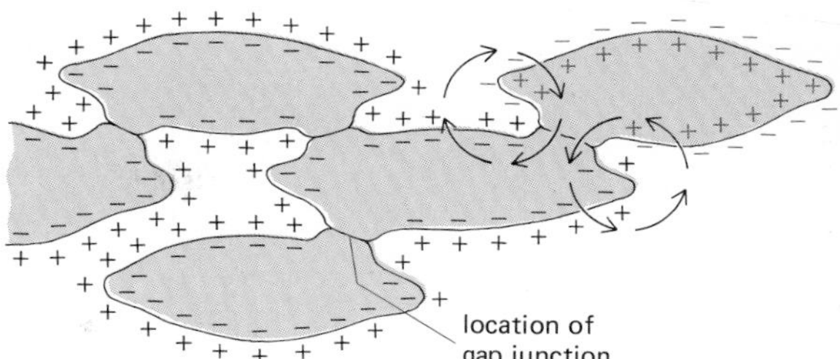

FIGURE 5-26 ***Action potentials from one smooth muscle cell can be conducted to others through gap junctions between adjacent cells.***

The term single-unit smooth muscle is applied to these cells because large numbers of them show a synchronized electric and mechanic activity and thus respond as if they were a single unit. What is responsible for coordinating the simultaneous activity of so many separate cells? The electron microscope has revealed that the membranes of adjacent single-unit smooth muscle cells are joined together to form gap junctions (Chap. 3). In the regions of the gap junctions the membranes of the adjacent cells join to form small channels directly linking the cytoplasms of the joined cells (Fig. 5-26). These gap junctions provide a low-resistance pathway for the conduction of electric activity from cell to cell. Thus an action potential in one cell is conducted from cell to cell through the gap junctions and the cells so joined together electrically respond as a single unit. As we shall see later, cardiac muscle cells are also linked together by gap junctions.

The basic electric event in the membranes of single-unit smooth muscle appears to be a spontaneous depolarization of the membrane which occurs at regular intervals and is known as a *pacemaker potential.* This pacemaker potential appears as a relatively slow depolarization of the membrane potential which, when it reaches the threshold potential, triggers off an action potential. Following repolarization of the action potential, the membrane again begins to undergo depolarization leading to a second action potential, and so on (Fig. 5-27). In a population of cells joined together by gap junctions, all the cells do not show spontaneous pacemaker potentials. Cells which do not have pacemaker activity, however, fire action potentials at the same rate as the pacemaker cells because the action potential from the pacemaker cell has been propagated throughout the population of cells through the gap junctions (Fig. 5-27). Thus the pacemaker cell sets the pace at

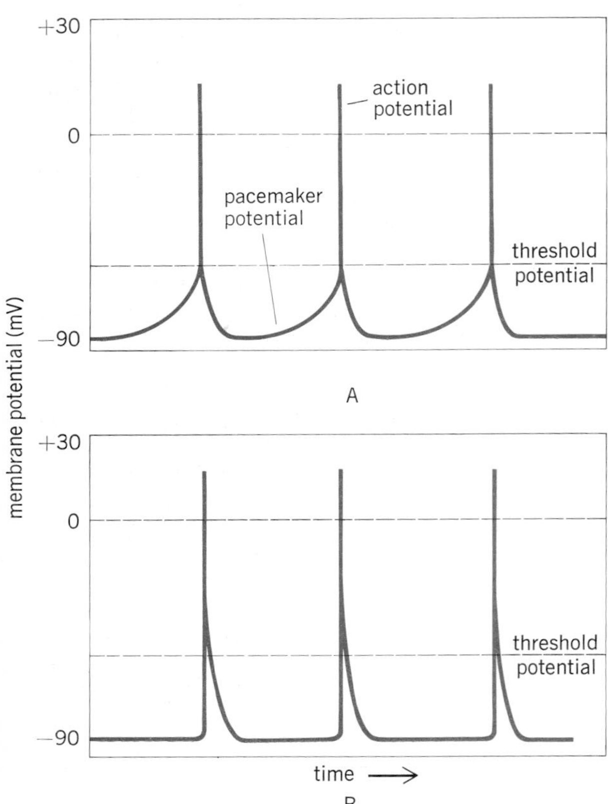

FIGURE 5-27 (A) ***Spontaneous action potentials in pacemaker cells.*** (B) ***Action potentials in non-pacemaker cells connected to a pacemaker cell by gap junctions.***

which all the cells in the population fire action potentials.

In addition to the pacemaker potentials, much slower waves of depolarization and repolarization are often observed upon which are superimposed bursts of action potentials (Fig. 5-28A). The origin of these slow waves of activity is still unclear. They may reflect the electric summation of a number of interconnected pacemaker cells which fail to trigger propagated action potentials. When the mechanical activity of the muscle is recorded simultaneously with the electric activity, the contractions of the muscle are seen to correspond to the periods when action potentials are occurring in the muscle, and the waves of contraction and relaxation correspond to the slow-wave frequency (Fig. 5-28B). Furthermore, the intensity of the mechanical response increases with the frequency of the action potentials. This is equivalent to the mechanical summation of activity in skeletal muscle that results from multiple action potentials.

Although the responses we have discussed thus far result from spontaneous electric activity within the muscle itself, this activity can be modified by external agents such as nerves, hormones, metabolic intermediates, mechanical stretch, and a variety of drugs. All these agents appear to act upon the smooth muscle membrane to produce either a depolarization or a hyperpolarization of the membrane potential. If the membrane becomes depolarized, so that it is nearer to threshold, the frequency of spontaneous action potentials increases and the resulting mechanical activity of the muscle increases. Stretching the muscle leads to depolarization of the membrane, which increases the

FIGURE 5-28 (A) ***Slow-wave oscillations in a membrane potential triggering bursts of action potentials.*** (B) ***The oscillation in tension corresponds to the frequency of slow-wave potential changes; the magnitude of the tension corresponds to action-potential frequency.***

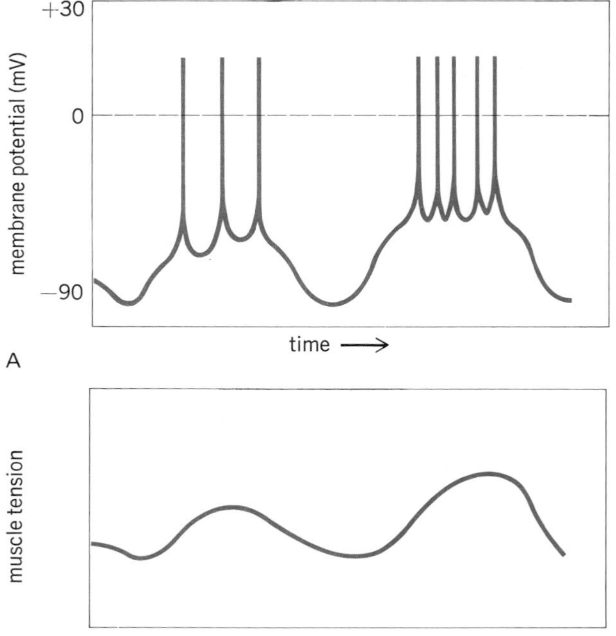

frequency of action potentials, which produces a contraction which tends to oppose further stretch of the muscle. This provides a form of negative feedback which tends to keep the length of the muscle constant. In contrast, stretching a multiunit innervated smooth muscle usually does not elicit a contraction. Thus, this response to stretch is a property of single-unit smooth muscle and probably reflects the instability inherent in their membranes. Agents which lead to a hyperpolarization of the membrane, on the other hand, tend to decrease the frequency of spontaneous action potentials, inhibiting contraction and leading to smooth muscle relaxation.

As was indicated earlier, the specific properties of smooth muscles show great diversity between different tissues. One form of this diversity is illustrated in Fig. 5-29, which shows the opposite response of vascular and intestinal single-unit smooth muscle to sympathetic stimulation. The same neurotransmitter, norepinephrine, is released from the sympathetic ending in each tissue, yet the vascular smooth muscle becomes depolarized and contracts whereas the intestinal smooth muscle becomes hyperpolarized and relaxes. How can the same chemical agent produce such opposite effects? The molecular mechanism is unknown but it seems clear that the different responses are not produced by the chemical effects of norepinephrine directly. The neurotransmitter merely acts as a trigger at the receptor site which sets in motion a set of permeability changes which were initially built into the structure of the cell membranes of these two different cells. The mechanism necessary to produce a given type of permeability change must already be built into the membrane and norepinephrine acts only to release that built-in mechanism, not to produce the permeability change itself.

CARDIAC MUSCLE

Cardiac (heart) muscle has properties similar to those of both skeletal and single-unit smooth muscle. It is a striated muscle having myofibrils with thick and thin filaments. The sliding-filament type of contraction is found in cardiac muscle, which has a length-tension relationship similar to that shown by skeletal muscle.

Cardiac muscle also has a well-developed sarcoplasmic reticulum. Action potentials in the cardiac muscle membrane lead to the release of calcium from the sarcoplasmic reticulum and thereby to the activation of the actomyosin contractile system.

The metabolism of cardiac muscle is designed for endurance rather than speed or strength. A continuous supply of oxygen must be maintained to the heart muscle if it is to continue to supply ATP to the contractile machinery. Cardiac cells deprived of oxygen for as little as 30 s cease to contract, and heart failure ensues.

Cardiac muscle most resembles single-unit smooth muscle in its spontaneous activity and the presence of gap junctions. The properties of cardiac muscle will be developed more fully as they apply to the functioning of the cardiovascular system discussed in Chap. 14.

FIGURE 5-29 ***Different responses of vascular smooth muscle and intestinal smooth muscle to norepinephrine released from a sympathetic nerve ending.***

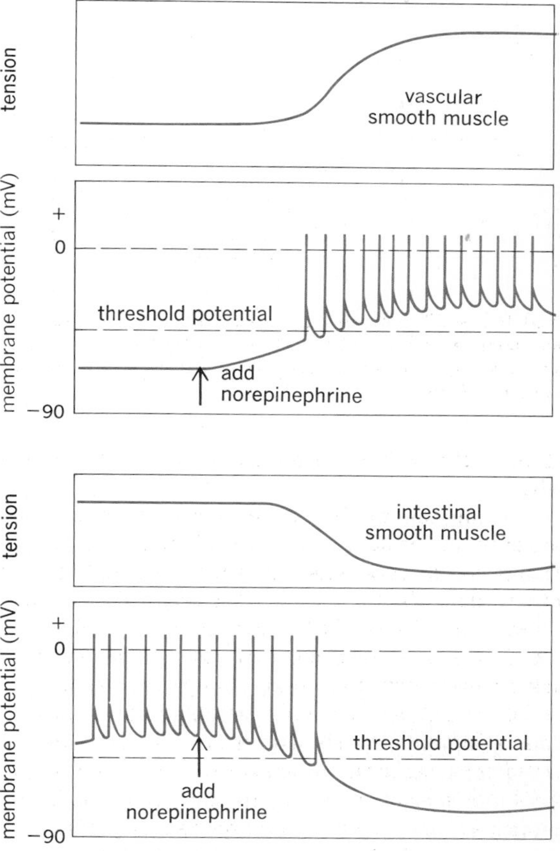

CHAPTER 6
Homeostasis and Control Systems

HOMEOSTASIS

As described in Chap. 2, Claude Bernard was the first to recognize the central importance of maintaining a stable internal environment (extracellular fluid). This concept was further elaborated and supported by the American physiologist W. B. Cannon, who emphasized that such stability could be achieved only through the operation of carefully coordinated physiologic processes which he termed *homeostatic*. The activities of tissues and organs must be regulated and integrated with each other in such a way that any change in the internal environment automatically initiates a reaction to minimize the change. *Homeostasis* denotes the stable conditions which result from these compensating regulatory responses. Some changes in the composition of the internal environment do occur, of course, but the fluctuations are minimal and are kept within narrow limits through the multiple coordinated homeostatic processes, descriptions of which constitute the bulk of the remaining chapters.

Concepts of regulation and relative constancy have already been introduced in the context of a single cell. In Chap. 1, we described how metabolic pathways within a cell are regulated by the principle of mass action and by changes in enzyme activity so as to maintain the concentrations of the various metabolites within the cell. Chapter 1 also described the control of protein synthesis and the mechanisms by which the genetic apparatus maintains a constant species line. In human beings, these basic intracellular primitive regulators still remain, so that each individual cell exhibits some degree of self-regulation, but the existence of a multitude of different cells organized into specialized tissues, which are further combined to form organs, obviously imposes the need for overall regulatory mechanisms to coordinate and integrate the activities of all cells. For this, intercellular communication over relatively long distances is essential. Such communication is accomplished by means of nerves and the blood-borne chemical messengers known as *hormones*.

The mechanisms by which these two communications systems operate are the subject of Chaps. 4 and 12, but their overall role and the basic characteristics of homeostatic processes can be appreciated only in terms of *control systems*, to which we now turn.

GENERAL CHARACTERISTICS OF CONTROL SYSTEMS

We shall define a homeostatic, or control, system as a collection of interconnected components which functions to keep a physical or chemical parameter of the body relatively constant.

Let us first analyze a nonbiological control system (Fig. 6-1) designed to maintain the temperature of a water bath at approximately 30°C despite fluctuations in room temperature from 25 to 10°C. Since the water temperature is always to be higher than the room temperature, there is a continuous loss of heat to the room from the water. Moreover, the lower the room temperature, the greater is this heat loss. Accordingly, the water must be continuously heated in order to offset the loss, and the degree of heating must be altered whenever the room temperature changes. This adjustment of heat input to heat loss so that water temperature remains approximately 30°C is the job of the control system.

The first system component, known as a *sensor*, is the temperature-sensitive instrument *A*. It generates an electric current the magnitude of which is inversely proportional to the water temperature; in other words, the higher the temperature, the less the current flow. (This current is always so small that its magnitude per se does not affect the water temperature.) The current flows through wire *B* into the control box *C* constructed in such a way that the amount of current flowing out of the box in wire *D* is directly proportional to that entering along *B*. The current from the box is fed to the heating unit *E* within the water bath. Its activity and therefore the amount of heat it produces per unit time is directly proportional to the strength of the signal to it, i.e., the current flow in *D*.

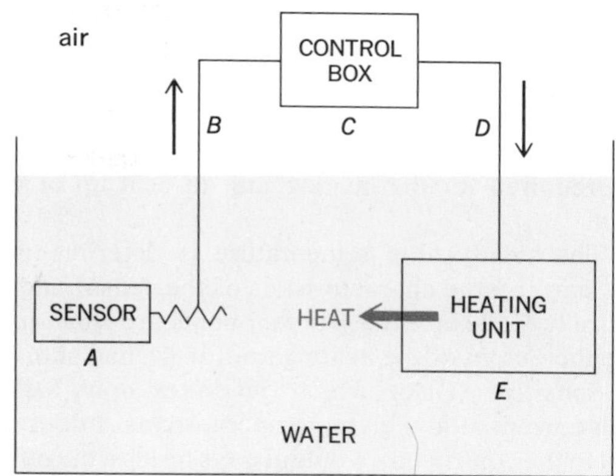

FIGURE 6-1 ***Components of a control system for regulating the temperature of a water bath.***

We fill the bath with water at room temperature, close the switches, and allow the system to operate (Fig. 6-2). Current generated in the sensor *A* controls the output of the heating unit by way of the control box. Because of the initially low water temperature, the magnitude of current flow from *A* is large and the heating unit is running full blast. This heats the water rapidly, but as the water temperature rises, two opposing events occur: (1) More heat is lost from the bath to the room and (2) the signal from *A* decreases and results in a decreased input to the heating unit, thereby decreasing the amount of heat it produces. The system ultimately stabilizes at a particular water temperature when heat

FIGURE 6-2 ***Effects of filling a water bath with water at room temperature and allowing the system to operate. Shown are the initial state, i.e., just after the water was added, and the ultimate steady state achieved. Note the difference in output of the sensor and the heating unit in the two states.***

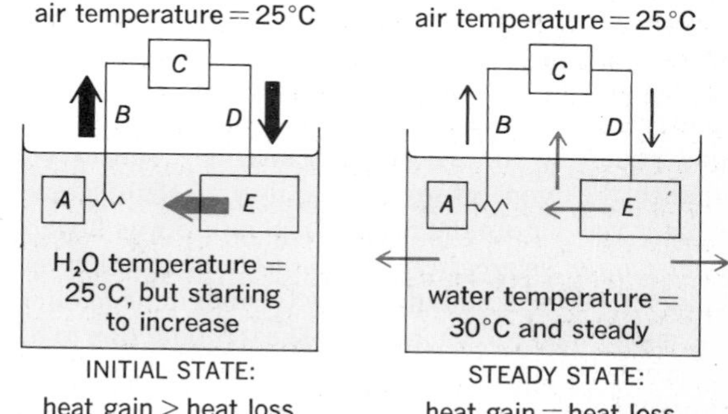

loss to the room exactly equals heat gain from the unit. At this point, the system is said to be in a *steady state;* input equals output, and the temperature remains steady. Actually there is always some oscillation around the steady-state temperature because of the time required for the heating unit to heat up or cool down.

The steady-state temperature is determined, in large part, by the characteristics of the sensor and the control box C, since these components are what determine the output of the heating unit. If we had chosen a heat-sensitive sensor which generates only half as much current at any given temperature as A does, the input to the control box would always be less, the output from the control box would be less, and the heating unit would always be generating less heat. Therefore, the steady-state temperature of the bath would be lower. Similarly, by altering the transforming function, i.e., the relationship between current in and current out, of the control box, we alter the steady-state temperature ultimately reached. In any control system, the actual steady state, or so-called operating point of the system, depends upon the characteristics of the individual components of the system. Our components were chosen to achieve an operating point of 30°C.

Most important is that this type of system resists any changes from the operating point. In our example the control system automatically prevents any significant deviation of the water temperature from the steady state established. Suppose that after steady state (30°C) has been reached, the room temperature is suddenly lowered (and kept low) so that the loss of heat from water to room is increased. This loss unbalances heat loss and heat gain, and the water temperature falls. But the decrease in water temperature immediately increases the current generated in A; therefore, more current flows out of the control box C and the heating unit increases its activity, thereby raising the water temperature back toward its original value. A new steady state will be reached when heat loss once again equals heat gain, both having been increased by a proportionate amount. What is the new steady-state temperature at this point? If the system is extremely sensitive to change, the temperature will be only very slightly below what it was before the room temperature was lowered *but it cannot be precisely the same*. Compensation is incomplete because the new steady state depends upon the maintenance of an increased heat production to balance the increased heat loss, and this increased heat production is due to the increased signal coming from A. The reason A generates more current is the slightly lower water temperature. If the temperature actually returned completely to normal, the signal from A would return to normal, heat production would return to normal, and the water temperature would immediately decrease as heat loss again became greater than heat gain (recall that the room temperature is being maintained lower than normal). Thus, control systems of this type cannot absolutely prevent changes from occurring in the physical or chemical variable being regulated, but they do keep such changes within very narrow limits, dependent upon the sensitivity of the system.

We can now summarize the basic characteristics of a control system in general terms. There must be a component which is sensitive to the variable being regulated and which changes its output as the variable changes. There must be a continuous flow of information from this sensor to an integrating control box, from which, in turn, the so-called command signal flows to an apparatus which responds to the signal by altering its rate of output (heat, in our example).

There remains one more concept which we have described but not named: *feedback*. The ultimate effects of the change in output by the system must somehow be made known (fed back) to the sensor which initiated the sequence of events. Our example illustrated perhaps the commonest form of feedback: When the water temperature is reduced, the sensor A detects this change and relays the information to the control box, which in turn signals the heating unit to increase its output; sensor A is "informed" of this change in output by the resulting rise in water temperature; accordingly, its current generation is again altered, which, in turn, results in an alteration of input to the control box and thereby the heating unit. The water temperature acts as a continuous link between the sensor and the heating unit (without feedback, the sensor's signal would be unrelated to the heating-unit output and the system would be unable to maintain constancy). This type of feedback, in which an increase in the output of the system results in a decrease in the input, is known as *negative feedback*. It clearly leads to stability of a system and is crucial to the efficient operation of homeostatic mechanisms.

Note that a major characteristic of negative-feedback control systems is that they *restore* the regulated variable toward normal after its initial displacement, but they cannot *prevent* the initial displacement. Suppose, however, that we add another component to our temperature-control system of Fig. 6-2, namely, a thermosensor on the *outside* of the water bath which can detect changes in room temperature. Now, when the room temperature is lowered, as in our example, this information is immediately relayed to the control box which causes the heating unit to increase its output. In this manner, additional heat can be supplied to the water bath *before* the water temperature begins to fall. Thus, the use of the external sensor permits the system to *anticipate* a pending fall in water temperature and begin to take action to counteract the change before it occurs. This provides *feedforward* information which has the net effect of minimizing fluctuations in the level of the parameter being regulated. We shall see that the body frequently makes use of feedforward control in conjunction with negative-feedback systems.

There is, however, another type of feedback known as *positive feedback* in which an initial disturbance in a system sets off a train of events which increases the disturbance even further. Generally, such cycles occur explosively; they usually lead to instability and are quite uncommon. However, several important positive-feedback relationships occur in the body, blood clotting being an example.

COMPONENTS OF LIVING CONTROL SYSTEMS

Reflex Arc

Homeostatic control systems in living organisms manifest virtually the same characterstics as those just described, but some of the terminology is different. The components are shown in Fig. 6-3 and are completely analogous to those of Fig. 6-1.

A *stimulus* is defined as a detectable change in the environment, such as a change in temperature, potassium concentration, pressure, etc. A *receptor* is the component which receives the stimulus, i.e., detects the environmental changes (it is identical to the sensor of the previous section). The stimulus acts upon the receptor to alter the signal emitted by the receptor, and this signal is the information relayed to the control box, or *integrating center*. The pathway between the receptor and the integrating center is known as the *afferent pathway*.

The integrating center usually receives input from many receptors, some of which may be responding to quite different types of stimuli. Thus, the output of the integrating center reflects the net effect of the total afferent input; i.e., it represents an integration of numerous and frequently conflicting bits of information.

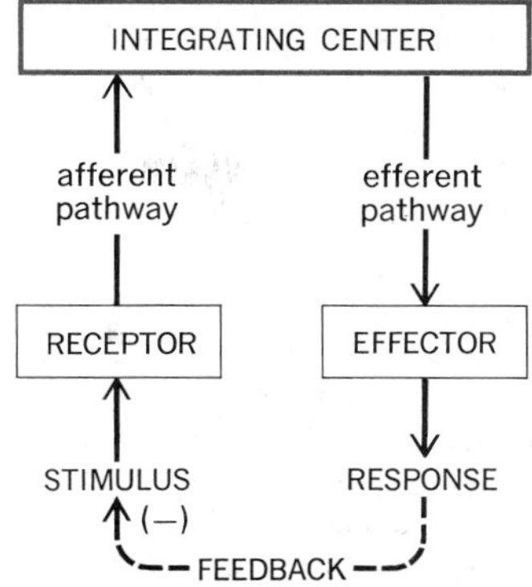

FIGURE 6-3 ***General components of a reflex arc with negative feedback. The response of such a system has the effect of counteracting or eliminating the original stimulus. This phenomenon of negative feedback is emphasized by the minus sign in the feedback loop.***

The output of the integrating center is then relayed to the last component of the system, the device whose change in activity constitutes the overall response of the system. This component, known as an *effector*, is analogous to the heating unit of our previous example. The information going from the integrating center to the effector is like a command directing the effector to alter its activity. The pathway along which this information travels is known as the *efferent pathway*.

As a result of the effector's response, the original stimulus (environmental change) which triggered this entire sequence of events may be counteracted (at least in part), just as in our example the increased heat production by the heating unit caused the lowered water temperature to return toward normal. As the stimulus is diminished by the effector's response, the activity of the receptor is diminished, so that the flow of information from receptor to integrating center returns toward the original level and, in turn, the effector's activity is turned toward its previous rate. Thus, the counteracting of the stimulus by the effector's response constitutes the *negative feedback*.

Most biological control systems belong to the general category of stimulus-response sequences known as reflexes. A *reflex* is the sequence of events elicited by a stimulus, though we may be aware only of the final event in the sequence, the *reflex response* (pulling one's hand away from a hot stove, for example). The entire reflex, including the response, often occurs without any awareness on the part of the person. The pathway mediating the reflex is known as the *reflex arc*, and its components, as described above, are:

1 Receptor
2 Afferent pathway
3 Integrating center
4 Efferent pathway
5 Effector

As an example, let us again take a thermoregulatory system, this time the homeostatic control system which maintains internal body temperature relatively constant. The receptors are neuronal endings in the brain which generate action potentials at a rate determined by their temperature. This information is relayed by neurons (the afferent pathway) to a specific part of the brain (which acts as the integrating center) which in turn influences the rate of firing of the neurons to skin blood vessels, sweat glands, and skeletal muscle. The latter neurons are the efferent pathway, and the structures they innervate, the effectors. Rates of sweat production, skin blood flow, and muscle contraction are altered to raise or lower heat loss and production and thus restore body temperature to its operating point.

Sometimes the term "reflex" is restricted to situations in which the first four components are all parts of the nervous system. However, the afferent and efferent information can be carried by nervous or hormonal pathways. In any case, two different components must serve as afferent and efferent pathways; thus the input to, or output from, the integrating center may both be neural, but they must be two different nerve fibers. Of course, they can also both be hormonal, or one neural and the other hormonal. Depending on the specific nature of the reflex, the integrating center may reside in either the nervous system or an endocrine gland.

Finally, we must identify the effectors, the cells whose outputs constitute the ultimate responses of the reflexes. Actually, most cells of the body act as effectors in that their activity is subject to control by nerves or hormones. Muscle and gland, however, comprise the major effectors of biological control systems.

To summarize, most biological control systems function to keep a physical or chemical parameter of the body relatively constant. One may analyze any such system by answering a series of questions: (1) What is the parameter (blood glucose, body temperature, blood pressure, etc.) which is being maintained constant in the face of changing conditions? (2) Where are the receptors which detect changes in the state of this parameter? (3) Where is the integrating center to which these receptors send information and from which information is sent out to the effectors, and what is the nature of these afferent and efferent pathways? (4) What are the effectors and how do they alter their activities so as to restore the regulated parameter toward normal (i.e., the operating point of the system)?

Some Semantic Problems There are problems which arise when one attempts to categorize the components of some complex reflex arcs according to the five terms listed above. For example, in the reflex arc shown in Fig. 6-4, the stimulus, receptor, and final effector (muscle) are quite clear, as is the designation of the first nerve (*A*) in the chain and the last hormone (*B*) as afferent and efferent pathways. But what about the brain, nerve *B*, hormone *A*, and the two endocrine glands? Actually, one is dealing here with a chain of small reflex arcs all subserving the overall reflex arc. Thus, endocrine gland *A* may be considered to be an effector whose response is the secretion of hormone *A*, or it may be viewed as an integrating center; in the latter case, nerve *B* becomes its afferent input and hormone *A* becomes its efferent output. It is fruitless to assign rigid terms to the interior components of such complex reflex chains; it is more important to appreciate the sequence of events rather than worry about the labels.

Another problem is that some reflex arcs lack the usual afferent pathway. This may seem puzzling, but the following example may help (Fig. 6-5). Parathor-

FIGURE 6-4 ***Complex reflex arc composed of multiple hormones and nerve fibers.***

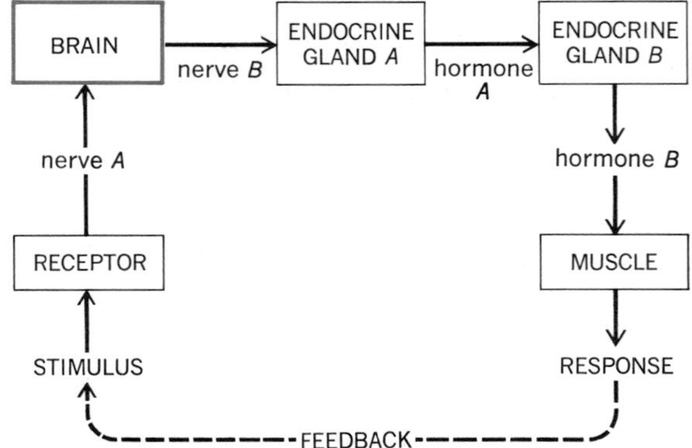

mone is a hormone which is secreted by the parathyroid glands and which acts upon bone, causing it to increase its release of calcium into the blood. When the blood calcium concentration is decreased for any reason, an increased amount of parathormone is secreted into the blood; the blood-borne parathormone reaches bone throughout the body and induces it to release more calcium into the blood. Clearly, in this reflex, blood-borne parathormone acts as the efferent pathway and bone is the effector. But we have made no mention yet of the receptors or the afferent pathway involved. Actually, the parathyroid gland cells are themselves sensitive to the calcium concentration of the blood supplying them; thus, the same cells which produce the hormone act as the receptors for this reflex. Clearly, there is no afferent arc in such a situation since the receptors and integrating center (i.e., hormone-producing cells) are one and the same.

Finally, we must point out a problem to which we will return in several subsequent chapters. In the most narrow sense of the word, a reflex is an involuntary, unpremeditated, unlearned response to a stimulus; the pathway over which this chain of events occurs is "built in" to all members of a species. Examples of such basic reflexes would be pulling one's hand away from a hot stove or shutting one's eyes as an object rapidly approaches the face. However, there are also many responses which appear to be automatic and stereotyped but which actually are the result of learning and practice. For example, experienced drivers perform many complicated acts in operating their cars; to them these motions are, in large part, automatic, stereotyped, and unpremeditated, but they occur only because a great deal of conscious effort was spent to learn them. We shall refer to such acts as *learned* or *acquired.* In general, most reflexes, no matter how basic they may appear to be, are subject to alteration by learning; i.e., there is often no clear distinction between a basic reflex and one with a learned component.

FIGURE 6-5 ***Homeostatic reflex by which plasma calcium concentration is controlled. Note the absence of an afferent pathway.***

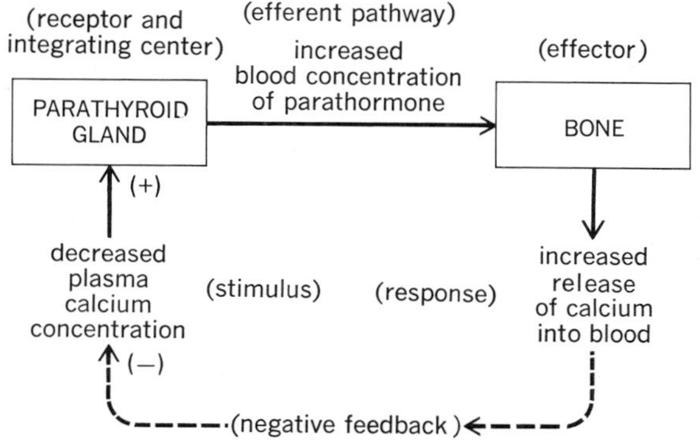

Local Homeostatic Responses

Besides reflexes, another group of biological responses is of immense importance for homeostasis. We shall call them *local responses.* Local responses are initiated by a change in the external or internal environment, i.e., a stimulus, which acts upon cells in the immediate vicinity of the stimulus inducing an alteration of cell activity with the net effect of counteracting the stimulus. Thus, a local response is, like a reflex, a sequence of events proceeding from stimulus to response, but, unlike a reflex, the entire sequence occurs only in the area of the stimulus, no hormones or nerves being involved.

Two examples should help clarify the nature and significance of local responses: (1) Damage to an area of skin causes the release of certain chemicals from cells in the damaged area which help the local defense against further damage; (2) an exercising muscle liberates chemicals into the extracellular fluid which act locally to dilate the blood vessels in the area, thereby permitting the required inflow of additional blood to the muscle. The great significance of such local responses is that they provide individual areas of the body with mechanisms for local self-regulation.

Chemical Mediators

It should be evident that the *sine qua non* of reflexes is the ability of cells to communicate with one another, i.e., the capacity of one cell to alter the activity of another. When a hormone is involved in a reflex, it is clear that the communication between cells, i.e., between endocrine gland cell and effector, is accomplished by a chemical agent, the hormone (the blood, of course, acting as the delivery service). As described in Chap. 4, virtually all nerve fibers in the body also communicate with each other or with effectors by means of chemical agents or mediators. Thus, one neuron alters the activity of the next neuron in a reflex chain by releasing a substance from its ending; this *chemical transmitter* diffuses across the very narrow space separating the two neurons and acts upon the second, altering its activity. Similarly, chemical transmitters released from the ends of the neurons going to effectors constitute the immediate signal, or input, to the effector cells.

The detailed physiology of these chemical transmitters and of neuron-neuron or neuron-effector communication were described in Chaps. 4 and 5. We mention them here to emphasize that chemical mediators, whether they are secreted by endocrine gland cells or released from neuronal endings, constitute the ultimate messages by which one cell signals another to alter its activity. This is true not only for reflexes but for local responses, as the examples illustrate.

In future chapters, we shall describe the roles of a considerable number of chemical transmitters, but one aspect of their physiology is quite striking (and confusing), namely, the fact that the same chemical may act as a transmitter in many different sites with widely differing effects. For example, *acetylcholine* is one of the transmitters for communication between many neurons, neurons and skeletal muscle, neurons and heart muscle, neurons and gland cells, neurons and smooth muscles. *Epinephrine* and *norepinephrine* also have a wide spectrum of functions. These three have received the most study in the past but others (*dopamine, histamine, serotonin,* etc.) are rapidly gaining on them. Recently, an entirely new set of substances has gained the limelight—the family of fatty acids called *prostaglandins*. They were originally discovered in semen but are now known to be present in most tissues. Their precise physiologic roles are yet to be established, but they may well be involved in the function of smooth muscle, nerves, the liver, adipose tissue, the circulation, and the reproductive organs. We shall discuss these possibilities in subsequent chapters where relevant.

Receptor Sites

The mechanisms by which chemical mediators act to alter cellular activity represent one of the most intensively studied areas in physiology today. Present thinking centers around the concept of *membrane receptor sites*.[1] It is believed that the first step in the action of most, if not all, chemical mediators is their combination with certain specific molecules, i.e., receptor sites, within the cell-membrane structure. This combination of chemical mediator and membrane component somehow alters the membrane structure. In some cases it leads to changes in membrane permeability and the rate at which a particular substance is transported across the membrane; in the main, however, the precise mechanisms by which this combination induces an alteration in the cell's activity, i.e., a response, are unknown.

One extremely important characteristic of chemical mediation that is understandable in terms of membrane receptor sites is *specificity*. A chemical mediator—hormone, neural transmitter, or locally released chemical agent—influences only certain cells and not others, although there is considerable variation in the degrees of specificity manifested by the different mediators. The likeliest explanation is that membranes of different cell types all differ; accordingly, only certain cell types, frequently just one, possess the precise membrane receptor site required for combination with a given chemical mediator. Conversely, the different molecular structures of the mediators explain why only certain mediators, but not others, can influence any given cell type.

[1] This term is frequently shortened is scientific literature to receptor, a usage that unfortunately can cause confusion because receptor sites are totally different from afferent receptors, described earlier. The reader must carefully distinguish between these terms.

THE BALANCE CONCEPT AND CHEMICAL HOMEOSTASIS

One of the most important concepts in the physiology of control system is that of balance. Our example of the water bath was really a study of heat balance within the water bath, i.e., the control system functioned to maintain a precise balance between the rates at which heat was added to and left the bath. Almost every homeostatic system in the body can be studied in terms of balance; some regulate the balance of a physical parameter (heat, pressure, flow, etc.), but most are concerned with the balance of a chemical component of the body. This section is intended to provide a foundation of general principles upon which can be built the study of any specific chemical substance.

Figure 6-6 is a generalized schema of the possible pathways involved in the balance of a chemical substance. The *pool* occupies a position of central importance in the balance sheet; it is the body's readily available quantity of the particular substance and is frequently identical to the amount present in the extracellular fluid. The pool functions as "middleman," receiving from and contributing to all the other pathways.

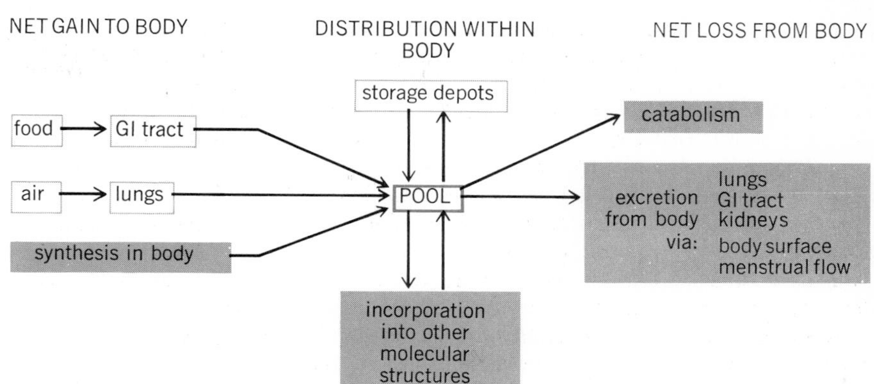

FIGURE 6-6 ***Balance diagram for a chemical substance.***

The pathways on the left of the figure are sources of *net gain* to the body. A substance may be ingested and then absorbed from the gastrointestinal (GI) tract. It is important to realize that all that is ingested may not be absorbed; some may either fail to be absorbed or be consumed by the bacteria residing in the gut. The lungs offer another site of entry to the body for gases (O_2) and airborne chemicals. Finally, the substance may be synthesized by cells within the body itself. (Bacteria in the GI tract may also synthesize nutrients which can be used by the body.)

The pathways to the right of the figure are sources of *net loss* from the body. A substance may be excreted in the urine, feces, expired air, and menstrual flow, as well as from the surface of the body (skin, hair, nails, sweat, tears, etc.). The substance may be catabolized or transformed within the body to some other chemical; this fate—the opposite of synthesis—represents net loss of the substance.

The central portion of the figure illustrates the *distribution* of the substance within the body. From the readily available pool, it may be taken up by storage depots; conversely, material may leave the storage depots to reenter the pool. Finally, the chemical may be incorporated into some other molecular structure (fatty acids into membranes, iodine into thyroxine, etc.). This process is reversible in that the substance is liberated again whenever the more complex molecule is broken down. For this reason, this pathway differs from catabolism or transformation by which the substance is irretrievably lost. This pathway is also distinguished from storage in that the latter has no function other than the passive one of storage, whereas the incorporation of the substance into other molecules is done to fulfill an active function of the substance (e.g., the function of iodine in the body is to provide an essential component of the thyroxine molecule).

Of course, it should be recognized that every pathway of this generalized schema is not applicable to every substance; for example, the mineral electrolytes cannot be synthesized or catabolized by the body.

The orientation of the figure illustrates two important generalizations concerning the balance concept: (1) The total body balance depends upon the rates of total body net gain and net loss; (2) the pool concentration depends not only upon total body losses and gains but upon exchanges of the substance within the body.

It should be apparent that, with regard to total body balance, three states are possible: (1) Loss exceeds gain, the total amount of the substance in the body decreases, and the person is said to be in *negative balance;* (2) gain exceeds loss, the total amount of the substance in the body increases, and the person is said to be in *positive balance;* (3) gain equals loss, and balance is stable. Physiology is, in large part, concerned with the homeostatic mechanisms which match gain with loss to achieve a stable balance. Pool size, too, tends to be maintained relatively constant as a result of these homeostatic mechanisms as well as those operating on the pathways for internal exchange. Clearly a stable balance can be upset by alteration of the magnitude of any single pathway in the schema, e.g., severe negative water balance can occur in the presence of increased sweating. Conversely, stable balance can be restored by homeostatic control of the pathways. Therefore, much research has been concerned with determining which are the key homeostatically controlled pathways for each substance, what are the specific mechanisms involved, and what are the limits of intake and output beyond which balance cannot be achieved.

Essential Nutrients: an Example

There are many substances which are required for normal or optimal body function but which are synthesized by the body either not at all or in amounts inadequate to achieve balance. They are known as *essential nutrients*. Because they are all excreted or catabolized at some finite rate, a continuous new supply must be provided by the diet. Approximately 50 in number, they are water, 8 amino acids, several unsaturated fatty acids, approximately 20 vitamins, and a similar number of inorganic minerals.

It should be reemphasized that the term essential nutrient is reserved for substances that fulfill *two* criteria: They not only must be essential for good health but must not be synthesized by the body in adequate amounts. Thus, glucose, although "essential" for normal metabolism, is not classified as an essential nutrient because the body normally can synthesize all it needs.

The physiology of each essential nutrient can be studied in terms of the schema of Fig. 6-6, i.e., by analyzing each pathway relevant for that nutrient. There is considerable variation in the relative importances of the pathways for the homeostatic regulation of the different nutrients. Thus, in Chap. 18 we shall see that ingestion and urinary excretion are the main controlled variables for water. In contrast, control of iron balance, as described in Chap. 13, is dependent largely upon the control of iron absorption by the gastrointestinal tract. Balance of essential amino acids is achieved in still another way; the rate of catabolism of these amino acids (specifically the loss of the NH_2 group from the amino acid) is reduced in the presence of amino acid deficiency (and increased in the presence of excess).

The reason for placing so much emphasis on the pathways which are homeostatically controlled is that alteration of the nutrient flow via these pathways constitutes the mechanism by which balance is achieved wherever a primary change occurs in any of the other pathways. For example, iron balance can be upset either by a primary change in intake or excretion; in either case, balance can be reestablished by a compensating change in the rate of absorption.

The key event in triggering off the homeostatic response is a change in the body content of the nutrient, manifested frequently as a change in pool content, i.e., a change in the internal environment. Therefore, it is important to realize that nothing in the body is maintained *absolutely* constant; relatively small deflections from normal are continuously occurring and constitute the signals for the control systems which then operate to limit the extent of deviation.

Let us look more closely at the spectrum of conditions possible as we alter primary intake of an essential nutrient, assuming all the other pathways are functioning normally. At zero intake, balance is impossible since excretion or catabolism, or both, cannot be reduced to zero; accordingly negative balance persists as the body stores of the nutrient are progressively depleted. Thus, the minimal combined rate of excretion and catabolism sets the lower limit for achievement of balance. At the other end of the spectrum, the maximal combined rate of excretion and catabolism sets the upper limit for achievement of balance, i.e., the maximal intake of nutrient compatible with balance. If intake is greater than this, body stores will continuously increase, with the potential for toxic effects. This occurs for certain of the fat-soluble vitamins (A, D, and K) and is a problem for other nutrients as well.

Between the extremes there obviously exists a range of intakes at which balance can be maintained without overt manifestations of either deficiency or toxicity. How wide this range is depends upon how much the rates of excretion or catabolism can be altered. (For example, sodium balance can be achieved readily at intakes between 0.3 and 25 g/day.) It must be reemphasized that, despite the fact that balance is stable at all intakes in this range, there are small differences in body content and pool size over the range, since these differences are required to drive the homeostatic control systems.

The ultimate aim of nutrition is to determine which intake in this range is the *optimal* intake for each nutrient. The critical question may be stated as follows: Does ingestion of more than the minimal amount of a nutrient produce greater health, growth, intelligence, etc., i.e., is there an amount of nutrient which is *optimal* for health rather than merely adequate for avoiding disease? But this question raises the further question: Optimal for what? For example, most Americans ingest very large quantities of protein (which supplies essential amino acids), and this almost certainly has contributed to our increased body height. This would seem desirable, but in experimental animals it has been found that protein intakes which are optimal for maximal total body growth enhance the development of cancers and arteriosclerosis as well. Clearly, "optimal for what?" cannot be answered without a clearly definable "what" and a careful consideration of many factors, including existing environmental influences and life style.

BIOLOGICAL RHYTHMS

A striking characteristic of many bodily functions is the rhythmic changes they manifest. Body temperature, for example, fluctuates considerably during a normal 24-h period, as do the concentrations of many hormones. Such daily rhythms are called *circadian* (literally, around the day). Most likely, they occur as a result of changes in the operating points of the control systems regulating them, but we generally are uncertain as to their precise mechanisms or significance. We do know that modern life, with its stepped-up pace, rapid changes, and creation of artificial environments, may frequently disrupt them with, as yet, unknown results. Many of the body's rhythms are much longer than 24 h. The menstrual cycle is the best-known longer cycle, but there may well be others with even greater time spans.

Central to this field is the problem of "biological clocks" within the body which set the rhythms. This question has received particular attention by physiologists of reproduction, but it is now recognized as a critical area in the biology of growth, aging, and many other fields relevant to human physiology.

PART TWO

INTEGRATED SYSTEMS OF THE BODY

CHAPTER 7

The Musculoskeletal System

Section A. Basic Characteristics of the Musculoskeletal System

Previous chapters described the characteristics of bone and muscle as tissues. This chapter discusses the specific bones and skeletal muscles of the body and the ways in which they function together.

TYPES OF BONES

Individual bones vary in shape and are divided into four categories (Fig. 7-1): long, short, flat, and irregular. *Long bones* occur in the upper and lower limbs, varying in size from the large thigh bone (femur) to the distal bone of the little finger. As described in Chap. 3, they are tube-shaped and have a long shaft (diaphysis); their ends (epiphyses) are expanded. The shaft of long bones contains a central marrow cavity. The *short bones,* such as the bones of the wrists or ankles, are roughly cuboidal. The small, rounded *sesamoid bones* are a type of short bone. These bones, of which the patellas (or knee caps) are the most familiar example, are almost always near joints or embedded in tendons where the tendons make a sharp bend around a bony surface. With the obvious exception of the patella, the sesamoid bones are usually associated with joints in the hands and feet. The *flat bones* include the bones of the upper portion of the skull and the shoulder blades. They are formed of an inner layer of trabecular bone (the *diploe*) between two layers of compact bone. Bones such as the vertebrae and some of the bones of the face, which fit none of the above categories, are classified as *irregular*.

The surfaces of virtually all bones are characterized by structural variations such as grooves, protrusions, holes, and shallow depressions (Table 7-1): the

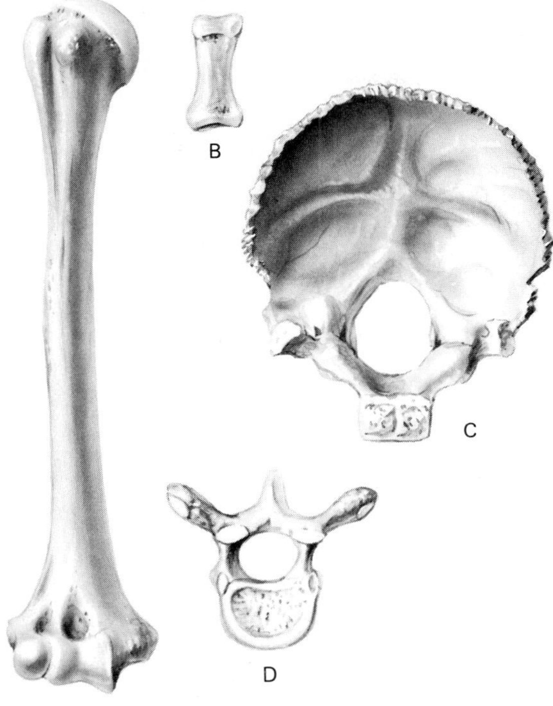

FIGURE 7-1 ***Four categories of bone: (A) long, (B) short, (C) flat, and (D) irregular.***

grooves and holes provide passageways for nerves and blood vessels; the large protrusions at the ends of bones form parts of joints; shallow depressions and ridges often serve as the attachment points for fibrous tissue.

TABLE 7-1
SURFACE MARKINGS OF BONE

	Definition
Protrusions (processes):	
Condyle	A relatively large, rounded protuberance which forms articulations
Epicondyle	A prominence above or on a condyle
Crest	A prominent ridge or border
Head	A rounded projection separated from the main part of the bone by a constricted region (the *neck*)
Tubercle	A small, rounded projection
Tuberosity	A large, rounded process, often with a roughened surface
Trochanter	A large, blunt process found only on the femur
Spine	A sharp, pointed process
Line	A low ridge
Flat surfaces and depressions:	
Facet	A flat or shallow surface for articulations
Fovea	A small, shallow depression
Groove or sulcus	A furrow that accommodates a blood vessel, nerve, tendon, or other soft structure
Fossa	A depression or groove
Notch	A deep indentation in the edge of a bone
Neck	The narrow section of bone between the head and the shaft
Sinus	A cavity within a bone
Fissure	A groove or narrow cleft passage
Foramen	A hole or perforation through a bone
Meatus	A tubelike passageway or canal within a bone

JOINTS

A *joint* or *articulation* is defined as the meeting place of two or more bones. Some bones meet at movable joints, others at only slightly movable joints, and still others at immovable joints. The bones involved in an articulation do not actually touch each other but are separated by various types of tissue. The type of connection between the articulating bones determines the type and range of movement, if any, between them.

The classification of joints into three groups depends upon the type of connecting tissue and the mobility of the joint. The fibrous or fixed joints are *synarthroses,* the cartilaginous or slightly movable joints are *amphiarthroses,* and the synovial or freely movable joints are *diarthroses* (*arthro,* joint). (Some joints mentioned below do not fit easily into this classification system. For example, syndesmoses, classified as fibrous joints because of their structure, are actually slightly movable whereas synchondroses classified by structure as cartilaginous joints, are immobile.)

Fibrous Joints (Synarthroses)

The tissue between the bones is fibrous connective tissue in fibrous joints. These are usually tight joints which allow little or no movement. They include sutures and syndesmoses, which are distinguished from each other, in part, by the length of the connecting fibers. *Sutures,* which have short connections, occur between the flat bones in the upper part of the skull where the bones, each growing from its own ossification center, contact each other and trap fibrous connective tissue between them (e.g., see Fig. 7-15B). In the early stages of growth before the opposing bones of the skull have joined completely, only regions of connective tissue exist where the bones will eventually be. These, the *fontanelles* of the infant skull (Fig. 7-2), usually close by the eighteenth month of life. In *syndesmoses,* the fibers of the connective tissue within the joint are longer and more plentiful and some movement is possible. The interosseous membrane between the radius and ulna (see Fig. 7-31) is a syndesmosis. In both types of fibrous joints the connective-tissue fibers blend with the periosteum of the bones.

Cartilaginous Joints (Amphiarthroses)

Cartilaginous joints are *symphyses,* which retain a permanent cartilage or modified cartilage between the articulating bones, and *synchondroses,* in which cartilage appears temporarily and is eventually replaced by bone. Limited movement is possible at symphyses, but as mentioned above, synchondroses allow little or no movement. As described in Chap. 3, the normal process of bone elongation in bones with epiphyses involves synchondroses between the shaft and ends of the bone; after growth is complete, this region ossifies. A second example of a synchondrosis occurs between the sternum and first rib. In symphyses, a layer of hyaline cartilage is retained on the surface of each of the articulating bones; the cartilages of the two bones are then joined by fibrous connective tissue or a fibrocartilage pad plus strong ligaments. Together these can be compressed or displaced and therefore offer some movement. Vertebrae are joined by symphyses, and the cartilaginous discs between them give considerable flexibility to the spinal column. Another example of a symphysis is the joint between the pubic bones, the symphysis pubis (Fig. 7-3).

Synovial Joints (Diarthroses)

Synovial joints work on an entirely different principle from the fibrous and cartilaginous joints and have a mechanism that allows free movement. The major parts of the articulating surfaces of the bones involved are not

TABLE 7-2
MOVEMENTS AT SYNOVIAL JOINTS

Movement	*Definition*
Flexion	A movement that decreases the angle between two bones, bending a joint
Extension	A movement that increases the angle between two bones, straightening a joint
Dorsiflexion	Flexion of the foot at the ankle joint, the foot and toes are turned upward as in standing on the heel
Plantar flexion	Extension of the foot at the ankle joint, the foot and toes are turned downward toward the sole of the foot as in standing on tiptoe
Hyperextension	Continuation of extension beyond the anatomical position, as in bending the head backward
Abduction	Movement of a bone away from the midline of the body or body part as in raising the arm or spreading the fingers
Adduction	Movement of a bone toward the midline of the body or part
Rotation	Movement of a bone around its own axis as in moving the head to indicate "no" or turning the palm of the hand up and then down
Circumduction	Movement of a bone in a circular direction so the distal end scribes a circle while the proximal end remains stationary as "winding up" to throw a ball
Inversion	Turning inward, movement of the foot at the ankle joint so the sole faces inward
Eversion	Turning outward, movement of the foot at the ankle joint so the sole faces outward
Protraction	Movement of the clavicle (collar bone) or mandible (lower jaw bone) forward on a plane parallel to the ground
Retraction	Movement of the clavicle or mandible backward on a plane parallel to the ground
Supination	Rotation of forearm so that the palm faces forward or upward, movement of the whole body so that the face and abdomen are upward
Pronation	Rotation of the forearm so that the palm faces backward or downward, movement of the whole body so the face and abdomen are downward

FIGURE 7-2 ***Skull of a newborn infant showing the fontanelles.***

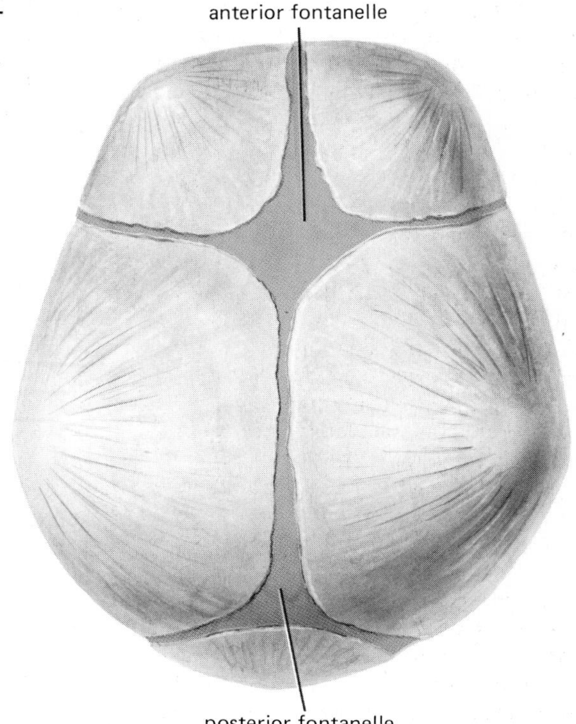

anterior fontanelle
sphenoid fontanelle
mastoid fontanelle

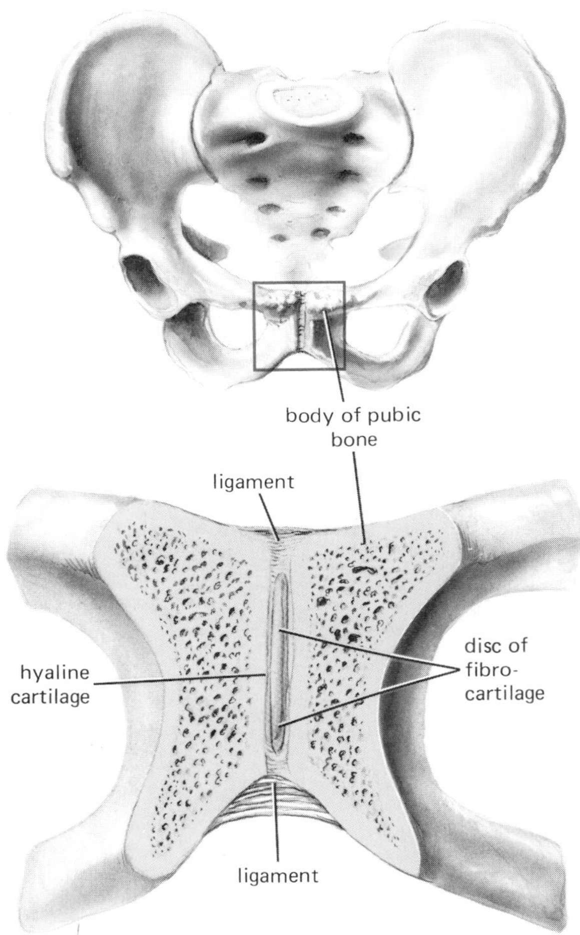

FIGURE 7-3 ***Section through the pubic region to show the structure of a symphysis.***

directly connected, although they are indirectly linked by a fibrous *articular capsule* (*joint capsule*) (Fig. 7-4) and often by *accessory ligaments* as well. These ligaments may be inside or outside the capsule, and they may be separate or applied to it. They serve to limit the types of movement and help hold the articulating bones in place.

The bony surfaces involved in the joint are covered by a thin layer of hyaline cartilage, and it is actually the two cartilaginous surfaces that slide past each other during movement. Their movement is facilitated by a viscous *synovial fluid,* which lubricates the cartilages, cushions shocks, and provides a nutrient source. This fluid is formed by a sheet of specialized tissue, the *synovial membrane*. The membrane lines the entire joint capsule, thus forming a cavity (the *articular,* or *joint, cavity*) which contains the synovial fluid. The cavity extends along the sides of the bones under the capsule and ends as the capsule fibers attach to the bone. In some synovial joints a pad of fibrous tissue or fibrocartilage lies in the joint cavity (Fig. 7-4B). The variety of synovial joints and the movements they permit (Table 7-2 and Fig. 7-5) are considerable, and only a few are listed here.

1 The *ball and socket joint* offers the greatest degree of movement, allowing flexion-extension, abduction-adduction, medial and lateral rotation, and circumduction, the last being a combination of flexion-extension and abduction-adduction. The hip and shoulder joints are of this type (Figs. 7-30 and 7-38).

2 *Condyloid joints,* such as the wrist, allow movement in two planes at right angles to each other. Thus, such joints permit flexion-extension, abduction-adduction, and circumduction, but prevent rotation.

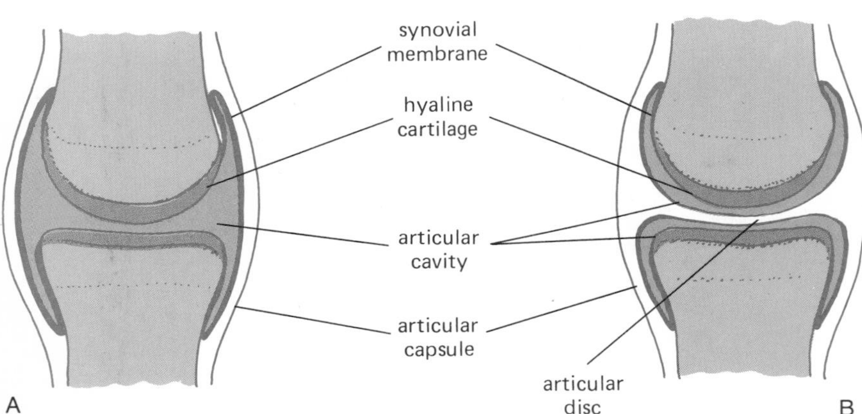

FIGURE 7-4 ***Diagrammatic sections through (A) a simple synovial joint, and (B) a synovial joint with an articular disc. (From R. Warwick and P. L. Williams, "Gray's Anatomy," 35th Brit. ed., Saunders, Philadelphia 1973.)***

FIGURE 7-5 ***Types of movements possible at various synovial joints.***

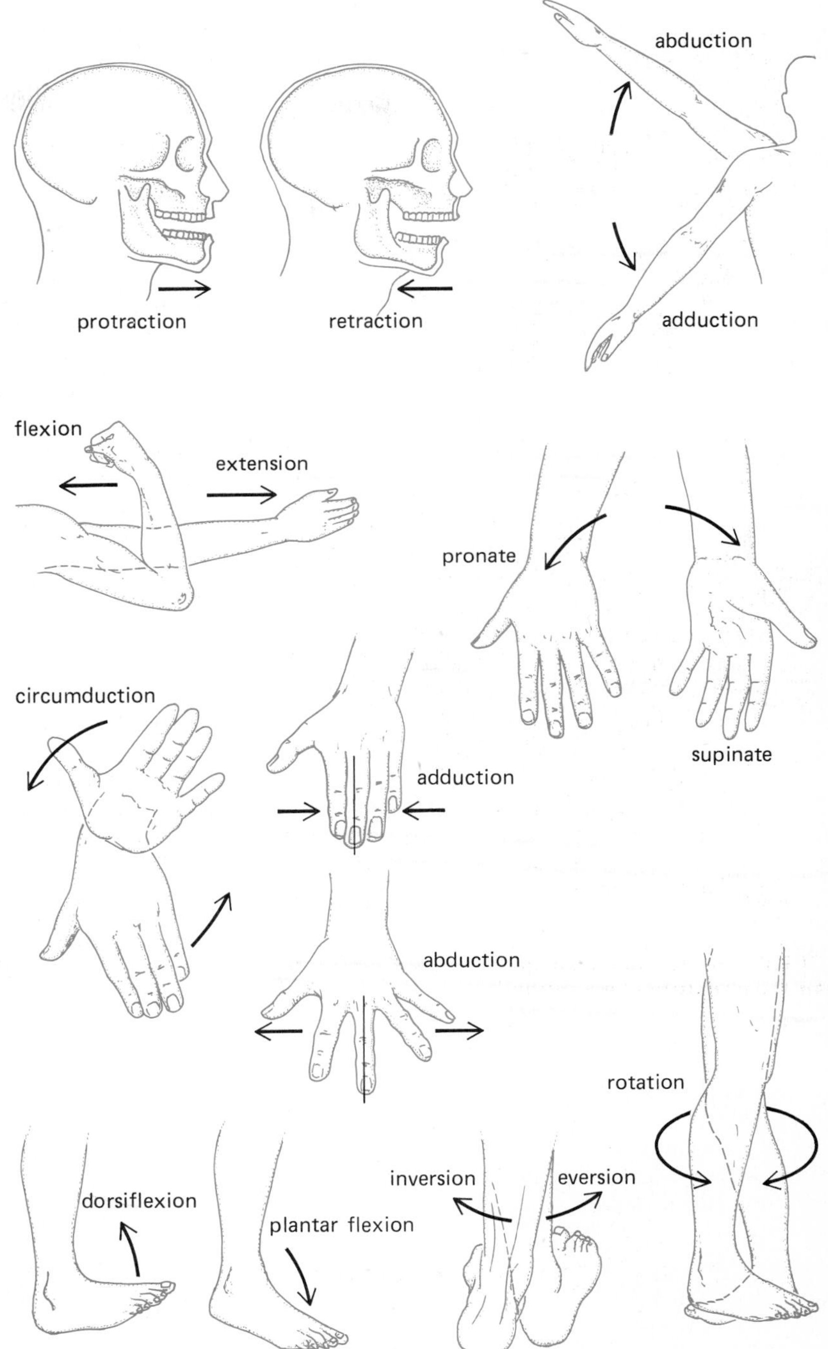

3 *Hinge joints,* which occur in the elbow (Fig. 7-32) and between the segments of the fingers, allow movement around a single axis at right angles to the bones and permit only flexion-extension.

4 *Gliding joints* provide simple sliding or gliding movements. They exist between the articular processes of the vertebrae, and between some of the bones of the wrist, hand, ankle, and foot.

Synovial membranes sometimes form cavities, such as *bursas* and *synovial sheaths,* that are not parts of joints. They occur in situations where two structures that must move relative to each other are in tight opposition; such situations are present where tendons are deflected around bone or where the skin must move freely over bony tissue, such as at the knee and elbow. The bursas in these locations are flattened sacs of synovial membranes supported by dense connective tissue; the membranes are separated by a thin film of synovial fluid and permit relatively frictionless movement (Fig. 7-6). Synovial sheaths surround some tendons (Fig 7-7): They are composed of a very narrow synovial cavity, which contains a thin film of fluid and is enclosed within synovial membranes. The cavity and its enclosing membranes encircle the tendon, forming a sleeve or sheath around it. They occur in the wrists and ankles where tendons pass over bones.

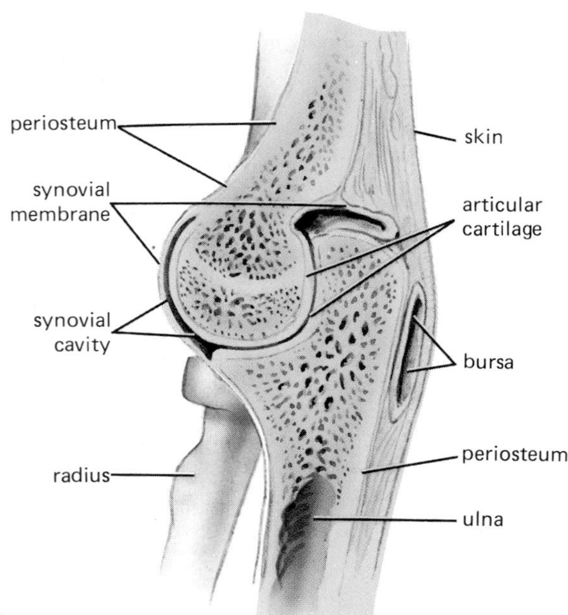

FIGURE 7-6 ***Bursa in the elbow joint.***

FIGURE 7-7 ***Diagram of a tendon sheath.***

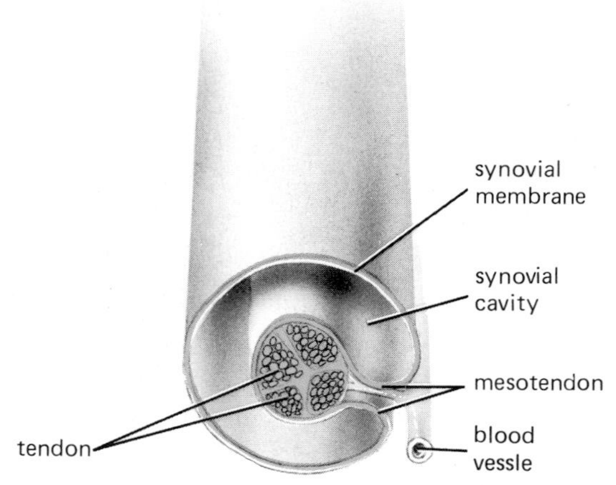

Disorders of Joints

The most common injuries to joints are dislocations and sprains. A *dislocation* is the actual displacement of the articulating surfaces of the bones; realignment may be difficult if the surrounding connecting tissues have been damaged or if the skeletal muscles attaching to the bones are in spasm. In a *sprain,* the bones are not displaced, but the ligaments attached to the bones may be torn or otherwise damaged. *Arthritis* is defined as inflammation of a joint. Any of the joint tissues may be involved, and the results are pain, swelling, and impaired movement. Arthritis can be caused by a large number of diseases; the cause is sometimes known (as in gouty arthritis, the result of accumulation of uric acid crystals in the joints), but in most cases the cause remains unclear.

MUSCLES: ATTACHMENTS AND FIBER ARRANGEMENTS

The contraction of skeletal muscles causes the movement of bones at joints. The two ends of a skeletal muscle are usually attached via tendons to the bones which meet at the joint. In general, contraction of the muscle produces relatively more movement of one bone than of the other. The muscle attachment to the bone of lesser movement is called the muscle's *origin;* that to the bone of greater movement is the *insertion.* Attachment via tendons, a *fibrous attachment,* allows the force of the muscle contraction to be concentrated on a small region of bone. A second type of attachment, the *fleshy attachment,* spreads the muscle's force over a wider area of bone; in this type, the muscle fibers seem to attach directly to bone but the attachment is actually via short connective-tissue elements.

The *force* that a muscle generates via its attachment depends on the number and thickness of its fibers, while the *maximum range* (i.e., amount of shortening) depends on the fibers' lengths. The fiber arrangement within the muscle is also important. Fascicles within a muscle may be parallel, oblique, or spiral relative to the muscle's line of pull (Fig. 7-8). Muscles with parallel fascicles may be flat, short, and rectangular (e.g., the thyrohyoid, Fig. 7-22B) or straplike (the sternohyoid, Fig. 7-22B, or sartorius, Fig. 7-42A). Straplike muscles, which are long and thin, are particularly useful in situations requiring a large range of motion and little power, as in stretching. When the fascicles are oblique to the muscle's line of pull, the muscles are classed as *triangular* (temporalis, Fig. 7-16B, or adductor longus, Fig. 7-24A) or *pennate* (featherlike). Examples of pennate muscles of different complexities are the flexor pollicus longus (Fig. 7-35A), rectus femoris (Fig. 7-42A), and deltoid (Fig. 7-34A). Pennate muscles frequently have a short range of motion, but the large number of fibers can generate considerable force. Spiral muscles (trapezius, Fig. 7-23) apply rotational forces to the bones to which they are attached. Many muscles have components that fit more than one of these categories.

FIGURE 7-8 ***Types of muscle based on their general form and arrangement of fascicles. Note that not all muscles have demonstrable tendons of origin and insertion. (From R. Warwick and P. L. Williams, "Gray's Anatomy," 35th Brit. ed., Saunders, Philadelphia 1973.)***

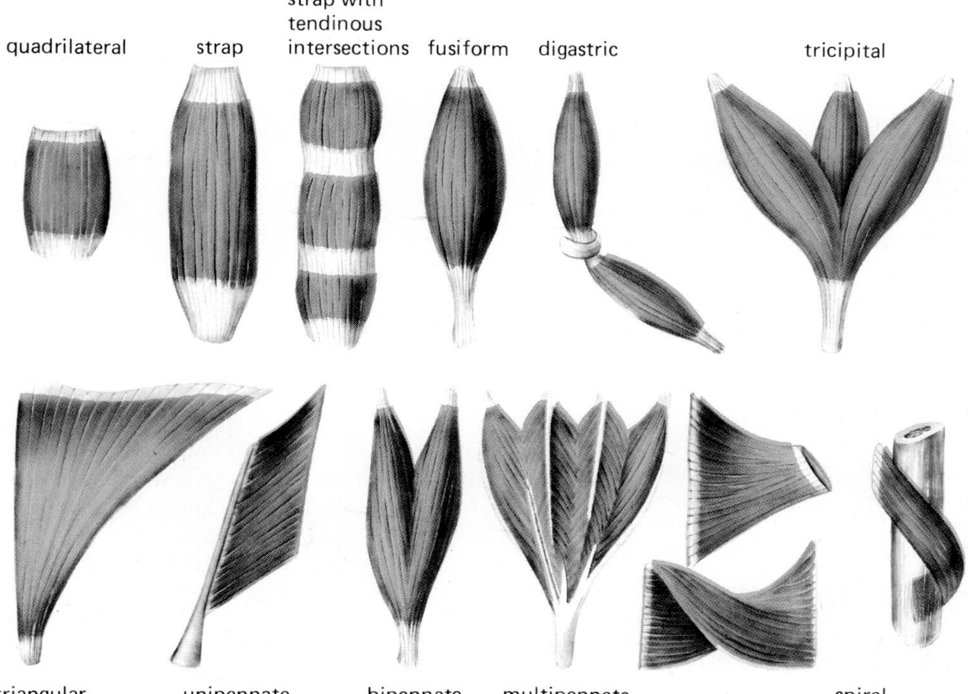

MECHANICS OF MOVEMENTS AT JOINTS

A contracting muscle exerts a force on bones through its connecting tendons. When the force is great enough, the bone moves as the muscle shortens. A contracting muscle exerts only a pulling force; as the muscle shortens, the bones attached to it are pulled toward each other. *Flexion* of a limb is its bending or movement toward the body and *extension* is straightening or movement away from the body. These motions require at least two separate muscles, one to cause flexion and the other extension. From Fig. 7-9 it can be seen how contraction of the biceps causes flexion of the forearm and contraction of the triceps causes its extension. Both muscles exert a pulling force upon the forearm when they contract. Groups of muscles which produce oppositely directed movements of a limb are known as *antagonists*. Other sets of antagonistic muscles are required to cause side-to-side movements or rotation of a limb. In some muscles contraction leads to two types

FIGURE 7-9 ***Antagonistic muscles for flexion and extension of the forearm.***

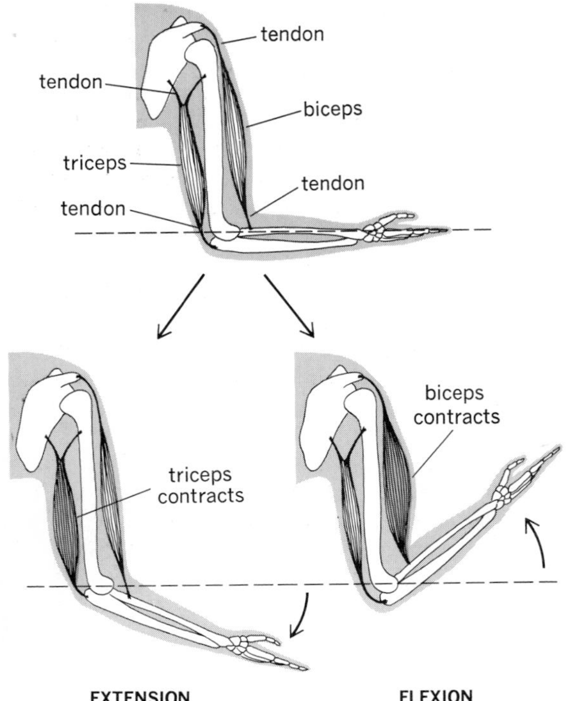

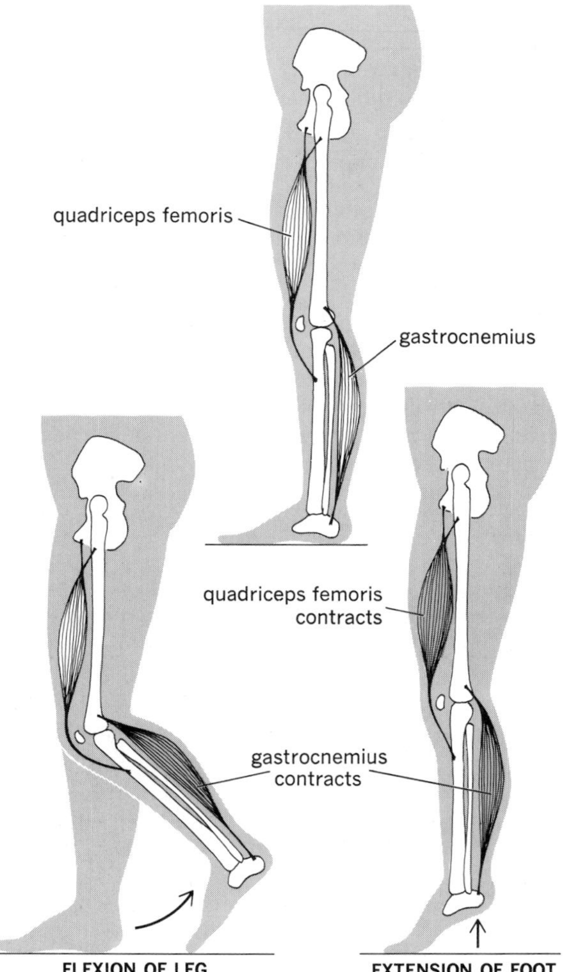

FIGURE 7-10 ***Flexion of the leg or extension of the foot follows contraction of the gastrocnemius muscle, depending on the activity of the quadriceps femoris muscle.***

of limb movement. In Fig. 7-10 contraction of the gastrocnemius muscle in the calf causes foot extension and flexion of the lower leg, as in walking. Contraction of the gastrocnemius at the same time as that of the quadriceps femoris (which causes extension of the lower leg) prevents the knee joint from bending, leaving only the ankle joint capable of moving; the foot is extended, and the body rises on tiptoe.

The arrangements of the muscles, bones, and joints in the body form lever systems. The basic principle of

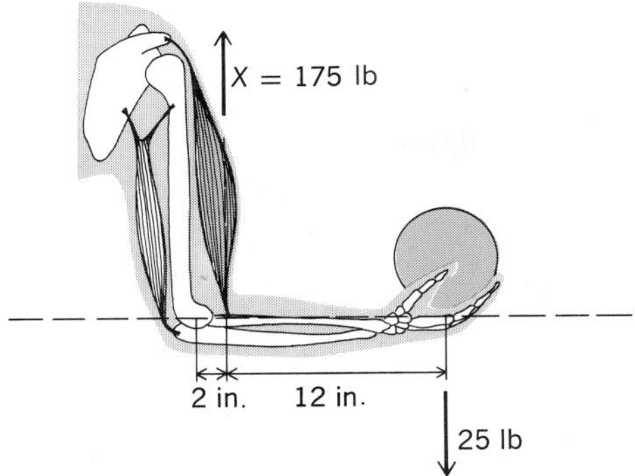

FIGURE 7-11 ***Mechanical equilibrium of forces acting on the forearm while supporting a 25-lb load.***

a lever can be illustrated by the flexion of the forearm by the biceps muscle (Fig. 7-11). The biceps exerts an upward pulling force on the forearm about 2 in away from the elbow. A 25-lb weight held in the hand exerts a downward force of 25 lb about 14 in from the elbow. It can be demonstrated according to the laws of physics that a rigid body (the forearm) is in mechanical equilibrium (not accelerating) if the product of the downward force (25 lb) and its distance from the elbow (14 in) is equal to the product of the upward force exerted by the muscle (X) and its distance from the elbow (2 in). From this relationship one can calculate the force the biceps must exert on the forearm to support a 25-lb load: $25 \times 14 = 2X$, thus $X = 175$ lb. Such a system is working at a mechanical disadvantage since the force exerted by the muscle is considerably greater than the load it is supporting.

Contracting muscles exert much more pressure on the bones than the physical weight of the body does. During quiet standing, the forces exerted by the muscles on the hip joint are six times greater than the forces applied by the body's weight. In powerful exertion, the forces exerted by the muscles are even larger. In athletes with very powerful muscles, the large forces exerted by contracting muscles under conditions of maximum exertion sometimes tear the tendon away from the muscle or bone, or in rare cases break the bone.

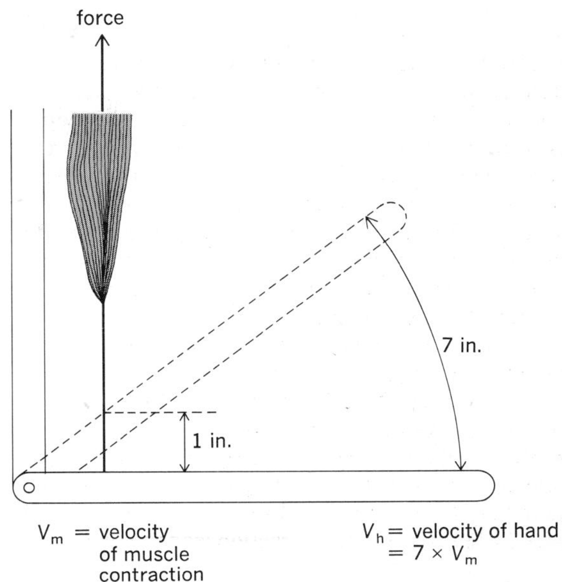

FIGURE 7-12 ***Small movements of the biceps muscle are amplified by the lever system of the arm, producing large movements of the hand.***

The mechanical disadvantage under which the muscles operate is offset by increased maneuverability. In Fig. 7-12, when the biceps shortens 1 in, the hand moves through a distance of 7 in. Since the muscle shortens 1 in in the same amount of time that the hand moves 7 in, the velocity at which the hand moves is seven times faster than the rate of muscle contraction. The lever system of the arm amplifies the movements of the muscle. Short, relatively slow movements of the muscle produce longer and faster movements of the hand. Thus, a pitcher can throw a baseball at 100 mi/hr even though his muscles shorten at only a fraction of this velocity. Skeletal muscles shorten at the rate of about 5 to 10 muscle lengths per second. Thus, the longer the muscle, the faster is its velocity of shortening.

This completes our analysis of the general characteristics of the musculoskeletal system. We now describe the specific bones, muscles, and joints of the body.

SECTION B
Atlas of the Musculoskeletal System

The Musculoskeletal System of the Shoulder Girdle and Arm

The Musculoskeletal System of the Pelvic Girdle and Leg

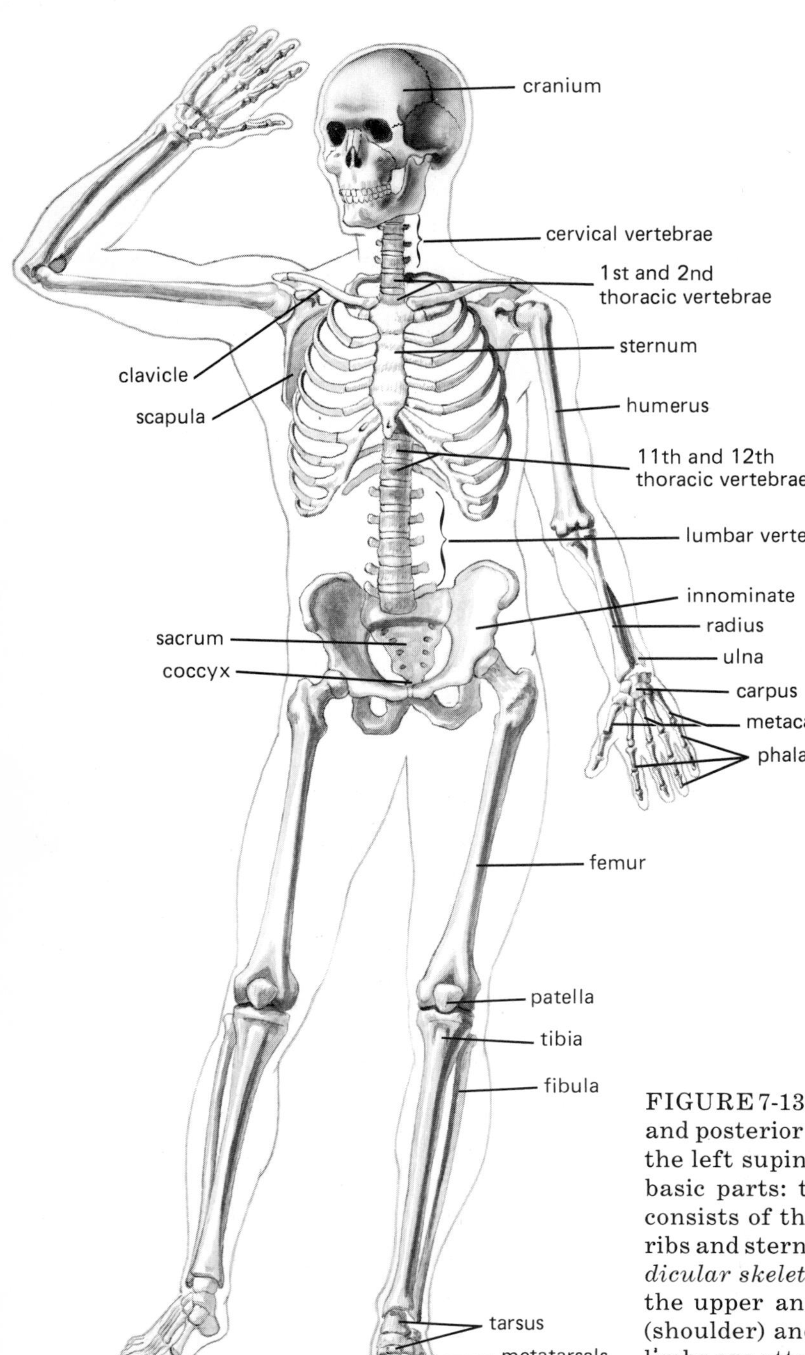

FIGURE 7-13 The human skeleton, anterior (A) and posterior (B) views. The right hand is prone, the left supine. The skeleton is divided into two basic parts: the primary *axial skeleton*, which consists of the skull, vertebral column, and the ribs and sternum (or breastbone); and the *appendicular skeleton*, which consists of the bones of the upper and lower limbs and of the pectoral (shoulder) and pelvic (hip) girdles by which the limbs are attached to the axial skeleton. The 206 bones of the human skeleton are listed in Table 7-3.

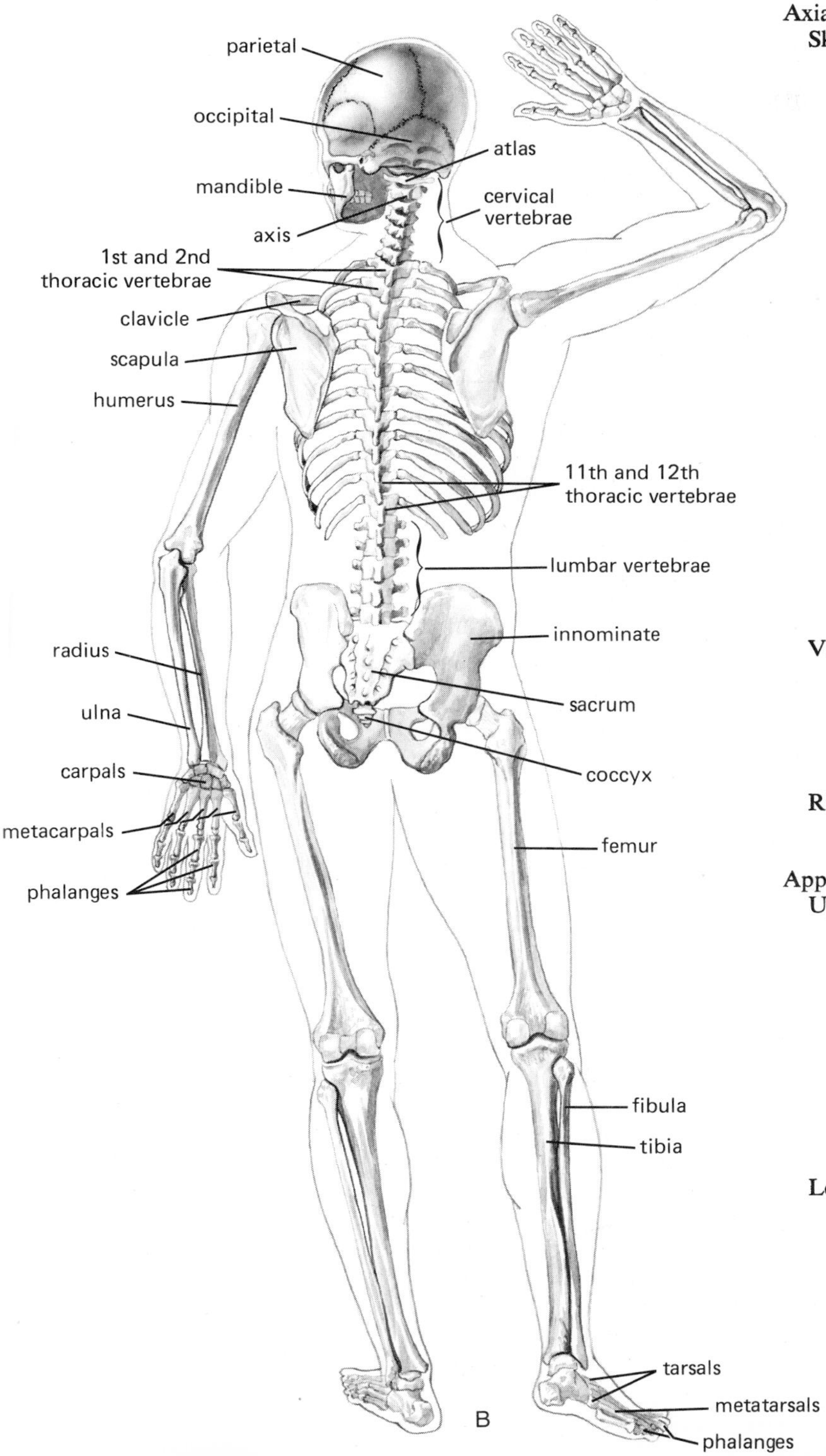

TABLE 7-3

BONES OF THE HUMAN SKELETON

	Number of bones			
Axial skeleton				80
Skull			29	
Cranium (brainbox)		8		
Parietal	2			
Temporal	2			
Frontal	1			
Occipital	1			
Sphenoid	1			
Ethmoid	1			
Face		14		
Maxilla	2			
Zygomatic	2			
Nasal	2			
Lacrimal	2			
Palatine	2			
Inferior nasal conchae	2			
Mandible	1			
Vomer	1			
Others		7		
Auditory ossicles				
Malleus	2			
Incus	2			
Stapes	2			
Hyoid	1			
Vertebral column			26	
Cervical vertebrae		7		
Thoracic		12		
Lumbar		5		
Sacral (5 fused)		1		
Coccygeal (usually 4 fused)		1		
Rib cage			25	
Ribs		24		
Sternum (3 fused)		1		
Appendicular skeleton				126
Upper limb			64	
Pectoral girdle		4		
Clavicle	2			
Scapula	2			
Arm		60		
Humerus	2			
Radius	2			
Ulna	2			
Carpal	16			
Metacarpal	10			
Phalanges	28			
Lower limb			62	
Pelvic girdle (3 fused)		2		
Leg		60		
Femur	2			
Patella	2			
Tibia	2			
Fibula	2			
Tarsal	14			
Metatarsal	10			
Phalanges	28			
Total				206

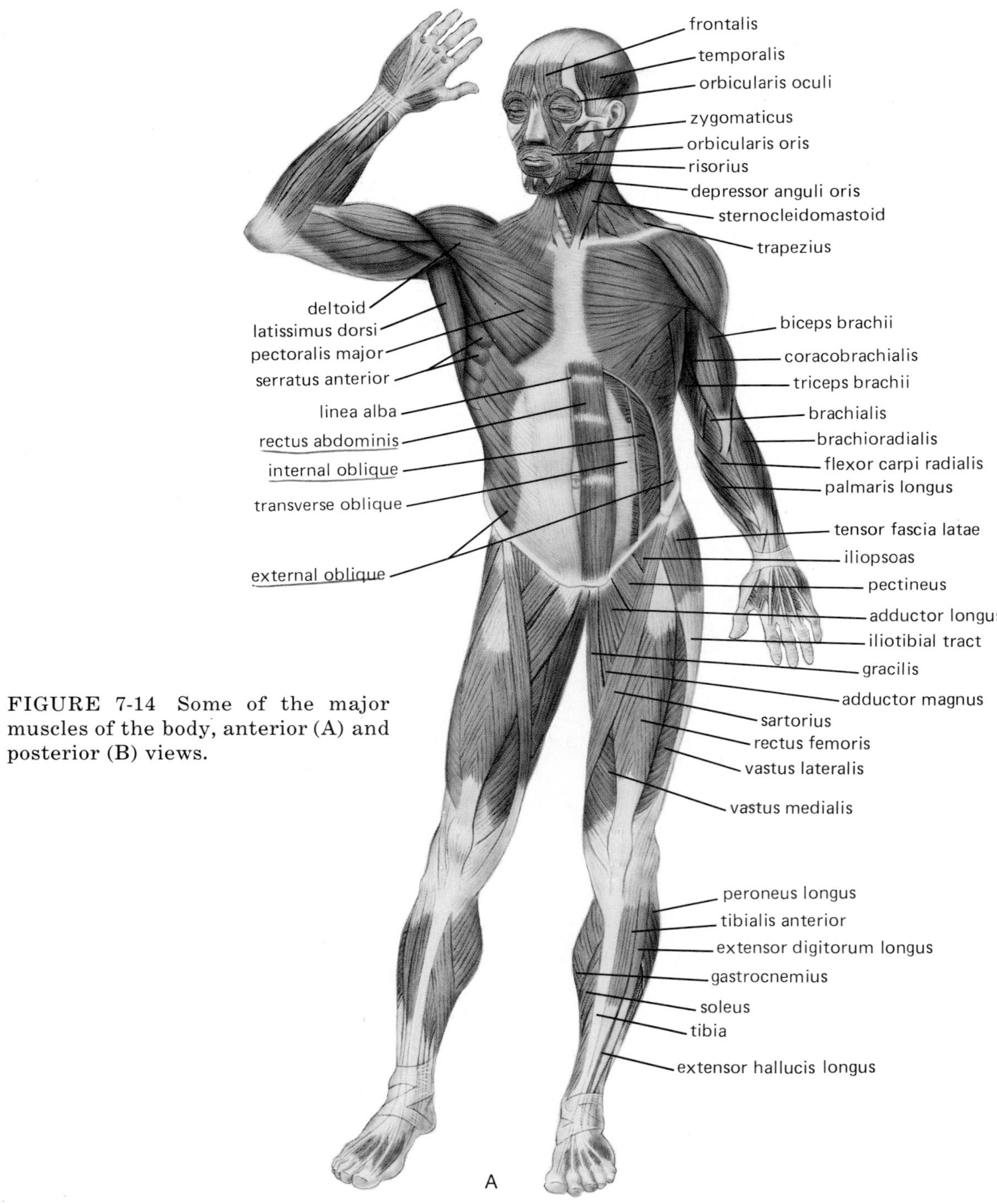

FIGURE 7-14 Some of the major muscles of the body, anterior (A) and posterior (B) views.

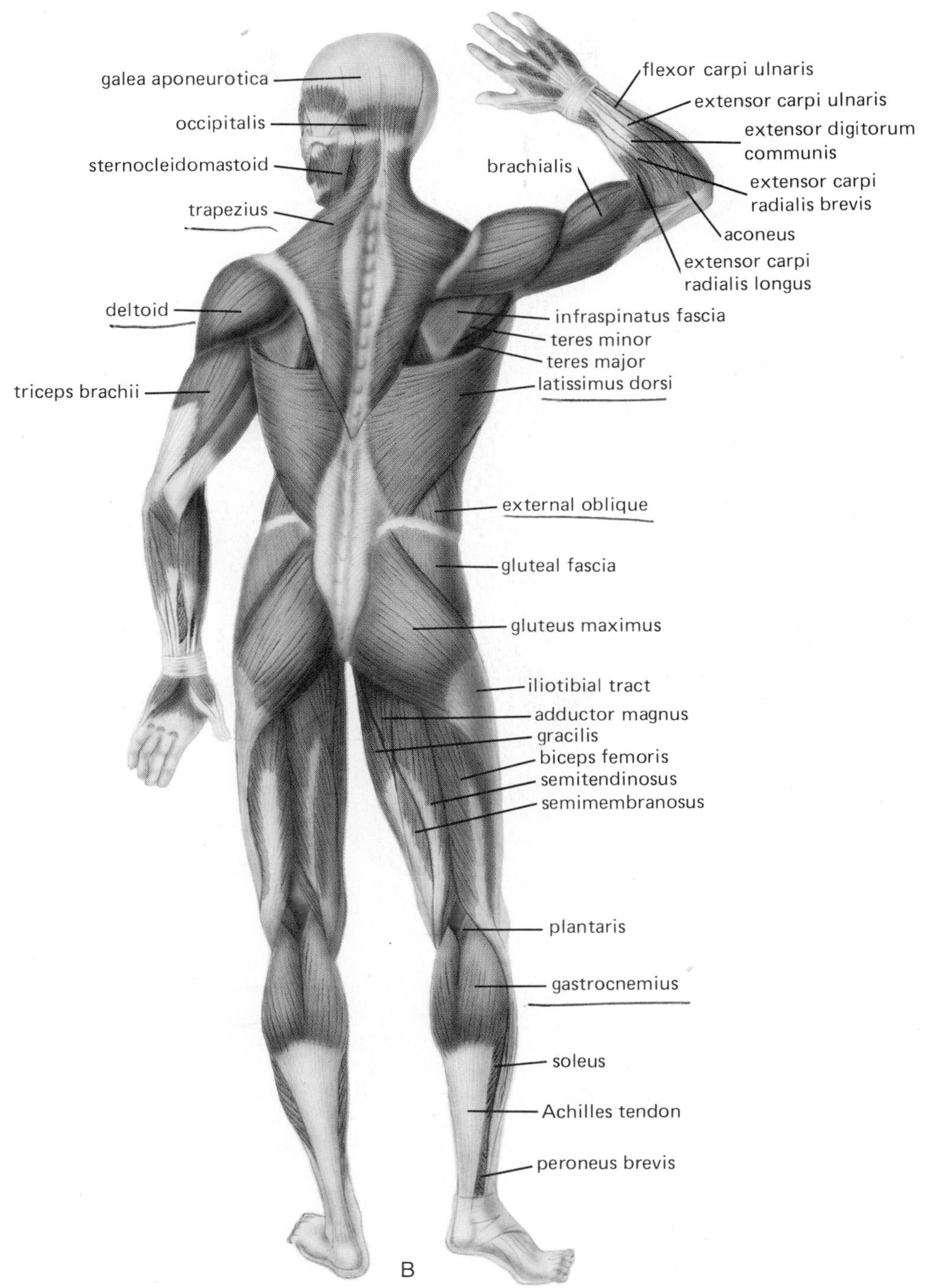
galea aponeurotica
occipitalis
sternocleidomastoid
trapezius
deltoid
triceps brachii
flexor carpi ulnaris
extensor carpi ulnaris
extensor digitorum communis
brachialis
extensor carpi radialis brevis
aconeus
extensor carpi radialis longus
infraspinatus fascia
teres minor
teres major
latissimus dorsi
external oblique
gluteal fascia
gluteus maximus
iliotibial tract
adductor magnus
gracilis
biceps femoris
semitendinosus
semimembranosus
plantaris
gastrocnemius
soleus
Achilles tendon
peroneus brevis

B

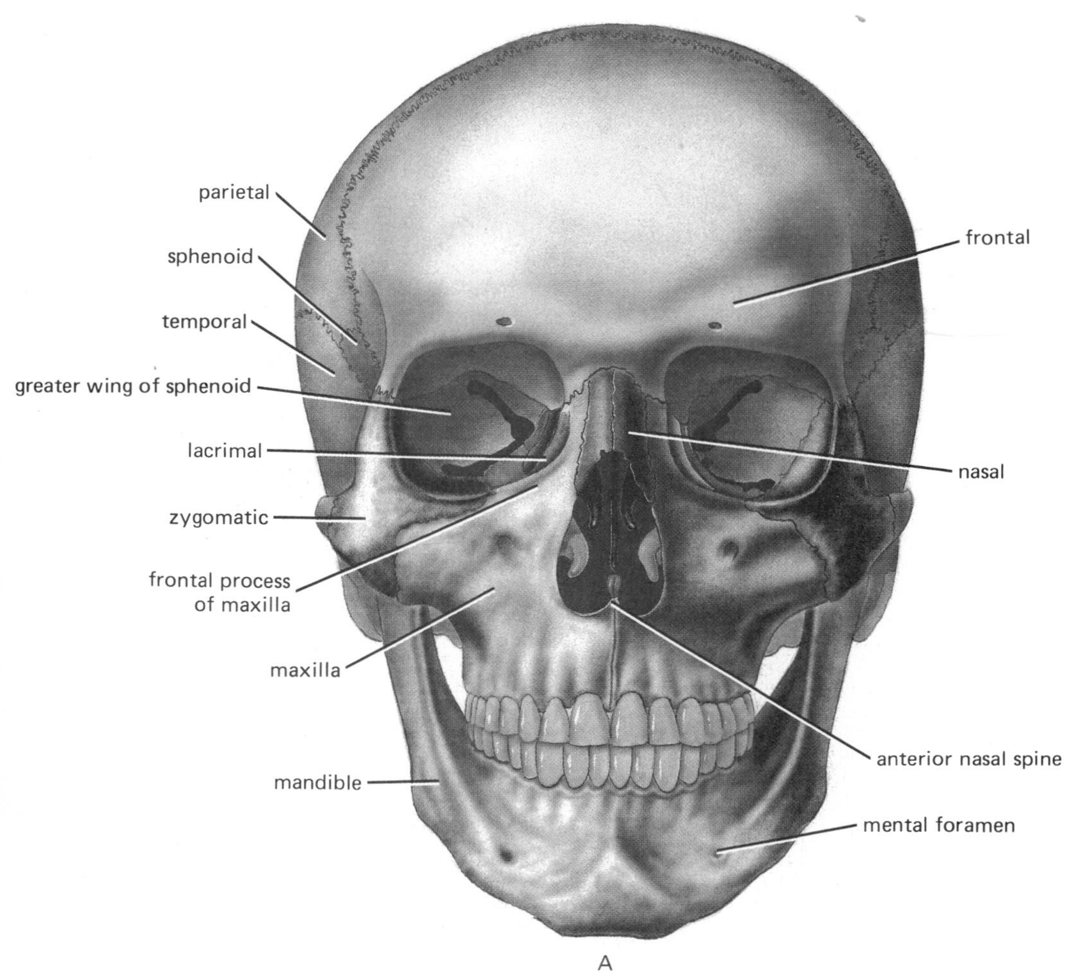

FIGURE 7-15 The skull. Views of the anterior (A), left lateral (B), and superior (C) aspects of the skull and an internal (D) and external (E) view of the base of the skull.

The skull is a complex of bones adapted to support and protect the brain and several of the special senses, to isolate the cerebral blood vessels from extracranial variations in pressure, and to get and process food. To serve these functions the skull has two regions: a 'brainbox', called the *calvaria* or *neurocranium*, and a facial skeleton.

The anterior part of the neurocranium (Fig. 7-15A) is the forehead, formed by the *frontal bone* which passes back over the top of the skull as far as the *coronal suture* (Fig. 7-15B and C) where it meets the right and left *parietal bones*. These two bones meet in the midline at the top of the skull forming the *sagittal suture* and extend back to meet the *occipital bone* at the *lambdoid suture*. The parietal bones extend laterally to contact the greater wing of the *sphenoid bone* and the squamous part of the *temporal bone* (Fig. 7-15B).

When the skull is viewed from the side (Fig. 7-15B), the lateral portion of the *mandible* can be seen. The *condylar process* at the posterior portion of the bone fits into the articular fossa on the under surface of the squamous portion of the

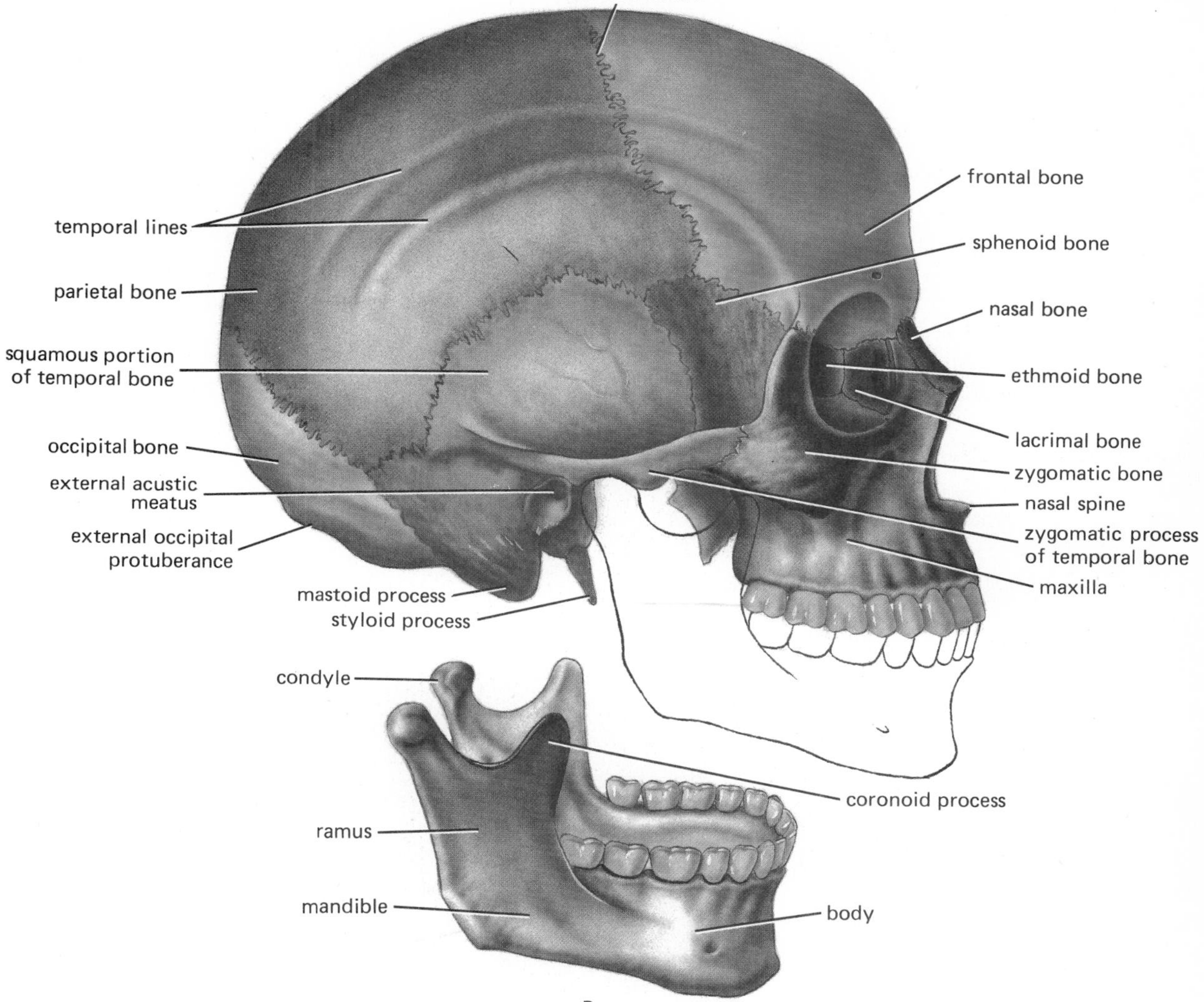

coronal suture
temporal lines
parietal bone
squamous portion of temporal bone
occipital bone
external acustic meatus
external occipital protuberance
mastoid process
styloid process
frontal bone
sphenoid bone
nasal bone
ethmoid bone
lacrimal bone
zygomatic bone
nasal spine
zygomatic process of temporal bone
maxilla
condyle
coronoid process
ramus
mandible
body

B

temporal bone. Just behind the condylar process in the temporal bone is the *external acoustic meatus*, or ear passage. Above and in front of the acoustic meatus the *zygomatic process* of the temporal bone arches forward to meet the *zygomatic*, or cheek, *bone*, the two together forming the *zygomatic arch*. If the mandible is removed, a small bony projection can be seen extending downward from the sphenoid bone; this is the *pterygoid hamulus*.

When the top of the calvaria and the brain are removed to expose the floor of the skull (Fig. 7-15D), one can see three distinct shallow depressions—the *anterior*, *middle*, and *posterior cranial fossae*. The floor of the anterior fossa forms the roof of the eye sockets (orbits) and nasal cavity. The frontal bone forms most of the orbital roofs, but a narrow strip of perforated bone, the *cribriform plate of the ethmoid bone*, forms the roof of the nasal cavity. The olfactory nerve fibers pass through these holes on their way from the nose to the brain. The rest of the floor of the anterior fossa is formed by the sphenoid bone. The middle fossa has a narrow central region but expands on each side. The central part, formed by the sphenoid bone, contains a hollow (the *sella turcica*) which houses the pituitary gland. The lateral parts are formed by the

FRONTAL
BONE
coronal suture
PARIETAL
BONE
sagittal suture
lambdoid suture
OCCIPITAL
BONE

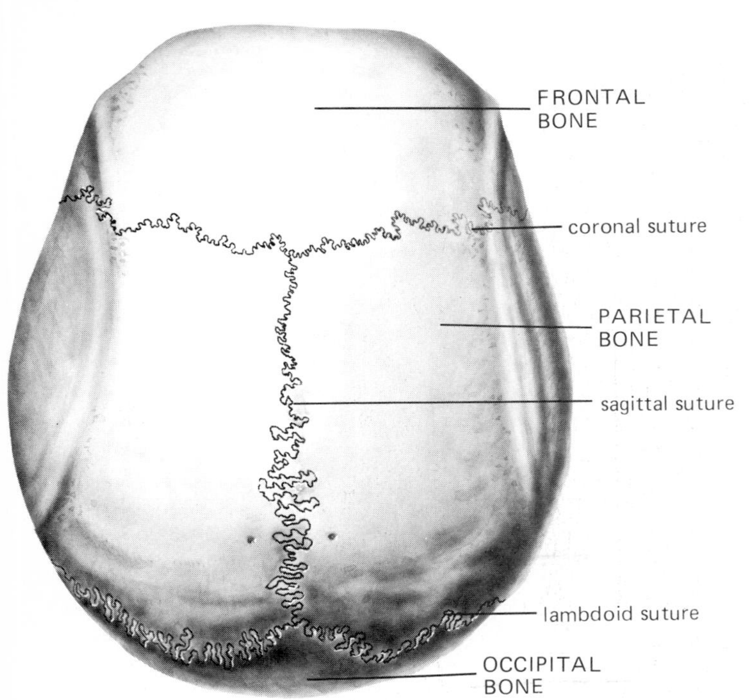

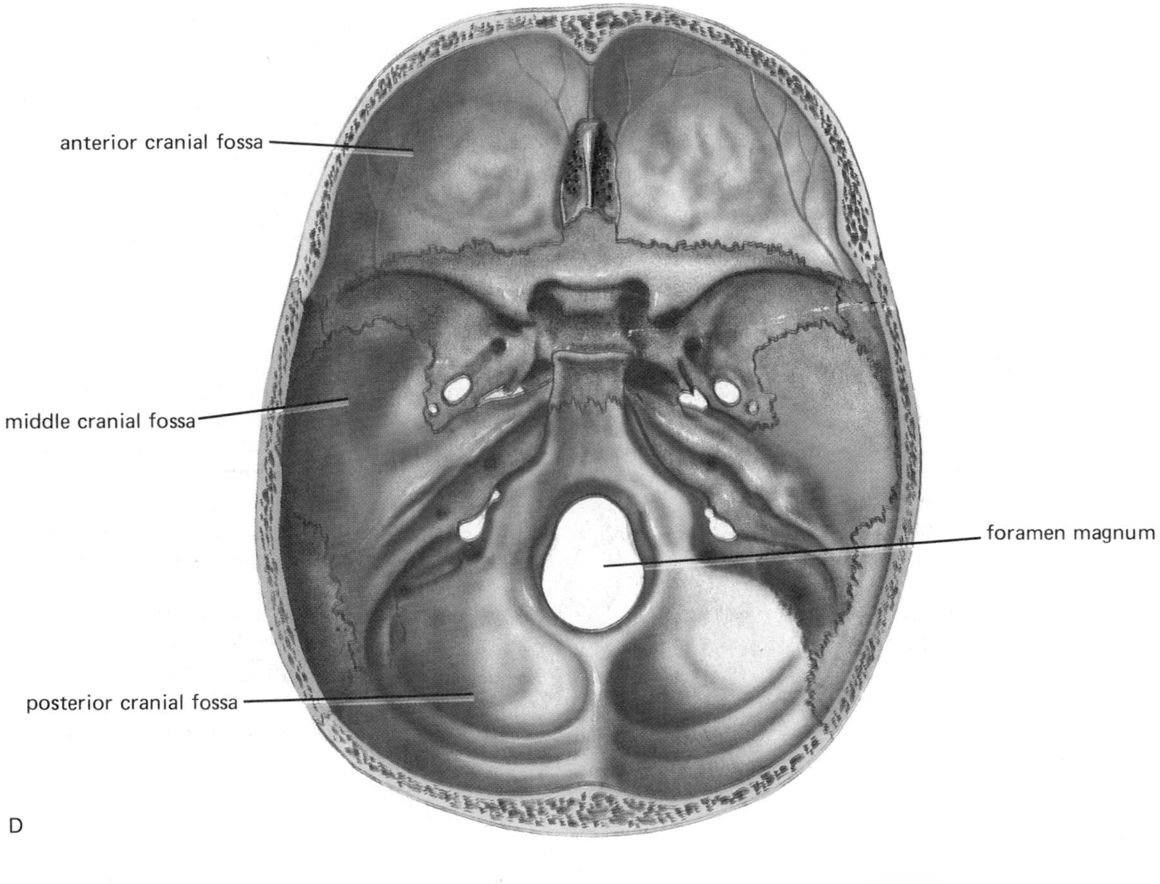

anterior cranial fossa
middle cranial fossa
foramen magnum
posterior cranial fossa

D

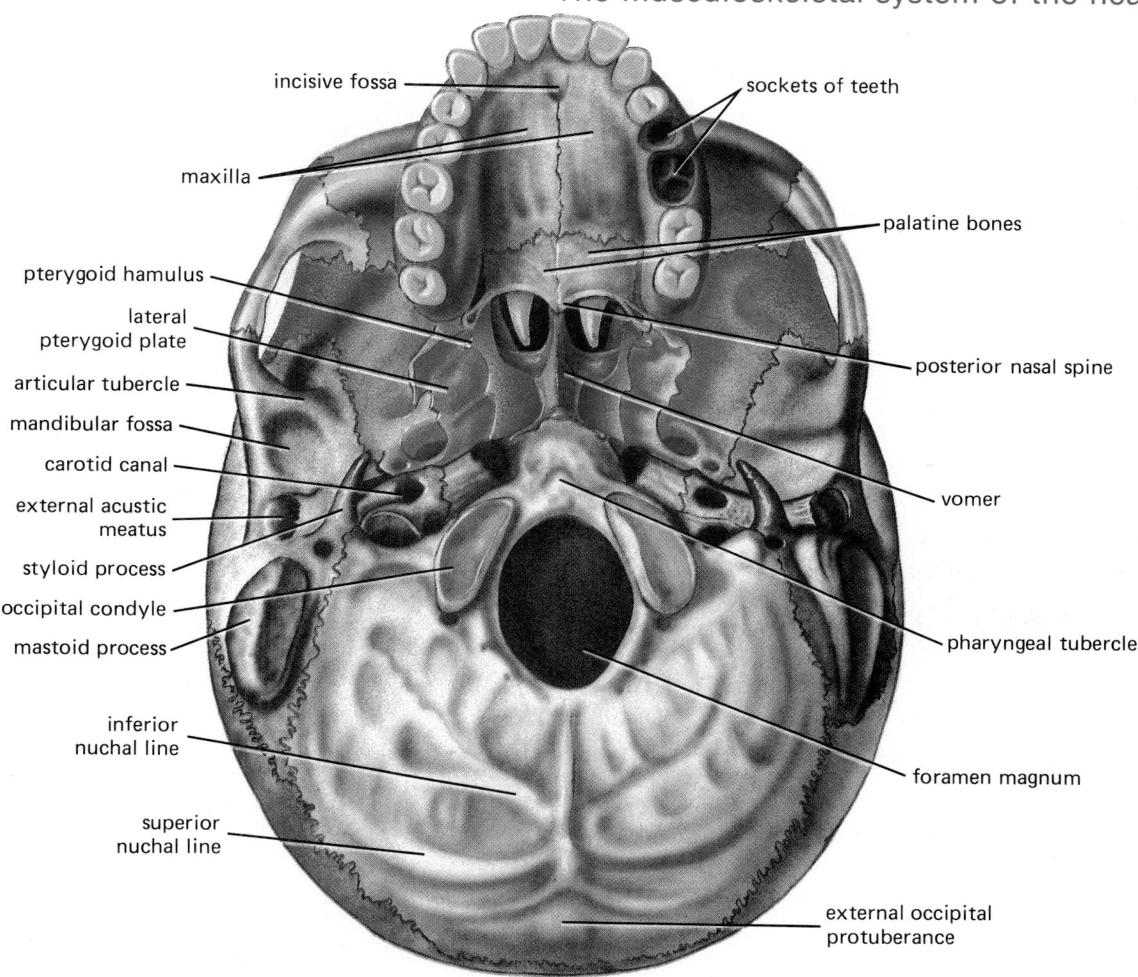

E

sphenoid bone and petrous parts of the temporal bones. The sphenoid curves up along the sides of the skull to meet the parietal bones (Fig. 7-15B). The floor of the posterior cranial fossa is formed to a large extent by the occipital bone (Fig. 7-15B). The large opening in this fossa, the *foramen magnum*, lies entirely within this bone. The posterior fossa accommodates the cerebellum, pons, and medulla of the brain.

The inferior, or *base*, of the skull (Fig. 7-15E) is very irregular. The occipital bone forms the largest part of it. On each side of the foramen magnum is an *occipital condyle*, which articulates with the lateral mass of the atlas (first cervical vertebra). Lateral to the condyles, the occipital bone articulates with the petrous portion of the temporal bones. The posterior part of the occipital bone forms the *external occipital crest*, to which is attached the upper end of the *ligamentum nuchae*. Two ridges are apparent on the bone, the *superior* and *inferior nuchal lines*, to which several muscles are attached. In the anterior part of the base of the skull, the *bony palate* can be seen (the mandible has been removed). It is made up of four bones, two maxillae and two palatine bones. The anterior portion of the palate is formed by the *palatine processes* of the two *maxillary bones* and the posterior portion by the *horizontal plates* of the *palatine bones*.

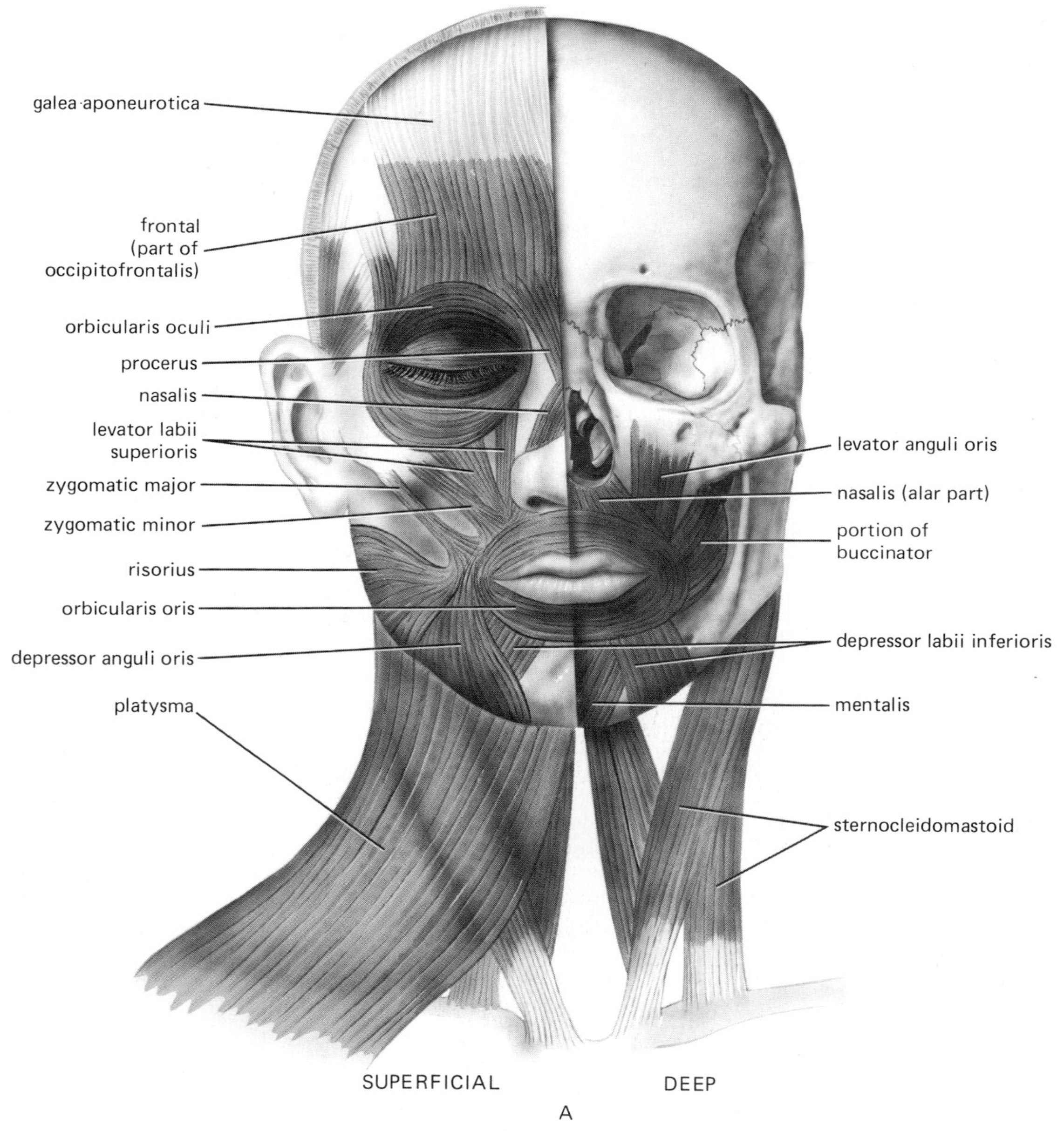

FIGURE 7-16 Muscles of the face, anterior (A) and right lateral (B) views. The superficial muscles are shown on the left half of part A, the deep muscles on the right. Many of the muscles of the face serve facial expressions, one of the most revealing outward signs of inner emotional states. Others move the eyes and control the intricate lip and cheek movements associated with speech, eating, and drinking. The muscles of facial expression differ from most skeletal mus-

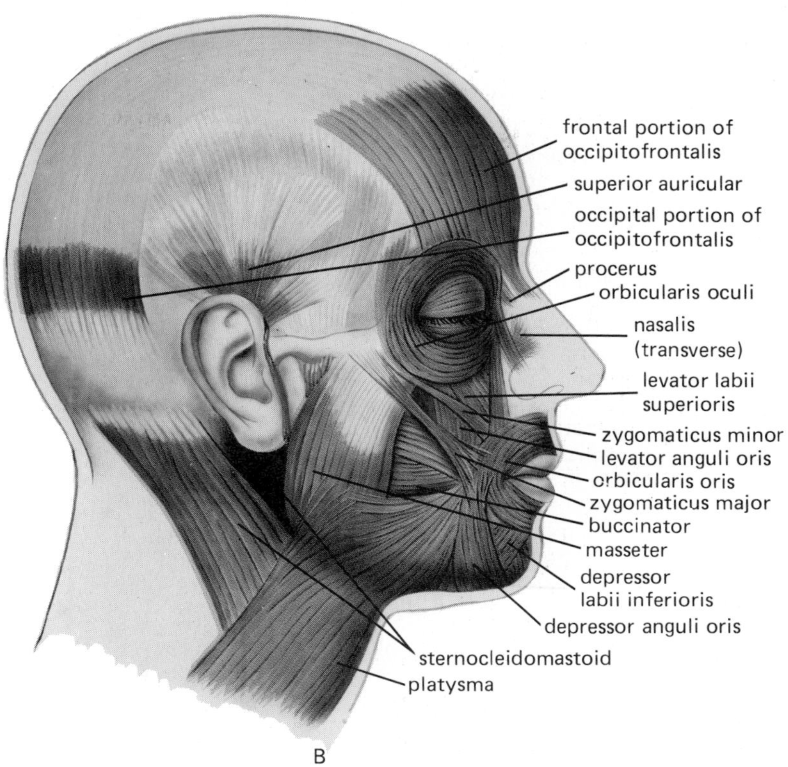

cles in that while their origin is most frequently on bone, their insertion is on skin and its underlying connective tissue. The muscles of the face are listed in Table 7-4 according to their location. The muscles that control the eyeball are listed in Table 9-1. The muscle of the scalp (*occipitofrontalis*, or *epicranius*) contains two frontal and two occipital parts in the forehead and occipital regions, respectively. They are joined by a strong membranous sheet, the *galea aponeurotica*. The shape of the mouth and position of the lips are controlled by elevators and retractors of the upper lip, depressors and retractors of the lower lip, and a sphincter, which surrounds the mouth opening.

TABLE 7-4
MUSCLES OF THE FACE

Muscle	*Figure*	*Nerve*	*Origin*	*Insertion*	*Action and comments*
Muscle of the scalp: Occipitofrontalis (Epicranius)					
Occipital parts	7-16B	Posterior auricular branch of VII	Occipital bone, mastoid process of temporal bone	Galea aponeurotica	Draws scalp backward
Frontal parts	7-16A and B	Temporal branch of VII	Galea aponeurotica	Skin of the eyebrows, and the root of the nose	Elevates the eyebrows, wrinkles forehead
Muscles of the eyelids:					
Orbicularis oculi	7-16A and B	Temporal and zygomatic branches of VII	Frontal, maxillary, and zygomatic bones and medial palpebral ligament	Skin around eye, lateral palpebral ligament	An elliptical muscle which occupies the eyelids, surrounds the orbit, and spreads on to the temporal region and cheek; closes eyelids
Corrugator	Not shown	Temporal branch of VII	Brow ridge of frontal bone	Skin of eyebrow	Draws eyebrows together
Muscles of the nose:					
Procerus	7-16A and B	Buccal branches of VII	Lower part of nasal bone, lateral nasal cartilage	Skin between the eyebrows	Wrinkles bridge of nose
Nasalis	7-16A and B	Upper buccal branch of VII	Maxilla next to incisor and canine teeth	Bridge and side of nose	Has transverse and alar parts. Widens anterior nasal aperture, especially in deep inspiration

TABLE 7-4
MUSCLES OF THE FACE *(Continued)*

Muscle	*Figure*	*Nerve*	*Origin*	*Insertion*	*Action and comments*
Muscles of the mouth:					
Levator labii superioris	7-16A and B	Buccal branches of VII	Maxilla and zygomatic bone	Muscles in upper lip	Raises upper lip and turns it outward
Zygomaticus					
Minor	7-16A and B	Buccal branches of VII	Zygomatic bone	Muscles in upper lip	Elevates upper lip
Major	7-16A and B	Zygomatic branches of VII	Zygomatic bone	Muscles at angle of mouth	Draws angle of mouth upward and outward
Levator anguli oris	7-16A and B	Buccal branches of VII	Canine fossa of maxilla	Muscles at angle of mouth	Raises angle of mouth
Mentalis	7-16A	Mandibular marginal branch of VII	Incisor fossa of mandible	Skin of chin	Raises and protrudes lower lip, wrinkles skin of chin
Depressor labii inferioris	7-16A and B	Mandibular marginal branch of VII	Mandible	Skin and muscles of lower lip	Draws lower lip downward and a little laterally
Depressor anguli oris	7-16A and B	Mandibular marginal branch of VII	Mandible	Muscles at angle of mouth	Draws angle of mouth downward and laterally
Buccinator	7-16A and B, 7-22C	Buccal branches of VII	Mandible and maxilla in region near molars	Muscles at angle of mouth	Compresses cheeks against teeth, provides a stable lateral wall to oral cavity for pressure in speech, sucking, and mastication
Orbicularis oris	7-16A and B	Buccal branches of VII	Muscle fibers surrounding mouth	Fibers encircle mouth; some attach to skin and muscles at angle of mouth	Closes lips, presses lips against teeth, protrudes lips, and shapes lips in speech
Risorius	7-16A and B	Buccal branches of VII	Fascia of masseter muscle	Skin at angle of mouth	Retracts angle of mouth

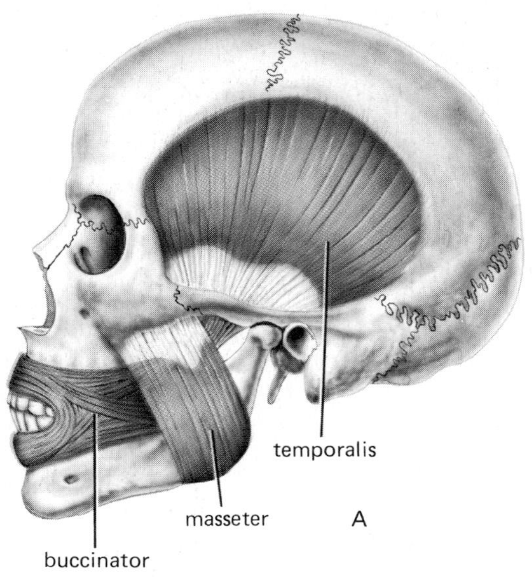

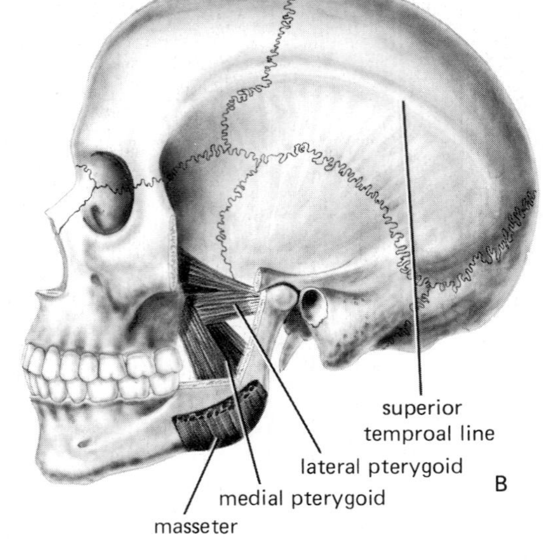

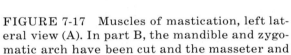

FIGURE 7-17 Muscles of mastication, left lateral view (A). In part B, the mandible and zygomatic arch have been cut and the masseter and temporalis partially or completely removed to show the underlying muscles.

TABLE 7-5

MUSCLES OF MASTICATION (CHEWING)

Muscle	*Figure*	*Nerve*	*Origin*	*Insertion*	*Action and comments*
Masseter	7-17	Mandibular division of V	Maxilla, zygomatic arch (temporal and zygomatic bones)	Lateral surface of angle of mandible	Elevates mandible, closing jaw; small effect in lateral movements or protrusion
Temporalis	7-16B, 7-17	Mandibular division of V	Temporal fossa	Coronoid process and ramus of mandible	Elevates mandible, closing jaw; draws mandible backward after protrusion, assists in lateral movements
Lateral pterygoid	7-17	Mandibular division of V	Greater wing of the sphenoid bone and lateral pterygoid plate	Disc and neck of mandible	Assists in opening mouth, protrusion of jaw, and grinding movements
Medial pterygoid	7-17	Mandibular division of V	Pterygoid plate, palatine bone, and maxilla	Ramus and medial surface of angle of mandible	Assists elevation and protrusion of mandible and side-to-side movements of jaw

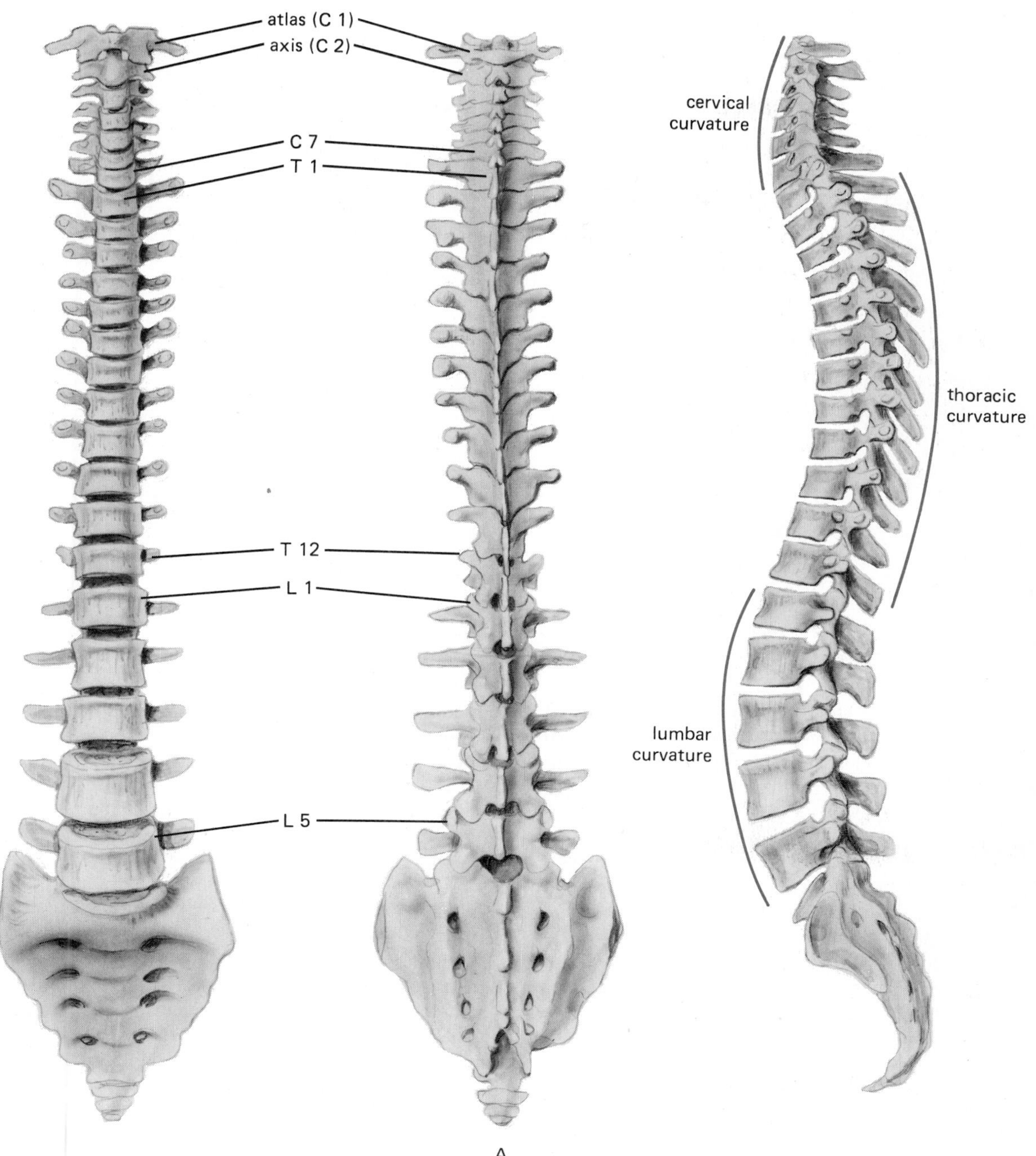

FIGURE 7-18 (A) The vertebral column: anterior, posterior, and lateral views. There are 26 vertebrae: 7 cervical, 12 thoracic, 5 lumbar, 1 sacrum (formed of 5 fused sacral vertebrae), and 1 coccyx (usually consists of 4 fused rudimentary vertebrae, but the number may be 3 or 5).

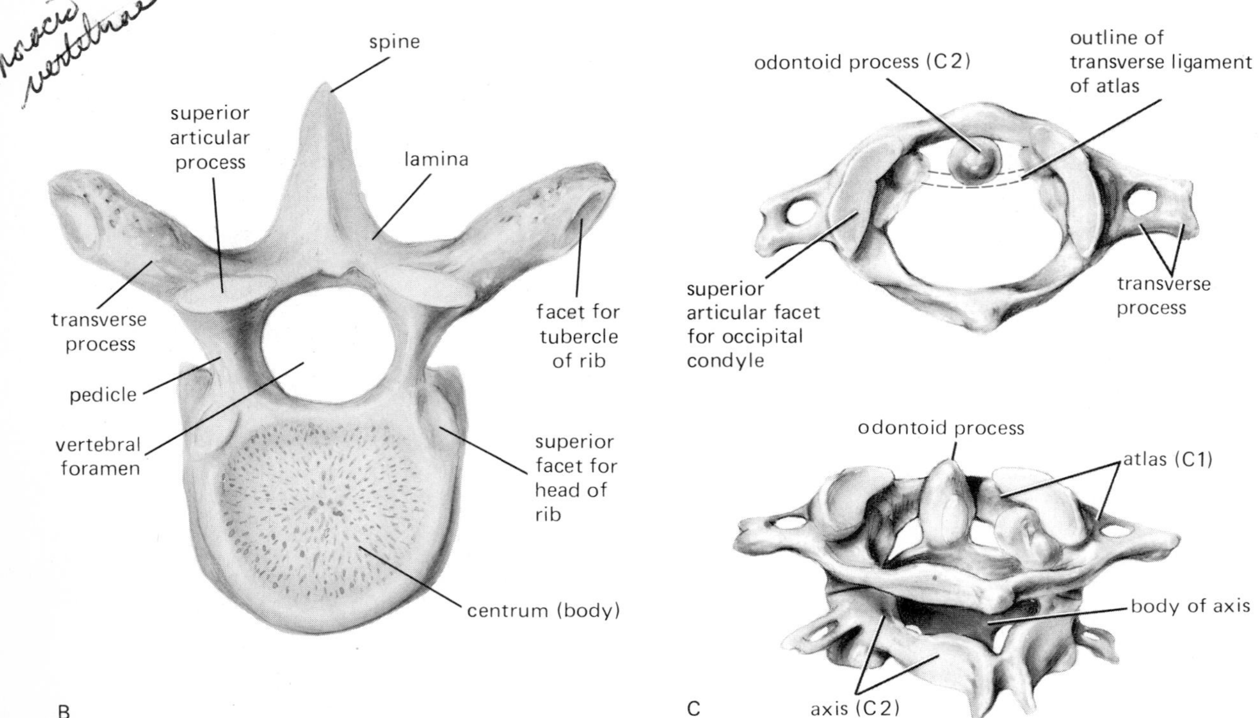

(B) Superior view of a characteristic vertebra. It consists basically of a large *body*, which lies ventral to the spinal cord, and an *arch*, which extends from the back of the body, enclosing a space called the *vertebral foramen* in which the spinal cord with its associated coverings and blood vessels is situated. The arch has two sections: lateral *pedicles* through which it joins the body, and behind these, the *lamina*. Where the two lamina fuse together a midline *spinous process* occurs; it is these "spines" that can be felt or seen protruding along the midline of the back. Extending somewhat to the sides of the spinous process are pairs or *transverse* and *superior and inferior articular processes*.

When the vertebrae are in place in the column, the inferior articular processes, which bend downward, meet the upward jutting superior processes of the vertebrae just below. Between these joined processes and the main column of the vertebral bodies and discs is an opening, the *intervertebral foramen* (Fig. 7-18A), through which the peripheral nerves pass. Most of the vertebrae are similar enough to our "characteristic vertebra" that they do not need separate discussions. However, the atlas, axis, sacrum, and coccyx deserve mention. The *atlas* (Fig. 7-18C) is the first cervical vertebra; it received its common name because it literally supports the "globe" of the head. It is unusual in that it lacks a vertebral body, the space of the missing body being filled in by an upward projection from the body of the second cervical vertebra, the *axis*. The odontoid process of the axis forms, in fact, a pivot around which the atlas and head rotate. The condyles of the occipital bone of the skull sit on the facets of the two lateral masses of the atlas, rocking in the facets during nodding movements of the head.

The *sacrum* (Fig. 7-18D) is a large triangular structure formed by the fusion of the five sacral vertebrae, the wide upper part articulating with the fifth lumbar vertebra and the narrow lower end with the coccyx. The bony characteristics of

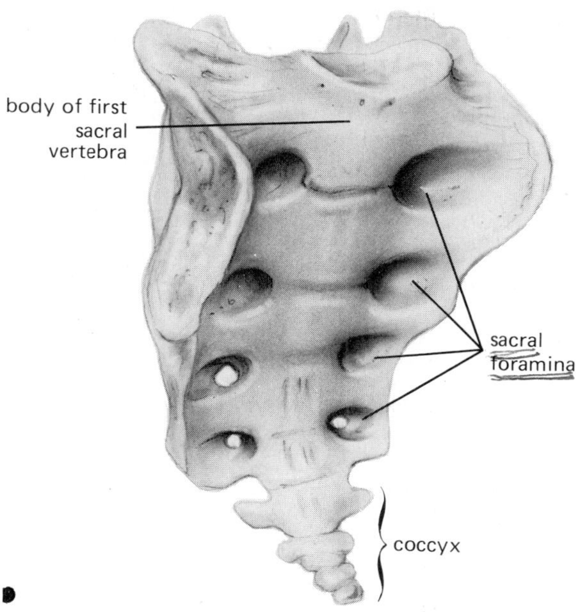

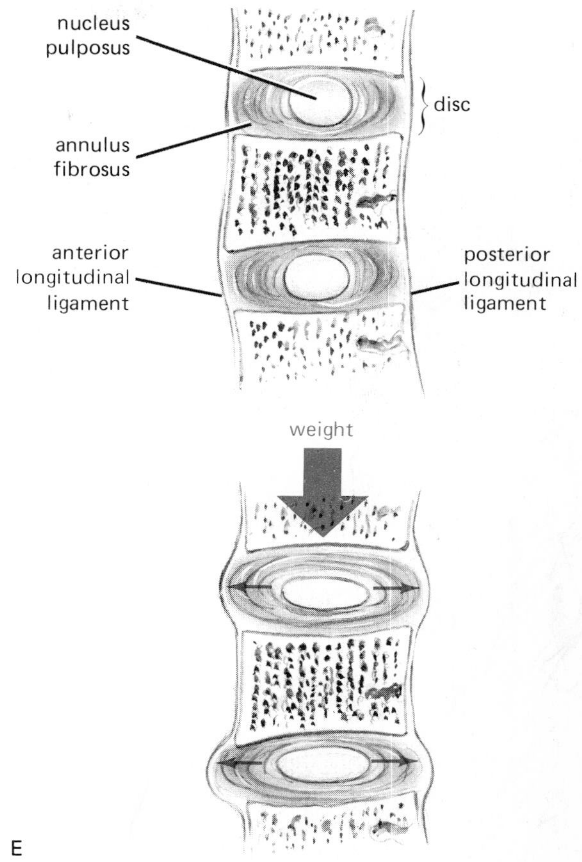

the sacrum reflect the fusion of the bodies, processes, and laminae of the individual sacral vertebrae. The *sacral canal*, formed by the vertebral foramina of the individual vertebrae, passes through the sacrum. Typically the sacra of males and females differ, that of the female being shorter and wider and having a deeper concavity than that of the male.

The *coccyx* (Fig. 7-18D) is a small triangular bone, consisting of from three to five rudimentary vertebrae, the second, third, and fourth coccygeal vertebrae being simply tiny nodules of bone. All the coccygeal vertebrae may be fused together, but the first often remains separate. The coccygeal vertebrae are solid, lacking a central canal.

The upper and lower surfaces of adjacent vertebral bodies are strongly bound to each other by fibrocartilage *intervertebral discs*, which lend a certain degree of flexibility to the complete vertebral column and act as shock absorbers, becoming temporarily flatter and broader and bulging from between the vertebrae when they are compressed (Fig. 7-18E). This is facilitated by the presence of a semifluid fibrogelatinous mass (the *nucleus pulposus*) within each disc. However, sudden heavy strains can cause the nucleus to break through the ring of cartilaginous fibers (the *annulus fibrosus*) surrounding it; this is called a *ruptured disc*. Discs occur between all the vertebrae down to the space between the last lumbar and first sacral vertebrae; they account for one-fourth of the total length of the spinal column.

Anterior and *posterior longitudinal ligaments* join the bodies of adjacent vertebrae while strong *ligamenta flava* join the laminae. *Supraspinal ligaments* join the spinous processes below the seventh cervical vertebra, but above that level they give way to the *ligamentum nuchae*, which is a triangular, sheet-like ligament extending up to attach to the skull.

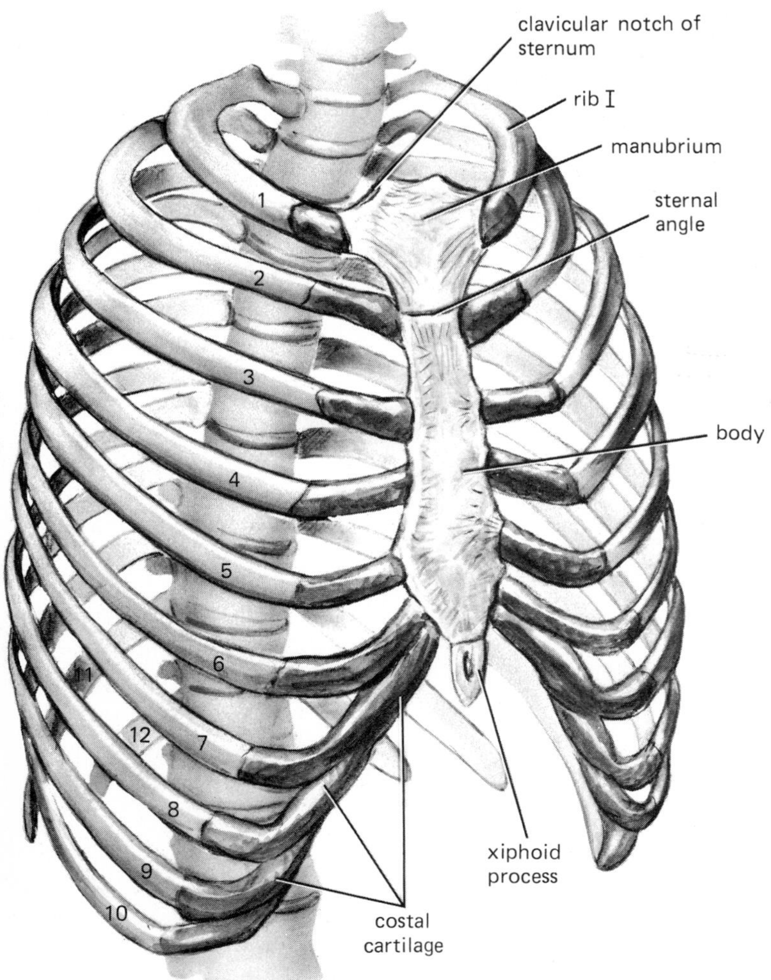

FIGURE 7-19 A three-quarter view of the skeleton of the chest. The skeleton of the chest, comprising the *sternum* (or breastbone), 12 pairs of ribs, and the 12 thoracic vertebrae, forms the bony part of the thoracic cage. It protects the heart and lungs and provides attachments for muscles of the upper arms, back, and abdomen as well as the thorax itself. Moreover, the red marrow of the ribs and sternum is one of the main sites of red blood cell formation in the adult. The sternum has three parts: the *manubrium, body,* and *xiphoid process.* The first seven pairs of ribs articulate with the lateral borders of the sternum, and the *clavicles* (or collarbones) articulate with its superior angles. An indentation, the *jugular notch,* lies between the clavicular articulations.

FIGURE 7-20 (A) Position of the right first rib and clavicle as they articulate with the sternum. (B) The superior surface (a) and inferior surface (b) of the right clavicle. The two clavicles extend almost horizontally across the upper part of the thorax as far as the shoulder. They lie just under the skin and can be easily felt. They are an important part of the shoulder girdle (Fig. 7-27) because they serve as struts for the upper limbs, propping the shoulders out from the chest so the arms can swing freely, but they are also the attachment site of many neck muscles and therefore must be introduced here. The lateral end of each clavicle is flattened and articulates with the *acromion* of the scapula; the medial end is enlarged and articulates with the *clavicular notch* on the manubrium portion of the sternum. The clavicle is an "S"-shaped bone, the curves enabling the bone to withstand the shocks it transmits from the shoulder to the axial skeleton; nevertheless, it is the most frequently fractured bone in the body.

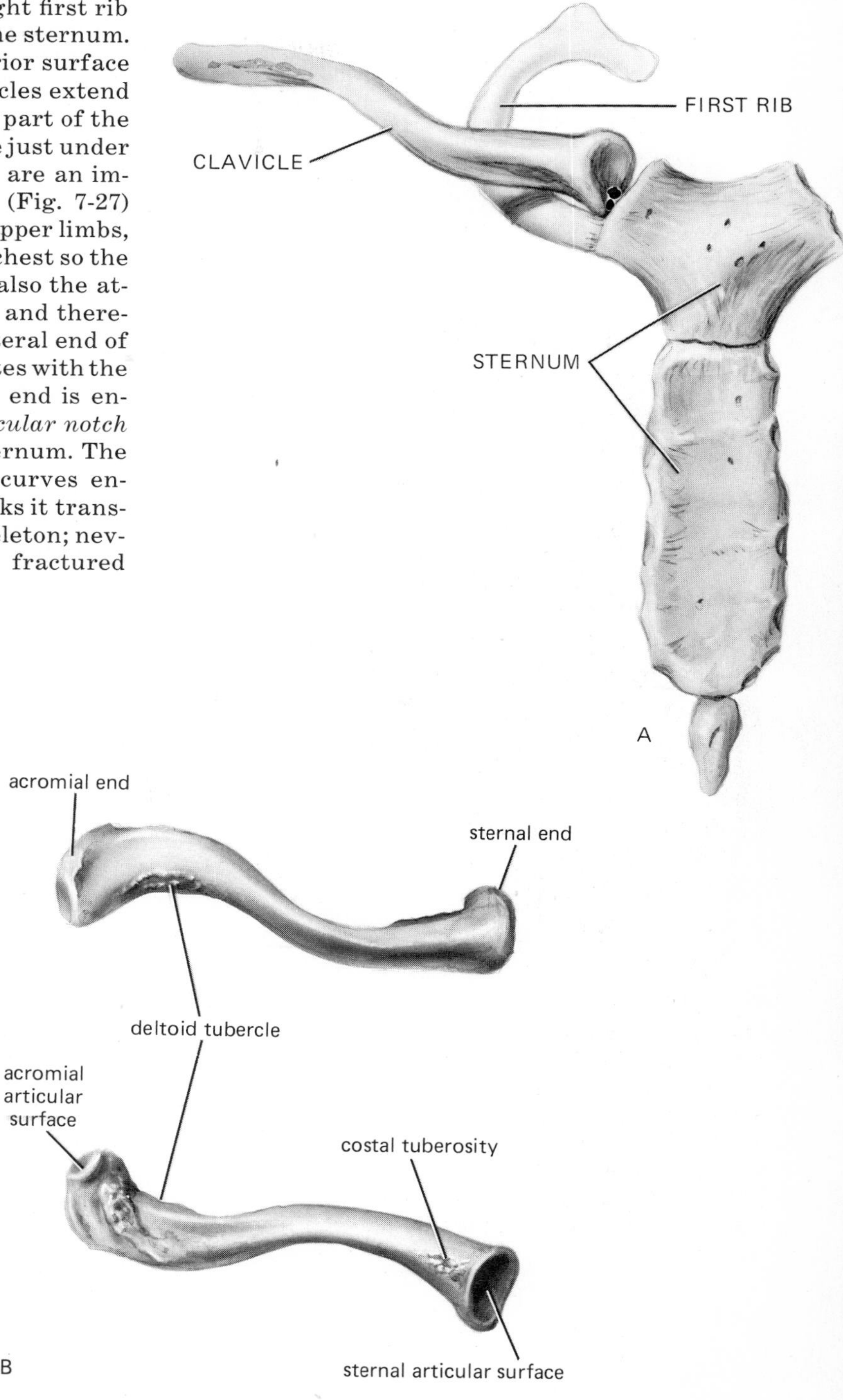

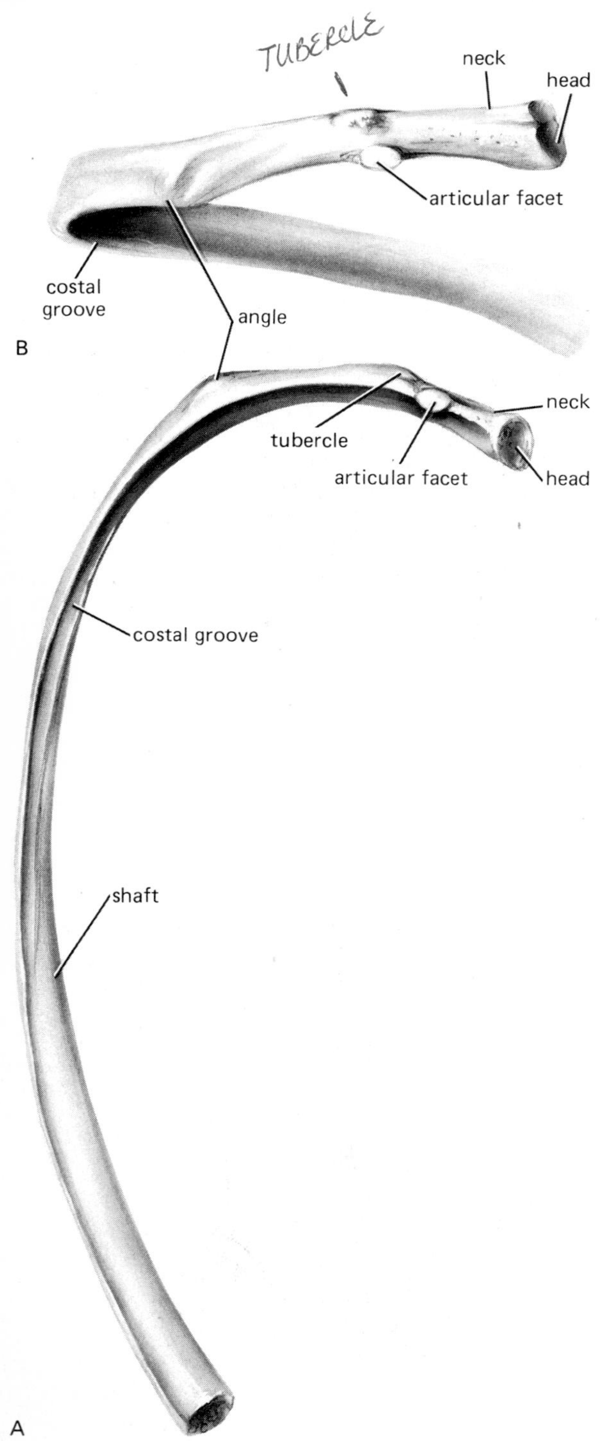

FIGURE 7-21 Inferior (A) and posterior (B) views of a typical rib. There are usually 12 pairs of ribs (some people have 11 or 13 pair) numbered from above downward, the first pair lying just beneath the clavicles. The main parts of a typical rib are the *anterior end, shaft, head, neck,* and *tubercle.* The head and tubercle, which are at the posterior, or vertebral, end of the rib, bear facets for articulation with the vertebrae, the head attaching to the body of a vertebra, and the neck and tubercle attaching to the transverse process. *Costal cartilages,* pieces of hyaline cartilage, lie between the anterior ends of the ribs and the sternum (Fig. 7-19) and contribute significantly to the mobility and elasticity of the walls of the thorax. The costal cartilages of the first seven pairs of ribs attach to the sternum; those of the 8th, 9th, and 10th pairs attach to the cartilages of the ribs immediately above them; and those of the 11th and 12th pairs do not attach to anything (thus, these last ribs are called *floating ribs*). The first seven pairs of ribs, being attached to the sternum, are the *true ribs* and pairs 8-12 are the *false ribs.*

FIGURE 7-22 Muscles of the neck. (A) Right lateral view of the more superficial muscles of the neck. (B) The supra- and infrahyoid muscles. The clavicle, sternohyoid, and sternocleidomastoid have been cut on the right side. Acting together, the two groups of muscles help flex the head and neck. The suprahyoid muscles, while elevating the hyoid bone, elevate the larynx as well because it is attached to the hyoid by ligaments; this action is important in swallowing because it helps prevent food from passing into the lungs.

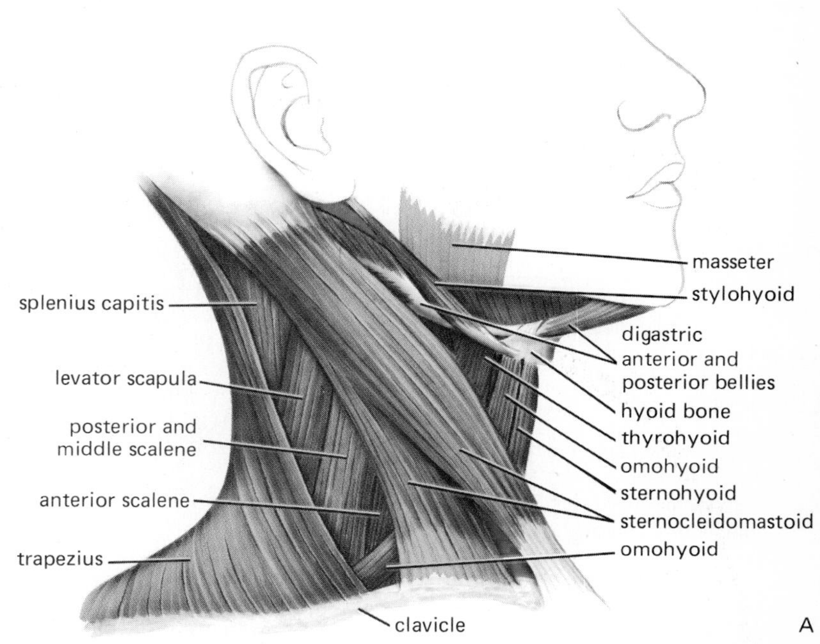

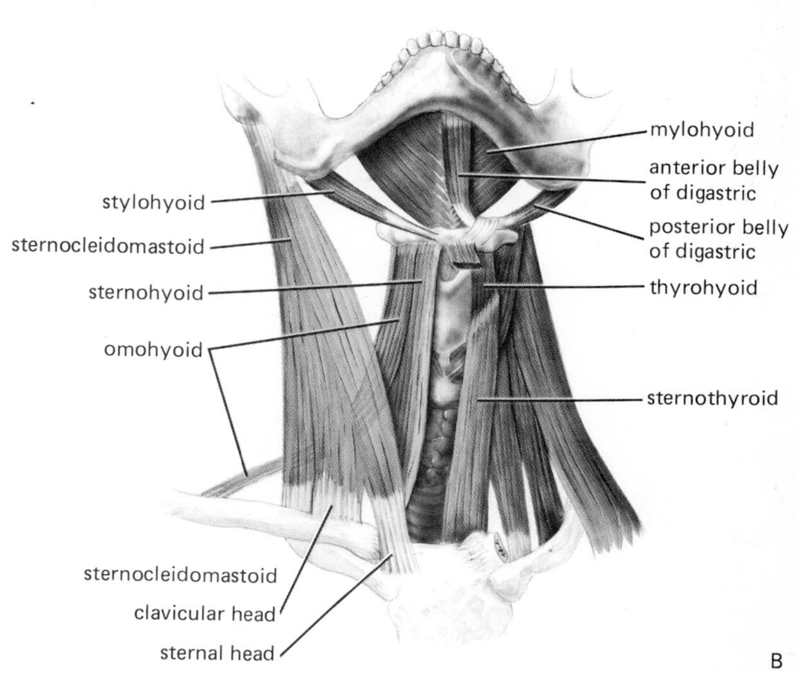

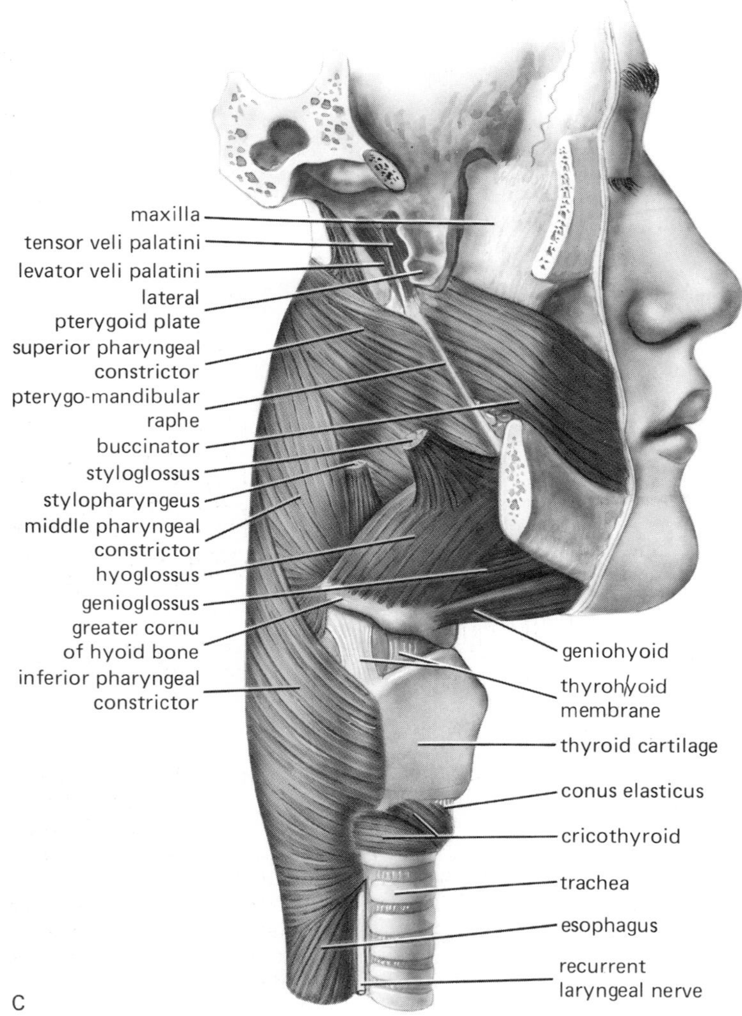

(C) Muscles of the pharynx. Parts of overlying bones (the zygomatic arch and mandible) and muscles (the masseter, temporalis, pterygoids, stylopharyngeus, styloglossus, hyoglossus, and all the infrahyoid group) have been removed.

Not shown in Fig. 7-22 are the deep anterior vertebral muscles which pass from one vertebra to another or from the vertebral column to the skull. These muscles flex the head on the neck or, contracting on only one side, tip the head toward that side.

TABLE 7-6
MUSCLES OF THE NECK

Muscle	*Figure*	*Nerve*	*Origin*	*Insertion*	*Action and comments*
Superficial and lateral cervical muscles:					
Platysma	7-16A and B	Cervical branch of VII	Fascia covering pectoralis major and deltoid muscles	Body of mandible and skin and subcutaneous tissue of lower part of face	Wrinkles skin in neck, assists in lowering mandible, draws down lower lip and angle of mouth
Trapezius	7-22A	Described with scapular muscles; see Table 7-12	See Table 7-12	See Table 7-12	See Table 7-12
Sterno-cleido-mastoid	7-22A and B, 7-23	Spinal accessory nerve, second and possibly third cervical nerves	Manubrium of sternum and clavicle	Mastoid process of temporal bone	Singly: tilts head toward shoulder of same side, rotates head to face opposite side; together: draw head forward, flex cervical part of vertebral column; if head is fixed, they help raise the thorax in forced inspiration
Suprahyoid muscles:					
Digastric	7-22A and B	Branch of the mandibular division of V and a branch of VII	Mastoid portion of temporal bone, symphysis of mandible	Hyoid bone, occasionally mandible	Depresses mandible, can elevate hyoid bone
Stylo-hyoid	7-22A and B	Branch of VII	Styloid process of temporal bone	Hyoid bone	Elevates the hyoid bone and draws it back, elongating the floor of the mouth. Can stabilize hyoid bone for tongue muscle action
Mylo-hyoid	7-22A to C	Branch of the mandibular division of V	Mylohyoid line on mandible	Hyoid bone and median raphe	Elevates floor of mouth in the first phase of swallowing; elevates hyoid bone or depresses mandible
Genio-hyoid	7-22C	Branch from XII containing fibers from first cervical spinal nerve	Inferior mental spine on mandible	Body of hyoid bone	Elevates hyoid bone and draws it forward; depresses mandible when hyoid bone is fixed
Infrahyoid muscles:					As a group, these are antagonists to the suprahyoid muscles
Sterno-hyoid	7-22A and B	Ansa cervicalis	Manubrium of the sternum	Body of hyoid bone	Depresses hyoid bone after it has been raised as in swallowing
Sterno-thyroid	7-22B	Ansa cervicalis	Manubrium of sternum and cartilage of first rib	Lamina of thyroid cartilage	Draws larynx downward after it has been elevated as in swallowing

TABLE 7-6
MUSCLES OF THE NECK *(Continued)*

Muscle	*Figure*	*Nerve*	*Origin*	*Insertion*	*Action and comments*
Thyro-hyoid	7-22A and B	First cervical spinal nerve	Lamina of the thyroid cartilage	Body of hyoid bone	Depresses hyoid bone or raises larynx
Omo-hyoid	7-22A and B	Ansa cervicalis	Upper border of scapula	Body of hyoid bone	Depresses hyoid bone after it has been elevated
Anterior vertebral muscles:					
Longus colli	Not shown	C2-6 spinal nerves	(See individual entries below)	(See individual entries below)	Bends neck forward, flexes it laterally, and rotates it to the opposite side
Lower oblique part	Not shown	(See Longus colli)	Bodies of T1-3 vertebrae	Transverse processes of C5-6 vertebrae	(See Longus colli)
Upper oblique part	Not shown	(See Longus colli)	Transverse processes of C3-5 vertebrae	Anterior arch of atlas	(See Longus colli)
Vertical part	Not shown	(See Longus colli)	Bodies of C5-7 and T1-3 vertebrae	Bodies of C2-4 vertebrae	(See Longus colli)
Longus capitis	Not shown	C1-3 spinal nerves	Transverse processes of C3-6 vertebrae	Basilar part of occipital bone	Flexes the head
Rectus capitis anterior	Not shown	C1-2 spinal nerves	Lateral mass and transverse process of atlas	Basilar part of occipital bone	Flexes the head
Rectus capitis lateralis	Not shown	C1-2 spinal nerves	Transverse process of atlas	Jugular process of occipital bone	Bends the head to the same side
Lateral vertebral muscles:					
Anterior scalene	7-22A	C4-6 spinal nerves	Transverse processes of C3-6 vertebrae	First rib	When the first rib is stabilized, it bends cervical portion of vertebral column forward and laterally and rotates it in the opposite direction; when the cervical vertebrae are stabilized, it assists elevation of the first rib in respiration
Middle scalene	7-22A	C3-8 spinal nerves	Transverse processes of C2-7 vertebrae	First rib	When the first rib is stabilized, bends cervical part of vertebral column to same side; when the cervical vertebrae are stabilized, helps raise first rib in respiration
Posterior scalene	Not shown	C5-8 spinal nerves	Transverse processes of C4-6 vertebrae	Second rib	Bends the lower part of the cervical portion of the vertebral column to the same side if second rib is fixed; when the cervical vertebrae are stabilized, helps elevate the second rib in respiration

TABLE 7-6
MUSCLES OF THE NECK *(Continued)*

Muscle	*Figure*	*Nerve*	*Origin*	*Insertion*	*Action and comments*
Intrinsic muscles of the larynx					
Crico-thyroid	7-22C	External branch of superior laryngeal	Outer surface of the cricoid cartilage	Inferior horn and lower border of the thyroid cartilage	Regulates tension of vocal cords
Crico-arytenoid					
Posterior	17-7	Recurrent laryngeal	Lamina of the cricoid cartilage	Arytenoid cartilage	Opens the glottis, separates the vocal folds, increases tension on the vocal folds
Lateral	17-7	Recurrent laryngeal	Arch of the cricoid cartilage	Arytenoid cartilage	Closes the glottis, approximates the vocal processes
Arytenoid					
Transverse	17-7	Recurrent laryngeal	Arytenoid cartilage	The opposite arytenoid cartilage	An unpaired muscle that bridges the space between the two arytenoid cartilages, approximating them and thus closing the glottis
Oblique	Not shown	Recurrent laryngeal	Arytenoid cartilage	The opposite arytenoid cartilage	Acts as a sphincter to the inlet of the larynx
Thyro-arytenoid	17-7	Recurrent laryngeal	Angle of the thyroid cartilage and the crico-thyroid ligament	Arytenoid cartilage	Shortens and relaxes the vocal ligaments; the lower, deeper fibers of this form a bundle known as the *vocalis*

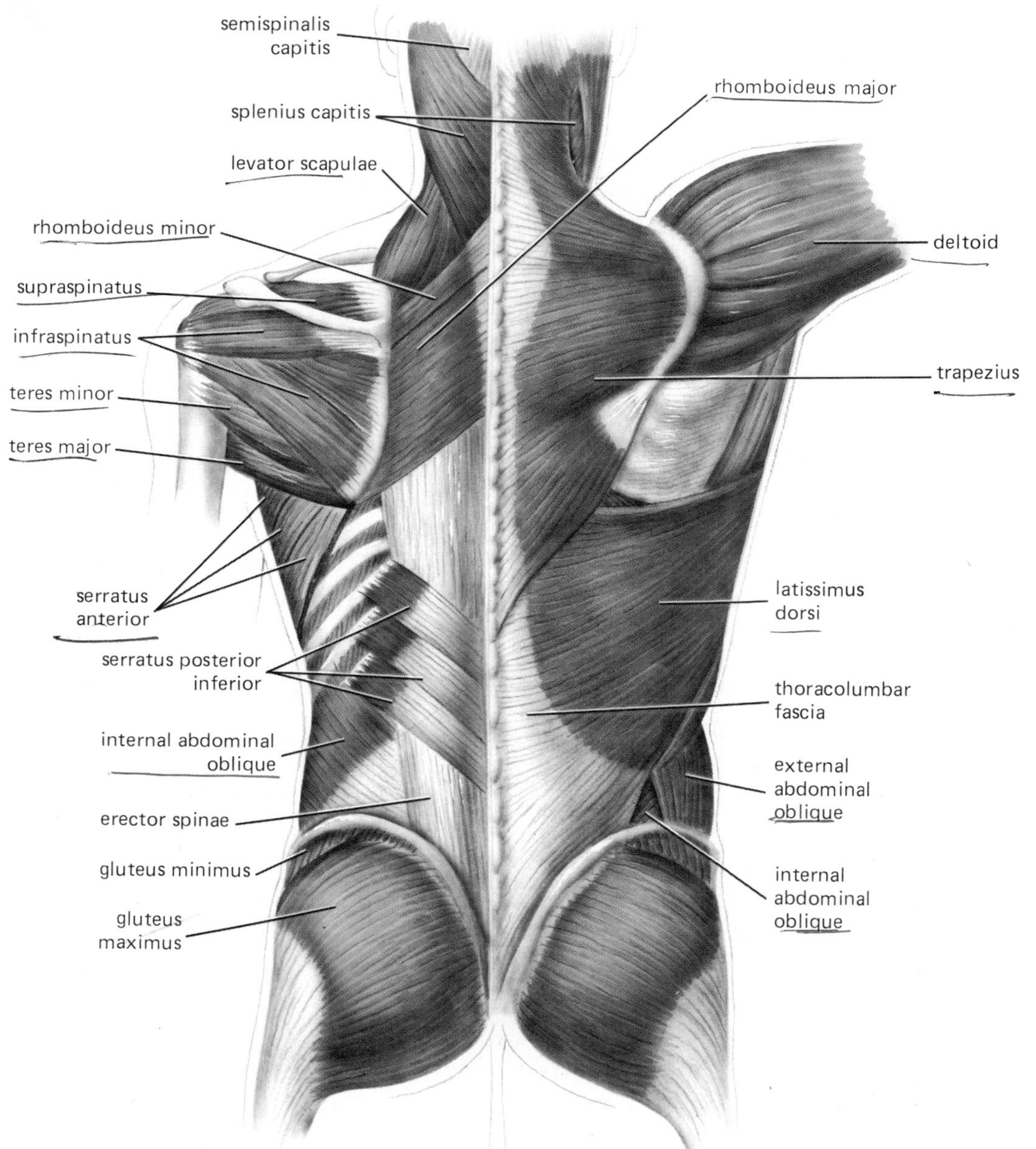

FIGURE 7-23 Superficial muscles of the back of the trunk and neck. The sternocleidomastoid, trapezius, latissimus dorsi, deltoid, and external oblique have been removed on the left side. Muscles connecting the pectoral (shoulder) girdle to the vertebral column (the *trapezius*, *latissimus dorsi*, *rhomboideus major* and *minor*, and *levator scapulae*) are located superficially in the back and arise from the vertebrae and ligamentum nuchae. (The *ligamentum nuchae* is a connective-tissue sheet in the upper back and back of the neck; it runs from the superior nuchal line and external occipital crest of the skull down to the spine of the seventh cervical vertebra and attached to the spines of the intervening vertebrae. It provides sites for muscle attachments that cannot be accommodated by the shortened spines of the cervical vertebrae.)

The scapula moves freely over the posterior wall of the thorax. Muscles connecting the scapula with the vertebral column are able to move the scapula but more often serve to stabilize it during movements of the shoulder and arm.

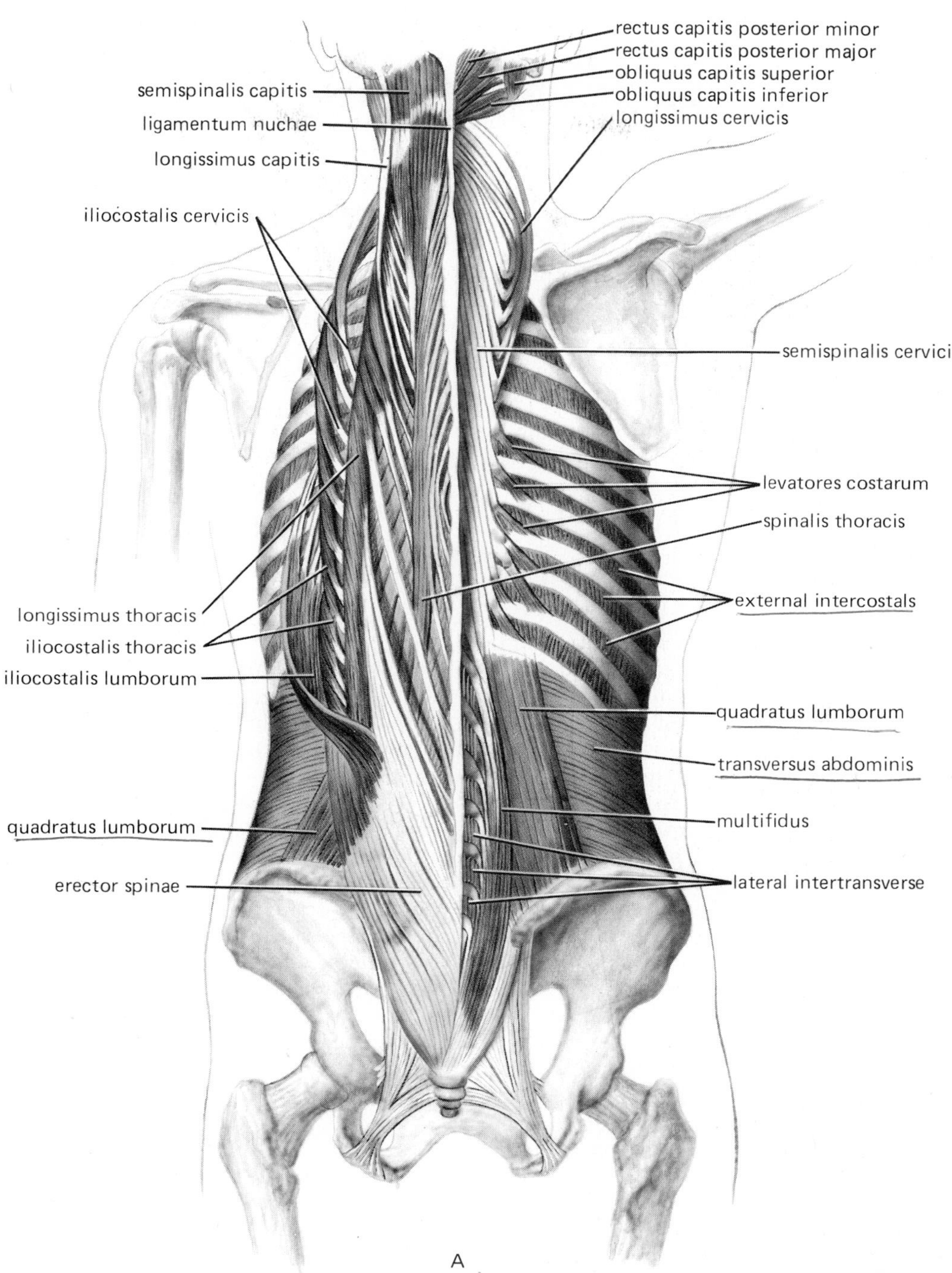

FIGURE 7-24 (A) Deep muscles of the back. The semispinalis capitis and erector spinae and some of its upward continuations have been removed on the right. The longissimus cervicis has been displaced laterally. This complex group of deep muscles extends from the pelvis to

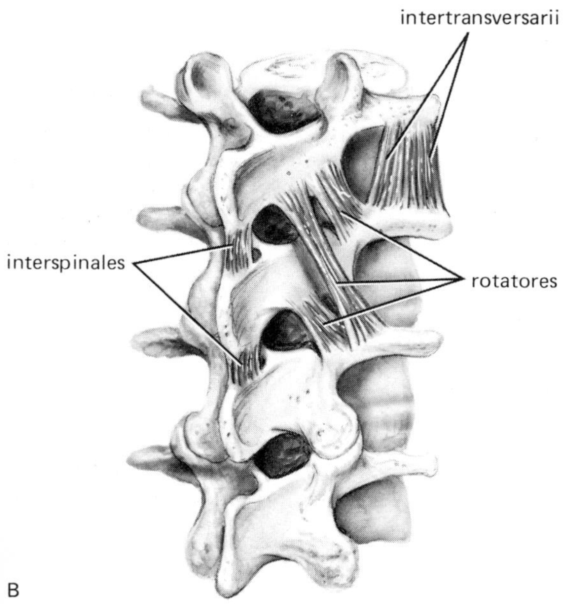

the skull. It includes extensors and rotators of the head and neck and the short segmental, extensor, and rotator muscles that control the vertebral column. The deep muscles of the back are covered by the *vertebral fascia* in the thorax, *prevertebral (nuchae) fascia* in the neck, and *thoracolumbar (lumbar) fascia* in the lumbar area. One of the major muscles of the back is the *erector spinae (sacrospinalis)* which lies in the angles between the spine and transverse processes of the vertebrae. As it ascends through upper lumbar regions, it breaks into three columns, the *iliocostalis*, *longissimus*, and *spinalis muscles*. When the muscles on both sides act together, they straighten or extend the vertebral column and hold it upright; acting singly, they twist the column. The suboccipital muscles extend the head at the atlanto-occipital joints and rotate the head and atlas on the axis. The muscles of the thorax connect the ribs to each other or to the vertebrae. They are all involved in movements of the ribs and are active during respiration.

(B) Section of vertebral column showing the short muscles of the back and their relation to the vertebrae. Short muscles (multifidus, rotatores, interspinales, and intertransversarii) function for the most part as postural muscles; they control the movement of the vertebrae relative to one another so segments of the column can be stabilized during movements of the whole column. In this way, the short muscles of the back ensure the efficient action of the long muscles.

TABLE 7-7
DEEP MUSCLES OF THE BACK

Muscle	*Figure*	*Nerve*	*Origin*	*Insertion*	*Action and comments*
Splenius	(See individual entries below)	(See individual entries below)	(See individual entries below)	(See individual entries below)	Acting together: draws head backward; separately: draws head to one side and slightly rotates it, turning the face to the same side
Splenius capitis	7-22A and 7-23	Middle cervical spinal nerves	Ligamentum nuchae, spine of C7 and T1-4 vertebrae	Mastoid process of temporal bone	(See Splenius)
Splenius cervicis	Not shown	Lower cervical spinal nerves	Spines of T3-6 vertebrae	Transverse processes of C1-3 vertebrae	(See Splenius)
Iliocostalis	7-24A	All three muscles in this group are innervated by the lower cervical, thoracic, and lumbar spinal nerves	(See individual entries below)	(See individual entries below)	Extends the vertebral column and flexes it laterally
Iliocostalis lumborum	7-24A	(See Iliocostalis)	Iliac crest and sacrospinal aponeurosis	Lower six or seven ribs	(See Iliocostalis)
Iliocostalis thoracis	7-24A	(See Iliocostalis)	Lower seven ribs	Upper seven ribs and transverse process of C7 vertebra	(See Iliocostalis)
Iliocostalis cervicis	7-24A	(See Iliocostalis)	3-6th ribs	Transverse processes of C4-6 vertebrae	(See Iliocostalis)
Longissimus	7-24A	Lower cervical, thoracic, and lumbar spinal nerves	(See individual entries below)	(See individual entries below)	(See individual entries below)
Longissimus thoracis	7-24A	(See Longissimus)	Transverse processes of the L1-2 and T6-12 vertebrae and the thoracolumbar fascia as far as the crest of the ilium and sacrum	Transverse processes of thoracic and lumbar vertebrae and lower borders of ribs	Bends vertebral column backward and laterally
Longissimus cervicis	7-24A	(See Longissimus)	Transverse processes of T1-5 vertebrae	Transverse processes of C2-6 vertebrae	Bends vertebral column backward and laterally
Longissimus capitis	7-24A	(See Longissimus)	Transverse processes of T1-5 and articular processes of C4-7 vertebrae	Mastoid process of the temporal bone	Extends the head and turns the face toward the same side

TABLE 7-7
DEEP MUSCLES OF THE BACK *(Continued)*

Muscle	*Figure*	*Nerve*	*Origin*	*Insertion*	*Action and comments*
Spinalis thoracis	7-24A	Lower cervical and thoracic spinal nerves	Spines of T11-12 and L1-2 vertebrae	Spines of upper thoracic vertebrae	Extends the vertebral column; blends with longissimus thoracis and semispinalis thoracis
Semispinalis	(See individual entries below)	Cervical and thoracic spinal nerves	(See individual entries below)	(See individual entries below)	(See individual entries below)
Semispinalis thoracis	Not shown	(See Semispinalis)	Transverse processes of T6-10 vertebrae	Spines of T1-4 and C6-7 vertebrae	Extends thoracic portion of vertebral column and rotates it toward the opposite side
Semispinalis cervicis	Not shown	(See Semispinalis)	Transverse processes of T1-6 vertebrae	Spines of C2-5 vertebrae	Extends cervical portion of vertebral column and rotates it toward the opposite side
Semispinalis capitis	7-24A	(See Semispinalis)	Transverse processes of T1-7 and C7 vertebrae and the articular processes of C4-6	Occipital bone between superior and inferior nuchal line	Extends the head and turns the face slightly toward the opposite side
Multifidus	7-24A	Spinal nerves	Processes of vertebrae from C4 to sacrum	Spinous processes of vertebrae two to four segments above origin	Extends vertebral column and rotates it toward opposite side
Rotatores	7-24B	Spinal nerves	Processes of vertebrae from C4 to the sacrum	Spinous process of vertebra adjacent to or second from origin	Extends vertebral column and rotates it toward opposite side
Interspinales (minor)	7-24B	Spinal nerves	Spinous processes of C and L vertebrae	Spinous process of vertebra adjacent to origin	Extends vertebral column; steadies vertebrae during movement of the vertebral column as a whole
Intertransversarii (minor)	7-24B	Spinal nerves	Transverse processes of C and L vertebrae	Transverse process of adjacent vertebra	Flexes vertebral column laterally. Steadies vertebrae during movement of the vertebral column as a whole

TABLE 7-8
SUBOCCIPITAL MUSCLES

Muscle	*Figure*	*Origin*	*Insertion*	*Action and comments*
(All are supplied by the first spinal nerve)				
Rectus capitis posterior:				
Major	7-24A	Spine of the axis	Inferior nuchal line of the occipital bone	Extends the head and turns the face toward the same side
Minor	7-24A	Posterior arch of the atlas	Below the inferior nuchal line of the occipital bone	Extends the head
Obliquus capitis:				
Inferior	7-24A	Spine and lamina of the axis	Transverse process of the atlas	Turns the face toward the same side
Superior	7-24A	Transverse process of the atlas	Occipital bone between the superior and inferior nuchal lines	Bends the head backward and to the same side

TABLE 7-9
MUSCLES OF THE THORAX

Muscle	*Figure*	*Nerve*	*Origin*	*Insertion*	*Action and comments*
Intercostals:	(See individual entries below)	Adjacent intercostal nerves	(See individual entries below)	(See individual entries below)	Thin layers of muscle and tendon which occupy the spaces between the ribs; external intercostals are more superficial than internal intercostals
External	7-24A, 7-25	(See Intercostals)	Lower border of one rib	Upper border of the rib below the rib of origin	Eleven pairs; possibly elevate the ribs
Internal	7-25	(See Intercostals)	Costal groove of one rib	Upper border of the rib below the rib of origin	Eleven pairs; oriented obliquely with their fibers at right angles to those of the external intercostals; probably depress the ribs; maintain integrity of the intercostal space and reduce the space between the ribs in respiration
Subcostals	Not shown	Adjacent intercostal nerve	Inner surface of one rib	Inner surface of the second or third rib below the rib of origin	Their fibers are parallel to the internal intercostals; like them they probably depress the ribs; well developed only in lower thorax
Transversus thoracis (sternocostalis)	Not shown	Adjacent intercostal nerve	Body of the sternum, the xiphoid process and costal cartilages of lower ribs	Costal cartilages of the second to sixth ribs	Draws down the costal cartilage to which it is attached

TABLE 7-9
MUSCLES OF THE THORAX *(Continued)*

Muscle	*Figure*	*Nerve*	*Origin*	*Insertion*	*Action and comments*
Levatores costarum	7-24A	Corresponding thoracic nerve	Transverse process of C7 and T1-11 vertebrae	Upper edge of the rib below the vertebrae of origin	Elevate the ribs, but their role in respiration is disputed; also rotate and flex vertebral column
Serratus posterior:					
Superior	Not shown	Second through fifth intercostal nerves	Ligamentum nuchae, spines of C7 and T1-3 vertebrae, and supraspinous ligament	Second through fifth ribs	Can elevate ribs, but their role in respiration is not clear
Inferior	7-23	T9-12 spinal nerves	Spines of T11-12 and L1-3 vertebrae and the supraspinous ligament	Lower four ribs	Draws lower ribs downward and backward, but possibly not in respiration; stabilizes lower ribs in respiration
Diaphragm	14-2	The motor supply is via the phrenic nerve, the sensory via the lower intercostal nerves	Xiphoid process, cartilages and lower six ribs on each side, the lumbocostal arches (arcuate ligaments), and the lumbar vertebrae	Central tendon of the diaphragm	Dome-shaped sheet which separates the thoracic and abdominal cavities; principal muscle of inspiration, the contracting fibers draw the central tendon downward

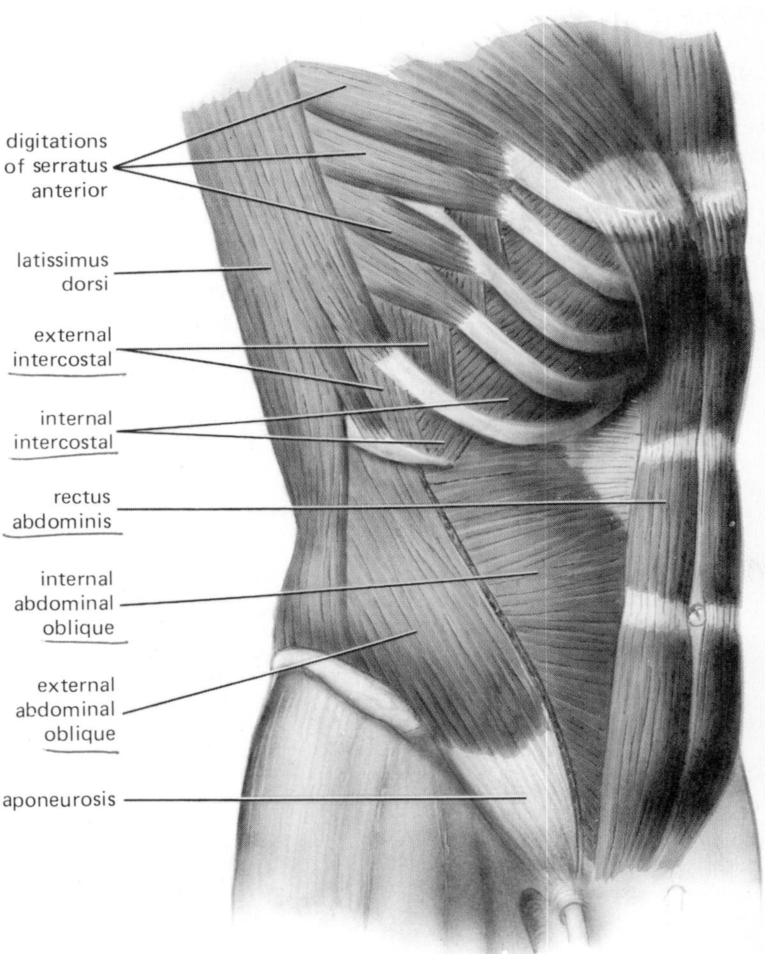

FIGURE 7-25 Muscles of the right side of the trunk. The digitations from the external abdominal oblique to the ribs are intact but most of the muscle has been removed. The abdominal muscle group consists of four flat muscular sheets (the internal and external abdominal oblique and the transversus and rectus abdominis) and two smaller muscles (the cremaster and pyramidalis, which suspend the testes and tense the abdominal fascia). The *linea alba* is a tough connective-tissue band between the right and left rectus abdominis muscles; it is formed from the fascia coverings of the external and internal abdominal oblique and transversus abdominis muscles from the two sides and extends from the xiphoid process of the sternum to the symphysis pubis. The *inguinal ligament* is formed from the lower border of the aponeurosis of the external abdominal oblique muscle. It passes between the anterior superior iliac spine and the pubic tubercle and marks the separation between the abdominal wall and the leg. Above and parallel to the inguinal ligament is the *inguinal canal,* a slanted opening in the lower abdominal wall, through which passes the spermatic cord in the male or the round ligament of the uterus in the female.

FIGURE 7-26 The muscles of the perineum of the male (A) and female (B). The *perineum* is the outlet of the pelvic girdle (see Fig. 7-36). The entire space is somewhat diamond shaped, the anterior half forming the *urogenital triangle* and the posterior half the *anal triangle*. A *superficial perineal space* in the urogenital triangle houses the root of the penis in the male and of the clitoris in the female. A deeper compartment, the *deep perineal space*, contains the membranous portion of the urethra and the bulbourethral glands in the male and the urethra and a portion of the vagina in the female as well as the muscles, nerves, and blood vessels associated with them. Two of the major muscles of the pelvis, the levator ani and coccygeus, form the *pelvic diaphragm*, a sheet of muscle which supports the pelvic viscera, particularly during periods of increased intraabdominal pressure. The anal canal, urethra, and, in the female, the vagina, pass through the pelvic diaphragm. A fibromuscular node, the *perineal body*, is an important point of attachment for many of the perineal muscles.

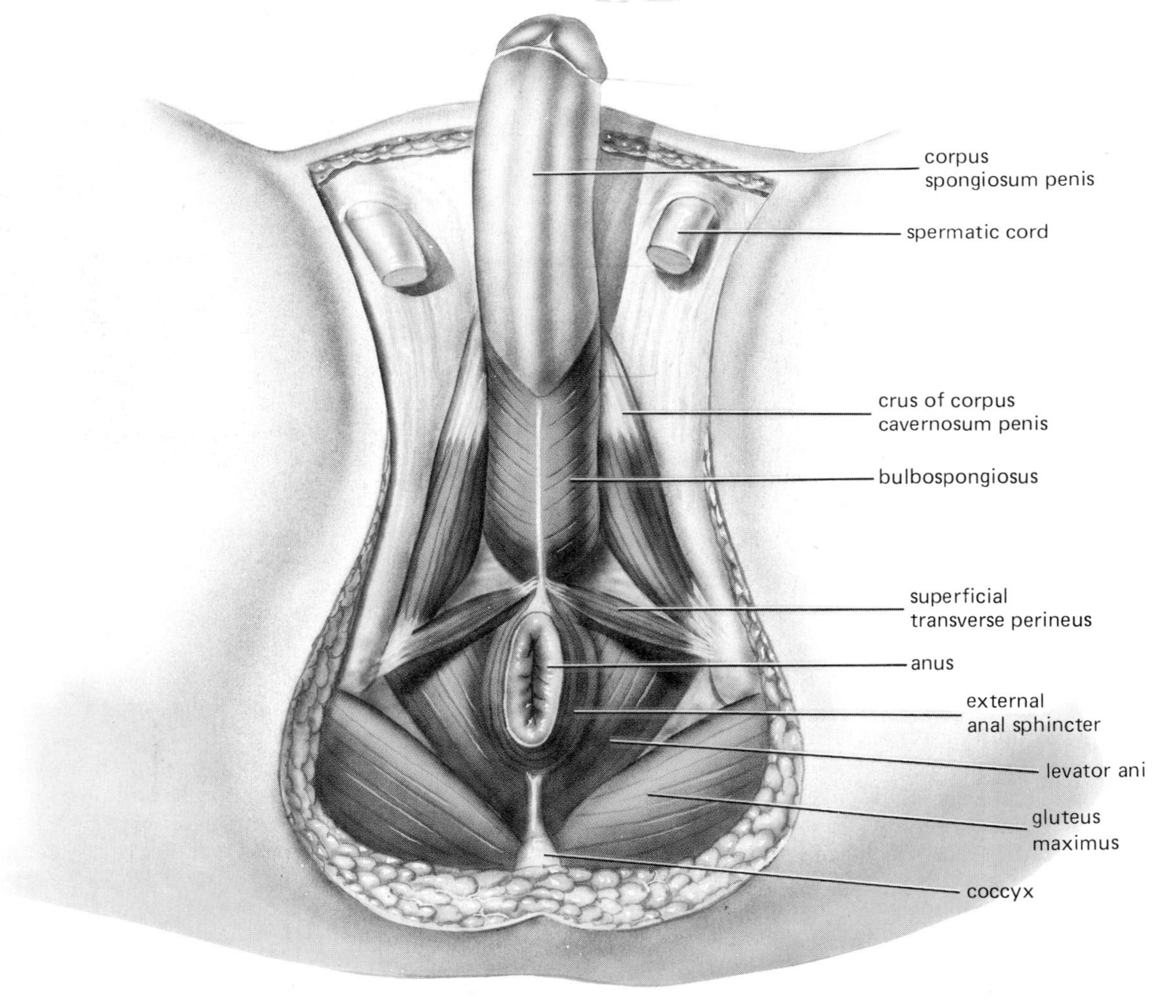

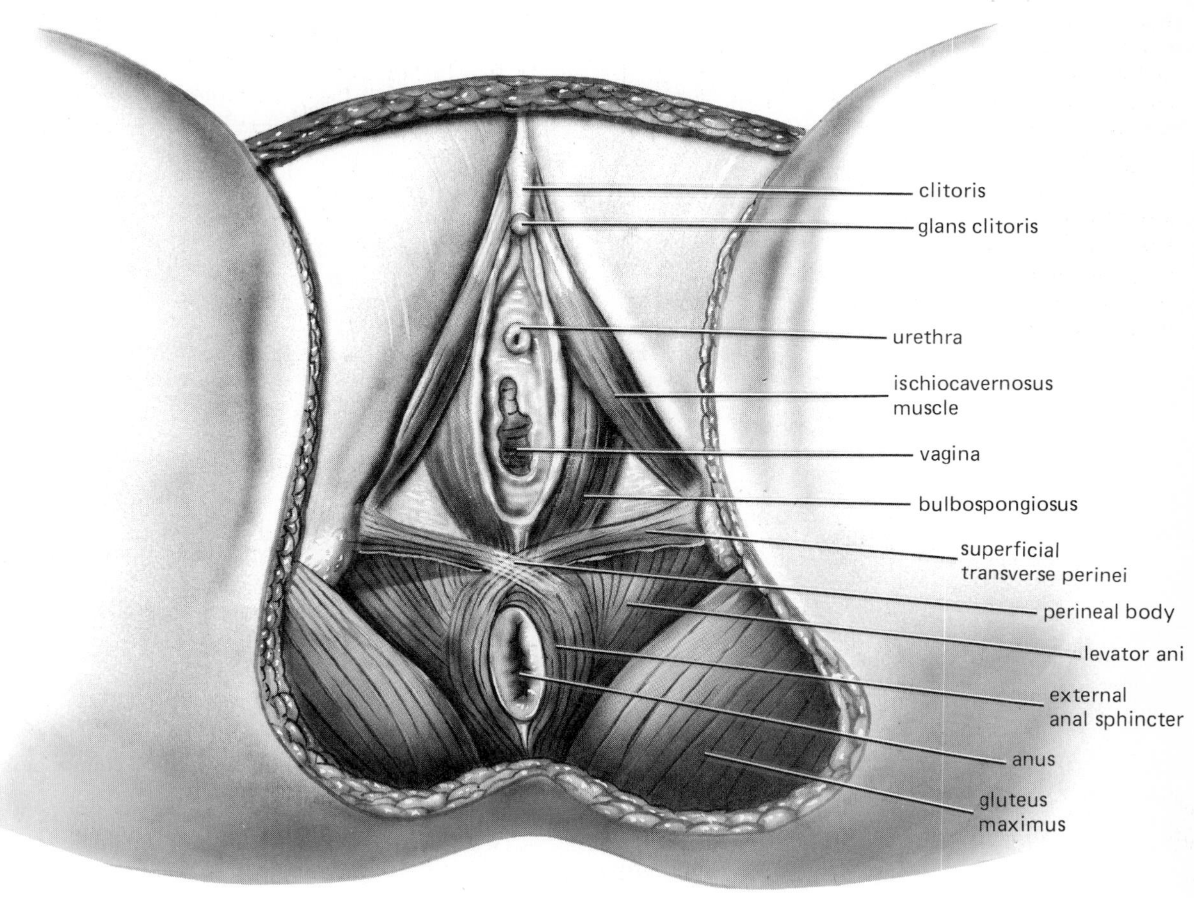
clitoris
glans clitoris
urethra
ischiocavernosus muscle
vagina
bulbospongiosus
superficial transverse perinei
perineal body
levator ani
external anal sphincter
anus
gluteus maximus

B

TABLE 7-10
MUSCLES OF THE ABDOMEN

Muscle	*Figure*	*Nerve*	*Origin*	*Insertion*	*Action and comments*
Abdominal oblique:					
External	7-23, 7-25	T6-12 spinal nerves	Lower eight ribs	Iliac crest and spine; pubic tubercle and crest; interdigitates with muscle of opposite side (linea alba)	Rotates and flexes vertebral column; tenses abdominal wall
Internal	7-23, 7-25	T6-12 and L1 spinal nerves	Inguinal ligament, iliac crest, and thoracolumbar fascia	Lower three or four ribs, aponeurosis; pubic crest; interdigitates with same muscle of opposite side (linea alba)	Rotates and flexes vertebral column; tenses abdominal wall
Cremaster	Not shown	L1-2 spinal nerves, genitofemoral nerve	Inguinal ligament, internal oblique and transversus muscles	Pubic tubercle	Pulls the testes upward; although its fibers are striated, it is usually not under voluntary control
Transversus abdominis	7-24A	T6-12 and L1 spinal nerves	Inguinal ligament, iliac crest, and the thoracolumbar fascia, ribs 7-12	Interdigitates with muscle of opposite side (linea alba), conjoint tendon to pubis	Supports abdominal viscera
Rectus abdominis	7-25	T6-12 spinal nerves	Xiphoid process and fifth to seventh costal cartilage	Pubic crest and symphysis pubis	Flexes vertebral column, tenses abdominal wall
Quadratus lumborum	7-24A	T12 and L1-4 spinal nerves	Iliac crest, transverse processes of L1-4 vertebrae	Twelfth rib, transverse processes of L1-4 vertebrae	Stabilizes the last rib, steadies the origin of the diaphragm; may flex vertebral column if pelvis is fixed

TABLE 7-11
MUSCLES OF THE PELVIS AND PERINEUM

Muscle	*Figure*	*Nerve*	*Origin*	*Insertion*	*Action and comments*
Muscles of the pelvis:					
Piriformis	See Table 7-19	See Table 7-19	See Table 7-19	See Table 7-19	See Table 7-19
Obturator internus	See Table 7-19	See Table 7-19	See Table 7-19	See Table 7-19	See Table 7-19
Levator ani	7-26A and B	Branch of S4 and perineal branch of pudendal nerve	Dorsal surface of the pubis	Spine of the ischium	Forms great part of floor of the pelvic cavity; constricts the lower end of the rectum and, in the female, of the vagina
Coccygeus	7-26A	Branch of S4 and 5	Spine of the ischium and supraspinous ligament	Margin of the coccyx and side of the sacrum	Draws the coccyx forward and raises the pelvic floor
External anal sphincter	7-26A and B	Perineal branch of S4 and the inferior rectal nerves	Anteriorly, it ends over the bulbospongiosus muscle	Posteriorly, it ends in subcutaneous tissue over the coccyx	The *superficial portion* lies just under the skin of the anus, the *deep portion* circles the anal canal and internal anal sphincter; composed of skeletal muscle tissue
Male urogenital region:					
Transverse perineus					
Superficial	7-26A	Perineal branch of the pudendal nerve	Tuberosity of the ischium	Perineal body	Fixes central tendon, thereby stabilizing muscles inserting on it
Deep	Not shown	Perineal branch of the pudendal nerve	Fascia over ischial ramus	Perineal body	Compresses urethra
Bulbospongiosus (bulbocavernosus)	7-26A	Perineal branch of the pudendal nerve	Median raphe and perineal body	Fascia of erectile tissue of the penis	Compresses urethra and can stop urination; aids erection by compressing veins draining penis
Ischiocavernosus	7-26A	Perineal branch of the pudendal nerve	Tuberosity of the ischium	Aponeurosis of bulb of penis	Aids erection by compressing bulb of penis
Sphincter urethrae	Not shown	Perineal branch of the pudendal nerve	Perineal ligament and fascia of pudendal vessels; inferior pubic ramus	Central tendinous raphe	Compresses urethra, essential for the voluntary control of urination

TABLE 7-11
MUSCLES OF THE PELVIS AND PERINEUM *(Continued)*

Muscle	*Figure*	*Nerve*	*Origin*	*Insertion*	*Action and comments*
Female urogenital region:					
Transverse perineus					
Superficial	7-26B	Perineal branch of the pudendal nerve	Tuberosity of the ischium	Perineal body	Stabilizes central tendon and muscles which anchor upon it
Deep	Not shown	Perineal branch of the pudendal nerve	Fascia over the ramus of the ischium	Joins muscle of opposite side, some fibers join wall of vagina	Compresses urethra and vagina
Bulbospongiosus	7-26B	Perineal branch of the pudendal nerve	Perineal body	Blends with external anal sphincter	Surrounds the vaginal opening and acts as a sphincter of vaginal opening
Ischiocavernosus	7-26B	Perineal branch of the pudendal nerve	Tuberosity and ramus of the ischium	Aponeurosis of clitoris	Compresses veins draining clitoris, thereby aiding erection
Sphincter urethrae	Not shown	Perineal branch of the pudendal nerve	Transverse perineal ligament	Joins muscle of opposite side or wall of vagina	Compresses urethra and vagina, essential for the voluntary control of urination

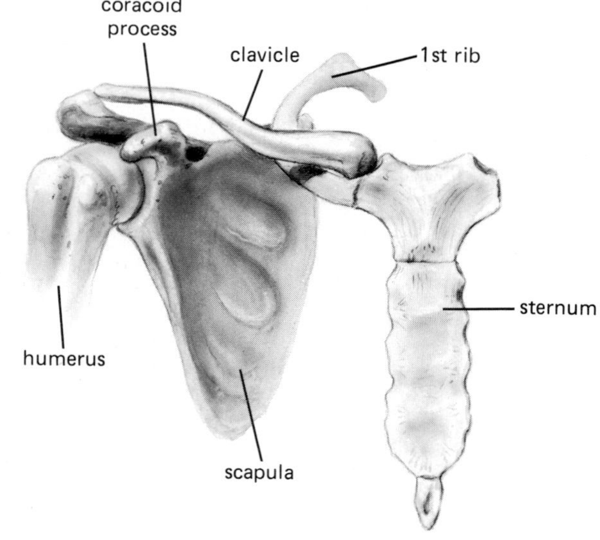

FIGURE 7-27 The pectoral, or shoulder, girdle. The upper limbs consist of the shoulders, arms, forearms, wrists, and hands. The framework of the arm is attached to the axial skeleton (vertebral column) by the *pectoral*, or *shoulder*, girdle. The bones of the pectoral girdle are the two *scapulae*, or "angel wings," on the posterior wall of the thoracic cage, and the two *clavicles*, or collar bones. The single point of bony contact between the upper limb and the trunk is at the *sternoclavicular joint*, a small and mobile articulation; there is no direct articulation between the skeleton of the upper limb and the vertebral column. The pectoral girdle articulates with the arms at freely movable ball-and-socket joints between the *glenoid fossa* of each scapula and *head* of the humerus.

FIGURE 7-28 Dorsal (A) and lateral (B) views of the right scapula. The scapula is a large, flattened bone which lies against the posterior wall of the thorax, extending from the level of the second to the seventh ribs. The *body* of the bone is thin and even translucent, but in the region of the glenoid fossa, acromion, and coracoid process the bone is weighty. The *spine* of the scapula projects from the dorsal surface, extending from the medial (vertebral) edge of the bone to the *acromion* at the point of the shoulder where the acromion projects forward to articulate with the clavicle. A shallow depression of the lateral surface of the bone, the *glenoid fossa*, forms the shoulder joint with the head of the humerus.

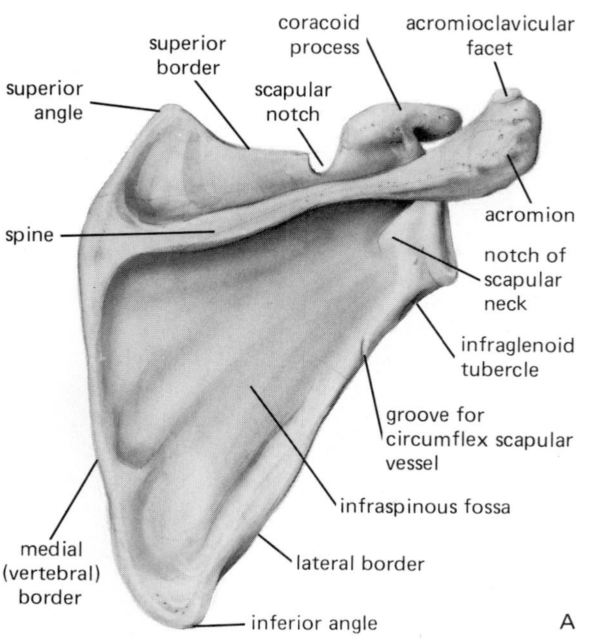

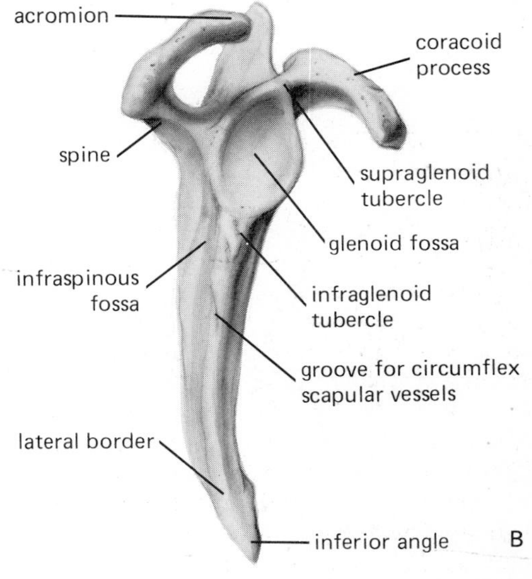

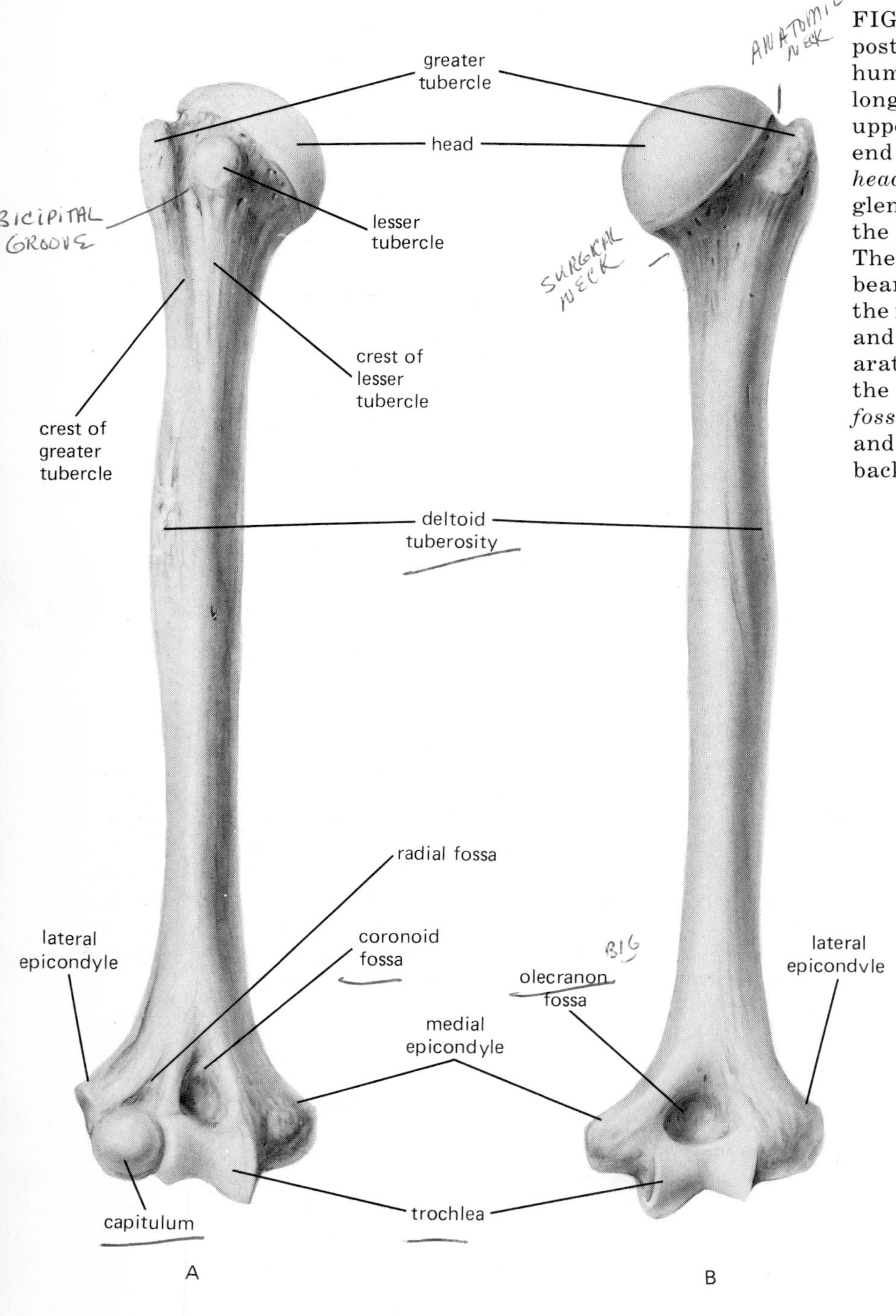

FIGURE 7-29 Anterior (A) and posterior (B) views of the right humerus. The humerus is the longest and largest bone of the upper limb. The expanded upper end of the bone has a rounded *head*, which articulates with the glenoid fossa of the scapula, and the *greater* and *lesser tubercles*. The lower end is more irregular, bearing articular surfaces for the radius and ulna. The lateral and medial epicondyles are separated by the *trochlea*. Above the trochlea are the *coronoid fossa* on the front of the bone and the *olecranon fossa* on the back.

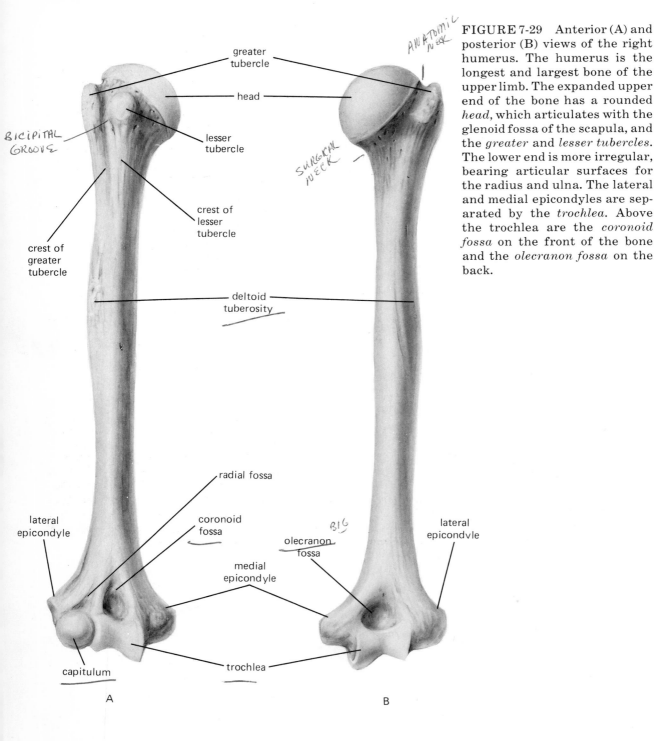

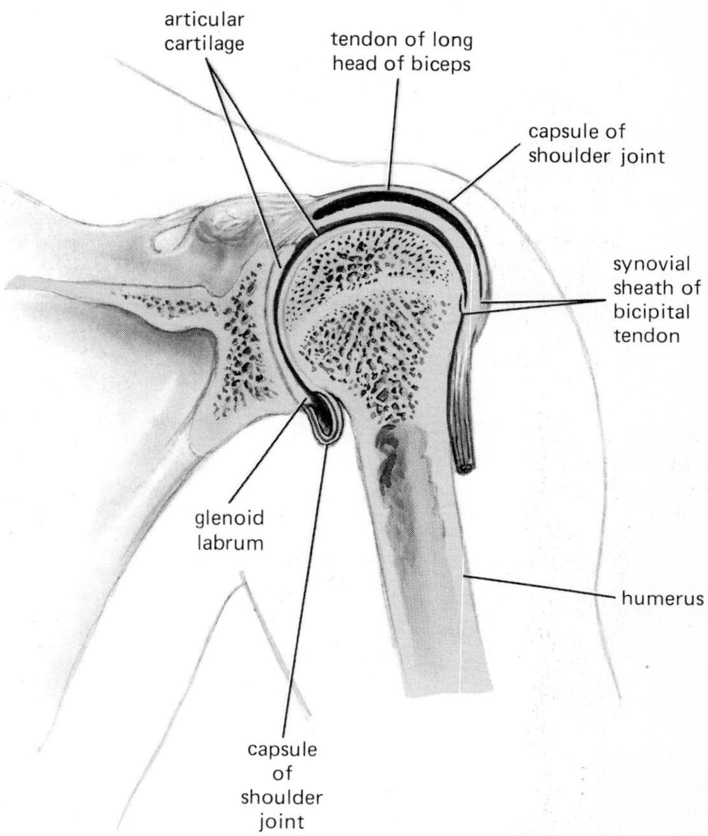

FIGURE 7-30 A section through the shoulder joint. The roughly spherical head of the humerus and shallow glenoid fossa of the scapula form the shoulder joint, which is of the ball-and-socket type. Both articular surfaces are covered by a layer of hyaline articular cartilage. The shallowness of the glenoid fossa allows considerable range of motion but little security, and the joint is supported by the muscles which surround it and a connective-tissue ring, the *glenoid labrum*, which surrounds the fossa and deepens it. In addition to the glenoid labrum, the ligamentous structures of the joint include the fibrous capsule, and the glenohumeral, coracohumeral, and transverse humeral ligaments.

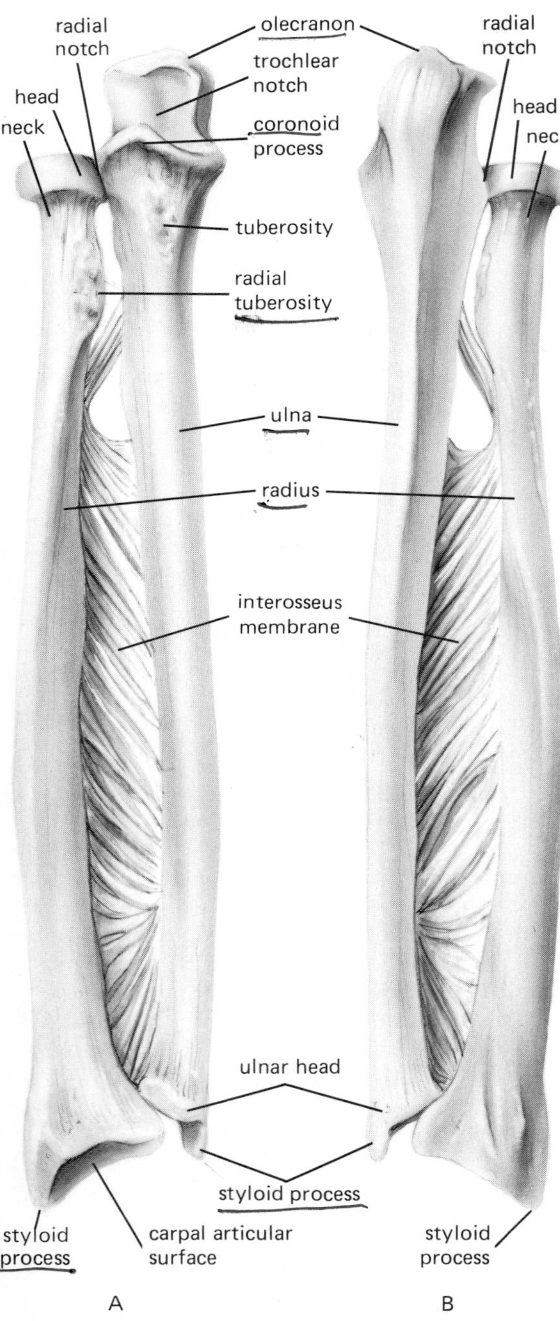

FIGURE 7-31 Anterior (A) and posterior (B) views of the right radius and ulna. The radius and ulna are the two bones of the forearm; the radius is on the lateral (thumb) side and the ulna on the medial side. The two bones articulate with each other at the upper and lower ends of the forearm and are connected throughout their length by a flexible connective-tissue *interosseous membrane*. The radius is better developed at its lower end and plays a larger role in the articulations of the wrist than in the elbow joint, while the ulna is just the opposite. The upper end of the ulna bears *olecranon* and *coronoid processes*, which act like stationary jaws to grasp the trochlea of the humerus. Its lower end makes little contact with the wrist. The radius articulates with the humerus (and ulna) at its upper end and with the wrist (and ulna) at its lower end. The lower end extends as a *styloid process*, which can be felt when the hand is relaxed.

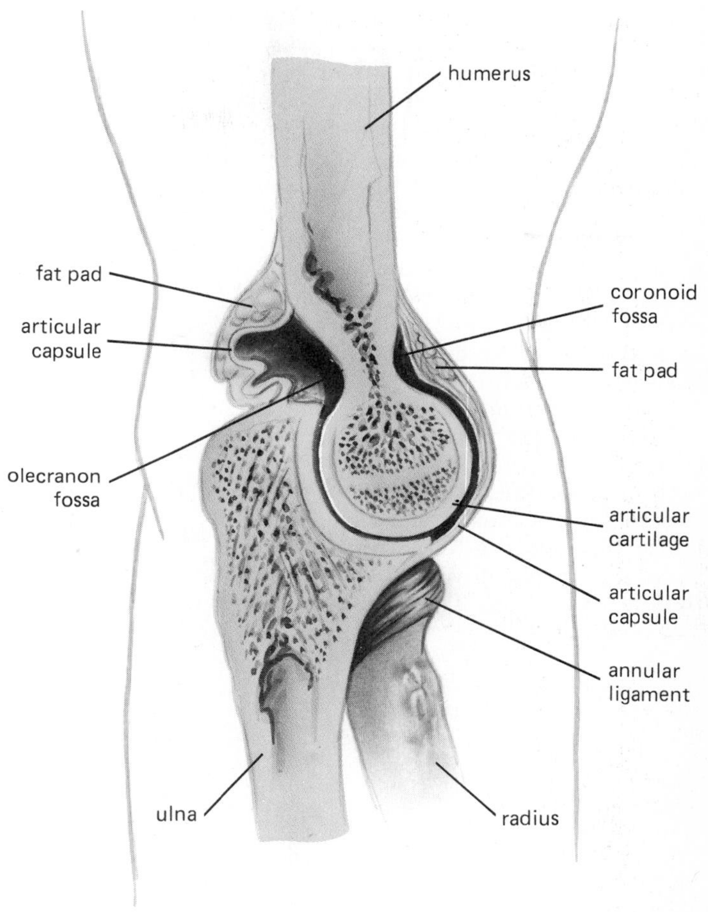

FIGURE 7-32 A sagittal section through the left elbow joint. The elbow joint includes two articulations—one between the capitulum of the humerus and head of the radius *(humeroradial articulation)* and the other between the trochlea of the humerus and the trochlear notch of the ulna (Figs. 7-29 and 7-31); thus, it is a compound joint. It acts as a hinge, the radius and ulna swinging on the capitulum and trochlea of the humerus during flexion and extension of the joint. There is also a slight degree of rotation of the forearm during these movements. Flexion of the joint is accomplished chiefly by the brachialis, biceps, and brachioradialis muscles and is limited by apposition of the soft tissues of the arm and forearm. Extension is achieved by the triceps and aconeus muscles and is limited by tension of joint capsule and muscles on the front of the joint.

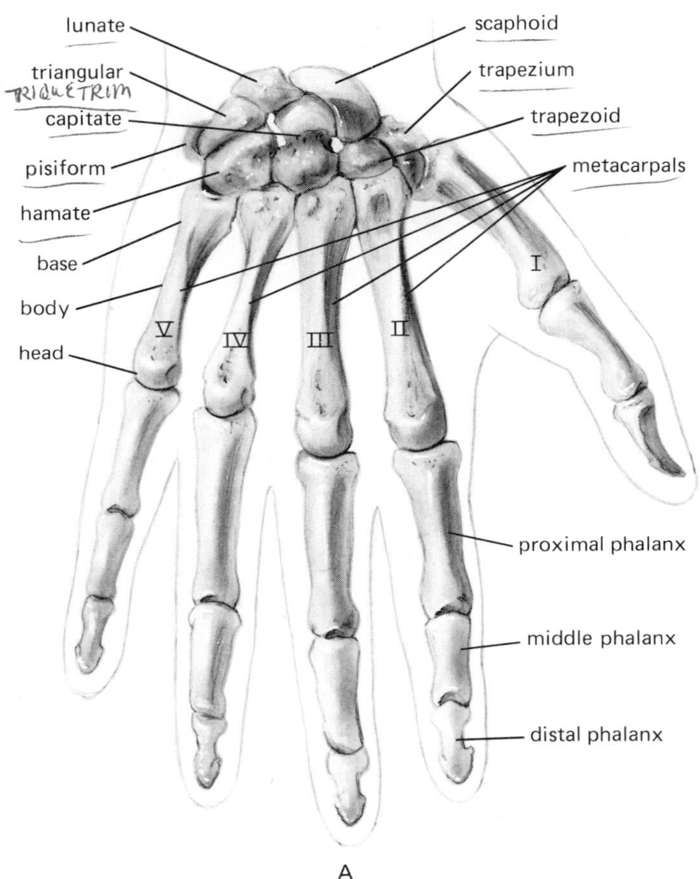

FIGURE 7-33 Dorsal (A) and ventral (B) views of the bones of the right wrist and hand. The skeleton of the hand has three parts: bones of the wrist *(carpals)*, palm *(metacarpals)*, and fingers *(phalanges)*. The eight carpal bones are arranged in two rows, each with four bones. Except for the small *pisiform* bone, all the carpal bones articulate with the bones adjacent to them. At the wrist joint the *scaphoid*, *lunate*, and *triangular* carpal bones articulate with the lower end of the radius and with the articular disc within the joint between the radius and ulna. There are five metacarpal bones in each hand; their bases articulate with the outer row of carpal bones and their heads with the proximal phalanges. The knuckles are formed by the heads of the metacarpal bones. Each of the four fingers contains three phalanges; designated as *proximal*, *middle* and *distal*. The thumb has only two phalanges.

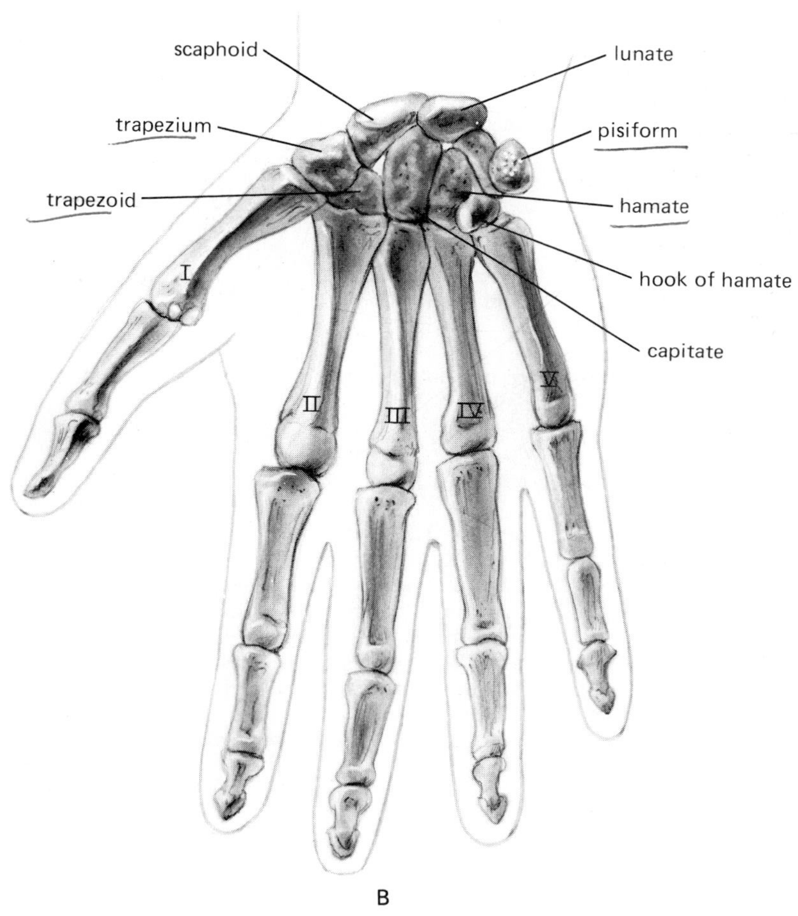
scaphoid
lunate
trapezium
pisiform
trapezoid
hamate
hook of hamate
capitate
I
II
III
IV
V

B

TABLE 7-12
MUSCLES CONNECTING THE UPPER LIMB AND VERTEBRAL COLUMN

Muscle	*Figure*	*Nerve*	*Origin*	*Insertion*	*Action and comments*
Trapezius	7-22A, 7-23	Spinal accessory nerve and C3-4 spinal nerve	Superior nuchal line of the occipital bone, the occipital protuberance, the ligamentum nuchae, the spinous processes of C7 and T1-12 vertebrae, and the corresponding supraspinous ligaments	Clavicle, the acromion and spine of the scapula	Steadies the scapula and controls its position and movement during use of the arm; acting with other muscles, it elevates, rotates, or retracts the scapula; when the shoulder is fixed, it draws the head backward and laterally
Latissimus dorsi	7-23, 7-25, 7-34A to C	Thoracodorsal nerve from the posterior chord of the brachial plexus	Spines of T6-12, lumbar, and sacral vertebrae, the supraspinous ligaments, crest of the ilium, and the lower ribs	Intertubercular sulcus of the humerus	Active in adduction, extension, and medial rotation of the humerus; aids in depressing the arm against resistance and backward swinging of the arm; pulls trunk upward and forward during climbing
Rhomboideus	7-23	Dorsal scapular nerve from C4-5	(See individual entries below)	(See individual entries below)	Helps control position and movement of the scapula during use of the arm; acting with other muscles, retracts and rotates the scapula and braces back the shoulder
Major	7-23	(See Rhomboideus)	Spines of T2-5 vertebrae and the supraspinous ligaments	Medial border of the scapula	(See Rhomboideus)
Minor	7-23	(See Rhomboideus)	Ligamentum nuchae and spines of C7 and T1 vertebrae	Medial border of scapula at the spine	(See Rhomboideus)
Levator scapulae	7-23	C3-5 spinal nerves	Transverse processes of the atlas, axis, and C3-4 vertebrae	Medial border of scapula above the spine	Helps control the position and movement of the scapula during use of the arm; when the cervical part of the vertebral column is fixed, helps elevate the scapula; when the shoulder is fixed, bends the neck to the same side

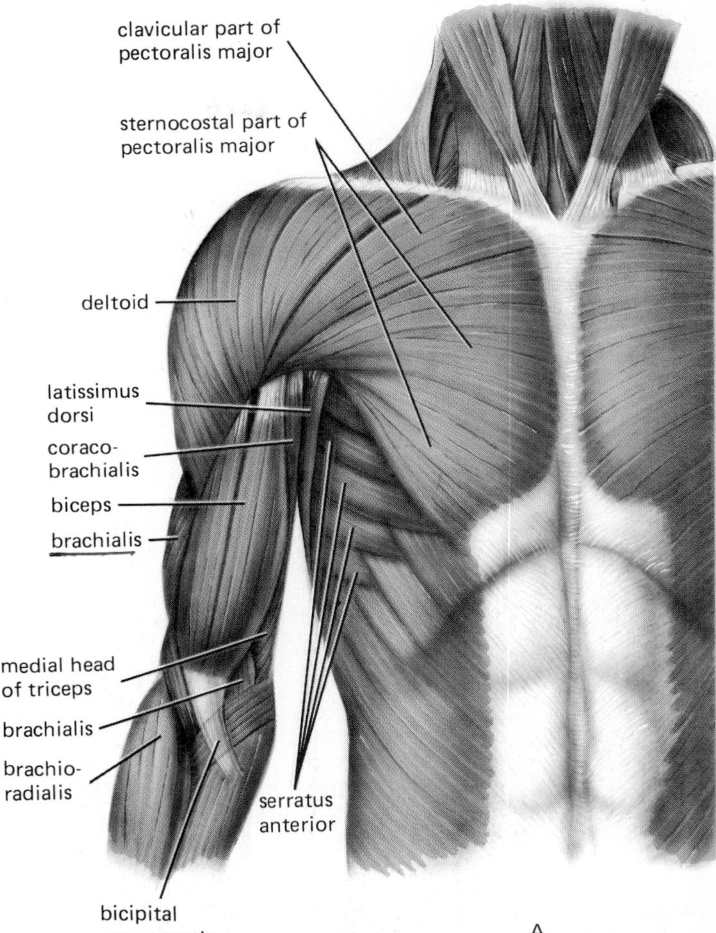

FIGURE 7-34 Superficial (A) and deep (B) muscles of the front of the chest and arm. Only the right side is shown. (C) The triceps and dorsal scapular muscles in a posterior view of the right shoulder. Part of the scapula has been removed. The muscles of the upper arm include the *coracobrachialis*, which acts only on the shoulder joint; the *biceps* and *triceps*, which act on both the shoulder and elbow joints, and the *brachialis*, which acts on the elbow joint.

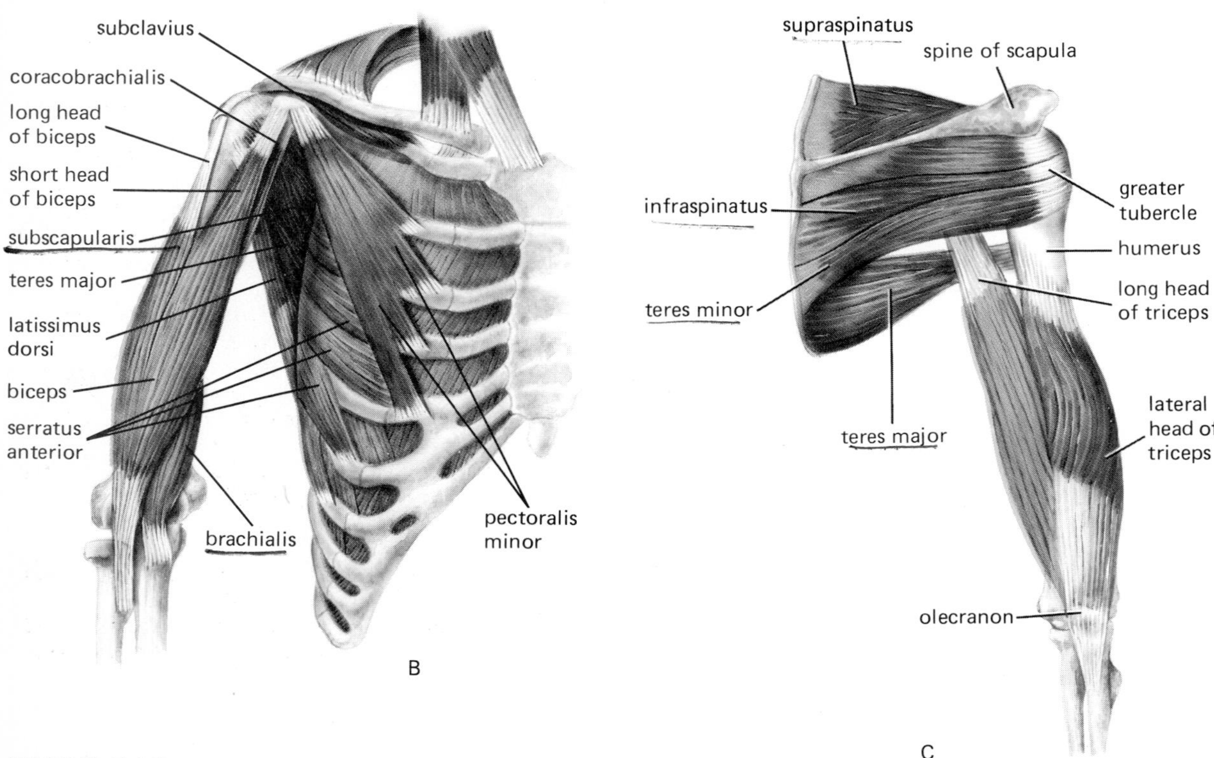

TABLE 7-13

MUSCLES CONNECTING THE UPPER LIMB AND THORACIC WALL

Muscle	*Figure*	*Nerve*	*Origin*	*Insertion*	*Action and comments*
Pectoralis:					
Major	7-34A	Lateral and medial pectoral nerves	Clavicle, anterior surface of the sternum, cartilages of the true ribs, and aponeurosis of the external abdominal oblique muscle	Crest of greater tubercle of the humerus	Adduction and medial rotation of the humerus; when the arm is extended, pectoralis major draws it forward and medially; in climbing, draws trunk forward and upward
Minor	7-34A and B	Medial pectoral nerves	3-5 ribs and intercostal fascia	Coracoid process of the scapula	Aids in drawing scapula forward around the chest wall and rotating scapula to depress point of shoulder
Subclavius	7-34B	Branch of brachial plexus from C5-6	Junction of first rib and its costal cartilage	Clavicle	Pulls point of shoulder down and forward and steadies the clavicle during movements of the shoulder
Serratus anterior	7-23, 7-25, 7-34A and B	Long thoracic nerve from C5-7	Upper nine ribs and intercostal fascia	Medial border of the scapula	Helps draw scapula forward in reaching and pushing movements; with the trapezius, rotates the scapula to elevate the point of the shoulder

TABLE 7-14
MUSCLES OF THE SHOULDER

Muscle	*Figure*	*Nerve*	*Origin*	*Insertion*	*Action and comments*
Deltoid	7-23, 7-34A and C	Axillary nerve from C5-6	Clavicle, acromion and spine of the scapula	Shaft of the humerus	Aids in drawing arm forward or backward and rotating it medially or laterally, depending upon which of its fibers are active; principal abductor of the humerus
Subscapularis	7-34B	Subscapular nerves from C5-6	Subscapular fossa of scapula	Lesser tubercle of humerus	Steadies head of humerus, rotates arm medially
Supraspinatus	7-23, 7-34C	Suprascapular nerve from C5-6	Supraspinous fossa of scapula	Greater tubercle of the humerus	Steadies head of humerus, initiates abduction of the arm
Infraspinatus	7-23, 7-34C	Suprascapular nerve from C5-6	Infraspinous fossa of scapula	Greater tubercle of the humerus	Steadies head of humerus, rotates arm laterally
Teres					
Minor	7-23, 7-34	Axillary nerve from C5-6	Dorsal surface of scapula	Greater tubercle of the humerus	Steadies head of humerus, rotates arm laterally
Major	7-23, 7-34B and C	Subscapular nerve from C6-7	Inferior angle of scapula	Crest of the lesser tubercle of the humerus	Draws humerus medially and backward and rotates it medially

TABLE 7-15
MUSCLES OF THE UPPER ARM

Muscle	*Figure*	*Nerve*	*Origin*	*Insertion*	*Action and comments*
Coracobrachialis	7-34A and B	Musculocutaneous nerve from C5-7	Coracoid process of scapula	Shaft of the humerus	Draws arm forward and medially
Biceps brachii	7-34A and B	Musculotendonous nerve from C5-6	Short head: coracoid process of scapula; long head: supraglenoid tubercle	Tuberosity of radius	A powerful supinator of the arm, flexes elbow joint, stabilizes head of humerus
Brachialis	7-34A and B	Musculocutaneous nerve from C5-6 and radial nerve from C7	Lower half of the humerus and intermuscular septa	Tuberosity and coronoid process of the ulna	Flexes the elbow joint
Triceps	7-34A and C	Radial nerve from C6-8	Long head: infraglenoid tubercle of scapula; lateral head: shaft of the humerus; medial head: shaft of the humerus	Olecranon process of ulna	Extends the arm at the elbow joint; slightly extends and adducts humerus and stabilizes head of humerus

7-35 Muscles of the forearm. The [illegible] muscles (A and B) generally lie on the anterior side of the forearm whereas the extensors are on the posterior surface. The origins of these muscles are on the sides of the elbows (their bulk would interfere with the joints' motion if they were on the front or back). A common tendon of origin of several of the superficial flexors attaches to the medial epicondyle of the humerus (Fig. 7-29) whereas the lateral epicondyle and supracondylar ridge are important origins for the superficial extensors. The origins of the deep flexors and extensors are from the radius and ulna and the interosseous membrane. The muscles of the forearm can be divided into six functional groups: muscles which rotate the radius on the ulna (pronator teres, pronator quadratus, and supinator); muscles which flex the hand at the wrist (flexor carpi radialis, flexor carpi ulnaris, and palmaris longus); muscles which flex the fingers and thumb (flexor digitorum superficialis, flexor digitorum profundus, and flexor pollicis longus); muscles which extend the hand at the wrist (extensor carpi radialis longus, extensor carpi radialis brevis, and extensor carpi ulnaris); muscles which extend the fin-

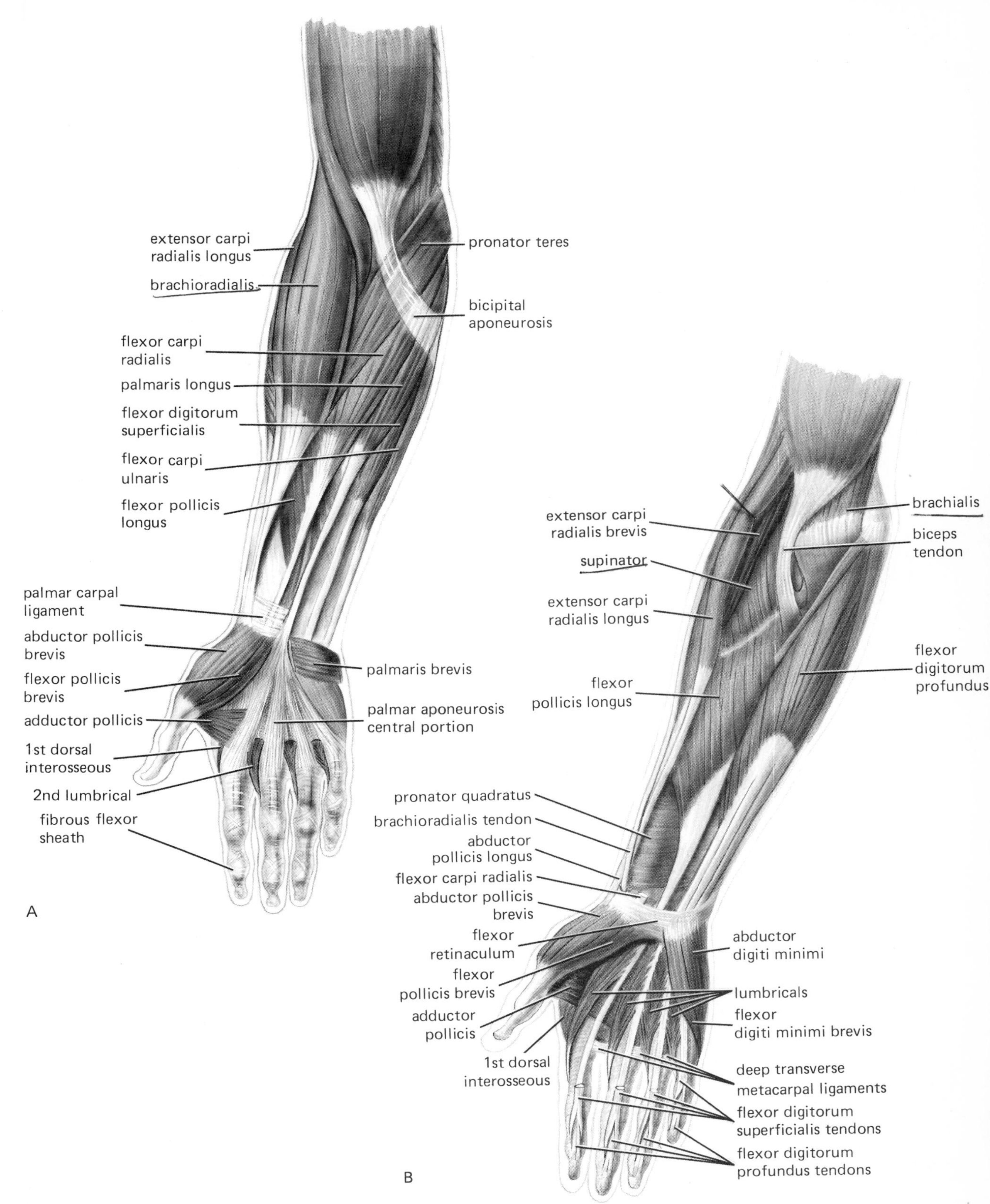
extensor carpi radialis longus
brachioradialis
flexor carpi radialis
palmaris longus
flexor digitorum superficialis
flexor carpi ulnaris
flexor pollicis longus
palmar carpal ligament
abductor pollicis brevis
flexor pollicis brevis
adductor pollicis
1st dorsal interosseous
2nd lumbrical
fibrous flexor sheath
pronator teres
bicipital aponeurosis
palmaris brevis
palmar aponeurosis central portion
A
extensor carpi radialis brevis
supinator
extensor carpi radialis longus
flexor pollicis longus
brachialis
biceps tendon
flexor digitorum profundus
pronator quadratus
brachioradialis tendon
abductor pollicis longus
flexor carpi radialis
abductor pollicis brevis
flexor retinaculum
flexor pollicis brevis
adductor pollicis
1st dorsal interosseous
abductor digiti minimi
lumbricals
flexor digiti minimi brevis
deep transverse metacarpal ligaments
flexor digitorum superficialis tendons
flexor digitorum profundus tendons
B

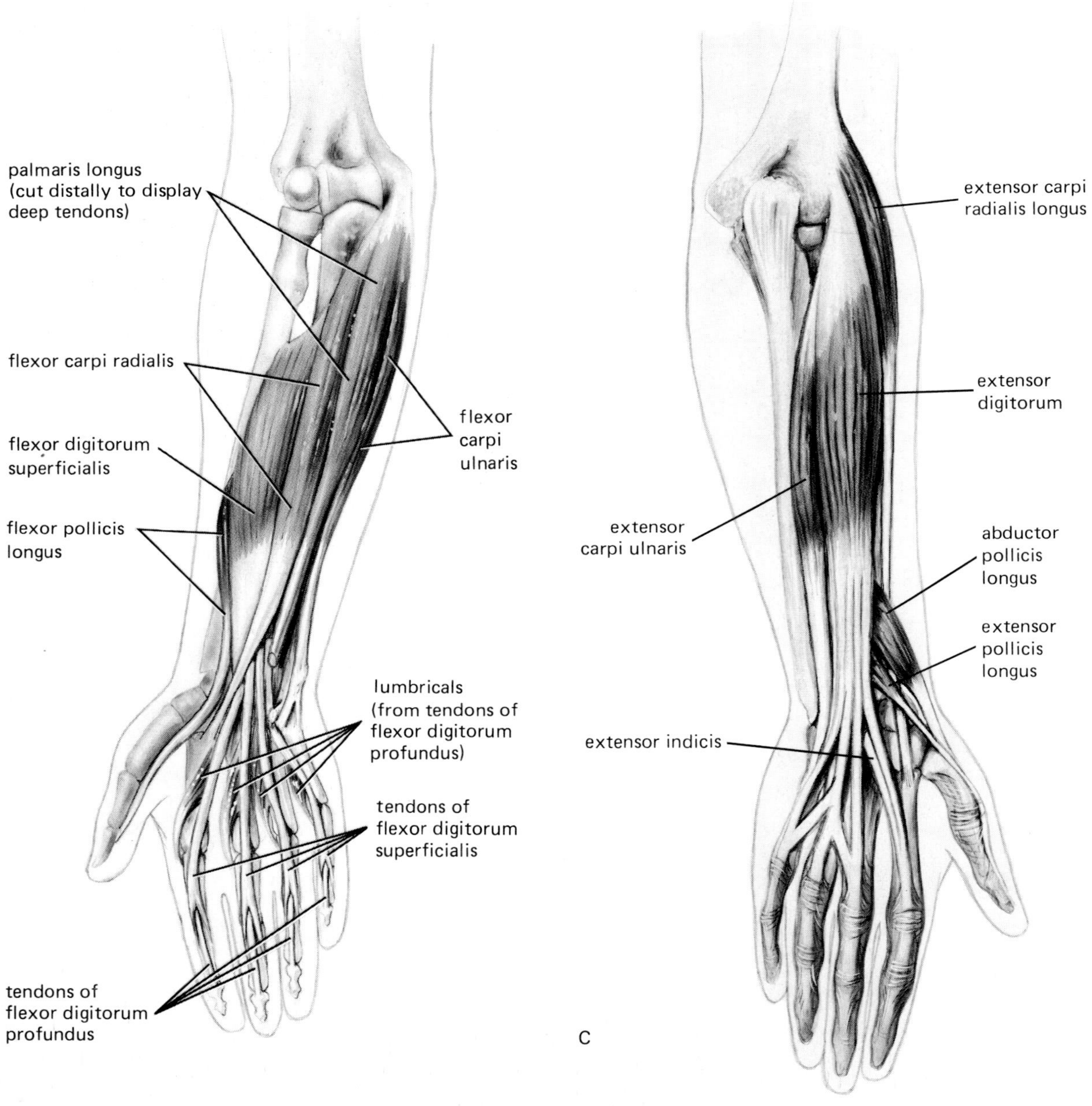
palmaris longus
(cut distally to display
deep tendons)
flexor carpi radialis
flexor digitorum
superficialis
flexor pollicis
longus
flexor
carpi
ulnaris
lumbricals
(from tendons of
flexor digitorum
profundus)
tendons of
flexor digitorum
superficialis
tendons of
flexor digitorum
profundus
extensor carpi
radialis longus
extensor
digitorum
extensor
carpi ulnaris
abductor
pollicis
longus
extensor
pollicis
longus
extensor indicis

C

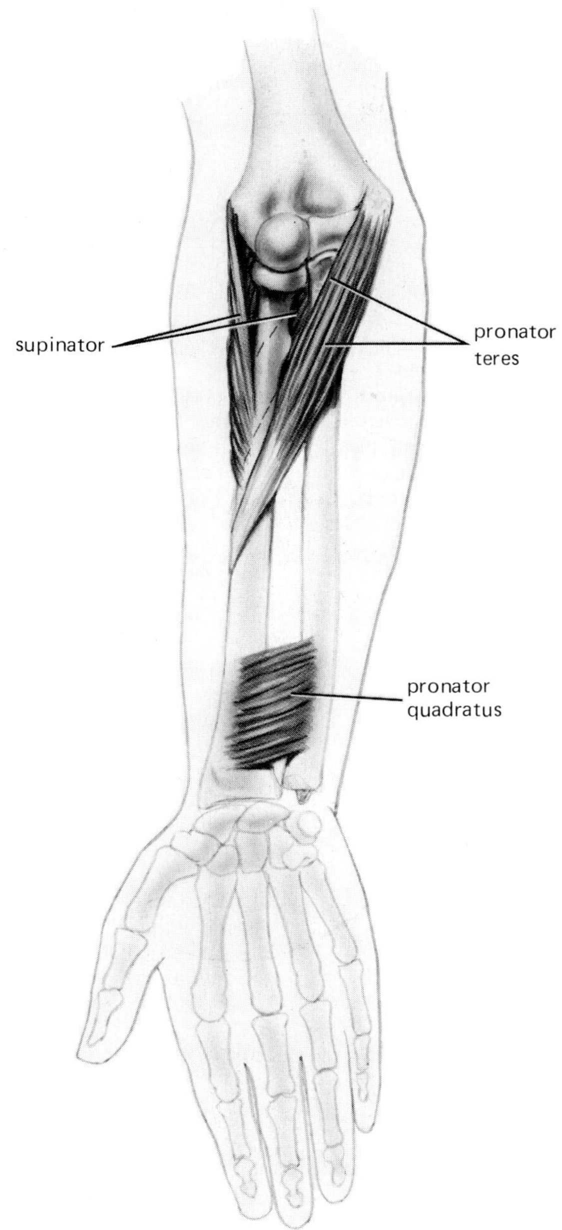

gers (extensor digitorum, extensor indicis, and extensor digiti minimi); and muscles which extend the thumb (abductor pollicis longus, extensor pollicis brevis, and extensor pollicis longus). Brachioradialis has not been mentioned; it is an elbow flexor and has no action on the hand or fingers. The muscles of the forearm are enclosed in a sheath of fascia. This sheath is reinforced on the dorsal side of the wrist by connective tissue fibers running between the radius and the ulna, triangular bone, and pisiform bone; this thickening forms the *extensor retinaculum.* A similar reinforcement, the *palmar carpal ligament,* occurs on the anterior side of the wrist. Part C shows more complete dissections of some of the forearm muscles.

TABLE 7-16
MUSCLES OF THE FOREARM

Muscle	*Figure*	*Nerve*	*Origin*	*Insertion*	*Action and comments*
Superficial flexor muscles:					
Pronator teres	7-35A and C	Median nerve from C6-7	Humeral head: medial epicondyle of humerus; ulnar head: coronoid process of the ulna	Shaft of radius	Rotates the radius upon the ulna turning the palm backward (i.e., it pronates the forearm)
Flexor carpi radialis	7-35A to C	Median nerve from C6-7	Medial epicondyle of humerus	Base of second and third metacarpal bones	Aids in flexing wrist and abducting the hand
Palmaris longus	7-35A	Median nerve from C7-8	Medial epicondyle of humerus	Palmar aponeurosis	Flexes the wrist and tenses palmar fascia
Flexor carpi ulnaris	7-35A to C	Ulnar nerve from C7-8	Medial epicondyle of the humerus and the olecranon process of the ulna	Pisiform, hamate, and fifth metacarpal bones	Aids in flexing the wrist and adducting the hand
Flexor digitorum superficialis	7-35A and C	Median nerve from C7-8 and T1	Medial epicondyle of the humerus, medial ulna, anterior border of the radius	Middle phalanges of digits 2 to 5	Flexes the middle and then the proximal phalanges and flexes the wrist
Deep flexor muscles:					
Flexor digitorum profundus	7-35B	Ulnar and median nerves from C8 and T1	Upper three-fourths of the ulna and adjacent interosseus membrane	Distal phalanges of digits 2 to 5	Flexes distal phalanges and aids in flexing the wrist
Flexor pollicis longus	7-35A to C	Median nerve from C8 and T1	Radius, adjacent interosseus membrane	Distal phalanx of the thumb	Flexes the phalanges of the thumb
Pronator quadratus	7-35B and C	Median nerve from C8 and T1	Distal end of shaft of ulna	Distal end of radius	Pronates the forearm, opposes separation of the lower ends of the radius and ulna
Superficial extensor muscles:					
Brachioradialis	7-34A and B, 7-35A	Radial nerve from C5-7	Lateral supracondylar ridge of the humerus	Styloid process of radius	Flexes the elbow joint

TABLE 7-16
MUSCLES OF THE FOREARM *(Continued)*

Muscle	*Figure*	*Nerve*	*Origin*	*Insertion*	*Action and comments*
Extensor carpi radialis					
Longus	7-35A to C	Radial nerve from C6-7	Lateral supracondylar ridge of humerus	Base of the second metacarpal bone	Aids in extending the wrist and abducting the hand
Brevis	7-35A	Deep radial nerve from C7-8	Lateral epicondyle of humerus	Base of second and third metacarpal bone	Aids in extending the wrist and abducting the hand
Extensor digitorum	7-35C	Deep radial nerve from C7-8	Lateral epicondyle of the humerus	Phalanges of second to fifth fingers	Extends fingers, hand, and forearm
Extensor digiti minimi	Not shown	Deep radial nerve from C7-8	Common extensor tendon and intermuscular septa	Phalanges of little finger	Extends the little finger and wrist
Extensor carpi ulnaris	7-35C	Deep radial nerve from C7-8	Lateral epicondyle of the humerus, posterior border of ulna	Base of fifth metacarpal bone	Aids in extending and adducting the wrist
Anconeus	Not shown	Radial nerve from C7-8 and T1	Lateral epicondyle of the humerus	Olecranon process and shaft of the ulna	Extends elbow joint
Deep extensor muscles:					
Supinator	7-35B and C	Deep radial nerve from from C5-6	Lateral epicondyle of the humerus and ligaments of the elbow joint and from supinator crest of ulna	Upper shaft of radius	Supinates hand and forearm
Abductor pollicis longus	7-35B and C	Posterior interosseus nerve from C7-8	Shaft of the radius and ulna and the interosseus membrane	First metacarpal bone	Abducts and extends the thumb
Extensor pollicis brevis	Not shown	Posterior interosseus nerve from C7-8	Posterior surface of the radius and the interosseus membrane	Base of the proximal phalanx of the thumb	Extends proximal phalanx of the thumb and helps extend metacarpal bone
Extensor pollicis longus	7-35C	Posterior interosseus nerve from C7-8	Shaft of the ulna and the interosseus membrane	Base of the distal phalanx of the thumb	Extends the metacarpal and proximal and distal phalanges of the thumb; rotates extended thumb laterally
Extensor indicis	Not shown	Posterior interosseus nerve from C7-8	Posterior surface of the ulna and interosseus membrane	Extensor expansion of index finger	Extends the index finger, aids in extending the wrist

TABLE 7-17
MUSCLES OF THE HAND

Muscle	*Figure*	*Nerve*	*Origin*	*Insertion*	*Action and comments*
Abductor pollicis brevis	7-35A and B	Median nerve from C8 and T1	Flexor retinaculum, tubercles of scaphoid bone and trapezium	Base of the proximal phalanx of the thumb	Abducts thumb and rotates it medially
Opponens pollicis	Not shown	Median nerve from C8 and T1	Tubercle of the trapezium and the flexor retinaculum	Palmar surface of the metacarpal bone of the thumb	Flexes metacarpal bone of thumb and rotates it medially, bringing the tip of the thumb in contact with the palmar surface of the fingers
Flexor pollicis brevis	7-35A and B	Median and ulnar nerves from C8 and T1	Flexor retinaculum, tubercle of the trapezium	Base of the proximal phalanx of the thumb	Flexes the proximal phalanx of the thumb and flexes the metacarpal bone and rotates it medially
Adductor pollicis	7-35A and B	Ulnar nerve from C8 and T1	Capitate bone, bases of the second and third metacarpal bones	Base of the proximal phalanx of the thumb	Approximates the thumb to the palm of the hand
Palmaris brevis	7-35A	Ulnar nerve from C8 and T1	Flexor retinaculum and palmar aponeurosis	Skin of the ulnar side of the hand	Wrinkles the skin on the ulnar side of the palm of the hand and deepens the hollow of the palm
Abductor digiti minimi	7-35B	Ulnar nerve from C8 and T1	Pisiform bone, tendon of flexor carpi ulnaris	Base of the proximal phalanx of the little finger	Abducts the little finger away from the fourth
Flexor digiti minimi brevis	7-35B	Ulnar nerve from C8 and T1	Hamate bone and flexor retinaculum	Base of the proximal phalanx of the little finger	Flexes the little finger
Opponens digiti minimi	Not shown	Ulnar nerve from C8 and T1	Hamate bone and flexor retinaculum	Fifth metacarpal bone	Draws the fifth metacarpal bone forward and rotates it laterally
Lumbricals	7-35A to C	Median and ulnar nerves from C8 and T1	Tendons of the flexor digitorum profundis	Extensor expansion covering the dorsal surface of each finger	Flex digits at joint between metacarpals and phalanges; extend terminal phalanges
Dorsal and palmar interossei	7-35A and B	Deep ulnar nerve from C8 and T1	Metacarpal bones	Proximal phalanges and extensor expansion	Abduct and adduct digits 2 to 5, aid in extending terminal phalanges

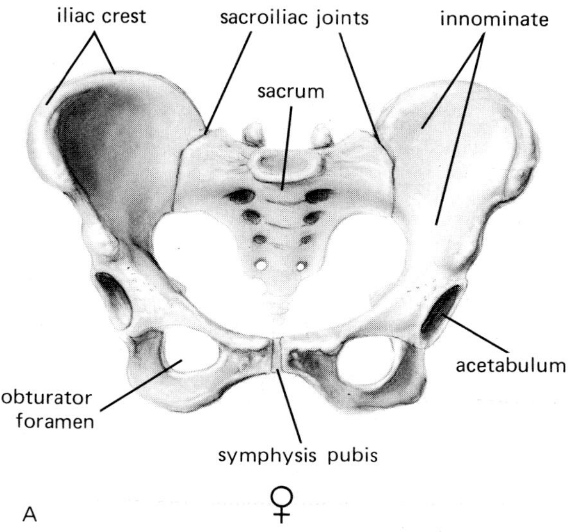

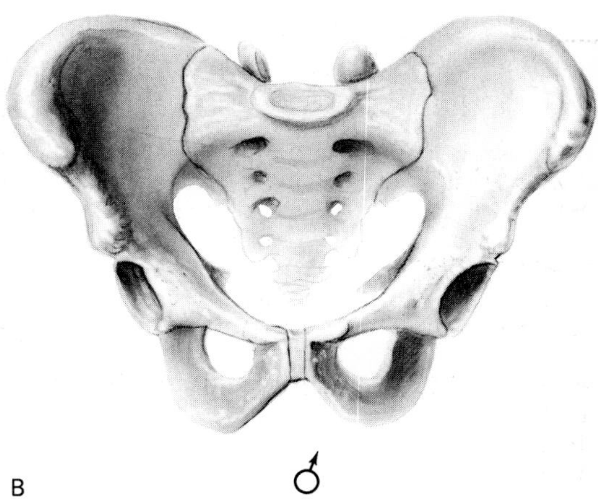

FIGURE 7-36 Anterior view of a female (A) and male (B) pelvis. The bony *pelvis* is a ring, formed by the two innominate, or hip, bones and the sacrum, which attaches the bones of the legs to the vertebral column. The innominate bones articulate with each other in front at the *symphysis pubis* and with the sacrum in back at the *sacroiliac joints*. (The term "pelvis" also applies to the general region where the trunk joins the legs and to the cavity within the bony pelvis.) The bony pelvis provides the most marked skeletal difference between the female and male; these differences are closely related to the body weight that must be supported in upright posture and locomotion and, in the female, the process of childbirth. In the male the pelvis is more massive and the markings for muscle and ligament attachments are more pronounced whereas in the female the pelvis is shallower and the openings broader.

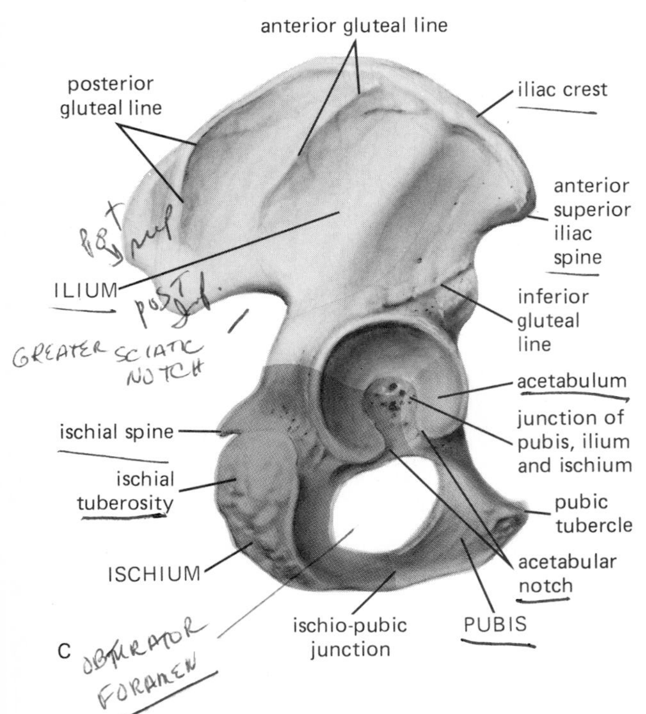

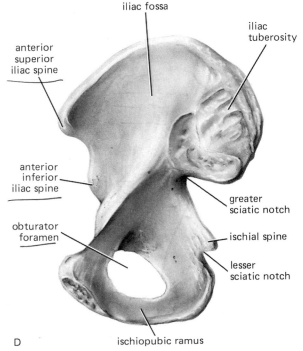

iliac fossa
iliac tuberosity
anterior superior iliac spine
anterior inferior iliac spine
greater sciatic notch
obturator foramen
ischial spine
lesser sciatic notch
ischiopubic ramus
D

Lateral (external) (C) and medial (D) aspects of the left innominate bone. The innominate bones are also called the *os coxae;* each consists of three parts: the *ilium*, *ischium*, and *pubis*, which are joined by cartilage in youth but fused together in the adult. The innominate bones are large and irregularly shaped with a constriction in the middle. The lateral surface of each contains a deep, cup-shaped hollow, the *acetabulum*, which articulates with the round head of the femur to form the hip joint. Above the acetabulum, the bone expands into a fan-shaped projection, the *iliac crest;* below is a large opening in the bone, the *obturator foramen.*

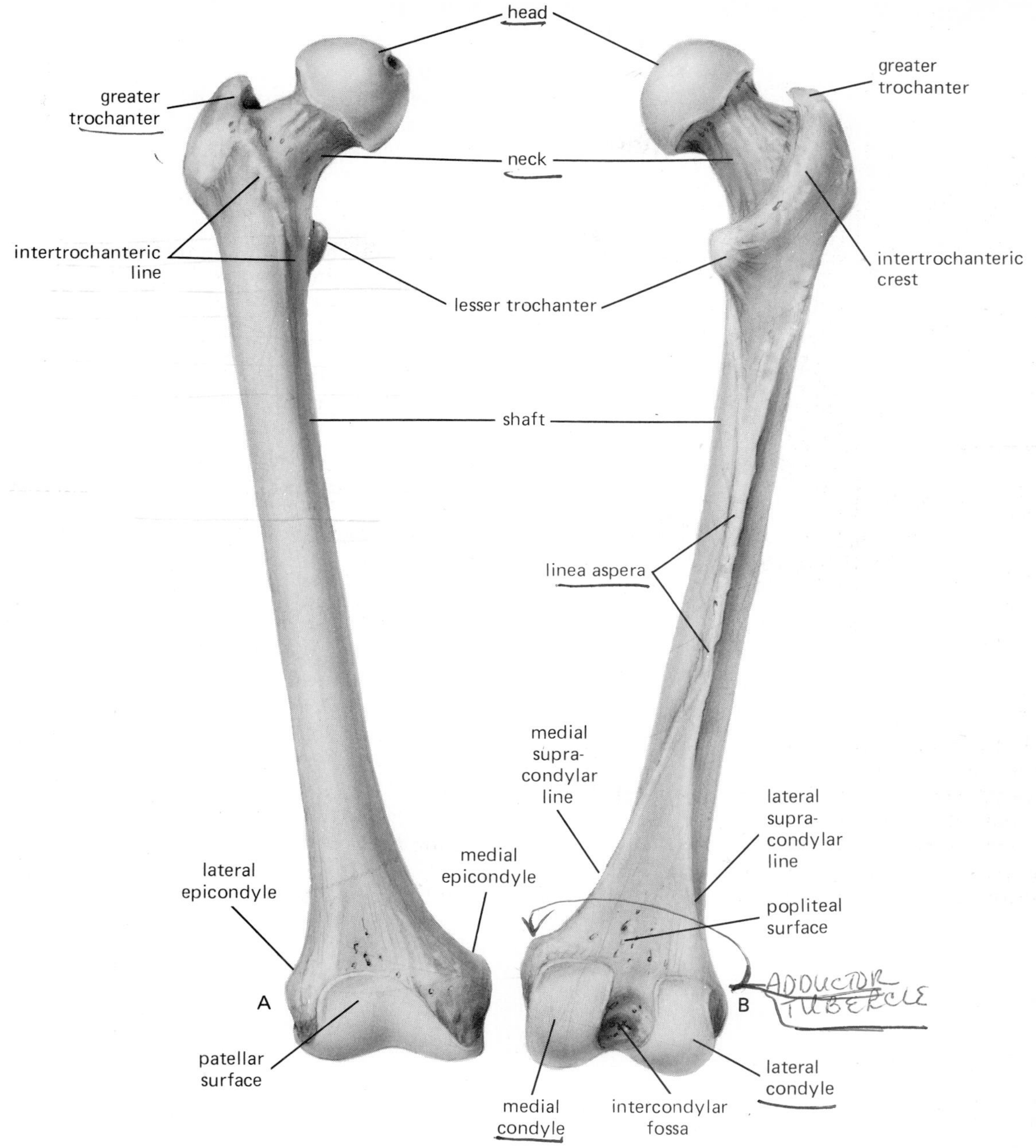

FIGURE 7-37 Anterior (A) and posterior (B) views of the right femur. The *femur,* or thigh bone, is the longest and strongest bone in the body, the length reflecting the length of man's stride and the strength reflecting the weight and muscular forces it must withstand. The rounded *head* at the upper end of the bone articulates with the acetabulum of the innominate bone to form the hip joint. The *neck* and *greater* and *lesser trochanters* are other major landmarks of the upper end. The two knuckle-like *condyles* at the lower end of the bone articulate with the tibia.

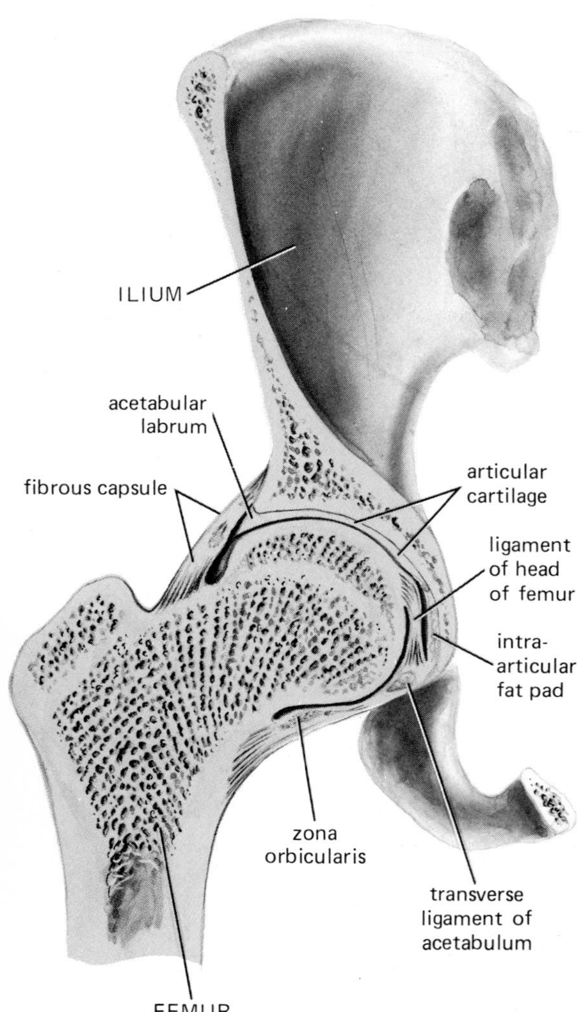

FIGURE 7-38 A section through the hip joint. The head of the femur fits into the shallow cuplike fossa of the acetabulum of the innominate bone to form the hip joint. The articular surfaces of both bones are almost completely covered with cartilage except at the point of attachment of the ligaments. The ligamentous structures associated with the joint are the fibrous capsule, the cartilaginous ring around the acetabulum (the acetabular labrum), the ligament of the head of the femur, the transverse ligament of the acetabulum, and other ligaments connecting the femur with the innominate bone. The joint is of the ball-and-socket type and is capable of flexion-extension, adduction-abduction, and circumduction (a combination of the previous four) as well as lateral and medial rotation. The major hip *flexors* are psoas major and iliacus; they are assisted by pectineus, rectus femoris, sartorius, and the adductors. The major hip *extensors* are gluteus maximus and the hamstring muscles (see Fig. 7-42). *Abduction* is achieved by gluteus medius and minimus with tensor fascia latae and sartorius, and *adduction* by the adductors longus, brevis, and magnus with pectineus and gracilis. *Medial rotation*, a weak movement, is achieved by tensor fascia latae and anterior fibers of the gluteus medius and minimus. *Lateral rotation* is strong, performed chiefly by the obturator muscles, gemelli, quadratus femoris, and gluteus maximus.

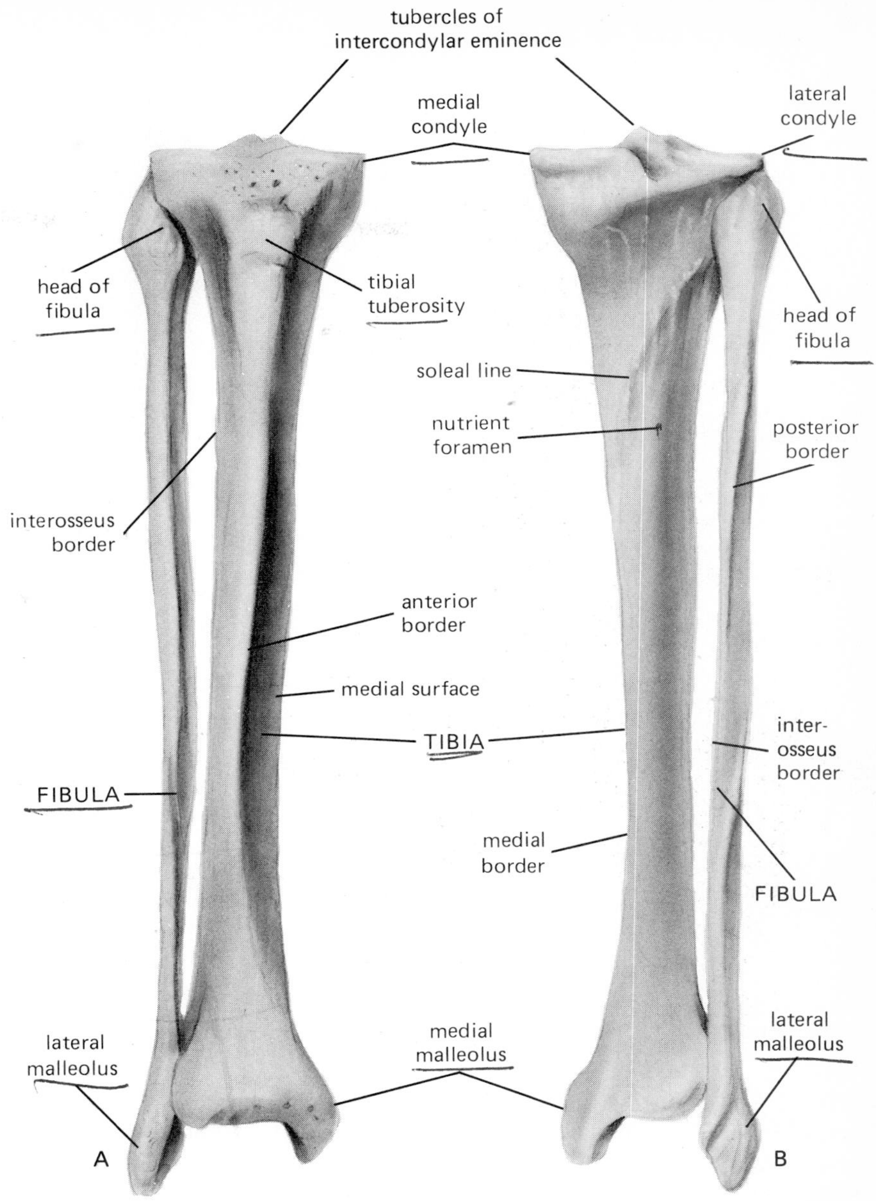

FIGURE 7-39 Anterior (A) and posterior (B) views of the right tibia and fibula. The *tibia*, the weight-bearing bone of the leg, is the more medial and stronger of the two. Its upper end has two prominent masses, the *medial* and *lateral condyles*, which articulate with the corresponding condyles of the femur to form the knee joint; its lower end is smaller and bears on its medial surface a large *medial malleolus*. The shaft of the tibia lies just under the skin and, over most of its length, forms a sharp crest, the *shin*. The *fibula*, the lateral bone of the leg, is more slender than the tibia and serves mainly for muscle attachments and to complete the lateral part of the ankle joint. The bone consists of a head, shaft, and lower end; the head articulates with the tibia at the inferior surface of the lateral condyle, and the lower end articulates with the talus bone in the ankle and again with the tibia. Tibiofibular articulations and interosseous membrane which connects their shafts allow only very slight movements of the bones relative to each other, a distinct difference between these two bones and the radius and ulna of the forearm.

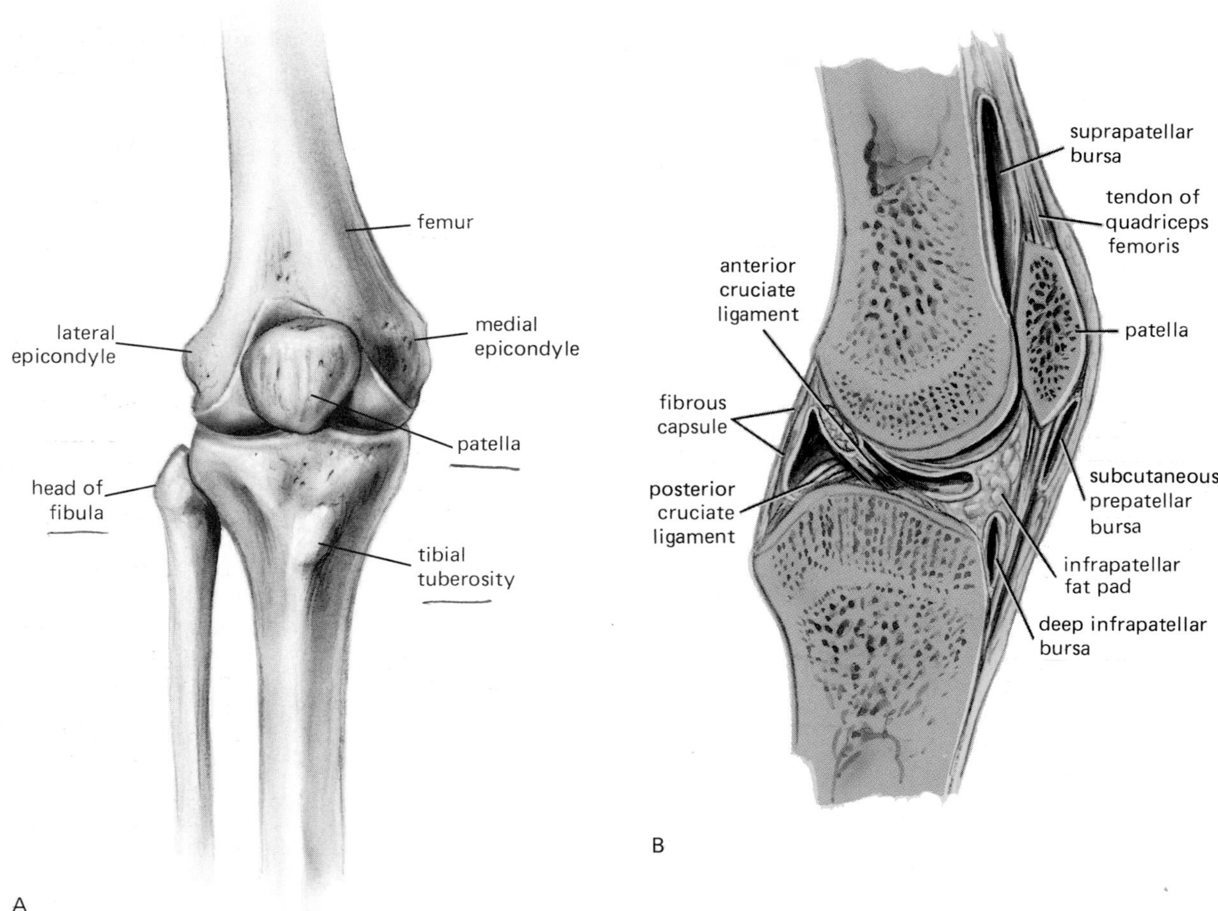

FIGURE 7-40 (A) Anterior view of the right knee joint. (B) Sagittal section through the knee joint. The *knee*, the largest joint in the body, is composed of two articulations between the opposed medial and lateral condyles of the femur and tibia. Full extension of the knee involves a screw-type action, the lateral condyle of the femur sliding forwards and the medial condyle backwards, the collateral ligaments tightening to hold the bones against one another. As the knee is flexed, the obliquely-oriented popliteus muscle rotates the femur laterally and draws the lateral condyle of the femur backward, "unlocking" the joint. The support of the body on the opposed ends of two long bones is not basically stable, but the stability is improved by strong ligaments and a strong capsule reinforced by tendons and aponeuroses. The knee is functionally a hinge joint, but it also allows a small amount of rotation, particularly when in the flexed position. Flexion is produced chiefly by biceps femoris, semitendinosus and semimembranosus; extension by quadriceps femoris; medial rotation of the leg by popliteus, semimembranosus, and semitendinosus; and lateral rotation of the leg by the biceps femoris alone. Associated with the knee and sharing a common compartment with the knee joint is the *patella*, a sesamoid (i.e., more or less oval, seedlike) bone in the front of the knee.

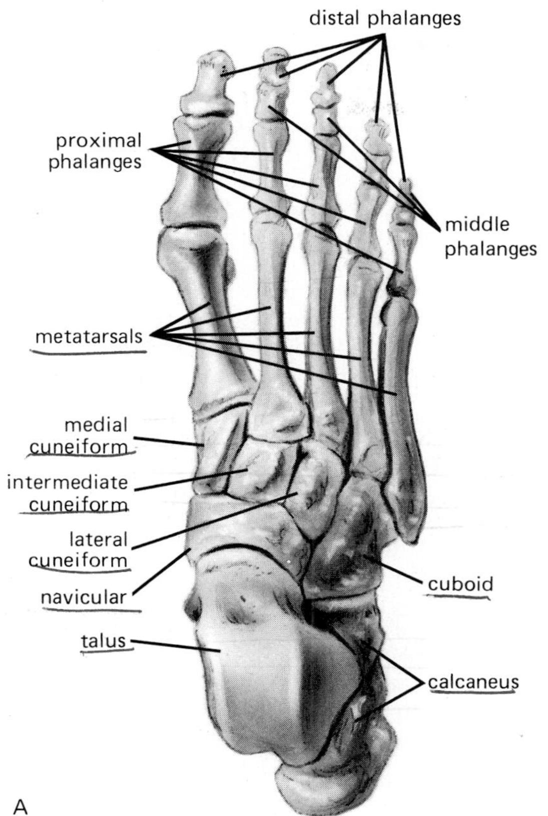

FIGURE 7-41 Dorsal (A) and lateral (B) views of the bones of the right foot. The skeleton of the foot has three divisions: the *tarsal* bones of the ankle, the *metatarsal* bones of the foot proper, and the *phalanges* of the toes. The seven tarsal bones comprise the skeleton of the posterior half of the foot. Like the carpal bones of the wrist, they are arranged in two rows, although the proximal row (i.e., that nearer the heel) is irregular, the *talus* lying above the *calcaneus*, and the *navicular* bone lying between the talus and bones of the second row. This arrangement allows the weight of the body to be transmitted from the talus to the other tarsal and metatarsal bones over the arched form of the foot rather than directly downward through the talus. The talus articulates with the tibia, but the calcaneus is the largest of the tarsal bones. It projects posteriorly to form the "heel bone" and receives the attachments of the calf muscles via the calcaneal, or Achilles, tendon. Five metatarsal bones connect the tarsus with the phalanges, two phalanges in the great toe and three in each of the other four toes. Unlike the thumb, which is rotated even in its resting position, the great toe is in the same plane as the other toes with its flexor surface facing the ground. Though utilizing bones and arragements similar to the hand, the foot comprises one of the most highly specialized features of human beings.

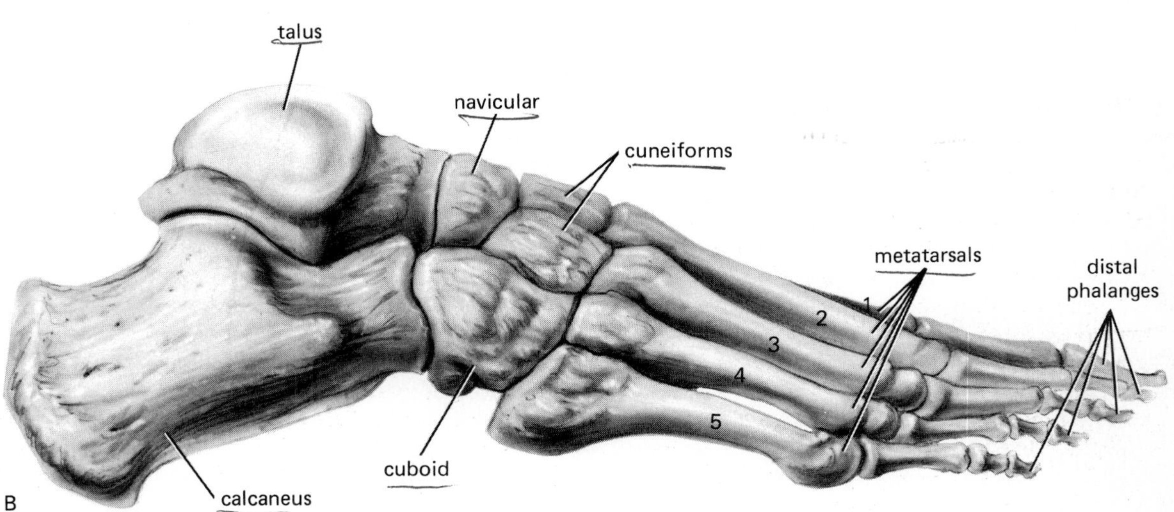

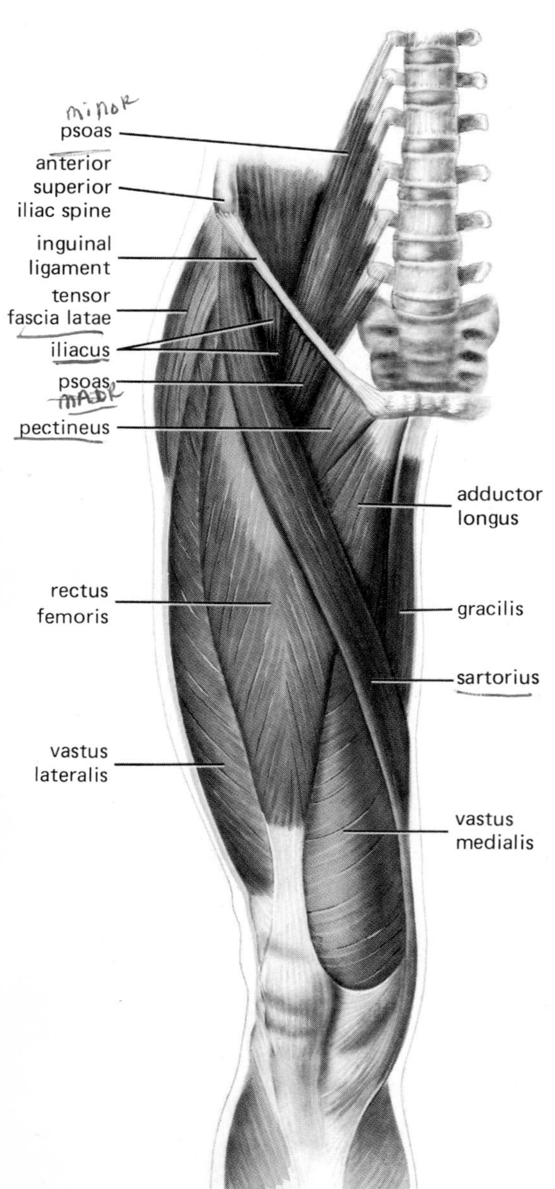

FIGURE 7-42 Anterior (A) and posterior (B) views of the muscles of the right hip and thigh. Specific muscles are shown in part C. The muscles of the thigh may be conveniently divided into three groups separated by inward extensions from the *fascia lata*, a sheet of fascia which forms an almost stocking-like covering of the tissues of the thigh. Muscles of the anterior group (see Table 7-19) occupy the entire anterior one-half of the thigh and almost completely surround the femur. They powerfully extend the leg at the knee and, in a weaker action, flex the thigh at the hip. (The major flexors of the thigh are the iliacus and psoas major—sometimes collectively called the iliopsoas, Table 7-18.) Of the anterior group, the rectus femoris and the three muscles of the vastus group act on the knee through a common tendon and are considered as one muscle, the *quadriceps femoris*. The medial group includes the muscles which lie in the gluteal region and abduct and laterally rotate the thigh as well as the thigh adductors. The posterior group includes the thigh flexors *(biceps femoris, semitendinosus,* and *semimembranosus)*, which span the hip and knee joint and integrate extension of the hip with flexion of the knee and are often grouped together as the "*hamstring muscles.*"

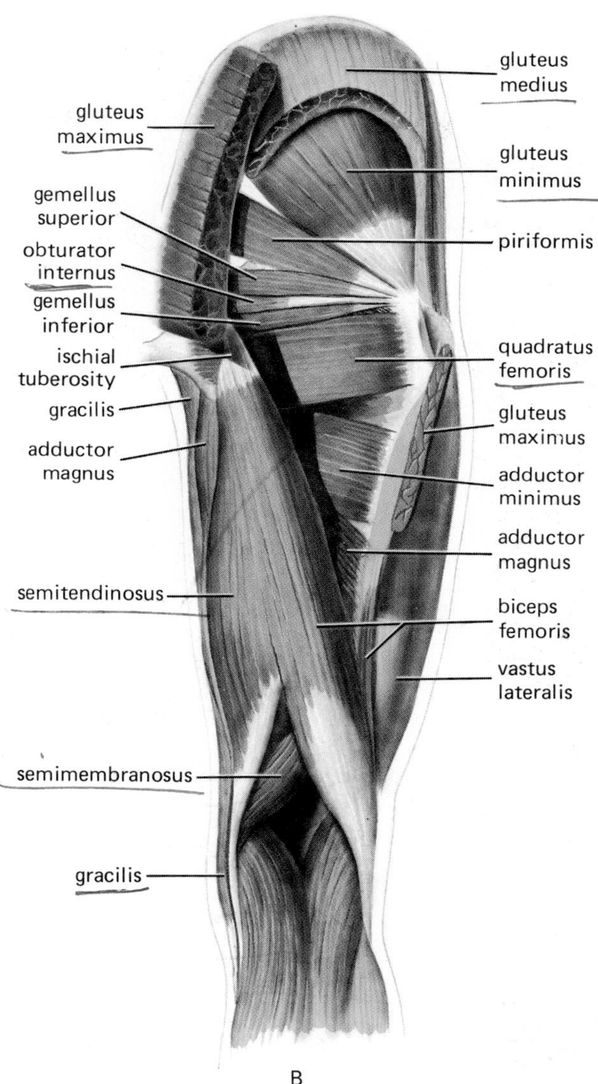
gluteus medius
gluteus maximus
gluteus minimus
gemellus superior
obturator internus
piriformis
gemellus inferior
ischial tuberosity
quadratus femoris
gracilis
gluteus maximus
adductor magnus
adductor minimus
adductor magnus
semitendinosus
biceps femoris
vastus lateralis
semimembranosus
gracilis

B

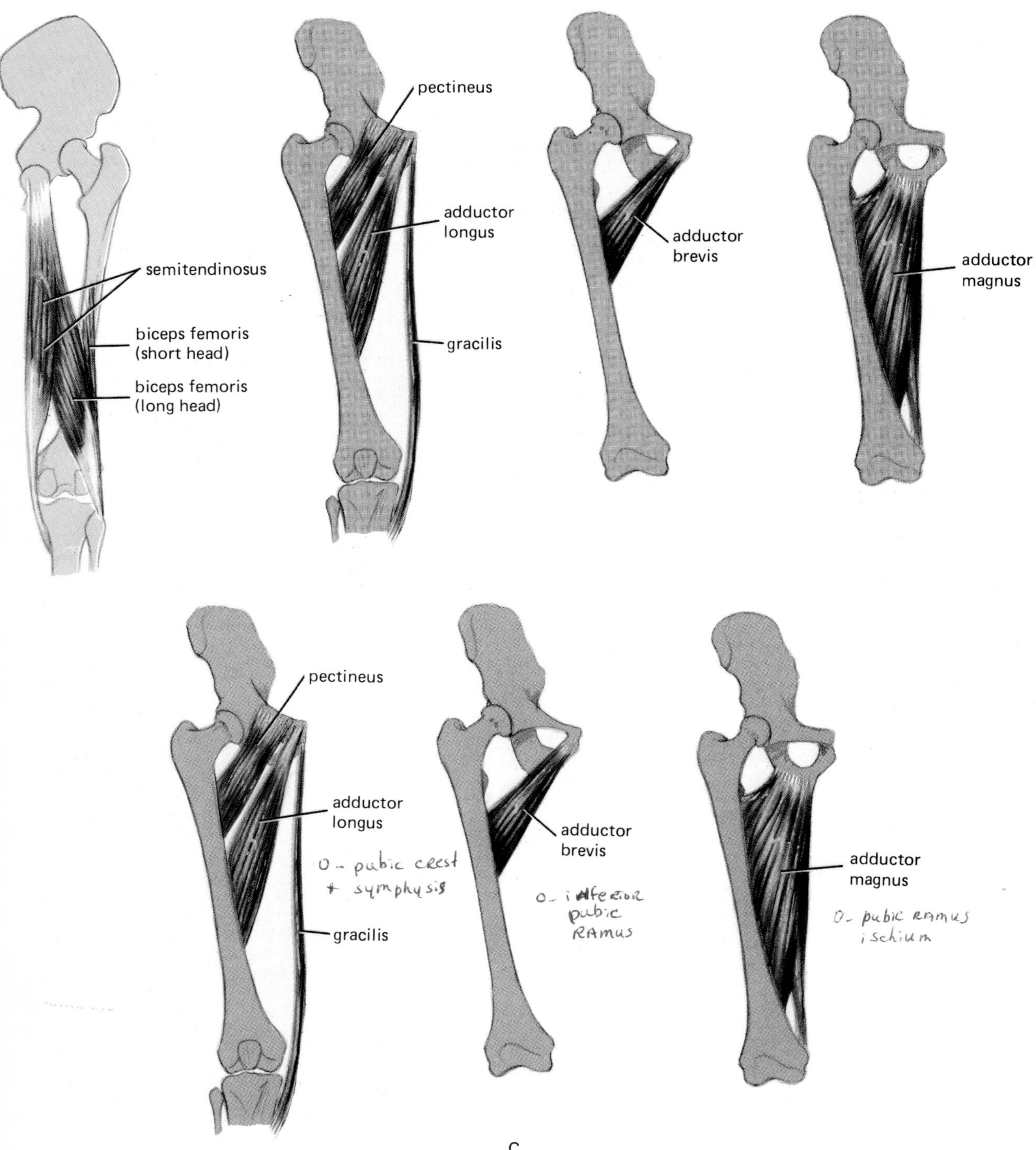
semitendinosus
biceps femoris (short head)
biceps femoris (long head)
pectineus
adductor longus
gracilis
adductor brevis
adductor magnus
pectineus
adductor longus
O – pubic crest + symphysis
gracilis
adductor brevis
O – inferior pubic ramus
adductor magnus
O – pubic ramus ischium

C

TABLE 7-18
MUSCLES OF THE ILIAC REGION

Muscle	*Figure*	*Nerve*	*Origin*	*Insertion*	*Action and comments*
Psoas					Flexes thigh upon the pelvis
Major	7-42A	L1-3 spinal nerves	Transverse processes of the lumbar vertebrae; the vertebral bodies and intervertebral discs	Inserts with the fibers of the iliacus on the lesser trochanter of the femur	Acts with the iliacus (see below)
Minor	Not shown	L1 spinal nerve	Bodies of T12 and L1 vertebrae and the disc between them	Iliopectinal eminence and iliac fascia	Weak flexor of the trunk
Iliacus	7-42A	Branches of the femoral nerve from L2-3	Upper two-thirds of the concavity of the iliac fossa, inner lip of the iliac crest, lateral part of the sacrum, and adjacent ligaments	Tendon of psoas major and femur below lesser trochanter	Assists in flexion of the thigh; when acting with psoas major and the femur is stabilized, bends the trunk and pelvis forward, as in raising the trunk from a lying to a sitting posture

TABLE 7-19
MUSCLES OF THE THIGH AND GLUTEAL REGION

Muscle	*Figure*	*Nerve*	*Origin*	*Insertion*	*Action and comments*
Anterior femoral muscles:					
Tensor fasciae latae	7-42A	Superior gluteal nerve from L4-5	Outer lip of the iliac crest, anterior superior iliac spine, and fascia lata	Iliotibial part of fascia lata, lateral femoral condyle and femur via lateral septum	Extends the knee with lateral rotation of the leg; when the knee is stabilized, steadies the head of the femur in the pelvis and the condyles of the femur on the tibia and thus helps maintain upright posture
Sartorius	7-42A and B, 7-43B	Femoral nerve from L2-3	Anterior superior iliac spine and notch	Medial surface of the tibia	The longest muscle in the body; assists in flexing the leg on the thigh and the thigh on the pelvis; helps abduct the thigh and rotate it laterally
Quadriceps femoris	7-42A to C	Femoral nerve from L2-4	(See individual entries below)	The tendons of the four parts of the quadriceps unite to form a single tendon which attaches to the base of the patella and tubercle of the tibia	The great extensor muscle of the leg; covers almost all the front and sides of the femur; can be divided into four parts

TABLE 7-19
MUSCLES OF THE THIGH AND GLUTEAL REGION *(Continued)*

Muscle	*Figure*	*Nerve*	*Origin*	*Insertion*	*Action and comments*
Rectus femoris	7-42A	(See Quadriceps femoris)	Arises by two heads: from the inferior iliac spine and from the groove above the acetabulum and capsule of the hip joint	Base of the patella	Extends leg at the knee; also helps flex the thigh on the pelvis; can flex hip and extend knee simultaneously
Vastus lateralis	7-42A	(See Quadriceps femoris)	Intertrochanteric line, borders of the greater trochanter, gluteal tuberosity, and linea aspera of the femur	Lateral border of the patella and the quadriceps femoris tendon	Extends leg at knee; stabilizes knee joint
Vastus medialis	7-42A 7-43B	(See Quadriceps femoris)	Intertrochanteric line, spiral line, linea aspera, and medial supracondylar line of the femur, tendons of the adductor longus and magnus	Medial border of the patella and the quadriceps femoris tendon	Extends leg at knee; stabilizes patella and knee joint
Vastus intermedius	Not shown	(See Quadriceps femoris)	Upper two-thirds of the femoral shaft and the lateral intermuscular septum	Quadriceps femoris tendon, lateral border of the patella, and lateral condyle of the tibia	Extends leg at knee
Articularis genu	Not shown	Femoral nerve from L2-4	Lower part of the shaft of the femur	Upper part of the synovial membrane of the knee joint	Pulls the synovial membrane of the knee joint upward during extension of the leg
Medial femoral muscles:					
Gracilis	7-42A to C, 7-43B	Obturator nerve	Lower half of the body and inferior ramus of the pubis and ramus of the ischium	Upper part of the medial surface of the tibia	Flexes the leg and rotates it medially; may also adduct the thigh
Pectineus	742A and C	Femoral nerve from L2-3 and accessory obturator nerve from L3	Pecten of the pubis and adjacent fascia	Femur between the lesser trochanter and linea aspera	Adducts the thigh and flexes it on the pelvis

TABLE 7-19
MUSCLES OF THE THIGH AND GLUTEAL REGION *(Continued)*

Muscle	*Figure*	*Nerve*	*Origin*	*Insertion*	*Action and comments*
Adductor					Adducts thigh; aids in control of gait and posture
Longus	7-42A and C	Obturator nerve from L2-4	Front of the body of the pubis	Linea aspera of the middle third of the femur	Probably a medial rotator of the thigh
Brevis	7-42C	Obturator nerve from L2-4	Body and inferior ramus of the pubis	Femur between the lesser trochanter and linea aspera	
Magnus	7-42B and C	Obturator nerve and tibial division of the sciatic form L2-4	Inferior ramus of the pubis, ramus of the ischium and ischial tuberosity	Gluteal tuberosity, linea aspera, medial supracondylar line, and the medial condyle of the femur	Probably a medial rotator of the thigh
Gluteus					Extensors and abductors of the hip joint
Maximus	7-42B	Inferior gluteal nerve from L5 and S1	Posterior gluteal line of the ilium and area above and behind it, lower part of the sacrum, side of the coccyx, and sacrotuberous ligament	Iliotibial tract of the fascia lata and gluteal tuberosity of the femur	When the pelvic girdle is stabilized, extends the flexed thigh; when the hip joint is stabilized, maintains upright posture; lateral rotator of thigh; active in raising trunk after stooping; steadies the femur on the tibia; largest muscles mass in the body
Medius	7-42B	Superior gluteal nerve from L5 and S1	Outer surface of the ilium and adjacent fascia	Greater trochanter of the femur	When the pelvic girdle is stabilized, abducts thigh and rotates it medially, aids in maintaining upright posture during stepping. Stabilizes pelvis when one foot is raised
Minimus	7-42B	Superior gluteal nerve from L5 and S1	Outer surface of the ilium	Greater trochanter of the femur	When the pelvic girdle is stabilized, abducts the thigh and rotates it medially; aids in maintaining upright posture during stepping. Stabilizes pelvis when one foot is raised
Piriformis	7-42B	Branches from L5 and S1-2	Front of the sacrum, ilium near posterior iliac spine, and capsule of sacroiliac joint	Greater trochanter of the femur	Rotates the extended thigh laterally; abducts the flexed thigh

TABLE 7-19
MUSCLES OF THE THIGH AND GLUTEAL REGION *(Continued)*

Muscle	*Figure*	*Nerve*	*Origin*	*Insertion*	*Action and comments*
Obturator internus	Not shown	Branch from L5 and S1	Wall of obturator foramen of hip bone	Greater trochanter of the femur	Rotates extended thigh laterally; abducts flexed thigh
Gemellus					
Superior	7-42B	Nerve to the obturator internus from L5 and S1	Spine of the ischium	Trochanter fossa with the tendon of the obturator internus	Rotates the extended thigh laterally; abducts the flexed thigh
Inferior	7-42B	Nerve to quadratus femoris and inferior gemellus from L5 and S1	Tuberosity of the ischium	Trochanter fossa with the tendon of the obturator internus	Rotates the extended thigh laterally; abducts the flexed thigh
Quadratus femoris	7-42B	Nerve to quadriceps femoris and inferior gemellus from L5 and S1	Tuberosity of the ischium	Trochanteric crest of the femur	Rotates thigh laterally
Obturator externus	Not shown	Branch of obturator nerve from L5 and S1	Ramus of the pubis and ischium on medial side of obturator foramen and obturator membrane	Trochanteric fossa of the femur	Rotates thigh laterally

TABLE 7-19
MUSCLES OF THE THIGH AND GLUTEAL REGION *(Continued)*

Muscle	*Figure*	*Nerve*	*Origin*	*Insertion*	*Action and comments*
Posterior femoral muscles:					
Biceps femoris	7-42B and C	Sciatic nerve from L5 and S1-2. Long head: common perineal divisions; short head: tibial division	Long head: ischial tuberosity and sarcotuberous ligament; short head: linea aspera, lateral supracondylar line, and lateral intermuscular septum	Head of the fibula, fibular collateral ligament, and lateral condyle of the tibia	When the hip joint is stabilized, flexes the leg on the thigh; when the knee is stabilized, extends the hip joint, drawing the trunk upright after stooping; rotates semiflexed knee laterally; rotates thigh laterally when hip is extended; the tendon of the biceps femoris is called the *lateral hamstring*
Semitendinosus	7-42B and C, 7-43B	Tibial division of the sciatic nerve from L5 and S1-2	Ischial tuberosity and adjacent aponeurosis	Upper part of the medial surface of the tibia	When the hip joint is stabilized, flexes the leg on the thigh; when the knee is stabilized, extends the hip joint and draws trunk upright; rotates thigh medially when hip is extended
Semimembranosus	7-42B, 7-43B	Tibial division of the sciatic nerve from L5 and S1-2	Ischial tuberosity	Posterior border of the medial condyle of tibia	When the hip joint is stabilized, flexes the leg on the thigh; when the knee is stabilized, extends the hip joint and draws the trunk upright; rotates thigh medially when hip is extended

FIGURE 7-43 Anterior (A), medial (B), and posterior (C) views of the superficial muscles of the right leg. The fascia lata of the thigh continues into the leg as the *crural fascia. Intermuscular septa*, which pass from the fascia deep into the leg, and the interosseous membrane separate the muscles into anterior, lateral, and posterior compartments. The large superficial muscles of the posterior group (gastrocnemius, plantaris, and soleus) form the calf of the leg. Their contraction produces plantar flexion of the foot; their size is directly related to man's upright posture and method of walking. The deep muscles of the posterior group (tibialis posterior, flexor digitorum longus, and flexor hallucis longus) act in flexion of the toes, plantar flexion, and medial rotation (inversion) of the foot. The anterior-compartment muscles (tibialis anterior, extensor hallucis longus, and extensor digitorum longus) extend the toes and dorsiflex the foot, whereas the muscles of the lateral compartment (peroneus longus and brevis) evert and abduct the foot and assist in plantar flexion.

The crural fascia is reinforced in the ankle region by transverse and oblique fibers which form thick bands, or *retinacula* (views A, B, and C). These hold the tendons which cross the ankle close to the bone and prevent their bowing out when under tension. There are extensor, flexor, and peroneal retinacula.

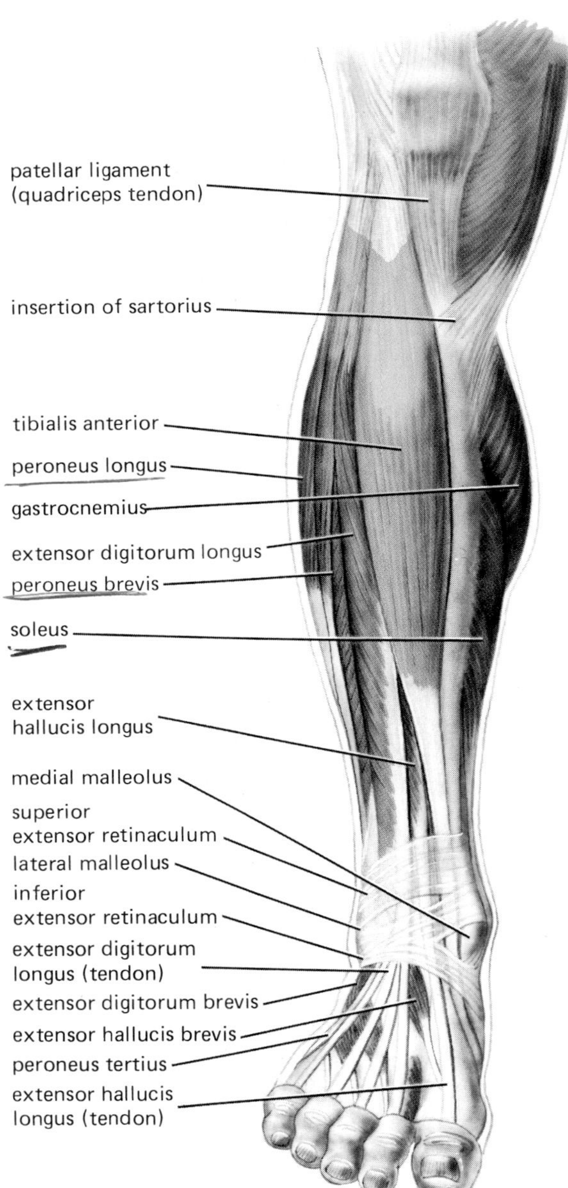

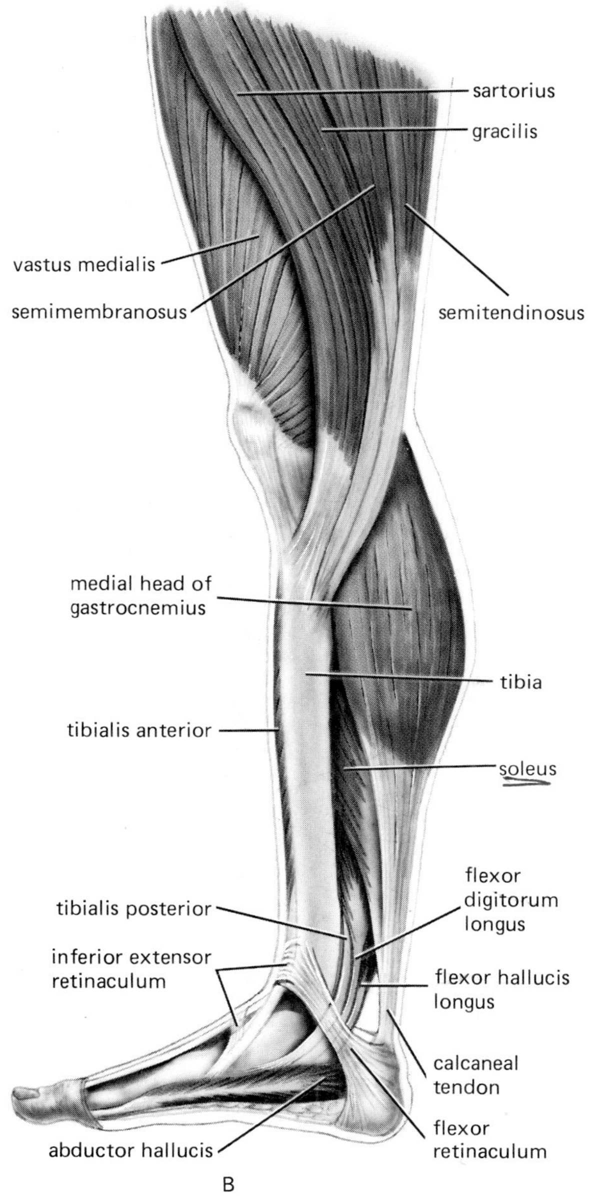
sartorius
gracilis
vastus medialis
semimembranosus
semitendinosus
medial head of gastrocnemius
tibia
tibialis anterior
soleus
flexor digitorum longus
tibialis posterior
inferior extensor retinaculum
flexor hallucis longus
calcaneal tendon
abductor hallucis
flexor retinaculum

B

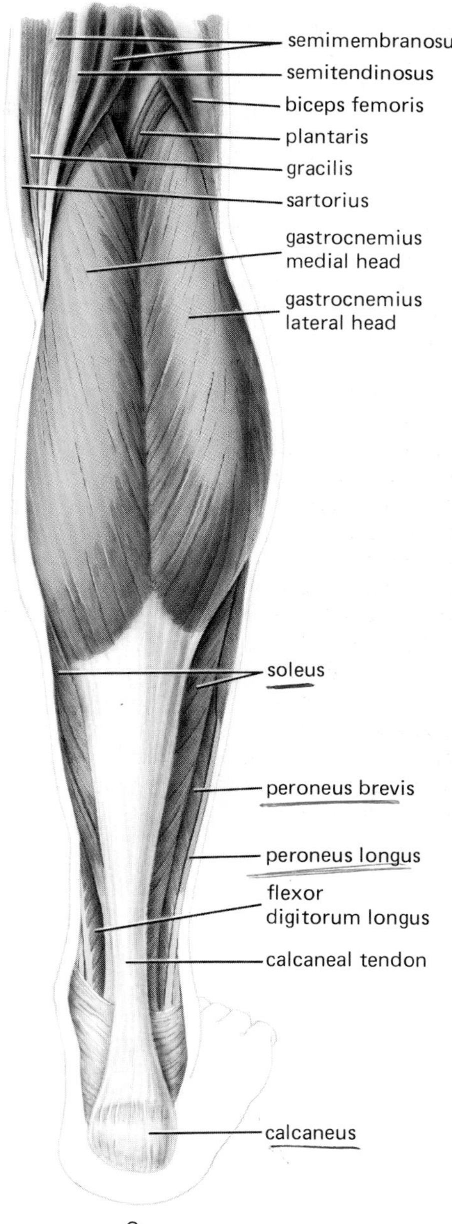
semimembranosus
semitendinosus
biceps femoris
plantaris
gracilis
sartorius
gastrocnemius medial head
gastrocnemius lateral head
soleus
peroneus brevis
peroneus longus
flexor digitorum longus
calcaneal tendon
calcaneus

C

TABLE 7-20
MUSCLES OF THE LEG

Muscle	*Figure*	*Nerve*	*Origin*	*Insertion*	*Action and comments*
Anterior crural muscles:					
Tibialis anterior	7-43A and B	Deep peroneal nerve from L4-5	Lateral condyle and upper half of shaft of tibia, interosseous membrane, and crural fascia	Medial cuneiform and base of the first metatarsal bone	Dorsiflexes and inverts the foot; if ankle is stabilized, can draw the body forward
Extensor hallucis longus	7-43A	Deep peroneal nerve from L5 and S1	Medial surface of the fibula and interosseous membrane	Base of the distal phalanx of the great toe	Extends the phalanges of the great toe and dorsiflexes the foot
Extensor digitorum longus	7-43A	Deep peroneal nerve from L5 and S1	Lateral condyle of the tibia, medial surface of the fibula, interosseous membrane, and crural fascia	Divides into four slips on the dorsum of the foot; joined by tendons of the extensor digitorum brevis to form *dorsal digital expansion;* attaches to middle and distal phalanx	Extends the toes and aids in dorsiflexing the foot
Peroneus tertius	7-43A	Deep peroneal nerve from L5 and S1	Medial surface of the fibula, interosseous membrane, and intermuscular septum	Base of the fifth metatarsal bone	Dorsiflexes the foot, acting as part of the extensor digitorum longus
Lateral crural muscles:					
Peroneus longus	7-43A and C	Superficial peroneal nerve from L5 and S1-2	Head and lateral surface of the fibula, crural fascia, and intermuscular septum	Base of the first metatarsal bone and the medial cuneiform	Everts and plantar flexes the foot, maintains the concavity of the foot in the early phase of stepping and in tiptoeing
Peroneus brevis	7-43C	Superficial peroneal nerve from L5 and S1-2	Lateral surface of the fibula and middle two-thirds of the intermuscular septum	Base of the fifth metatarsal bone	Participates in eversion of the foot; may help steady the leg on the foot and prevent overinversion of the foot
Posterior crural muscles:					
Gastrocnemius	7-43A to C	Tibial nerve from S1-2	Arises by two heads from the condyles of the femur and adjacent parts of the joint capsule	Posterior surface of the calcaneus as part of the calcaneal (Achilles) tendon	Forms the belly of the calf of the leg; one of the chief plantar flexors of the foot; flexes the knee; provides propelling force in walking, running, and leaping

TABLE 7-20
MUSCLES OF THE LEG *(Continued)*

Muscle	*Figure*	*Nerve*	*Origin*	*Insertion*	*Action and comments*
Soleus	7-43A to C	Tibial nerve from S1-2	Back of the head and posterior surface of the fibula and medial border of the fibula, and a fibrous band between the tibia and fibula	Posterior surface of the calcaneus as part of the calcaneal tendon	An important plantar flexor of the foot, steadies leg on foot for postural stability
Plantaris	7-43C	Tibial nerve from S1-2	Lateral supracondylar line of the femur	Posterior surface of the calcaneus as part of the calcaneal tendon	Accessory to the gastrocnemius
Popliteus	Not shown	Tibial nerve from L4-5 and S1	Lateral condyle of the femur and a popliteal ligament	Posterior surface of the tibia and the tendinous expansion covering it	If the tibia is stabilized, it rotates the femur laterally; "unlocks" the fully extended knee at the beginning of flexion
Flexor hallucis longus	7-43B	Tibial nerve from S2-3	Posterior surface of the fibula, interosseous membrane, and intermuscular septum	Base of the distal phalanx of the great toe	Flexes the phalanges of the great toe when the foot is off the ground and plantar flexes the foot; when the foot is on the ground, aids in stabilizing the metatarsal bones and ball of the foot. Is the push-off muscle in walking
Flexor digitorum longus	7-43B and C	Tibial nerve from S2-3	Posterior surface of the tibia and adjacent fascia	Forms long flexor tendons of the sole of the foot; ends on bases of distal phalanges	When the foot is off the ground, flexes phalanges of toes as plantar flexes foot; when the foot is on the ground, aids in stabilizing the metatarsal bones and ball of the foot
Tibialis posterior	7-43B	Tibial nerve from L4-5	Posterior surface of the tibia and interosseous membrane and posterior surface of the fibula, adjacent fascia, intermuscular septum	Tubercle of the navicular, medial, and intermediate cuneiform, and bases of the 2-4 metatarsals	Principal invertor of the foot, affects degree of flattening of foot during walking

FIGURE 7-44 Muscles of the right foot. The muscles of the foot often parallel those of the hand, particularly on the plantar surface (sole) where they are similar in name and number. (However, the term *hallucis* (hallux = great toe) is substituted for the term *pollicis* (pollex = thumb).) The muscles of the foot serve quite different purposes than those of the hand because the foot is adapted for stability, bearing weight, and providing propulsion for walking, running, and jumping and the hand is adapted for grasping.

The muscles on the plantar surface of the foot are covered by a tough thickening of the fascia, the *plantar aponeurosis*. (The dorsal surface has few muscles.) The plantar aponeurosis begins at the calcaneus bone and spreads over the sole of the foot, splitting into five forks as it approaches the toes. It helps sustain the arched form of the foot and gives origin to some of the intrinsic muscles (muscles which lie entirely within the foot). The plantar muscles may be divided into three groups: medial, lateral, and intermediate. The medial group contains the abductor (abductor hallucis) and flexor (flexor hallucis brevis) of the great toe whereas the lateral group contains the abductor and flexor of the small toe, i.e., the "digiti minimi." The central group, which is deep to the plantar aponeurosis, contains the lumbricals, interossei, and short flexors of the toes.

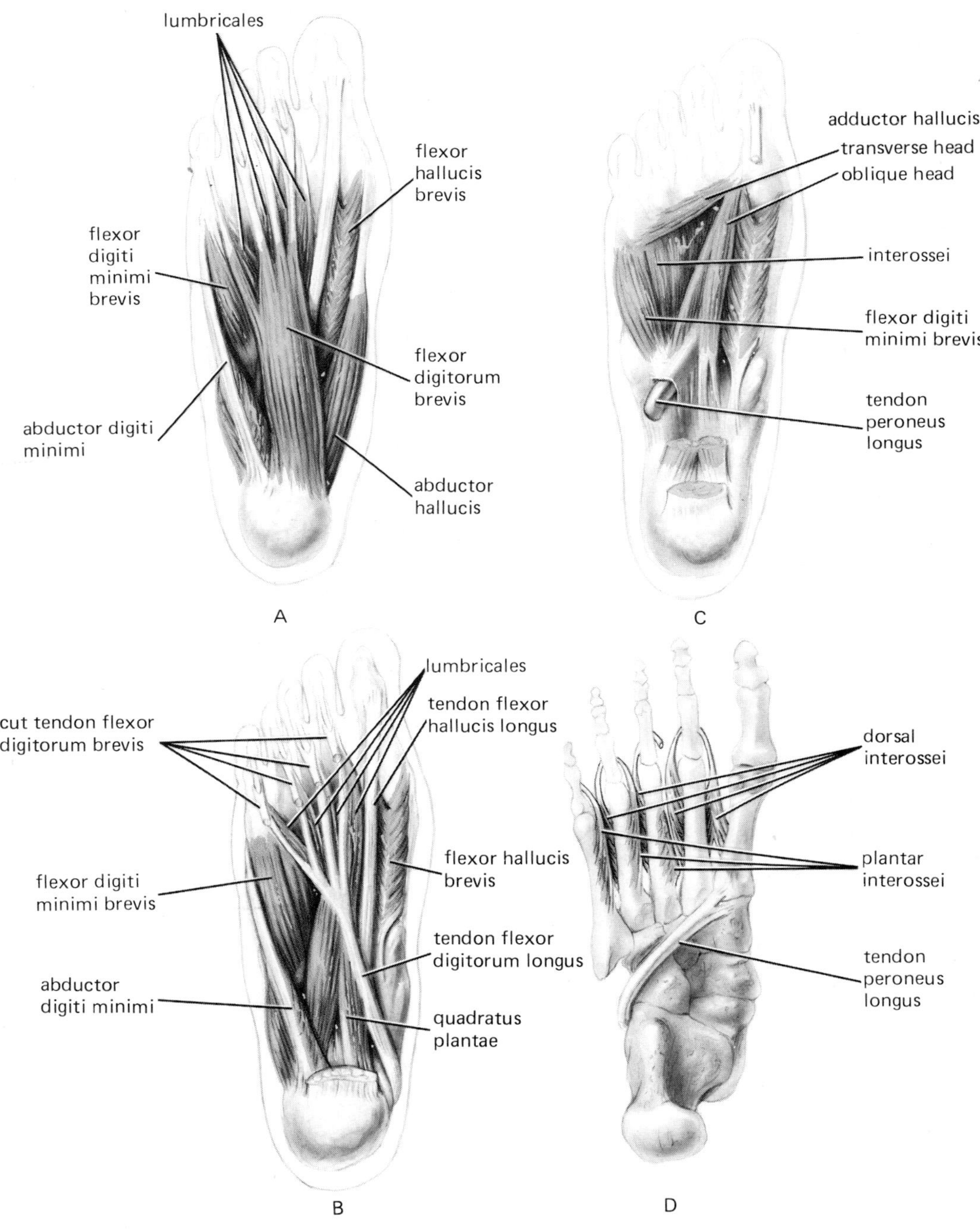
lumbricales
flexor hallucis brevis
flexor digiti minimi brevis
flexor digitorum brevis
abductor digiti minimi
abductor hallucis
A
adductor hallucis
transverse head
oblique head
interossei
flexor digiti minimi brevis
tendon peroneus longus
C
lumbricales
tendon flexor hallucis longus
cut tendon flexor digitorum brevis
flexor hallucis brevis
flexor digiti minimi brevis
tendon flexor digitorum longus
abductor digiti minimi
quadratus plantae
B
dorsal interossei
plantar interossei
tendon peroneus longus
D

TABLE 7-21
MUSCLES OF THE FOOT

Muscle	*Figure*	*Nerve*	*Origin*	*Insertion*	*Action and comments*
Dorsal muscle of the foot:					
Extensor digitorum brevis	7-43A	Branch of deep peroneal nerve from S1-2	Calcaneus, interosseous ligament, and extensor retinaculum	Base of the proximal phalanx of the great toe, tendons of extensor digitorum longus	Aids in extension of the phalanges of the middle three toes
Plantar muscles of the foot:					
Abductor hallucis	7-43B, 7-44A and C	Medial plantar nerve from S2-3	Calcanean tuberosity, flexor retinaculum, plantar aponeurosis, and adjacent intermuscular septum	Base of the proximal phalanx of the great toe	Aids in maintaining the concavity of the foot
Flexor digitorum brevis	7-44A	Medial plantar nerve from S2-3	Calcanean tuberosity, plantar aponeurosis, adjacent intermuscular septum	Shaft of the intermediate phalanges	Aids in maintaining the concavity of the foot
Abductor digiti minimi	7-44A and B	Lateral plantar nerve from S2-3	Calcanean tuberosity, plantar aponeurosis, adjacent intermuscular septum	Base of the proximal phalanx of the fifth toe	Aids in maintaining the concavity of the foot
Lumbricals	7-44A and B	Medial or lateral plantar nerves from S2-3	Tendons of flexor digitorum longus	Proximal phalanges	Four small muscles which flex the proximal phalanges
Flexor hallucis brevis	7-44A to C	Medial plantar nerve from S2-3	Cuboid and lateral cuneiform bones, tendon of tibialis posterior	Base of the proximal phalanx of the great toe	Flexes the proximal phalanx of the great toe
Adductor hallucis	7-44C	Branch of lateral plantar nerve from S2-3	Oblique head: bases of 2-4 metatarsal bones; transverse head: plantar ligaments of 3-5 toes	Sesamoid bone and base of first phalanx of great toe	Adducts the great toe and aids in maintaining the arches of the foot
Flexor digiti minimi brevis	7-44A to C	Branch of lateral plantar nerve from S2-3	Base of the fifth metatarsal and adjacent muscle sheaths	Base of the proximal phalanx of the fifth toe	Flexes the proximal phalanx of the small toe
Dorsal interossei	7-44D	Branch of lateral nerve from S2-3	Four muscles, each arising by two heads from the sides of the adjacent metatarsal bones	Bases of the proximal phalanges	Abduct the toes
Plantar interossei	7-44D	Branch of the lateral plantar nerve from S2-3	Three muscles, each arising from the base and side of the 3-5 metatarsal bones	Base of the proximal phalanges of the same toe	Adduct the toes

CHAPTER 8

The Nervous System: I. Structure

SUPPORTING STRUCTURES OF THE NERVOUS SYSTEM

Skull • Vertebral Column • Meninges

GLIA

SPINAL CORD

BRAIN

Brainstem • Cerebellum • Cerebrum • ***Diencephalon • Cerebral Hemispheres***

PERIPHERAL NERVOUS SYSTEM

Autonomic Nervous System • Spinal Nerves • Cranial Nerves

BLOOD SUPPLY OF THE CENTRAL NERVOUS SYSTEM

VENTRICULAR SYSTEM AND CEREBROSPINAL FLUID

The various parts of the nervous system are interconnected, but for convenience they can be divided into the *central nervous system,* composed of the brain and spinal cord, and the *peripheral nervous system,* consisting of the nerves extending from the brain and spinal cord (Fig. 8-1). The anatomy of the nervous system is extremely complex, and its basic features can best be approached in terms of its embryologic development.

The central nervous system develops from a column of ectoderm cells on the dorsal surface of the embryo. As the cells divide, increasing in number, a *neural groove* forms along the length of the column, with a ridge along each side (Fig. 8-2). The groove dips deeper and deeper into the mesoderm layer of the embryo and eventually seals over at the top, forming the *neural tube*. The head, or rostral, end of the neural tube becomes the brain, changing during the fourth and fifth weeks of development from a simple to a complex tube with five distinct regions, or *vesicles,* each corresponding to one of the five basic regions of the adult brain (Fig. 8-3). The caudal (tailward) part of the neural tube develops into the spinal cord. The hollow cavities within the neural tube are filled with fluid. As the brain develops, the shapes of the cavities change but they remain connected with each other through narrow channels (*aqueducts*); these interconnected fluid-filled cavities form the *ventricular system* of the mature central nervous system.

The caudal end of the primitive neural tube is lined with cells which divide many times to form the billions

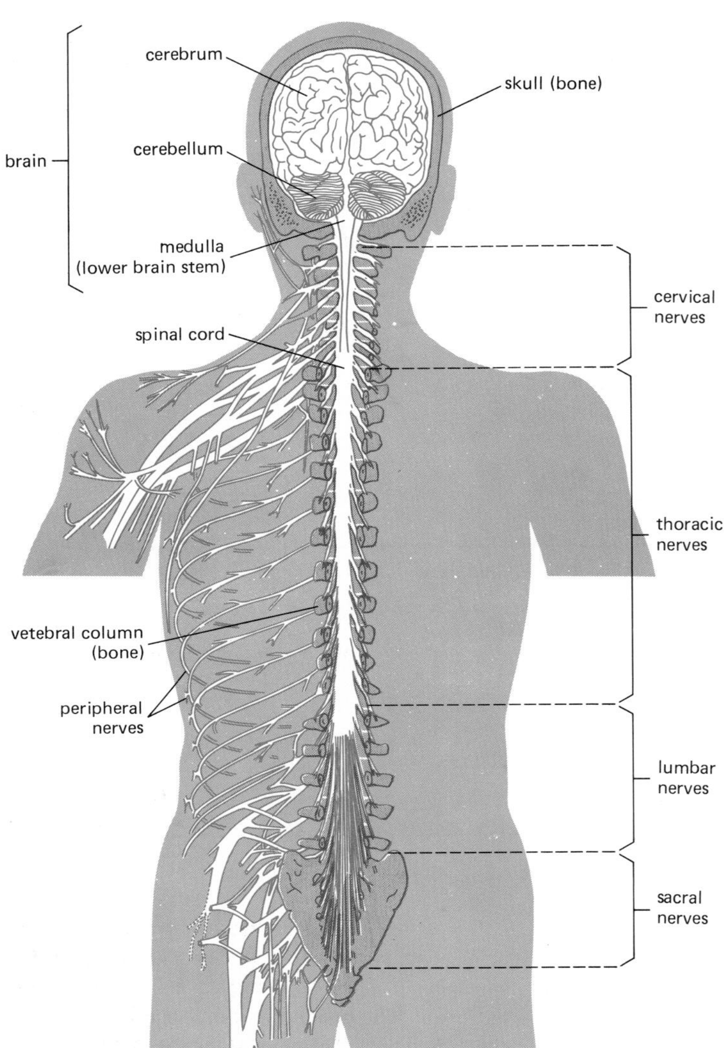

FIGURE 8-1 ***Nervous system viewed from behind. The back of the skull and backs of the bony vertebrae have been removed to expose the brain and spinal cord. The cranial nerves are not shown.*** **(Adapted from Woodburne.)**

of neurons of the spinal cord. This cell division leads to the formation of the central *gray matter,* which is composed of neuronal cell bodies (and neuroglia). Clusters of neurons in the gray matter are called *nuclei,* or *cell columns*. Processes (primarily axons) grow out from these cell bodies to form a layer on the outer surface of the tube; this is the early *white matter* (Fig. 8-4).

Some of the cells within the neural tube send axons out to the primitive muscles and glands, becoming one group of efferent neurons of the peripheral nervous system. Cells from two small columns, the *neural crest* columns, become the afferent neurons. These cells send out a process which branches, one branch migrating outward with the peripheral nerve to connect with or become a sensory receptor, the other migrating centrally to enter the brain or spinal cord where it synapses with other neurons; the important result is that the cell bodies of all afferent neurons lie outside the central nervous system. These cell bodies of the afferent columns form the *sensory ganglia* (known also as *dorsal root*

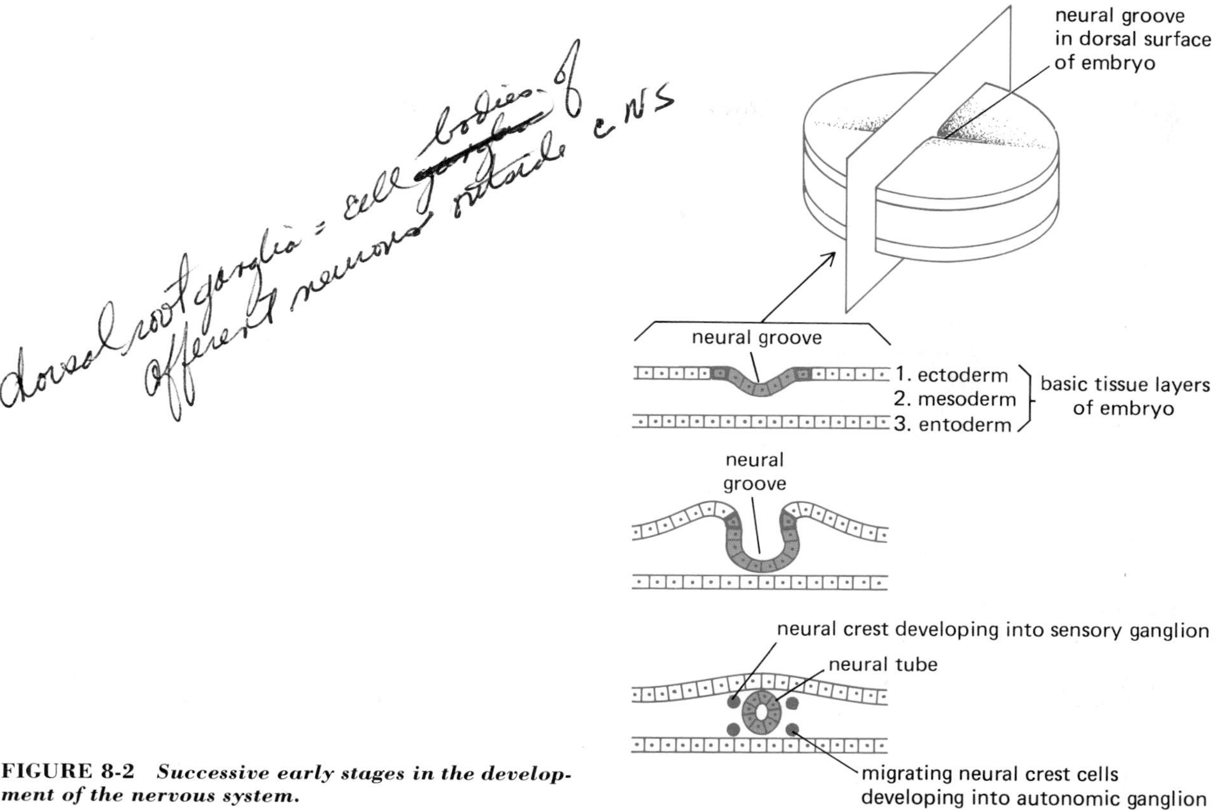

FIGURE 8-2 ***Successive early stages in the development of the nervous system.***

ganglia at the level of the spinal cord). Other neural crest cells migrate away from the original columns and differentiate into a second type of ganglia, those for the autonomic neurons.

SUPPORTING STRUCTURES OF THE NERVOUS SYSTEM

The supporting structures of the nervous system develop in the embryo from the mesoderm tissue that immediately surrounds the neural tube. The outer layer is bony: the *skull* and *vertebral column*. Between the bone and the nervous tissue are three membranes, or *meninges*.

Skull

Certain bones of the skull are adapted to support and protect the brain and several of the special senses and to isolate the cerebral blood vessels from extracranial variations in pressure. These bones form the *neurocranium* or *calvaria;* they are shown in Fig. 7-15. When the top of the skull and the brain are removed to expose the floor of the neurocranium, three distinct shallow depressions are exposed: the *anterior, middle,* and *posterior cranial fossae* (Fig. 7-15D); these support the base of the brain. The posterior cranial fossa, which accommodates the cerebellum, pons, and medulla, is perforated by a large opening, the *foramen magnum,* which encircles the central nervous system at the transition between the medulla and spinal cord. Blood vessels and connective-tissue structures also pass through it.

Vertebral Column

The spinal cord is surrounded in its entire length by the bony vertebral column, which was discussed in de-

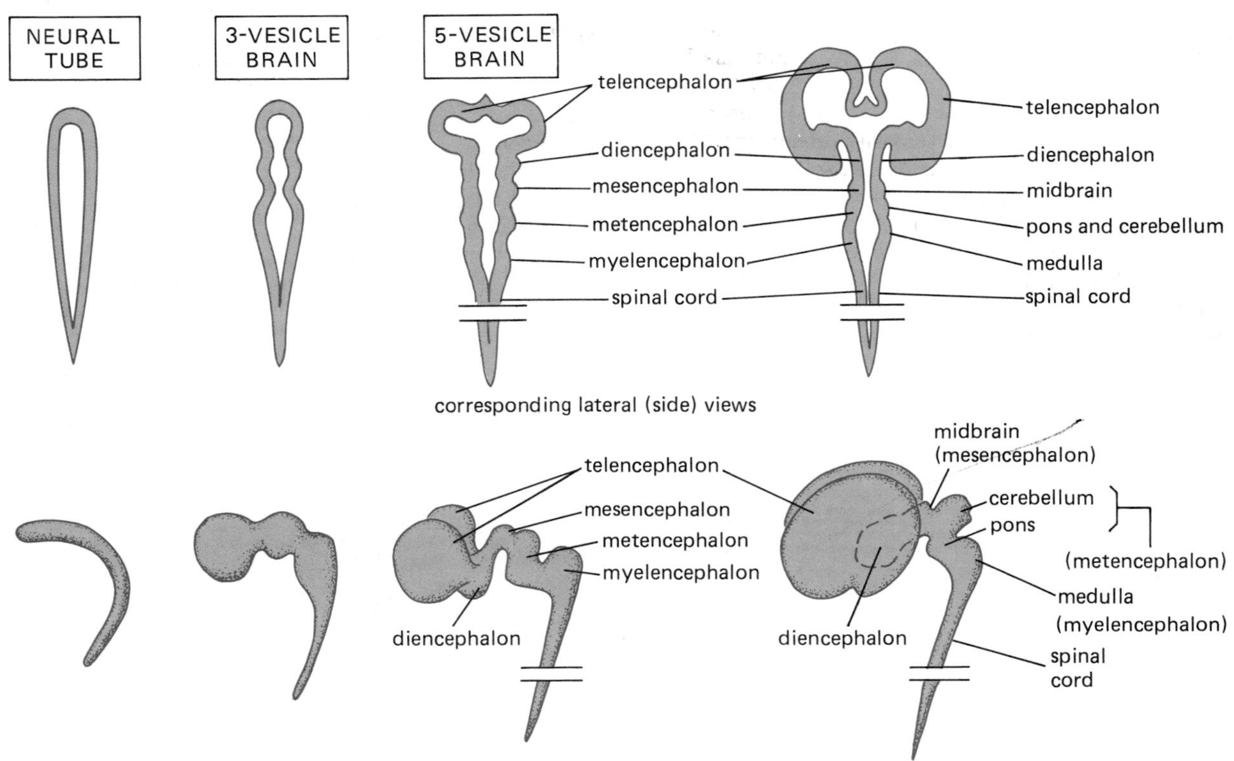

FIGURE 8-3 ***Schematic dorsal*** **(top row)** ***and lateral*** **(bottom row)** ***views of the developing and adult brains. The hollow center of the neural tube becomes the ventricular system in the adult brain.***

tail in Chap. 7. We are concerned here only with its relationship to the spinal cord. The spinal cord extends from the point where it joins the brain at the foramen magnum down to the region of the second lumbar vertebra, a length of about 45 cm (18 in). It lies within the *vertebral canal,* a passage through the successive *vertebral foramens* of the interconnected vertebrae (Fig. 8-5). Rather than ending abruptly, the lower part of the spinal cord narrows, forming the *conus medullaris.*

Meninges

Within the skull and vertebral column, respectively, the brain and spinal cord are enclosed in three membranous coverings, or *meninges* (Fig. 8-6): the *dura mater* (next to the bone), the *arachnoid* (in the middle), and the *pia mater,* which is next to the nervous tissue. There is a space between the pia and arachnoid, which is filled with cerebrospinal fluid; this is the *subarachnoid space.* The pia follows the contours of the nervous tissue more closely than the arachnoid does, and in some places the two layers are quite widely separated. These spaces, called *cisterns,* contain large pools of cerebrospinal fluid. A *subdural space* separates the dura and arachnoid, but it is so narrow that it is better termed a potential space.

The spinal cord is anchored to the dura by the *den-*

tate ligaments, which are narrow, fibrous bands lying on each side of the spinal cord (Fig. 8-6). They extend from the pia mater, along which they form a continuous band running the length of the spinal cord, out to the dura mater where they are attached at intervals between the exiting nerves.

As the spinal cord ends, the pia mater becomes continuous with a connective-tissue thread, the *filum terminale,* which descends in the midst of a bundle of nerve roots passing from the spinal cord down through the vertebral canal to their point of exit to the periphery. These nerve roots resemble a horsetail and have, therefore, been named the *cauda equina* (Fig. 8-5C). The filum terminale descends past the point of exit of the nerve roots, finally becoming closely associated with the dura and with it attaching to the back of the coccyx.

The subarachnoid space surrounding the filum is fairly wide, and samples of cerebrospinal fluid may be obtained from it by passing a needle through the midline of the back, between the spines of the third and fourth (or fourth and fifth) lumbar vertebrae, and through the dura mater into the subarachnoid space, care being taken to avoid damaging the nerve roots of the cauda equina. This procedure is called a *lumbar,* or *spinal, puncture.*

The dura surrounding the spinal cord is attached to bone in relatively few places; over most of its surface it is separated from the periosteum and ligaments of the vertebral canal by an *extradural* (or *epidural*) *space,* which contains a small amount of fatty tissue and a few blood vessels.

The meninges surrounding the brain (*cerebral meninges*) are continuous with those around the spinal cord. The cerebral dura serves the twofold purpose of periosteum for the cranial cavity and of supporting

FIGURE 8-4 ***Early stages in the formation of the spinal cord and afferent and efferent neurons. (B) Cross sections cut at level of the spinal cord in (A).***

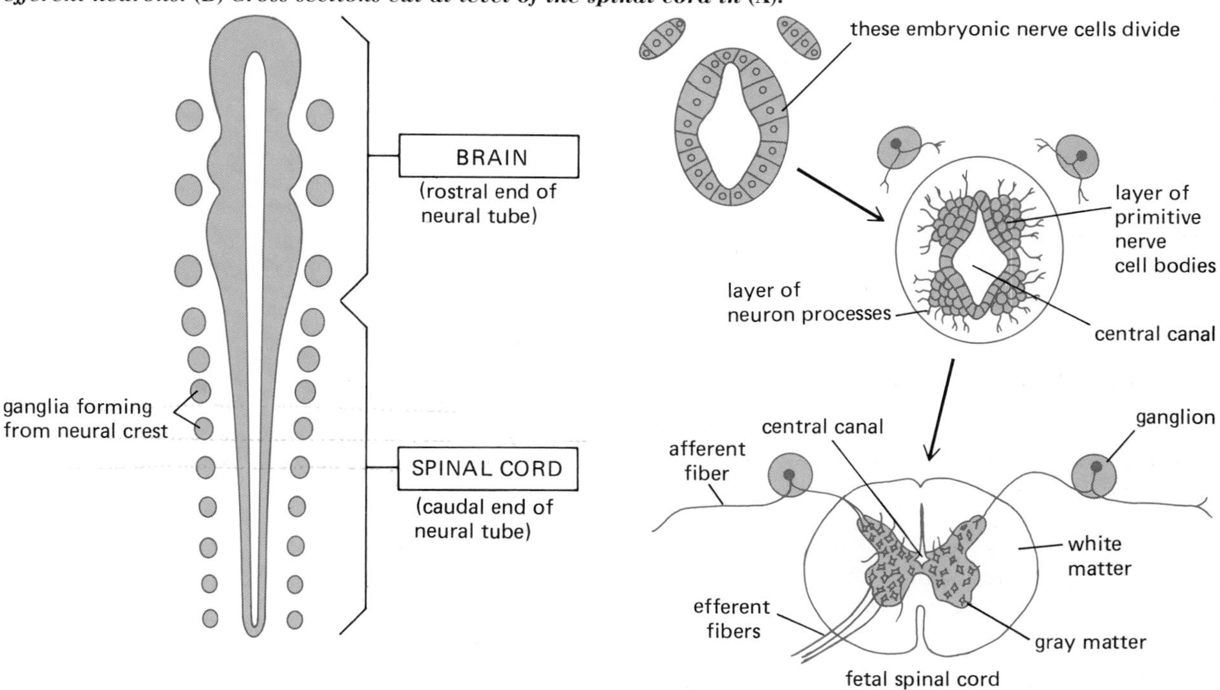

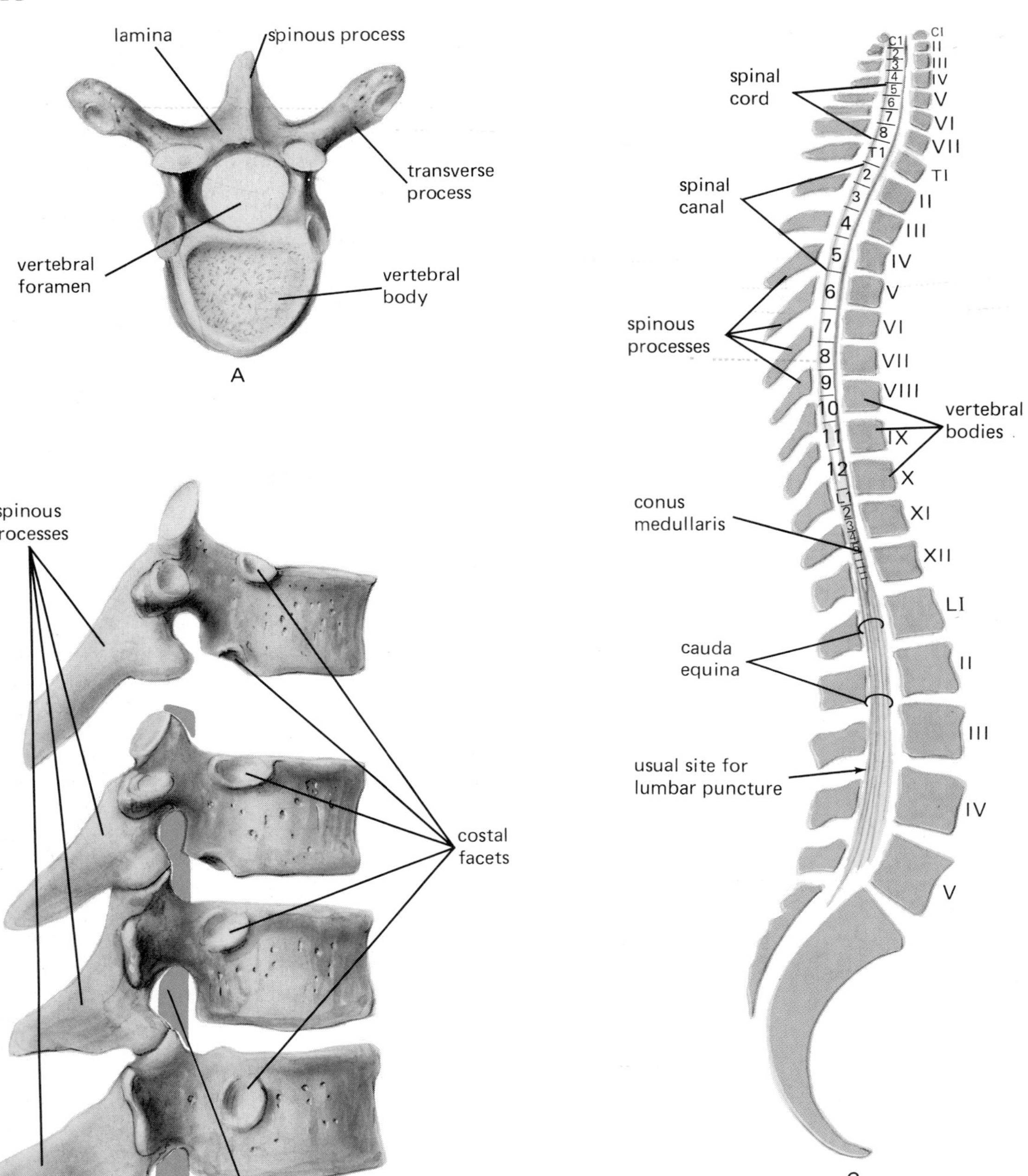

FIGURE 8-5 (A) *Typical vertebra (superior view).* (B) *Location of the vertebral canal relative to several vertebrae* (lateral view). (C) *Position of the spinal cord within the vertebral canal.*

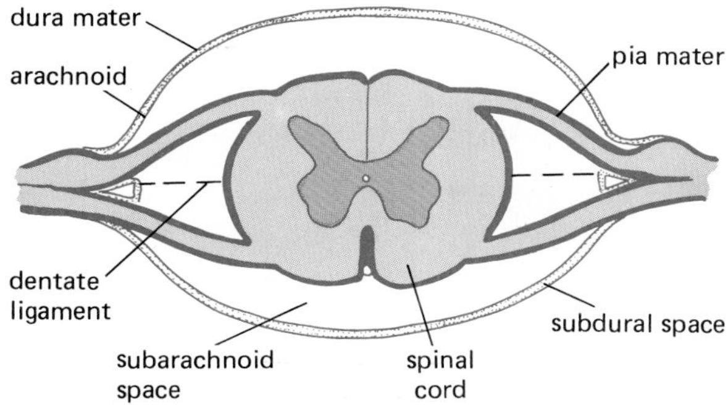

FIGURE 8-6 ***Cross section through the spinal cord and its membranes.*** **(Adapted from Warwick and Williams.)**

membrane for the brain. It contains the large blood vessels (*venous sinuses*) which drain the blood from the brain. At the openings, or *foramens,* in the skull, the dura follows the edges of the bone, passing from the interior of the skull to its outer surface, where it blends with the periosteum of the skull and the connective-tissue coat of the peripheral nerves. (At the foramen magnum, it is continuous with the spinal dura.)

GLIA

Only 15 percent or so of the cells in the central nervous system are neurons; the remainder are *glial cells,* or *neuroglia.* Glial cells do not give action-potential-like responses, and, except for the few cases mentioned below, their precise functional relation to neurons and their role in the physiology of the central nervous system are unknown. Glial cell bodies are generally smaller than the cell bodies of neurons, and the processes are shorter although they may be numerous and highly branched.

Processes of one type of glial cell, the *astrocyte* (Fig. 8-7), are tightly packed around the cell bodies and dendritic processes of neurons, sometimes forming a particularly tight covering over the regions where the endings of one neuron contact the cell bodies or dendrites of another (synapses). Astrocytes may affect the composition of the extracellular fluid in their immediate vicinity by providing a membrane across which ions released from active neurons can distribute, and it has been suggested that they play a role in the developing brain by guiding neurons as they migrate to their final adult locations and by directing the growing neuronal processes. They may affect the composition of the extracellular fluid as a whole by regulating the passage of some substances between the blood and the extracellular fluid of the central nervous system.

Oligodendrocytes (Fig. 8-7), a second type of neuroglia, are sometimes found around nerve cell bodies, but they also occur along bundles of axons in the brain and spinal cord. Some of the oligodendrocytes surrounding nerve fibers within the central nervous system form the myelin coating of the axons; the functions of the others are not known. *Microglia,* a type of small glial cell, are numerous only in disease states in which they possibly act as scavengers. *Ependymal cells,* the final type of glial cells, line the brain ventricles. There are no glial cells in the peripheral nervous system, but the Schwann cells and satellite cells in the ganglia there are similar to glia and perform some of the glial functions, e.g., myelin formation.

SPINAL CORD

The spinal cord in the adult is a slender cylinder about as big around as the little finger. Figure 8-8 shows the basic division of the internal structures of the cord into central gray and peripheral white regions. The gray portion is butterfly-shaped, the two halves of the "butterfly" being joined by a central *gray commissure.* Each "wing" has a *dorsal* and *ventral horn* and, in some segments of the cord, a *lateral horn* as well. The gray region is surrounded by white matter, which consists largely of bundles of myelinated nerve fibers (*tracts*) running longitudinally through the cord, some descending to convey information from the brain to the spinal cord (or from upper to lower levels of the cord), others ascending to transmit in the opposite direction. The fiber bundles are so organized that a given tract or pathway contains fibers transmitting one type of information. For example, the fibers which transmit information from light-touch receptors in the skin travel in the same pathway. Some of these spinal-cord pathways are illustrated in Fig. 8-9; their specific features will be described in Chaps. 9 and 10.

Afferent fibers enter the sides of the spinal cord via the *dorsal roots,* which contain the dorsal root ganglia (the cell bodies of the afferent neurons) (Fig. 8-8). All efferent fibers leave via the *ventral roots;* their cell bodies are in the ventral or lateral horns of the spinal gray matter.

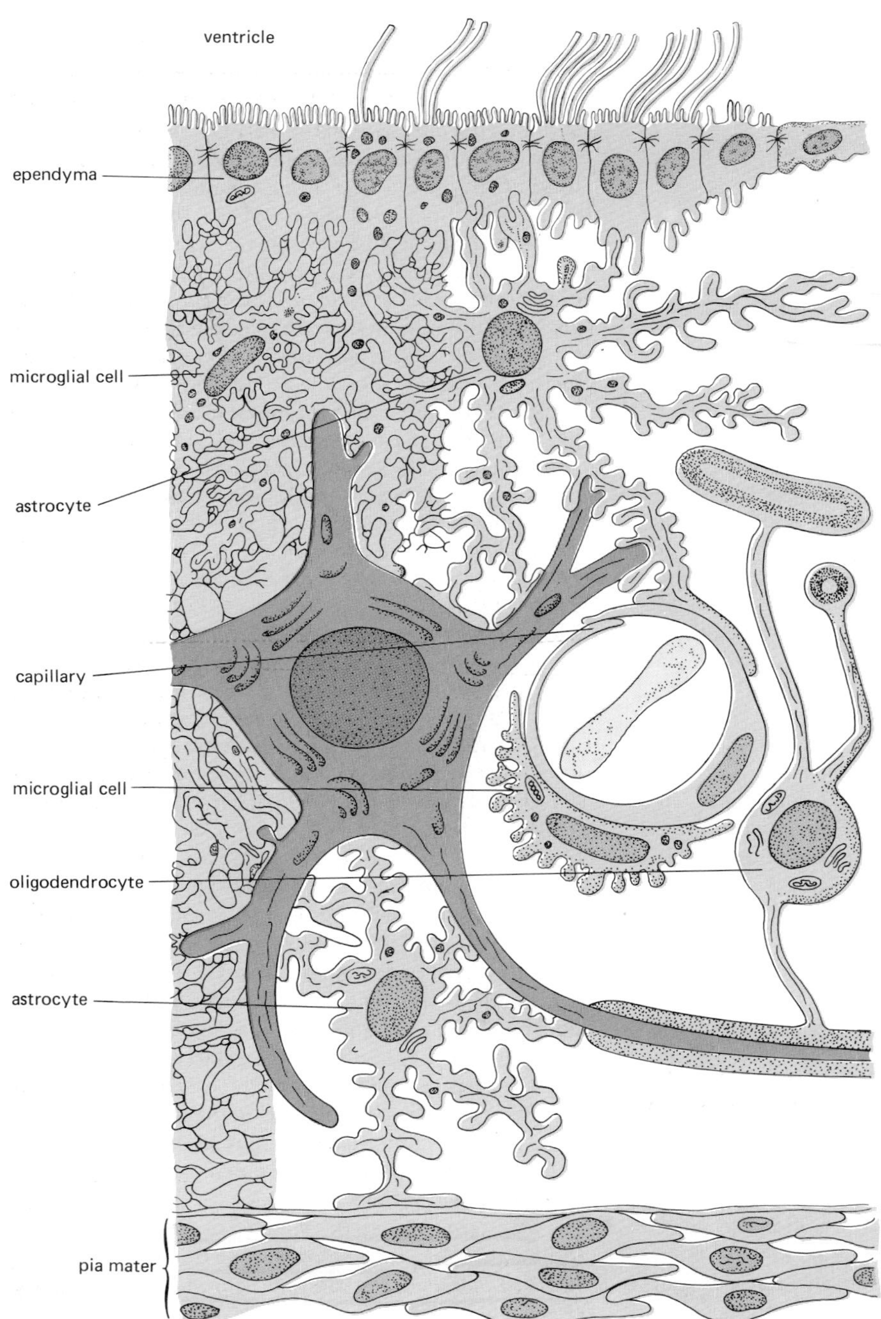

FIGURE 8-7 ***The four types of neuroglia: astrocytes, oligodendrocytes, microglia, and ependymal cells.*** **(Redrawn from R. E. M. Moore in Warwick and Williams.)**

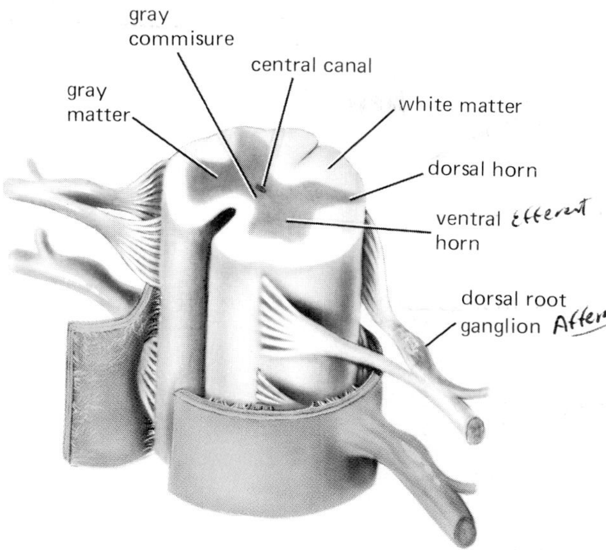

FIGURE 8-8 ***Diagrammatic cross section of the spinal cord.***

FIGURE 8-9 ***Major tracts of the spinal cord. Descending pathways are shown on the left and ascending on the right, but in fact the spinal cord is symmetrical.***

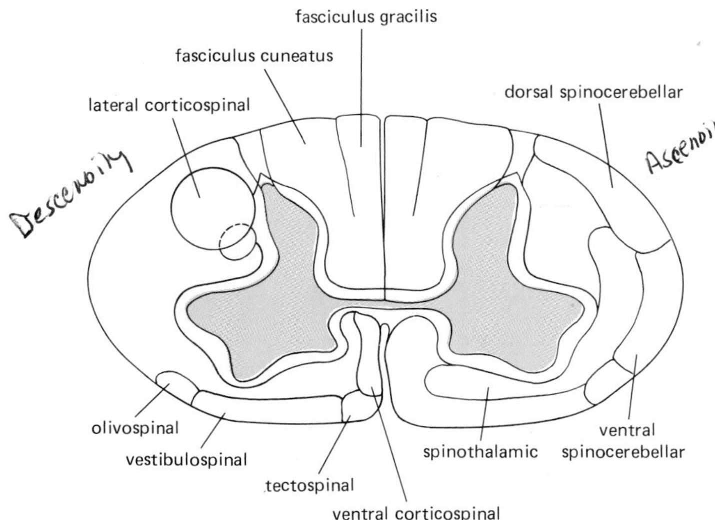

BRAIN

The adult brain is composed of six subdivisions: the *telencephalon, diencephalon, midbrain, pons, medulla,* and *cerebellum* (Fig. 8-3). The midbrain, pons, and medulla together form the *brainstem;* the telencephalon (cerebrum) and diencephalon together constitute the *forebrain.*[1]

Brainstem

The brainstem (Fig. 8-10) is literally the stalk of the brain, through which pass all the nerve fibers relaying signals of afferent input and efferent output between the spinal cord and higher brain centers. In addition, the brainstem contains the cell bodies of the efferent division of the cranial nerves; the axons of these neurons go out to the periphery to control the head's skeletal muscles, smooth muscles, and glands, the heart, and the smooth muscles and glands of most thoracic and abdominal viscera. The brainstem also receives many afferent fibers from the head and visceral cavities via the cranial nerves. In contrast to the distinct white and gray areas of the spinal cord, the tracts and nuclei of the brainstem are intermingled.

The medulla (sometimes called *medulla oblongata*) is the section of the brainstem continuous with the spinal cord below and the pons above. Its junction with the cord reflects a gradual change from the external tracts and internal columns of nuclei that exist at the upper levels of the cord. Efferent axons emerging from the medulla via cranial nerves VIII, IX, X, XI, and XII control areas of the mouth, throat, neck, thorax, and abdomen.

The pons is both wider and thicker than the medulla and is easily distinguished by a band of fibers running across its ventral surface (see Fig. 8-23A). These fibers converge at each side of the pons into bundles called the *middle cerebellar peduncles,* one of the three pairs of fiber bundles carrying information between the brainstem and cerebellum. The afferent and efferent components of cranial nerves V, VI, and VII connecting with the pons are from the head.

The midbrain is a relatively short part of the brainstem and is somewhat constricted in comparison with the pons. It is traversed by a huge number of axons subserving the corticospinal and spinocortical pathways. It contains the major nuclei associated with eye movements.

[1] Some neuroanatomists classify the diencephalon with the brainstem rather than with the forebrain.

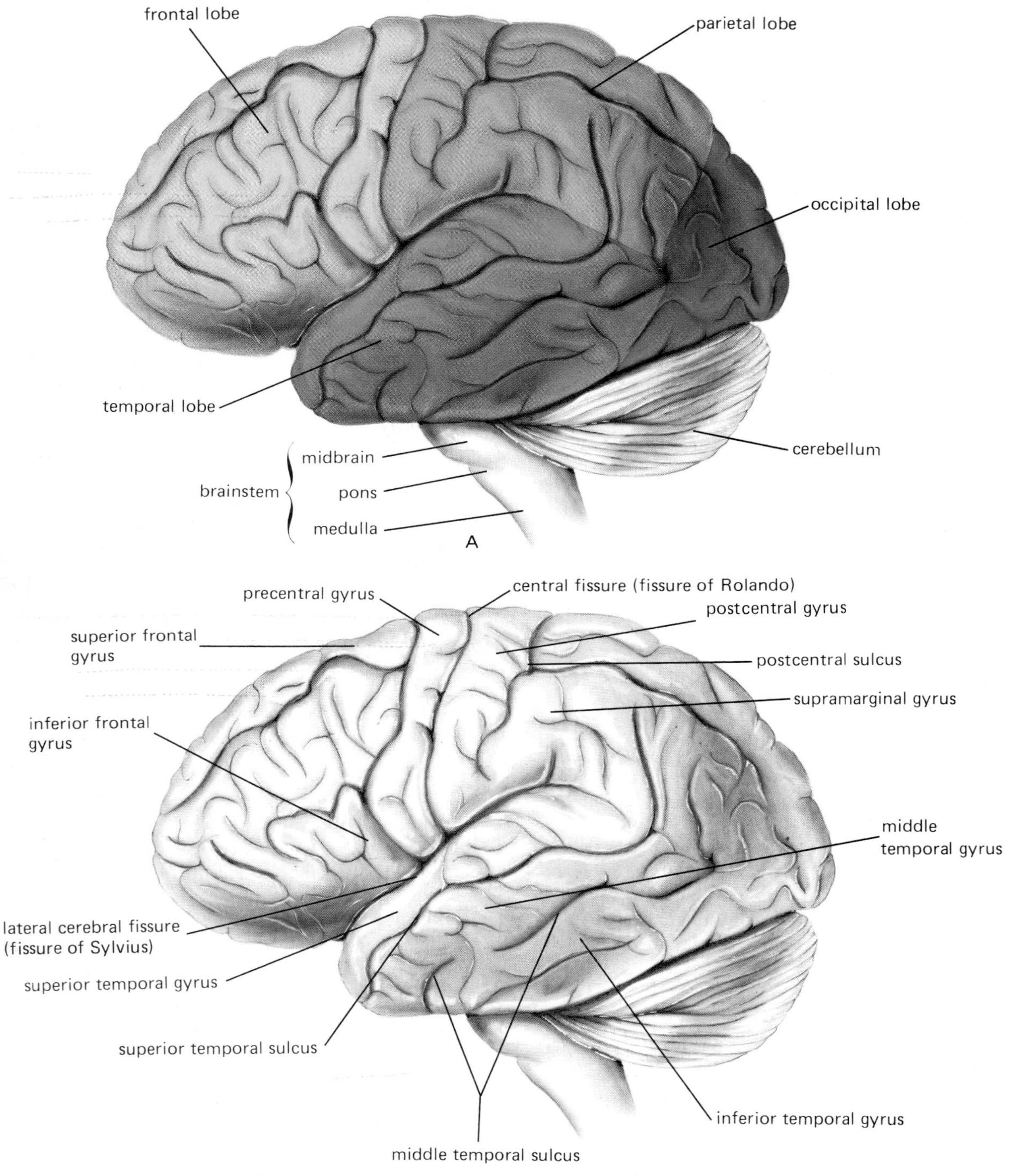

FIGURE 8-10 (A) ***Regions of the surface of the adult human brain.*** (B) ***View of the lateral surface of the brain.***

Forming the core of the brainstem is the *reticular formation,* which is composed of a diffuse collection of small, many-branched neurons. These neurons receive and integrate information from afferent fibers of the cranial nerves and pathways from the spinal cord and other regions of the brain. Reticular-formation neurons are clustered together, forming certain of the brainstem nuclei and "centers," such as the cardiovascular, respiratory, swallowing, and vomiting centers.

The output of the reticular formation can be divided functionally into descending and ascending systems. The descending components influence the function of both somatic and autonomic efferent neurons in the cranial and spinal nerves, and the ascending components affect such things as wakefulness and the direction of attention to specific events.

Cerebellum

The cerebellum is chiefly involved with muscle functions; it helps to maintain balance and provide smooth, directed movements. It consists of paired *cerebellar hemispheres* joined by a median *vermis.* The surface of the hemispheres is thrown into folds called *folia.* Parts of the cerebellum are connected with each other by association fibers and with other parts of the brain and spinal cord by projection fibers. (*Association fibers* connect parts within a given brain structure whereas *projection fibers* connect different structures. Association and projection fibers are found not only in the cerebellum but in other parts of the brain as well.) The many projection fibers of the cerebellum are grouped into three large bundles, or peduncles, on each side; the cerebellum is connected to the midbrain by the *superior cerebellar peduncles,* to the pons by the *middle cerebellar peduncles,* and to the medulla by the *inferior cerebellar peduncles.* The location of the cerebellum relative to the brainstem and cerebrum can be seen in Figs. 8-1 and 8-10.

Cerebrum

The large part of the brain remaining when the brainstem and cerebellum have been excluded is the cerebrum. It consists of a central core, the *diencephalon,* and right and left *cerebral hemispheres.* The hemispheres are connected to each other by fiber bundles known as *commissures,* the *corpus callosum* (Fig. 8-11) being the largest. Areas within a single hemisphere are connected to each other by association fibers (Fig. 8-12).

Diencephalon The diencephalon is rostral to the brainstem (Fig. 8-3) and is hidden beneath the cerebral hemispheres (Fig. 8-11). It is divided by a thin, vertical, fluid-filled space, the *third ventricle* (Fig. 8-13). The largest part of the diencephalon is the *thalamus,* which comprises two large egg-shaped masses of gray matter, one on each side of the third ventricle. Each is divided into several major parts and each part into distinct nuclei, which are the way stations and important integrating centers for most sensory inputs on their way to the cortex. In addition to their connections with the cortex, the thalamic nuclei have connections with each other, with neighboring nonthalamic masses of gray matter, and with the long ascending and descending paths in the brainstem and spinal cord. Through its connections with the hypothalamus, the thalamus is involved in a wide range of activities involving hormones, smooth muscle, and glands.

The *hypothalamus,* which is part of the diencephalon lying below the thalamus, is a tiny region whose volume is about 5 to 6 cm^3 (Fig. 8-13); it is responsible for the integration of many basic behavioral patterns which involve correlation of autonomic, endocrine, and somatic functions. Indeed, the hypothalamus appears to be the single most important control area for regulation of the internal environment. It is also one of the brain areas associated with emotions; stimulation of some hypothalamic areas leads to behavior interpreted as rewarding or pleasurable, and stimulation of other areas is associated with unpleasant feelings. Neurons of the hypothalamus are affected by a variety of hormones and other circulating chemicals.

At the top of the diencephalon is the *epithalamus,* in which is found the medial, unpaired *pineal body,* whose function in human beings is still unknown (Fig. 8-11).

Cerebral Hemispheres The surface of each hemisphere is divided into several parts, or *lobes* (*frontal, parietal, occipital,* and *temporal;* Fig. 8-10A), whose functions will be discussed in later chapters. When viewing the hemispheres, however, it is not the major lobes one notices; rather it is the numerous ridges (*gyri,* or *convolutions*) and furrows (*sulci,* or *fissures*) into which the surfaces are folded (Fig. 8-10B). These give room for considerable expansion of the outer layers of the hemispheres.

This outer portion of the cerebral hemispheres, the *cerebral cortex* (Fig. 8-13), is a shell of gray matter, a cellular layer about 3 mm ($\frac{1}{4}$ in) thick containing billions

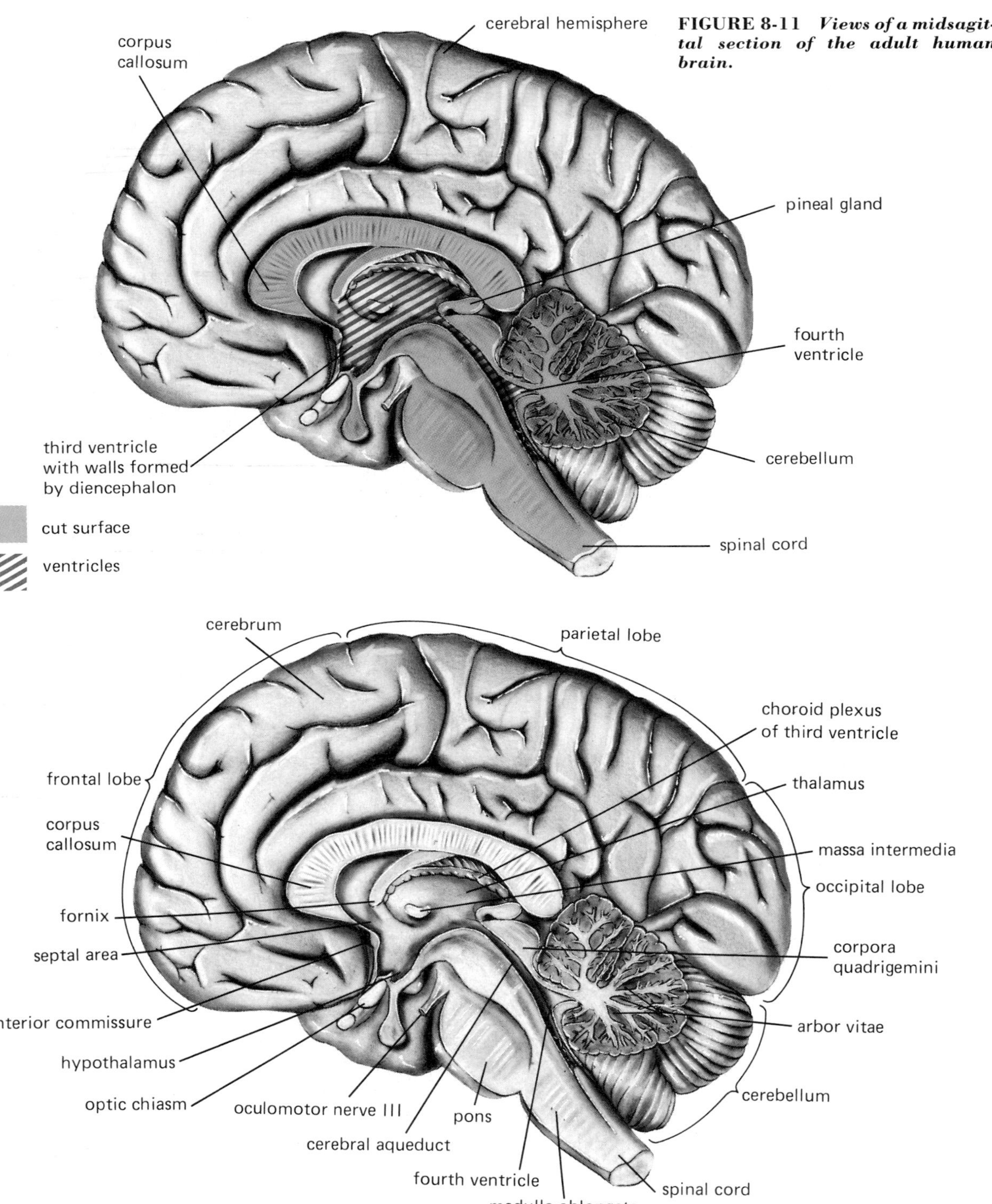

FIGURE 8-11 ***Views of a midsagittal section of the adult human brain.***

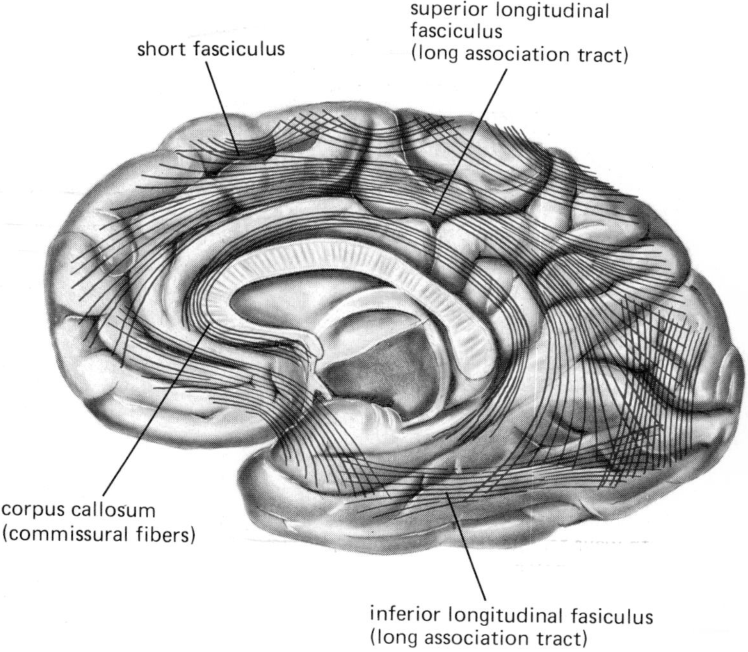

FIGURE 8-12 ***Association pathways of the cerebral hemispheres.***

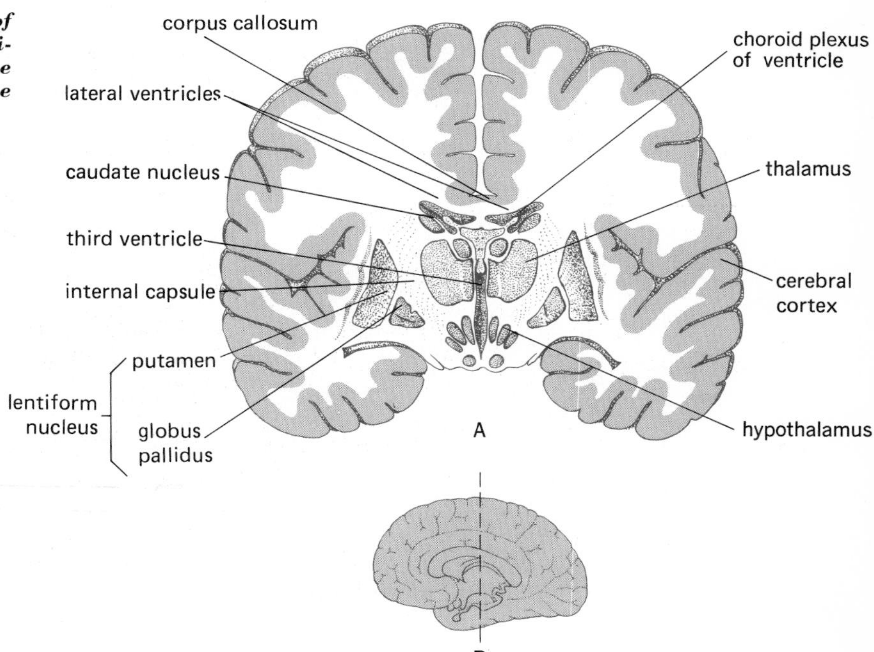

FIGURE 8-13 ***Coronal section of the brain. The dashed line AB indicates the location of the section. The caudate and lentiform nuclei are part of the basal ganglia.***

of neurons; it covers the entire surface of the cerebrum. Only a third of the cortex is visible from the surface, the remainder being hidden deep within the sulci. Popular opinion calls the cortex the "site of the mind and the intellect"; scientific opinion considers it to be an integrating area necessary for the bringing together of basic afferent information into complex perceptual images and ultimate refinement of control over both the autonomic and somatic systems; somatic reactions are more severely affected by cortical damage than are autonomic activities.

In other parts of the cerebrum, nerve-fiber tracts predominate, their whitish myelin coating distinguishing them as white matter. Lying in the midst of this white matter, between the cerebral cortex and cell clusters of the thalamus, are two large regions of gray matter known as the *basal ganglia,* each ganglion really a heterogeneous collection of subcortical nuclei with different structures, functions, and names (Fig. 8-13). They are part of the motor system and will be discussed in Chap. 10. The bands of white matter separating the basal ganglia from the nuclei of the diencephalon (Fig. 8-13) are the *internal capsules,* radiations of fibers passing between the cortex and other parts of the central nervous system.

PERIPHERAL NERVOUS SYSTEM

The peripheral nervous system consists of the spinal and cranial nerves extending from the spinal cord and brain, respectively. Each nerve contains many individual fibers (i.e., processes from many distinct neurons) grouped into *fasciculi;* a single fasciculus may contain relatively few fibers or hundreds of them (Fig. 8-14). The size and total number of fasciculi within a nerve vary greatly from one nerve to another and along the length of any one nerve. The fasciculi are bound together into a nerve by a connective-tissue sheath, the *epineurium,* which, along with arteries, veins, lymphatics, and fat cells, fills in the spaces between the fasciculi. A similar but less fibrous connective-tissue sheath, the *perineurium,* directly surrounds each fasciculus. The perineurium serves not only the protective and supportive functions of the other connective-tissue coverings;

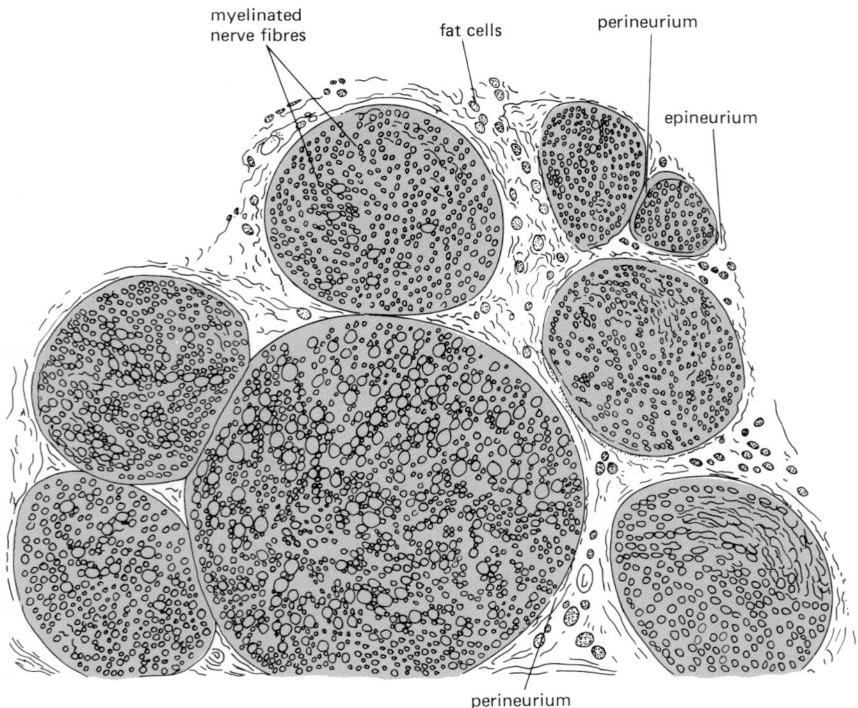

FIGURE 8-14 ***Cross section of a peripheral nerve showing the connective-tissue sheaths of the nerves and nerve fibers.***

it also performs barrier functions, isolating to some extent the enclosed nerve fibers from the outside environment. The individual nerve fibers are covered by a delicate connective-tissue *endoneurium*.

Each peripheral nerve fiber is surrounded by Schwann cells much as the central fibers are surrounded by oligodendroglia. The Schwann cells have been likened to elongated beads strung on a slender thread, each bead representing a Schwann cell and the thread representing the axon. The narrow spaces between the beads represent the nodes of Ranvier, described in Chap. 4. The Schwann cells lie beneath the endoneurium, separating it from the nerve fiber itself. Some of the axons of the peripheral nervous system are wrapped in layers of Schwann-cell membranes, which form the fiber's myelin coat. Other fibers are unmyelinated and are simply tucked into invaginations of the Schwann cell, one Schwann cell often enclosing several unmyelinated fibers (Fig. 8-15).

Because most nerves contain both afferent and efferent fibers, the peripheral nervous system can be separated into *afferent* and *efferent divisions*. Afferent neurons convey information from receptors in the periphery to the central nervous system; as described earlier, their cell bodies all lie outside the central nervous system. These neurons are sometimes called *first-order neurons* because they are the first cells activated in the synaptically linked chains of neurons which handle incoming information. They are also frequently called sensory neurons, but we hesitate to use this term because it implies that the information transmitted by these neurons is destined to reach consciousness, and this is not always true. For example, we have no conscious awareness of our blood pressure even though we have receptors sensitive to this variable.

The efferent division is more complicated than the afferent, being subdivided into *somatic* and *autonomic*. Although this separation is justified by many anatomic and physiologic differences, the simplest distinction between the two is that somatic efferents innervate skeletal muscle and the autonomics innervate smooth and cardiac muscle and glands (other differences are listed in Table 8-1).

The cell bodies of the somatic efferent neurons are located in groups within the brain or spinal cord; their large-diameter, myelinated axons leave the central nervous system and pass directly, i.e., without any synapses, to skeletal muscle cells. The transmitter substance released by these neurons is *acetylcholine*. Because activity of somatic efferent neurons causes contraction of the innervated skeletal muscle cells, these neurons are often called *motor neurons*. Motor neurons can be activated by local reflex mechanisms, or they can be activated by pathways which descend from higher brain centers, but, in either case, their excitation always leads to *contraction* of skeletal muscle cells; there are no motor neurons which, when active, inhibit skeletal muscle fibers.

TABLE 8-1
DIFFERENCES BETWEEN SOMATIC AND AUTONOMIC EFFERENT FIBERS

Somatic fibers:
1 Are a one-neuron pathway to the effector; i.e., they do not synapse once they have left the central nervous system.
2 Innervate skeletal muscle.
3 Always lead to excitation of the muscle.

Autonomic fibers:
1 Are a two-neuron pathway to the effector; i.e., they synapse once in ganglia after they have left the central nervous system.
2 Innervate smooth or cardiac muscle or gland cells.
3 Can lead to excitation or inhibition of the effector cells.

Autonomic Nervous System

The term "autonomic nervous system" is somewhat unfortunate because it conjures up another "nervous system" distinct from the central and peripheral nervous systems. Keep in mind that the term refers sim-

FIGURE 8-15 ***Schwann cell enclosing several unmyelinated nerve fibers.*** **(Adapted from Warwick and Williams.)**

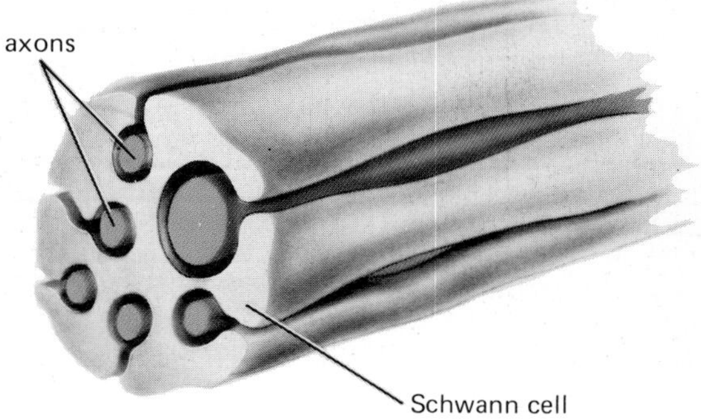

ply to one of the efferent divisions of the peripheral nervous system. Fibers of the autonomic division of the peripheral nervous system innervate cardiac or smooth muscle cells or glands. Anatomic and physiologic differences within the autonomic nervous system are the basis for its further subdivision into *sympathetic* and *parasympathetic* components (Table 8-2). The cell bodies of the first neurons in the two divisions are

TABLE 8-2
DIVISIONS OF THE PERIPHERAL NERVOUS SYSTEM

I. Afferent fibers
II. Efferent fibers
   A. Somatic
   B. Autonomic
      1. Sympathetic
      2. Parasympathetic

located in different areas of the central nervous system, and their fibers leave at different levels, the sympathetic from the thoracic and lumbar regions of the spinal cord and the parasympathetic from the brain and the sacral portion of the spinal cord (Fig. 8-16). Thus, the sympathetic division is also called the *thoracolumbar division,* and the parasympathetic is called the *craniosacral division.* Although the two divisions leave at different levels, the heart and many glands and smooth muscles are innervated by both sympathetic and parasympathetic nerve fibers; i.e., they receive *dual innervation.* (Do not confuse this with reciprocal innervation, which is the inhibition of an active muscle's antagonist.)

The fibers of the autonomic nervous system synapse once after they have left the central nervous system and before they arrive at the neuroeffector junctions (Fig. 8-17). These synapses outside the central nervous system occur in cell clusters called *ganglia.* The two divisions of the autonomic nervous system differ with respect to the locations of their ganglia. Many of the sympathetic ganglia lie close to the spinal cord, forming the paired chains known as the *sympathetic trunks* (Fig. 8-16); others lie approximately midway between the spinal cord and the innervated organ. In contrast, the parasympathetic ganglia lie close to or within the walls of the effector organ. The fibers passing between the central nervous system and the ganglia are the *preganglionic* autonomic fibers; those passing between the ganglia and the effector organ are the *postganglionic* fibers.

In both sympathetic and parasympathetic divisions, the chemical transmitter for the ganglionic synapse between pre- and postganglionic fibers is *acetylcholine.* The chemical transmitter at the junction between the *parasympathetic* postganglionic fiber and the effector cell is also acetylcholine. Fibers that release acetylcholine are called *cholinergic* fibers. The transmitter between the *sympathetic* postganglionic fiber and the effector cell is *norepinephrine*[2], a member of the chemical family of catecholamines (Fig. 8-17). Early experiments on the function of the sympathetic nervous system were done using epinephrine rather than the closely related norepinephrine; moreover, the epinephrine was called by its British name, adrenaline, so that fibers which release norepinephrine came to be called *adrenergic* fibers.

Many drugs stimulate or inhibit the various components of the autonomic nervous system. Those whose actions "mimic" the actions of the sympathetic nervous system are called *sympathomimetic* drugs. Amphetamine and phenylephrine are drugs of this class. Conversely, several choline compounds and the mushroom poison muscarine mimic parasympathetic actions and are *parasympathomimetic* drugs. Drugs which block the actions of the atuonomic nervous system are *sympatholytic* or *parasympatholytic*. The plant belladonna yields atropine, which is a parasympatholytic drug. In fact, the plant received its common name because extracts of the leaf caused dilatation of the pupil of the eye (by blocking the parasympathetic constriction of the iris sphincter muscle) and transformed the users into "beautiful women." (The plant's other name, deadly nightshade, arose because a lethal overdose of the extract was a favorite poison in the Middle Ages.)

The actions of the autonomic nervous system depend upon the nature of not only the chemical released by the postganglionic cell but also the effector cell's receptor sites and intracellular machinery. Four classes of receptor sites have been identified.

[2] There are a few exceptions to this statement; they will be identified where appropriate.

FIGURE 8-16 ***Diagrammatic summary of autonomic nervous system. Only one of the two sympathetic trunks is shown. Not shown are the fibers passing to the glands and smooth muscle cells in the body walls.*** **(Adapted from Langley et al., 1974.)** ►

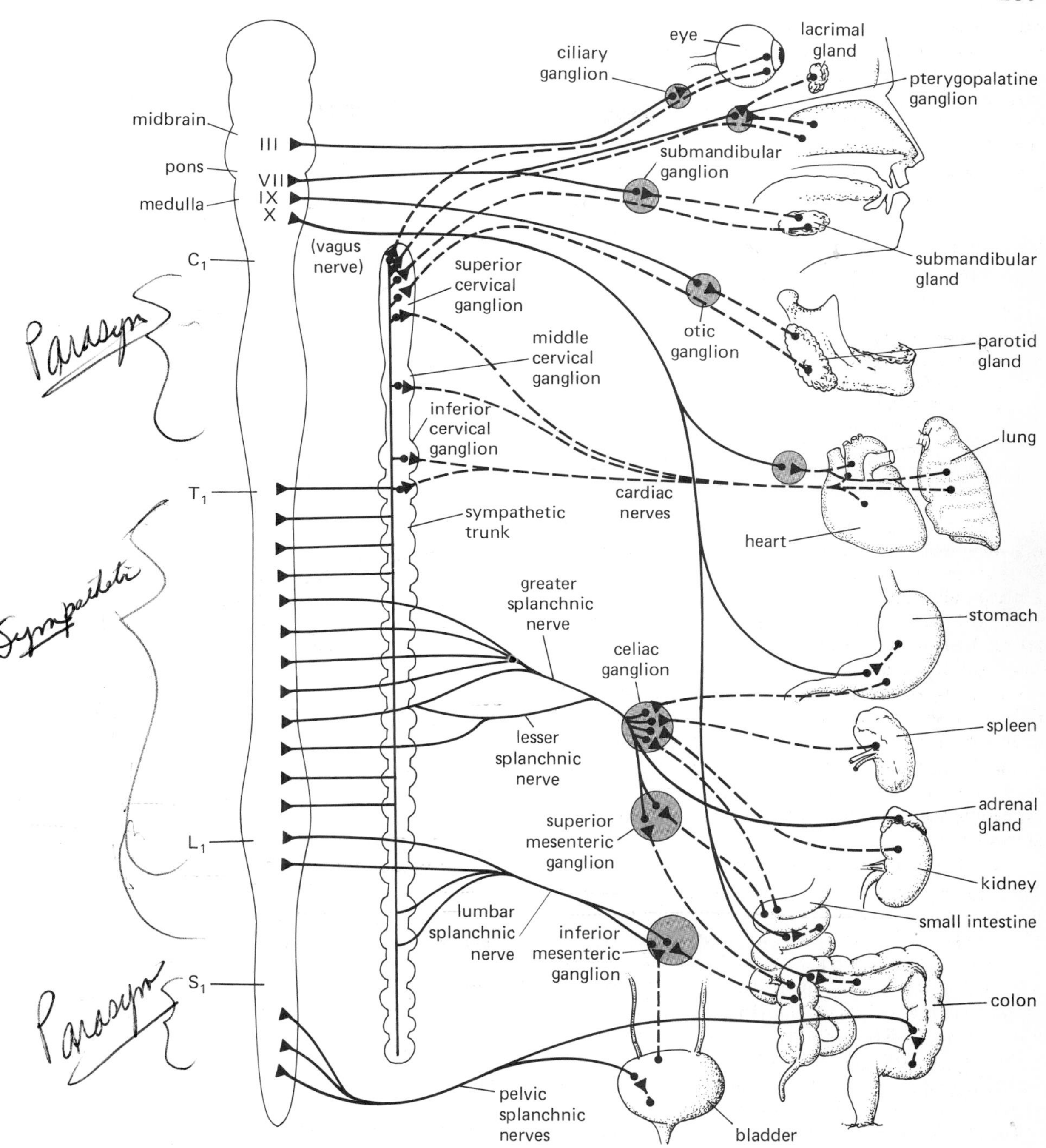
eye
lacrimal gland
ciliary ganglion
pterygopalatine ganglion
midbrain
III
pons
VII
IX
X
medulla
submandibular ganglion
submandibular gland
(vagus nerve)
C_1
superior cervical ganglion
otic ganglion
parotid gland
middle cervical ganglion
inferior cervical ganglion
lung
T_1
cardiac nerves
sympathetic trunk
heart
greater splanchnic nerve
stomach
celiac ganglion
lesser splanchnic nerve
spleen
adrenal gland
superior mesenteric ganglion
L_1
kidney
lumbar splanchnic nerve
inferior mesenteric ganglion
small intestine
S_1
colon
pelvic splanchnic nerves
bladder

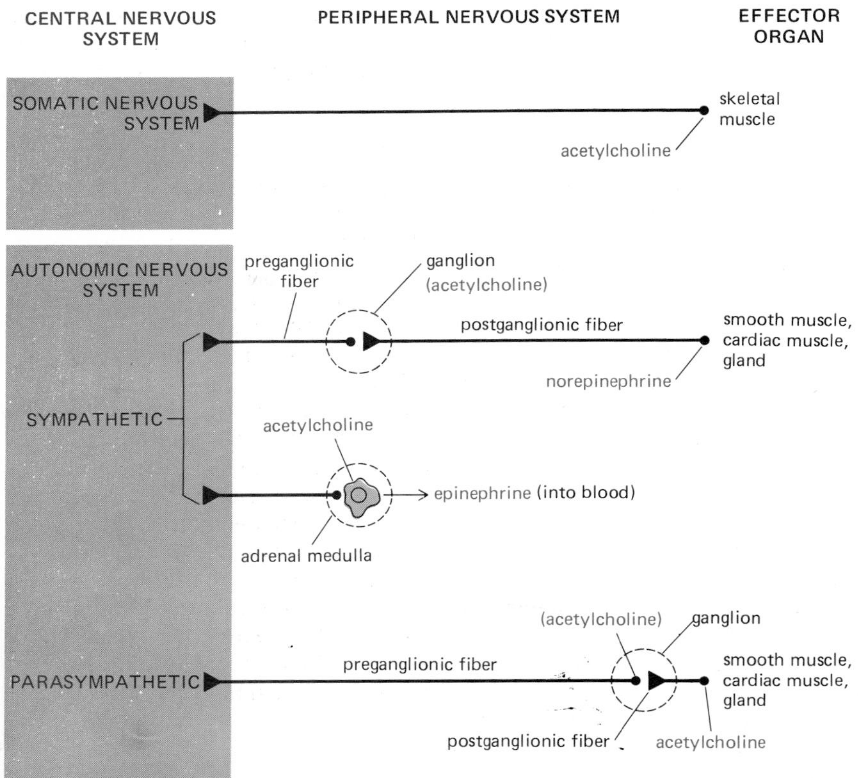

FIGURE 8-17 ***Efferent divisions of the peripheral nervous system.***

One "ganglion" in the sympathetic nervous system never developed long postganglionic fibers; instead, upon activation of its preganglionic nerves, the cells of this "ganglion" discharge their transmitters into the bloodstream. This "ganglion," called the *adrenal medulla,* is therefore an endocrine gland and will be described in more detail in Chap. 12. The fact that is important here is that it releases a mixture of about 80 percent epinephrine and 20 percent norepinephrine. These substances, properly called hormones rather than neurotransmitters, are transported via the blood and interstitial fluid to receptor sites on effector cells sensitive to them. The receptor sites may be the same ones that sit beneath the terminals of the sympathetic postganglionic neurons and are normally activated by the transmitter delivered directly to them, or they may be other, noninnervated receptor sites.

An important characteristic of autonomic neurons is *dual innervation* of most effector organs, i.e., innervation of the same organ by both divisions. Whatever one division does to the effector organ, the other division frequently does just the opposite. For example, action potentials over the sympathetic neurons to the heart increase the heart rate; action potentials over the parasympathetic fibers decrease it. In the intestine, activation of the sympathetic fibers reduces contraction of the smooth muscle in the intestinal wall; the parasympathetics increase contraction. Dual innervation with fibers of opposite action provides for a very fine degree of control over the effector organ. It is like equipping a car with both an accelerator and a brake. With only an accelerator, one could slow the car simply by decreasing the pressure on the accelerator, but the combined effects of releasing the accelerator and applying the brake provide faster and more accurate control. To prevent the sympathetic and parasympathetic systems' opposing effects from conflicting with each other, the two systems are usually activated reciprocally; i.e., as the activity of one system is enhanced, the activity of the other is depressed.

Skeletal muscle cells, in contrast, do not receive dual innervation; they are activated by somatic motor neurons which are only excitatory.

Because glands, smooth muscles, and the heart participate as effectors in almost all bodily functions, it follows that the autonomic nervous system has an extremely widespread and important role in the homeostatic control of the internal environment. It exerts a wide array of effects which can be very difficult to remember. Some common denominators are that, in general, the sympathetic system helps the body to cope with challenges from the outside environment, and the parasympathetic seems to be more responsible for internal housekeeping, such as digestion, defecation, and urination. The sympathetic system is utilized in situations involving stress or strong emotions such as fear or rage, whereas the parasympathetic system is most active during recovery or at rest. The sympathetic nervous system provides the responses to a situation leading to "fight or flight." For example, the sympathetic system increases blood flow to exercising muscles and sustains blood pressure in case of severe blood loss; it decreases activity of the gastrointestinal tract, increases the metabolic production of energy, and increases sweating, changes which provide energy utilization most appropriate to the emergency. These and many other functions of the sympathetic nervous system (and parasympathetic nervous system) will be discussed in relevant places throughout this book.

Autonomic responses usually occur without conscious control or awareness as though they were indeed autonomous (in fact, the autonomic nervous system has been called the involuntary nervous system). However, it is wrong to assume that this need always be the case, for recent experiments have indicated that discrete visceral and glandular responses can be learned. For example, to avoid an electric shock, a rat can learn to selectively increase or decrease its heart rate and a rabbit can learn to constrict the vessels in one ear while dilating those in the other. The implications of such voluntary control of autonomic functions in human medicine are enormous.

These experiments are also important in showing that small segments of the autonomic response can be regulated independently; thus, overall autonomic responses, made up of many small components, are quite variable. Rather than being gross, undiscerning discharges, they are finely tailored to the specific demands of any given situation.

This completes our general description of the afferent and efferent divisions of the peripheral nervous system. We now turn to the specific ways in which these components are organized in the spinal and cranial nerves which constitute the peripheral nervous system.

Spinal Nerves

Shortly after leaving the spinal cord, the ventral root (carrying efferent fibers) and the dorsal root (carrying afferent fibers) from the same side at the same level combine to form a *spinal nerve;* thus, all spinal nerves contain afferent and efferent fibers. The 31 pairs of spinal nerves are grouped as follows: 8 cervical, 12 thoracic, 5 lumbar, 5 sacral, and 1 coccygeal, the individual nerves being commonly named for the intervertebral foramens through which they pass. Thus, the *first cervical nerve* leaves between the occipital bone of the skull and the first cervical vertebra (the atlas), the *second cervical nerve* leaves the vertebral canal between the atlas and the second cervical vertebra, the *eighth cervical nerve* leaves between the seventh cervical and the first thoracic vertebrae, etc. (Fig. 8-5C).

Shortly after emerging from the intervertebral foramen, each spinal nerve divides into a *dorsal* and a *ventral branch* (Fig. 8-18). The dorsal branches supply the muscles and skin of the back; the ventral branches supply the limbs and front and sides of the trunk. Figure 8-19 shows the segmental distribution of the spinal

FIGURE 8-18 ***Distribution of a typical spinal nerve.***

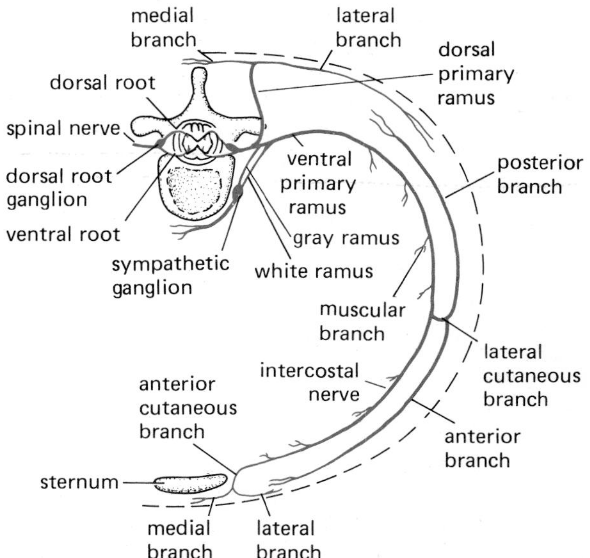

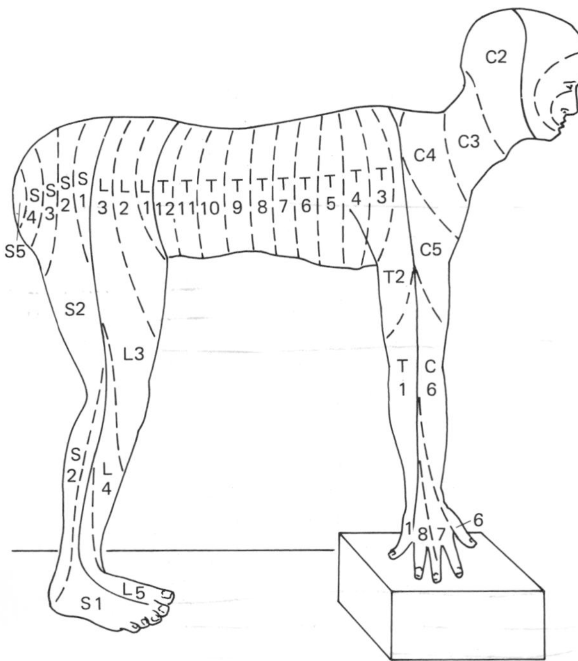

FIGURE 8-19 ***Segmental distribution of the spinal nerves. C = cervical, T = thoracic, L = lumbar, and S = sacral.***

nerves as they travel to the periphery. The dorsal branches remain separate; i.e., they do not join the dorsal branches of other nerves. The ventral branches from the thoracic region also remain separate, but the ventral branches from other regions of the spinal cord intermingle with fibers from neighboring nerves, forming *nerve plexuses*. Thus, a single peripheral nerve leaving a plexus may contain fibers from several different levels of the spinal cord. The four great plexuses, the *cervical, brachial, lumbar*, and *sacral plexuses*, can be seen in Fig. 8-1; their compositions are shown more clearly in Figs. 8-20 and 8-21. The cervical plexuses receive sympathetic fibers from the *superior cervical ganglion* of the sympathetic trunk, and the brachial plexuses receive their sympathetics from the *middle* and *inferior (cervical) ganglia* of the sympathetic trunk (Fig. 8-16). (Sometimes the inferior cervical and first thoracic ganglia are combined to form a single ganglion known as the *stellate ganglion*.) The peripheral ramifications of the plexuses are shown in Figs. 8-20B and C and 8-21B and C.

A potentially confusing aspect of spinal-nerve anatomy is the relationship of the nerves to the sympathetic trunks. Recall that the ventral roots of the thoracic and first few lumbar nerves carry sympathetic preganglionic fibers (whose cell bodies of origin are in the lateral horn of the spinal cord). A short distance after the ventral and dorsal roots have joined and the resulting nerve has emerged from the intervertebral foramen, all the preganglionic sympathetic fibers contained in that nerve leave by way of a branch called the *white ramus communicans*, which carries them to the sympathetic trunk.

On entering this ganglionic chain, a preganglionic fiber may pursue one of three courses: (1) It may synapse immediately with postganglionic neurons of the ganglion it first enters. (2) It may travel up or down the trunk before synapsing with postganglionic neurons in ganglia belonging to more superior or inferior segments. This explains how it is that the sympathetic trunk contains ganglia adjacent to all segments of the spinal cord despite the fact that the trunk receives preganglionic fibers only from the thoracic and upper lumbar segments. (3) It may not synapse in the sympathetic trunk at all but may pass directly through the trunk into the *splanchnic nerves* to reach the farther-removed sympathetic ganglia (or the adrenal medulla) where it synapses with postganglionic neurons.

How do the axons of postganglionic neurons, i.e., the second neurons in the pathways described in routes 1 and 2 above, leave the sympathetic trunk? Many of them leave by way of the *gray rami communicantes* which carry them to the spinal nerves with which they then travel to their destinations in the glands and smooth muscle of the body wall; every spinal nerve receives a gray ramus from the trunk. (As shown in Fig. 8-22B, an anatomic curiosity is that the gray rami enter the spinal nerve at a point closer to the intervertebral foramen than the site of exit of the white rami; thus these sympathetic fibers form a loop as they pass from the spinal cord to the periphery.) The remaining postganglionic neurons of the trunk, those whose axons do not leave the trunk via the gray rami, leave instead by fiber bundles which supply the heart and other thoracic structures; these come mainly from the cervical and thoracic segments of the sympathetic trunk.

Cranial Nerves

The 12 pairs of cranial nerves are listed in Table 8-3; the origins of all but the first (the olfactory nerve, which joins the olfactory bulb) are shown in Fig. 8-23A.

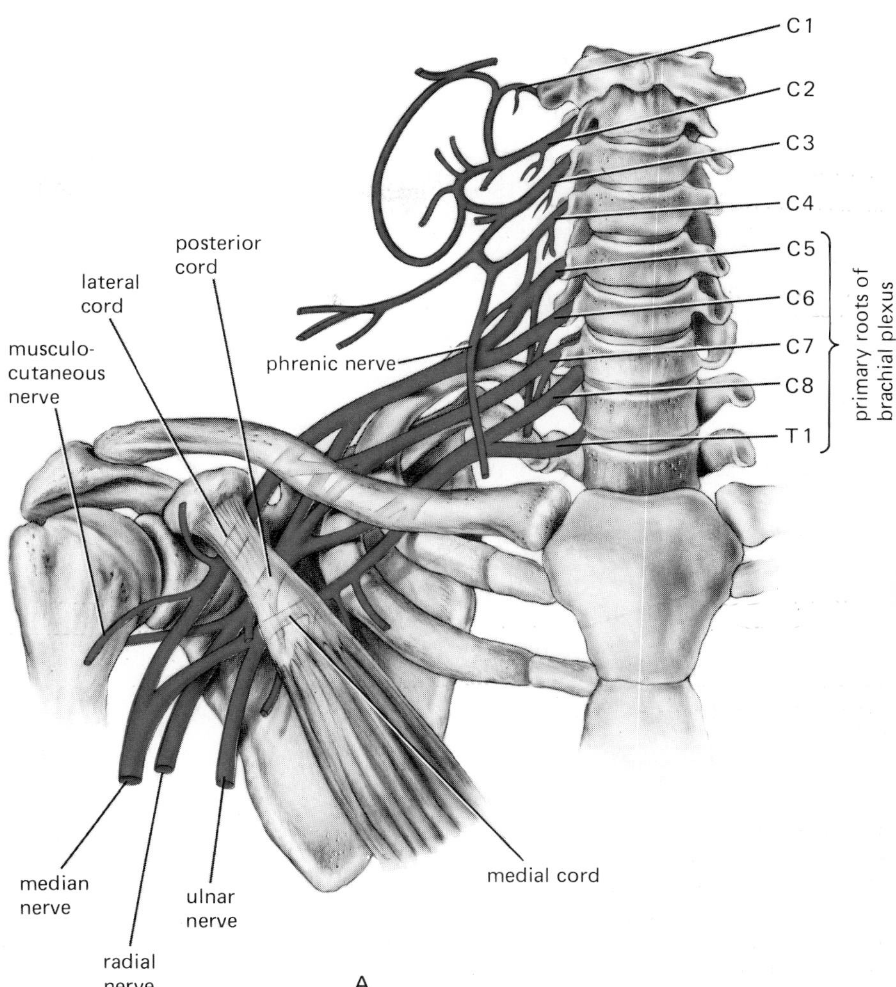

FIGURE 8-20 (A) ***Major branches of the cervical and brachial plexuses. Major nerves to the anterior*** **(B)** ***and posterior*** **(C)** ***surface of the arm.***

The nerves pass from the brain out to the periphery through holes, or *foramens,* in the skull; these are shown in Fig. 8-23B.

Not all the cranial nerves have both afferent and efferent components, and not all of those with efferent components have both autonomic and somatic parts. (Autonomic fibers that do occur belong to the parasympathetic, or *craniosacral,* division of the autonomic nervous system.) Efferent fibers arise within the brainstem from cell bodies contained within the nucleus of that nerve. For example, in the brainstem (from which most of the cranial nerves arise) there are nuclei of the oculomotor, trochlear, abducens, and facial nerves.

Afferent fibers arise outside the brain from cell bodies situated either in ganglia occurring along the nerve as it passes to the periphery or within a sense organ such as the eye, ear, or nose. The central processes of the afferent neurons enter the brain and synapse in a given nucleus of termination.

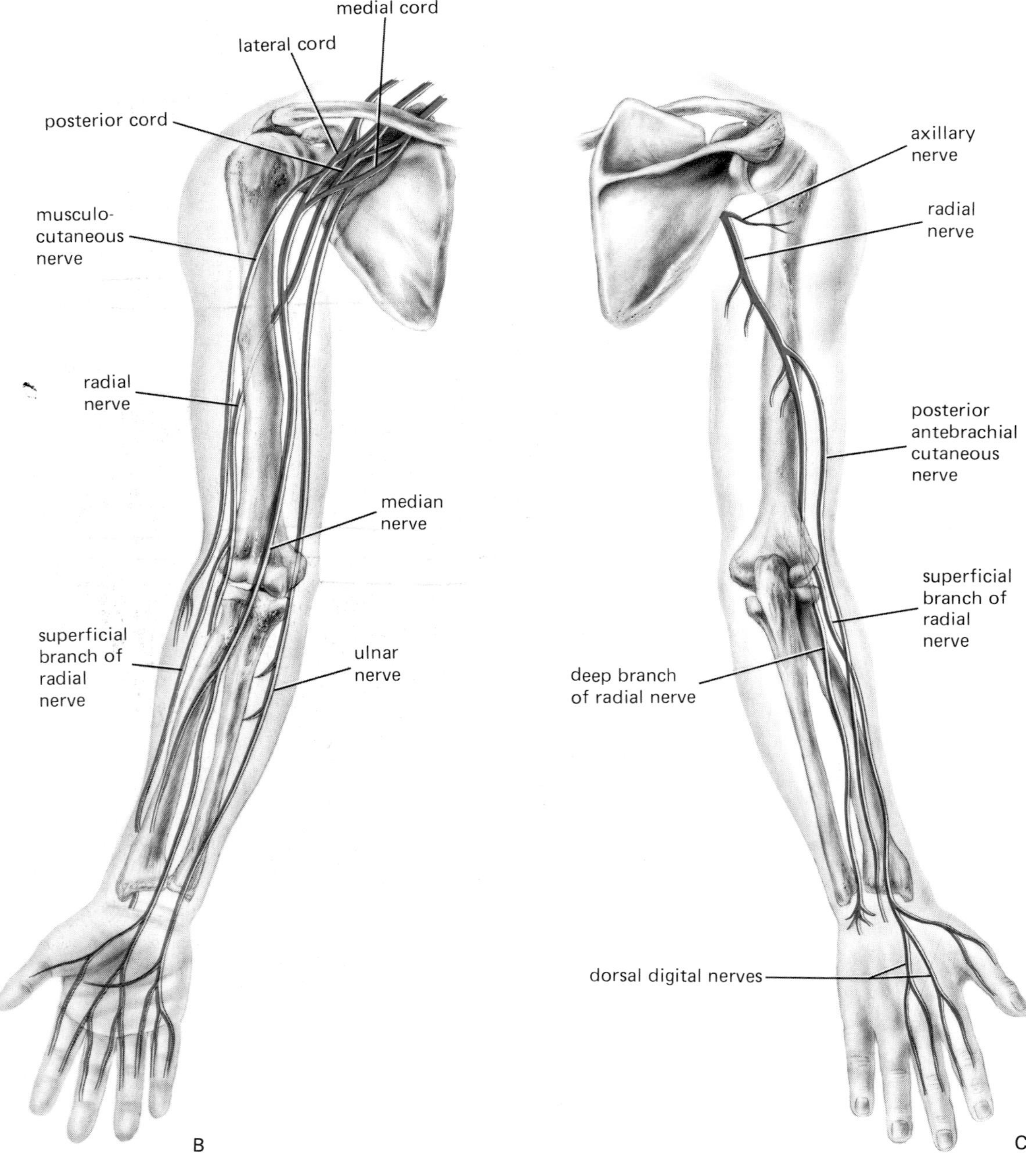
medial cord
lateral cord
posterior cord
musculo-cutaneous nerve
radial nerve
median nerve
superficial branch of radial nerve
ulnar nerve
B
ANTERIOR
axillary nerve
radial nerve
posterior antebrachial cutaneous nerve
superficial branch of radial nerve
deep branch of radial nerve
dorsal digital nerves
C
POSTERIOR

FIGURE 8-20 ***(Continued)***

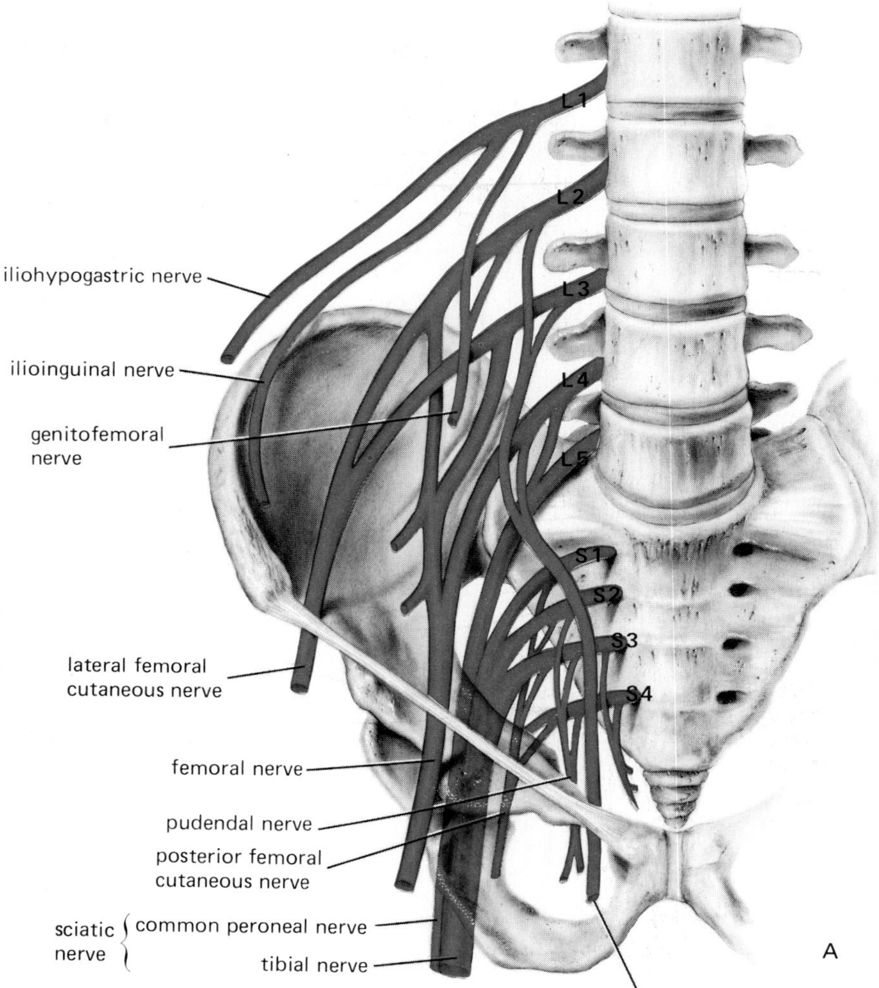

FIGURE 8-21 (A) ***Major branches of the lumbar and sacral plexuses. Major nerves to the anterior*** (B) ***and posterior*** (C) ***surface of the leg.***

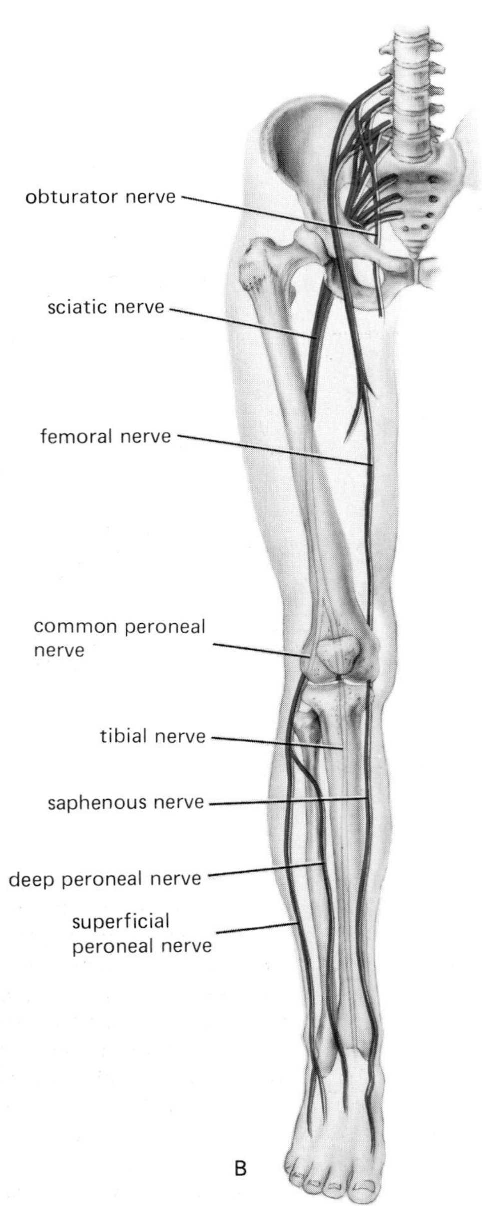
obturator nerve
sciatic nerve
femoral nerve
common peroneal nerve
tibial nerve
saphenous nerve
deep peroneal nerve
superficial peroneal nerve
B

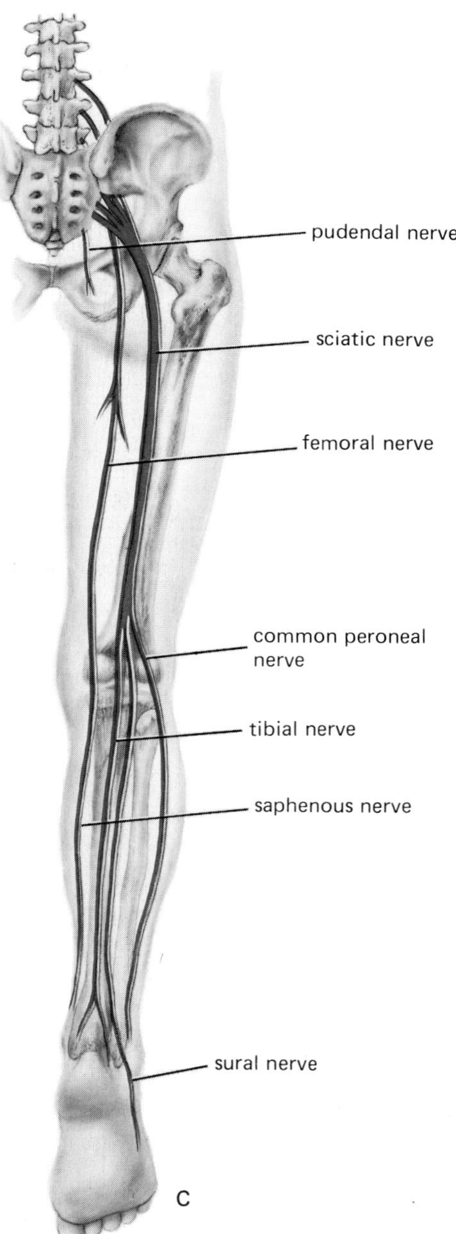
pudendal nerve
sciatic nerve
femoral nerve
common peroneal nerve
tibial nerve
saphenous nerve
sural nerve
C

FIGURE 8-21 ***(Continued)***

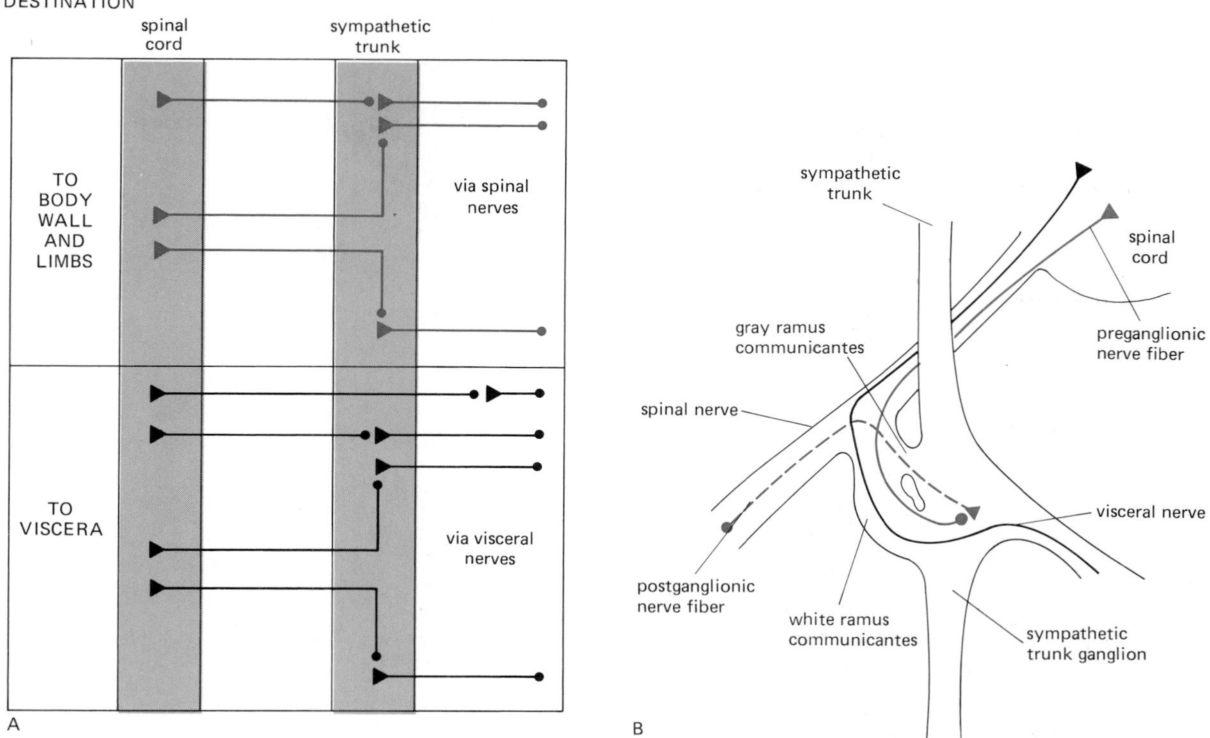

FIGURE 8-22 ***Possible relations of the sympathetic nerves and sympathetic trunk.***

BLOOD SUPPLY OF THE CENTRAL NERVOUS SYSTEM

Glucose is normally the only substrate that can be metabolized sufficiently rapidly by the brain to supply its energy requirements, and most of the energy from glucose is transferred to high-energy ATP molecules during its oxidative breakdown. The glycogen stores of the brain are negligible; thus the brain is completely dependent upon a continuous blood supply of glucose and oxygen. Although the adult brain is only 2 percent of the body weight, it receives 15 percent of the total blood supply at rest to support the high oxygen utilization. If the oxygen supply is cut off for 4 to 5 min or if the glucose supply is cut off for 10 to 15 min, brain damage will occur. In fact, the most common cause of brain damage is stoppage of the blood supply (a *stroke*). The cells in the region deprived of nutrients cease to function and die.

The central nervous system receives a rich blood supply, but the exchange of substances between the blood and the brain or spinal cord is handled differently from the somewhat unrestricted movement of substances from capillaries in other organs. Dye injected into a vein stains all the organs of the body except the brain and spinal cord. A complex group of *blood-brain barrier* mechanisms closely controls both the kinds of substances which enter the extracellular space of the brain and the rate at which they enter. The barrier probably comprises both anatomic structures and physiologic transport systems which handle different classes of substances in different ways. The blood-brain barrier mechanisms precisely regulate the chemical composition of the extracellular space of the brain and prevent harmful substances from reaching neural tissue.

The arterial blood supply to the brain comes from two sources, the *internal carotid* and *vertebral* arteries (see Figs. 15-5 and 8-24), the two vertebral arteries

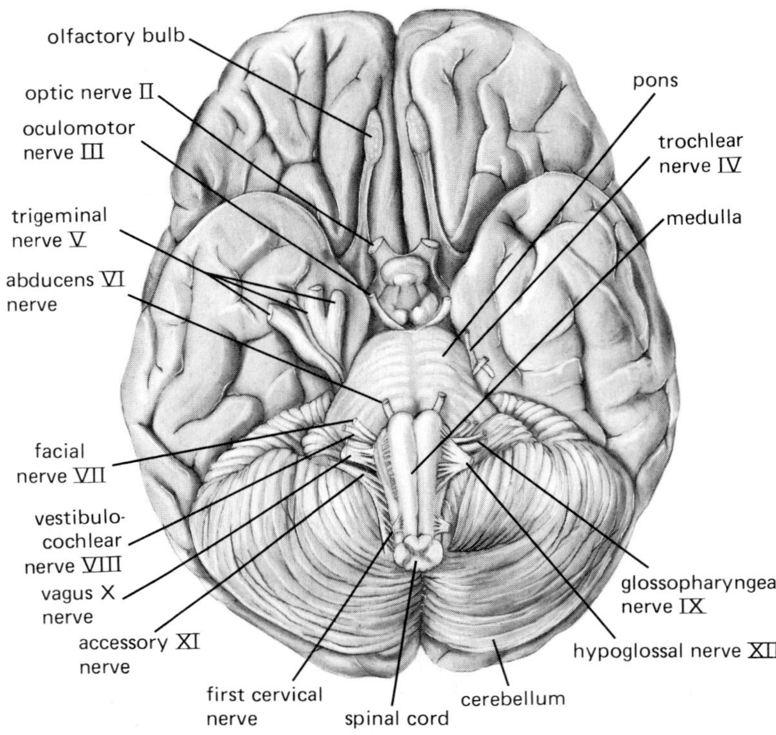

FIGURE 8-23 (A) ***View of the base of the brain showing the cranial nerves except for the olfactory nerve (I). The olfactory nerves synapse with neurons in the olfactory bulbs.*** (B) ***Base of the cranial cavity, which supports the base of the brain, and the foramens through which the cranial nerves (and other structures) pass.***

supplying the posterior circulation of the brain and the internal carotids supplying the anterior. After entering the skull through the foramen magnum, the vertebral arteries run through the subarachnoid space and pass upward along the brainstem to send branches to the cerebellum and medulla and downward along the anterior surface of the spinal cord to feed the *anterior spinal artery* (Fig. 8-24).

As they reach the level of the pons, the two vertebral arteries join to form the *basilar artery* (Fig. 8-24), which sends branches to the pons, midbrain, and parts of the cerebellum before ending at the upper boundary of the pons. As it ends, the basilar artery branches, forming two *posterior cerebral arteries,* which supply the occipital lobes of the cerebrum and parts of the temporal lobes.

The blood supply to the anterior part of the brain is furnished by the two internal carotid arteries, which enter the cranial cavity through the carotid canal of the temporal bone and almost immediately branch into the *anterior* and *middle cerebral arteries* (Fig. 8-24). The anterior branch supplies blood to the medial surface of the cerebral hemispheres, and the middle branch supplies principally the lateral surfaces. The distribution of the anterior and posterior cerebral arteries can be seen in Fig. 8-25.

These two blood supplies to the brain (the vertebrals and internal carotids) are connected by *posterior communicating arteries,* and the two anterior cerebral arteries are connected by an anterior communicating artery (Fig. 8-24). These connecting arteries complete a circle, the *circle of Willis,* which provides an important safeguard to ensure the supply of blood to all parts of the brain despite the blockage of either the vertebral or internal carotid arteries.

The veins of the brain, comprising the cerebral, cerebellar, and brainstem veins, have very thin walls and no valves. As they leave the substance of the brain, they pierce the arachnoid membrane, cross the subarachnoid space, and enter the dura, within which they

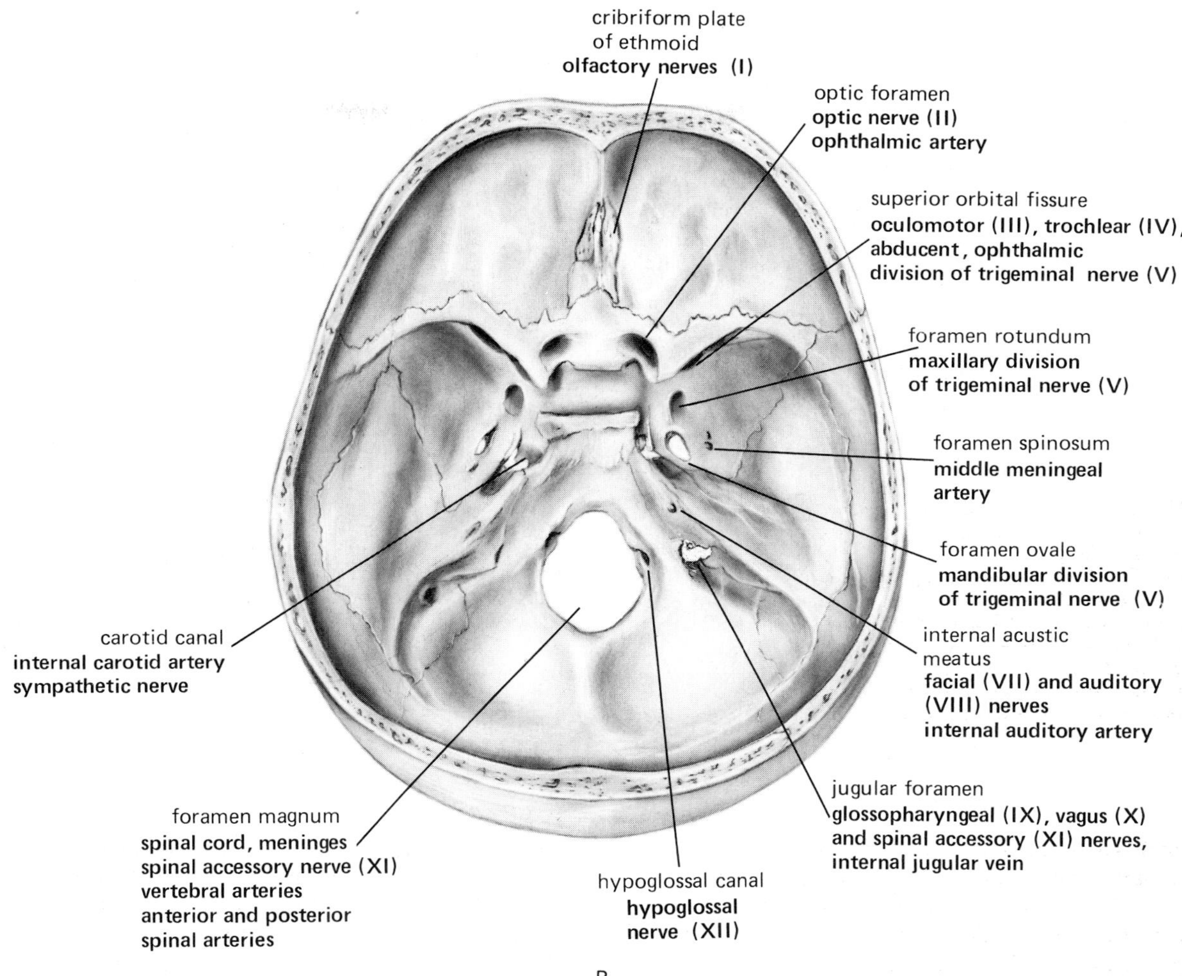

B

TABLE 8-3

CRANIAL NERVES (MAJOR COMPONENTS)

No.	*Name*	*Components*	*Central connections or origins*	*Peripheral distribution*	*Function*
I	Olfactory	Afferent	Olfactory bulb	Mucous membrane of olfactory region of nose	Smell
II	Optic	Afferent	Nucleus in thalamus (*lateral geniculate*)	Retina of eye	Vision
III	Oculomotor	Efferent	Nuclei in midbrain (*oculomotor*)	Muscles (*superior, inferior,* and *medial rectus; inferior oblique;* and *levator palpebrae*)	Eye movements, raise upper eyelid

TABLE 8-3
CRANIAL NERVES (MAJOR COMPONENTS) *(Continued)*

No.	*Name*	*Components*	*Central connections or origins*	*Peripheral distribution*	*Function*
			Nucleus in midbrain (*Edinger-Westphal*)	*Pupillary constrictor* and *ciliary* muscles of eye	Regulation of pupil size, accommodation of lens
IV	Trochlear	Efferent	Nucleus in midbrain (*trochlear nucleus*)	*Superior oblique* muscle	Eye movements
V	Trigeminal	Afferent	Nucleus in pons and medulla (*chief sensory nucleus of* V and *nucleus of spinal tract of* V)	Face, greater part of scalp, teeth, mouth, nasal cavity	Sensory from face, nose, and mouth
		Efferent	Nucleus in pons (*motor nucleus of* V)	Muscles (*medial and lateral pterygoids, masseter, temporalis*)	Chewing movements
VI	Abducens	Efferent	Nucleus in pons (*abducent nucleus*)	*Lateral rectus* muscle	Eye movements
VII	Facial	Afferent	Nucleus in pons (*nucleus of tractus solitarius*)	Anterior part of tongue and soft palate	Taste
		Efferent	Nucleus in pons (*facial nucleus*)	Muscles of face, scalp, and outer ear	Facial expression
			Nucleus in pons (*superior salivatory nucleus*)	Submandibular and sublingual salivary glands, lacrimal glands, glands of nasal and palatine mucosa	Secretions
VIII	Acoustic	Afferent	Nucleus in pons (*cochlear* and *vestibular nuclei*)	Hair cells of organ of Corti, semicircular canals, maculae, and saccule	Hearing, balance, change of rate of motion of head
IX	Glossopharyngeal	Afferent	Nucleus in medulla (*nucleus of tractus solitarius*)	Back of tongue, pharynx, carotid sinus, and body	Taste and other sensations of tongue, changes in levels of blood pressure and gases
		Efferent	Nucleus in medulla (*nucleus ambiguus*)	*Stylopharyngeus* muscle	Swallowing movements
			Nucleus in medulla (*inferior salivatory nucleus*)	Parotid salivary gland	Secretions
X	Vagus	Afferent	Nuclei in medulla (*sensory nucleus of vagus, nucleus of tractus solitarius*)	Taste buds on epiglottis; larynx, trachea, pharynx, heart, lungs, esophagus, small intestine, part of colon	Pain, stretching, changes in levels of blood pressure and gases, taste
		Efferent	Nuclei in medulla (*dorsal motor nucleus, nucleus ambiguus*)	Smooth muscle of bronchi, esophagus, stomach, and intestines; heart; striated muscles of pharyngeal constrictors and intrinsic muscles of larynx	Movement and secretion by the organs supplied
XI	Accessory	Efferent	Spinal and medullary nucleus (*accessory nucleus*)	Muscles of larynx, *trapezius, sternocleidomastoid*	Shoulder movements, turning head, voice production
XII	Hypoglossal	Efferent	Nucleus in medulla (*hypoglossal nucleus*)	Muscles of tongue	Tongue movements

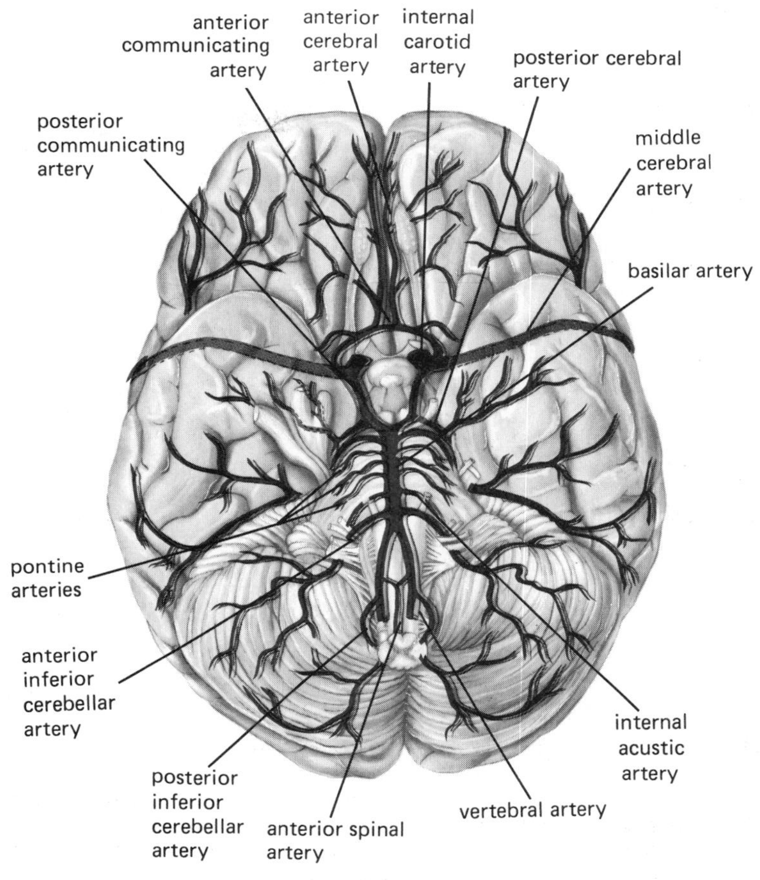

FIGURE 8-24 ***Major arteries on the base of the brain.***

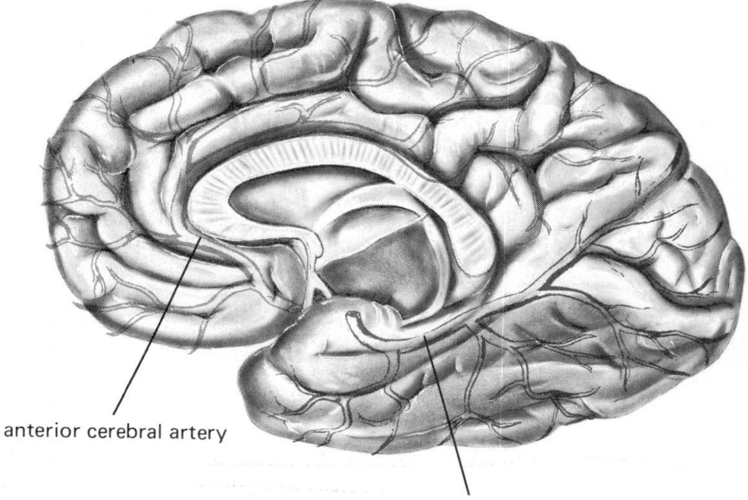

FIGURE 8-25 ***Distribution of the anterior and posterior cerebral arteries over the midline surface of the cerebral hemisphere.***

form large *venous sinuses,* which drain eventually into the *internal jugular veins* (Figs. 8-26 and 15-6).

The spinal cord is nourished by several arteries, including the *spinal arteries,* which run along the anterior and posterior surfaces of the cord. The blood leaves the cord via the *anterior* and *posterior spinal veins.*

VENTRICULAR SYSTEM AND CEREBROSPINAL FLUID

The central nervous system is perfused not only by its blood supply but by a second fluid, the *cerebrospinal fluid.* This clear fluid surrounds the outer surface of the

FIGURE 8-26 ***Major veins and venous sinuses of the brain.***

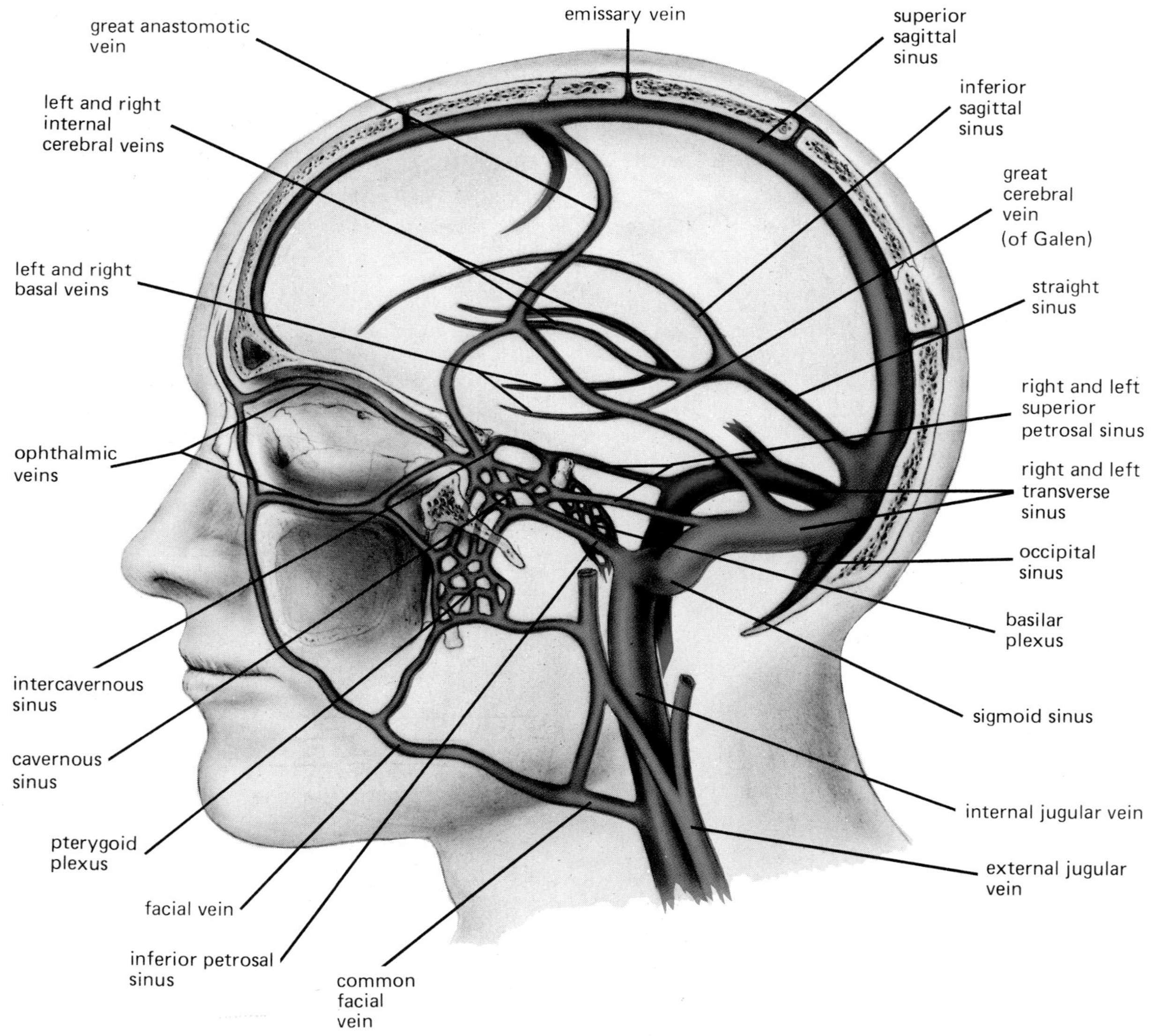

brain and spinal cord so that the central nervous system literally floats in a cushion of cerebrospinal fluid. Since the brain is a soft and delicate tissue about the consistency of jelly, it is thus protected from sudden and jarring movements of the head. Cerebrospinal fluid also fills the four ventricles in the brain. The first and second ventricles (called the *lateral ventricles;* Fig. 8-27) are deep inside the two cerebral hemispheres. They meet in the midline and connect there with the *third ventricle,* which dips down vertically between the two halves of the diencephalon. The third ventricle, in turn, connects through a narrow channel, the *cerebral aqueduct,* with the *fourth ventricle,* an expanded space on the dorsal surface of the pons and part of the medulla (Fig. 8-11B). The fourth ventricle contracts at its posterior end to form the fine *central canal,* which runs through the caudal part of the medulla and on through the full length of the spinal cord. The central canal is usually occluded in the adult.

Cerebrospinal fluid is secreted into the ventricles by highly vascular tissues, the *choroid plexuses.* A barrier is present between the capillaries of the choroid plexus and the cerebrospinal fluid. Consistent with the barrier mechanisms, cerebrospinal fluid is a selective secretion, not a simple filtrate of plasma. For example, protein, potassium, and calcium concentrations are lower in cerebrospinal fluid than in a filtrate of plasma, whereas sodium and chloride are higher. The mechanisms responsible for this selective formation of spinal fluid are not known.

As the cerebrospinal fluid flows from its origin at the choroid plexuses within the ventricles, substances diffuse between it and the extracellular space of brain tissue since the walls of the ventricles are permeable to most substances. This exchange across the ventricular walls allows nutrients to enter brain tissue from cerebrospinal fluid and the end products of brain metabolism to leave. (However, the major sites of nutrient and end-product exchange are the cerebral capillaries.) The cerebrospinal fluid moves from its origin, back through the interconnected ventricular system to the brainstem, where it passes through three openings in the roof of the fourth ventricle (a single *median* and two *lateral apertures*) out to the subarachnoid space on the surface of the brain. This is a continuation of the cerebrospinal-fluid-filled subarachnoid space surrounding the spinal cord.

Aided by circulatory, respiratory, and postural pressure changes, the cerebrospinal fluid finally flows to the top of the outer surface of the brain, where it enters the venous sinuses through one-way valves. If the path of flow is obstructed at any point between its site of formation and its final reabsorption into the vascular system, cerebrospinal fluid builds up, causing *hydrocephalus,* or "water on the brain." The pressure against which cerebrospinal fluid continues to be secreted is quite high—high enough to damage the brain so that mental retardation accompanies severe, untreated cases.

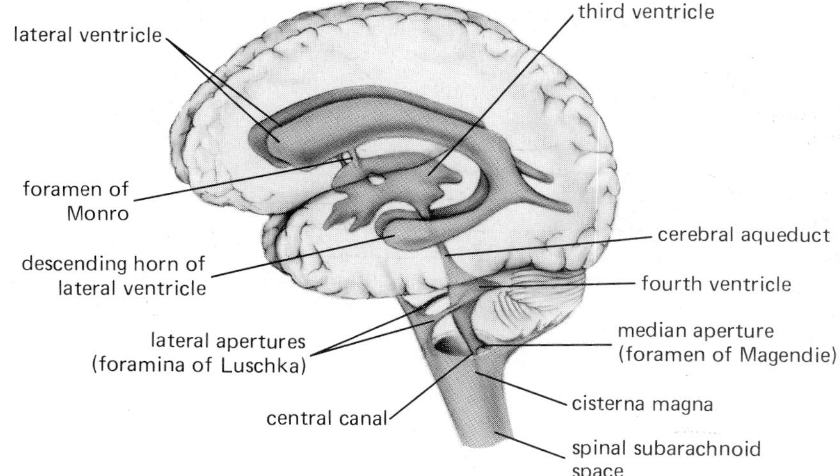

FIGURE 8-27 ***Cerebral ventricles and their connections with the subarachnoid space of the spinal cord and base of the brain. The continuation of the subarachnoid space over the surface of the brain is not shown. The two lateral ventricles (one in each hemisphere) are also called the first and second ventricle.***

CHAPTER 9

The Nervous System: II. Sensory Systems

Man's awareness of the world is determined by the physiologic mechanisms involved in the processing of afferent information, including such steps as the conversion of stimulus energy into coded neural activity indicating the quality, intensity, location, and duration of the stimulus. The action potentials are coded in different temporal patterns along different nerve fibers. This code represents the information from the external world even though, as is frequently the case with symbols, it differs vastly from the information it represents. The coded afferent information may or may not have a conscious correlate; i.e., it may or may not be incorporated into a conscious awareness of the physical world. Afferent information which does have a conscious correlate is called *sensory information* and, for the purposes of this book, that conscious experience of objects and events of the external world which we acquire from the neural processing of afferent information is called *perception*.

Intuitively, it might seem that sensory systems operate like electric equipment, but this is true only up to a point. As an example, let us compare telephone transmission with our auditory sensory system. The telephone changes sound waves into electric impulses, which are then transmitted along wires to the receiver; thus far the analogy holds. (Of course, the mechanisms by which electric currents and action potentials are transmitted are quite different, but this does not affect our argument.) The telephone then changes the coded electric impulses *back into sound waves*. Here is the crucial difference, for our brain does not physically translate the code into sound; rather the coded information itself or some correlate of it is what we perceive as sound. At present there is absolutely no understanding how coded action potentials or composites of them can be associated with conscious sensations.

BASIC CHARACTERISTICS OF SENSORY CODING

It is worthwhile to restate that all information transmitted by the nervous system over distances greater than a few millimeters is signaled in the form of action potentials traveling over specific neural pathways. Several different kinds of information must be relayed by this code: stimulus quality, intensity, and localization.

Stimulus Quality

As described in Chap. 4, receptors possess differential sensitivities; i.e., each receptor type responds more readily to one form of energy than to others. Therefore, the type of receptor activated by a stimulus constitutes the first step in the coding of different types (*modalities*) of stimuli. If, however, the differential sensitivity of receptors is to play an important role in the separation of the various sensory modalities, the afferent nerve fibers and at least some of the ascending spinal cord and brain pathways activated by the receptors must retain the same degree of specificity, carrying information that pertains to only one sensory modality. As expected, therefore, there are specific pathways ("labeled lines," as it were) for the different modalities. In tracing these pathways we shall begin at the receptor.

A single afferent neuron plus all the receptors it innervates make up a *sensory unit*. In a few cases the afferent neuron innervates a single receptor, but generally the peripheral end of an afferent neuron divides into many fine branches, each terminating at a receptor (Fig. 9-1). When the sensory unit contains more than one receptor, all the receptors are differentially sensitive to the same stimulus energy form. The *receptive field* of a neuron is that area which, if stimulated, leads to activity in the neuron (Fig. 9-1).

Central processes of afferent neurons terminate in the central nervous system, often diverging to terminate on several (or many) interneurons (Fig. 9-2A). The central processes of the afferent neurons also overlap, so that the processes of many afferent neurons converge upon a single interneuron (Fig. 9-2B). If the interneuron is to retain the specificity of the receptors, sensory units converging upon it must be of the same modality.

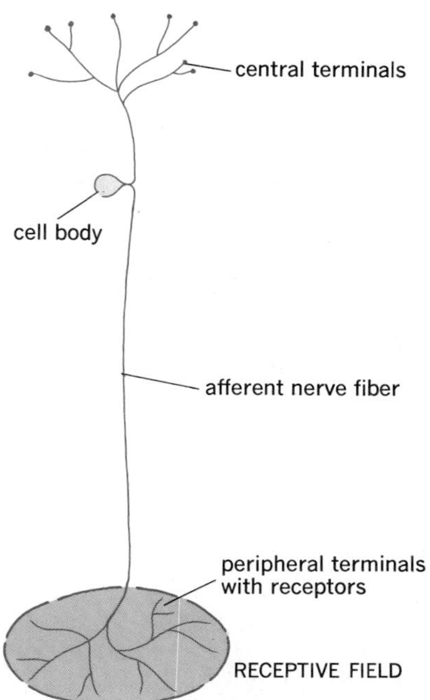

FIGURE 9-1 ***Sensory unit and receptive field.***

The parallel chains of interneurons, which are grouped together to form the ascending pathways of the central nervous system, are of two kinds, *specific* and *nonspecific*. Each chain of neurons conveying specific imformation consists of three to five synaptically connected neuronal links. Several sensory units may converge upon a given specific chain of neurons, but all these sensory units respond to the same stimulus energy, and specificity is maintained. The specific pathways (except for the olfactory pathways) pass to the thalamus of the brain and synapse there with neurons which go to the cerebral cortex.

In contrast to the specific pathways, chains of neurons conveying nonspecific information are activated by sensory units of several different modalities and therefore convey only general information about the level of excitability; i.e., they indicate that *something* is happening, usually without specifying just what (or where). The nonspecific pathways feed into areas of the brain, such as the brainstem reticular formation, which are not highly discriminative but are important in determining states of consciousness such as sleep and wakefulness. Some of the tracts which ascend in the

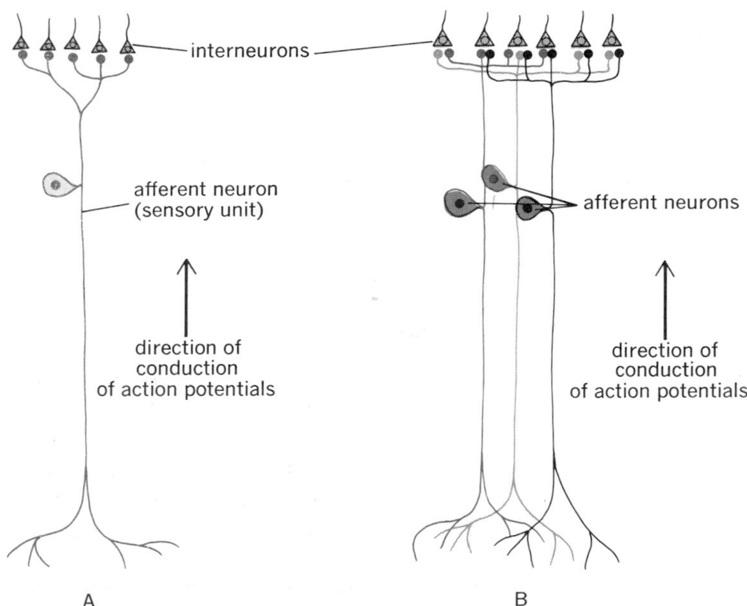

FIGURE 9-2 ***(A) Divergence of afferent-neuron (sensory unit) terminals. (B) Convergence of afferent neurons onto single interneurons.***

spinal cord, e.g., the *spinothalamic pathways,* contain both specific and nonspecific fibers.

Pathways carrying specific information about the different sensory modalities go to different areas of the cortex (Fig. 9-3). The fibers subserving the somatic sensory modalities (touch, temperature, etc.) synapse at cortical levels in the *somatosensory cortex,* a strip of cortex on the postcentral gyrus, which lies in the parietal lobe just behind the boundary between the parietal and frontal lobes (the central sulcus, Fig. 9-3). The specific pathways which originate in receptors of the taste buds, after synapsing in brainstem and thalamus, probably pass to cortical areas adjacent to the face region of the somatosensory strip. However, the specific pathways from the ears, eyes, and nose do not pass to the somatosensory cortex but go to other primary receiving areas (Fig. 9-3). The pathways subserving olfaction are different from all the others in that they do not pass directly through the thalamus but pass instead into the parts of the limbic system (see Fig. 11-3).

Thus stimulus quality is indicated by the specific sensitivity of individual receptors and the pathways conveying the information to the primary sensory areas of the brain. It is not known how the activation of neurons in different parts of the cortex results in the different sensations.

FIGURE 9-3 ***Primary sensory areas of the cerebral cortex. The central sulcus is also called the central fissure.***

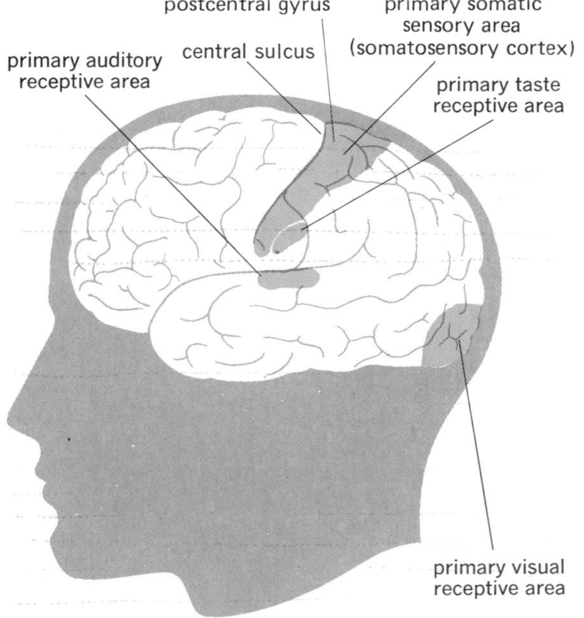

Stimulus Intensity

The second kind of information contained in the code is stimulus intensity or quantity. As described in Chap. 4, one important mechanism for signaling intensity is the number of sensory units activated; the greater the intensity of the stimulus, the greater is the number of sensory units activated. A second mechanism is the frequency at which the sensory unit fires. Action-potential frequency correlates with stimulus intensity for afferent pathways leading to the sensory experiences of touch, temperature, limb position (joint extension or flexion), taste, sound (loudness), and light (brightness). Figure 9-4 shows the afferent impulses in a single cold fiber as the cutaneous receptor is gradually cooled from 34 to 26°C (normal body temperature is close to 37°C). The action-potential frequency in temperature-sensitive sensory units is linearly related to the perceived intensity, but in other sensory systems the relation between stimulus intensity and perceived intensity is more complicated.

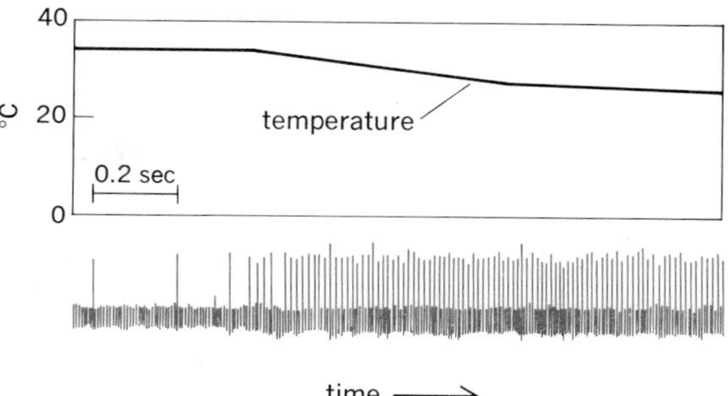

FIGURE 9-4 ***Single cold receptor signals a drop in temperature from 34 to 26°C with an increase in firing rate of action potentials in its afferent nerve fiber.*** **(Adapted from Hensel and Bowman.)**

Stimulus Localization

A third factor to be relayed in the code is the *location of the stimulus*. The specific pathways channel the afferent information in a relatively unmixed manner and indicate stimulus location as well as stimulus quality. Since only sensory units from a restricted area converge upon any one interneuron, the specific afferent pathway which begins with that particular interneuron transmits exclusively information about that restricted area. Before discussing the terminations of the specific afferent pathways, let us examine more closely how information is actually fed into them. The branching peripheral terminals of the afferent neuron spread over areas of variable size (2 to 200 mm^2 in skin). The thresholds of the receptors of a single sensory unit vary within the area covered by the peripheral terminals (the receptive field of the afferent neuron), usually being lowest at the geometric center. Thus, a stimulus of a given intensity causes more activity in an afferent neuron if it occurs at the center of the receptive field (point *A*, Fig. 9-5) than at the periphery (point *B*). Since the peripheral terminations of afferent neurons also overlap to a great extent (Fig. 9-6), the placement of a stimulus determines not only the rate at which a single afferent nerve fiber fires but also the balance of activity within a group of sensory units. In the example in Fig. 9-6, neurons *A* and *C*, stimulated near the edge of their receptive fields, where the thresholds are higher, fire at a lower frequency than neuron *B*, stimulated at the center of its receptive field. Because of this gradient of sensitivity across the receptive field, the information content of the pattern of activity in a population of afferent neurons is great.

As we have seen above, stimulus strength is related to the firing frequency of the afferent neuron, but a high frequency of impulses in the single afferent fiber of Fig. 9-5 could mean either that a stimulus of moderate intensity was applied at the center of the

FIGURE 9-5 ***Two stimulus points, A and B, in the receptive field of a single afferent neuron.***

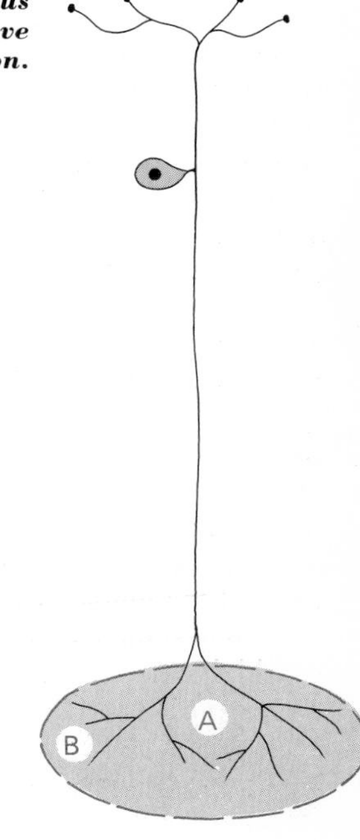

receptive field (point *A*) or that a strong stimulus was applied at the periphery (point *B*). Neither the intensity nor the localization of the stimulus can be detected precisely. But in a group of sensory units (Fig. 9-6), a high frequency of action potentials in neuron *B* arriving simultaneously with a lower frequency of action potentials in neurons *A* and *C* permits accurate localization of the stimulus. Once the location of the stimulus within the receptive field of neuron *B* is known, the firing frequency of neuron *B* can be taken as a meaningful measure of stimulus intensity.

The precision with which a stimulus can be localized and differentiated from an adjacent stimulus depends on the size of the receptive field covered by a single afferent neuron and the amount of overlap of nearby receptive fields. For example, the ability to discriminate between two adjacent mechanical stimuli to the skin is greatest on the thumb, fingers, lips, nose, and cheeks, where the sensory units are small and overlap considerably. The localization of visceral sensations is less precise than that of somatic stimuli because there are fewer afferent fibers and each has a larger receptive field.

We have given examples stressing the role played by receptive-field size and distribution in accurate stimulus localization; now we demonstrate the importance of functional interaction between the neurons of the afferent pathways. In one important type of such interaction the afferent neurons and interneurons themselves inhibit other parallel afferent components. As mentioned above, the neurons at the center of the receptive field (where the stimulus application is most intense) fire most rapidly. They relay information about stimulus quality, intensity, and location to the brain; in addition, via their axon collaterals they inhibit other parallel afferent fibers which are from the periphery of the stimulated area and therefore fire more slowly. Thus, a localized area of sensation is surrounded by an area of inhibition and decreased sensitivity, which serves to refine and clarify the information about stimulus localization on its way to higher levels of the central nervous system and to intensify contrasts. Such inhibition can be demonstrated in the following way. While pressing the tip of a pencil against the finger with one's eyes

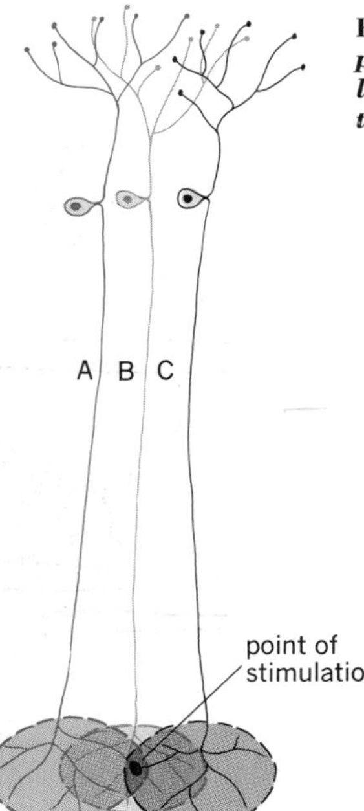

FIGURE 9-6 ***Stimulus point falls within the overlapping receptive fields of three afferent neurons.***

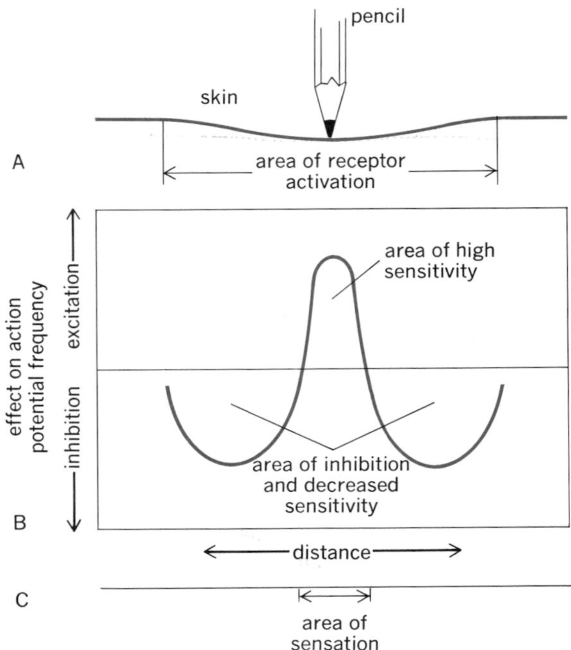

FIGURE 9-7 **(A)** ***Pencil tip pressed against skin depresses surrounding tissues. Receptors are activated under the pencil tip and in the adjacent tissue.*** **(B)** ***Because of lateral inhibition, the central area of excitation is surrounded by an area of inhibition.*** **(C)** ***The sensation is localized to a more restricted region than that in which mechanoreceptors were actually activated.***

closed, one can localize the pencil point quite precisely, even though the region around the pencil tip is also indented and mechanoreceptors within this entire area are activated (Fig. 9-7A). This potential information is discarded by inhibitory mechanisms, and the pencil tip is accurately localized (Fig. 9-7C). Such inhibition occurs in the pathways of virtually all sensory modalities and is of great importance for the detection and emphasis of contrast, serving to lessen the weaker responses and collect the stronger ones into a common pathway.

We now turn to the receptor mechanisms and specialized patterns of coding of the specific sensory systems.

SPECIFIC SENSORY SYSTEMS

Receptors are classified in various ways. For example, they may be classified according to their modality (*thermoreceptors,* which sense changes in temperature; *mechanoreceptors,* which are sensitive to pressure changes; *chemoreceptors,* which are sensitive to chemical changes in the surrounding fluid; etc.). Alternatively, receptors may be classified according to their locations in the body and their functions in sensory awareness. *Exteroceptors* are located in the skin and organs of the special senses and respond to stimuli from the external environment. *Proprioceptors* respond to stimuli in the deeper tissues, particularly the joints, muscles, and tendons, and, with some of the exteroceptors, provide information about the body's position in space. *Interoceptors,* which are generally sensitive to stretch, include the receptors in the walls of the viscera and blood vessels. The afferent neurons having interoceptors at their peripheral ends are *visceral* afferent neurons; those having exteroceptors or proprioceptors are called *somatic* afferent neurons. In this book, however, we shall group them together and refer simply to *afferent neurons.*

Somatic Sensation

Somatic in this context refers to the framework or outer walls of the body, including skin, skeletal muscle, tendons, and joints, as opposed to the viscera (organs located in the thoracic and abdominal cavities).[1] Somatic receptors respond to mechanical stimulation of the skin or hairs and underlying tissues, rotation or bending of joints, temperature changes, and possibly some chemical changes. Their activation gives rise to the sensations of touch, pressure, heat, cold, the awareness of the position and movement of the parts of the body, and pain. Recalling that by *receptor* we mean the afferent-nerve fiber ending plus any specialized nonneural cells associated with it, we can say that probably each of these sensations is associated with a specific type of receptor; i.e., there are distinct receptors for heat, cold, touch, pressure, joint position, and pain.

The somatic receptors are classified as either *free nerve endings* or *encapsulated endings.* Free nerve endings, which have neither myelin nor Schwann-cell coatings (although they may be part of an afferent neuron which is myelinated), divide and ramify through many tissues of the body, including the epidermis and connective tissue of the skin. These endings are often implicated in the transduction of painful or thermal stimuli. Encapsulated endings have specialized nonneural cells associated with them, and, as the name implies, the nerve terminal and its surrounding cells are enclosed in a capsule. One type of encapsulated ending, the pacinian corpuscle, was introduced in Chap. 4. Another is the *Meissner corpuscle* (Fig. 9-8), in which the nerve terminal winds back and forth between extensions of a specialized lamellar cell in the core of the corpuscle. Like the pacinian corpuscle, the Meissner corpuscle is presumed to be a rapidly adapting mechanoreceptor. Other encapsulated endings have been described, but their functions and even their identities are often not clear.

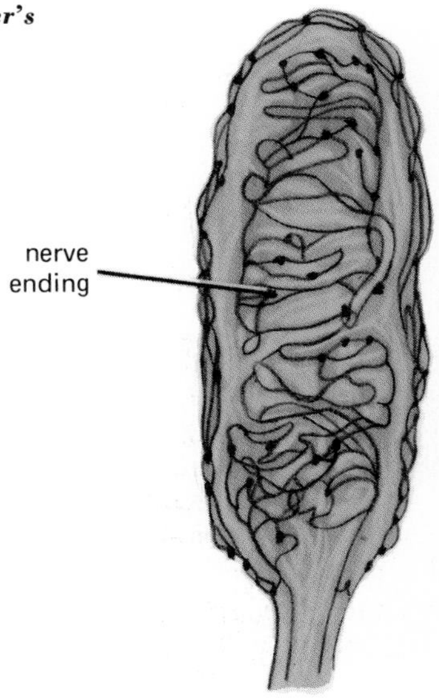

FIGURE 9-8 ***Meissner's corpuscle.***

[1] Somatic also includes the eye and ear but not taste and smell, which are visceral.

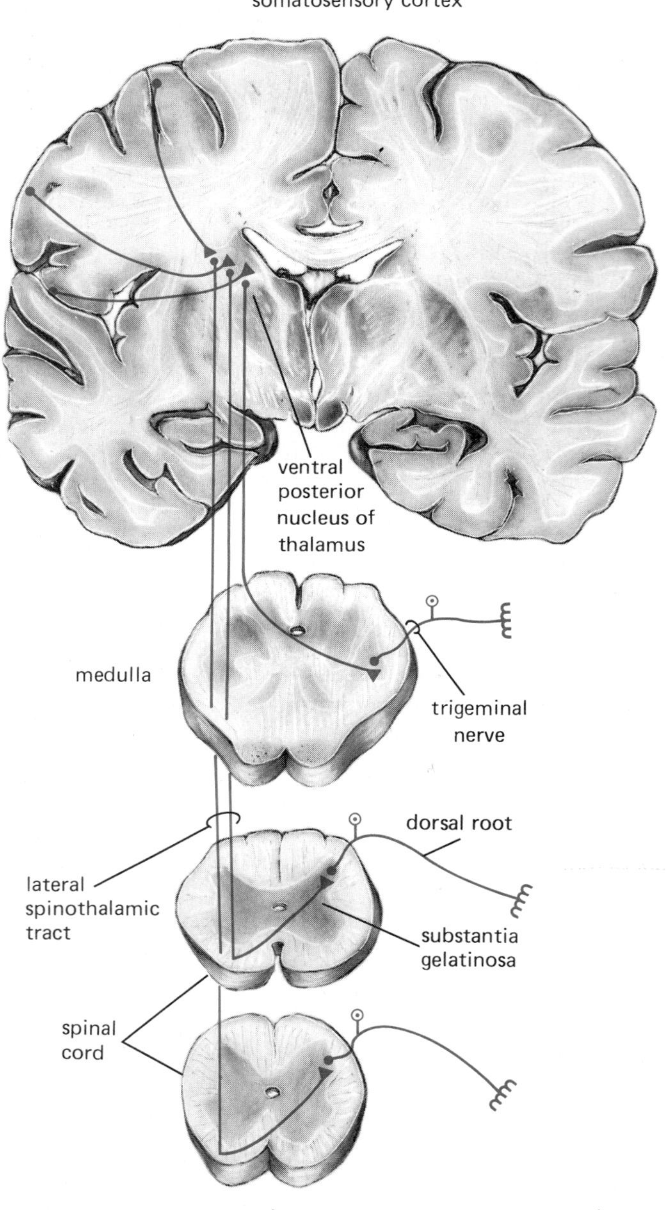

FIGURE 9-9 ***Examples of ascending pathways. (A) The lateral spinothalamic tract, which conveys pain and temperature information. (B) The posterior columns (fasciculus gracilis and fasciculus cuneatus) and spinocerebellar tracts, which convey proprioceptive information. Axons in the posterior columns are the central processes of afferent (first order) neurons; they terminate in the nuclei gracilis and cuneatus.***

After entering the central nervous system, the afferent neurons from the somatic receptors synapse with interneurons which enter pathways passing through the brainstem and thalamus to the somatosensory cortex. The specific pathways cross from their side of entry to the opposite side of the central nervous system in the spinal cord or brainstem; thus the sensory pathways from receptors on the left side of the body go to the somatosensory strip of the right cerebral hemisphere and vice versa (Fig. 9-9). In the somatosensory cortex,

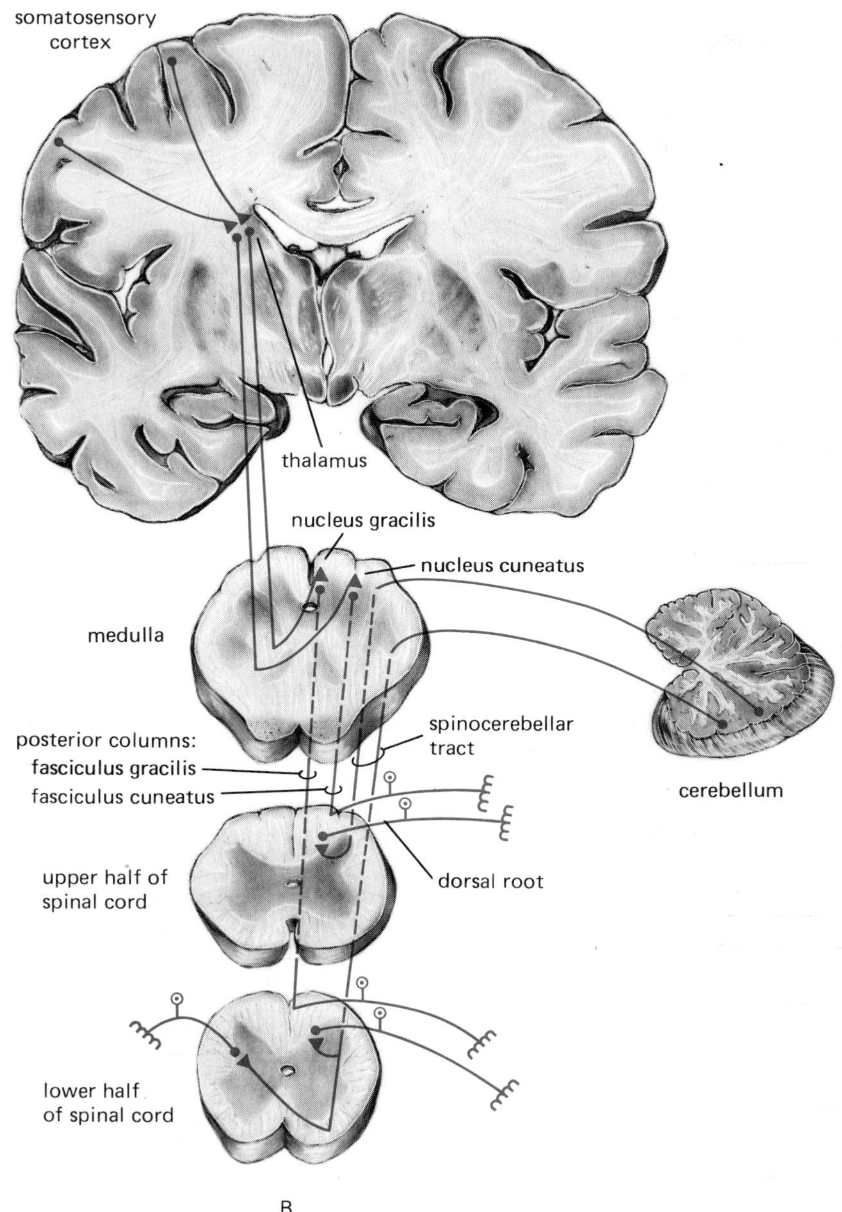

B

the terminations of the individual components of the specific somatic pathways are grouped according to the location of the receptors. The pathways which originate in the foot end nearest the longitudinal dividing line between the two cerebral hemispheres. Passing laterally over the surface of the brain, one finds the terminations of the pathways from leg, trunk, arm, hand, face, tongue, throat, and viscera (Fig. 9-10). The parts with the greatest sensitivity (fingers, thumb, and lips) are represented by the largest areas of somato-

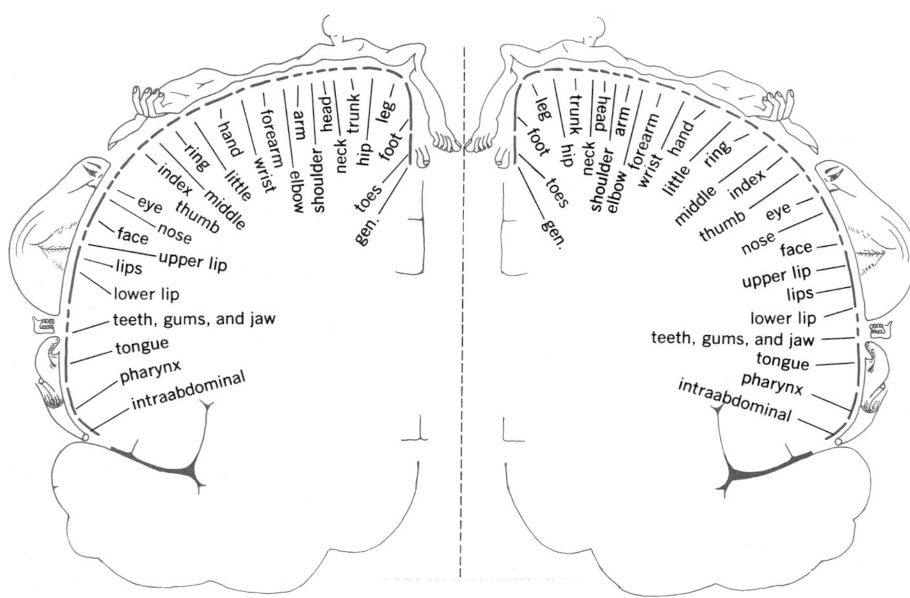

FIGURE 9-10 ***Location of pathway terminations for different parts of the body in the somatosensory cortex. The left half of the body is represented on the right hemisphere of the brain, and the right half of the body is represented on the left cerebral hemisphere.***

sensory cortex. The sensory strip of one cerebral hemisphere is duplicated in the opposite hemisphere.

Touch-Pressure One of the best examples of how receptor specificity is determined by the characteristics of the surrounding tissue is the pacinian corpuscle, discussed in Chap. 4. The nerve terminal of the pacinian corpuscle is surrounded by alternating layers of cells and extracellular fluid in such a way that a mechanical stimulus reaches the afferent-nerve terminal only after displacing the layers of the surrounding capsule (see Fig. 4-18). Rapidly applied pressures are transmitted to the nerve terminal without delay and give rise to a generator potential, presumably by the mechanisms described in Chap. 4. The energy of slowly administered or sustained forces is, in part, absorbed by the elastic-tissue components of the capsule and is therefore partially dissipated before it reaches the nerve terminal at the core. Thus, the receptor fires only at the fast onset—and perhaps again at the release—of the mechanical stimulus but not under sustained pressure. What the pacinian corpuscle signals is not pressure but *changes* in pressure with time. It can effectively discriminate stimuli vibrating up to frequencies of 300 hertz (Hz). Other mechanoreceptors which adapt more slowly than the pacinian corpuscles provide information about both the *rate* of stimulus application and the stimulus *intensity*.

Joint Position Joint capsules contain encapsulated and free nerve endings. Activation of these receptors (as well as those in the associated ligaments and tendons) and their pathways through the nervous system gives rise to the conscious awareness of the position and movement of the joints and of stresses acting upon them. Input from these receptors is integrated with visual and other information to provide awareness of the position of the body in space. The proprioceptive pathways also provide input to the cerebellum and other brain areas involved in the unconscious coordination of movements and maintenance of balance. Four such pathways, two leading to somatosensory cortex and two leading to the cerebellum, are illustrated in Fig. 9-9B. The different receptors in the joints and ligaments are activated by mechanical stimuli such as stretching, twisting, or compressing or by painful stimuli caused by damage. Their combined sensitivities signal movement and final position of the joint.

Temperature Free nerve endings are commonly associated with temperature sensitivity. Several hypothetical receptor mechanisms have been proposed for such receptors. Those responding to high-temperature stimuli might work in the following way: Increased temperature and the associated thermal agitation of molecular bonds cause configurational changes in protein molecules in the nerve ending; these alter the membrane permeability, causing a generator potential, which leads to the formation of an action potential. Although this mechanism has not been established, it is the most plausible of those currently proposed. The so-called cold receptors are activated by lower temperatures; their mechanism is unknown.

Pain A stimulus which causes or is on the verge of causing tissue damage often elicits a sensation of pain and a reflex escape or withdrawal response as well as a gamut of physiologic changes which resemble the effects of activation of the sympathetic nervous system in fear, rage, or fight or flight. These physiologic changes usually include faster heart rate, higher blood pressure, greater secretion of epinephrine into the bloodstream, increased blood sugar, less gastric secretion and motility, decreased blood flow to the viscera and skin, dilated pupils, and sweating. Moreover, the experience of pain includes an emotional component of fear, anxiety, and sense of unpleasantness as well as information about the stimulus's location, intensity, and duration. And probably more than any other type of sensation, the experience of pain can be altered by past experiences, suggestion, emotions (particularly anxiety), and the simultaneous activation of other sensory modalities.

This complex nature of pain can be accounted for by saying that the stimuli which give rise to pain result in a sensory experience *plus* a reaction to it, the reaction including the emotional response (anxiety, fear) and behavioral response (withdrawal or other defensive behavior). Both the sensation and the reaction to the sensation must be present for tissue-damaging stimuli to cause suffering. The sensation of pain can be dissociated from the emotional and behavioral reactive component by drugs, e.g., morphine, or by selective brain operations which interrupt pathways connecting the frontal lobe of the cerebrum with other parts of the brain. When the reactive component is no longer associated with the sensation, pain is felt, but it is not necessarily disagreeable; the patient does not mind as much. Thus, satisfactory pain relief can be obtained even though the perception of painful stimuli is not reduced.

The receptors whose stimulation gives rise to pain are high-threshold receptors at the free endings of certain small unmyelinated or lightly myelinated afferent neurons. These receptors fire specifically in response to tissue-damaging pressure, intense heat, or irritating chemicals, their firing frequency increasing as the severity of the stimulus rises.

The central connections of these afferent "pain fibers" are not well understood, but it is postulated that the afferent information is transmitted via two classes of ascending pathways simultaneously. The specific pathways (in the *lateral spinothalamic tracts,* Fig. 9-9A), which go to the thalamus and cerebral cortex, are involved in the perception of pain; i.e., they convey information about where, when, and how strongly the stimulus was applied. The nonspecific pathways, which go to the brainstem reticular formation and a part of the thalamus different from that supplied by the specific pathways, arouse the aversive, reactive response to the stimulus. These same neurons which are activated by the nonspecific pathway are interconnected with the hypothalamus and other areas of the brain which play major roles in integrating autonomic and endocrine stress responses and the behavioral patterns of aggression and defense.

Descending pathways capable of altering the transmission of information in the afferent neurons, spinal pathways, or brain centers are known to exist in most sensory systems, but they are particularly important in pain. They are thought to be one means by which emotions, past experiences, state of attention, etc., can alter sensitivity to pain. When the descending pathways reduce the activity in the pain pathways, the unpleasant emotions and response behavior, as well as the specific pain perception, are diminished.

Vision

The Eye The eyes are embedded in a fatty cushion within the orbits of the skull. The exposed surface of the eyeball is kept moist by a continual flow of fluid from the *lacrimal gland,* located in the superolateral corner of the orbit (Fig. 9-11). The fluid drains from the surface of the eyeball into *lacrimal canaliculi* which join to form the *nasolacrimal duct* which, in turn, empties into the nasal cavity. The lacrimal fluid not only keeps the eyeball moist and clean but also protects it against infection by means of antibacterial chemicals contained in it. The lacrimal glands are innervated by autonomic neurons which stimulate the increased secretion (tears) associated with emotions. The eyeballs are also protected by a thin layer of epithelium, the *conjunctiva,* which covers their anterior surface except for the cornea.

Each eye consists of a three-layered membranous sac and its contents: the *aqueous humor, vitreous body,* and *lens.* The external layer of the eye is the opaque

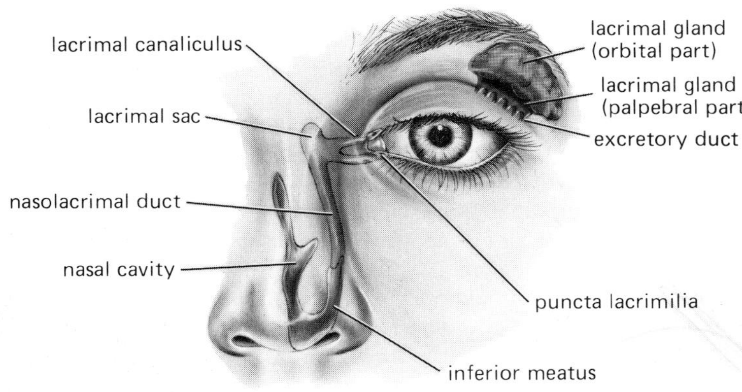

FIGURE 9-11 ***Lacrimal apparatus, which consists of the lacrimal gland and its excretory ducts and the lacrimal canaliculi, lacrimal sac, and nasolacrimal duct.***

sclera in the posterior part of the eye and the *cornea* in the anterior part (Fig. 9-12). The transparent cornea, which projects as a flattened dome from the front of the sclera, has neither blood nor lymph vessels, but it has numerous free nerve endings. The highly vascular middle layer from the back of the eye forward is made up of the *choroid, ciliary body,* and *iris.* The iris, which is behind the cornea and in front of the lens, is a circular diaphragm with a central opening, the *pupil.* Light can enter the eye only through the pupil because the iris itself is pigmented and light cannot pass through it. The iris exerts important control over the amount of light entering the eye. Near the border of the pupil the iris contains a ring of smooth muscle cells, the *pupillary sphincter,* innervated by the parasympathetic division of the autonomic nervous system. Activation of the sphincter muscle decreases the diameter of the pupil and decreases the amount of light entering the eye. Radially arranged smooth muscle fibers in the iris, the *pupillary dilator* muscles, are innervated by sympathetic nerve fibers; when activated, they enlarge the pupil.

The ciliary body supports the lens via *suspensory,* or *zonular, ligaments.* It also contains the *ciliary muscles,* which change the shape of the lens for near or far vision, and it secretes a fluid (the aqueous humor) which fills the anterior chamber of the eye. The aqueous humor provides a path for the metabolites of the avascular lens and cornea, and it also maintains pressure within the eyeball so that the latter's dimensions remain constant. It is drained from the anterior chamber via the *canal of Schlemm* (Fig. 9-12B). An increased aqueous-humor pressure, often due to obstruction of the canal of Schlemm, is associated with the condition *glaucoma.* The inner, nervous layer of the eye is the *retina,* which contains the receptors (*rods* and *cones*) for light. The cup formed by the sclera, choroid, and retina is filled by a transparent gel, the *vitreous body.*

Light The receptors of the eye are sensitive to only that tiny portion of the vast spectrum of electromagnetic radiation which we call light (Fig. 9-13). Radiant energy is described in terms of wavelengths and frequencies. The *wavelength* is the distance between two successive wave peaks (Fig. 9-14) and varies from several kilometers long at the top of the spectrum to minute fractions of a millimeter at the bottom end. Those wavelengths capable of stimulating the receptors of the eye are between 400 and 700 nm (a nanometer is one-billionth of a meter). Light of different wavelengths is associated with different color sensations; for example, light having a wavelength of about 540 nm gives rise to the sensation of green and light having a wavelength of about 565 nm gives rise to the sensation of red.

The Optics of Vision Light can be represented most simply by a ray or line drawn in the direction in which the wave is traveling. Light waves are propagated in all directions from every point of a light source. These divergent light waves must pass through an optical system which focuses them back into a point before an accurate image of the light source is achieved. In the eye itself, the image of the object being viewed must be focused upon the retina where the light-sensi-

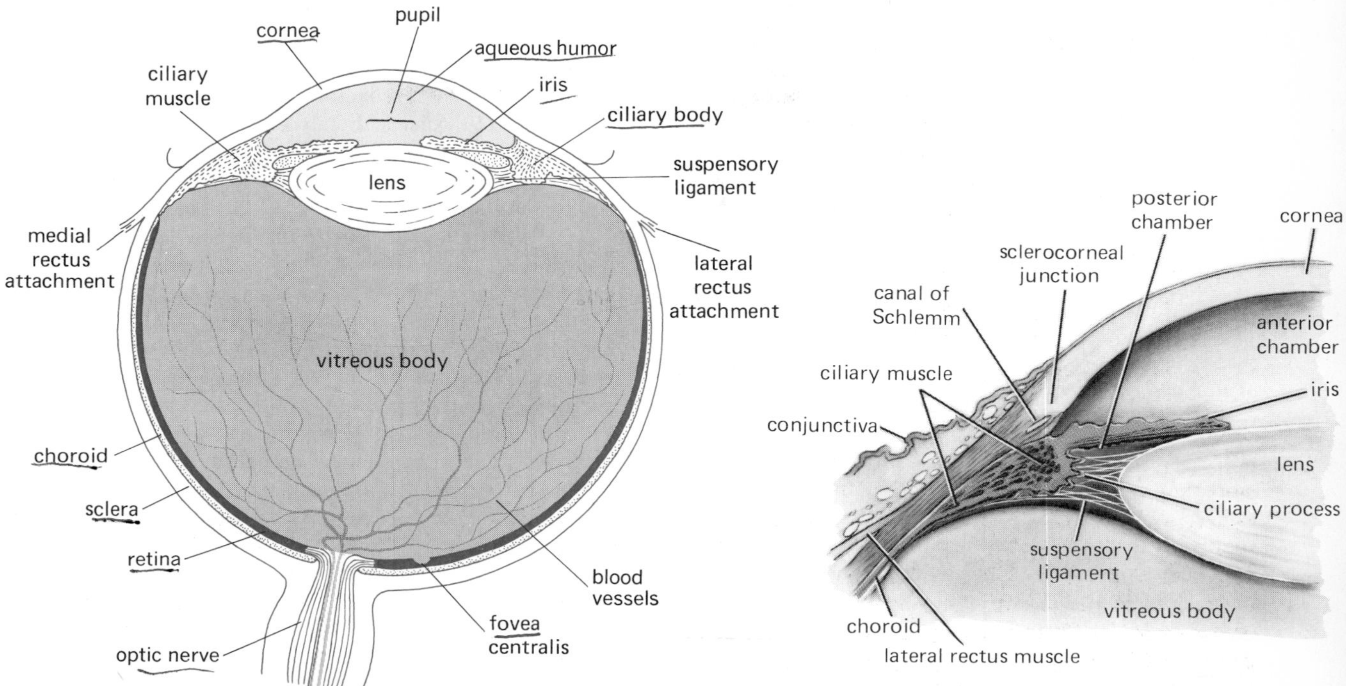

FIGURE 9-12 ***Section through the right human eye. The blood vessels depicted run along the back of the eye between the retina and the vitreous humor.***

tive receptor cells of the eye are located. The lens and cornea of the eye (Fig. 9-12) are the optical systems which focus the image of the object upon the retina. At a boundary between two substances, such as the cornea of the eye and the air outside it, the rays are bent so that they travel in a new direction. The degree of bending depends in part upon the angle at which the light enters the second medium. The cornea plays a larger role than the lens in focusing the image because light rays are bent more in passing from air into the cornea than in passing into and out of the lens.[2]

The surface of the cornea is curved so that light rays coming from a single point source hit the cornea at different angles and are bent different amounts, but all in such a way that they are directed to a point after emerging from the lens (Fig. 9-15). Notice what happens to the image when the object being viewed has

[2] All transparent structures of the eye participate in the refraction of light, but the cornea and lens are the most important.

FIGURE 9-13 ***Electromagnetic spectrum.***

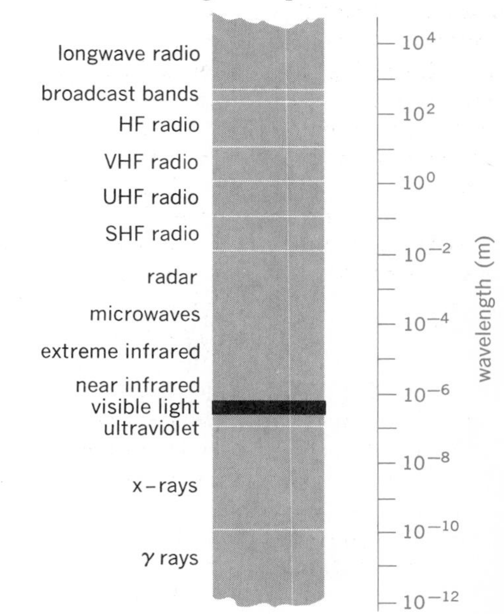

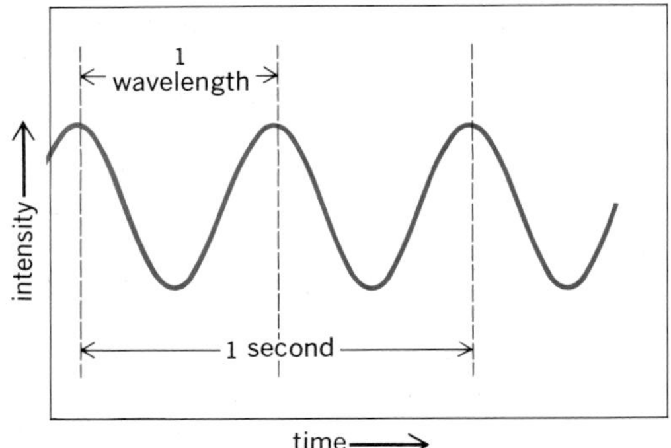

FIGURE 9-14 ***Properties of a wave. The frequency of this wave is 2 Hz.***

more than one dimension (Fig. 9-15B); the image on the retina is upside down relative to the original light source. It is also reversed left to right.

The shape of the cornea and lens and the length of the eyeball determine the point where light rays reconverge. Although the cornea performs the greater part quantitatively of focusing the visual image on the retina, all adjustments for distance are made by changing the shape of the lens. Such changes are called *accommodation*. The shape of the lens is controlled by the ciliary muscles and the tension they apply to the suspensory ligaments of the lens. The lens is flattened when distant objects are to be focused upon the retina and allowed to assume a more spherical shape to provide additional bending of the light rays when near objects are viewed (Fig. 9-16). The ciliary muscles are controlled by parasympathetic nerve fibers.

Cells are added to the lens throughout life but only to the outer surface. This means that cells at the center of the lens are both the oldest and the farthest away from the nutrient fluid which bathes the outside of the lens (if capillaries ran through the lens, they would interfere with its transparency). These central cells age and die first, and with death they become stiff, so that accommodation of the lens for near and far vision becomes more difficult. This impairment in vision is known as *presbyopia* and is one reason why many people who never needed glasses before start wearing them in middle age.

Cells of the lens can also become opaque so that detailed vision is impaired; this is known as *cataract*. The defective lens can usually be removed surgically from persons suffering from cataract, and with the addition of compensating eyeglasses, effective vision can be restored.

FIGURE 9-15 ***Refraction (bending) of light by the lens system of the eye. The light source is (A) a point and (B) an object consisting of many point sources.***

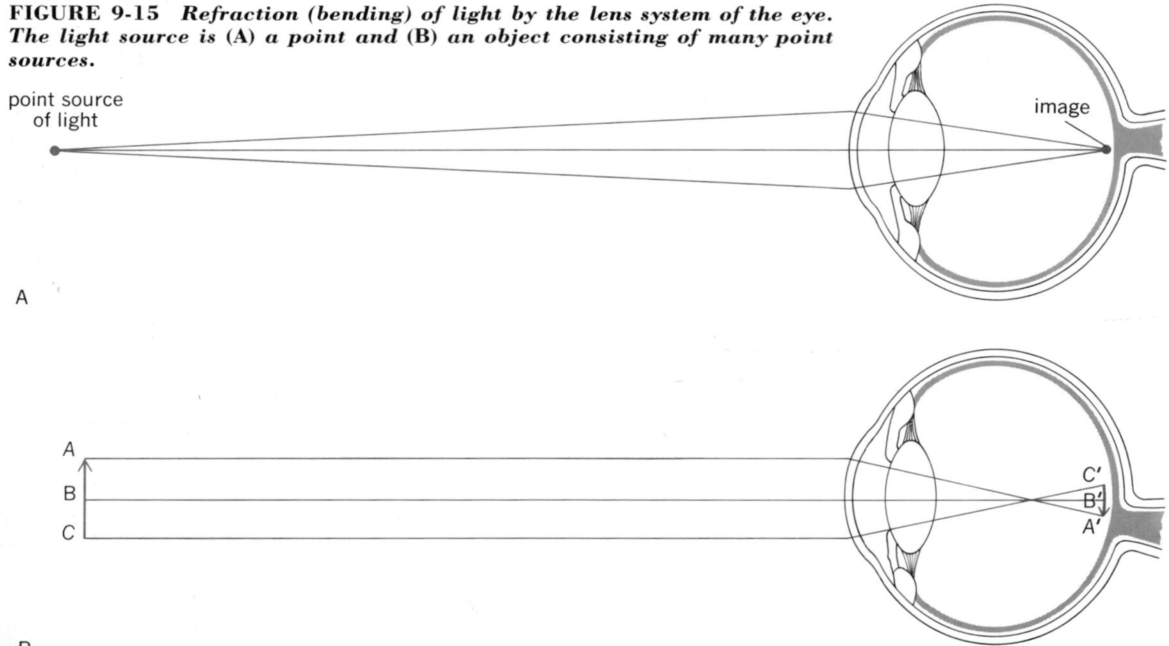

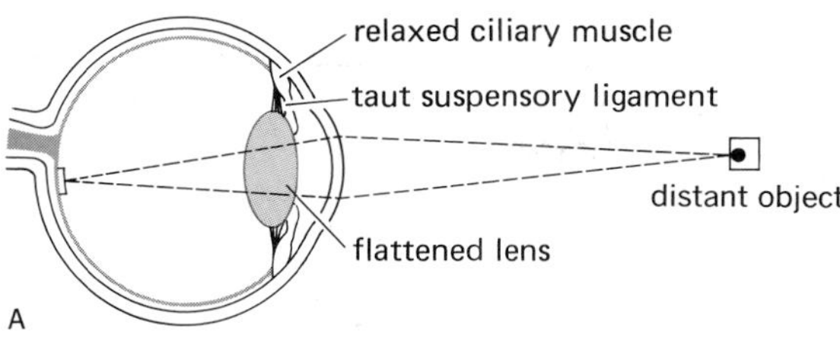

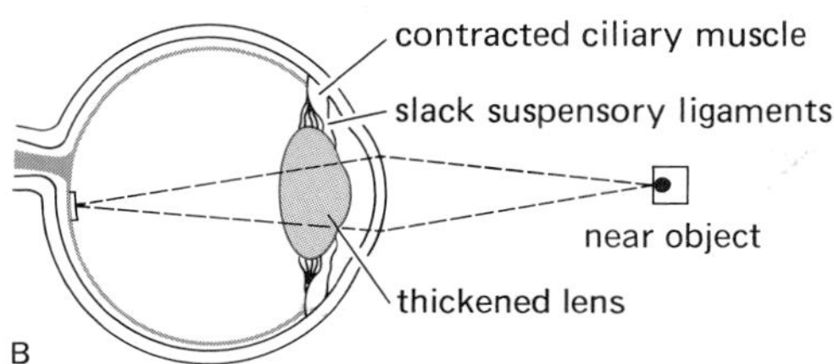

FIGURE 9-16 ***Accommodation for distant and near vision by the pliable lens.* (A) *The lens is stretched for distant vision so that it adds the minimum amount of focusing power.* (B) *The lens thickens for near vision to provide greater focusing power.***

Defects in vision occur if the eyeball is too long in relation to the lens size, for then the images of near objects fall on the retina but the images of far objects are focused in front of the retina. This is a *nearsighted,* or *myopic,* eye, which is unable to see distant objects clearly. If the eye is too short for the lens, distant objects are focused on the retina and near objects are focused behind it (Fig. 9-17); this eye is *farsighted,* or *hyperopic,* and near vision is poor. Defects in vision also occur where the lens or cornea does not have a smoothly spherical surface. The improperly shaped eyeball or irregularities in the cornea (*astigmatism*) or lens can usually be compensated for by eyeglasses (Fig. 9-17).

As mentioned above, the amount of light entering the eye is controlled by the iris. The pupillary sphincter muscle of the iris reflexly contracts in bright light, decreasing the diameter of the pupil; this not only reduces the amount of light entering the eye but also directs the light to the central and most optically accurate part of the lens. Conversely, the sphincter relaxes in dim light, when maximal sensitivity is needed.

Receptor Cells The receptor cells in the retina are called either *rods* or *cones* because of their microscopic appearance (Fig. 9-18). Both cell types contain light-sensitive molecules called *photopigments,* whose prime function is to absorb light. Light energy causes the photopigments to change their molecular configuration which, in turn, alters the properties of the receptor-cell membranes in which they are situated. Unlike other receptor cells that have been studied, in response to stimulation the membrane *decreases* its permeability to sodium ions, which *hyperpolarizes* the receptor-cell membrane. This seems to release other neurons in the visual pathway from inhibition.

There are four kinds of photopigments: one, *rhodopsin,* which is very sensitive to low levels of illumination, and three, *erythrolabe, chlorolabe,* and *cyanolabe,* which are sensitive to light wavelengths of the three primary colors, red, green, and blue, respectively. All four photopigments are made up of a protein (*opsin*) bound to a *chromophore* molecule. The chromophore is always the same slight variant of vitamin A, but the opsin differs in each of the four photopigments and confers the specific light sensitivities upon it, i.e., determines whether it responds to all light or selectively to red, blue, or green. The photic energy (light) acts upon the chromophore, which then splits away from the opsin, changing the molecular configuration. After this breakdown of the photopigment in the presence of light, the chromophore molecule is rearranged and rejoined to opsin to restore the photopigment. Thus, the only action of light in vision is to change the chromophore; everything else in the sequence leading to vision—whether chemical, physiologic, or psychologic—is a "dark" consequence of this one light reaction.

Because the rod receptor cells contain rhodopsin, they are very sensitive, being able to detect very small amounts of light and acting as the photoreceptors during conditions of poor illumination and for night vision. Their responses do not indicate color, showing only shades of gray; they do indicate brightness. Their *acuity,* i.e., their ability to distinguish one point in space from another nearby point, is very poor. Rods are most numerous in the peripheral retina, i.e., that part closest to the lens, and are absent from the very center of the retina (the *fovea*) (Fig. 9-12A). There are three types of cones, each containing one of the three photopigments for color vision. Cones operate only at high levels of illumination and are the photoreceptors for day vision. Cone visual acuity is very high, and because cones are concentrated in the center of the retina, it is that part which we use for finely detailed vision.

Each receptor cell in the retina (whether rod or cone) synapses upon a second neuron (a *bipolar* cell, Fig. 9-19A) which in turn synapses upon a *ganglion* cell.

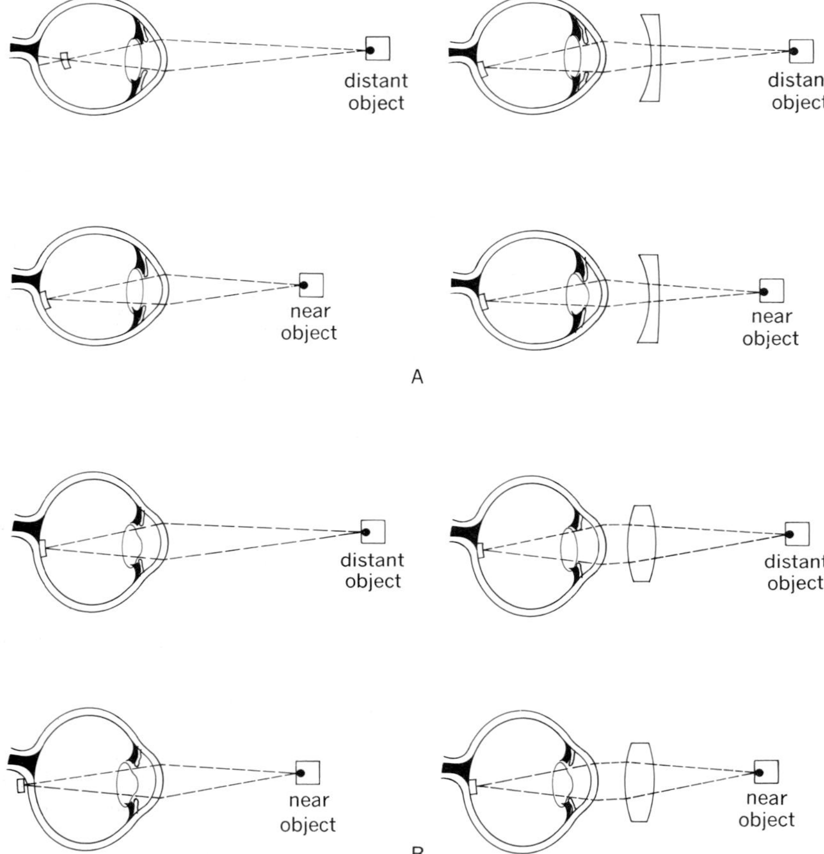

FIGURE 9-17 **(A)** ***In the nearsighted eye, light rays from a distant source are focused in front of the retina. A concave lens placed before the eye bends the light rays out sufficiently to move the focused image back onto the retina. When near objects are viewed through concave lenses, the eye accommodates to focus the image on the retina.*** **(B)** ***The farsighted eye must accommodate to focus the image of of distant objects upon the retina. (The normal eye views distant objects with a flat, stretched lens.) The accommodating power of the lens of the eye is sufficient for distant objects, and these objects are seen clearly. The lens cannot accommodate enough to keep images of near objects focused on the retina, and they are blurred. A convex lens converges light rays before they enter the eye and allows the eye's lens to work in a normal manner.***

The axons of the ganglion cells form a bundle called the *optic nerve* (Fig. 9-12) which passes directly into the brain. Generally, cone receptor cells have relatively direct lines to the brain; i.e., each bipolar cell receives synaptic input from relatively few cones, and each ganglion cell receives synaptic input from relatively few bipolar cells. This relative lack of convergence provides precise information about the area of the retina that was stimulated, but it offers little opportunity for the summation of subthreshold events to fire the ganglion cell. Conversely, many rod cells converge on bipolar and ganglion cells, and, although acuity is poor, opportunities for spatial and temporal summation are good. Therefore, a relatively low-intensity light stimulus that would cause only a subthreshold response in a cone ganglion cell can cause an action potential in a rod ganglion cell. Thus, the difference in acuity and light sensitivity between rod and cone vision is due, at least in part, to the anatomic wiring patterns of the retina.

These differences explain why objects in a darkened theater are indistinct and appear only in shades of gray; with such low illumination the cones do not reach threshold and fail to fire, so that all vision is supplied by the more sensitive but less accurate rod vision. The loss of visual acuity in dim light is due in part to the shift from cone to rod receptors.

The sensitivity of the eye improves after being in the dark for some time, due to *dark adaptation*. The modern theory of dark adaptation still has many unsolved problems but, in general, states that the excitability of the rod visual pathways depends on the number of intact rhodopsin molecules in the rods. In bright light so many rod rhodopsin molecules are broken down that the rods are ineffective, and vision is chiefly due to cone activation. When one moves from bright light to a darkened room, there are relatively few intact rhodop-

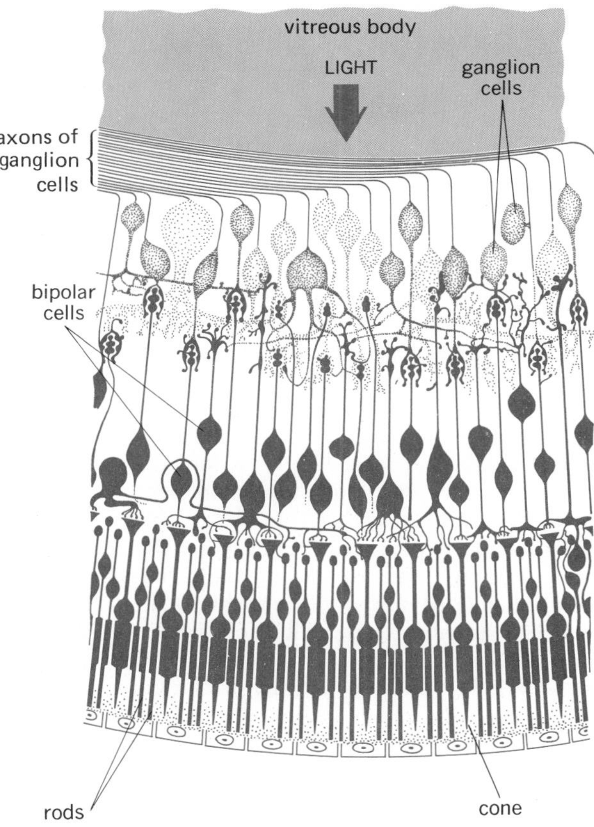

FIGURE 9-18 ***Human retina. Light entering the eye must pass through the fibers and cells of the retina before reaching the sensitive tips of the rods and cones. After leaving the eyeball, the axons of the ganglion cells form the optic nerve.*** **(Adapted from Gregory.)**

sin molecules, but as the rhodopsin slowly regenerates in the dark, visual sensitivity improves.

Visual-system Pathway and Coding In experiments on visual-system coding, simple visual shapes such as white bars against a black background were projected onto a screen in front of an anesthetized animal while the activity of single cells in the visual system was recorded. (This discussion is based on research done chiefly on frogs, cats, and monkeys, but it almost certainly applies to people as well.) Different parts of the retina could be stimulated by varying the position of the bar on the screen. We shall follow the information processing elucidated by these experiments through the various stages of the visual pathway, starting at the level of the ganglion cells (Fig. 9-19A).

It is found that even within the retina an amazing amount of data processing has occurred. The retinal ganglion cells discharge spontaneously; i.e., they fire in the absence of any light stimulus. This spontaneous activity gives the cell an important second signal with which to work; it can either increase or decrease its rate of firing. Each receptor-cell–bipolar-cell–ganglion-cell chain is synaptically connected to other similar chains by cells which conduct laterally through the retina. These interconnections, which occur at both bipolar- and ganglion-cell levels, provide the pathways by which many receptor cells converge upon a single ganglion cell. The greater the degree of convergence, the larger is the area of the retina that can influence the ganglion cell.

The receptive fields of the ganglion cells are circular; i.e., any light falling within a specific circular area of the retina influences the activity of a given ganglion cell. The response of the ganglion cell varies markedly, depending on the region of the receptive field stimulated. Some ganglion cells speed up their rate of firing when a spot of light is directed at the center of their receptive field and slow down their firing when the periphery is stimulated. Such a cell is said to have an *on* center. The activity of a ganglion cell of this type is shown in Fig. 9-20A. Other ganglion cells, such as cell 2, have just the opposite response, decreasing activity when the center of the receptive field is stimulated (Fig. 9-20B) and increasing activity when the light stimulates the periphery (Fig. 9-20C). Notice the spontaneous activity of the *off*-center cell and the abrupt inhibition of its activity when the light is turned on. Often when the intermediate region of the receptive field is stimulated, the cell responds both when the light is turned on and again when it is turned off; i.e., it has an on-off response. Therefore, one ganglion cell can give an on, off, or on-off response, depending upon which region of its receptive field is stimulated.

Moreover, the basic pattern of ganglion-cell activity can be greatly modified. An *on* response increases, i.e., the frequency of firing action potentials increases, if the intensity of the light spot is greater, if the diameter of the spot is larger, or if the spot is moved. The *on* response decreases if the diameter of the spot becomes so much larger that it encroaches upon adjacent *off* regions or if a second spot is shown simultaneously on a nearby *off* region.

The axons of the ganglion cells form the optic nerve, which passes to the brain. The optic nerves from the two eyes meet at the *optic chiasm* (optic crossing) where some of the fibers cross over to the opposite side

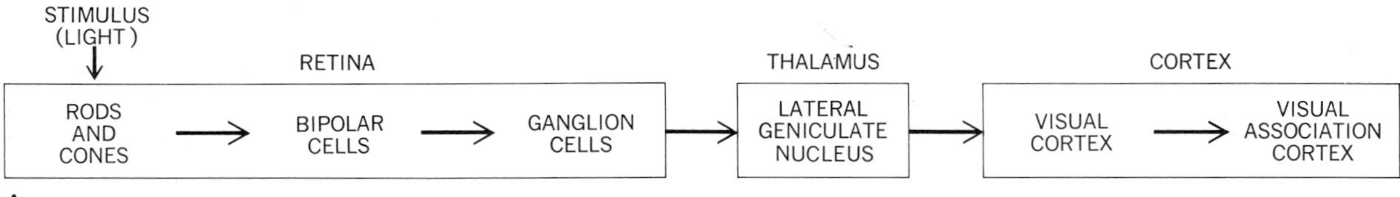

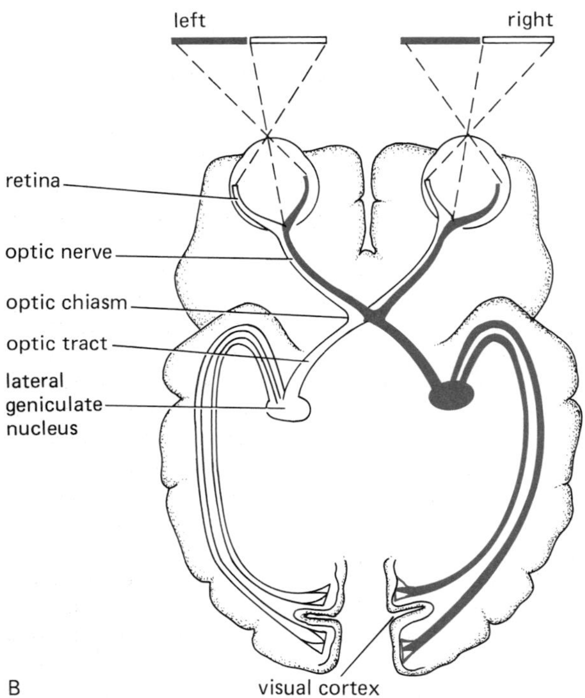

FIGURE 9-19 ***(A) Diagrammatic representation of the cells in the visual pathway. (B) The visual pathway.***

of the brain (Fig. 9-19B). This partial crossover provides both cerebral hemispheres with input from both eyes. After entering the brain, the visual pathways pass to the *lateral geniculate nucleus* in the thalamus.

The response of a single lateral geniculate cell resembles that of a retinal ganglion cell in being concentric with either an on-center–off-periphery pattern or vice versa. Movement of the stimulus across the receptive field of a cell always produces a stronger response than a stationary stimulus does, but the increase does not depend on the direction of movement.

The partially processed visual information is transmitted along the axons of the lateral geniculate neurons to the *primary visual cortex,* where the processing continues. Although the receptive fields of retinal ganglion cells and lateral geniculate neurons are usually concentric, with on or off centers, the receptive fields of cells in the visual cortex vary widely in organization. They are divided into *off* and *on* regions, but the divisions are no longer concentric (Fig. 9-21), having a side-by-side arrangement of excitatory and inhibitory areas with straight boundaries rather than circular ones. Diffuse light over the entire receptive field generally gives little or no response because the effects of the simultaneously stimulated *on* and *off* areas cancel out. The most effective stimulus is one which covers the *on* area but does not encroach upon the *off* area, e.g., long, narrow slits of light; dark, rectangular bars against a light background (lines); or straight-line borders between areas of different brightness. The orientation of the optimum stimulus varies from cell to cell. Some cortical

FIGURE 9-20 ***Recordings of the activity of a single ganglion cell. (A) Response of an on-center ganglion cell (cell 1) which increased its activity when light stimulated the center of its receptive field. (B) Activity of an off-center cell (cell 2) is suppressed when the center of its receptive field is stimulated. (C) The same off-center cell increases its activity when the light is restricted to the periphery.*** **(Adapted from Hubel and Wiesel, *J. Physiol.*, 154.)**

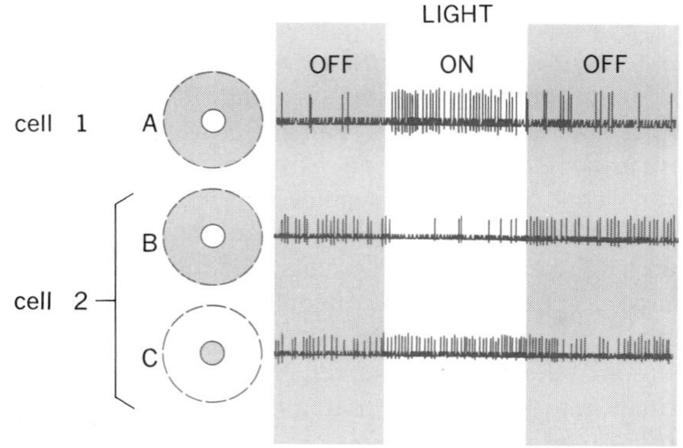

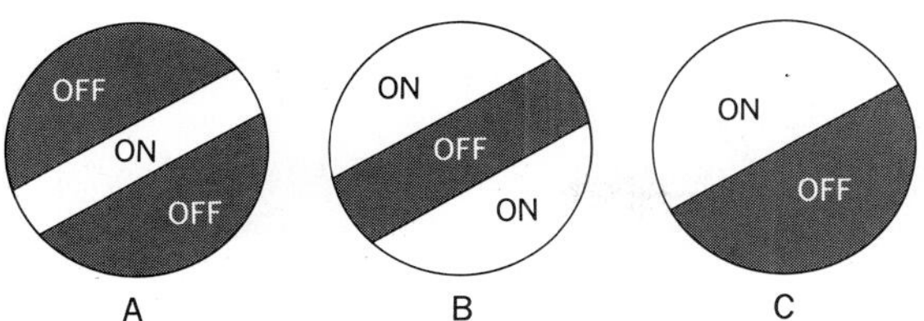

FIGURE 9-21 ***Retinal receptive fields of simple cortical cells are no longer arranged concentrically but are organized to provide information about lines and borders.*** **(Adapted from Hubel and Wiesel, *J. Physiol.*, 160.)**

cells respond only when the stimulus is in motion across the visual field (Fig. 9-22). Most cells at cortical levels can be influenced from either eye with the most effective stimulus form, orientation, and rate of movement similar for both eyes. The cortical response increases when the two eyes are stimulated simultaneously.

The multiple interconnections in the visual pathways provide for active data processing rather than the simple transmission of action potentials. By means of these intricate cellular hookups, cells of the visual pathways respond only to selected features of the visual world. They are organized to handle information about line, contrast, movement, and color, but they are not very good intensity detectors. Note, also, that the visual system does not form a picture in the brain but through the simultaneous activation of many neurons forms a specifically coded electrical statement.

FIGURE 9-22 ***Complex cortical cells increase their rate of firing action potentials only when a bar of light moves across the visual field in a specific direction (e.g., vertically, as in A) and not in any other direction (e.g., horizontally, as in B). Recorded activity of the cells is shown in traces at bottom.***

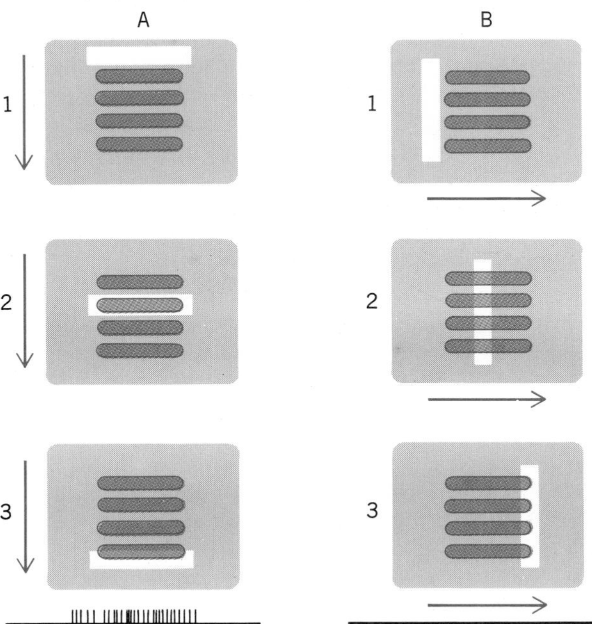

Color Vision Light is the source of all colors. Pigments, such as those mixed by a painter, serve only to reflect, absorb, or transmit different wavelengths of light, yet the nature of the pigments determines how light of different wavelengths will react. For example, an object appears red because all wavelengths other than those of 565 nm are absorbed by the material; light of 565 nm is reflected to excite the red-catching photopigment of the retina. Light perceived as white is a mixture of all wavelengths, and black is the absence of all light. Sensation of any color can be obtained by the appropriate mixture of three lights, red, blue, and green (Fig. 9-23). Light and pigments are properties of the physical world, but color exists only as a sensation in the mind of the beholder. The problem for the scientist is to discover how the perception of a brilliantly colored world results from packets of photic energy of varying wavelengths.

Color vision begins with the activation of the photopigments in the cone receptor cells. Normal human retinas, as we have seen, have cones which contain either red-, green-, or blue-sensitive photopigments, responding optimally to light of 565-, 540-, and 435-nm wavelengths, respectively. Although each type of cone is excited most effectively by light of one particular wavelength, it responds to other wavelengths as well; thus, for any given wavelength, the three cone types are excited to different degrees. For example (Fig. 9-24), in response to a light of 540-nm wavelength, the green cones fire maximally, the red cones at about two-thirds of their maximum rate, and the blue cones not at all. Our sensation of color depends upon the ratios of these three cone outputs. The cells processing color vision follow the pathways described earlier for the line-contrast processors (Fig. 9-19). The cones synapse upon bipolar cells, and the bipolar cells synapse upon ganglion cells, etc.

The fact that there are three different kinds of cone cells explains the various types of color blindness. Most people (over 90 percent of the male population and over

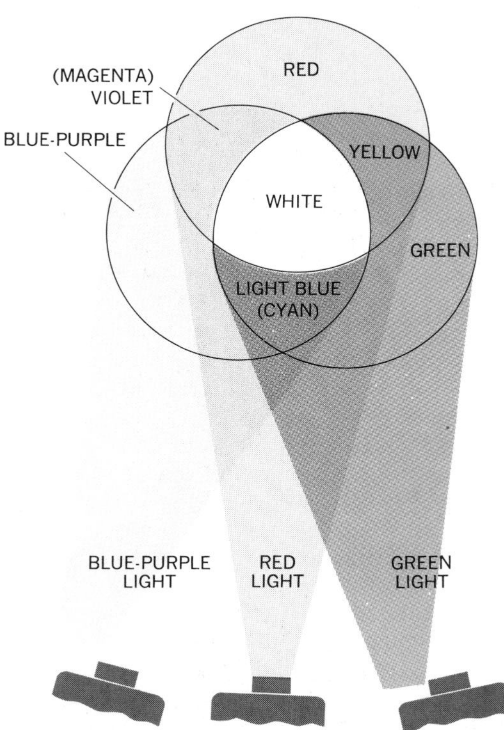

FIGURE 9-23 ***All known colors can be produced by different combinations of the three primary wavelengths of light, those giving rise to the color sensations of red, blue, and green.*** **(Adapted from Gregory.)**

FIGURE 9-24 ***Response of the three photopigments to light of different wavelengths.*** **(Adapted from Gregory.)**

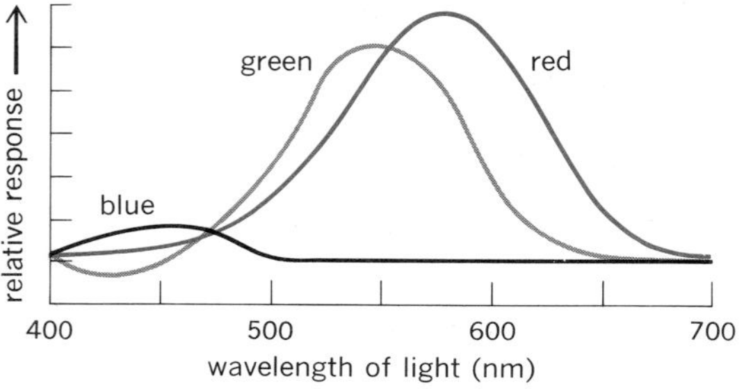

99 percent of the female population) have normal color vision; i.e., their color vision is determined by the differential activity of the three types of cones. Most color-blind (or better, color-defective) people appear to lack one of the three photopigments, and their color vision is therefore formed by the differential activity of the remaining two types of cones. For example, people with green-defective vision see as if they have only red- and blue-sensitive cones.

Notice that a push-pull type of behavior, in which paired neurons respond in opposite directions, is a common feature of the visual coding system. One cell responds when the light goes on, the other when it is turned off; one cell is sensitive to illumination at the center of the receptive field, the other to light at the periphery; one is sensitive to red, the other to green; one is excitatory, the other inhibitory. Pairs of neurons having opposing behaviors emphasize *contrasts* in the visual stimuli; much of our visual perception depends on such contrasts.

Eye Movement The cones are highly concentrated in a specialized area of the retina known as the *fovea,* and images focused there are seen with the greatest acuity. In order to keep the visual image focused on the fovea, six muscles (the extraocular muscles; Fig. 9-25) are attached to each eyeball. They are capable of rotating the eyeball in any direction (Table 9-1). The four recti pass from their attachment at the back of the orbit to the sclera, their positions of attachment on the eye being designated by their names. The superior oblique also originates from the back of the orbit, but before attaching to the eye it passes through a cartilaginous loop, the *trochlea,* which allows a change in the direction of the muscle's pull much as a pulley changes the direction of pull by a rope. The inferior oblique originates on the nasal side of the orbit and passes under the eyeball to attach to its lateral surface (Fig. 9-25).

The lateral and medial recti are antagonists, one muscle relaxing as the other contracts. They can either sweep the two eyes horizontally, both eyes turning to the right or left simultaneously, or they can turn the eyes medially (convergence) for near vision or laterally (divergence) for distant vision. They cannot raise or lower the eyes; these movements are done by the inferior and superior recti and the two oblique muscles.

There are four main classes of eye movements:

1 Search for visual targets. This type of movement is a small rapid jerk called a *saccade.* This pattern of periods of steady fixation interrupted by quick changes in fixation occurs, for example, while examining an object or reading. In addition to search of the visual field, saccades move the visual image over the receptors, preventing adaptation. In fact, if such movements of the eye are stopped, all color and most detail fade away in a matter of seconds. Saccades are among the fastest movements in the body. These movements also occur during certain periods of sleep when the eyes are closed. Perhaps then they are associated with "watching" the visual imagery of dreams.

2 **Tracking of visual objects.** These smooth movements cause the eyes to follow an object if it moves throughout the visual field.

3 **Compensation for movements of the head.** If a stationary visual object is focused on the fovea and the head is moved to the left, the eyes must be moved an equal distance to the right if the object's image is to remain focused on the fovea; if the head moves up, the eyes must move down.

4 **Convergence and Divergence.** These types of movements are used to track a visual object in depth through the visual field, turning the eyes inward (convergence) as the object comes closer and outward (divergence) as it moves farther away.

The four movement types seem to be controlled by separate neurologic systems, yet they cooperate in almost all eye movements. Except for those eye movements which compensate for movements of the head, the control systems depend upon information from the retina. The compensating movements obtain their information about the movement of the head from the semicircular canals of the vestibular system, which will be described below.

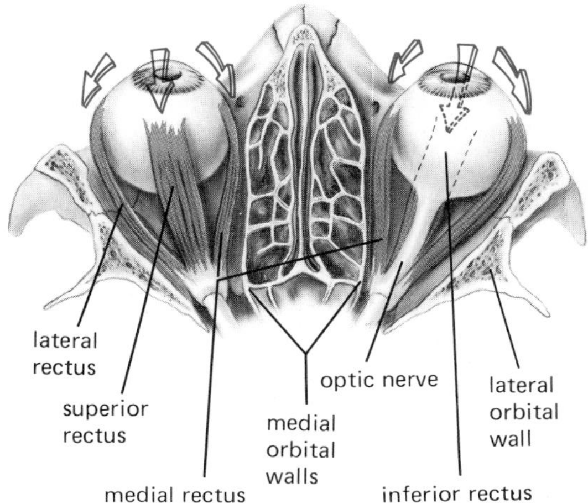

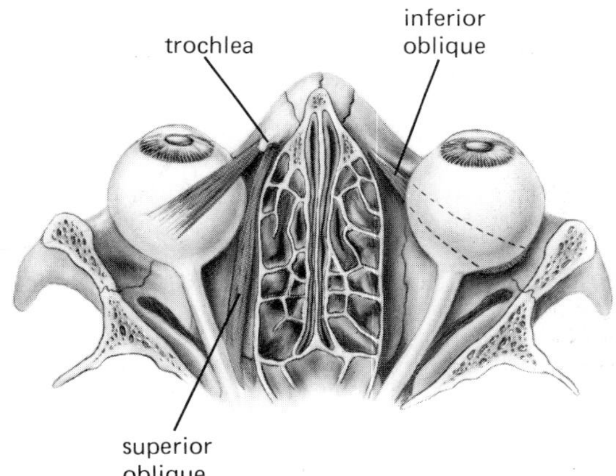

FIGURE 9-25 ***Extraocular muscles.***

TABLE 9-1

EXTRAOCULAR MUSCLES

Muscle	*Action*		*Innervation*
Superior rectus	Rotates eye medially, elevates eye.	Convergence	Oculomotor (III)
Inferior rectus	Rotates eye medially, depresses eye.	Convergence	Oculomotor (III)
Medial rectus	Rotates eye medially.	Convergence	Oculomotor (III)
Lateral rectus	Rotates eye laterally.	Divergence	Abducens (VI)
Superior oblique	Rotates eye downward and lateralward.	Divergence	Trochlear (IV)
Inferior oblique	Rotates eye upward and lateralward.	Divergence	Oculomotor (III)

Hearing

Sound energy is transmitted through air as a disturbance of air molecules. When there are no air molecules, as in a vacuum, there can be no sound. The disturbance of air molecules that makes up a sound wave consists of regions of compression, in which the air molecules are close together and the pressure is high, alternating with areas of rarefaction, where the molecules are farther apart and the pressure is lower. Anything capable of creating such disturbances can serve as a sound source. A tuning fork at rest emits no sound (Fig. 9-26), but if it is struck sharply, it gives rise to a pure tone. As the arms of the tuning fork move, they push air molecules ahead of them, creating a zone of compression, and pull apart the molecules behind them, leaving a zone of rarefaction (Fig. 9-26B). As they move in the opposite direction, they again create pressure waves of compression and rarefaction (Fig. 9-26C). The molecules in an area of compression, pushed together by the vibrating prong of the tuning fork, bump into the molecules ahead of them, push them together, and create a new region of compression. Individual molecules travel only short distances, but the disturbance passed from one molecule to another can travel many miles; it is in these disturbances (sound waves) that sound energy is transmitted. The sound dies out only when so much of the original sound energy has been dissipated that one sound wave can no longer disturb the air molecules around it. The tone emitted by the tuning fork is said to be *pure* because the waves of rarefaction and compression are regularly spaced. The waves of speech and many other common sounds are not regularly spaced but are complex waves made up of many frequencies of vibration.

The sounds heard most keenly by human ears are those from sources vibrating at frequencies between 1,000 and 4,000 Hz, but the entire range of frequencies audible to human beings extends from 20 to 20,000 Hz. The *frequency* of vibration of the sound source is related to the pitch we hear; the faster the vibration, the higher the pitch. We can also detect loudness and tonal quality, or timbre, of a sound. The difference between the packing (or pressure) of air molecules in a zone of compression and a zone of rarefaction, i.e., the *amplitude* of the sound wave, is related to the loudness of the sound that we hear. The number of sound frequencies in addition to the fundamental tone, i.e., the degree of *purity* of the sound wave, is related to the quality or timbre of the sound. We can distinguish some 400,000 different sounds. We can distinguish the note A played on a piano from the same note played on a violin, and we can identify voices heard over the telephone. We can also selectively *not* hear sounds, tuning out the babel of a party to concentrate on a single voice. How is this ac-

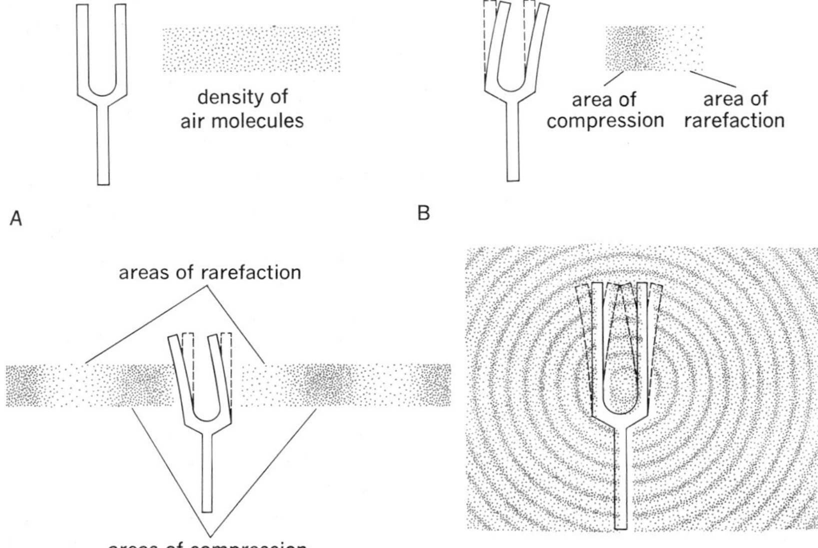

FIGURE 9-26 ***Formation of sound waves.***

complished by an apparatus small enough to fit into a teacup?

The first step in hearing is usually the entrance of pressure waves into the *ear canal* (Fig. 9-27). The waves reverberate from the side and end of the ear canal so that it is filled with the continuous vibrations of pressure waves. The *tympanic membrane* (*eardrum*) is stretched across the end of the ear canal. The air molecules, under slightly higher pressure during a wave of compression, push against the membrane, causing it to bow inward. The distance the membrane moves, although always very small, is a function of the force and velocity with which the air molecules hit it and is therefore related to the loudness of the sound. During the following wave of rarefaction, the membrane returns to its original position. The exquisitely sensitive tympanic membrane responds to all the varying pressures of the sound waves, vibrating slowly in response to low-frequency sounds and rapidly in response to high tones. It is sensitive to pressures to which the most delicate touch receptors of the skin are totally insensitive.

The tympanic membrane separates the ear canal from the *middle-ear cavity* (Fig. 9-27). The pressures in these two air-filled chambers are normally equal, but a difference can be produced with sudden changes in altitude, as in an elevator or airplane. This difference distorts the tympanic membrane and causes pain. The outer ear canal is, of course, normally at atmospheric pressure. The middle ear is exposed to atmospheric pressure only through the *eustachian tube,* which connects the middle ear to the pharynx and nose or mouth. The slitlike ending of the eustachian tube in the pharynx is normally closed; but during yawning, swallowing, or sneezing, when muscle movements of the pharynx open the entire passage, the pressure in the middle ear equilibrates with atmospheric pressure.

The second step in hearing is the transmission of sound energy from the tympanic membrane, through the cavity of the middle ear, and then to the receptor cells in the *inner ear,* which are surrounded by liquid. The major function of the middle ear (Fig. 9-27) is to transfer movements of the air in the outer ear to the liquid-filled chambers of the inner ear. The liquid in the inner ear is more difficult to move than air; thus, the pressure transmitted to the inner ear must be increased. This is achieved by a movable chain of three small middle-ear bones (the *incus,* or anvil; the *malleus,* or hammer; and the *stapes,* or stirrup), which couple the tympanic membrane to a membrane-covered opening (the *oval window*), which separates the middle and inner ear. The total force on the tympanic membrane is transferred to the much smaller oval window. The *total* force on the oval window is the same as that on the tympanic membrane, but because the oval window is so much smaller, the *force per unit area* (i.e., pressure) is increased 15 to 20 times. Additional advantage is gained through the lever action of the three middle-ear bones. Thus, the tiny amounts of energy involved are transferred to the inner ear with relatively small loss. The amount of energy transmitted to the inner ear can be modified by the contraction of two small muscles in the middle ear which alter the tension of the tympanic membrane and the position of the third middle-ear bone (*stapes*) in the oval window. These muscles protect the delicate receptor apparatus from intense sound stimuli and possibly aid intent listening over certain frequency ranges.

Thus far, the entire system has been concerned with the transmission of the sound energy into the inner ear, where the receptors are located. The inner ear consists of two parts: a series of cavities (the *bony labyrinth*) within the petrous portion of the temporal bone and a series of communicating membranous sacs (the *membranous labyrinth*), which is contained within the bony cavities. The bony labyrinth is made up of the *vestibule, semicircular canals,* and *cochlea;* these cavities are filled with a clear fluid (*perilymph*) in which the membranous labyrinth is suspended. The membranous labyrinth is filled with *endolymph.* The canal of the cochlea, which is concerned with hearing, spirals for $2\frac{3}{4}$ turns around a bony central pillar (the *modiolus*). A shelf of bone (the *osseous spiral lamina*) projects from the modiolus and partially divides the cochlear canal into two passageways, or scalae: the *scala vestibuli* and the *scala tympani* (see Fig. 9-29). The division is completed by the *basilar membrane,* which extends from the osseous spiral lamina to the opposite side of the cochlea. These scalae, being portions of the bony labyrinth, are filled with perilymph. A portion of the membranous labyrinth, the *cochlear duct,* lies within the cochlea.

As the pressure wave in the ear canal pushes in on the tympanic membrane, the chain of middle-ear bones rocks the footplate of the stapes against the membrane covering the oval window, causing it to bow into the scala vestibuli (Figs. 9-28 and 9-29). The wall of the scala vestibuli is largely bone, but there are two paths by which the pressure waves can be dissipated. One

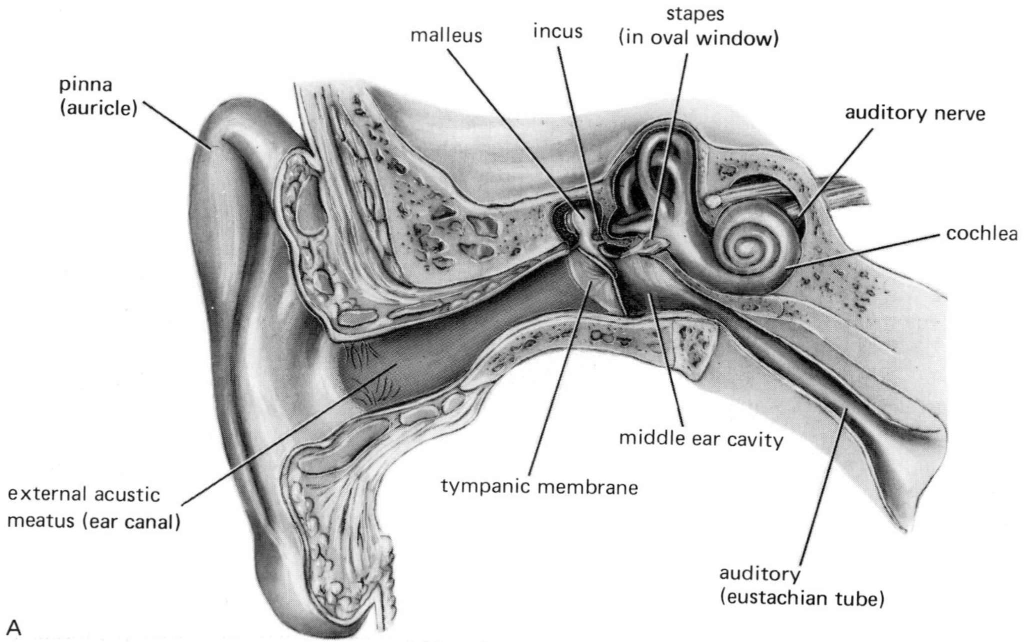

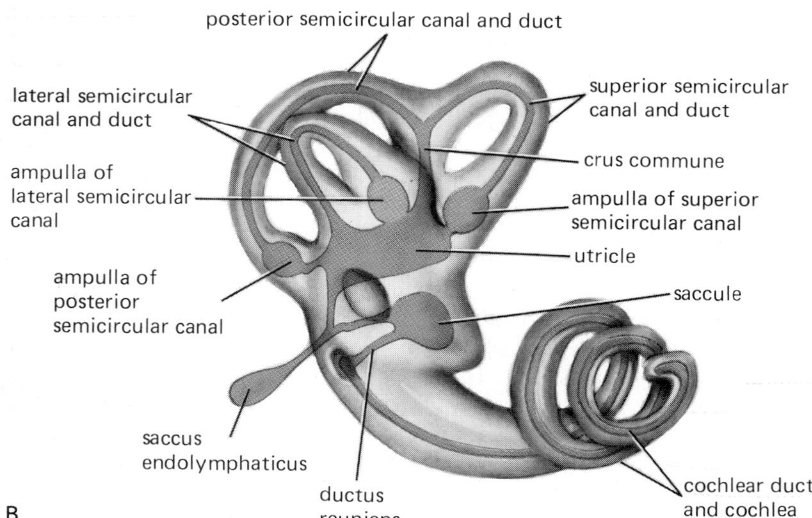

FIGURE 9-27 **(A)** ***Anatomy of the human ear. The malleus, incus, and stapes are the middle ear bones.*** **(B)** ***The membranous labyrinth within the bony labyrinth. The saccus endolymphaticus is a blind pouch of the membranous labyrinth.***

path is to the end of the scala vestibuli, where the waves pass around the end of the cochlear duct into the scala tympani, and back to another membrane-covered window, the round window, which they bow out into the middle-ear cavity. However, most of the pressure waves are transmitted to the cochlear duct and thereby to the basilar membrane, which is deflected into the scala tympani.

The pattern by which the basilar membrane is deflected is important because this membrane contains

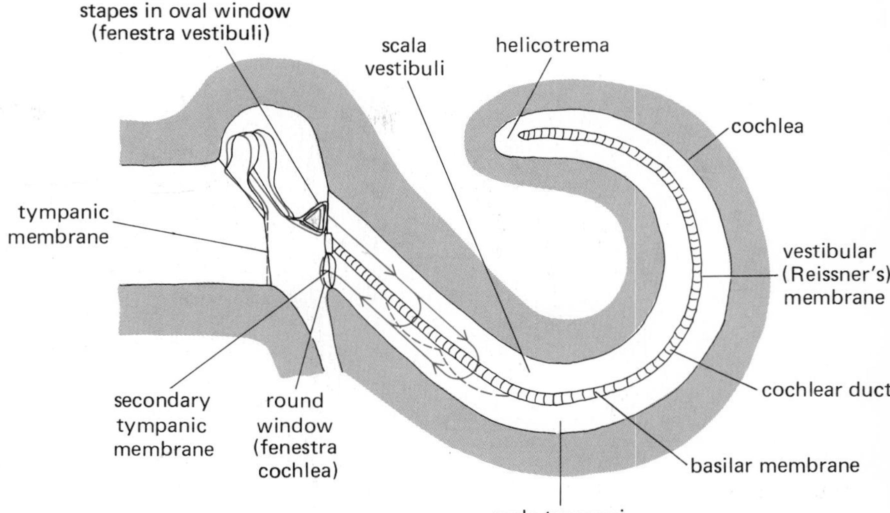

FIGURE 9-28 ***Auditory portions of the middle and inner ear. The position of the membranes and middle-ear bones is shown at rest*** **(solid lines)** ***and following the inner displacement of the tympanic membrane by a sound wave*** **(dashed line).** ***Arrows show the different paths of two sound pressure waves of different frequency. The cochlea has been partially unwound.***

the spiral *organ of Corti* and its sensitive receptor cells which transform sound energy, i.e., the pressure wave, into action potentials. At the end of the cochlea closest to the middle-ear cavity, the basilar membrane is narrow and relatively stiff, but it becomes wider and more elastic as it extends throughout the length of the cochlear spiral. The stiff end nearest the middle-ear cavity vibrates immediately in response to the pressure changes transmitted to the scala vestibuli, but the responses of the more distant parts are slower. Thus, with each change in pressure in the inner ear, a wave of vibrations is made to travel down the basilar membrane.

The region of maximal displacement of the basilar membrane varies with the frequency of vibration of the sound source. The properties of the membrane nearest the oval window and middle ear are such that this region resonates best with high-frequency tones and undergoes the greatest amplitude of vibration when high-pitched tones are heard. The traveling wave soon dies out once it is past this region. Lower tones also cause the basilar membrane to vibrate near the middle-ear cavity, but the vibration wave travels out along the membrane for greater distances. The more distant regions of the basilar membrane vibrate maximally in response to low tones. Thus the frequencies of the incoming sound waves are in effect sorted out along the length of the basilar membrane (Fig. 9-30).

Where the displacement of the basilar membrane is a maximum, the stimulation of the receptors (the *hair cells* of the organ of Corti; Fig. 9-29) which ride upon the membrane is the greatest. The fine hairs on the top of the receptor cells are in contact with the overhanging *tectorial membrane,* which projects inward from the side of the osseous spiral lamina. As the basilar membrane is displaced by pressure waves in the scala vestibuli and cochlear duct, the hair cells move in relation to the tectorial membrane, and, consequently, the hairs are displaced. In this interaction, the incoming sound energy is transformed from the vibrating molecules of pressure waves to electric events in the hair cells, for in some way (the precise mechanism is not known) movements of the hairs cause a depolarization of the hair cells which is similar to the generator potential of the receptors. The hair cells are easily damaged by exposure to high-intensity noises such as the typical live amplified rock music concerts and engines of jet planes and revved-up motorcycles. The damaged sensory hairs form giant, abnormal hair structures or are lost altogether and, in cases of long exposure to loud sounds, areas of hair cells and their supporting cells completely degenerate (Fig. 9-31).

In normal hearing, the generator potentials formed by the activated hair cells lead ultimately to the production of action potentials in the peripheral endings of the afferent nerves, the resulting action potentials being

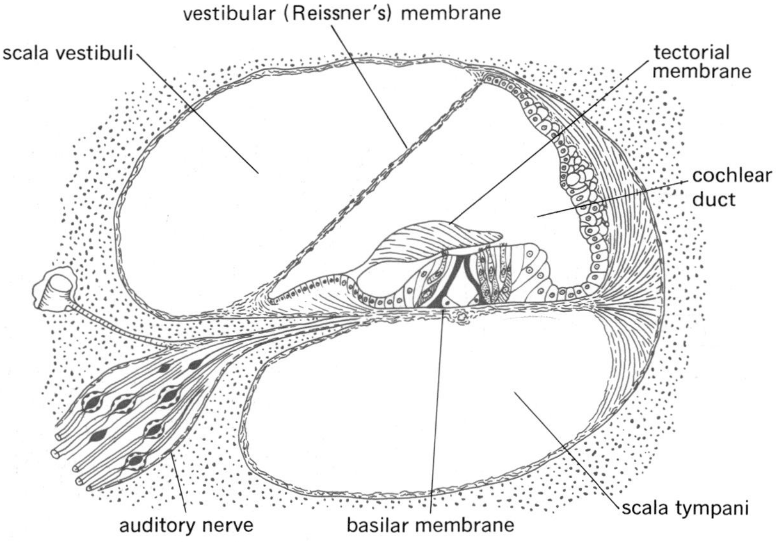

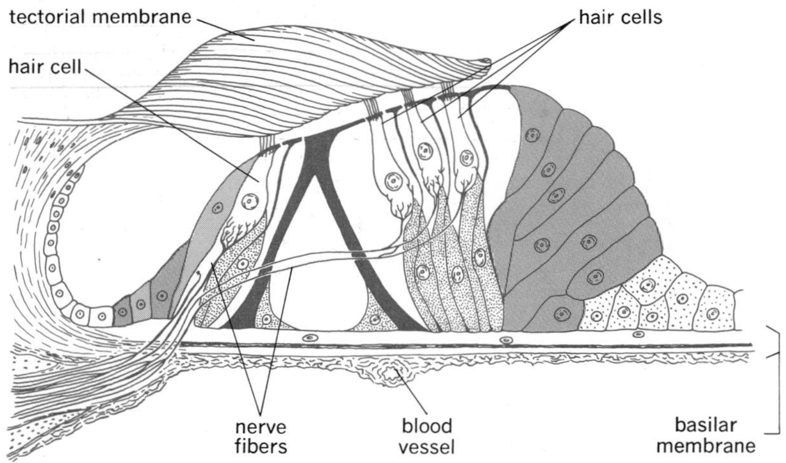

FIGURE 9-29 ***Cross section of the membranes and compartments of the inner ear with a detailed view of the spiral organ of Corti with its hair cells and other structures upon the basilar membrane.*** **(Adapted from Rasmussen.)**

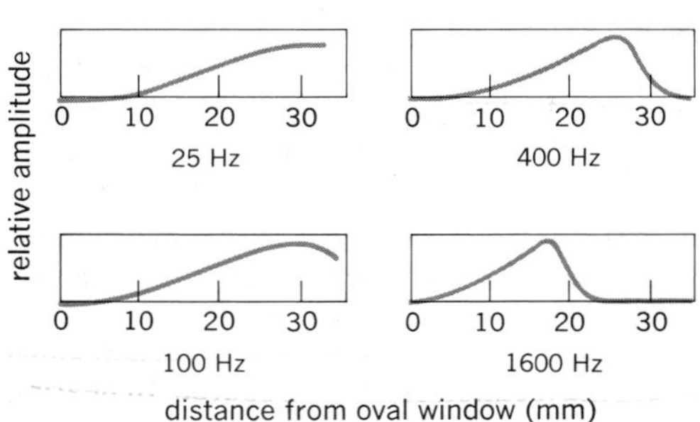

FIGURE 9-30 ***Point along the basilar membrane where the traveling wave peaks is different with different sound frequencies. The region of maximal displacement of the basilar membrane occurs near the end of the membrane for low-pitched (low-frequency) tones and near the oval window and middle ear for high-pitched tones*** **(Adapted from von Békésy and Rosenblith.)**

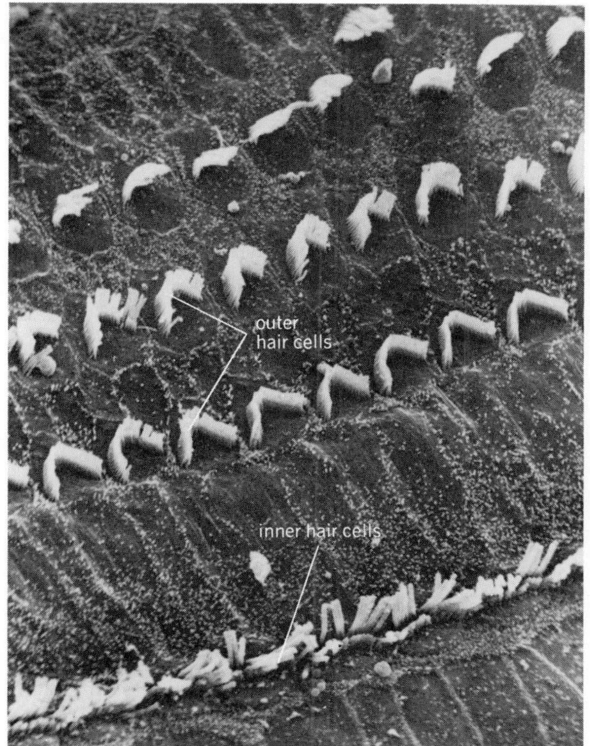

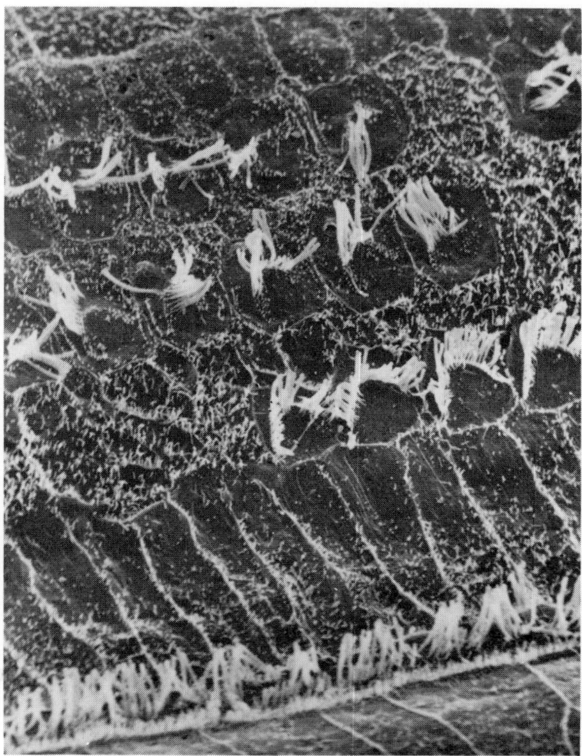

FIGURE 9-31 ***Injury to the inner ear by intense noise. (A) Normal organ of Corti (guinea pig) showing the three rows of outer hair cells and single row of inner hair cells. (B) Injured organ of Corti after 24-h exposure to noise levels typical of very loud rock music (2,000-Hz-octave band at 120 dB). Several outer hair cells are missing, and the cilia of others no longer form the orderly W pattern of the normal ear. Note also the increased number and size of small villi on the cell surfaces.*** **(Scanning electron micrograph by Robert E. Preston. Courtesy Joseph E. Hawkins, Kresge Hearing Research Institute.)**

transmitted into the central nervous system. The greater the energy of the sound wave (loudness), the greater is the movement of the basilar membrane, the greater the amplitude of the generator potential, and the greater the frequency of action potentials in the afferent nerve.

This completes our analysis of the auditory receptor mechanisms. We should like to be able to follow the coding of information from the organ of Corti to auditory cortex, as we did for vision. However, the precise mechanisms are considerably less clear than is the present case for vision and we therefore will not do so, except to describe sound localization.

Times of occurrence of sounds are important clues used to detect the location of a sound source. In hearing (and in vision and smell) there is the problem in stimulus localization of projecting the stimulus to an external source. Although the receptors stimulated by sound are located in the cochlea of the inner ears, the source of the sound is perceived to be the phonograph speaker across the room. Comparison of the times of onset and intensities of sounds at each of the two ears are the two most important clues in finding a sound source. In fact, the head is turned when localizing sounds to emphasize the difference in this information. The difference in time of onset helps particularly to localize low-frequency sounds (low pitch). A sound originating on the left stimulates the left ear slightly before it stimulates the right (Fig. 9-32), periods of compression and rarefaction of each sound wave occurring slightly earlier on the left.

The sound is also of slightly greater intensity on the left, and the difference in sound intensity becomes an increasingly important clue as the sound frequency increases and pitch rises. The sound is localized to the side where it is louder.

A simple experiment can demonstrate the role of inequality of sound intensity upon sound localization. A quietly hummed tone is localized at the middle of the head. If, while the humming continues, one ear is lightly plugged, the sound intensity in that ear increases because of the greater sound reverberations in the blocked ear canal. As the tone increases in loudness, it becomes localized in the closed ear. Even this slight imbalance in intensity or loudness is sufficient to cause large changes in localization of the stimulus. The first step in this comparison occurs in the brainstem at the level of entry of the auditory nerves before there has been much chance for synaptic alteration and delay to modify the information, but in human beings the cerebral cortex is necessary for actual sound localization.

FIGURE 9-32 ***Difference in input between the two ears helps to localize sound sources.***

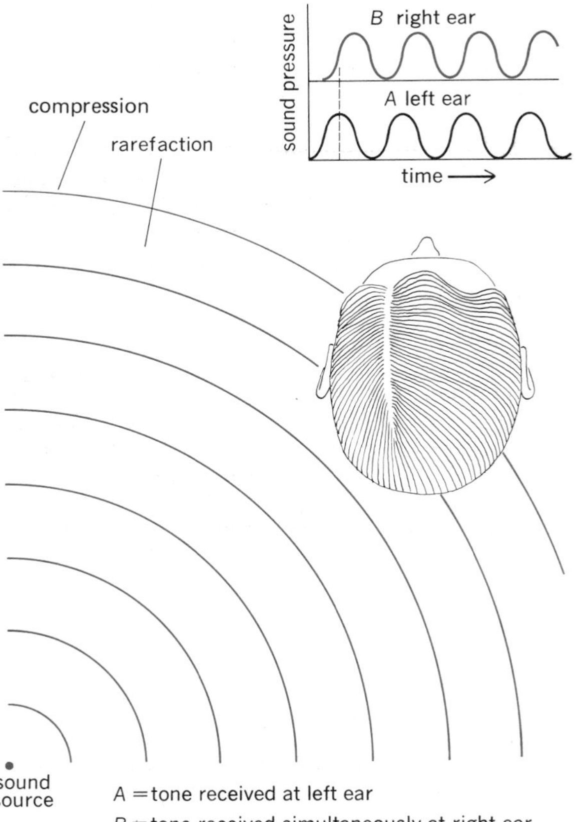

Vestibular System

The vestibular system contains mechanoreceptors specialized to detect changes in both the motion and position of the head. The receptors are part of the *vestibular apparatus* which is part of the membranous labyrinth; it consists of three *semicircular ducts* enclosed within the semicircular canals, and a *utricle* and *saccule* within the vestibule (Fig. 9-27B), all of which are filled with a specialized fluid, the *endolymph.*

The three semicircular canals on each side of the skull are arranged at right angles to each other (Fig. 9-33). The actual receptors of the semicircular canals are hair cells which sit at the ends of the afferent neurons. The sensory hairs are closely ensheathed by a gelatinous mass, the *cupula,* which almost blocks the channel of the semicircular duct at that point.

The receptor system in the semicircular canals works in the following way. Whenever the head is moved, the bony-tunnel wall, its enclosed membranous semicircular duct, and the attached bodies of the hair cells, of course, turn with it. The endolymph fluid filling the duct, however, is neither attached to the skull nor automatically pulled with it; instead, because of inertia, the fluid tends to retain its original position, i.e., to be "left behind." As the bodies of the hair cells move with the skull, the hairs are pulled against the relatively stationary column of endolymph and are bent (Fig. 9-34). The speed and magnitude of the movement of the head determine the degree to which the hairs are bent and the hair cells stimulated. As the inertia is overcome, the hairs slowly return to their resting position; for this reason, the hair cells are stimulated only during *changes* in rate of motion, i.e., during acceleration of the head. During motion at a constant speed, stimulation of the hair cells ceases.

The transducing mechanisms of these receptor cells, by which bending of the hairs gives rise to action potentials in the afferent nerve, are not known. Although the junction between the hair cell and afferent nerve fiber has the anatomic features of a chemically mediated synapse, the mechanism of synaptic transmission and the nature of the transmitter substance are unknown.

Even when the head is motionless, the afferent nerve fibers are activated at a relatively low resting fre-

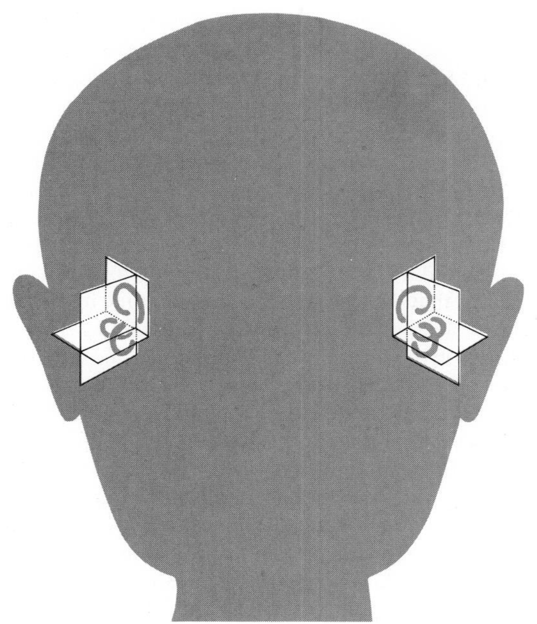

FIGURE 9-33 ***Relationship of the two sets of semicircular canals.***

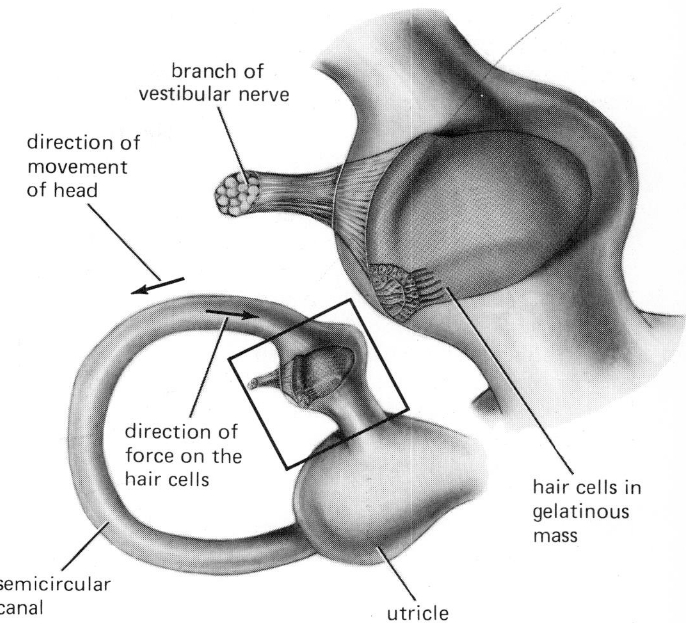

FIGURE 9-34 ***Diagram of a semicircular canal.***

quency (Fig. 9-35A). The explanation for this resting activity is unknown—perhaps it is due to the spontaneous leak of chemical transmitter from the receptor cell—but it allows the receptor cells to signal information by either increasing or decreasing the frequency of action potentials in the afferent nerve fiber.

Thus, in some way, the shearing force bending the hairs on the receptor cells is related to the frequency of action potentials in the afferent nerve; when the hairs are bent one way, the rate of firing speeds up; when the hairs are bent in the opposite direction, the firing frequency slows down (Fig. 9-35B and C).

Whereas the semicircular canals signal the rate of change of motion of the head, the utricle and saccule contain the receptors of the vestibular system which provide information about the position of the head relative to the direction of the forces of gravity. The receptor cells here, too, are mechanoreceptors sensitive to the movement of projecting cilia, or hairs. The hair cells of the utricle and saccule are collected into groups from which the hairs protrude into a gelatinous substance. In the utricle and saccule tiny calcium carbonate stones, or *otoliths,* are embedded in the gelatinous covering of the hair cells, making the gelatinous substance heavier than the surrounding endolymph. When the head is tipped, the gelatinous-otolith material changes its position, pulled by gravitational forces to the lowest point in the utricle or saccule. The shearing forces of the gelatinous-otolith substance against the hair cells bend the hairs and stimulate the receptor cells.

FIGURE 9-35 ***Relation between position of hairs and activity in afferent nerve.* (A) *Resting state;* (B) *movement in one direction;* (C) *opposite movement.* (Adapted from Wersall, Gleisner, and Lundquist.)**

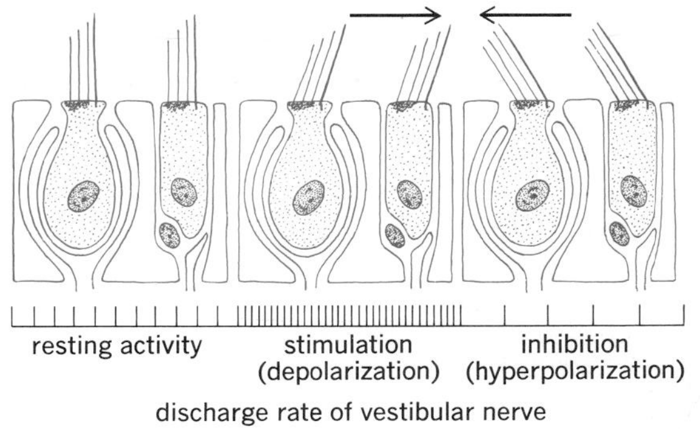

The information from the vestibular apparatus is used for two purposes. The first is to control the muscles which move the eyes so that, in spite of changes in the position of the head, the eyes remain fixed on the same point. As the head is turned to the left, the balance of afferent input from the vestibular apparatus on each side is altered. Impulses from the vestibular nuclei activate the ocular muscles which turn the eyes to the right and inhibit their antagonists. The eyes turn toward the right as the head turns toward the left, and the net result is that the eyes remain fixed on the point of interest.

Vestibular information is also utilized in reflex mechanisms for maintaining upright posture. In monkeys, cats, and dogs the vestibular apparatus plays a definite role in the postural fixation of the head, orientation of the animal in space, and reflexes accompanying locomotion. However, in human beings very few postural reflexes are known to depend primarily on vestibular input, despite the fact that the vestibular organs are sometimes called the sense organs of balance.

Taste and Smell

Receptors sensitive to certain chemicals in the environment are *chemoreceptors*. There are chemoreceptors which respond to chemical changes in the internal environment (e.g., the oxygen and hydrogen receptors in certain of the large blood vessels) or in the external environment (e.g., the receptors for the sense of taste and the sense of smell). In human beings, these external chemoreceptors are much less important than the receptors for vision or hearing although that is certainly not true for most animals. Although taste and smell affect a person's appetite, the initiation of digestion, and the avoidance of harmful substances, they do not exert strong or essential influence.

Taste The specialized receptor organs for the sense of taste are the 10,000 or so *taste buds*. Some are located in the walls of the *papillae,* small elevations on the tongue which are visible to the naked eye (Fig. 9-36), but most are scattered over the surface of the tongue, roof of the mouth, pharynx, and larynx. The latter receptors are not activated by substances in the mouth but do fire during swallows. Inside the taste buds the receptor cells are arranged like segments of an orange with the multifolded upper surfaces of the receptor cells extending into a small pore at the surface of the taste bud, where they are bathed by the fluids of the mouth (Fig. 9-36C).

Taste sensations are traditionally divided into four basic groups: sweet, sour, salt, and bitter, but different types of taste buds or receptor cells which would support this specificity have not been identified. In fact, a single receptor cell can respond in varying degrees to many different chemical substances falling into more than one of the basic categories.

What makes a receptor cell respond? The mechanisms by which the taste receptors are stimulated and action potentials generated are not known. It has been suggested that the first step is a loose binding of the individual molecules of the chemical substance with specific sites on the receptor cell membrane. The fact that a single receptor can be responsive to more than one basic taste quality could be explained if a single receptor cell had several different sites, each capable of binding with a different type of molecule. The theory goes on to suggest that the chemical-substance–receptor-site combination alters the cell membrane, forming pores through which ions move to change the membrane potential of the cell. It is known that the membrane of the receptor cells depolarizes when the cell is stimulated chemically.

Beneath the taste buds lie the nerve fibers which enter the buds to end at the receptor cells. The receptors on the anterior two-thirds of the tongue and soft palate are innervated by the facial nerve (VII), those on the posterior third of the tongue and region of the vallate papillae by the glossopharyngeal nerve (IX), and those on the pharynx and larynx by the vagus nerve (X). One nerve fiber may innervate several receptor cells, and one receptor cell may be innervated by several different neurons. There is clearly no one-to-one relationship by which each receptor cell has a single line into the central nervous system. How can we distinguish so many different taste sensations when the receptor cells lack specificity both in terms of the kind of chemical to which they will respond and the way in which they are connected to the brain? The frequency of action potentials in single nerve fibers increases in response to increasing concentrations of the chemical stimulant; therefore, frequency signals quantity, but what signals the quality? The afferent fibers involved in taste show different firing patterns in response to different substances; e.g., one fiber may fire very rapidly when the stimulatory substance is salt but only sporadically when it is sugar, and another fiber may have just the opposite reaction. This variation in relative sensitivity makes the pattern of firing within a group of neurons meaningful, and awareness of the specific taste of a sub-

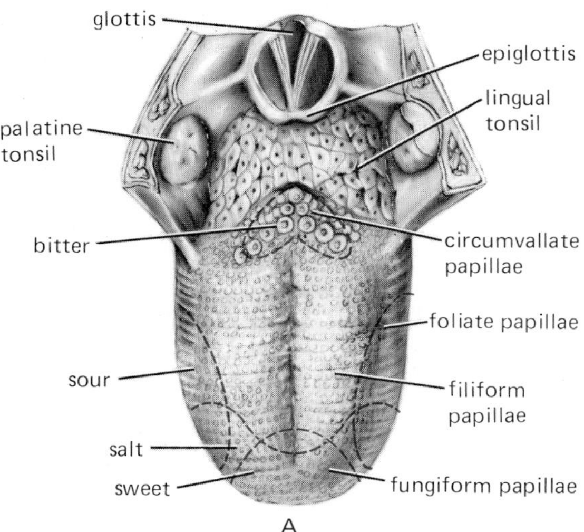

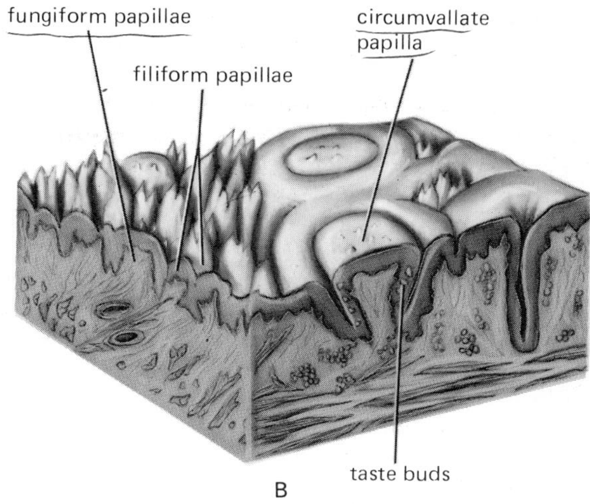

FIGURE 9-36 (A) ***Dorsal surface of the tongue. Major structures and the areas of specific taste sensitivities are indicated.*** (B) ***Cross section of the tongue surface showing three of the four types of papillae.*** (C) ***Structure and innervation of a taste bud.***

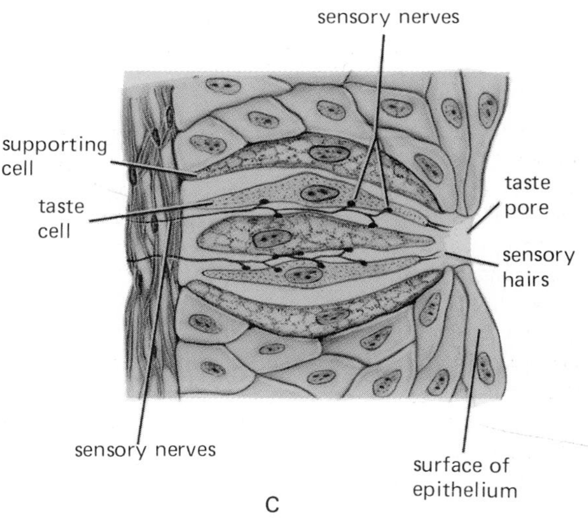

stance probably depends upon the relative activity in a number of different neurons rather than that in a specific neuron. Identification of the substance is aided by information about its temperature and texture which is transmitted to the central nervous system from receptors on the tongue and surface of the oral cavity. The odor of the substance clearly helps, too, as is attested by the common experience that food lacks taste when one has a stuffy head cold.

Smell The olfactory receptors which give rise to the sense of smell lie in a small patch of mucosa, i.e., membrane which secretes mucus, in the upper part of the nasal cavity (Fig. 9-37A). Because the *olfactory mucosa* is above the path of the main air currents that enter the nose with inspiration, the odorous molecules must either diffuse up to the receptor cells or be drawn up by changes in respiration such as sniffing. The receptor cells (Fig. 9-37B) are bipolar modified neurons having two processes; one passes toward the brain, forming the *olfactory nerve*, and the other bears many fine cilia and extends out to the surface of the olfactory mucosa.

Before an odorous substance can be detected, it must first release molecules which diffuse into the air and pass into the nose to the region of the olfactory mucosa, dissolve in the layer of mucus covering the receptors, establish some sort of relation with the receptor, and depolarize the membrane enough to initiate an action potential in the afferent nerve fiber. Perhaps the molecule combines with a receptor site on the membrane in such a way that the membrane perme-

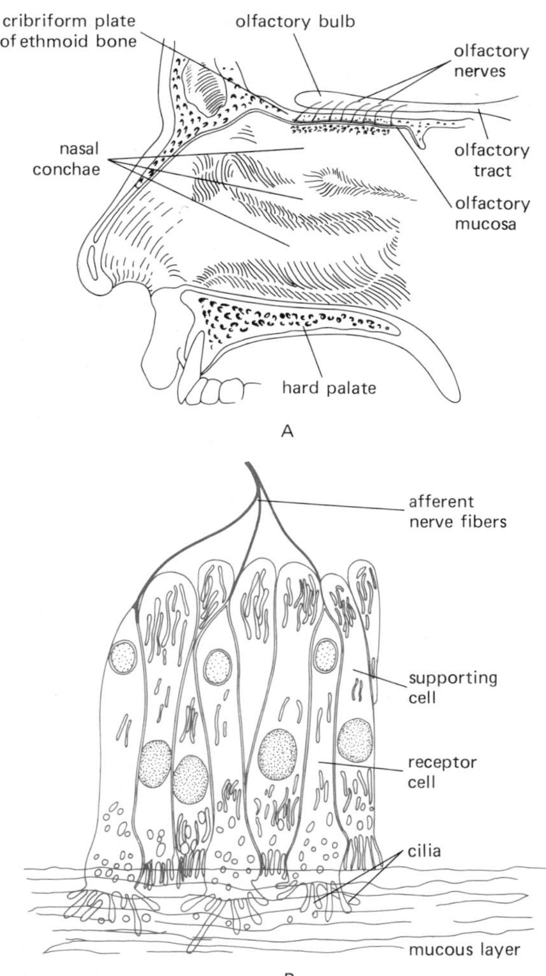

FIGURE 9-37 ***Location and structure of the olfactory receptors.***

ability changes and ions move across the membrane to depolarize the cell, for upon stimulation of the olfactory mucosa with odorous substances, a generator potential can be recorded which changes with stimulus quality and intensity.

The physiologic basis for discrimination between the tens of thousands of different odor qualities is speculative. There are no apparent differences in receptor cells, at least on a microscopic level, to account for it, but meaningful differences between receptors are thought to exist at molecular levels. One theory is that odor discrimination depends upon a large number of different types of interactions between the odor-substance molecules and receptor sites. This theory is based on the supposition that the receptor cells probably have 20 or 30 different types of receptor sites, each type capable of interacting with many different odor molecules but responding best to a molecule with a particular size, polarity, and shape. The receptor-site populations vary from one receptor cell to another. For example, one receptor cell may have mainly receptor sites of types A, B, and C, whereas another cell has types B, D, and E. The molecules of the odor substance combine with most types of receptor sites, but the "fit" varies. A good contact occurs when the odor molecule and receptor sites' size, shape, and polarity match; in such a case the resulting depolarization of the receptor cell is large. In cases of poorer fit, the receptor cell still depolarizes but to a smaller degree. All depolarizations occurring together within one cell are summed, forming a generator potential, which determines the firing rate in the afferent nerve fiber. Receptor cells with different receptor-site populations respond to the same odor substance with different firing rates. Thus, it is the simultaneous yet differential stimulation of many receptor cells that provides the basis for the discrimination. Moreover, olfactory discrimination depends only partially upon the action-potential pattern generated in the different afferent neurons; it also varies with attentiveness, state of the olfactory mucosa (acuity decreases when the mucosa is congested, as in a head cold), hunger (sensitivity is greater in hungry subjects), sex (women in general have keener olfactory sensitivities than men), and smoking (decreased sensitivity has been repeatedly associated with smoking). And just as the awareness of the taste of an object is aided by other senses, so the knowledge of the odor of a substance is aided by the stimulation of other receptors. This is responsible for the description of odors as pungent, acrid, cool, or irritating.

FURTHER PERCEPTUAL PROCESSING

Our actual perception of the events around us often involves areas of the brain other than the primary sensory cortex. It is apparent that the sensory information of the primary sensory areas achieves further elaboration through the neural activity in *cortical association areas*. These brain areas lie outside the classic primary cortical sensory or motor areas but are connected to them by association fibers (Fig. 9-38). Although it often

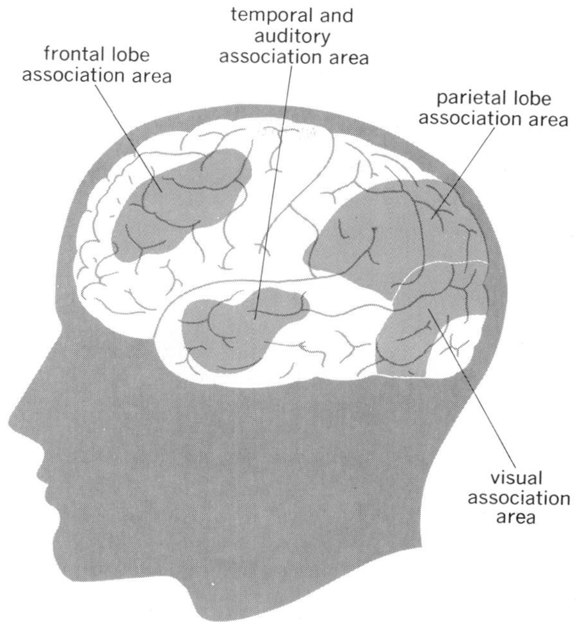

FIGURE 9-38 ***Areas of association cortex.***

has not been possible to elucidate the specific roles performed by the association areas, they are acknowledged to be of the greatest importance in the maintenance of higher mental activities in human beings. This elaboration of sensory information is demonstrated in the following experiments. For example, if areas of the primary visual cortex are stimulated (when the brain surface is exposed under local anesthesia during neurosurgical procedures and the patient is awake), the patient "sees" a flash of light. Upon stimulation of the association areas surrounding the visual cortex, the patient reports seeing more elaborate visual sensations such as brilliant colored balloons floating around in an infinite sky. Upon stimulation of association areas still farther from the primary visual cortex, the patient might report visual memories which seem to be reenacted before his eyes. Another example of the embellishment of the sensory experience provided by association areas can be found in persons who have undergone removal of parts of cortex because of tumors or accidents. A person who has no primary visual cortex is blind; if a chair is placed in her path, she walks into it because she does not see it. In contrast, a person who has a functional primary visual cortex but has no visual association areas sees the chair and therefore does not walk into it but is not able to say that the object in his path is a chair or to explain its function. Similarly, a patient with damaged auditory association areas may hear a spoken word but not comprehend its meaning. Patients with damaged parietal association areas can feel a small, cool object in their hands but know neither that it is called a key nor that it is used to open doors. As mentioned earlier, the mechanisms responsible are unknown.

Such further perceptual processing involves arousal, attention, learning, memory, language, and emotions, and it involves comparing the information presented via one sensory modality with that of another. For example, we may hear a growling dog, but our perception of the actual event taking place varies markedly, depending upon whether our visual system detects that the sound source is an angry animal or a loudspeaker.

Another step in the processing of any sensory information is testing the appropriateness of our interpretation. For example, when we see a can on the supermarket shelf whose label contains a picture of pears or tomatoes, we understand that we are looking at cans containing these items. However, if the label pictures a green giant, we do not think that the can contains the flesh of this bizarre creature since we know that that is impossible. We reject that notion and look for further clues to tell us of the can's contents.

Sometimes the objects we view are ambiguous and have more than one logical interpretation. Such a figure is the drawing entitled "My wife and my mother-in-law," illustrated in Fig. 9-39. When first trying to identify such a figure, we pick out one detail, say the curved line at the midportion of the left side of the drawing. If it reminds us of a bent nose, we immediately seek to verify our impression by looking for the expected eyes, mouth, hair, etc. An old woman's face is perceived, but as we look at it, the image suddenly shifts. The line that was a nose is now a chin; the image of a stylish young woman appears. Since both images are equally plausible, our interpretation of the image shifts back and forth between them. It is an interesting property of our perceptual mechanisms that we see either one or the other of the images; it is impossible to see them both as plausible at the same time.

We put great trust in our sensory-perceptual processes, despite the inevitable modifications we know to exist. Some factors known to distort our perceptions of the real world are as follows:

FIGURE 9-39 ***Ambiguous figure, "My wife and my mother-in-law," in which the young girl's chin is the woman's nose, created by cartoonist W. E. Hill in 1915.***

1 Afferent information is distorted by receptor mechanisms and by its processing along afferent pathways, e.g., by accommodation.
2 Emotions, personality, and social background can influence perceptions so that two people can witness the same events and yet perceive them differently.
3 Not all information entering the central nervous system gives rise to conscious sensations. Actually, this is a very good thing because many unwanted signals, generated by the extreme sensitivity of our receptors, are canceled out. The afferent systems are very sensitive. Under ideal conditions the rods of the eye can detect the flame of a candle 17 mi away. The hair cells of the ear can detect vibrations of an amplitude much lower than that caused by the flow of blood through the vascular system and can even detect molecules in random motion bumping against the tympanic membrane. Olfactory receptors respond to the presence of only four to eight odorous molecules. It is possible to detect one action potential generated by a pacinian corpuscle. If no mechanisms existed to select, restrain, and organize the barrage of impulses from the periphery, life would be unbearable. Information in some receptors' afferent pathways is not canceled out; it simply lacks the capacity for expression of a conscious correlate. For example, stretch receptors in the muscles detect changes in the length of the muscles, but activation of these receptors does not lead to a conscious sense of anything. Similarly, stretch receptors in the walls of some of the blood vessels effectively monitor both absolute blood pressure and its rate of change, but we have no conscious awareness of our blood pressure.
4 We lack suitable receptors for many energy forms. For example, we can have no direct information about radiation and radio or television waves until they are converted to an energy form to which we are sensitive. Many regions of the body are insensitive to touch, pressure, and pain because they lack the appropriate receptors. The brain itself has no pain or pressure receptors, and brain operations can be performed painlessly on patients who are still awake, provided that the cut edges of the sensitive brain coverings are infused with local anesthetic.

However, the most dramatic examples of a clear difference between the real world and our perceptual world can be found in illusions and drug- and disease-induced hallucinations when whole worlds can be created and mistaken for reality but can be proved false by physical measurements.

Any sense organ can give false information; e.g., pressure on the closed eye is perceived as light in darkness, and electric stimulation of any sense organ produces the sensory experience normally arising from the activation of that receptor. Why do such illusions appear? The two bits of false information just mentioned arise because most afferent fibers transmit information about one modality, and action potentials in a given afferent pathway going to a certain area of the brain signal the kind of information normally carried in that pathway. The association of the signal with the particular sensory experience has been in part built into the system during development and in part acquired through learning. The explanations of other illusions vary.

Levels of Perception

The two processes of transmitting data through the nervous system and interpreting it cannot be separated. Information is processed at each synaptic level of the afferent pathways. There is no one point along the afferent pathways or one particular level of the central nervous system below which activity cannot be a conscious sensation and above which it is a recognizable, definable sensory experience. Perception has many levels, and it seems that the many separate stages are arranged in a hierarchy, with the more complex stages receiving input only after it is processed by the more elementary systems. Every synapse along the afferent pathways adds an element of organization and contributes to the sensory experience.

CHAPTER 10

The Nervous System: III. Motor Control

The execution of a coordinated movement is a complicated process. Consider reaching to pick up an object. The fingers are extended and then flexed, the degree of extension depending upon the size of the object to be grasped, and the force of flexion depending upon the weight and consistency of the object. Simultaneously, the wrist, elbow, and shoulder are extended, and the trunk is inclined forward, the exact movements depending upon the distance of the object and the direction in which it lies. The shoulder must be stabilized to support the weight first of the arm and then of the object. Upright posture must be maintained in spite of the body's continually shifting center of gravity.

The building blocks for this seemingly simple action—as for all movements—are active motor units, each comprising one motor neuron together with all the skeletal muscle cells it innervates (Chap. 5). Neural inputs from many sources converge upon the motor neurons to control their activity; these interrelating systems are the subject of this chapter.

Each of the myriad coordinated body movements is characterized by a set of motor-unit activities which take place in both space and time. The ordered recruitment of specific motor neurons, i.e., the "program" required to achieve a given movement, is the function of the cerebral cortex, basal ganglia, cerebellum, brainstem, and spinal-cord nuclei.

A currently accepted hypothesis is that a general command, such as "pick up sweater" or "write signature" or "answer telephone," is sent from the cerebral cortex to the cerebellum and basal ganglia, where the actual motor-neuronal recruitment patterns are programmed. The programs are developed from stored patterns of motor-neuron activity relevant to the movement and from afferent information about the current (starting) position of the joints to be moved. The stored patterns arise from genetically determined connections between neurons and from learned behaviors. The program arrived at is sent via nuclei in the thalamus back to the motor cortex where it changes the firing patterns of the motor-cortex cells. The axons of these cells descend to alter the activity of the appropriate motor units; via axon collaterals, they also inform the basal ganglia and cerebellum of the commands that are being sent. This starts the cycle over again, the basal ganglia and cerebellum comparing the new directives with current information about the limb position and computing adjustments, which are sent back to the motor cortex. Thus, the basal ganglia and cerebellum receive a constant stream of information from the cortex about what actions are supposed to be taking place while they receive reports from the periphery about the actions that actually are taking place. Any discrepancies between the intended and actual movements are detected, and program corrections are sent to the motor cortex. This cycle from the motor cortex to the cerebellum and basal ganglia and back again continues throughout the course of the movement. For very rapid movements that do not provide enough time for continual correction, the entire course of the action is preprogrammed. The cerebellum deals more with these movements, and the basal ganglia are more involved with slow, continuous movements.

However, it should be recognized that the system is incomplete as described, for it does not take into account the mechanisms by which the "decision" to make a particular movement is reached. What neural events actually occur in the brain to cause one to "decide" to pick up an object in the first place? Presently we have no insight on this question.

Given such a system, it is difficult to use the word "voluntary" with any real precision. We shall use *voluntary* to refer to those actions which are characterized as follows:

1 These are actions we think about. The movement is accompanied by a conscious awareness of what we are doing and why we are doing it rather than the feeling that it "just happened," a feeling that often accompanies reflex responses.

2 Our attention is directed toward the action or its purpose.

3 The actions are the result of learning. Actions known to have disagreeable consequences are less likely to be performed voluntarily.

In the previous example of reaching to pick up an object, the activation of some of the motor units, such as those actually involved in grasping the object, can be classified clearly as voluntary; but most of the muscle activity associated with the act is initiated without any conscious, deliberate effort. In fact, almost all motor behavior involves both conscious and unconscious components, and the distinction between the two cannot be made easily.

Even a highly conscious act such as threading a needle involves the unconscious postural support of the hand and arm and inhibition of antagonistic muscles (those muscles whose activity would oppose the intended action, in this case the finger extensor muscles which straighten the fingers), and unconscious basic reflexes such as dropping a hot object can be influenced by conscious effort. If the hot object is something that took a great deal of time and effort to prepare, one probably would not drop it but would try to inhibit the reflex, holding on to the object until it could be put down safely. Most motor behavior is neither purely voluntary nor purely involuntary but falls at some point on a spectrum between these two extremes. But even this statement is of little help because patterned muscle movements shift along the spectrum according to the frequency with which they are performed.

For example, when a person first learns to drive a car with standard transmission, stopping is a fairly complicated process involving the accelerator, clutch, and brake. The sequence and force of the various operations depend on the speed of the car, and correct implementation requires a great deal of conscious attention. With practice the same actions become automatic. If a child darts in front of the car of an experienced driver, he does not have to think about the situation and decide to remove his foot from the accelerator and depress the brake and clutch. Upon seeing the child, he immediately and automatically stops the car. A complicated pattern of muscle movements is shifted from the highly conscious end of the spectrum over toward the involuntary end by the process of learning.

Whether activated voluntarily or involuntarily, given motor units are frequently called upon to serve many different functions. For example, one demand upon the muscles of the limbs, trunk, and neck is made by postural mechanisms; these muscles must support the weight of the body against gravity, control the position of the head and different parts of the body relative to each other to maintain equilibrium, and regain stable, upright posture after accidental or intentional shifts in position. Superimposed upon these basic postural requirements are the muscle movements associated with locomotion. For these purposes, the muscles must be capable of transporting the body from one place to another under the coordinated commands of neural mechanisms for alternate stepping movements and shifting the center of gravity. And added to the requirements of posture and locomotion can be the highly skilled movements of a ballerina or hockey player. The motor units are activated and the sometimes conflicting demands are settled, usually without any conscious, deliberate effort.

We now turn to an analysis of the individual components of the motor control system. We begin with local control mechanisms because their activity serves as a base upon which the descending pathways frequently exert their influence.

It should be realized that most information on the motor system has been obtained by planned experiments on frogs, cats, and monkeys. Similar information about human beings has been provided chiefly by accidental damage or disease of various parts of the nervous system; only rarely can experiments be performed during neurosurgical procedures. The information on motor control in human beings is not as explicit as that obtained from animal experiments, but it shows that data from animals can be applied to human beings with some reservations. The same neuroanatomic apparatus is present in human beings, cats, and monkeys, but the emphasis sometimes varies. Whenever possible, the physiology of neuromuscular control in human beings will be discussed.

LOCAL CONTROL OF MOTOR NEURONS

Much of the synaptic input to the motor neurons arises from neurons at the same level of the central nervous system as the motor neurons. Indeed, some of these neurons are activated by receptors in the very muscles controlled by the motor neurons and in other nearby muscles as well as the tendons associated with the muscles. These receptors monitor muscle length and tension and pass this information into the central nervous system via afferent nerve fibers. This input forms the afferent component of purely local reflexes which provide negative-feedback control over muscle length and tension. In addition, it is transmitted to the cerebral cortex, cerebellum, and basal ganglia where it can be integrated with input from other types of receptors.

Length Monitoring Systems and the Stretch Reflex

Embedded within skeletal muscle are stretch receptors which are made up of afferent-nerve endings wrapped around modified muscle cells, both partially enclosed in a fibrous capsule. The entire structure is called a *muscle spindle*. The modified muscle fibers within the spindle are known as *spindle* (or *intrafusal;* fusal = spindle) *fibers;* the typical skeletal muscle cells outside the spindle are the *skeletomotor* (or *extrafusal*) *fibers* (Fig. 10-1).

The muscle spindles are positioned in the muscle in such a way that passive stretch of the entire muscle pulls on the spindle fibers, stretching them and activating their receptors. Conversely, contraction of the skeletomotor fibers and the resultant shortening of the muscle release tension on the muscle spindle and slow down the rate of firing of the stretch receptor. The muscle spindle is complicated in that there are different kinds of spindle receptors, one responding to the magnitude of the stretch, another probably to both the absolute magnitude of the stretch and the speed with which it occurs. The importance of the kind of information relayed by the first receptor is apparent: It tells the central nervous system about the length of the muscle. However, by indicating the rate of change of the muscle length, the second type of receptor allows the central nervous system to anticipate the magnitude of the stretch. If the rate of stretch is increasing very rapidly, the stretch itself cannot stop immediately, and an additional change in length can be predicted. Although these receptors are separate anatomic and physiologic entities, they will be referred to collectively as the *muscle-spindle stretch receptors*.

When the afferent nerves from the muscle spindle enter the central nervous system, they divide into branches which can take several different paths. One group of terminals (*A* in Fig. 10-2) directly forms excitatory synapses upon the motor neurons going back to the muscle that was stretched, thereby completing a reflex arc known as the *stretch reflex*. This reflex is probably most familiar in the form of the knee jerk, which is tested as part of routine medical examinations. The physician taps on the patellar ligament, which is the lower part of the tendon that stretches over the knee and connects the quadriceps femoris muscle in the thigh to the tibia in the foreleg. As the tendon is

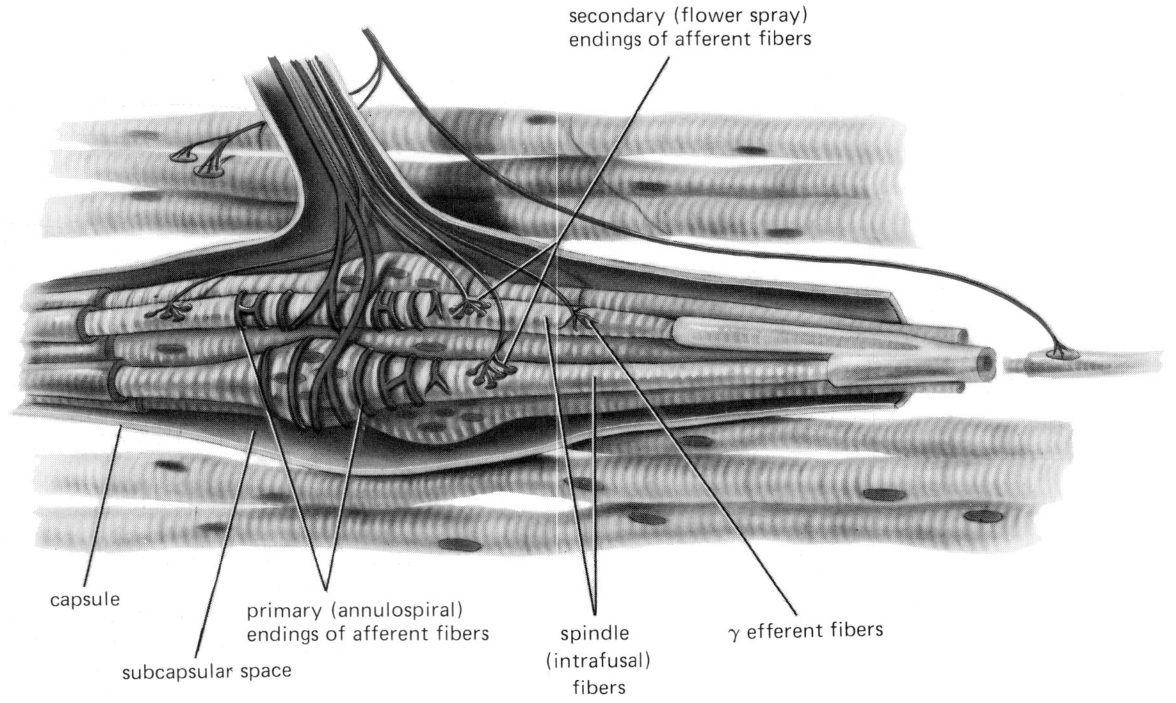

spindle capsule
afferent nerve fiber
skeletomotor (extrafusal)
muscle fiber
spindle (intrafusal)
muscle fiber

B

C

FIGURE 10-1 **(A)** ***Diagram of a muscle spindle.*** **(B)** ***Contraction of the skeletomotor fibers removes tension on the spindle stretch receptors and lowers the rate of firing in the afferent nerve.*** **(C)** ***Passive stretch of the skeletomotor fibers activates the spindle stretch receptors and causes a higher rate of firing in the afferent nerve. Arrows in*** **B** ***and*** **C** ***indicate the direction of force on the muscle spindles.*** **[Part A redrawn from P. A. Merton, How We Control the Contractions of Our Muscles, *Sci. Am.*, 226: 30, May (1972).]**

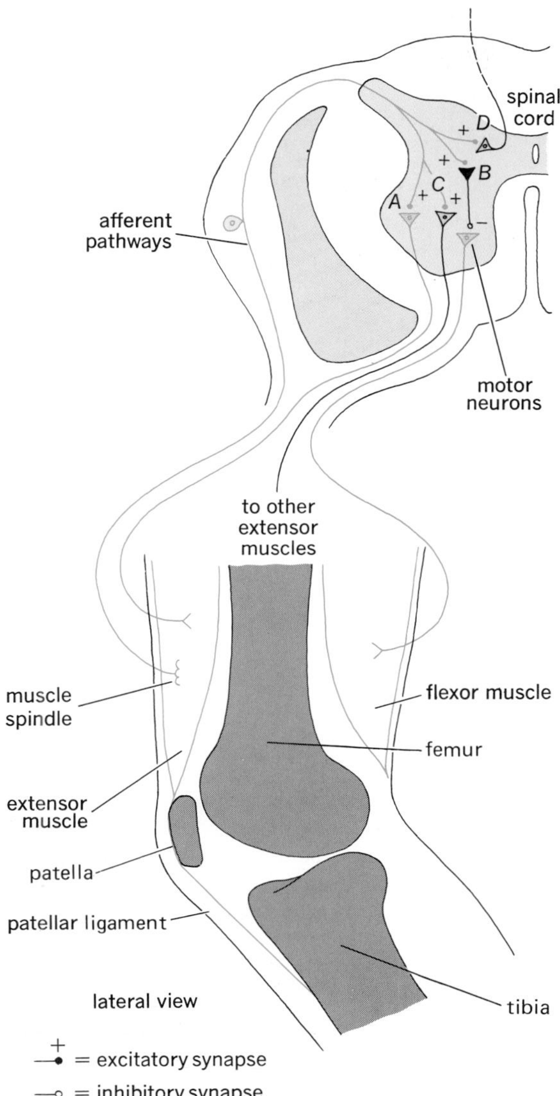

FIGURE 10-2 ***Terminals of the afferent fiber from the muscle spindle involved in the knee jerk.***

depressed, the quadriceps femoris is stretched, and the receptors in its muscle spindles are activated. Coded information about the change in length of the muscle is fed back to the motor neurons controlling the same muscle. The motor units are excited, and the patient's foreleg is raised to give the familiar knee jerk. The proper performance of the knee jerk tells the physician that the afferent limb of the reflex, the balance of synaptic input to the motor neuron, the motor neuron itself, the neuromuscular junction, and the muscle are functioning normally. The knee jerk nicely illustrates a major physiologic function of the stretch reflex; it permits the muscle to resist any passively induced change in its length and is a good example of a negative-feedback control system.

Because the group of afferent terminals mediating the stretch reflex synapses directly with the motor neurons without the interposition of any interneurons, the stretch reflex is called *monosynaptic*. Stretch reflexes are the only known monosynaptic reflex arcs in human beings; all other reflex arcs are *polysynaptic,* having at least one interneuron (and usually many) between the afferent and efferent pathways.

A second group of afferent terminals (*B* in Fig. 10-2) ends on interneurons which, when excited, inhibit the motor neurons controlling antagonistic muscles, whose contraction would interfere with the reflex response. For example, the normal response to the knee-jerk reflex is straightening of the knee to extend the foreleg. The antagonists to these extensor muscles are a group of flexor muscles which, when activated, draw the foreleg back and up against the thigh. If both opposing groups of muscles are activated simultaneously, the knee joint is immobilized and the leg becomes a stiff pillar. This is certainly what is required in some situations, but if the foreleg is to be extended from a flexed position, the motor neurons which activate the flexor muscles must be inhibited as the motor neurons controlling the extensor muscles are activated. The excitation of one muscle and the simultaneous inhibition of its antagonistic muscle is called *reciprocal innervation.*

A third group of terminals (*C* in Fig. 10-2) ends on interneurons which, when excited, activate *synergistic muscles,* i.e., muscles whose contraction assists the reflex motion. For example, in the knee jerk, interneurons facilitate motor neurons which control other leg extensor muscles.

A fourth group of afferent terminals (*D* in Fig. 10-2) synapses with interneurons which convey information about the muscle length to areas of the brain dealing with coordination of muscle movement. Although the muscle stretch receptors initiate activity in pathways eventually reaching the cerebral cortex, information relayed by these action potentials does not have a conscious correlate; rather the conscious aware-

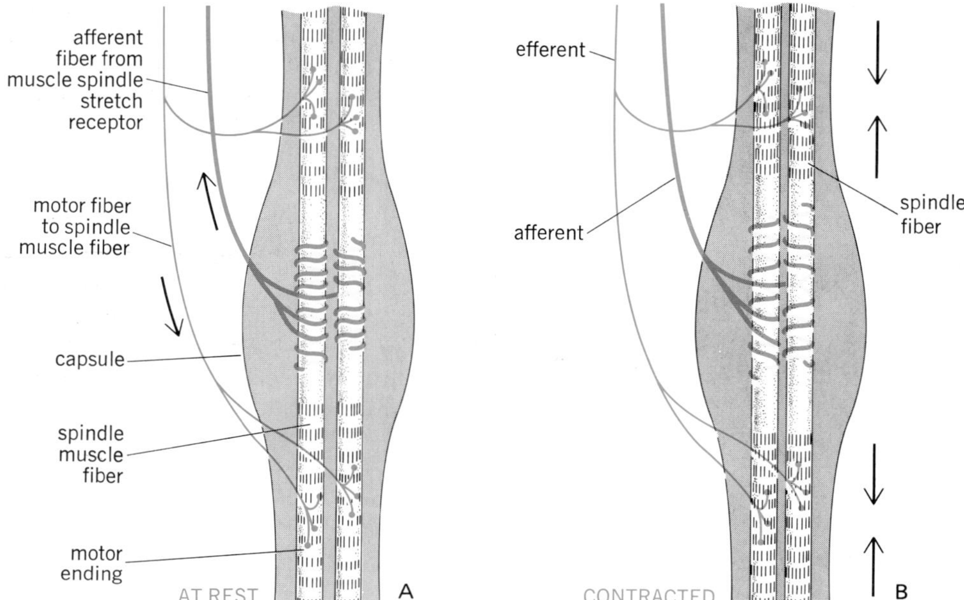

FIGURE 10-3 (A) ***Diagram of muscle-spindle innervation.*** (B) ***As the two striated ends of a spindle fiber contract, they pull on the center of the fiber and stretch the receptor, which is located there.*** **[From P. A. Merton, How We Control the Contractions of Our Muscles, *Sci. Am.*, 226: 30, May (1972).]**

ness of the position of a limb or joint comes from the joint, ligament, and skin receptors (Chap. 9).

Alpha-Gamma Coactivation The muscle spindles are structurally parallel to the large skeletomotor muscle fibers so that stretch on them is removed when the skeletomotor fibers contract (Figs. 10-1 and 10-3). If the spindle stretch receptors were permitted to shorten at this time they would stop firing action potentials and this important afferent information would be lost. To prevent this, the spindle muscle fibers themselves are frequently made to contract during the shortening of the skeletomotor fibers, thus maintaining tension in the spindle and firing in the receptors. The spindle fibers are not large and strong enough to shorten whole muscle and move joints; their sole job is to produce tension on the spindle stretch receptors. The muscle fibers in the spindles shorten in response to motor neuron activity (Fig. 10-3). The motor neurons which activate the spindle fibers are not, however, the same motor neurons which activate the skeletomotor muscle fibers. The motor neurons controlling the skeletomotor muscle fibers are larger and are classified as *alpha motor neurons;* the smaller neurons whose axons innervate the spindle fibers are known as the *gamma motor neurons*. The latter neurons are activated primarily by synaptic input from descending pathways.

In many voluntary and involuntary movements alpha and gamma motor neurons are *coactivated,* i.e., fired at almost the same time. To understand the usefulness of alpha-gamma coactivation, consider picking up a book whose weight is unknown. Suppose that the initially programmed strength of alpha motor-unit firing is not sufficient to lift the book. The skeletomotor muscle fibers will be unable to shorten but the spindle fibers, activated simultaneously by the gamma motor neurons, will shorten and the spindle receptors will be stretched. By way of the stretch reflex, the excitatory synaptic input to the alpha motor units will increase, causing summation of contraction, recruitment of additional motor units, and greater muscle tension. The overall route—descending pathway, gamma motor neuron, spindle muscle fiber, stretch receptor and afferent neuron, alpha motor neuron—is known as the *gamma loop* (Fig. 10-4). Thus the gamma loop provides a mechanism by which motor commands and muscle performance can be compared at the local level and compensation brought about on the spot. Moreover, information that the spindles are longer than expected (i.e., that the skeletomotor fibers have not shortened enough) will be transmitted to those higher brain centers involved in programming and controlling motor behavior so that they can alter their output as well.

Alpha-gamma coactivation works the other way too. If the initial program caused too intense alpha motor-unit activity, the book would be lifted too rapidly. The faster-than-expected shortening of the spindle fibers would remove tension from the spindles, stopping the receptors' firing. This would reflexly remove a component of excitatory input from the alpha motor neurons, automatically slowing the muscle movement to a more desirable rate. Thus, coactivation of alpha and gamma motor neurons can lead to fine degrees of regulation of muscle activity.

Tension Monitoring Systems

A second component of the local motor-control apparatus monitors tension rather than length. The receptors employed in this system are the *Golgi tendon organs,* which are encapsulated structures located in the tendon near its junction with the muscle. Endings of afferent-nerve fibers are wrapped around collagen bun-

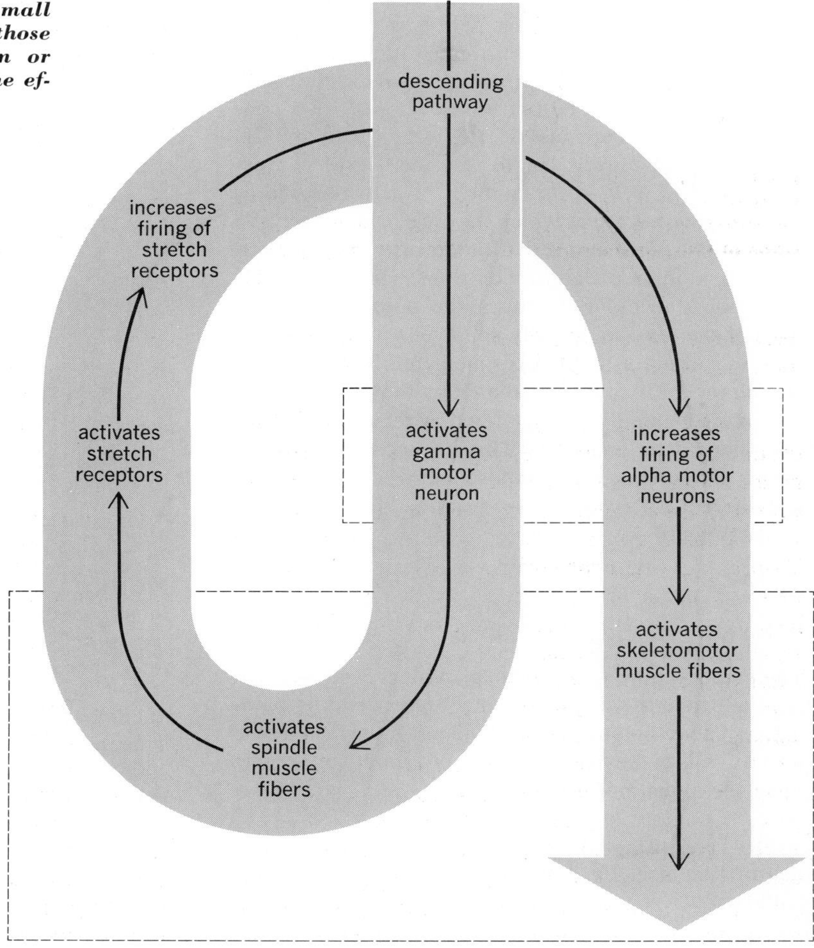

FIGURE 10-4 ***Gamma loop. The small rectangle at the center indicates those effects occurring in the brainstem or spinal cord; the large rectangle, the effects in the muscle.***

dles of the tendon, which are slightly bowed in the resting state. When the skeletomotor fibers of the attached muscle contract, they pull on the tendon, straightening the collagen bundles and distorting the receptor endings of the afferent nerves. The receptors fire in proportion to the increasing force or tension generated by a contracting muscle. Their activity results in the reflex inhibition of the motor neurons of the contracting muscle (Fig. 10-5). Some of the Golgi tendon organs have high thresholds and respond only when the tension is very high. These high-threshold receptors may function as safety valves, inhibiting the muscle when the force it generates is great enough to damage the limb. The remainder of the Golgi tendon organs have lower thresholds, comparable to the receptors of the muscle spindle, and they supply the motor control systems with continuous information about the tension generated. This information is necessary for effective movement because a given input to a group of motor neurons does not always provide the same amount of tension. The tension developed by a contracting muscle depends on the velocity of muscle shortening, the muscle length, and the degree of muscle fatigue as well as the number of activated motor neurons and the rate at which they are firing (Chap. 5). Because one set of inputs to the motor neurons can lead to a large number of different tensions, a feedback of information is necessary to inform the motor control systems of the tension actually achieved.

As is true for the reflexes mediated by the spindle stretch receptors, those mediated by the Golgi tendon organs are generally characterized by reciprocal effects on antagonistic muscles. This is accomplished by means of interneurons juxtaposed between the afferent and efferent pathways just as described for the stretch reflex (Fig. 10-5).

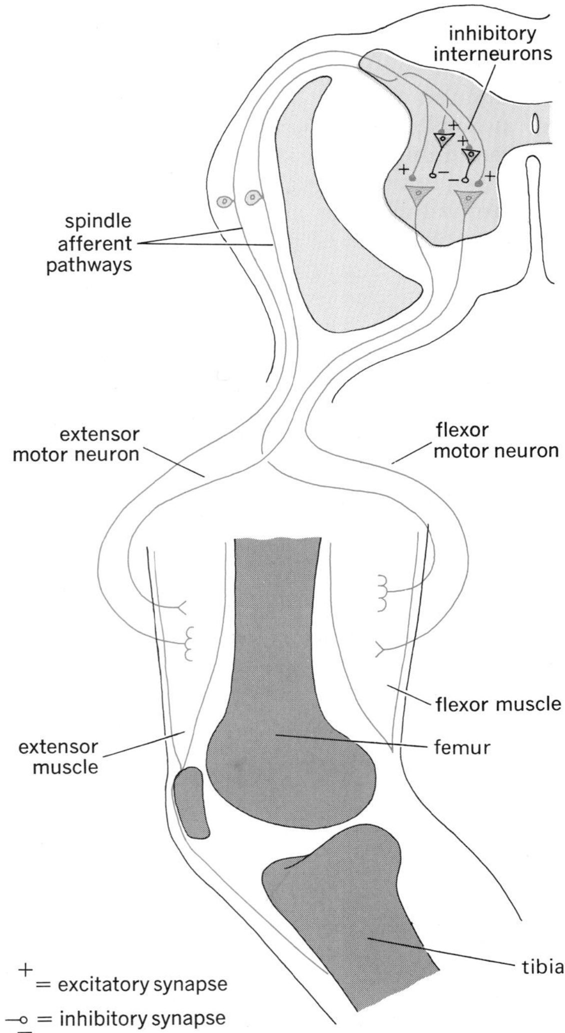

FIGURE 10-5 ***Golgi tendon organ component of the local control system.***

DESCENDING PATHWAYS AND THE BRAIN CENTERS WHICH CONTROL THEM

The cerebral cortex, subcortical centers, and cerebellum influence the motor neurons by descending pathways. There are three mechanisms by which these pathways alter the balance of synaptic input converging upon the alpha motor neurons:

1 By synapsing directly upon the alpha motor neurons themselves. This has the advantage of speed and specificity.

2 By synapsing on the gamma motor neurons, which, via the gamma loop, influence the alpha motor neurons. This pathway and number 1 above usually operate together (alpha-gamma coactivation). As described earlier, this has the advantage of maintaining output from the stretch receptors and providing a means for local on-the-spot compensation.

3 By synapsing on interneurons, often the same ones subserving the local reflexes. Although this route is not as fast as directly influencing the motor neurons, it has the advantage of the coordination built into the interneuron network as described earlier (e.g., reciprocal innervation).

The degree to which each of these three mechanisms is employed varies, depending upon the nature of the descending pathway, of which there are two major categories: the *corticospinal pathway* and the *multineuronal pathway*.

Corticospinal Pathway

The fibers of the corticospinal pathway, as the name implies, have their cell bodies in the cerebral cortex. The axons of these cortical neurons pass through the internal capsule of the cerebrum, through the pons, and into the medulla where about two-thirds of the fibers cross the midline and turn to descend on the opposite side of the spinal cord as the *lateral corticospinal tract*. Thus, the skeletal muscles on the left side of the body are controlled largely by neurons in the right half of the brain, and vice versa. The fibers that do not cross in the medulla pass straight into the spinal cord, forming the *ventral corticospinal tract*. (Most of these fibers eventually cross at their level of termination in the spinal cord.) Once they have left their cells of origin in the cortex, the fibers of the corticospinal pathway pass directly and without any additional synapsing to end in the immediate vicinity of the motor neurons (Fig. 10-6). The corticospinal pathways are also called the *pyramidal tracts* or *pyramidal system*, perhaps because they form an elevation (the pyramids) in the brainstem just before they cross to the opposite side or because they were formerly thought to arise solely from the giant pyramidal neurons of the cortex. The group of fibers controlling muscles of the eye, face, tongue, and throat branch away from these descending pathways in the brainstem as the *corticobulbar pathways*, which contact motor neurons whose axons travel out with the cranial nerves.

The corticospinal pathway is the major mediator of fine, intricate movements. However, it is not the sole mediator of such movements, since surgical section of this pathway in human beings does not completely eliminate them although it does make them weaker, slower, and less well coordinated. Clearly, the multineuronal pathway must also contribute to the performance of delicate movements.

The fibers of the corticospinal pathway end in all three of the ways described, i.e., on alpha motor neurons, gamma motor neurons, and interneurons. In

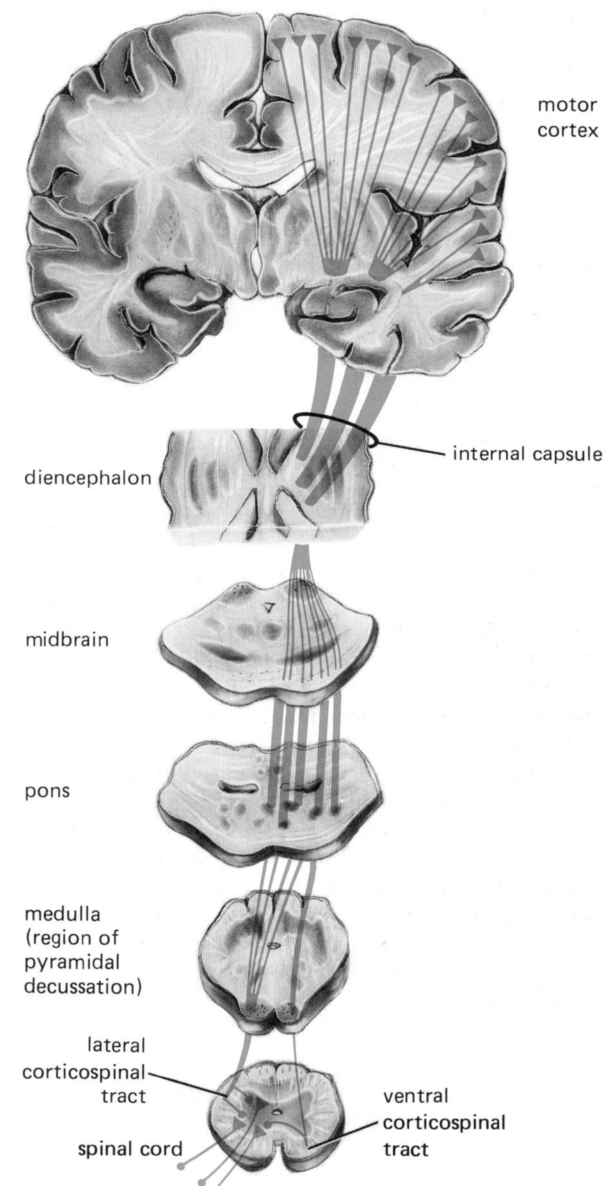

FIGURE 10-6 ***Diagram of the corticospinal pathway. In the medulla most of its fibers cross to descend in the opposite side of the spinal cord.*** **(From J. C. Eccles, "The Understanding of the Brain," p. 107, McGraw-Hill, New York 1973.**

addition they end presynaptically on the central terminals of afferent neurons and, through collateral branches, on neurons of the ascending afferent pathways. The overall effect of their input to afferent systems is to limit the area of skin, muscle, or joints allowed to influence the cortical neurons, thereby sharpening the focus of the afferent signal and improving the contrast between important and unimportant information. The collaterals also convey the information that a certain motor command is being delivered and possibly give rise to the sense of effort. Because of this descending (motor) control over ascending (sensory) information, there is clearly no real functional separation of these two systems.

Multineuronal Pathway

Some of the neurons of the cerebral cortex do not go directly to the region of the motor neurons; rather, they form the first link in the multineuronal pathway which ultimately goes to motor neurons of the brainstem and spinal cord. These cortical neurons synapse in many of the nuclei of the cerebrum and brainstem. These nuclei are often those which provide input to the basal ganglia and cerebellum or receive output from them. The names of the nuclei which finally give rise to the fibers which descend to the levels of the motor neurons are used in the names of the specific tracts of the pathway. For example, the multineuronal pathways include *vestibulospinal tracts* from the vestibular nuclei, *reticulospinal tracts* from the reticular formation, etc. The pathway is thus made up of a chain of functionally related neurons, all integrating and transmitting information about control of the skeletal muscles. The final neuron in the chain ends on either interneurons or the gamma motor neurons.

Despite the distinctions between the corticospinal and multineuronal pathways, it is wrong to imagine a complete separation of function. All movements, whether automatic or voluntary, require the continual coordinated interaction of both pathways.

Cerebral Cortex

Many areas of cerebral cortex give rise to the two descending pathways described above, but a large number of the fibers come from the posterior part of the frontal lobe, which is therefore called the *motor cortex* (Fig. 10-7). The function of the neurons in the motor cortex varies with position in the cortex. As one starts at the top of the brain and moves down along the side (*A* to *B* in Fig. 10-7), the cortical neurons lie in such a way

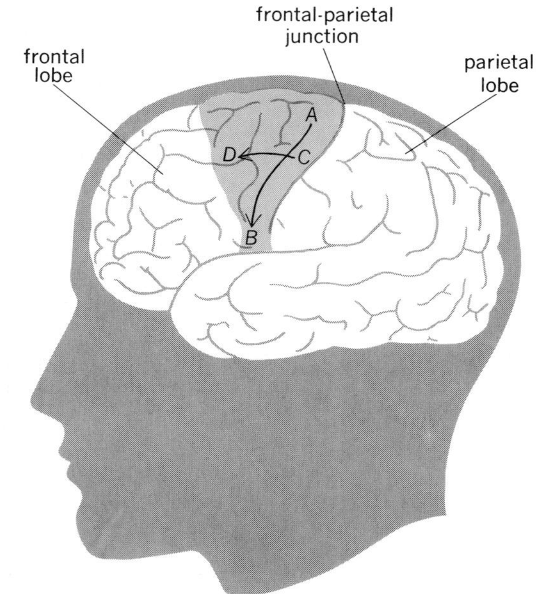

FIGURE 10-7 ***Regions of the motor cortex.***

that neurons affecting movements of the toes and feet are at the top of the brain, followed (as one moves laterally along the surface of the brain) by neurons controlling leg, trunk, arm, hand, fingers, neck, and face. The size of each of the individual body parts in Fig. 10-8 is proportional to the amount of cortex devoted to its control; clearly, the cortical areas representing hand and face are the largest. The great number of cortical neurons for innervation of the hand and face is one of the factors responsible for the fine degree of motor control that can be exerted over those parts.

The neurons also change character as one explores from the back of the motor cortex forward (*C* to *D* in Fig. 10-7). Those in the back portion, i.e., closest to the junction of the frontal and parietal lobes, mainly contribute to the corticospinal pathway. Moving anteriorally (forward) in the motor cortex, this zone of neurons gradually blends into the group which forms the first link in the multineuronal pathway. However, none of these cortical neurons functions as an isolated unit. Rather, they are interconnected so that those cortical neurons controlling motor units having related functions fire together. The more delicate the function of any skeletal muscle, the more nearly one-to-one is the relationship between cortical neurons and the muscle's motor units.

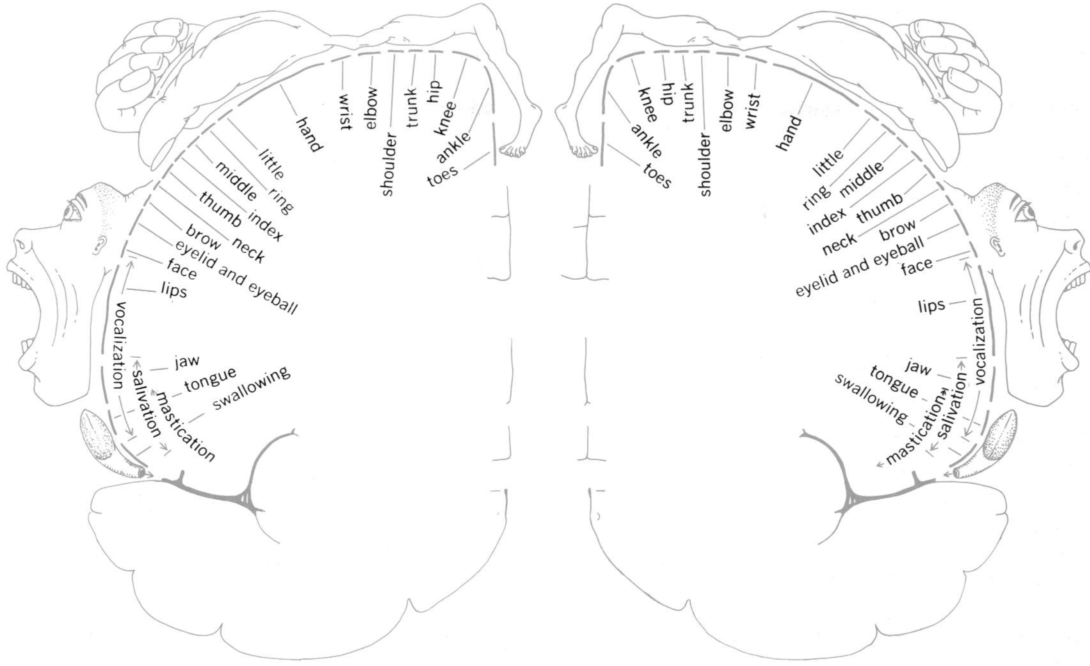

FIGURE 10-8 ***Arrangement of the motor cortex.***

Throughout this description we have presented the cerebral cortex as the origin of the descending pathways to motor neurons. But how are the action potentials which travel in these descending pathways initiated? As in most other neurons, the firing pattern of these cells is determined solely by the balance of synaptic activity impinging upon them (there is no evidence for the spontaneous generation of action potentials in these cells). What, then, are the inputs to these cortical cells? First, they receive a large amount of information from the receptors listed earlier in this chapter. Given this input, one might visualize the cortical cells as the integrators of long reflex arcs initiated by the stimulation of receptors, and, indeed, some movement is explainable in these terms. Yet such explanations do not seem to account for the initiation of movement by conscious intention. For example, if the neurons of the motor cortex are stimulated in conscious patients during surgical exposure of the cortex, the patient moves but not in a purposefully organized way. The movement made depends upon the part of the motor cortex stimulated. The patient is aware of the movement but says, "You made me do that," recognizing that it is not a voluntary movement. Stimulation of parts of cortex in the parietal lobe sometimes causes patients to say that they want to make a certain movement, but the movement does not occur.

Somehow, in the brain, and it is not known how, immediate afferent information must be associated with data acquired from past experiences and with the neural activity of present feelings (I want to, I feel like doing, etc.). It is the output of this synthesis which ultimately commands the motor cortex. Electrodes placed on the skull to record the electric activity of the brain during the "decision-making period" begin to pick up distinctive activity patterns about 800 ms (0.8 s) before the movement begins. This *readiness potential* is recorded over wide areas of association cortex of both hemispheres. This widespread activity feeds into the motor cortex and also into the cerebellum and basal ganglia; 50 to 60 ms prior to the movement the electric activity changes to a sharper *motor potential* localized over the motor-cortex region of the hemisphere controlling the involved muscles. Thus, the motor cortex is not the prime initiator of the movement; it is simply a tremendously important final relay station of the cortex.

Finally, the input to the cortex neurons not only must be directed toward some purpose but must select from a variety of ways of achieving that purpose. For a simple example, a rat trained to press a lever whenever a signal light flashes will do so quite consistently but, depending upon its original position in the cage, its movements can be quite varied. If the rat is to the left of the lever, it moves to the right; if it is to the right, it moves to the left. If its paw is on the floor, it raises the paw; if its paw is above the lever, it lowers the paw. The only consistent act is pressing the lever. The movements performed to achieve the purpose are variable and seem almost inconsequential. Yet, although it is the end result that matters, only the intervening acts or movements can be programmed by the nervous system. How a given program is selected is not known.

Subcortical Centers: Basal Ganglia, Brainstem Nuclei, and Brainstem Reticular Formation

The multineuronal motor pathway descending from the cortex synapses in the basal ganglia, brainstem nuclei, and brainstem reticular formation. These neuronal clusters, or centers, have important roles in the control of postural mechanisms and coordination of the many simultaneous movements of locomotion. They also serve to correlate fine, detailed voluntary movements with the appropriate postural mechanisms upon which these movements are superimposed. In addition, the basal ganglia serve a special role in the voluntary control of slow, smooth movements. These structures send information back to the motor cortex, which is involved with both fast and slow movements, to modulate its output. These functions become apparent when the basal ganglia are damaged or diseased (for example, Parkinson's disease), as the patient exhibits excessive and disorganized movements and has a marked defect in the voluntary production of smooth motions of different speeds.

Cerebellum

The cerebellum sits on top of the brainstem, as can be seen in Fig. 8-10. It consists of two *cerebellar hemispheres* joined by a median *vermis;* its surface is marked by numerous parallel fissures. The *cerebellar cortex* is a convoluted shell of gray matter over a core of nerve fibers ascending to and descending from it. Purkinje-cell neurons of the cerebellar cortex provide the sole output pathway from the cortex; they synapse in deep nuclei in the white matter at the base of the cerebellum where final program adjustments are made before the cerebellar output is transmitted to specific nuclei in the brainstem and diencephalon. The neural connections between the cerebellum and the rest of the brain pass through the three pairs of cerebellar peduncles (Chap. 8). The cerebellum does not initiate movement but acts by influencing other regions of the brain responsible for motor activity. Destruction of the cerebellum does not cause the loss of any specific movement; instead it is associated with a general inadequacy of that movement. The disturbance in coordinated movement is on the same side of the body as the cerebellar damage; i.e., the abnormal movement is on the left side if the cerebellar damage is on the left. The main problems of persons with cerebellar damage are as follows:

1 They cannot perform movements smoothly. If they try to grasp an object with the hand, the movement is jerky and is accompanied by oscillating, to-and-fro tremors which become more marked as they approach the object. When their limbs are at rest, they are steady and motionless; but if they try to move them for any reason, even to help maintain balance, they go into oscillations which are sometimes quite wild. This oscillating tremor is also known as *intention tremor* and is classically associated with cerebellar damage. It is opposite to the resting tremor that occurs with basal ganglion disease.

2 Persons suffering from cerebellar damage walk awkwardly with the feet well apart. They have such difficulty maintaining their balance that their gait is reeling and drunken.

3 They cannot start or stop movements quickly or easily and, if asked to rotate the wrist back and forth as rapidly as possible, their motions are slow and irregular.

4 They may not be able to combine the movements of several joints into a single, smooth, coordinated motion. To move the arm, they might first move the shoulder, then the elbow, and finally the wrist.

In cases of severe cerebellar damage, the combined difficulties of poor balance and unsteady movements may become so great that the person is incapable of walking or even standing alone. Since speech depends on the intricately timed coordination of many muscle movements, cerebellar damage is accompanied by speech disturbances. The person with cerebellar damage shows no evidence whatsoever of sensory or intellectual deficits.

From this discussion, the reader will deduce that

the cerebellum is involved in the control of muscles utilized both in maintaining steady posture and in effecting coordinated, detailed movements. It receives input from both cortex and subcortical centers with information about what the muscles *should* be doing and from many afferent systems with information about what the muscles *are* doing. If there is a discrepancy between the two, an error signal is sent from the cerebellum to the cortex and subcortical centers where new commands are initiated to decrease the discrepancy and smooth the motion.

The cerebellum plays a special role in the control of rapid movements (and thus serves as the counterpart to the basal ganglia, which influence slow movements). Rapid movements are largely "preprogrammed" in their entirety, rather than modified during their course as slower movements are. Such preprogramming requires calculation of the time needed for the movement (taking into account the particular amount of muscle force necessary in any special situation) and integration of these data with information about the moved structure's resting position and final location. The cerebellum performs this function and, like the basal ganglia, sends the information to the motor cortex.

The afferent inputs to the cerebellum come from the vestibular system, eyes, ears, skin, muscles, joints, and tendons, i.e., from the major receptors affected by movement. Inputs from receptors in a single small area of the body end in the same region of the cerebellum as do the inputs from the higher brain centers controlling the motor units in that same area. Thus information from the muscles, tendons, and skin of the arm arrive at the same area of cerebellar cortex as the motor commands for "arm" from the cerebral cortex. This permits the cerebellum to compare motor commands with muscle performance.

This completes our analysis of the components of motor control systems. We now analyze the interactions of these components in two situations: maintenance of upright posture and walking.

MAINTENANCE OF UPRIGHT POSTURE AND BALANCE

The skeleton supporting the body is a system of long bones and a many-jointed spine which cannot stand alone against the forces of gravity. Even when held together with ligaments and covered with flesh, it cannot stand erect unless there is coordinated muscular activity. This applies not only to the support of the body as a whole but also to the fixation of segments of the body on adjoining segments, e.g., the support of the head, which is held erect without any conscious effort and without fatigue in a normal awake person. A decrease in postural fixation of the head is seen when someone falls asleep sitting up; the head nods until the chin is on or near the chest.

Added to the problem of supporting one's own weight against gravity is that of maintaining equilibrium. A human being is a very tall structure balanced on a relatively small base, and the center of gravity is quite high, being situated just below the small of the back. For stability, the center of gravity must be kept within the small area determined by the vertical projection of that base. Yet human beings are almost always in motion, swaying back and forth and side to side even when standing still. Clearly, they often operate under conditions of unstable equilibrium and would be toppled easily by physical forces in the environment if their equilibrium were not protected by reflex postural mechanisms.

The maintenance of posture and balance is accomplished by means of complex counteracting reflexes, all the components of which we have met previously. The efferent arc of the reflexes is, of course, the alpha motor neurons to the skeletal muscle. The major coordinating centers are the basal ganglia, brainstem nuclei, and reticular formation which influence the motor neurons mainly via the descending multineuronal pathway. What is the source of afferent input? One might predict the existence of "center-of-gravity" receptors but, in fact, no such receptors exist. Rather, information about the location of the center of gravity is given by the integration of all the afferent signals from muscles, joints, skin, vestibular system, and eyes. This integration provides the subcortical coordinating centers with a "map" of the position of the whole body in space.

In such a system it is extremely difficult to assign a certain percentage of importance to any one afferent system, even when the exact conditions are specified. However, it does seem that in human beings, under conditions which permit vision, visual information is probably most important. Yet, so influential are the other inputs and so adaptable is the overall system that a blind person maintains balance quite well, with only slight loss of precision. Moreover, with changing circumstances, the balance of importance between different afferent inputs may change considerably.

Let us take the vestibular system as an example. Despite the fact that the vestibular organs are called the sense organs of balance, persons whose vestibular mechanisms have been completely destroyed may have very little disability in everyday life (one such man was even able to ride a motorcycle). Such persons are not seriously handicapped as long as their visual system, joint-position receptors, and cutaneous receptors are functioning. However, they do have difficulty walking in darkness over uneven ground or walking down stairs, where they cannot see a point immediately in front of their feet for visual reference. Thus, in a normal person vestibular input must increase in importance under such circumstances. Finally, vestibular information provides the only clue to orientation with respect to gravity when one is swimming under water, where visual and skin and joint input is confused.

Receptors in the joints and associated ligaments and tendons and their pathways through the nervous system play an important role in the unconscious control of posture and movement and also give rise to the conscious awareness of the position and movement of the joints (Chap. 9). Recall that joint receptors are accurate indicators of movement and position. Skin receptors sensing contact of the body with other surfaces also play a role in regulation of body posture. That both cutaneous and joint-position receptors contribute to postural mechanisms can be shown by the following tests.

Walking in the dark, one's gait is halting and uncertain; stability and confidence are greatly improved by the simple act of running a fingertip along a wall. The fingertip certainly provides no physical support, but it adds significant afferent information to the coordinating centers in the subcortical centers and cerebellum. The joint-position receptors add information too. When a blindfolded person lacking a vestibular system is tilted from an upright position, the trunk does not make appropriate motions to counteract the tilt, but the limbs do so as a result of stimulation of the joint-position receptors in the moving joints. Moreover, the movements are similar to those a normal person would make (but they are not as vigorous; they do not always affect all limbs and may not prevent falling).

Several more examples will further illustrate how appropriate responses usually require the integration of more than one input. Normal subjects show large postural reflex responses to instability of the base upon which they are seated or standing; but under these conditions, both vestibular receptors and joint-position receptors are excited. If the vestibular system alone is stimulated by moving the head sharply in any direction, there is no reflex movement of the trunk or limbs. It seems, therefore, that the postural reflexes occur only when the *body* is unstable. In the central nervous system the integration of synaptic information determines whether the stimuli from the vestibula are to excite the postural mechanisms or not, a process which at one time allows the vestibular system to excite these large reflex movements and at another time when they would be inappropriate shuts them out.

The two figures in Fig. 10-9 illustrate this point. Man A is clearly in danger of falling and is in need of muscle movements to keep him upright. Consider the synaptic input to the central nervous system: The vestibular inputs from the two sets of semicircular canals are different, the inputs from the joints and ligaments of the limbs on the two sides of his body are different, and the joint receptors from the section of vertebral column that runs through his neck indicate that his neck is not bent. The final integration of all these synaptic inputs leads to the initiation of postural reflex responses. Man B is not in danger of falling even though the input from his vestibular systems is exactly the same as that of man A, because the total synaptic input to the central ner-

FIGURE 10-9 ***Interaction of the vestibular and skin, joint, and tenson receptors.***

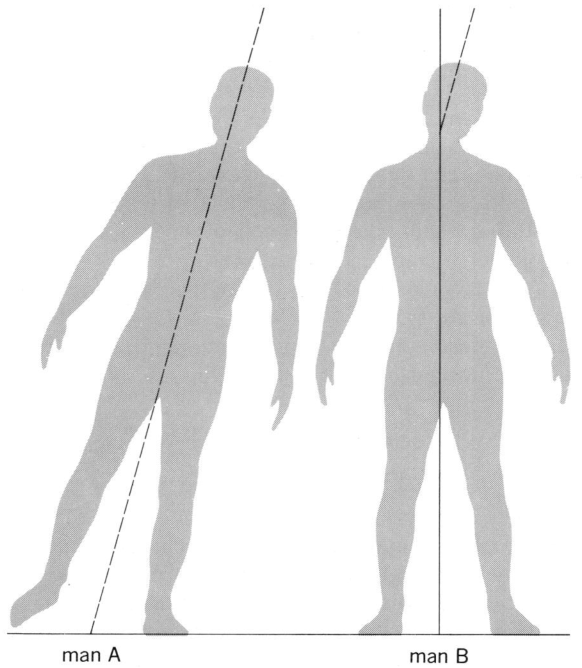

vous system is quite different. There is no imbalance of input from his limb joint receptors, but there is imbalance from his neck joint receptors. The information in man A indicates instability; that in man B does not. The reactions which would be initiated by vestibular stimulation do not occur unless the input from joint receptors indicates that the body is unstable.

Finally, we must point out the special contribution of the muscle-spindle (stretch) receptors to the maintenance of upright posture against the force of gravity. These length-monitoring receptors provide information, via afferent and ascending pathways, to the higher brain centers but, in addition, they initiate the locally occurring stretch reflex. Imagine a woman standing upright; gravity causes her knees to begin to buckle. As this occurs the patellar tendon and extensor muscles are stretched and the stretch reflex is elicited, resulting in increased contraction of the extensor muscles to straighten the knee and prevent her from falling.

From the above description, it might seem (incorrectly) that the stretch reflexes, by themselves, could maintain upright position against the forces of gravity. In fact, their input is effective in enhancing motor activity only where there is a simultaneously occurring facilitation of the alpha motor neurons by descending pathways. Indeed, the providing of such facilitation is a major function of the multineuronal pathway. As might be predicted, this pathway generally facilitates preferentially the motor neurons to the so-called antigravity muscles, for example, the extensor muscles of the legs.

Finally, what happens if all the postural reflexes, once initiated, are unsuccessful and the person loses her balance? If a person is tilted, the arm on the lower side is pulled toward the body, but if the tilting goes so far that she is in danger of losing her balance, the arm on the lower side is quickly extended to break the fall. The first reflex pattern in response to tilt breaks up and a different set of reflexes takes over as soon as there is no hope of restoring balance. Indeed, it is just such a transformation that accounts for stepping in walking.

WALKING

Postural fixation of the body is intimately related to the problems of locomotion and maintaining equilibrium in the course of movement. In considering locomotion, the need for some structure or mechanism capable of carrying the body along is added to the basic requirement of antigravity support of the body. In walking, the human body is balanced on the very small base provided by one foot. The weight of the body is supported on each leg alternately, and to accomplish this the body moves from side to side in such a way that the center of gravity is alternately poised over the right and then the left leg. Only when the center of gravity is shifted over the right leg can the left foot be raised from the ground and advanced. As the left foot is lifted, the trunk of the body sways to the right to counterbalance the weight of the left leg (Fig. 10-10). It must be apparent that strict and delicate control of the center of gravity is essential to permit these movements without loss of equilibrium.

Effective stepping can be evoked in newborn infants if the child is held with the feet set on a surface and the legs supporting the weight of the body. The stepping movements can be initiated by simply tilting the child's body forward and rocking it slightly from side to side. Thus, small shifts in the center of gravity start the coordinated, alternate flexion and extension of the leg involved in purposeful stepping.

The stimulus necessary to trigger the stepping is a slight forward tilt of the body. When a child or adult takes a step, the weight of the body is shifted to one foot, and the opposite foot is lifted from the ground. The body is allowed to fall forward and loses its equilibrium until it is caught on the leg which has swung forward. During this process, the center of gravity has moved both sideways and forward from its original position over one leg to a similar position over the other leg. This action is repeated rhythmically, and its continuation depends on both components of the shift in the center of gravity, the forward shift, which causes the fall forward and the sideways shift, which allows one foot to be lifted and advanced. Thus, there are four necessary components for locomotion: antigravity support of the body, stepping, control of the center of gravity to provide equilibrium, and a means of acquiring forward motion. All four of these components must be present simultaneously and continuously. It would obviously be futile to apply forward motion without an adequate stepping mechanism. An interesting point is that the forward fall serves to both provide the advance in position and evoke stepping. In fact, if a person leans forward beyond a certain point, he must either take a step or fall, and in this case the stepping is part of a protective reflex.

In persons with diseases of the basal ganglia, disturbances in locomotion often occur, and occasionally the disorders may become so great that the patients become immobilized. Such patients can stand and can make rhythmic, alternate stepping movements, but

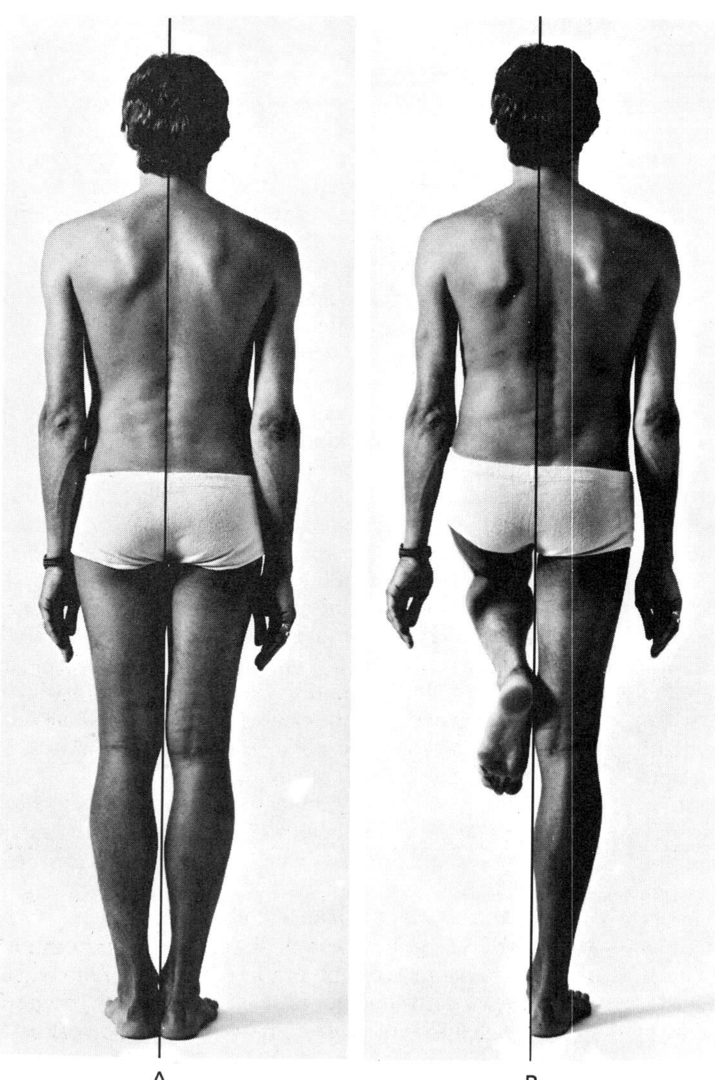

FIGURE 10-10 ***Postural changes with stepping.*** **(A)** ***Normal standing posture.*** **(B)** ***As the left foot is raised, the whole body leans toward the right so that the center of gravity is vertically above the right foot.***

they cannot walk. One such patient with diseased basal ganglia, who walks only poorly, is able to get along very well if she carries a 14-lb chair in front of her. Her disorder is in the basal ganglion mechanisms which control the forward shift in the center of gravity, and she cannot lean forward to acquire the forward progression necessary for locomotion. The weight of the chair has the effect of bringing the center of gravity forward. Her disorder is only in the ability to tilt her body forward; she does not lack the ability to rock her body from side to side. Other patients have good front-to-back control but lack adequate lateral control.

CHAPTER 11

The Nervous System: IV. Consciousness and Behavior

Despite the fact that the word behavior is commonly used to refer to those actions which are external, readily visible events, it really denotes the response of individuals to their total environment, both external and internal. Thus behavior may be defined simply as anything a person does, including both somatic and autonomic events.

CONSCIOUSNESS

The term *consciousness* includes two distinct concepts, *states of consciousness* and *conscious experience*. The second concept refers to those things of which a person is aware—thoughts, feelings, perceptions, ideas, dreams, reasoning—during any of the states of consciousness. The state of consciousness, i.e., whether awake, asleep, drowsy, etc., is defined both by behavior, covering the spectrum from coma to maximum attentiveness, and by the pattern of brain activity that can be recorded electrically, usually as the electric-potential difference between two points on the scalp. This record is the *electroencephalogram* (EEG).

The wavelike pattern of the EEG changes in frequency and amplitude as behavior changes from attentive alertness through quiet resting and drowsiness to sleep (Fig. 11-1). The EEG is produced by the intermittent synchronization of the electric activity of small groups of neurons of the cerebral cortex. The basic units of this electric activity are thought to be individual synaptic potentials (or groups of them) rather than action potentials. These active synapses are driven by some subcortical center, possibly the thalamus. Cortical neurons other than those synchronized at the moment also have fluctuating membrane potentials due

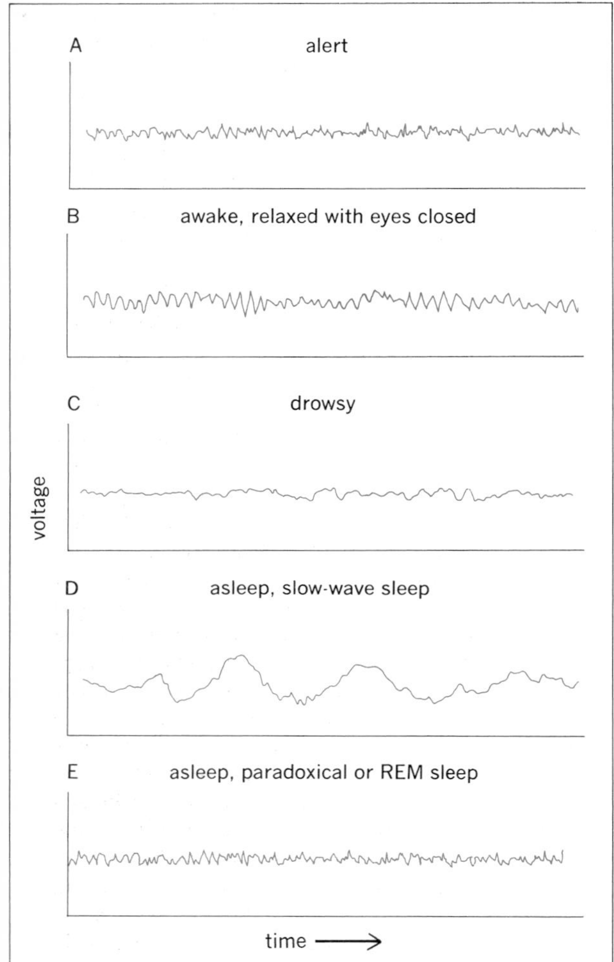

FIGURE 11-1 ***EEG patterns corresponding to various states of consciousness.***

to synaptic input, but their activity is not synchronized and, therefore, the potentials tend to cancel each other out. Periodically, new groups of neurons are synchronized.

The EEG is a useful clinical tool because the normal patterns are altered over brain areas that are diseased or damaged. It is also useful in defining states of consciousness, but it is not known what function, if any, this electric activity serves in the brain's task of information processing. We do not know whether these electric waves actually influence brain activity or whether they are merely epiphenomena. (An epiphenomenon is a phenomenon which occurs with an event but is not causally related to it, for example, the sound of a baseball bat striking a ball. The sound results from the impact but does not influence how far the ball will travel.)

States of Consciousness

Waking State and Arousal Behaviorally, the waking state is far from homogeneous, comprising the infinite variety of things one can be doing. The prominent EEG wave pattern of an awake, relaxed adult whose eyes are closed is a slow oscillation of 8 to 13 Hz, known as the *alpha rhythm* (Fig. 11-1B). Each individual has a characteristic pattern of alpha rhythm, and so does each region of the brain. The alpha rhythms are nearly always larger at the back of the head over the area of visual cortex, and they are also larger when people are not thinking.

When people are attentive to an external stimulus (or are thinking hard about something), the alpha rhythm is replaced by lower, faster oscillations (Fig. 11-1A). This transformation is known as *EEG arousal* and is associated with the act of attending to stimuli rather than with the perception itself; for example, if people open their eyes in a completely dark room and try to see, EEG arousal occurs. Also, with decreasing attention to repeated stimuli, the EEG pattern reverts to the alpha rhythm. Thus, the alpha rhythm with its larger and more regular oscillations is associated with decreased levels of attention.

When alpha rhythms are being generated, subjects commonly report that they feel relaxed and happy. A high degree of alpha rhythm is also associated with meditational states. However, people who normally experience high numbers of alpha episodes have not been shown to be psychologically different from others with lower levels, and the relation between brain-wave activity and subjective mood is obscure. People have been trained to increase the amount of alpha brain rhythms by providing them with a feedback signal such as a tone whenever alpha rhythm appears in their EEG. Increasing the number of alpha episodes has been used for the control of some kinds of chronic pain. This is successful in some cases, possibly because (1) the alpha training distracts attention from the pain to the feedback signal and inner feelings, (2) the patient believes that the method works, (3) the relaxation associated with alpha episodes decreases the patient's anxiety, or (4) awareness of the possibility of control over the pain changes its meaning and, therefore, the response to it.

Sleep Although average people spend about one-third of their lives sleeping, we know little of the functions served by it. We do know that sleep is an active process and not a mere absence of wakefulness. Moreover, it is not a single simple phenomenon; there are two distinct states of sleep characterized by different EEG and behavior patterns.

The EEG pattern changes profoundly in sleep. As a person becomes drowsy, the alpha rhythm is gradually replaced by irregular, low-voltage potential differences (Fig. 11-1C), and as sleep deepens, the EEG waves become slower, larger, and more irregular (Fig. 11-1D). This *slow-wave sleep* is periodically interrupted by episodes of *paradoxical sleep,* during which the subject still seems asleep but has an EEG pattern similar to that of EEG arousal, i.e., an awake, alert person (Fig. 11-1E).

Paradoxical sleep and slow-wave sleep are differentiated by behavioral criteria as well. During slow-wave sleep these criteria are not clear-cut, and it is difficult to tell precisely when a person passes from drowsiness into slow-wave sleep. There is considerable tonus in postural muscles and only a small change in cardiovascular or respiratory activity. The sleeper can be awakened fairly easily during slow-wave sleep, and if awakened, rarely reports dreaming. Slow-wave sleep has a characteristic kind of mentation but is described by subjects as "thoughts" rather than "dreams." The thoughts are more plausible and conceptual, and more concerned with recent events of everyday life and more like waking-state thoughts than are true dreams.

During episodes of paradoxical sleep, on the other hand, the behavioral criteria are precise. At the onset of paradoxical sleep, there is an abrupt and complete inhibition of tone in the postural muscles, although periodic episodes of twitching of the facial muscles and limbs and rapid eye movements behind the closed lids occur. (Paradoxical sleep is therefore also called *rapid-eye-movement* or *REM sleep*.) Respiration and heart rate are irregular, and blood pressure may go up or down. When awakened during paradoxical sleep, 80 to 90 percent of the time subjects report that they have been dreaming.

Continuous recordings show that the two states of sleep follow a regular 30- to 90-min cycle, each episode of paradoxical sleep lasting 10 to 15 min. Thus, slow-wave sleep constitutes about 80 percent of the total sleeping time in adults, and paradoxical sleep about 20 percent. The time spent in paradoxical sleep increases toward the end of an undisturbed night. Normally it is not possible to pass directly from the waking state to an episode of paradoxical sleep; it is entered only after at least 30 min of slow-wave sleep. Thus, subjects awakened at the beginning of every period of paradoxical sleep can be prevented from spending much time in that state although their total sleeping time remains approximately normal. After being deprived of paradoxical sleep for several nights, all subjects spend a greater than usual proportion of time in paradoxical sleep the next time they sleep. Thus, the total number of hours spent in paradoxical sleep tends to remain constant.

What is the functional significance of sleep? What happens to the brain during sleep? We now know that the brain, as a whole, does not rest during sleep and there is no generalized inhibition of activity of cerebral neurons. On the contrary, there is a considerable amount of neuronal activity during slow-wave sleep, and many areas of the brain are more active during paradoxical sleep than they are during waking. The blood flow and oxygen consumption of the brain, signs of its metabolic activity, do not decrease in sleep.

However, during sleep there is a change in distribution or reorganization of neuronal activity, some individual neurons being less active than during waking although the brain as a whole remains relatively active. Although sleep is not a period of generalized rest for the whole brain, it may represent a period of rest for certain specific elements, during which they can replenish substrates necessary for their generation of action potentials. Yet, when isolated neural tissue is exposed to extreme rates of stimulation far exceeding those occurring under physiologic circumstances, neurons recover within a period of minutes. Alternatively, it has been suggested that the functional significance of sleep lies not in short-term recovery but in the relatively long-term chemical and structural changes that the brain must undergo to make learning and memory possible.

Role of Reticular Formation The prevailing state of consciousness is the resultant of the interplay between three neuronal systems, one causing arousal and the other two sleep, all three of which are parts of the reticular formation. The reticular formation lies in the central core of the brainstem in the midst of the neural pathways ascending and descending between the brain and spinal cord, and the neurons of the reticular formation receive a continuous sample of the neural activity in these pathways, including information from (1) areas of the cerebrum (cerebral cortex, basal ganglia, limbic system, and other regions deep within the cerebrum); (2) spinal cord; and (3) cerebellum and other brainstem structures (Fig. 11-2A). The output of the reticular formation neurons is determined by these inputs as well as by spontaneous activity generated in the reticular formation itself. The reticular formation projects to (1) the spinal cord, (2) the cerebellum, and (3) the subcortical and cortical areas of the cerebrum.

FIGURE 11-2 (A) ***Convergence of descending, local, and ascending influences upon reticular formation*** (shaded area). (B) ***Projections from reticular formation to spinal cord, brainstem and cerebellum, and cerebrum.*** (Adapted from Livingston.)

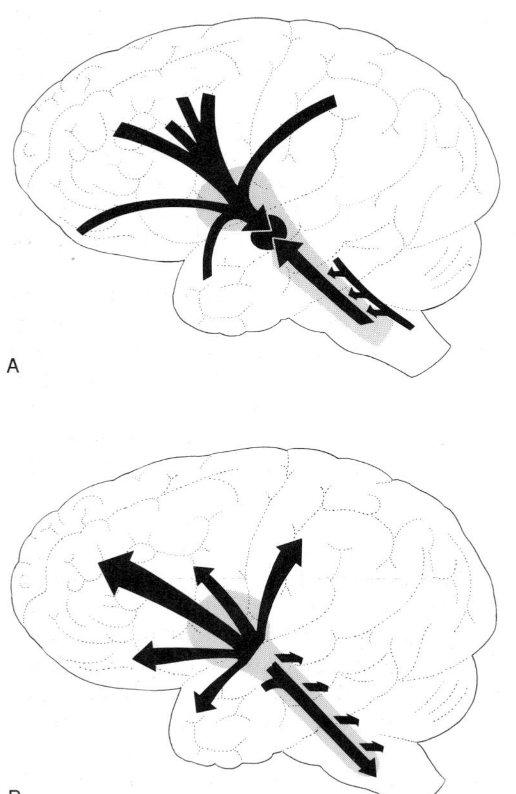

There are also a great many synaptic endings in the brainstem, so that the reticular formation acts upon itself (Fig. 11-2B). Thus, the reticular formation influences and is influenced by virtually all areas of the central nervous system.

However, the reticular formation is not homogeneous, and discrete areas frequently have specific functions: It helps to coordinate skeletal muscle activity (Chap. 10); it contains the primary cardiovascular and respiratory control centers (Chaps. 16 and 17); it monitors the huge number of messages ascending and descending through the central nervous system (Chaps. 9 and 10). In this chapter we are concerned with its role in determining states of consciousness.

Reticular Activating System In 1934, it was discovered that the EEG of a cerebrum surgically isolated from the spinal cord and lower three-fourths of the brainstem loses the wave patterns typical of an awake animal, indicating that some neural structures within the separated brainstem or spinal cord are essential for the maintenance of a waking EEG. These neural structures lie within the brainstem reticular formation. Electric or chemical stimulation of this area causes EEG arousal, whereas its destruction produces coma and the EEG characteristic of the sleeping state.

This component of the reticular formation is called the *reticular activating system* (RAS). As they pass from the brainstem into the central core of the cerebrum, the neurons of the RAS activate the *diffuse thalamic projection system.* These thalamic neurons synapse in the cortex, but unlike the thalamic projections described in Chap. 9, they are not involved in the transmission of information about specific sensory modalities; rather, they carry on the functions of the RAS and maintain the EEG and behavioral characteristics of the awake state.

Single neurons in the reticular formation may be activated by any afferent modality—a flash of light, a ringing bell, a touch on the skin—to "arouse" the brain. Human beings are able to perceive a stimulus only when the nervous system is oriented and appropriately receptive toward it, and it is the neurons of the reticular activating system which arouse the brain and facilitate information reception by the appropriate neural structures. However, the sensitivity of this system is selective. A mother may awaken instantly at her baby's faintest whimper whereas she can sleep peacefully through the roar of a jet plane passing overhead.

One of the phenomena leading to selectivity is known as *habituation.* Presentation of a novel stimulus to an awake animal usually leads to an orienting response, during which the animal stops whatever it is doing and looks around or listens intently and the EEG switches from the quiet resting alpha rhythm to that characteristic of arousal. On the other hand, the monotonous repetition of a stimulus of constant strength leads to a progressive decrease in response (habituation). For example, when a loud bell is sounded for the first time, it may evoke a startle response in the animal; but after several ringings, the animal makes progressively less response and eventually may ignore the bell altogether. An extraneous stimulus of another modality or the same stimulus at a different intensity restores the original response (*dishabituation*). Habituation is not due to receptor fatigue or adaptation but is mediated at least in part by the reticular formation.

Although the mechanism of action of centrally acting anesthetics is not known, it has been postulated that some act by interfering with the transmission of neural activity in the reticular formation rather than by blocking the direct transmission of afferent information to the cortex. Also, some stimulating drugs (e.g., the amphetamines) work by enhancing the transmission of nerve impulses through the reticular system.

Parts of the brain other than the reticular system are also important for wakefulness and the alert state. For example, the cortex is necessary for sustained wakefulness, and maintenance of the alert state seems to involve an interplay between cortex and the reticular formation. Electric stimulation of some regions of the cortex activates the reticular formation and awakens or alerts an animal as surely as afferent stimulation does. Finally, certain hypothalamic areas are implicated in the EEG and behavioral aspects of the waking state.

Sleep Centers The control of sleep is exerted by two neuronal systems which oppose the tonic activity of the RAS. The two neuronal clusters, one in the central core of the brainstem, the other in the pons, are also part of the reticular formation. The brainstem-core neurons tonically release the transmitter 5-hydroxytryptamine (5-HT, also known as *serotonin*); when serotonin levels become high enough, the neurons of the RAS are inhibited. This results in the loss of awake conscious behavior and its EEG manifestations and the replacement of these by the behavior and EEG characteristics of slow-wave sleep.

These brainstem-core neurons also facilitate the sleep center of the pons, whose activity induces paradoxical sleep. Pathways ascending from this paradoxical-sleep center of the pons establish the low-voltage, fast EEG pattern and activate the muscles of the eyes; descending pathways inhibit motor-neuronal activity, which results in loss of muscle tone.

Of great importance is the fact that, while instituting the EEG and behavioral manifestations of paradoxical sleep, the neurons of the sleep center in the pons also influence the central core of the brainstem in a true feedback fashion. We have said that the release of serotonin by these central core neurons causes slow-wave sleep and facilitates the paradoxical-sleep center. It is thought that the paradoxical-sleep-center neurons feed back and stimulate the *uptake* of serotonin by the endings of the brainstem-core neurons. The decrease in free (i.e., extracellular) serotonin concentration lessens the inhibition on the RAS and permits a return of the awake state. Waking continues until sleep is again triggered by the release of sufficient serotonin to inhibit the RAS. Thus, cycling of sleeping and waking states of consciousness is due, at least in part, to slow accumulation and dissipation of chemical transmitters.

As presented above, sleep is basically the result of the cyclic inhibition of the RAS by brainstem-core neurons. However, it should be reemphasized that this inhibition can be overridden by input from afferent pathways or other brain centers so that RAS activity is maintained sufficiently high to keep one awake or interrupt sleep. In fact, the waking mechanisms seem to be more easily activated than those causing sleep. An example familiar to all parents is that it is much easier to arouse a sleeping child than to get an alert, attentive child to sleep.

When deprived of sleep, we sleep longer at the next sleep cycle to "catch up." However, this recovery period is not directly proportional to the period of sleep deprivation; for example, a good night's sleep of 14 to 16 h by one subject was sufficient to repay at least the slow-wave component of 300 h of sleep deprivation (the paradoxical-sleep debt took longer to make up).

Conscious Experience

All subjective experiences are popularly attributed to the workings of the mind. This word conjures up the image of a nonneural "me," a phantom interposed between afferent and efferent impulses, with the implication that mind is something more than action potentials and synapses. The truth of the matter is that physiologists and psychologists have absolutely no idea of the mechanisms which give rise to conscious experience. Nor are there even any scientifically meaningful hypotheses concerning the problem.

Conscious experiences are difficult to investigate because they can be known only by verbal report. Such studies lack scientific objectivity and must be limited to human beings. In an attempt to bypass these difficulties scientists have studied the behavioral correlates of mental phenomena in other animals. For example, a rat deprived of water performs certain actions to obtain it. These actions are the behavioral correlates of thirst. But it must be emphasized that we do not know whether the rat consciously experiences thirst; this can only be inferred from the fact that human beings are conscious of thirst under the same conditions.

However, one crucial question which cannot be investigated in experimental animals is whether conscious experiences actually influence behavior. Although, intuitively, it might seem absurd to question this, the fact is that the answer is crucial for the development of one's concept of human beings. It is possible that conscious experience is an epiphenomenon.

Consider the following sequence of events: The ringing of a telephone reminds a student that she had promised to call her mother; she finishes the page she has been reading and makes the call. What causes her to do so? The epiphenomenon view holds that the conscious awareness accompanies but does not influence the passage of information from afferent to motor pathways. Thus, in this view, behavior occurs automatically in response to a stimulus, memory stores supplying the direct link between afferent and efferent activity. In contrast, the processing of the afferent information (the sound of the bell), acting through memory stores, could result in the conscious awareness of her promise, which, in turn, leads to the relevant activity in motor pathways descending from the cortex. There is no way of choosing between these two views at present.

In contrast to this presently unapproachable question, some aspects of conscious experience have yielded, at least in part, to experimentation in human beings. In general, these can be described by two questions: What is the relationship between the conscious experience and information arriving over the principal afferent pathways? Is there an anatomically distinct area of the central nervous system involved in conscious experience as opposed to those areas of the brain engaged in the unconscious processing of information or the execution of automatic movements?

The first question has been approached by studies performed in conscious human subjects whose brains are exposed for neurosurgery. Tiny electrodes lowered into different areas of the thalamus record the activity of single cells. In certain thalamic cells a given somatic stimulus, such as movement of a joint, regularly produces a given response which contains in coded form the precise location and intensity of the stimulus. The electric response recorded is unchanged regardless of whether the patients are keenly aware of each stimulus or whether their attention is diverted so that they are unaware that they have been stimulated; i.e., the electric activity is the same whether or not information about the stimulus is incorporated into conscious experience. The thalamic cells which respond in this manner are in nuclei known to form part of the specific afferent pathway. Thus, information relayed in the specific ascending pathways does not necessarily become part of conscious experience.

There are other neurons in nonsensory parts of the thalamus which have quite different properties and patterns of activity which are more closely related to conscious experience. Each of these cells, called *novelty detectors,* responds to input from many parts of the body, but it rapidly *ceases* responding to repeated stimuli of the same kind as the person's attention to them wanes. The firing pattern is more closely related to the degree to which the person is aware of a given stimulus than to precise information about a certain sensory modality.

Other evidence also suggests that conscious experience is determined by central structures as well as by peripheral stimuli. Clinical examples of conscious sensory experiences existing in the absence of neural input are not unusual. For example, after a limb has been amputated, the patient sometimes feels as though it were still present. The nonexistent limb, called a phantom limb, can be the "site" of severe pain.

In this context, it is interesting to study the conscious experiences of persons undergoing periods of sensory deprivation. Student volunteers lived 24 h a day in as complete isolation as possible—even to the extent that their movements were greatly restricted. External stimuli were almost completely absent, and stimulation of the body surface was relatively constant. At first the students slept excessively, but soon they began to be disturbed by vivid hallucinations which sometimes became so distorted and intense that the students refused to continue the experiment. The neural bases of these hallucinations are poorly understood, but it has been suggested that central structures may generate patterns of activity corresponding to those normally elicited by peripheral stimuli when varied sensory input is absent and that conscious experience is not solely dependent upon the senses.

There seems to be an optimal amount of afferent stimulation necessary for the maintenance of the normal, awake consciousness. Levels of stimulation greater or less than this optimal amount can lead to trances, hypnotic states, hallucinations, "highs," or other altered states of consciousness. In fact, alteration of sensory input is commonly used to induce intentionally such experiences.

The answer to the second question: Is there a specific brain area in which conscious experience resides? can be gleaned from conscious persons undergoing neurosurgical procedures and from persons who have accidental or disease-inflicted damage to parts of the brain; but the answer is still far from clear. We have mentioned that it is possible that conscious experience may be just another aspect of the neural activity in those brain centers which receive and process afferent information. Or, on the other hand, conscious experience may depend upon the transmission of the processed afferent information to special parts of the brain whose function is to arrive at and release the contents of conscious experience. The only available clues as to how or where this might be done have been gained from inference. For example, with evolution comes (we assume) greater complexity of conscious experience. Since the brain of human beings is distinguished anatomically from that of other mammals by a greatly increased volume of cerebral cortex, this is a logical place to look for the seat of conscious experience. The cortex has been stimulated when patients on the operating table are fully alert and the brain is exposed. If a certain area of association cortex in temporal lobe is stimulated, the subject may report one of two types of changes in his conscious experience. Either he is aware of a sudden change in his interpretation of the present situation, i.e., what he is seeing or hearing suddenly becomes familiar or strange or frightening or coming closer or going away, or he has a sudden flashback or awareness of an earlier experience. Although he is still aware of where he is, an earlier experience comes to him and repeats itself in the same order and detail as the original experience. It may have been a particular occasion when he was listening to music. If asked to do so, he can hum an accompaniment to the music. If in the past he thought the music beautiful, he thinks so again. During such electric stimulation, visual or auditory experiences are recalled only if the patient was attentive to them when they originally occurred. Other experiences which are also part of conscious experience have never been produced by such stimulation. No one has ever reported periods when he or she was trying to make a decision or solve a problem or add up a row of figures. It is also interesting that no brain area other than temporal association cortex has been found from which complete memories have been activated. However, one cannot assume that association cortex is the site of the stream of consciousness, because stimulation of these cortical neurons causes the propagation of action potentials to many other parts of the brain.

It is best to take a different view of the matter; for, in the words of one famous neurosurgeon, "Consciousness is not something to be localized in space." It is a function of the integrated action of the brain. Sensations and perceptions form part of conscious experience, and yet there is no one point along the ascending pathways or one particular level of the central nervous system below which activity cannot be a conscious sensation and above which it is a recognizable, definable sensory experience. Every synapse along the ascending pathways adds an element of meaning and contributes to the sensory experience.

On the other hand, consciousness and, we presume, the accompanying conscious experiences, are inevitably lost when the function of regions deeper within the cerebrum or of the nerve fibers passing to association cortex from the reticular formation are interrupted by injury. Although the matter is far from settled, evidence suggests that neuronal systems in the reticular formation of the brainstem and regions deep within the cerebrum are involved in brain mechanisms necessary for perceptual awareness. It has been suggested that this system of fibers has widespread interactions with various areas of the cortex and somehow determines which of these functional areas is to gain temporary dominance in the ongoing stream of the conscious experience.

The concept we want to leave as an answer to the question of where the conscious experience resides is perhaps best presented in the following analogy. In an attempt to say which part of a car is responsible for its controlled movement down a highway, one cannot specify the wheels or axle or engine or gasoline. The final performance of an automobile is achieved only through the coordinated interaction of many components. In a similar way, the conscious experience is the result of the coordinated interaction of *many* areas of the nervous system. One neuronal system would be incapable of creating a conscious experience without the effective interaction of many others.

MOTIVATION AND EMOTION

Motivation

Motivation is presently undefinable in neurophysiologic terms, but it can be defined in behavioral terms as the processes responsible for the goal-directed quality of behavior. Much of this behavior is clearly related to homeostasis, i.e., the maintenance of a stable internal environment, an example being putting on a sweater when one is cold. In such homeostatic goal-directed behavior specific bodily needs are being satisfied, the word "needs" having a physicochemical correlate. Thus, in our example the correlate of need is a drop in body temperature, and the correlate of need satisfaction is return of the body temperature to normal. The neurophysiologic integration of much homeostatic goal-directed behavior will be discussed later (thirst and drinking, Chap. 18; food intake and temperature regulation, Chap. 20; reproduction, Chap. 21).

However, many kinds of motivated behavior, e.g., the selection of a particular sweater on the basis of style, have little if any apparent relation to homeostasis. Clearly, much of the behavior of human beings fits this latter category. Nonetheless, the generalization that motivated behavior is induced by needs and is sustained until the needs are satisfied is a useful one despite the inability to understand most needs in physicochemical terms.

A concept inseparable from motivation is that of *reward* and *punishment,* rewards being things that organisms work for or things which strengthen behavior leading to them, and punishments being the opposite. They are related to motivation in that rewards may be said to satisfy needs. Many psychologists believe that rewards and punishments constitute the incentives for learning. Because virtually all behavior is shaped by learning, reward and punishment become crucial factors in directing behavior. Although some rewards and punishments have conscious correlates, many do not. Accordingly, much of human beings' behavior is influenced by factors (rewards and punishments) of which they are unaware.

We have thus far described motivation without regard to its neural correlates. As was true for conscious experience, nothing is known of the mechanisms which underlie the subjective components of this phenomenon, nor is it known how rewards and punishments influence learning and behavior. Present knowledge is limited to recognition of some of the brain areas (and their interconnecting pathways) which are important in motivated behavior.

It should not be surprising that the brain area most important for the integration of motivated behavior related to homeostasis is the hypothalamus, since it contains the integrating centers for thirst, food intake, temperature regulation, and many others. Much information concerning the reinforcing effects of rewards and punishments on hypothalamic function has been obtained through *self-stimulation* experiments, in which an unanesthetized experimental animal regulates the rate at which electric stimuli are delivered through electrodes previously implanted in discrete brain areas. The animal is placed in a box containing a lever it can press. If no stimulus is delivered to the animal's brain when the bar is pressed, it usually presses it occasionally, perhaps out of boredom or curiosity. However, if a stimulus is delivered to the brain as a result of the bar press, a different behavior can result, depending upon the location of the electrodes. If the animal increases its bar-pressing rate above control, the electric stimulus is, by definition, rewarding; if it decreases it, the stimulus is punishing. Thus, the rate of bar pressing is a measure of the effectiveness of the reward (or punishment). Bar pressing that results in self-stimulation of the sensory and motor systems produces response rates not significantly different from the control rate. Brain stimulation through electrodes implanted in certain areas of hypothalamus serves as a positive reward. Animals with electrodes in these areas bar-press to stimulate their brains from 500 to 5,000 times per hour. In fact, electric stimulation of some areas of the hypothalamus is more rewarding than external rewards; e.g., hungry rats often ignore available food for the sake of electrically stimulating their brains.

This rewarding effect of self-stimulation is not found in all areas of the hypothalamus but is most closely associated with those areas which normally mediate highly motivated behavior, e.g., feeding, drinking, and sexual behavior. Consistent with this is the fact that the animal's rate of self-stimulation in some areas increases when it is deprived of food; in other areas, it is decreased by castration and restored by administration of sex hormones. Thus, it appears that neurons controlling homeostatic goal-directed behavior are themselves intimately involved in the reinforcing effects of reward and punishment.

Although it has been generally assumed that such feeding and drinking behaviors are associated with the underlying feelings of hunger and thirst, it is puzzling to determine why an animal would self-stimulate in order to experience "thirst" or "hunger." An explanation of the paradox is that the reinforcing effects of self-

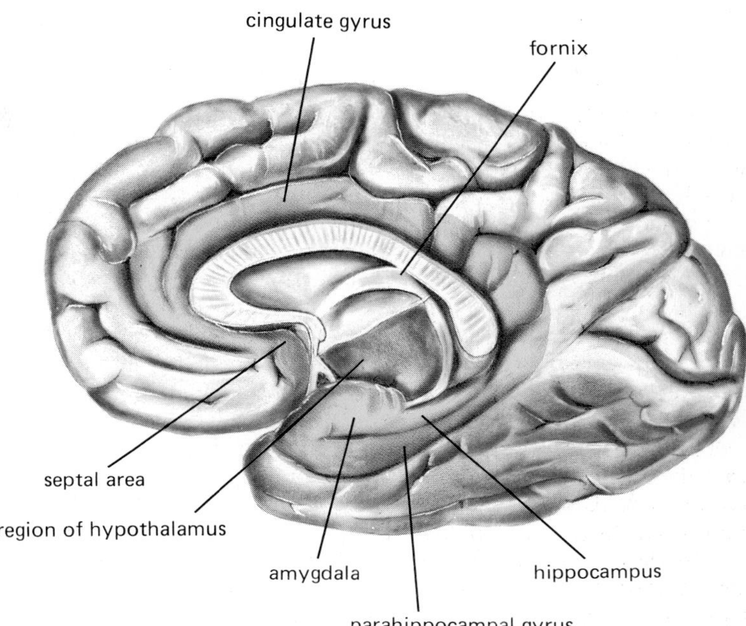

FIGURE 11-3 ***Structures of the limbic system are in the shaded area on this midsagittal section through the brain.***

stimulation are not thirst or hunger sensations per se; rather that stimulation (i.e., activation) of the neural pathways underlying these behaviors is in itself reinforcing and can, therefore, provide the motivation to engage in self-stimulation behavior.

Emotion

Related to motivation are the complex phenomena of *emotion*. Scientists are presently trying to understand the operation of the chain of events leading from the perception of an emotionally toned stimulus, i.e., the subjective feelings of fear, love, anger, joy, anxiety, hope, etc., to the complex display of emotional behavior, and they are beginning to find some answers.

Most experiments point to the involvement of the *limbic system*, which is an interconnected group of brain structures within the cerebrum, including portions of the frontal-lobe cortex, temporal lobe, thalamus, and hypothalamus as well as the circuitous neuron pathways connecting all parts together (Fig. 11-3). Besides being connected with each other, the parts of the limbic system have connections with many other parts of the central nervous system. For example, it is likely that information from all the different afferent modalities can influence activity within the limbic system, whereas activity of the limbic system can result in a wide variety of autonomic responses and body movements. This should not be surprising since many emotional feelings are accompanied by autonomically mediated responses such as sweating, blushing, and heart-rate changes, and by somatic reponses such as laughing and sobbing.

The limbic system has been studied in experimental animals, using electric stimulation of specific areas within it. The physiologic results of these procedures vary markedly but justify the hypothesis that three distinct neural systems (septal area–hippocampus, septal area–hypothalamus, amygdala) mediate the various emotional behaviors. Of course, in these experiments there was no way to assess the subjective emotional feelings of the animals; instead they were observed for behaviors which usually are associated with emotions in human beings. As different areas of the limbic system were stimulated in awake animals (the stimulus is a small electric current delivered through electrodes previously implanted while the animal was anesthetized), three types of behavior resulted.

After stimulation of one area the animal actively approaches a situation as though expecting a reward. Stimulation of a second area causes the animal to stop the behavior it is performing, as though it knew it would lead to punishment. Stimulation of a third area of the limbic system causes the animal to arch its back, puff out its tail, hiss, snarl, bare its claws and teeth, flatten its ears, and strike. Simultaneously, its heart rate, blood pressure, respiration, salivation, and concentrations of plasma epinephrine and fatty acids all increase. Clearly, this behavior typifies that of an enraged or threatened animal.

Limbic areas have also been stimulated in awake human beings undergoing neurosurgery. These patients, relaxed and comfortable in the experimental situation, report vague feelings of fear or anxiety during periods of stimulation to certain areas even though they are not told when the current is on. Stimulation of other areas induces pleasurable sensations which the subjects find difficult to define precisely.

Surgical damage to parts of the limbic system in experimental animals is another commonly used tool; it leads to a great variety of changes in behavior, particularly that associated with emotion. Destruction of a nucleus in the tip of the temporal lobe produces docility in an otherwise savage animal, whereas surgical damage to an area deep within the brain produces vicious rage in a tame animal; and the rage caused by this lesion can be counteracted by a lesion in the tip of the temporal lobe. A rage response can also be caused by destruction of part of the hypothalamus. Lesioned animals sometimes manifest bizarre sexual behavior in which they attempt to mate with animals of other species; females frequently assume male positions and attempt to mount other animals.

Self-stimulation experiments have shown the presence of reward and punishment responses in various parts of the limbic system. When the electrodes are in certain midline areas, the animal presses the bar once and never goes back, indicating that stimulation of these brain areas has a punishing effect. These are the same brain areas which, when stimulated, give rise to behavioral activity signifying avoidance, rage, or escape. Conversely, self-stimulation of other limbic areas has a strong rewarding effect.

Stimulation of certain hypothalamic areas (like the stimulation of other limbic structures described above) elicits behavior that *seems* to have a strong subjective emotional component; yet if the hypothalamus is isolated from the other portions of the limbic system, the emotional component is lacking. For example, stimulation of a cat's brain can cause enraged, aggressive behavior complete with attack directed at any available object, but as soon as the stimulation ends, the animal immediately reverts to its usual friendly behavior. It seems as though the actions lacked emotionality and purpose, representing only the motor component of the behavior. We use the word "purpose" but could have said that, except for the experimentally induced stimulus, the behavior lacked "motivation."

What is the relationship between the hypothalamus and the rest of the limbic system? The three components of the limbic system mentioned above converge in the medial hypothalamus, which acts as an integrating center. For example, in cats the medial hypothalamus exerts a tonic inhibition on the neural pathways which lead to fight-or-flight behavior. However, upon receipt of appropriate environmental stimuli, the nuclei of the temporal lobe inhibit the medial hypothalamus, thus decreasing its inhibitory influence over the fight-or-flight system and allowing activity in that system to increase. The resulting emotional behavior, then, results from the balance of input to the medial hypothalamic integrating centers. Notice that, although the structures involved in the control of emotional behavior are predominantly located in the limbic system, and (as described in previous chapters) the main controlling centers for consummatory behavior related to homeostasis are located in the hypothalamus, the two meet and interact at the level of the medial hypothalamus.

Finally the subjective aspects, or feelings, that make up part of an emotional experience possibly also involve the cortex, particularly cortex of the frontal lobes, which is implicated because changes in emotional states frequently occur following damage there. These alterations in mood and character are described as fear, aggressiveness, depression, rage, euphoria, irritability, or apathy. There are indications that frontal regions may exert inhibitory influences upon the hypothalamus and other areas of the limbic system. There may be facilitatory frontal regions as well. Anatomic connections between frontal cortex and hypothalamus exist to support the suggested interrelationship of these two areas in motivated and emotional behaviors. Excitatory and inhibitory influences from the limbic system and possibly from nonlimbic areas of cortex are of great importance in determining and patterning the level of excitability of hypothalamic and brainstem neurons. How activity in the limbic system is initiated and influences other brain areas is poorly understood.

Chemical Mediators for Emotion-Motivation

Norepinephrine and other amines play important roles in the mediation of some emotional and behavioral states. For example, norepinephrine is believed to be a neurotransmitter in the pathways subserving rage, for drugs which enhance the effect of norepinephrine also increase rage and those which block it diminish rage behavior. Norepinephrine also seems to operate in the active-approach system. Thus, an animal given a drug that depletes brain norepinephrine stores manifests a decrease in active avoidance of a punishing shock and in rates of self-stimulation, whereas drugs enhancing norepinephrine release have the opposite effect.

The biogenic amines, particularly norepinephrine and dopamine, are also associated with subjective mood. Decreased norepinephrine is associated with depression and increased norepinephrine with elevated mood; furthermore, the antidepressant drugs are thought to work by increasing brain amine concentration.

The biogenic amines are also implicated in the networks subserving learning. This association is not unexpected since, as we have just stated, they are involved in the neural systems underlying reward and punishment, and many psychologists believe that rewards and punishments constitute the incentives for learning.

NEURAL DEVELOPMENT AND LEARNING

Evolution of the Nervous System

During one of its earliest stages of evolution, the nervous system was probably a simple three-neuron system with a limited number of interneurons interposed between the afferent and efferent nerve cells. The interneuronal component expanded rapidly until it came to be by far the largest part. The interneurons formed networks of increasing complexity, at first involved mainly with the stability of the internal environment and position of the body in space. Those cells with increasing specialization of function came to be localized at one end of the primitive nervous system, and the brain began to evolve.

The brainstem, which is the oldest part of the brain in this evolutionary sense, retains today many of the anatomic and functional characteristics typical of those most primitive brains. With continued evolution, newer, increasingly complex structures were added on top of (or in front of) the older ones. They developed as paired symmetric tissues, the cerebral hemispheres, and reached their highest degree of sophistication with the formation of the cerebral cortex.

The newer structures served in part to elaborate, refine, modify, and control already existing functions. For example, it is possible to perceive the somatic stimuli of pain, touch, and pressure with only a brainstem, but the stimulus cannot be localized without a functioning cerebral cortex. Perhaps even more important is that these newer, more anterior parts of the brain came to be involved in the perception of goals, the ordering of goal priorities, and the patterning of behaviors to serve in pursuit of these goals.

Development of the Individual Nervous System

Although all nerve cells are present at birth, their rate of growth in size, number of dendrites, degree of myelinization, etc., varies considerably. Notice the changes in visual cortex cells in just the first 3 months of life (Fig. 11-4) and the gradual change in electric activity (Fig. 11-5). These developmental patterns are reflected in behavioral changes. It is believed that the cortex is relatively nonfunctional at birth, a notion supported by the fact that infants born without a cortex have almost the same behavior and reflexes as shown by a normal infant. As the cortex develops, it seems gradually to exert an inhibitory control over the lower (and phylogenetically older) structures.

With cortical development, some of the reflexes whose integrating centers are in subcortical structures come under at least a degree of cortical control. Examples are the rooting and sucking reflexes present in infancy. (The stimulus is a touch on the infant's cheek; in response, the head turns toward the stimulus, the mouth opens, the stimulating object is taken into the mouth, and sucking begins.) The reflex is essential for the survival of the young, but it is superseded by other eating behaviors. As the cerebral hemispheres develop and autonomy increases, allowing the elaboration of hand feeding, the primitive rooting and sucking reflexes are inhibited. The basic reflex arc does not disappear; it is simply inhibited. For example, while examining a patient with damaged frontal lobes of the brain, a physician standing behind the patient and out of his view quietly reached over and touched the patient's cheek. In response, the patient nuzzled the physician's finger until it was in his mouth and even began sucking on it. Upon realizing what he had unconsciously and automatically done, he was quite embarrassed; but his behavior, released from its normal inhibition because of the brain damage, serves as a perfect illustration of the above point.

An adequate nutritive environment is required for normal brain development. The level of hormones is also important; for example, decreased thyroid hormone concentrations during development cause a type

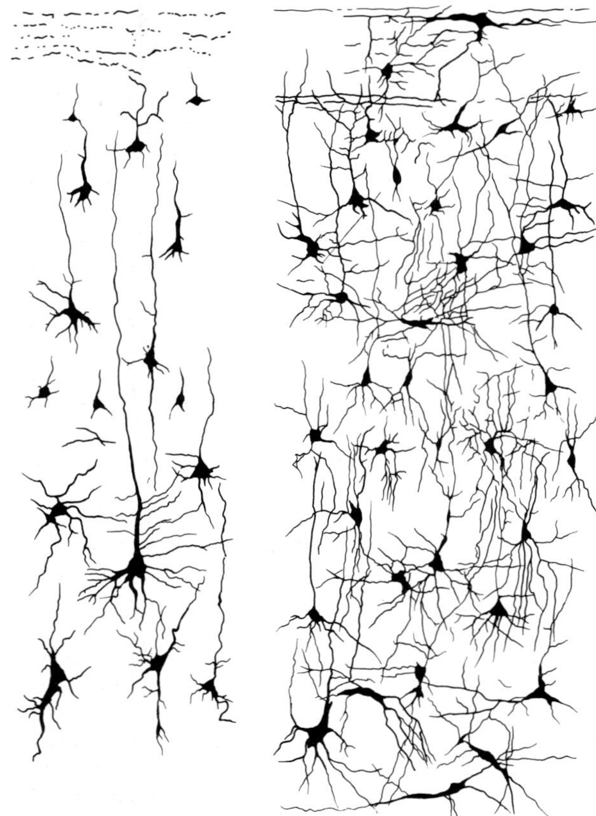

FIGURE 11-4 ***Visual cortex of a newborn*** **(left)** ***and a 3-month-old child*** **(right). (Redrawn from J. L. Conel, "The Postnatal Development of the Human Cerebral Cortex," vol. 1, The Cortex of the Newborn, Harvard University Press, Cambridge, 1939.)**

of mental retardation known as *cretinism*. Although it was formerly believed that the developing fetus "took what it needed" nutritionally from the mother and that, in cases of malnutrition after birth, the brain was "spared" at the expense of the rest of the body, there is increasing evidence that nutritional deprivation both before and after birth has serious and irreversible effects on the chemical and structural maturation of the brain. Intermittent deprivation affects those cells that are still developing at that time more than cells already mature.

Despite evidence indicating the selective growth and interconnectivity of nerve cells, the nature-nurture controversy, i.e., how much of development is due to genetically determined patterns and how much to experience and subsequent learning, must also be considered, for, in fact, the course of neural development can be altered.

However, the period during which this alteration can occur is genetically built into the neurons' developmental timetable. The modifiability of some neurons is limited to so-called *critical periods* early in development but in others it persists longer. Thus, some neuronal organizational patterns are highly specified early in development and are unmodifiable thereafter. These periods of maximal structural and functional

FIGURE 11-5 ***EEGs from the same individual, showing the onset of the alpha rhythm at 4 months, the attainment of adult wave frequencies at 10 years of age, and little change thereafter.*** **(Redrawn from D. B. Lindsley, Attention, Consciousness, Sleep, and Wakefulness, in "Handbook of Physiology; Neurophysiology," vol. 3, pp. 1553–1593, American Physiological Society, Washington, D.C., 1960.)**

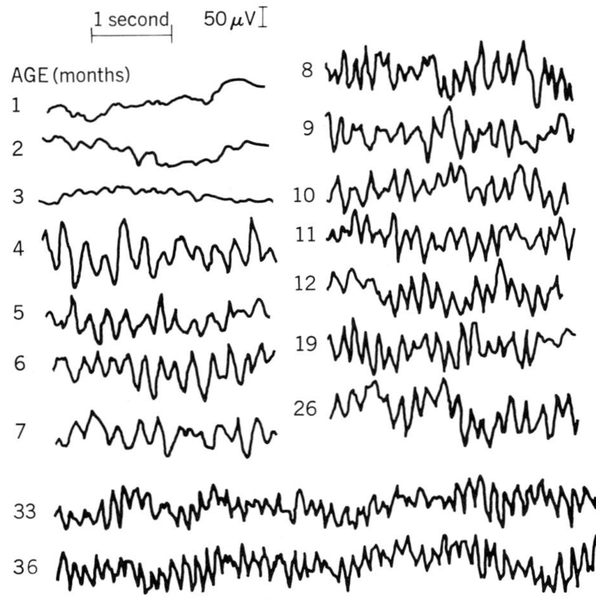

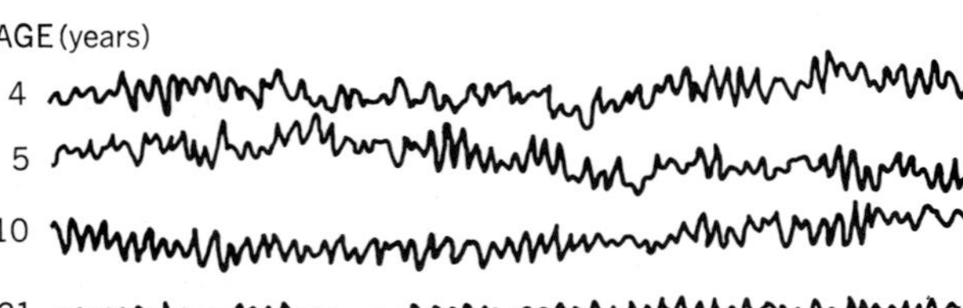

growth depend upon the availability of proper internal and external environmental conditions for their full development. As an example of such a critical period it was found that the visual systems of sheep dogs deprived of sight in the first 5 weeks of life are anatomically, biochemically, and electrophysiologically retarded; this does not occur if the deprivation occurs from 5 to 10 weeks of age or during a 5-week period in adult dogs.

Other neurons remain uncommitted and are modifiable by function and experience until late in development. But in the case of either limited or lengthy periods of modifiability, it can occur only within the constraints imposed by the neurons' genetic code and its experiential history. The genes specify the capacity whereas the environmental stimuli determine its specific expression and content.

Learning and Memory

Learning is the increase in the likelihood of a particular response to a stimulus as the consequence of experience. Rewards or punishments, as mentioned earlier, are crucial ingredients of learning, as is contact with, and manipulation of, the environment. A variety of hypotheses have been advanced recently to explain how individually acquired information may be stored by the brain. The postulated neural correlate of memory is called the *memory trace*.

The processes involved in laying down the memory trace occur in a matter of minutes or hours, during which the memory is known as *short-term memory*. After this somewhat labile formative period, the memory is stored as *long-term memory*. Short-term memory involves the cerebral cortex, and long-term memory involves the limbic system; however, there is no exclusive site for memory storage because removal of various parts of the brain does not remove specific memories.

Short-term Memory Short-term memory is a limited-capacity storage process which serves as the initial depository of information. Generally, as new items enter, older ones are displaced, suggesting that the information is organized in a temporal sequence. Information that has entered short-term memory may be forgotten, recycled back through short-term memory by actively rehearsing the information, or transferred to a more durable storage mode. Recycling keeps the item at the front of the temporal sequence so that it will not be forgotten and increases the probability that it will be transferred to long-term storage.

One theory suggests that the memory trace during the early phases of learning is a reverberating neural circuit in which electric activity passes around and around in closed neuronal loops. There is certainly evidence for the existence of such neuronal pathways in the brain, and activity once started in such loops could be maintained to keep the "memory" of the input for some time. That such reverberating circuits could be responsible for the temporary storage of acquired information in short-term memory is supported by evidence that conditions such as coma, deep anesthesia, electroconvulsive shock, and insufficient blood supply to the brain, which interfere with the electric activity of the brain, also interfere with the retention of recently acquired information. These same states do not interfere with long-term memory. When persons become unconscious from a blow on the head, they often cannot remember anything that happened for about 30 min before they were hit. This phenomenon is called *retrograde amnesia*. The loss of consciousness in no way interferes with memories of experiences that were learned before the period of amnesia.

Long-term Memory The existence of at least two stages of memory, short- and long-term memory, is borne out by the behavior of patients with specific amnesias. Such patients can learn perfectly well and recall the information immediately after it has been presented, but they lose the material much faster than normal people do, particularly if it involves verbal clues. It is as though they had difficulty transferring the information from short- to long-term storage.

Behavioral investigators and common experience indicate that memories of past events and well-learned behavior patterns normally can have very long lifespans. They may be changed or suppressed by other experiences, but, contrary to popular opinion, memories do not usually fade away or decay with time. This stability and durability combined with the fact that removing parts of the brain does not remove specific memories suggest that memory is stored in widespread chemical form or that the memory trace is an alteration in structure of some elements of the brain.

Retrieval of items from short-term memory seems to be much faster than retrieval from long-term memory, perhaps because the stores in long-term memory are so much larger. There are two major theories to explain long-term memory.

Molecular Theories One theory states that large, stable molecules within neurons are changed during learning and that information is stored in the specific configuration of these molecules. The molecules most frequently implicated are RNA and protein. We are already familiar with information storage mechanisms of this kind, e.g., the coding of genetic information by the nucleotide sequences in chromosomal DNA and the transfer of this information to RNA and proteins (Chap. 1). Present evidence indicates that these macromolecules are also involved in the laying down of long-term memory. For learning, the question becomes: Are there chemical processes that in some way constitute the store of learned material in a coded form?

Morphologic Theories These theories suggest that changes in the relationship between cells occur during the formation of a memory trace. One theory suggests that modifications of the glial cells surrounding neurons provide the storage of the memory trace. A second theory suggests that new synaptic relationships are established or old synapses become more efficient when new information is transmitted to the brain. These modifications of synaptic relationships could occur with the swelling or shrinking of nerve processes as the result of use or disuse, an increase or decrease in the concentration of synaptic vesicles, changes at the pre- or postsynaptic membranes, etc.

Relationships between cells could also change with the interposition of small, newly formed nerve cell processes between the input and output elements of the nervous system. Indeed, the maturing brain undergoes radical structural transformations after birth: a great outgrowth of dendritic processes of the nerve cells representing an increase in the potential connectivity of neurons; formation of many glial cells, whose processes become interdigitated among the neurons; continuing formation of myelin around neuronal axons; and increased density of blood vessels in the brain. These morphologic changes in the developing brain are to a large extent genetically determined maturational processes. However, studies have suggested that the structural organization of the maturing brain is sensitive to conditions of the physical and social environment of the animal.

Forgetting Our understanding of the mechanisms of forgetting, like that of the mechanisms of learning, is still at the level of theories. One such idea is the *interference theory of forgetting,* which states that forgetting results from competition between responses at the time of recall. Competition comes from conflicting information stored both before and after the storage of the particular item that is being recalled. Thus we carry with us (in the form of prior learning) the source of much of our forgetting.

Summary of Theories of Learning The molecular and morphologic theories are certainly not mutually exclusive because a change in structural interrelationships can occur only through a change in macromolecules such as RNA and protein. However, the steps by which a particular experience results in a specific alteration in RNA and protein are obscure. Hypotheses to suggest how macromolecules can produce neuronal and, ultimately, behavioral changes are also lacking. It is probable that more than one mechanism will be found to be involved in the processes of learning and remembering.

The physical mechanisms ultimately accepted to explain learning must be able to account for the following phenomena:

1 Learning can occur very rapidly. In fact, under some situations, learning can occur in one trial.

2 Learning must be translated from an initial form involving action potentials to some permanent form of storage that can survive deep anesthesia, trauma, or electroconvulsive shock, which disrupt the normal patterns of neural conduction in the brain.

3 Information can be retained over long periods of time during which most components of the body have been renewed many times. Learning and memory must therefore reside either in systems that do not turn over rapidly or in systems which are self-perpetuating.

4 Information can be retrieved from memory stores after long periods of disuse. The common notion that memory, like muscle, always atrophies with lack of use is largely wrong.

5 Learning opposing responses interferes with the memory of initial responses to the same stimulus.

6 When learning a specific task, after an initial period of rapid learning, the process seems to slow down and proceed at a rate that offers diminishing returns.

LANGUAGE

As demonstrated by *aphasias,* i.e., specific language deficits not due to mental defects, language is separable into two components, conceptualization and expression. In forms of aphasia related to conceptualization, patients cannot understand spoken or written language even though their hearing and vision are unimpaired. In expressive aphasias, patients are unable to carry out the coordinated respiratory and oral movements necessary for language even though they can move their lips and tongue, understand spoken language, and know what they want to say. Expressive aphasias are often associated with an inability to write.

Different areas of cortex are related to specific aspects of language. Areas in the frontal lobe near motor cortex are involved in the articulation of speech, whereas areas in the parietal and temporal lobe are involved in sensory functions and language interpretation. These cortical specializations are not present at birth but are established gradually in childhood during language acquisition. Why language functions localize in the left hemisphere in 96 percent of the population is not known, because during early childhood both hemispheres have language potential. Accidental damage to the left hemisphere of children under 2 years causes no impediment of future language development, and language develops in the intact right hemisphere. Even if the left hemisphere is traumatized after the onset of language, language is reestablished in the right hemisphere after transient periods of loss. The prognosis becomes rapidly worse as the age at which damage occurs increases, so that after the early teens, language is interfered with permanently. The dramatic change in the possibility of establishing language (or a second language) in the teens is possibly related to the fact that the brain attains its final structural, biochemical, and functional maturity at that time. Apparently, with maturation of the brain, language functions are irrevocably assigned and the utilization of language propensities of the right hemisphere is no longer possible.

Recent evidence that language functions reside in the left hemisphere has been obtained from studies on patients whose main commissures (nerve fiber bundles) joining the two cerebral hemispheres have been cut to relieve uncontrollable epilepsy (neuronal activity, which starts at a cluster of abnormal cortical neurons, spreads throughout adjacent cortex, and gives rise to seizures and convulsions). Essentially, this operation leaves two separate cerebral hemispheres with a single brainstem, two separate mental domains within one head. Events experienced, learned, and remembered by one hemisphere remain unknown to the other because the memory processing of one hemisphere is inaccessible to the other. Because complex language resides in the left hemisphere, only that hemisphere can communicate orally or in writing about its conscious experience. If vision is limited so that only that part of the retina whose fibers pass to the right hemisphere is excited

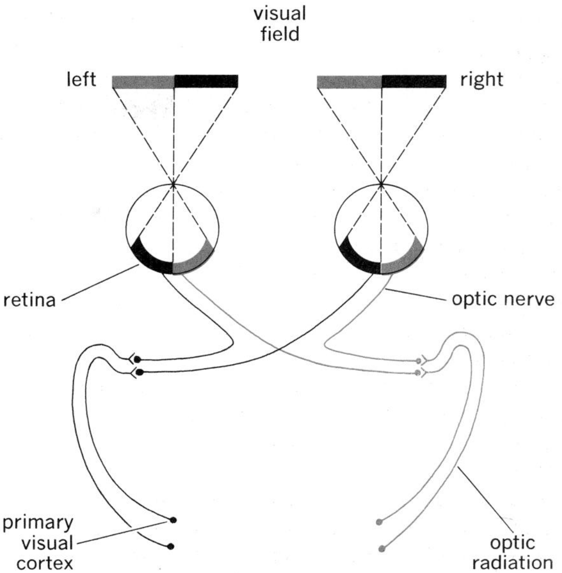

FIGURE 11-6 ***Visual pathways. Information about objects in the left half of the visual field is projected to the right cerebral hemisphere and vice versa.***

(Fig. 11-6), the left hemisphere, which controls speech, is unaware of the visual experience and the patient is unable to describe it either orally or in writing. On the other hand, if the portion of retina projecting to the left hemisphere is activated, the patient can describe the experience without difficulty. The right silent hemisphere does understand language, however, for if the word for an object is flashed only to that hemisphere and the patient is told to point to the object or demonstrate what the object is used for, the patient complies, responding, of course, with the left hand since the right hemisphere controls the muscles of the opposite side of the body. Therefore the right hemisphere does have linguistic functions but not those involved in graphic or vocal expression.

CONCLUSION

Until recently it was commonly thought that control of most complex behavior such as thinking, remembering, learning, etc., was handled almost exclusively by the cerebral cortex. Actually, damage of cortical areas outside of motor and sensory areas produces behavioral results that are subtle rather than obvious (the one exception, language, is highly sensitive to cortical damage), and stimulation of the cortex causes little change in the orientation or level of excitement of the animal.

In general, it is best to consider that particular behavioral functions are not controlled exclusively by any one area of the nervous system but that the control is shared or influenced by structures in other areas. Cortex and the subcortical regions – particularly limbic and reticular systems – form a highly interconnected system in which many parts contribute to the final expression of a particular behavioral performance. Moreover, the nervous system is so abundantly interconnected that it is difficult to know where any particular subsystem begins or ends.

In the early seventeenth century Descartes taught that all things in nature, including human beings, are machines. The brain's mode of operation was compared to that of a clock. When computers became widely used, the brain was compared to a computer. The most recent analogy compares the brain to a hologram, a photographic process which records specially processed lightwaves themselves, rather than the image of an object. These widely divergent analogies only emphasize how little we know of how the brain really functions.

CHAPTER 12 The Endocrine System

The endocrine system constitutes the second great communications system of the body, the hormones serving as blood-borne messengers which regulate cell function. The endocrine system consists of the hormone-producing endocrine glands. As described in Chap. 3, the term *endocrine* broadly denotes a gland that secretes its product into the interstitial fluid from where it diffuses into the blood or lymph. However, the term is usually used in a more restricted sense to include only those glands whose secretory products are hormones; thus, for example, the liver, which secretes nonhormonal materials (glucose and other metabolites) into the blood is excluded from this category.

We shall define a *hormone* as a chemical substance synthesized by a specific organ or tissue and secreted into the blood, which carries it to other sites in the body, where its actions are exerted. The word specific in the definition is important to distinguish true hormones from another class of substances, the so-called parahormones, which are metabolic products produced by *many* organs of the body that exert effects on distant sites, e.g., the hydrogen ion. In terms of chemical struc-

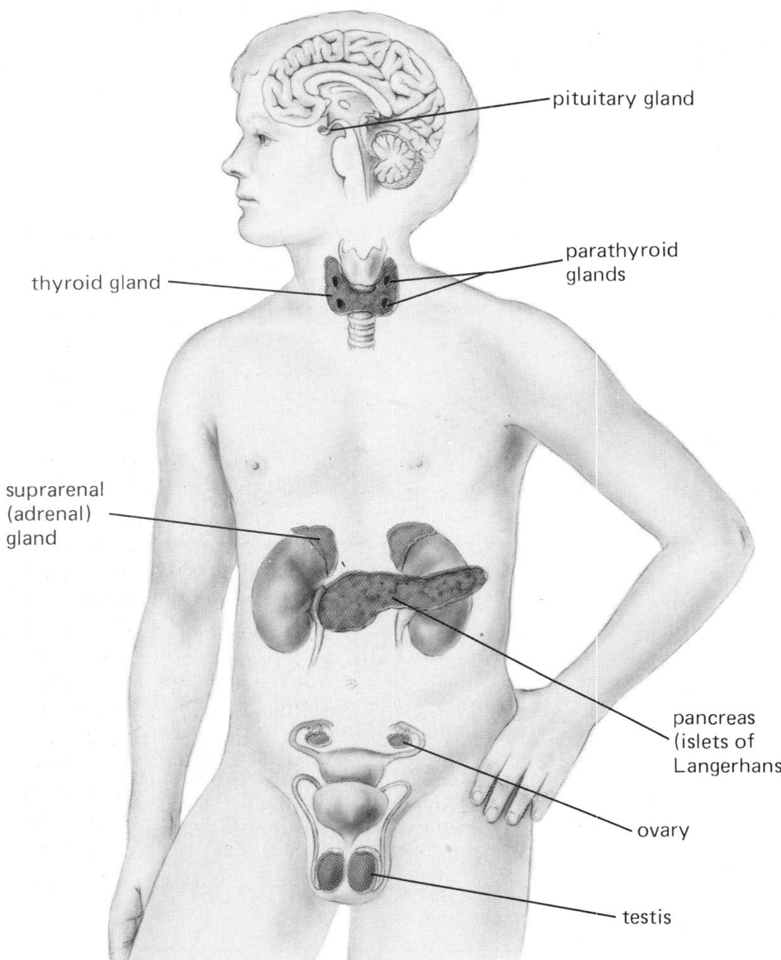

FIGURE 12-1 ***Location of the major endocrine glands. Note that the parathyroid glands actually lie on the posterior surface of the thyroid. Also note that this hypothetical bisexual figure has both ovaries and testes; this does not occur in normal human beings.***

ture, hormones generally fall into two categories: steroids and amino acid derivatives, the latter ranging in size from small molecules containing single amine groups to very large proteins.

Figure 12-1 illustrates the locations of the major endocrine glands. Clearly, the endocrine system differs from most of the other systems of the body in that the various glands are not in anatomic continuity with each other. However, they do form a system in the functional sense. It should also be noted that some of the glands are completely distinct organs (the pituitary, for example) whereas others are found within larger organs having nonendocrine functions as well (the gonads, for example).

Hormones serve to control and integrate many bodily functions: reproduction (Chap. 21), organic metabolism and energy balance (Chap. 20), and mineral metabolism (Chap. 18). The ability to reproduce is absolutely dependent upon a normally functioning endocrine system; in contrast, no other bodily function absolutely requires hormonal control, and, for this reason, the endocrine system is not strictly essential for life. However, we must quickly point out that such a life would be extremely precarious and abnormal; individuals would be unable to adapt to environmental alteration or stress, and their physical and mental abilities would be drastically impaired. They would require the constant attention bestowed upon a hothouse plant.

In recent years, it has become clear that the nervous and endocrine systems actually function as a single interrelated system. The central nervous system, particularly the hypothalamus, plays a crucial role in controlling hormone secretion; and conversely, hormones markedly alter neural function and strongly influence many types of behavior. These interrelationships form the area of study known as *neuroendocrinology*.

Table 12-1 summarizes the physiology of the major endocrine glands and the hormones they secrete. As emphasized previously, the hormones function as components of the body's control systems; accordingly, the detailed physiologic roles of the various hormones will be described in later chapters in the context of the control systems in which they participate. The aim of this chapter is to provide the foundation for these later descriptions by presenting: (1) the general characteristics and principles which apply to almost all hormones; (2) the anatomy of the simple endocrine glands (the anatomy of the hormone-secreting cells of the organs having multiple functions – gonads, pancreas, kidneys, gastrointestinal tract, and thymus – is best described with the overall anatomy of these organs); and (3) the types of direct input which act upon endocrine-secreting cells to cause production and release of the hormones into the blood.

HORMONE-TARGET-ORGAN CELL SPECIFICITY

Hormones travel in the blood and are therefore able to reach virtually all tissues. This is obviously very different from the efferent nervous system, which can send messages selectively to specific organs. Yet, the body's response to hormones is not all-inclusive but highly specific, in some cases involving only one organ or group of cells. In other words, despite the ubiquitous distribution of a hormone via the blood, only certain cells are capable of responding to the hormone; they are known as *target-organ cells*. By unknown evolutionary mechanisms, cells have become differentiated so as to respond in a highly characteristic manner only to certain hormones, this ability to respond depending upon specific receptor sites on cell components. Specialization of target-organ receptor sites explains the specificity of action of hormones; e.g., thyroid-stimulating hormone is produced by the anterior pituitary and affects significantly only the thyroid gland and no other tissue, and insulin causes increased glucose uptake by many cells but not by all, the brain cells being one of the important exceptions.

GENERAL FACTORS WHICH DETERMINE THE BLOOD CONCENTRATIONS OF HORMONES

Rate of Secretion

With few exceptions, hormones are not secreted at constant rates. As emphasized previously, a regulatory system must be capable of altering its output. Normally, some secretion is always occurring; therefore, the rate can be increased or decreased. This pattern is completely analogous to the phenomena of tonic activity, facilitation, and inhibition manifested by the nervous system.

Secretion encompasses two processes, intracellular synthesis and release into the blood. For very short periods of time, release can occur in the absence of synthesis because all endocrine cells store some finished product, but over any prolonged period synthesis must obviously keep pace with release. Too little information is available to permit dissociation of the inputs which control these distinct processes, and so the two are usually incorporated into the general term secretion. Hormone deficiencies can be caused by failure of synthesis due to lack of an essential chemical needed for production of the particular hormone. Thus, hormone production requires a supply of chemical precursors either from the diet or other body cells (Fig. 12-2), and the manner in which the rest of the body metabolizes these precursors becomes of great importance.

Rates of Inactivation and Excretion

The concentration of a hormone in the plasma depends not only upon the rate of secretion but also upon the rate of removal from the blood. Sometimes the hormone is inactivated by the cells upon which it acts, but for most hormones, the pathway of removal from the blood is the liver or the kidneys. Accordingly, patients with kidney or liver disease may suffer from excess of certain hormones solely as a result of reduced hormone inactivation. In any case, it is essential to realize that all hormones are continuously removed by excretion or inactivation, so that maintenance of blood concentrations requires continual secretion.

Because, for many hormones, the rate of urinary excretion is directly proportional to the rate of glandular secretion, physiologists often use the rate of excretion as an indicator of secretory rate.

TABLE 12-1

SUMMARY OF THE MAJOR HORMONES

Gland	*Hormone*	*Major function/Control of:*
Hypothalamus	Releasing hormones	Secretions of the anterior pituitary
	Oxytocin	(See posterior pituitary)
	Antidiuretic hormone	(See posterior pituitary)
Anterior pituitary	Growth hormone (somatotropin, STH)†	Growth; organic metabolism
	Thyroid-stimulating hormone (TSH)	Thyroid gland
	Adrenocorticotropic hormone (ACTH)	Adrenal cortex
	Prolactin	Breasts (milk formation)
	Gonadotropic hormones:	Gonads
	Follicle-stimulating hormone (FSH)	
	Luteinizing hormone (LH)	
Posterior pituitary‡	Oxytocin	Milk secretion; uterine motility
	Antidiuretic hormone (ADH, Vasopressin)	Water excretion
Adrenal cortex	Cortisol	Organic metabolism; response to stress
	Androgens	Growth and, in women, sexual activity
	Aldosterone	Sodium and potassium excretion
Adrenal medulla	Epinephrine	Organic metabolism; cardiovascular function; response to stress
	Norepinephrine	
Thyroid	Thyroxine (T-4)	Energy metabolism; growth and development
	Triiodothyronine (T-3)	
	Calcitonin	Plasma calcium
Parathyroids	Parathyroid hormone (parathormone, PTH, PH)	Plasma calcium and phosphate
Gonads		
Female: ovaries	Estrogen	Reproductive system; growth and development; breasts
	Progesterone	
Male: testes	Testosterone	Reproductive system; growth and development
Pancreas	Insulin	Organic metabolism; plasma glucose
	Glucagon	
Kidneys	Renin	Adrenal cortex; blood pressure
	Erythropoietin	Erythrocyte production
	1,25-Dihydroxycholecalciferol	Calcium balance
Gastrointestinal tract	Gastrin	Gastrointestinal tract; liver; pancreas; gallbladder
	Secretin	
	Cholecystokinin	
Thymus	Thymus hormone (thymosin)	Lymphocyte development
Pineal	Melatonin	? Sexual maturity

† The names and abbreviations in parentheses are synonyms.

‡ The posterior pituitary stores and secretes these hormones; they are synthesized in the hypothalamus.

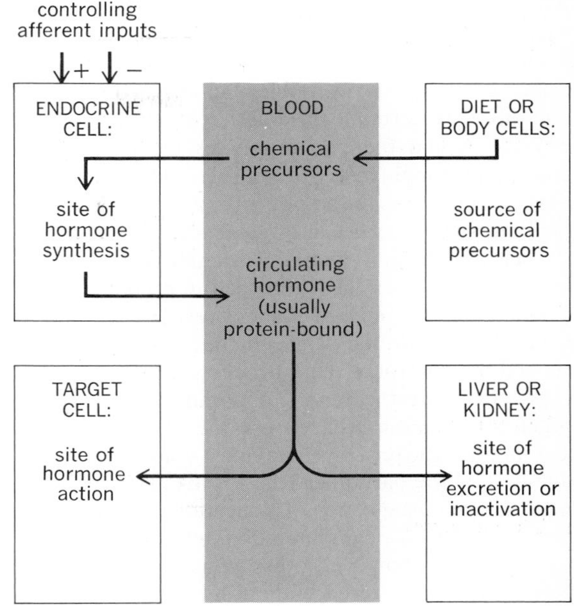

FIGURE 12-2 ***General factors determining blood concentrations of hormones.***

Transport in the Blood

Many of the hormone molecules which circulate in the blood are bound to various plasma proteins; the free moiety is usually quite small and is in equilibrium with the bound fraction:

$$\text{“Free” hormone} + \text{protein} \rightleftharpoons \text{hormone} - \text{protein}$$

It is important to realize that only the free hormone can exert effects on the target-organ cells.

MECHANISMS OF HORMONE ACTION

Common Denominators of Hormonal Effects

Hormones exert their effects by altering the rates at which specific cellular processes proceed. It must be emphasized that hormones never initiate a process;[1] they merely alter its rate. For example, the absence of insulin results in markedly reduced glucose uptake by cells but not absolute cessation. The specific cell processes accelerated or decelerated by hormones are numerous and varied, but most of them fit into one of two general categories (both of which require the combination of hormone with a specific receptor site), namely, alteration of the activity of a crucial enzyme, and alteration of the rate of membrane transport of a substance.

Alteration of Enzyme Activity What is meant by a "crucial enzyme" in this context? The reader should review the section on "interacting metabolic pathways" in Chap. 1. Recall that most metabolic reactions are truly reversible whereas others proceed generally in one direction under the influence of one set of enzymes and in the reverse direction under the influence of a second set of enzymes. The enzymes which catalyze these "one-way" reactions are the primary ones regulated by various hormones. Let us consider, as an example, the relationship between glucose and glycogen in the liver:

$$\text{Glucose} \underset{\text{enzyme B}}{\overset{\text{enzyme A}}{\rightleftharpoons}} \text{glycogen}$$

[1] This statement refers to the primary biochemical effect of the hormone (see below) rather than to the biological events ultimately resulting from these effects. For example, the pituitary hormone LH induces ovulation, and ovulation will not occur in its absence. Accordingly, LH "initiates" ovulation. However, the biochemical reactions (RNA and protein synthesis) which ultimately lead to ovulation are not "all-or-none" events, and their rates are altered, not initiated, by LH. There is no contradiction since it is clear that a certain level of change in the biochemical reaction rates is required to trigger off an essentially "all-or-none" event—ovulation.

Although there are actually multiple steps in both pathways, each catalyzed by a different enzyme, two enzymes (A + B) are particularly critical since they catalyze the major irreversible reactions in opposing directions. The hormone insulin increases the activity of enzyme A and thereby stimulates the formation of glycogen from glucose. In contrast, the hormone epinephrine increases the activity of enzyme B and thereby facilitates the catabolism of glycogen to glucose.

How is the activity of a particular type of enzyme increased? One way is for the cell to produce more of the enzyme, a prominent effect of certain hormones. In Chap. 1 we described the mechanics of protein synthesis and the control of these processes. Hormones may exert effects on the genetic apparatus of their target-organ cells to induce (or repress) the synthesis of RNA and, in turn, the proteins (enzymes) whose synthesis is directed by the particular RNA. This may well be the major biochemical action of most steroid hormones (although it is by no means limited to steroid hormones).

There are, however, other ways by which enzyme activity may be altered without a change in the total number of enzyme molecules in the cells, i.e., with no change in enzyme synthesis. Many enzymes exist within a cell in both active and inactive forms; thus, the number of active enzymes can be increased by converting some of the inactive molecules into active ones. This appears to be the common denominator for a number of hormones.

It is of great interest that certain hormones induce both the synthesis of new enzyme molecules and an increased activity of the enzyme molecules already present in the cell. The advantages of this dual effect are considerable: Induction of new enzyme synthesis requires hours to days, whereas the activation of molecules already present can occur within minutes. Thus, the hormone simultaneously exerts a rapid effect and sets into motion a long-term adaptation.

Alteration of Membrane Transport The effect of many hormones is to facilitate or inhibit the transport of substances into the cell. For example, glucose enters most cells by carrier-mediated, facilitated diffusion, and insulin somehow affects cell membranes so as to increase the rate of glucose transport (this is quite distinct from the insulin effect on enzyme activity described above). Other hormones inhibit glucose transport. A similar type of pattern involving hormonally mediated inhibition and stimulation also operates for the membrane transport of amino acids and other organic metabolites. Finally, the transport of ions and water by kidney cells and others is also influenced by hormones. The inhibition or stimulation of membrane transport is without question one of the major modes of action of many hormones, but we do not know the specific chemical or physical mechanism by which any of these actions is exerted.

Direct and Indirect Effects

Another important consideration is that of direct versus indirect hormone effects. Because intracellular chemical reactions are so closely interrelated, and because all cells of the body are interconnected by the blood, it should be evident that a single effect may set into motion an extensive chain of subsequent events, which may well be more important than the initial event for both the single cell and the total body.

Indirect Effects within a Cell Again we take insulin as our example. As just described, one of its major effects is to increase the transport of glucose into cells, increasing the cellular concentration of glucose and, by mass action, the rates of the many intracellular chemical reactions in which glucose participates. For example, looking again at the synthesis of glycogen (Eq. 12-1), it should be evident that an increased glucose concentration drives this reaction to the right, resulting in synthesis of more glycogen. The important generalization to be derived from this example is that the direct effect of a hormone may initiate multiple indirect effects within the cell.

Indirect Effects on Other Cells When insulin is injected into a person, the blood concentration of glucose rapidly decreases because glucose is leaving the blood and being taken up by cells all over the body. Conversely, insulin deficiency causes the blood glucose to rise because of deficient uptake. When the blood glucose becomes very high, large quantities of glucose, sodium, and water appear in the urine (the mechanisms will be described in Chap. 18). These urinary losses are not due to any direct effect of insulin on the kidney but result indirectly from the high blood glucose. Despite its indirect nature, this urinary loss is one of the major causes of sickness and death in patients with insulin deficiency (diabetes mellitus).

Cyclic AMP A major goal of endocrinologists is to identify the precise initial biochemical actions of hormones on their target-organ cells. For example, we have pointed out that an effect of epinephrine is to activate the crucial enzyme (enzyme B in Eq. 12-1) leading to glycogen breakdown. But exactly how does it do this? Does it act directly on the enzyme itself or does it act several steps removed from the enzyme? To take another example, just what does insulin do to cells that leads to the increased transport of glucose? In recent years it has become apparent that a large number of hormones actually have *identical* initial biochemical actions, namely, activation of the enzyme adenyl cyclase, which is found in cell membranes throughout the body. This has become known as the *second-messenger* or *cyclic AMP* system which is summarized in Fig. 12-3. Adenyl cyclase is a membrane-bound enzyme which, when activated, catalyzes the transformation of cell ATP (on the inner side of the cell membrane) to another molecule known as cyclic AMP. The sole action of the hormone is to interact with a receptor site on the cell membrane so as to activate adenyl cyclase. The cyclic AMP generated as a result then acts within the cell as a "second messenger" to produce the alteration of cell function associated with that hormone. Note that the hormone itself does not gain entry to the cell; this explains how those protein hormones which cannot penetrate cell membranes can still alter cell function.

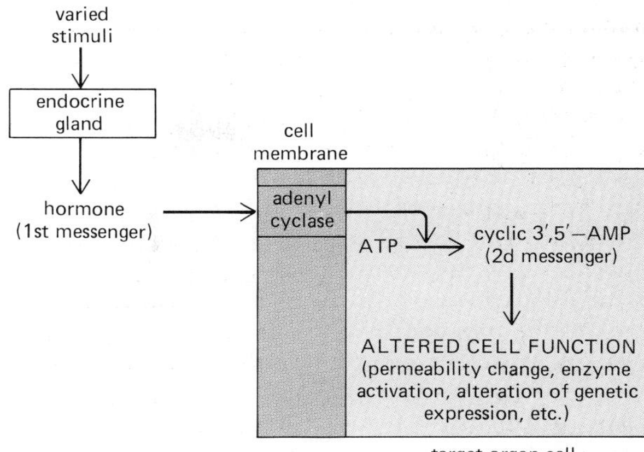

FIGURE 12-3 ***Second-messenger or cyclic-AMP mechanism of hormone action.***

At least 12 hormones have been shown to exert their biochemical actions by stimulating the intracellular synthesis of cyclic AMP. Several others may reduce cell concentrations of cyclic AMP by inhibiting adenyl cyclase. Let us take as an example the action of the hormone epinephrine on glucose formation from glycogen in the liver. Epinephrine activates adenyl cyclase which then catalyzes the formation of cyclic AMP. In turn, cyclic AMP stimulates the conversion of an inactive enzyme to an active form which catalyzes the critical reaction leading to the breakdown of glycogen.

For his elucidation of this beautiful unifying principle, Earl W. Sutherland was awarded the Nobel Prize for medicine and physiology in 1971. Moreover, cyclic AMP plays important roles in nonendocrine regulatory mechanisms as well. Some examples are the control of antibody production and the regulation of vision, and the list will surely grow rapidly as further experiments are done. However not all hormones act via cyclic AMP, the most notable exceptions being the various steroid hormones. It is also unlikely that all the actions of all nonsteroid hormones can be explained by cyclic AMP.

The perceptive reader may have detected an apparent inconsistency in this description of cyclic AMP; if the generation of cyclic AMP is the common biochemical action of many hormones, why do not all these hormones produce identical effects in the body? The answer is that the adenyl cyclase systems in different target-organ cells differ in their abilities to be activated by different hormones. This is best explained by qualitative differences in membrane receptor sites in different tissues. Thus the adenyl cyclase receptor sites in liver cells are able to interact with epinephrine but not parathyroid hormone, whereas just the reverse holds true for bone.

There is a second problem relating to cyclic AMP and specificity: How is it that several hormones, all of which influence cyclic AMP, can have different effects on the same cell? The most likely explanation is that there are qualitatively different receptor sites in the cell, some of which respond to one hormone, some to another. This hypothesis goes on to postulate a compartmentalization of cyclic AMP within the cell, i.e., cyclic AMP generated as a result of the interaction between hormone A and its specific receptor site does not gain access to the same intracellular site on which cyclic AMP generated from the interaction of hormone B and its receptor site does (and vice versa).

Mechanism of Steroid-hormone Action As mentioned above, the common denominator of most (if not all) of the effects of the steroid hormones is an increased synthesis of proteins (enzymes, structural proteins, etc.) by their specific target-organ cells. This increased protein synthesis is the result of hormone-induced stimulation of the synthesis of RNA. The actual mechanism by which steroid hormones influence nuclear constituents so as to alter the rate of RNA synthesis is presently the subject of intense study. The first step is the passage of the steroid across the outer cell membrane into the cell cytoplasm (steroid hormones, unlike protein hormones and amines, are quite lipid-soluble and can readily cross cell membranes). Once in the cell, the hormone binds to a hormone-specific soluble cytoplasmic protein. Accordingly, this protein, which must contain specific molecular receptor sites for the particular steroid hormone, is known as the "receptor." The hormone-protein complex then moves into the nucleus and combines with a chromatin protein (i.e., a protein associated with DNA). It is this molecular combination that triggers off the specific RNA synthesis, i.e., the transcription of the relevant DNA sequences. This system has come to be known as the "mobile-receptor model" since it involves the movement of the hormone-receptor complex from the cytoplasm into the nucleus. In contrast, the cyclic AMP system described earlier utilizes fixed receptor sites on the outer cell membrane of target-organ cells.

Hormone Interactions on Target-organ Cells

Because virtually all hormones are always being secreted at some rate, finite blood concentrations of all hormones exist at all times. These concentrations may vary over wide ranges in response to stimuli, but since the blood contains *some* of each hormone, cells are constantly exposed to the simultaneous effects of many hormones. This allows for complex hormone-hormone interactions on the target-organ cells, the most important phenomenon being that of *permissiveness*. In general terms, frequently hormone A must be present for the full exertion of hormone B's effect. In essence, A is "permitting" B to exert its action. Generally, only a very small quantity of the permissive hormone is required. For example (Fig. 12-4), the hormone epinephrine causes marked release of fatty acids from adipose tissue only in the presence of thyroid hormone. Many of the defects seen when an endocrine gland is removed or ceases to function because of disease actually result from loss of the permissive powers of the hormone secreted by that gland.

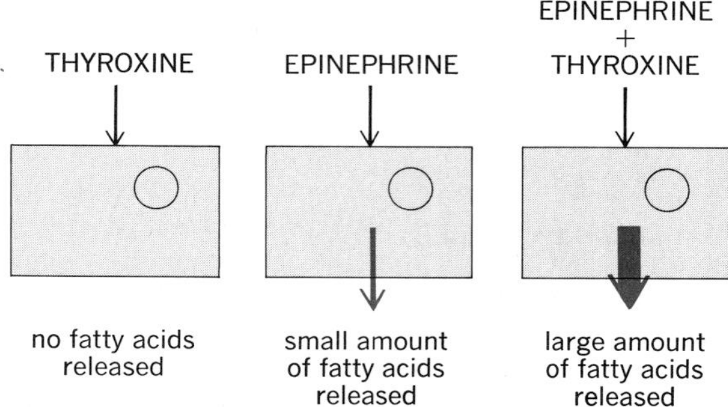

FIGURE 12-4 ***Ability of thyroxine to permit epinephrine-induced liberation of fatty acids from adipose-tissue cells.***

Pharmacologic Effects

Administration of very large quantities of a hormone may have results which are never seen in a normal person, although these so-called *pharmacologic effects* sometimes occur in endocrine diseases when excessive amounts of hormone are secreted. These effects are of great importance in medicine since hormones in pharmacologic doses are used as therapeutic agents. Perhaps the most famous example is that of the adrenal hormone cortisol, which is highly useful in suppressing allergic and inflammatory reactions. Mental changes, including outright psychosis, may also be induced by large quantities of cortisol and are frequently a striking symptom of patients suffering from hyperactive adrenal glands.

THE ENDOCRINE GLANDS: CONTROL OF SECRETION AND ANATOMY

The anatomy of each endocrine gland is unique, as are the immediate inputs which control the secretion of hormones by the glands. However, these inputs fall into one of five categories, and an appreciation of them should greatly facilitate the understanding of each specific hormone as it is discussed in subsequent chapters. Table 12-2 summarizes the categories to be discussed; it must be reemphasized that all the material contained in this table will be presented again in later chapters where relevant. It should also be noted that the table does not constitute a complete inventory of all hormones or of the immediate inputs for any given hormone but includes only those which help to illustrate the common patterns.

TABLE 12-2

SUMMARY OF THE CONTROL OF HORMONE SECRETION†

Hypothalamic neurons directly release:
- Oxytocin } From posterior pituitary
- Antidiuretic hormone } From posterior pituitary
- Hypothalamic releasing factors

Hypothalamic releasing factors directly control the release of:
- Growth hormone
- Thyroid-stimulating hormone (TSH)
- Adrenocorticotropic hormone (ACTH)
- Gonadotropic hormones (FSH and LH)
- Prolactin

(All from anterior pituitary)

Anterior pituitary hormones directly control the release of:
- Thyroid hormone
- Cortisol (from adrenal cortex)
- Gonadal hormones
  - (female: estrogen and progesterone)
  - (male: testosterone)

Autonomic neurons directly control the release of:
- Epinephrine and norepinephrine (from adrenal medulla)
- Renin (from kidney)
- Insulin and glucagon (from pancreas)
- Gastrointestinal hormones
- ? Others

Plasma concentrations of ions or nutrients directly control the release of:
- Parathyroid hormone
- Insulin and glucagon (from pancreas)
- Aldosterone (from adrenal cortex)
- Calcitonin

† As described in the text, this table does not necessarily list all the controls of each hormone.

The Pituitary

It is evident from Table 12-1 that the *pituitary gland* (or *hypophysis cerebri*) is of major importance in hormone secretion. This gland lies in the *hypophyseal fossa,* a pocket in the sphenoid bone just below the hypothalamus (Fig. 12-5) to which it is connected by a stalk (the *infundibulum*) containing neurons and small blood vessels. It is composed of three lobes, the *anterior, intermediate,* and *posterior lobes,* each of which is a more or less distinct gland. The term *adenohypophysis* refers to the anterior and intermediate lobes and their part of the stalk; *neurohypophysis* refers to the posterior lobe and its part of the stalk. In human beings, the intermediate lobe is rudimentary, and its function is unclear. It contains two substances called melanocyte-stimulating hormones (MSH) which are known to cause skin darkening in lower vertebrates; however, their function in human beings is unknown.

The arteries of the pituitary arise from the right and left internal carotid arteries. The *superior hypophyseal arteries* (Fig. 12-6), before entering the hypothalamus and stalk, give off a *trabecular artery,* which descends to supply the lower part of the stalk. The *inferior hypophyseal arteries* supply the posterior pituitary and anastomose with the superior hypophyseal arteries, but they do not enter the secretory portions of the anterior pituitary; that blood supply will be described below. The veins draining the pituitary enter neighboring venous sinuses in the dura mater.

The Anterior Pituitary Hormones Despite its close proximity to the brain, the anterior pituitary is not neural but is composed of true glandular tissue, which produces at least six different protein hormones. Secretion of each of the six hormones may occur independently of the others; i.e., the anterior pituitary comprises, in effect, six endocrine glands anatomically associated in a single structure. Until recently, the cells of the anterior pituitary were grouped mainly according to their propensities for being stained by certain dyes; those cells stained strongly by acid dyes are known as *acidophils,* or *α cells,* and those stained strongly by basic dyes are *basophils,* or *β cells* (Fig. 12-7). These two groups comprise the chromophilic or "color-attracting" cells, in contrast to a third group, the *chromophobes,* which have little affinity for either dye. However, more recent evidence using histologic techniques specific for the different hormones has documented that these cell types are not homogeneous; i.e., there are distinct subgroups within the categories, each group responsible for the secretion of one (or, at most, two) hormones. Thus, different groups of acidophils secrete growth hormone, prolactin, or ACTH, and groups of basophils secrete TSH, FSH, and LH. The precise function of the chromophobes is unknown; they may represent nonsecretory phases of the other cell types.

The major function of two of the anterior pituitary hormones is to stimulate the secretion of other hormones: (1) *Thyroid-stimulating hormone* (TSH) induces secretion of thyroid hormone from the thyroid.

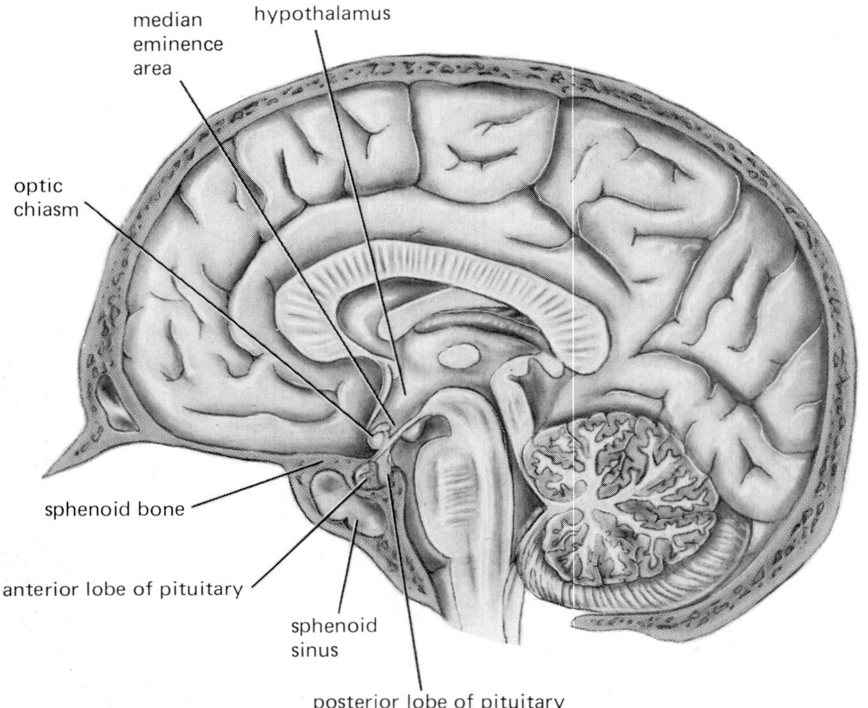

FIGURE 12-5 ***Relation of the pituitary gland to the brain and hypothalamus.***

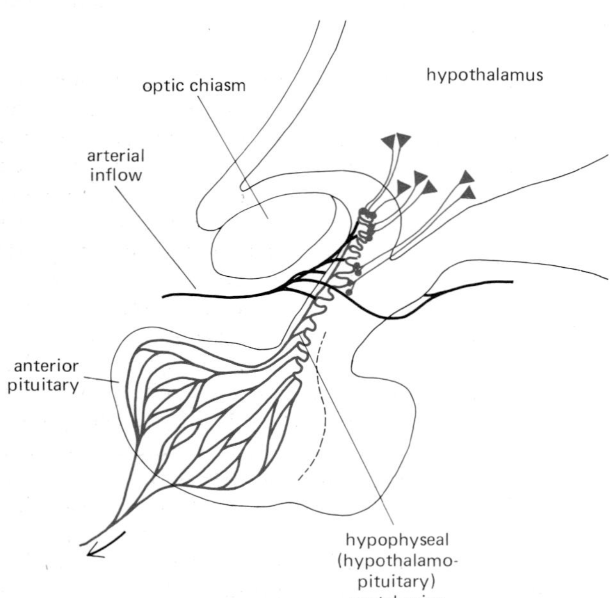

FIGURE 12-6 ***Hypothalamus–anterior-pituitary vascular connections. The hypothalamic neurons, which secrete releasing factors, end on the capillary loops of the primary capillary plexus of the portal system carrying blood from the hypothalamus to the anterior pituitary.***

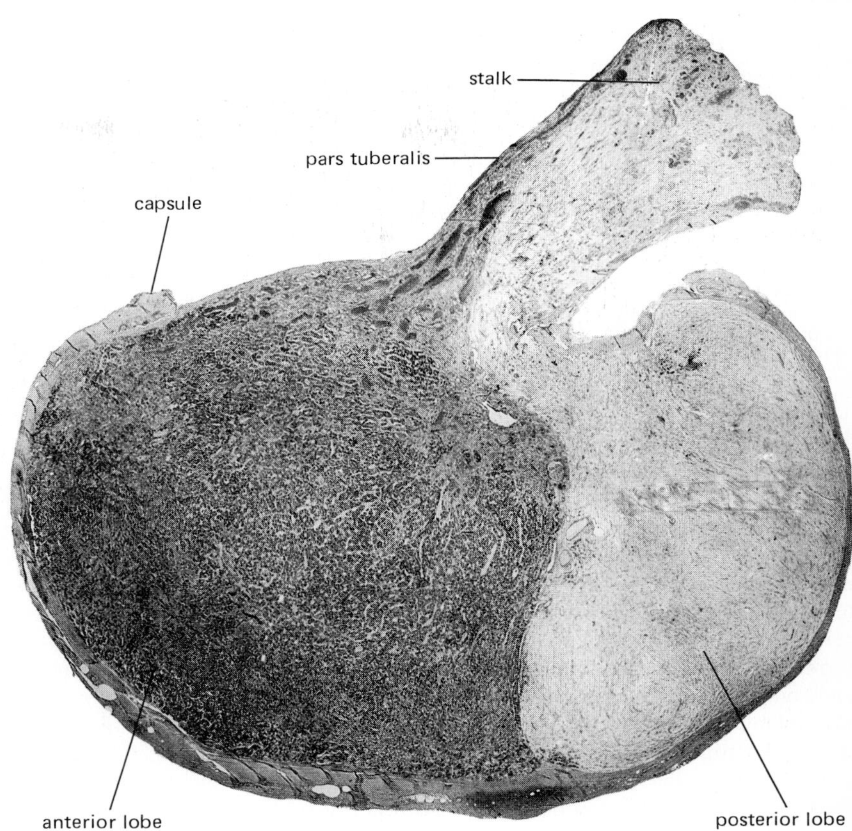

FIGURE 12-7 ***Histologic preparation of a sagittal section through the pituitary gland.*** **(From Wm. F. Windle, "Textbook of Histology," 5th ed., McGraw-Hill, New York, 1976.)**

(2) *Adrenocorticotropic hormone* (ACTH), meaning "hormone which stimulates the adrenal cortex," is responsible for stimulating the secretion of *cortisol.* Thus, the important target organs for TSH and ACTH are the thyroid and adrenal cortex, respectively.

Two other anterior pituitary hormones, the *gonadotropic hormones, follicle-stimulating hormone* (FSH) and *luteinizing hormone* (LH), primarily control the secretion of sex hormones (*estrogen, progesterone,* and *testosterone*) by the gonads. The gonadotropins differ from TSH and ACTH in that, besides controlling the secretion of other hormones, they have a second major role, the growth and development of the reproductive cells, the sperm and ova. The gonads are the sole target organs for the anterior pituitary gonadotropins.

It should now be clear why the anterior pituitary is frequently called the master gland; it secretes six hormones itself and controls the secretion of three or four (depending upon the person's sex) other hormones.

What about the two remaining anterior pituitary hormones; do they also control the secretion of some other hormones? The answer is no. *Prolactin*'s major target organs are the breasts, and *growth hormone* exerts multiple metabolic effects upon many organs and tissues. The target organs and functions of the anterior pituitary hormones are summarized in Fig. 12-8.

Let us now return to the target organs of the tropic hormones. We have said that the secretion rates of thyroid hormone, cortisol, and the gonadal sex hormones are stimulated by anterior pituitary tropic hormones. Are they also controlled by other types of input? The answer is no in normal (and nonpregnant) persons. The sole control of these hormones is via the pituitary. This can easily be proved by observing them after surgical removal of the anterior pituitary; the secretion of thyroid hormone, cortisol, and sex hormones ceases almost completely. The thyroid gland,

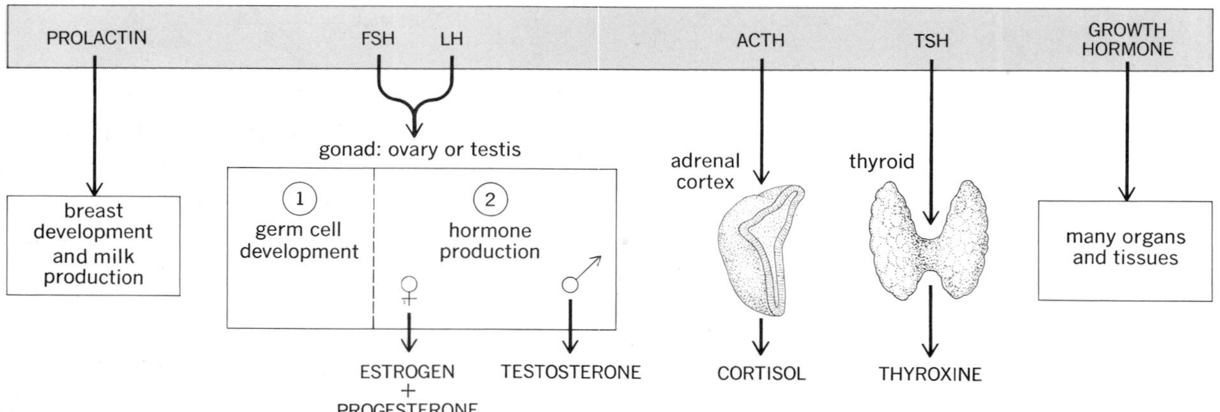

FIGURE 12-8 ***Target organs and functions of the six anterior pituitary hormones.***

most of the adrenal cortex, and the gonads greatly decrease in size and take on a nonfunctioning appearance; these observations clearly show that the tropic hormones control not only the secretion of their target-gland hormones but the growth and development of the target glands themselves. The reason the adrenal cortex does not atrophy completely is that the cells which secrete the second major cortical hormone, *aldosterone,* are primarily controlled not by ACTH but by other inputs to be described in Chap. 18.

Control of Anterior Pituitary Hormone Secretion What direct inputs control secretion of the anterior pituitary hormones? One important type of input for the tropic hormones is the target-gland hormone itself, a beautiful example of negative feedback (Chap. 6). Thus, ACTH stimulates cortisol secretion and increases the blood concentration of cortisol, which acts upon the anterior pituitary to inhibit ACTH release. In this manner, any increase in ACTH secretion is partially prevented by the resultant increase in cortisol secretion, as illustrated in Fig. 12-9. This same pattern of negative feedback is exerted by thyroxine and the sex hormones on their respective pituitary tropic hormones (the sex hormone effects are actually more complex than those shown in the figure, as will be discussed in Chap. 21). It is evident that such a system is highly effective in damping hormonal responses, i.e., limiting the extremes of hormone secretory rates. However, if this negative-feedback relationship were the sole source of anterior pituitary control, there would be no way of altering anterior pituitary output; some unchanging equilibrium blood concentrations of pituitary and target-gland hormone would always be maintained. Obviously, there must be some other type of input to the anterior pituitary. In reality, this other input is the major controller of anterior pituitary function.

The major inputs controlling release of anterior pituitary hormones are a group of so-called releasing factors produced in the hypothalamus. An appreciation of the anatomic relationships between the hypothalamus and anterior pituitary is essential for understanding this process. Although the anterior pituitary lies just below the hypothalamus, there are no important neural connections between the two, but there is an unusual capillary-to-capillary connection (Fig. 12-6). The superior hypophyseal arteries end in the base of the hypothalamus (the *median eminence*) as intricate capillary tufts which recombine into the *hypophyseal* (or hypothalamopituitary) *portal veins* (the term "portal" denotes veins which connect two distinct capillary beds). These pass down the stalk and into the secretory portion of the anterior pituitary where they break into a second capillary bed, the *anterior pituitary capillaries,* which provide most of the vascular supply to that organ.

Thus the hypophyseal portal veins offer a local route for flow of capillary blood from hypothalamus to anterior pituitary. The axons of neurons which origi-

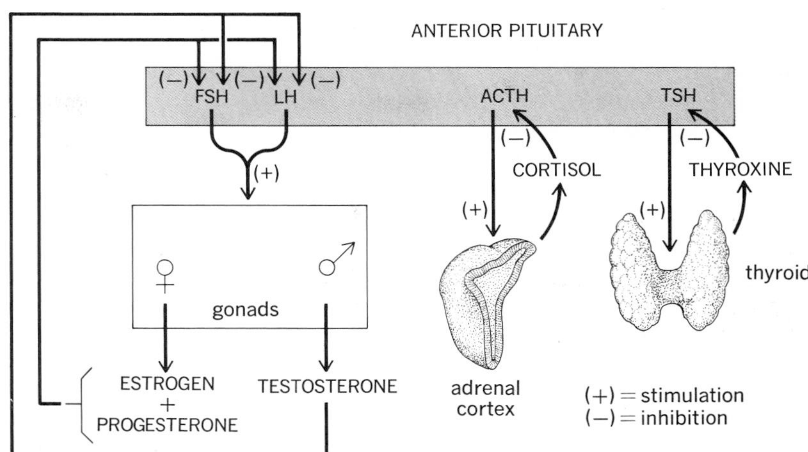

FIGURE 12-9 ***Negative feedback of target-organ hormones on their respective anterior pituitary tropic hormones.***

nate in diverse areas of the hypothalamus terminate in the median eminence around the hypothalamic capillary origins of the portal vessels. These neurons secrete into the capillaries substances which are carried by the portal vessels to the anterior pituitary, where they act upon the various pituitary cells to control hormone secretion (Fig. 12-10). We are dealing with multiple discrete substances, each controlling the release of only one (or, at most, two) type of pituitary hormone. Most of these substances stimulate release of their relevant hormones and are therefore termed *hypothalamic releasing factors* (TSH-releasing factor, ACTH-releasing factor, etc.). At least one, that which controls prolactin secretion, inhibits rather than stimulates prolactin release and is termed *prolactin-inhibiting factor* (PIF). Moreover, the system is even more complex than this in that some hormones are controlled by dual systems of hypothalamic substances, one inhibitory and the other stimulatory.

Each factor appears to be secreted only by neurons in a discrete portion of the hypothalamus, i.e., one group of neurons secretes PIF, a different group secretes TSH-releasing factor, etc. Regardless of the origin of the hypothalamic neurons, the releasing factors are all secreted into the hypothalamopituitary portal vessels. The alert reader will have recognized that these substances fulfill all the criteria for our definition of a hormone and really should be called that instead of factor. This change in nomenclature is presently under way. It should also be noted that the "releasing" factors

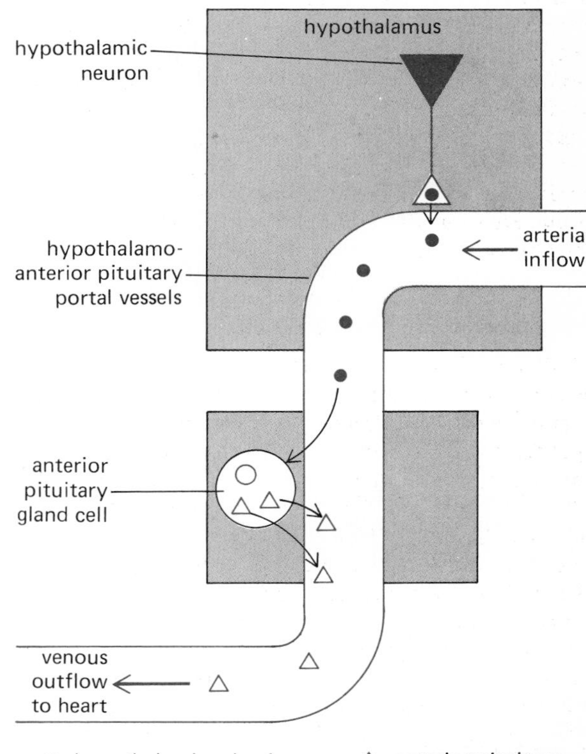

FIGURE 12-10 ***Control of anterior pituitary secretion by a hypothalamic releasing factor.***

control not only the release of their respective pituitary hormones, but their synthesis as well.

These relationships form much of the foundation of neuroendocrinology. We stated earlier that the nervous and endocrine systems actually function as a single interrelated system; the anterior pituitary may be the master gland, but its function is primarily controlled by the hypothalamus via the releasing factors. It now appears that many diseases characterized by inadequate secretion of one or more pituitary hormones really are due to hypothalamic malfunction rather than primary pituitary disease.

Our analysis has pushed the critical question one step further: The hypothalamic releasing factors control anterior pituitary function, but what controls secretion of the releasing factors? The answer is neural and hormonal input to the hypothalamic neurons which secrete the releasing factors. The hypothalamus receives neural input, both facilitory and inhibitory, from virtually all areas of the body; the specific type of input which controls the secretion rate of the individual releasing factors will be described in future chapters when we discuss the relevant anterior pituitary or target-gland hormone. It suffices for now to point out that endocrine disorders may be generated, via alteration of hypothalamic activity, by all manner of neural activity, such as stress, anxiety, etc. An example is sterility caused by severe emotional upsets.

Hormonal influences upon the hypothalamus are also important. Some of the negative-feedback effect of thyroxine, cortisol, and sex hormones upon pituitary tropic-hormone secretion is actually mediated via the hypothalamus, i.e., by inhibition of releasing-factor secretion. For example, cortisol acts not only directly upon the anterior pituitary to inhibit ACTH secretion but also upon the hypothalamus to inhibit ACTH-releasing-factor secretion, an event which also reduces ACTH secretion. Presently, there is still controversy over the quantitative importance of these two negative-feedback sites for the various hormones, but the generalization that both sites are involved, albeit to varying extents for each hormone, seems likely. In addition, it is quite likely that growth hormone, one of the two anterior pituitary hormones which have no target-organ hormones, exerts a negative-feedback control, via the hypothalamus, over its own secretion. The situation for prolactin is presently too unclear to permit speculation. Our description of the interrelationships of the hypothalamus, anterior pituitary, and target glands is now complete, as shown in Fig. 12-11.

Hypothalamus—Posterior-Pituitary Function The *posterior pituitary* lies just behind the anterior pituitary in the same bony pocket in the sphenoid bone at the base of the hypothalamus, but its structure is totally different from its neighbor's. The posterior pituitary is actually an outgrowth of the hypothalamus and is true neural tissue. Two well-defined clusters of hypothalamic neurons, the *supraoptic* and *paraventricular nuclei,* send out nerve fibers which pass by way of the connecting stalk to end within the posterior pituitary in close proximity to capillaries (Fig. 12-12). The two hormones, *oxytocin* and *antidiuretic hormone* (ADH, vasopressin), released from the posterior pituitary are actually synthesized in the hypothalamic cells. Antidiuretic hormone is chiefly a product of the supraoptic nucleus and oxytocin of the paraventricular nucleus. After synthesis in the neuronal cell bodies, the hormones are enclosed in small vesicles which move slowly down the cytoplasm of the neuron axons to accumulate at the nerve endings. Release into the capillaries occurs in response to generation of an action potential within the nerve. Thus, these hypothalamic neurons secrete hormones in a manner quite analogous to that described previously for hypothalamic releasing factors, the essential difference being that the releasing factors are secreted into capillaries which empty directly into the anterior pituitary whereas the posterior pituitary capillaries drain primarily into the general body circulation.

It is evident, therefore, that the term posterior pituitary hormones is somewhat of a misnomer since the hormones are actually synthesized in the hypothalamus, of which the posterior pituitary is merely an extension. However, when oxytocin and antidiuretic hormone were discovered, this fact was not known, and both were believed to be synthesized in small posterior pituitary cell bodies now known to be connective-tissue cells. Indeed, the entire posterior pituitary can be surgically removed with only a temporary loss of oxytocin and ADH secretion; capillaries soon grow up along the connecting stalk, and hormone release returns to normal levels. We must therefore add two more names to our growing list of hormones under direct or indirect control of the hypothalamus.

The Adrenal Glands

The paired adrenal (or *suprarenal*) glands are small triangular organs, one on top of each kidney; each measures only approximately 5 by 3 by 1 cm. Each adrenal comprises two distinct endocrine glands, an

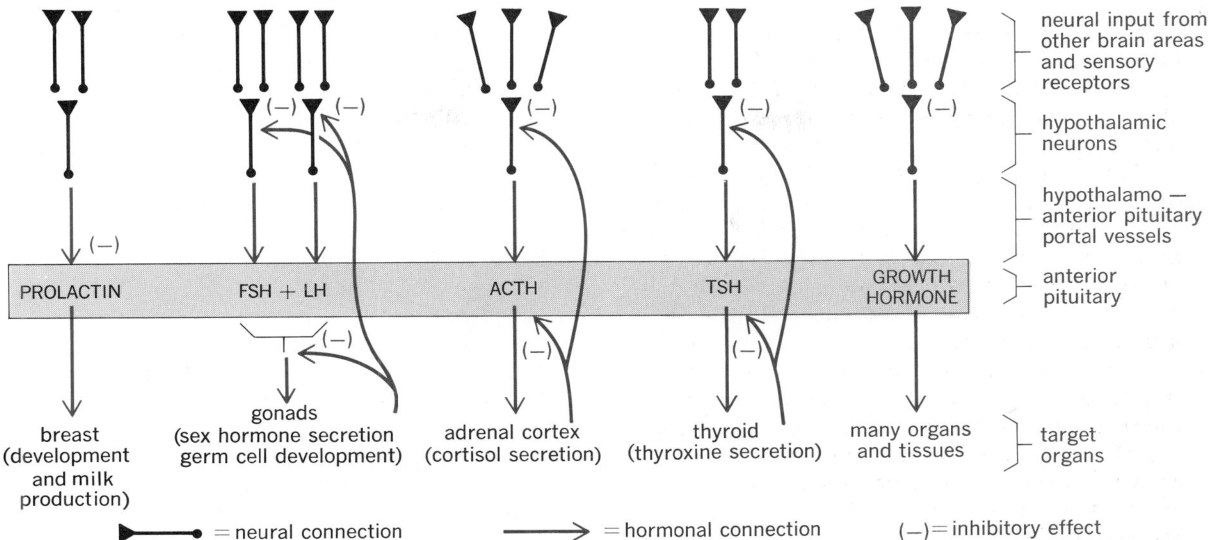

FIGURE 12-11 ***Summary of neural–anterior-pituitary–target-organ relationships. This model will almost certainly become more complicated as research progresses. For example, current evidence suggests the possible existence of dual systems of hypothalamic substances, one inhibitory and the other stimulatory for several of the hormones.***

inner *adrenal medulla* and its surrounding *adrenal cortex* (Fig. 12-13).

The cortex, which forms the larger part of the gland, secretes several steroid hormones, the most important of which are: (1) *aldosterone,* which is essential for the maintenance of sodium and potassium balance and is therefore known as a *mineralocorticoid;* (2) *cortisol,* known as a *glucocorticoid* because it has important effects, among others, on carbohydrate and protein metabolism; and (3) *sex steroids,* produced in small but significant quantities. The adrenal cortex is itself divided into three layers: a thin outer *zona glomerulosa,* a middle *zona fasciculata,* and an inner *zona reticularis* which abuts on the medulla (Fig. 12-13). The cells of all three layers contain the abundant lipids typical of steroid-secreting cells; the zona glomerulosa synthesizes aldosterone, and the other zones produce cortisol (and the sex steroids). As described above, the secretion of cortisol is totally under the control of ACTH from the anterior pituitary whereas aldosterone secretion is controlled by nonpituitary inputs discussed in Chap. 18.

The adrenal medulla, constituting only 10 percent of the total gland, contains *chromaffin cells* arranged in

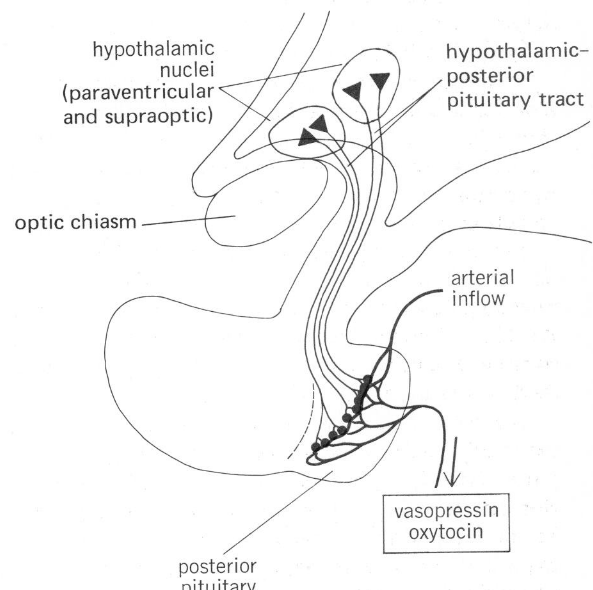

FIGURE 12-12 ***Relationship between the hypothalamus and posterior pituitary.*** **(Adapted from Guillemin and Burgus.)**

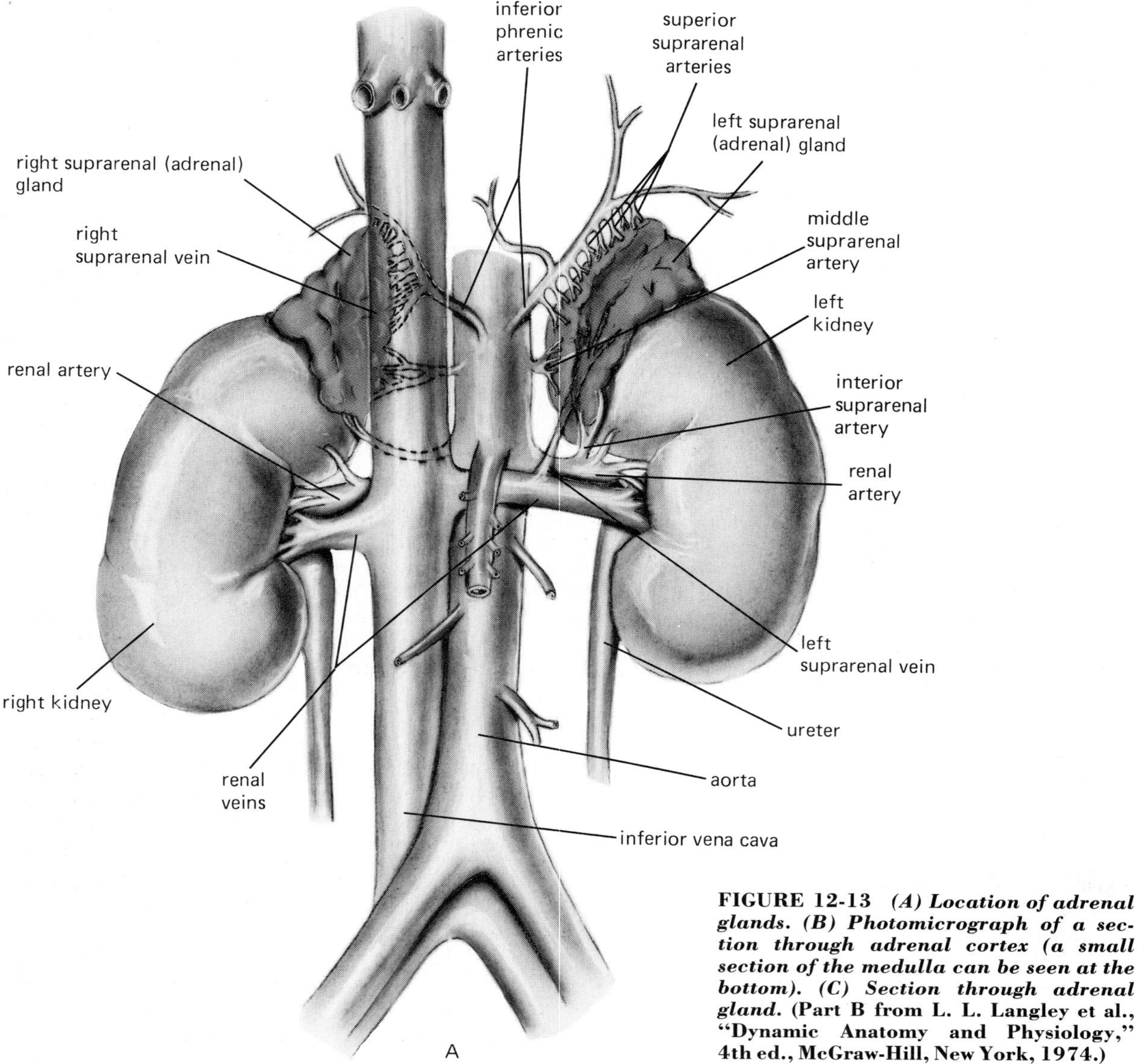

FIGURE 12-13 ***(A) Location of adrenal glands. (B) Photomicrograph of a section through adrenal cortex (a small section of the medulla can be seen at the bottom). (C) Section through adrenal gland.*** **(Part B from L. L. Langley et al., "Dynamic Anatomy and Physiology," 4th ed., McGraw-Hill, New York, 1974.)**

rows along the edges of wide venous sinuses. These cells are under the control of sympathetic preganglionic nerves and, in response to nerve stimulation, secrete their hormones into the extracellular space surrounding the venous sinuses; the hormones pass through the sinus walls and enter the circulation. In human beings, the hormone released by the medulla is for the most part the amino acid derivative *epinephrine* (a smaller amount of norepinephrine is also secreted), a substance closely related to norepinephrine but with several distinctly different functional properties. In controlling epinephrine secretion, the adrenal medulla behaves just like a sympathetic ganglion and is dependent upon stimulation by the sympathetic preganglionic fibers. Destruction of these incoming nerves causes marked reduction of epinephrine release and failure to increase

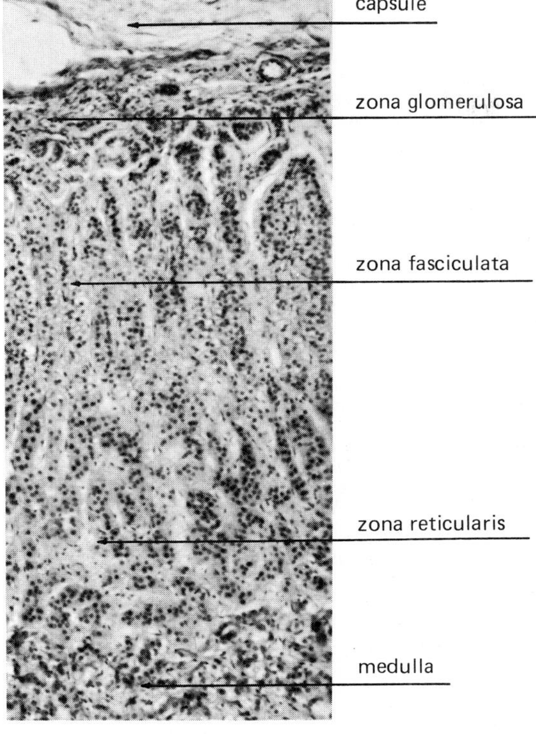

B

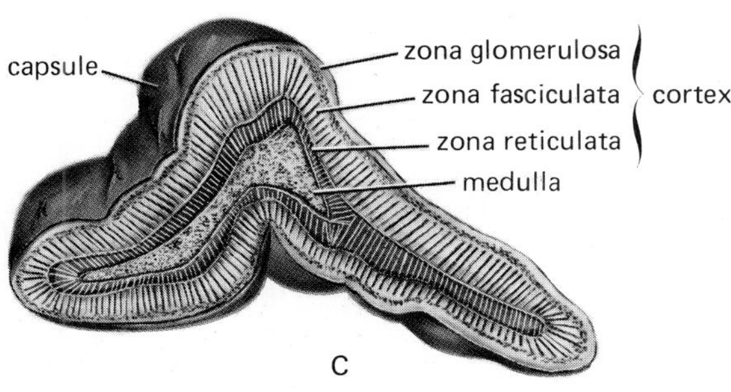

C

secretion in response to the usual physiologic stimuli.

The adrenal medulla is best viewed as a general reinforcer of sympathetic activity. Its secretion of epinephrine into the blood serves to increase the overall sympathetic functions of the body. We shall discuss in later chapters the specific reflexes which cause enhanced sympathetic activity and elicit epinephrine secretion; suffice it to say that these reflexes are under the strong control of higher brain centers, particularly the hypothalamus. Here is another hormone which is literally a product of the nervous system.

Role of the Autonomic Nervous System in Hormone Secretion

We have seen that one hormone—epinephrine (and norepinephrine)—is actually a product of the sympathetic nervous system. This is not the only influence the autonomic nervous system has on the endocrine system. Certain of the endocrine glands receive a rich supply of sympathetic and/or parasympathetic postganglionic fibers, and the activity of these neurons influences their rates of hormone secretion. As shown in Table 12-2, examples are the secretion of renin by the kidneys, insulin and glucagon by the pancreas, and the gastrointestinal hormones. Others will almost certainly be discovered. In no case yet studied, other than the adrenal medulla, does this autonomic input serve as the sole regulator of the hormone's secretion but is one of multiple inputs.

The Thyroid Gland

The thyroid gland secretes two iodine-containing amino acid derivatives, thyroxine and triiodothyronine (collectively known as *thyroid hormone*[2]), which have widespread effects on energy metabolism, growth and development, and brain function. The thyroid is located in the lower part of the neck in front of the trachea (Fig. 12-14A). It consists of right and left lobes connected by a narrow *isthmus,* all of which are covered by a thin fibrous connective-tissue capsule. The basic unit of the thyroid is a *follicle,* which is a sphere of epithelial *follicular cells* surrounding a semifluid material known as *colloid* (Fig. 12-14B). Surrounding the follicles is a loose connective tissue in which *parafollicular cells* [also called C, for clear, or light, cells] are embedded. These cells, which are the second major cell type of the thyroid, secrete still another hormone, *calcitonin.*

Thyroid physiology received its greatest stimulus when it was discovered that the enlarged thyroids (goiters) so common among inland populations were completely preventable by the administration of small quantities of iodine, as little as 4 g/year. The other ingredient for thyroxine synthesis is the amino acid *tyrosine,* which can be produced from a wide variety of substances within the body and therefore offers no supply problem.

[2] For simplicity, we shall refer in this book only to thyroxine which is produced in greater quantity.

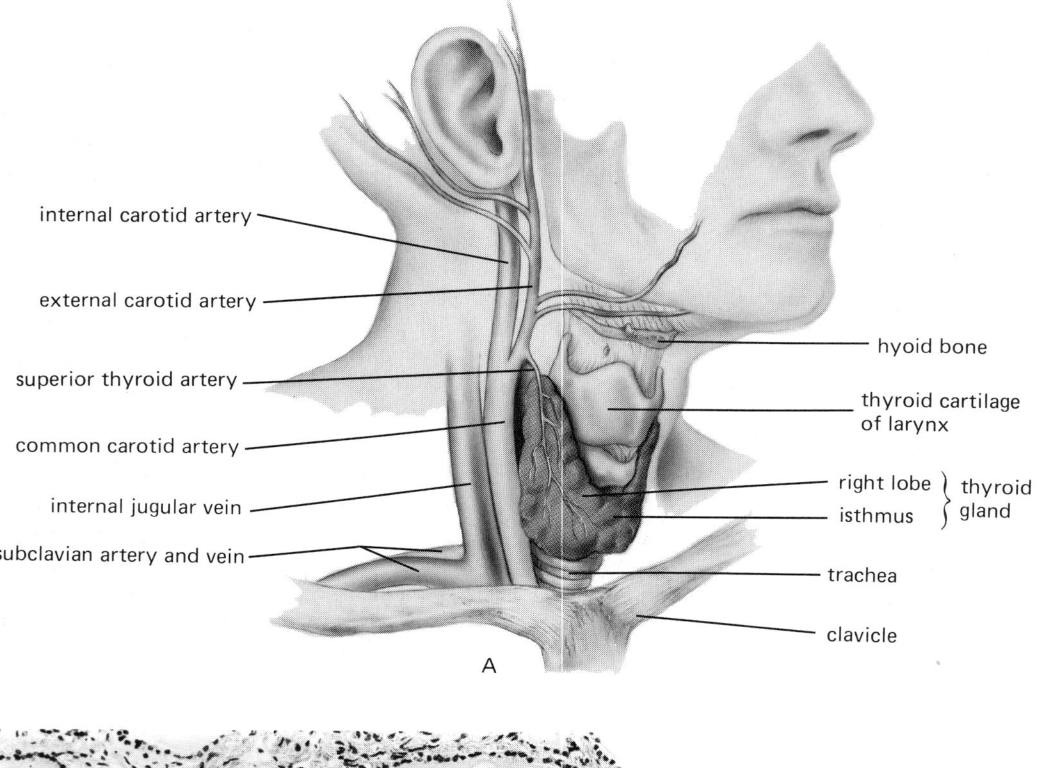

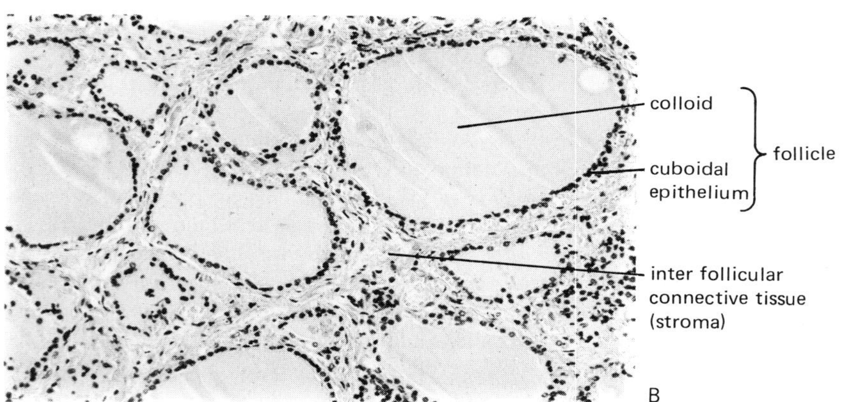

FIGURE 12-14 ***(A) Thyroid gland. (B) Histology of the thyroid gland.*** **(Part B from Wm. F. Windle, "Textbook of Histology," 5th ed., McGraw-Hill, New York, 1976.)**

Much of the ingested iodine absorbed by the gastrointestinal tract (which converts it to iodide, the ionized form of iodine) is removed from the blood by the follicular cells of the thyroid, which manifest a remarkably powerful active-transport mechanism for iodide. Once in the cells, the iodide is reconverted to an active form of iodine, which immediately combines with tyrosine. Two molecules of iodinated tyrosine then combine to give thyroxine. During these synthetic biochemical processes, the tyrosine and thyroxine molecules are bound to a polysaccharide-protein material known as *thyroglobulin,* which constitutes most of the colloid within the follicles. The normal gland may store several weeks' supply of thyroxine in this bound form.

Hormone release into the blood occurs by uptake of the thyroglobulin from the colloid by the follicular cells, enzymatic splitting of the thyroxine from the thyroglobulin, and the entry of this freed thyroxine into the blood. Finally, as described above, the overall process is controlled by a pituitary *thyroid-stimulating hormone,* which stimulates certain key rate-limiting steps and thereby alters the rate of thyroxine secretion. A variety of defects—dietary, hereditary, or disease-induced—may decrease the amount of thyroxine released into the blood. One such defect results from dietary iodine deficiency. However, this deficiency need not lead to permanent reduction of thyroxine secretion because the thyroid gland enlarges (goiter) so as to provide greater utilization of whatever iodine is available. This response is mediated by thyroid-stimulating hormone, which induces thyroid enlargement whenever the blood concentration of thyroxine decreases; the blood concentration of thyroid-stimulating hormone increases because of diminished feedback inhibition secondary to lowered blood thyroxine.

Once in the blood, most of the thyroxine becomes bound to certain plasma proteins and circulates in this form. This protein-bound thyroxine is in equilibrium with a much smaller amount of free thyroxine, the latter being the effective hormone. Figure 12-15 presents a summary of the thyroxine pathways. Since most of the iodine in the plasma is in the thyroid-hormone molecules, and since most of the thyroid hormone is bound to protein, a useful estimate of the level of circulating thyroid hormone is gained by measuring the plasma protein-bound iodine (PBI).

The Parathyroid Glands

The parathyroid hormone is one of the most important regulators of plasma calcium and phosphate concentrations (Chap. 18). The parathyroid glands are very small oval structures, which lie along the back of the thyroid gland between the gland and its capsule (Fig. 12-16A). Usually there are four parathyroids, two behind each lobe of the thyroid. Connective tissue surrounds the parathyroids, forming a thin *capsule,* and passes into the substance of the gland, carrying with it the large blood vessels, nerves, and lymphatics. The glandular cells themselves (the *principal* or *chief* cells), which secrete parathyroid hormone, are arranged in columns separated by a rich network of sinusoidal capillaries (Fig. 12-16B).

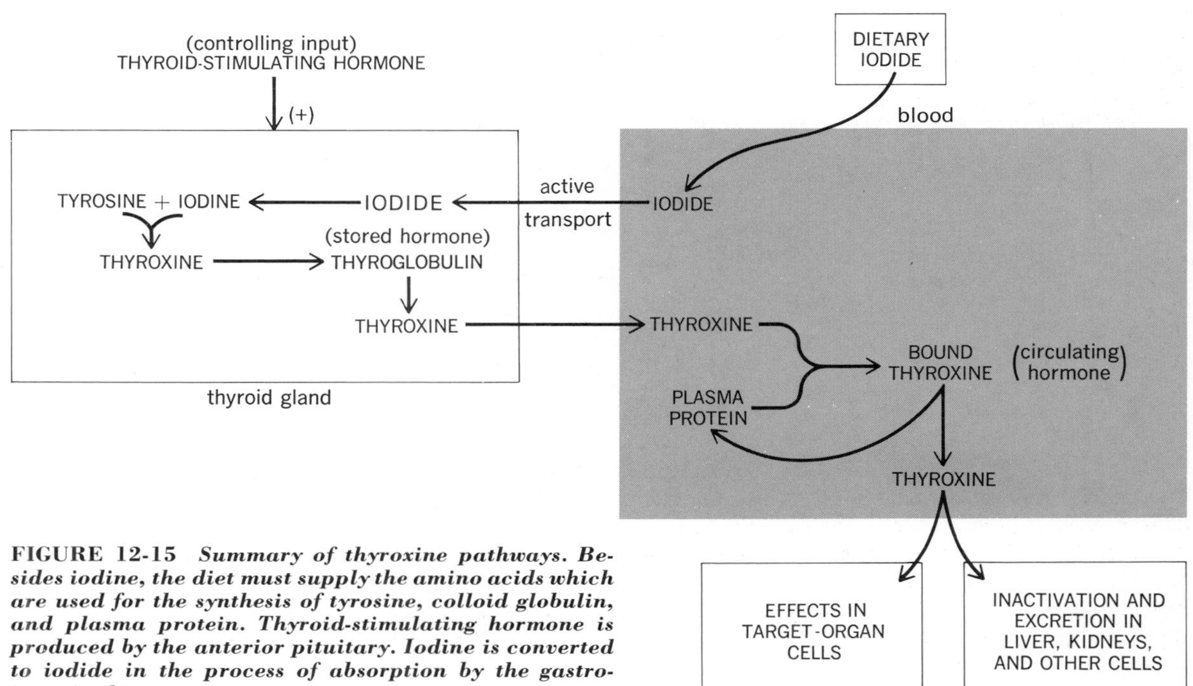

FIGURE 12-15 ***Summary of thyroxine pathways. Besides iodine, the diet must supply the amino acids which are used for the synthesis of tyrosine, colloid globulin, and plasma protein. Thyroid-stimulating hormone is produced by the anterior pituitary. Iodine is converted to iodide in the process of absorption by the gastrointestinal tract.***

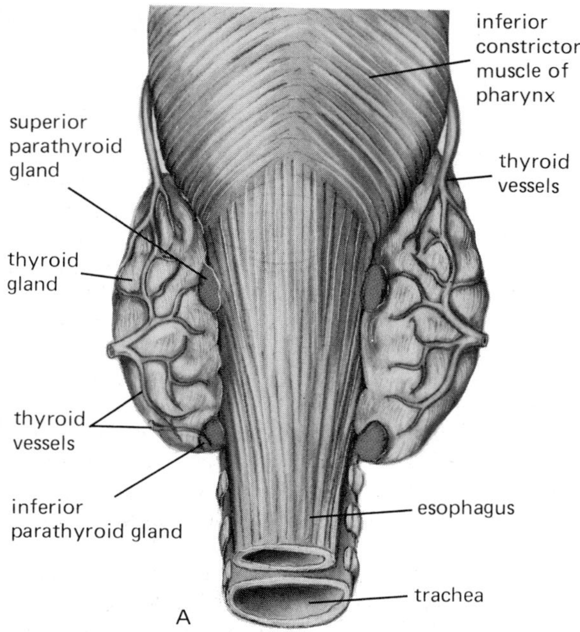

FIGURE 12-16 ***(A) Esophagus and thyroid gland viewed from the back. The parathyroid glands (usually four in number) are on the posterior surface of the thyroid. (B) Histology of the parathyroid gland.*** **(Part B from Wm. F. Windle, "Textbook of Histology," 5th ed., McGraw-Hill, New York, 1976.)**

Parathyroid hormone is the first hormone in our discussion the secretion of which does not primarily involve either the pituitary, hypothalamus, or sympathetic nervous system. Rather, the cells of the parathyroid gland respond directly to the calcium concentration of the blood supplying them. Since the major function of parathyroid hormone is to regulate plasma calcium concentration, this control mechanism is appropriate and remarkably simple: The hormone-secreting cells are themselves sensitive to the plasma substance their hormone regulates.

As shown in Table 12-2, this type of pattern exists not just for parathyroid hormone but for other hormones as well. Thus, the secretion of insulin and glucagon is analogous in that the pancreatic cells which produce these hormones are controlled, in large part, directly by the glucose concentration of the plasma supplying them; the adaptive rationale is also analogous in that a major function of insulin and glucagon is regulation of the plasma glucose concentration. Similarly, aldosterone helps to regulate plasma potassium concentration, and the adrenal cells which secrete it are directly controlled, at least in part, by the plasma potassium concentration.

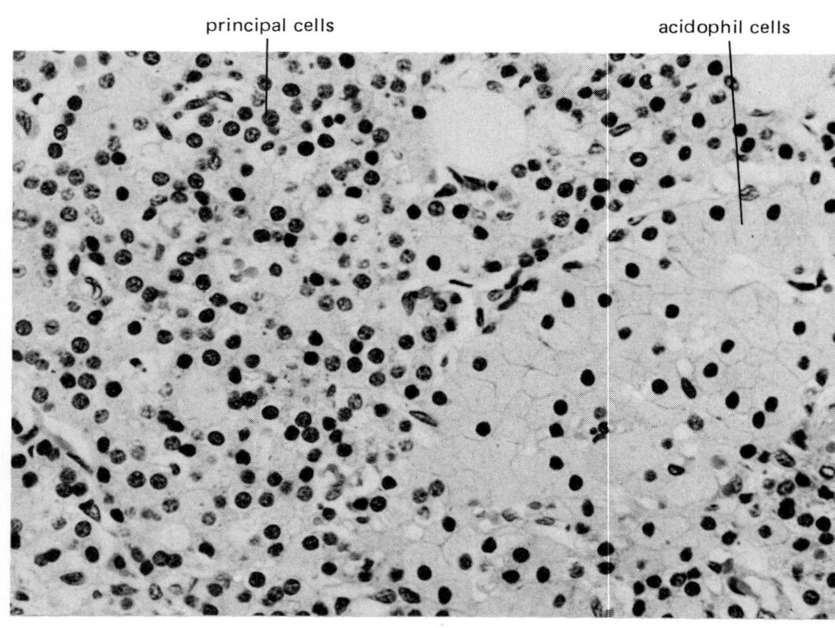

Summary

Table 12-1 should now appear less formidable. The control mechanisms for most of the hormones involve the direct or indirect participation of the hypothalamus and pituitary. The anterior pituitary hormones are controlled primarily by releasing factors secreted into the hypothalamopituitary portal vessels by neurons in the hypothalamus. In turn, the four anterior pituitary tropic hormones control hormone secretion by their target-organ glands, the thyroid, adrenal cortex, and gonads. These glands exert a negative-feedback control over their own secretion via the effects of their hormones on both the hypothalamus and anterior pituitary. The hypothalamus itself also produces two hormones, oxytocin and ADH, which are released from nerve endings in the posterior pituitary. And finally, the hypothalamus exerts profound control over the autonomic nervous system,[3] including the adrenal medulla. Thus, this small area of brain, which weighs 4 g in the adult human being, acts as a compact integrating center, receiving messages, both neural and hormonal, from all areas of the body and sending out efferent messages via both the nerves and hormones. As we shall see, it also regulates body temperature, food intake, water balance, and a host of other autonomic, endocrine, and behavioral activities.

Despite the central role of the hypothalamopituitary system and the autonomic nervous system, there are important hormones, including aldosterone (from the adrenal cortex), insulin and glucagon (from the pancreas), parathyroid hormone, the kidney hormones, and the gastrointestinal hormones, whose secretion is controlled, at least in part, by distinct mechanisms. In some cases, the gland cells which secrete these hormones are controlled by the very plasma constitutents regulated by the hormone, but there are still other specific types of control not shown in Table 12-2; for example, a dominant controller of the gastrointestinal hormones is the nutrient composition of the gastrointestinal contents. And finally it must be emphasized that the secretion of any given hormone may be controlled by multiple inputs; at present count the cells which secrete insulin are normally influenced by no less than eight distinct types of input, plasma glucose concentration being only one. Thus, many endocrine cells act as true "integrating centers."

[3] The hypothalamus is, of course, not the only brain area controlling autonomic outflow; many others are involved in determining autonomic influence over endocrine glands and other organs and tissues.

THE PROBLEM OF MULTIPLE HORMONE SECRETION

The phenomenon of multiple hormone secretion by a single gland is clearly evident in Table 12-1. In certain glands, such as the pancreas, the hormones are secreted by completely distinct cells; in other glands, it is likely that, although some separation of function exists, a single cell may secrete more than one hormone. Even in such cases, it is essential to realize that each hormone has its own unique control mechanism; i.e., there is no massive undifferentiated release of the multiple hormones.

The adrenal cortex and gonads offer a particularly vexing problem of multiple hormone secretion. The hormones secreted by these organs all have a common ringlike type of lipid structure known as a steroid. Subtle changes in the steroid molecule produce great alterations of physiologic activity (compare, for example, the structures of the male sex hormone, testosterone, and one of the female sex hormones, progesterone, in Fig. 12-17). The biochemical pathways leading to steroid synthesis are complex and overlapping so that the adrenals and gonads secrete similar steroids. Only the major ones are presented in Table 12-1, but the reader should also be aware of several facts:

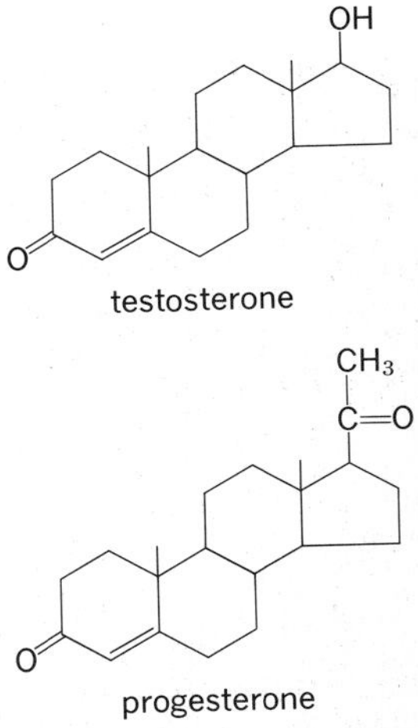

testosterone

progesterone

FIGURE 12-17 ***Structures of testosterone and progesterone.***

1 There are several closely related male sex hormones, all termed *androgens*. Testosterone is the dominant hormone of the group, and we shall usually use this name rather than the more general term, androgens, when referring to these hormones.

2 The name estrogen actually is a general term used to include several closely related, normally secreted female sex hormones.

3 The adrenal glands normally secrete significant quantities of androgen and estrogen.

4 The male and female gonads normally secrete very small quantities of estrogen and androgen, respectively, but the amounts are probably too small to be of significance (except as a reminder that there are no chemicals unique to males or females).

A frequent source of confusion concerning the endocrine system concerns not multiple hormone secretion by a single gland but the mixed general functions exhibited by the gonads, pancreas, kidneys, and gastrointestinal tract. All these organs contain endocrine gland cells but also perform other completely distinct nonendocrine functions. For example, most of the pancreas is concerned with the production of digestive enzymes; these are produced by exocrine glands, i.e., the secretory products are transported not into the blood but into ducts leading, in this case, into the intestinal tract. The endocrine function of the pancreas is performed by completely distinct nests of endocrine cells, the islets of Langerhans, which are scattered throughout the pancreas. This pattern is true for all the organs of mixed function; the endocrine function is always subserved by gland cells distinct from the other cells which constitute the organ.

CHAPTER 13

The Circulatory System: I. Blood

The circulatory system comprises a set of tubes, *blood vessels,* through which *blood* flows, and a pump, the *heart,* which produces this flow. Blood is composed of specialized cellular elements and a liquid, *plasma*, in which they are suspended. The cells are the *red blood cells,* or *erythrocytes;* the *white blood cells,* or *leukocytes;* and the *platelets* (which are really cell fragments). Ordinarily, the constant motion of the blood keeps the cells well dispersed throughout the plasma, but if a sample of blood is allowed to stand (clotting prevented), the cells slowly sink to the bottom. This process can be speeded up by centrifuging. By this means, the percentage of total blood volume which is cells, known as the *hematocrit,* can be determined. The normal hematocrit is approximately 45 percent. The total blood volume of an average man is approximately 8 percent of his total body weight. Accordingly, for a 70-kg man

Total blood weight $= 0.08 \times 70$ kg $= 5.6$ kg

One kilogram of blood occupies approximately 1 l; therefore

Total blood volume $= 5.6$ l

The hematocrit is 45 percent; therefore

Total cell volume[1] $= 0.45 \times 5.6$ l $= 2.5$ l
Plasma volume $= 5.6 - 2.5$ l $= 3.1$ l

PLASMA

Plasma is an extremely complex liquid. It consists of a large number of organic and inorganic subtstances dissolved in water (Table 13-1). The most abundant

[1] Since the vast majority of all blood cells are erythrocytes, the total cell volume is approximately equal to the erythrocyte volume.

TABLE 13-1
CONSTITUENTS OF PLASMA

Constituent	*Amount/Concentration*	*Major functions*
Water	90% of plasma	Medium for carrying all other constituents.
Electrolytes (inorganic)	Total = < 1% of plasma	Keep H_2O in extracellular compartment. Act as buffers. Function in membrane excitability.
Na^+	142 mEq/l (142 mmol/l)	
K^+	4 mEq/l (4 mmol/l)	
Ca^{2+}	5 mEq/l (2.5 mmol/l)	
Mg^{2+}	3 mEq/l (1.5 mmol/l)	
Cl^-	103 mEq/l (103 mmol/l)	
HCO_3^-	27 mEq/l (27 mmol/l)	
Phosphate (mostly HPO_4^{2-})	2 mEq/l (1 mmol/l)	
SO_4^{2-}	1 mEq/l (0.5 mmol/l)	
Proteins	6% of plasma (2.5 mmol/l)	Provide colloid osmotic pressure of plasma. Act as buffers. Bind other plasma constituents (lipids, hormones, vitamins, metals, etc.). Clotting factors. Enzymes, enzyme precursors. Antibodies (immune globulins). Hormones.
Albumins	4.5 g/100 ml	
Globulins	2.5 g/100 ml	
Fibrinogen	0.3 g/100 ml	
Gases		
CO_2	60 ml/100 ml plasma	
O_2	0.2 ml/100 ml	
N_2	0.9 ml/100 ml	
Nutrients		
Glucose and other carbohydrates	100 mg/100 ml	
Amino acids	40 mg/100 ml	
Lipids	500 mg/100 ml	
Cholesterol	150–250 mg/100 ml	
Vitamins		
Trace elements		
Waste products		
Urea	34 mg/100 ml	
Creatinine	1 mg/100 ml	
Uric acid	5 mg/100 ml	
Bilirubin	0.2–1.2 mg/100 ml	
Hormones		

solutes by weight are the proteins, which together compose approximately 7 percent of the total plasma weight. The *plasma proteins* vary greatly in their structure and function, but they can be classified, according to certain physical and chemical reactions, into three broad groups: the *albumins, globulins,* and *fibrinogen.* The albumins are three to four times more abundant than the globulines and usually are of smaller molecular weights. The plasma proteins, with notable exceptions, are synthesized by the liver, the major exception being the group known as *gamma globulins.* which are formed in the lymph nodes and other lymphoid tissues (Chap. 22). The plasma proteins serve a host of important functions which will be described in relevant chapters, but it must be emphasized that normally they are *not* taken up by cells and utilized as metabolic fuel. Accordingly, they must be viewed quite differently from most other organic constituents of plasma, such as glucose, which use the plasma as a vehicle for transport but function in cells. The plasma proteins function in the plasma itself or, under certain circumstances, in the interstitial fluid. Finally, plasma should be distinguished from *serum,* which is plasma after removal of fibrinogen has occurred as a result of clotting.

In addition to the organic solutes—proteins, nutrients, and metabolic end products—plasma contains a large variety of mineral electrolytes, the concentrations of which are shown in Table 13-1, along with that of protein. The value in millimoles per liter for protein may seem puzzling in view of the statement that protein is the most abundant plasma solute by *weight.* Remember, however, that molarity is a measure not of the weight but of the *number* of molecules or ions per unit volume. Protein molecules are so large in comparison with sodium ions that a very small number of them greatly outweighs a much larger number of sodium ions. The osmolarity (and, therefore, water concentration) of a solution depends upon the *number,* not the weight, of the solute particles present. Accordingly, sodium is the single most important determinant of total plasma osmolarity.

CELLULAR ELEMENTS OF THE BLOOD

Red Blood Cells

Each milliliter of blood contains approximately 5 billion erythrocytes. Since there is approximately 5,000 ml of blood in the average person, the total number of erythrocytes in the human body is about 25 trillion. The shape of these cells is a biconcave disk, i.e., thicker at the edge than in the middle, like a doughnut with a center depression on each side instead of a hole (Fig. 13-1). Their thickness is approximately 2 μm at the margin of the disk and 1 μm in the center, and their average diameter is slightly greater than 7 μm. This shape and these small dimensions have adaptive value in that oxygen and carbon dioxide can rapidly diffuse throughout the entire cell interior. The outstanding physiologic characteristic of erythrocytes is the presence of the iron-containing protein *hemoglobin,* which binds oxygen; hemoglobin constitutes approximately one-third of the total cell weight. Another erythrocyte substance of great importance is the enzyme *carbonic anhydrase,* which, as we shall see, facilitates the transportation of carbon dioxide.

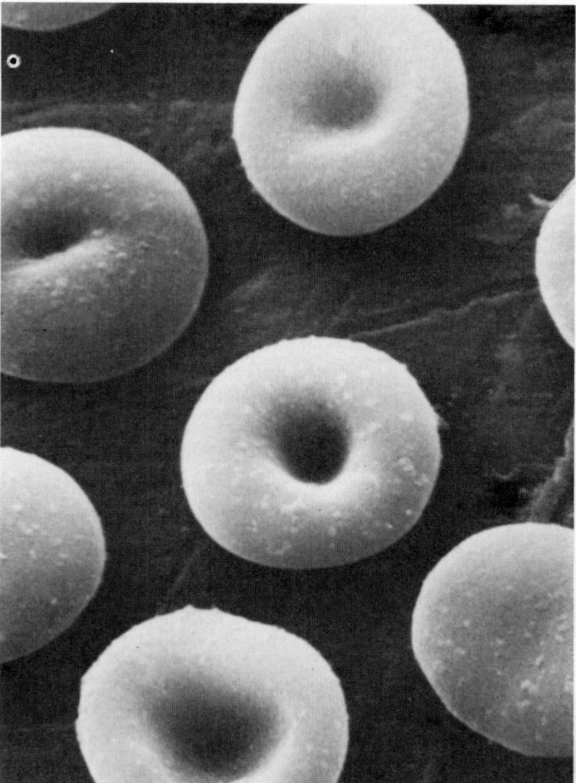

FIGURE 13-1 ***Scanning electron micrograph of red blood cells.*** **(From R. G. Kessel and C. Y. Shih, "Scanning Electron Microscopy in Biology," Springer-Verlag, New York, 1974, p. 265.)**

Erythrocyte and Hemoglobin Balance Erythrocyte volume and hemoglobin content are not fixed but are subject to physiologic control. Erythrocytes are incomplete cells in that they lack both nuclei and the metabolic machinery to synthesize new proteins. Thus, they can neither reproduce themselves nor maintain their normal structure for any length of time. As essential enzymes within them deteriorate and are not replaced, the cells age and ultimately die. Fortunately, their oxygen-carrying ability is not significantly diminished during the aging period. The average life of an erythrocyte is approximately 120 days, which means that almost 1 percent of the total erythrocytes in the body are destroyed every day. Destruction of erythrocytes is accomplished by a group of large cells, the *phagocytes,* found in liver, spleen, bone marrow, and lymph nodes, usually lining the blood vessels or lying close to them. These cells, the physiology of which will be described in detail in Chap. 22, ingest and destroy the erythrocytes by breaking down their large complex molecules with enzymes. The hemoglobin molecule is broken down and converted to a yellow molecule named *bilirubin*, which is released into the blood. The liver cells pick up this substance from the blood and add it to the bile, but if the liver is damaged or overloaded because of an abnormally high rate of erythrocyte destruction, the bilirubin may accumulate in the blood and give a yellow color to the skin. This is called *jaundice*.

Obviously, in a normal person, a quantity of erythrocytes equal to that destroyed must be simultaneously synthesized and released into the circulatory system. The site of erythrocyte production (*erythropoiesis*) is the *bone marrow*, a specialized, highly cellular form of connective tissue. Bone marrow occurs in two forms, *yellow* and *red marrow*, the yellow consisting largely of fat cells within a connective-tissue matrix. The red marrow is the site of erythrocyte production; its color is due to the numerous erythrocytes and their pigmented precursors that are there. Although red marrow occurs throughout the skeleton at birth, it is gradually replaced by yellow marrow so that, by 20 to 25 years of age, hemopoiesis occurs only in the bones of the skull, clavicles, ribs, vertebrae, sternum, pelvis, and upper ends of the humeri and femurs; active bone marrow constitutes approximately 5 percent of the total adult body weight. The erythrocytes are originally descended from large bone-marrow *stem cells* (*hemocytoblasts*) which contain no hemoglobin but do have nuclei and are therefore capable of cell division. After several cell divisions, cells emerge which are identifiable as immature erythrocytes because they contain hemoglobin. As maturation continues, these cells accumulate increased amounts of hemoglobin, and their nuclei become progressively smaller until they ultimately disappear completely. The mature erythrocyte leaves the bone marrow via the rich network of marrow capillaries and enters the general circulation, in which it will travel for some 120 days.

This growth process obviously requires a number of different raw materials. The formation of the erythrocyte itself requires the usual nutrients and structural materials: amino acids, lipids, and carbohydrates. In addition, certain growth factors, including vitamin B_{12} and folic acid, are essential for normal erythrocyte formation. Finally, the formation of an erythrocyte requires the materials which go into the making of hemoglobin: iron, amino acids, and the organic molecules which are incorporated into the protein portion of hemoglobin. A lack of any of these growth factors or raw materials results in the failure of normal *erythropoiesis* (erythrocyte formation) and a decreased quantity of effective circulating erythrocytes (*anemia*). The substances which are most commonly lacking are iron and vitamin B_{12}.

Iron Iron is obviously an essential component of the hemoglobin molecule since it is with this element that the oxygen actually combines. The balance of iron and its distribution within the body are shown schematically in Fig. 13-2. About 70 percent of the total body iron is in hemoglobin, and the remainder is stored primarily in the liver, spleen, and bone marrow. As erythrocytes are destroyed, most of the iron released from hemoglobin is returned to these depots. Small amounts of iron, however, are lost each day via the urine, feces, sweat, and cells sloughed from the skin. In addition, women lose a significant quantity of iron via menstrual blood. In order to remain in iron balance, the amount of this metal lost from the body must be replaced by ingestion of iron-containing foods. An upset of this balance results either in iron deficiency and inadequate hemoglobin production or in an excess of iron in the body and serious toxic effects. The control of iron balance is unusual in that it resides primarily in the intestinal epithelium, which absorbs the iron from ingested food. The intestine absorbs only a fraction of

FIGURE 13-2 ***Summary of iron balance. The sizes of the boxes represent the quantity of iron involved. The magnitude of exchange in a particular direction varies according to conditions.***

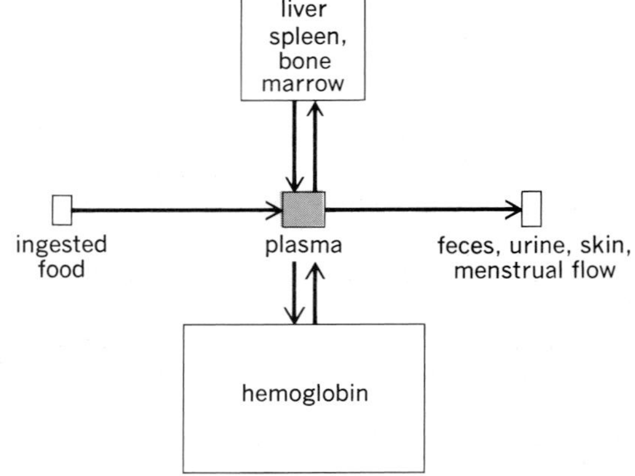

the ingested iron, and, what is more important, this fraction is increased or decreased depending upon the state of body iron balance. These fluctuations appear to be mediated by changes in the iron content of the intestinal epithelium itself. When body stores of iron are ample, the epithelial iron store is increased, and its presence somehow prevents the absorption of most of the iron ingested in food. When the body stores of iron drop (as a result of hemorrhage, for example), some of the iron stored in the intestinal epithelium enters the blood and is transported to the bone marrow. The resulting lowering of intestinal epithelial iron content permits increased intestinal absorption of dietary iron and returns total body iron to normal.

Vitamin B_{12} Normal erythrocyte formation requires extremely small quantities (one-millionth of a gram per day) of a cobalt-containing molecule, vitamin B_{12}, which, by mechanisms still unknown, permits final maturation of the erythrocyte. This substance is not synthesized within the body, which therefore depends upon its dietary intake. Absorption of vitamin B_{12} from the gastrointestinal tract into the blood requires a substance (still unidentified) normally secreted by the epithelial lining of the stomach. Its absence prevents normal absorption of vitamin B_{12} and causes deficient erythrocyte production and *pernicious anemia*.

Regulation of Erythrocyte Production In a normal person, the total volume of circulating erythrocytes remains remarkably constant. Such constancy is required both for delivery of oxygen to the tissues and for maintenance of the blood pressure. From the preceding paragraphs it should be evident that a constant number of circulating erythrocytes can be maintained only by balancing erythrocyte production and destruction (or loss). Wide variations in the latter result from erythrocyte-destroying diseases or hemorrhage, and the balance must therefore be obtained by controlling the rate of erythrocyte production. During periods of severe erythrocyte destruction or hemorrhage, the normal rate of erythrocyte production can be increased more than sixfold.

It is easiest to discuss the control of erythrocyte production by first naming the mechanisms *not* involved. In the previous section, we listed a group of nutrient substances, such as iron and vitamin B_{12}, which must be present for normal erythrocyte production. However, none of these substances actually *regulates* the rate of production.

The direct control of erythropoiesis is exerted by a hormone called *erythropoietin*. The best evidence now indicates that the cells producing erythropoietin are located in the kidneys, although other sites in the body may also contribute to its production. Normally, a small quantity of erythropoietin is circulating, which stimulates the bone marrow to produce erythrocytes at a certain basal rate. An increase in circulating erythropoietin further stimulates the bone marrow and increases erythropoiesis, whereas a decrease in circulating erythropoietin below the normal level decreases erythropoiesis.

At the present time, the components of the reflex pathway which controls erythropoietin production remain largely a mystery. The common denominator of changes which increase circulating erythropoietin is precisely what one would logically expect—a decreased oxygen delivery to the tissues. As will be described in Chap. 17, this can result either from decreased erythrocyte volume, decreased blood flow, or decreased oxygen delivery from the lungs into the blood. As a result, erythrocyte production is increased, the oxygen-carrying capacity of the blood is increased, and oxygen delivery to the tissues is returned toward normal. Unfortunately, neither the receptors which detect the change in oxygen delivery nor the afferent pathway which relays this information to the erythropoietin-producing cells has been discovered.

White Blood Cells

If one takes a drop of blood, adds appropriate dyes, and examines it under a microscope, the various cell types can be seen. Over 99 percent of all the cells are erythrocytes. The remaining cells, the *white blood cells*, are classified according to their structure and affinity for various dyes. The name *polymorphonuclear granulocytes* refers to the three types of cells with lobulated nuclei and abundant cytoplasmic granules. The granules of one group show no dye preference, and the cells are therefore called *neutrophils*. The granules of the second group take up the red dye eosin, thus giving the cells their name *eosinophils*. Cells of the third group have an affinity for a basic dye and are called *basophils*. All three types of granulocytes are produced in the bone marrow (Fig. 13-3) and released into the circulation. It should be understood that, unlike the erythrocytes, the major functions of leukocytes are exerted not within the blood vessels but in the interstitial fluid; i.e., leukocytes utilize the circulatory system only as the route for reaching a damaged or invaded area.

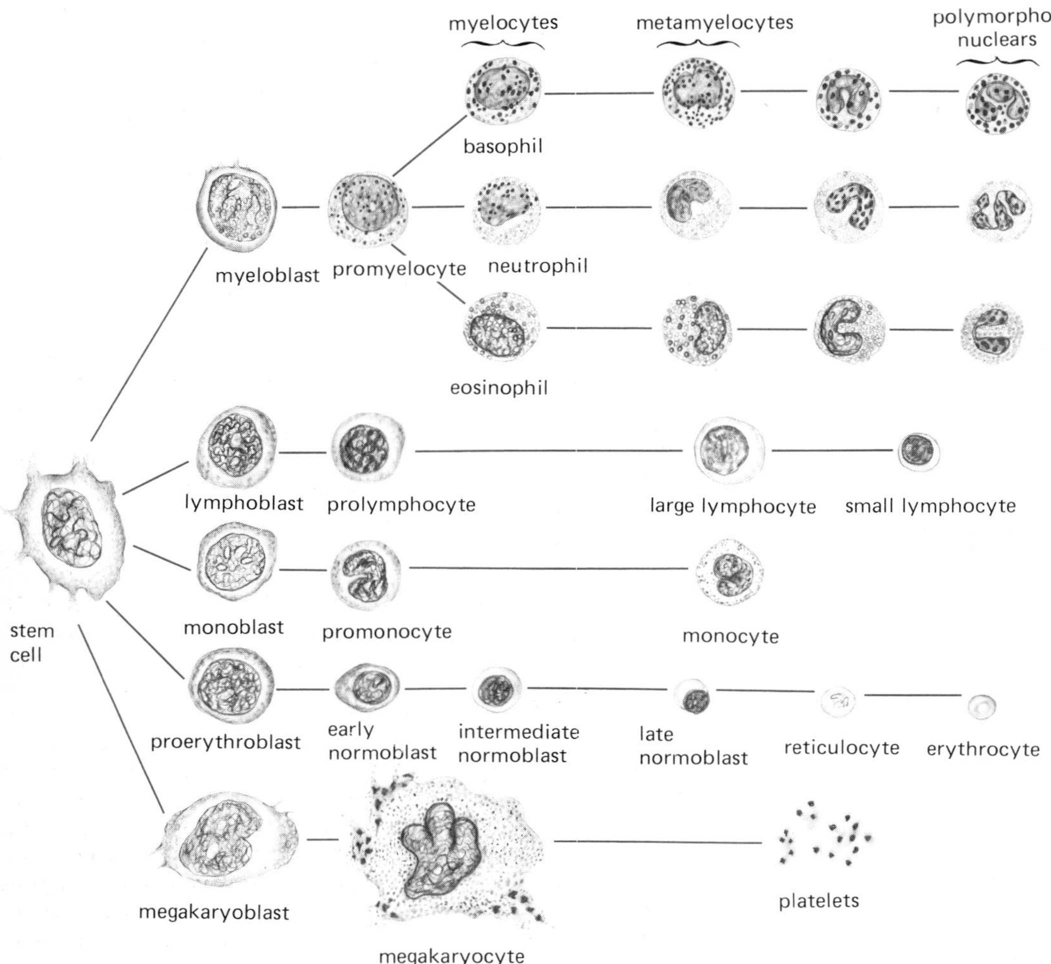

FIGURE 13-3 ***Development of the formed elements of the blood from bone marrow cells. Note that all the leucocyte cell forms are derived from one stem cell type. (Redrawn from Whitby and Britton.)***

Once there, they leave the blood vessels to enter the tissue and perform their functions.

The primary function of the neutrophil is *phagocytosis,* the ingestion and digestion of particulate material. It has a life-span of only a few days and is incapable of division, so that its supply must be continuously replenished by the bone marrow. The eosinophil is also involved in phagocytosis but its precise function is unknown. The basophil, in contrast, is not a phagocytic cell; rather it contains powerful chemicals, such as histamine, which it releases locally, a response which, as we shall see, may importantly contribute to tissue damage and allergy. The basophil is virtually identical to the group of cells known as *mast cells* which are found in connective tissue throughout the body and do not circulate; i.e., the basophil is, in essence, a circulating mast cell.

A second type of leukocyte quite different in appearance from the granulocytes is the *monocyte.* These cells, which are also produced by the bone marrow, are somewhat larger than the granulocytes, with a single oval or horseshoe-shaped nucleus and relatively few cytoplasmic granules. Upon entering an area which has been invaded or wounded, monocytes are transformed into *macrophages*. Thus, the function of circulating monocytes is to provide to the tissues a source of new macrophages, the function of which is described in Chap. 22.

TABLE 13-2
NUMBERS AND DISTRIBUTION OF ERYTHROCYTES AND WHITE CELLS IN NORMAL HUMAN BLOOD

Total erythrocytes = 5,000,000,000 cells per milliliter of blood
Total white cells = 7,000,000 cells per milliliter of blood
Percent of total white cells:
- Polymorphonuclear granulocytes
  - Neutrophils, 50–70
  - Eosinophils, 1–4
  - Basophils, 0.1
- Mononuclear cells
  - Monocytes, 2–8
  - Lymphocytes, 20–40

The final class of leukocyte is the *lymphocyte* (in a normal person the white blood cells have approximately the distribution shown in Table 13-2). Their outstanding structural features are a relatively large nucleus and scanty surrounding cytoplasm. The circulating pool of lymphocytes continually travels from the blood across capillary membranes to the lymph, thence to lymph nodes, and, finally, back to the blood. However, circulating lymphocytes constitute only a very small fraction of the total body lymphocytes, most of which are found at any instant in the lymphoid tissues. The life history of lymphocytes and the interactions between them and the various lymphoid organs are quite complex and will be described in Chap. 22. Suffice it here to point out that the lymphocytes (and a daughter-cell line, the *plasma cells*) are responsible for specific defenses against foreign invaders.

Platelets

In human beings, the circulating platelets (Fig. 13-4) are colorless corpuscles much smaller than erythrocytes and containing numerous granules; there are approximately 250 million per milliliter of blood (compare this with the number of white blood cells given in Table 13-2). The platelets are not complete cells since they lack nuclei. They originate from certain large cells (*megakaryocytes*) found in bone marrow and apparently are portions of cytoplasm of these cells which are pinched off and enter the circulation. The factors that control the rate of platelet formation are unknown.

HEMOSTASIS: THE PREVENTION OF BLOOD LOSS

All animals with a vascular system must be able to minimize blood loss consequent to vessel damage. In human beings, blood coagulation is one of several important mechanisms for hemostasis. The hemostatic mechanism which predominates varies, depending upon the kind and number of vessels damaged and the location of the injury.

We shall discuss the probable sequence of events in response to damage to small vessels—arterioles, capillaries, venules—because they are the most common source of bleeding in everyday life and because the hemostatic mechanisms are most effective in dealing with such injuries. In contrast, the bleeding from a severed artery of medium or large size is not usually controllable by the body and requires radical aids such as application of pressure and ligatures. Venous bleeding is less dangerous because of the vein's low hydrostatic pressure; indeed, the drop in hydrostatic pressure induced by simple elevation of the bleeding part may stop the hemorrhage. In addition, if the venous bleeding goes into the tissues, the accumulation of blood (*hematoma*) may increase interstitial pressure enough to eliminate the pressure gradient required for continued blood loss. The hemostatic events in small vessels (Table 13-3) do not occur in neat orderly sequence but overlap in time and are closely interrelated functionally.

TABLE 13-3
SYNOPSIS OF HEMOSTATIC EVENTS IN SMALL BLOOD VESSELS

1 Contraction of smooth muscle in the wall of a damaged vessel
2 Sticking together of injured endothelium
3 Clumping of platelets to form a plug
4 Facilitation of the initial vasoconstriction
5 Blood coagulation, i.e., formation of a fibrin clot
6 Retraction of the clot

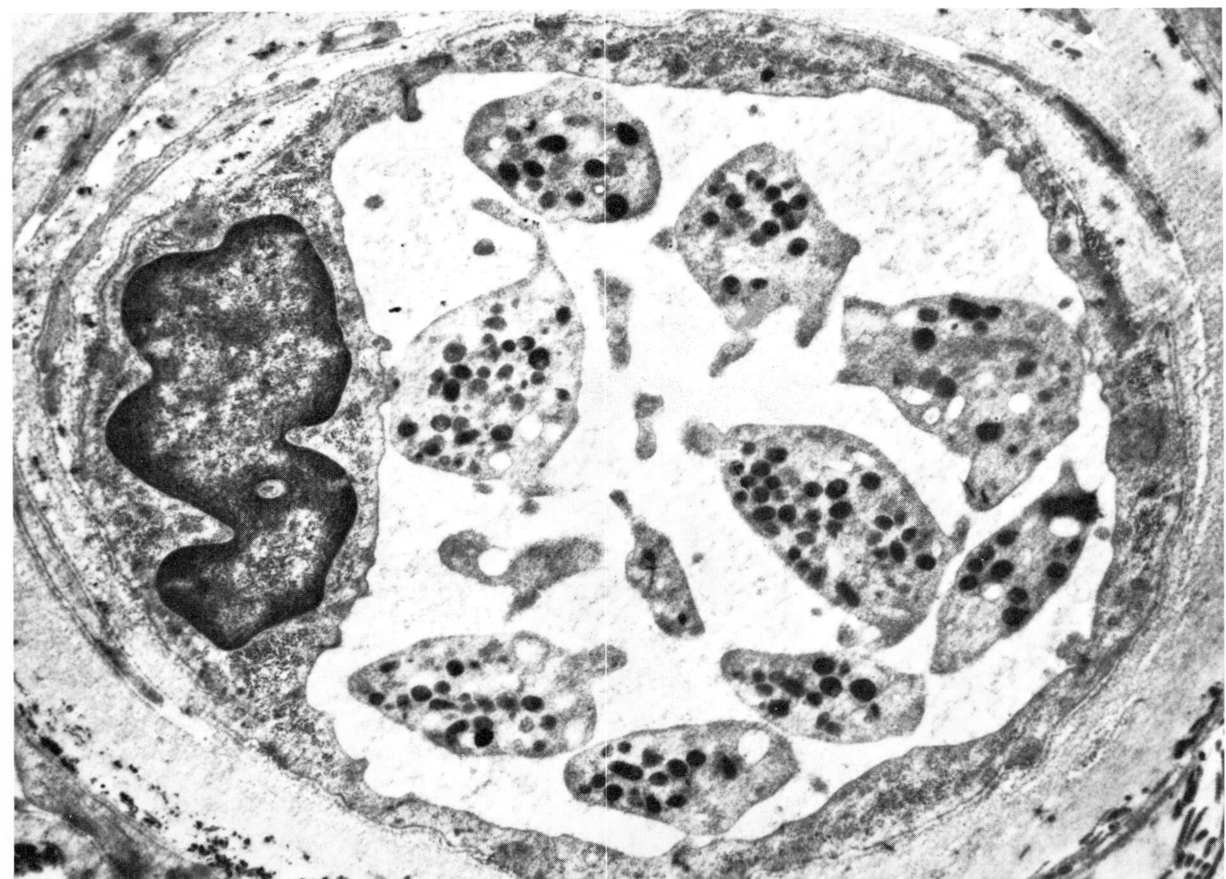

FIGURE 13-4 ***Electron micrograph of a capillary containing several blood platelets in its lumen. Note the numerous granules and lack of nuclei.*** **(From W. Bloom and D. W. Fawcett, "A Textbook of Histology," 9th ed., Saunders, Philadelphia, 1968.)**

Hemostatic Events Prior to Clot Formation

When a blood vessel is severed or injured, its immediate response is to constrict. The rupture of the vessel wall in some manner directly stimulates the smooth muscle or nerves supplying it. This spasm is particularly long-lasting in larger veins and arteries and may be intense enough to close the severed end completely. In addition, the spasm presses the opposed endothelial surfaces of the vessel together; this contact induces a stickiness capable of keeping them "glued" together against high pressures even after the active vasoconstriction has begun to wane.

The next step involves the platelets. The platelets play a critical role in the last four hemostatic components of our synopsis. Although platelets have a propensity for adhering to many foreign or rough surfaces, they do not adhere to the normal endothelial cells lining the blood vessels. However, injury to a vessel disrupts the endothelium and exposes the underlying connective tissue with its collagen molecules. Platelets adhere strongly to collagen, and this attachment somehow triggers the release from the platelets' granules of potent chemical agents, including *adenosine diphosphate* (*ADP*). This ADP then causes the surface of the adhered platelets to become extremely sticky so that new platelets adhere to the old ones and an aggregate or plug of platelets is rapidly built up by this self-perpetuating (positive-feedback) process. This platelet plug may occlude small vessels so as to slow or even stop bleeding. In addition, the aggregated platelets release *serotonin* and *epinephrine,* both of which are powerful vasoconstrictors. The first four events in hemostasis are summarized in Fig. 13-5.

Blood Coagulation: Clot Formation

Despite the participation of the four mechanisms just described, blood coagulation is usually the dominant hemostatic defense in human beings, as attested by the fact that, with few exceptions, abnormal bleeding is associated with some clotting defect. Pure vascular defects, interfering with the preclot hemostatic mechanisms, do occur but are only rarely the cause of abnormal bleeding.

The event transforming blood into a solid gel is the conversion of the plasma protein *fibrinogen* to *fibrin.* Fibrinogen is a soluble, large, rod-shaped protein (molecular weight approximately 340,000) produced by the liver and always present in the plasma of normal persons. Its conversion to fibrin is catalyzed by the enzyme *thrombin:*

$$\text{Fibrinogen} \xrightarrow{\text{thrombin}} \text{fibrin}$$

In this reaction, several small negatively charged polypeptides are split from fibrinogen, conferring upon the remaining large molecule a high degree of attraction for molecules like it. They join each other end to end and side by side to form fibrin (Fig. 13-6). This polymerization causes the fluid portion of the blood to gel, rather like a gelatin dessert. In addition, the cellular elements of the blood in the vicinity become entangled in the meshwork and contribute to its strength. It must be emphasized that the clot is basically due to fibrin and can occur in the absence of blood cells (except the platelets).

Since fibrinogen is always present in the blood, the enzyme thrombin must normally be absent and its for-

FIGURE 13-5 ***Summary of hemostatic mechanisms not dependent upon blood coagulation. The dashed line indicates the positive-feedback effect of ADP on platelet adhesion and agglutination.***

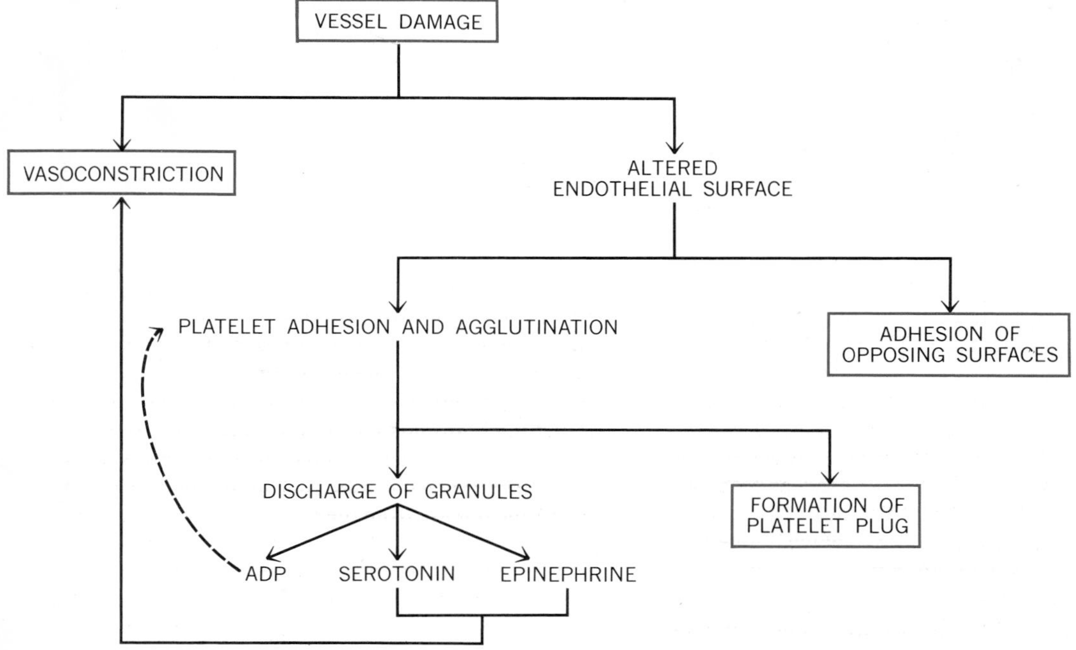

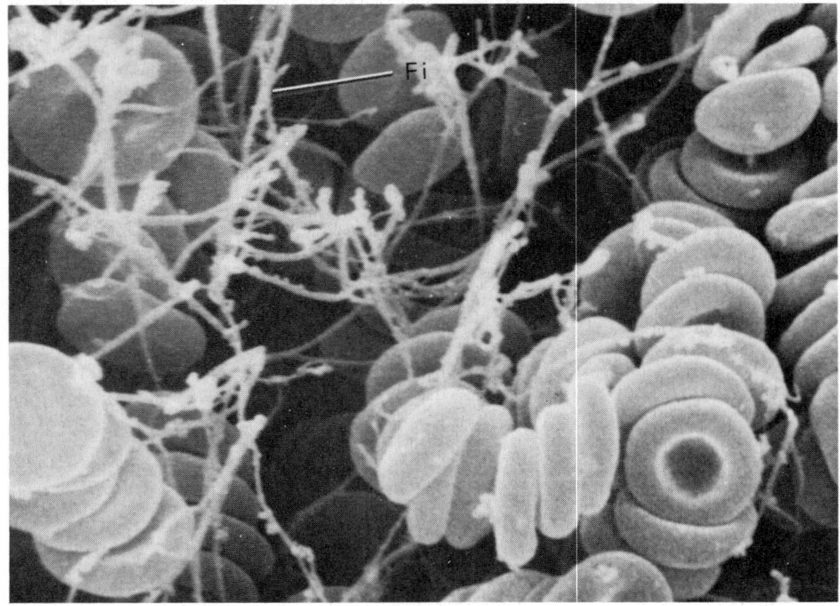

FIGURE 13-6 ***Scanning electron micrograph of fibrin.*** **(From R. G. Kessel and C. Y. Shih, "Scanning Electron Microscopy in Biology," Springer-Verlag, New York, 1974, p. 265.)**

mation must be triggered by vessel damage. The generation of thrombin follows the same general principle as fibrin, in that an inactive precursor, *prothrombin,* is produced by the liver and is normally present in the blood, being enzymatically converted to thrombin during clot formation:

$$\text{Prothrombin} \xrightarrow{?} \text{thrombin}$$
$$\text{Fibrinogen} \xrightarrow{\text{thrombin}} \text{fibrin}$$

We have now only pushed the essential question one step further back: What catalyzes the conversion of the prothrombin to thrombin? (This is the question mark in the equation.) The answer is that this reaction is catalyzed by a plasma factor which itself is activated by another plasma factor, which. . . . Thus, we are dealing with a *cascade* of plasma protein factors, each normally inactive until activated by the previous one in the sequence (Fig. 13-7). Ultimately, the final factor in the sequence is activated and in turn catalyzes the activation of prothrombin, i.e., its conversion to thrombin. The designations A and B in our model are arbitrarily chosen to demonstrate the general principle. The first factor in our diagram is called *Hageman factor* and, as will be described later, it has several other important functions in addition to initiating clotting.

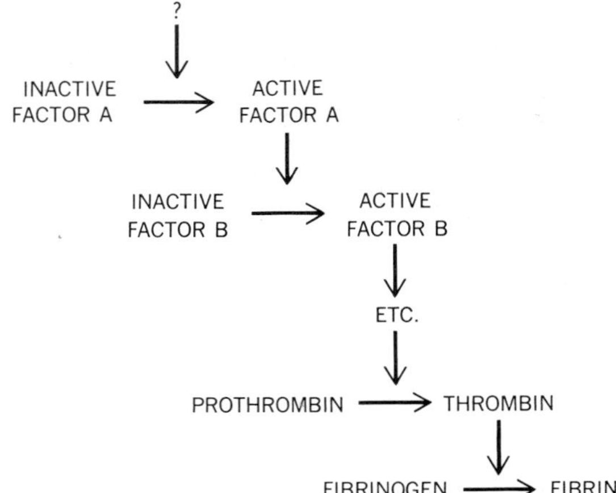

FIGURE 13-7 ***Cascade theory of blood clotting. Each substance left of an arrow is normally present in plasma but requires activation by the action of the previous substance in the sequence. The name of factor A is Hageman factor.***

In addition to these protein plasma factors, calcium is required as cofactor for several steps. However, calcium deficiency is never a cause of clotting defects in human beings since only very small concentrations are required.

Figure 13-7 reveals that we have still not answered the basic question of what *initiates* clotting. What activates the first protein (Hageman factor or factor A in the figure) in the catalytic sequence? The answer is *contact of this protein with a damaged vessel surface,* most likely with the collagen fibers underlying the damaged endothelium (as was true for platelet aggregation).

However, even contact activation of this first factor cannot produce clotting in the absence of platelets. A phospholipid substance exposed on the surface of platelets during their adhesion and agglutination is required as cofactor for several of the steps in the catalytic sequence. Thus, the critical event initiating clot formation is contact of the blood with a damaged surface for two reasons: (1) It activates the first factor in the activation sequence, and (2) it causes platelet adhesion and exposure of a phospholipid cofactor. One final detail is that thrombin markedly enhances the adhesion and agglutination of platelets; thus, once thrombin formation has begun, the overall reaction progresses explosively owing to the positive-feedback effect of thrombin. Our growing figure can now be completed (Fig. 13-8).

This entire process occurs only locally at the site of vessel damage. Each active component is formed, functions, and is rapidly inactivated without spilling over into the rest of the circulation. Otherwise, because of the chain-reaction nature of the response, the appearance in the *overall circulation* of platelet phospholipid and any single activated factor would induce massive widespread clotting throughout the body.

The liver plays several important indirect roles in the overall functioning of the clotting mechanism (Fig. 13-9). First, it is the site of production for many of the plasma factors, prothrombin, and fibrinogen, although the reflex mechanisms controlling their rates of synthesis are unknown. Second, the bile salts produced by the liver are required for normal gastrointestinal absorption of the fat-soluble *vitamin K,* which is an essential cofactor in the hepatic synthesis of prothrombin and the plasma factors. For these two reasons, patients with liver disease or defective gastrointestinal fat absorption frequently have serious bleeding problems.

Finally, a word must be said about the contribution of a tissue (rather than blood) factor to clotting. If one extracts almost any of the body's tissues and injects the extract into unclotted normal blood in a siliconized tube, clotting occurs within seconds. The explanation is that the tissues contain a substance known as *tissue thromboplastin* which can substitute for both platelet phospholipid and several of the plasma factors, and thus an abnormal surface is no longer required to initiate clotting. This is known as the *extrinsic clotting pathway* to distinguish it from the *intrinsic pathway* described above. Its quantitative contribution to normal intravascular clotting is unclear but it may play an

FIGURE 13-8 ***Summary of blood-clotting mechanism. The dashed line indicates the positive-feedback effect of thrombin on platelet adhesion (recall that ADP exerts a similar positive feedback).***

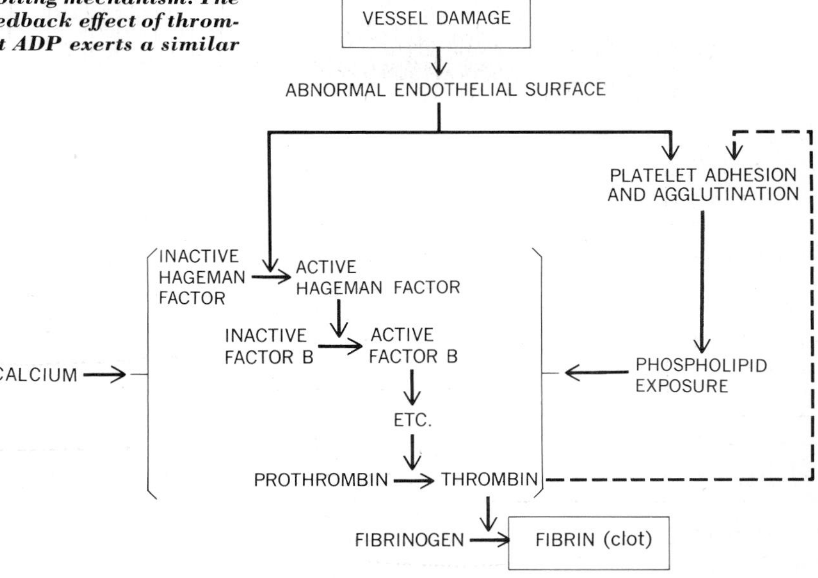

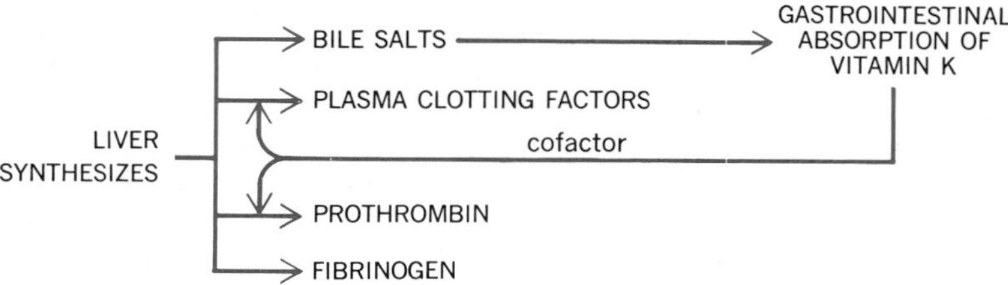

FIGURE 13-9 ***Role of the liver and vitamin K in the synthesis of plasma clotting factors, prothrombin, and fibrinogen.***

essential role in the response to many bacterial infections by initiating fibrin clots which may block further spread of the bacteria.

Clot Retraction

When blood is carefully collected and placed in a glass test tube, clotting usually occurs in several minutes (because the glass surface acts like a damaged vessel surface), the entire volume of blood appearing as a coagulated gel. However, during the next 30 min, a striking transformation occurs; the clot literally retracts, squeezing out the fluid which constituted a large fraction of the gel. The end result is a small hard clot at the bottom of the tube with a large volume of serum floating on top (plasma without fibrinogen is called serum). The fibrin meshwork with its entangled cells has become denser and stronger. This same process in the body is known as clot retraction. Besides increasing the strength of the clot, it has the advantage of pulling the vessel walls adhering to the clot closer together.

Clot retraction is due to the platelets. As fibrin strands form around them during clotting, the agglutinated mass of platelets sends out adhering pseudopods along them. The pseudopods then contract, pulling the fibrin fibrils together and squeezing out the serum. It has been shown that the platelets contain actomyosinlike contractile protein and that they split ATP during clot retraction. The various roles of platelets in hemostasis are summarized in Fig. 13-10.

The Anticlotting System

It has long been observed that clots frequently disappear after lengthy standing. Blood also fails to clot in a variety of special circumstances, including endometrial invasion by the fetal placental cells. This is due to a proteolytic enzyme called *plasmin,* which is able to decompose fibrin, thereby dissolving a clot. The physiology of plasmin bears some striking similarities to that of the coagulation factors in that it circulates in blood in an inactive form which is enzymatically converted to active plasmin by the action of activated Hageman factor as well as by other substances found in tissue fluid (Fig. 13-11). It may seem paradoxical that the same substance, Hageman factor, triggers off simultaneously the clotting and anticlotting systems. Yet, in fact, this makes sense since the generated plasmin becomes trapped within the newly formed clot and very slowly dissolves it, thereby contributing to tissue repair at a time when the danger of hemorrhage is past.

The anticlotting system no doubt has other functions as well. It may be that small amounts of fibrin are constantly being laid down throughout the vascular tree and that plasmin acts on this fibrin to prevent clotting. The lung tissue, for example, contains a substance which activates plasmin; this probably explains the lung's ability to dissolve the fibrin clumps which its capillaries filter from the blood. Moreover, the uterine wall is extremely rich in a similar activator, and thus normal menstrual blood generally does not clot.

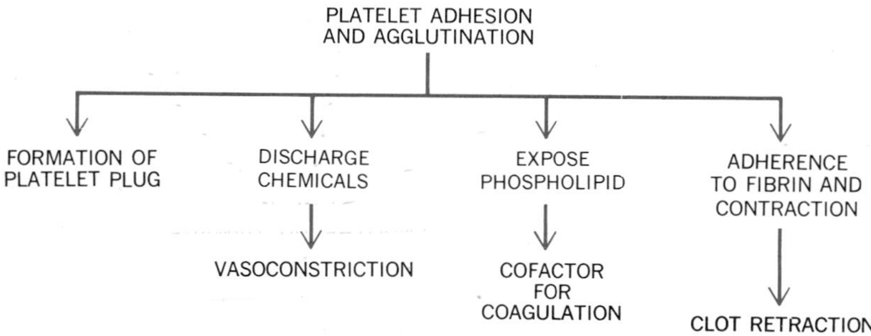

FIGURE 13-10 ***Summary of platelet functions in hemostasis.***

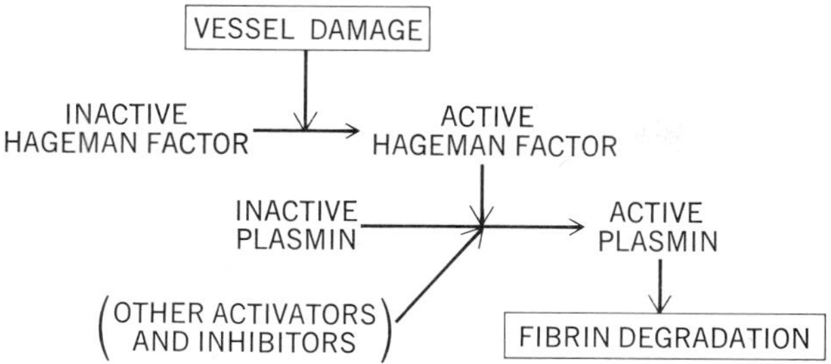

FIGURE 13-11 ***Summary of plasmin anticlotting system.***

A second naturally occurring anticoagulant is *heparin*. This substance, found in various cells of the body, especially mast cells, acts by interfering with the ability of thrombin to split fibrinogen. Despite its presence in the body and the fact that it is the most powerful anticoagulant known, there is really no good evidence to prove that it plays a physiologic role in clot prevention. On the other hand, heparin is widely used as an anticoagulant in medicine.

Excessive Clotting: Intravascular Thrombosis

Formation of a clot in a bleeding vessel is obviously a homeostatic physiologic response, but the formation of clots within intact vessels is pathologic. It may occur in the veins, the microcirculation, or arteries. Coronary arterial occlusion secondary to thrombosis (thrombus means clot) is one of the major killers in the United States today. The sequence of events leading to thrombosis is the subject of intensive study, and numerous theories have been proposed. A brief synopsis of some of them is in order because of the great importance of the subject and its illustration of the basic physiologic processes.

One of the dominant theories today postulates that the clotting mechanism in persons prone to thrombosis is hyperactive, as manifested by the reduced time it takes for withdrawn blood to clot in a test tube. Perhaps one of the plasma factors is present in excessive amounts, or a normally occurring anticoagulant is deficient. This theory emphasizes that the blood itself is the cause of excessive clotting. However, there seems little question that hypercoagulability is not always essential for thrombosis, since hemophiliacs have been known to suffer from coronary thrombosis.

A second category of theories puts the blame on the blood vessels. Since initiation of blood clotting is primarily dependent upon the state of the blood vessel lining, even minor transient alterations in the endothelial surface could trigger the autocatalytic sequence leading to clot formation. These vessel-oriented theories can explain many of the situations associated with vascular thrombosis (Fig. 13-12): (1) Stasis, i.e., decreased movement, of blood in veins, which occurs during quiet standing, valve malfunction, or cardiac insufficiency, may induce damage in the vein wall as a result of oxygen lack. (2) Inflammation of veins and other vessels caused by bacteria, allergic reactions, or toxic substances may cause vessel damage. (3) Deposition of lipids and connective tissue in arterial walls (atherosclerosis) causes marked thickening and irregularity of the arterial lining, although there is certainly no agreement as to which comes first, the deposits or the clot.

Thus, the three major conditions predisposing to clot formation are consistent with the damaged-lining theory. Association does not prove causality, however, and it should be noted that this concept and the hypercoagulability theory are not mutually exclusive. Both probably are valid, depending upon the circumstances.

Regardless of the initiating event, there is no question that a clot, no matter how small, provides a suitable surface upon which more clot can form. Thus the thrombus grows and may eventually occlude the entire vessel, thereby leading to damage of the tissue supplied or drained by the vessel. A second important factor in vessel closure during clot growth is the release of vasoconstrictors from freshly adhered platelets. Finally, the chances are greatly increased of clot fragments breaking off and being carried to the lungs (if from a vein) or other organs (if from an artery). These *emboli* not only plug the microcirculation but in the lung may induce totally inappropriate cardiovascular reflexes culminating in hypotension, disturbance of the cardiac rhythm, and death.

FIGURE 13-12 ***Damaged-vessel theory of abnormal intravascular clotting.***

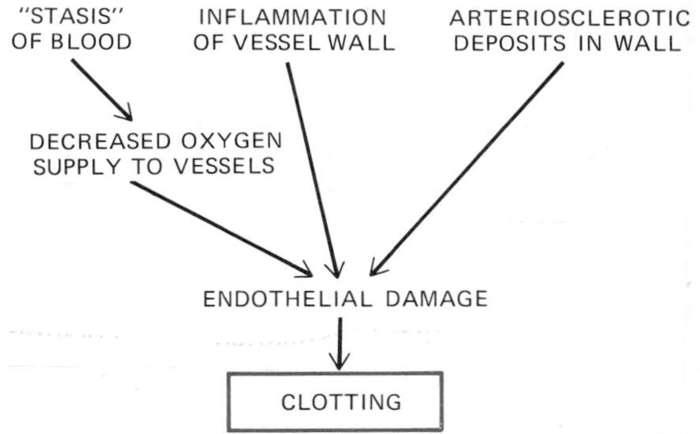

CHAPTER 14

The Circulatory System: II. The Heart

Physiology as an experimental science began in 1628, when William Harvey demonstrated that the entire circulatory system forms a circle, so that blood is continuously being pumped out of the heart through one set of vessels and returning to the heart via a different set. In human beings, as in all mammals, there are actually two circuits, both originating and terminating in the heart, which is divided longitudinally into two functional halves. Blood is pumped via one circuit (the *pulmonary circulation*) from the right half of the heart through the lungs and back to the left half of the heart. It is pumped via the second circuit (the *systemic circulation*) from the left half of the heart through all the tissues of the body, except, of course, the lungs, and back to the right half of the heart (Fig. 14-1). In both circuits, the vessels carrying blood away from the heart are called *arteries*, and the vessels carrying blood from the lungs and tissues back to the heart are called *veins*. In the systemic circuit, blood leaves the left half of the heart via a single large artery, the aorta; from the aorta, branching arteries conduct blood to the various organs and tissues. Blood returns to the heart via two large veins, the superior and inferior venae cavae.

In a normal person, blood can pass from the systemic veins to the systemic arteries only by first being pumped through the pulmonary circuit, thus oxygenating all the blood returning from the body tissues before it is pumped back to them. Normally the total volumes of blood pumped through the pulmonary and systemic circuits during a given period of time are equal. In other words, the right heart pumps the same amount of blood as the left heart. Only under unusual circumstances, such as malfunction of one half of the heart, do these volumes differ from each other, and then only tran-

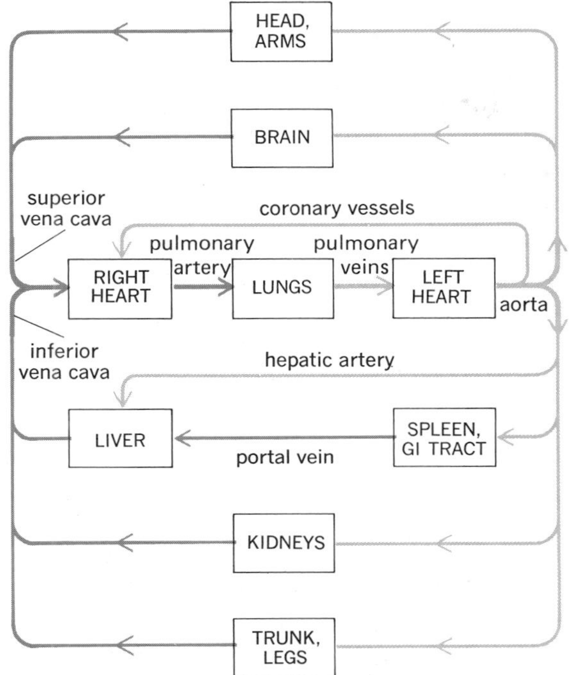

FIGURE 14-1 ***Diagrammatic representation of the cardiovascular system in the adult human being. Darker shading indicates blood with low oxygen content.***

siently. In a resting normal man, the amount of blood pumped simultaneously by each half of the heart is approximately 5 l/min. During heavy work or exercise, the volume may increase as much as fivefold to 25 l/min.

ANATOMY OF THE HEART

The heart is a muscular organ located in the chest (*thoracic*) cavity between the two lungs and covered by a fibrous sac, the *pericardium*. One-third of the heart lies to the right of the median plane, and two-thirds to the left (Fig. 14-2). The walls of the heart are composed primarily of muscle (*myocardium*), the structure of which is different from either skeletal or smooth muscle. The inner surface of the myocardium, i.e., the surface in contact with the blood within the heart chambers, is lined by a thin layer of cells (*endothelium*).

The human heart is divided longitudinally into right and left halves (Fig. 14-3), each consisting of two chambers, an *atrium* and a *ventricle*. The cavities of the atrium and ventricle on each side of the heart communicate with each other, but the right chambers do not communicate directly with those on the left. They are separated from each other by partitions called the *interatrial* and *interventricular septa*. Thus, right and left atria and right and left ventricles are distinct.

The tip of the left ventricle, which points downward, forward, and toward the left, forms the *apex* of the heart, and the upper surfaces, which face backward and toward the right, form the *base*. The walls of the left atrium form the major part of the base of the heart. A small earlike pouch extends from each atrium: These pouches are termed the *right* and *left auricles* (the term auricle is often wrongly used synonymously with atrium).

Perhaps the easiest way to picture the architecture of the heart is to begin with its fibrous skeleton, which comprises four rings of dense connective tissue joined together (Fig. 14-4). To the tops of these rings are anchored the muscle masses of the atria, pulmonary trunk, and aorta. To the bottoms are attached the muscle masses of the ventricles. The connective-tissue rings form the openings between the atria and ventricles and between the great arteries and ventricles. To these rings are attached four sets of valves (Figs. 14-4 and 14-5).

Between the cavities of the atrium and ventricle in each half of the heart are the *atrioventricular valves* (*AV valves*), which permit blood to flow from atrium to ventricle but not from ventricle to atrium. The right and left AV valves are called, respectively, the *tricuspid* and *mitral* valves. When the blood is moving from atrium to ventricle, the valves lie open against the ventricular wall, but when the ventricles contract, the valves are brought together by the increasing pressure of the ventricular blood, and the atrioventricular opening is closed. Blood is therefore forced into the pulmonary trunk (from the right venticle) and into the aorta (from the left ventricle) instead of back into the atria. To prevent the valves themselves from being forced upward into the atrium, they are fastened by thin fibrous strands, the *chordae tendineae*, to the *papillary muscles*, which are projections of the ventricular walls (Fig. 14-3). These muscular projections do *not* open or close the valves; they act only to limit the valves' movements and prevent them from being everted.

The openings of the ventricles into the pulmonary trunk and aorta are also guarded by valves, the *pulmonary* and *aortic valves*, which permit blood to flow into these arteries but close immediately, preventing reflux of blood in the opposite direction. There are no true

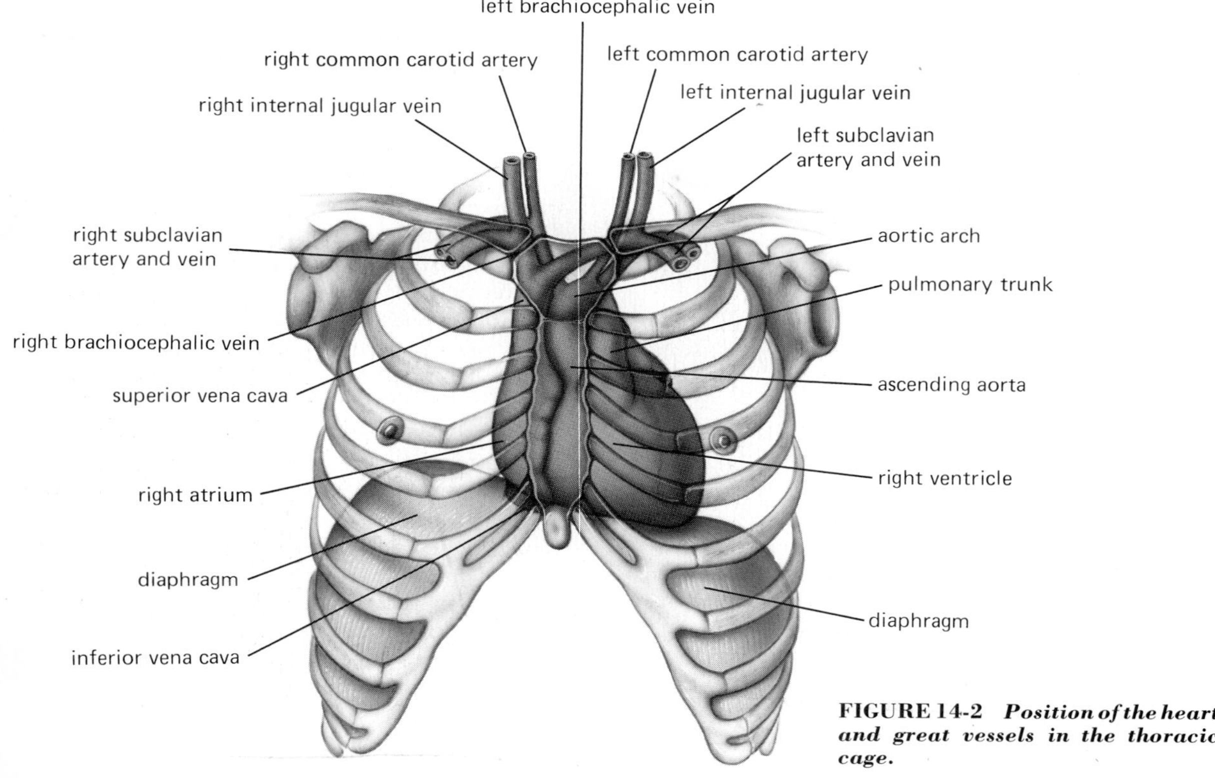

FIGURE 14-2 ***Position of the heart and great vessels in the thoracic cage.***

valves at the entrances of the venae cavae and pulmonary veins into the right and left atrium, respectively.

We can now list the structures through which blood flows in passing from the systemic veins to the systemic arteries: superior and inferior venae cavae; right atrium; tricuspid valve (right AV valve); right ventricle; pulmonary valve; pulmonary arteries, arterioles, capillaries, venules, veins; left atrium; mitral valve (left AV valve); left ventricle; aortic valve; aorta. The driving force for this flow of blood, as we shall see, comes solely from the active contraction of the cardiac muscle. The valves play no part at all in initiating flow and only prevent the blood from flowing in the opposite direction.

The cells constituting the heart walls do not exchange nutrients and metabolic end products with the blood within the heart chambers. The heart, like all other organs, receives its blood supply via arterial branches (the *right* and *left coronary arteries,* Fig. 14-6A and B) which arise from the aorta. Most of the coronary veins drain into the *coronary sinus,* which empties directly into the right atrium.

The walls of the atria and ventricles are composed of layers of cardiac muscle which are tightly bound together and completely encircle the blood-filled chambers. Thus, when the walls of a chamber contract, they come together like a squeezing fist, thereby exerting pressure on the blood they enclose. Cardiac muscle cells combine certain of the properties of smooth and skeletal muscle. The individual cell is striated (Fig. 14-7), containing both the thick myosin and thin actin filaments described for skeletal muscle. Cardiac cells are considerably shorter than the long, cylindrical skeletal fibers and have several branching processes. The processes of adjacent cells are joined end to end at structures known as intercalated disks, within which are points of membrane fusion, i.e., gap junctions, which allow action potentials to be transmitted from one car-

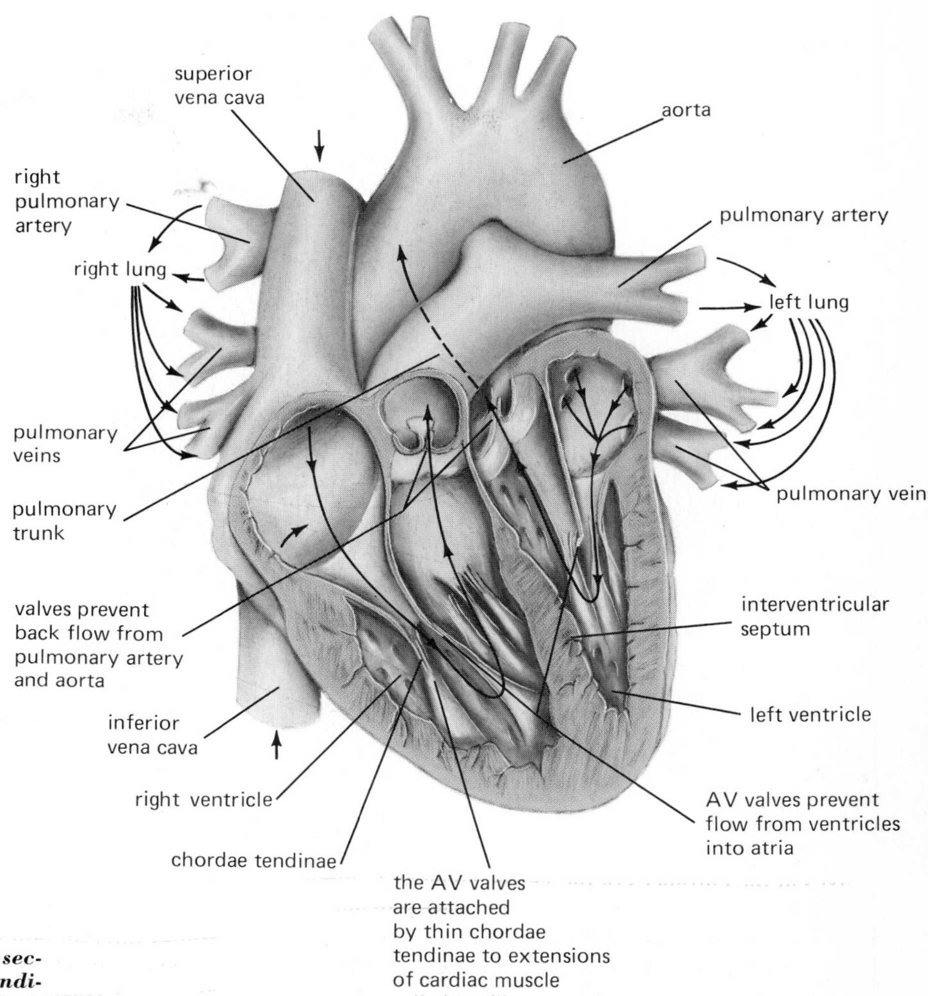

FIGURE 14-3 ***Diagrammatic section of the heart. The arrows indicate the direction of blood flow.***

diac cell to another, in a manner similar to that in smooth muscle.

Besides the usual type of cardiac muscle shown in Fig. 14-7, certain areas of the heart contain specialized muscle fibers called *nodal fibers* and *Purkinje fibers*, which have a different appearance and are essential for normal excitation of the heart. They constitute a network known as the *conducting system* of the heart and also are in contact with fibers of the usual cardiac muscle to form gap junctions which permit passage of action potentials from one cell to another.

HEARTBEAT COORDINATION

Contraction of cardiac muscle, like all other types, is triggered by depolarization of the muscle membrane. In dealing with the mechanisms by which membrane excitation is initiated and spread within the heart we may wonder what would happen if all the many muscle fibers in the heart were to contract in a random manner. One result would be lack of coordination between pumping by each corresponding atrium and ventricle, but this defect is dwarfed by the more serious lack of muscle coordination within the ventricles. The blood

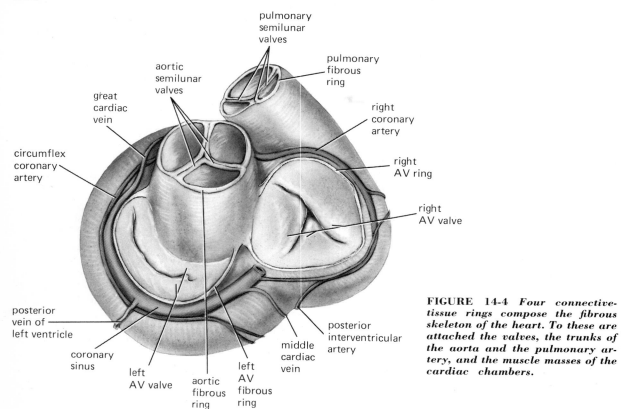

FIGURE 14-4 ***Four connective-tissue rings compose the fibrous skeleton of the heart. To these are attached the valves, the trunks of the aorta and the pulmonary artery, and the muscle masses of the cardiac chambers.***

FIGURE 14-5 ***Position of the heart valves and direction of blood flows during ventricular relaxation (left) and ventricular contraction (right). (Adapted from Carlson, Johnson, and Cavert.)***

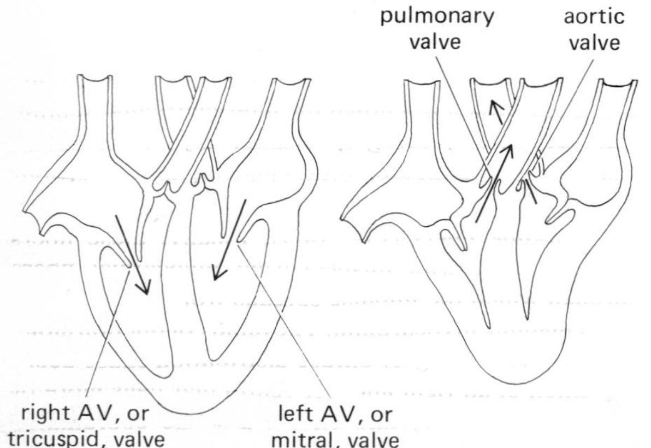

would be sloshed back and forth within the ventricular cavities instead of being ejected into the aorta and pulmonary trunk. In other words, the complex muscle masses which form the ventricular pumps must contract more or less simultaneously for efficient pumping.

Such coordination is made possible by two factors, already mentioned: (1) The gap junctions allow spread of an action potential from one fiber to the next so that excitation in one muscle fiber spreads throughout the heart; (2) the specialized conducting system within the heart facilitates the rapid and coordinated spread of excitation. Where and how does the action potential first arise, and what are the path and sequence of excitation?

Origin of the Heartbeat

Cardiac muscle cells, like certain forms of smooth muscle, are autorhythmic; i.e., they are capable of spontaneous, rhythmic self-excitation. When the individual cells of a salamander embryo heart are separated and placed in salt solution, the individual cells are seen beating spontaneously. But they are beating at different rates. Figure 14-8 shows recordings of membrane potentials from two such cells, the most important feature of which is the gradual depolarization causing the membrane potential to reach threshold, at which point an action potential occurs. Following the action potential, the membrane potential returns to the initial resting value, and the gradual depolarization begins. It should be evident that the slope of this depolarization, i.e., the rate of membrane potential change per unit time, determines how quickly threshold is reached and the next action potential elicited. Accordingly, cell *A* has a faster rate of firing than cell *B*. This capacity for autonomous depolarization toward threshold makes the rhythmic self-excitation of the muscle cells possible. It is due to a decreasing membrane permeability to potassium, but just how this change is generated "spontaneously" remains obscure.

In the course of the salamander experiment above, many individual cells form gap junctions. When such a gap junction is formed between two cells previously contracting autonomously at different rates, both the joined cells contract at the faster rate (Fig. 14-8). In other words, the faster cell sets the pace, causing the initially slower cell to contract at the faster rate. The mechanism is straightforward: The action potential generated by the faster cell causes depolarization, via the gap junction, of the second cell's membrane to threshold, at which point an action potential occurs in this second cell. The important generalization emerges that, because of the gap junctions between cardiac muscle cells, all the cells are excited at the rate set by the cell with the fastest autonomous rhythm.

Precisely the same explanation holds for the origination of the heartbeat in the intact heart; several areas of the adult mammalian heart demonstrate these same characteristics of autorhythmicity and pacemaking, the one with the fastest inherent rhythm being a small mass of specialized myocardial cells embedded in the right atrial wall near the entrance of the superior vena cava (Fig. 14-9). Called the *sinoatrial* (SA) *node*, it is the normal pacemaker for the entire heart. Figure 14-10 is an intracellular recording from an SA node cell; note the slow depolarization toward threshold which initiates the action potential. Compare this SA nodal action potential with that of unspecialized nonautorhythmic atrial cells, which fail to show the pacemaker potential. In unusual circumstances, if some other area of the heart becomes more excitable and develops a faster spontaneous rhythm than the SA node, the new area begins to determine the rhythm for the entire heart.

Sequence of Excitation

The cells of the SA node make contact with the surrounding atrial myocardial fibers (Figs. 14-9 and 14-11). From the SA node, the wave of excitation spreads throughout the right atrium along ordinary atrial myocardial cells, passing from cell to cell by way of the gap junctions. There are also specialized fiber bundles (*internodal tracts*) which conduct the impulse from the SA node directly to the left atrium and the base of the right atrium, thereby ensuring the virtually simultaneous contraction of both atria. How does the excitation spread to the ventricles? At the base of the right atrium very near the interventricular septum, the wave of excitation traveling along ordinary atrial fibers as well as the internodal tracts encounters a second small mass of specialized cells, the *atrioventricular* (AV) *node*. This node and the bundle of fibers leaving it constitute the only myocardial link between the atria and ventricles, all other areas being separated by nonconducting connective tissue. This anatomic pattern ensures that excitation will travel from atria to ventricles only through the AV node, but it also means that malfunction of the AV node may completely dissociate atrial and ventricular contraction. The AV node manifests one particularly important characteristic; the propagation of action potentials through the node is delayed for approximately 0.1 s, allowing the atria to contract and empty their contents into the ventricles before ventricular contraction.

After leaving the AV node, the impulse travels through a group of fibers known as the *bundle of His*, which in turn divides into *right and left bundle branches*, which run down the interventricular septum and then ramify along specialized myocardial fibers which run down the interventricular septum throughout much of the right and left ventricular myocardium as the *Purkinje fibers* (Fig. 14-9). Finally, these fibers make contact with unspecialized myocardial fibers through which the impulse spreads from cell to cell in the remaining myocardium. The rapid conduction along these fibers and highly diffuse distribution cause depolarization of all right and left ventricular cells more or less simultaneously and ensure a single coordinated contraction.

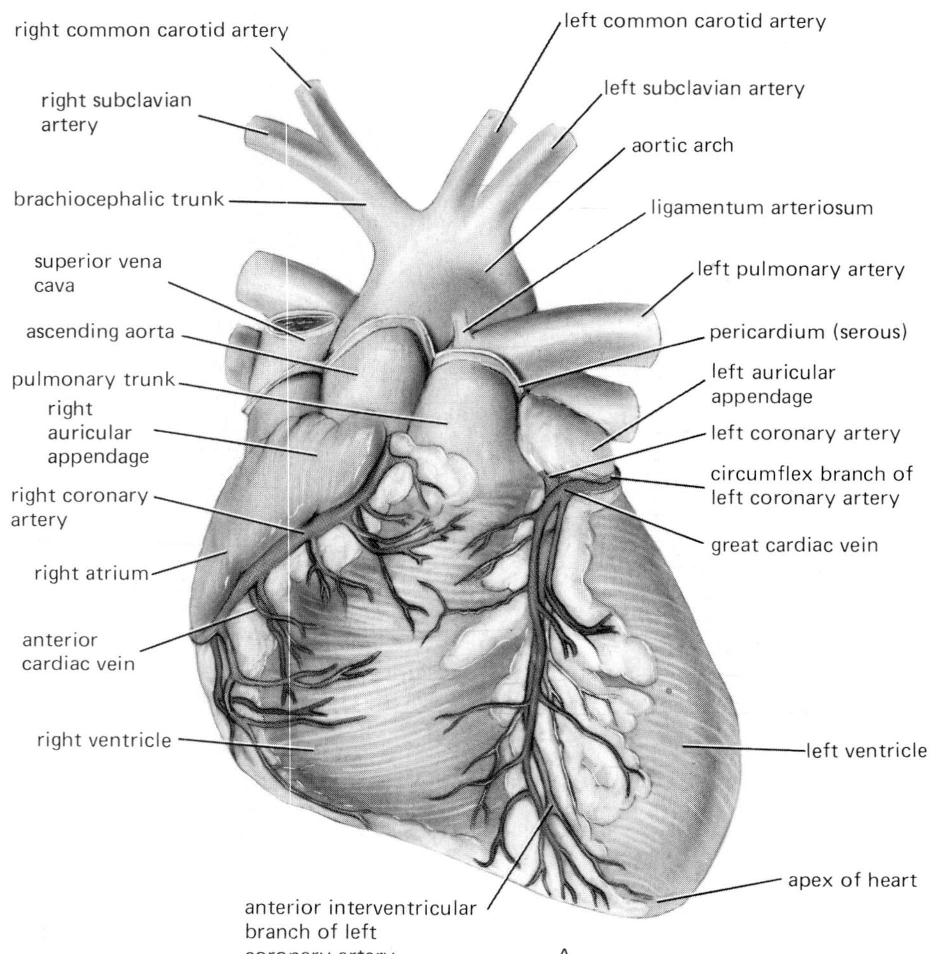

FIGURE 14-6 ***Anterior (A) and posterior (B) views of the heart showing the coronary blood vessels.***

Refractory Period of the Heart

The pumping of blood requires alternate periods of contraction and relaxation. Imagine the result of a prolonged tetanic contraction of cardiac muscle like that described for skeletal muscle. Obviously, pumping would cease and death would ensue. In reality such contractions never occur in the heart because of the long *refractory period* of cardiac muscle. Recall that in any excitable membrane an action potential is accompanied by a period during which the membrane is completely insensitive to a stimulus regardless of intensity. Following this absolute refractory period comes a second period during which the membrane can be depolarized again but only by a more intense stimulus. In skeletal muscle, the absolute refractory periods are very short (1 to 2 ms) compared with the duration of contraction, and a second contraction can be elicited before the first is over. In contrast, the absolute refractory period of cardiac muscle lasts almost as long as the contraction (250 ms), and the muscle cannot be excited in time to produce summation (Fig. 14-12).

A common situation explainable in terms of the refractory period is shown in Fig. 14-13. In many people, drinking several cups of coffee causes increased excitability of certain areas of the atria or ventricles due to the action of caffeine. When one of these areas (*ectopic foci*) fires just after completion of a normal contraction but before the next SA nodal impulse, a premature wave of excitation causes a contraction. As a result, the next normal SA nodal impulse occurs during the refractory period of the premature beat and is not propagated since the myocardial cells are not excitable (the SA

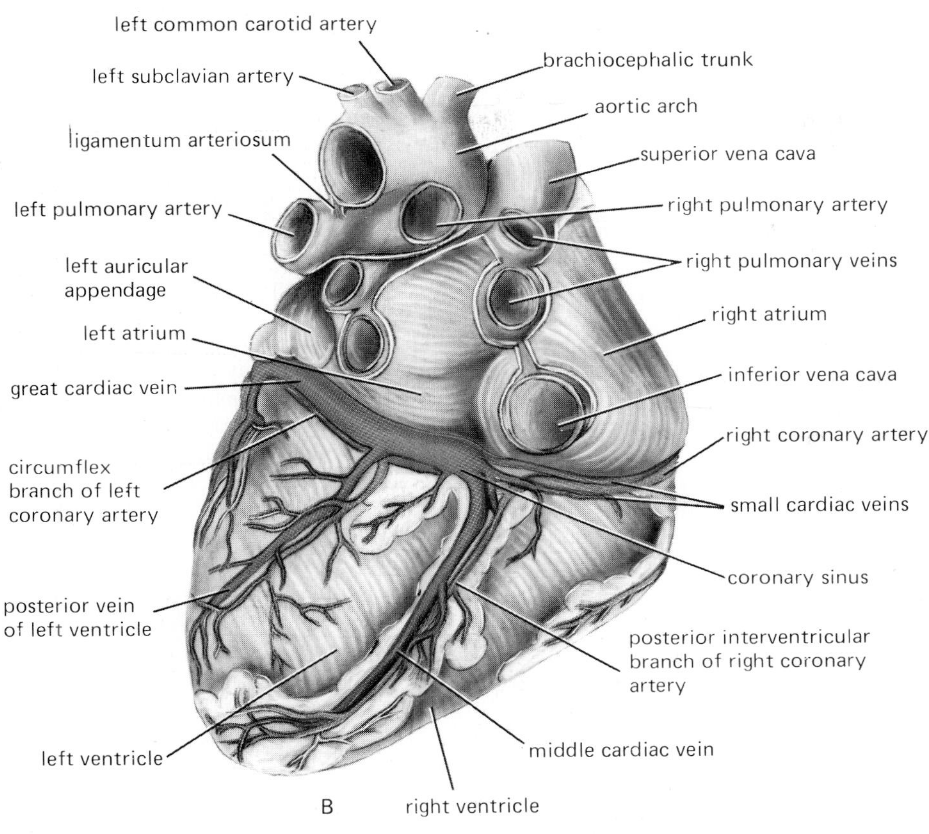

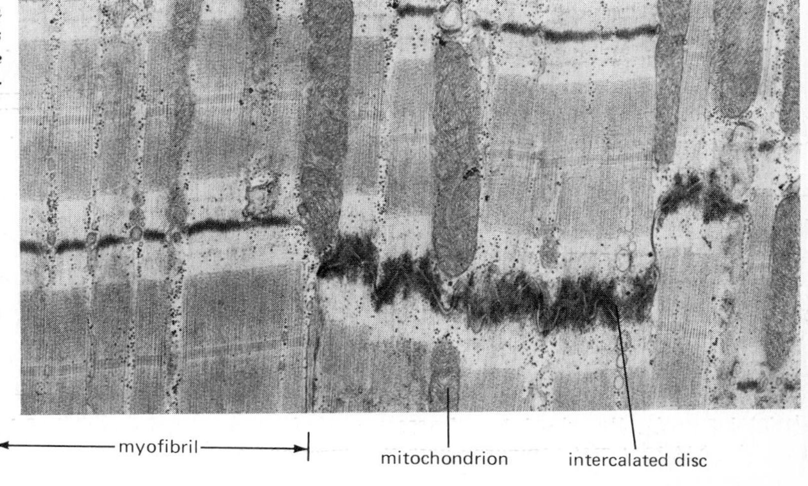

FIGURE 14-7 ***Electron micrograph of cardiac muscle. Note the striations similar to those of skeletal muscle. The wide dark bands indicate interdigitating areas called intercalated disks.***

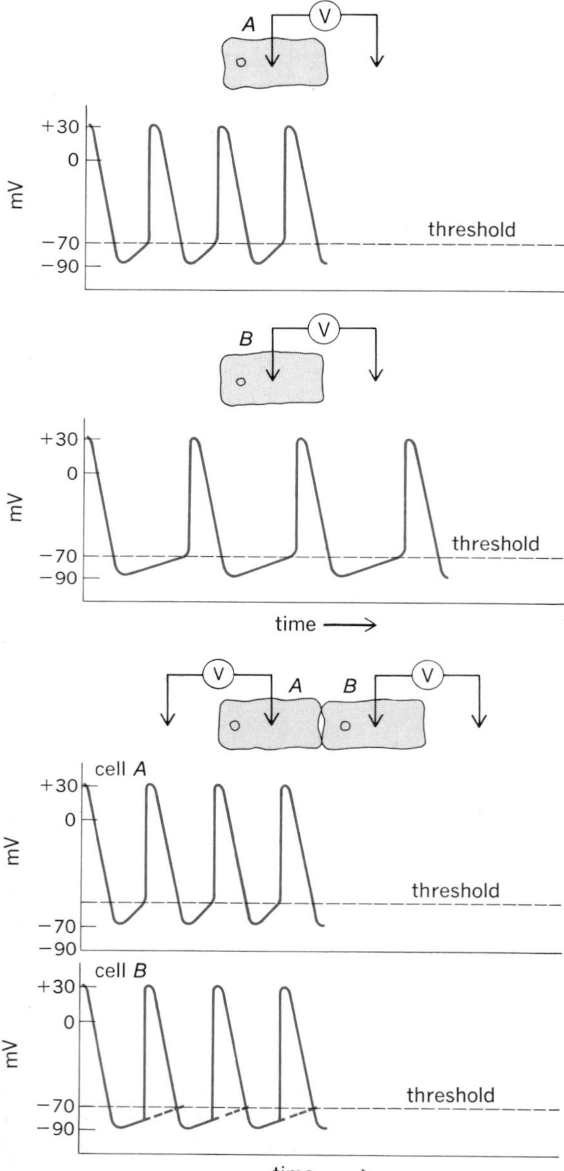

FIGURE 14-8 ***Transmembrane potential recordings from cardiac muscle cells grown in tissue culture. The dashed lines in the bottom recording of cell B indicate the course depolarization would have followed if the cells had not been joined.***

node still fires because it has a shorter refractory period). The second SA nodal impulse after the premature contraction is propagated normally. The net result is an unusually long delay between beats. The contraction after the delay is unusually strong, and the person is aware of his or her heart pounding. Moreover, if the hyperexcited area continued to discharge at a rate higher than that of the SA node, it might then capture the role of pacemaker and drive the heart at rates as high as 200 to 300 beats per minute, compared with a normal rate of 70.

Many similar examples are important clinically and help to clarify the normal physiologic process. For example, disease commonly damages human cardiac tissue and hampers conduction through the AV node. Frequently, only a fraction of the atrial impulses are transmitted into the ventricles; thus, the atria may have a rate of 80 beats per minute and the ventricles only 60. If there is complete block at the AV node, none of the atrial impulses get through but a portion of the ventricle just below the AV node usually begins to initiate excitation at its own spontaneous rate. This rate is quite slow, generally 25 to 40 beats per minute, and completely out of synchrony with the atrial contractions, which continue at the normal higher rate. Under such conditions, the atria are totally ineffective as pumps since they are usually contracting against closed AV valves, but atrial pumping, as we shall see, is relatively unimportant for cardiac functioning (except during relatively strenuous exercise). Some patients have transient recurrent episodes of complete AV block signaled by fainting spells (due to decreased brain blood flow). These spells result because the ventricles do not begin their own impulse generation immediately and cardiac pumping ceases temporarily.

Partial AV block need not be caused by disease and may frequently represent a normal life-saving adaptation. Imagine a patient with an ectopic area driving her atria at 300 beats per minute. Ventricular rates this high are very inefficient because of inadequate filling time. Fortunately, the long refractory period of the AV node may prevent passage of a significant fraction of the impulses and the ventricles beat at a slower rate.

Another group of abnormalities apparently is characterized by a prolonged or unusual conduction route so that the impulse constantly meets an area which is no longer refractory and keeps traveling around the heart in a so-called circus movement. This may lead to con-

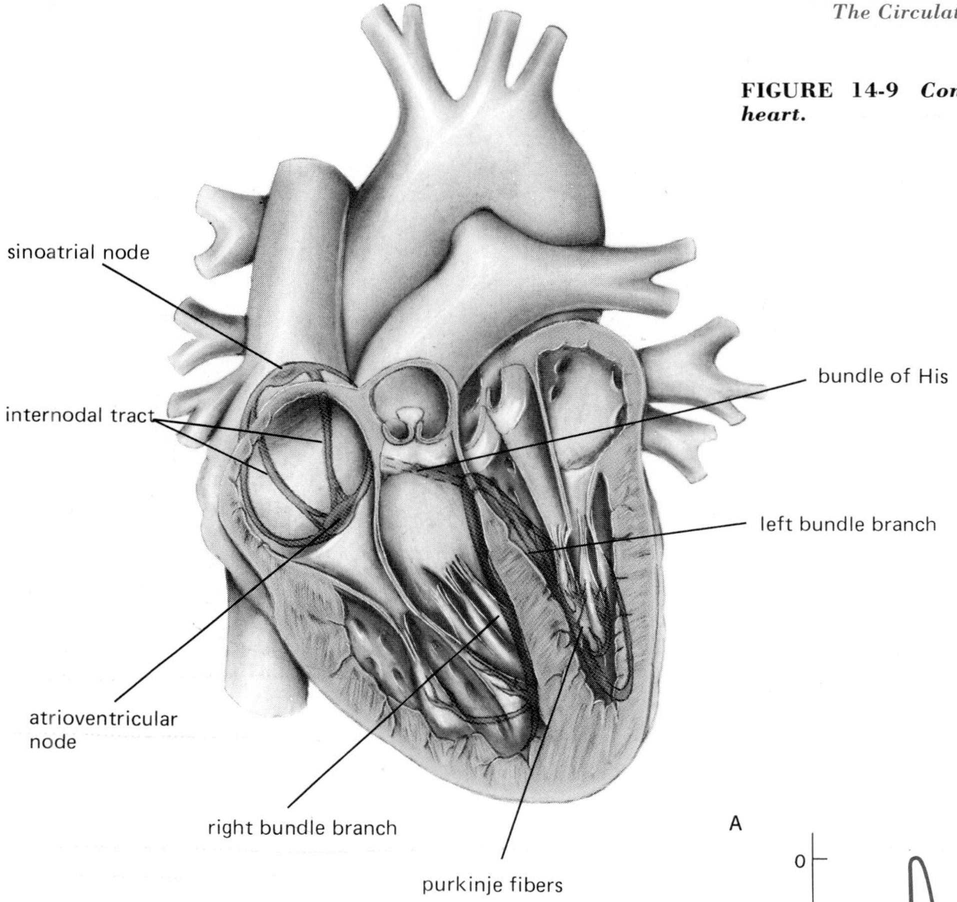

FIGURE 14-9 ***Conducting system of the heart.***

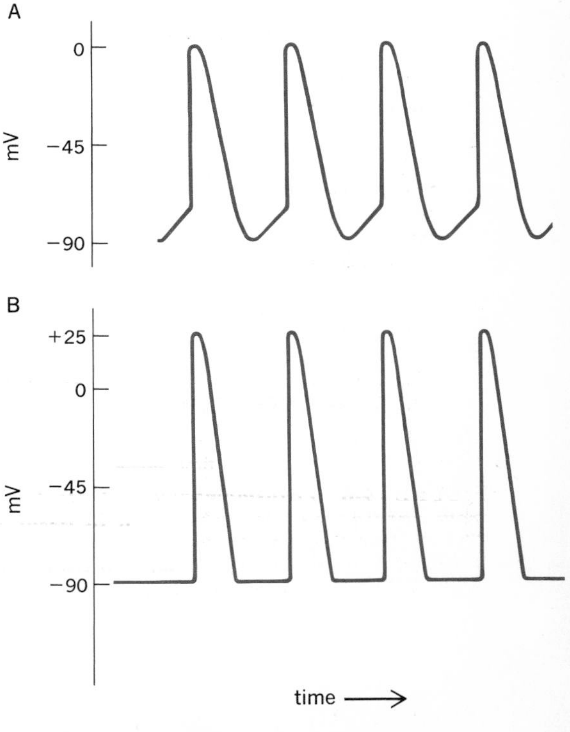

FIGURE 14-10 ***Transmembrane potential recordings from SA node cell (A) and atrial muscle fiber (B).***

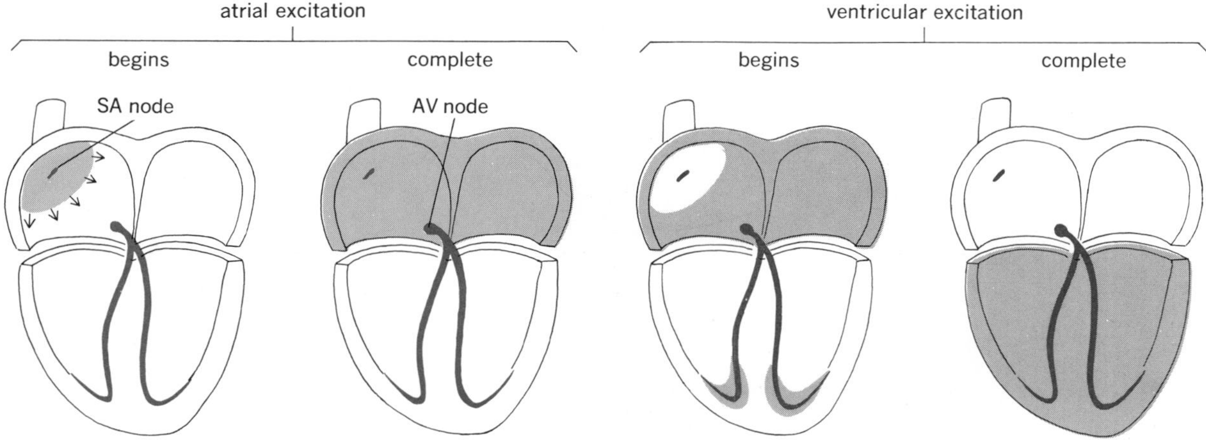

FIGURE 14-11 ***Sequence of cardiac excitation. Atrial excitation is complete before ventricular excitation begins because of the delay at the AV node. (Adapted from Rushmer.)***

FIGURE 14-12 ***Relationship between membrane potential changes and contraction in a single cardiac muscle cell. The refractory period lasts almost as long as the contraction.***

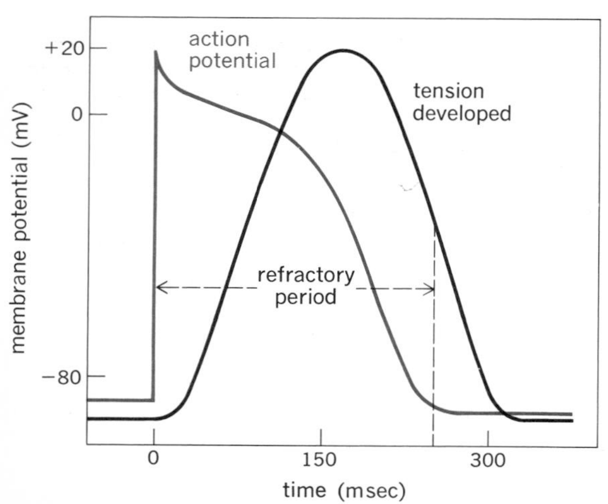

tinuous, completely disorganized contractions (*fibrillation*), which can cause death if they occur in the ventricles. Indeed, ventricular fibrillation is the immediate cause of death from electrocution.

The Electrocardiogram

The *electrocardiogram* (ECG)[1] is primarily a tool for evaluating the electric events within the heart. The action potentials of cardiac muscle can be viewed as batteries which cause current flow throughout the body fluids. These currents produce voltage differences at the body surface which can be detected by attaching small metal plates at different places on the body. Figure 14-14 illustrates a typical normal ECG recorded as the potential difference between the right and left wrists. The first wave *P* represents atrial depolarization. The second complex *QRS*, occurring approximately 0.1 to 0.2 s later, represents ventricular depolarization. The final wave *T* represents ventricular repolarization. No manifestation of atrial repolarization is evident because it occurs during ventricular depolarization and is masked by the *QRS* complex. Figure 14-15A gives one example of the clinical usefulness of the ECG. This patient is suffering from partial AV nodal block so that only one-half the atrial impulses are being transmitted. Note that every second *P* wave is not

[1] Traditionally, electrocardiogram has been abbreviated EKG, the K being derived from the Greek word for heart. ECG is now the preferred term.

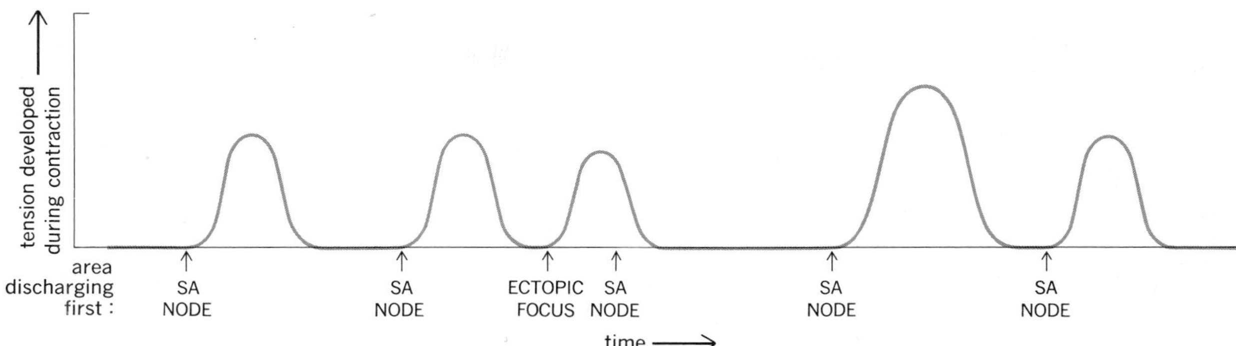

FIGURE 14-13 ***Effect of an ectopic discharge on ventricular contraction. The arrows indicate the times at which the SA node or ectopic focus fires. The premature beat induced by the ectopic discharge makes the ventricular muscle refractory at the time of the next SA-node impulse. The failure of this impulse to induce a contraction results in a longer period than normal before the next beat.***

followed by a *QRS* and *T*. Because many myocardial defects alter normal impulse propagation, and thereby the shapes of the waves, the ECG is a powerful tool for diagnosing heart disease.

FIGURE 14-14 ***Typical electrocardiogram. P, atrial depolarization; QRS, ventricular depolarization; T, ventricular repolarization.***

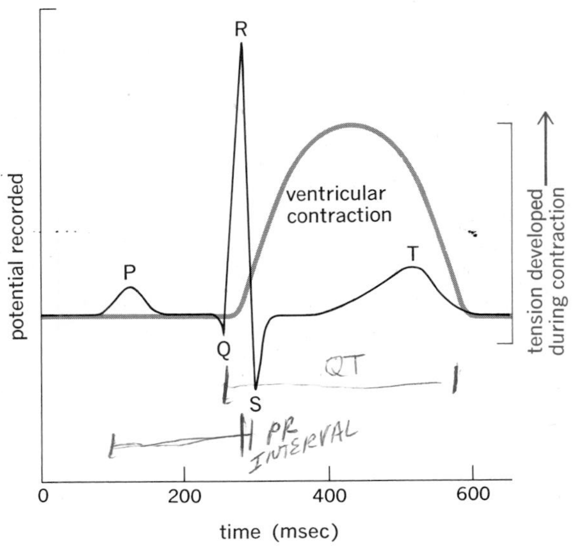

MECHANICAL EVENTS OF THE CARDIAC CYCLE

Fluid always flows from a region of higher pressure to one of lower pressure. (This important concept will be developed formally in Chap. 15; at present, we need deal with it only as an intuitively obvious phenomenon.) The sole function of the heart is to pump blood through the various organs of the body; the fluid pressure generated by cardiac contraction accomplishes the

FIGURE 14-15 ***Electrocardiograms from two persons suffering from atrioventricular block. (A) Partial block; one-half of the atrial impulses are transmitted to the ventricles. (B) Complete block; there is absolutely no synchrony between atrial and ventricular electric activities.***

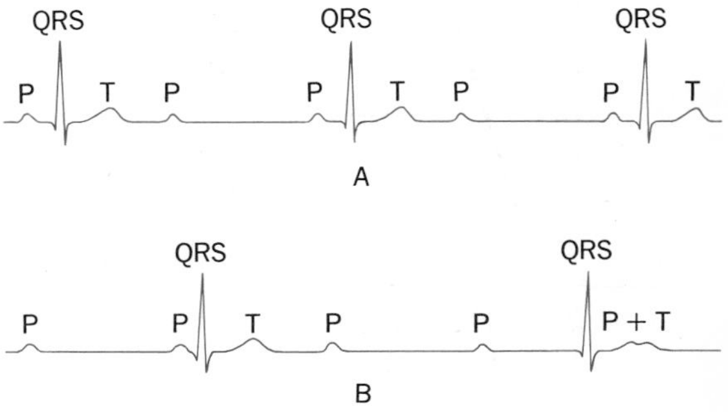

task, the heart valves serving only to direct the flow. The orderly process of depolarization described in the previous section triggers contraction of the atria followed rapidly by ventricular contraction. The flow and pressure changes induced by these contractions are summarized in Fig. 14-16. The reader should follow our analysis of the heart picture and the pressure profile in each phase carefully since an understanding of the *cardiac cycle* is essential. *Systole* is the name of the period of ventricular contraction; *diastole* is ventricular relaxation. We start our analysis with the events of late diastole, considering first only the left heart, events on the right being qualitatively identical.

Late Diastole

The left atrium and ventricle are both relaxed; left atrial pressure is very slightly higher than left ventricular (because blood is entering the atrium from the pulmonary veins); therefore, the AV valves are open, and blood is passing from atrium to ventricle. This is an important point: The ventricle receives blood from the atrium throughout most of diastole, not just when the atrium contracts. Indeed, at rest, approximately 80 percent of ventricular filling occurs before atrial contraction.[2] Note that the aortic valve (between aorta and left ventricle) is closed because the aortic pressure is higher than the ventricular pressure. The aortic pressure is slowly falling because blood is moving out of the arteries and through the vascular tree; in contrast, ventricular pressure is rising slightly because blood is entering from the atrium, thereby expanding the ventricular volume. At the very end of diastole, the SA node discharges, the atrium depolarizes (as shown by the *P* wave of the ECG), the atrium contracts (note the small rise in atrial pressure), and a small volume of blood is added to the ventricle. The amount of blood in the ventricle just prior to systole is called the *end-diastolic volume*.

Systole

The wave of depolarization passes through the ventricle (*QRS* complex) and triggers ventricular contraction. As the ventricle contracts, it squeezes the blood contained in it and ventricular pressure rises steeply. Almost immediately, this pressure exceeds the atrial pressure and closes the AV valve, thus preventing backflow into the atrium. Since for a brief period the aortic pressure still exceeds the ventricular, the aortic valve remains closed and the ventricle does not empty despite contraction. This early phase of systole is called *isovolumetric ventricular contraction* because ventricular volume is constant; i.e., the lengths of the muscle fibers remain approximately constant as in an isometric skeletal muscle contraction. This brief phase ends when ventricular pressure exceeds aortic, the aortic valve opens, and *ventricle ejection* occurs. The ventricular-volume curve shows that ejection is rapid at first and then tapers off. *The ventricle does not empty completely;* the amount remaining after ejection is called the *end-systolic volume*. As blood flows into the aorta, the aortic pressure rises with ventricular pressure. Atrial pressure also rises slowly throughout the entire period of ventricular ejection because of continued flow of blood from the veins. Note that peak aortic pressure is reached before the end of ventricular ejection; i.e., the pressure actually is beginning to fall during the last part of systole despite continued ventricular ejection. This phenomenon is explained by the fact that the rate of blood ejection during this last part of systole is quite small (as shown by the ventricular-volume curve) and is less than the rate at which blood is leaving the aorta (and other large arteries) via the arterioles; accordingly the volume and, therefore, the pressure within the aorta begin to decrease.

Early Diastole

When contraction stops, the ventricular muscle relaxes rapidly owing to release of tension created during contraction. Ventricular pressure therefore falls almost immediately below aortic pressure, and the aortic valve closes. However, ventricular pressure still exceeds atrial so that the AV valve remains closed. This phase of early diastole, obviously the mirror image of early systole, is called *isovolumetric ventricular relaxation*. It ends as ventricular pressure falls below atrial, the AV valves open, and ventricular filling begins. Filling occurs rapidly at first and then slows down as atrial pressure decreases. The fact that ventricular filling is almost complete during early diastole is of the greatest importance; it ensures that filling is not seriously impaired during periods of rapid heart rate, e.g., exercise, emotional stress, fever, despite a marked reduction in the duration of diastole. When rates of approximately 200 beats per minute or more are reached,

[2] It is for this reason that the conduction defects discussed above, which eliminate the atria as efficient pumps, do not seriously impair cardiac function, at least at rest. Many persons lead relatively normal lives for many years despite atrial fibrillation. Thus, in many respects, the atrium may be conveniently viewed as merely a continuation of the large veins.

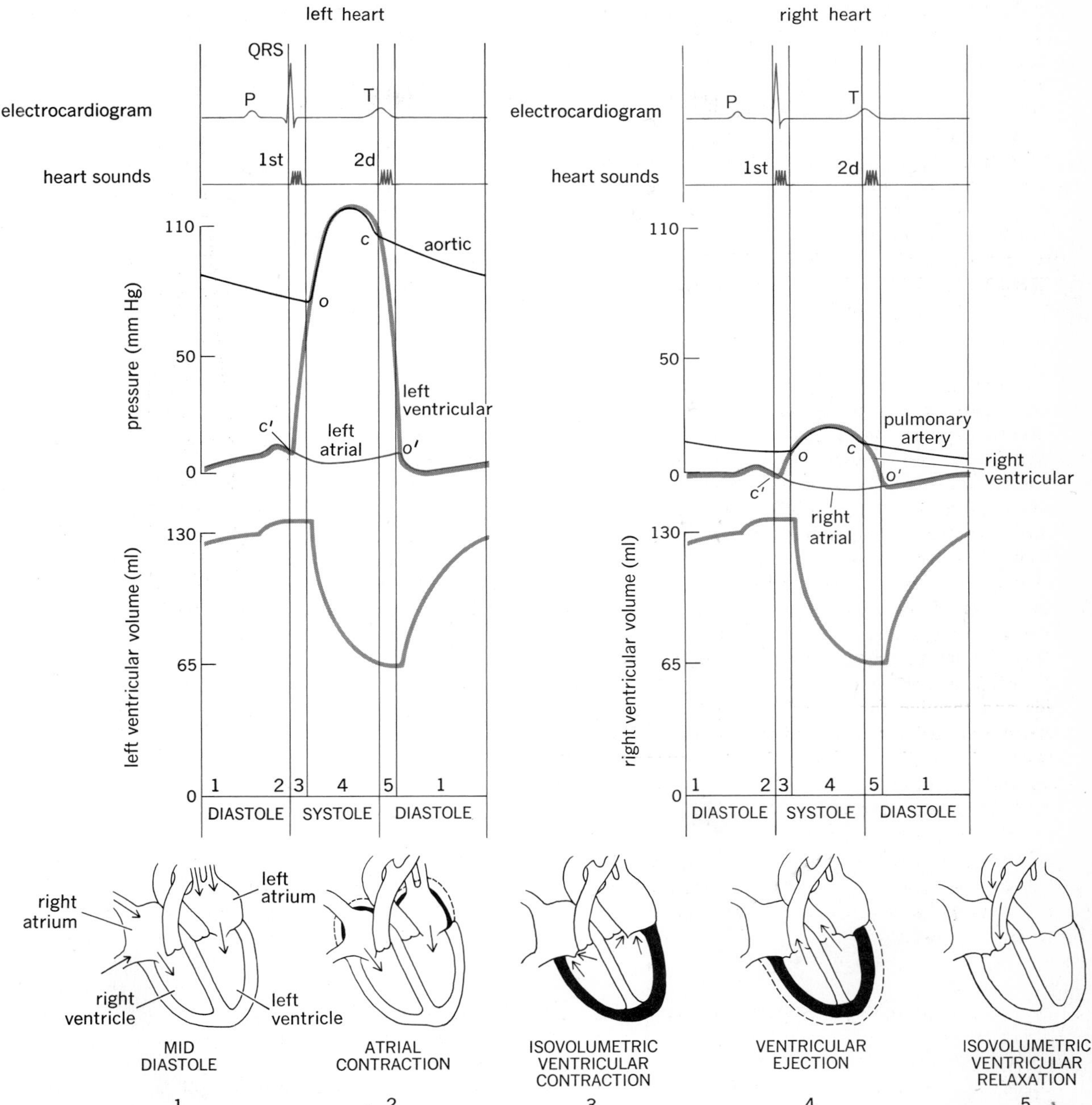

FIGURE 14-16 **(Left)** ***Summary of events in the left heart and aorta during the cardiac cycle. At c′ the AV valve closes; at o′ it opens. At o the aortic valve opens; at c it closes. The contracting portions of the heart are shown in black.*** **(Right)** ***Summary of events in the right heart and pulmonary arteries during the cardiac cycle. At c′ the AV valve closes; at o′ it opens. At o the pulmonary valve opens; at c it closes. (Adapted from Ganong.)***

filling time is inadequate, and cardiac pumping is impaired. Significantly, the AV node in normal adults does not conduct at rates greater than 200 to 250 per minute.

Pulmonary Circulation Pressures

Figure 14-16 summarizes the simultaneously occurring events in the right heart and pulmonary arteries, the patterns being virtually identical to those just described for the left heart. There is one striking quantitative difference: The ventricular and arterial pressures in the right heart are considerably lower during systole. The pulmonary circulation is a low-pressure system (for reasons to be described later). This difference is clearly reflected in the ventricular architecture, the right ventricular wall being much thinner than the left (Fig. 14-17). Note, however, that despite the lower pressure, the right ventricle ejects the same amount of blood as the left.

Heart Sounds

Two heart sounds are normally heard through a stethoscope placed on the chest wall (see Fig. 14-16). The first sound, a low pitched *lub*, is associated with closure of the AV valves at the onset of systole; the second, a high-pitched *dub*, is associated with closure of the pulmonary and aortic valves at the onset of diastole. These sounds, which result from vibrations caused by valvular closure, are perfectly normal, but heart murmurs are frequently (although not always) a sign of heart disease. When blood flows smoothly in a streamlined manner, i.e., layers of fluid sliding evenly over one another, it makes no sound, but turbulent flow pro-

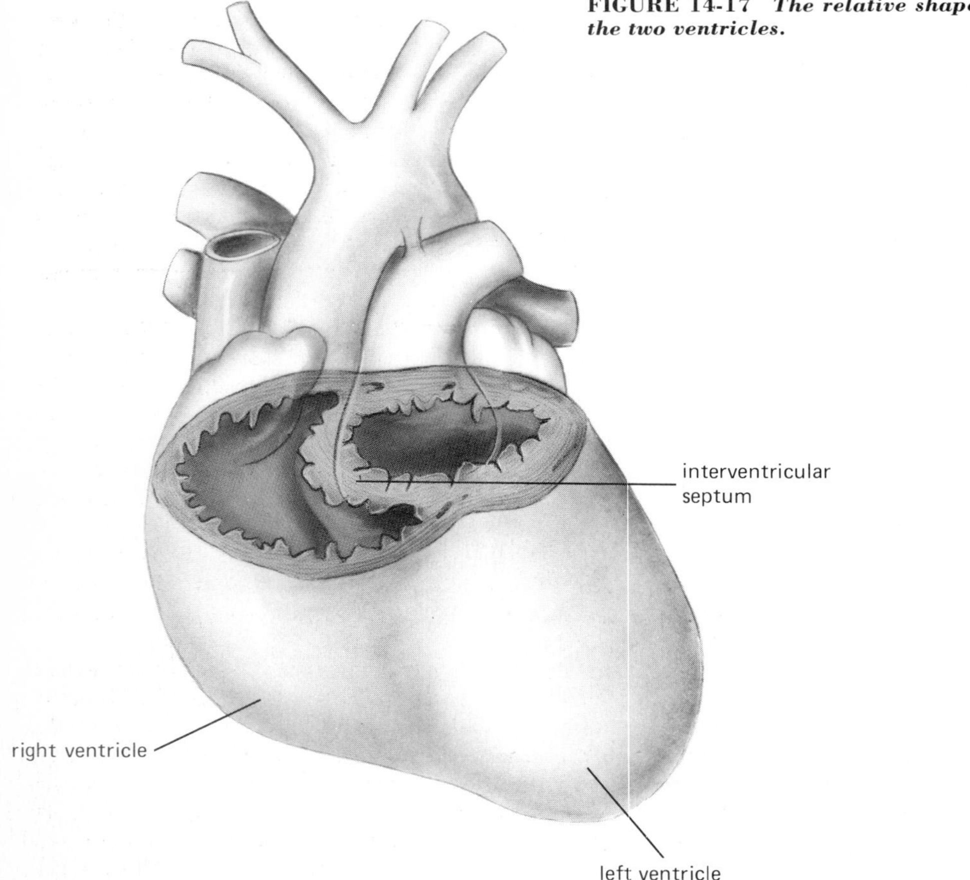

FIGURE 14-17 ***The relative shape and wall thickness of the two ventricles.***

duced by unusually high velocities makes a noise. This noise is heard as a murmur or sloshing sound. Turbulence can be produced by blood flowing rapidly in the usual direction through an abnormally narrowed valve, backward through a damaged leaky valve, or between the two atria or two ventricles via a small hole in the septum. The exact timing and location of the murmur provide the physician with a powerful diagnostic clue. For example, a murmur heard throughout systole suggests a narrowed pulmonary or aortic valve or a hole in the interventricular septum. The diagnosis can then be completed by the use of specialized techniques.

THE CARDIAC OUTPUT

The volume of blood pumped *by each ventricle* per minute is called the *cardiac output,* usually expressed as liters per minute. It must be remembered that the cardiac output is the amount of blood pumped by *each* ventricle, *not* the total amount pumped by both ventricles. The cardiac output is determined by multiplying the *heart rate* and the volume of blood ejected by each ventricle during each beat (*stroke volume*):

$$\underset{\text{l/min}}{\text{Cardiac output}} = \underset{\text{beats/min}}{\text{heart rate}} \times \underset{\text{l/beat}}{\text{stroke volume}}$$

For example, if each ventricle has a rate of 72 beats per minute and ejects 70 ml with each beat, what is the cardiac output?

$$\text{CO} = 72 \text{ beats/min} \times 0.07 \text{ l/beat} = 5.0 \text{ l/min}$$

These values are approximately normal for a resting adult. During periods of exercise, the cardiac output may reach 20 to 25 l/min. Obviously, heart rate or stroke volume or both must have increased. Physical exercise is but one of many situations in which various tissues and organs require a greater flow of blood; e.g., flow through skin vessels increases when heat loss is required, and flow through intestinal vessels increases during digestion. Some of the increased flow can be obtained merely by decreasing blood flow to some other organ, i.e., by redistributing the cardiac output, but most of the supply must come from a greater total cardiac output. The following description of the factors which alter the two determinants of cardiac output, heart rate and stroke volume, applies in all respects to both the right and left heart since stroke volume and heart rate are the same for both the right and left ventricles.

Control of Heart Rate

The rhythmic discharge of the SA node occurs spontaneously in the complete absence of any nervous or hormonal influences. However, it is under the constant influence of both nerves and hormones. A large number of parasympathetic and sympathetic fibers end on the SA node as well as on other areas of the conducting system. The parasympathetics to the heart are contained in the vagus nerves (Fig. 8-16). The *sympathetic cardiac nerves,* which contain the postganglionic sympathetic fibers, arise from the thoracic levels of the sympathetic trunk or the cervical ganglia (Fig. 8-16). The preganglionic sympathetic neurons lie in the upper thoracic levels of the spinal cord.

Stimulation of the parasympathetic nerves (or local application of acetylcholine) causes slowing of the heart and, if strong enough, may stop the heart completely for some time; cutting the parasympathetics causes the heart rate to increase. The effects of the sympathetic nerves are just the reverse; nerve stimulation (or local application of norepinephrine) increases the heart rate, whereas cutting the sympathetics slows the heart. Both the sympathetic and parasympathetic nerves normally discharge at some finite rate. Apparently, in the resting state, the parasympathetic influence is dominant since simultaneous removal of all nerves causes the heart rate to increase to approximately 100 beats per minute. This is the inherent autonomous discharge rate of the SA node.

Figure 14-18 illustrates the nature of the sympathetic and parasympathetic influence on SA node function. Sympathetic stimulation increases the slope of the pacemaker potential, which causes the cell to reach threshold more rapidly and is responsible for the rate change. Stimulation of the parasympathetics has the opposite effect; the slope of the pacemaker potential decreases, threshold is reached more slowly, and heart rate decreases. The underlying alterations of membrane permeability induced by the sympathetic mediator, norepinephrine, and the parasympathetic mediator, acetylcholine, are still uncertain.

Factors other than the cardiac nerves also can alter heart rate. Epinephrine, the hormone liberated from the adrenal medulla, speeds the heart; this is not surprising since epinephrine is a blood-borne sympathetic mediator similar in structure to norepinephrine (Chaps. 8 and 12).

The heart rate is also sensitive to many other factors, including temperature, plasma electrolyte concentrations, and hormones other than epinephrine. How-

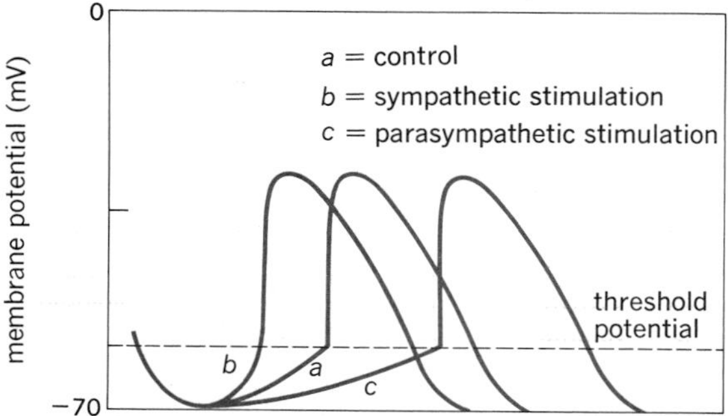

FIGURE 14-18 ***Effects of sympathetic and parasympathetic nerve stimulation on the slope of the pacemaker potential of an SA node cell. (Adapted from Hoffman and Cranefield.)***

ever, these are generally of lesser importance, and the heart rate is primarily regulated very precisely by balancing the slowing effects of parasympathetic discharge against the accelerating effects of sympathetic discharge, both operating on the SA node (Fig. 14-19).

Control of Stroke Volume

There are always considerable amounts of blood remaining in the ventricles after contraction. The volume of blood ejected during each ventricular contraction is obviously the difference between the volume of blood contained in the ventricle at the end of diastole (*end-diastolic volume*) and the volume remaining at the end of systole:

FIGURE 14-19 ***Summary of major factors which influence heart rate. All effects are exerted upon the SA node. The figure, as drawn, shows how heart rate is increased; conversely, heart rate is slowed when sympathetic activity and epinephrine are decreased and when parasympathetic activity is increased.***

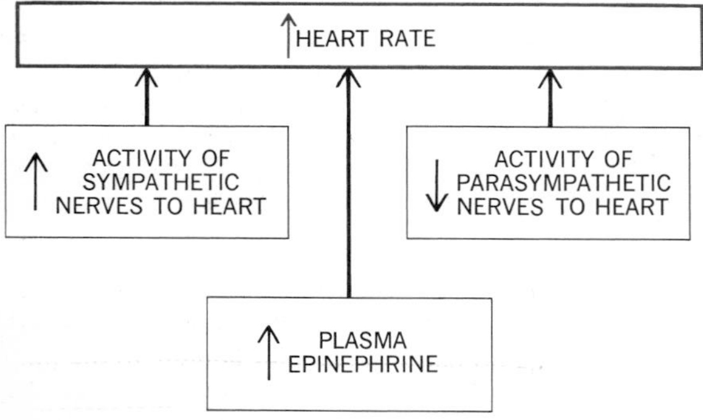

Stroke volume
= end-diastolic volume − end-systolic volume

An increased end-diastolic volume usually leads to an increased stroke volume. We shall refer to this as *intrinsic control* because it represents an inherent property of cardiac muscle. The stroke volume can also be increased by the nerves to the heart and several circulating hormones; we shall call this *extrinsic control.*

Intrinsic Control: The Relationship Between Stroke Volume and End-diastolic Volume It is possible to study the completely intrinsic adaptability of the heart by means of the so-called heart-lung preparation (Fig. 14-20). Tubes are placed in the heart and vessels of an anesthetized animal so that blood flows from the very first part of the aorta (just above the exit of the coronary arteries) into a blood-filled reservoir and from there into the right atrium. The blood then is pumped by the right heart as usual via the lungs into the left heart. The net effect is to nourish the heart and lungs normally and to deprive the rest of the animal's body of blood and cause death, thereby abolishing all nervous and hormonal activity. A key feature of this preparation is that the pressure causing blood flow into the heart can be altered simply by raising or lowering the reservoir. This is analogous to altering the venous and atrial pressures and causes changes in the quantity of blood entering the ventricles during diastole. When the reservoir is raised, the following sequence of events is observed:

1 Diastolic filling increases, thereby increasing end-diastolic volume.

2 Stroke volume increases as a result of a more forceful contraction but not enough to eject all the blood which entered during the previous diastole, and end-systolic volume increases.

3 For several more beats diastolic filling slightly exceeds systolic ejection despite progressively more forceful contractions, and the end-diastolic volume becomes progressively larger.

4 Ultimately the distended heart contracts forcefully enough so that the stroke output becomes equal to the diastolic filling.

The net result is a new steady state, in which the ventricle is distended and diastolic filling and stroke volume are both increased, but equally. The mechanism underlying this completely intrinsic adaptation is that cardiac muscle, like other muscle, increases its strength

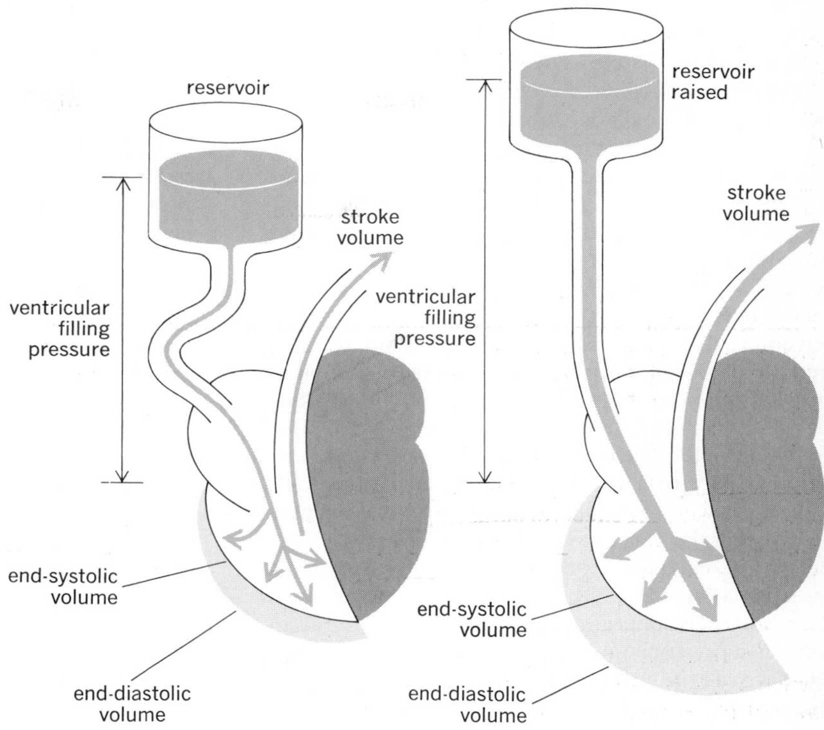

FIGURE 14-20 ***Demonstration of the intrinsic control of stroke volume (Starling's law of the heart). By raising the reservoir, the pressure causing ventricular filling is increased. The increased filling distends the ventricle, which responds with an increased strength of contraction.***

of contraction when it is stretched. Thus, in the experiment above, the increased diastolic volume stretches the ventricular muscle fibers and causes them to contract more forcefully. This relationship was expounded by the British physiologist Starling, who observed that there was a direct proportion between the diastolic volume of the heart, i.e., the length of its muscle fibers, and the force of contraction of the following systole. It is now referred to as *Starling's law of the heart.* A typical response curve obtained by progressively increasing end-diastolic volume is shown in Fig. 14-21. Note that marked overstretching causes the force of contraction, and thereby the stroke volume, to fall off. Thus, heart muscle manifests a length-tension relationship very similar to that described earlier for skeletal muscle (see Fig. 5-18) and explainable in terms of the sliding-filament mechanism of muscle contraction. However, unlike skeletal muscle, cardiac muscle length in the resting state is less than that which yields maximal tension during contraction so that an increase in length produces an increase in contractile tension.

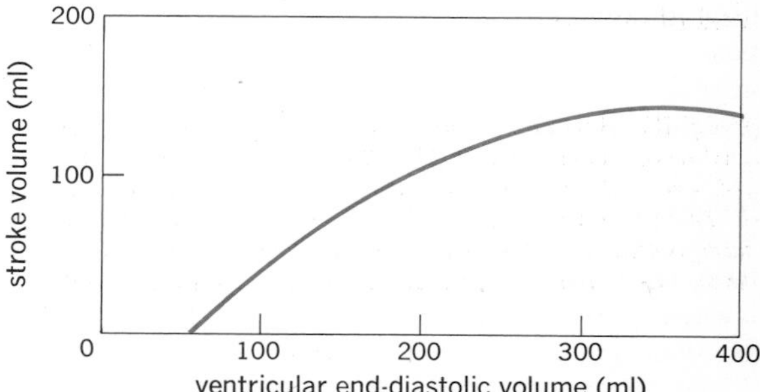

FIGURE 14-21 ***Relationship between ventricular end-diastolic volume and stroke volume (Starling's law of the heart). The data were obtained by progressively increasing ventricular filling pressure, as in Fig. 14-20.***

This intrinsic relationship between end-diastolic volume and stroke volume, originally demonstrated in the heart-lung preparation, applies equally to the intact human being. End-diastolic volume, therefore, becomes a crucial determinant of cardiac output.

What then are the factors which determine end-diastolic volume, i.e., the degree of ventricular distension just before systole? The simplest way to approach this question is to view the ventricle as an elastic chamber, much like a balloon. A balloon enlarges when one blows into it because the internal pressure acting upon the wall becomes greater than the external pressure. The more air blown in, the higher the internal pressure becomes. The degree of distension therefore depends upon the pressure difference across the wall and the distensibility of the wall. This is precisely the situation for the ventricle. Ignore the problem of ventricular distensibility (for the reason that it does not appear to be under physiologic control) and concentrate only on the *transmural,* or across-the-wall, difference in pressure.

What is the ventricular transmural pressure at the end of diastole? The internal pressure, of course, is the fluid pressure exerted by the blood against the walls. The external pressure surrounding the heart is the pressure within the chest cavity (thorax), or *intrathoracic pressure.* A typical end-diastolic ventricular blood pressure is 4 mm Hg. Physiologists state pressures with the atmospheric pressure, i.e., the pressure of the atmospheric air surrounding the body, given as zero. Thus, when we say the end-diastolic pressure is 4 mm Hg, we really mean 4 mm Hg greater than atmospheric pressure. If we assume the atmospheric pressure to be 760 mm Hg, the true internal pressure is 764 mm Hg. This distinction must be remembered, especially when one is discussing events within the thoracic cage, because, for reasons to be described in Chap. 17, the intrathoracic pressure of the fluid surrounding the heart, lungs, and all other intrathoracic structures is *less* than atmospheric. This subatmospheric intrathoracic pressure is frequently termed a "negative" pressure, but this terminology should be avoided since there is no such thing in nature as a negative pressure. The intrathoracic pressure averages approximately 5 mm Hg *less* than atmospheric pressure (or a true pressure of 755 mm Hg); accordingly, the pressure difference acting to distend the ventricles at the end of diastole is $4 + 5 = 9$ mm Hg, i.e., $764 - 755 = 9$ mm Hg.

We can now answer our original question. End-diastolic ventricular volume, i.e., distension, can be increased either by increasing the intraventricular blood pressure or by decreasing the intrathoracic pressure, or both. The latter always occurs during inspiration (see Chap. 17) and accounts, in part, for the increased stroke volume which characteristically occurs during inspiration. However, it is primarily by changes in the end-diastolic intraventricular pressure that end-diastolic volume is controlled. As can be seen from Fig. 14-16, the pressure within the ventricles during diastole closely approximates the atrial pressure, since the valves are open and the chambers are connected by the wide AV orifice through which blood is flowing into the ventricle. Accordingly, the diastolic ventricular blood pressure is determined by the atrial pressure. As we shall see, the atrial pressure, in turn, is determined by the rate of blood flow from the veins into the atria.

The significance of this mechanism should now be apparent; an increased flow of blood from the veins into the heart automatically forces an equivalent increase in cardiac output by distending the ventricle and increasing stroke volume, just like the heart-lung apparatus when we increased "venous return" by elevating the reservoir. This is probably the single most important mechanism for maintaining equality of right and left output. Should the right heart, for example, suddenly begin to pump more blood than the left, the increased blood flow to the left ventricle would automatically produce an equivalent increase in left ventricular output and blood would not be allowed to accumulate in the lungs. Another example of Starling's law has already been described above, namely, the pounding that occurs after a premature contraction. Recall that an unusually long period elapses between the premature contraction and the next contraction; the period for diastolic filling is increased, end-diastolic volume increases, and the force of contraction is increased (Fig. 14-13). It is this strong contraction, which may actually lift the heart against the chest wall, that the person is aware of. In summary, end-diastolic volume and stroke volume are generally increased whenever atrial pressure increases; the factors which determine atrial pressure will be described in Chap. 15.

Extrinsic Control: The Sympathetic Nerves Until recently, it was generally assumed that the mechanism described by Starling's law could explain almost all changes in stroke volume observed under physiologic conditions. During exercise, for example, it was believed that the increased venous return produced by various factors led to cardiac distension and increased stroke output. However, experiments have not borne out this hypothesis; x-ray photographs of exercising human beings (and other mammals) have clearly shown that the normal heart does not usually distend during exercise and may even decrease in size, despite the fact that stroke volume usually increases.

It must be stressed that the relationship described by Starling's law is not invalid but simply is not the sole determinant of ventricular strength of contraction. The other major factor is the sympathetic nerves, which are distributed not only to the SA node and conducting system but to all myocardial cells. The effect of the sympathetic mediator, norepinephrine, is to increase ventricular (and atrial) *contractility,* defined as the strength of contraction at any given initial muscle-fiber length, i.e., end-diastolic volume. Not only is the contraction more powerful but both it and relaxation occur more rapidly. These latter effects are quite important since, as described earlier, increased sympathetic activity to the heart also increases heart rate. As the heart rate increases, the time available for diastolic filling decreases, but the more rapid contraction and relaxation induced simultaneously by the sympathetic neurons partially compensate for this problem by permitting a larger fraction of the cardiac cycle to be available for filling. Moreover, because the ventricles relax so rapidly after a contraction the intraventricular pressure falls rapidly, thereby creating an enhanced pressure gradient for flow of blood into the ventricles. This is, in essence, a "sucking" effect which facilitates ventricular filling. The ability of these effects to maintain diastolic filling is, of course, not unlimited, and diastolic filling is significantly reduced at very high heart rates. The significance of this interplay between diastolic filling time, heart rate, and contractility will be analyzed further in the section on exercise in Chap. 16.

Circulating epinephrine produces changes in contractility similar to those induced by the sympathetic nerves to the heart. Moreover, a decreased contractility can be obtained by reducing the rate of sympathetic discharge below the usual tonic level. The mechanism by which norepinephrine and epinephrine increase contractility probably involves an increased liberation of calcium during excitation. In contrast to the sympathetic nerves, the parasympathetic nerves to the heart have relatively little effect on ventricular contractility.

The interrelationship between the intrinsic (Starling's law) and extrinsic (cardiac nerves) mechanisms as measured in a heart-lung preparation is illustrated in Fig. 14-22. The dashed line is the same as the line shown in Fig. 14-21 and was obtained by slowly raising ventricular pressure while measuring end-diastolic volume and stroke volume; the solid line was obtained similarly for the same heart but during sympathetic-nerve stimulation. Starling's law still applies, but during nerve stimulation the stroke volume is greater at any

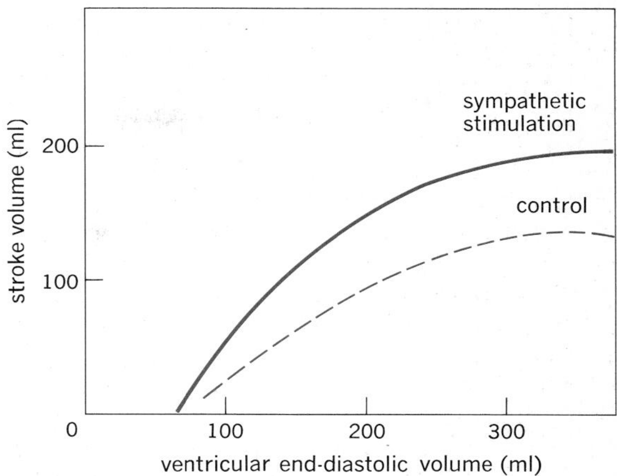

FIGURE 14-22 ***Effects on stroke volume of stimulating the sympathetic nerves to the heart.***

FIGURE 14-23 ***Major factors which influence stroke volume. The figure as drawn shows how stroke volume is increased; a reversal of all arrows in the boxes would illustrate how stroke volume is decreased. Refer to the text for details.***

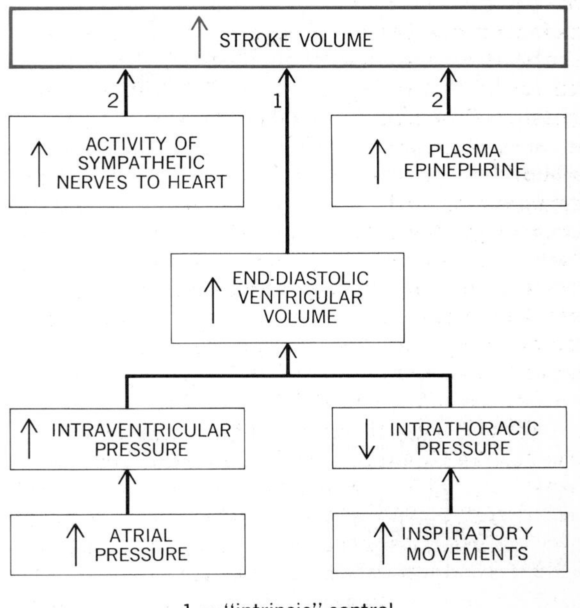

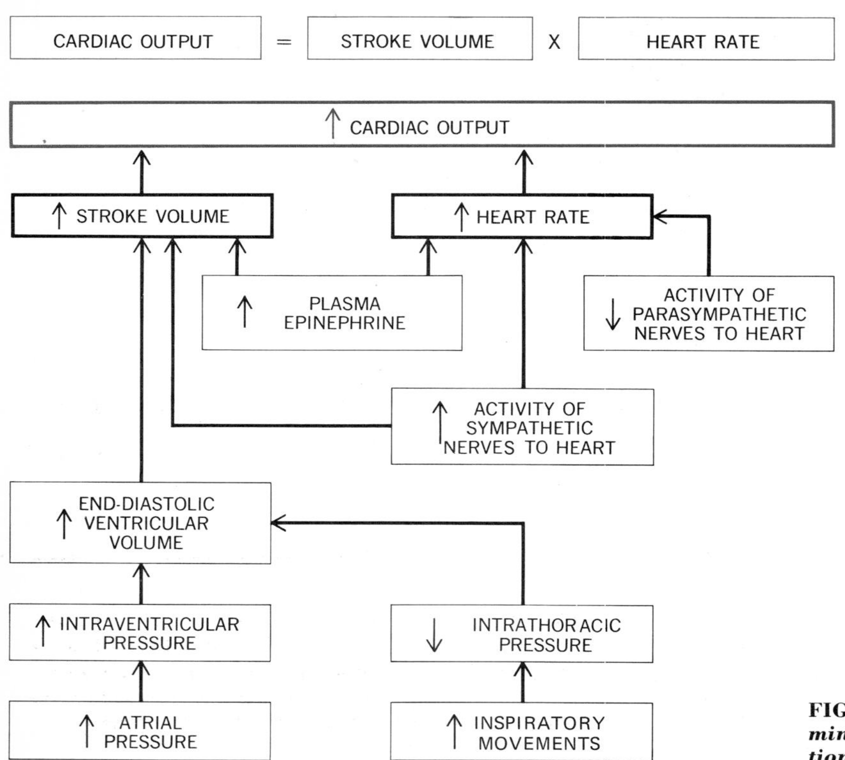

FIGURE 14-24 ***Major factors determining cardiac output (an amalgamation of Figs. 14-19 and 14-23).***

given end-diastolic volume. In other words, the increased contractility leads to *a more complete ejection* of the end-diastolic ventricular volume.

In summary (Fig. 14-23) stroke volume is controlled both by an intrinsic cardiac mechanism dependent only upon changes in end-diastolic volume and by an extrinsic mechanism mediated by the cardiac sympathetic nerves (and circulating epinephrine). The contribution of each mechanism in specific physiologic situations and the reflexes controlling the nerves will be described in Chap. 16.

Summary of Cardiac Output Control

A summary of the major factors which determine cardiac output is presented in Fig. 14-24, which combines the information of Fig. 14-19 (factors influencing heart rate) and Fig. 14-23 (factors influencing stroke volume).

CHAPTER 15

The Circulatory System: III. The Vascular System

DESIGN OF THE VASCULAR SYSTEM

As mentioned in the preceding chapter, in the systemic vascular circuit the blood leaves the left half of the heart via the aorta; from this single large artery branching arteries conduct blood to the various organs and tissues. These arteries divide in a highly characteristic manner into progressively smaller branches, much of the branching occurring within the specific organ or tissue supplied. The smallest branches, called *arterioles,* differ structurally and functionally from the arteries. Ultimately the arterioles branch into a huge number of very small, thin vessels, termed *capillaries.* The capillaries unite to form larger vessels (*venules*) which, in turn, unite to form fewer and still larger vessels, termed *veins* (Fig. 15-1). The veins from different organs and tissues unite to form two large veins, the *inferior vena cava* (from the lower portion of the body) and the *superior vena cava* (from the upper half of the body). By these two veins blood is returned to the right half of the heart. There is an exception to this general pattern of blood flow: *Portal systems* have two capillary beds interposed between the arteries and the veins which finally return blood to the heart. Portal systems serve specialized functions and occur between the hypothalamus and anterior pituitary (the hypothalamohypophyseal portal system, Chap. 12), between many organs of the digestive system and the liver (the hepatic portal system, Chaps. 15 and 19), and within the kidneys (Chap. 18).

The pulmonary circulation is similar to the systemic circuit. Blood leaves the right half of the heart via a single large artery, the *pulmonary artery*, which divides into two arteries, one supplying each lung. Within the lungs, the arteries continue to branch, ultimately forming arterioles, which then divide into capillaries. These capillaries unite to form small venules, which unite to form larger and larger veins. The blood leaves the lungs via the largest of these, the *pulmonary veins*, which empty into the left half of the heart. The blood flowing through the systemic veins, right half of the heart, and pulmonary arteries has a low oxygen content. As this blood flows through the lung capillaries, it picks up large quantities of oxygen; therefore, the blood in the pulmonary veins, left heart, and systemic arteries is high in oxygen. As this blood flows through the capillaries of tissues and organs throughout the body, much of the oxygen leaves the blood, resulting in the low oxygen content of systemic venous blood.

It should be evident that all the blood pumped by the right heart flows through the lungs; in contrast, only a fraction of the total left ventricular output flows through any single organ or tissue. In other words, the systemic circulation comprises numerous different pathways "in parallel." They all originate as large arteries branching off from the aorta. The only significant deviation from this pattern is the portal blood supply to the liver, much of which is not arterial but venous blood, which has just left the pancreas, spleen, and gastrointestinal tract.

Figures 15-2A and B show the major arteries and veins. Tables 15-1 to 15-4 and Figs. 15-3 to 15-10 summarize the distribution of these major vessels.

The functional and structural characteristics of the blood vessels change with successive branching, yet the entire cardiovascular system from the heart to the smallest capillary has one structural component in common, a smooth, low-friction lining of *endothelial cells*. All vessels larger than capillaries have layers of tissue surrounding the endothelium which provide supporting connective-tissue elements to counter the pressure of the contained blood, elastic elements to dampen pressure pulsations and minimize flow variations throughout the cardiac cycle, and muscle fibers to actively control the diameter of the vessel lumen.

These coats, or tunics, from within outward are named the *tunica intima, tunica media*, and *tunica adventitia*. In general, the tunica intima comprises a single layer of endothelial cells supported on its outer surface by a thin connective-tissue network. The tunica media of arteries is a fibromuscular coat with an *internal* and *external elastic lamina* at its inner and outer borders, and the tunica adventitia is largely a connective-tissue coat. The thickness and composition of the outer two layers vary with the type and size of vessel, reflecting its functional role. In general, arteries have thicker walls and smaller lumina than veins, and the tunica media is the thickest coat in arteries whereas the adventitia is the thickest coat in veins.

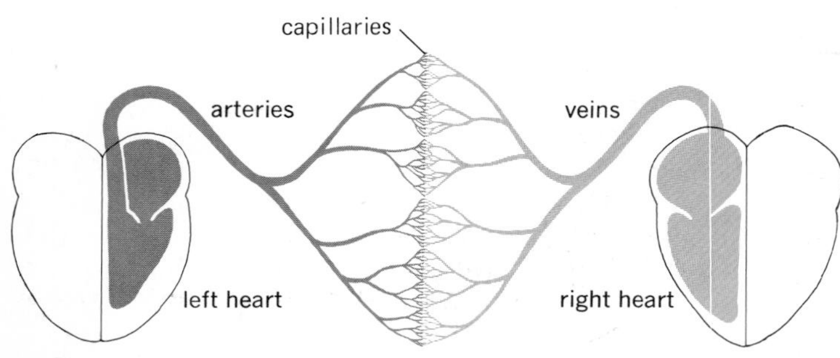

FIGURE 15-1 ***Systemic circulation as two trees connected by capillaries. As indicated by the color change, oxygen leaves the blood during passage through the capillaries.*** **(Adapted from Rushmer.)**

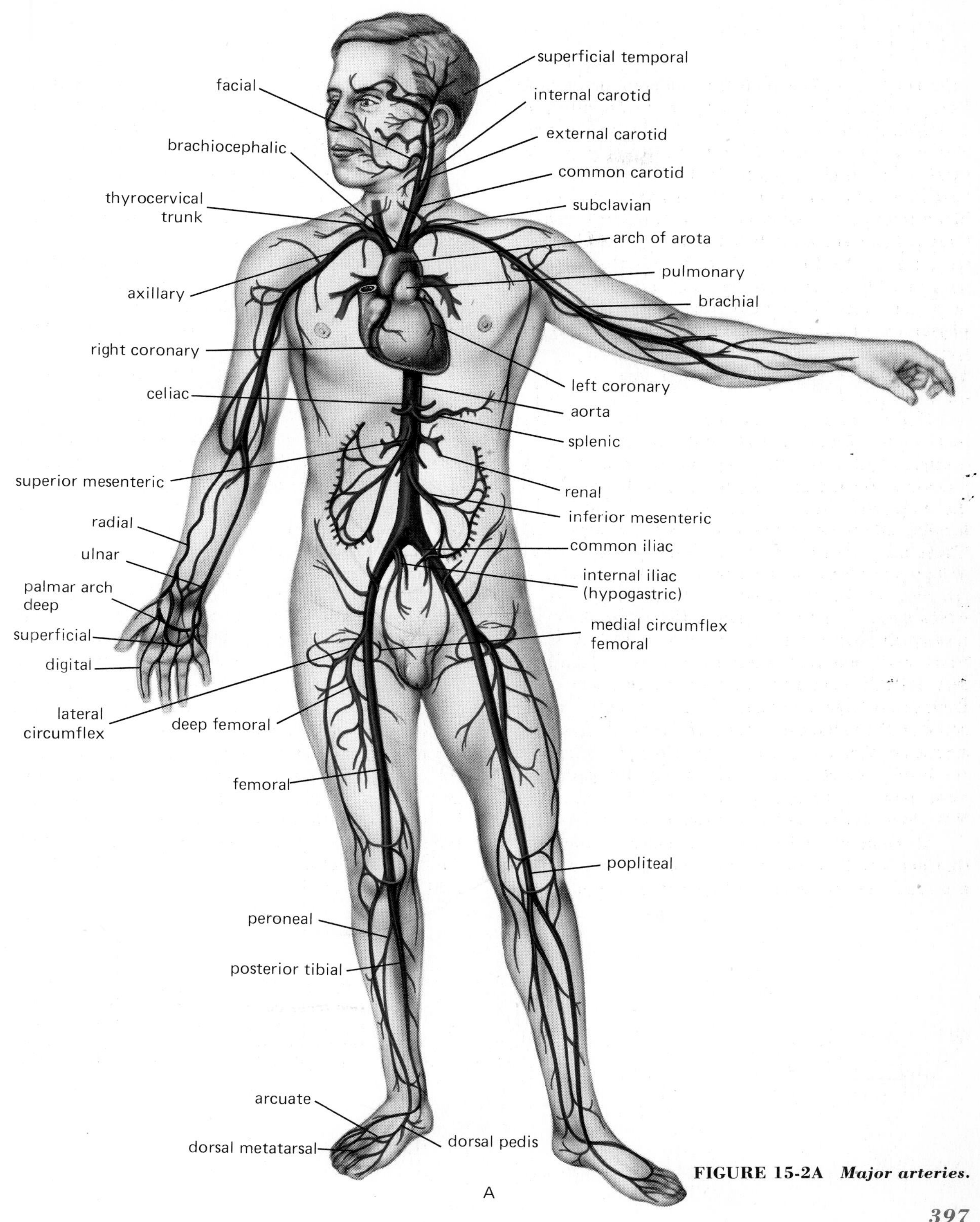
superficial temporal
facial
internal carotid
external carotid
brachiocephalic
common carotid
thyrocervical trunk
subclavian
arch of arota
pulmonary
axillary
brachial
right coronary
left coronary
celiac
aorta
splenic
superior mesenteric
renal
inferior mesenteric
radial
common iliac
ulnar
internal iliac (hypogastric)
palmar arch deep
superficial
medial circumflex femoral
digital
lateral circumflex
deep femoral
femoral
popliteal
peroneal
posterior tibial
arcuate
dorsal metatarsal
dorsal pedis
A

FIGURE 15-2A *Major arteries.*

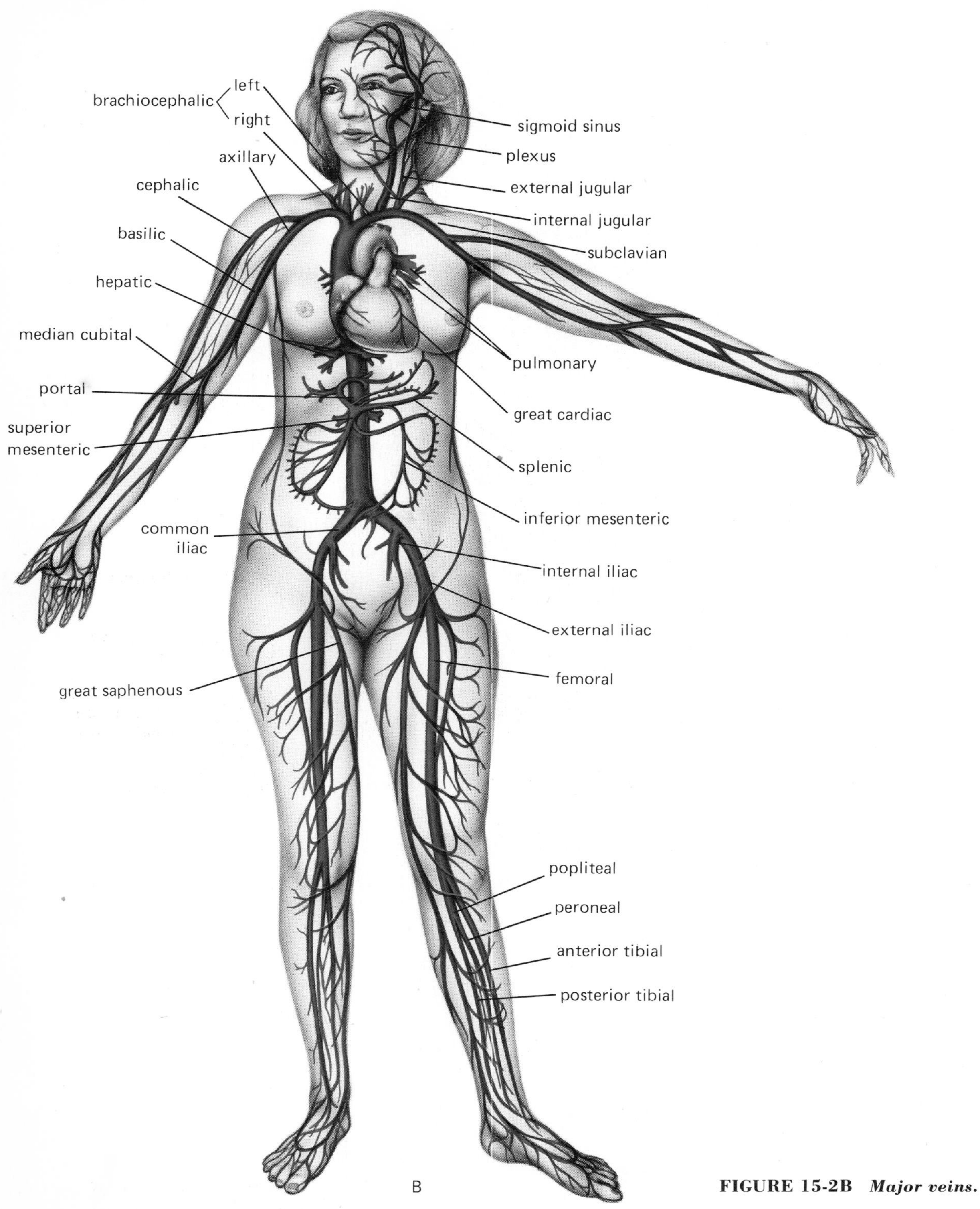
brachiocephalic
left
right
axillary
cephalic
basilic
hepatic
median cubital
portal
superior mesenteric
common iliac
great saphenous
sigmoid sinus
plexus
external jugular
internal jugular
subclavian
pulmonary
great cardiac
splenic
inferior mesenteric
internal iliac
external iliac
femoral
popliteal
peroneal
anterior tibial
posterior tibial
B

FIGURE 15-2B ***Major veins.***

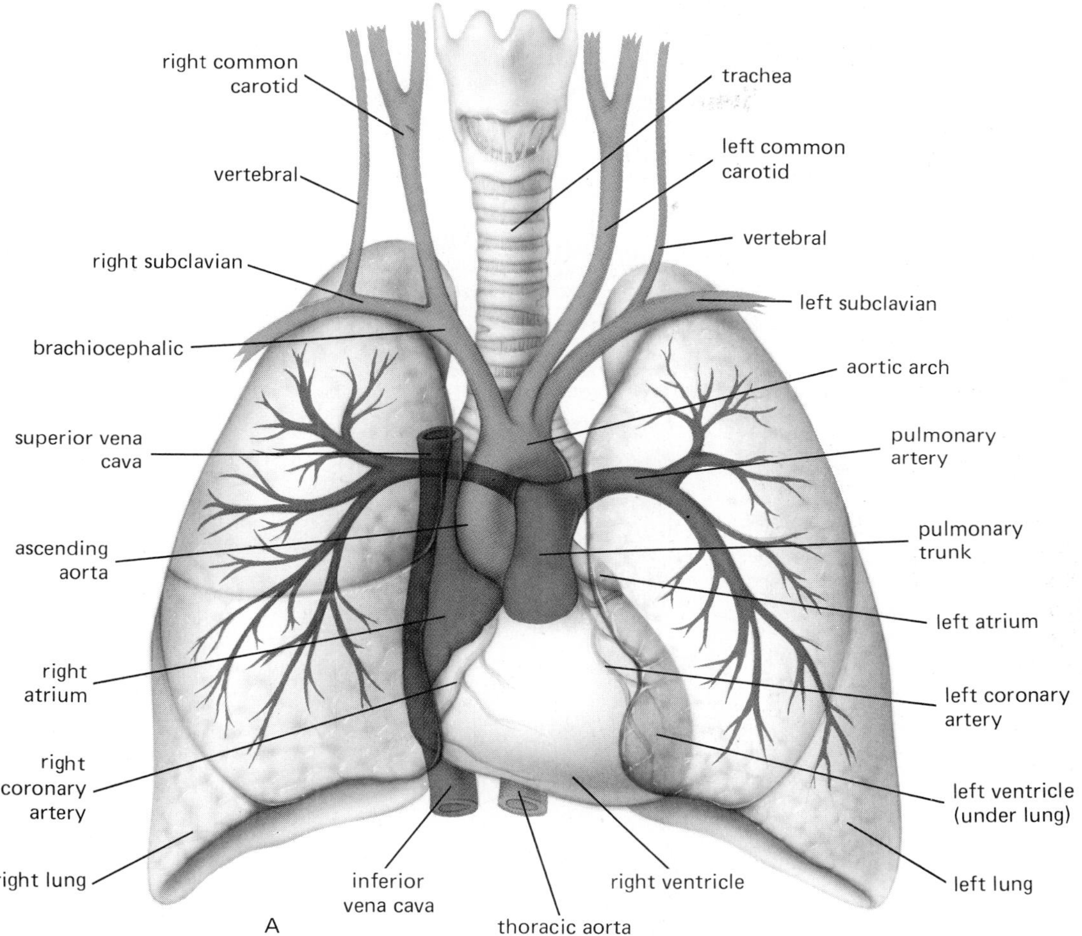

FIGURE 15-3A ***The aortic arch. The aorta is the main vessel carrying oxygenated blood from the heart to the tissues of the body.***

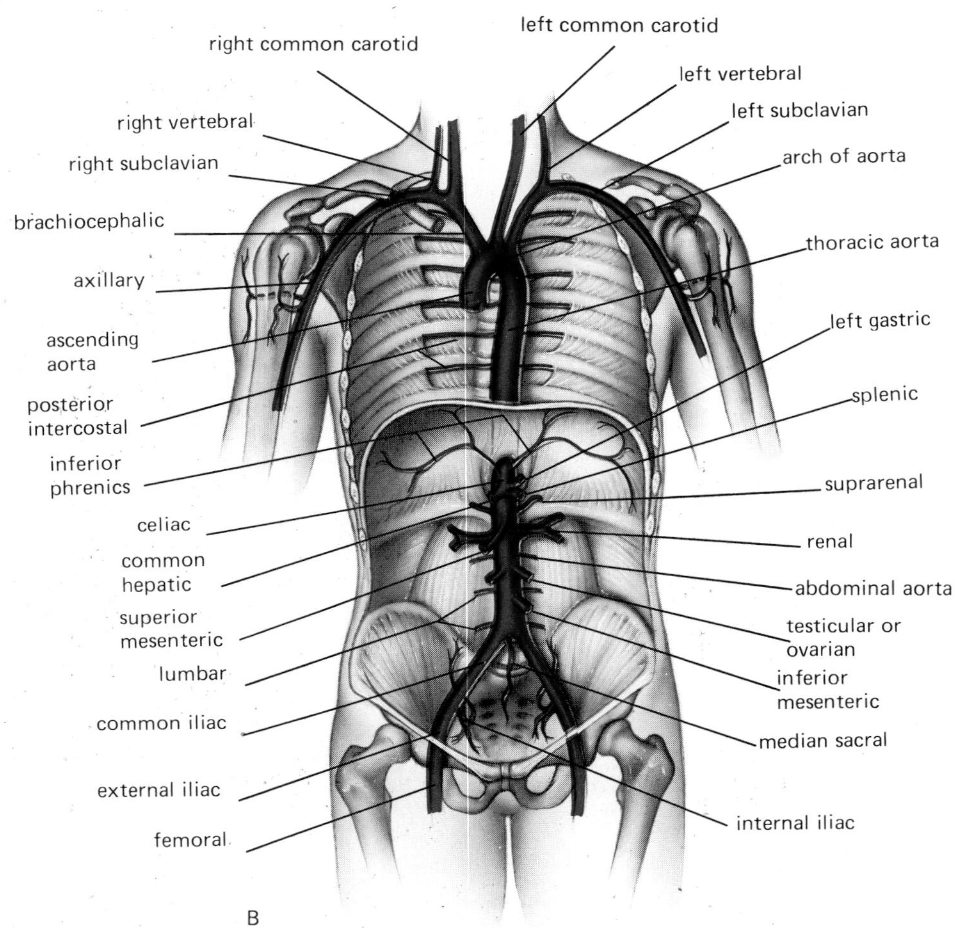

FIGURE 15-3B *Thoracic portion of the descending aorta and abdominal aorta. After leaving the left ventricle, the aorta passes upward and then, at the upper end of the sternum, arches backward and toward the left. It descends along the posterior body wall just in front of the thoracic vertebrae a little to the left of midline. As it descends, it gradually moves toward the midline. It passes from the thoracic cavity through the* aortic opening *(or* aortic hiatus*) in the diaphragm into the abdominal cavity. As it continues to descend, its diameter decreases rapidly as major vessels branch from it. The aorta ends at the level of the 4th lumbar vertebra by dividing into the* right *and* left common iliac arteries, *which supply tissues of the pelvic region and legs. The description of the aorta and its branches in Table 15-1 is divided into ascending aorta, aortic arch, thoracic portion of the descending aorta, and abdominal aorta.*

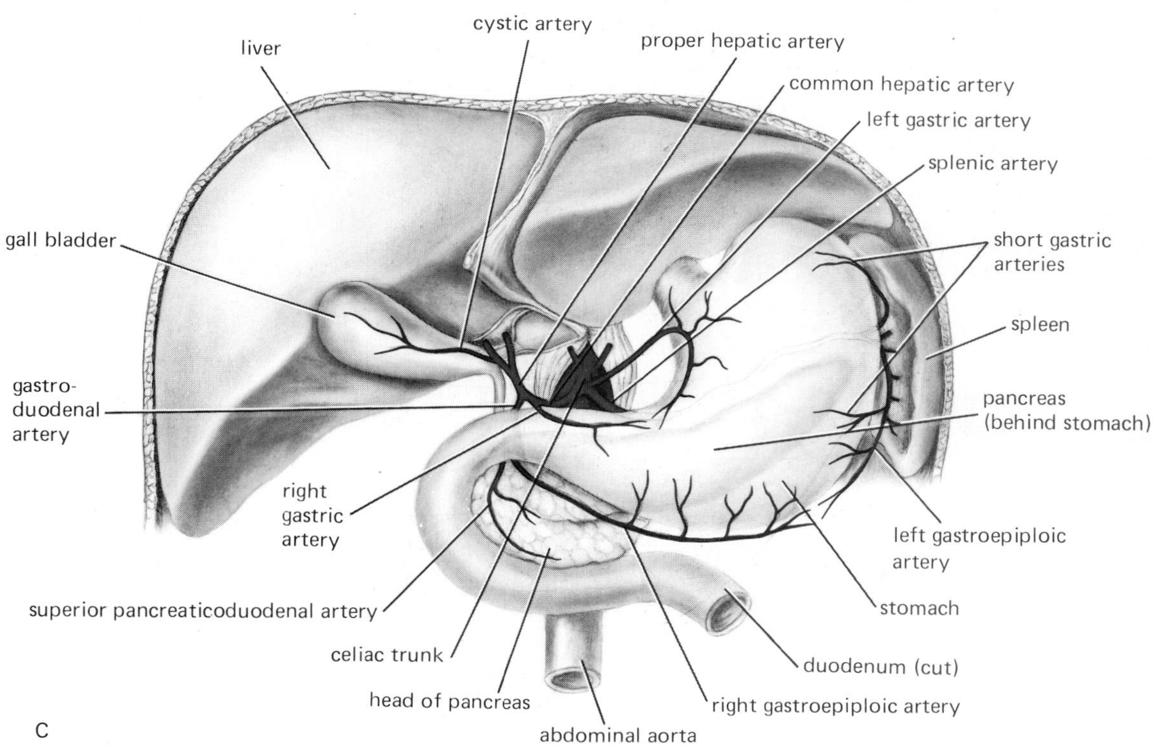

FIGURE 15-3C ***Celiac trunk. Just after the aorta passes through the diaphragm, a wide vessel, the* celiac trunk, *branches off. It divides into three main parts: the* left gastric, hepatic, *and* splenic arteries.**

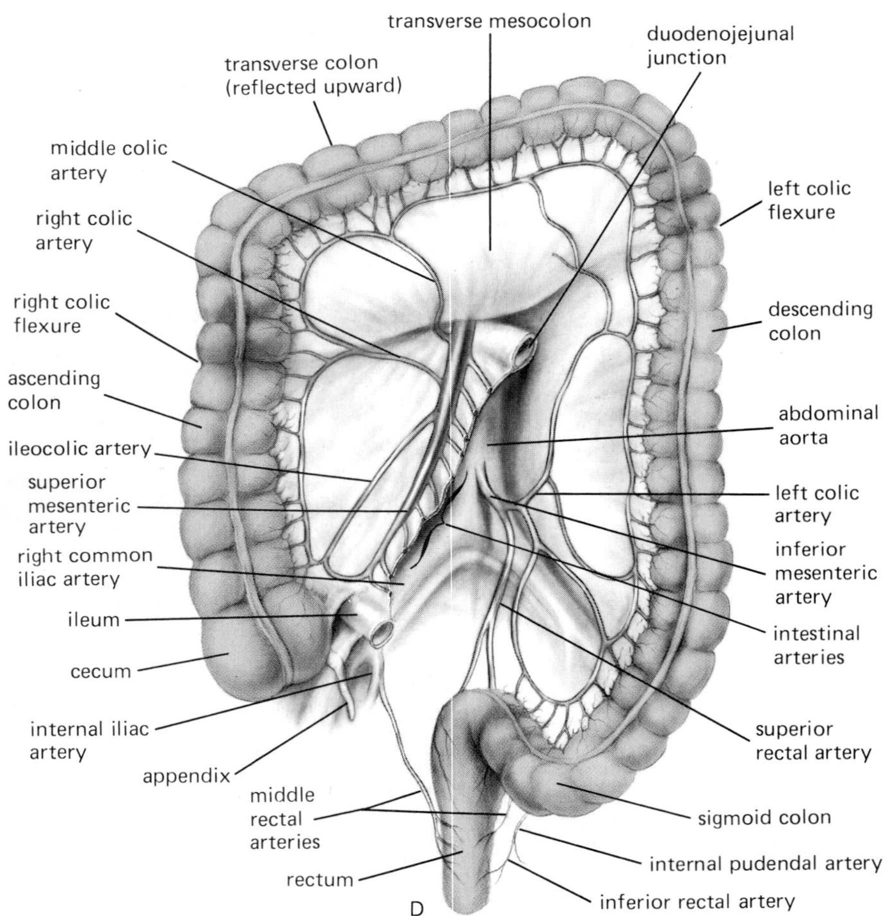

FIGURE 15-3D *Superior and inferior mesenteric arteries. The superior mesenteric artery leaves the aorta about 1 cm below the celiac trunk. It supplies the entire small intestine except for the superior end of the duodenum. It also supplies the cecum, ascending colon, and most of the transverse colon. The inferior mesenteric artery supplies the left one-third of the transverse colon and the entire descending colon, sigmoid colon, and rectum.*

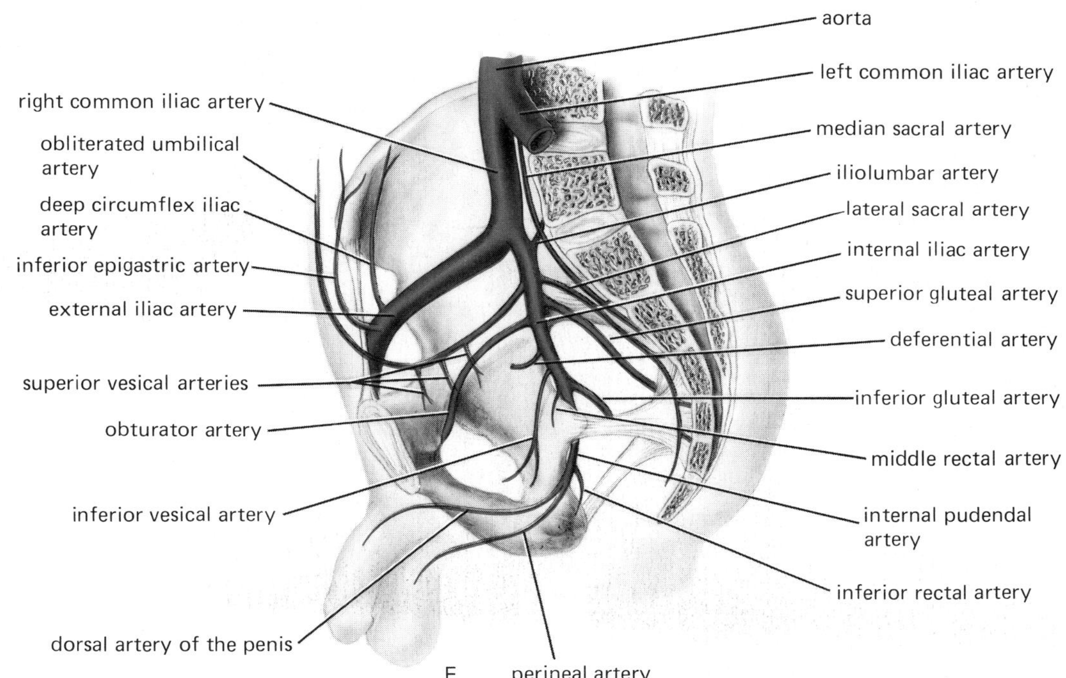

FIGURE 15-3E ***Internal iliac artery. From the internal iliac artery a*** **vesical branch** ***supplies the bladder, prostate and seminal vesicles, and ureter; a*** **rectal artery** ***supplies the lower rectum and prostate and seminal vesicles; a*** **uterine branch** ***supplies the uterus; a*** **vaginal branch** ***supplies the vagina; the*** **obturator artery** ***supplies muscles and bones of the pelvic region; the*** **internal pudendal** ***supplies the genitalia and perineum; a*** **gluteal artery** ***supplies the buttocks and back of thighs; a*** **sacral artery** ***supplies the sacral vertebrae and their contents; and the*** **iliolumbar artery** ***supplies the lumbar vertebrae, ilium, and nearby muscles.***

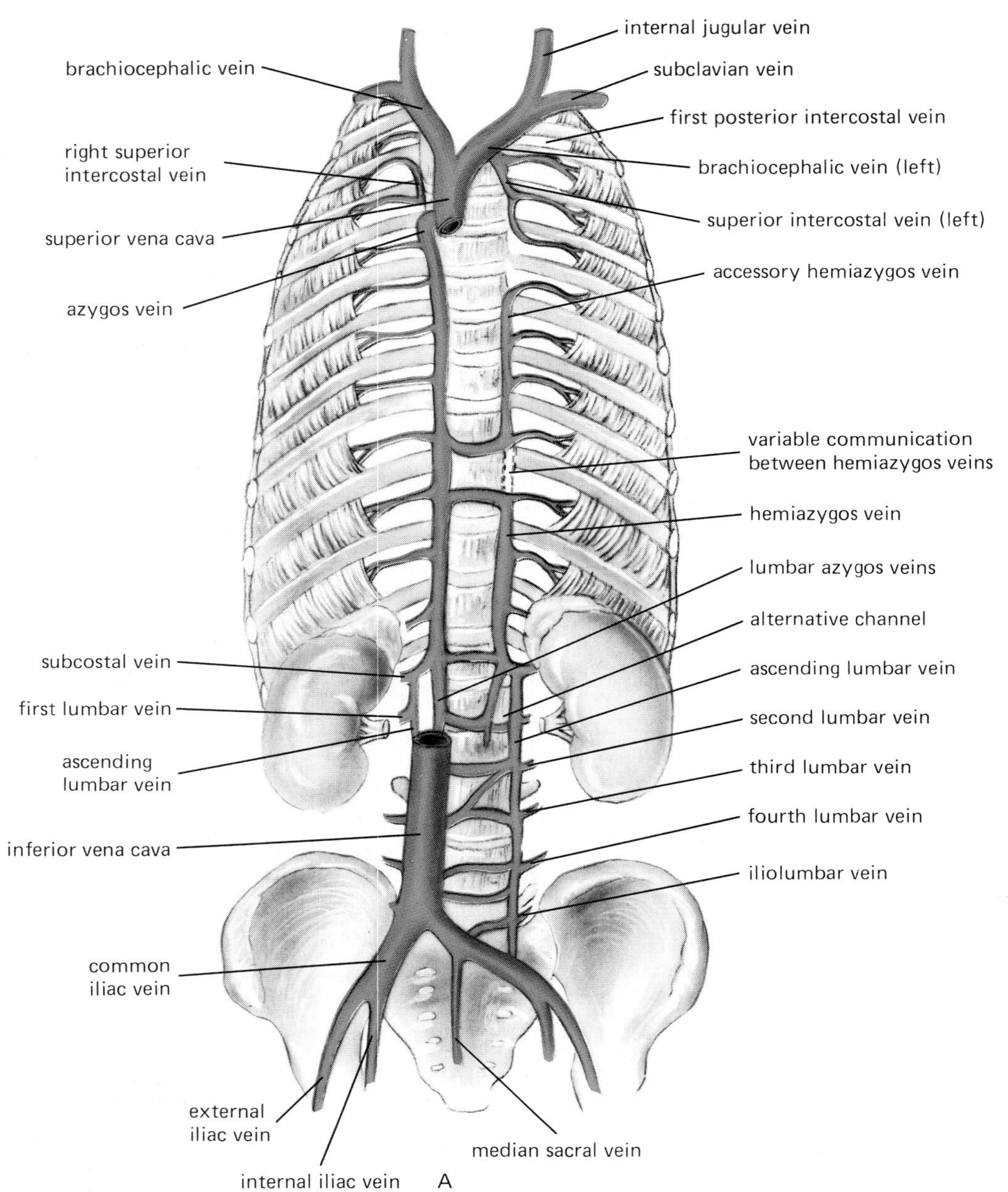

FIGURE 15-4 ***Major veins of the thorax and abdomen. All the blood returning to the right heart passes through the superior or inferior vena cavae, except for the blood from the cardiac veins, which drain directly into right atrium from the coronary sinus. The superior vena cava is the major vein of the thorax and upper extremities. It is formed by the union of the right and left brachiocephalic veins. The inferior vena cava is formed by the union of the common iliac veins; it passes upward through the abdominal cavity next to the aorta, receiving segmental veins which drain the tissues of the abdominal wall, and the lumbar, testicular (or ovarian), renal, right suprarenal (or adrenal), inferior phrenic, and hepatic veins. Thus, the inferior vena cava conveys blood from the body below the diaphragm to the right atrium.***

The segmental veins draining the thoracic wall do not drain directly into the vena cavae, because of the lack of continuity of the vena cavae as they join the right heart. The segmental veins drain instead into the azygos system of veins, which lies along the posterior wall of the thoracic cage. This system includes the azygos vein, which

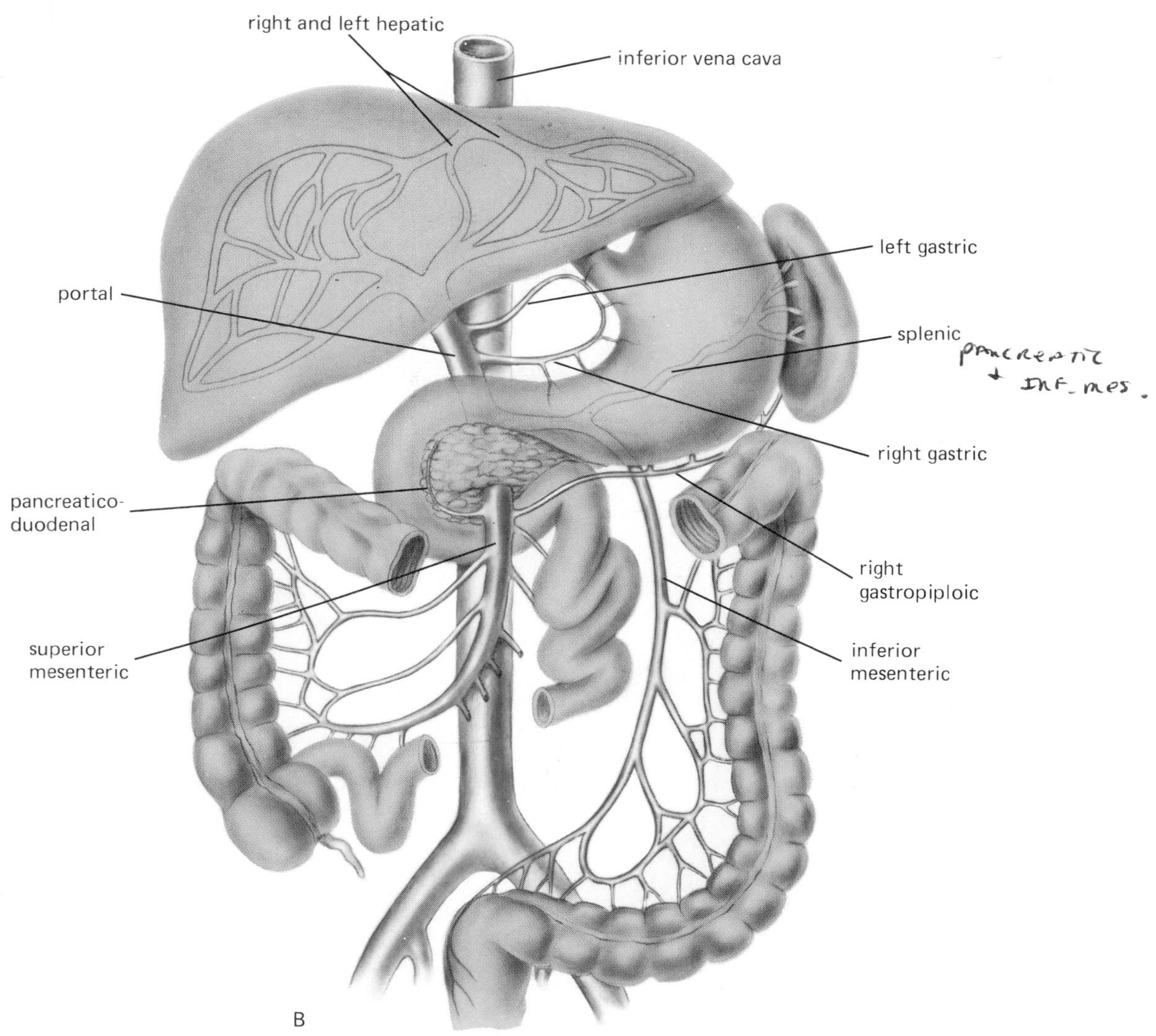

runs along the right side of the vertebral column, and the hemiazygos and accessory hemiazygos veins, which run along the left side. The azygos is a continuation of the right ascending lumbar vein, and the hemiazygos is a continuation of the left. The hemiazygos and accessory hemiazygos veins cross the midline to empty into the azygos vein, which drains into the superior vena cava. Because of its connections with the ascending lumbar veins (and with the inferior vena cava) the azygos system can serve as a bypass for the inferior vena cava if it should become obstructed. In addition to the intercostal veins, the esophageal, pericardial, and bronchial veins drain into the azygos system.

The hepatic vein, mentioned rather casually above, deserves special attention because through it the venous blood from most of the abdominal viscera drains into the inferior vena cava. The venous drainage from these viscera is not led directly back to the heart through successively larger veins, as is the case for most tissues; instead the venous blood passes through a second capillary bed, forming a portal system. Thus, venous blood from the abdominal viscera is collected into a portal vein, which passes to the liver where it again breaks into sinusoidal capillary beds. The portal blood and the blood from the liver's own arterial supply leave via the hepatic vein, which drains into the inferior vena cava. The portal vein is formed by the union of the superior mesenteric and splenic veins; the splenic vein receives blood from the pancreatic and inferior mesenteric veins. Even the coronary, pyloric, and gastroepiploic veins, which drain the stomach, empty into the hepatic portal vein. The hepatic circulation is described in greater detail in Chap. 19.

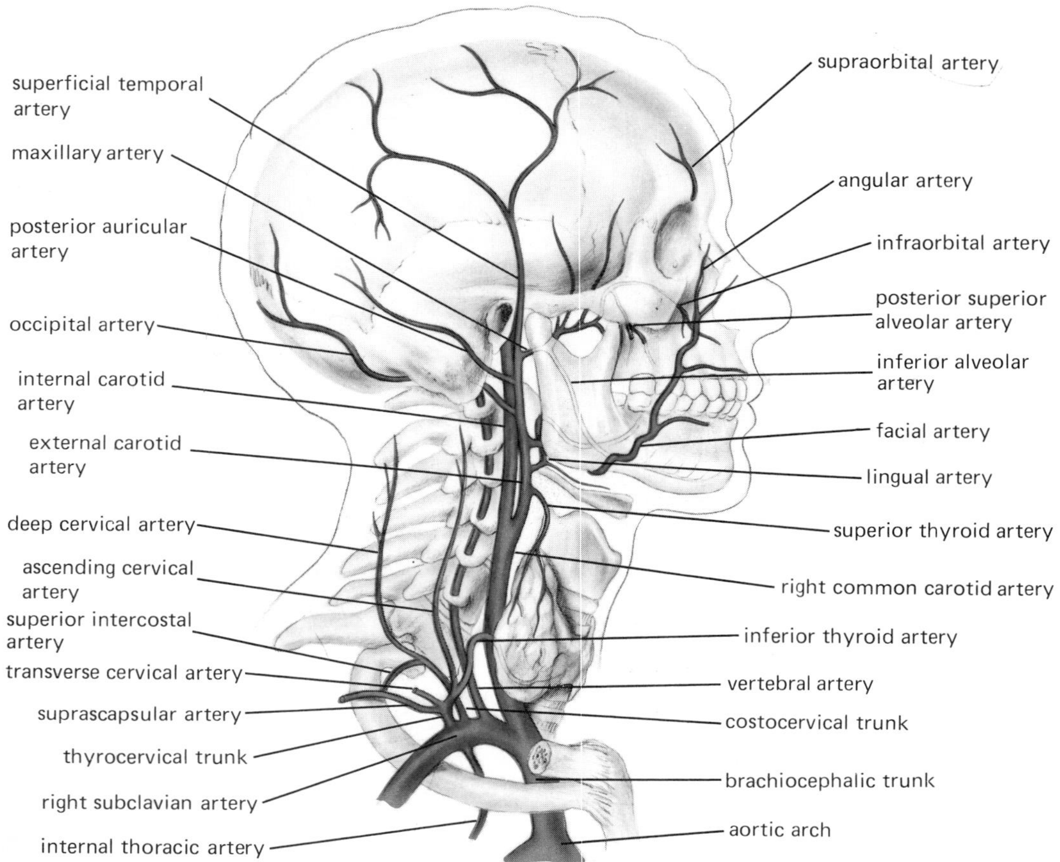

FIGURE 15-5 ***The common carotid arteries are the principal arteries of the head and neck. They have different origins, the right common carotid arising from the brachiocephalic trunk while the left arises directly from the aortic arch, but as they ascend in the neck, their paths become similar. (Figure 15-5 shows only the vessels on the right side.) High in the neck, at the level of the upper border of the thyroid cartilage, each common carotid divides into an external and internal carotid artery. The external branch supplies the exterior of the head, the face, and a large part of the neck, and the internal branch supplies the tissues within the cranial and orbital cavities, i.e., the brain and eye and the tissues associated with them. The branches of the external carotid and a few of the branches of the internal carotid are listed in Table 15-2; the branches of the internal carotid which supply the brain were described in Chapter 8.***

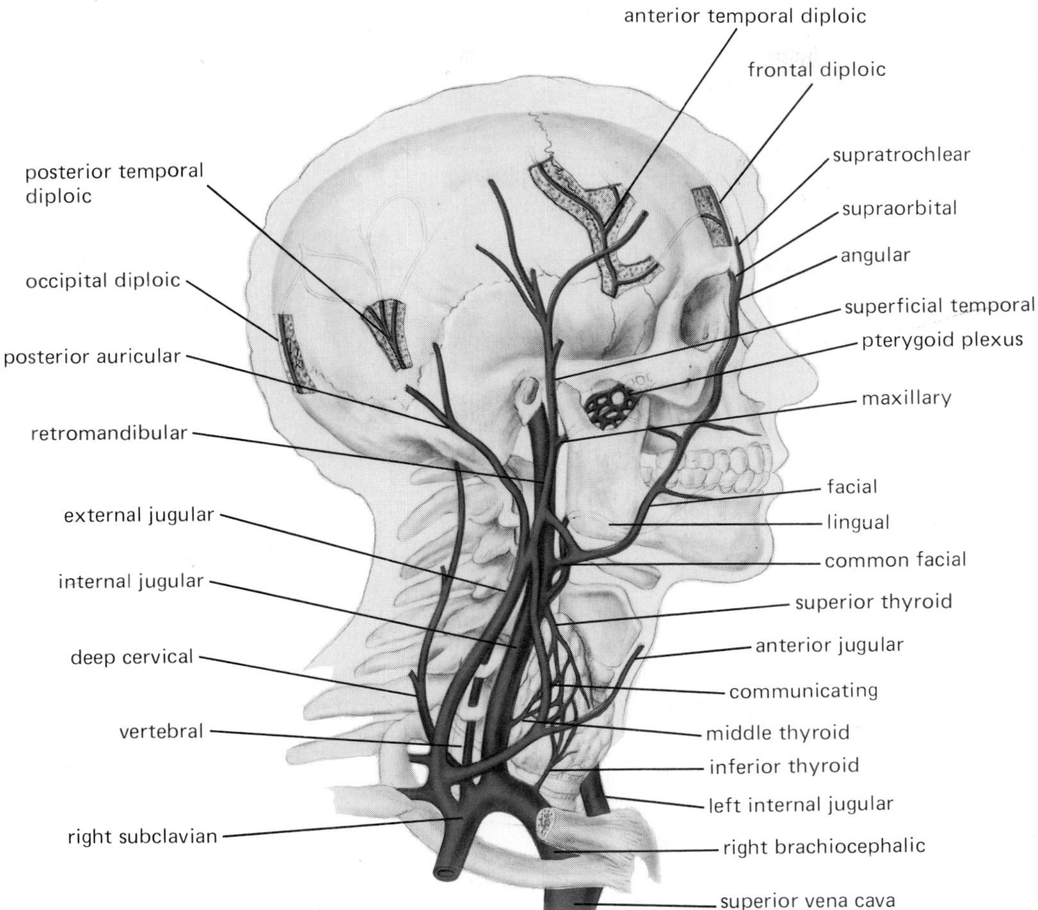

FIGURE 15-6 ***Veins of the head and neck. The head and neck are drained by the jugular veins. The superficial veins of the head and neck which drain into the external jugular accompany similarly-named arteries. Thus, venous blood in the anterior portion of the face, which is supplied by the facial artery, drains into the anterior facial vein; that behind the ear supplied by the posterior auricular artery, drains into the posterior auricular vein. The right and left external jugular veins empty into the subclavian vein of the same side. The internal jugular vein collects blood from the brain, parts of the face, and the neck. The venous drainage of the brain is discussed in Chap. 8.***

The right and left internal jugular veins arise as continuations of the sigmoid sinus within the cranial cavity (see Chap. 8). They drain blood from the brain and parts of the face and neck. As they pass behind the clavicle, they join the subclavian veins on the same side, forming the brachiocephalic veins. These in turn unite to form the superior vena cava.

The right and left vertebral veins drain the vertebral and spinal-cord regions and the deep muscles at the back of the neck. They descend along the path of the vertebral artery to empty into the brachiocephalic veins.

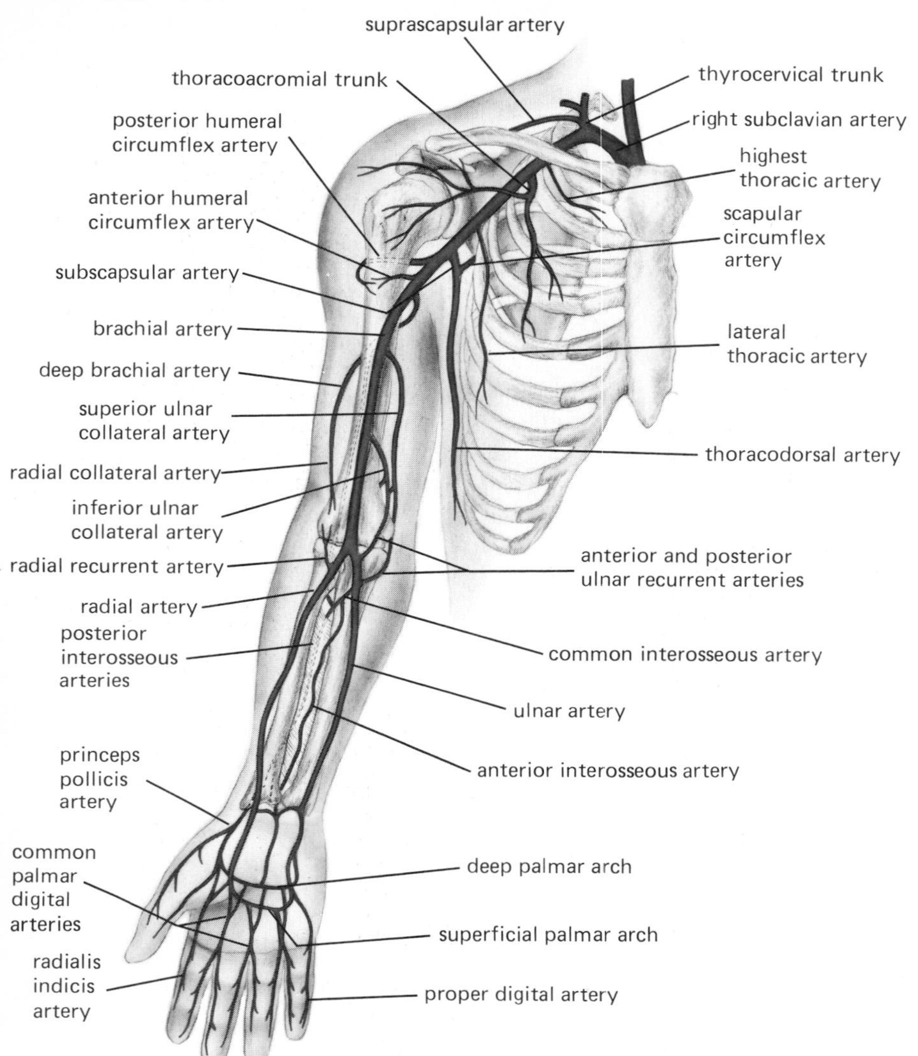

FIGURE 15-7 ***Major arteries of the arm. The major artery of the arm has three different names, depending upon the region it is crossing. From its origin to the outer border of the first rib, it is called the subclavian artery. The right and left subclavian arteries have different origins, the right artery arising from the brachiocephalic trunk while the left arises directly from the aortic arch, but they soon become similar. The subclavian arteries give rise to the three vessels which supply the head and neck (vertebral artery, thyrocervical trunk, and costocervical trunk, Table 15-3); the internal thoracic artery (or mammary artery), which supplies the chest wall and thoracic organs; and the dorsal scapular artery, which supplies muscles of the back and shoulder. At the outer border of the first rib, the subclavian artery becomes the axillary artery, which supplies structures of the lateral chest wall and the shoulder joint. The axillary artery becomes the brachial artery in the upper part of the arm. The brachial artery terminates just below the elbow joint by dividing into a radial and ulnar artery. The radial artery courses down the radial (or thumb) side of the forearm while the ulnar artery passes down the lateral side near the ulna. Terminal branches of the two arteries join, forming arches across the palm. Arteries project from these arches to supply the fingers.***

FIGURE 15-8 ***Two sets of veins drain the upper extremities; the deep veins, which accompany the arteries, and the superficial veins. The superficial and deep veins anastomose freely with each other, and both sets of veins have valves. The dorsal and palmar surfaces of the hand have complex venous networks which drain the fingers and metacarpals. The dorsal venous network coalesces into the cephalic vein, which passes from the back of the hand, around the radius, to the medial surface of the forearm. On the ulnar side, the dorsal venous network of the hand forms the basilic vein. The venous plexus on the palmar surface of the hand drains into the median vein of the forearm; this vein terminates near the elbow by joining the basilic vein or the median cubital (the vein commonly used for venipunctures). The median cubital vein flows into the basilic vein, which then continues upward, changing its name to the axillary vein in the upper part of the arm. The cephalic vein empties into the axillary vein just below the level of the clavicle. At the outer border of the first rib, the axillary vein becomes the subclavian vein, which unites with the internal jugular to form the brachiocephalic vein. The deep veins are venae comitantes and take the names of the arteries they accompany. Thus, deep and superficial palmar venous arches drain into radial and ulnar veins, which unite to form the brachial veins, paired veins on either side of the brachial artery. The brachial veins join the axillary vein.***

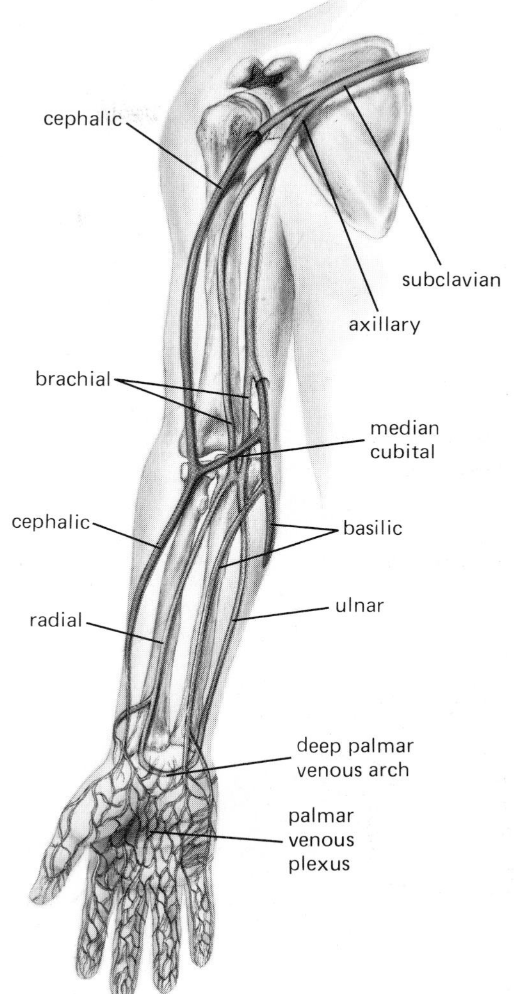

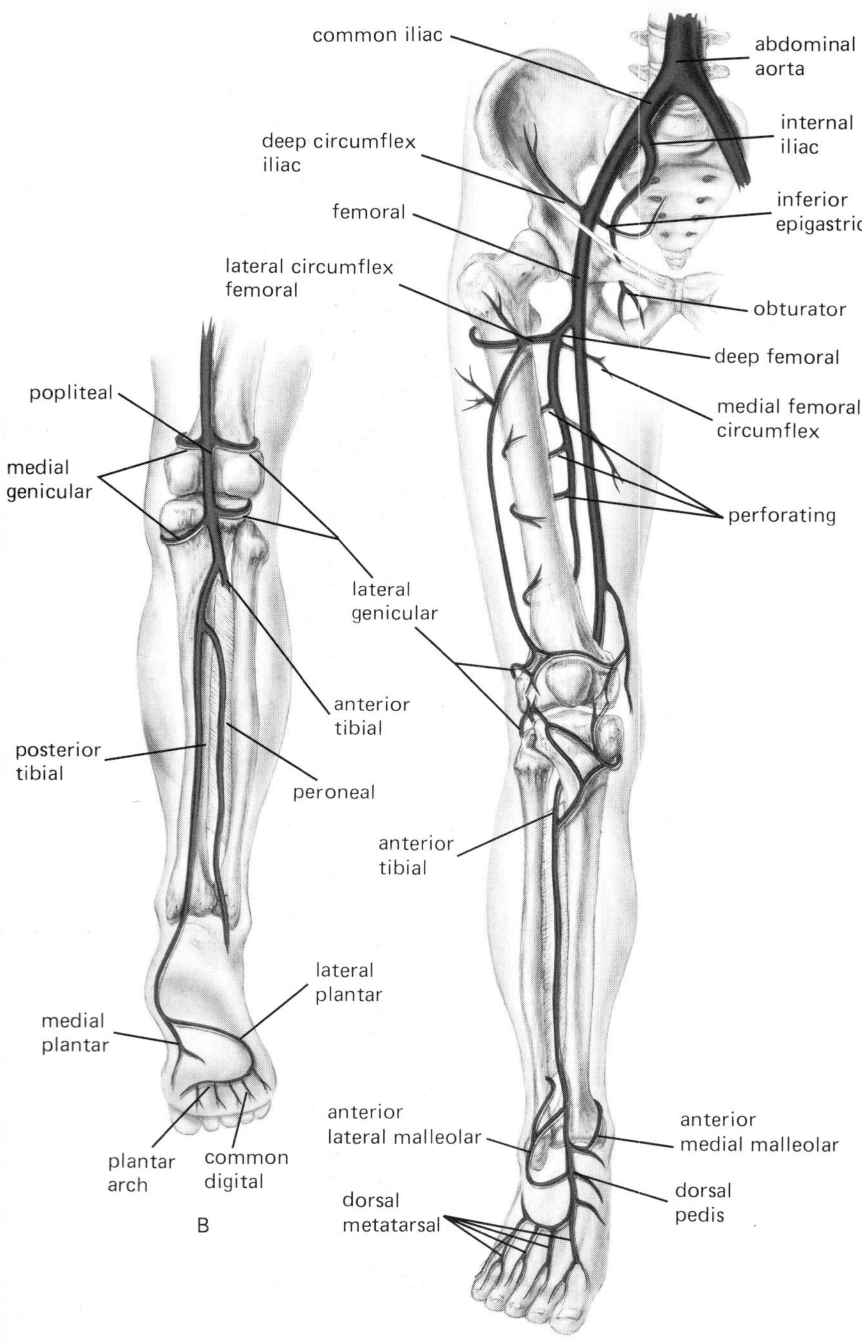

FIGURE 15-9 ***Arterial supply to the lower extremities. At the level of the fourth lumbar vertebra, the abdominal aorta divides into the right and left common iliac arteries, each of which soon divides into an external and internal iliac artery. The internal branch supplies the viscera and walls of the pelvis and the genitalia (Table 14-2) while the external branch supplies the leg. The external iliac artery enters the thigh behind the inguinal ligament and becomes the femoral artery. The femoral artery lies on the front and medial side of the thigh and then passes to the back of the leg; it ends about two-thirds of the way down the thigh when it becomes the popliteal artery. The popliteal artery branches just below the knee, forming the anterior and posterior tibial arteries. Branches from both these vessels anastomose freely and form an arterial network around the ankle. The anterior tibial ends as the dorsal artery of the foot (dorsalis pedis) and the posterior tibial ends as the medial and lateral plantar arteries. The lateral plantar artery and dorsal artery of the foot anastomose, forming the plantar arch. Branches from the plantar arch supply the toes.***

FIGURE 15-10 ***Venous drainage of the legs. The veins of the leg consist of superficial and deep veins; the superficial veins are immediately under the skin. The major superficial veins are the* great *and* small saphenous veins; *most of their branches are not named. The veins from the toes and the dorsal surface of the foot form a* dorsal venous arch, *which flows into an irregular venous network. Medial and lateral* marginal veins *from the sides and soles of the foot drain this network. The medial marginal veins joins other veins on the medial side of the foot to form the great saphenous vein, the longest vein in the body. The lateral marginal and other lateral veins form the small, or lesser, saphenous vein, which begins its ascent at the back of the leg. As it ascends, the small saphenous vein anastomoses frequently with the great saphenous and with the deep veins. The great saphenous joins the* femoral vein *just below the inguinal ligament. The small saphenous joins the* popliteal vein *behind the knee. The popliteal vein is formed by union of the* anterior tibial, peroneal, *and* posterior tibial veins, *deep veins which accompany the arteries. The popliteal vein becomes the femoral vein higher in the thigh. All the veins of the leg have numerous valves.***

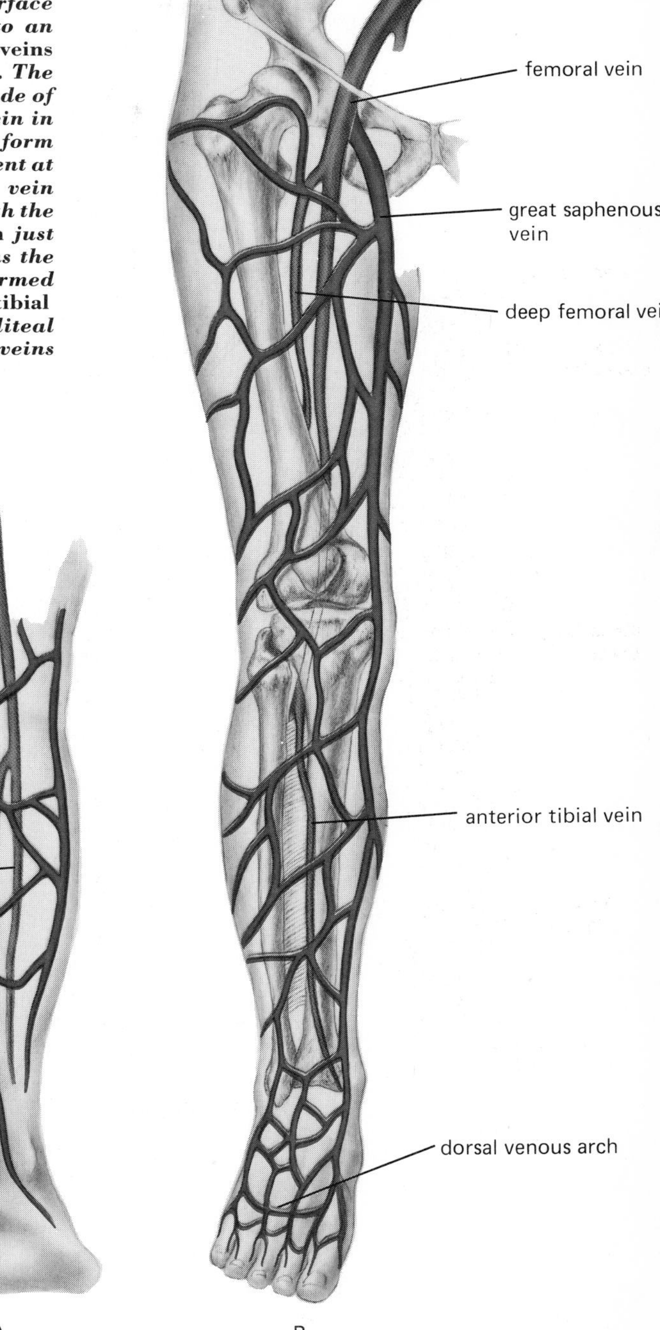

TABLE 15-1
MAJOR ARTERIES BRANCHING FROM VARIOUS SEGMENTS OF THE AORTA

Artery	*Major branches*	*Region supplied*	*Comments*
Ascending aorta			
Right and left coronary arteries		Heart (myocardium, epicardium, endocardium)	Arise behind the cusps of the aortic valve.
	Marginal branch (right) Posterior interventricular branch (right) Anterior interventricular branch (left) Circumflex (left)		
Arch of the aorta			Top of the arch is at the level of the manubrium of the sternum.
Brachiocephalic trunk			This is the largest branch of the aorta. Also called the *innominate artery*.
	Right common carotid artery	Head and neck	One of the two principal arteries of the head and neck; the other is the left common carotid.
	Right subclavian artery	Right arm	At the outer border of the first rib, the name changes from subclavian to *axillary*.
Left common carotid artery	(Branches of the common carotid arteries are listed in Table 15-2.)	Head and neck	
Left subclavian artery	(Branches of the subclavian arteries are listed in Table 15-3.)	Left arm	
Thoracic portion of the descending aorta			Runs from the 4th to the 12th thoracic vertebra where it passes through the diaphragm. The bronchial and esophageal branches are collectively called the *visceral branches;* those supplying the chest muscles, skin, and breasts are the *parietal branches*.
Bronchial arteries		Bronchial airways, tissues of lungs, esophagus, pericardium	
Esophageal arteries		Esophagus	
Mediastinal branches		Lymph nodes and fatty tissue of the mediastinum	
Superior phrenic branches		Diaphragm	
Posterior intercostal arteries		Intercostal spaces	Usually nine pairs.

TABLE 15-1
MAJOR ARTERIES BRANCHING FROM VARIOUS SEGMENTS OF THE AORTA *(Continued)*

Artery	*Major branches*	*Region supplied*	*Comments*
	Dorsal branch	Spinal cord, muscles, and skin of back	
	Muscular branch	Chest muscles	
	Cutaneous branch	Skin of chest	
	Mammary branch	Breasts	
Subcostal arteries			Similar to the intercostals but below the 12th rib.
Abdominal aorta			Branches divisible into four sets: those from the ventral and lateral surfaces of the aorta go to the viscera; the dorsal branches go to the body wall. The terminal branches (the common iliac arteries) supply the pelvis and legs.
Inferior phrenic arteries		Diaphragm and lower esophagus	
	Superior suprarenal branches	Suprarenal (adrenal) glands	
Celiac trunk (Fig. 15-3C)			
	Left gastric artery	Stomach, esophagus	Smallest branch of the celiac trunk.
	Common hepatic artery	Liver, stomach, pancreas, duodenum, gallbladder, and bile duct	
	Splenic artery	Pancreas, stomach, omentum, spleen	Largest branch of the celiac trunk.
Middle suprarenal arteries		Suprarenal (adrenal) gland	
Superior mesenteric artery (Fig. 15-3D)		Most of the small intestine, cecum, ascending colon, part of the transverse colon	
	Inferior pancreaticoduodenal artery	Head of pancreas, duodenum	
	Jejunal and ileal branches	Small intestine	There are usually 12 to 15 of these branches.
	Ileocolic artery	Ascending and part of the transverse colon, appendix, lower ileum	
	Right colic artery	Ascending colon	
	Middle colic artery	Transverse colon	
Renal arteries		Kidneys, ureters, surrounding tissue	
	Inferior suprarenal arteries	Suprarenal (adrenal) glands	
Testicular (or ovarian) arteries		Testes (or ovaries)	

TABLE 15-1

MAJOR ARTERIES BRANCHING FROM VARIOUS SEGMENTS OF THE AORTA *(Continued)*

Artery	*Major branches*	*Region supplied*	*Comments*
Inferior mesenteric artery (Fig. 15-3D)	Left colic artery	Transverse and descending colon	
	Sigmoid arteries	Descending and sigmoid colon	
	Superior rectal artery	Rectum	
Lumbar arteries		Muscles and skin of the back, vertebral canal, and its contents	4 to 5 pairs in series with the posterior intercostals.
Median sacral artery		Sacral vertebrae, rectum	
Common iliac arteries (Fig. 15-3E)		Pelvic region and legs	
	External iliac	Leg (see Table 15-4)	
	Internal iliac (Fig. 15-3E)	Viscera and walls of the pelvis, perineum, and gluteal region	

TABLE 15-2

MAJOR ARTERIES OF THE HEAD AND NECK

Artery	*Major branches*	*Region supplied*	*Comments*
External carotid artery			
	Superior thyroid artery	Thyroid gland and adjacent muscles	
	Lingual artery	Tongue and floor of the mouth, oropharynx, sublingual gland, and neighboring muscles	
	Facial artery	Muscles and tissues of the face below the level of the eyes, the submandibular gland, tonsil, and soft palate	Pulsations of this artery can be felt over the base of the mandible. Beyond the branch to the upper lip, the facial artery is called the *angular artery*.

TABLE 15-2

MAJOR ARTERIES OF THE HEAD AND NECK ***(Continued)***

Artery	*Major branches*	*Region supplied*	*Comments*
	Occipital artery	Muscles, skin, and other tissue in the region behind the ear and the back part of the scalp	
	Posterior auricular artery	Parotid gland, muscles, skin, and other tissues of the ear and posterior scalp regions	
	Superficial temporal artery	Parotid gland, temporomandibular joint, outer ear, forehead, temporal region of the scalp, adjacent muscles	The external carotid terminates by dividing into the superficial temporal and maxillary arteries in the region of the parotid gland.
	Maxillary artery	Upper and lower jaws, teeth, muscles of mastication, palate, nose, and dura mater	
Internal carotid artery			The internal carotid gives off no branches in the neck. It passes through the carotid canal of the temporal bone and, after giving off several branches, terminates by dividing into the anterior and middle cerebral arteries. They were discussed in Chap. 8.
	Hypophyseal artery	Pituitary gland	For further details see Chap. 12.
	Ophthalmic artery	Eye, lacrimal gland, ocular muscles, nasal cavity, forehead	The ophthalmic artery passes from the cranial to the orbital cavity via the optic canal.
Right and left subclavian arteries			These are the stem arteries of the arms but give off branches to the head and neck early in their course. The early branches are included here; the later branches are described in Table 15–3.
	Vertebral artery	Spinal cord, vertebrae and surrounding tissues, deep neck structures	Ascends through foramens in the transverse processes of the cervical vertebrae. Before entering the cranial cavity through the foramen magnum it gives off branches to neck structures. Within the cranial cavity, it joins the vertebral artery of the opposite side to form the basilar artery (see Chap. 8).
	Thyrocervical trunk	Thyroid gland, neck muscles, trachea, esophagus	
	Costocervical trunk	Muscles at the back of the neck, vertebral canal, first intercostal space	

TABLE 15-3

MAJOR ARTERIES OF THE ARM

Artery	*Major branches*	*Region supplied*	*Comments*
Subclavian artery	Vertebral artery	(See Table 15-2)	
	Internal thoracic (mammary) artery	Thoracic pleura, pericardium, fat and lymph nodes of mediastinum, sternum, intercostal muscles, muscles and skin of chest wall, breast, muscles and skin of abdominal wall	
	Thyrocervical trunk	(See Table 15-2)	
	Costocervical trunk	(See Table 15-2)	
	Dorsal scapular artery	Muscles of the upper back and shoulder	
Axillary artery			Continuation of the subclavian artery from the outer border of the first rib.
	Highest thoracic artery	Pectoralis muscles and chest wall	
	Thoracoacromial artery	Muscles of the chest and shoulder	
	Lateral thoracic artery	Muscles of the chest	
	Subscapular artery	Muscles of the chest and shoulder	
	Anterior and posterior circumflex humeral arteries	Head of the humerus, shoulder joint, muscles of shoulder	
Brachial artery			Continuation of the axillary artery. It ends about a centimeter below the elbow by dividing into the radial and ulnar arteries. Initially, it lies medial to the humerus, but it shifts until it lies on the anterior surface of the arm at its termination. Commonly used for blood pressure measurements.
	Deep brachial artery (profunda)	Muscles of the upper arm, humerus	
	Main nutrient artery	Muscles and tissues of the upper arm	
	Muscular branches	Muscles of the upper arm	
	Inferior and superior ulnar collateral arteries	Elbow and upper region of the forearm	
Radial artery			One of the terminal divisions of the brachial. Commonly used for taking the pulse. At its termination, unites with the deep branch of the ulnar artery to form the deep palmar arch.

TABLE 15-3

MAJOR ARTERIES OF THE ARM ***(Continued)***

Artery	*Major branches*	*Region supplied*	*Comments*
	Radial recurrent artery	Elbow joint and muscles of the forearm	
	Muscular branches	Muscles on the radial (thumb) side of the forearm	
	Palmar carpal branch	Bones and joints of the wrist	Passes medially across the wrist to join the palmar branch of the ulnar artery, forming a palmar carpal arch.
	Superficial palmar branch	Muscles of the thumb	
	Dorsal carpal branch	Lower ends of the radius and ulna; gives rise to arteries supplying metacarpals	Joins the dorsal carpal branch of the ulnar artery to form the dorsal carpal arch.
Ulnar artery		Muscles on the medial side of the forearm and hand	Other terminal branch of the brachial artery. Ends as the superficial palmar arch, which crosses the palm.
	Anterior ulnar recurrent and posterior arteries	Elbow joint and muscles of the upper forearm	
	Common interosseous artery	Wrist and forearm	
	Muscular branches	Muscles of the ulnar region of the forearm	
	Deep branch		With the deep radial artery, it completes the deep palmar arch.
	Common palmar digital arteries	Fingers	These are branches of the superficial palmar arch (the termination of the ulnar artery as it crosses the palm).

TABLE 15-4

MAJOR ARTERIES OF THE LEG

Artery	*Major branches*	*Region supplied*	*Comments*
Femoral artery			Continuation of the external iliac as it passes beneath the inguinal ligament into the thigh.
	Muscular branches	Muscles of the upper thigh	
	Deep femoral artery (profunda)	Principal arterial supply to the adductor, extensor, and flexor muscles	Forms many anastomoses with the iliac arteries above and the popliteal below. This later anastomosing branch is the *fourth perforating artery*. Branches of the profunda are the *lateral* and *medial circumflex femoral arteries,* numerous *muscular branches,* and the *perforating arteries,* so named because they pass through muscle insertions to reach the back of the thigh. The *descending genicular artery* supplies the knee.
Popliteal artery			After the femoral artery passes between the extensor and adductor muscles of the thigh to emerge on the back of the leg, it becomes the popliteal artery (the popliteal fossa is the region at the back of the knee). Just below the knee, it terminates by dividing into the anterior and posterior tibial arteries.
	Sural arteries	Gastrocnemius, soleus, and plantaris muscles	
	Superior, middle, and inferior genicular arteries	Knee joint	Form an intricate genicular anastomosis around the patella and ends of the femur and tibia.
Anterior tibial artery			One of the terminal branches of the popliteal. Arising at the back of the leg, it passes forward to run along the front. Continues on the dorsum (upper surface) of the foot as the dorsal artery of the foot (dorsalis pedis).
	Anterior and posterior tibial recurrent arteries	Knee joint, tibiofibular joint	
	Muscular branches	Adjacent muscles and skin	
	Anterior medial malleolar artery	Medial side of the ankle joint	Arteries around the ankle, including the medial and lateral malleolars, dorsal artery of the foot, and other branches of the anterior and posterior tibial arteries, anastomose freely, forming a ring around the ankle joint.
	Anterior lateral malleolar artery	Lateral side of the ankle	
Dorsal artery of the foot (dorsalis pedis)			Continuation of the anterior tibial distal to the ankle. Eventually turns downward to the sole of the foot to complete the *plantar arch.*

TABLE 15-4

MAJOR ARTERIES OF THE LEG *(Continued)*

Artery	*Major branches*	*Region supplied*	*Comments*
	Tarsal arteries	Tarsal region of foot	
	Arcuate artery		Passes laterally over the foot. Gives rise to the dorsal metatarsal arteries, which supply the toes.
Posterior tibial artery			Continuation of the popliteal artery.
	Circumflex fibular artery	Bone and joint structures	
	Peroneal artery	Muscles, fibula	
	Medial malleolar branches	Ankle joint	Joins anastomosing arteries of the ankle.
	Calcaneal branch	Heel	
	Medial plantar artery	Metatarsal region and toes	A terminal branch of the posterior tibial. Passes along medial side of the foot.
	Lateral plantar artery	Adjoining muscles, skin, and sole of the foot	Larger terminal branch of the posterior tibial. Passes along the lateral surface of the foot. Turns medially to unite with the dorsal artery of the foot to form the deep plantar arch. Branches from the deep plantar arch supply the toes.

BASIC PRINCIPLES OF PRESSURE, FLOW, AND RESISTANCE

Fluid flows through a tube in response to a difference (*gradient*) in pressure between the two ends of the tube. The total volume flowing per unit time is directly proportional to the pressure ΔP. It is not the absolute pressure in the tube which determines flow but the difference in pressure between the two ends. This direct proportionality of flow F and pressure difference ΔP can be written

$$F = k\,\Delta P$$

where k is the proportionality constant describing how much flow occurs for a given pressure difference. Knowing only the pressure difference between two ends of a tube is not enough to determine how much fluid will flow; one must also know the numerical value of k. This constant is simply a measure of the ease with which fluid will flow through a tube. Usually we do not deal directly with k but with the reciprocal of k, known as *resistance R:*

$$R = \frac{1}{k}$$

Thus, the larger k is, the smaller R is. R answers the question: How difficult is it for fluid to flow through a tube at any given pressure? Its name, resistance, is therefore quite appropriate. Substituting $1/R$ for k, the basic equation becomes

$$F = \frac{\Delta P}{R}$$

In words, flow is directly proportional to the pressure difference and inversely proportional to the resistance. This basic equation applies to the flow of fluids in general, not just in living systems.

Determinants of Resistance

Resistance (see Table 15-5) is essentially a measure of friction since it is basically the friction between tube wall and fluid and between the molecules of the fluid themselves which opposes flow. Resistance depends upon the nature of the fluid and the geometry of the tube.

TABLE 15-5
FACTORS WHICH DETERMINE RESISTANCE TO FLOW

$$\text{Resistance} \propto \begin{cases} \text{viscosity of fluid} \\ \text{tube length} \\ 1/(\text{tube radius})^4 \end{cases}$$

Nature of the Fluid Maple syrup flows less readily than water because the molecules of maple syrup slide over each other only with great difficulty. This property of fluids is called *viscosity*. As we shall see, changes in blood viscosity can have important effects on blood flow because they change the resistance to flow. Generally, however, the viscosity of blood is relatively constant and only a minor factor in determining resistance.

Geometry of the Tube Both the length and radius of a tube affect its resistance to flow (Fig. 15-11). Resistance is directly proportional to the length of the tube, but since the lengths of the blood vessels remain constant in the body, we shall not be concerned with this relationship. Because of complex relationships between the tube wall and the fluid, the resistance increases markedly as tube radius decreases. The exact relationship is given by the following formula, which states that the resistance is inversely proportional to the fourth power of the radius (the radius multiplied by itself four times):

$$R \propto \frac{1}{r^4}$$

The extraordinary dependence of resistance upon radius can be appreciated by the fact (shown in Fig. 15-11) that doubling the radius increases the flow sixteenfold. As we shall see, the radius of blood vessels can be changed significantly and constitutes the most important factor in the control of resistance to blood flow.

ARTERIES

The aorta and other large arteries have thick walls with a relatively thick tunica media (Figs. 15-12A and B), which is filled with repeating concentric layers of elastic membranes separated by layers of smooth muscle cells and connective-tissue fibers whereas, in comparison, the tunica adventitia is thin. Smaller arteries and arterioles have much more smooth muscle than large arteries. The walls of the larger arteries themselves receive blood vessels, called the *vasa vasorum,* but nutrients for the smaller arteries simply diffuse in from the blood that is within the vessel itself. The arterial walls are supplied with nerves, the majority of which are postganglionic sympathetic fibers. They course through the adventitia, contacting the smooth muscle cells through breaks in the external elastic lamina. The

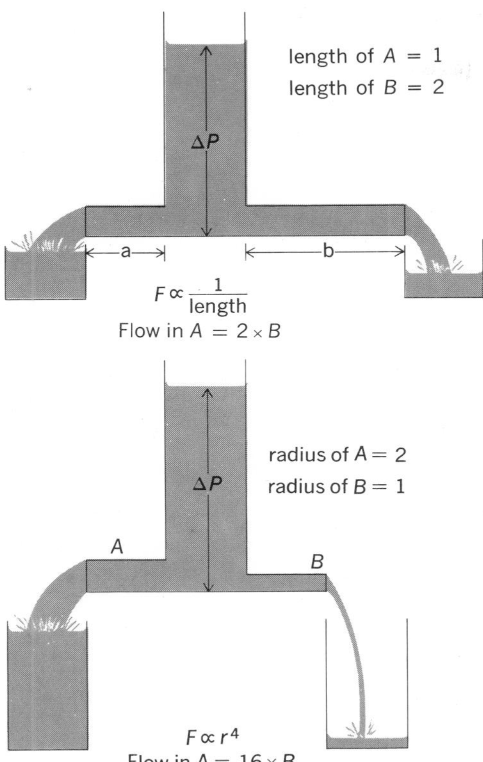

FIGURE 15-11 ***Effects of tube length and radius on resistance to flow.***

arteries are enclosed in thin connective-tissue sheaths, usually with an accompanying vein and nerve.

Despite the presence of smooth muscle in the vessel wall, there is relatively little fluctuation in the state of muscle activity. Because the arteries have large radii, they serve as low-resistance pipes conducting blood to the various organs. Their second major function is to act as a pressure reservoir driving blood through the tissues.

Arterial Blood Pressure

Recall once more the factors determining the pressure within an elastic container, e.g., a balloon filled with water. The pressure inside the balloon depends upon the volume of water within it and the distensibility of the balloon, i.e., how easily its walls can be stretched. If the walls are very stretchable, large quantities of water can go in with only a small rise in pressure; conversely, a small quantity of water causes a large pressure rise in a balloon which strongly resists stretching.

These principles can now be applied to an analysis of arterial function. The contraction of the ventricles ejects blood into the pulmonary and systemic arteries during systole. If a precisely equal quantity of blood were to flow simultaneously out of the arteries via arterioles, the total volume of blood in the arteries would remain constant and arterial pressure would not change. Such, however, is not the case. As shown in Fig. 15-13, a volume of blood equal to only about one-third the stroke volume leaves the arteries during systole. The excess volume distends the arteries and raises the arterial pressure. When ventricular contraction ends, the stretched arterial walls recoil passively (like a stretched rubber band upon release) and the arterial pressure continues to drive blood through the arterioles. As blood leaves the arteries, the pressure slowly falls, but the next ventricular contraction occurs while there is still adequate blood in the arteries to stretch them partially, so that the arterial pressure does not fall to zero. In this manner, the arterial pressure provides the immediate driving force for tissue blood flow. The aortic pressure pattern shown in Fig. 15-14 is typical of the pressure changes which occur in all the large systemic arteries. The pulmonary-artery pressure profile is similar but with all pressures smaller. The maximum pressure is reached during peak ventricular ejection and is called *systolic pressure*. The minimum pressure obviously occurs just before ventricular contraction and is called *diastolic pressure*. They are generally recorded as systolic/diastolic, that is, 125/75 mm Hg in our example. The pulse, which can be felt in an artery, is due to the difference between systolic and diastolic pressure. This difference $(125 - 75 = 50)$ is called the *pulse pressure*. The factors which alter pulse pressure are the following: (1) An increased stroke volume tends to elevate systolic pressure because of greater arterial stretching by the additional blood. (2) Decreased arterial distensibility, as in arteriosclerosis, may cause a marked increase in systolic pressure because the wall is stiffer; i.e., any given volume of blood produces a greater pressure rise.

It is evident from the figure that arterial pressure is constantly changing throughout the cardiac cycle and the average pressure (*mean pressure*) throughout the cycle is not merely the value halfway between systolic and diastolic pressure, because diastole usually lasts longer than systole. Actually, the true mean arterial pressure can be obtained only by complex methods, but for most purposes it is reasonably accurate to assume the mean pressure is the diastolic pressure plus one-third of the pulse pressure. Thus, in our example,

aorta
artery
arteriole
precapillary sphincter
capillary
venule
vein
vena cava
inside diameter 25 mm 4 mm 30 μm 35 μm 8 μm 20 μm 5 mm 2.0 mm
wall thickness 2 mm 1 mm 20 μm 30 μ 1 μ 2 μ 0.5 mm 1.5 mm
components of wall:
endothelium
elastic connective tissue
muscle
fibrous connective tissue
A

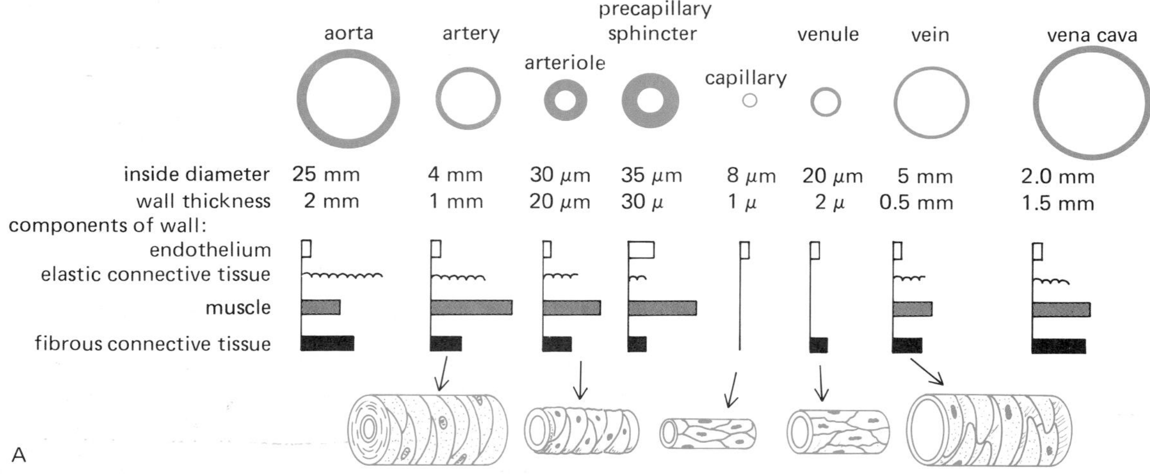

FIGURE 15-12A ***Comparison of the wall components of the different types of blood vessels.*** **(Adapted from Rushmer.)**

FIGURE 15-12B ***Blood and lymph vessels (transverse sections).***

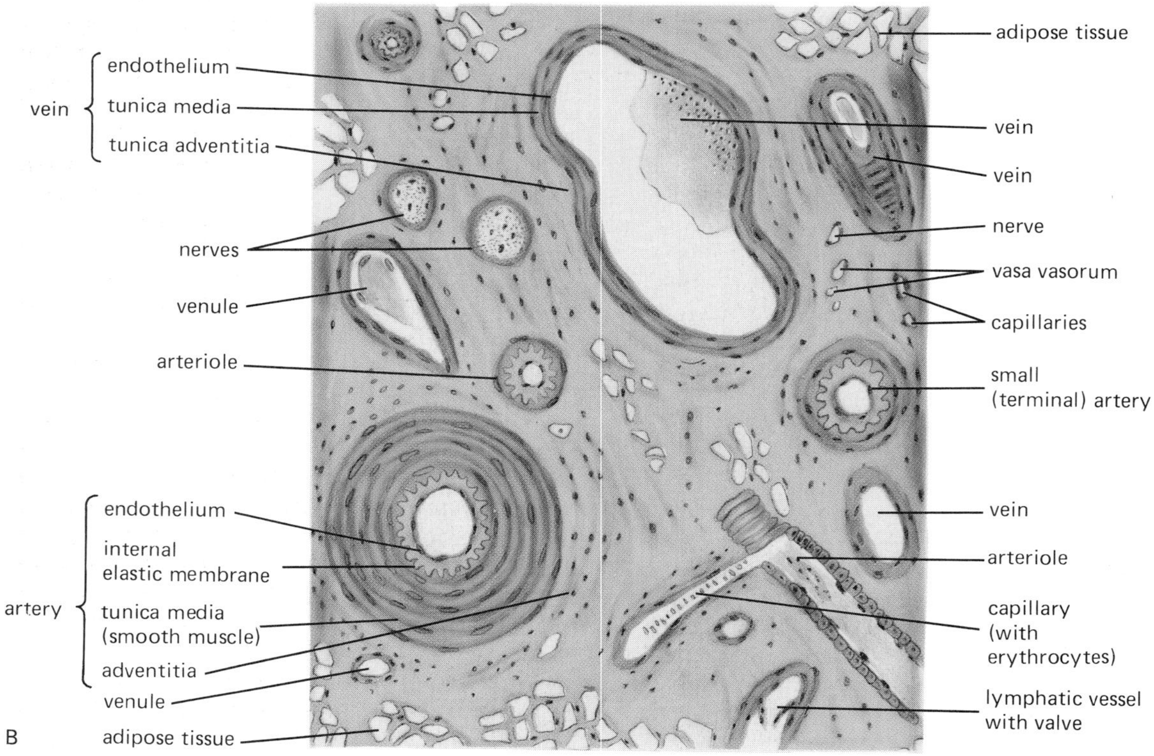

vein
endothelium
tunica media
tunica adventitia
nerves
venule
arteriole
artery
endothelium
internal elastic membrane
tunica media (smooth muscle)
adventitia
venule
adipose tissue
B
adipose tissue
vein
vein
nerve
vasa vasorum
capillaries
small (terminal) artery
vein
arteriole
capillary (with erythrocytes)
lymphatic vessel with valve

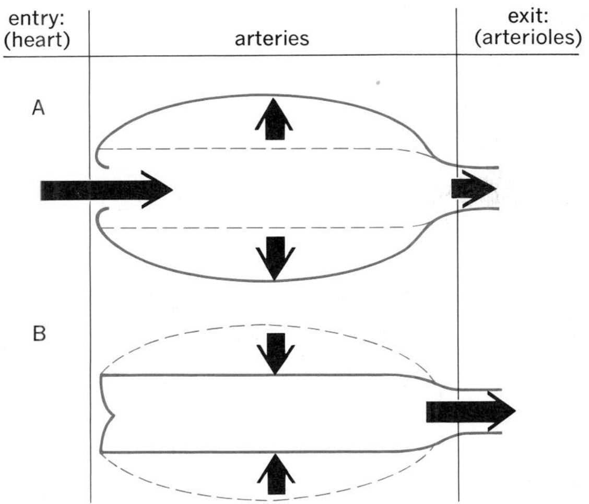

FIGURE 15-13 *Movement of blood into and out of the arteries during the cardiac cycle. The lengths of the arrows denote relative quantities. During systole A, less blood leaves the arteries than enters and the arterial walls stretch. During diastole B, the walls recoil psssively, driving blood out of the arteries.*

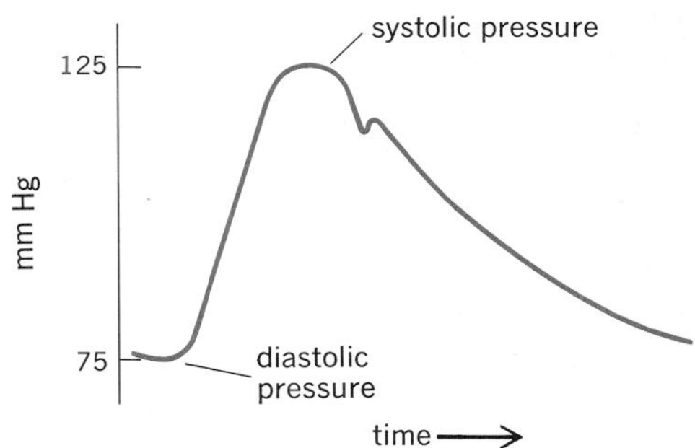

FIGURE 15-14 *Typical aortic pressure fluctuations during the cardiac cycle.*

$$\text{Mean pressure} = 75 + (\tfrac{1}{3} \times 50) = 92 \text{ mm Hg}$$

The mean arterial pressure is actually the most important of the pressures described because it is the *average* pressure driving blood into the tissues throughout the cardiac cycle. In other words, if the pulsatile pressure changes were eliminated and the pressure throughout the cardiac cycle were always equal to the mean pressure, the total flow would be unchanged. It is a closely regulated quantity; indeed, the reflexes which accomplish this regulation constitute the basic cardiovascular control mechanisms and will be described in detail in Chap. 16.

One last point: We can refer to "arterial" pressure without specifying to which artery we are referring because the aorta and other arteries have such large diameters that they offer only negligible resistance to flow and the pressures are therefore similar everywhere in the arterial tree.

Measurement of Arterial Pressure

Both systolic and diastolic blood pressure are readily measured in human beings with the use of a sphygmomanometer (Fig. 15-15). A hollow cuff is wrapped around the arm and inflated with air to a pressure greater than systolic blood pressure (Fig. 15-15A). The high pressure in the cuff is transmitted through the tissues of the arm and completely collapses the arteries under the cuff, thereby preventing blood flow to the lower arm. The air in the cuff is now slowly released, causing the pressure in the cuff and arm to drop. When cuff pressure has fallen to a point just below the systolic pressure (Fig. 15-15B), the arterial blood pressure at the peak of systole is greater than the cuff pressure, causing the artery to expand and allow blood flow for this brief time. During this interval, the blood flow through the partially occluded artery occurs at a very high velocity because of the small opening for blood passage and the large pressure gradient. The high-velocity blood flow produces turbulence and vibration, which can be heard through a stethoscope placed over the artery just below the cuff. The pressure measured on the manometer attached to the cuff at which sounds are first heard as the cuff pressure is lowered is identified as the systolic blood pressure. These first sounds are soft tapping sounds, corresponding to the peak systolic pressure reached during ejection of blood from the heart. As the pressure in the cuff is lowered further, the time of blood flow through the artery during each cycle becomes longer (Fig. 15-15C). The tapping sound becomes louder as the pressure is lowered. When the cuff pressure reaches the diastolic blood pressure, the sounds become dull and muffled, as the artery remains open throughout the cycle, allowing continuous turbulent flow (Fig. 15-15D). Just below diastolic pressure all sound stops as flow is now continuous and nonturbulent through the completely open artery. Thus, systolic pressure is measured as the cuff pressure at which sounds first appear and diastolic pressure as the cuff pressure at which sounds disappear.

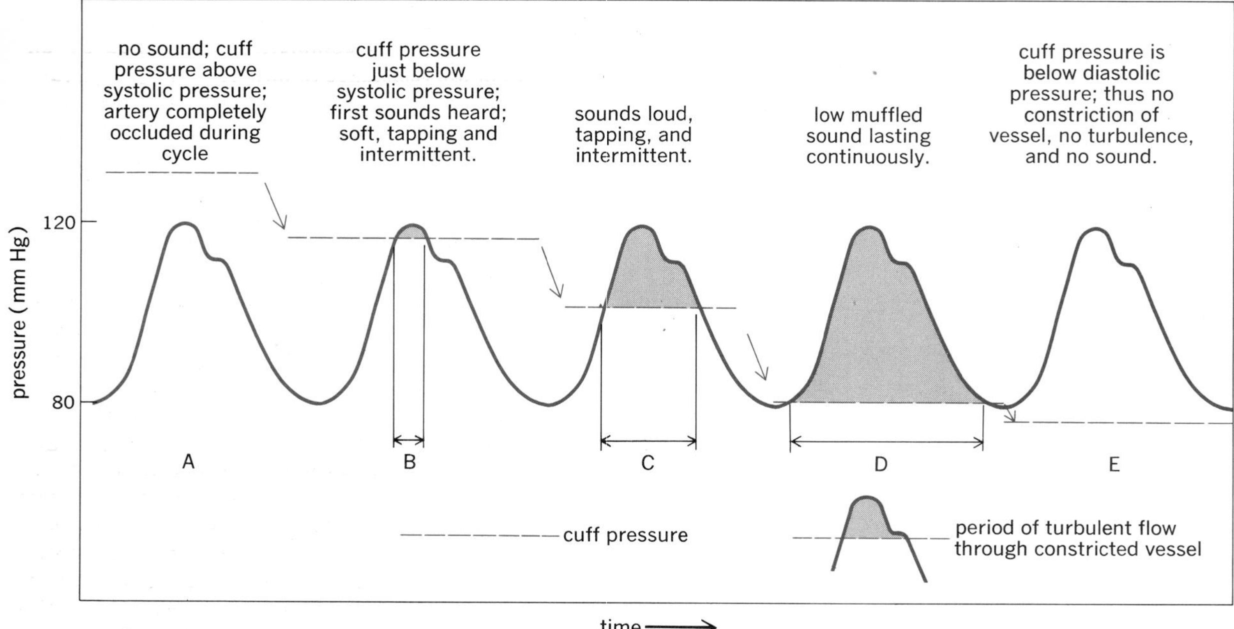

FIGURE 15-15 ***Sounds heard through a stethoscope while the cuff pressure of a sphygmomanometer is gradually lowered. Systolic pressure is recorded at B and diastolic pressure at the point of sound disappearance.***

ARTERIOLES

Each organ or tissue obviously receives only a fraction of the total left ventricular cardiac output. The typical distribution for a resting normal adult is given in Fig. 15-16. The digestive tract (including the liver), the kidneys, and the brain receive the largest supplies. Perhaps the most striking aspect of brain blood flow is its remarkable constancy. It really requires little, if any, additional energy to think; i.e., whether we are staring blankly into space or contemplating the theory of relativity, the total energy consumption of the brain remains virtually unchanged. In contrast, the energy consumption of muscular tissues of the body (heart, skeletal muscle, uterus, etc.) varies directly with the degree of muscle activity. Now compare the values at rest with those for exercise in Fig. 15-16. There is a large increase in blood flow to the exercising skeletal muscle and heart. Skin blood flow also has increased, kidney flow has decreased, and brain flow is unchanged. Obviously, the total cardiac output has increased; but, more important for our present purposes, the distribution of flow has greatly changed. Cardiac output is distributed to the various organs and tissues according to their functions and needs at any given moment. The remainder of this section describes the arterioles, that segment of the vascular tree primarily responsible for control of blood-flow distribution.

The arterioles begin as the terminal branches of the smallest arteries reach diameters of 100 μm or so. Their endothelial lining is separated from the smooth muscle cells of the tunica media by mere patches of elastic tissue, which become more and more scanty as the arterioles decrease in size until they finally disappear in the finest terminal arterioles. The smooth muscle cells are wound around the arteriole in a spiral (Fig. 15-17) and electrically coupled to each other by occasional gap junctions. There are several layers of these cells in the larger arterioles, but they too thin out until there is just a single layer in the finest terminal arterioles.

Figure 15-18 illustrates the major principles by

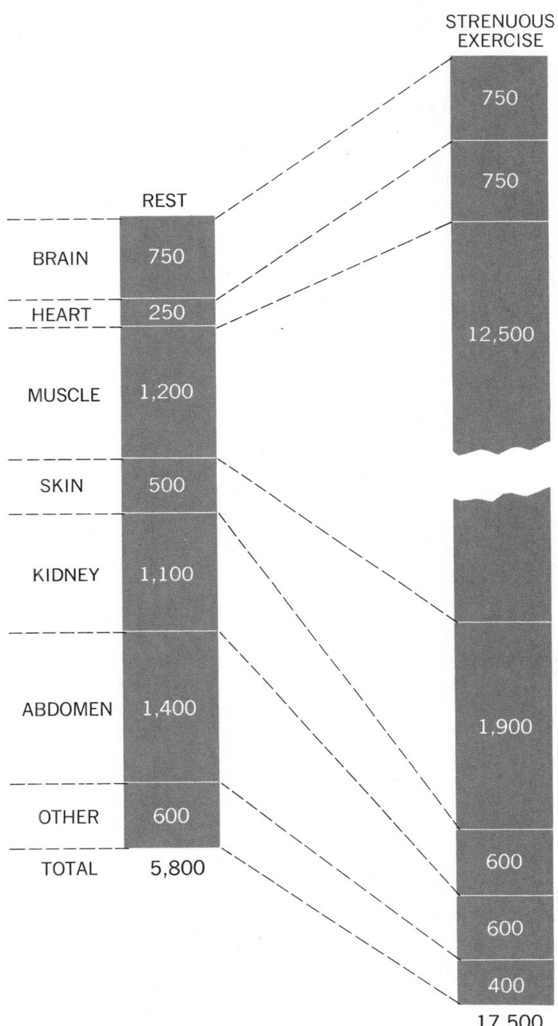

FIGURE 15-16 ***Distribution of blood flow to the various organs and tissues of the body at rest and during strenuous exercise. The numbers show blood flow in milliliters per minute.*** **(Adapted from Chapman and Mitchell.)**

which contraction or relaxation of the arteriolar smooth muscle and the resulting change in vessel diameter control blood-flow distribution. The principles are presented in terms of a simple model, a fluid-filled tank with a series of compressible outflow tubes. What determines the rate of flow through each exit tube? As always,

$$\text{Flow} = \frac{\Delta P}{R}$$

Since the driving pressure is identical for each tube, differences in flow are completely determined by differences in the resistance to flow offered by each tube. The lengths of the tubes are approximately the same, and the viscosity of the fluid is a constant; therefore, differences in resistance offered by the tubes are due solely to differences in their radii. Obviously, the widest tube has the greatest flow. If we equip each outflow tube with an adjustable cuff, we can obtain any combination of *relative* flows we wish.

This analysis can now be applied to the cardiovascular system. The tank is analogous to the arteries, which serve as a pressure reservoir, the major arteries themselves being so large that they contribute little resistance to flow. The smaller terminal arteries begin to offer some resistance, but it is relatively slight. Therefore, all the arteries of the body can be considered a single pressure reservoir. The arteries branch within each organ into the next series of smaller vessels, the arterioles, which are now narrow enough to offer considerable resistance. The arterioles are the major site of resistance in the vascular tree and are therefore analogous to the outflow tubes in the model. The smooth muscle in their walls is under precise physiologic control and can relax or constrict, thereby changing the radius of the inside (*lumen*) of the arteriole. *Thus, the pattern of blood-flow distribution depends primarily upon the degree of arteriolar smooth muscle constriction within each organ and tissue.*

The mechanisms controlling arteriolar constriction and dilatation fall into two general categories: (1) local controls, which serve the metabolic needs of the specific tissue in which they occur, and (2) reflex controls, which integrate and coordinate the needs of the whole body.

Local Controls

Active Hyperemia Certain organs and tissues, particularly the heart, skeletal muscle, and other muscular organs, manifest an increased blood flow (*hyperemia*) any time their metabolic activity is increased. For example, the blood flow to exercising skeletal muscle increases in direct proportion to the increased activity of the muscle. This phenomenon, known as *active hyperemia,* is the direct result of arteriolar dilatation within the more active organ. This vasodilatation does not depend upon the presence of nerves or hormones but is a locally mediated response. The adaptive value

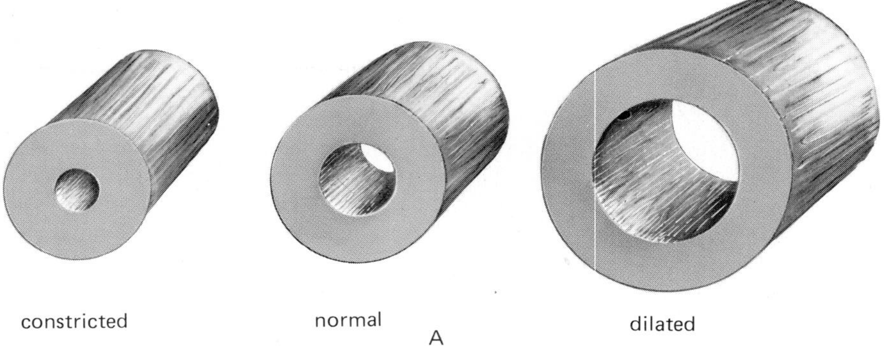

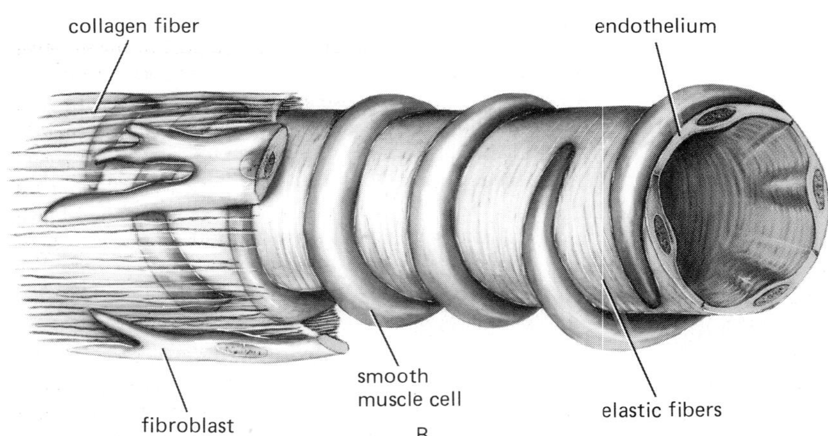

FIGURE 15-17 (A) ***Change in lumen diameter and wall thickness of a dilated and constricted arteriole as compared to normal.*** **(B)** ***Structure of an arteriole wall. The endothelium forms the internal coat, the smooth muscle fibers the middle coat, and the connective-tissue elements (fibroblasts and collagen fibers) the external coat.***

of the phenomenon should be readily apparent; an increased rate of activity in an organ automatically produces an increased blood flow to that organ by relaxing its arterioles.

What causes the arterioles to dilate? It is evident that the mechanism must involve local chemical changes which result from the increased cellular activity, but the relevant chemical changes have not yet been precisely identified. Moreover, the relative contributions of the various local factors thus far implicated seem to vary, depending upon the organs involved. At present, therefore, we can name only some of the factors which appear to be involved. A *decreased oxygen concentration* occurs locally as a result of increased utilization of oxygen by the more active cells. Conversely, local concentrations of *carbon dioxide* and *hydrogen ion* increase. Other *metabolites* also increase in concentration as a result of the greater metabolic activity. The concentrations of certain ions, particularly *potassium,* frequently increase (perhaps as a result of enhanced movement out of muscle cells during the more frequent action potentials). Local osmolarity also increases (i.e., water concentration decreases) as a result of the increased breakdown of high-molecular-weight substances. Changes in all these variables—decreased oxygen, increased carbon dioxide, ions, osmolarity, and metabolites—have been shown to cause arteriolar dilatation under controlled experimental conditions, and they probably all contribute, more or less, to the active hyperemia response. It must be emphasized that all these chemical changes act *directly* upon the arteriolar smooth muscle, causing it to relax (dilate), no nerves or hormones being involved.

It should not be too surprising that the phenomenon of active hyperemia is most highly developed in heart and skeletal muscle, which show the widest range

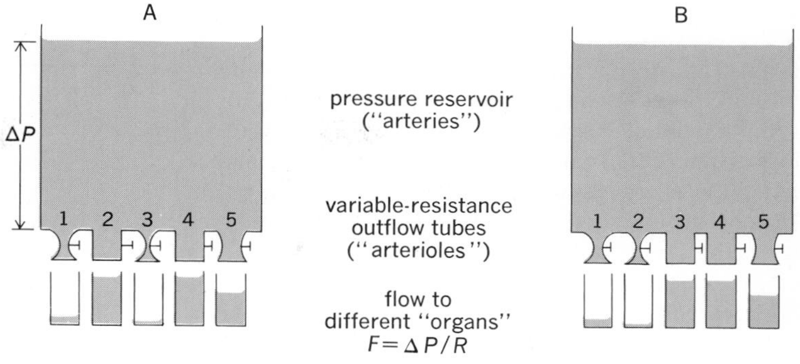

FIGURE 15-18 ***Physical model of the relationship between arterial pressure, arteriolar radius in different organs, and blood-flow distribution. Blood has been shifted from organ 2 to organ 3 (in going from A to B) by constricting the "arterioles" of 2 and dilating those of 3.***

of normal metabolic activities of any organs or tissues in the body. It is highly efficient, therefore, that their supply of blood be primarily determined *locally* by their rates of activity. The gastrointestinal tract also manifests a great capacity for active hyperemia in keeping with its relatively wide range of metabolic activity.

Reactive Hyperemia If a tourniquet is placed around the upper arm and tightened to shut off the arterial inflow, the tissues of the arm are deprived of blood (*ischemia*). When the tourniquet is loosened, the arm becomes red and very warm, signs of a greatly increased blood flow. Increased flow following a period of ischemia, called *reactive hyperemia,* occurs because the ischemia has produced local arteriolar dilatation. The explanation appears to be quite similar to that for active hyperemia. While the blood flow was reduced, the supply of oxygen to the tissue was diminished and the local oxygen concentration decreased. Simultaneously, the concentrations of carbon dioxide, hydrogen ion, and metabolites all increased because they were not removed by the blood as fast as they were produced. In other words, the events of active and reactive hyperemia are similar because both reflect an imbalance of blood supply and level of cellular metabolic activity. The adaptive value of reactive hyperemia is that a tissue which suffers ischemia, say as a result of partial occlusion of the artery supplying it, automatically tends to maintain its blood supply because of local arteriolar dilatation.

Histamine and the Response to Injury Injury to the skin (and probably other tissues as well) causes local release from cells of a chemical substance known as *histamine*. It makes arteriolar smooth muscle relax and is probably a major cause of vasodilatation in an injured area. This phenomenon, a part of the general process known as inflammation, is described in detail in Chap. 22 and is mentioned here only to point out that histamine is *not* the vasodilator substance responsible for active or reactive hyperemia. In addition to histamine, several other locally released chemicals alter arteriolar tone in response to tissue injury or blood-vessel damage and are also described in Chap. 22.

Reflex Controls

Sympathetic Nerves Most arterioles in the body receive a rich supply of sympathetic postganglionic nerve fibers. These nerves (with one major exception) release norepinephrine, which acts upon vascular smooth muscle to cause vasoconstriction. The only organs whose arterioles are not significantly influenced by these constrictor fibers are the brain and the heart.[1] (The adaptive significance for these exceptions will be clear after the role of these sympathetic nerves has been explained.) If almost all the nerves to arterioles are constrictor in action, how can reflex arteriolar dilatation be achieved? Since the sympathetic nerves are seldom completely quiescent but discharge at some finite rate, which varies from organ to organ, the nerves always cause some degree of tonic constriction; from this basal position, further constriction is produced by increased sympathetic activity, whereas dilatation can be achieved by decreasing the rate of

[1] The vasculature of the brain and heart receives sympathetic neurons the activity of which may be altered under certain circumstances. However, present evidence indicates that these neurons are usually of negligible importance when compared with the local control of these vascular beds.

sympathetic activity below the basal level. The skin offers an excellent example of these processes: Skin arterioles of a normal unexcited person at room temperature are already under the influence of a high rate of sympathetic discharge; an appropriate stimulus (fear, loss of blood, etc.) causes reflex enhancement of this activity; the arterioles constrict further, and the skin pales. In contrast, an increased body temperature reflexly inhibits the sympathetic nerves to the skin, the arterioles dilate, and the skin flushes. This generalization cannot be stressed too strongly: *Control of the sympathetic constrictor nerves to arteriolar smooth muscle can accomplish either dilatation or constriction.*

In contrast to the processes of active and reactive hyperemia, the primary functions of these nerves are concerned *not* with the coordination of *local* metabolic needs and blood flow but with reflexes that help maintain an adequate blood supply at all times to vital organs such as the brain and heart. As their common denominator these reflexes have the regulation of arterial blood pressure; they will be described in detail in Chap. 16.

There is, as we said, one exception to the generalization that sympathetic nerves to arterioles release norepinephrine. A group of sympathetic (*not* parasympathetic) nerves to the arterioles in skeletal muscle instead releases acetylcholine, which causes arteriolar dilatation and increased blood flow. It still must be emphasized that *most* sympathetic nerves to skeletal muscle arterioles release norepinephrine; thus skeletal muscle arterioles receive a dual set of sympathetic nerves. The only known function of the vasodilator fibers is in the response to exercise or stress and will be described in Chap. 16; the vasoconstrictor fibers mediate all other situations involving neural control of skeletal muscle arterioles.

Parasympathetic Nerves With but one major exception (the blood vessels of certain areas in the genital tract), there is no significant parasympathetic innervation of arterioles. It is true that stimulation of the parasympathetic nerves to certain glands is associated with an increased blood flow, but this may be secondary to the increased metabolic activity induced in the gland by the nerves, with resultant local active hyperemia.

Hormones Several hormones cause constriction or dilatation of arteriolar smooth muscle, one of the most important being epinephrine, the hormone released from the adrenal medulla. In most vascular beds, epinephrine, like the sympathetic nerves, causes vasoconstriction; surprisingly, in other vascular beds, epinephrine may induce vasodilatation. However, it is likely that the effects of circulating epinephrine on arterioles are quantitatively of little significance when compared with those exerted by norepinephrine released from sympathetic-nerve endings. Angiotensin, a hormone to be discussed in Chap. 18, may also strongly constrict arterioles under certain conditions.

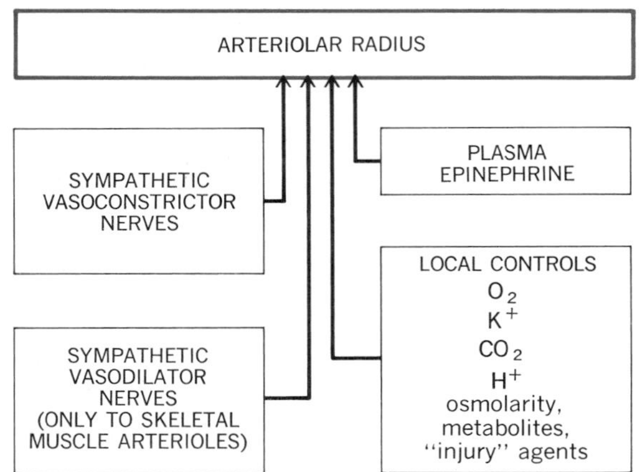

FIGURE 15-19 ***Major factors affecting arteriolar radius.***

Figure 15-19 summarizes the factors which determine arteriolar radius.

CAPILLARIES

At any given moment, approximately 5 percent of the total circulating blood is flowing through the capillaries. Yet this 5 percent is really the only blood in the cardiovascular system which is performing the ultimate function of the entire system, namely, the exchange of

FIGURE 15-20 ***Structure of the capillary wall.***

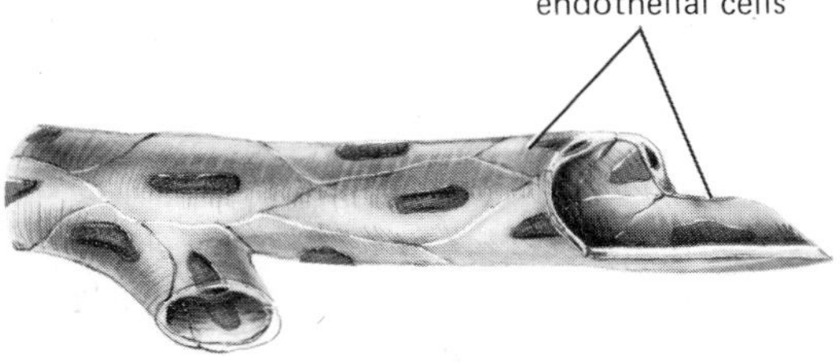

nutrients and metabolic end products. All other segments of the vascular tree subserve the overall aim of getting adequate blood flow through the capillaries. The capillaries permeate every tissue of the body; no cell is more than 0.15 mm from a capillary. Therefore, diffusion distances are very small, and exchange is highly efficient. There are thousands of kilometers of capillaries in an adult person, each individual capillary being only about 1 mm long.

Capillaries throughout the body vary somewhat in structure, but the typical capillary (Fig. 15-20) is a thin-walled tube of endothelial cells without elastic tissue, connective tissue, or smooth muscle to impede transfer of water and solutes. The squamous cells which constitute the endothelial lining interlock like pieces of a jigsaw puzzle. This thin capillary membrane behaves as though it were perforated by small pores through which water and solute particles smaller than proteins readily move. Sinusoids differ from other capillaries in that they are much wider and have irregular lumens, and their walls, particularly in the liver, spleen, and bone marrow, contain fixed macrophages. The permeability of capillaries varies throughout the body, liver sinusoids being the "leakiest" and brain capillaries the "tightest."

Anatomy of the Capillary Network

Figure 15-21 illustrates diagrammatically the general anatomy of the small vessels which constitute the so-called microcirculation. Blood enters the capillary network from the arterioles. Most tissues appear to have two distinct types of capillaries: "true" capillaries and thoroughfare channels. The thoroughfare channels connect arterioles and venules directly. From these channels exit and reenter the network of true capillaries across which materials actually exchange. The site at which a true capillary exits is protected by a ring of smooth muscle, the *precapillary sphincter,* which continually opens and closes so that flow through any given capillary is usually intermittent. Generally, the more active the tissue, the more precapillary sphincters are open at any moment. The sphincters are best visualized as functioning in concert with arteriolar smooth muscle to regulate not only the total flow of blood through the tissue capillaries but the number of functioning capillaries as well.

Resistance of the Capillaries

Since a capillary is very narrow, it offers a considerable resistance to flow, but for two reasons, the resistance is not of critical importance for cardiovascular function: (1) Despite the fact that the capillaries are actually narrower than the arterioles, the huge total number of capillaries provides such a great cross-sectional area for flow that the *total* resistance of *all* the capillaries is considerably less than that of the arterioles. (2) Because capillaries have no smooth muscle, their radius (and, therefore, their resistance) is not subject to active control and simply reflects the volume of blood delivered to them via the arterioles (and the volume leaving via the venules).

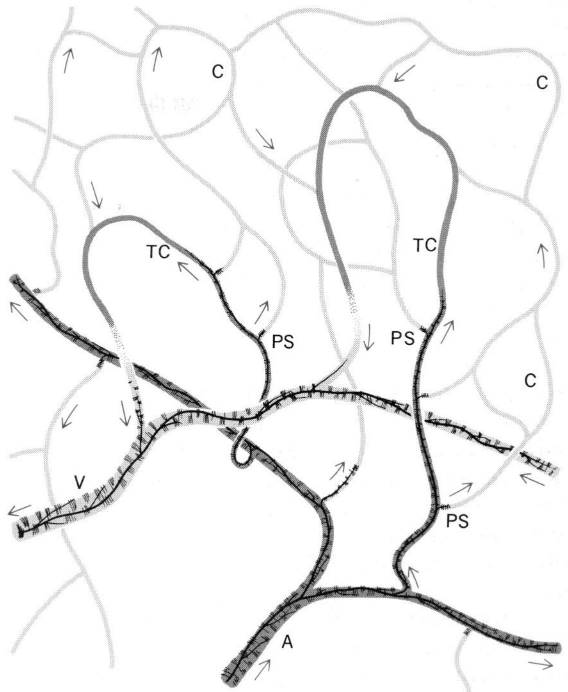

FIGURE 15-21 ***Diagram of microcirculation. Note the thinning of the smooth muscle coat in the thoroughfare channels and its complete absence in the true capillaries. The black lines on the surface of the vessels are nerve fibers leading to smooth muscle cells.*** **(Adapted from Zweifach.)**

Velocity of Capillary Blood Flow

Figure 15-22 illustrates a simple mechanical model of a series of 1-cm-diameter balls being pushed down a

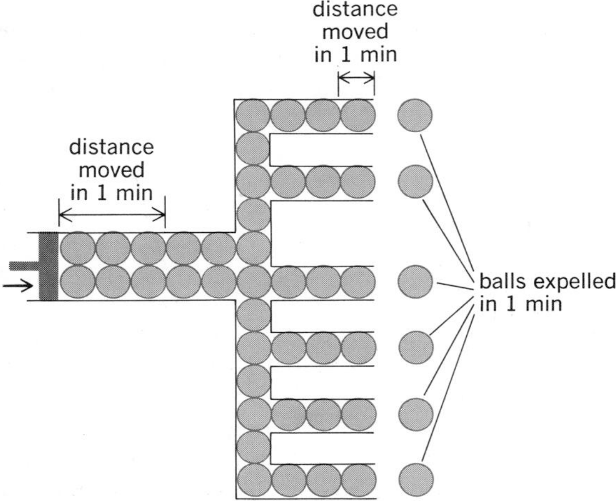

FIGURE 15-22 ***Relationship between cross-sectional area and velocity of flow. The total cross-sectional area of the small tubes is three times greater than that of the large tube. Accordingly, velocity of flow is one-third as great in the small tubes.***

single tube which branches into narrower tubes. Although each tributary tube has a smaller cross section than the wide tube, the sum of the tributary cross sections is much greater than the area of the wide tube. Let us assume that in the wide tube each ball moves 3 cm/min. If the balls are 1 cm in diameter and they move two abreast, six balls leave the wide tube per minute and enter the narrow tubes. Obviously, then, six balls must be leaving the narrow tubes per minute. At what speed does each ball move in the small tubes? The answer is 1 cm/min. This example illustrates the following important generalization: When a continuous stream moves through consecutive sets of tubes, the velocity of flow decreases as the sum of the cross-sectional areas of the tubes increases. This is precisely the case in the cardiovascular system (Fig. 15-23); the blood velocity is very great in the aorta, progressively slows in the arteries and arterioles, and then markedly slows as it passes through the huge cross-sectional area of the capillaries (600 times the cross-sectional area of the aorta). The speed then progressively increases in the venules and veins because the cross-sectional area decreases. The adaptive significance of this phenomenon is very great; blood flows through the capillaries so slowly (0.07 cm/s) that there is adequate time for exchange of nutrients and metabolic end products between the blood and tissues.

Diffusion Across the Capillary Wall: Exchanges of Nutrients and Metabolic End Products

There is no active transport of solute across the capillary wall, materials crossing primarily by simple passive diffusion.[2] As described in the next section, there is some movement of fluid by bulk flow, but it is of negligible importance for the exchange of nutrients and metabolic end products. Because fat-soluble substances penetrate cell membranes easily, they probably

[2] The only probable exception to this statement is the movement of small amounts of protein by pinocytosis.

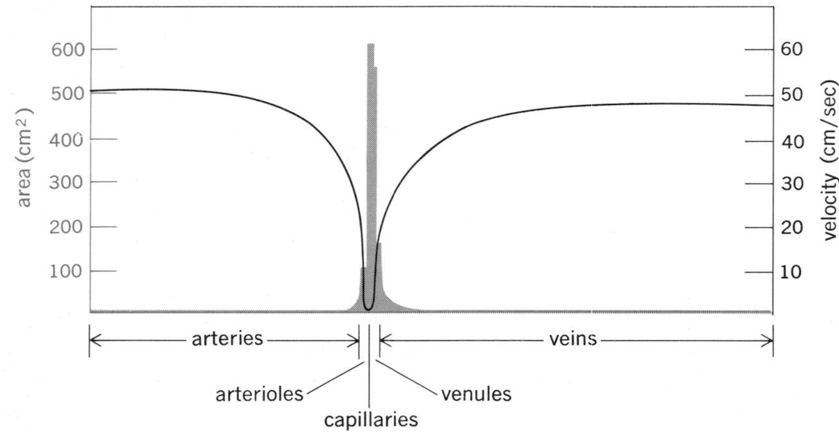

FIGURE 15-23 ***Relation between cross-sectional area and velocity of flow in the systemic circulation. The values are those for a 30-lb dog. Velocity of blood flow through the capillaries is about 0.07 cm/sec.*** **(Adapted from Rushmer.)**

pass directly through the endothelial capillary cells. Many ions and molecules are poorly soluble in fat and so pass through pores between adjacent endothelial cells. In any case, nearly all nutrients and metabolic end products diffuse across the capillary with great speed.

What is the sequence of events involved in capillary-cell transfers? Tissue cells do not exchange material *directly* with blood; the interstitial fluid always acts as middleman. Thus, nutrients diffuse across the capillary wall into the interstitial fluid, from which they gain entry to cells. Conversely, metabolic end products move first across cell membranes into interstitial fluid, from which they diffuse into the plasma. Thus, two membrane-transport processes must always be considered, that across the capillary wall and that across the tissue cell membrane. The cell-membrane step may be by diffusion or by carrier-mediated transport, but, as described above, the transcapillary movement is always by diffusion. Since, to achieve *net* transport of any substance by diffusion, a concentration gradient is required, transcapillary diffusion of nutrients and metabolic end products proceeds primarily in one direction because of diffusion gradients for these substances between the blood and interstitial fluid.

How do these diffusion gradients arise? Let us take two examples, those of glucose and carbon dioxide transcapillary movement in muscle. Glucose is continuously consumed after being transported from interstitial fluid into the muscle cells by carrier-mediated transport mechanisms; this removal from interstitial fluid lowers the interstitial-fluid glucose concentration below that of plasma and creates the gradient for glucose diffusion out of the capillary. Carbon dioxide is continuously produced by muscle cells, thereby creating an increased intracellular carbon dioxide concentration, which causes diffusion of carbon dioxide into the interstitial fluid; in turn, this causes the interstitial carbon dioxide concentration to be greater than that of plasma and produces carbon dioxide diffusion into the capillary. We chose these particular examples to emphasize the fact that net movement of a substance between interstitial fluid and cells is the event which establishes the transcapillary plasma-interstitial diffusion gradients but that it does not matter whether the substance moves across the cell membrane by diffusion or carrier-mediated transport.

When a tissue increases its rate of metabolism, it must obviously obtain more nutrients from the blood and eliminate more metabolic end products. One important mechanism for achieving this is active hyperemia, described above, which increases the blood flow to the tissue. A second important and quite simple mechanism involves alterations of the plasma-interstitial concentration gradient. Let us return to our example of glucose and muscle. When the muscle increases its activity, it also increases its uptake of glucose, thereby lowering the interstitial glucose concentration below normal. This sets up an increased plasma-interstitial glucose concentration gradient which causes the *net* diffusion of glucose out of the capillary to be increased. In other words, this change in concentration gradient has allowed a greater fraction of the total blood glucose to be extracted from the blood as it flows through the capillaries. Thus, there need not be an absolutely strict correlation between a tissue's activity and its blood supply, at least within moderate limits. Similar changes in diffusion gradients permit a tissue to obtain nutrients and eliminate metabolic end products adequately in spite of modest reductions of blood flow to it.

It should now be clear how the interstitial fluid functions as the true immediate environment for the body's cells and constitutes the body's *internal environment*.

Bulk Flow Across the Capillary Wall: Distribution of the Extracellular Fluid

Since the capillary wall is highly permeable to water and to almost all the solutes of the plasma with the exception of the plasma proteins, it behaves like a porous filter through which protein-free plasma moves by bulk flow, known as *ultrafiltration,* under the influence of a hydrostatic pressure gradient. The magnitude of the bulk flow is directly proportional to the hydrostatic pressure difference between the inside and outside of the capillary, i.e., between the capillary blood pressure and the interstitial-fluid pressure. Normally, the former is much larger than the latter, so that a considerable hydrostatic pressure gradient exists to drive the filtration of protein-free plasma out of the capillaries into the interstitial fluid. Why then does all the plasma not filter out into the interstitial space instead of remaining in the capillaries? The explanation was first elucidated by Starling (the same scientist who expounded the law of the heart which bears his name) and depends upon the principles of osmosis.

In Chap. 1 we described how a net movement of water occurs across a semipermeable membrane from a solution of high water concentration to a solution of low water concentration. Recall that the concentration of water depends upon the concentration of solute molecules or ions dissolved in the water. When two solutions A and B, which are separated by a semipermeable membrane, have identical concentrations of all solutes, the water concentrations are identical and no net water movement occurs. When, however, a quantity of a nonpermeating substance is added to solution A, the water concentration of A is reduced below that of solution B and a net movement of water will occur by osmosis from B into A. Of great importance is that osmotic flow of water (solvent) "drags" along with it any dissolved solutes to which the membrane is highly permeable. Thus, a difference in water concentration can result in the movement of both water and permeating solute in a manner virtually indistinguishable from the bulk flow produced by a hydrostatic pressure difference. The difference in water concentration resulting from the presence of the nonpenetrating solute can therefore be expressed in units of pressure (millimeters of mercury).

This analysis can now be applied to capillary fluid movements. The plasma within the capillary and the interstitial fluid outside it contain large quantities of low-molecular-weight solutes (crystalloids), e.g., sodium, chloride, or glucose. Since the capillary lining is highly permeable to all these crystalloids, they all have almost identical concentrations in the two solutions. There are small concentration differences occurring for substances consumed or produced by the cells, but these tend to cancel each other, and, accordingly, no significant water-concentration difference is caused by the presence of the crystalloids. In contrast, the plasma proteins can diffuse across the capillary wall only very slightly and therefore have a very low interstitial-fluid concentration. This difference in protein concentration between plasma and interstitial fluid means that the water concentration of the plasma is lower than that of interstitial fluid, inducing an osmotic flow of water from the interstitial compartment into the capillary. Along with the water are carried all the different types of crystalloids dissolved in the interstitial fluid. Thus, osmotic flow of fluid, like bulk flow, does not alter the concentrations of the low-molecular-weight substances of plasma or interstitial fluid.

In summary, two opposing forces act to move fluid across the capillary: (1) The hydrostatic pressure difference between capillary blood pressure and interstitial-fluid pressure favors the filtration of a protein-free plasma out of the capillary; (2) the water-concentration difference between plasma and interstitial fluid, which results from the protein-concentration differences, favors the osmotic movement of interstitial fluid into the capillary. Accordingly, the movements of fluid depend directly upon four variables: the capillary hydrostatic pressure, interstitial hydrostatic pressure, plasma protein concentration, and interstitial-fluid protein concentration.

We may now consider quantitatively how these variables act to move fluid across the capillary wall (Fig. 15-24). Much of the arterial blood pressure has already been dissipated as the blood flows through the arterioles, so that pressure at the beginning of the capillary is 35 mm Hg. Since the capillary also offers resistance to flow, the pressure continuously decreases to 15 mm Hg at the end of the capillary. The interstitial pressure is so close to zero that it can be ignored. The difference in protein concentration between plasma and interstitial fluid causes a difference in water concentration (plasma water concentration less than interstitial-fluid water concentration), which induces an osmotic flow of fluid into the capillary equivalent to that produced by a hydrostatic pressure difference of 25 mm Hg. It is evident that in the first portion of the capillary the hydrostatic pressure difference is greater than the osmotic forces and a net movement of fluid out of the capillary occurs; in the last portion of the capillary, however, a net force causes fluid movement into the capillary (termed *absorption*). The net result is that the early and late capillary events tend to cancel each other out, and there is little overall net loss or gain of fluid (Fig. 15-24A). In a normal person there is a small net filtration; as we shall see, this is returned to the blood by lymphatics.

The analysis of capillary fluid dynamics in terms of different events occurring at the arterial and venous ends of the capillary is oversimplified. It is likely that any given capillary manifests either net filtration or net absorption along its entire length because the arteriole supplying it is either so dilated or so constricted as to yield a capillary hydrostatic pressure above or below 25 mm Hg along the entire length of the capillary. This does not alter the basic concept that, taken as a unit, a capillary bed manifests net absorption or filtration, depending upon the average levels of hydrostatic pressures within the individual capillaries constituting the bed.

Figure 15-24B and C illustrates, however, the effects on this equilibrium of changing capillary pressure. In Figure 15-24B, the arterioles in the organ have been dilated, and the capillary pressure therefore increases since less of the arterial pressure is dissipated in the passage through the arterioles. Outward filtration now

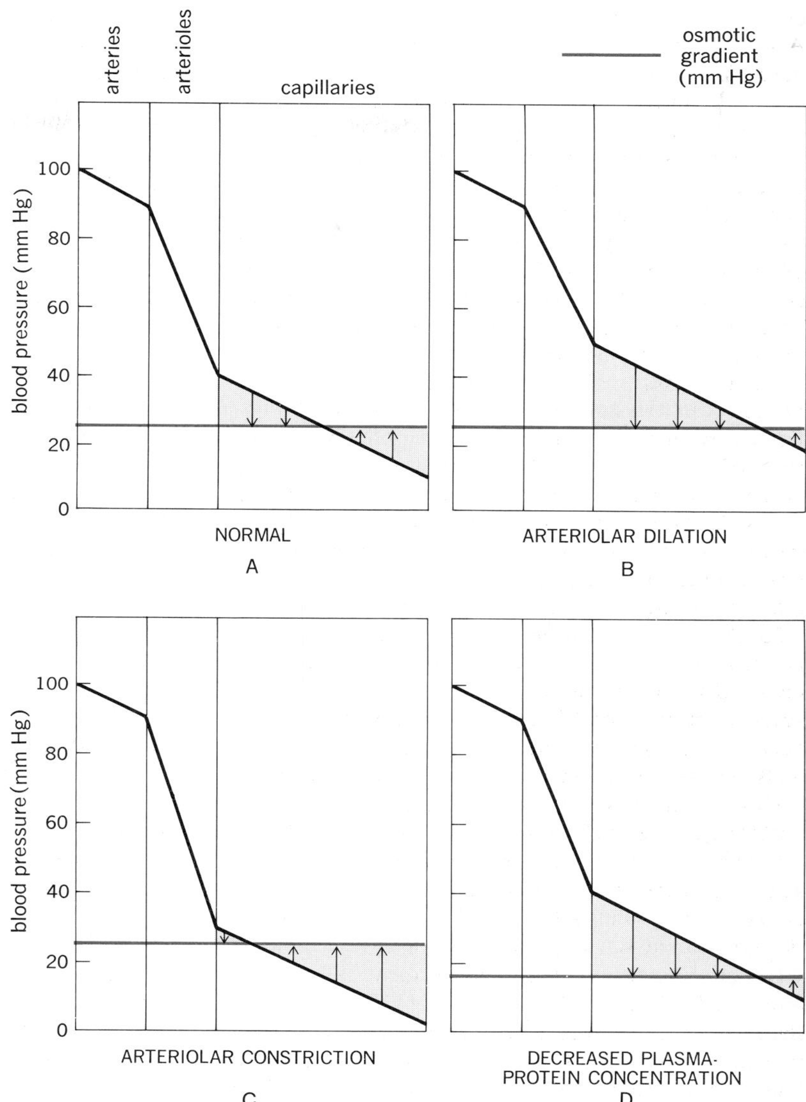

FIGURE 15-24 ***Relevant filtration-absorption forces acting across the capillary wall in several situations. Arrows down indicate filtration out of the capillary. Arrows up indicate fluid movement from interstitium into capillary. The shaded areas denote relative magnitudes of the fluid movements.***

predominates, and some of the plasma enters the interstitial fluid. In contrast, marked arteriolar constriction (Fig. 15-24C) produces decreased capillary pressure and net movement of interstitial fluid into the vascular compartment. Figure 15-24D shows how net absorption or filtration can be produced in the absence of capillary pressure changes whenever plasma protein concentration is altered. Thus, in liver disease, protein synthesis decreases, plasma protein concentration is reduced, plasma water concentration is increased, net filtration occurs, and fluid accumulates in the interstitial space.

The transcapillary protein-concentration difference can also be decreased by a quite different event, namely, the leakage of protein across the capillary wall into the interstitium whenever the capillary lining is damaged. This eliminates the protein-concentration difference, and local edema occurs as a result of the unchecked hydrostatic pressure difference still acting across the capillary. The fluid accumulation in a blister is an excellent example.

The major function of this capillary filtration-absorption equilibrium should now be evident: *It determines the distribution of the extracellular-fluid volume between the vascular and interstitial compartments.* Obviously, the ability of the heart to pump blood depends upon the presence of an adequate volume of blood within the system. Recall that the interstitial-fluid volume is three to four times larger than the plasma volume; therefore, the interstitial fluid serves as a reservoir which can supply additional fluid to the circulatory system or draw off excess. The important role this equilibrium plays in the physiologic response to many situations, such as hemorrhage, will be described in Chap. 16.

It should be stressed again that capillary filtration and absorption do not alter *concentrations* of any substance (other than protein) since movement is by bulk flow; i.e., everything in the plasma (except protein) or the interstitial fluid moves together. The reason this process of filtration plays no significant role in the exchange of nutrients and metabolic end products between capillary and tissues is that the total quantity of a substance (such as glucose or oxygen) moving into or out of the capillary during bulk flow is extremely small in comparison with the quantities moving by diffusion. For example, during a single day approximately 20,000 g of glucose crosses the capillary into the interstitial fluid by diffusion but only 20 g enters by bulk flow. Of course, only a small fraction of this glucose is utilized by the cells, the remainder moving back into the blood, again almost entirely by diffusion.

VEINS

Most of the pressure imparted to the blood by the heart is dissipated as blood flows through the arterioles and capillaries, so that pressure in the small venules is only approximately 15 mm Hg and only a small pressure remains to drive blood back to the heart. One of the major functions of the veins is to act as low-resistance conduits for blood flow from the tissues back to the heart. This function is performed so efficiently that the total pressure drop from venule to right atrium is only about 10 mm Hg, the right atrial pressure being 0 to 5 mm Hg. The resistance is low because the veins have a large diameter.

The walls of the veins are composed of three coats, the main difference between the veins and the arteries being in the relative thinness of the middle coat in the veins (Fig. 15-12A and B). The tunica media of the veins has less smooth muscle and elastic connective tissue than the same layer in arteries, features related to the lower blood pressures they encounter.

The veins perform a second extremely important function which has only recently been appreciated: They adjust their total *capacity* to accommodate variations in blood volume. The veins are the last set of tubes through which the blood must flow on its trip to the heart. The force immediately driving this venous return is the venous pressure (more precisely, the pressure gradient between the veins and atria). In turn, the rate of venous return, i.e., inflow to the atria, is one of the most important determinants of atrial pressure. In a discussion of the control of cardiac output in Chap. 14, we emphasized that the atrial pressure was the major determinant of ventricular end-diastolic volume and thereby of intrinsic control of stroke volume. Combining these two statements, we now see that venous pressure is a crucial determinant of stroke volume via the intermediation of atrial pressure and ventricular end-diastolic volume.

This completes our description of the pressure changes throughout the vascular tree. The normal pressure profiles for the systemic and pulmonary circulation are given in Fig. 15-25. Note that the pulmonary pressures are considerably smaller than the systemic pressures for reasons shortly to be described. Note also that the resistance offered by the arterioles effectively damps the pulse; by doing so, the arterioles convert the pulsative arterial flow into a continuous capillary flow.

Determinants of Venous Pressure

The factors determining pressure in any elastic tube, as we know, are the volume of fluid within it and the distensibility of its wall. Accordingly, total blood volume is one important determinant of venous pressure. The relatively thin walls of the veins are much more distensible than arterial walls and can accommodate large volumes of blood with little increase of internal pressure. This is illustrated by comparing Fig. 15-26 with Fig. 15-25; approximately 60 percent of the total blood volume is present in the systemic veins at any given moment, but the venous pressure averages less than 10 mm Hg. In contrast, the systemic arteries contain less than 15 percent of the blood at a pressure of approximately 100 mm Hg. This pressure-volume rela-

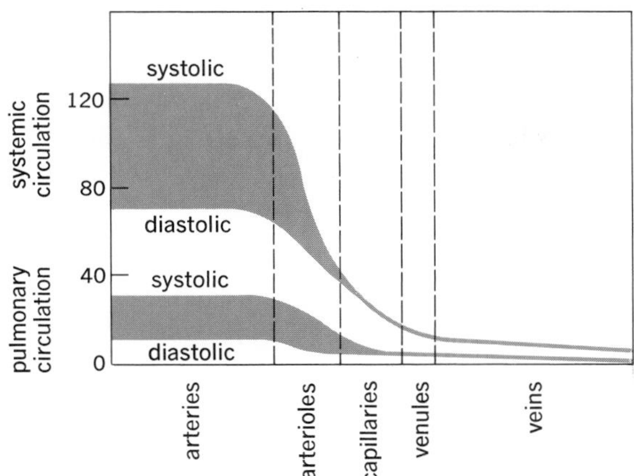

FIGURE 15-25 ***Summary of pressures in the vascular system.***

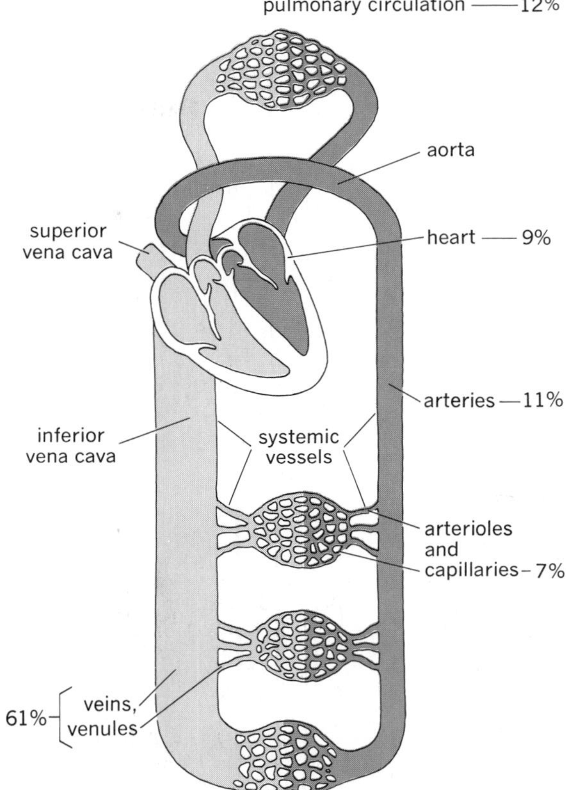

FIGURE 15-26 ***Distribution of blood in the different portions of the cardiovascular system. Compare this distribution of blood volumes with the pressures for the relevant areas shown in Fig. 15-25.*** **(Adapted from Guyton.)**

tionship of the veins allows them to act as a reservoir for blood. The walls of the veins contain smooth muscle richly innervated by sympathetic vasoconstrictor nerves, stimulation of which causes venous constriction, thereby increasing the stiffness of the wall, i.e., making it less distensible, and raising the pressure of the blood within the veins. Increased venous pressure drives more blood out of the veins into the right heart. Thus, venous constriction exerts precisely the same effect on venous return as giving a transfusion.

The great importance of this effect can be visualized by the example in Fig. 15-27. A large decrease in total blood volume initially reduces the pressures everywhere in the circulatory system, including the veins; venous return to the heart decreases, and cardiac output decreases. However, reflexes to be described cause increased sympathetic discharge to the venous smooth muscle, which contracts, thereby returning venous pressure toward normal, restoring venous return and cardiac output.

Two other mechanisms can decrease venous capacity, increase venous pressure, and facilitate venous return; these are the skeletal muscle "pump" and the effects of respiration upon thoracic and abdominal veins (respiratory "pump"). During skeletal muscle contraction, the veins running through the muscle are partially compressed, thereby reducing their diameter and decreasing venous capacity. As will be described in Chap. 17, during inspiration, the diaphragm descends, pushes on the abdominal contents, and increases abdominal pressure. The large veins which pass through the abdomen are partially compressed by this increased pressure; this facilitates movement of blood, but only toward the heart because venous valves in the legs prevent backflow. Simultaneously, the pressure in the chest (thorax) decreases, and this decrease is transmitted passively to the intrathoracic veins and right atrium. The net effect is to increase the pressure gradient between the right atrium and veins outside the thorax; accordingly, venous return to the heart is enhanced.

In summary (Fig. 15-28), the effects of the venous muscle tone, skeletal muscle pump, and respiratory pump are to facilitate return of blood to the heart. The

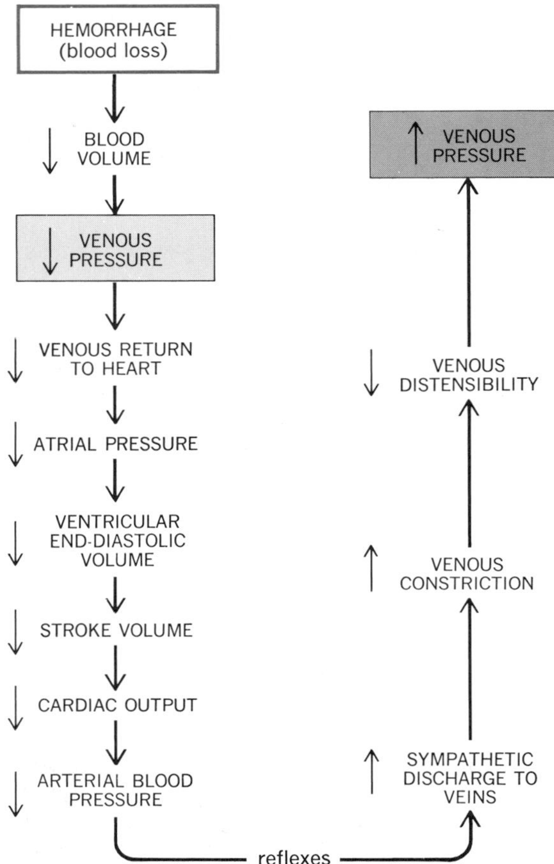

FIGURE 15-27 ***Role of venoconstriction in maintaining venous pressure during blood loss. The increased venous smooth muscle constriction returns the decreased venous pressure toward, but not to, normal. The reflexes involved in this response are described in Chap. 16.***

net result is that atrial pressure and, thereby, cardiac output are determined in large part by these factors.

Effects of Venous Constriction on Resistance to Flow

We have seen that decreasing the diameter of the veins increases venous pressure, which increases venous return to the heart. However, this decreased diameter also increases resistance to flow, a phenomenon which would retard venous return if the effect of venous constriction upon resistance were not so slight as to be negligible. The veins have such large diameters that a slight decrease in size (which has great effects on venous capacity) produces little increase in resistance. This is just the opposite of the arterioles, which are so narrow that they contain little blood at any moment (Fig. 15-26), further decrease having little effect on blood displacement toward the heart but even a slight decrease in diameter producing a marked increase in resistance to flow. Flow back to the heart, therefore, tends to be impaired by arteriolar constriction and enhanced by venous constriction. It should be stressed, however, that abnormally great increases in venous resistance, say from an internal blood clot or a tumor compressing from the outside, may markedly impair blood flow. Under such conditions, blood accumulates behind the lesion, pressures in the small veins and capillaries drained by the occluded vein increase, capillary filtration increases, and the tissue becomes edematous, i.e., swollen with excess interstitial fluid.

Venous Valves

Many veins in the body, particularly in the limbs, have valves which close so as to allow flow only toward

FIGURE 15-28 ***Major factors determining venous pressure and thereby atrial pressure. The figure as drawn shows how venous and atrial pressures are increased; reversing the arrows in the boxes indicates how these pressures can be reduced-***

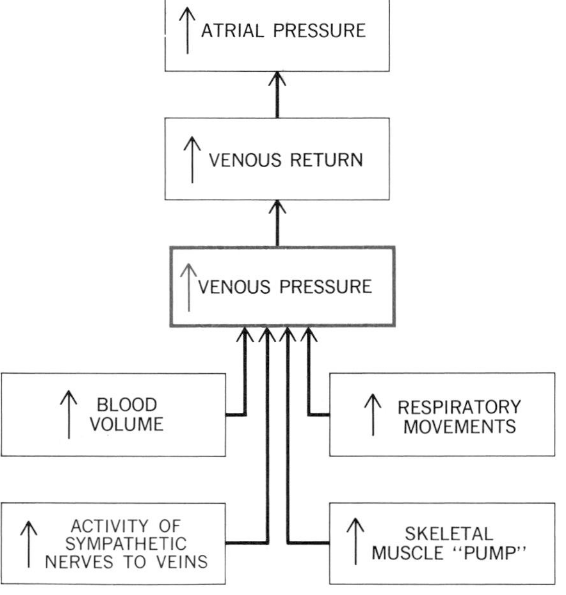

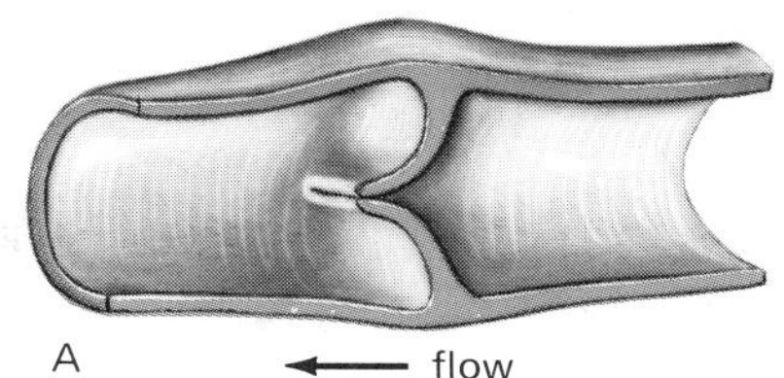

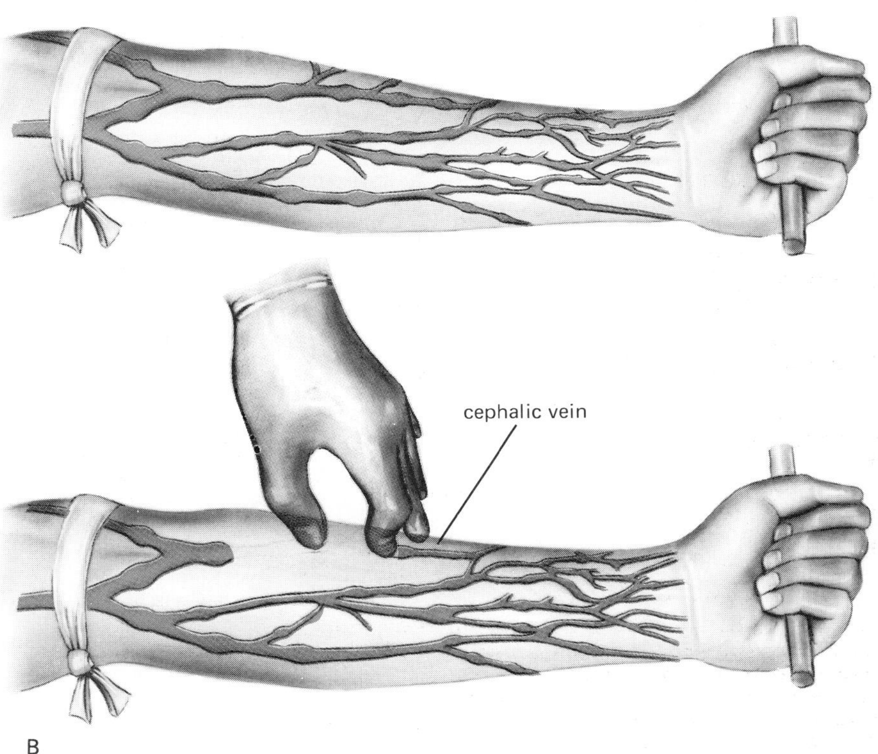

FIGURE 15-29 (A) ***Normal venous valve. Any tendency toward retrograde flow would immediately push the valve leaflets together.*** (B) ***Demonstration of the location of venous valves.***

the heart (Fig. 15-29). Why are these valves necessary if the pressure gradient created by cardiac contraction always moves blood toward the heart? We have seen how two other forces, the muscle pump and inspiratory movements, facilitate flow of venous blood. When these forces squeeze the veins, blood would be forced in both directions if the valves were not there to prevent backward flow. As we shall see, valves also play a critical role in counteracting the effects of upright posture.

LYMPHATICS

The lymphatics are not part of the circulatory system per se but constitute a one-way route from the tissue spaces to blood. The lymphatic system in human beings constitutes an extensive network of thin-walled vessels resembling the veins. It arises as a group of blind-end lymph capillaries which are present in almost all organs of the body. These capillaries are apparently permeable to virtually all interstitial-fluid constituents (including

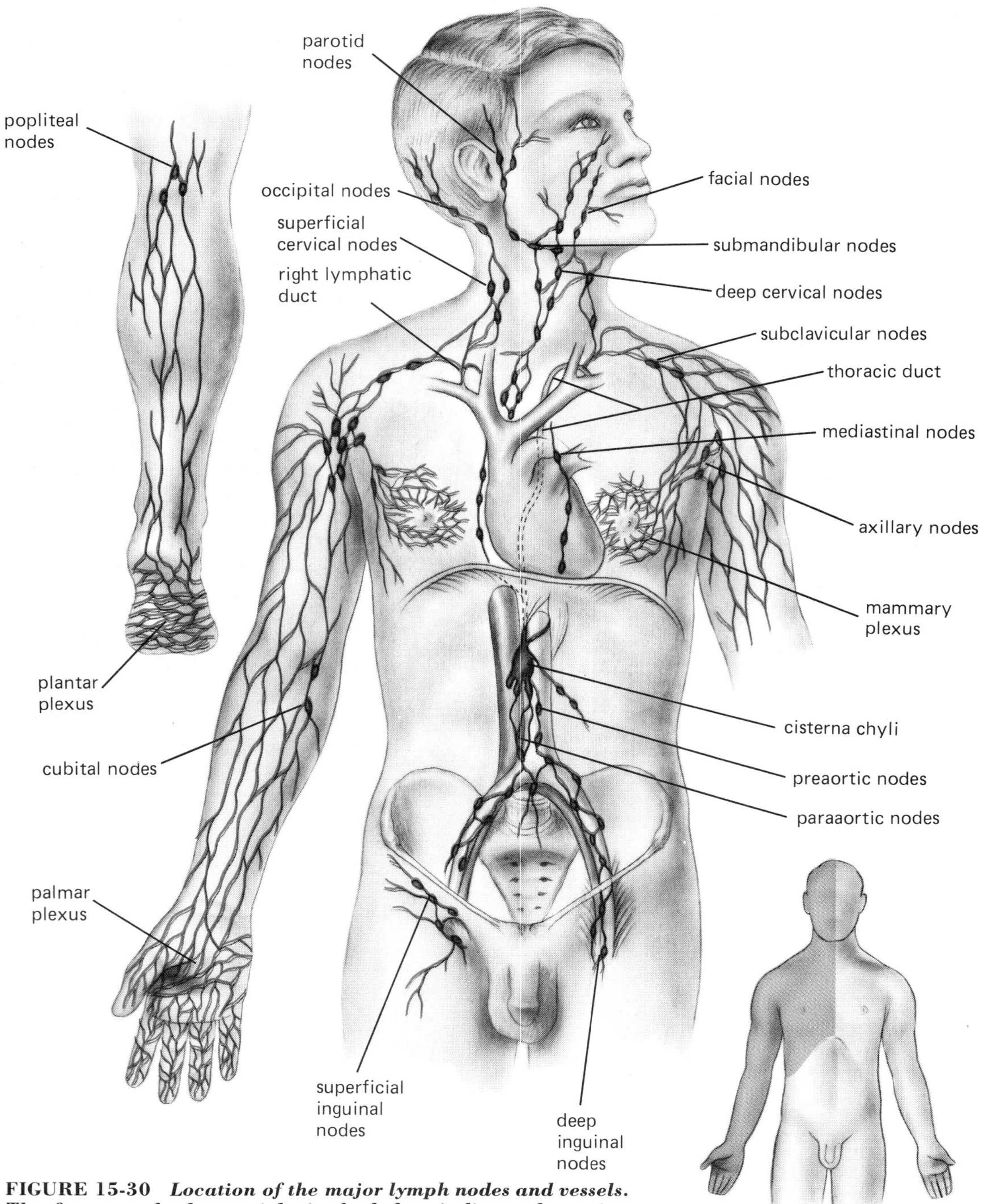

FIGURE 15-30 ***Location of the major lymph nodes and vessels. The figure at the lower right is shaded to indicate the region drained by the right lymphatic duct.***

protein), which either diffuse or filter into them. The pale lymph then drains into larger vessels called *lymphatics*. At some point before returning to the venous bloodstream, the lymph filters through chains of *lymph nodes* (Chap. 22). Eventually the lymph is collected into the major lymphatic ducts, the *thoracic duct* and the *right lymphatic duct,* which empty their lymph into the left and right brachiocephalic veins, respectively (Fig. 15-30). Thus, all fluid drained from the tissue spaces is eventually returned to the venous circulation.

Functions of the Lymphatic System

Return of Excess Filtered Fluid In a normal person, the fluid filtered out of the capillaries each day slight exceeds that reabsorbed. This excess is returned to the blood via lymphatics. Partly for this reason, lymphatic malfunction leads to increased interstitial fluid, i.e., edema.

Return of Protein to the Blood Most capillaries in the body have a slight permeability to protein, and, accordingly, there is a small steady loss of protein from the blood into the interstitial fluid. The protein returns via the lymphatics. The breakdown of this cycle is without question the most important cause of the marked edema seen in patients with lymphatic malfunction. Because protein (in small amounts) is normally lost from the capillaries, failure of the lymphatics to remove it allows the interstitial protein concentration to increase to that of the plasma. This failure reduces or eliminates the protein-concentration difference and thus the water-concentration difference across the capillary wall and permits the net movement of quantities of fluid out of the capillary into the interstitial space.

Specific Transport Functions In addition to these nonspecific transport functions, the lymphatics also provide the pathway by which certain specific substances reach the blood. The most important is fat absorbed from the gastrointestinal tract via lymphatic capillaries called *lacteals*. It is likely that certain high-molecular-weight hormones reach the blood via the lymphatics.

Lymph Nodes Besides its transport functions, the lymphatic system plays a critical role in the body's defenses against disease. This function, which is mediated by the lymph nodes located along the larger lymphatic vessels, will be described in Chap. 22.

Mechanism of Lymph Flow

How does the lymph move with no heart to push it? The best explanation at present is that lymph flow depends primarily upon forces external to the vessels, e.g., the pumping action of the muscles through which the lymphatics flow and the effects of respiration on thoracic-cage pressures. Since the lymphatics have valves similar to those in veins, external pressures would permit only unidirectional flow.

CHAPTER 16

The Circulatory System: IV. Integration of Cardiovascular Function in Health and Disease

In Chap. 6 we described the fundamental ingredients of all reflex control systems: (1) the internal-environmental variable being regulated, i.e., maintained relatively constant, and the receptors sensitive to it; (2) afferent pathways passing information from the receptors to (3) a control center, which integrates the different afferent inputs; (4) efferent pathways controlling activity of (5) effector organs, whose output raises or lowers the level of the regulated variable. The control and integration of cardiovascular function will be described in these terms. *The major variable being regulated is the systemic arterial blood pressure.* The central role of arterial pressure and the adaptive value of keeping it relatively constant should be apparent from the ensuing discussion.

ARTERIAL PRESSURE, CARDIAC OUTPUT, AND ARTERIOLAR RESISTANCE

Adequate blood flow through the vital organs (brain and heart) must be completely maintained at all times; the brain, for example, suffers irreversible damage within 3 min of ischemia. In contrast, many areas of the body, e.g., the gastrointestinal tract, the kidney, skeletal muscle, and skin, can withstand moderate reductions of blood flow for longer periods of time or even severe reductions if only for a few minutes. The mean arterial blood pressure is the driving force for blood flow through all the organs. The distribution of flow, i.e., the actual flow through the various organs at any given arterial pressure, depends primarily upon the radii of the arterioles in each vascular bed. A critical relationship has not been emphasized before, although it is implicit in the basic pressure-flow equation; these two factors, arterial pressure and arteriolar resistance, are *not* independent variables; *arteriolar resistance is one of the major determinants of arterial pressure.* This can be illustrated by the simple mechanical model shown in Fig. 16-1. A pump pushes fluid into a cylinder at the rate of 1 l/min; at steady state, fluid leaves the cylinder via the outflow tubes at 1 l/min, and the height of the fluid column, which is the driving pressure for outflow, remains stable. Assuming that the radii of the adjustable outflow tubes are all equal so that the flows through them are equal, we disturb the steady state by loosening the cuff on the outflow tube 1, thereby increasing its radius, reducing its resistance, and increasing its flow. The total outflow for the system is now greater than 1 l/min, more fluid leaves the reservoir than enters via the pump, and the height of the fluid column begins to decrease. In other words, a change in outflow resistance must produce changes in the pressure of the res-

FIGURE 16-1 ***Model to illustrate the dependency of arterial blood pressure upon arteriolar resistance, showing the effects of dilating one arteriolar bed upon arterial pressure and organ blood flow if no compensatory adjustments occur. The middle panel is a transient state before the new equilibrium occurs. In one respect, the illustration of the model is misleading in that the arterial reservoir is shown containing very large quantities of blood. In fact, as we have seen, the volume of blood in the arteries is quite small.***

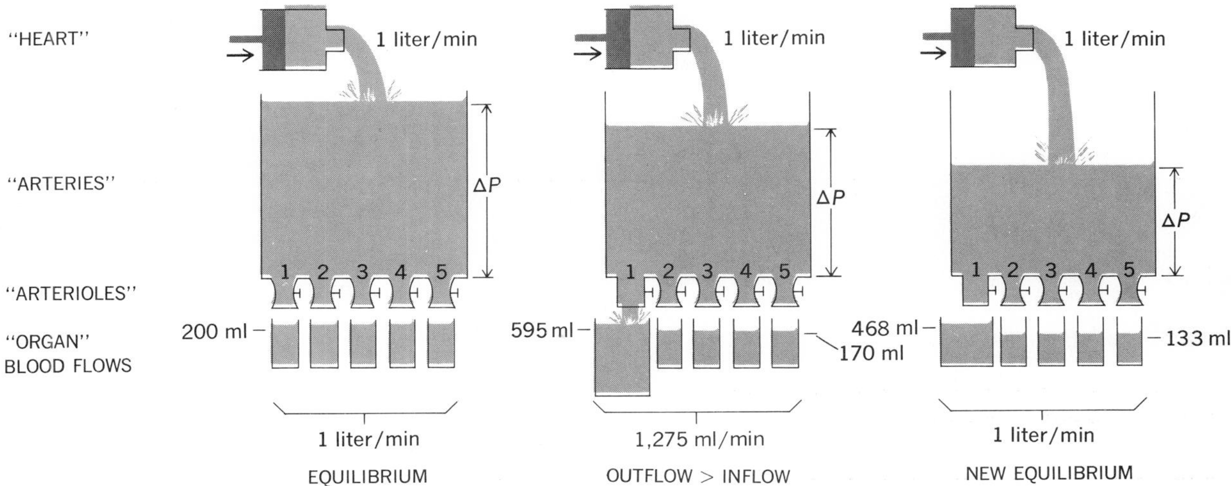

ervoir (unless some compensatory mechanism is brought into play). As the pressure falls, the rate of outflow via all tubes decreases. Ultimately, in our example, a new steady state is reached when the reservoir pressure is low enough to cause only 1 l/min outflow despite the decreased resistance of tube 1.

This analysis can be applied to the cardiovascular system by equating the pump with the heart, the reservoir with the arteries, and the outflow tubes with various arteriolar beds. An analogy to opening outflow tube 1 is exercise; during exercise, the skeletal muscle arterioles dilate, primarily because of active hyperemia, thereby decreasing resistance. If the cardiac output and the arteriolar diameters of all other vascular beds remain unchanged, the increased runoff through the skeletal muscle arterioles causes a decrease in arterial pressure. This, in turn, decreases flow through all other organs of the body, including the brain and heart. Indeed, even the exercising muscles themselves suffer a lessening flow (below that seen immediately after they dilated) as arterial pressure falls. Thus, the only way to guarantee the essential flow to the vital organs and the additional flow to the exercising muscle is to prevent the arterial pressure from falling.

This can be accomplished by changing the radii of the other arteriolar vascular beds or cardiac output or both. Figure 16-2 demonstrates the first major possibility, simultaneously tightening one or more of outflow tubes 2 to 5. This partially compensates for the decreased resistance of tube 1, and the total outflow resistance of all tubes can be shifted back toward normal. Therefore, the total outflow remains near 1 l/min; of course, the distribution of flow is such that flow in tube 1 is increased and all the others are decreased. If, for some reason, tube 5 is declared a vital pathway the flow through which should never be altered, the adjustments can always be made in tubes 1 to 4.

Applied to the body, this process is obviously analogous to control of total vascular resistance. When the skeletal muscle arterioles dilate during exercise, the total resistance of all vascular beds can still be maintained if arterioles constrict in other organs, such as the kidneys and gastrointestinal tract, which can readily suffer moderate flow reductions for at least short periods of time. In contrast, the brain and heart arterioles remain unchanged, thereby assuring constant brain blood supply. (The importance of the absence of significant activation of sympathetic vasoconstrictor fibers to the arterioles of brain and heart should now be apparent.)

This type of resistance juggling, however, can compensate only within limits. Obviously if tube 1 opens very wide, even total closure of the other tubes cannot compensate completely. Moreover, if the closure is prolonged, absence of flow will cause severe tissue damage. There must therefore be a second compensatory mechanism: increasing the inflow by increasing the activity of the pump (Fig. 16-3). When tube 1 widens and total outflow increases, the reservoir column can be completely maintained by simultaneously increasing the inflow from the pump. Thus, at the new

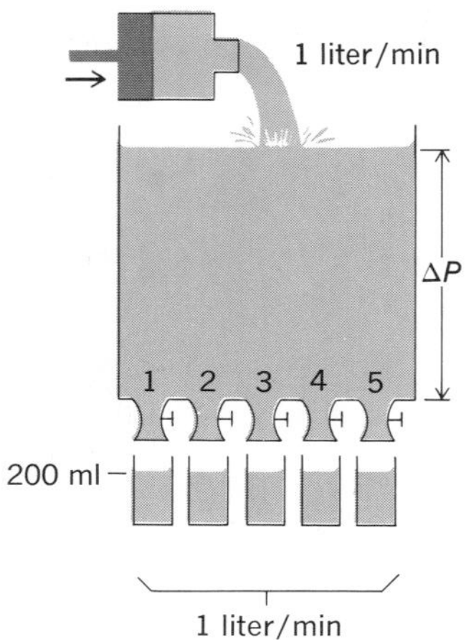

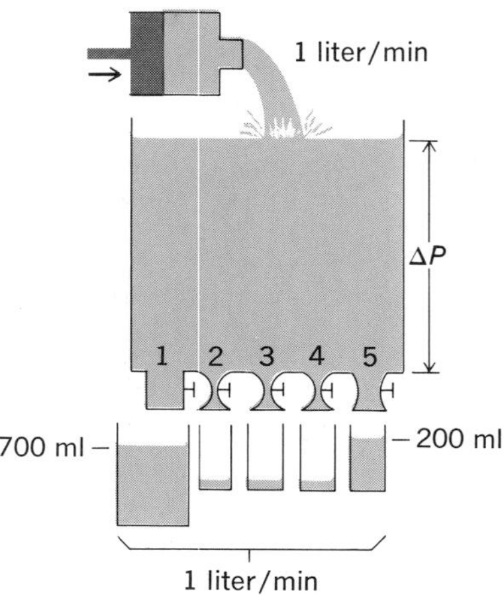

FIGURE 16-2 ***Compensation for dilation in one bed by constriction in others. When outflow tube 1 is opened, outflow tubes 2 to 4 are simultaneously tightened so that the total outflow resistance remains constant, total rate of runoff remains constant, and reservoir pressure remains constant.***

equilibrium, the total outflow and inflow are still equal, the reservoir pressure is unchanged, outflow through tubes 2 to 5 is unaltered, and the entire increase in outflow occurs through tube 1. Applied to the body, it should be evident that, when the blood vessels dilate, arterial pressure can be maintained constant by stimulating the heart to increase cardiac output. *Thus, the regulation of arterial pressure not only assures blood supply to the vital organs but provides a means for coordinating cardiac output with total tissue requirements.* In summary, the regulation of arterial blood pressure is accomplished both by control of cardiac output and arteriolar resistance.

It should now be possible to formalize these qualitative relationships for the entire cardiovascular system, using our basic pressure-flow equation. Flow through any tube is directly proportional to the pressure gradient between the ends of the tube and inversely proportional to the resistance.

$$\text{Flow} = \frac{\Delta P}{R}$$

Rearranging terms algebraically

$$\Delta P = \text{flow} \times R$$

This is simply another way of looking at the same equation, a way which clearly shows the dependence of pressure upon flow and resistance, which we have just described, using our models. Because the vascular tree is a continuous closed series of tubes, this equation holds for the entire system, i.e., from the very first portion of the aorta to the last portion of the vena cava just at the entrance to the heart. Therefore

Flow = cardiac output
ΔP = mean aortic pressure − late vena cava pressure
R = total resistance

where total resistance means the sum of the resistances of all the vessels in the systemic vascular tree. This usually is termed *total peripheral resistance.*

Since the late vena cava pressure is very close to 0 mm Hg, the formula

$$\Delta P = \text{mean aortic pressure} - \text{late vena cava pressure}$$

becomes

$$\Delta P = \text{mean aortic pressure} - 0$$

or

$$\Delta P = \text{mean aortic pressure}$$

Moreover, since the mean pressure is essentially the same in the aorta and all large arteries, the pressure term in the equation becomes

$$\Delta P = \text{mean arterial pressure}$$

The pressure-flow equation for the entire vascular tree now becomes

$$\text{Mean arterial pressure} = \text{cardiac output} \times \text{total peripheral resistance}$$

Recall that the arteries and veins are so large that they contribute very little to the total peripheral resistance.

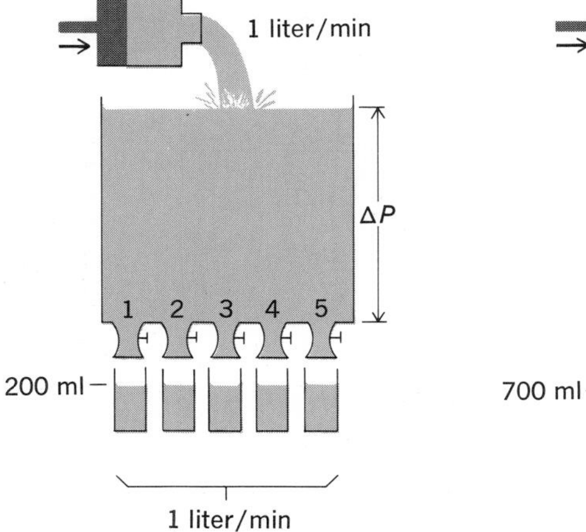

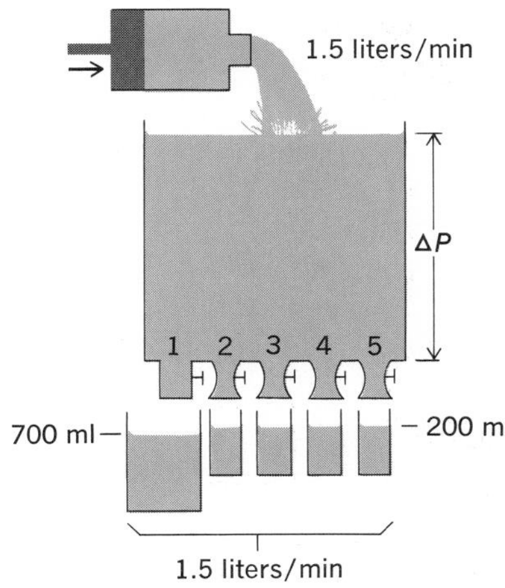

FIGURE 16-3 ***Compensation for dilation by increasing pump output. When outflow tube 1 is dilated, the total resistance decreases and total rate of runoff increases. The pump output is simultaneously increased by precisely the same amount, so that reservoir pressure remains constant.***

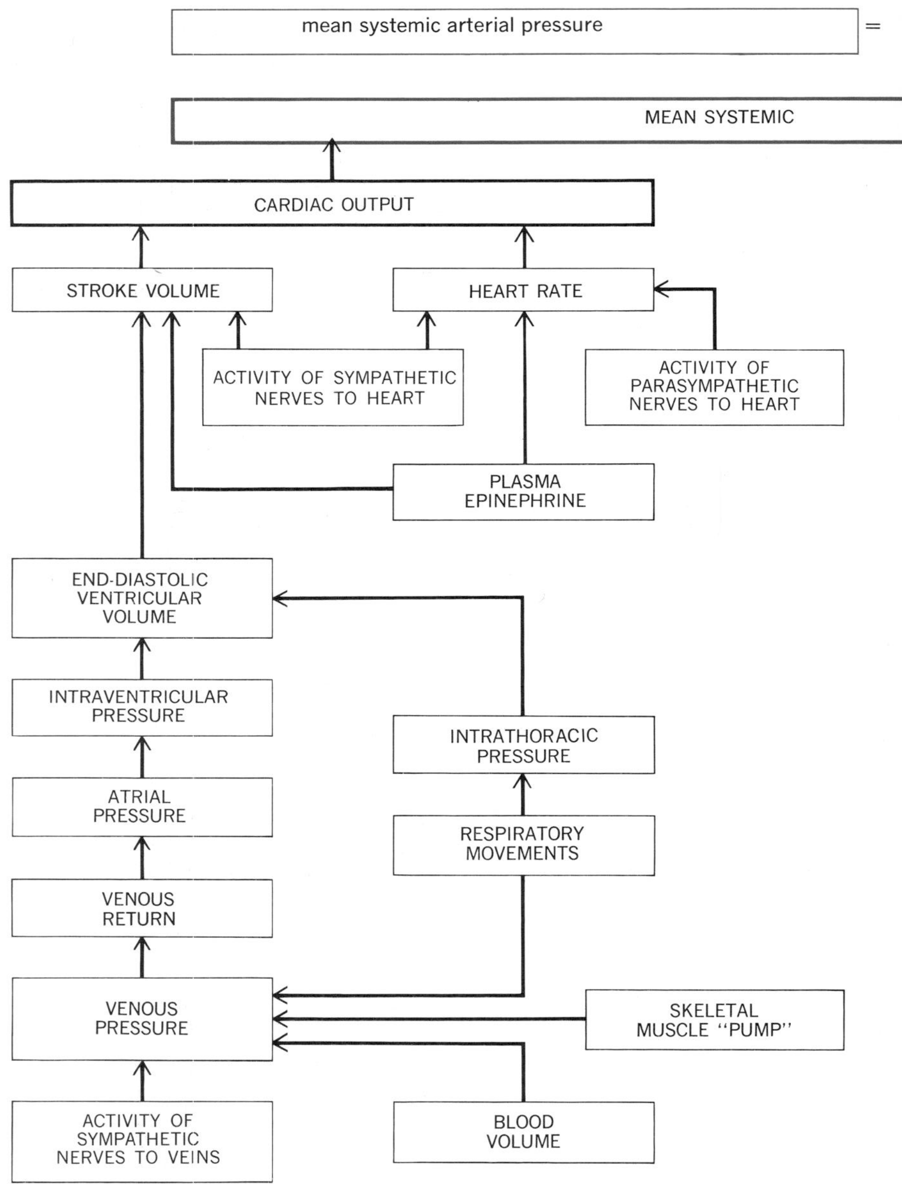
mean systemic arterial pressure
=
MEAN SYSTEMIC
CARDIAC OUTPUT
STROKE VOLUME
HEART RATE
ACTIVITY OF SYMPATHETIC NERVES TO HEART
ACTIVITY OF PARASYMPATHETIC NERVES TO HEART
PLASMA EPINEPHRINE
END-DIASTOLIC VENTRICULAR VOLUME
INTRAVENTRICULAR PRESSURE
INTRATHORACIC PRESSURE
ATRIAL PRESSURE
RESPIRATORY MOVEMENTS
VENOUS RETURN
VENOUS PRESSURE
SKELETAL MUSCLE "PUMP"
ACTIVITY OF SYMPATHETIC NERVES TO VEINS
BLOOD VOLUME

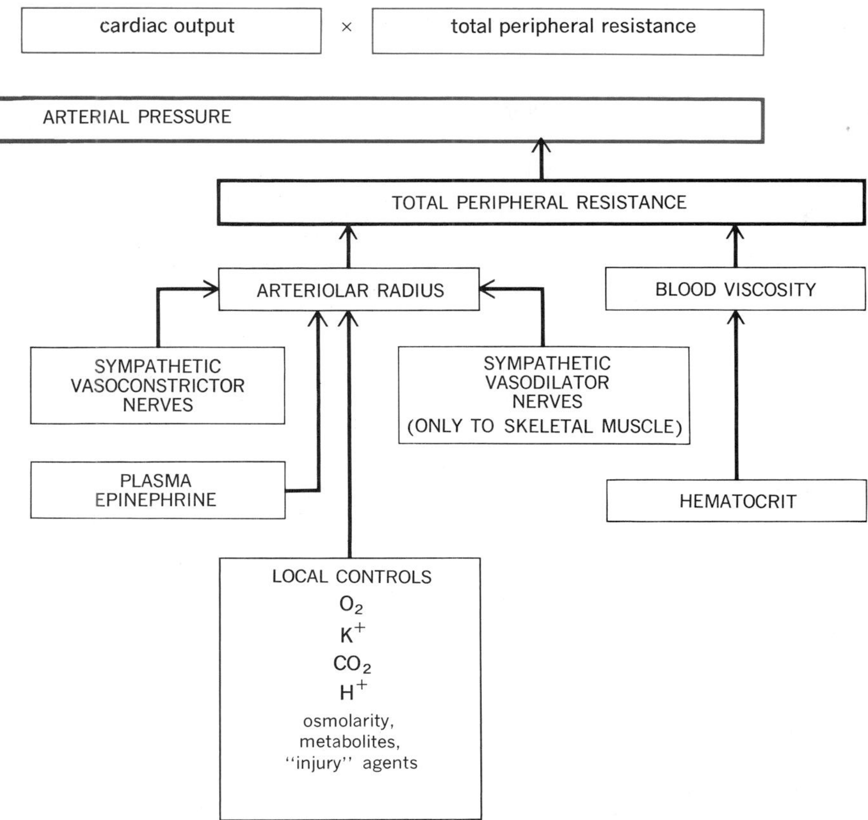

FIGURE 16-4 *Summary of effector mechanisms and efferent pathways which regulate systemic arterial pressure, an amalgamation of Figs. 14-19, 14-23, 15-19, and 15-28, with the addition of the effect of hematocrit on resistance.*

The major sites of resistance are the arterioles. The capillaries also offer significant resistance, but they have no muscle and their diameter reflects primarily the diameter of the arterioles supplying them. For these reasons, it is convenient to consider changes in arteriolar radius as virtually the only determinant of variations in total peripheral resistance.[1] Thus, the equation formally and quantitatively states the basic relationships described earlier, namely, that *arterial blood pressure can be increased, either by increasing cardiac output or total peripheral resistance.*

This equation is the fundamental equation of cardiovascular physiology. Given any two of the variables, the third can be calculated. For example, we can now explain why pulmonary arterial pressure is much lower than the systemic arterial pressure (see Fig. 14-16). The blood flow per minute, i.e., cardiac output through the pulmonary and systemic arteries, is, of course, the same; therefore, the pressures can differ only if the resistances differ. Thus, the pulmonary arterioles must be wider and offer much less resistance to flow than the systemic arterioles. In other words, the total pulmonary vascular resistance is lower than the total systemic peripheral resistance. This permits the pulmonary circulation to function as a low-pressure system.

Figure 16-4 (page 445) presents the grand scheme of effector mechanisms and efferent pathways which regulates systemic arterial pressure. None of this information is new, all of it having been presented in previous figures. The reader can now appreciate how the function of the heart and various vascular segments are coordinated to achieve this. A change in any single variable shown in the figure will, all others remaining constant, produce a change in mean arterial pressure by altering either cardiac output or total peripheral resistance. Conversely, any such deviation in mean arterial pressure can be eliminated by the reflex alteration of some other variable. It should be evident from the figure that the reflex control of cardiac output and peripheral resistance involves primarily (1) sympathetic nerves to heart, arterioles, and veins, (2) parasympathetic nerves to the heart, and (3) release of epinephrine from the adrenal medulla. In one sense, we have approached arterial pressure regulation backward, in that the past chapters have described the effector sites (heart, arterioles, veins) and motor pathways (autonomic nervous system). Now we must complete the reflexes by describing the monitoring systems (receptors and afferent pathways to the brain) and the control centers in the brain.

CARDIOVASCULAR CONTROL CENTERS IN THE BRAIN

The primary cardiovascular control center is in the medulla, the first segment of brain above the spinal cord. The axons of the neurons which comprise this center make synaptic connections with the autonomic neurons and via these connections exert dominant influence over them. The medullary cardiovascular center is absolutely essential for blood-pressure regulation. The relevant medullary neurons are sometimes divided into cardiac and vasomotor centers, which are then further subdivided and classified, but because these areas actually constitute diffuse networks of highly interconnected neurons, we prefer to call the entire area the *medullary cardiovascular center.*

An important aspect of its function is that of reciprocal innervation. The synaptic distribution of the medullary axons and the input to their cell bodies are such that when the parasympathetic nerves to the heart are stimulated, the sympathetic nerves to the heart, as well as to the arterioles and the veins, are usually simultaneously inhibited (Fig. 16-5). Conversely, parasympathetic inhibition and sympathetic stimulation are usually elicited simultaneously.[2] This pattern is important because there is always some continuous discharge of the autonomic nerves. Therefore, the heart can be slowed by two simultaneous events: inhibition of the sympathetic activity to the SA node and enhancement of the parasympathetic activity to the SA node. The converse is also true for accelerating the heart. In contrast, only sympathetic fibers significantly innervate the ventricular muscle itself and the arteriolar and venous smooth muscle. However, the muscle activity can still be decreased below normal by inhibiting the basal sympathetic activity.

Other areas of the brain, particularly in the hypothalamus, have an important influence on blood pressure, but there is good reason to believe that most of them exert their effects via the medullary centers; i.e., nerve impulses from them descend to the medulla and

[1] As described in Chap. 15, changes in blood viscosity can also contribute to change in flow resistance.

[2] These generalizations are oversimplifications. There is a considerable degree of separateness in the control of sympathetic and parasympathetic discharge, depending upon the precise circumstances eliciting the reflexes.

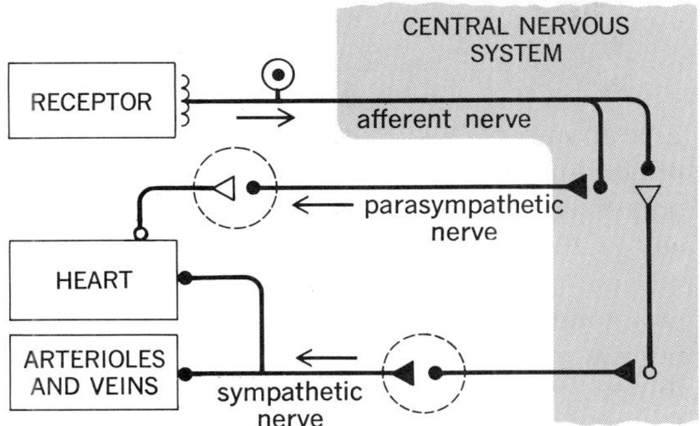

FIGURE 16-5 ***Reciprocal innervation in the control of the cardiovascular system. Afferent input, which stimulates the parasympathetic nerves to the heart, simultaneously inhibits the sympathetic nerves to the heart, arterioles, and veins.***

through synaptic connections alter the discharge of the primary medullary neurons. It is through these pathways that factors such as pain, anger, body temperature, and many others can alter blood pressure. There is one major exception: The sympathetic vasodilator fibers to skeletal muscle arterioles are apparently not controlled by the medullary centers but are under the direct influence of neuronal pathways originating in the cerebral cortex and hypothalamus. This pathway and the sympathetic vasodilators are activated only during exercise and stress and play no role in any of the many other cardiovascular responses.

RECEPTORS AND AFFERENT PATHWAYS

We have now to discuss the last link in arterial pressure regulation, namely, the receptors and afferent pathways bringing information into the medullary centers. The most important of these are the *arterial baroreceptors*.

Arterial Baroreceptors

It is only logical that the reflexes which homeostatically regulate arterial pressure originate primarily with receptors within the arteries which are pressure-sensitive. High in the neck each of the carotid arteries supplying the brain divides into the external and internal carotid arteries. At this bifurcation, the wall of the artery is thinner than usual and contains a large number of branching, vinelike nerve endings (Fig. 16-6). This small portion of the artery is called the *carotid sinus*. There the nerve endings are apparently highly sensitive to stretch or distortion; since the degree of wall stretching is directly related to the pressure within the artery, the carotid sinus actually serves as a pressure receptor (*baroreceptor*). The nerve endings come together to form afferent neurons which travel with the glossopharyngeal (IXth cranial) nerve to the medulla, where they eventually synapse upon the neurons of the cardiovascular center. An area functionally similar to the carotid sinuses found in the *arch of the aorta* constitutes a second important arterial baroreceptor. The afferent fibers from these receptors travel with the vagus (Xth cranial) nerve.

Action potentials recorded in single afferent fibers from the carotid sinus (Fig. 16-6) demonstrate the pattern of response by these receptors. In Fig. 16-6 the arterial pressure within the carotid sinus was artificially controlled. At a steady nonpulsatile pressure of 100 mm Hg, there is a tonic rate of discharge by the nerve. This rate of firing can be decreased or increased by lowering or raising the arterial pressure, respectively. Note that the fiber shows no fatigue or adaptation. The arterial baroreceptors are responsive not only to the mean arterial pressure but to the pulse pressure as well, an increased pulse pressure producing an increased firing. This responsiveness adds a further degree of sensitivity to blood-pressure regulation since small changes in certain important factors (such as blood volume) cause changes in pulse pressure before they become serious enough to affect mean pressure.

Our description of the major blood-pressure-regulating reflex is now complete; an increase in arterial pressure increases the rate of discharge of the carotid sinus and aortic arch baroreceptors; these impulses travel up the afferent nerves to the medulla and, via appropriate synaptic connections with the neurons of the medullary cardiovascular centers, induce (1) slowing of the heart because of decreased sympathetic discharge and increased parasympathetic discharge, (2) decreased myocardial contractility because of decreased sympathetic activity, (3) arteriolar dilatation because of decreased sympathetic discharge to arteriolar smooth muscle, and (4) venous dilatation because of decreased sympathetic discharge to smooth muscle. The net result is a decreased cardiac output (decreased heart rate and stroke volume), decreased peripheral resistance, and return of blood pressure toward normal.

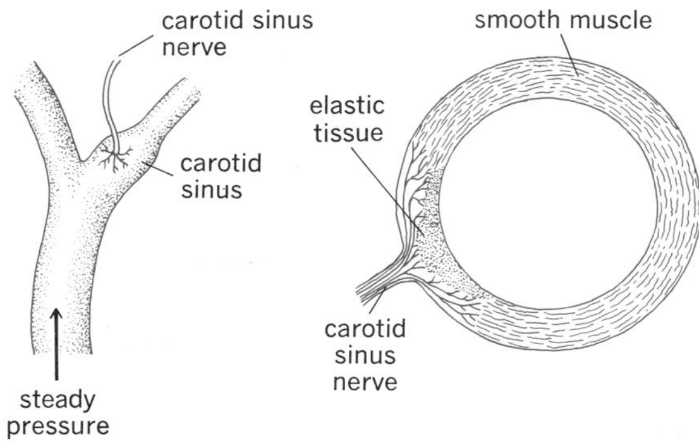

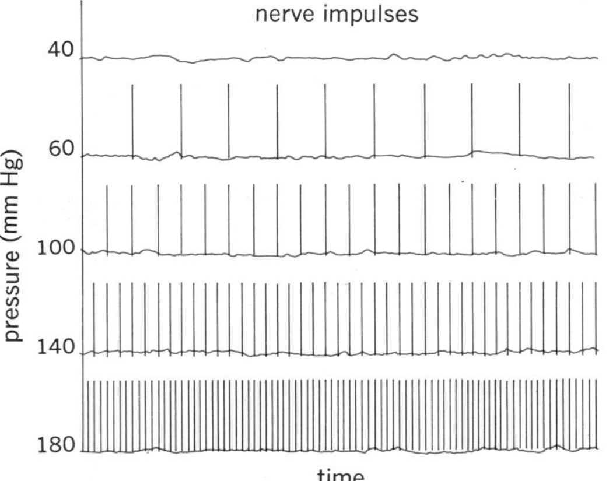

FIGURE 16-6 ***Action potentials recorded from the carotid sinus nerve during nonpulsatile, i.e., steady, perfusion of an isolated carotid artery. The baroreceptors discharge at greater and greater frequencies as the pressure is increased.*** **(Adapted from Rushmer.)**

Other Baroreceptors

Other portions of the vascular tree contain nerve endings sensitive to stretch, namely, other large arteries, the large veins, and the cardiac walls themselves. They seem to function like the carotid sinus and aortic arch in that the few from which electric activity has been recorded show increased rates of discharge with increasing pressure. By means of these receptors, the medulla is kept constantly informed about the venous and atrial pressure, and a further degree of sensitivity is gained. Thus, a slight decrease in atrial pressure begins to facilitate the sympathetic nervous system even before the change becomes sufficient to lower cardiac output and arterial pressure far enough to be detected by the arterial baroreceptors. As we shall see in Chap. 18, the atrial baroreceptors are particularly important for the control of body sodium and water.

Chemoreceptors

The aortic and carotid arteries contain specialized structures sensitive primarily to the concentrations of arterial oxygen but also to those of carbon dioxide and hydrogen ion. Since these receptors are far more important for the control of respiration, they are described in Chap. 17, but they also send information to the medullary cardiovascular centers, the result being that blood pressure tends to be reflexly increased by decreased arterial oxygen. Changes in carbon dioxide and hydrogen-ion concentrations also alter blood pressure reflexly, but the effects are small and the pathways quite complex.

Summary

The medullary cardiovascular control centers are true integrating centers, receiving a wide variety of information from baroreceptors, chemoreceptors, peripheral sensory receptors of all kinds (pain, cold, etc.), and many higher brain centers, particularly the hypothalamus, Therefore, it is not surprising that at every moment arterial pressure reflects the resultant response to all these inputs. Sudden anger increases the pressure; fright may actually cause hypotension (low blood pressure) severe enough to cause fainting, but this complexity should not obscure the important generalization that the primary regulation of arterial pressure is exerted by the baroreceptors, particularly those in the carotid sinus and aortic arch. Other inputs may alter the pressure somewhat from minute to minute, but the mean arterial pressure in a normal person is maintained by the baroreceptors within quite narrow limits.

HEMORRHAGE AND HYPOTENSION

The decrease in blood volume caused by bleeding produces a drop in blood pressure (*hypotension*) by the sequence of events previously shown in Fig. 15-27. The most serious consequences of the lowered blood pressure are the reduced blood flow to the brain and cardiac muscle. Compensatory mechanisms restoring arterial pressure toward normal are summarized in Fig. 16-7; their effects can best be appreciated from the data of Table 16-1. Kidney flow is even lower 5 min after the hemorrhage, despite the improved arterial pressure, but

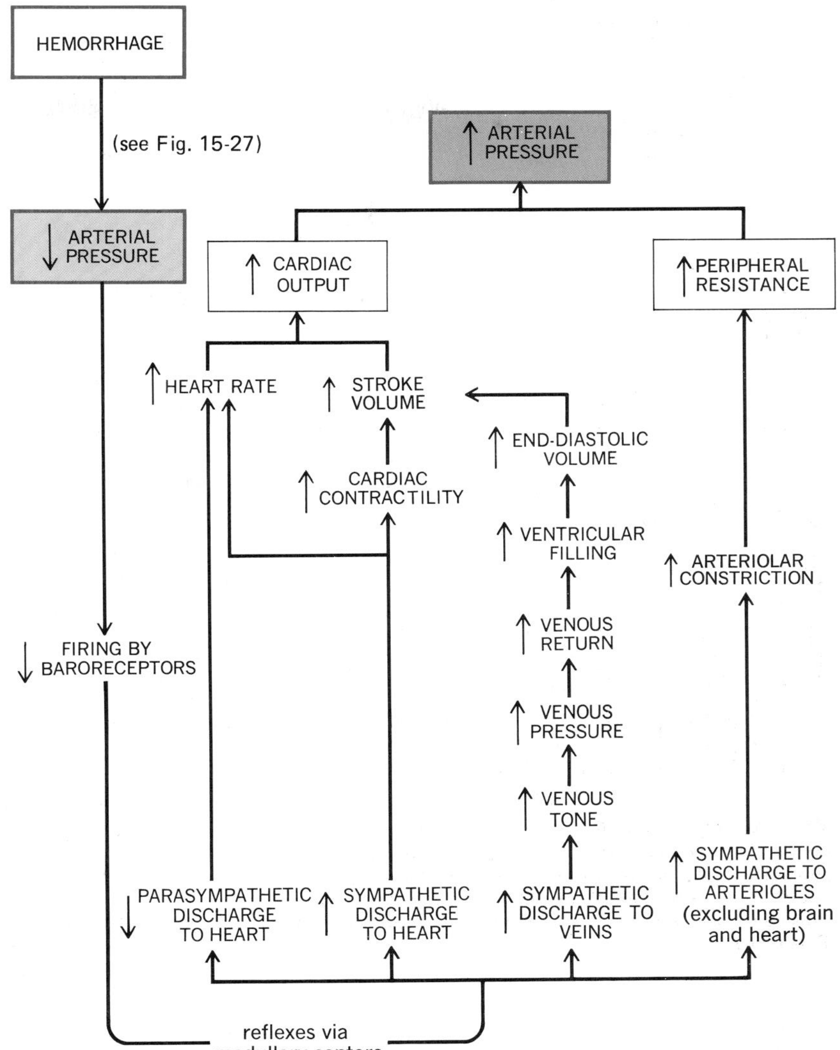

FIGURE 16-7 ***Reflex mechanisms by which lower arterial pressure following blood loss is brought back toward normal. The compensatory mechanisms do not restore arterial pressure completely to normal.***

we recall that one of the important compensatory mechanisms is increased arteriolar constriction in many organs; thus, kidney blood flow is reduced in order to maintain arterial blood pressure and thereby brain and heart blood flow.

A second important compensatory mechanism involving capillary fluid exchange results from both the decrease in blood pressure and the increase in arteriolar constriction (Fig. 16-8). Thus, the initial event—blood loss and decreased blood volume—is in large part compensated for by the movement of interstitial fluid into the vascular system. Indeed, as shown in Table 16-2, several hours after a moderate hemorrhage, the blood volume may be virtually restored to normal.

TABLE 16-1
CARDIOVASCULAR EFFECTS OF HEMORRHAGE

	Prehemorrhage	*Posthemorrhage Immediate*	*Posthemorrhage 5 min*
Arterial pressure, mm Hg	125/75	80/55	115/75
Left atrial pressure, mm Hg	4	2	2.5
End-diastolic volume, ml	150	75	90
Stroke volume, ml	75	40	53
Heart rate, beats/min	70	70	91
Cardiac output, ml/min	5,250	2,800	4,775
Kidney blood flow, ml/min	1,300	1,000	850
Brain blood flow, ml/min	1,300	1,000	1,275

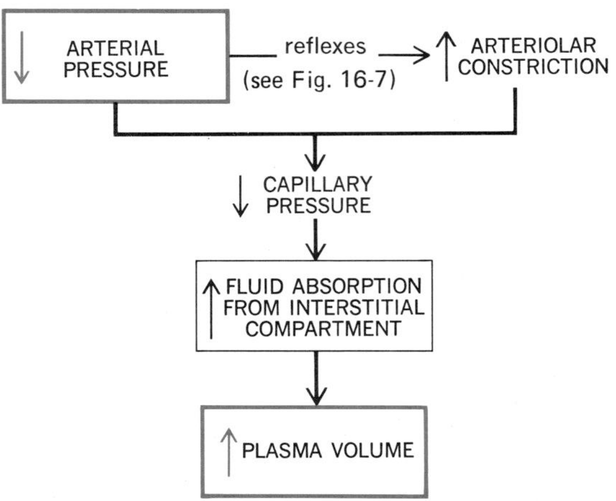

FIGURE 16-8 ***Mechanisms compensating for blood loss by movement of interstitial fluid into the capillaries. This response is diagramed in Fig. 15-24C.***

The entire compensation is due to expansion of the plasma volume; replacement of the lost erythrocytes requires days. Note that much of the albumin lost in the hemorrhage has already been replaced by synthesis of new protein (by the liver). This phenomenon is of great importance for expansion of plasma volume, as can be seen by considering the capillary filtration-absorption equilibrium (Fig. 15-24). As capillary hydrostatic pressure decreases as a result of the hemorrhage, interstitial fluid enters the plasma; this fluid, however, contains virtually no protein, so that its entrance dilutes the plasma proteins and increases the plasma water concentration. The resulting reduction of the water-concentration difference between the capillaries and the interstitial fluid hinders further fluid reabsorption and would prevent the full compensatory expansion if it were not that the rapid synthesis of new plasma protein minimizes this fall in protein concentration and movement of interstitial fluid into the plasma can continue. It is not known what stimulates the liver to synthesize new protein. We must emphasize that this capillary mechanism has only *redistributed* the extracellular fluid; ultimate replacement of the plasma lost from the body involves the control of fluid ingestion and kidney function, both described in Chap. 18.

TABLE 16-2
FLUID SHIFTS AFTER HEMORRHAGE

	Normal	*Immediately after hemorrhage*	*3 h after hemorrhage*
Total blood volume, ml	5,000	4,000 (↓20%)	4,900
Erythrocyte volume, ml	2,300	1,840 (↓20%)	1,840
Plasma volume, ml	2,700	2,160 (↓20%)	3,060
Plasma albumin mass, g	135	108 (↓20%)	125

The compensatory mechanisms just described are highly efficient; losses of as much as 1 to 1.5 l of blood (approximately 20 percent of total blood volume) can be sustained with only slight reduction of mean arterial pressure. When greater losses occur, severe hypotension (*shock*) may be precipitated. If prolonged for several hours, shock becomes irreversible; the person dies even after blood transfusions and other appropriate therapy. Irreversible shock is manifested when the compensatory mechanisms are overridden; thus, a person who has been in mild shock for several hours may suddenly show a marked worsening of the hypotension. Apparently, the major event is a decrease in cardiac contractility; the mechanism of deterioration leading to irreversible shock is unknown.

Loss from the body of large quantities of extracellular fluid (rather than whole blood) can also cause hypotension. It may occur via the skin, as in severe sweating or burns, the gastrointestinal tract, as in diarrhea or vomiting, or unusually large urinary losses. Regardless of the route, the loss decreases circulating blood volume and produces symptoms and compensatory phenomena similar to those seen in hemorrhage.

Hypotension may be caused by events other than blood or fluid loss. Fainting in response to strong emotion is a common form of hypotension. Somehow the higher brain centers involved with emotions act upon the medullary cardiovascular centers to inhibit sympathetic activity and enhance parasympathetic activity (Fig. 16-9), resulting in decreased arterial pressure and brain flow. Fortunately, this whole process is usually transient, with no aftereffects, although a weak heart may suffer damage during the period of reduced blood flow to the cardiac muscle.

Other important causes of hypotension seem to have a common denominator in the liberation within the body of chemicals which relax arteriolar smooth muscle. There the cause of hypotension is clearly excessive arteriolar dilatation and reduction of peripheral resistance, an important example being the hypotension which occurs during severe allergic responses.

It may be of interest to point out the physiologic reasons for not treating a patient in shock in ways commonly favored by the uninformed, namely, administering alcohol and covering the person with mounds of blankets. Both alcohol and excessive body heat, by actions on the central nervous system, cause profound dilatation of skin arterioles, thus lowering peripheral resistance and decreasing arterial blood pressure still further. As shown below, the worst possible thing is to try to get the person to stand up.

It should be noted that the simple act of getting out of bed and standing up is equivalent to a mild hemorrhage, because the changes in the circulatory system in going from a lying, horizontal position to a standing, vertical position result in a decrease in the effective circulating blood volume. Effective blood volume upon standing is reduced by two factors (Fig. 16-10), both resulting from the increase in gravitational blood pressure in the legs and feet and the distensibility of blood vessels, namely, increased distension of the veins and increased filtration of fluid from the capillaries into the interstitial space of the legs. The ensuing decrease in arterial pressure causes reflex compensatory adjustments similar to those shown in Fig. 16-7 for hemorrhage.

Perhaps the most effective compensation is the contraction of skeletal muscles of the leg, which compresses the veins, thereby diminishing the degree of venous distension and pooling, and causes a marked reduction in capillary hydrostatic pressure, thereby reducing the rate of fluid filtration out of the capillaries. Muscular contraction produces intermittent, complete emptying of veins within the upper leg, so that uninterrupted columns of venous blood from the heart to the feet no longer exist. Thus the skeletal muscle pump reduces both venous pooling and capillary filtration. A common example of the importance of this compensation is when soldiers faint after standing very still, i.e., with minimal contraction of the abdominal and leg muscles, for long periods of time. Here the fainting may be considered a useful compensatory mechanism in that venous and capillary pressure changes induced by gravity are eliminated once the person is prone, the pooled venous blood is mobilized, and the previously filtered fluid is reabsorbed into the capillaries. Thus, the wrong thing to do to anyone who has fainted is to hold him upright.

EXERCISE

In order to maintain muscle activity during exercise, a large increase in blood flow is required to provide the oxygen and nutrients consumed and to carry away the carbon dioxide and heat produced. Thus, cardiac output may increase from a resting value of 5 l/min to the maximal values of 35 l/min obtained by trained athletes. The increased skeletal muscle blood flow results from marked dilatation of the skeletal muscle arterioles mediated by the sympathetic vasodilator fibers (stimulated by descending pathways from the hypothalamus) and by local factors associated with active hyperemia. In a person just about to begin exercising, the skeletal muscle flow actually increases before the onset of muscular activity and therefore of active hyperemia. This anticipatory response providing a rapid initial supply of blood to the muscle can, in large part, be blocked by cutting the sympathetic nerves or by administering drugs which inhibit the actions of acetylcholine. It must be stressed, however, that once exercise has begun, these sympathetic vasodilator nerves are of little importance and active hyperemia plays the primary role in producing vasodilatation.

The cardiovascular response to exercise has already been shown in our series of models (Figs. 16-1 to 16-3). The decrease in peripheral resistance resulting

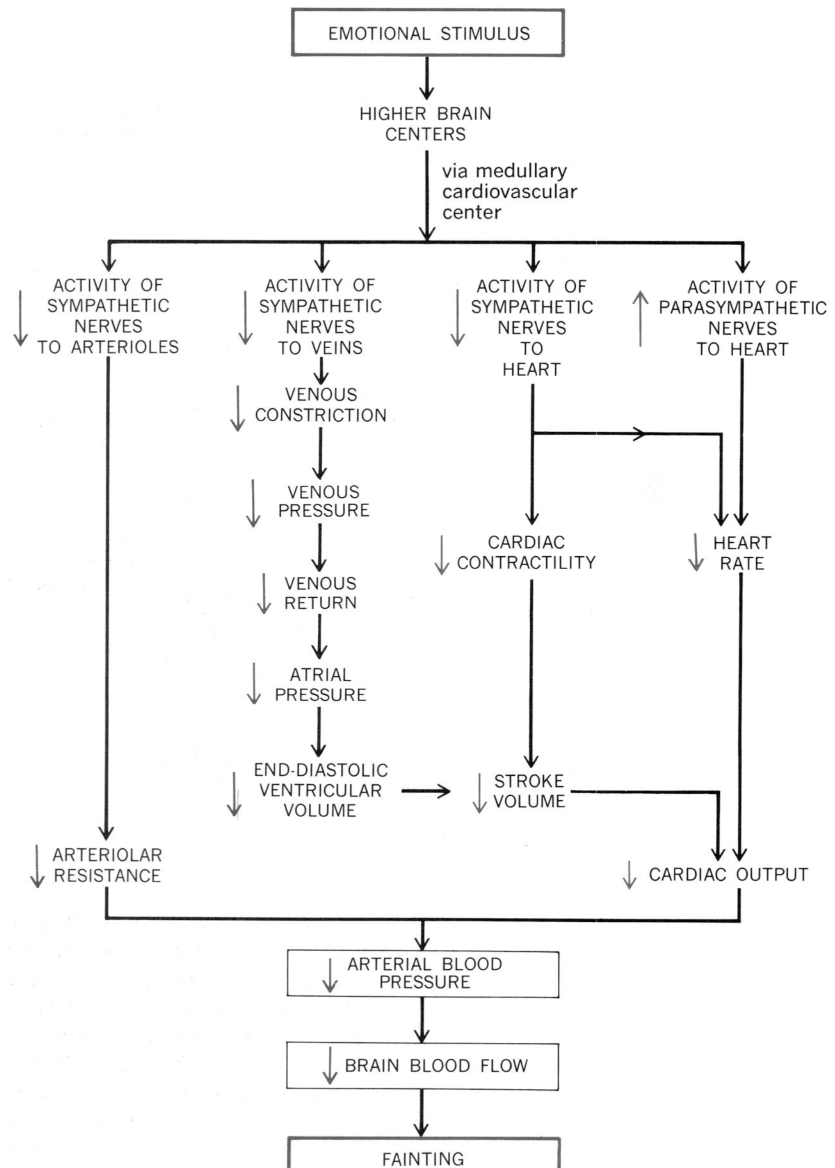

FIGURE 16-9 ***Mechanisms inducing fainting in response to a strong emotional stimulus.***

from dilatation of skeletal muscle arterioles is partially offset by constriction of arterioles in other organs, particularly the gastrointestinal tract and kidneys. However, the "resistance juggling" is quite incapable of compensating for the huge dilatation of the muscle arterioles, and the net result is a marked decrease in total peripheral resistance.

The cardiac output increase during exercise is associated with greater sympathetic activity and less parasympathetic activity to the heart. Thus, heart rate

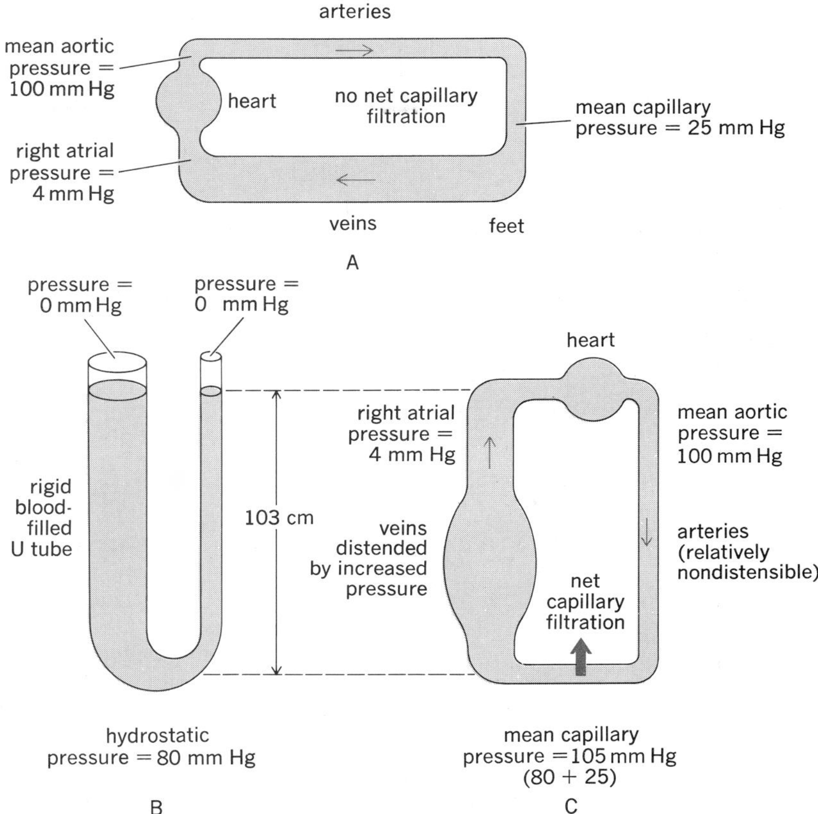

FIGURE 16-10 ***Effects of standing upon blood pressure, blood flow, and capillary filtration.***

and stroke volume both rise, causing an increased cardiac output. The heart-rate changes are usually much greater than stroke-volume changes. Note (Fig. 16-11) that, in our example, the increased stroke volume occurs without change in end-diastolic ventricular volume; accordingly, the former is ascribable completely to the increased contractility induced by the cardiac sympathetic nerves. The stability of end-diastolic volume in the face of reduced time for ventricular filling (increased heart rate) is, in part, attributable to the fact, described earlier, that the sympathetic nerves increase the speed of contraction and relaxation as well as inducing a "suctionlike" effect which facilitates filling.

However, it would be incorrect to leave the impression that enhanced sympathetic activity to the heart completely accounts for the elevated cardiac output which occurs in exercise, for such is not the case. The fact is that cardiac output could not be increased to high levels unless the venous return to the heart were not simultaneously facilitated to the same degree, for otherwise end-diastolic volume would fall and stroke volume would decrease (because of Starling's law). Therefore, factors promoting venous return during exercise are extremely important. They are (1) the marked activity of the skeletal muscle pump, (2) the increased "respiratory-pump" activity resulting from increased respiratory movements, (3) sympathetically mediated increase in venous tone, and (4) the ease with which blood flows from arteries to veins through the dilated skeletal muscle arterioles. These factors may be so powerful that venous return is enhanced enough to cause an increase in end-diastolic ventricular volume; under such conditions, stroke volume (and, thereby, cardiac output) is further enhanced.

What happens to arterial blood pressure during exercise? As always, the mean arterial pressure depends

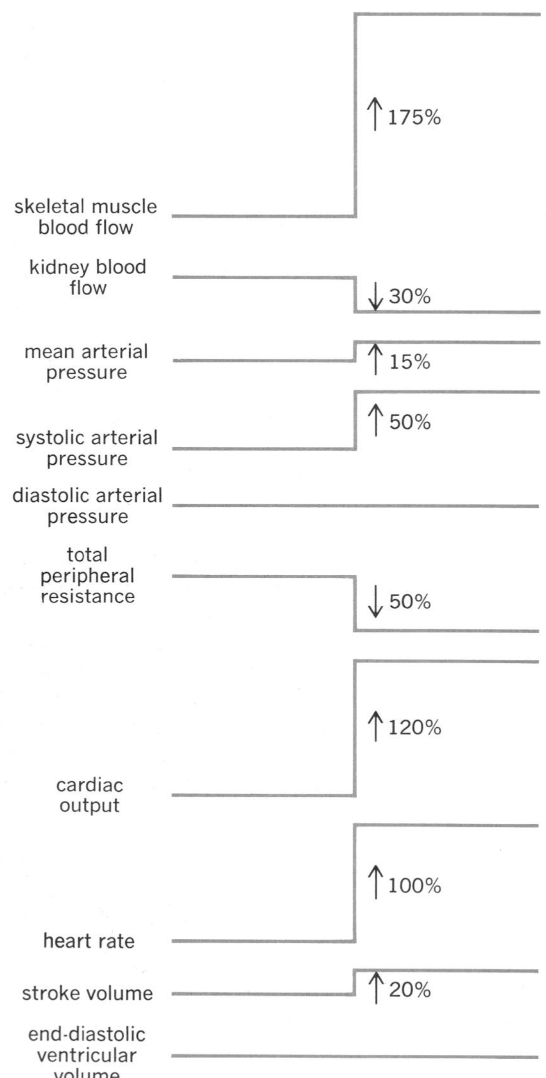

FIGURE 16-11 ***Summary of cardiovascular changes during mild exercise.***

only upon the cardiac output and peripheral resistance. During most forms of exercise, the cardiac output tends to increase somewhat more than the peripheral resistance decreases so that mean arterial pressure usually increases slightly. However, the pulse pressure may show a marked increase because of greater systolic pressure and a relatively constant diastolic pressure. The former is due primarily to the faster ejection. These changes are all shown schematically in Fig. 16-11.

It is evident from this description that the sympathetic nerves (and inhibition of the parasympathetics) play an important role in the cardiovascular response to exercise. However, a problem arises when we try to understand the mechanisms which control the autonomic nervous system during exercise. The pulsatile (and mean) arterial pressure tends to be elevated, which should cause the arterial baroreceptors to signal the medullary centers to decrease cardiac output and dilate arterioles of the abdominal organs. Obviously, then, the arterial baroreceptors not only cannot be the origin of the cardiovascular changes in exercise but actually oppose these changes. Similarly, other possible inputs, such as oxygen and carbon dioxide, can be eliminated since they show little, if any, change (Chap. 17). We shall meet this same problem again when we describe (Chap. 17) increased respiration during exercise. A major clue is the finding that electric stimulation of certain hypothalamic areas in resting unanesthetized dogs produces all the cardiovascular changes usually observed during exercise. On the basis of these and other experiments, the best present working hypothesis is that a center in the hypothalamus acts, via descending pathways, upon the medullary centers (and directly upon the sympathetic vasodilators to skeletal muscle arterioles) to produce the changes in autonomic function so characteristic of exercise. What is the input to the hypothalamus? We do not know for certain, but it is probably information coming from the same motor areas of the cerebral cortex which are responsible for the skeletal muscle contraction. It is likely that, as these fibers descend from cerebral cortex to the spinal-cord motor neurons, they give off branches to the relevant hypothalamic centers. This system would nicely coordinate the skeletal muscle contraction with the blood supply needed to support it.

HYPERTENSION

Hypertension (high blood pressure) is defined as a chronically increased arterial pressure. In general, the dividing line between normal pressure and hypertension is taken to be 140/90 mm Hg, although systolic pressures above 140 are frequently not associated with ill effects. However, the diastolic pressure is really the most important index of hypertension. Hypertension is one of the major causes of illness and death in America today. It is estimated that approximately 6 million people suffer from this disease, the lethal end result of which may be heart failure (see below), brain stroke (occlusion or rupture of a cerebral blood vessel), or kidney damage, all caused by prolonged hypertension and its attendant strain on the various organs.

Theoretically, hypertension could result from an increase in cardiac output or peripheral resistance or both. In fact, at least in well-established hypertension, the major abnormality is increased peripheral resistance due to abnormally reduced arteriolar diameter. What causes arteriolar narrowing? In most cases, we do not know. This has led to the strange label of "essential hypertension," meaning hypertension of unknown cause. In a small fraction of cases, the cause of hypertension is known: (1) Certain tumors of the adrenal medulla secrete excessive amounts of epinephrine; (2) certain tumors of the adrenal cortex secrete excessive amounts of hormones which lead to hypertension by as yet unknown mechanisms; (3) many diseases which damage or decrease the blood supply to the kidneys are associated with hypertension, but despite intensive efforts by many investigators, the actual factor(s) mediating *renal hypertension* remains unknown. In Chap. 18, we shall describe the renin-angiotensin system as the prime controller of aldosterone secretion; long before this effect was recognized, it was known that angiotensin is a powerful constrictor of arterioles, and most researchers in the field believed it to be the basis for renal hypertension. Now, the story seems much less certain; regardless of mechanism, however, there is no question that the kidneys frequently play a critical role in hypertension.

Besides the three known causes of hypertension described above and several others, we are left with the vast majority of patients in the essential, or unknown, category. Many hypotheses have been proposed for increased arteriolar constriction but none proved. At present, much evidence seems to point to excessive sodium ingestion or retention within the body as a common denominator of renal, adrenal, and even essential hypertension, but how the sodium is involved in the increased arteriolar constriction remains unknown. In any case, sodium restriction has become a major form of therapy in hypertension. Virtually all other forms of therapy involve drugs which act upon some aspect of autonomic function to produce arteriolar dilatation. This does not mean that excessive sympathetic tone was the original cause of the hypertension but only that, whatever the cause, anything which dilates the arterioles reduces the blood pressure.

The perceptive reader might also wonder why the arterial baroreceptors do not, by way of the reflexes they initiate, return the blood pressure to normal. The reason seems to be that, in chronic hypertension, the baroreceptors are "reset" at a higher level; i.e., they regulate blood pressure but at a greater pressure. We have no explanation for this phenomenon.

CONGESTIVE HEART FAILURE

The heart may become weakened for many reasons; regardless of cause, however, the failing heart induces a similar procession of signs and symptoms grouped under the category of *congestive heart failure*. Patients with early mild heart disease may show at rest no significant abnormalities because of the great safety factor, or reserve, in cardiac function. However, the ability to perform exercise is impaired, as evidenced by shortness of breath and early fatigue. Ultimately, the cardiac reserve becomes inadequate to supply normal amounts of blood even at rest, and the patient becomes bedridden. Finally, the cardiac output may become too low to support life.

The basic defect in heart failure is a decreased contractility of the heart, but the molecular mechanism is unknown. As shown in Fig. 16-12, the failing heart shifts downward to a lower Starling curve. How can this be compensated for? Increased sympathetic stimulation would help to increase contractility, and this does occur. However, an even more striking compensation is increased ventricular end-diastolic volume. The failing heart is generally engorged with blood, as are the

FIGURE 16-12 ***Relationship between end-diastolic ventricular volume and stroke volume in normal and failing hearts. The normal curves are those shown previously in Figs. 14-21 and 14-22. The failing heart can still eject an adequate stroke volume if the sympathetic activity to it is increased or if the end-diastolic volume increases, i.e., if the ventricle becomes more distended.***

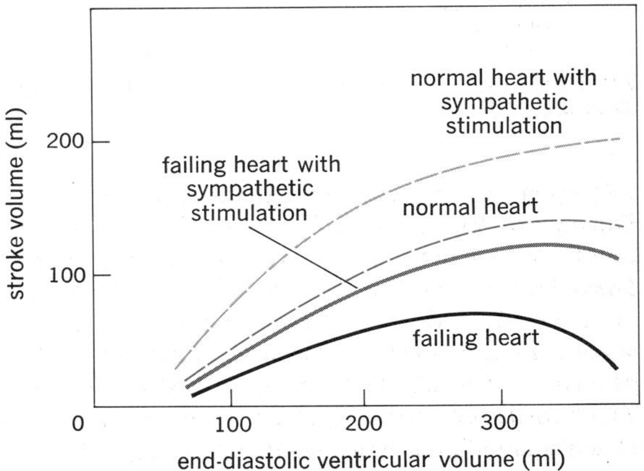

veins and capillaries, the major cause being an increase (sometimes massive) in plasma volume. The sequence of events is still not completely understood; decreased cardiac output in some manner (perhaps by reducing the blood flow to the kidneys) leads to failure of normal sodium and water excretion by the kidneys. The retained fluid then causes expansion of the extracellular volume, increasing venous pressure, venous return, and end-diastolic ventricular volume and thus tending to restore stroke volume toward normal.

Another result of elevated venous and capillary pressure is increased filtration out of the capillaries, with resulting edema. This accumulation of tissue fluid may be the chief symptom of ventricular failure; the legs and feet are usually most prominently involved (because of the additional effects of gravity), but the same engorgement is occurring in other organs and may cause severe malfunction. The most serious result occurs when the left ventricle fails; in this case, the excess fluid accumulates in the lung air sacs (*pulmonary edema*) because of increased pulmonary capillary pressure, and the patient may actually drown in his own fluid. This situation usually worsens at night; during the day, because of the patient's upright posture, fluid accumulates in the legs, but it is slowly absorbed when he lies down at night, the plasma volume expands, and an attack of pulmonary edema is precipitated.

Thus, what began as a useful compensation becomes potentially lethal because the tension-length relationship for muscle holds only up to a point, beyond which further stretching of the muscle may actually cause decreased strength of contraction. Thus, expansion of plasma volume may so increase end-diastolic volume as to decrease contractility (Fig. 16-12) and produce a rapidly progressing downhill course.

The treatment for congestive heart failure is easily understood in these terms: The precipitating cause should be corrected if possible; contractility can be increased by a drug known as digitalis; excess fluid should be eliminated by the use of drugs which increase excretion of sodium and water by the kidneys; the patient should be kept at rest so as to reduce the cardiac output required to fulfill the body's metabolic needs.

"HEART ATTACKS" AND ATHEROSCLEROSIS

We have seen that the myocardium does not extract oxygen and nutrients from the blood within the atria and ventricles but depends upon its own blood supply via the coronary vessels. The coronary arteries exit from the aorta just above the aortic valves and lead to a branching network of small arteries, arterioles, capillaries, venules, and veins similar to those in all other organs. The rate of blood flow depends primarily upon the arterial blood pressure and the resistance offered by the coronary vessels. The degree of arteriolar constriction, or dilatation, is almost entirely determined by local metabolic control mechanisms, there being little if any neural control. This is just what one would expect in an organ with varying metabolic requirements which must be met at all times to insure survival of the entire organism. Insufficient coronary blood flow leads to myocardial damage and, if severe enough, to death of the myocardium (*infarction*), a so-called *heart attack*. This may occur as a result of decreased arterial pressure but is more commonly due to increased vessel resistance following coronary atherosclerosis.

Atherosclerosis is a disease characterized by a thickening of the arterial wall with connective tissue and deposits of cholesterol. The mechanism by which thickening occurs is not clear, but it is known that smoking, obesity, inactivity, nervous tension, and a variety of other factors markedly predispose one to this disease of aging. The suspected relationship between atherosclerosis and blood concentrations of cholesterol and saturated fatty acids has probably received the most widespread attention, and many studies are presently trying to evaluate the likely hypothesis that high blood concentrations of these lipids increase the rate and the severity of the atherosclerotic process. Cholesterol is an important physiologic substance because it is the precursor of certain hormones and the bile acids (Chap. 19). Since it is found only in animals, ingestion of animal fats (including egg yolks) constitutes the major dietary source. However, the liver (and other body cells) is capable of producing large quantities of cholesterol, so that even profound reductions of dietary intake frequently do not lower blood cholesterol concentration significantly because the liver responds by producing more. Indeed, it may well be their high content of saturated fatty acids, rather than cholesterol, which causes the ingestion of animal fat to predispose one to atherosclerosis. Vegetable fat, in contrast, contains primarily unsaturated fatty acids and may actually lower blood cholesterol.

The incidence of coronary atherosclerosis in the United States is extraordinarily great; it is estimated to cause 500,000 deaths per year. The mechanism by which atherosclerosis reduces coronary blood flow is quite simple; the fat deposits and fibrous thickening narrow the vessels and increase resistance to flow. This is usually progressive, leading often ultimately to complete occlusion. Acute coronary occlusion may occur because of sudden formation of a clot on the roughened vessel surface or breaking off of a deposit, which then lodges downstream, completely blocking a smaller vessel. If, on the other hand, the atherosclerotic process causes only gradual occlusion, the heart may remain uninjured because of the development, over time, of new accessory vessels supplying the same area of myocardium. It should be stressed that before complete occlusion many patients experience recurrent transient episodes of inadequate coronary blood flow, usually during exertion or emotional tension. The pain associated with this is termed *angina pectoris.*

The cause of death from coronary occlusion and myocardial infarction may be either severe hypotension resulting from weakened contractility or disordered cardiac rhythm resulting from damage to the cardiac conducting system, but in addition the severe hypotension which may be associated with a heart attack is frequently due to reflex inhibition of the sympathetic nervous system and enhancement of the parasympathetics. The origin of these totally inappropriate and frequently lethal reflexes is not known. Finally, should the patient survive an acute coronary occlusion, the heart may be left permanently weakened, and a slowly progressing heart failure may ensue. On the other hand, many people lead quite active and normal lives for many years after a heart attack.

We do not wish to leave the impression that atherosclerosis attacks only the coronary vessels, for such is not the case. Most arteries of the body are subject to this same occluding process. For example, cerebral occlusions (*strokes*) are extremely common in the aged and constitute an important cause of sickness and death (200,000 per year). Wherever the atherosclerosis becomes severe, the resulting symptoms always reflect the decrease in blood flow to the specific area. In recent years, synthetic materials have been developed from which tubes can be made and surgically substituted for a diseased segment of artery.

THE FETAL CIRCULATION AND CHANGES IN THE CIRCULATORY SYSTEM AT BIRTH

Blood is carried from the placenta to the fetus by an *umbilical vein,* which enters the abdomen of the fetus through the *umbilicus,* or *navel,* and goes immediately to the liver (Fig. 16-13). This blood, of course, carries the oxygen and nutrients required by the fetus. From the liver, the blood enters the inferior vena cava where it mixes with the blood from the inferior half of the fetus. (One of the vessels carrying blood from the liver to the inferior vena cava is the *ductus venosus,* which is obliterated and present as the *ligamentum venosum* in the adult.)

The blood from the inferior vena cava enters the right atrium. Rather than flowing into the right ventricle, most of this blood passes through an opening in the interatrial septum (the *foramen ovale*) directly into the left atrium where it mixes with the small amount of blood returning to the heart from the pulmonary veins (see below). From the left ventricle the blood is pumped into the aorta.

Some of the blood from the inferior vena cava and most of the blood from the superior vena cava, upon entering the right atrium, passes into the right ventricle and on into the pulmonary trunk. However, the resistance to blood flow through the lungs is very high in the fetus and only a small portion of the blood in the pulmonary trunk actually enters the lung tissue. Most of it is shunted through the *ductus arteriosus,* a fetal vessel which carries blood from the pulmonary trunk to the aorta (Fig. 16-13). The ductus arteriosus enters the descending portion of the aortic arch beyond the sites where the coronary arteries and the arteries to the upper limbs and head exit. Therefore, most of the blood from the inferior vena cava (which has passed through the foramen ovale and left heart) flows to the heart muscle, upper limbs, and head of the fetus while the blood from the superior vena cava (which has passed through the right ventricle, pulmonary trunk, and ductus arteriosus) is shunted to the lower limbs, abdominal organs, and the two *umbilical arteries,* which carry it back to the placenta. The significance of this distribution lies in the fact that inferior vena cava blood is much better oxygenated than superior vena cava blood because much of the former has returned from the placenta, the site of the fetus' oxygen uptake; accordingly, the heart and brain receive better-oxygenated blood.

In order for blood to flow from the right atrium into the left atrium or from the pulmonary trunk through the ductus arteriosus into the aorta, the pressures in the right heart must exceed those in the left heart and systemic circulation, a condition that reverses shortly after birth.

At birth the muscular walls of the umbilical vessels constrict in response to various stimuli such as changes in oxygen tension, temperature, and the degree of stretch on the vessels; this constriction markedly in-

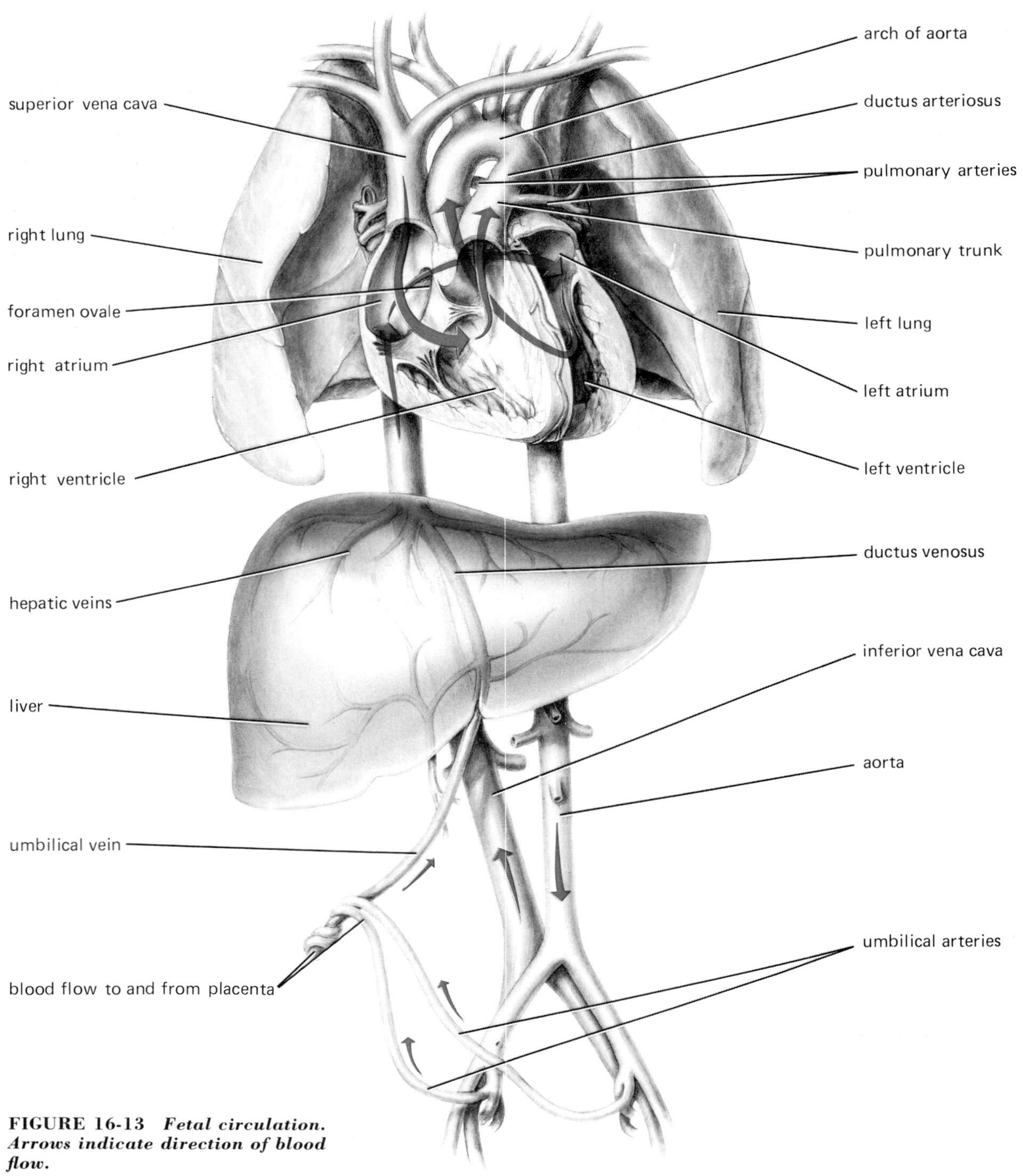

FIGURE 16-13 ***Fetal circulation. Arrows indicate direction of blood flow.***

creases the total systemic peripheral resistance in the fetal circulation and also decreases venous return to the heart. Simultaneously the infant's first breath is stimulated by temperature or tactile stimulation during birth or, perhaps, by the decrease in blood oxygen content resulting from the infant's separation from the placenta. The mechanical actions of the chest movements associated with the first breath and the smooth muscle relaxation induced by the increased blood oxygen resulting from the breath cause the resistance in the pulmonary vessels to decrease markedly. This decrease, coupled with the decreased venous return to the heart and the increased total peripheral resistance, lower the pressure in the right heart and raise it in the left heart, thereby reversing the fetal pressure gradient. This reversal, in turn, abruptly closes the valve over the foramen ovale, and the leaflets of the valve physically fuse within several days. Simultaneously, the decreased pressure in the pulmonary vessels and the increased pressure in the aorta cause a reversal of flow through the ductus arteriosus; this is short-lived, however, because the walls of the ductus begin to constrict and the vessel lumen becomes obliterated, thereby eliminating all flow through it. Closure of the ductus is stimulated by the increased oxygen content of the blood flowing through it and probably by hormones as well. Failure of either the ductus arteriosus or foramen ovale to close after birth leads to some of the more common congenital heart defects; fortunately, these can usually be corrected surgically.

CHAPTER 17 The Respiratory System

ORGANIZATION OF THE RESPIRATORY SYSTEM

Nose and Pharynx • Larynx • Trachea and Main Bronchi • Lungs • *Intrapulmonary Airways • Site of Gas Exchange in the Lungs: The Alveoli* **• Respiratory Mucosa • Blood Supply of the Respiratory System • Thoracic Cage •** *Movements of the Thorax • Relation of the Lungs to the Thoracic Cage*

INVENTORY OF STEPS INVOLVED IN RESPIRATION

EXCHANGE OF AIR BETWEEN ATMOSPHERE AND ALVEOLI: VENTILATION

Concept of Intrapleural Pressure • Inspiration • Expiration • Quantitative Relationship Between Atmosphere-Intraalveolar Pressure Gradients and Air Flow: Airway Resistance • Control of Pulmonary Arterioles • Lung-Volume Changes during Breathing • Air Distribution within the Lungs • *Anatomic Dead Space • Alveolar Dead Space* **• Work of Breathing**

EXCHANGE AND TRANSPORT OF GASES IN THE BODY

Basic Properties of Gases • Behavior of Gases in Liquids • Pressure Gradients of Oxygen and Carbon Dioxide within the Body • Transport of Oxygen in the Blood: The Role of Hemoglobin • *Effect of* P_{O_2} *on Hemoglobin Saturation • Effect of Acidity on Hemoglobin Saturation • Effect of Temperature on Hemoglobin Saturation • Effect of DPG on Hemoglobin Saturation* **• Functions of Myoglobin • Transport of Carbon Dioxide in the Blood**

CONTROL OF RESPIRATION

Neural Generation of Rhythmic Breathing • Control of Ventilatory Volume • *Control of Ventilation by Oxygen • Control of Ventilation by Carbon Dioxide and Hydrogen Ion* **• Control of Ventilation during Exercise •** *Decreased* P_{O_2} *as the Stimulus? • Increased* P_{CO_2} *as the Stimulus? • Increased Hydrogen Ion as the Stimulus?* **• Control of Respiration by Other Factors •** *Temperature • Input from the Cerebral Cortex • Epinephrine • Reflexes from Joints and Muscles • Baroreceptor Reflexes • Protective Reflexes • Pain • Emotion • Voluntary Control of Breathing*

Most cells in the human body obtain the bulk of their energy from chemical reactions involving oxygen. In addition, cells must be able to eliminate the major end product of these oxidations, carbon dioxide. A unicellular organism can exchange oxygen and carbon dioxide directly with the external environment, but this is obviously impossible for most cells of a complex organism like the human body, since only a small fraction of the total cells (skin, gastrointestinal lining, respiratory lining) is in direct contact with the external environment. In order to survive, large animals have had to develop specialized systems for the supply of oxygen and elimination of carbon dioxide. These systems are not the same in all complex animals since evolution often follows several pathways simultaneously. The organs of gas exchange with the external environment in fish are gills; those in human beings are *lungs*. Specialized blood components have also evolved which permit the transportation of large quantities of oxygen and carbon dioxide between the lungs and cells.

In a human being at rest, the body's cells consume approximately 200 ml of oxygen per minute. Under conditions of high oxygen need, e.g., exercise, the rate of oxygen consumption may increase as much as thirtyfold. Equivalent amounts of carbon dioxide are simultaneously eliminated. It is obvious, therefore, that mechanisms must exist which coordinate breathing with metabolic demands. We shall see in Chap. 18 that the control of breathing also plays an important role in the regulation of the acidity of the extracellular fluid.

Before describing the basic processes of oxygen supply, carbon dioxide elimination, and breathing control, we must first define the terms *respiration* and *respiratory system*. Respiration has two quite different meanings: (1) the metabolic reaction of oxygen with carbohydrate and other organic molecules and (2) the exchange of gas between the cells of an organism and the external environment. The various steps of the second process form the subject matter of this chapter; the first process was described in Chap. 1. The term respiratory system refers only to those structures which are involved in the exchange of gases between the blood and external environment; it does not include the transportation of gases in the blood or gas exchange between blood and the tissues. Admittedly, this definition is arbitrary since it includes only half of the processes involved in respiration, but it has become firmly established by long usage. The respiratory system consists of the lungs, the series of passageways leading to the lungs, and the chest structures responsible for movement of air in and out of the lungs.

ORGANIZATION OF THE RESPIRATORY SYSTEM

In order for air to reach the lungs, it must first pass through a series of air passages connecting the lungs to the nose and mouth (Fig. 17-1) and then through the continuation of these air passages in the lungs themselves (Fig. 17-2). All these air passages together are termed the *conducting portion* of the respiratory system. Their walls are supported by bone or cartilage to ensure that the airways remain open. The smallest tubes of the conducting portion of the lung lead into the *respiratory portion* where the passages contain on their walls tiny blind sacs, the *alveoli,* which are the sites of gas exchange within the lungs. The system of air passages terminates in clusters of alveoli (see Fig. 17-9A). All portions of the air passageways and alveoli receive a rich supply of blood by way of blood vessels, which constitute a large portion of the total lung substance (Fig. 17-3). Between the air tubes and blood vessels of the lungs are large quantities of elastic connective tissue, which plays an important role in breathing.

The alveoli are such a great distance from the external environment that diffusion could not possibly provide the amount of oxygen and carbon dioxide exchange necessary for metabolism; therefore, a pressure gradient capable of generating bulk flow between the two regions is needed. This pressure gradient is developed by changes in lung volume, the pressure in the air spaces of the lungs decreasing as their volume increases, and vice versa. However, the lungs themselves lack muscle and are therefore passive elastic containers with no inherent ability to increase their volume; lung expansion is accomplished instead by the action of the intercostal muscles, which move the ribs, and the diaphragm muscle.

Nose and Pharynx

Air can enter the respiratory passages via either the nose or mouth, although the nose is the normal route. The *nasal cavity* is an irregularly shaped space extending from the bony palate, which separates the nose and mouth cavities, upward to the base of the cranial cavity (Fig. 17-4). It is divided into right and left nasal cavities by the *nasal septum*. The front of each cavity opens onto the face through the *nostril;* the back opens into the nasal pharynx through the *posterior nasal aperture,* or *choana*.

The slight enlargement of each cavity just inside the nostril is the *vestibule*. It is lined with coarse hairs, which tend to trap foreign substances carried into the

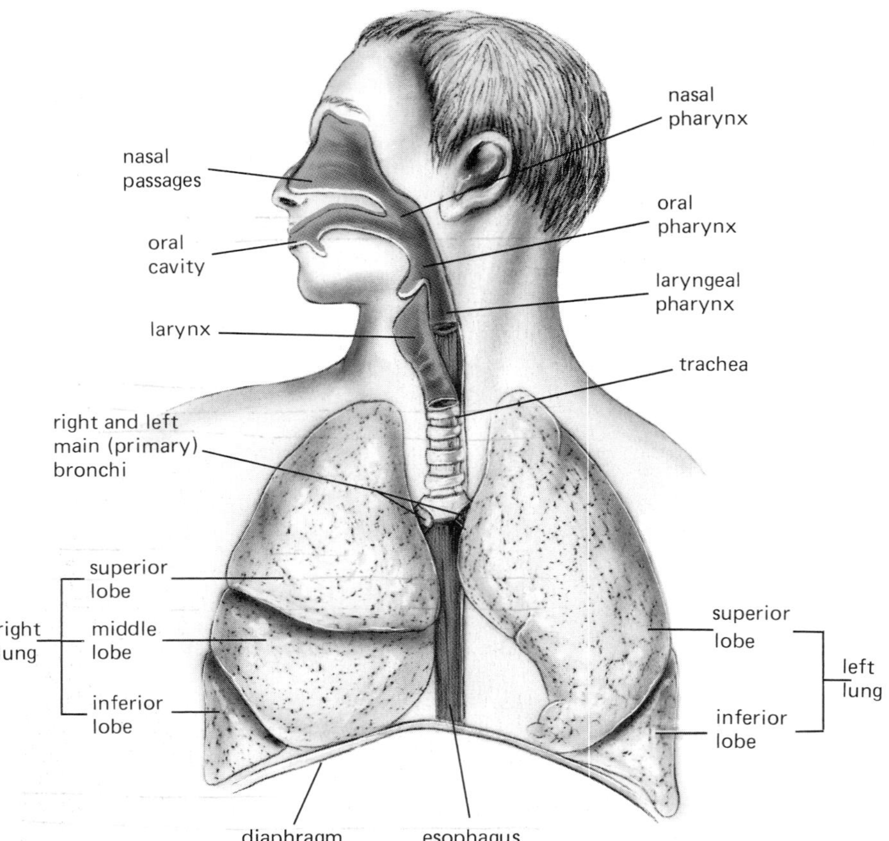

FIGURE 17-1 ***Organization of the respiratory system.***

nose with the inspired air. The roof of the nasal cavity contains the receptors for the sense of smell and is called the *olfactory region.* The remaining *respiratory region* extends from the midline nasal septum to the lateral walls, which have three bony projections, the *superior, middle,* and *inferior conchae* (Fig. 17-4). Under each concha is an air space, the *superior, middle,* and *inferior meatus,* which connects freely with the main air passage of the cavity. The lacrimal ducts (tear ducts) empty into the nasal cavities.

The bony framework of the nasal cavity is formed by the fusion of several bones (the *nasal, ethmoid, frontal, lacrimal,* and *palatine bones, maxillae,* and *vomer;* Figs. 7-15 and 17-5). The cartilaginous part of the framework is formed by the *septal* and *nasal cartilages* (Fig. 17-5). The frontal, ethmoid, sphenoid, and maxillary bones contain *paranasal sinuses,* large air pockets which are lined with mucous membranes (Fig. 17-4). The mucus secreted by these membranes drains through small openings into the nasal cavities. The sinuses add resonance to the voice and decrease the weight of the skull, but these functions are relatively unimportant. The sinuses are probably simply manifestations of the growth patterns of the bones in which they occur.

After passing through the nose (or mouth), air enters the *pharynx* (throat), a passage common to the routes followed by air and food. The pharynx is a muscular tube lined with mucous membrane; it extends from the base of the skull downward behind the nasal cavity, mouth, and larynx where it becomes continuous with the esophagus (Fig. 17-4).

The upper, *nasal part* of the pharynx lies behind the nasal cavities and communicates with them through

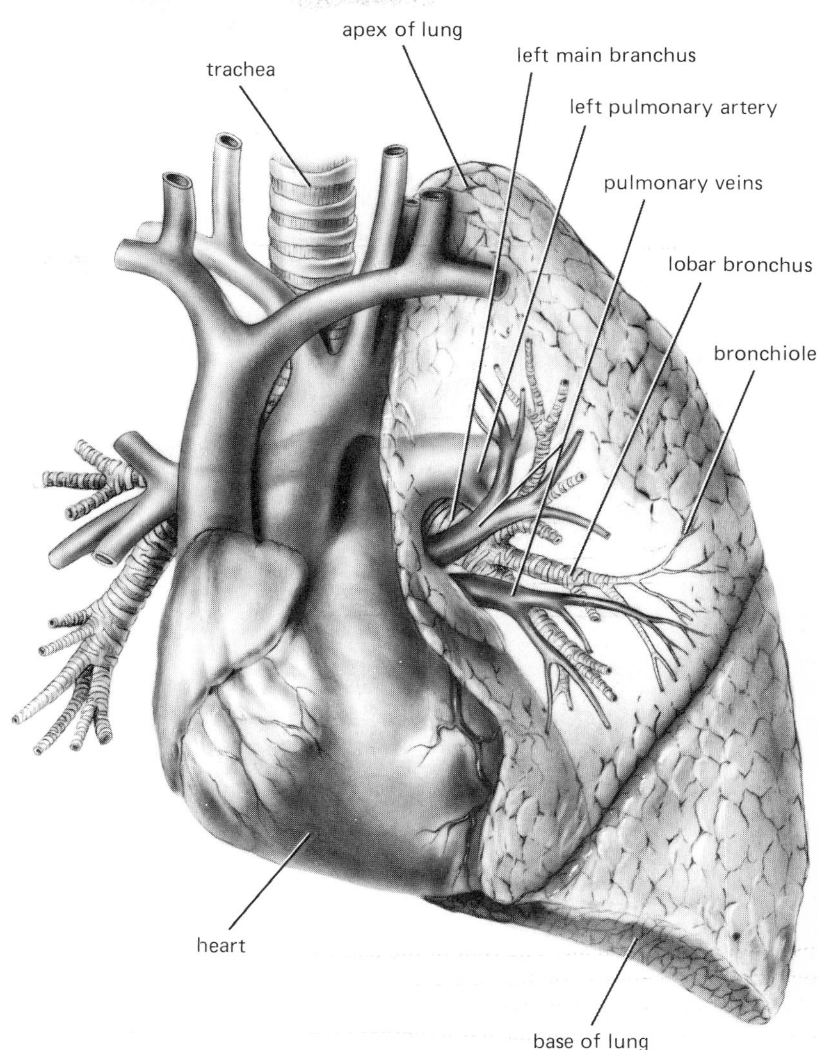

FIGURE 17-2 ***Airways of the adult human lung, starting with the trachea and continuing with its many branches in the lungs. The substance of the lungs appears transparent so the relationships between the airways and blood vessels can be seen.***

the posterior nasal apertures. The nasal pharynx also communicates with the middle-ear cavities through the *auditory,* or *eustachian, tubes* (Figs. 9-27A and 17-4). Near these openings patches of lymphoid tissue, the *pharyngeal tonsils,* lie in the mucous membrane. (Hypertrophied pharyngeal tonsils are *adenoids.*) At the level of the soft palate, the nasal region of the pharynx ends and the *oral pharynx* begins; it extends downward to the level of the epiglottis (Fig. 17-4). Anteriorly, the oral pharynx opens into the oral (mouth) cavity. The oral pharynx contains the *palatine tonsils,* which with the pharyngeal tonsils form a partial ring of lymphoid tissue around the opening of the digestive and respiratory tubes; the ring is completed by the *lingual tonsils.* (The histology and function of the tonsils will be discussed in Chap. 22.) The pharynx continues downward as the *laryngeal pharynx,* which opens into the larynx.

Larynx

The larynx, which lies in the front part of the neck, serves three functions: It is the air passageway between the pharynx and lungs; it acts as a protective sphincter

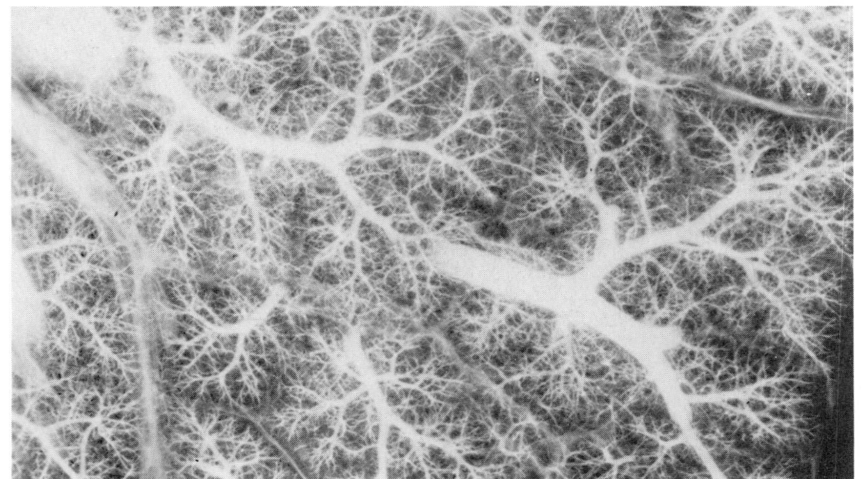

FIGURE 17-3 ***Autoradiogram of the blood vessels in a portion of a mammalian lung. A radioactive substance was injected into the vascular system and the lung subsequently removed. The lung tissue was laid on a photographic plate, and the energy associated with radioactive decay of the substance in the blood vessels exposed the plate in the same way that light exposes film. Development of the exposed plate produced the autoradiogram shown here.*** **(Courtesy of Dr. Robert E. Heitzman.)**

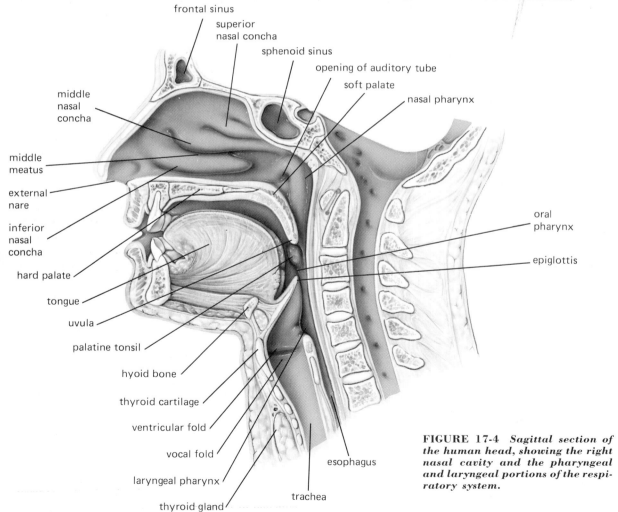

frontal sinus
superior nasal concha
sphenoid sinus
opening of auditory tube
soft palate
nasal pharynx
middle nasal concha
middle meatus
external nare
inferior nasal concha
hard palate
tongue
uvula
palatine tonsil
hyoid bone
thyroid cartilage
ventricular fold
vocal fold
laryngeal pharynx
thyroid gland
oral pharynx
epiglottis
esophagus
trachea

FIGURE 17-4 ***Sagittal section of the human head, showing the right nasal cavity and the pharyngeal and laryngeal portions of the respiratory system.***

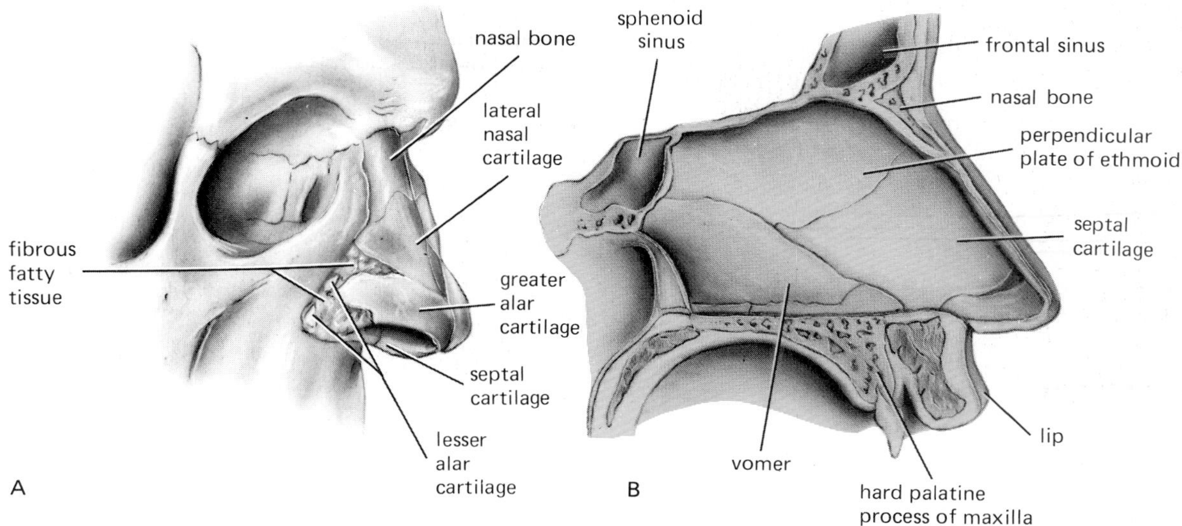

FIGURE 17-5 ***Nasal cartilages:* (A) *the external nose;* (B) *the nasal septum.***

to prevent solids and liquids from passing into the bronchi and lungs; and it is involved in the production of sounds. Its framework consists of nine cartilages and the ligaments, membranes, and muscles connecting them (Fig. 17-6). The largest cartilage of the larynx, the *thyroid cartilage,* forms the hard bump in the front of the neck known as the *laryngeal prominence,* or Adam's apple. This prominence is hardly visible in children or adult females, but it is quite marked in males after puberty. The inlet to the larynx lies in the anterior wall of the pharynx; it is bounded by the epiglottis, arytenoid cartilages, and the aryepiglottic folds (Fig. 17-6). Continuing from the inlet, the cavity of the larynx expands into a wide *vestibule,* which ends below at the *true vocal folds.* These are stretched between the thyroid cartilage anteriorly and the arytenoid cartilages posteriorly (Fig. 17-7). The free margins of the true vocal folds contain the vocal ligaments. The vocal folds and the space between them are called the *glottis* and are the part of the larynx most directly involved with the production of sounds. The actual space between the vocal folds and the arytenoid cartilages is the *glottic aperture;* its width and shape vary with the movements of the arytenoid cartilages and vocal folds. The muscles involved in these movements are listed in Table 7-26. Above the vocal folds are two thick, membranous *vestibular folds.* These, often called the false vocal folds, plus the true vocal folds are commonly termed the *vocal cords.*

Trachea and Main Bronchi

The *trachea* constitutes the tubular air passageway from the larynx to the lungs (Figs. 17-1 and 17-2), where at the *carina* it divides into two *main* (primary) *bronchi* (sing.: *bronchus*). The main bronchi, after traveling only a short distance beyond this tracheal division, enter the lungs.

The trachea and bronchi are kept open by irregular rings of cartilage embedded in their walls. The tracheal rings are C-shaped, almost completely circling the air passageway, with their openings at the back where the trachea lies against the esophagus (Fig. 17-8). The gaps are bridged by smooth muscle and connective-tissue bands, which cause changes in the diameter of the tracheal lumen when they contract or are stretched. Adjacent rings are joined by connective tissue and smooth muscle. The structure of the main bronchi is similar to that of the trachea.

Lungs

The *lungs* are located in the thoracic cavity, one on each side, separated from each other by the mediastinum and its contents. They lie free in their respective

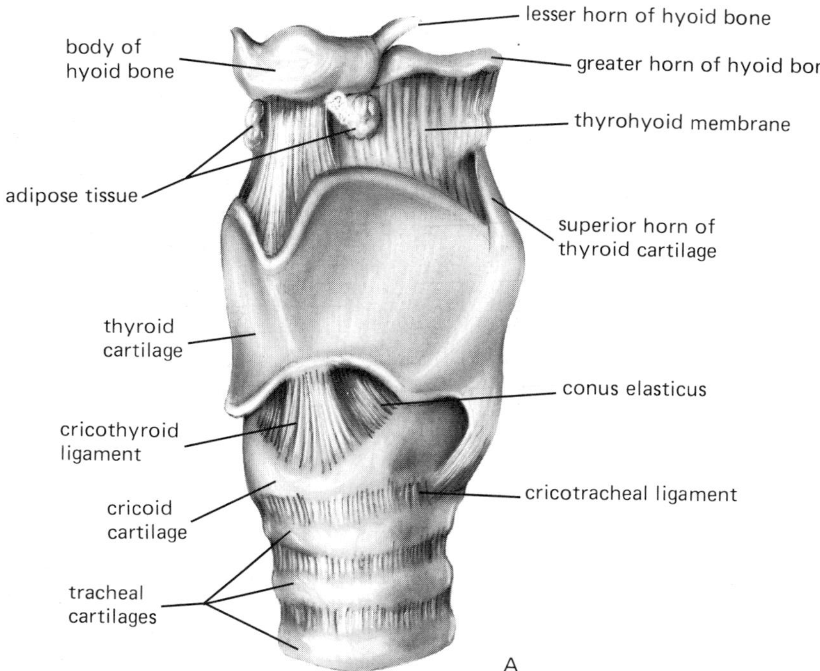

FIGURE 17-6 **(A)** ***Anterior view of the laryngeal portion of the respiratory system.*** **(B)** ***Sagittal section through the larynx. The anterior part of the larynx is toward the left.***

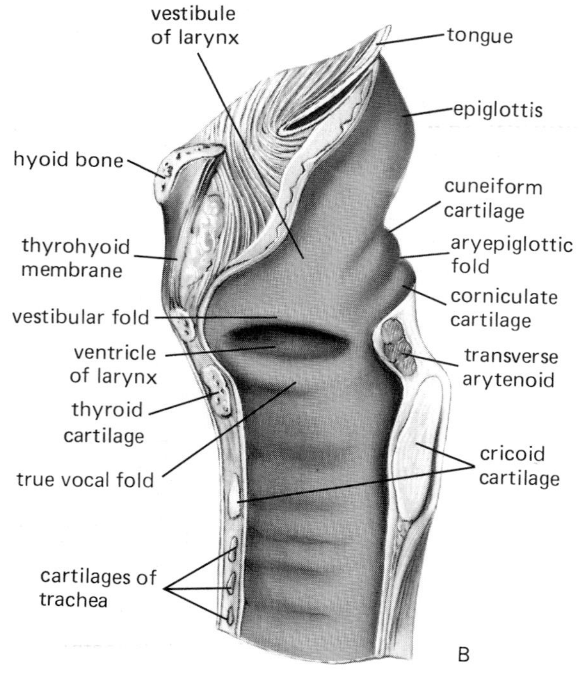

halves of the thoracic cavity, except for attachments to the heart and trachea. The heart, which lies in the mediastinum slightly to the left of midline, occupies part of the thoracic cavity on that side, and the left lung is slightly smaller than the right. The left lung is divided into two *lobes,* the right into three. The medial surface of each lung has a slight depression surrounding the place where the main bronchus and its accompanying nerves and blood vessels enter the lung; this depression is the *hilum* of the lung, with the bronchus, nerves, and vessels forming the *root* of the lung. The upper part of the lung, which lies against the top of the thoracic cavity, is the *apex,* and the lower part lying against the diaphragm is the *base.*

The substance of the lungs is light and spongy, and their color, at least at birth, is rose-pink. By adulthood the color changes to a dark slaty gray mottled with patches that can be almost black because of inhaled particulate matter that has been retained and deposited in the lung tissues.

Intrapulmonary Airways The lungs are not simply hollow balloons but have a highly organized structure consisting of air-containing tubes, blood vessels, and elastic connective tissue. The airway system

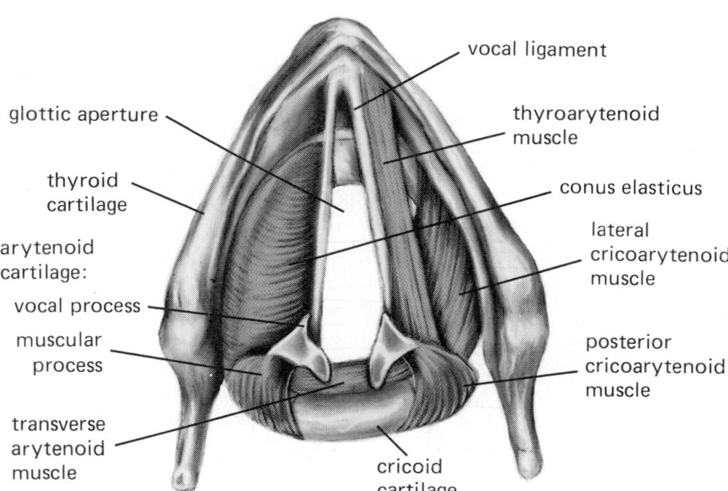

FIGURE 17-7 ***Diagrammatic cross section of the larynx through the region of the glottis. The superficial muscles have been removed on the left.***

of a lung has been compared to a tree because the main bronchus entering it undergoes many divisions, the resulting branches becoming smaller and closer together with each division (Fig. 17-2). The large branches are called *bronchi.* Their walls contain cartilage which decreases in amount as the bronchi successively branch. As the cartilage declines, the muscle layer increases, and rings of smooth muscle completely surround the bronchi. The two main bronchi divide into *lobar,* or *secondary, bronchi,* one of which enters each of the lobes of the lungs where they divide into *tertiary bronchi,* of which there are 10 in each lung. Major branches such as these are termed *segmental bronchi* if they enter a self-contained, functionally independent unit of the lung. Such independent segments of lung tissue are known as *bronchopulmonary segments.*

Where the cartilage disappears and the diameter is reduced to about 1 mm, the bronchi become *bronchioles.* The rings of smooth muscle persist and are joined together by diagonal muscle fibers. Like the trachea and main bronchi, the smooth muscle of the intrapulmonary airways is richly innervated and sensitive to certain circulating hormones, e.g., epinephrine. Contraction or relaxation of this muscle alters resistance to air flow. The bronchioles continue to divide, their last branches being the *terminal bronchioles.* This marks the end of the *conducting portion* of the respiratory system and the beginning of the thinner-walled *respiratory portion* where the actual gas exchange occurs.

Site of Gas Exchange in the Lungs: The Alveoli Beyond the terminal bronchioles, the airway walls contain small outpocketings of eipthelium. As the airway continues to divide, these outpocketings, or *alveoli,* increase in frequency until the airway ends in grapelike clusters of alveoli (Fig. 17-9). The walls of the alveoli consist of a loose mesh of elastic connective-tissue fibers lined with a thin layer of squamous

FIGURE 17-8 ***Cross section through the trachea.*** **(Redrawn from Bloom and Fawcett.)**

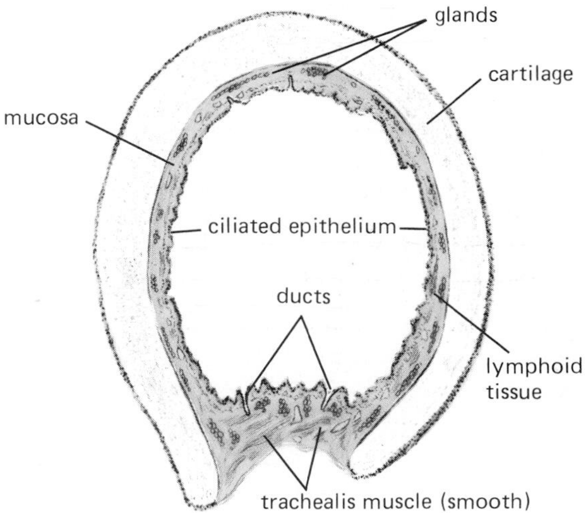

epithelium. Running in these walls are numerous capillaries, the endothelial lining of which lies snugly against the epithelium lining the alveoli (Fig. 17-10) separated from it only by a thin layer of interstitium. Thus the blood in a capillary is separated from the air in an alveolus only by an extremely thin barrier (0.2 μm compared with 7 μm, which is the diameter of an average red blood cell). The total area of alveoli in contact with capillaries is 70 m^2 (the size of a badminton court) in human beings. This immense area combined with the thin barrier permits the rapid exchange of large quantities of oxygen and carbon dioxide. In addition to its predominant squamous cells, the alveolar epithelium also contains phagocytic cells and specialized cells, *type II cells,* which produce the critical substance, pulmonary surfactant, to be discussed below. Finally, there are pores in the alveolar membranes which permit some flow of air between alveoli (Figs. 17-9 and 17-10). This "collateral ventilation" can be very important when the duct leading to an alveolus is occluded by disease, since some air can still enter this alveolus by way of pores between it and adjacent alveoli.

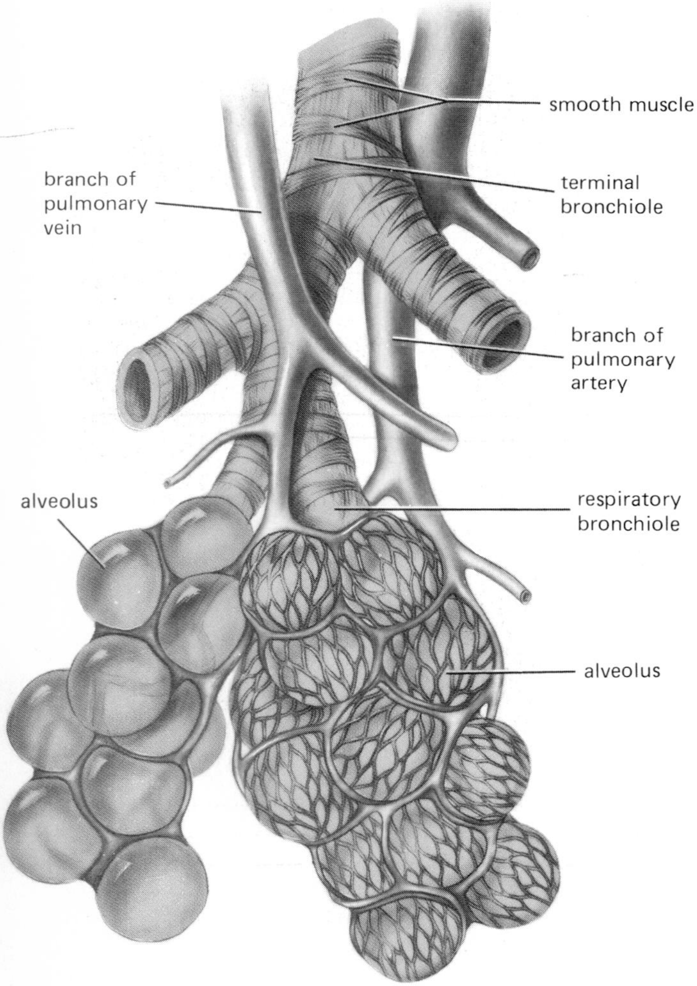

FIGURE 17-9 ***Clusters of alveoli form alveolar sacs, which are connected to the air passageways of the lung. The intimate relationship between the respiratory system and blood vessels at the level of the alveoli can be seen. The respiratory bronchioles branch to form* alveolar ducts *from which* alveolar sacs *open. The walls of all these structures contain alveoli; those opening from the alveolar sacs are distinguished as* terminal alveoli.**

Respiratory Mucosa

The entire airway is lined with epithelial tissue the characteristics of which vary, depending upon the area of the respiratory system (Table 17-1). The oral part of

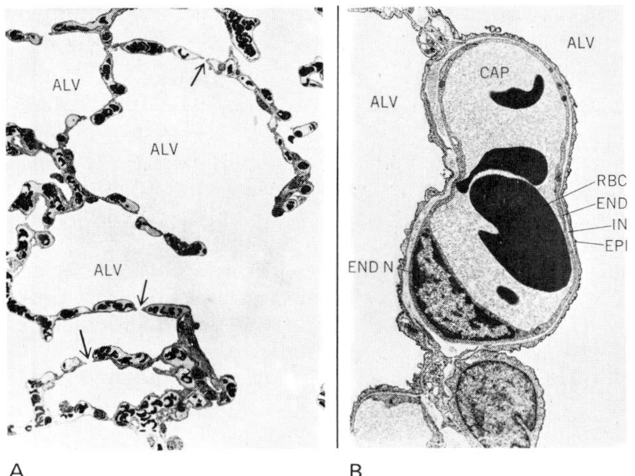

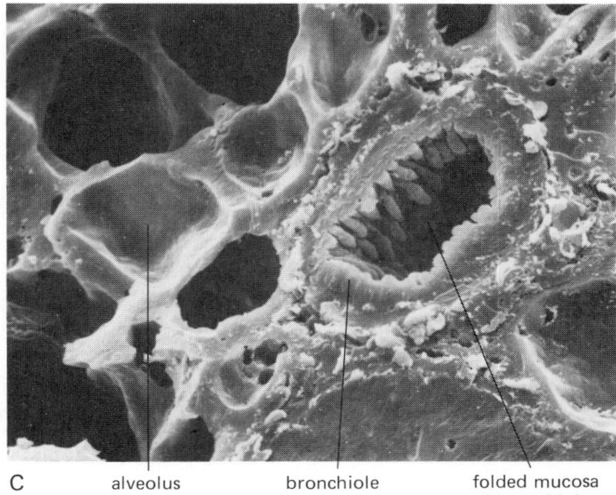

FIGURE 17-10 (A) ***Low-power electron micrograph of dog lung alveoli*** **(ALV).** ***Note the capillaries in the walls between alveoli (the dark disklike objects are erythrocytes). The arrows denote pores in the alveolar walls.*** **(B)** ***Higher magnification of a portion of an alveolar wall, showing a single capillary*** **(CAP)** ***surrounded by an alveolar epithelial cell. The nucleus*** **(END N)** ***and cytoplasm*** **(END)** ***of a capillary endothelial cell are visible, as are the nucleus and cytoplasm*** **(EPI)** ***of an alveolar epithelial cell. Note that the blood in the capillary is separated from air in the alveoli*** **(ALV)** ***only by the thin membrane consisting of endothelium, interstitial fluid*** **(IN),** ***and epithelium.*** **(C and D)** ***Scanning electron micrographs of rat and mouse lung.*** **(Parts A and B, courtesy of E. R. Weibel. From *Physiol. Rev.*, 53:424 (1973). Parts C and D from R. G. Kessel and C. Y. Shih, "Scanning Electron Microscopy in Biology," p. 293, Springer-Verlag, Berlin, 1974.)**

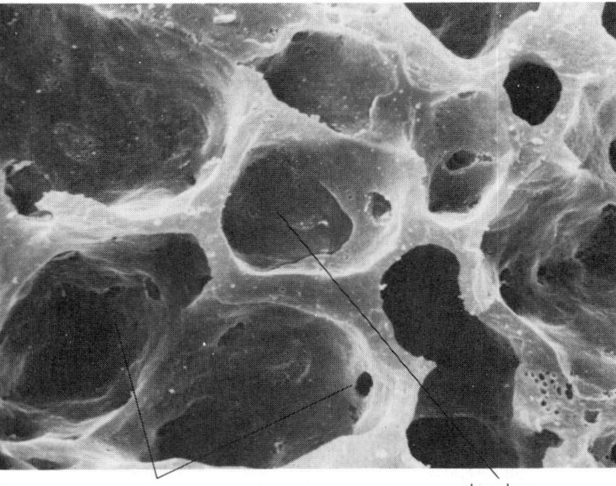

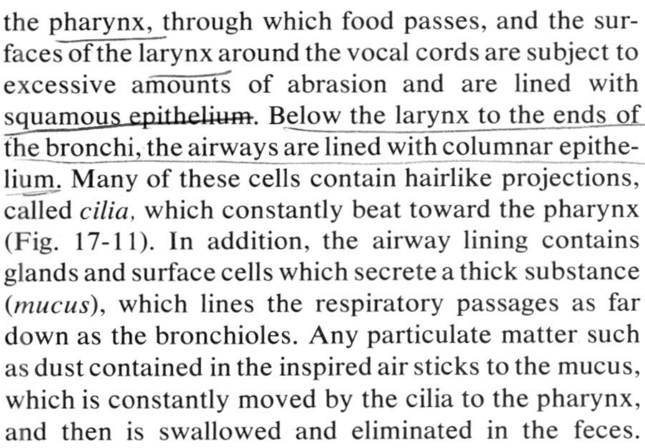

the pharynx, through which food passes, and the surfaces of the larynx around the vocal cords are subject to excessive amounts of abrasion and are lined with squamous epithelium. Below the larynx to the ends of the bronchi, the airways are lined with columnar epithelium. Many of these cells contain hairlike projections, called *cilia,* which constantly beat toward the pharynx (Fig. 17-11). In addition, the airway lining contains glands and surface cells which secrete a thick substance (*mucus*), which lines the respiratory passages as far down as the bronchioles. Any particulate matter such as dust contained in the inspired air sticks to the mucus, which is constantly moved by the cilia to the pharynx, and then is swallowed and eliminated in the feces. Besides keeping the lungs clean, this mechanism is important in the body's total defenses against bacterial infection, since many bacteria enter the body on dust particles. A major cause of lung infection is probably paralysis of ciliary activity by noxious agents, including substances in cigarette smoke. This, coupled with the stimulation of mucus secretion induced by these same agents, may result in partial or complete airway obstruction by the stationary mucus. (A smoker's early-morning cough is the attempt to clear this obstructive mucus from the airways.) A second protective mechanism is provided by the phagocytic cells, which are present in the respiratory-tract lining in great numbers. These cells, which engulf dust, bacteria, and debris, are

TABLE 17-1
COMPOSITION OF THE RESPIRATORY AIRWAYS

Component	*Predominant cell types in airway lining*	*Cartilage?*	*Muscle?*
Conducting portion			
Trachea	Pseudostratified tall columnar ciliated epithelial cells and goblet cells, which secrete mucus; mucus-secreting exocrine glands are present.	Yes, 16–20 incomplete rings.	Yes, bands bridge the gaps between ends of cartilages.
Bronchi	Large bronchi are similar to trachea, but as the bronchi get smaller, the height of the cells decreases.	Yes, present as large irregular often helical plates.	Yes, forms a complete layer in the larger bronchi; present as isolated bundles in the smaller bronchi.
Bronchioles and terminal bronchioles	Cuboidal epithelium with some ciliated cells; mucus-secreting elements are few or absent; the number of ciliated cells gradually decreases in the terminal bronchioles and a new cell type (Clara cell) appears.	No	Yes, present as isolated bundles rather than as a complete layer.
Respiratory portion			
Respiratory bronchioles	Low cuboidal epithelium changing to flat transitional cells at the entrance of the occasional alveoli which are present.	No	Yes, present as isolated bundles
Alveolar ducts	Lining is so extensively interrupted by alveoli that there is essentially no airway wall between the rings of cells which surround the alveolar entrances.	No	Yes, muscle forms a sphincter at the outlet of the last alveolar duct.
Alveoli	Squamous epithelium (alveolar type I cells), alveolar type II cells, and macrophages ("dust cells").	No	No

also injured by cigarette smoke and other air pollutants. Air is also warmed and moistened by contact with the epithelial lining as it flows through the respiratory passages.

Blood Supply of the Respiratory System

The lungs receive blood from two different arterial systems: from the *pulmonary arteries,* which carry deoxygenated (venous) blood to the lungs to pick up oxygen and release carbon dioxide, and from the *bronchial arteries,* which carry arterial blood to nourish the lung tissues.

The two pulmonary arteries arise from the pulmonary trunk shortly after it leaves the right ventricle, one artery entering each lung. The arteries branch, generally following the branches of the bronchi, and finally end in the dense capillary networks within the alveolar walls where the oxygen and carbon dioxide of the blood equilibrate with the alveolar air. Blood that is now oxygenated empties from the capillaries into branches of the pulmonary veins, which also receive a small amount of deoxygenated blood directly from arteriovenous shunts. The branches of the veins gradually merge until all the blood is collected into the four large *pulmonary veins,* two from each lung, which pass to the left atrium.

The bronchial arteries (branches of the aorta) form

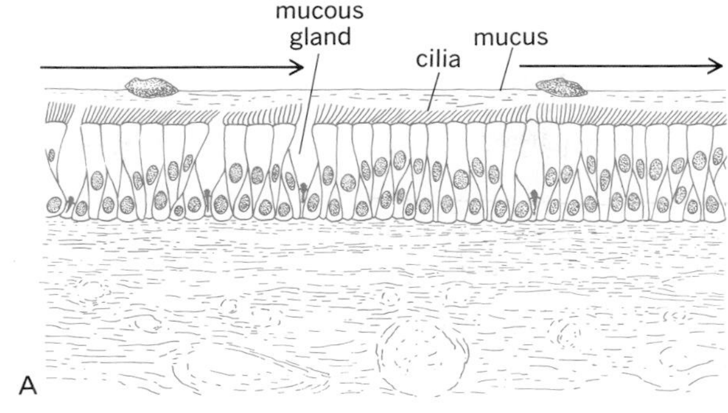

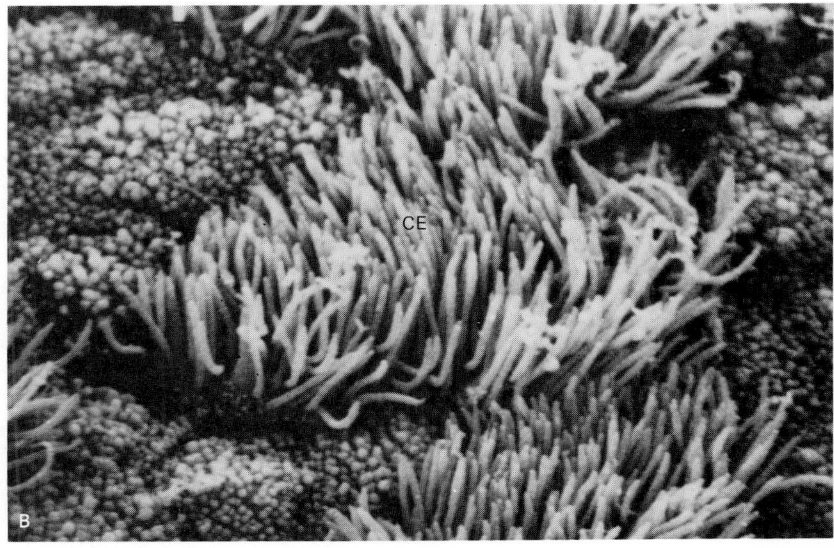

FIGURE 17-11 (A) ***Epithelial lining of the respiratory tract. The arrows indicate the upward direction in which the cilia move the overriding layer of mucus, to which foreign particles are stuck.*** (B) ***Scanning electron micrograph of the mucosa of a mammalian trachea. The two cell types of the mucosa, the ciliated epithelial cell*** (CE) ***and mucus-secreting goblet cells can be seen.*** **(Part B from R. G. Kessel and C. Y. Shih, "Scanning Electron Microscopy in Biology," p. 291, Springer-Verlag, Berlin, 1974.)**

capillary plexuses in the large connective-tissue structures of the lungs and the airway walls. Some of the blood from the bronchial capillary plexuses drains into the pulmonary veins, mixing with the arterial blood there; the rest empties into the *bronchial veins,* which eventually return the blood to the superior vena cava.

Thoracic Cage

The thoracic cage is a closed compartment. It is bounded at the neck by muscles and connective tissue, and it is completely separated from the abdomen by the diaphragm. The outer walls of the thoracic cage are formed by the sternum (breastbone), 12 pairs of ribs, and the intercostal muscles, which lie between the ribs. These walls also contain large amounts of elastic connective tissue.

The *diaphragm* is the principal muscle of respiration, but fibers of both external and internal intercostal muscles are also active during the various phases of the respiratory cycle; those firing during inspiration are called the *inspiratory intercostals,* and those active during forced expiration are the *expiratory intercostals.* Additional muscles can be recruited during forced respiration; some of the *accessory muscles of respiration* are the scalenes, sternocleidomastoids, posterior neck and back muscles (including the trapezius), muscles of the palate, tongue, and cheeks, and the abdominal muscles.

Movements of the Thorax The thoracic cage is quite mobile, and this mobility is ultimately responsible for respiration. Its dimensions increase during inspiration and return to resting size during normal expiration. The changes during inspiration are the result of contraction of various muscles, notably the diaphragm and inspiratory intercostals. Expiration is usually passive and occurs simply as a result of cessation of the inspiratory contractions.

During inspiration, the dimensions of the thoracic cage are increased in all three directions. Movement of the ribs, mainly by the inspiratory intercostal muscles, is responsible for the lateral and anteroposterior changes. Each rib can be regarded as a lever whose fulcrum lies near the vertebral column so that a slight movement of the rib at its vertebral end is much magnified at its sternal end. When a rib is elevated by muscle contraction, its shaft moves outward from the medial plane, increasing the lateral dimensions of the thoracic cage; simultaneously, it pushes the sternum forward, increasing the anteroposterior dimensions (Fig. 17-12).

FIGURE 17-12 ***Position of a rib at end inspiration*** **(dashed line)** ***and at end expiration*** **(solid line).**

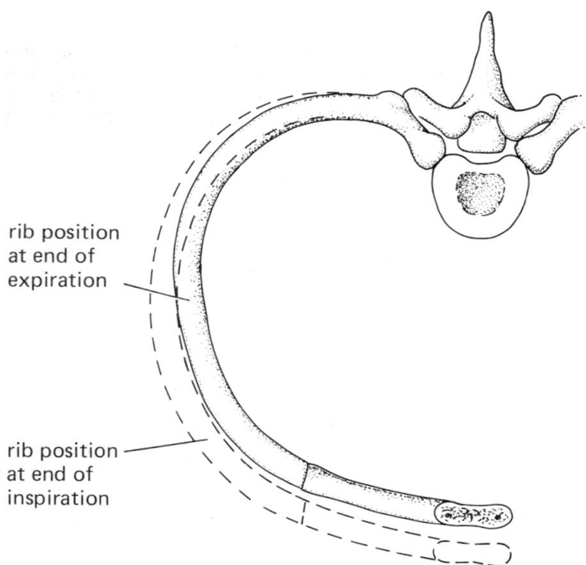

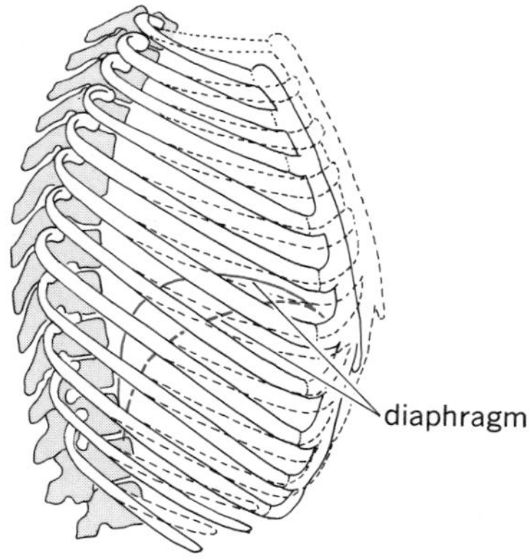

FIGURE 17-13 ***Movements of chest wall and diaphragm during breathing. The contracting intercostal muscles move the ribs upward and outward during inspiration while the contracting diaphragm moves downward.*** **(Adapted from McNaught and Callender.)**

The vertical (craniocaudal) dimension of the thoracic cavity is increased mainly by contraction of the diaphragm (Fig. 17-13). As the muscle fibers of the diaphragm contract, the central tendinous dome moves downward, pushing on the abdominal viscera, whose movements down and outward are made easier by the simultaneous relaxation of the muscles of the abdominal wall.

Relation of the Lungs to the Thoracic Cage The preceding section described how the dimensions of the thoracic cage are altered during normal respiration. To understand how these changes produce similar changes in the dimensions of the lungs, the relationship of the lungs to the thoracic cage must be appreciated.

Firmly attached to the entire interior of the thoracic cage is the *pleura,* a thin layer of loose connective tissue covered by a layer of mesothelium. The pleura forms two completely enclosed sacs in the thoracic cage, one on each side of the midline. The relationship between the lungs and pleura can be visualized by imagining what happens when one punches a fluid-filled balloon (Fig. 17-14). The arm represents the main bronchus leading to the lung, the fist is the lung, and the balloon is the pleural sac. The outer portion of the fist becomes coated by one surface of the balloon. In addition, the balloon is pushed back upon itself so that its surfaces lie close together. This is precisely the relation between the lung and pleura except that the pleural surface coating the lung is firmly attached to the lung sur-

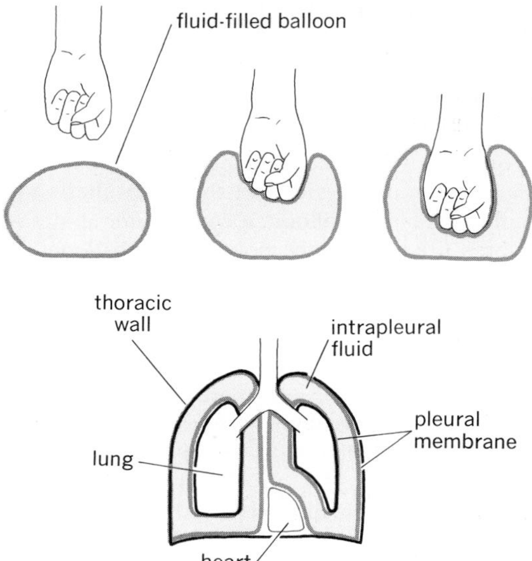

FIGURE 17-14 ***Relationship of lungs, pleura, and thoracic cage, analogous to pushing one's fist into a fluid-filled balloon. Note that there is no communication between the right and left intrapleural fluids. The volume of intrapleural fluid is greatly exaggerated; normally it consists of an extremely thin layer of fluid between the pleural membrane lining the inner surface of the thoracic cage and the pleural membrane lining the surface of the lungs.***

face. This layer of pleura (called the *visceral,* or *pulmonary, pleura*) and the outer layer which lines the interior thoracic wall (the *parietal pleura*) are so close to each other that they are virtually in contact, being separated only by a very thin layer of *intrapleural fluid.* In a normal person, these two surfaces always maintain this intimate relationship, which is of great importance for breathing.

INVENTORY OF STEPS INVOLVED IN RESPIRATION

1 Exchange of air between the atmosphere (external environment) and alveoli. This process includes the movement of air in and out of the lungs and the distribution of air within the lungs. Not only must a large volume of new air be delivered to the alveoli but it must be distributed proportionately to the millions of alveoli within each lung. This entire process is called *ventilation* and occurs by bulk flow.

2 Exchange of oxygen and carbon dioxide between alveolar air and lung capillaries by diffusion. The volume and distribution of the pulmonary (lung) blood flow are extremely important for normal functioning of this process.

3 Transportation of oxygen and carbon dioxide by the blood.

4 Exchange of oxygen and carbon dioxide between the blood and tissues of the body by diffusion as blood flows through tissue capillaries.

EXCHANGE OF AIR BETWEEN ATMOSPHERE AND ALVEOLI: VENTILATION

Like blood, air moves by bulk flow from a high pressure to a low pressure. We have seen (Chap. 1) that bulk flow can be described by the equation

$$F = k(P_1 - P_2)$$

That is, flow is proportional to the pressure difference between two points, k being the proportionality constant. For air flow, the two relevant pressures are the *atmospheric pressure* and the *intraalveolar pressure,* gas flow into or out of the lungs thus being given by

$$F = k(P_{atm} - P_{alv})$$

At sea level the atmospheric pressure is 760 mm Hg and is obviously not subject to control, short of putting a person in a spacesuit or diving bell. Since atmo-

FIGURE 17-15 ***Relationships required for breathing. When the alveolar pressure P_{alv} is less than atmospheric pressure, air enters the lungs. Flow F is directly proportional to the pressure difference, k being the proportionality constant.***

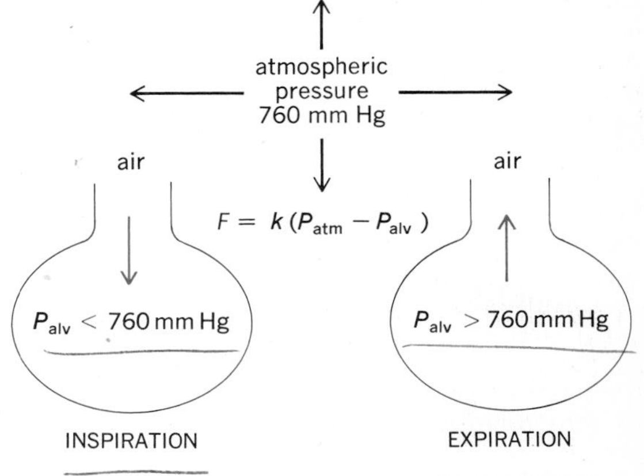

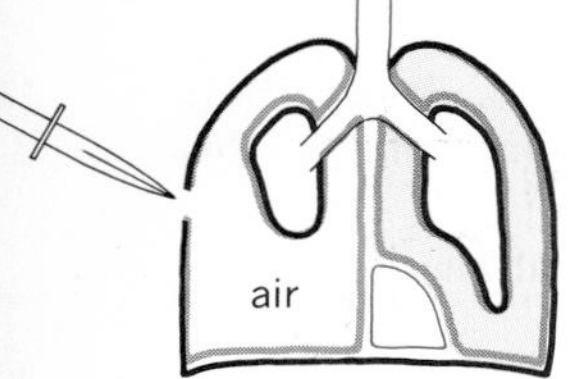

FIGURE 17-16 ***Lung collapse caused by a stab wound piercing the thoracic cage. Note that the air in the pleural space did*** **not** ***come from the lungs since the lung wall is still intact.***

spheric pressure remains relatively constant, if air is to be moved in and out of the lungs, the air pressure in the lungs, i.e., the intraalveolar pressure, must be made alternately less than and greater than atmospheric pressure (Fig. 17-15).

Concept of Intrapleural Pressure

If one cuts open the chest of an animal, being careful to cut only the thoracic wall but not the lung, the lung collapses immediately (Fig. 17-16). This is precisely what happens when a person is stabbed in the chest. Normally the highly elastic lungs are stretched within the intact chest, and the force responsible is eliminated when the chest is opened. This force is the subatmospheric pressure in the pleural fluid.

In a newborn child, the lungs and thoracic cage have approximately the same dimensions when unstretched. After birth, the thoracic cage grows more rapidly than the lungs; therefore, the thoracic wall tends to move away from the outer lung surface, but separation is prevented by the presence of the intrapleural fluid. An analogy (Fig. 17-17) may help illustrate the mechanism. Imagine two balloons of slightly different size, one inside the other. Since the inner, smaller balloon is open at the top so that there is free communication between its interior and the atmospheric air surrounding the larger balloon, it contains air at atmospheric pressure. The space between the balloons is completely filled with water. The forces acting upon the wall of the inner balloon are the inner air pressure and the water pressure surrounding it. Initially, as shown in the left half of Fig. 17-17, these pressures are approximately equal, and there is no tension in the wall of the inner balloon. When we enlarge the outer balloon by pulling on it in all directions, the inner balloon expands by an almost equal amount and its walls become highly stretched and taut. *This occurs because of a decrease in the fluid pressure of the water surrounding it.* Water is highly *indistensible;* i.e., any attempt to expand or compress a completely water-filled space causes a marked decrease or increase, respectively, of the fluid pressure in the space (it is much more difficult to compress a water-filled balloon than an air-filled balloon of the same size). Thus the pull on the external balloon produces a drop in the fluid pressure surrounding the inner balloon, which now becomes less than the air pressure in the inner balloon, and a transmural pressure gradient pushes out the wall of the inner balloon. As the inner wall moves outward, the internal air pressure falls slightly, but atmospheric air immediately enters through the opening so that atmospheric pressure is maintained. Thus, the external fluid pressure remains lower than the internal air pressure and the inner balloon expands until the force of its elastic recoil becomes great enough to balance this distending pressure difference. The balloon is behaving just like a stretched spring. The crucial event induced by expanding the outer balloon is a reduction of the fluid pressure between the balloons; this, in turn, causes the inner balloon to expand.

FIGURE 17-17 ***How fluid between two balloons causes the inner balloon to expand whenever the outer one does. (1) Outer balloon is expanded by an outside force. (2) Expansion of the fluid space causes the fluid pressure to decrease. (3) Fluid pressure is less than internal air pressure; therefore, the wall of the inner balloon is pushed out. (4) As the inner balloon expands, its internal air pressure decreases and air moves in from the atmosphere.***

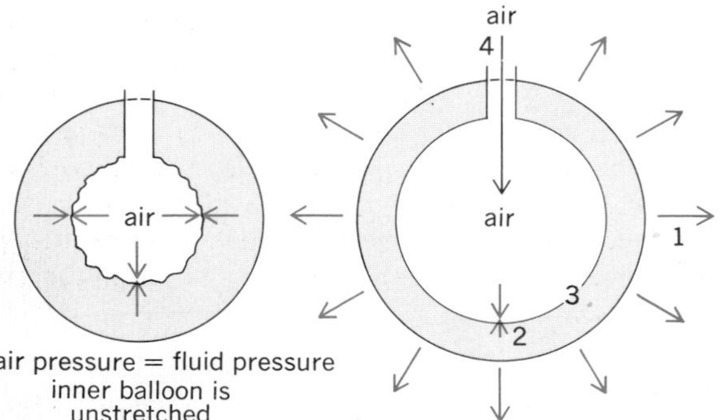

We can apply this analogy to the lungs (air-filled inner balloon), thoracic cage (outer balloon), and intrapleural fluid (the water between the balloons). As the thoracic cage expands during growth, it pulls slightly away from the outer surface of the lungs. This drops the *intrapleural fluid pressure* below that of the intraalveolar air pressure, a pressure difference that forces the lungs to distend. The lungs must expand to virtually the same degree as the thoracic cage, and their elastic walls become greatly stretched. The tendency for the lungs to recoil as a result of this stretch is balanced by the difference between the intraalveolar air pressure and the intrapleural fluid pressure.

Why the lung collapses when the chest wall is opened should now be apparent. The low intrapleural pressure is significantly less than the pressure of the atmospheric air outside the body; i.e., it is subatmospheric. When the chest wall is pierced, atmospheric air rushes into the intrapleural space, the pressure difference across the lung wall is eliminated, and the stretched lung collapses. Air in the intrapleural space is known as a *pneumothorax.*

The subatmospheric pressure of the intrapleural fluid, which is generated by the different growth rates of lung and thoracic cage, is maintained throughout life.[1] Regardless of whether the person is inspiring, expiring, or not breathing at all, the intrapleural pressure is always lower than the air pressure in the lungs, and the lungs are considerably stretched.[2] However, the gradient between intrapleural and intraalveolar pressures does vary during breathing and directly causes the changes in lung size which occur during inspiration and expiration. Recall that (Chap. 14) subatmospheric intrathoracic pressure was mentioned in describing the forces which determine end-diastolic ventricular volume.

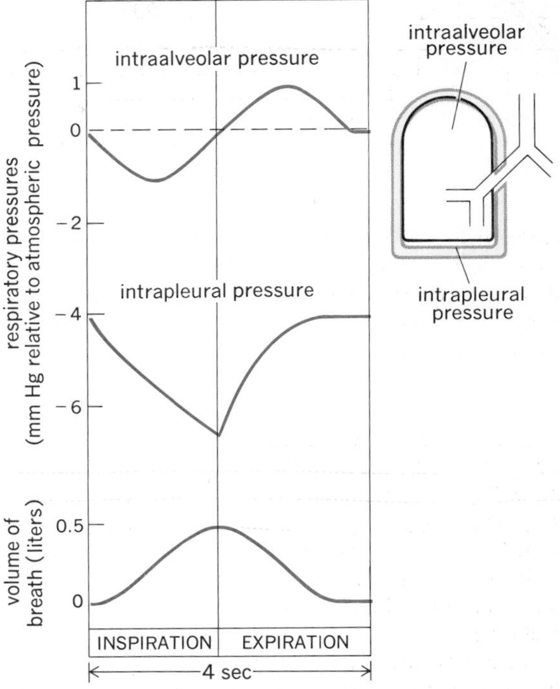

FIGURE 17-18 ***Summary of intraalveolar and intrapleural pressure changes and air flow during inspiration and expiration of 500 ml of air. Note that, on the pressure scale at the left, normal atmospheric pressure (760 mm Hg) has a scale value of zero.***

Inspiration

The left half of Fig. 17-18 summarizes events which occur during inspiration. Just before the inspiration begins, i.e., at the conclusion of the previous expiration, the respiratory muscles are relaxed and no air is flowing. The intrapleural pressure is subatmospheric (for reasons described above). The intraalveolar pressure, i.e., the air pressure in the alveoli, is exactly atmospheric because the alveoli are in free communication with the atmosphere via the airways. Inspiration is initiated by the contraction of the diaphragm and inspiratory intercostal muscles which, as described above, enlarge the volume of the thoracic cage. This expansion of the thoracic cage is merely a speeded-up version of the events described for growth. As the thoracic cage begins to move away from the lung surface, the intrapleural fluid pressure abruptly decreases, i.e., becomes even more subatmospheric. This increases the difference between the intraalveolar and intrapleural pressures, and the lung wall is pushed out. Thus, when the inspiratory muscles increase the thoracic dimensions, the lungs are also forced to enlarge because of the changes in intrapleural pressure.[3] This further stretching of the lung causes an increase in the volumes of all the air-containing passages and alveoli in the lung. As the alveoli enlarge, the air pressure in them drops to less than atmospheric, causing bulk flow of air from the at-

[1] Only during a forced expiration does intrapleural pressure exceed atmospheric pressure.

[2] Since intrapleural pressure is transmitted throughout the intrathoracic fluid surrounding not only the lungs but the heart and other intrathoracic structures as well, it is frequently termed the *intrathoracic pressure*.

[3] Another function of the intercostal muscles is to stiffen the intercostal spaces so that the tissues there are not drawn inward with the decrease in intrathoracic pressure.

mosphere through the airways into the alveoli until their pressure again equals atmospheric. Thus, air is literally sucked into the expanding lungs.

Expiration

The expansion of thorax and lungs produced during inspiration by active muscular contraction stretches both lung and thoracic wall elastic tissue. When inspiratory-muscle contraction ceases and these muscles relax, the stretched tissues recoil to their original length since there is no force left to maintain the stretch. An obvious analogy is the snap of a stretched rubber band when it is released. The tissue recoil causes a rapid and complete reversal of the inspiratory process, as shown in the right side of Fig. 17-18. The thorax and lung spring back to their original sizes, alveolar air becomes temporarily compressed so that its pressure exceeds atmospheric, and air flows from the alveoli through the airways out into the atmosphere. Normal expiration is thus completely passive, depending only upon the cessation of inspiratory-muscle activity and the relaxation of these muscles. Under certain conditions, however, particularly when resistance to air flow is abnormally high, expiration can be facilitated by the contraction of expiratory intercostal muscles and abdominal muscles, which actively decrease thoracic dimensions. The abdominal muscles help by increasing intraabdominal pressure to force the diaphragm up higher into the thorax and by depressing the lower ribs and flexing the trunk.

It should be noted that the analysis of Fig. 17-18 treats the lungs as a single alveolus. The fact is that there are significant regional differences in both intraalveolar and intrapleural pressures throughout the lungs and thoracic cavity. These differences are due, in part, to the effects of gravity and to local differences in the elasticity of the chest structures. They may be of great importance in determining the pattern of ventilation in disease states.

Quantitative Relationship Between Atmosphere-Intraalveolar Pressure Gradients and Air Flow: Airway Resistance

What is the quantitative relationship between the atmosphere-intraalveolar pressure gradients and the volume of air flow? It is expressed by precisely the same equation[4] given for the circulatory system by

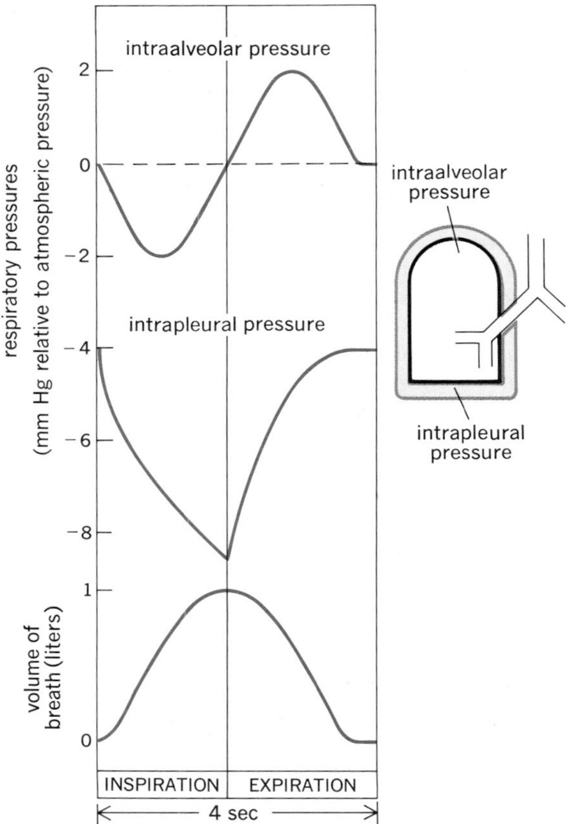

FIGURE 17-19 ***Summary of intraalveolar and intrapleural pressure changes and air flow during inspiration of 1,000 ml of air. Compare these values with those given in Fig. 17-18.***

Flow = pressure gradient/resistance

The volume of air which flows in or out of the alveoli is directly proportional to the pressure gradient between the alveoli and atmosphere and inversely proportional to the *resistance* to flow offered by the airways. Normally the magnitude of this pressure gradient is increased by taking deeper breaths, i.e., by increasing the strength of contraction of the inspiratory muscles. This causes, in order, an increased expansion of the thoracic cage, a greater drop in intrapleural pressure, increased expansion of the lungs, lower intraalveolar pressure, and increased flow of air into the lungs (Fig. 17-19).

What factors determine airway resistance? Resistance is (1) directly proportional to the magnitude of

[4] Note that, as for blood flow, we have merely transformed the basic equation, flow $= k(P_1 - P_2)$, by using resistance, which is the reciprocal of k.

the interactions between the flowing gas molecules, (2) directly proportional to the length of the airway, and (3) inversely proportional to the fourth power of the airway radius. These factors, of course, are counterparts of the factors determining resistance in the circulatory system; and in the respiratory tree, just as in the circulatory tree, resistance largely depends upon the radius of the conducting tubes.

The airway diameters are normally so large that they actually offer little total resistance to air flow, and interaction between gas molecules is also usually negligible, as is the contribution of airway length. Therefore, the total resistance remains so small that minute pressure gradients suffice to produce large volumes of air flow. As we have seen (Fig. 17-18), the average pressure gradient during a normal breath at rest is less than 1 mm Hg; yet, approximately 500 ml of air is moved by this tiny gradient.

The diameter of the airways becomes critically important in certain conditions such as asthma, which is characterized by severe bronchiolar smooth muscle constriction and plugging of the airways by bronchial secretions. Resistance to air flow may become great enough to prevent air flow completely, regardless of the atmosphere-alveolar pressure gradient.

Airway size and resistance may be altered by physical, nervous, or chemical factors. The most important normal physical factor is simply expansion of the lungs; during inspiration, airway resistance decreases because the overall enlargement of the lungs pulls on the airways and widens them. Conversely, during expiration, airway resistance increases. For this reason patients with abnormal airway resistance, as in asthma, have less difficulty inhaling than exhaling, with the result that air may be trapped in the lungs and the lung volume progressively increased.

Nervous regulation of airway size is mediated by the autonomic nervous system, the sympathetic neurons causing relaxation of the airway smooth muscle (decreased resistance) and the parasympathetics causing smooth muscle contraction (increased resistance). These reflexes are important in causing airway constriction upon inhalation of chemical irritants but their precise contribution to control of airway resistance under normal conditions is unclear.

As would be expected from knowledge of the effects of the sympathetic nerves on airway resistance, circulating epinephrine also causes airway dilatation. This is a major reason for administering epinephrine or epinephrine-like drugs to patients suffering from airway constriction, as in an asthmatic attack. In contrast, histamine and certain other chemical mediators cause bronchiolar constriction (and increased mucus secretion as well) and may be the cause of the airway constriction observed in allergic attacks. This explains the use of antihistamines to relieve the respiratory symptoms of allergies.

It is interesting that contraction of certain skeletal muscles can also influence the airway diameter. For example, the *alae nasi* widen the nostrils, and the laryngeal muscles alter the diameter of the larynx.

So far we have been discussing *total* airway resistance, but discrete *local* changes in airway resistance are important in promoting efficient gas exchange. The bronchioles contain smooth muscle highly responsive to the carbon dioxide concentration of the medium surrounding them. High carbon dioxide increases bronchodilatation and low carbon dioxide increases bronchoconstriction. These effects are exerted locally directly upon the smooth muscle and are independent of nerves or hormones. What is the significance of this sensitivity? The lungs are composed of approximately 300 million discrete alveoli, each receiving carbon dioxide from the pulmonary capillary blood. To be most efficient, the right proportion of alveolar air and capillary blood should be available to each alveolus, a pattern local changes in bronchiolar tone help maintain. In Fig. 17-20, for example, the left alveolus is receiving too much air for its blood supply and vice versa on the right. Because of the deficient blood flow, the left alveolus receives little carbon dioxide and the bronchiole supplying it is exposed to a low carbon dioxide concentration and constricts. Conversely, the area around the poorly ventilated right alveolus has a high carbon dioxide concentration, and the bronchiole supplying it dilates. By this completely local mechanism, the ventilation and blood supply are matched.

Control of Pulmonary Arterioles

Chapter 16 described in detail how the systemic arterioles control the distribution of blood to the various organs and tissues. The pulmonary arterioles perform an analogous function in controlling the distribution of blood to different alveolar capillaries, thus providing a second mechanism for matching air flow and blood flow. This control is based, in large part, on the great sensitivity to oxygen of the pulmonary arteriolar smooth muscle; a decreased oxygen concentration causes arteriolar constriction, whereas an increased oxygen concentration causes vasodilatation. In addition, the pulmonary vessels are also sensitive to their local hydrogen-ion concentration; an increased hydrogen-ion concentration causes constriction whereas a decreased concentration causes vasodilatation. (As will be emphasized subsequently, the hydrogen-ion concentration reflects, in large part, the partial pressure of carbon dioxide.) These purely local (no nerves or hormones) effects of oxygen and hydrogen ion on pulmonary arterioles are precisely the opposite of those exerted by them on systemic arterioles (Chap. 16); no

FIGURE 17-20 ***Adjustment of air flow by means of carbon dioxide–induced changes in bronchiolar constriction. Increased carbon dioxide causes bronchiolar dilatation; conversely, decreased carbon dioxide results in bronchiolar constriction.*** **(Top panel)** ***Initial state;*** **(bottom panel)** ***after compensation.***

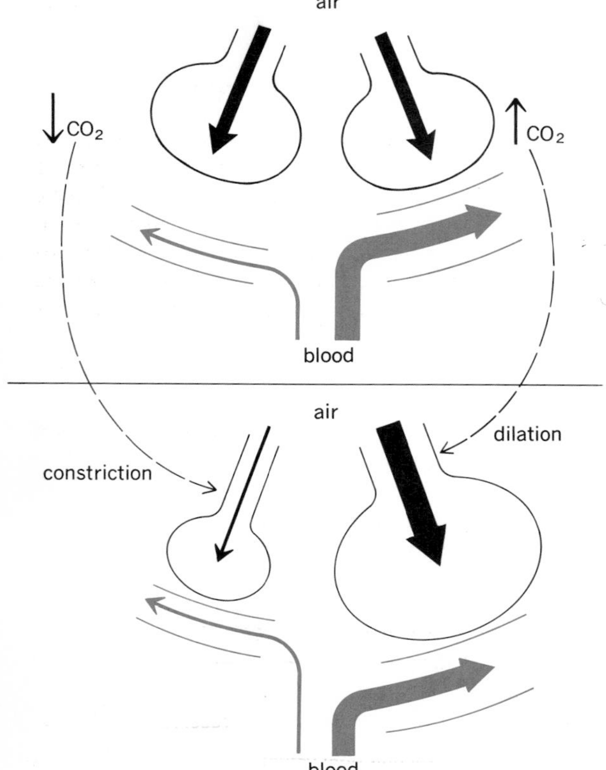

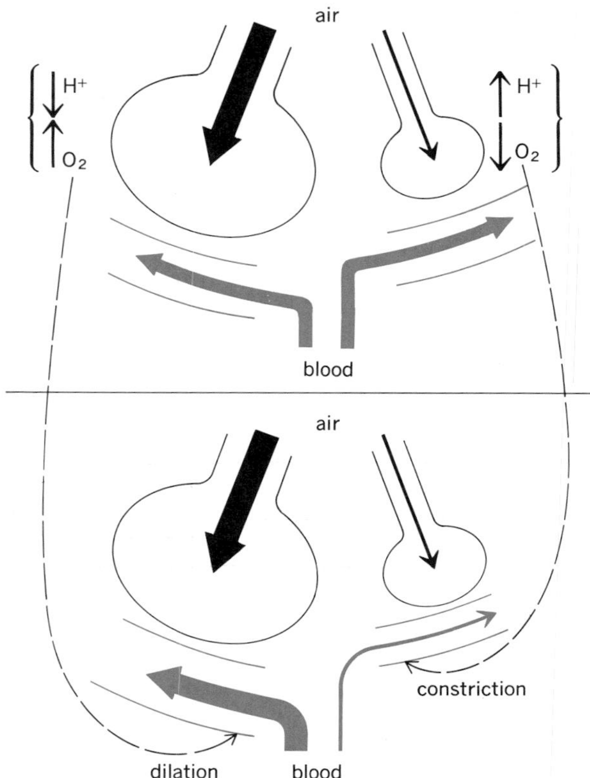

FIGURE 17-21 ***Adjustment of blood flow to air flow by means of hydrogen-ion– and oxygen-induced changes in arteriolar constriction. Increased oxygen and decreased hydrogen ion cause arteriolar dilatation; conversely, decreased oxygen and increased hydrogen ion result in arteriolar constriction.*** **(Top panel)** ***Initial state;*** **(bottom panel)** ***after compensation.***

chemical explanation for this striking difference is known. In Fig. 17-21, with the same uneven distribution pattern as Fig. 17-20, the left alveolar area has high oxygen and low hydrogen ion because it receives much air but little blood; conversely, the right alveolar area has low oxygen and high hydrogen ion. This causes vasoconstriction of the right arteriole and dilatation of the left and shunts blood away from the poorly ventilated alveolus and to the richly ventilated one. Of course, the events of Figs. 17-20 and 17-21 occur simultaneously (Table 17-2) so that the final situation lies somewhere in between. The net result is efficient

TABLE 17-2
MAJOR FACTORS WHICH CONTROL AIRWAY AND PULMONARY VASCULAR RESISTANCE

	Airways	*Arterioles*
Constricted by:	Histamine Parasympathetic nerves $\downarrow CO_2$	$\downarrow O_2$ $\uparrow H^+$
Dilated by:	Epinephrine Sympathetic nerves $\uparrow CO_2$	$\uparrow O_2$ $\downarrow H^+$

matching of air flow and blood flow in 300 million alveoli.

Before leaving the subject of the pulmonary circulation, we wish to reemphasize the fact, described in Chap. 15, that the pulmonary circulation is a low-pressure circuit. The normal pulmonary capillary pressure is only 15 mm Hg. This is the major force favoring movement of fluid out of the pulmonary capillaries into the interstitium and alveoli. It is well below the major force favoring absorption, namely, the plasma colloid osmotic pressure of 25 mm Hg. Accordingly, the alveoli normally remain dry, a feature essential for normal gas exchange. Should pulmonary capillary pressure increase greatly, as occurs when the left ventricle "fails," fluid accumulates in the alveoli (pulmonary edema) with serious consequences for gas exchange across the alveolar wall.

Lung-Volume Changes during Breathing

The volume of air entering or leaving the lungs during a single breath is called the *tidal volume* (Fig. 17-22). For a breath under resting conditions, this is approximately 500 ml. We are all aware that the resting thoracic excursion is small compared with a maximal breathing effort. The volume of air which can be inspired over and above the resting tidal volume is called the *inspiratory reserve* and amounts to 2,500 to 3,500 ml of air. At the end of a normal expiration, the lungs still contain a large volume of air, part of which can be exhaled by active contraction of the expiratory muscles; it is called the *expiratory reserve* and measures approximately 1,000 ml of air. Even after a maximal expiration, some air (approximately 1,000 ml) still remains in the lungs and is termed the *residual volume*.

What, then, is the maximum amount of air which can be moved in and out during a single breath? It is the sum of the normal tidal, inspiratory-reserve, and expiratory-reserve volumes. This total volume is called the *vital capacity*. During heavy work or exercise, a person uses part of both the inspiratory and expiratory reserves (particularly the former) but rarely uses more than 50 percent of his or her total vital capacity, because deeper breaths than this require exhausting activity of the inspiratory and expiratory muscles. The greater depth of breathing during exercise greatly increases pulmonary ventilation, and still larger increases are produced by speeding the rate of breathing as well.

The *total pulmonary ventilation* per minute (*minute ventilatory volume*) is determined by the tidal volume times the respiratory rate (expressed as breaths per minute). For example, at rest, a normal person moves approximately 500 ml of air in and out of the lungs with each breath and takes 10 breaths each minute. The total minute ventilatory volume is therefore 500 ml × 10 = 5,000 ml of air per minute. However, as we shall see, not all this air is available for exchange with the blood.

Air Distribution within the Lungs

Anatomic Dead Space The respiratory tract, as we have seen, is composed of conducting airways and the alveoli. Within the lungs, exchanges of gases with the blood occur only in the alveoli and not in the conducting airways, the total volume of which is approximately 150 ml. Picture, then, what occurs during expiration: 500 ml of air is forced out of the alveoli and

FIGURE 17-22 ***Lung volumes.*** **(Adapted from Comroe.)**

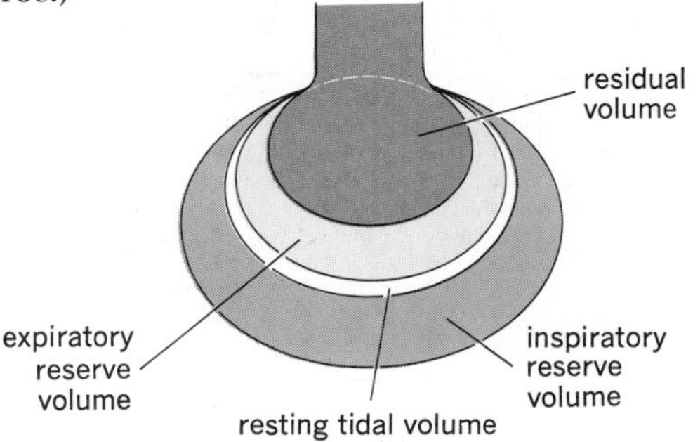

through the airways. Approximately 350 ml of this air is exhaled at the nose or mouth, but approximately 150 ml still remains in the airways at the end of expiration. During the next inspiration, 500 ml of air flows into the alveoli, but the first 150 ml of entering air is not atmospheric but the 150 ml of alveolar air left behind. Thus, only 350 ml of new atmospheric air enters the alveoli during the inspiration. At the end of inspiration, 150 ml of fresh air also fills the conducting airways, but no gas exchange with the blood can occur there. At the next expiration, this fresh air will be washed out and again replaced by old alveolar air, thus completing the cycle. The end result is that 150 ml of the atmospheric air entering the respiratory system during each inspiration never reaches the alveoli but is merely moved in and out of the airways. Because these airways do not permit gas exchange with the blood, the space within them is termed the *anatomic dead space*.

The volume of fresh atmospheric air entering the alveoli during each inspiration, then, equals the volume of air in the anatomic dead space subtracted from the total tidal volume. Thus, for a normal breath

Tidal volume = 500 ml
Anatomic dead space = 150 ml
Fresh air entering alveoli = 350 ml

This total is called the *alveolar ventilation*. The term is somewhat confusing, because it seems to indicate that only 350 ml air enters and leaves the alveoli with each breath. This is not true—the total is 500 ml of air, but only 350 ml is fresh air.

What is the significance of the anatomic dead space and alveolar ventilation? Total minute ventilatory volume, as explained above, is equal to the tidal volume of each breath multiplied by the number of breaths per minute. But since only that portion of inspired atmospheric air which enters the alveoli, i.e., the alveolar ventilation, is useful for gas exchange with the blood, the magnitude of the alveolar ventilation is of much greater significance than the magnitude of the total pulmonary ventilation, as can be demonstrated readily by the data in Table 17-3.

In this experiment subject A breathes rapidly and shallowly, B normally, and C slowly and deeply. Each subject has exactly the same total ventilatory volume; i.e., each is moving the same amount of air in and out of the lungs each minute. Yet, when we subtract the anatomic-dead-space ventilation from the total ventilatory volume, we find marked differences in alveolar ventilation. Subject A has no alveolar ventilation (and would become unconscious in several minutes), whereas C has a considerably greater alveolar ventilation than B, who is breathing normally. The important deduction to be drawn from this example is that increased depth of breathing is far more effective in elevating alveolar ventilation than an equivalent increase of breathing rate. Conversely, a decrease in depth can lead to a critical reduction of alveolar ventilation. This is because a fraction of *each* tidal volume represents anatomic-dead-space ventilation. If the tidal volume decreases, this fraction increases until, as in subject A, it may represent the entire tidal volume. On the other hand, any increase in tidal volume goes entirely toward increasing alveolar ventilation. Alveolar ventilation is calculated as follows:

Alveolar ventilation (ml/min)
= frequency (breaths/min)
× [tidal volume (ml) − dead space (ml)]

These concepts have important physiologic implications. Most situations, such as exercise, which necessitate an increased oxygen supply (and carbon dioxide elimination) reflexly call forth a relatively greater increase in breathing depth than rate. Indeed, well-

TABLE 17-3
EFFECT OF BREATHING PATTERNS ON ALVEOLAR VENTILATION

Subject	*Tidal volume, ml/breath*	×	*Frequency, breaths/min*	=	*Total pulmonary ventilation, ml/min*	*Dead space ventilation, ml/min*	*Alveolar ventilation, ml/min*
A	150		40		6,000	150 × 40 = 6,000	0
B	500		12		6,000	150 × 12 = 1,800	4,200
C	1,000		6		6,000	150 × 6 = 900	5,100

trained athletes can perform moderate exercise with very little increase, if any, in respiratory rate. The mechanisms by which rate and depth of respiration are controlled will be described in a later section of this chapter.

Alveolar Dead Space Some inspired air is not useful for gas exchange with the blood even though it reaches the alveoli because some alveoli, for various reasons, receive too little blood supply for their size. Air which enters these alveoli during inspiration cannot exchange gases efficiently because there is insufficient blood. The inspired air in this *alveolar dead space* must be distinguished from that in the anatomic dead space. It is quite small in normal persons but may reach lethal proportions in several kinds of lung diseases, e.g., emphysema. As we have seen, it is minimized by the local mechanisms which match air and blood flows.

Work of Breathing

During inspiration, active muscular contraction provides the energy required to expand the thorax and lungs. What determines how much work these muscles must perform in order to provide a given amount of ventilation? First, there is simply the stretchability of the thorax and lungs. To expand these structures they must be stretched. The easier they stretch, the less energy is required for a given amount of expansion. Much of the work of breathing goes into stretching the elastic tissue of the lung, but an even larger fraction goes into stretching a different kind of "tissue" – water itself! The air in each alveolus is separated from the alveolar membranes by an extremely thin layer of fluid; in a sense, therefore, the alveoli may be viewed as air-filled bubbles lined with water. At an air-water interface, the attractive forces between water molecules cause them to squeeze in upon the air within the bubble (Fig. 17-23). This force, known as *surface tension,* makes the water lining very like highly stretched rubber which constantly tries to shorten and resists further stretching. Thus inspiration requires considerable energy to expand the lungs because of the difficulty of distending these alveolar bubbles. Indeed, the surface tension of pure water is so great that lung expansion would require exhausting muscular effort and the lungs would tend to collapse. It is extremely important, therefore, that specialized alveolar cells, the type II cells, produce a phospholipoprotein complex, known as *pulmonary surfactant,* which intersperses with the water molecules and markedly reduces their cohesive force,

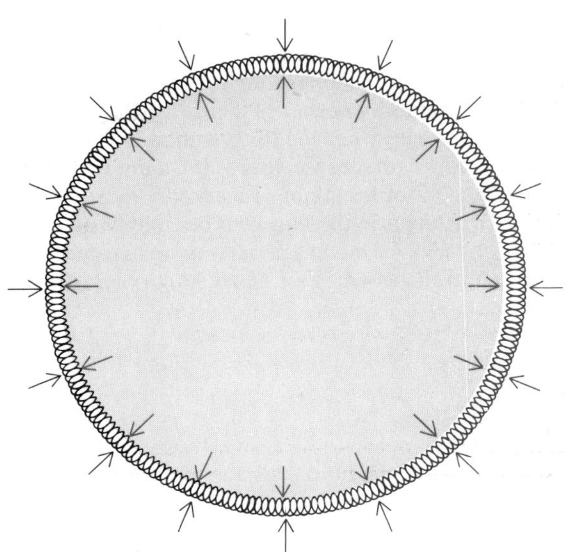

FIGURE 17-23 ***Forces acting on the surface of a bubble. The springs and dark arrows represent the surface tension resulting from the cohesive forces of water molecules at the air-water interface. This tension is opposed by the air pressure within the bubble*** **(colored arrows).**

thereby lowering the surface tension. Surfactant is continuously replenished by these alveolar cells, and normal ventilation of the lungs seems to be the stimulus for its production. A striking example of what occurs when insufficient surfactant is present is provided by the disease known as "respiratory-distress syndrome of the newborn," which frequently afflicts premature infants in whom the surfactant-synthesizing cells are too immature to function adequately. The infant is able to inspire only by the most strenuous efforts which may ultimately cause complete exhaustion, inability to breathe, lung collapse, and death. The recent discovery that the administration of adrenal steroids markedly enhances the maturation process of the surfactant-synthesizing cells may provide an important means of combating this disease.

The second factor determining the degree of muscular work required for a certain amount of ventilation is the magnitude of the airway resistance. When airway resistance is increased by bronchiolar constriction or secretions (as in asthma), the usual pressure gradient does not suffice for adequate air inflow and a deeper breath is required to create a larger pressure gradient.

One might imagine from this discussion (and from observing an athlete exercising hard) that the work of breathing uses up a major portion of the energy spent by the body. Not so; in a normal person, even during heavy exercise, the energy needed for breathing is only about 3 percent of the total expenditure. It is only in disease, when the work of breathing is markedly increased by structural changes in the lung or thorax, by loss of surfactant, or by an increased airway resistance, that breathing itself becomes an exhausting form of exercise.

EXCHANGE AND TRANSPORT OF GASES IN THE BODY

We have completed our discussion of alveolar ventilation, but this is only the first step in the total respiratory process. Oxygen must move across the alveolar membranes into the pulmonary capillaries, be transported by the blood to the tissues, leave the tissue capillaries, and finally cross cell membranes to gain entry into cells. Carbon dioxide must follow a similar path in reverse (Fig. 17-24). At rest, during each minute, body cells consume approximately 200 ml of oxygen and produce approximately the same amount of carbon dioxide. The relative amounts depend primarily upon what nutrients are being used for energy; e.g., when glucose is utilized, one molecule of carbon dioxide is produced for every molecule of oxygen consumed:

$$C_6H_{12}O_6 + 6O_2 \longrightarrow 6CO_2 + 6H_2O + \text{energy}$$

The ratio (CO_2 produced)/(O_2 consumed) is known as the *respiratory quotient* (RQ) and accordingly is 1 for glucose. When fat is utilized, only 7 molecules of carbon dioxide are produced for every 10 molecules of oxygen consumed, and RQ = 0.7. For simplicity, Fig. 17-24 assumes that the carbon dioxide and oxygen amounts are equal and the total volumes of air inspired and expired therefore identical.

At rest, the total pulmonary ventilation equals 5 l of air per minute. Since only 20 percent of atmospheric air is oxygen (most of the remainder is nitrogen), the total oxygen input is $20\% \times 5\,l = 1\,l$ of O_2 per minute. Of this inspired oxygen, 200 ml crosses the alveoli into the pulmonary capillaries, and the remaining 800 ml is exhaled. This 200 ml of oxygen is carried by 5 l of blood, which is the pulmonary blood flow (cardiac output) per minute. Note, however, that blood entering the lungs already contains large quantities of oxygen, to which this 200 ml is added. This blood is then pumped by the left ventricle through the tissue capillaries of the body, and 200 ml of oxygen leaves the blood to be taken up and utilized by cells. Because only a fraction of the total blood oxygen actually leaves the blood, some oxygen remains in the blood when it returns to the heart and lungs. It is obvious but important that the quantities of oxygen added to the blood in the lungs and removed in the tissues are identical. As shown by Fig. 17-24, the story reads in reverse for carbon dioxide. As we shall see, most of the blood carbon dioxide is actually in the form HCO_3^-, but we have shown it as CO_2 for simplicity.

The pumping of blood by the heart obviously propels oxygen and carbon dioxide between the lungs and tissues, but what forces induce the net movement of these molecules across the alveolar, capillary, and cell membranes? The answer is *diffusion. There is no active membrane transport for oxygen or carbon dioxide; they move solely by passive diffusion.* As described in Chap. 1, diffusion can effect the net transport of a substance only when a concentration gradient exists for it. Understanding the mechanisms involved depends upon familiarity with some basic chemical and physical properties of gases, to which we now turn.

Basic Properties of Gases

A gas consists of individual molecules constantly moving at great speeds. Since rapidly moving molecules bombard the walls of any vessel containing them, they therefore exert a *pressure* against the walls. The magnitude of the pressure is increased by anything which increases the bombardment. The pressure a gas exerts is proportional to (1) the temperature, because heat increases the speed at which molecules move, and (2) the concentration of the gas, i.e., the number of molecules per unit volume. In other words, when a certain number of molecules are compressed into a smaller volume, there are more collisions with the walls.

The pressure of a gas is therefore a measure of the concentration and speed of its molecules. Gases move from a region of higher pressure to a region of lower pressure, and diffusion is thus the result of the continuous movement of gas molecules. Some molecules, of course, are moving against the gradient, but many more are moving from the region of higher pressure to that of lower, and we should really speak of *net* diffusion.

Of great importance is the relationship between different gases, i.e., different kinds of molecules, such as oxygen and nitrogen, in the same container. In a mix-

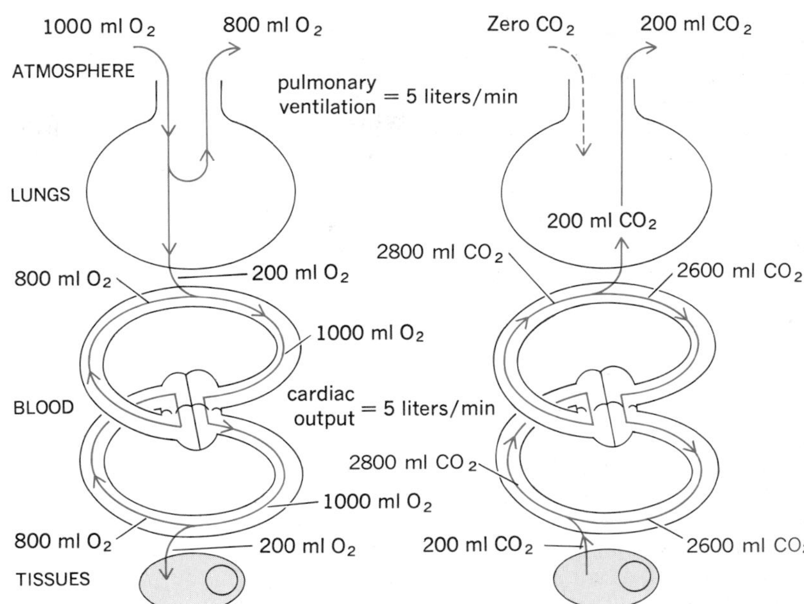

FIGURE 17-24 ***Summary of exchanges between atmosphere, lungs, blood, and tissues during 1 min. RQ is assumed to be 1.***

ture of gases, the pressure exerted by each gas is independent of the pressure exerted by the others because gas molecules are normally so far apart that they do not interfere with each other. Since each gas behaves as though the other gas were not present, the total pressure of a mixture of gases is simply the sum of the individual pressures. These individual pressures, termed *partial pressures*, are denoted by a *P* in front of the symbol for the gas, the partial pressure of oxygen thus being represented by P_{O_2}. Gas pressures are usually expressed in millimeters of mercury, the same units used for the expression of hydrostatic pressure.

Behavior of Gases in Liquids

Several factors determine the uptake of gases by liquids and the behavior of gases dissolved in liquids. When a free gas comes into contact with a liquid, the number of gas molecules which dissolve in the liquid is directly proportional to the pressure of the gas. This phenomenon is clear from the basic definition of pressure. Suppose, for example, that oxygen is placed in a closed vessel half full of water. Oxygen molecules constantly bombard the surface of the water, some entering the water and dissolving. Since the number of molecules striking the surface is directly proportional to the pressure of the oxygen gas P_{O_2}, the number of molecules entering the water is also directly proportional to P_{O_2}. How many entering molecules actually stay in the water? Since the dissolved oxygen molecules are also constantly moving, some of them strike the water surface from below and escape into the free oxygen above. The rate of escape from the water and the rate of entry into the water are equal when the rates of bombardment are equal, i.e., when the pressures of the oxygen in the free gas and in the water become identical. Thus, we come back to our earlier statement: The number of gas molecules which will dissolve in a liquid is directly proportional to the pressure of the gas. When the free-gas pressure is higher than the pressure of the gas in a liquid, a number of molecules must dissolve in the liquid for the pressure of the dissolved gas to equal the pressure of the free gas. Conversely, if a liquid containing a dissolved gas at high pressure is exposed to that same free gas the pressure of which is lower, gas molecules leave the liquid and enter the free gas until the free- and dissolved-gas pressures become equal. These are precisely the phenomena occurring between alveolar air and pulmonary capillary blood.

It should also be apparent that dissolved gas molecules diffuse *within* the liquid from a region of higher gas pressure to a region of lower pressure, an effect which underlies the exchange of gases between cells, tissue fluid, and capillary blood throughout the body.

This discussion has been in terms of proportionalities rather than absolute amounts. The number of gas molecules which will dissolve in a liquid is *proportional* to the gas pressure, but the *absolute* number also depends upon the *solubility* of the gas in the liquid. Thus, if a liquid is exposed to two different gases at the same pressures, the numbers of molecules of each gas which are dissolved at equilibrium are not necessarily identical but reflect the solubilities of the two gases. Nevertheless, doubling the gas pressures doubles the number of gas molecules dissolved.

Pressure Gradients of Oxygen and Carbon Dioxide within the Body

With these basic gas properties as foundation, we can discuss the diffusion of oxygen and carbon dioxide across alveolar, capillary, and cell membranes. The pressures of these gases in atmospheric air and in various sites of the body are given in Fig. 17-25 for a resting person at sea level. The rest of this section is devoted to an elaboration of this figure.

Atmospheric air consists primarily of nitrogen and oxygen with very small quantities of water vapor, carbon dioxide, and inert gases such as argon. The sum of the partial pressures of all these gases is termed *atmospheric pressure* or *barometric pressure*. It varies in different parts of the world as a result of differences in altitude, but at sea level it is 760 mm Hg. Since air is 20 percent oxygen, the Po_2 of inspired air is $20\% \times 760 = 152$ mm Hg.

The first question suggested by Fig. 17-25 is why the partial pressures of the constituents of expired air are not identical to those of alveolar air. Recall that approximately 150 ml of the inspired atmospheric air during each breath never gets down into the alveoli but remains in the airways (dead space). This air does not exchange carbon dioxide or oxygen with blood and is expired along with alveolar air during the subsequent expiration. Therefore, the Po_2 and Pco_2 of the total expired air are higher and lower, respectively, than those of alveolar air.

The next question concerns the alveolar gas pressures themselves. One might logically reason that the alveolar gas pressures must vary considerably during the respiratory cycle, since new atmospheric air enters only during inspiration. In fact, however, the variations in alveolar Po_2 and Pco_2 during the cycle are so small as to be negligible because, as explained in the section on lung volumes, a large volume of gas is always left in the lungs after expiration. This remaining alveolar gas contains large quantities of oxygen and carbon dioxide, and when the new air enters, it mixes with the alveolar air already present, lowering its Pco_2 and raising its Po_2, but only by a small amount. For this reason, the alveolar-gas partial pressures remain *relatively* constant throughout the respiratory cycle, and we may use the single alveolar pressures shown in Fig. 17-25 in our subsequent analysis of alveolar-capillary exchange, ignoring the minor fluctuations.

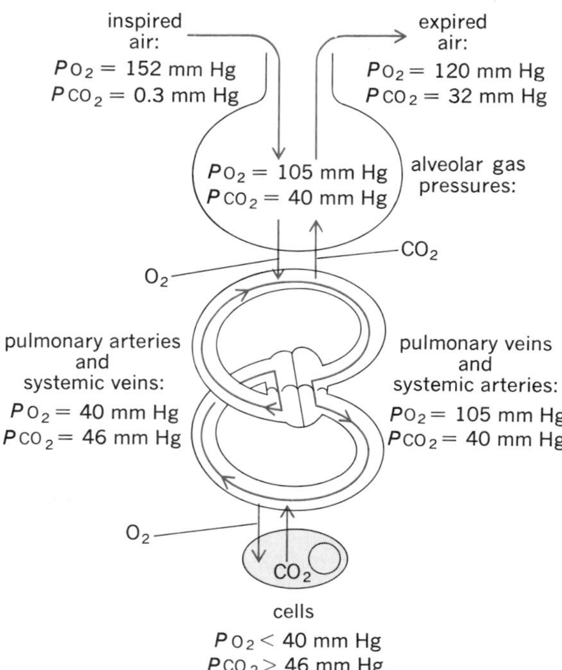

FIGURE 17-25 ***Summary of carbon dioxide and oxygen pressures in the inspired and expired air and various places within the body.***

Our next question concerns the exchange of gases between alveoli and pulmonary capillary blood. The blood which enters the pulmonary capillaries is, of course, systemic venous blood pumped to the lungs via the pulmonary arteries. Having come from the tissues, it has a high Pco_2 (46 mm Hg) and a low Po_2 (40 mm Hg). As it flows through the pulmonary capillaries, it is separated from the alveolar air only by an extremely thin layer of tissue. The differences in the partial pressures of oxygen and carbon dioxide on the two sides of this alveolar-capillary membrane result in the net diffusion of oxygen into the blood and of carbon dioxide into the alveoli. As this diffusion occurs, the capillary blood Po_2 rises above its original value and the Pco_2 falls. The net diffusion of these gases ceases when the alveolar and capillary partial pressures become equal. In a normal person, the rates at which oxygen and carbon dioxide diffuse are so rapid and the blood flow through the capillaries so slow that complete equilibrium is always reached (at rest, a red blood cell takes 0.75 s to pass through the pulmonary capillaries). Thus, the blood that leaves the lungs to return to the heart and be pumped into the arteries has the same Po_2 and Pco_2 as alveolar air. (Actually, the Po_2 of arterial blood is slightly less than that of alveolar air, but we shall ignore this discrepancy in our analysis.)

The diffusion of gases between alveoli and capillaries may be impaired in a number of ways. The disease emphysema, which is intimately related to cigarette smoking, is characterized by the breakdown of the alveolar walls with the formation of fewer but larger alveoli. The result is a reduction in the total area available for diffusion. In a different kind of defect, caused by membrane thickening or by pulmonary edema, the area available for diffusion is normal but the molecules must travel a greater distance. Finally, without becoming thicker the alveolar walls may become denser and less permeable, as for example when beryllium is inhaled and deposited on the walls.

As the arterial blood enters capillaries throughout the body, it becomes separated from the interstitial fluid only by the thin, highly permeable capillary membrane. The interstitial fluid, in turn, is separated from intracellular fluid by cell membranes which are also quite permeable to oxygen and carbon dioxide. Metabolic reactions occurring in these cells are constantly consuming oxygen and producing carbon dioxide. Therefore, as shown in Fig. 17-25, intracellular Po_2 is lower and Pco_2 higher than in blood. As a result, a net diffusion of oxygen occurs from blood to cells, and a net diffusion of carbon dioxide from cells to blood. In this manner, as blood flows through capillaries, its Po_2 decreases and its Pco_2 increases until, by the end of the capillaries, equilibrium has been reached. This accounts for the venous blood values shown in Fig. 17-25. Venous blood returns to the right ventricle and is pumped to the lungs, where the entire process begins again.

In summary, no active transport mechanisms are required to explain the exchange of gases in the lungs and tissues. The consumption of oxygen in the cells and the supply of new oxygen to the alveoli create Po_2 gradients which produce net diffusion of oxygen from alveoli to blood in the lungs and from blood to cells in the rest of the body. Conversely, the production of carbon dioxide by cells and its elimination from the alveoli via expiration create Pco_2 gradients which produce net diffusion of carbon dioxide from blood to alveoli in the lungs and from cells to blood in the rest of the body.

Transport of Oxygen in the Blood: The Role of Hemoglobin

Table 17-4 summarizes the oxygen content of arterial blood. Each liter of arterial blood contains the same number of oxygen molecules as 200 ml of gaseous oxygen. Oxygen is present in two forms: (1) physically dissolved in the blood water and (2) chemically bound to *hemoglobin* molecules. The amount of oxygen which can be physically dissolved in blood is directly proportional to the Po_2 of the blood, but because oxygen is relatively insoluble in water, only 3 ml of oxygen can be dissolved in 1 l of blood at the normal alveolar and arterial Po_2 of 100 mm Hg. In contrast, in 1 l of blood, 197 ml of oxygen, more than 98 percent of the total, is carried in the red blood cells chemically bound to hemoglobin.

TABLE 17-4
OXYGEN CONTENT OF ARTERIAL BLOOD

1 l arterial blood contains:	
	3 ml O_2 physically dissolved
	197 ml O_2 chemically bound to hemoglobin
Total	200 ml O_2
Cardiac output = 5 l/min	
O_2 carried to tissues/min = 5 × 200 = 1,000 ml	

The chemical reaction between oxygen and hemoglobin is usually written

$$O_2 + Hb \rightleftharpoons HbO_2 \qquad (17\text{-}1)$$

Hemoglobin combined with oxygen (HbO_2) is called *oxyhemoglobin;* not combined (Hb), it is called *reduced hemoglobin* or deoxyhemoglobin. Because the number of sites on the hemoglobin molecule which bind oxygen is limited, only a certain number of oxygen molecules can be combined with the hemoglobin molecule. When hemoglobin has been converted completely to HbO_2, it is said to be *fully saturated.* When hemoglobin exists as mixed Hb and HbO_2, it is said to be *partially saturated.*

The *percentage saturation* of hemoglobin is a measure of the fraction of total hemoglobin which is combined with oxygen, i.e., in the form of HbO_2.

What factors determine the extent to which oxygen will combine with hemoglobin? By far the most important is the P_{O_2} of the blood, i.e., the concentration of physically dissolved oxygen. The blood hydrogen-ion concentration and temperature also play significant roles as do certain chemicals produced by the red blood cells themselves.

Effect of $\mathbf{P_{O_2}}$ ***on Hemoglobin Saturation*** From inspection of Eq. 17-1 and the law of mass action, it is obvious that raising the P_{O_2} of the blood should increase the combination of oxygen with hemoglobin. The experimentally determined quantitative relationship between these variables is shown in Fig. 17-26. When a sample of blood is placed in a flask and exposed to a large volume of free oxygen, a net diffusion of oxygen occurs from the free gas into the blood until the blood and free-gas partial pressures become equal. If

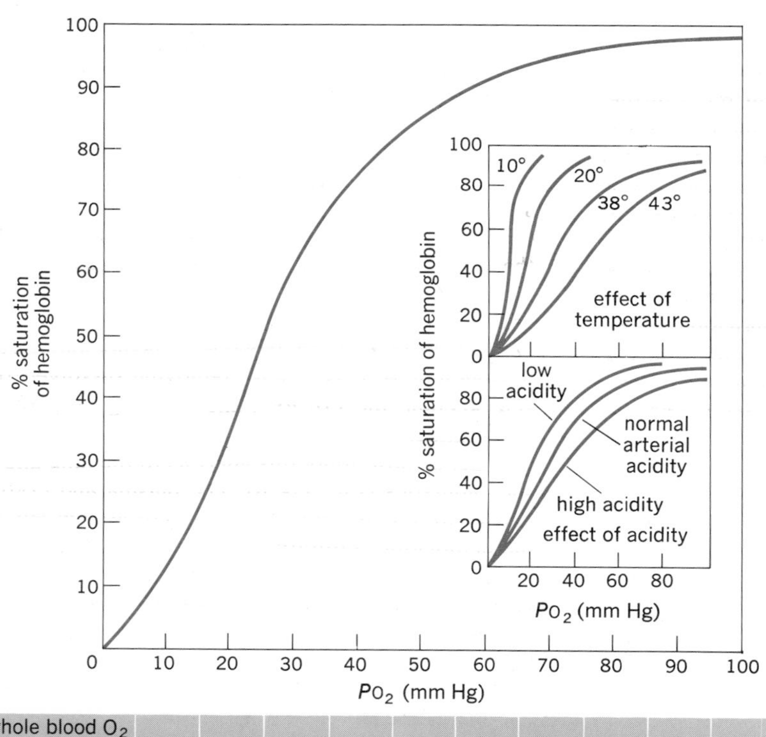

FIGURE 17-26 ***Hemoglobin-oxygen dissociation curve. The large curve applies to blood at 38°C and the normal arterial hydrogen-ion concentration (acidity). The inset curves illustrate the effects of altering temperature and acidity on the relationship between*** $\mathbf{P_{O_2}}$ ***and hemoglobin saturation with oxygen.*** **(Adapted from Comroe.)**

A whole blood O_2 ml O_2/100 ml blood	2.73	7.06	11.49	15.12	16.85	17.98	18.75	19.14	19.57	19.80
B dissolved O_2 ml O_2/100 ml blood	0.03	0.06	0.09	0.12	0.15	0.18	0.21	0.24	0.27	0.30
C O_2 combined with Hb (A–B)	2.70	7.00	11.40	15.00	16.70	17.80	18.54	18.90	19.30	19.50
D % saturation of Hb	13.5	35	57	75	83.5	89	92.7	94.5	96.5	97.5

the volume of free gas is extremely large compared with the volume of blood, the final equilibrium Po_2 is very close to that of the original free gas. Using such a procedure, therefore, we can achieve any flask blood Po_2 we wish. By this means, we produce in 10 different samples of blood 10 different oxygen partial pressures ranging from 10 to 100 mm Hg and analyze the effect of Po_2 on hemoglobin saturation by measuring the fraction of hemoglobin combined with oxygen in each flask. Data from such an experiment are plotted in Fig. 17-26, which is called an oxygen-hemoglobin dissociation, i.e., saturation, curve. It is an S-shaped curve with a steep slope between 10 and 60 mm Hg Po_2 and a flat portion between 70 and 100 mm Hg Po_2. In other words, the extent to which hemoglobin combines with oxygen increases very rapidly from 10 to 60 mm Hg so that, at a Po_2 of 60 mm Hg, 90 percent of the total hemoglobin is combined with oxygen. From this point on, a further increase in Po_2 produces only a small increase in oxygen uptake. The adaptive importance of this plateau at higher Po_2 values is very great for the following reason. Many situations (severe exercise, high altitudes, cardiac or pulmonary disease) are characterized by a moderate reduction of alveolar and arterial Po_2; even if the Po_2 fell from the normal value of 100 to 60 mm Hg, the total quantity of oxygen carried by hemoglobin would decrease by only 10 percent, since hemoglobin saturation is still close to 90 percent at a Po_2 of 60 mm Hg. The plateau therefore provides an excellent safety factor in the supply of oxygen to the tissues.

We now retrace our steps and reconsider the movement of oxygen across the various membranes, this time including hemoglobin in our analysis. It is essential to recognize that the oxygen which is chemically bound to hemoglobin does not contribute to the Po_2 of the blood. Only gas molecules which are *free in solution,* i.e., physically dissolved, can participate in the bombardment which creates a gas pressure. Therefore, the diffusion of oxygen is directly governed only by that portion which is dissolved, a fact which permitted us to ignore hemoglobin in discussing transmembrane pressure gradients. However, the presence of hemoglobin plays a critical role in determining the *total amount* of oxygen which will diffuse, as illustrated by a simple example (Fig. 17-27). Two solutions separated by a semipermeable membrane contain equal quantities of oxygen, the gas pressures are equal, and no net diffusion occurs. Addition of hemoglobin to compartment B destroys this equilibrium because much of the oxygen combines with hemoglobin. Despite the fact that the

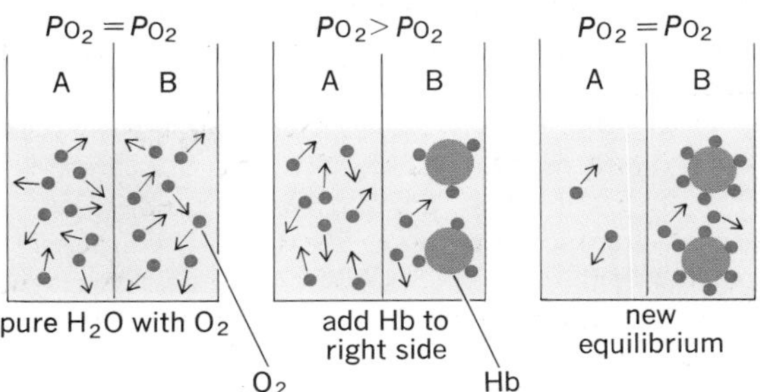

FIGURE 17-27 ***Effects of adding hemoglobin on the distribution of oxygen between two compartments separated by a semipermeable membrane. At the new equilibrium, the Po_2 values are equal to each other but lower than before the hemoglobin was added; however, the total oxygen, i.e, both dissolved and chemically combined with hemoglobin, is much higher on the right side of the membrane.*** **(Adapted from Comroe.)**

total quantity of oxygen in compartment B is still the same, the number of molecules *dissolved* has decreased; therefore, the Po_2 of compartment B is less than that of A, and net diffusion of oxygen occurs from A to B. At the new equilibrium, the oxygen pressures are once again equal, but almost all the total oxygen is in compartment B combined with hemoglobin. Thus, hemoglobin acts as a sink, removing dissolved oxygen, keeping the Po_2 low, and allowing net diffusion to continue.

Let us now apply this analysis to the lung and tissue capillaries (Fig. 17-28). The plasma and erythrocytes entering the lungs have a Po_2 of 40 mm Hg, and the hemoglobin saturation is 75 percent. Oxygen diffuses from the alveoli because of its higher Po_2 (100 mm Hg) into the plasma; this increases plasma Po_2 and induces diffusion of oxygen into the erythrocytes, elevating erythrocyte Po_2 and causing increased combination of oxygen and hemoglobin. Thus, the vast preponderance of the oxygen diffusing into the blood from the alveoli does not remain dissolved but combines with hemoglobin. In this manner, the blood Po_2 remains less than that of the alveolar Po_2 until hemoglobin is virtually completely saturated, and the diffusion gradient favoring oxygen movement into the blood is maintained despite the very large transfer of oxygen. In the tissue capillaries, the procedure is reversed: As the blood enters the capillaries, plasma Po_2 is greater than interstitial fluid Po_2 and net oxygen diffusion occurs across the capillary membrane; plasma Po_2 is now lower than

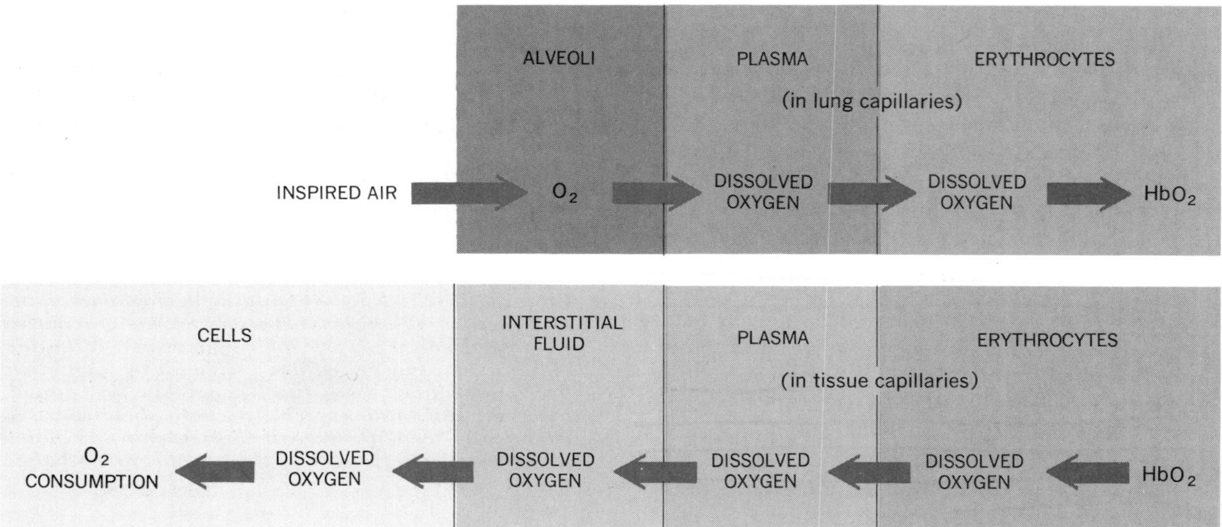

FIGURE 17-28 ***Oxygen movements in the lungs and tissues. Movement of air into the alveoli is by bulk flow; all movements across membranes are by passive diffusion.***

erythrocyte Po_2, and oxygen diffuses out of the erythrocyte into the plasma. The lowering of erythrocyte Po_2 causes the dissociation of HbO_2, thereby liberating oxygen; simultaneously, the oxygen which diffused into the interstitial fluid is moving into cells along the concentration gradient generated by cell utilization of oxygen. The net result is a transfer of large quantities of oxygen from HbO_2 into cells purely by passive diffusion.

The fact that hemoglobin is still 75 percent saturated at the end of tissue capillaries under resting conditions underlies an important automatic mechanism by which cells can obtain more oxygen whenever they increase their activity. An exercising muscle consumes more oxygen, thereby lowering its intracellular Po_2; this increases the overall blood-to-cell Po_2 gradient and the diffusion of oxygen out of the blood; in turn, the resulting reduction of erythrocyte Po_2 causes additional dissociation of hemoglobin and oxygen. An exercising muscle can thus extract virtually all the oxygen from its blood supply. This process is so effective because it takes place in the steep portion of the hemoglobin dissociation curve. Of course, an increased blood flow to the muscle also contributes greatly to the increased oxygen supply.

Effect of Acidity on Hemoglobin Saturation

The large oxygen-hemoglobin dissociation curve illustrated in Fig. 17-26 is for blood having a specific hydrogen-ion concentration equal to that found in arterial blood. When the same experiments are performed at a different level of acidity, the curve changes significantly. A group of such curves is shown in the inset of Fig. 17-26 for different hydrogen-ion concentrations. It is evident that, regardless of the existing acidity, the percentage saturation of hemoglobin is still determined by the Po_2. However, a change in acidity causes highly significant shifts in the center curve. An increased hydrogen-ion concentration moves the curve downward and to the right, which means that, at any given Po_2, hemoglobin has less affinity for oxygen when the acidity is high. For reasons to be described later, the hydrogen-ion concentration in the tissue capillaries is greater than in arterial blood. Blood flowing through tissue capillaries becomes exposed to this elevated hydrogen-ion concentration and therefore loses even more oxygen than if the decreased Po_2 had been the only factor involved. Conversely, the hydrogen-ion concentration is lower in the lung capillaries than in the systemic venous blood, so that hemoglobin picks up more oxygen in the lungs than if only the Po_2 were involved. Finally, the more active a tissue is, the greater its hydrogen-ion concentration; accordingly, hemoglobin releases even more oxygen during passage through these tissue capillaries, thereby providing the more active cells with additional oxygen. The hydrogen ion exerts this effect on the affinity of hemoglobin for oxygen by combining with the hemoglobin and altering its molecular structure. The importance of this combination for the regulation of extracellular acidity will be described in Chap. 18.

Effect of Temperature on Hemoglobin Saturation The effect of temperature on the oxygen-hemoglobin dissociation curve (inset of Fig. 17-26) resembles that of an increase in acidity. The implication is similar: Actively metabolizing tissue, e.g., exercising muscle, has an elevated temperature, which facilitates the release of oxygen from hemoglobin as blood flows through the muscle capillaries.

Effect on DPG on Hemoglobin Saturation Red cells contain large quantities of the substance 2,3-diphosphoglycerate (DPG) which is present in only trace amounts in other mammalian cells. DPG, which is produced by the red cells during glycolysis, binds reversibly with hemoglobin, causing it to change its conformation and release oxygen. Therefore, the effect of increased DPG is to shift the curve downward and to the right (just as does an increased temperature or hydrogen-ion concentration). The net result is that whenever DPG is increased there is enhanced unloading of oxygen as blood flows through the tissues. Such an increase is triggered by a variety of conditions associated with decreased oxygen supply to the tissues and helps to maintain oxygen delivery.

In summary, we have seen that oxygen is transported in the blood primarily in combination with hemoglobin. The extent to which hemoglobin binds oxygen is dependent upon the Po_2, hydrogen-ion concentration, temperature, and DPG. The first three of these factors cause the release of large quantities of oxygen from hemoglobin during blood flow through tissue capillaries and virtually complete conversion of reduced hemoglobin to oxyhemoglobin during blood flow through lung capillaries. An active tissue increases its extraction of oxygen from the blood because of its lower Po_2 and higher hydrogen-ion concentration and temperature. DPG also plays a role when oxygen supply is low.

Functions of Myoglobin

Myoglobin, an iron-containing protein found in cardiac and skeletal muscle cells, resembles hemoglobin in that it binds oxygen reversibly. Its major function is to act as an intracellular carrier which facilitates the diffusion of oxygen throughout the muscle cell. In addition, it provides a store of oxygen which the cell can call upon during sudden changes in activity.

Transport of Carbon Dioxide in the Blood

As is true for oxygen, the quantity of carbon dioxide which can physically dissolve in blood at physiologic carbon dioxide partial pressures is quite small, certainly much smaller than the large volume of carbon dioxide which must be constantly transported from the tissues to the lungs.

Carbon dioxide can undergo the reaction

$$CO_2 + H_2O \rightleftharpoons \underset{\text{Carbonic acid}}{H_2CO_3}$$

which goes quite slowly unless it is catalyzed by the enzyme carbonic anhydrase. The quantities of both dissolved carbon dioxide and carbonic acid are directly proportional to the Pco_2 of the solution. The actual amount of carbonic acid in blood is small because carbonic acid almost completely ionizes according to the equation

$$H_2CO_3 \rightleftharpoons HCO_3^- + H^+$$

Combining these two equations, we find

$$CO_2 + H_2O \xrightleftharpoons{\text{carbonic anhydrase}} H_2CO_3 \rightleftharpoons HCO_3^- + H^+ \qquad \textbf{(17-2)}$$

Thus, the addition of carbon dioxide to a liquid results ultimately in bicarbonate and hydrogen ions. Carbon dioxide can also react directly with proteins, particularly hemoglobin, to form *carbamino* compounds.

$$CO_2 + Hb \rightleftharpoons HbCO_2 \qquad \textbf{(17-3)}$$

When arterial blood flows through tissue capillaries, oxyhemoglobin gives up oxygen to the tissues and carbon dioxide diffuses from the tissues into the blood, where the following processes occur (Fig. 17-29):

1 A small fraction (8 percent) of the carbon dioxide remains physically dissolved in the plasma and red blood cells.

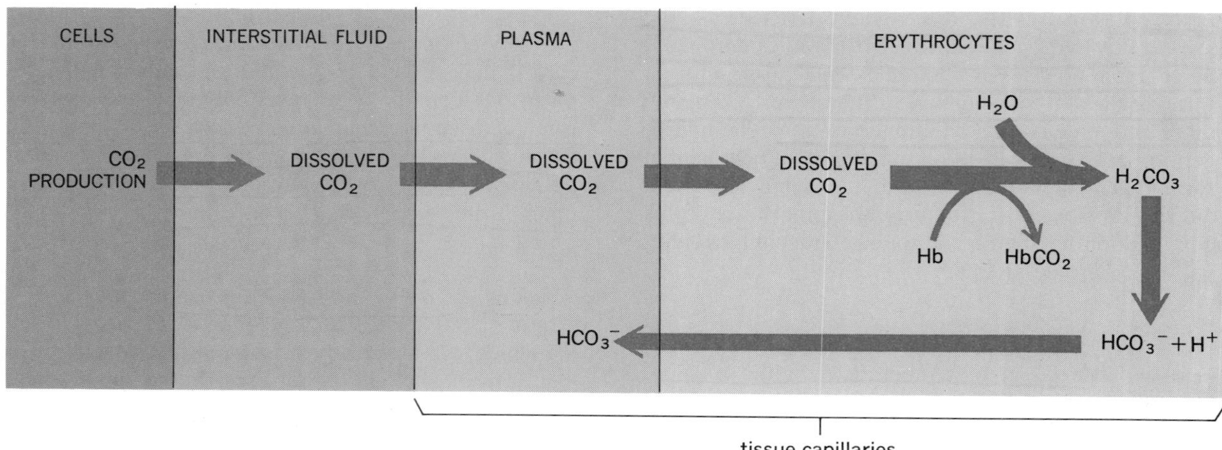

FIGURE 17-29 ***Summary of carbon dioxide movements and reactions as blood flows through tissue capillaries. All movements across membranes are by passive diffusion. Note that most of the CO_2 ultimately is converted to HCO_3^-; this occurs almost entirely in the erythrocytes (because the carbonic anhydrase is located there), but most of the HCO_3^- then diffuses out of the erythrocytes into the plasma.***

2 The largest fraction (67 percent) of the carbon dioxide undergoes the reactions described in Eq. 17-2 and is converted into bicarbonate and hydrogen ions. This occurs primarily in the red blood cells because they contain large quantities of the enzyme carbonic anhydrase but the plasma does not. This explains why tissue capillary hydrogen-ion concentration is higher than that of the arterial blood and increases as metabolic activity increases. The fate of these hydrogen ions will be discussed in Chap. 18. Bicarbonate, in contrast to carbon dioxide, is extremely soluble in blood.

3 The remaining fraction (25 percent) of the carbon dioxide reacts directly with hemoglobin to form $HbCO_2$, as in Eq. 17-3.

Since these are all reversible reactions, i.e., they can proceed in either direction, depending upon the prevailing conditions, why do they all proceed primarily to the right, toward generation of HCO_3^- and $HbCO_2$, as blood flows through the tissues? In any chemical reaction, increasing the concentration of any of the reacting substances on the left side of the equation drives the reaction toward the right. The converse, of course, is also true. Once again, the answer is provided by the law of mass action: It is the increase in carbon dioxide concentration which drives these reactions to the right as blood flows through the tissues.

Obviously, a sudden lowering of blood P_{CO_2} has just the opposite effect. HCO_3^- and H^+ combine to give H_2CO_3, which generates carbon dioxide and water. Similarly, $HbCO_2$ generates hemoglobin and free carbon dioxide. This is precisely what happens as venous blood flows through the lung capillaries (Fig. 17-30). Because the blood P_{CO_2} is higher than alveolar, a net diffusion of carbon dioxide from blood into alveoli occurs. This loss of carbon dioxide from the blood lowers the blood P_{CO_2} and drives these chemical reactions to the left, thus generating more dissolved carbon dioxide. Normally, as fast as this carbon dioxide is generated from HCO_3^- and H^+ and from $HbCO_2$, it diffuses into the alveoli. In this manner, all the carbon dioxide delivered into the blood in the tissues now is delivered into the alveoli, from which it is expired and eliminated from the body.

CONTROL OF RESPIRATION

In dealing with the mechanisms by which the basic respiratory processes are controlled, we shall be concerned primarily with two questions: By what mechanisms are rhythmic breathing movements generated? What factors control the rate and depth of breathing, i.e., the total ventilatory volume?

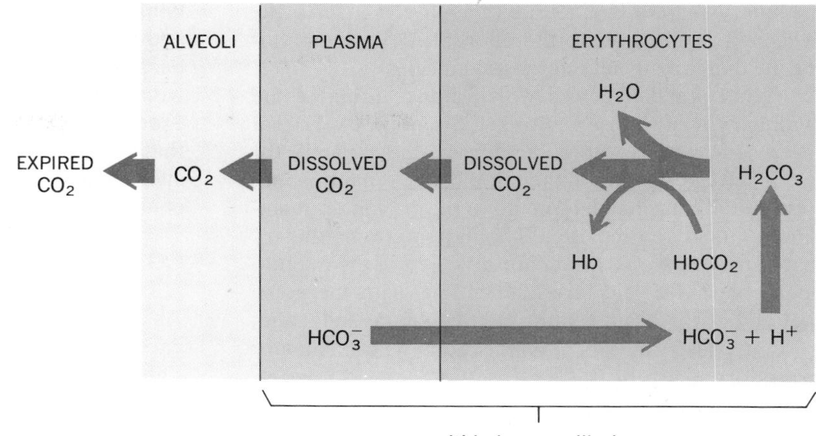

FIGURE 17-30 ***Summary of carbon dioxide movements and reactions as blood flows through the lung capillaries. All movements across membranes are by passive diffusion. The plasma-erythrocyte phenomena are simply the reverse of those occurring during blood flow through the tissue capillaries, as shown in Fig. 17-29. The breakdown of H_2CO_3 is catalyzed by carbonic anhydrase.***

Neural Generation of Rhythmic Breathing

Like cardiac muscles, the inspiratory muscles normally contract rhythmically; however, the origins of these contractions are quite different. Cardiac muscle has automaticity; i.e., it is capable of self-excitation. The nerves to the heart merely alter this basic inherent rate and are not actually required for cardiac contraction. On the other hand, the diaphragm and intercostal muscles consist of skeletal muscle, which cannot contract unless stimulated by nerves. Thus, breathing depends entirely upon cyclic respiratory muscle excitation by the *phrenic nerve* (to the diaphragm) and the *intercostal nerves* (to the intercostal muscles). The two phrenic nerves arise from the 3rd, 4th, and 5th cervical spinal nerves, one on each side of the vertebral column. They pass downward in the mediastinum and perforate the diaphragm, each dividing to supply the various areas of that muscle. The 11 pairs of intercostal nerves are formed from the 1st to 11th thoracic spinal nerves. Destruction of the phrenic and intercostal nerves or the spinal cord areas from which they originate (as in poliomyelitis, for example) results in complete paralysis of the respiratory muscles and death, unless some form of artificial respiration can be rapidly instituted.

At the end of expiration, when the chest is at rest, a few impulses are still passing down these nerves. Like other skeletal muscles, therefore, the respiratory muscles have a certain degree of resting tonus. This muscular contraction is too slight to move the chest but plays a role in maintaining normal posture. Inspiration is initiated by an increased rate of firing of these inspiratory motor units. More and more new motor units are recruited, and thoracic expansion increases. In addition, the firing frequency of the individual units increases. By these two measures, the force of inspiration increases as it proceeds. Then almost all these units stop firing, the inspiratory muscles relax, and expiration occurs as the elastic lungs recoil. In addition, when expiration is facilitated by contraction of expiratory muscles, the nerves to these muscles, having been quiescent during inspiration, begin firing during expiration.

By what mechanism are nerve impulses to the respiratory muscles alternately increased and decreased? Control of this neural activity resides primarily in neurons with cell bodies in the lower portion of the brainstem, *the medulla,* which also contains the cardiovascular control centers. If the spinal cord is cut at any point between the medulla and the areas of the spinal cord from which the phrenic and intercostal nerves originate, breathing ceases. This experiment demonstrates that these efferent nerves are controlled by synaptic connections with neurons which descend in the spinal cord from the medulla.

By means of tiny electrodes placed in various parts of the medulla to record electric activity, neurons have been found which discharge in perfect synchrony with inspiration and smaller numbers of other neurons which discharge synchronously with expiration, called *inspiratory* and *expiratory neurons,* respectively. The respiratory cycle is established primarily by these medullary inspiratory neurons which give rise to the descending pathways to the efferent nerves innervating the diaphragm and intercostal muscles.

What are the factors which induce firing of the medullary inspiratory neurons? It is quite likely that these neurons have, to some degree, the capacity for cyclic self-excitation; on the other hand, synaptic input from other neurons also seems to play a role in maintaining their rhythmicity. The inputs to medullary inspiratory neurons which are involved in the maintenance of rhythmicity are connections with the pons, the area of the brainstem just above the medulla, and, in some cases, reciprocal connections with medullary expiratory neurons.

Control of Ventilatory Volume

In the preceding section, we were concerned with the mechanisms that generate rhythmic breathing. It is obvious, however, that the actual respiratory rate is not fixed but can be altered over a wide range. Similarly, the depth and force of breathing movements can also be altered. As we have seen, these two factors, rate and depth, determine the alveolar ventilatory volume. Generally, rate and depth change in the same direction, although there may be important quantitative differences. For simplicity, we shall describe the control of total ventilation without attempting to discuss whether rate or depth makes the greatest contribution to the change.

Depth of respiration depends upon the number of motor units firing and their frequency of discharge, whereas respiratory rate depends upon the length of time elapsing between the bursts of motor unit activity. As described above, the respiratory motor units are directly controlled by descending pathways from the medullary respiratory centers. The efferent pathways for control of ventilation are therefore clear-cut, and the critical question becomes: What is the nature of the afferent input to these centers? In other words, what variables does the control of ventilatory volume regulate?

This may seem a ridiculously complex way of phrasing a question when the answer seems so intuitively obvious. After all, respiration ought to supply oxygen as fast as it is consumed and ought to excrete carbon dioxide as fast as it is produced. *But* how do the respiratory centers "know" what the body's oxygen requirements are? The logical way to approach this question is to ask what detectable changes would result from imbalance of metabolism and ventilation. Certainly the most obvious candidates are the plasma P_{O_2} and P_{CO_2}. Inadequate ventilation would lower the P_{O_2} because consumption would get ahead of supply and would elevate the P_{CO_2} because production would exceed elimination. Less obvious, perhaps, is the fact that arterial hydrogen-ion concentration also is exquisitely sensitive to changes in ventilation. Recall the equilibrium between H^+, HCO_3^-, and CO_2.

$$CO_2 + H_2O \rightleftharpoons H_2CO_3 \rightleftharpoons HCO_3^- + H^+$$

As described by the law of mass action, any increase in carbon dioxide concentration drives this reaction to the right, thereby liberating additional hydrogen ions. *Any increase or decrease in plasma* P_{CO_2} *is accompanied by changes in plasma hydrogen-ion concentration.*

However, these statements of fact in no way prove that any of these plasma concentrations actually are involved in respiratory regulation. The next step is to ask whether there are, indeed, receptors which can detect the levels of these variables in plasma and transmit the information to the respiratory centers. If so, where are these receptors located and what contribution do they make to overall control of ventilation? We shall describe the answers to these questions first for oxygen and then for carbon dioxide and hydrogen ion.

Control of Ventilation by Oxygen The rationale of the experiments to be described is quite simple: Alteration of inspired P_{O_2} produces changes in plasma P_{O_2}; if plasma P_{O_2} is important in controlling ventilation, we should observe definite changes in ventilation. If a normal person takes a single breath of 100 percent oxygen (without being aware of it), after a latent period of about 8 s, a transient reduction in ventilation occurs of approximately 10 to 20 percent. Such studies have demonstrated that the normal sea-level plasma P_{O_2} of 100 mm Hg exerts a tonic stimulatory effect adequate to account for approximately 20 percent of total ventilation. Thus, the increase in plasma P_{O_2} produced by the 100 percent oxygen removed this tonic stimulation and reduced ventilation.

Conversely, does a reduction in plasma P_{O_2} below normal cause a further increase in ventilation? Figure 17-31 illustrates the average response of a group of healthy subjects who breathed low P_{O_2} gas mixtures for 8 min. Note that no significant stimulation of respiration was observed until the oxygen content of the inspired air was reduced to one-half of normal air. Even at this point, some of the subjects failed to increase their ventilation despite an arterial P_{O_2} of 40 mm Hg and hemoglobin saturation of 75 percent. Only when the oxygen lack was severe did all subjects manifest a sustained large increase in ventilation.

More recent studies, however, have demonstrated that our sensitivity to low P_{O_2} is actually greater than was originally concluded from data like those in Fig. 17-31. In the earlier experiments, a second important variable, the plasma P_{CO_2}, was not controlled. As we shall see in the subsequent section, voluntarily increasing one's ventilation causes a reduction of plasma P_{CO_2}, and this chemical change acts as a profound respiratory depressant. What apparently happened to the subjects of Fig. 17-31 was that, although even mild reductions of P_{O_2} actually raised the ventilation at first, this increase lowered their plasma P_{CO_2}, which inhibited ventilation and thereby counteracted the stimulatory effects of the reduced P_{O_2}. When this type of experiment was repeated with the P_{CO_2} held constant, all subjects manifested a considerably greater increase in ventilation when exposed to low P_{O_2} gas mixtures. The new experiments emphasize that human beings are indeed sensitive to a reduction of plasma P_{O_2} but the respiratory stimulation may be overridden by other simultaneously occurring changes to which they are even more sensitive, particularly changes in P_{CO_2} and hydrogen-ion concentration. The concept that respiration is controlled at any instant by multiple factors is of great importance.

The receptors stimulated by low P_{O_2} are located at the bifurcation of the common carotid arteries and in the arch of the aorta, quite close to, but distinct from, the baroreceptors described in Chap. 16 (Fig. 17-32). Known as the *carotid* and *aortic bodies*, they are composed of epithelial-like cells and neuron terminals in intimate contact with the arterial blood. The nerve fibers from the carotid body are derived from the glossopharyngeal (IXth cranial) nerve and those from the aortic arch are derived from the vagus (X); they enter the brainstem, where they synapse ultimately with the neurons of the medullary centers. A low P_{O_2} increases the rate at which the receptors discharge, resulting in an increased number of action potentials traveling up the afferent nerves and a stimulation of the medullary inspiratory neurons. If an animal is exposed to a low P_{O_2} after the afferent nerves from the carotid and aortic bodies have been cut, the usual rapid increase in alveolar ventilation is not observed, demonstrating

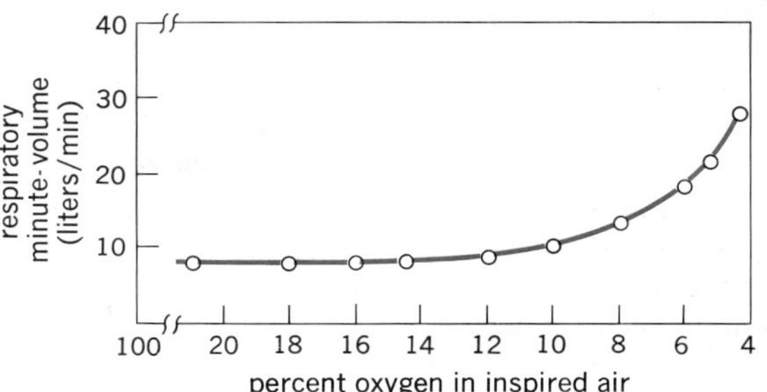

FIGURE 17-31 ***Effects of altering the oxygen content of inspired air on ventilation in a normal person. The points are the average values for all subjects.*** **(Adapted from Comroe.)**

that the immediate stimulatory effects of a low P_{O_2} are completely mediated via this pathway.

What is the precise stimulus to these chemoreceptors? It is most likely their own internal P_{O_2}, that is, the concentration of dissolved oxygen within them. Normally, their blood supply is so huge relative to their utilization of oxygen that their internal P_{O_2} is virtually identical to the arterial P_{O_2}. For this reason we can state that, in effect, they monitor arterial P_{O_2}. Thus, any time arterial P_{O_2} is reduced by lung disease, hypoventilation, or high altitude, these chemoreceptors are stimulated and call forth a compensatory increase in ventilation. On the other hand, the oxygen deficiency of anemia (decreased hemoglobin content of the blood) does not stimulate ventilation because, although total oxygen content of the blood is reduced, the arterial P_{O_2} is quite normal. This also explains why there is no respiratory stimulation in carbon monoxide poisoning. Carbon monoxide, a gas which reacts with the same sites on the hemoglobin molecule as oxygen, has such a remarkable affinity for these sites that even small amounts reduce the ability of oxygen to combine with hemoglobin. Since it does not affect the amount of oxygen which can physically dissolve in blood, the P_{O_2} is unaltered, the carotid and aortic bodies are not stimulated,[5] and the patient faints and dies of oxygen lack without ever increasing his or her ventilation.

[5] Unfortunately, although there is general agreement that carbon monoxide does not stimulate ventilation, recent experiments have suggested that it may, in fact, stimulate the chemoreceptors. An explanation of this discrepancy has not been forthcoming.

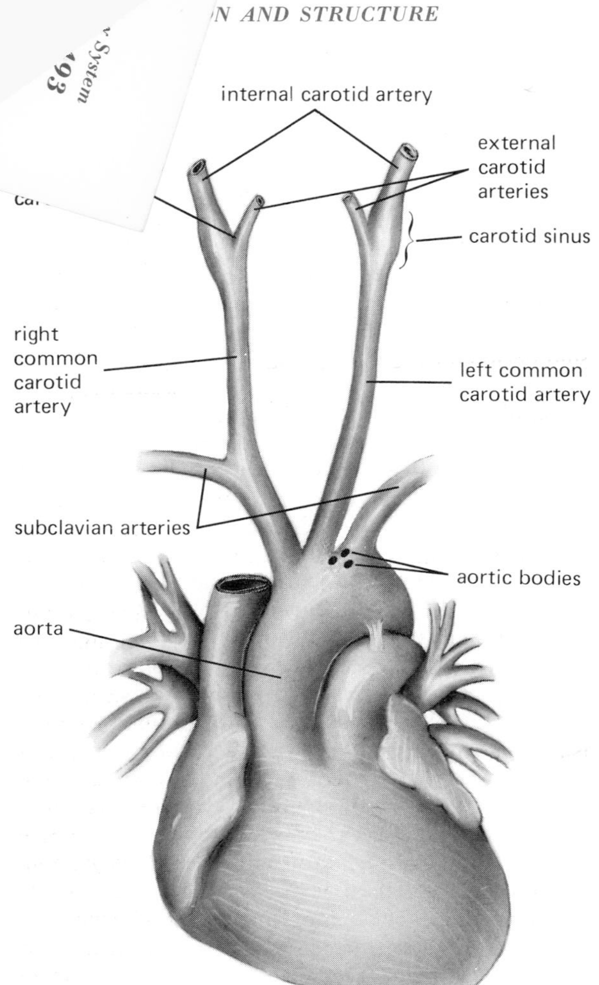

FIGURE 17-32 ***Location of the carotid and aortic bodies. Each common carotid bifurcation contains a carotid sinus and a carotid body.***

Control of Ventilation by Carbon Dioxide and Hydrogen Ion The preponderant role that CO_2 plays in the control of ventilation can be demonstrated by a few relatively simple experiments. (1) Figure 17-33 illustrates the effects on respiratory volume of increasing the P_{CO_2} of inspired air. Normally, atmospheric air contains virtually no carbon dioxide. In the experiment illustrated, the subject breathed from bags of air containing variable quantities of carbon dioxide. The presence of carbon dioxide in the inspired air causes an elevation of alveolar P_{CO_2} and thereby an elevation of arterial P_{CO_2} as well. This increased P_{CO_2} markedly stimulated ventilation, an increase of 5 mm Hg in alveolar P_{CO_2} causing a 100 percent increase in ventilation. Obviously, ventilation is acutely dependent upon the P_{CO_2}. (2) A subject is asked to breathe as rapidly and deeply as she can. When the period of voluntary hyperventilation is over, the subject is told merely to breathe naturally. All subjects manifest markedly reduced breathing for the next few minutes, and a few stop breathing completely (apnea), often for 1 to 2 min (Fig. 17-34). Ventilation is inhibited because during the period of hyperventilation carbon dioxide was blown off faster than it was produced; accordingly, at the moment when the subject ceased to hyperventilate, her plasma P_{CO_2} was lower than normal, and ventilation was inhibited until plasma P_{CO_2} returned toward normal as a result of accumulation of metabolically produced carbon dioxide. Note in Fig. 17-34 that the subject began breathing again, at least intermittently, before the P_{CO_2} was completely back to normal; this is due to stimulation of the carotid and aortic bodies by the extremely low P_{O_2}.

The opposite of hyperventilation is holding one's breath. Why can a person hold his breath for only a relatively short time? The lack of ventilation causes an accumulation of carbon dioxide and increased plasma P_{CO_2}; the ability of this increased P_{CO_2} to stimulate the respiratory center is so powerful that it overcomes the voluntary inhibition of the respiratory center (the latter mediated by descending pathways from the cerebral cortex). Unfortunately, underwater swimmers have misguidedly made use of these facts. They voluntarily hyperventilate for several minutes before submerging and are therefore able to hold their breaths for long periods of time. This is a very dangerous procedure, particularly during exercise, when oxygen consumption is high; because of our relative insensitivity to oxygen deficits, a rapidly decreasing P_{O_2} may cause fainting and drowning before ventilation is stimulated.

The experiments we have discussed strongly support the hypothesis that P_{CO_2} is the major determinant of respiratory center activity. An increase in P_{CO_2} stimulates ventilation and thereby promotes the excretion of additional carbon dioxide. Conversely, a decrease in P_{CO_2} below normal inhibits ventilation and thereby allows metabolically produced carbon dioxide to accumulate and return the P_{CO_2} to normal. In this manner, the arterial P_{CO_2} is stabilized at the normal value of 40 mm Hg.

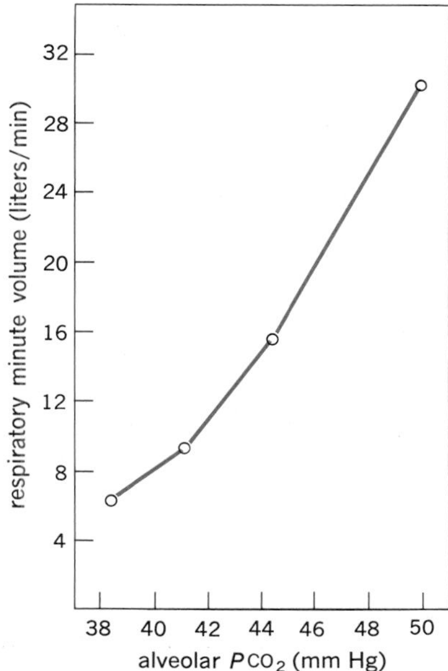

FIGURE 17-33 ***Effects on respiration of increasing alveolar*** **$P\text{CO}_2$** ***by adding carbon dioxide to inspired air. The increase in respiratory volume is due to an increase in both tidal volume and respiratory rate. This study was performed on normal human subjects.*** **(Adapted from Lambertsen.)**

FIGURE 17-34 ***Variations in alveolar gas pressures after voluntary overbreathing for 2 min. The actual breathing movements are shown by the jagged lines at the bottom of the figure.*** **(Adapted from Douglas and Haldane.)**

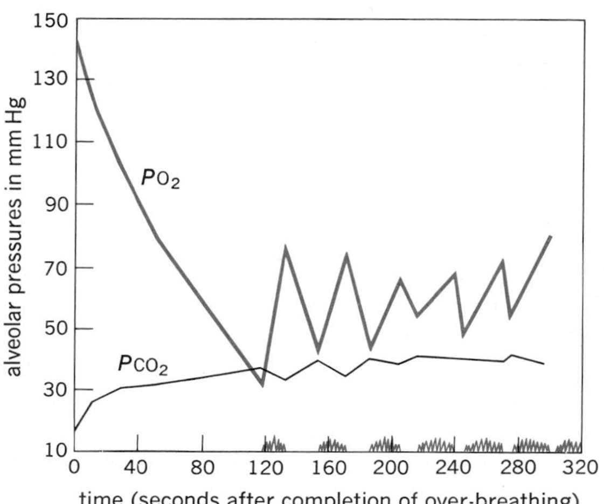

Much evidence indicates that the effects of carbon dioxide on ventilation are due not to carbon dioxide itself but to the associated changes in hydrogen-ion concentration. For example, the stimulant effects of breathing mixtures containing large amounts of carbon dioxide are probably due not to the effects of molecular carbon dioxide but to those of the increased hydrogen-ion concentration resulting from the chemical reactions described above. This is why we stressed the relationship between $P\text{CO}_2$ and hydrogen-ion concentration.

The critical hydrogen-ion concentration appears to be not that of the arterial blood but rather that of brain extracellular fluid, i.e., cerebrospinal fluid. An increased hydrogen-ion concentration increases the rate of discharge of the inspiratory neurons by acting upon nearby chemosensitive cells (*central chemoreceptors*) having synaptic input to them; conversely, a decreased hydrogen-ion concentration inhibits their discharge. One is left with the rather startling conclusion that the control of breathing (at least during rest) is aimed primarily at the regulation of *brain hydrogen-ion concentration!* However, this actually makes perfectly good sense, in terms of survival of the organism. As emphasized above, the relationship between carbon dioxide and hydrogen ion ensures that the regulation of hydrogen-ion concentration also produces relative constancy of $P\text{CO}_2$ as well. To continue the chain, the close relationship between oxygen consumption and carbon dioxide production ensures that ventilation adequate to maintain $P\text{CO}_2$ constant by excreting carbon dioxide as fast as it is produced also suffices to supply adequate oxygen (except in certain kinds of lung disease). Moreover, the shape of the hemoglobin dissociation curve minimizes any minor deficiency of oxygen supply, and any major oxygen deficit induces reflex respiratory stimulation via the carotid and aortic bodies. In contrast, brain function is extremely sensitive to changes in hydrogen-ion concentration so that even small increases or decreases in brain hydrogen-ion concentration could induce serious malfunction.

It should be noted that the central chemoreceptors are not the only receptors sensitive to hydrogen ion. The carotid and aortic bodies, responsible for the low-oxygen reflexes, are also sensitive to changes in plasma hydrogen-ion concentration, and these peripheral chemoreceptors can also function as receptors in reflexes which alter ventilation in response to altered hydrogen ion concentration. They are particularly important in situations in which the deviation in plasma hydrogen-ion concentration is caused by events other than changes in $P\text{co}_2$; for example, patients with severe diabetes synthesize increased amounts of organic acids, which raise plasma hydrogen-ion concentration and result in ventilatory stimulation mediated, at least in part, by the peripheral chemoreceptors. The relationship between ventilation and hydrogen-ion concentration is only one aspect of the regulation of extracellular-fluid hydrogen-ion concentration described in Chap. 18.

Throughout this section we have described the stimulatory effects of carbon dioxide on ventilation. It should also be noted that very high levels of carbon dioxide depress the entire central nervous system, including the respiratory centers, and may therefore be lethal. Closed environments, such as submarines and space capsules, must be designed so that carbon dioxide is removed as well as oxygen supplied.

Control of Ventilation during Exercise

During heavy exercise, the alveolar ventilation may increase ten- to twentyfold to supply the additional oxygen needed and excrete the excess carbon dioxide produced. On the basis of our three variables—$P\text{o}_2$, $P\text{co}_2$, and hydrogen-ion concentration—it might seem easy to explain the mechanism which induces this increased ventilation. Unhappily, such is not the case.

Decreased $P\text{o}_2$ as the Stimulus? It would seem logical that, as the exercising muscles consume more oxygen, plasma $P\text{o}_2$ would decrease and stimulate respiration. But, in fact, arterial $P\text{o}_2$ is *not* significantly reduced during exercise (Fig. 17-35). The alveolar ventilation increases in exact proportion to the oxygen consumption; therefore, $P\text{o}_2$ remains constant. Indeed, in exhausting exercise, the alveolar ventilation may actually increase relatively more than oxygen consumption, resulting in an *increased* $P\text{o}_2$.

Increased $P\text{co}_2$ as the Stimulus? This is virtually the same story. Despite the marked increase in carbon dioxide production, the precisely equivalent increase in alveolar ventilation excretes the carbon dioxide as rapidly as it is produced, and arterial $P\text{co}_2$ remains constant (Fig. 17-35). Indeed, for the same reasons given above for oxygen, the arterial $P\text{co}_2$ may actually decrease during exhausting exercise.

Increased Hydrogen Ion as the Stimulus? Since the arterial $P\text{co}_2$ does not change (or decreases) during exercise, there is no accumulation of excess hydrogen ion as a result of carbon dioxide accumulation. Although there is an increase in arterial hydrogen-ion concentration for quite a different reason, namely, generation and release into the blood of lactic acid and other acids during exercise, the changes in hydrogen-ion concentration are not nearly great enough, particularly in only moderate exercise, to account for the increased ventilation.

We are left with the fact that, despite intensive study for more than 100 years by many of the greatest respiratory physiologists, we do not know what input stimulates ventilation during exercise. Our big three—$P\text{o}_2$, $P\text{co}_2$, and hydrogen ion—appear presently to be inadequate, but many physiologists still believe that they will ultimately be shown to be the critical inputs. They reason that the fact that $P\text{co}_2$ remains constant during moderate exercise is very strong evidence that ventilation is actually controlled by $P\text{co}_2$. In other words, if $P\text{co}_2$ were not the major controller, how else could it remain unchanged in the face of the marked increase in its production? They reason that there may be a change in sensitivity of the chemoreceptors to CO_2 or hydrogen ion so that changes undetectable by experimental methods might be responsible for the stimulation of ventilation. Of course, this view is based presently on the theoretical grounds; it remains to be seen whether experiments can validate it.

Moreover, the problem is even more complicated; there is an abrupt increase (within seconds) in ventilation at the onset of exercise and an equally abrupt decrease at the end. Clearly, these changes occur too rapidly to be explained by alteration in some chemical constituent of the blood.

Control of Respiration by Other Factors

Temperature An increase in body temperature frequently occurs as a result of increased physical activity and contributes to stimulation of alveolar ventilation. This facilitation of the respiratory centers is probably due both to a direct physical effect of increased temperature upon the respiratory-center neurons and to stimulation via pathways from thermoreceptors of the hypothalamus.

Input from the Cerebral Cortex In Chap. 16, we described how the cardiovascular responses to ex-

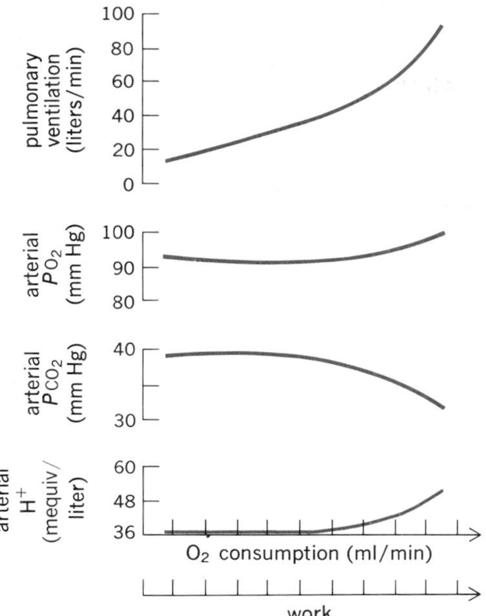

FIGURE 17-35 ***Relation of ventilation, arterial gas pressures, and hydrogen-ion concentration to the magnitude of muscular exercise.*** **(Adapted from Comroe.)**

ercise were probably mediated by input to the hypothalamus from branches of the neuron pathways descending from the cerebral cortex to the motor nerves. It is likely that these fibers send branches into the medullary respiratory centers as well. However, the magnitude of this contribution to stimulation of ventilation is not known.

Epinephrine Injection of epinephrine into a person produces stimulation of respiration by unknown mechanisms. Epinephrine secretion is uniformly increased during heavy exercise and probably contributes to the increased ventilation.

Reflexes from Joints and Muscles Many receptors in joints and muscles can be stimulated by the physical movements which accompany muscle contraction. It is quite likely that afferent pathways from these receptors play a significant role in stimulating the respiratory centers during exercise. Thus, the mechanical events of exercise help to coordinate alveolar ventilation with the metabolic requirements of the tissues.

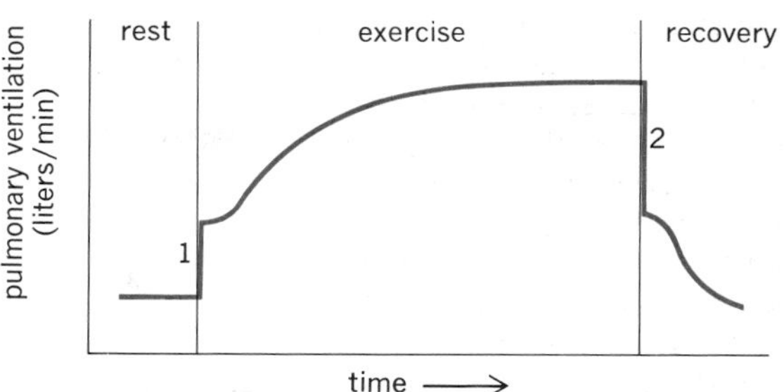

FIGURE 17-36 ***Ventilation changes in time during exercise. Note the abrupt increase (1) at the onset of exercise and the equally abrupt but larger decrease (2) at the end of exercise.***

Baroreceptor Reflexes We have already discussed the carotid sinus and aortic arch reflexes as the primary controllers of cardiovascular function. These reflexes can also alter alveolar ventilation, but their effect in human beings is usually minor.

Protective Reflexes A group of responses, most familiar being the *cough* and the *sneeze,* protect the respiratory tract against irritant materials. These reflexes originate in receptors which line the respiratory tract. When they are excited, the result is stimulation of the medullary respiratory centers (via afferent nerves) in such a manner as to produce a deep inspiration and a violent expiration. In this manner, particles can be literally exploded out of the respiratory tract. Another example of a protective reflex is the immediate cessation of respiration which is frequently triggered when noxious agents are inhaled.

Pain Painful stimuli anywhere in the body can produce reflex stimulation of respiration. This is why one formerly spanked newborn infants to start their respirations.[6] Another common example is the deep inspiration induced by a sudden shock such as entering a cold shower.

Emotion Emotional states are often accompanied by marked stimulation of respiration, as

[6] The mucus and fluids present in a newborn's respiratory tract are now usually aspirated from the back of the baby's throat through a tube. The presence of this tube is stimulus enough to increase respiration. Now too, drugs are used more cautiously; lower doses of sedatives to the mother before delivery cause less depression of the baby's respiration.

evidenced by the rapid breathing rate which characterizes fright and many similar emotions. In addition, the movement of air in or out of the lungs is an absolute requirement for such involuntary expressions of emotion as laughing and crying. In such situations, the respiratory centers must be primarily controlled by descending pathways from higher brain centers.

Voluntary Control of Breathing Although we have discussed in detail the involuntary nature of most respiratory reflexes, it is quite obvious that we retain considerable voluntary control of respiratory movements. This is accomplished by descending pathways from the cerebral cortex to the neurons supplying the intercostal muscles and diaphragm. As we have seen, this voluntary control of respiration cannot be maintained when the involuntary stimuli, such as an elevated $P\text{CO}_2$ or hydrogen-ion concentration, become intense. Besides the obvious forms of voluntary control, e.g., breath holding, respiration must also be controlled during the production of complex voluntary actions such as speaking and singing.

CHAPTER 18
The Urinary System

The urinary system is responsible for the formation of urine and its conveyance out of the body. Urine is formed by the *kidneys.* It then drains from each kidney via a large duct, the *ureter,* which leads to the *urinary bladder,* the reservoir for urine (Fig. 18-1). The *urethra,* the last component of the urinary system, carries urine from the bladder to the outside. In males, it also serves as the passageway for semen.

The major task achieved by the kidneys' formation of urine is regulation of the composition of the internal environment. A cell's function depends not only upon receiving a continuous supply of organic nutrients and eliminating its metabolic end products but also upon the existence of stable physicochemical conditions in the extracellular fluid bathing it. Among the most important substances contributing to these conditions are water, sodium, potassium, calcium, and hydrogen ion. This chapter is devoted to a discussion of the mechanisms by which the total amounts of these substances in the body and their concentrations in the extracellular fluid are maintained relatively constant, mainly as a result of the kidneys' activities.

A substance appears in the body either as a result of ingestion or as a product of metabolism. Conversely, a substance can be excreted from the body or consumed in a metabolic reaction. Therefore, if the quantity of any substance in the body is to be maintained at a constant level over a period of time, the total amounts ingested and produced must equal the total amounts excreted and consumed. This is a general statement of the *balance concept* described in Chap. 6. For water and hydrogen ion, all four possible pathways apply. However, for the mineral electrolytes balance is simpler since they are neither synthesized nor consumed by cells, and their total body balance thus reflects only ingestion versus excretion.

As an example, let us describe the balance for total body water (Table 18-1). It should be recognized that these are average values, which are subject to considerable variation. The two sources of body water are metabolically produced water, resulting largely from the oxidation of carbohydrates, and ingestion of water in liquids and so-called solid food (a rare steak is approximately 70 percent water). There are four sites from which water is lost to the external environment: skin, lungs, gastrointestinal tract, and kidneys. The loss of water by evaporation from the cells of the skin and the lining of respiratory passageways is a continuous process, often referred to as *insensible loss* because the person is unaware of its occurrence. Additional water can be made available for evaporation from the skin by the production of sweat. The normal gastrointestinal loss of water (in feces) is quite small but can be severe in vomiting or diarrhea.

TABLE 18-1
NORMAL ROUTES OF WATER GAIN AND LOSS IN ADULTS

	Milliliters per day
Intake:	
Drunk	1,200
In food	1,000
Metabolically produced	350
Total	2,550
Output:	
Insensible loss (skin and lungs)	900
Sweat	50
In feces	100
Urine	1,500
Total	2,550

Under normal conditions, as can be seen from Table 18-1, water loss exactly equals water gain, and no net change of body water occurs. This is obviously no accident but the result of precise regulatory mechanisms. The question then is: Which processes involved in water balance are controlled to make the gains and losses balance? The answer, as we shall see, is voluntary intake (*thirst*) and urinary loss. This does not mean that none of the other processes is controlled but that their control is not primarily oriented toward water balance. Carbohydrate catabolism, the major source of the water of oxidation, is controlled by mechanisms directed toward regulation of energy balance. Sweat production is controlled by mechanisms directed toward temperature regulation. Insensible loss (in human beings) is truly uncontrolled, and fecal water loss is generally quite small and unchanging.

The mechanism of thirst is certainly of great importance, since body deficits of water, regardless of cause, can be made up only by ingestion of water, but it is also true that our fluid intake is often influenced more by habit and sociological factors than by the need to regulate body water. The control of urinary water loss is the major mechanism by which body water is regulated.

By similar analyses, we find that the body balances of most of the ions determining the properties of the extracellular fluid are regulated primarily by the kidneys.

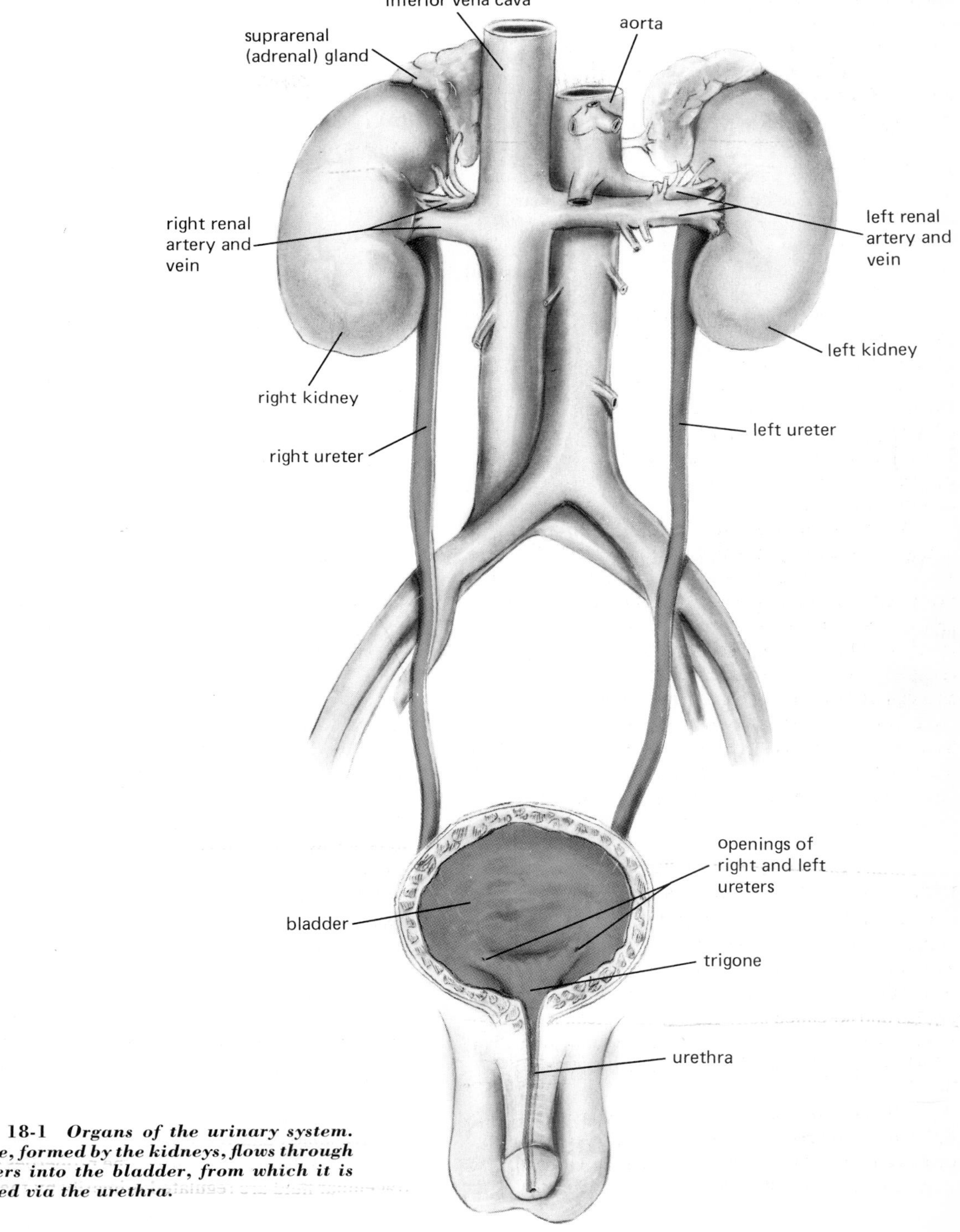

FIGURE 18-1 ***Organs of the urinary system. The urine, formed by the kidneys, flows through the ureters into the bladder, from which it is eliminated via the urethra.***

To appreciate the importance of these kidney regulations and the fact that severe kidney malfunction is rapidly fatal, one need only make a partial list of the more important simple inorganic substances which constitute the internal environment and which are regulated by the kidney: water, sodium, potassium, chloride, calcium, magnesium, sulfate, phosphate, and hydrogen ion. It is worth repeating that normal biological processes depend on the constancy of this internal environment, the implication being that the amounts of these substances must be held within very narrow limits, regardless of large variations in intake and abnormal losses resulting from disease (hemorrhage, diarrhea, vomiting, etc.). Indeed, the extraordinary number of substances which the kidney regulates and the precision with which these processes normally occur accounted for the kidney's being the last stronghold of the nineteenth-century vitalists, who simply would not believe that the laws of physics and chemistry could fully explain renal function. By what mechanism does urine flow rapidly increase when a person ingests several glasses of liquid? How is it that the patient on an extremely low salt intake and the heavy salt eater both urinate precisely the amounts of salt required to maintain their sodium balance? What mechanisms decrease the urinary calcium excretion of children deprived of milk?

This regulatory role is obviously quite different from the popular conception of the kidneys as glorified garbage disposal units which rid the body of assorted wastes and "poisons." It is true that several of the complex chemical reactions which occur within cells result ultimately in end products collectively called waste products (primarily because they serve no known biological function in human beings); e.g., the catabolism of protein produces approximately 30 g of urea per day. Other waste substances produced in relatively large quantities are uric acid (from nucleic acids), creatinine (from muscle creatine), and the end products of hemoglobin breakdown. There are many others, not all of which have been completely identified. Most of these substances are eliminated from the body as rapidly as they are produced, primarily by way of the kidneys. Many of these waste products are harmless, although the accumulation of certain of them within the body during periods of renal malfunction accounts for some of the disordered body functions which eventually kill the patient suffering from severe kidney disease. However, most of the problems which occur in renal disease are due simply to disordered water and electrolyte metabolism.

The kidneys have another excretory function which is presently assuming increased importance, namely, the elimination from the body of foreign chemicals (drugs, pesticides, food additives, etc.). A final kidney function is to act as endocrine glands secreting at least two hormones: erythropoietin (Chap. 13) and renin (to be discussed later in this chapter).

Section A. Basic Principles of Renal Physiology

STRUCTURE OF THE KIDNEY AND URINARY SYSTEM

The kidneys are paired organs which lie in the posterior abdominal wall behind the lining of the abdominal cavity (peritoneum), one on each side of the vertebral column at the level of the T12 to L3 vertebrae. The right kidney usually lies a little lower than the left, possibly because the bulk of the liver is on the right side just above the right kidney. Each kidney is surrounded by a mass of fatty connective tissue, the *perirenal fat.* The medial border of the kidney is indented by a deep fissure called the *hilum,* through which pass the renal vessels and nerves and in which lies the funnel-shaped continuation of the upper end of the ureter, the *renal pelvis* (Fig. 18-2). The hilum leads into a shallow cavity (the *renal sinus*). The outer convex border of the renal pelvis is divided into *major calyxes,* each of which subdivides into several *minor calyxes.* The minor calyxes are cupped around the projecting apexes of the renal papillae.

The interior of each kidney is divided into two zones: an inner *renal medulla* and an outer *renal cortex.* The medulla consists of a number of cone-shaped masses (the *renal pyramids*) whose apexes form the papillae (Fig. 18-2). Each pyramid, topped by a region of the renal cortex, forms a *lobe* of the kidney. Approximately eight such lobes are present in each kidney. That part of the renal cortex which dips down between adjacent pyramids forms the *renal columns.*

In human beings, each kidney is composed of approximately 1 million tiny, closely packed units, all similar in structure and function, bound together by small amounts of connective tissue containing blood vessels, nerves, and lymphatics. One such unit, or *nephron,* is shown in Fig. 18-3A and B. The nephron consists of a *vascular component (the glomerulus)* and a *tubular component.* The mechanisms by which the kidneys perform their functions depend on the relationship

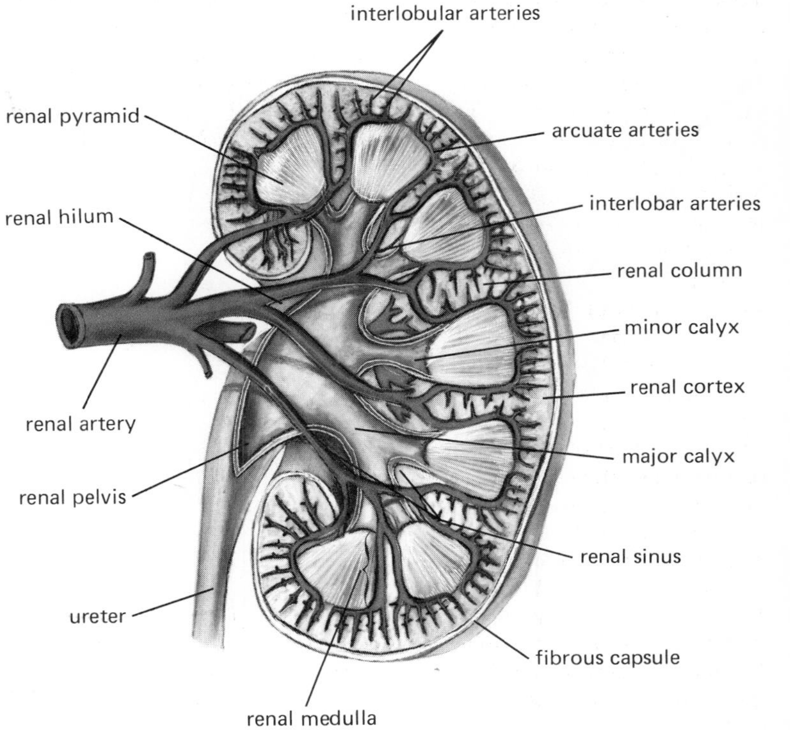

FIGURE 18-2 ***Coronal section through the kidney.***

between these two components.

Throughout its course, the tubule is composed of a single layer of epithelial cells which differ in structure and function from portion to portion. It originates as a blind sac, known as *Bowman's capsule,* which is lined with thin epithelial cells. On one side, Bowman's capsule is intimately associated with the glomerulus; on the other it opens into the first portion of the tubule, which is highly coiled and is known as the *proximal convoluted tubule.* The next portion of the tubule is a sharp hairpinlike loop, called *Henle's loop.* The tubule once more becomes coiled (the *distal convoluted tubule*) and finally runs a straight course as the *collecting duct.* From the glomerulus to the beginning of the collecting duct, each of the million tubules is completely separate from its neighbors, but there the tiny collecting ducts from separate tubules join to form larger ducts, which in turn form even larger ducts, which finally empty into the large central cavities of the renal pelvis (Fig. 18-3B). The urine is not altered after it leaves the collecting ducts. From the renal pelvis on, the remainder of the urinary system serves simply as plumbing.

To return to the other component of the nephron: What are the origin and nature of the glomerulus? Blood is supplied to the kidneys through the *renal arteries,* which arise on each side of the aorta. Each renal artery divides before entering the substance of the kidney, forming the *interlobar arteries.* These ascend within the renal columns between the pyramids until they reach the bases of the pyramids, where they give off many *arcuate arteries,* which run horizontally along the region of the corticomedullary junction (Fig. 18-4). The *interlobular arteries* arise from the arcuates, and each of these gives off, at right angles to itself, a series of arterioles, the *afferent arterioles,* each of which leads to a compact tuft of capillaries. This tuft of capillaries is the glomerulus, which protrudes into Bowman's capsule and is essentially floating in the fluid within the capsule (Fig. 18-5). The functional significance of this anatomic arrangement is that blood in the glomerulus is separated from the space within Bowman's capsule only by a thin layer of tissue composed of (1) the single-celled capillary lining, (2) a layer of basement membrane, and (3) the single-celled lining of Bowman's capsule. This thin barrier permits the *filtration* of fluid from the capillaries into Bowman's capsule.

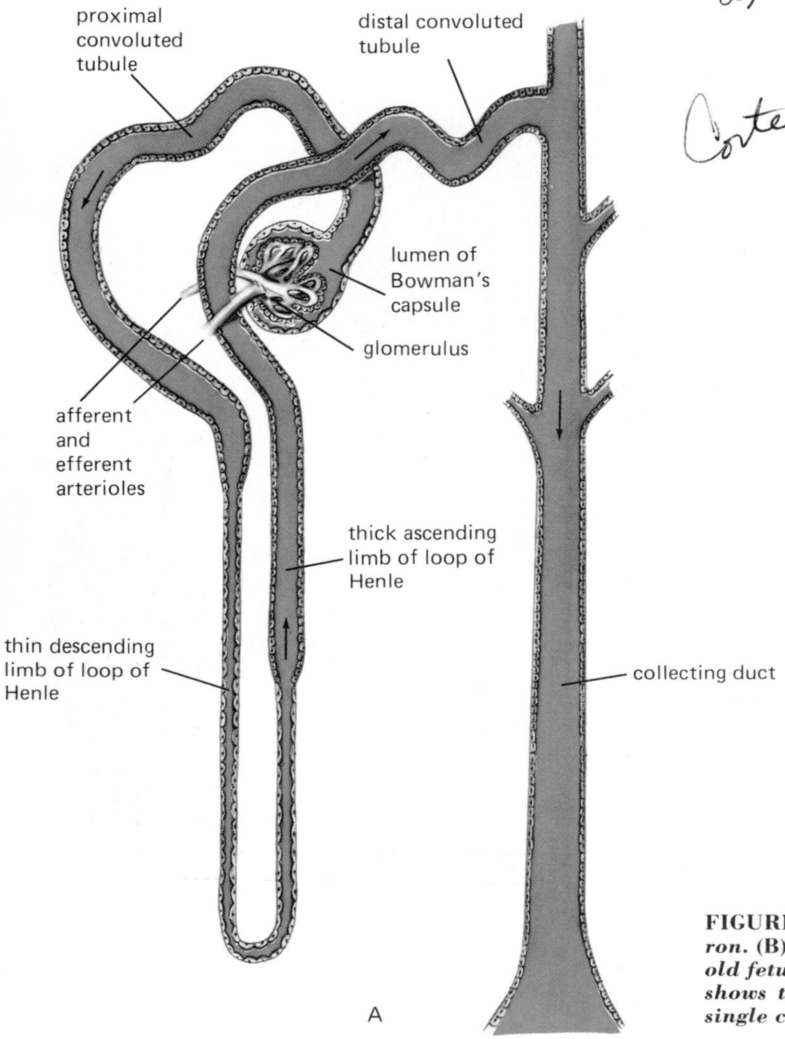

proximal convoluted tubule
distal convoluted tubule
lumen of Bowman's capsule
glomerulus
afferent and efferent arterioles
thick ascending limb of loop of Henle
thin descending limb of loop of Henle
collecting duct
A

FIGURE 18-3 (A) ***Basic structure of a nephron.*** (B) ***One lobe of a kidney from a 6-month-old fetus. The diagram in the center of the lobe shows the location of two nephrons around a single collecting duct.***

The glomerular capillaries are part of a portal system; instead of combining to form veins, they recombine to form another set of arterioles, the *efferent arterioles*. Thus, blood leaves the glomerulus through an arteriole which soon subdivides into a second set of capillaries (Fig. 18-4). These *peritubular capillaries* are profusely distributed to, and intimately associated with, all the remaining portions of the tubule. They rejoin to form the *interlobular veins* which end in the *arcuate veins* (Fig. 18-4). These drain through the *interlobar veins* into the *renal veins*, by which blood ultimately leaves the kidney.

There are important regional differences in the location of the various tubular and vascular components, as shown in Fig. 18-3B. The cortex contains all the glomeruli, proximal and distal convoluted tubules, and the first portions of the loops of Henle and collecting ducts. In human beings, approximately 85 percent of the nephrons originate in glomeruli located in the outer, or *superficial*, cortex and have relatively

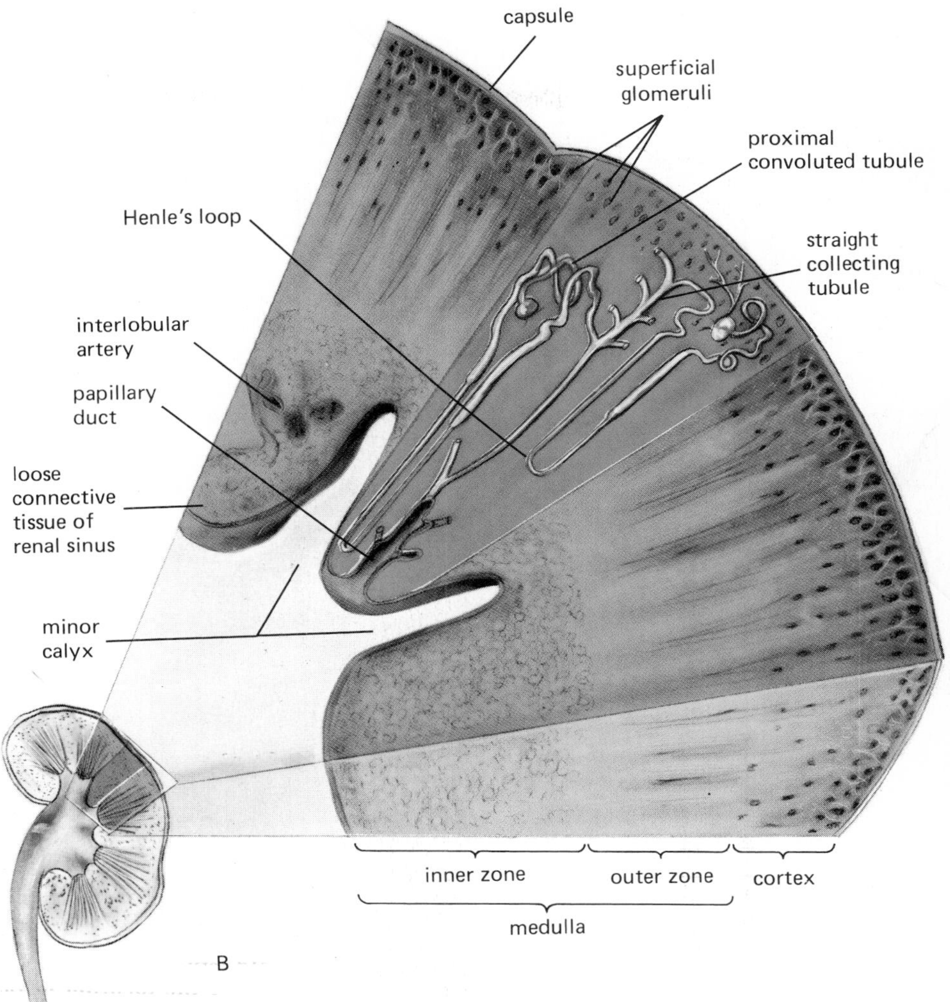

short loops of Henle, which extend only into the outer medulla. The remaining 15 percent originate in glomeruli located in the innermost, or *juxtamedullary*, cortex (i.e., "cortex just adjacent to the medulla"). These *juxtamedullary nephrons*, in contrast to the *superficial nephrons*, have long loops which extend deep into the medulla, where they run parallel to the collecting ducts. The vascular structure supplying the juxtamedullary nephrons also differs in that its efferent arterioles drain not only into the usual peritubular-capillary network of the cortex but also into thin hairpin-loop vessels (*vasa recta*), which run parallel to the loops of Henle and collecting ducts in the medulla. This arrangement has considerable significance for renal function, as will be described later.

One last anatomic feature should be pointed out. Note that, as the loop ascends into the cortex to become the distal tubule, the tubule contacts the arterioles supplying its nephron of origin (Fig. 18-3A). This area of contact is marked by unique structural

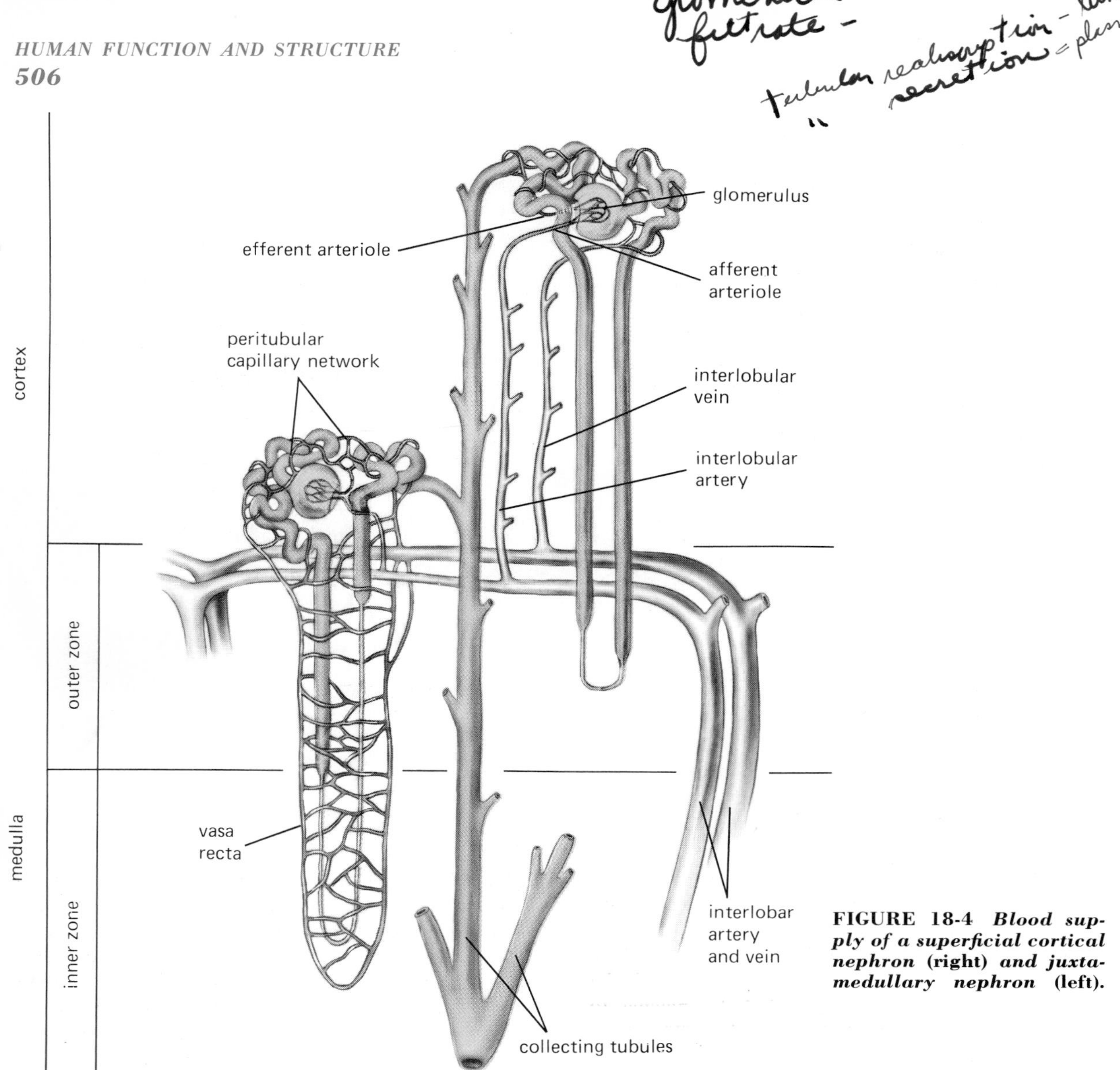

FIGURE 18-4 ***Blood supply of a superficial cortical nephron* (right) *and juxtamedullary nephron* (left).**

changes in both the arterioles and tubules and is known as the *juxtaglomerular apparatus*. Its structure and function will also be discussed later.

BASIC RENAL PROCESSES

Urine formation begins with the filtration of essentially protein-free plasma through the glomerular capillaries into Bowman's capsule. This *glomerular filtrate* contains all crystalloids (low-molecular-weight substances) in virtually the same concentrations as in plasma. The final urine which enters the renal pelvis is quite different from the glomerular filtrate because, as the filtered fluid flows from Bowman's capsule through the remaining portions of the tubule, its composition is altered. This change occurs by two general processes, *tubular reabsorption* and *tubular secretion*. The tubule is at all points intimately associated with the peritubular

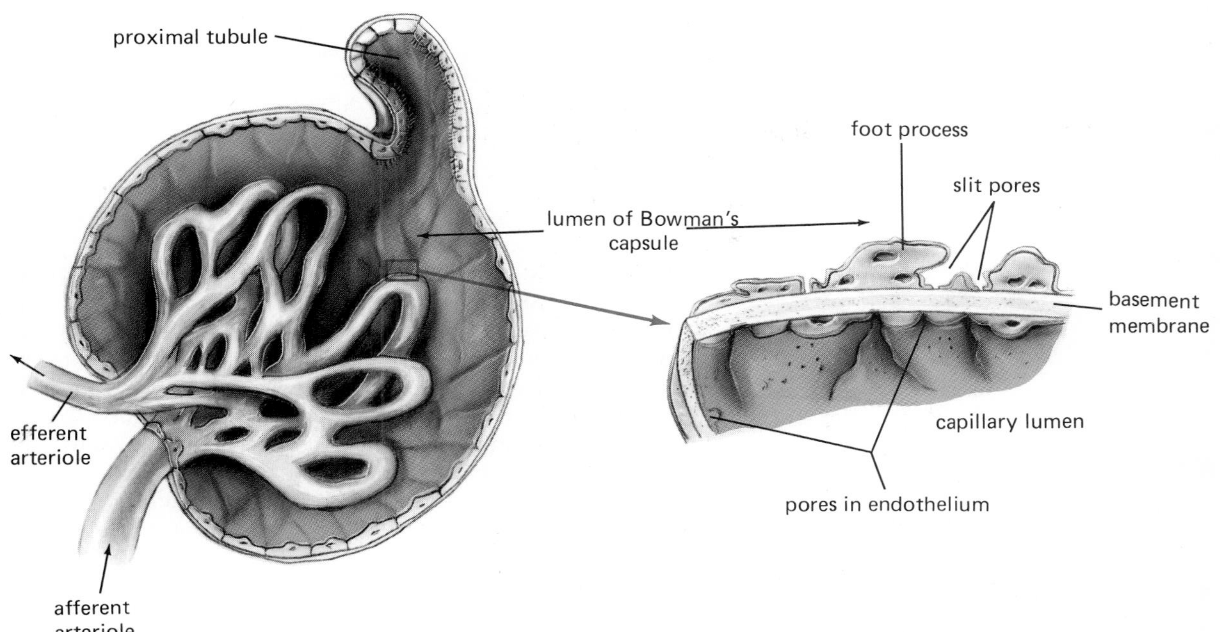

FIGURE 18-5 ***Anatomy of the glomerulus. The drawing on the right shows the glomerular membrane: tubular epithelium (only foot processes are shown), basement membrane, and capillary endothelium. The foot processes come together at the cell body like the tentacles of an octopus.***

capillaries, a relationship that permits transfer of materials between peritubular plasma and the inside of the tubule (*tubular lumen*). When the direction of transfer is from tubular lumen to peritubular capillary plasma, the process is called tubular reabsorption. Movement in the opposite direction, i.e., from peritubular plasma to tubular lumen, is called tubular secretion. (This term must not be confused with *excretion;* to say that a substance has been excreted is to say only that it appears in the final urine.) These relationships are illustrated in Fig. 18-6.

The most common relationships between these basic renal processes—glomerular filtration, tubular reabsorption, and tubular secretion—are shown in Fig. 18-7. Plasma containing substances X, Y, and Z enters the glomerular capillaries. A certain quantity of protein-free plasma containing these substances is filtered into Bowman's capsule, enters the proximal tubule, and begins its flow through the rest of the tubule. The remainder of the plasma, also containing X, Y, and Z, leaves the glomerular capillaries via an efferent arteriole and enters the peritubular capillaries. The cells composing the tubular epithelium can actively transport X (not Y or Z) from the peritubular plasma into the tubular lumen, but not in the opposite direction. By this combination of filtration and tubular secretion all the plasma which originally entered the renal artery is cleared of substance X, which leaves the body via the urine, thus reducing the amount of X remaining in the body. If the tubule were incapable of reabsorption, the Y and Z originally filtered at the glomerulus would also leave the body via the urine, but the tubule can transport Y and Z from the tubular lumen back into the peritubular plasma. The amount of reabsorption of Y is small, so that most of the filtered material does escape from the body, but for Z the reabsorptive mechanism is so powerful that virtually all the filtered material is transported back into the plasma, which flows through the renal vein back into the vena cava. Therefore no Z is lost from the body. Hence the processes of filtration

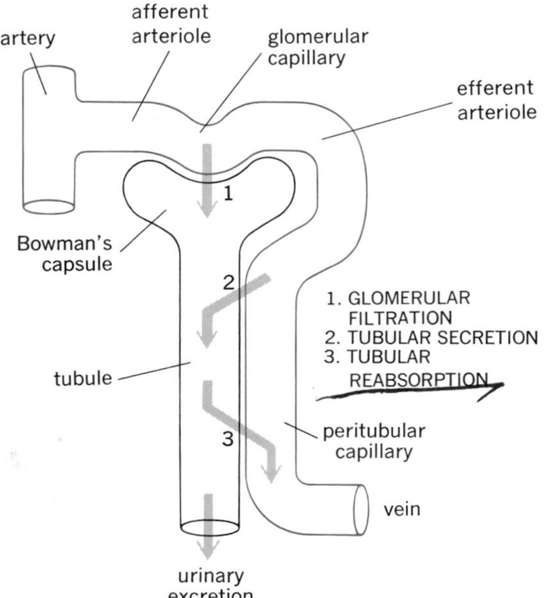

FIGURE 18-6 ***Three basic components of renal function.***

and reabsorption have canceled each other out, and the net result is as though Z had never entered the kidney at all.

The kidney works only on plasma (the erythrocytes supply oxygen to the kidney but serve no other function in urine formation). Each substance in plasma is handled in a characteristic manner by the nephron, i.e., by a particular combination of filtration, reabsorption, and secretion. The critical point is that *the rates at which the relevant processes proceed for many of these substances are subject to physiologic control.* What is the effect, for example, if the Y filtration rate is increased or its reabsorption rate decreased? Either change means that more Y is lost from the body via the urine. By triggering such changes in filtration or reabsorption whenever the plasma concentration of Y rises above normal, homeostatic mechanisms regulate plasma Y.

Glomerular Filtration

As described in Chap. 15, the capillaries of the body are freely permeable to water and solutes of small molecular dimension (crystalloids). They are relatively impermeable to large molecules (colloids), the most important of which are the plasma proteins. The glomerulus thus behaves qualitatively like any other capillary. This may seem surprising since the glomerular barrier is structurally different from most other capillaries in that it is composed of three layers: capillary endothelium, basement membrane, and a single-celled layer of epithelial cells. These epithelial cells (*podocytes*) have an unusual octopuslike structure in that they have a large number of extensions, or foot processes, which rest on the basement membrane (Fig. 18-5). Slits exist between adjacent foot processes and probably constitute the path through which the filtrate, once past the endothelial cells and basement membrane, travels to enter Bowman's capsule. The presence of this epithelial layer may account for certain quantitative differences between the glomerulus and other capillaries.

The mean blood pressure in the large arteries of the body is approximately 100 mm Hg, but this is not the hydrostatic pressure of the blood within the glomerular capillaries, since pressure is dissipated as the blood passes through the arterioles connecting the renal artery branches to the glomeruli. The glomerular capillary pressure is usually about 50 mm Hg. This is about half of mean arterial pressure and is considerably higher than in other capillaries of the body because the afferent arterioles leading to the glomeruli are wider than most arterioles and therefore offer less resistance to flow. The capillary hydrostatic pressure favoring filtration is not completely unopposed. There is, of course, fluid within Bowman's capsule which results in a capsular hydrostatic pressure of 10 mm Hg resisting further filtration into the capsule. A second opposing force results from the presence of protein in the plasma and its absence in Bowman's capsule. As in other capillaries, this unequal distribution of protein causes the water concentration of the plasma to be less than that of the fluid in Bowman's capsule. Again, as in other capillaries, the water-concentration difference is due completely to the plasma protein since all the crystalloids have virtually identical concentrations in the plasma and Bowman's capsule. The difference in water concentration induces an osmotic flow of fluid (water plus all the crystalloids) from Bowman's capsule into the capillary, a flow that opposes filtration. Its magnitude is equivalent to the bulk flow produced by a hydrostatic pressure difference of 30 mm Hg.

As can be seen from Table 18-2, the net filtration pressure is approximately 10 mm Hg. This pressure initiates urine formation by forcing an essentially protein-

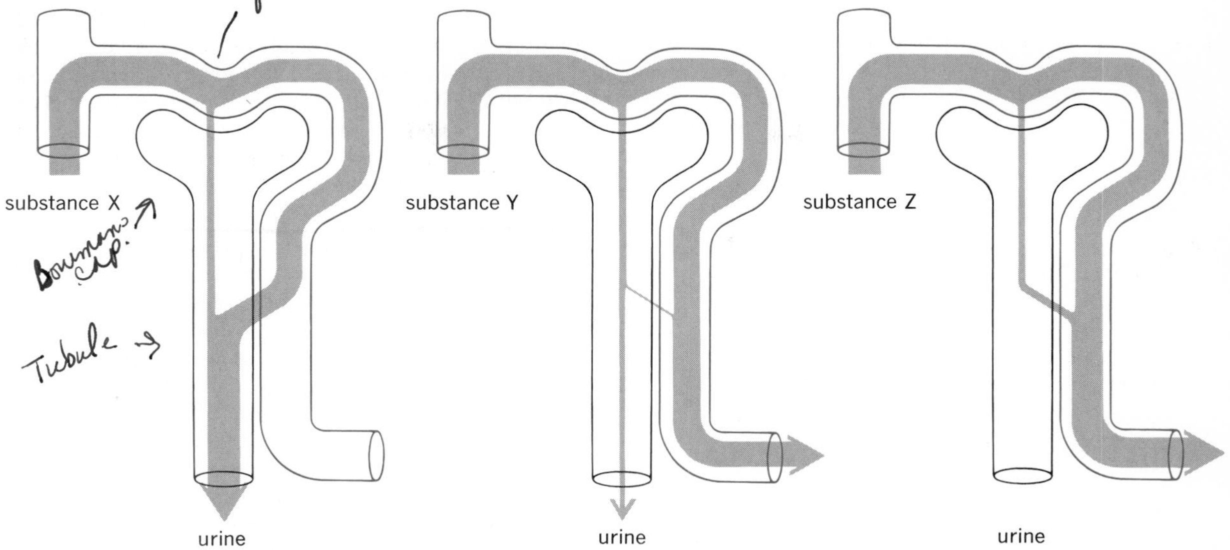

substance X
substance Y
substance Z
urine
urine
urine

FIGURE 18-7 ***Renal manipulation of three substances X; Y, and Z. X is filtered and secreted but not reabsorbed. Y is filtered, and a fraction is then reabsorbed. Z is filtered and is completely reabsorbed.***

TABLE 18-2
FORCES INVOLVED IN GLOMERULAR FILTRATION

Forces	*Millimeters of mercury*
Favoring filtration:	
Glomerular capillary blood pressure	50
Opposing filtration:	
Fluid pressure in Bowman's capsule	10
Osmotic gradient (water-concentration difference due to protein)	30
Net filtration pressure	10

free filtrate of plasma through glomerular pores into Bowman's capsule and thence down the tubule into the renal pelvis. It should be evident from this description that the glomerular membranes serve only as a filtration barrier and play no active, i.e., energy-requiring, role. The energy producing glomerular filtration is the energy transmitted to the blood (as hydrostatic pressure) when the heart contracts.

Before leaving this topic, we must point out the reason for use of the term "essentially protein-free" in describing the glomerular filtrate. In reality there is a very small amount of protein in the filtrate since the glomerular membranes are not perfect sieves for protein. Normally, less than 1 percent of serum albumin and virtually no globulin are filtered; whatever protein is filtered is normally completely reabsorbed so that no protein appears in the final urine. However, in diseased kidneys, the glomerular membranes may become much more leaky to protein so that large quantities are filtered and some of this protein appears in the urine.

Rate of Glomerular Filtration In human beings, the average volume of fluid filtered from the plasma into Bowman's capsule is 180 l/day (approximately 45 gal)! The implications of this remarkable fact are extremely important. When we recall that the average total volume of plasma in human beings is approximately 3 l, it follows that the entire plasma volume is filtered by the kidneys some 60 times a day. It is, in part, this ability to process such huge volumes of plasma that enables the kidneys to excrete large quantities of waste products and to regulate the constituents of the internal environment so precisely. The second implication concerns the magnitude of the reabsorptive process. The average person excretes between 1 and 2 l of urine per day. Since 180 l of fluid is filtered, approxi-

TOTAL CARDIAC output

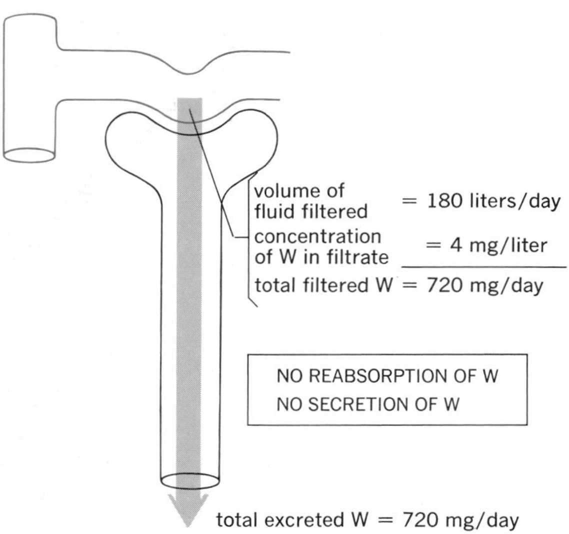

FIGURE 18-8 ***Measurement of glomerular filtration. W is filtered but is neither reabsorbed nor secreted.***

mately 99 percent of the filtered water must have been reabsorbed into the peritubular capillaries, the remaining 1 percent escaping from the body as urinary water.

How is the rate of glomerular filtration measured? To answer this question we use another imaginary substance W (Fig. 18-8). Over a 24-h period (or any other convenient period of time) the subject's urine is collected. The amount of W in this volume of urine is measured and found to be 720 mg. How did this amount of W get into the urine? From other experiments it has been determined that W is filtered at the glomerulus but is not secreted. Therefore, W can get into the urine only by filtration. Moreover, it also has been previously determined that W is not reabsorbed at all by the tubules. Thus, all the filtered W appears in the final urine. Therefore, in order to excrete 720 mg of W in 24 h, the subject must have filtered 720 mg of W during the same 24-h period. The question now becomes: How much glomerular filtrate contains 720 mg of W? Since W is a molecule of small dimensions, it must have the same concentration in the glomerular filtrate as in plasma. Several samples of the person's blood are obtained during the same 24-h period, and the plasma W concentration is found to be 4 mg/l of plasma. Since 1 l of plasma contains 4 mg of W, then 720/4, or 180, l contain 720 mg of W. In other words, 180 l of plasma having a W concentration of 4 mg/l must have been filtered during the 24-h period in order to account for the appearance of 720 mg of W in the final urine. This example is illustrated in Fig. 18-8. The validity of this analysis depends upon the fact that W is freely filterable at the glomerulus and is neither reabsorbed nor secreted by the tubules. A polysaccharide called *inulin* (not insulin) completely fits this description and is used for just such determinations in human beings and experimental animals.

This type of experiment, called *clearance technique,* has proved invaluable for gaining information about tubular reabsorption and secretion. For example, suppose we were interested in learning whether there is tubular reabsorption of phosphate. Using inulin, we would determine the glomerular filtration rate (GFR). (This must be repeated for every experiment because the GFR is not fixed but varies significantly.) The GFR in this particular experiment is found to be 165 l/day. The plasma concentration of phosphate is 1 mmol/l. Since phosphate is completely filterable at the glomerulus, 165 l/day × 1 mmol/l = 165 mmol/day is filtered. Finally, the amount of phosphate in the urine during this same 24-h period is found to be 40 mmol. Therefore, 165 − 40 mmol = 125 mmol of phosphate must have been reabsorbed by the tubules per 24 h. The generalization emerging from this example is that whenever the quantity of a substance excreted in the urine is *less* than the amount filtered during the same period of time, tubular reabsorption must have occurred. Conversely, if the amount excreted in the urine is *greater* than the amount filtered during the same period of time, tubular secretion must have occurred.

To complete this discussion of glomerular filtration we must consider the magnitude of the *total renal blood flow.* We have seen that none of the red blood cells and only a portion of the plasma which enters the glomerular capillaries are filtered into Bowman's capsule, the remainder passing via the efferent arterioles into the peritubular capillaries. Normally, the glomerular filtrate constitutes approximately one-fifth of the total plasma entering the kidney. Thus the total renal plasma flow is equal to 5 × 180 = 900 l/day or 0.610 l/min. Since plasma constitutes approximately 55 percent of whole blood, the total renal blood flow, i.e., erythrocytes plus plasma, must be approximately 1.1 l/min. Thus, the kidneys receive one-fifth to one-fourth of the total cardiac output (5 l/min) although their combined weight is less than 1 percent of the total body weight! These relationships are illustrated in Fig. 18-9.

Tubular Reabsorption

Many filterable plasma components are either completely absent from the urine or present in smaller quantities than were originally filtered at the glomerulus. This fact alone is sufficient to prove that these substances undergo tubular reabsorption. An idea of the magnitude and importance of these reabsorptive mechanisms can be gained from Table 18-3, which summarizes data for a few plasma components, all of which are handled by filtration and reabsorption.

TABLE 18-3

AVERAGE VALUES FOR SEVERAL COMPONENTS HANDLED BY FILTRATION AND REABSORPTION

Substance	*Amount filtered per day*	*Amount excreted per day*	*Percent reabsorbed*
Water, l	180	1.8	99.0
Sodium, g	630	3.2	99.5
Glucose, g	180	0	100
Urea, g	54	30	44

FIGURE 18-9 ***Magnitude of glomerular filtration rate, total renal plasma flow, and total renal blood flow. Only 20 percent of the plasma entering the kidneys is filtered from the glomerulus into Bowman's capsule. The remaining 80 percent flows through the glomerulus into the efferent arteriole and thence into the peritubular capillaries.***

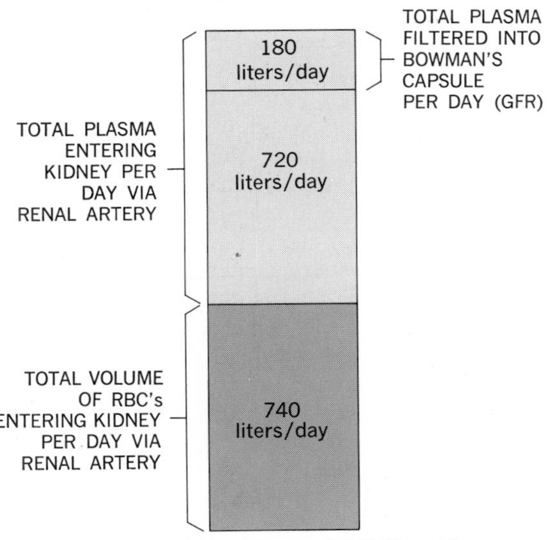

These are typical values for a normal person on an average diet. There are at least three important conclusions to be drawn from this table: (1) The quantities of material entering the nephron via the glomerular filtrate are enormous, generally larger than their total body stores; e.g., if reabsorption of water ceased but filtration continued, the total plasma water would be urinated within 30 min. (2) The quantities of waste products, such as urea, which are excreted in the urine are generally sizable fractions of the filtered amounts; thus, in mammals coupling a large glomerular filtration rate with a limited urea reabsorptive capacity permits rapid excretion of the large quantities of this substance produced constantly as a result of protein breakdown. (3) In contrast to urea and other waste products, the amounts of most useful plasma components, e.g., water, electrolytes, and glucose, which are excreted in the urine represent quite smaller fractions of the filtered amounts because of reabsorption. The rates at which many of these substances are reabsorbed (and therefore the rates at which they are excreted) are constantly subject to physiologic control.

Types of Reabsorption A bewildering variety of ions and molecules is found in the plasma. With the exception of the proteins (and a few ions tightly bound to protein), these materials are all present in the glomerular filtrate, and most are reabsorbed, to varying extents. It is essential to realize that tubular reabsorption is a qualitatively different process than glomerular filtration. The latter occurs by bulk flow in which water and all dissolved free crystalloids move together. In contrast, tubular reabsorption of various substances is by more or less discrete tubular transport mechanisms and by diffusion, although in many cases a single reabsorptive system transports several different components if they are similar in structure. For example, many of the simple carbohydrates are reabsorbed by the same system. As described in Chap. 1, transport processes can be categorized broadly as *active* or *passive,* and there are many examples of each in the kidney. The process is passive if no cellular energy is directly and specifically involved in the transport of the substance, i.e., if the substance moves downhill by simple or facilitated diffusion as a result of an electric or chemical concentration gradient. Active reabsorption, on the other hand, can produce net movement of the substance uphill against its concentration or electric gradient and therefore requires energy expenditure by the transporting cells.

Transport of any substance across the renal tubule involves a sequence of steps. In Chap. 1, when we described the basic characteristics of transport processes, we were dealing only with transport across a single membrane, i.e., from the outside of the cell to the inside, or vice versa. However, to cross the renal tubule, a substance must traverse not just one but a sequence of membranes. For example, to be reabsorbed, a sodium ion must gain entry to the tubular cell by crossing the cell membrane lining the lumen. It must then move through the cell's cytoplasm and cross the opposite cell membrane to enter the interstitial fluid. Finally, it must cross the capillary membranes to enter the plasma. The entire process is known as *transepithelial transport* and occurs not only in the kidney but in the gastrointestinal tract and other epithelial linings of the body.

In transepithelial transport the overall process is called *active* if one or more of the individual steps in the sequence is active. Sodium ions, for example, diffuse across the first tubular cell membrane and the cytoplasm of the cell. They are then actively transported out of the cell into the interstitial fluid; water accompanies the sodium, and the entire fluid then gains entry into the capillary by the bulk-flow process typical of all capillaries. Thus, three of the four steps are passive, but the crucial step is mediated by an active carrier process, and the overall process of sodium reabsorption is therefore said to be active. In this chapter we shall ignore the fact that multiple membranes lie between the lumina of the tubule and capillary and treat the tubular epithelium as a single membrane separating tubular fluid and plasma.

Many of the active reabsorptive systems in the renal tubule can transport only limited amounts of material per unit time, primarily because the membrane carrier responsible for the transport becomes saturated. The classic example is the tubular transport process for glucose. As we know, normal persons do not excrete glucose in their urine because tubular reabsorption is complete; but it is possible to produce urinary excretion of glucose in a completely normal person merely by administering large quantities of glucose directly into one of his veins. Recall that the filtered quantity of any freely filterable plasma component such as glucose is equal to its plasma concentration multiplied by the glomerular filtration rate. If we assume that, as we give our subject intravenous glucose, the glomerular filtration remains constant, the filtered load of glucose will be directly proportional to the plasma glucose concen-

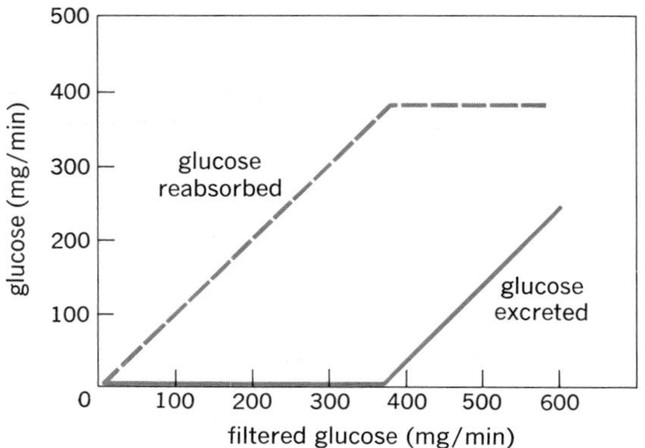

FIGURE 18-10 ***Saturation of the glucose transport system. Glucose is administered intravenously to a person so that plasma glucose and, thereby, filtered glucose are increased.***

tration (Fig. 18-10). We shall find that even after the plasma glucose concentration has doubled, the urine will still be glucose-free, indicating that his *maximal tubular capacity, T_m,* for reabsorbing glucose has not yet been reached. But as the plasma glucose and the filtered load continue to rise, glucose finally appears in the urine. From this point on any further increase in plasma glucose is accompanied by a directly proportionate increase in excreted glucose, because the T_m has now been reached. The tubules are now reabsorbing all the glucose they can, and any amount filtered in excess of this quantity cannot be reabsorbed and appears in the urine. This is precisely what occurs in the patient with diabetes mellitus. Because of a deficiency in pancreatic production of insulin, the patient's plasma glucose may rise to extremely high values. The filtered load of glucose becomes great enough to exceed the T_m and glucose appears in the urine. There is nothing wrong with the tubular transport mechanism for glucose, which is simply unable to reabsorb the huge filtered load.

Except for our experimental subject receiving intravenous glucose, the plasma glucose in normal persons never becomes high enough to cause urinary excretion of glucose because the reabsorptive capacity for glucose is much greater than necessary for normal filtered loads. However, for certain other substances, e.g., phosphate, the reabsorptive T_m is very close to the normal filtered load.

Just as glucose and phosphate provide excellent examples of actively transported solutes, urea provides an example of passive transport. Since urea is filtered at the glomerulus, its concentration in the very first portion of the tubule is identical to its concentration in peritubular capillary plasma. Then, as the fluid flows along the tubule, water reabsorption occurs, increasing the concentration of any intratubular solute not being reabsorbed at the same rate as the water, with the result that intratubular urea concentration becomes greater than the peritubular plasma concentration. Accordingly, urea is able to diffuse passively down this concentration gradient from tubular lumen to peritubular capillary. Urea reabsorption is thus a passive process and completely dependent upon the reabsorption of water, which establishes the diffusion gradient. In human beings, urea reabsorption varies between 40 and 60 percent of the filtered urea, the lower figure holding when water reabsorption is low and the higher when it is high.

Passive reabsorption is also of considerable importance for many foreign chemicals. The renal tubular epithelium acts in many respects as a lipid barrier; accordingly, lipid-soluble substances, like urea, can penetrate it fairly readily. Recall from Chap. 1 that one of the major determinants of lipid solubility is the polarity of a molecule—the less polar, the more lipid-soluble. Many drugs and environmental pollutants are nonpolar and, therefore, highly lipid-soluble. This makes their excretion from the body via the urine quite difficult since they are filtered at the glomerulus and then reabsorbed, like urea, as water reabsorption causes their intratubular concentrations to increase. Fortunately, the liver transforms most of these substances to progressively more polar metabolites which, because of their reduced lipid solubility, are poorly reabsorbed by the tubules and can therefore be excreted (polarity does not influence glomerular filtration which is bulk flow, not diffusion).

Tubular Secretion

Tubular secretory processes, which transport substances from peritubular capillaries to tubular lumen, i.e., in the direction opposite to tubular reabsorption, constitute a second pathway into the tubule, the first being glomerular filtration. Like tubular reabsorption processes, secretory transport may be either active or passive. Of the large number of different substances transported into the tubules by tubular secretion, only a few are normally found in the body, the most important being hydrogen ion and potassium. We shall see that most of the excreted hydrogen ion and potassium enters the tubules by secretion rather than filtration. Thus, renal regulation of these two important substances is accomplished primarily by mechanisms which control the rates of their tubular secretion. The kidney is also able to secrete a large number of foreign chemicals, thereby permitting their excretion from the body; penicillin is an example.

In the remainder of this chapter we shall see how the kidney acts as effector organ in a variety of homeostatic processes and how renal function is coordinated with that of other organs which also serve important regulatory roles as effector organs. Before turning to the individual variables being regulated, we complete our basic story by describing the mechanisms for eliminating urine from the body.

MICTURITION (URINATION)

From the kidneys, urine flows to the bladder through the ureters (Fig. 18-1), propelled by peristaltic contractions of the smooth muscle which makes up the ureteral wall and which has inherent rhythmicity, like other smooth muscles. The ureters, which are 25 to 30 cm long and about 3 mm in diameter, pass from the renal pelvis downward behind the peritoneum, approach the posterior wall of the bladder, pass obliquely through a gap in the bladder's muscular wall, and open into the bladder interior. The walls of the ureters have three coats: an inner *mucosa* of connective-tissue fibers covered by transitional epithelium; a middle *muscular coat* of circular and longitudinal layers of smooth muscle fibers and an outer connective-tissue *fibrous coat,* which blends with the kidney capsule and bladder wall at the two ends of the ureters. Under normal rates of urine formation, the downward-progressing peristaltic contractions of these muscles propel several jets of urine each minute into the bladder where it is temporarily stored.

The bladder is a balloonlike structure whose size, shape, and location vary with the amount of urine it contains (and with the degree of distension of nearby viscera such as the large intestine and uterus).

Like the ureters, the bladder has a three-layered wall: an inner *mucosa,* lined with transitional epithelium and continuous with the mucosa of the ureters above and the urethra below, which is thrown into folds when the bladder is empty; a middle muscular layer, which forms the *detrusor muscle* (Fig. 18-11); and an external fibrous layer. The detrusor muscle consists of three layers of smooth muscle fibers, an external and internal layer of longitudinally arranged fibers, and a middle layer of fibers oriented in a circular

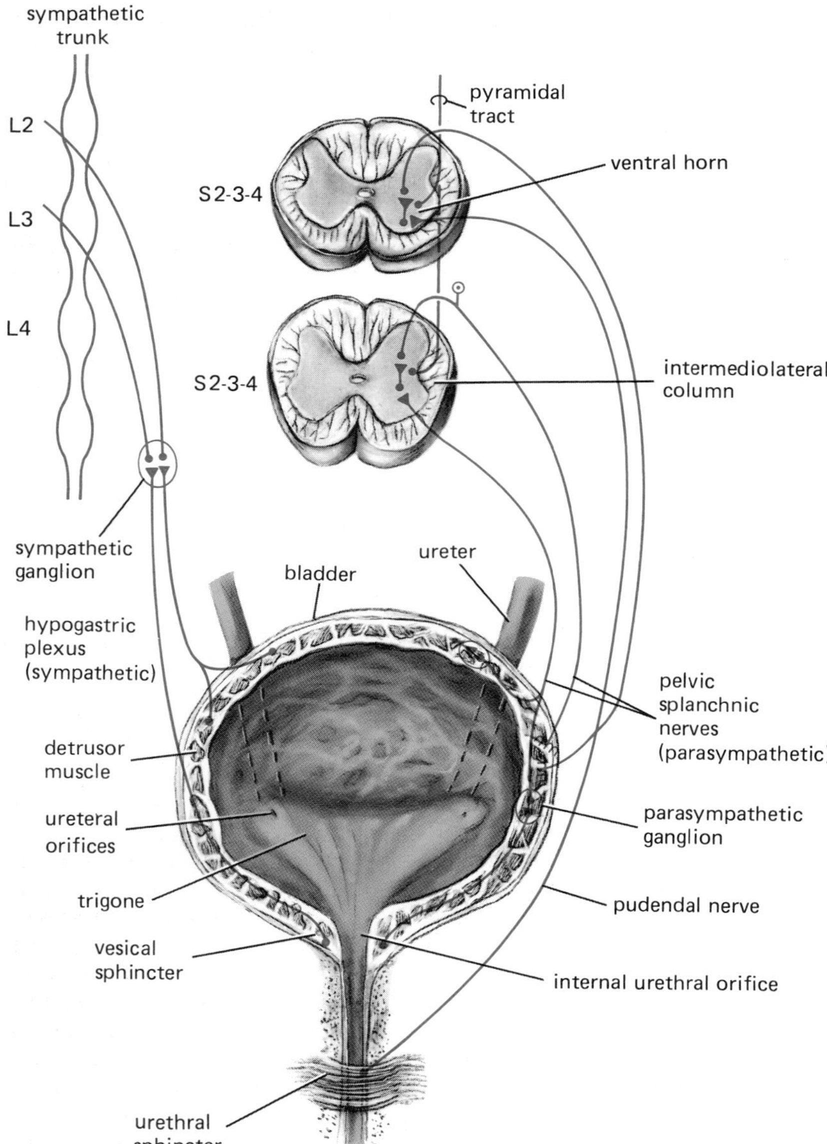

FIGURE 18-11 ***Basic structure and nerve supply of the bladder and external urethral sphincter. Stimulation of the mechanoreceptors in the bladder walls causes reflex stimulation of the parasympathetic nerves to the bladder and inhibition of the motor nerves to the external sphincter. The result is bladder contraction and sphincter relaxation. Higher-center input allows voluntary initiation or delay of micturition.***

direction (although there is considerable intermingling of the three layers). Upon contraction, the walls squeeze inward, increasing the pressure of the urine in the bladder. The smooth muscle layer at the base of the bladder is called the *vesical,* or *internal, sphincter.* It is not a distinct muscle but the last portion of the bladder; however, when the bladder is relaxed, this circular muscle is closed and functions as a sphincter. As the bladder contracts, this sphincter is pulled open simply by changes occurring in bladder shape during contraction. In other words, no special mechanism is required for its relaxation.

Another muscle important in the process of *micturition* (elimination of urine from the bladder) is a circular layer of skeletal muscle which surrounds the urethra farther down from the base of the bladder. When contracted, this *urethral,* or *external, sphincter* can hold the urethra closed against even strong bladder contractions.

Micturition is basically a local spinal reflex which can be influenced by higher brain centers. The bladder muscle receives a rich supply of parasympathetic fibers via the *pelvic splanchnic nerves,* the stimulation of which causes bladder contraction (Fig. 18-11). The external urethral sphincter is innervated by somatic motor nerves just like any other skeletal muscle. These fibers are in the *pudendal nerve,* which is a branch of the sacral plexus. The bladder wall contains many stretch receptors whose afferent nerves enter the spinal cord and eventually synapse with these parasympathetic and somatic motor neurons, stimulating the former and inhibiting the latter. Via descending pathways, higher brain centers synaptically facilitate and inhibit these motor pathways.

The following sequence of events leads to bladder emptying in an infant, in whom higher centers have only minor influence. When the bladder contains only small amounts of urine, its internal pressure is low, there is little stimulation of the bladder stretch receptors, the parasympathetics are relatively quiescent, and the somatic nerves are discharging at a moderately rapid rate. As the bladder fills with urine, it becomes distended, and the stretch receptors are gradually stimulated until their output becomes great enough to contract the bladder while simultaneously relaxing the external sphincter. Thus, the entire process is quite analogous to any other spinal reflex.

This process describes micturition adequately in the infant, but it is obvious that adults have the capacity either to delay micturition or to induce it voluntarily. In an adult, the volume of urine in the bladder required to initiate the spinal reflex for bladder contraction is approximately 300 ml. Delay is accomplished via descending pathways from the cerebral cortex which inhibit the bladder parasympathetics and stimulate the motor nerves to the external sphincter, thereby overriding the opposing synaptic input from the bladder stretch receptors. Voluntary initiation of micturition is just the opposite: Descending pathways from the cerebral cortex stimulate the bladder parasympathetics and inhibit the motor nerves to the sphincter, thereby summating with the afferent input from the stretch receptors and initiating micturition. It is by learning to control these pathways that a child achieves the ability to control the timing of micturition.

Section B. Regulation of Sodium and Water Balance

Table 18-1 is a typical balance sheet for water; Table 18-4 for sodium chloride. As with water, the excretion of sodium via the skin and gastrointestinal tract is normally quite small but may increase markedly during severe sweating, vomiting, or diarrhea. Hemorrhage, of course, can result in loss of large quantities of both sodium and water.

TABLE 18-4
NORMAL ROUTES OF SODIUM CHLORIDE INTAKE AND LOSS

	Grams per day
Intake:	
Food	10.5
Output:	
Sweat	0.25
Feces	0.25
Urine	10.0
Total output	10.5

Control of the renal excretion of sodium and water constitutes the most important mechanism for the regulation of body sodium and water. The excretory rates of these substances can be varied over an extremely wide range; e.g., a gross consumer of salt may ingest 20 to 25 g of sodium chloride per day whereas a patient on a low-salt diet may ingest only 50 mg. The normal kidney can readily alter its excretion of salt over this range. Similarly, urinary water excretion can be varied physiologically from approximately 400 ml/day to 25 l/day, depending upon whether one is lost in the desert or participating in a beer-drinking contest.

BASIC RENAL PROCESSES FOR SODIUM, CHLORIDE, AND WATER

None of these substances undergoes tubular secretion; each is freely filterable at the glomerulus and approximately 99 percent is reabsorbed as it passes down the tubules (Table 18-3). Indeed, the vast majority of all renal energy production must be used to accomplish this enormous reabsorptive task. The tubular mechanisms for reabsorption of these substances can be summarized by three generalizations: (1) The reabsorption of sodium is an active process, i.e., it is carrier-mediated, requires an energy supply, and can occur against an electrochemical gradient; (2) the reabsorption of chloride is by passive diffusion and depends upon the active reabsorption of sodium[1]; (3) the reabsorption of water is also by passive diffusion (osmosis) and also depends upon the active reabsorption of sodium.

Thus, active tubular sodium reabsorption is the primary force, which results in reabsorption of chloride and water as well. The relationship is explained as follows: Since the concentrations of all crystalloids are virtually identical in plasma and in Bowman's capsular fluid, in the very first portion of the proximal tubule no significant transtubular concentration gradients exist for sodium, chloride, or water. As the fluid flows down the tubule, sodium is actively reabsorbed into the peritubular capillaries.[2] Since sodium is a positively charged ion, its movement attracts a negatively charged chloride ion. In other words the sodium ion "pulls" the chloride ion with it as a result of its electric charge, thus obviating the need for a separate active-transport system for chloride.[3] (The remainder of this discussion thus implicitly includes chloride when it refers to sodium reabsorption.) Finally, what happens to the water concentration of the tubular fluid as a result of sodium reabsorption? Obviously this removal of solute lowers osmolarity, i.e., raises water concentration, below that of plasma. Thus, a water concentration gradient is created between tubular lumen and peritubular plasma which constitutes a driving force for water reabsorption via osmosis. If the water permeability of the tubular epithelium is very high, water molecules are reabsorbed passively almost as rapidly as the actively transported sodium ions, so that the tubular fluid is only slightly more dilute than plasma. In this manner almost all the filtered sodium and water could theoretically be reabsorbed and the final urine would still have approximately the same osmolarity as plasma. However, this reabsorption of water can occur only if the tubular epithelium is highly permeable to water. No matter how great the water concentration gradient, water cannot move if the epithelium is impermeable to it. In reality, the permeability of the last portions of the tubules (the distal tubules and collecting ducts) to water is subject to physiologic control. The major determinant of this permeability is a hormone known as *vasopressin* or *antidiuretic hormone* (ADH), the second name being preferable because it describes the effect of the hormone's action, antidiuresis, i.e., against a high urine volume. In the absence of ADH the water permeability of the distal tubule and collecting duct is very low, sodium reabsorption proceeds normally because ADH *has no effect on sodium reabsorption,* but water is unable to follow and thus remains in the tubule to be excreted as a large volume of urine. On the other hand, in the presence of ADH, the water permeability of these last nephron segments is very great, water reabsorption is able to keep up with sodium reabsorption, and the final urine volume is small.

[1] Recent evidence suggests that in one part of the tubule, the ascending loop of Henle, generalizations 1 and 2 may be reversed, i.e., chloride may be the actively transported ion with sodium following passively. The end result is the same.

[2] As mentioned earlier, we are treating the tubular epithelium as a single membrane separating tubular fluid and plasma.

[3] This description is, by necessity, highly simplified. The nature of the electric coupling between sodium and chloride movements is quite complex and still controversial.

CONTROL OF SODIUM EXCRETION: REGULATION OF EXTRACELLULAR VOLUME

The renal compensation for increased body sodium is excretion of the excess sodium. Conversely, a deficit in body sodium is prevented by reducing urinary sodium to an absolute minimum, thus retaining within the body the amount usually lost via the urine.

Since sodium is freely filterable at the glomerulus and actively reabsorbed but not secreted by the tubules, the amount of sodium excreted in the final urine represents the resultant of two processes, glomerular filtration and tubular reabsorption.

$$\text{Sodium excretion} = \text{sodium filtered} - \text{sodium reabsorbed}$$

It is possible, therefore, to adjust sodium excretion by controlling one or both of these two variables (Fig. 18-12). For example, what happens if the quantity of filtered sodium increases (as a result of a higher GFR) but the rate of reabsorption remains constant? Clearly, sodium excretion increases. The same final result could be achieved by lowering sodium reabsorption while holding the GFR constant. Finally, sodium excretion could be raised greatly by elevating the GFR and simultaneously reducing reabsorption. Conversely, sodium excretion could be decreased below normal levels by lowering the GFR or raising sodium reabsorption or both. Control of GFR and sodium reabsorption is therefore the mechanism by which renal regulation of sodium balance is accomplished.

The reflex pathways by which changes in total body sodium balance lead to changes in GFR and sodium reabsorption include (1) "volume" receptors (the reasons for the use of quotation marks will soon be apparent) and the afferent pathways leading from them to the central nervous system and endocrine glands; (2) efferent neural and hormonal pathways to the kidneys; (3) renal effector sites: the renal arterioles and tubules.

FIGURE 18-12 ***Sodium excretion is increased by increasing the GFR* (B), *by decreasing reabsorption* (C), *or by a combination of both* (D). *The arrows indicate relative magnitudes of filtration, reabsorption, and excretion.***

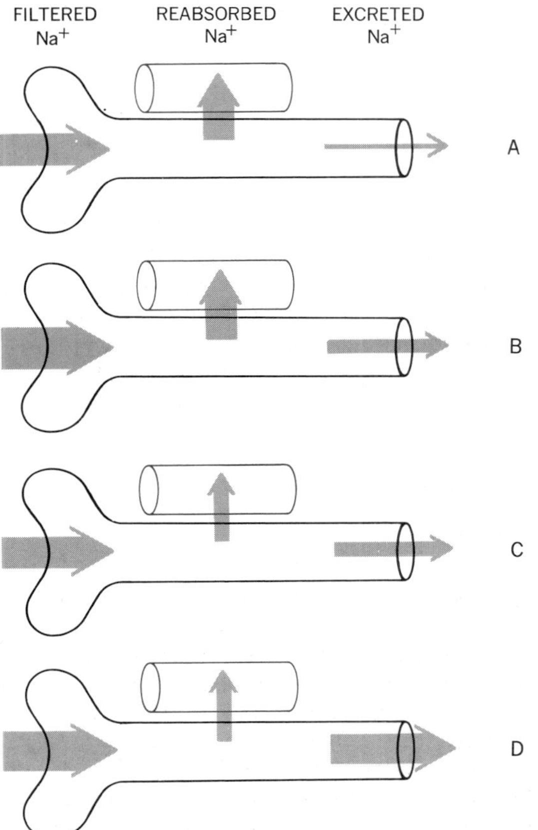

That cardiovascular volume receptors are appropriate receptors for regulating total body sodium is shown by the following analysis. What happens, for example, when a person ingests a liter of isotonic sodium chloride, i.e., a solution of salt with exactly the same osmolarity as the body fluids? It is absorbed from the gastrointestinal tract; all the salt and water remain in the extracellular fluid (plasma and interstitial fluid) and none enters the cells. Because of the active sodium "pumps" in cell membranes, sodium is effectively barred from the cells and therefore remains in the extracellular fluid. The water, too, remains since only the volume and not the osmolarity of the extracellular compartment has been changed; i.e., no osmotic gradient exists to drive the ingested water into cells. This example is one of many which could be used to illustrate the important generalization that total extracellular-fluid volume depends primarily upon the mass of extracellular sodium, which in turn correlates directly with total body sodium. However, rather than extracellular volume, per se, the variables monitored in sodium-regulating reflexes are closely correlated derivative values of this volume, specifically intravascular and intracardiac pressure or volume. For example, a decrease in plasma volume generally tends to lower the hydrostatic pressures within the veins, cardiac chambers, and arteries. The changes are detected by receptors within these structures (for example, the carotid sinus) and initiate the reflexes leading to renal sodium retention which helps to restore the plasma volume toward normal.

Control of GFR

Figure 18-13 summarizes the reflex pathway by which negative sodium balance (as, for example, in diarrhea) causes a decrease in GFR. Note that this is simply the basic baroreceptor reflex described in Chap. 16, where it was pointed out that decrease in cardiovascular pressures causes reflex vasoconstriction in many areas of the body. Conversely, an increased GFR can result from increased plasma volume and contribute to the increased renal sodium loss which returns extracellular volume to normal.

Control of Tubular Sodium Reabsorption

Present evidence indicates that, so far as long-term regulation of sodium excretion is concerned, the control of tubular sodium reabsorption is probably more important than that of GFR. For example, patients with chronic marked reductions of GFR usually maintain

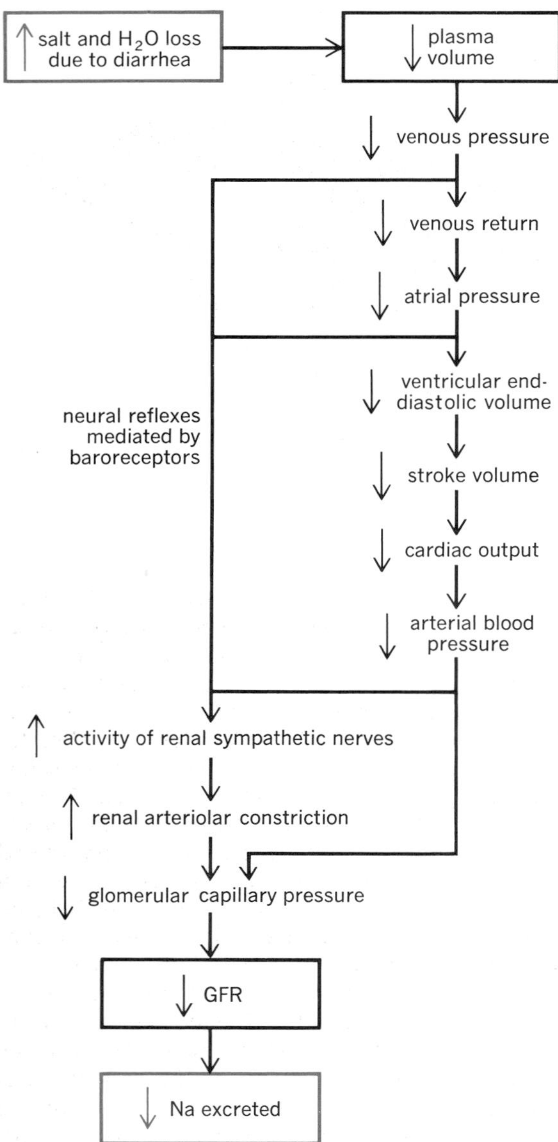

FIGURE 18-13 ***Pathway by which the GFR is decreased when plasma volume decreases. The baroreceptors which initiate the sympathetic reflex are probably located in large veins and the walls of the heart, as well as in the carotid sinuses and aortic arch.***

normal sodium excretion by decreasing tubular sodium reabsorption. The two major controllers of tubular sodium reabsorption are aldosterone and so-called third factor.

Aldosterone The adrenal cortex produces a hormone, called *aldosterone,* which stimulates sodium reabsorption, specifically by the distal tubules. In the complete absence of this hormone, the patient may excrete 25 g of salt per day, whereas excretion may be virtually zero when aldosterone is present in large quantities. In a normal person, the amounts of aldosterone produced and salt excreted lie somewhere between these extremes. It is interesting that aldosterone also stimulates sodium transport by other epithelia in the body, namely, by sweat and salivary glands and the intestine. The net effect is the same as that exerted on the renal tubules—a movement of sodium out of the luminal fluid into the blood. Thus aldosterone is an "all-purpose" stimulator of sodium retention.

Aldosterone secretion (and thereby tubular sodium reabsorption) is controlled by reflexes involving the kidneys themselves. Specialized cells (the *juxtaglomerular cells,* Fig. 18-14) lining the arterioles within the kidney synthesize and secrete into the blood a protein known as *renin* (not rennin), an enzyme catalyzing the reaction in which a small polypeptide *angiotensin* splits off from a large plasma protein *angiotensinogen* (Fig. 18-15). Angiotensin is a profound stimulator of aldosterone secretion and constitutes the primary input to the adrenal gland controlling production and the release of this hormone.

Angiotensinogen is synthesized by the liver and is always present in the blood; therefore, the rate-limiting factor in angiotensin formation is the concentration of plasma renin which, in turn, depends upon the rate of renin secretion by the kidneys. The critical question now becomes: What controls the rate of renin secretion? The answer to this, the first link in the reflex chain, is presently uncertain, mainly because there seem to be multiple inputs to the renin-secreting cells and it is not yet possible to assign quantitative roles to each of them. It is likely that the renal sympathetic nerves (and circulating epinephrine) constitute the single most important input in normal persons; this makes excellent sense, teleologically, since a reduction in body sodium and extracellular volume triggers off increased sympathetic tone to the kidneys (as shown in Fig. 18-13), thereby setting off the hormonal chain of events which restores sodium balance and extracellular volume to normal (Fig. 18-16).

Third Factor Until recently most renal physiologists believed that the control of sodium excretion could be explained completely in terms of change in GFR and aldosterone-dependent tubular sodium reab-

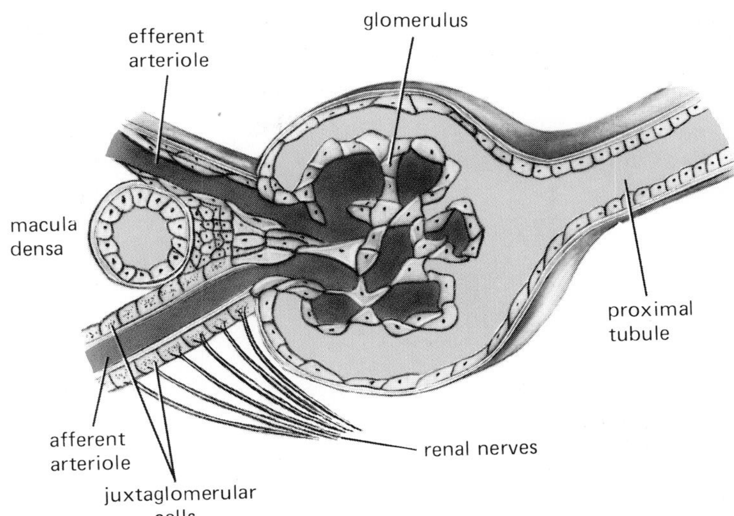

FIGURE 18-14 ***Juxtaglomerular apparatus.***

sorption. It is now clear that these two factors do not suffice, since there are circumstances in which GFR is low and aldosterone is high, yet sodium excretion is normal or increased. Therefore, there must exist a *third factor* which importantly influences tubular sodium reabsorption independently of aldosterone. However, the precise identity of this third factor has not yet been determined.

ADH Secretion and Extracellular Volume

Although we have discussed extracellular-volume regulation only in terms of the control of sodium excretion, it is clear that, to be effective in altering extracellular volume, the changes in sodium excretion must be accompanied by equivalent changes in water

FIGURE 18-15 ***Summary of the renin-angiotensin-aldosterone system. The plus sign denotes the stimulatory effect of angiotensin on aldosterone secretion.***

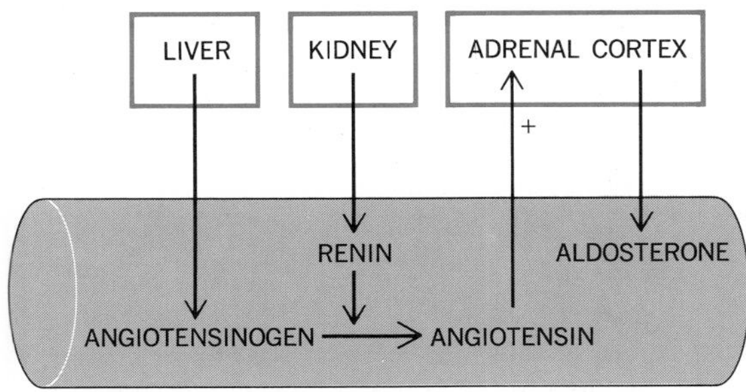

excretion. We have already pointed out that the ability of water to follow when sodium is reabsorbed depends upon ADH. Accordingly, a decreased extracellular volume must reflexly call forth increased ADH production as well as increased aldosterone secretion. What is the nature of this reflex? As described in Chap. 12, ADH is produced by a discrete group of hypothalamic neurons whose axons terminate in the posterior pituitary, from which ADH is released into the blood. These hypothalamic cells receive input from several vascular baroreceptors, particularly a group located in the left atrium. The baroreceptors are *stimulated* by *increased* atrial blood pressure, and the impulses resulting from this stimulation are transmitted via the vagus nerves and ascending pathways to the hypothalamus, where they *inhibit* the ADH-producing cells. Conversely, decreased atrial pressure causes less firing by the baroreceptors and stimulation of ADH synthesis and release (Fig. 18-17). The adaptive value of this baroreceptor reflex, one more in our expanding list, should require no comment.

RENAL REGULATION OF EXTRACELLULAR OSMOLARITY

We turn now to the renal compensation for pure water losses or gains, e.g., a person drinking 2 l of water, where no change in total salt content of the body occurs, only total water. The most efficient compensatory mechanism is for the kidneys to excrete the excess

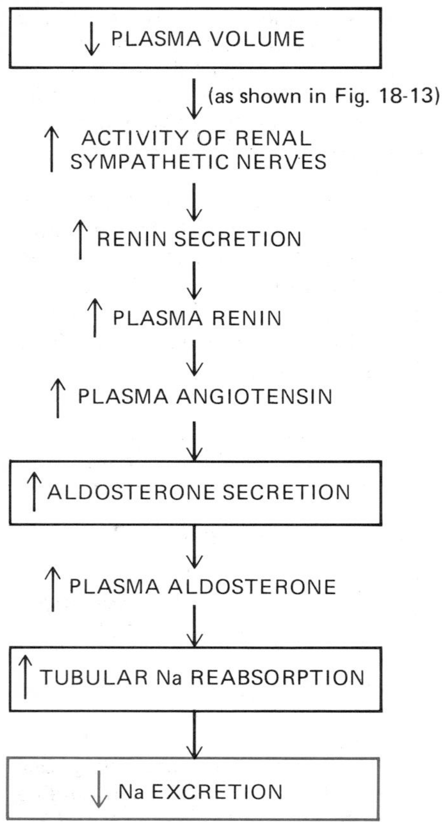

FIGURE 18-16 ***Pathway by which aldosterone secretion is increased when plasma volume is decreased. (The pathways by which the GFR is reduced are shown in Fig. 18-13.)***

water without altering their usual excretion of salt, and this is precisely what they do. ADH secretion is reflexively inhibited, as will be described below, tubular water permeability of the distal tubules and collecting ducts becomes very low, sodium reabsorption proceeds normally but water is unable to follow, and a large volume of extremely dilute urine is excreted. In this manner, the excess pure water is eliminated.

Thus it is easy to see how the kidneys produce a final urine having the same osmolarity as that of plasma or one having a lower osmolarity than plasma (hypoosmotic urine), the latter occurring whenever water reabsorption lags behind solute reabsorption, i.e., when plasma ADH is reduced. Clearly the formation of a hypoosmotic urine is a good compensation for an excess of water in the body.

But how can the kidneys produce a hyperosmotic urine, i.e., a urine having an osmolarity greater than that of plasma? For this to occur, does not water reabsorption have to "get ahead" of solute reabsorption? How can this happen if water reabsorption is always secondary to solute (particularly sodium) reabsorption? It would seem that, if our three generalizations are not to be violated, the kidneys cannot produce a hyperosmotic urine. Yet they do. Indeed, the final urine may

FIGURE 18-17 ***Pathway by which ADH secretion is increased when plasma volume decreases.***

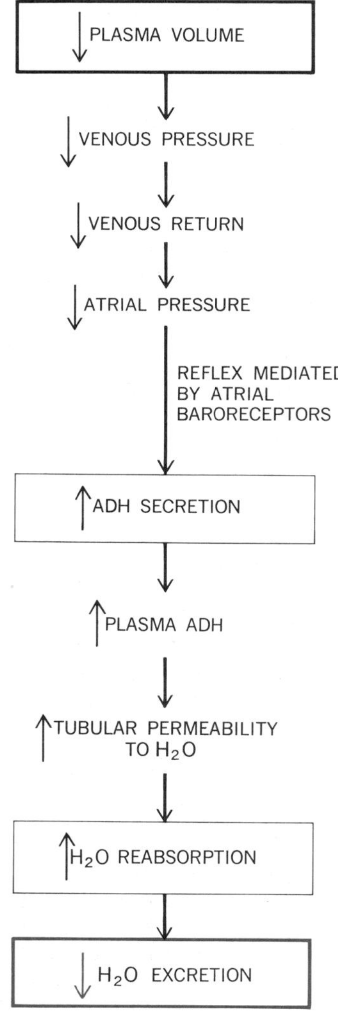

be as concentrated as 1,400 mosmol/l compared with a plasma osmolarity of 300 mosmol/l. Moreover, this concentrated urine is produced without violating the three generalizations.

Urine Concentration: The Countercurrent System

The ability of the kidneys to produce concentrated urine is not merely an academic problem. It is a major determinant of one's ability to survive without water. The human kidney can produce a maximal urinary concentration of 1,400 mosmol/l. The urea, sulfate, phosphate, and other waste products (plus the smaller number of nonwaste ions) which must be excreted each day amount to approximately 600 mosmol; therefore, the water required for their excretion constitutes an obligatory water loss and equals

$$\frac{600 \text{ mosmol/day}}{1{,}400 \text{ mosmol/l}} = 0.444 \text{ l/day}$$

As long as the kidneys are functioning, excretion of this volume of urine will occur, despite the absence of water intake. In a sense, persons lacking access to water may literally urinate to death (due to fluid depletion). If the body could produce a urine with an osmolarity of 6,000 mosmol/l then only 100 ml of water need be lost obligatorily each day and survival time would be greatly expanded. A desert rodent, the kangaroo rat, does just that; this animal never drinks water, the water produced in its body by oxidation of foodstuffs being ample for its needs.

The kidneys produce concentrated urine by a complex interaction of events involving the so-called countercurrent multiplier system residing in the loop of Henle. Recall that the loop of Henle, which is interposed between the proximal and distal convoluted tubules, is a hairpin loop extending into the renal medulla. The fluid flows in opposite directions in the two limbs of the loop, thus the name countercurrent. Let us list the critical characteristics of this loop.

1 The ascending limb of the loop of Henle (i.e., the limb leading to the distal tubule) *actively* transports sodium chloride out of the tubular lumen into the surrounding interstitium. (As mentioned earlier, in the ascending loop of Henle it is possible that chloride is the actively transported ion with sodium following passively; for simplicity, we shall refer to the process as "sodium chloride transport.") It is relatively impermeable to sodium chloride and water, so that passive fluxes into or out of it are small.

2 The descending limb of the loop of Henle (i.e., the limb into which drains fluid from the proximal tubule) does *not actively* transport sodium chloride; it is the only tubular segment that does not. Moreover, it is relatively permeable to both ions and water, so that *passive* fluxes into or out of it are large.

Given these characteristics, imagine the loop of Henle filled with a stationary column of fluid supplied by the proximal tubule. At first, the concentration everywhere would be 300 mosmol/l, since fluid leaving the proximal tubule is isosmotic to plasma, equivalent amounts of sodium chloride and water having been reabsorbed by the proximal tubule.

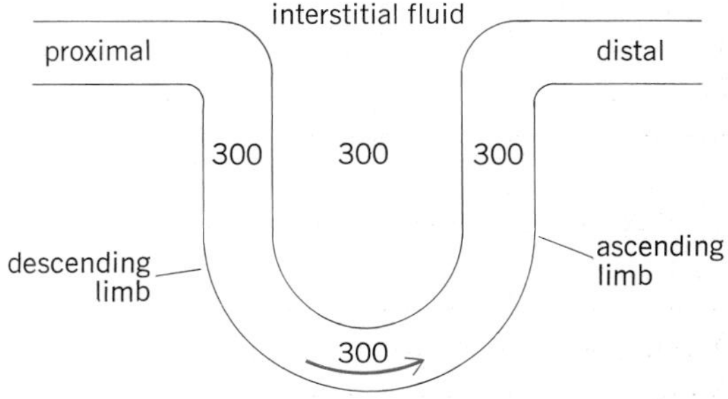

Now let the active pump in the ascending limb transport sodium chloride into the interstitium until a limiting gradient (say 200 mosmol/l) is established between ascending-limb fluid and interstitium.

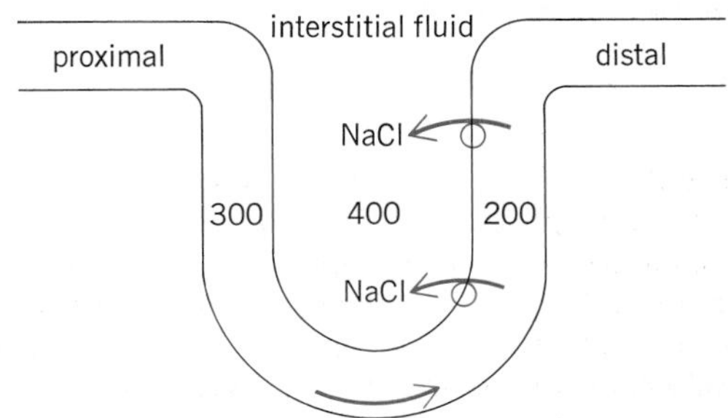

A limiting gradient is reached because the ascending limb is not completely impermeable to sodium and chloride; accordingly, passive backflux of ions into the lumen counterbalances active outflux, and a steady-state limiting gradient is established.

Given the relatively high permeability of the descending limb to sodium, chloride, and water, what net fluxes now occur between interstitium and descending limb? There is a net diffusion of sodium and chloride into the descending limb and a net diffusion of water out until the osmolarities are equal. The interstitial osmolarity is maintained at 400 mosmol/l during this equilibration because of active sodium transport out of the ascending limb.

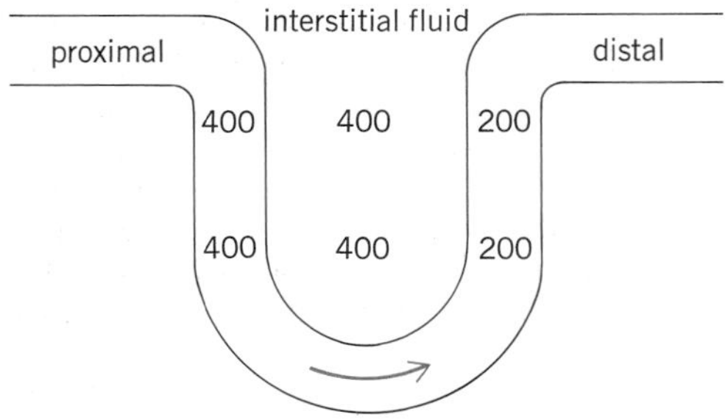

Note that the osmolarities of the descending limb and interstitium are equal and both are higher than that of the ascending limb. So far we have held the fluid stationary in the loop, but of course it is actually continuously flowing. Let us look at what occurs under conditions of flow (Fig. 18-18). Note that the fluid is progressively concentrated as it flows down the descending limb and then is progressively diluted as it flows up the ascending limb. Although only a 200-mosmol/l gradient is maintained across the ascending limb at any given *horizontal level* in the medulla, there exists a much larger osmotic gradient from the top of the medulla to the bottom (300 mosmol/l versus 1,400 mosmol/l). In other words, the 200-mosmol/l gradient established by active sodium chloride transport has been *multiplied* because of the *countercurrent* flow within the loop (i.e., flow in opposing directions through the two limbs of the loop). It should be emphasized that the active sodium chloride transport mechanism within the ascending limb is the essential component of the entire system; without it, the countercurrent flow would have no effect whatever on concentrations.

The highest concentration achieved at the tip of the loop depends upon many factors, particularly the length of the loop (the kangaroo rat has extremely long loops) and the strength of the sodium chloride pump. In human beings, the value reached is 1,400 mosmol/l, which, you will recall, is also the maximal concentration of the

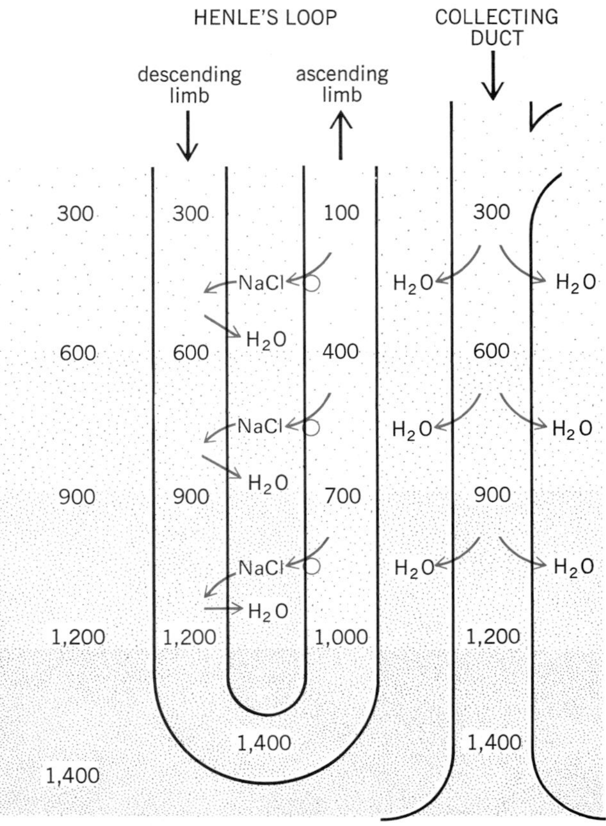

FIGURE 18-18 ***Operation of countercurrent multiplication of concentration in the formation of hypertonic urine.*** **(Adapted from R. F. Pitts.)**

excreted urine. But what has this system really accomplished? Certainly, it concentrates the loop fluid to 1,400 mosmol/l, but then it immediately redilutes the fluid so that the fluid entering the distal tubule is actually more dilute than the plasma. Where is the *final urine* concentrated and how?

The site of final concentration is in the collecting ducts. Recall that the collecting ducts course through the renal medulla parallel to the loops of Henle and are bathed by the interstitial fluid of the medulla. In the presence of maximal levels of ADH, fluid leaves the distal tubules isosmotic to plasma (that is, 300 mosmol/l) because it has reequilibrated with peritubular plasma. As this fluid then flows through the collecting ducts it equilibrates with the ever-increasing osmolarity of the interstitial fluid. Thus, the real function of the loop countercurrent multiplier system is to concentrate the *medullary interstitium*. Under the influence of ADH, the collecting ducts are highly permeable to water which diffuses out of the collecting ducts into the interstitium as a result of the osmotic gradient (Fig. 18-18). The net result is that the fluid at the end of the collecting duct has equilibrated with the interstitial fluid at the tip of the medulla. By this means, the final highly concentrated urine contains relatively less of the filtered water than solute, which is precisely the same as adding pure water to the extracellular fluid, and thereby compensating for a pure water deficit.

In contrast, in the presence of low plasma–ADH concentration, the collecting ducts, like the distal tubules, become relatively impermeable to water and the interstitial osmotic gradient is ineffective in inducing water movement out of the collecting ducts; therefore a large volume of dilute urine is excreted, thereby compensating for a pure water excess.

Osmoreceptor Control of ADH Secretion

To reiterate, pure water deficits or gains are compensated for by partially dissociating water excretion from that of salt through changes in ADH secretion. What receptor input controls ADH under such conditions? The answer is changes in extracellular osmolarity. The adaptive rationale should be obvious, since osmolarity is the variable most affected by pure water gains or deficits. What are the pathways by which osmolarity controls the hypothalamic ADH-producing cells? If osmolarity is the parameter being regulated, it follows that receptors must exist which are sensitive to extracellular osmolarity. These osmoreceptors are located in the supraoptic and paraventricular nuclei of the hypothalamus (Chap. 12), but the mechanism by which they detect changes in osmolarity are unknown. The hypothalamic cells which secrete ADH receive neural input from these osmoreceptors. Via these connections an increase in osmolarity stimulates them and increases their rate of ADH secretion, and, conversely, decreased osmolarity inhibits ADH secretion (Fig. 18-19).

We have now described two different afferent pathways controlling the ADH-secreting hypothalamic cells, one from baroreceptors and one from osmoreceptors. These hypothalamic cells are therefore true integrating centers whose rate of activity is determined by the total synaptic input. To add to the complexity, these cells receive synaptic input from many other brain areas, so that ADH secretion (and therefore urine flow) can be altered by pain, fear, and a variety of other factors. However, these effects are usually short-lived and should not obscure the generalization that ADH secretion is determined primarily by the states of extracellular volume and osmolarity. Alcohol is a powerful inhibitor of ADH release and probably accounts for much of the large urine flow accompanying the ingestion of alcohol.

The disease *diabetes insipidus,* which is distinct from diabetes mellitus, i.e., sugar diabetes, illustrates what happens when the ADH system is disrupted. This disease is characterized by the constant excretion of a large volume of highly dilute urine (as much as 25 l/day). In most cases, the flow can be restored to normal by the administration of ADH. These patients apparently have lost the ability to produce ADH, usually as a result of damage to the hypothalamus. Thus, renal

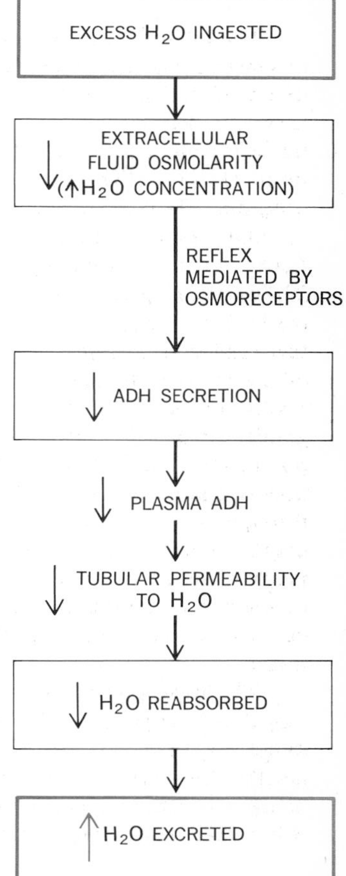

FIGURE 18-19 ***Pathway by which ADH secretion is lowered and water excretion raised when excess water is ingested.***

FIGURE 18-20 ***Pathways by which sodium and water excretion are decreased in response to severe sweating. This figure is an amalgamation of Figs. 18-13, 18-16, and 18-17 and the converse of Fig. 18-19. "Third factor" is not shown in the figure.***

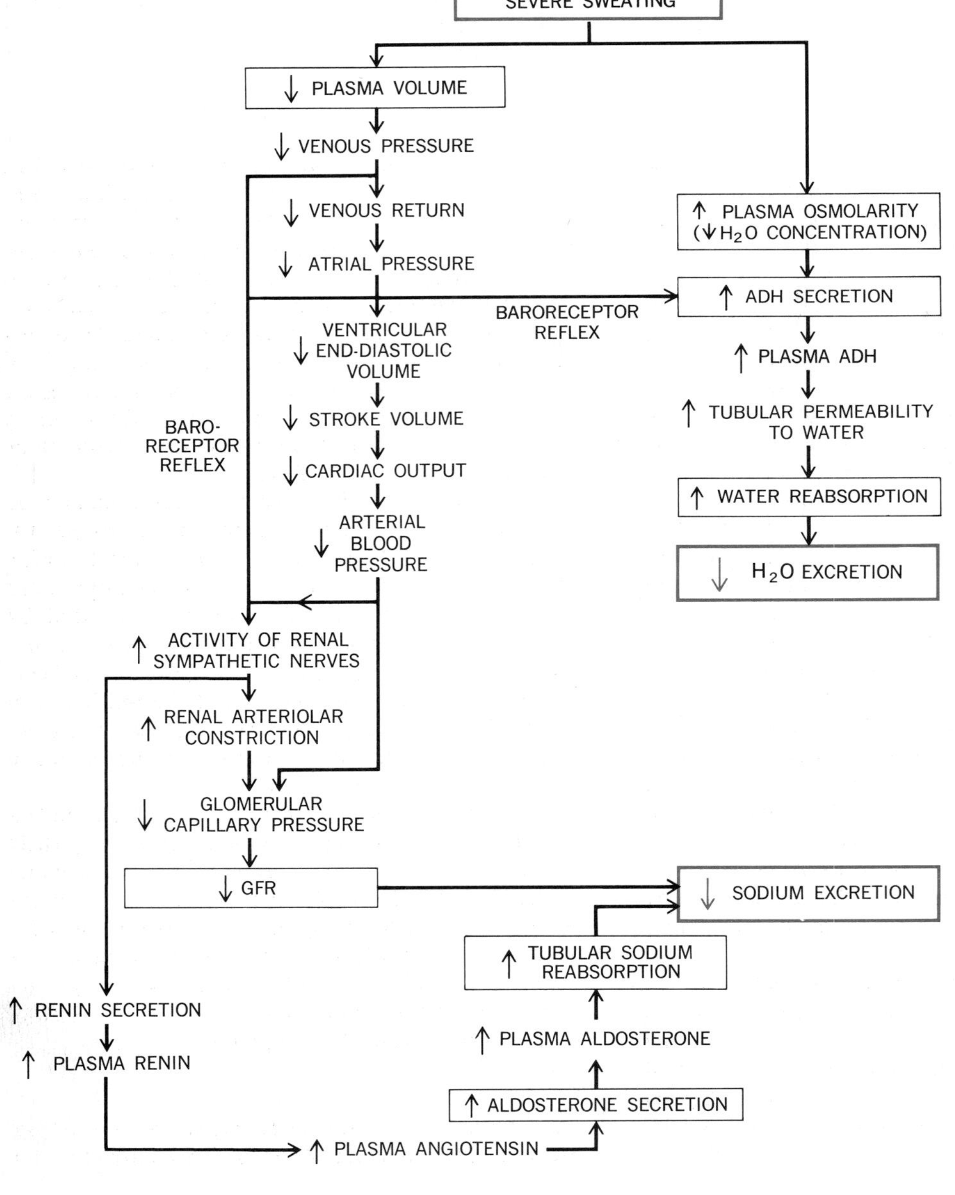

tubular permeability to water is low and unchanging regardless of extracellular osmolarity or volume. The very thought of having to urinate (and therefore to drink) 25 l of water per day underscores the importance of ADH in the control of renal function and body water balance.

Figure 18-20 shows all these factors which control renal sodium and water excretion in operation in response to severe sweating, as in exercise; the renal retention of fluid helps to compensate for the water and salt lost in the sweat.

THIRST

Now we must turn to the other component of the balance, control of intake. It should be evident that large deficits of salt and water can be only partly compensated by renal conservation and that ingestion is the ultimate compensatory mechanism. The subjective feeling of thirst, which drives one to obtain and ingest water, is stimulated both by a lower extracellular volume and a higher plasma osmolarity, the adaptive significance of both being self-evident. Note that these are precisely the same changes which stimulate ADH production. Indeed, the centers which mediate thirst are also located in the hypothalamus very close to those areas which produce ADH. Damage to these centers abolishes water intake completely. Conversely electric stimulation of them may induce profound and prolonged drinking (these water-intake centers are very close to, but distinct from, the food-intake centers to be described in a later chapter). Because of the similarities between the stimuli for ADH secretion and thirst, it is tempting to speculate that the receptors (osmoreceptors and atrial baroreceptors) which initiate the ADH-controlling reflexes are identical to those for thirst. Much evidence indicates that this is, indeed, the case.

A recent finding of considerable interest is that angiotensin stimulates thirst by a direct effect on the brain. Thus, the renin-angiotensin system is not only an important regulator of sodium balance but of water balance as well and constitutes one of the pathways by which thirst is stimulated when extracellular volume is decreased.

There are also other pathways controlling thirst. For example, dryness of the mouth and throat causes profound thirst, which is relieved by merely moistening them. It is fascinating that animals such camels (and human beings, to a lesser extent) which have been markedly dehydrated will rapidly drink just enough water to replace their previous losses and then stop; what is amazing is that when they stop, the water has not yet had time to be absorbed from the gastrointestinal tract into the blood. Some kind of "metering" of the water intake by the gastrointestinal tract has occurred, but its nature remains a mystery.

Section C. Regulation of Potassium, Calcium, and Hydrogen-ion Concentrations

POTASSIUM REGULATION

The potassium concentration of extracellular fluid is a closely regulated quantity. The importance of maintaining this concentration in the internal environment stems primarily from the role of potassium in the excitability of nerve and muscle. Recall that the resting-membrane potentials of these tissues are directly related to the ratio of intracellular to extracellular potassium concentration. Raising the external potassium concentration lowers the resting-membrane potential, thus increasing cell excitability. Conversely, lowering the external potassium hyperpolarizes cell membranes and reduces their excitability.

Since most of the body's potassium is found within cells, primarily as a result of active ion-transport systems located in cell membranes, even a slight alteration of the rates of ion transport across cell membranes can produce a large change in the amount of extracellular potassium. Unfortunately, very little is known about the physiologic control of these transport mechanisms, and obviously our understanding of the regulation of extracellular potassium concentration will remain incomplete until further data are obtained on this critical subject.

The normal person remains in total body potassium balance (as is true for sodium balance) by daily excreting an amount of potassium equal to the amount ingested minus the small amounts eliminated in the feces and sweat. Normally potassium losses via sweat and the gastrointestinal tract are small (although large quantities can be lost by the latter during vomiting or diarrhea). Again the control of renal function is the major mechanism by which body potassium is regulated.

Renal Regulation of Potassium

Potassium is completely filterable at the glomerulus. The amounts of potassium excreted in the urine are generally a small fraction (10 to 15 percent) of the filtered quantity, thus establishing the existence of tubular potassium reabsorption. However, it has also been demonstrated that under certain conditions the excreted quantity may actually exceed that filtered, thus establishing the existence of tubular potassium secretion. The subject is therefore complicated by the fact that potassium can be both reabsorbed and secreted by the tubule. Present evidence suggests, however, that normally almost all the filtered potassium is reabsorbed (by active transport) regardless of changes in body potassium balance. In other words, the reabsorption of potassium does not seem to be controlled so as to achieve potassium homeostasis. The important result of this phenomenon is that *changes* in potassium *excretion* are due to *changes* in potassium *secretion* (Fig. 18-21). Thus, during potassium depletion when the homeostatic response is to reduce potassium excretion to a minimal level, there is no significant potassium secretion, and only the small amount of potassium escaping reabsorption is excreted. In all other situations, to this same small amount of unreabsorbed potassium is added a variable amount of secreted potassium. Thus, in describing the homeostatic control of potassium excretion we may ignore changes in GFR or reabsorption and focus only on the factors which alter the rate of tubular potassium secretion.

FIGURE 18-21 ***Basic renal processing of potassium. Since virtually all the filtered potassium is reabsorbed, potassium excreted in the urine results from tubular secretion.***

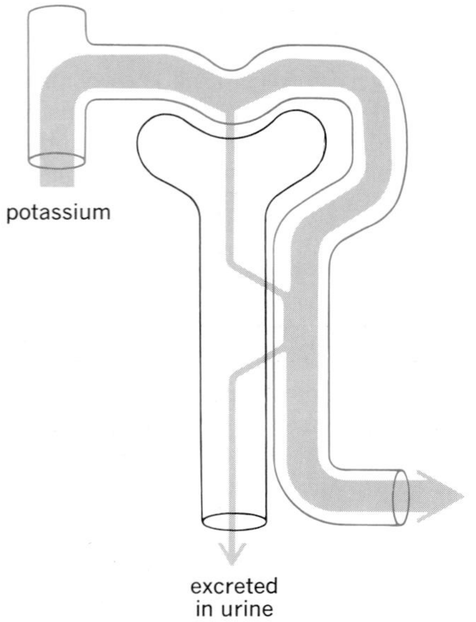

One of the most important of these factors is the potassium concentration of the renal tubular cells themselves. When a high-potassium diet is ingested, potassium concentration in most of the body's cells increases, including the renal tubular cells. This higher concentration facilitates potassium secretion into the lumen and raises potassium excretion. Conversely, a low-potassium diet or a negative potassium balance, e.g., from diarrhea, lowers renal-tubular-cell potassium concentration; this reduces potassium entry into the lumen and decreases potassium excretion, thereby helping to reestablish potassium balance.

A second important factor controlling potassium secretion is the hormone aldosterone, which besides assisting tubular sodium reabsorption enhances tubular potassium secretion. The reflex by which changes in extracellular volume control aldosterone production is completely different from the reflex initiated by an excess or deficit of potassium. The former constitutes a complex pathway, involving renin and angiotensin; the latter, however, seems to be much simpler (Fig. 18-22). The aldosterone-secreting cells of the adrenal cortex are apparently themselves sensitive to the potassium concentration of the extracellular fluid bathing them. For example, an increased intake of potassium leads to an increased extracellular potassium concentration, which in turn directly stimulates aldosterone production by the adrenal cortex. This extra aldosterone circulates to the kidney, where it increases tubular potassium secretion and thereby eliminates the excess potassium from the body. Conversely, a lowered extracellular potassium concentration would decrease aldosterone production and thereby inhibit tubular potassium secretion. Less potassium than usual would be excreted in the urine, thus helping to restore the normal extracellular potassium concentration. Again, complete compensation depends upon the ingestion of additional potassium.

The control and renal tubular effects of aldosterone are summarized in Fig. 18-23.

CALCIUM REGULATION

Extracellular calcium concentration is also normally held relatively constant, the requirement for precise regulation stemming primarily from the profound effects of calcium on neuromuscular excitability. A low

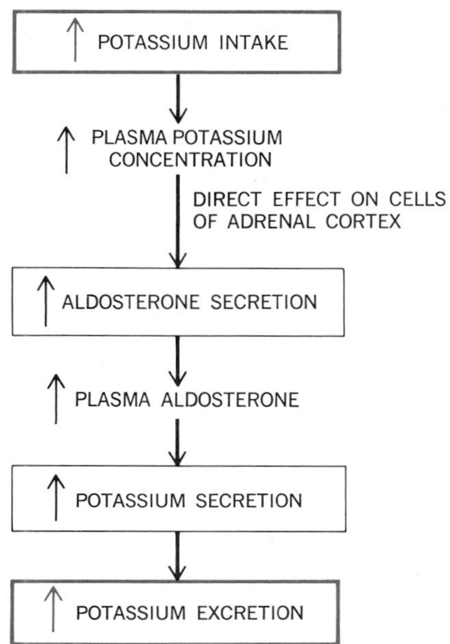

FIGURE 18-23 ***Summary of the control of aldosterone and its functions.***

calcium concentration increases the excitability of nerve and muscle cell membranes so that patients with diseases in which low calcium occurs suffer from *hypocalcemic tetany,* characterized by skeletal muscle spasms which can be severe enough to cause death by asphyxia. Calcium is also important in blood clotting, but low calcium is never a cause of abnormal clotting clinically because the levels required for this function are considerably below those which produce fatal tetany. Hypercalcemia is also dangerous in that it causes cardiac arrhythmias as well as depressed neuromuscular excitability.

Effector Sites for Calcium Homeostasis

At least three effector sites are involved in the regulation of extracellular calcium concentration: bone, the kidney, and the gastrointestinal tract.

Bone Approximately 99 percent of total body calcium is contained in bone, which consists primarily of a framework of organic molecules upon which calcium phosphate crystals are deposited. Contrary to popular opinion, bone is not an absolutely fixed, unchanging tissue but is constantly being remolded and, what is more important, is available for either the withdrawal or deposit of calcium from extracellular fluid.

Gastrointestinal Tract We have previously commented on the indiscriminate nature of most gastrointestinal absorptive processes. This is certainly not true for calcium absorption, which is subject to quite precise hormonal control.

FIGURE 18-23 ***Summary of the control of aldosterone and its functions.***

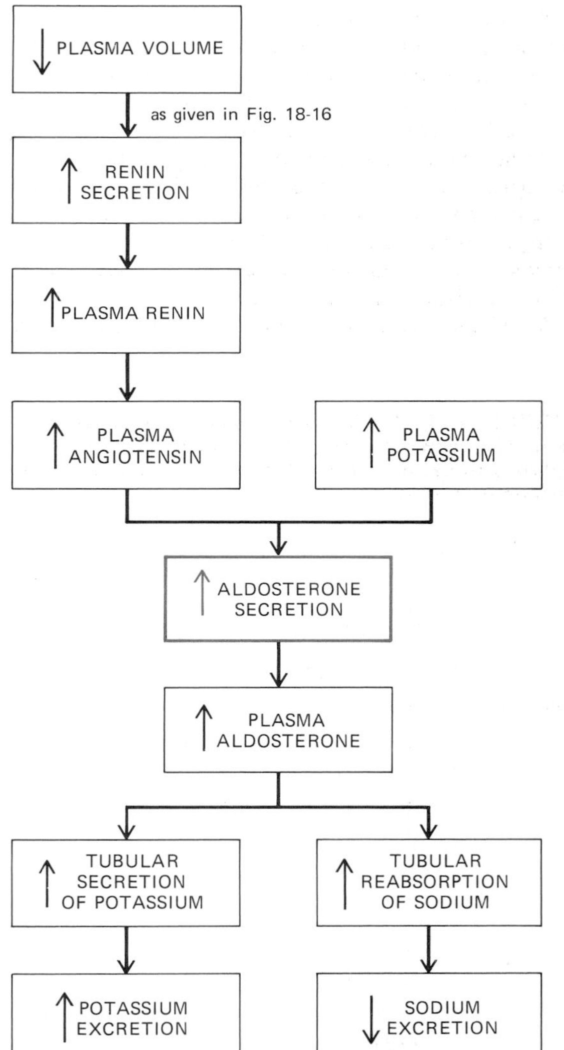

Kidney The kidney handles calcium by filtration and reabsorption. In addition, as we shall see, the renal handling of phosphate also plays an important role in the regulation of extracellular calcium.

Parathormone

All three of the effector sites described above are subject to control by a protein hormone, called *parathormone* (also called parathyroid hormone), produced by the parathyroid glands. Parathormone production is controlled directly by the calcium concentration of the extracellular fluid bathing the cells of these glands. Lower calcium concentration stimulates parathormone production and release, and a higher concentration does just the opposite. It should be emphasized that extracellular calcium concentration acts directly upon the parathyroids (just as was true of the relation between extracellular potassium and aldosterone production) without any intermediary hormones or nerves.

Parathormone exerts at least four distinct effects on the sites described earlier (Fig. 18-24):

1 It increases the movement of calcium (and phosphate) from bone into extracellular fluid, making available this immense store of calcium for the regulation of the extracellular calcium concentration. The mechanisms involved are still controversial.

2 It increases gastrointestinal absorption of calcium by stimulating the active-transport system which moves the ion from gut lumen to blood. This is an important mechanism for elevating plasma calcium concentration since, under normal conditions, considerable amounts of ingested calcium are not absorbed from the intestine but are eliminated via the feces.

3 It increases renal-tubular calcium reabsorption, thus decreasing urinary calcium excretion.

4 It reduces the renal-tubular reabsorption of phosphate, thus raising its urinary excretion and lowering extracellular phosphate concentration.

The adaptive value of the first three should be obvious: They all result in a higher extracellular calcium concentration, thus compensating for the lower concentration which originally stimulated parathormone production. Conversely, an increase in extracellular calcium concentration inhibits normal parathormone production, thereby producing increased urinary and fecal calcium loss and net movement of calcium from extracellular fluid into bone (Fig. 18-24).

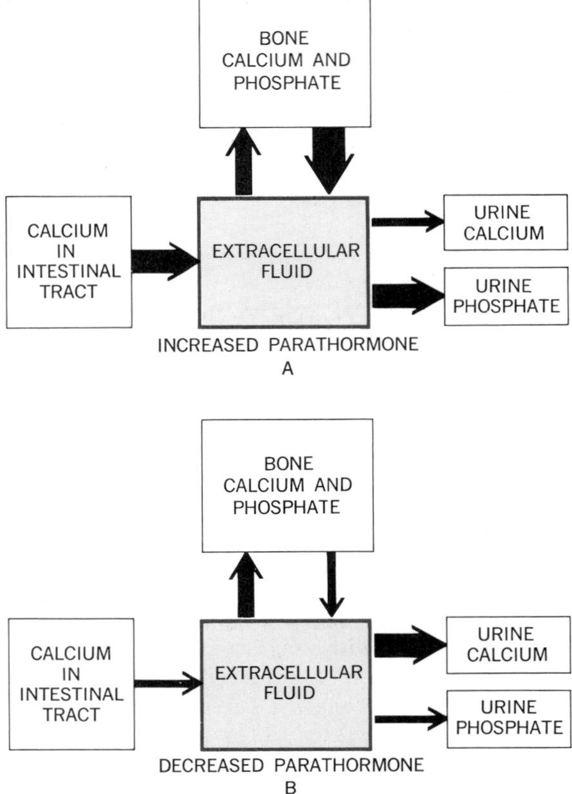

FIGURE 18-24 ***Effects of parathormone on the gastrointestinal tract, kidneys, and bone, the arrows signifying relative magnitudes. Note that when parathormone is decreased there is net movement of calcium and phosphate into bone, urine calcium is raised, and gastrointestinal absorption of calcium is reduced.***

The adaptive value of the fourth effect requires further explanation. Because of the solubility characteristics of undissociated calcium phosphate, the extracellular concentrations of ionic calcium and phosphate bear the following relationship to each other: The product of their concentrations, i.e., calcium times phosphate, is approximately a constant. In other words, if the extracellular concentration of phosphate increases, it forces the deposition of some extracellular calcium in bone, lowering the calcium concentration and keeping the calcium phosphate product a constant. The converse is also true. Imagine now what happens when parathormone releases calcium from bone. Both calcium and phosphate are released into the extracellular fluid. If the phosphate concentration is allowed to increase, further movement of calcium from bone is retarded; but in addition to this effect on bone, we have seen that parathormone also decreases tubular reabsorption of phosphate, thus permitting the excess phosphate to be eliminated in the urine. Indeed, extracellular phosphate may actually be reduced by this mechanism, which would allow even more calcium to be mobilized from bone.

Parathormone has other functions in the body, notably its role in milk production, but the four effects discussed above constitute the major mechanisms by which it integrates various organs and tissues in the regulation of extracellular calcium concentration.

Vitamin D

Vitamin D plays an important role in calcium metabolism, as attested by the fact that its deficiency results in poorly calcified bones. Vitamin D really should be called a hormone since it can be produced by the skin in the presence of sunlight, but since people's clothing prevents this reaction and they are dependent on dietary intake for its supply, it is classed as a vitamin. The major action of vitamin D is to stimulate active calcium absorption by the intestine. It is of greater importance in this regard than parathormone; indeed the action of parathormone is, itself, dependent upon the presence of adequate quantities of vitamin D. Thus, the major event in vitamin D deficiency is decreased gut calcium absorption, resulting in decreased plasma calcium. In children, the newly formed bone protein matrix fails to be calcified normally because of the low plasma calcium, leading to the disease *rickets*.

The activity of vitamin D in the body is controlled ultimately by plasma calcium. The vitamin D which is ingested or produced by the skin is relatively inactive and requires biochemical alteration first by the liver and then by the kidneys before it is fully able to stimulate gut calcium absorption. This activation is stimulated by parathormone; accordingly, a decreased plasma calcium stimulates the secretion of parathormone which, in turn, enhances the activity of vitamin D which, by its actions, helps to restore plasma calcium to normal. Thus, the feedback loops for parathormone and vitamin D activity are closely intertwined.

Calcitonin

A hormone known as calcitonin (also known as thyrocalcitonin) has recently been discovered which has significant effects on plasma calcium. Calcitonin is secreted by the parafollicular (or clear) cells within the thyroid gland which surround but are completely distinct from the thyroxine-secreting follicles (for anatomy of the thyroid gland, see Chap. 12). Calcitonin lowers plasma calcium primarily by inhibiting the release of calcium from bone to its surrounding fluids. Its secretion is controlled directly by the calcium concentration of the plasma supplying the thyroid gland; increased calcium causes increased calcitonin secretion. Thus, this system constitutes a second feedback control over plasma calcium concentration, one that is opposed to the parathormone system. However, its overall contribution to calcium homeostasis is minor compared with that of parathormone.

HYDROGEN-ION REGULATION

Most metabolic reactions are exquisitely sensitive to the hydrogen-ion concentration[4] of the fluid in which they occur. This sensitivity is due primarily to the marked influence on enzyme function exerted by the hydrogen ion. Accordingly, the hydrogen-ion concentration of the extracellular fluid is one of the most critical and closely regulated chemical quantities in the entire body.

Basic Definitions

The hydrogen ion is an atom of hydrogen which has lost its only electron. When dissolved in water, many compounds dissociate reversibly to produce negatively charged ions (anions), and hydrogen ions, e.g.,

$$\text{Lactic acid} \rightleftharpoons H^+ + \text{lactate}^-$$

$$\underset{\text{Carbonic acid}}{H_2CO_3} \rightleftharpoons H^+ + \underset{\text{Bicarbonate}}{HCO_3^-}$$

The double arrows mean that the reaction can proceed in either direction, depending upon conditions. Any compound capable of liberating a hydrogen ion in this manner is called an *acid*. Conversely, any substance which can accept a hydrogen ion is termed a *base*. Thus, in the reactions above, lactate and bicarbonate are bases since they can bind hydrogen ions. The hydrogen-ion concentration often is referred to in terms

[4] Hydrogen-ion concentration is frequently expressed in terms of pH, which is defined as the negative logarithm to the base 10 of the hydrogen-ion concentration: pH = $-\log H^+$. This can be confusing for several reasons, not the least of which is that pH decreases as H^+ increases. We have chosen not to use pH in this text.

of *acidity:* the higher the hydrogen-ion concentration, the greater the acidity (it must be understood that the hydrogen-ion concentration of a solution is a measure only of the hydrogen ions which are *free* in solution).

Strong and Weak Acids

Strong acids dissociate completely when they dissolve in water. For example, hydrochloric acid added to water gives

$$\underset{\text{Hydrochloric acid}}{HCl} \longrightarrow H^+ + Cl^-$$

Virtually no HCl molecules exist in the solution, only free hydrogen ions and chloride ions. On the other hand, a large number of *weak acids* do not dissociate completely when dissolved in water. For example, when dissolved, only a fraction of lactic acid molecules dissociate to form lactate and hydrogen ions, the other molecules remaining intact. This characteristic of weak acids forms the basis of an important chemical and physiologic phenomenon, *buffering*.

Buffer Action and Buffers

Figure 18-25 pictures a solution made by dissolving lactic acid and sodium lactate in water. The sodium lactate dissociates completely into sodium ions and lactate ions, but only a very small fraction of the lactic acid molecules dissociate to generate hydrogen ion. Accordingly, the solution has a relatively low concentration of hydrogen ion and relatively high concentrations of undissociated lactic acid molecules, sodium ions, and lactate ions. Lactic acid, hydrogen ion, and lactate are in equilibrium with each other:

$$\text{Lactic acid} \rightleftharpoons \text{lactate}^- + H^+$$

By the mass-action law, an increase in the concentration of any substance on one side of the arrows forces the reaction in the opposite direction. Conversely, a decrease in the concentration of any substance on one side of the arrows forces the reaction toward that side, i.e., in the direction which generates more of that substance. What happens if we add hydrochloric acid to this solution? Hydrochloric acid, a strong acid, completely dissociates and liberates hydrogen ions. This

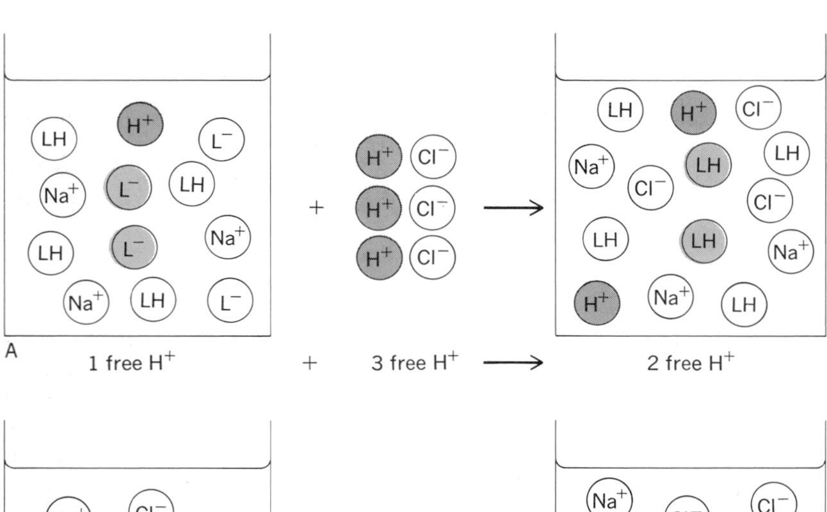

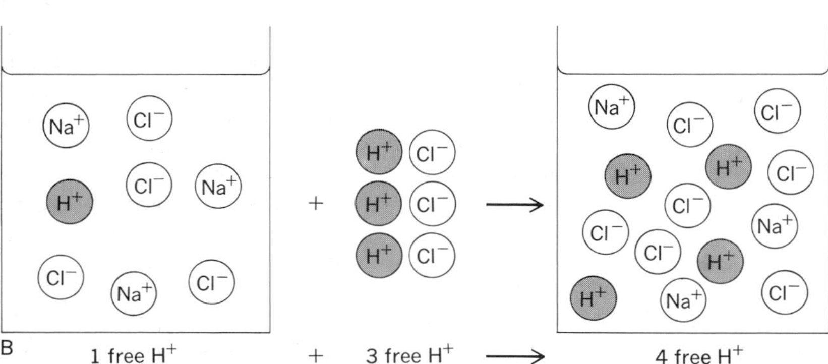

FIGURE 18-25 **(A)** ***Example of a buffer system.*** **LH** ***= lactic acid,*** **I$^-$** ***= lactate. When HCl is added to the beaker, two of the hydrogen ions react with lactate to give lactic acid; therefore, only one of the three added hydrogen ions remains free in solution. Contrast this to*** **B*****, in which no buffer is present and all three of the added hydrogen ions remain free in solution.***

excess of hydrogen ions drives the lactic acid reaction to the left, and many of the hydrogen ions liberated from the dissociation of hydrochloric acid thus combine with lactate to give undissociated lactic acid. As a result, many of the hydrogen ions generated by the dissociation of hydrochloric acid do not remain free in solution but become bound in lactic acid molecules. The final hydrogen-ion concentration is therefore smaller than if the hydrochloric acid had been added to pure water (Fig. 18-25).

Conversely, if we add a chemical which *removes* hydrogen ions instead of adding them, the lactic acid reaction is driven to the right, lactic acid molecules dissociate to generate more hydrogen ions, and the original fall in hydrogen-ion concentration is minimized. This process preventing large changes in hydrogen-ion concentration when hydrogen ion is added or removed from a solution is termed *buffering*, and the chemicals (in this case lactate and lactic acid) are called *buffers* or *buffer systems*.[5]

Generation of Hydrogen Ions in the Body

There are three major sources of hydrogen ion in the body:

1 Phosphorus and sulfur are present in large quantities in many proteins and other biologically important molecules. The catabolism of these molecules releases phosphoric and sulfuric acids into the extracellular fluid. These acids, to a large extent, dissociate into hydrogen ions and anions (phosphate and sulfate). For example,

$$H_2PO_4^- \longrightarrow H^+ + HPO_4^{2-}$$
$$H_2SO_4 \longrightarrow 2\ H^+ + SO_4^{2-}$$

2 Many organic acids, e.g., fatty acids and lactic acid, are produced as end products of metabolic reactions and also liberate hydrogen ion by dissociation.

3 We have already described (Chap. 17) the major source of hydrogen ion, namely, liberation of hydrogen ion from metabolically produced carbon dioxide via the reactions

$$CO_2 + H_2O \longrightarrow H_2CO_3 \longrightarrow H^+ + \underset{\text{Bicarbonate}}{HCO_3^-}$$

[5] Of necessity, this description of acids and buffering is oversimplified and incomplete; e.g., no mention has been made of the ionization of water itself or of the importance of the hydroxyl ion. The interested reader can find excellent descriptions of acid-base chemistry in any textbook of biochemistry.

As described in Chap. 17, the lungs normally eliminate carbon dioxide from the body as rapidly as it is produced in the tissues. As blood flows through the lung capillaries and carbon dioxide diffuses into the alveoli, the chemical reactions which originally generated HCO_3^- and H^+ from carbon dioxide and water in the venous blood are reversed:

$$CO_2 + H_2O \longleftarrow H_2CO_3 \longleftarrow H^+ + HCO_3^-$$

As a result, all the hydrogen ion generated from carbonic acid is completely reincorporated into water molecules. Therefore, there is *normally* no net gain or loss of hydrogen ion in the body from this source, but what happens when lung disease prevents adequate elimination of carbon dioxide? The retention of carbon dioxide within the body means an elevated extracellular P_{CO_2}. Some of the hydrogen ion generated in the venous blood from the reaction of this carbon dioxide with water would also be retained in the body and would raise the extracellular hydrogen-ion concentration. This hydrogen ion must be eliminated from the body (via the kidneys) in order for hydrogen-ion balance to be maintained.

Buffering of Hydrogen Ions within the Body

Between the generation of hydrogen ions in the body and their excretion (to be described) what happens? The hydrogen-ion concentration of the extracellular fluid is extremely small, approximately 0.00004 mmol/l. What would happen to this concentration if just 2 mmol of hydrogen ion remained free in solution following its dissociation from an acid? Since the total extracellular fluid is approximately 12 l, the hydrogen-ion concentration would increase by 2/12, or approximately 0.167 mmol/l. Since the original concentration was only 0.00004, this would represent more than a 4,000-fold increase. Obviously, such a rise does not really occur, for we have already observed that the extracellular hydrogen ion is kept remarkably constant. Therefore, of the 2 mmol of hydrogen ion liberated in our problem, only an extremely small portion can have remained free in solution. The vast majority has been bound (buffered) by other ions. The kidneys ultimately eliminate excess hydrogen ions produced in the body, but it is buffering which minimizes hydrogen-ion-concentration changes until excretion occurs.

The most important body buffers are bicarbonate–CO_2, large anions such as plasma protein and intracellular phosphate complexes, and hemoglobin. Recall that only free hydrogen ions contribute to the acidity of a solution. These buffers all act by binding hydrogen ions according to the general reaction

$$\text{Buffer}^- + \text{H}^+ \rightleftharpoons \text{H-buffer}$$

It is evident that H-buffer is a weak acid in that it can exist as the undissociated molecule or can dissociate to buffer$^-$ + H^+. When hydrogen-ion concentration increases, the reaction is forced to the right and more hydrogen ion is bound. Conversely, when hydrogen-ion concentration decreases, the reaction is forced to the left and hydrogen ion is released. In this manner, the body buffers stabilize hydrogen-ion concentration against changes in either direction.

Bicarbonate - CO_2 In this section and in Chap. 17 we have already described the relationships between HCO_3^-, H^+, and CO_2. Let us once more write the pertinent equations in their true forms as reversible equations:

$$H^+ + HCO_3^- \rightleftharpoons H_2CO_3 \rightleftharpoons H_2O + CO_2$$

The basic mechanism by which this system acts as a buffer should be evident: An increased extracellular hydrogen-ion concentration drives the reaction to the right, H^+ and HCO_3^- combine, hydrogen ion is thereby removed from solution, and the hydrogen-ion concentration turns toward normal. Conversely, a decreased extracellular hydrogen-ion concentration drives the reaction to the left, CO_2 and H_2O combine to generate hydrogen ion, and this additional hydrogen ion turns hydrogen-ion concentration toward normal. One reason for the importance of this buffer system is that the extracellular bicarbonate concentration is normally quite high and is closely regulated by the kidney. A second and even more important reason stems from the relationship between extracellular hydrogen-ion concentration and carbon dioxide elimination from the body.

When additional hydrogen ion is added to the extracellular fluid, i.e., when the H^+ combines with HCO_3^-, the extent to which this reaction can restore hydrogen-ion concentration to normal depends upon precisely how much additional H^+ actually combines with HCO_3^- and is thereby removed from solution. Complete compensation (which never actually occurs) can be obtained only if the HCO_3^-–CO_2 reaction proceeds to the right until all the additional H^+ has combined with HCO_3^-. However, this reaction obviously generates CO_2, which seriously hinders the further buffering ability of HCO_3^- because, as can be seen from the equation, any increase in the concentration of CO_2, by the mass-action law, tends to drive the reaction back to the left, thus preventing further net combination of $H^+ + HCO_3^-$. In reality, however, this expected increase in extracellular CO_2 does not occur. Indeed, during periods of increased body hydrogen-ion production from organic, phosphoric, or sulfuric acids, the extracellular CO_2 is actually *decreased!* What causes this? We have already studied the mechanism in Chap. 17: A greater extracellular hydrogen-ion concentration stimulates the respiratory centers to increase alveolar ventilation and thereby causes greater elimination of carbon dioxide from the body (Fig. 18-26). Thus, although an increased combination of H^+ and HCO_3^- generates more carbon dioxide, respiratory stimulation produced by the higher extracellular hydrogen-ion concentration results in the elimination of carbon dioxide even faster than it is generated. As a result, the extracellular carbon dioxide decreases, and, by the mass-action law, the further combination of H^+ and HCO_3^- is actually facilitated. It is this control of carbon dioxide elimination exerted by the extracellular hydrogen-ion concentration that allows the HCO_3^-–CO_2 system to function so efficiently as a buffer.

A lower extracellular hydrogen-ion concentration resulting from either decreased hydrogen-ion production or increased hydrogen-ion loss from the body is compensated for by just the opposite buffer reactions: (1) The lower hydrogen-ion concentration drives the HCO_3^-–CO_2 reaction to the left, carbon dioxide and water combine to generate $H^+ + HCO_3^-$, and this additional hydrogen ion turns hydrogen-ion concentration toward normal; (2) the lower hydrogen-ion concentration decreases alveolar ventilation and carbon dioxide elimination by inhibiting the medullary respiratory centers. The elevated extracellular carbon dioxide resulting from this process also serves to drive the HCO_3^-–CO_2 reaction to the left, thus allowing further generation of hydrogen ion.

When lung disease is itself the cause of the increased hydrogen-ion concentration, the efficiency of this buffer system is obviously greatly impaired. Actually a large fraction of any hydrogen-ion excess or deficit is always buffered by the other buffers listed.

Hemoglobin as a Buffer In Chap. 17 we pointed out that most metabolically produced carbon dioxide is carried from the tissues to the lungs in the form of HCO_3^-. These bicarbonate ions are generated by the hydration of CO_2 to form H_2CO_3 and the dissoci-

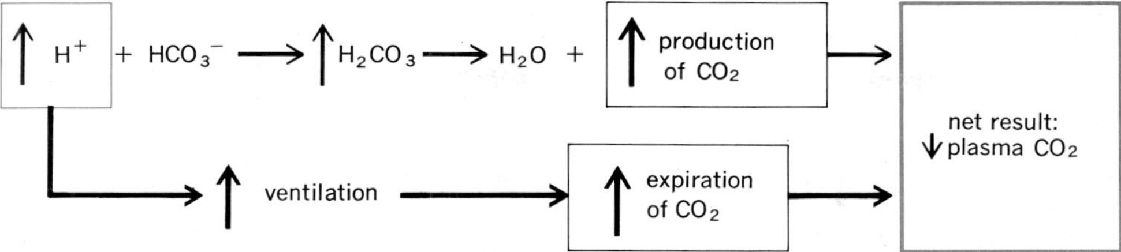

FIGURE 18-26 ***Effects of excess hydrogen ions on plasma carbon dioxide. The direct effect, by mass action, is to increase the production of carbon dioxide, but the indirect effect is to lower carbon dioxide by reflexly stimulating breathing. Since the latter effect predominates, the net effect is a reduction of plasma carbon dioxide.***

ation of H_2CO_3 to HCO_3^- and H^+. As blood flows through the lungs, the reaction is reversed and H^+ and HCO_3^- recombine. However, these hydrogen ions must be buffered while they are in transit from the tissues to the lungs, a function performed primarily by hemoglobin. Its suitability for this role depends upon a remarkable characteristic of the hemoglobin molecule: Reduced hemoglobin has a much greater affinity for hydrogen ion than oxyhemoglobin does. As blood flows through the tissues, a fraction of oxyhemoglobin loses its oxygen and is transformed into reduced hemoglobin. Simultaneously, a large quantity of carbon dioxide enters the blood and undergoes (primarily in the red blood cells) the reactions which ultimately generate HCO_3^- and H^+. Because reduced hemoglobin has a strong affinity for hydrogen ion, most of these hydrogen ions become bound to hemoglobin (Fig. 18-27). In this manner only a few hydrogen ions remain free, and the acidity of venous blood is only slightly greater than that of arterial blood. As the venous blood passes through the lungs, all these reactions are reversed. Hemoglobin becomes saturated with oxygen, and its ability to bind hydrogen ions decreases. The hydrogen ions are released, whereupon they react with HCO_3^- to give CO_2, which diffuses into the alveoli and is expired.

In Chap. 17 we described how the hydrogen-ion concentration of the blood is an important determinant of the ability of hemoglobin to bind oxygen. Now we have shown how the presence of oxygen is an important determinant of hemoglobin's ability to bind hydrogen ion. The adaptive value of these phenomena is enormous. Their combination in one molecule marks hemoglobin as a remarkable evolutionary development.

It should be emphasized that none of these buffer systems eliminates hydrogen ion from the body, instead causing the hydrogen ion to be bound in some molecule (H_2O, hemoglobin, etc.). Binding thus removes the hydrogen ion from solution, preventing it from contributing to the free-hydrogen-ion concentration; the actual elimination of hydrogen ion from the body is normally performed only by the kidneys.

Renal Regulation of Extracellular Hydrogen-ion Concentration

In a normal person the quantity of phosphoric, sulfuric, and organic acids formed depends primarily upon the type and quantity of food ingested. A high-protein diet, for example, results in increased protein breakdown and release of large quantities of sulfuric acid. The average American diet results in the liberation of 40 to 80 mmol of hydrogen ion each day. If hydrogen-

FIGURE 18-27 ***Buffering of hydrogen ions by hemoglobin as blood flows through tissue capillaries.***

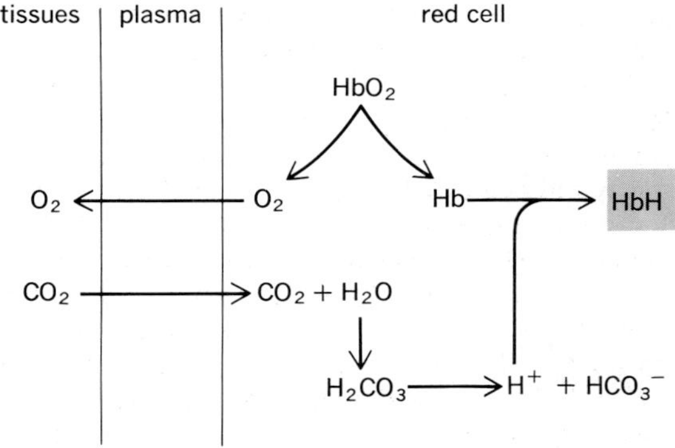

ion balance is to be maintained, the same quantity must be eliminated from the body each day. This loss occurs via the kidneys. In addition, the kidneys must be capable of *altering* their hydrogen-ion excretion in response to changes in body hydrogen-ion production, regardless of whether the source is carbon dioxide (in lung disease) or phosphoric, sulfuric, or organic acids. The kidneys must also be able to compensate for any gastrointestinal loss or gain of hydrogen ion resulting from disease.

Virtually all the hydrogen ion excreted in the urine enters the tubules via tubular secretion, the mechanism and its control being quite complex. Suffice it to say that the presence of excess acid in the body induces an increased urinary hydrogen-ion excretion. Conversely, the kidney responds to a decreased amount of acid in the body by lowering urinary hydrogen-ion excretion. The controlling effect of the acid appears to be primarily exerted directly upon the tubular cells, with no nerve or hormone intermediates. Obviously, such a system is effective in stabilizing hydrogen-ion concentration at the normal value.

The ability of the kidneys to excrete H^+ depends both upon tubular hydrogen-ion secretion and upon buffers in the urine. The major urinary buffers are HPO_4^{2-} and ammonia NH_3:

$$HPO_4^{2-} + H^+ \longrightarrow H_2PO_4^-$$
$$NH_3 + H^+ \longrightarrow NH_4^+$$

The HPO_4^{2-} in the tubular fluid has been filtered and not reabsorbed. In contrast, the ammonia of the tubular fluid is formed by the tubular cells themselves by the deamination of certain amino acids transported into the renal tubular cells from the peritubular capillary plasma. From the cells the ammonia diffuses into the lumen. Of great importance is that the amount which *remains* in the lumen depends upon the rate of hydrogen-ion accumulation there, since the tubular cell membrane is quite permeable to ammonia but not to NH_4^+. Accordingly, the more hydrogen ion there is in the lumen, the greater the conversion of NH_3 to NH_4^+, increased secretion of hydrogen ion automatically inducing an increased excretion of NH_3 in the form of NH_4^+. Another important feature of this system is that, by unknown mechanisms, the rate of ammonia production by the renal tubular cells increases whenever extracellular hydrogen-ion concentration remains elevated for more than 1 to 2 days; this extra ammonia provides the additional buffers required for increased hydrogen-ion secretion.

KIDNEY DISEASE

The term kidney disease is no more specific than car trouble, since many diseases affect the kidneys. Bacteria cause kidney infections, most of which are collectively called *pyelonephritis*. A common type of kidney disease, *glomerulonephritis*, results from an allergy incident to throat infection by a specific group of bacteria. Congenital defects, stones, tumors, and toxic chemicals are possible sources of kidney damage. Obstruction of the urethra or a ureter may cause injury due to a buildup of pressure and may predispose the kidneys to bacterial infection.

Disease can attack the kidney at any age. Experts estimate that there are at present more than 3 million undetected cases of kidney infection in the United States and that 25,000 to 75,000 Americans die of kidney disease each year.

Early symptoms of kidney disease depend greatly upon the type of disease involved and the specific part of the kidney affected. Although many diseases are self-limited and produce no permanent damage, others progress if untreated. The end state of progressive diseases, regardless of the nature of the damaging agent (bacteria, toxic chemical, etc.), is a shrunken, nonfunctioning kidney. Similarly, the symptoms and signs of profound renal malfunction are independent of the damaging agent and are collectively known as *uremia*, literally "urine in the blood."

The severity of uremia depends upon how well the impaired kidneys are able to preserve the constancy of the internal environment. Assuming that the patient continues to ingest a normal diet containing the usual quantities of nutrients and electrolytes, what problems arise? The key fact to keep in mind is that the kidney destruction has markedly reduced the glomerular filtration rate. Accordingly, the many subtances which gain entry to the tubule primarily by filtration are filtered in diminished amounts. Of the substances described in this chapter, this category includes sodium chloride, water, calcium, and a number of waste products. In addition, the excretion of potassium, hydrogen ion, and certain other substances is impaired because the diseased kidneys have a diminished capacity for tubular secretion. The buildup of these substances in the blood causes the symptoms and signs of uremia.

Artificial Kidney

The artificial kidney is an apparatus that eliminates the excess ions and wastes which accumulate in the blood when the kidneys fail. Blood is pumped from one of the patient's arteries through tubing which is bathed by a large volume of fluid. The tubing then conducts the blood back into the patient by way of a vein. The tubing is generally made of a cellophane, which is highly permeable to most solutes but relatively impermeable to protein—characteristics quite similar to those of capillaries. The bath fluid, which is constantly replaced, is a salt solution similar in ionic concentrations to normal plasma. The basic principle is simply that of dialysis, or diffusion. Because the cellophane is permeable to most solutes, as blood flows through the tubing, solute concentrations tend to equilibrate in the blood and bath fluid. Thus, if the plasma potassium concentration of the patient is above normal, potassium diffuses out of the blood into the bath fluid. Similarly, waste products and excesses of other substances diffuse across the cellophane tubing and thus are eliminated from the body.

Until recently patients were placed on the artificial kidney only when there was reason to believe that their kidney damage was only temporary and that recovery would occur if the patient could be kept alive during temporary renal failure. However, in recent years, technical improvements have permitted many patients to utilize the artificial kidney several times a week for unlimited periods, and thus patients with permanent kidney failure can be kept alive and functioning, many of them performing the dialysis in their homes.

The other major hope for patients with permanent renal failure is kidney transplantation. Although great strides have been made, the major problem remains the frequent rejection of the transplanted kidney by the recipient's body (see Chap. 22).

CHAPTER 19 The Gastrointestinal System

The gastrointestinal system includes the gastrointestinal tract (mouth, esophagus, stomach, small and large intestines), salivary glands, and portions of the liver and pancreas (Fig. 19-1). Its function is to transfer food and water from the external environment to the internal environment, where they can be distributed to the cells of the body by the circulatory system. Most food is taken into the mouth as large pieces of matter, consisting of high-molecular-weight substances, such as proteins and polysaccharides, which are unable to cross cell membranes. Before these substances can be absorbed, they must be broken down into smaller molecules, such as amino acids and monosaccharides. This breaking-down process, *digestion*, is accomplished by the action of acid and enzymes secreted into the gastrointestinal tract. The small mole-

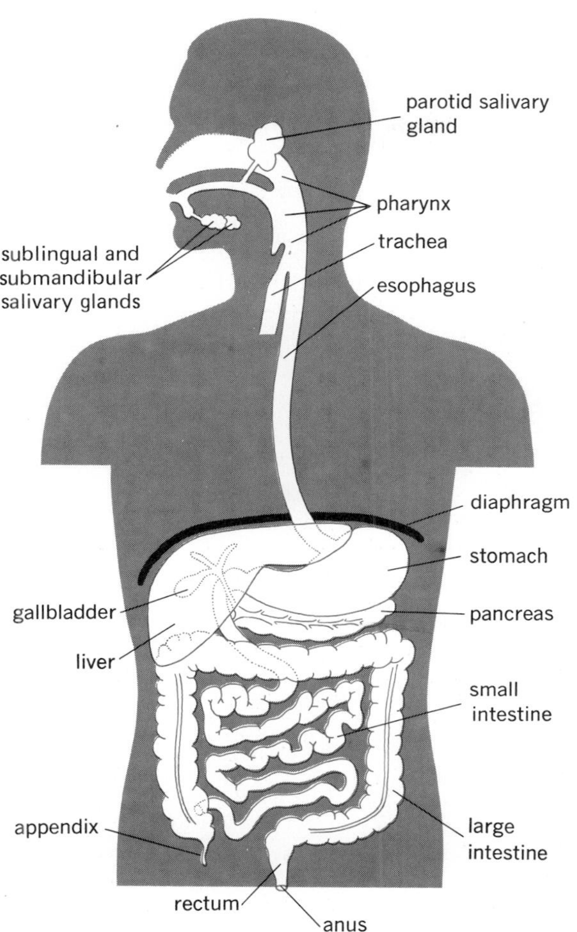

FIGURE 19-1 ***Anatomy of the gastrointestinal system.***

cules resulting from digestion cross the cell membranes of the intestinal cells and enter the blood and lymph, a process known as *absorption*. While these processes are taking place, contractions of the smooth muscle lining the walls of the gastrointestinal tract move the luminal contents through the tract. Contrary to popular belief, the gastrointestinal system is not a major excretory organ for eliminating wastes from the body. It is true that small amounts of some end products, such as the breakdown products of hemoglobin, are normally eliminated in the feces, but the elimination of most metabolic end products (wastes) from the internal environment is performed not by the gastrointestinal tract but by the lungs and kidneys. Feces co
bacteria and ingested material wh
digested and absorbed during its pass
trointestinal tract, i.e., material tha
never in the internal environment of the body.

In this chapter we examine the anatomy of the gastrointestinal system and four aspects of gastrointestinal function: (1) secretion, (2) digestion, (3) absorption, and (4) motility (Fig. 19-2). In Chap. 20 the mechanisms which control the distribution and utilization of absorbed food products as well as hunger and thirst will be considered.

STRUCTURE OF THE GASTROINTESTINAL TRACT

The gastrointestinal tract consists of a tube of variable diameter, 6.5 m in length, running through the body from mouth to anus. The lumen of this tube is continuous with the external environment, which means that its contents are technically outside of the body. This fact is relevant to an understanding of some of the tract's properties. For example, the lower portion of the intestinal tract is inhabited by millions of living bacteria, most of which in this location are harmless and even beneficial, but if the same bacteria enter the spaces of the abdominal cavity or the blood, as may happen as a result of a ruptured appendix, they are extremely harmful and even lethal.

Although there is some regional variation, the gastrointestinal tract throughout most of its length has the same general structure as the segment of intestine illustrated in Fig. 19-3. From the esophagus to the anus, its wall is made up of a series of layers which, beginning with the innermost, are as follows: (1) epithelium, (2) lamina propria, (3) muscularis mucosa, (4) submucosa, (5) muscularis externa, the main muscle coat, and (6) adventitia.

The first two layers constitute the *mucosa*. The lamina propria is simply a connective-tissue layer which underlies the epithelium. The *epithelium* provides the interface between ingested food and the body, and its characteristics vary according to the function of a particular region. For example, the pharynx, esophagus, and lower anal canal, which are subject to abrasion by relatively large particles, are lined with stratified squamous epithelium interspersed with mucous glands. The mucus serves as a lubricant to facilitate the passage of the material and, at the same time, protects the walls of the intestinal tract. The epithelium in other regions is of the simple columnar type

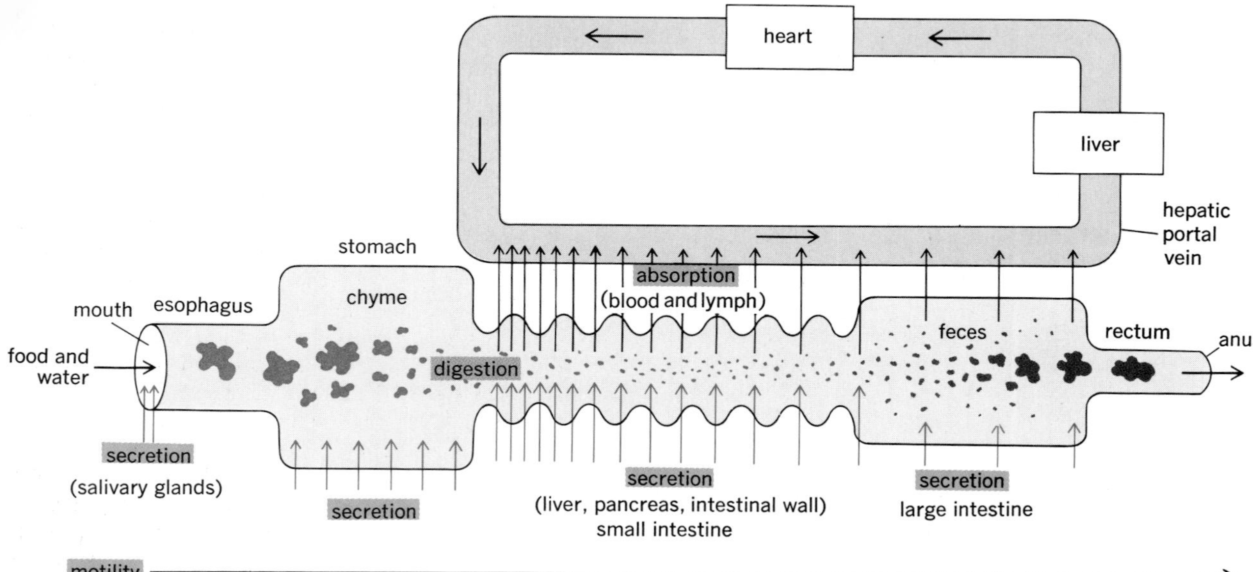

FIGURE 19-2 ***Summary of gastrointestinal activity involving motility, secretion, digestion, and absorption.***

FIGURE 19-3 ***Anatomy of a segment of the small intestine.***

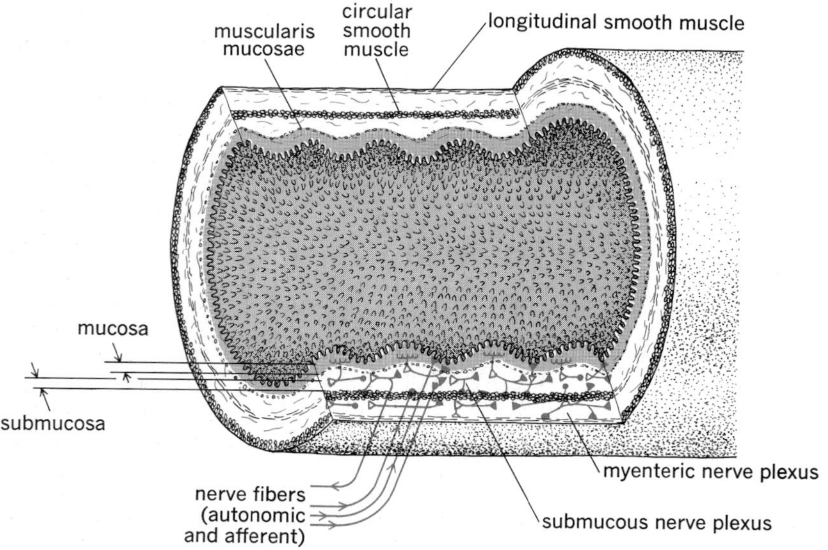

and consists primarily of two cell types: absorptive cells, which move material from the lumen into the blood and lymph vessels which are in the mucosal and submucosal layers, and secretory cells, which secrete mucus (the goblet cells) and digestive enzymes. The secretory cells are often clustered into exocrine glands, some dipping into other layers of the wall. Other exocrine glands secrete material through ducts which empty into the lumen of the gastrointestinal tract; these secretions not only contain the digestive enzymes but also provide the chemical environment necessary for optimal function of these enzymes.

The mucosa also contains endocrine cells which secrete hormones into the blood flowing through the capillaries in the walls of the gastrointestinal tract; the hormones are carried by the venous blood back to the heart and return to the gastrointestinal tract by way of the arterial system. Thus these hormones can affect smooth muscle activity or gland secretion in an area of the gastrointestinal tract far removed from the site where they are produced. The three major gastrointestinal hormones are *gastrin* (secreted by the stomach), *secretin,* and *cholecystokinin* (the latter two secreted by the first segment of the small intestine).

Below the mucosa is the *muscularis mucosa,* a layer of smooth muscle fibers whose contractions throw the overlying mucosa into ridges. The *submucosa* is another connective-tissue layer, rich in blood vessels, nerves, and lymph vessels. It is surrounded by the *muscularis externa,* which in most regions of the gastrointestinal tract consists of two layers: an outer layer with fibers running almost longitudinally along the tube and an inner layer where the muscles lie with a more circular orientation. Contractions of the longitudinal layer shorten the gastrointestinal tract, and contractions of the circular layer constrict its lumen, exerting pressure on the contents of the lumen and causing them to flow. The *adventitia* is yet another connective-tissue layer which blends with the surrounding structures and those supporting the gastrointestinal tract.

Support of the Abdominal Portion of the Gastrointestinal Tract

The stomach and intestines are suspended within the peritoneal cavity by thin sheets of connective tissue attached to the body wall. The sheets which attach the stomach to the body wall or to other organs are the *greater* and *lesser omenta* (singular: *omentum*); those which attach the intestines to the body wall are the *mesenteries*. (The mesentery for the large intestine is the *mesocolon.*) The omenta and mesenteries have numerous fat deposits, blood vessels, lymphatics, and nerves. They and the organs to which they attach are covered by peritoneum which is continuous with that lining the abdominal wall.[1]

The peritoneum, a serous membrane like the pleura of the thoracic cavity, can be viewed as a membranous bag that has been invaginated by organs. The invaginating organs in the thorax are, of course, the lungs; those in the abdominal, or peritoneal, cavity are the stomach and intestines. The uninvaginated part of the peritoneum adhering to the abdominal wall is the *parietal peritoneum,* and the part pushed inward by the organs (and, therefore, adhering to their surfaces) is the *visceral peritoneum* (Figs. 17-14 and 19-4). The parietal and visceral layers are in contact with each other so that the space between them, the *abdominal,* or *peritoneal, cavity,* is actually only a potential space. The free surface of the peritoneum (facing the abdominal cavity) is lined with flattened cells that look like squamous epithelium but are called *mesothelium* because of their embryonic origin from mesoderm. The surface is kept moist and lubricated by a small amount of serous fluid which allows the viscera to move past each other and the walls of the cavity.

REGULATION OF THE GASTROINTESTINAL SYSTEM

Gland secretion and smooth muscle contraction in the gastrointestinal system are regulated by nerves and hormones. The gastrointestinal tract has in its walls its own local nervous system in the form of two major nerve plexuses: the *myenteric plexus* (between the longitudinal and circular layers of smooth muscle) and the *submucous nerve plexus* (in the submucosa; Fig. 19-3). These two plexuses are found throughout the length of the gastrointestinal tract from esophagus to anus. They are composed of neurons forming synaptic junctions with other neurons in the plexus or ending in the regions around smooth muscles and glands. Most axons leave the myenteric plexus and synapse with neurons in the submucous plexus and vice versa, so that neural activity in one plexus influences the activity in the other. The axons in both plexuses branch profusely, and electric

[1] Knowledge of how the organs are held in place is actually very rudimentary. Pressure from other organs, tonus of the abdominal muscles, and the continuity of blood vessels, nerves, etc., certainly also play a role.

stimulation at one point in the plexus is found to lead to electric activity that is conducted both up and down the gastrointestinal tract. Thus, activity initiated in the plexus in the upper part of the small intestine may affect smooth muscle and gland activity in the stomach as well as in the lower part of the intestinal tract.

Nerve fibers from both the sympathetic and parasympathetic branches of the autonomic nervous system enter the intestinal tract and synapse with neurons in the internal nerve plexuses. Thus, these autonomic fibers (which we shall refer to as *external fibers* to distinguish them from the internal plexuses) exert many of their effects upon the glands and muscles of the gastrointestinal tract through the internal plexuses. Frequently, however, the external fibers (particularly the sympathetic fibers) bypass the plexuses and end in the immediate vicinity of gland cells and vascular smooth muscle. The major autonomic nerve supplying the gastrointestinal tract is the *vagus nerve,* which sends branches to the stomach, small intestine, and upper portion of the large intestine. This nerve is composed of efferent parasympathetic fibers and many afferent fibers from receptors and the nerve plexuses in the gastrointestinal wall.

The afferent pathways of the gastrointestinal tract are also complicated by the presence of the internal plexuses. The walls of the tract contain numerous sensory receptors, many of which are the dendritic endings of the internal plexus neurons, so that action potentials arising in them are conducted directly to the internal plexuses. In some cases the situation is similar to that of other afferent neurons in the body in that the receptors

FIGURE 19-4 ***Transverse section through the abdomen. The cut edges of the peritoneum and the abdominal cavity are shown in color. Normally the organs are virtually in contact with each other.***

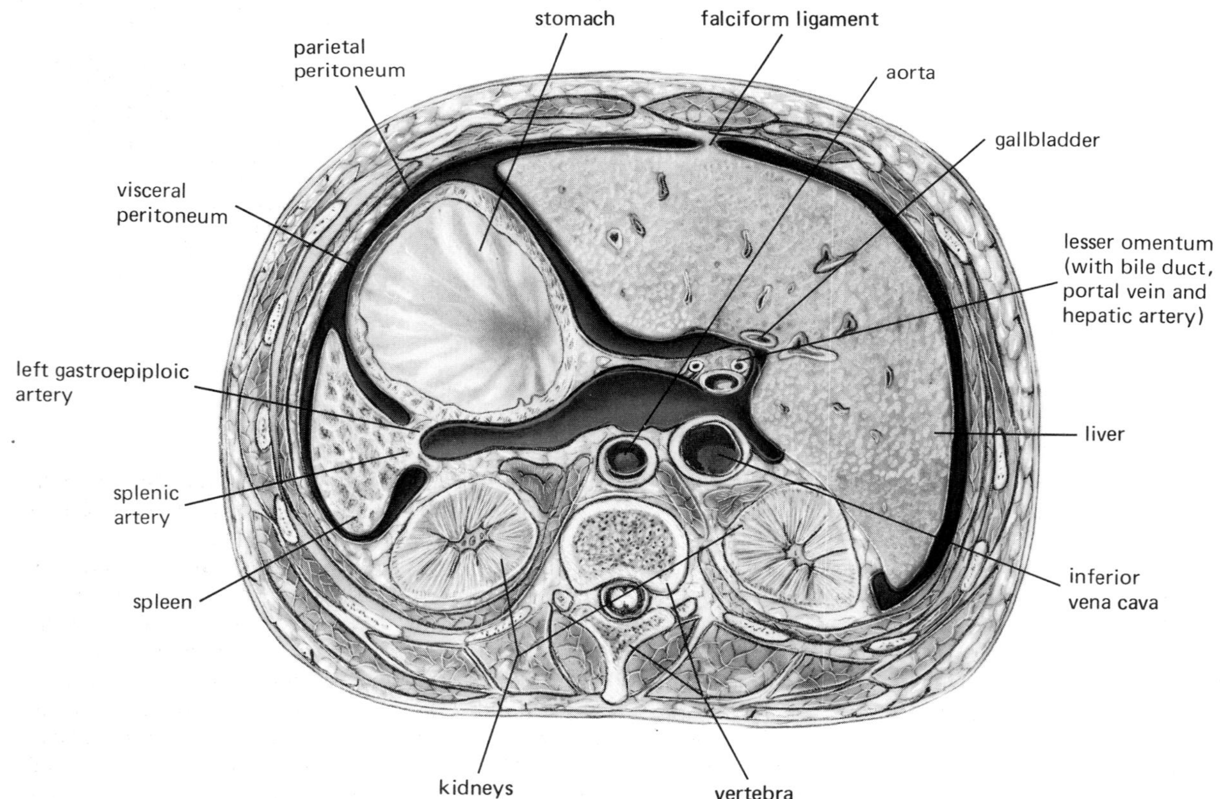

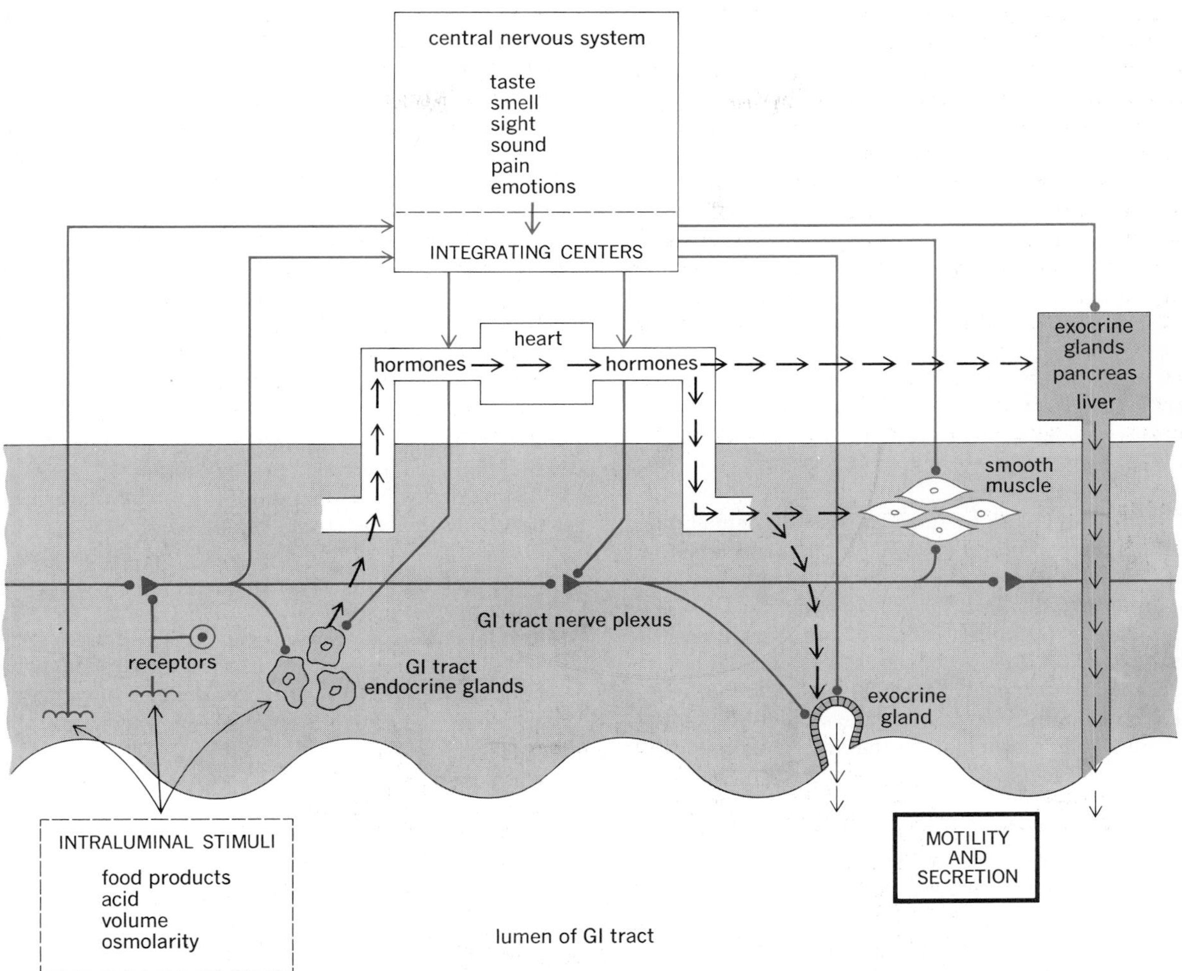

FIGURE 19-5 ***Summary of pathways regulating gastrointestinal activity.***

are the endings of neurons whose cell bodies are outside the gastrointestinal tract in ganglia. In these cases, the afferent information bypasses the internal plexuses and is conducted directly to the central nervous sytem via the vagus nerve.

The general pattern of gastrointestinal control systems can now be summarized (Fig. 19-5). The smooth muscle and exocrine glands are directly influenced by three efferent pathways: (1) external autonomic nerves, (2) internal plexus nerve fibers, and (3) hormones secreted by the gastrointestinal tract itself. The endocrine gland cells which secrete these hormones are controlled not only by these same three types of input but respond directly to changes in the composition of the gastrointestinal contents as well.

These neural and hormonal inputs constitute the efferent pathways of reflexes initiated by receptors in the gastrointestinal tract itself or in other parts of the body. The gastrointestinal receptors respond to changes in the chemical composition of the luminal contents and to the degree of wall distension. As noted above, the information from these receptors is conducted to the internal plexuses as well as to the central nervous system. In the former case, all the components of the reflex arc are located completely within the walls of the gastrointestinal tract; these so-called *short* reflexes confer a considerable degree of self-regulation upon the tract. The point should not be overemphasized, however, since most gastrointestinal reflexes initiated by receptors in the tract have some degree of central nervous system control. Of course, whenever the reflexes are initiated by receptors outside the tract, e.g., by the sight of food, the central nervous system must be involved in the response. Similarly, complex behavioral influences, e.g., emotion, operate through the central nervous system.

THE ORAL CAVITY, PHARYNX, AND ESOPHAGUS

The Oral Cavity

The *oral cavity* (mouth) consists of a small outer portion, the *vestibule,* between the teeth (and gums) and the lips, and a larger portion behind the teeth and gums; the latter extends back to the *oropharyngeal isthmus* where the oral cavity joins the oral portion of the pharynx (Fig. 17-4).

The lips are highly muscular folds covered externally by skin and internally by the stratified squamous epithelial mucous membrane that lines the oral cavity. The midpoint of each lip is connected to the corresponding gum by a fold of mucous membrane called a *frenulum.*

The roof of the mouth consists of a *hard* and a *soft palate.* The hard palate is formed by the *maxilla* and *palatine bones,* and the soft palate is a flexible sheet of densely packed collagen fibers projecting back from the hard palate and dipping downward. The chief function of the soft palate is to close off the opening between the respiratory and oral regions of the pharynx (the *pharyngeal isthmus*) during swallowing, sucking, blowing, and the production of some speech sounds. This closure is achieved by the action of muscles (the *levator veli palatini*), which elevate the soft palate and draw it toward the posterior wall of the pharynx as the posterior wall is being drawn forward by contraction of the *palatopharyngeal muscles.*

The *uvula* hangs down from the soft palate. Two folds of mucous membrane pass from the sides of the uvula to the pharyngeal walls and tongue; they form the *palatoglossal arch* (so named because it contains the palatoglossus muscle) and the *palatopharyngeal arch* (Fig. 19-6). The two arches together form the *isthmus of the fauces* and contain the *palatine tonsil.* (The palatoglossal arch alone forms the lateral borders of the oropharyngeal isthmus.)

The Tongue

The tongue is a muscular organ associated with chewing, swallowing, speech, and taste. Its anterior two-thirds is in the oral cavity, and the posterior one-third lies in the pharynx, where the *root* of the tongue is attached to the hyoid bone and mandible. The surface of the tongue is dotted with *papillae,* elevations in the epithelium and underlying connective tissue which give the tongue its rough appearance. Four types of papillae are distinguished: circumvallate, fungiform, filiform, and foliate (see Fig. 9-36B); they often contain taste buds within their walls. The inferior surface of the tongue is attached to the floor of the oral cavity by the *frenulum of the tongue.*

The right and left halves of the tongue contain extrinsic muscle fibers (the *genioglossus, hyoglossus, chondroglossus,* and *palatoglossus*), which have their attachments outside the tongue, and intrinsic fibers, which are contained totally within it; these fibers together constitute the *lingual muscles.* The muscle fibers cross each other in intricate patterns which account for the tongue's extremely varied and delicate movements. Lingual glands provide mucus and watery (serous) secretions.

The Teeth

The teeth are the hardest and most chemically stabilized tissues in the body. A tooth consists of two parts, the *crown,* which is covered by hard, whitish *enamel,* and the *root,* which is covered by yellowish, bonelike cement (Fig. 19-7). Most of the tooth substance consists of *dentin* which contains a central *pulp cavity* expanded in the crown region into a *pulp chamber,* narrowed in the root into a *pulp canal,* and open at the *apical foramen.* Dental pulp, from which the cavity takes its name, is a connective tissue which contains the vascular and nervous supply for the dentin.

The teeth are seated in their sockets in the *maxilla* (upper jaw) or *mandible* (lower jaw) bone within a relatively soft tissue, the *periodontal ligament* (Fig. 19-7), which allows slight movements of the teeth in response to forces generated in chewing. The periodontal tissues contain receptors which provide information concerning the pressures generated during chewing to the brain centers controlling chewing movements. Near the junction of the root and crown, the periodontal ligament is covered by the *gingiva,* or *gum,* which is continuous with the *oral mucosa* lining most of the oral cavity.

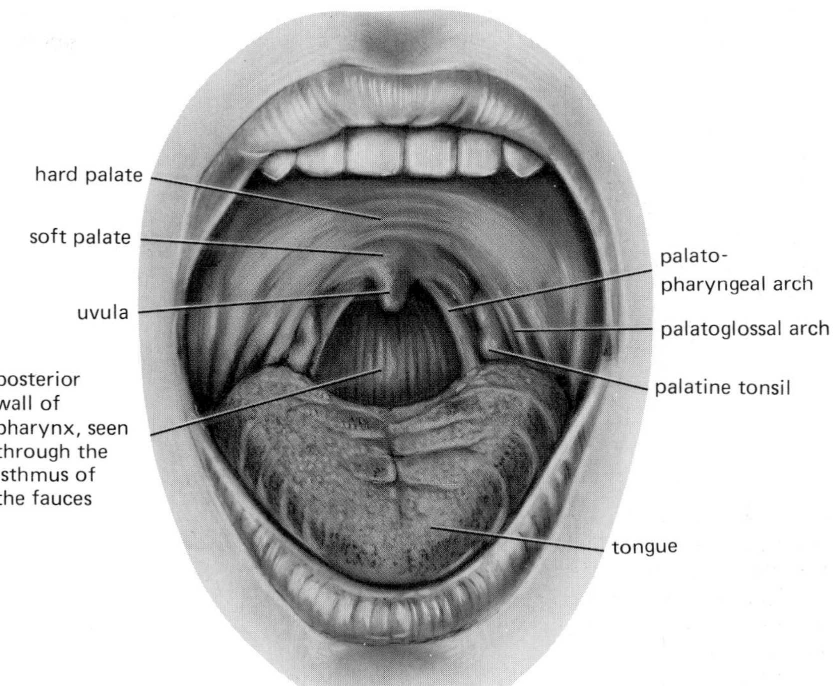

FIGURE 19-6 ***Cavity of the mouth.***

People, like other mammals, have two sets of teeth during their lifetime, a set of small *deciduous teeth* (the "baby teeth") and a set of larger *permanent teeth.* There are 20 deciduous teeth (10 in the upper and 10 in the lower jaw) and 32 permanent teeth. The latter consist of eight *incisors,* so named because they are used to incise, or bite, food (Fig. 19-8), and four *cuspids,* each bearing a single bump, or cusp. (These four teeth are sometimes called the canine teeth.) There are also eight *bicuspids* and twelve *molars.*

Chewing

The primary function of the teeth is to bite off and grind down chunks of food into pieces small enough to be swallowed. The incisors of an adult man can exert forces of 25 to 50 lb, and the molars the 200 lb required to crack a walnut or support a trapeze artist by his teeth. The rhythmic act of chewing is a combination of voluntary and reflex activation of the skeletal muscles of the mouth and jaw. In animals whose cerebral cortex has been destroyed, the reflex rhythmic movements of chewing still occur when food is placed in the mouth. Prolonged chewing of food, so characteristic of human beings, does not appear to be essential to the digestive process (many animals, such as the dog and cat, swallow their food almost immediately). Although chewing prolongs the subjective pleasure of taste, it does not appreciably alter the rate at which the food is digested and absorbed. On the other hand, attempting to swallow a particle of food too large to enter the esophagus may lead to choking, if the particle lodges over the trachea, blocking the entry of air into the lungs. A surprising number of preventable deaths occur each year from choking, the symptoms of which are often confused with the sudden onset of a heart attack so that no attempt is made to remove the obstruction from the airway.

The Salivary Glands and Secretion of Saliva

Saliva is secreted by three pairs of exocrine glands, the parotid, the submandibular, and the sublingual (Fig. 19-1), as well as by numerous small glands distributed throughout the surface of the mouth. The *parotid glands,* which are the largest of the salivary glands, lie

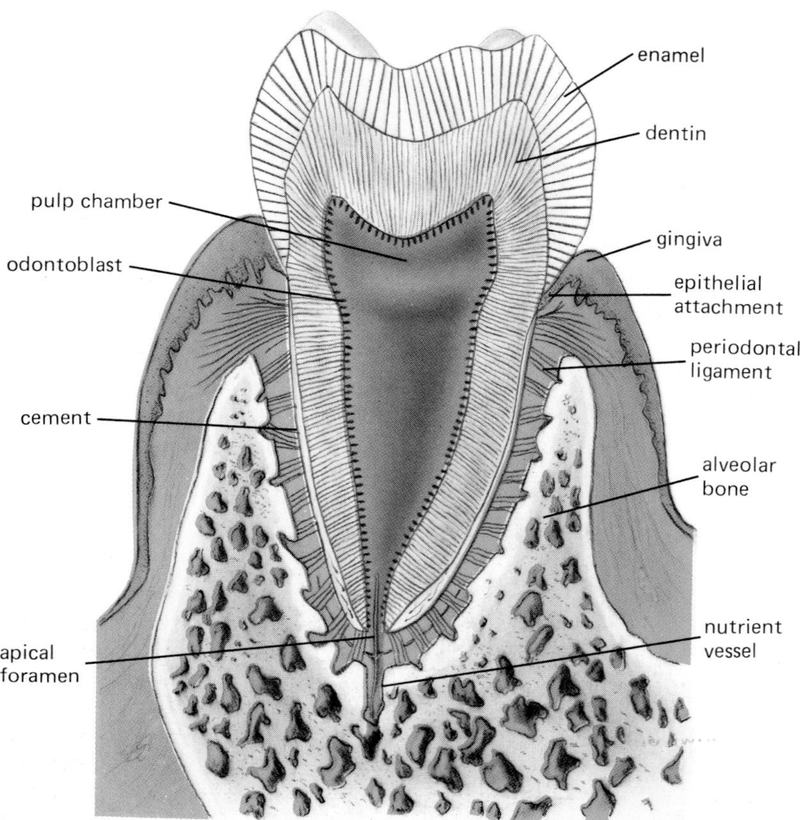

FIGURE 19-7 ***Section of a tooth in situ.***

on the sides of the face, in front of the ears; their ducts *Stensen's* open into the mouth opposite the second molar teeth. The *submandibular glands* lie in front of the angles of the mandible, and their ducts *Wharton's* empty at the sides of the frenulum of the tongue. The *sublingual glands,* the smallest of the three pairs, are beneath the mucous membrane of the floor of the mouth. Their secretions flow through many ducts *(of Rivinus),* most of which open onto the floor of the mouth.

Typical of all exocrine glands, these glands contain secretory cells arranged as acini at the ends of ducts. (The salivary glands are classified as compound tubuloacinar glands.) The epithelial cells, which form the acini, are of two types: serous cells (which produce a watery secretion) and mucous cells. The parotid glands have acini with only serous cells, whereas the sublingual and submandibular acini are mixed, containing cells of both types. The glands also contain myoepithelial cells, the contractions of which help to move the secretions along the ducts. Water accounts for 99 percent of the secreted fluid, the remaining 1 percent consisting of various salts and a few proteins. The major proteins of saliva are the *mucins,* which contain small amounts of carbohydrate attached to the amino acid side chains of the protein. When mixed with water, the mucins form a highly viscous solution known as *mucus.* Mucins are also secreted by gland cells located throughout the gastrointestinal tract. In the mouth the watery solution of mucus moistens and

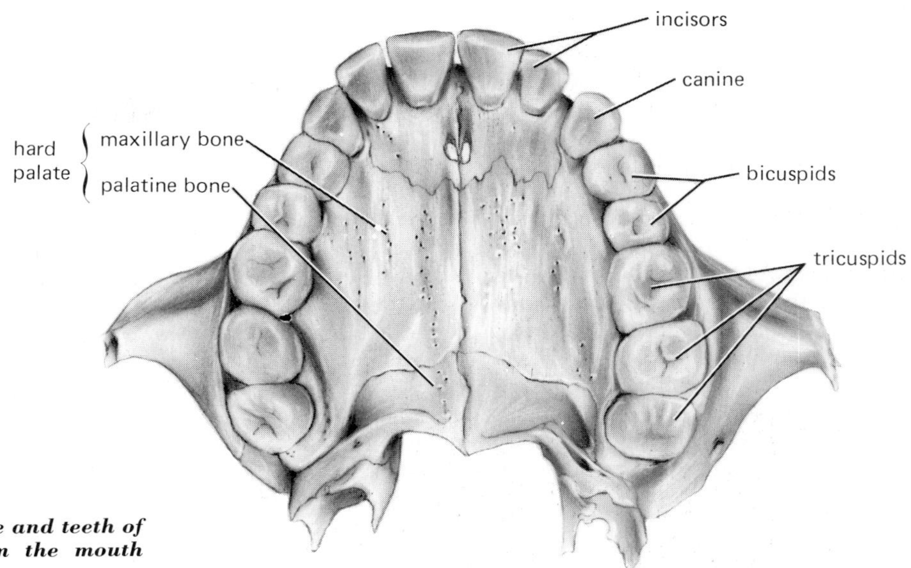

FIGURE 19-8 ***Bony palate and teeth of the upper jaw. View from the mouth cavity.***

lubricates the food particles, allowing them to slide easily along the esophagus into the stomach. (Imagine what it would be like trying to swallow a dry cracker.)

Another protein secreted by the salivary glands is the enzyme *amylase,* which catalyzes the breakdown of polysaccharides into disccharides. Although salivary amylase starts the digestive process in the mouth, food does not remain there long enough for much digestion to occur; however, amylase continues its digestive activity in the stomach until inhibited by the hydrochloric acid there.

A third function of saliva is to dissolve some of the molcules in the food particle; only in this dissolved state can they react with the chemoreceptors in the mouth, giving rise to the sensation of taste (Chap. 9).

The secretion of saliva is controlled by sympathetic and parasympathetic autonomic nerves to the glands. However, unlike their antagonistic activity in most organs, both systems stimulate secretion although the parasympathetic branch causes by far the greatest increase in fluid volume. During sleep very little saliva is secreted. In the awake state a basal rate of about 0.5 ml/min keeps the mouth moist. Food in the mouth increases the rate of salivary secretion. This response is mediated by nerve fibers from chemoreceptors and pressure receptors in the walls of the mouth and tongue. Chewing tasteless wax or marbles increases the rate of salivary secretion even though they have no taste or nutritional value since they activate the pressure receptors. The receptors send fibers via cranial nerves to the brainstem medulla, which contains the integrating center controlling salivary secretion. The most potent stimuli for salivary secretion are acid solutions, e.g., fruit juices and lemons, which may lead to a maximal secretion of 4 ml of saliva per minute. Salivation initiated by the sight, sound, or smell of food is very slight in human beings in contrast to the marked increase produced by these stimuli in dogs.

During the course of a day, between 1 and 2 l of saliva is secreted, most of which is swallowed. The proteins in saliva are broken down into amino acids by the digestive enzymes in the stomach and intestinal tract, and the amino acids, salts, and water absorbed into the circulation. This is a typical pattern for most of the secretions of the gastrointestinal tract. Although large amounts of fluid may be secreted into the tract during the course of a day, most of it, together with its salt and protein content, is digested and reabsorbed. If these secretions are not reabsorbed, large quantities of fluid and salt can be lost from the body through the gastrointestinal tract.

The Pharynx

The pharynx is a muscular tube, 12 to 14 cm long, which lies behind the nasal cavities, oral cavity, and larynx. It is discussed in Chap. 17.

Swallowing

Swallowing is a complex reflex initiated when the tongue forces a bolus of food into the rear of the mouth, and pressure receptors in the walls of the pharynx are stimulated. They send afferent impulses to the swallowing center in the medulla, which coordinates the swallowing process via efferent impulses to the 25 different skeletal muscles in the pharynx, larynx, and upper esophagus and the smooth muscles of the lower esophagus. As the bolus of food moves into the pharynx, the soft palate is elevated and lodges against the back wall of the pharynx, sealing off the nasal cavity and preventing food from entering this area (Fig. 19-9). The swallowing center generates a precisely timed sequence of nerve impulses which inhibit respiration and draw the pharyngeal walls upward and inward as the tongue pushes the bolus to the back of the pharynx. The larynx is raised and the glottis (the opening between the vocal cords) is closed, keeping food from getting into the trachea. Successive contractions of the pharyngeal constrictor muscles and gravity (when the body is upright) force the bolus past the epiglottis into the lower part of the pharynx and on into the esophagus. As the bolus passes the epiglottis, it tips the epiglottis backward to cover the closed glottis. It is the closure of the glottis, however, and not the folding of the epiglottis which is primarily responsible for preventing food from entering the trachea.

Once swallowing has been initiated, it cannot be stopped voluntarily even though it involves skeletal muscles. This reflex is a stereotyped all-or-none response, the entire coordination of which resides in the swallowing center in the medulla. The swallowing reflex is an example of a triggered reflex in which multiple responses occur in a regular temporal sequence that is predetermined by the synaptic connections in the coordinating center.

The Esophagus

The esophagus is a flaccid, empty tube about 25 cm long. Its layers blend with those of the pharynx above and the stomach below. Its epithelium is similar to that of the pharynx, being of the stratified squamous type. Mucous glands occur in the mucosa and deeper in the submucosa. The upper third of the esophagus in human beings is surrounded by skeletal muscle, the lower third by smooth muscle, and the middle third by both types of muscle fibers. At rest, the opening into the esophagus is closed, this region of the esophagus forming the *hypopharyngeal,* or *superior esophageal, sphincter.* The skeletal muscles in this region are so arranged that, when they are relaxed, the sphincter is closed by passive elastic tensions in the walls. The swallowing center initiates contraction of these muscles, opening the hypopharyngeal sphincter and allowing the bolus to pass into the esophagus. Immediately after the bolus has passed, the sphincter muscles relax, the sphincter closes, the glottis opens, and breathing resumes. The pharyngeal phase of swallowing lasts about 1 s.

FIGURE 19-9 ***Movement of a bolus of food through the pharynx and upper esophagus during swallowing.***

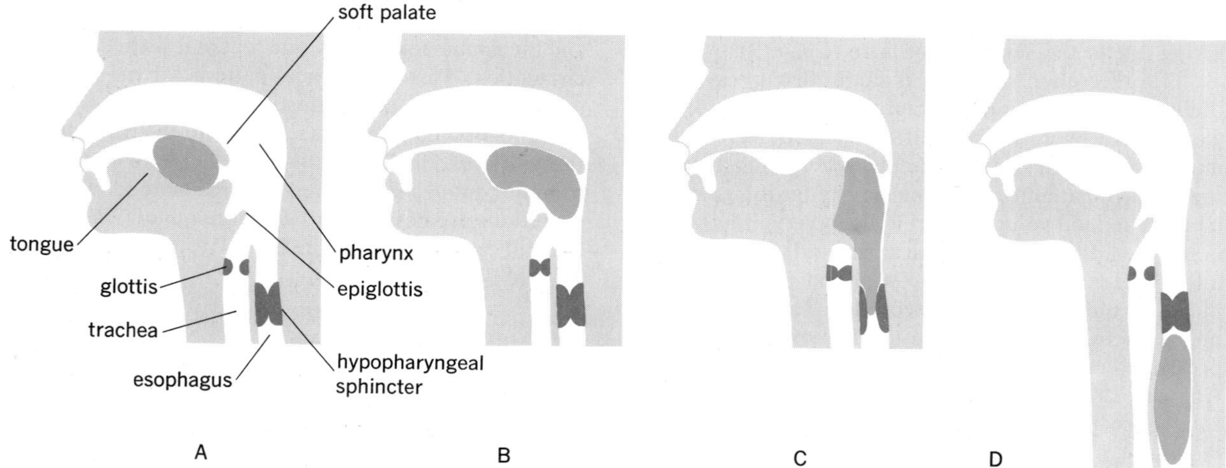

Once within the esophagus, the bolus is moved along it by a wave of sequential constrictions that pass along the walls of the esophagus and move toward the stomach. Such waves of contractile activity in the muscle layers surrounding a tube are known as *peristaltic waves* and are found in various parts of the body, including other portions of the gastrointestinal tract. A peristaltic wave takes a total of about 9 s to travel the length of the esophagus. The progression of the wave is controlled by the vagus nerves whose activity patterns are coordinated by the swallowing center in the medulla, unlike the peristaltic waves in other portions of the gastrointestinal tract, which are coordinated primarily through the internal nerve plexuses. If a large or sticky bolus of food, e.g., peanut butter, becomes stuck so that it is not carried along the esophagus by the initial wave, the distension of the esophagus triggers a second more forceful peristaltic wave, which begins without movement of the mouth or pharynx. Distension of the esophagus also reflexly stimulates increased salivary secretion. The combination of increased salivary fluid and repeated peristaltic waves helps dislodge the trapped bolus. Swallowing can occur while a person is standing on his head since it is not primarily gravity but the peristaltic wave which moves the contents of the esophagus toward the stomach.

The last 4 cm of the esophagus before it enters the stomach is known as the *gastroesophageal*, or *cardiac, sphincter*. Anatomically, this region appears no different from adjacent regions, but its functional activity is distinct. As the peristaltic wave begins in the esophagus, the gastroesophageal sphincter relaxes, allowing the bolus upon its arrival to enter the stomach. When swallowing is not taking place, this sphincter remains tonically contracted, forming a barrier between the contents of the stomach and the esophagus. Most of the esophagus lies in the thoracic cavity and is subject to a subatmospheric intrathoracic pressure of about −5 to −10 mm Hg. The stomach, however, lying below the diaphragm, has an internal pressure slightly above atmospheric, +5 to +10 mm Hg, due to its compression by the contents of the abdominal cavity. Thus, without the gastroesophageal sphincter, the pressure gradient from the stomach to the esophagus would tend to force the contents of the stomach into the esophagus.

The ability of the gastroesophageal sphincter to maintain a barrier between the stomach and esophagus is helped by the fact that its last portion lies below the diaphragm and is subject to the same pressures in the abdominal cavity as the stomach. Thus, if the pressure of the abdominal cavity is raised, e.g., during cycles of respiration or by contraction of the abdominal muscles, the pressures of both the stomach contents and this terminal segment of the esophagus are raised together and there is no change in the pressure gradient between the two. During pregnancy the growth of the fetus increases the pressure on the abdominal contents and displaces the terminal segment of the esophagus through the diaphragm into the thoracic cavity. The gastroesophageal barrier is therefore no longer assisted by changes in abdominal pressure, and during the last 5 months of pregnancy there is a tendency for the increased pressures in the abdominal cavity to force some of the contents of the stomach up into the esophagus. The hydrochloric acid from the stomach contents irritates the walls of the esophagus and causes contractile spasms of the smooth muscle, both of which are associated with the pain known as *heartburn* because the sensation appears to be located in the region over the heart. Hearburn often subsides in the last weeks of pregnancy as the uterus descends prior to delivery, decreasing the pressure on the abdominal organs. A newborn child has no intraabdominal segment of the esophagus and thus has a tendency to regurgitate.

In the abnormal condition called *achalasia,* the gastroesophageal sphincter fails to relax, and a whole meal may become lodged in the esophagus, entering the stomach very slowly. Distension of the esophagus results in pain in the chest which is often confused with pain originating from the heart. Achalasia appears to be the result of damage to, or absence of, the myenteric nerve plexus in the region of the gastroesophageal sphincter.

During swallowing, air is trapped in the pharynx ahead of the bolus, and most of this passes into the trachea before the glottis closes. Air mixed with the bolus or saliva is swallowed and then expelled by belching, passed on into the intestine, or absorbed into the bloodstream and removed via the lungs.

THE STOMACH

The stomach is a chamber located between the end of the esophagus and the beginning of the small intestine. Figure 19-10 indicates the major parts of the stomach and the folds, or *rugae,* present on its inner surface when it is empty. (These gradually flatten out as the stomach becomes filled with food.)

At the gastroesophageal junction the stratified squamous epithelium of the esophagus changes

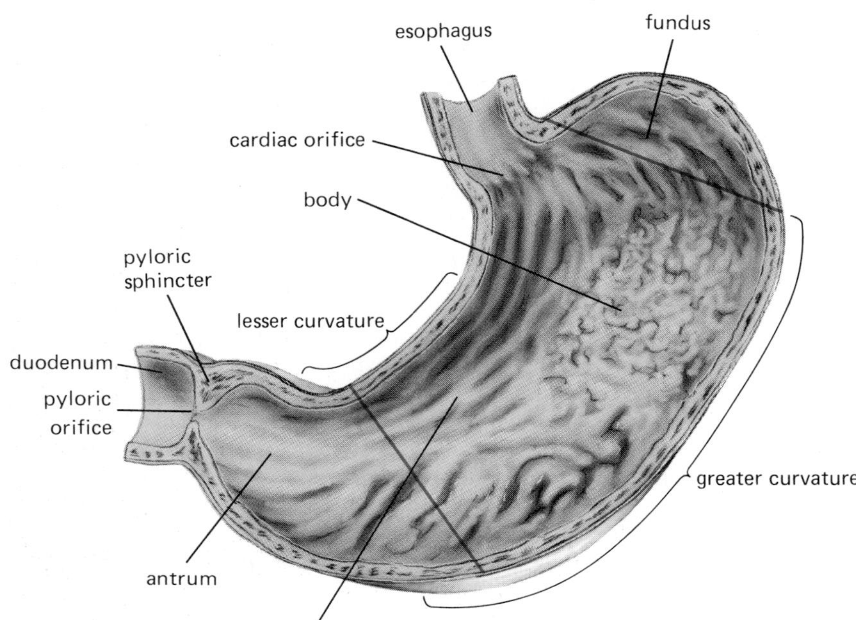

FIGURE 19-10 ***Interior of the stomach.***

abruptly to the simple columnar epithelium characteristic of the stomach. These cells secrete mucus, which lubricates the stomach wall and protects it to some extent from the abrasive contents of the meal and from the digestive enzymes and acid in the gastric juice. The cells have a very short life-span and are replaced every 3 days or so by the division of undifferentiated epithelial cells in the stomach lining. The lining of the stomach is indented by numerous *gastric pits* which are the openings to the *gastric glands* lying within the mucosa (Fig. 19-11). The glands are of three types: *cardiac glands* near the gastroesophageal junction; *pyloric glands* near the pylorus; and the main gastric glands, or *oxyntic glands,* of the body and fundus of the stomach. The most common cell type in the cardiac and pyloric glands secretes mucus, but cells of the main gastric glands secrete a very strong acid, hydrochloric acid, and several enzymes which, along with salivary amylase, begin the process of digestion.

Digestion in the stomach is limited to the breakdown of large lumps of food into a solution of individual molecules and fragments of large molecules, most of which are still too large to be absorbed. This mixture of fluids, partially digested food particles, and enzymes has the consistency of a thick soup and is known as *chyme.* The most important function of the stomach is to regulate the rate at which chyme enters the small intestine, where most of the process of digestion and absorption takes place. In the absence of the stomach, a normal-sized meal moves so rapidly through the small intestine that only a fraction of the food has time to be digested and absorbed.

Gastric Motility

The empty stomach has a volume of about 50 ml, and its lumen is little larger than that of the small intestine. The surface of the interior of the stomach is highly folded into ridges. Upon being filled with food, the stomach relaxes and the folds get smaller; thus the wall tension and intraluminal pressure change only slightly.

The muscular layer in the stomach wall has three layers: an external longitudinal layer, a circular layer, and an internal oblique layer. These muscle layers are relatively thin around the fundus and body but are thicker and more powerful in the lower portion of the stomach, the *antrum* (see Fig. 19-10). Emptying the stomach contents depends primarily upon the contractile activity of these smooth muscle layers.

The membrane potential of the longitudinal smooth muscle cells undergoes rhythmic oscillations of

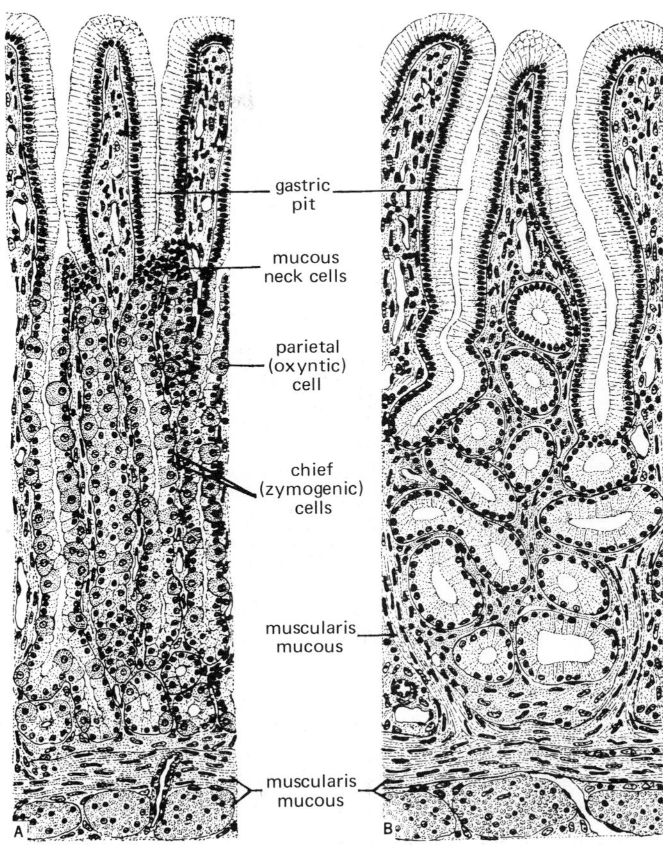

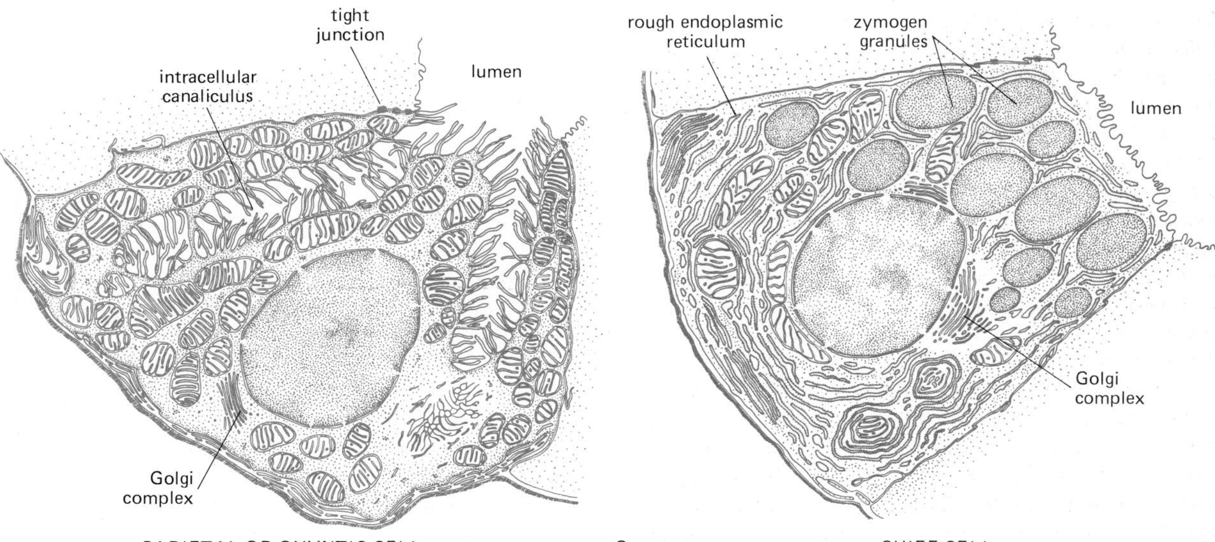

FIGURE 19-11 **(A and B)** ***Mucosa from the fundic and pyloric regions of the stomach.*** **(C)** ***Structure of the parietal and chief cells of the gastric glands.*** **(Part A from W. F. Windle, "Textbook of Histology," 5th ed., p. 384, McGraw-Hill, New York, 1976. Part B adapted from Ito.)**

about 10 to 15 mV, repeating at a rate of about three per minute. Known as the *basic electrical rhythm,* its magnitude and rate of propagation control the mechanical activity of the stomach wall. The oscillation in membrane potential is initiated by pacemaker cells in the region of the cardiac orifice. These potential changes are then conducted through gap junctions along the longitudinal layer of smooth muscle. When membrane depolarization is greatest during these cycles, action-potential spikes often occur and are associated with muscle contraction. The propagation of the slow-wave electric activity along the longitudinal smooth muscle produces a peristaltic wave of contraction which spreads along the stomach from the esophagus to the small intestine. The circular smooth muscle cells do not show a basic electrical rhythm, but since action potentials appear in these cells coincident with action potentials in the longitudinal layer, there must be some form of electrical connection between the layers, possibly through the internal nerve plexuses.

The basic electrical rhythm is continuously active in the empty stomach, but only occasionally is the depolarization great enough to give rise to action-potential spikes and a resulting wave of peristaltic contraction. Moreover, when these contractions do occur, they cause only a slight ripple in the wall. As fasting is prolonged the magnitude of the stomach contractions becomes greater and greater. These changes appear to be mediated by parasympathetic input via the vagus nerve which increases the sensitivity of the smooth muscle cells to the basic electrical rhythm. Attempts have been made to correlate these strong stomach contractions with the sensation of hunger pangs, but the correlation is generally poor.

During the first half hour after a meal, peristaltic activity in the stomach is very weak; thereafter, the waves increase in intensity and slightly in frequency. The waves start at the esophagus and proceed as a weak ripple over the body of the stomach at about 1 cm/s. As the wave approaches the large mass of muscle in the walls of the antrum, it speeds up, reaching velocities of 3 to 4 cm/s. The higher conduction velocity in the antrum means that a large part of the smooth muscle in this region undergoes an intense contraction almost simultaneously. This strong antral contraction is responsible for emptying material from the stomach into the *duodenum* (the first segment of the small intestine). Because of the size of the stomach and the frequency of contractions, two to three waves may be proceeding over the surface of the stomach simultaneously (Fig. 19-12).

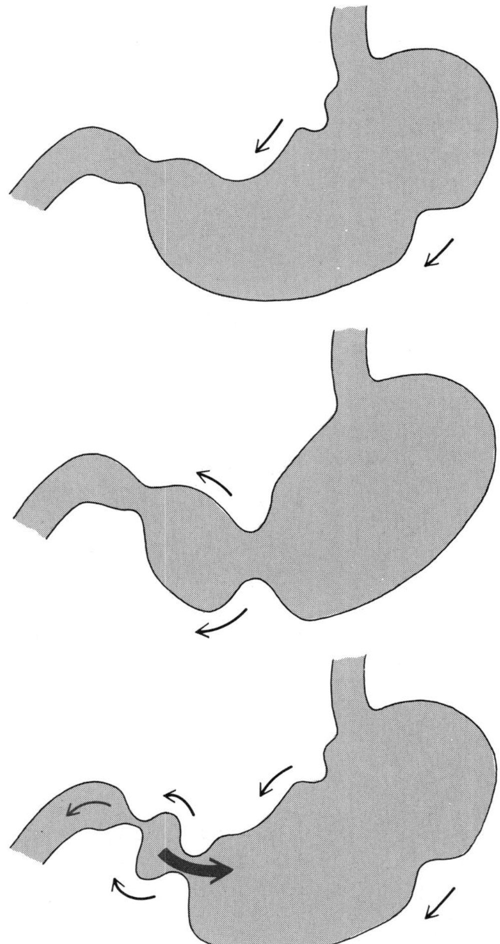

FIGURE 19-12 ***Peristaltic waves passing over the stomach empty a small amount of material into the duodenum. Most of the material is forced back into the antrum.***

The *pyloric sphincter* is a ring of smooth muscle and connective tissue between the terminal antrum and the duodenum (Fig. 19-10). Since at rest the pressure in the lumen of the duodenum is equal to or slightly greater than the pressure in the stomach, there is normally no pressure gradient to cause material to move from the stomach into the duodenum. The pyloric sphincter exerts only a slight pressure at the junction and is normally open most of the time. When a strong peristaltic wave arrives at the antrum, the pressure of the antral contents is increased, forcing some of the material into the duodenum. However, the antral contraction also closes the pyloric sphincter, with the result that the pressure in the antrum forces most of the antral contents back into the body of the stomach and only small amounts pass into the duodenum. The strong contractions of the antral region are the primary force acting to mix the contents of the stomach; little mixing occurs in the body of the stomach, which is subjected only to weak peristaltic waves.

Control of Gastric Emptying The amount of material forced into the duodenum from the stomach depends upon the strength with which the antral muscles contract. One of the factors which influences this contraction is the amount of food in the stomach. The stomach empties at a rate proportional to the volume of material in it at any given time. This effect may be mediated by the internal nerve plexus or may be a direct effect of stretching the smooth muscle, which partially depolarizes the membrane, thus requiring less depolarization by the basic electrical rhythm to reach threshold and fire an action potential that will lead to contraction. In addition, stretch receptors present in the walls of the stomach influence smooth muscle activity through both short and long reflex pathways.

The most important factor controlling gastric emptying is not gastric volume, however, but the chemical composition and amount of chyme in the duodenum. When the duodenum contains fat, acid, or hypertonic solutions, or when it is distended, gastric motility is reflexly inhibited. These reflexes are initiated by chemoreceptors, osmoreceptors, and pressure receptors located in the walls of the duodenum (Fig. 19-13) and mediated by external nerves, the internal nerve plexuses, and a group of hormones secreted by cells in the wall of the duodenum. Two of these hormones, *secretin* and *cholecystokinin* (abbreviated CCK), have been isolated and found to be small proteins. As will be described in subsequent sections, secretin and CCK affect the contractile activity of smooth muscles and glandular secretion, not just in the stomach but in many other locations of the gastrointestinal system. Their effects on the gastrointestinal system are not identical; some effectors respond much more strongly to one or the other of the two hormones. These two hormones are secreted by separate cells distributed throughout the duodenum. The stimulus for the release of secretin and CCK is the chemical composition of the chyme entering the intestine from the stomach. The only effective chemical stimulus for the release of secretin appears to be acid. In contrast, the products of protein and triglyceride digestion—certain essential amino acids and long-chain fatty acids—are the most potent stimuli for CCK release. Carbohydrates are ineffective in releasing either hormone. Finally, many of the stimuli which release the duodenal hormones also stimulate nerve cells in the internal nerve plexus which lead to reflex neural inhibition of gastric motility. This completely neural pathway is known as the *enterogastric reflex*. Let us look at each of these hormonal and neural reflexes more closely.

Fat in the duodenum is the most potent stimulus for the inhibition of gastric motility. Part of this inhibitory effect is mediated by CCK, but other, as yet unidentified, duodenal hormones appear to be released, since the relatively weak inhibitory activity of cholecystokinin cannot account for the magnitude of the inhibition that occurs in the presence of fat. When gastric emptying is slowed down, less fat enters the intestine per unit time, giving more time for digestion and absorption. The high fat content of eggs and milk may inhibit gastric emptying to the point where some of the meal may be found in the stomach after 6 h. In contrast, a meal of meat and potatoes (protein and carbohydrate) may empty in 3 h.

Hydrochloric acid, secreted by the stomach, empties into the duodenum, where, as we shall see, it is neutralized by sodium bicarbonate secreted by the pancreas. Unneutralized acid in the duodenum inhibits the emptying of more acid by way of the enterogastric reflex as well as by the release of secretin.

As molecules of protein and starch are digested in the duodenum to form large numbers of molecules of amino acids and glucose, the osmolarity of the solution rises. If the rate of absorption does not keep pace with the rate at which osmotically active molecules are formed by digestion, the osmolarity of the chyme becomes greater than blood and large volumes of water may enter the intestine by osmosis, causing significant lowering of blood volume. It is important, therefore, that reflex inhibition of gastric emptying by hypertonic solutions in the duodenum decrease the rate at which osmotically active molecules are formed in the intestine by reducing the amount of material available for digestion. This response is mediated primarily by the enterogastric reflex.

In addition to the control of gastric emptying by the gastric volume and duodenal contents, motility is influenced by sympathetic and parasympathetic nerve fibers to the stomach. Emotions such as sadness, depression, and fear tend to decrease motility; aggression or anger tends to increase it. These relationships

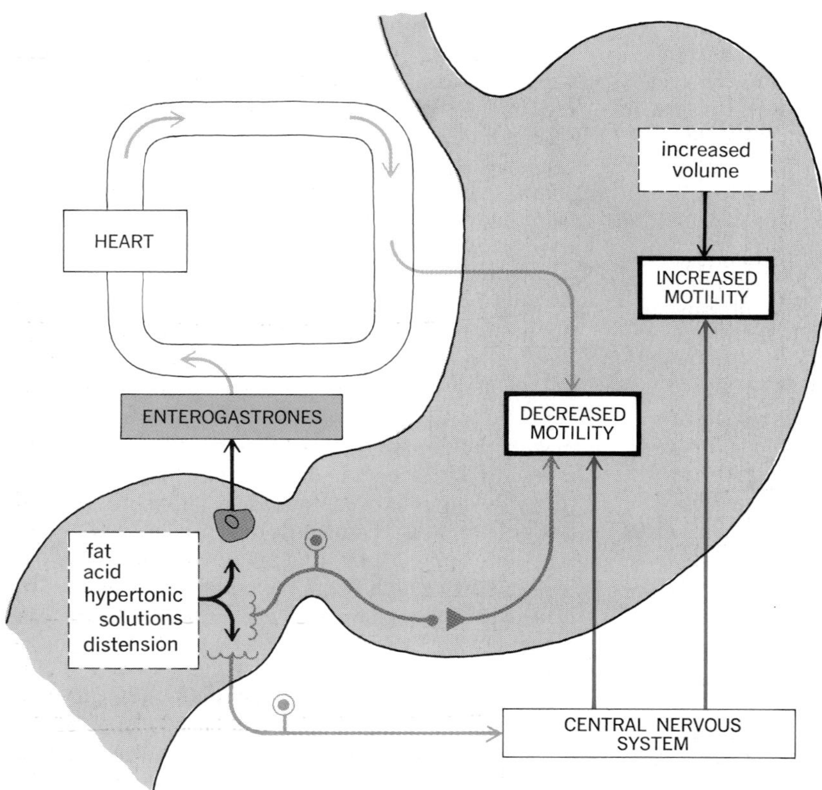

FIGURE 19-13 ***Summary of pathways controlling gastric motility.***

are not always predictable, however, and different people show different responses to apparently similar emotional states. Intense pain from any part of the body tends to inhibit gastric motility. These effects mediated through external nerves from higher centers in the nervous system usually affect not only the stomach but the entire gastrointestinal tract. Inhibition of motility is mediated by decreased parasympathetic activity and increased sympathetic activity.

Vomiting Vomiting is the forceful expulsion of the contents of the stomach and upper intestinal tract through the mouth. It is a complex reflex coordinated by the vomiting center in the brainstem medulla and is usually preceded by increased salivation, sweating, faster heart rate, and feelings of nausea—all characteristic of a general discharge of the autonomic nervous system. Vomiting begins with a deep inspiration, closure of the glottis, and elevation of the soft palate. The abdominal and thoracic muscles contract, raising the intraabdominal pressure, which is transmitted to the contents of the stomach; the gastroesophageal sphincter relaxes, and the high abdominal pressure forces the contents of the stomach into the esophagus. When the pressure is great enough, the contents of the esophagus are forced through the hypopharyngeal sphincter into the mouth. Vomiting is also accompanied by strong contractions of the upper portion of the small intestine which tend to force some of the intestinal contents back into the stomach.

Input to the vomiting center in the medulla comes from a number of receptors throughout the body. The primary stimuli are tactile stimulation of the back of the throat, e.g., sticking a finger in the back of the throat; excessive distension of the stomach or duodenum; high pressures within the skull; rotating movements of the head producing dizziness; and intense pain as well as certain chemical agents which act on chemoreceptors in the brain and upper parts of the gastrointestinal tract. Excessive vomiting can lead to large losses of secreted fluids, H^+, and salts from the body which would normally be reabsorbed. This can result in severe dehydration, upset the acid-base and salt balance of the body, and produce circulatory problems due to a decrease in plasma volume.

Gastric Secretions

HCl Secretion The normal human stomach secretes about 2 l of hydrochloric acid solution a day. Hydrochloric acid is a strong acid which completely dissociates in water into hydrogen and chloride ions. The concentration of hydrogen ions in the lumen of the stomach may reach 150 mM (isotonic HCl) compared with the concentration of hydrogen ion in the blood, 0.00004 mM. The secretion of acid is not absolutely essential to gastrointestinal function, since digestion and absorption occur in the small intestine in the absence of a stomach; however, the acid performs several functions which assist digestion. A high acidity denatures proteins and breaks intermolecular bonds, thereby breaking up connective tissue and cells, releasing ionized molecules into solution. Hydrochloric acid thus continues the process begun by chewing, namely, reducing large particles of food to smaller particles and individual molecules. However, the acid has relatively little ability to break down proteins and polysaccharides into amino acids and glucose. A second function is to kill most of the bacteria that enter along with food. This process is not 100 percent effective, and some live bacteria enter the intestinal tract where they may continue to multiply, especially in the large intestine. Many of these bacteria are harmless; others may release disease-producing toxins which reach the blood even after the bacteria themselves have been killed (Chap. 22). A third function of hydrochloric acid is to activate some of the enzymes secreted by the stomach.

Hydrochloric acid is secreted by the *parietal,* or *oxyntic, cells* lying in gastric pits in the body of the stomach (Fig. 19-11). These cells have many large mitochondria, and their cell membranes are indented at the luminal surface to form minute intracellular channels (*canaliculi*) which penetrate deep into the cell. The canaliculi greatly enlarge the surface area of the cell available for secretion, and their intracellular distribution brings them into close association with the mitochondria which produce the ATP necessary for operating the active-transport systems involved in acid secretion.

The molecular mechanism of acid secretion is unknown. Hydrogen ions must be moved against a very large concentration gradient, the concentration of hydrogen ions in the lumen of the stomach being as much as 3 million times greater than in the blood. Chloride ions are moved actively but against a much smaller concentration gradient, since their concentration in the stomach lumen is only about 1.5 times that of the blood. The two ions appear to be actively transported by separate pumps in the luminal membrane of the parietal cell. The enzyme carbonic anhydrase, similar to the enzyme in red blood cells, is present in the gastric mucosa and is believed to be involved in the secretion of acid, probably by way of the general pathway illustrated in Fig. 19-14. Whenever an acid is formed, an equivalent amount of base is also formed. Here the base is the bicarbonate ion, which enters the blood. The venous blood leaving the stomach is therefore more alkaline than the arterial blood entering it because of its higher bicarbonate concentration and lower hydrogen-ion concentration. Whenever there is considerable loss of the acid contents of the stomach through vomiting, the blood may become progressively more alkaline, requiring adjustments in the acid-base balance of the body by the kidneys and lungs. Under normal conditions acid secretion produces little change in the acid-base balance of the body because an amount of bicarbonate equal to that released into the blood by the stomach is secreted by the pancreas into the duodenum.

Enzyme Secretion and Digestion Although several different enzymes have been detected in gastric secretions, the only ones with any significant effect upon the normal digestive process are the proteolytic enzymes known as *pepsins.* At least seven separate pepsins have been isolated from human gastric mucosa. These enzymes are similar in their activity and will be referred to simply as pepsin. Pepsin is secreted by the *chief* (or *zymogenic*) cells, which contain numerous zymogen granules (Fig. 19-11). Acid is secreted only by the glands in the body of the stomach, but pepsin is secreted by both the gastric glands in the body and the pyloric glands in the antrum.

Pepsin is secreted in an inactive form known as *pepsinogen,* which is converted to pepsin by splitting off a small fragment of the molecule. The hydrochloric acid in the stomach initiates this process, and once pepsin is formed, it can act upon other molecules of pepsinogen to form more pepsin. Thus a small amount of pepsin acts autocatalytically to activate more pepsin.

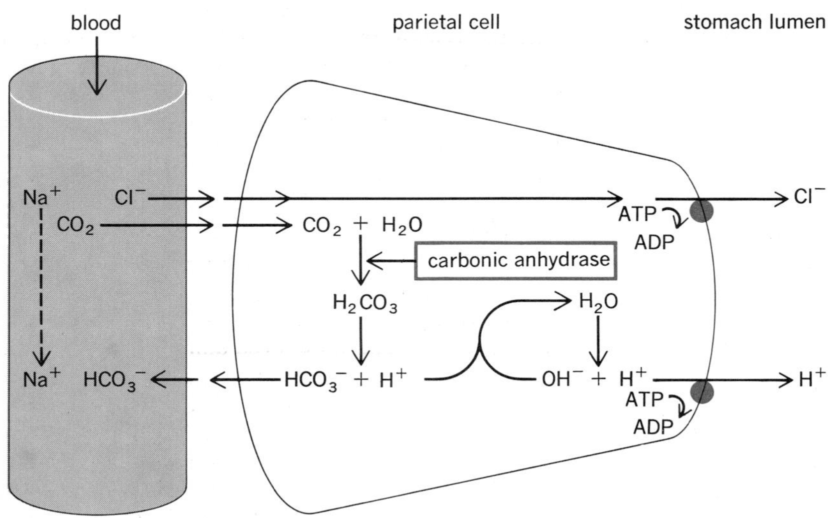

FIGURE 19-14 ***Secretion of HCl by parietal cells and the release of bicarbonate ions into the blood.***

Pepsin catalyzes the splitting of bonds between particular types of amino acids in protein chains. The products of pepsin digestion are primarily small peptide fragments composed of several amino acids. As we shall see, these fragments are one of the major chemical stimuli controlling hydrochloric acid production and pancreatic secretion. Pepsin is active only in an environment having the high hydrogen-ion concentration provided by the hydrochloric acid in the stomach. In the duodenum, where the stomach acid is neutralized by bicarbonate ions from the pancreas, pepsin is inactive.

The ability of acid and pepsin to break down particles of food depends upon their coming in contact with the food in the stomach. There is very little mixing of solid food in the stomach until it reaches the antrum, where strong peristaltic waves compress and mix the contents. Food in the body of the stomach remains as a semisolid mass which is attacked by acid and pepsin only at its surface, the interior of the mass being free of acid. On the other hand, although salivary amylase is inactivated by hydrochloric acid, it may continue to act upon starch for long periods of time in the stomach if it is inside the bolus, where acid has not yet reached. Bacteria in the same location may also escape the sterilizing action of hydrochloric acid.

Control of Gastric Secretion Acid and enzyme secretions in the stomach generally vary in a parallel manner. Most of the factors described below which stimulate the secretion of acid also stimulate enzyme secretion.

When no food is in the stomach, acid secretion occurs at a basal rate of about 0.5 ml/min. After a meal the rate of acid secretion increases considerably, reaching maximal values of about 3 ml/min. For reasons to be discussed shortly, the concentration of acid in the stomach decreases immediately following a meal and then rises as the rate of acid secretion declines.

Cephalic Phase Enhanced acid secretion begins even before food reaches the stomach. This is known as the *cephalic phase* of gastric secretion. Sight, smell, and taste all contribute to the stimulation, mediated by parasympathetic nerve fibers traveling to the stomach in the vagus nerve. Central nervous system factors other than those normally associated with the process of eating (such as emotions) may also stimulate secretion by this pathway.

Gastric Phase Food placed directly in the stomach through a tube elicits an increase in acid and enzyme secretion, the amount depending upon the chemical composition of the food and its volume. The chemical stimulation of gastric secretion is mediated by a hormone, *gastrin,* released by cells located in the walls of the terminal segment of the stomach.

Gastrin is the third major gastrointestinal hormone and, like secretin and CCK, has multiple sites of action throughout the gastrointestinal system. Because the last five amino acids in CCK are identical to those in gastrin, they are both able to bind to the same receptor sites; therefore, CCK shows weak gastrinlike activity and gastrin has weak CCK-like activity. However, because the affinity of the hormones for a given receptor site is determined by the remaining amino acids in their respective structures, the magnitude of the response produced at any given site differs between the two. For example, as we shall see, CCK is a more potent stimulator of pancreatic secretion than is gastrin.

In addition to gastrin's ability to stimulate hydrochloric acid secretion in the stomach, it also stimulates the contraction of the smooth muscle in the gastroesophageal sphincter. Thus, at the same time that gastrin is stimulating acid secretion, it helps to prevent the entry of acid into the esophagus. Patients who have had the lower portion of their stomach removed (e.g., because of cancer) often have an increased incidence of heartburn (irritation of the esophagus by acid) because of the loss of gastrin and its ability to keep the esophageal sphincter closed.

The major chemical stimulus for the release of gastrin is the presence of protein in the food entering the stomach. Intact protein, however, is much less effective than the peptide fragments produced by the action of pepsin upon these proteins. In addition, alcohol and caffeine (present in coffee, tea, and cola drinks) are potent stimuli for gastrin release and thus acid secretion. A cocktail before dinner and a bowl of soup, i.e., protein broth, before the main course both stimulate acid secretion and prepare the stomach for the main meal. On the other hand, the intake of large quantities of alcohol and coffee in the absence of food produces the secretion of very concentrated acid, which can irritate the linings of the esophagus, stomach, and duodenum. In addition to these chemical stimuli, gastrin is also released in response to stimulation of the parasympathetic fibers in the vagus nerves to the stomach.

It should be emphasized that gastrin is not the unique final common pathway for stimulating acid secretion. Distension of the terminal part of the stomach or stimulation of the portion of the vagus supplying that region enhances acid secretion independently of gastrin release. A summary of the pathways involved in controlling the gastric phase of acid secretion is given in Fig. 19-15.

Is there any inhibitory influence on acid secretion during the gastric phase? Obviously a lessening of any of the stimuli causing acid secretion will reduce the secretion. All the stimuli mentioned above lessen as the stomach empties its contents into the duodenum. In addition, acid itself is a powerful inhibitor of acid secretion. A high concentration of hydrogen ions directly inhibits the release of gastrin by the antrum and thus inhibits gastric secretion by reducing the amount of gastrin available to stimulate acid secretion. Before food enters the stomach, the hydrogen-ion concentration is high, inhibiting gastrin release. When food enters the stomach, the hydrogen-ion concentration drops, removing the inhibition of gastrin release and thereby acid secretion. Why does the hydrogen-ion concentration drop and remain low for some time in spite of increased acid secretion? Food, particularly protein, buffers the hydrogen ions, which combine with the ionized carboxyl groups and un-ionized amino groups on the protein molecules, thus removing them from solution and lowering the concentration of free hydrogen ions. Acid secretion continues as long as gastrin is not inhibited by a rise in hydrogen-ion concentration. As more and more acid is secreted, the buffering capacity of the proteins in food is exceeded and the hydrogen-ion concentration begins to rise, inhibiting gastrin release and turning off acid secretion. Thus, for two reasons, the total amount of acid secreted during a meal is directly proportional to the amount of protein in the meal: (1) The greater the amount of protein, the greater the buffering capacity, and thus the more acid can be secreted without appreciably raising the hydrogen-ion concentration and inhibiting gastrin release; (2) the greater the protein content of the meal, the more peptide fragments are formed, which directly stimulate the release of gastrin.

Intestinal Phase Like gastric motility, gastric secretion is inhibited by reflexes initiated in the duodenum. Thus distension, hypertonic solutions, fatty acids, amino acids, and acid in the duodenum all inhibit gastric secretion. The pathways (Fig. 19-16) are the same as those inhibiting motility: the internal and external nerves and the duodenal hormones secretin and CCK.[2] Thus, during the intestinal phase of gastric activity, as chyme passes into the duodenum it initiates reflexes which decrease acid secretion and gastric emptying, allowing time for digestion and absorption in the intestine.

Ulcers Considering the high concentration of acid and proteolytic enzymes secreted by the stomach, it is natural to wonder why the stomach does not digest

[2] CCK inhibits gastric secretion, in spite of its chemical similarity to gastrin, because when it is bound to the receptor sites for acid secretion in the stomach it prevents the more potent hormone, gastrin, from combining with the same receptor sites.

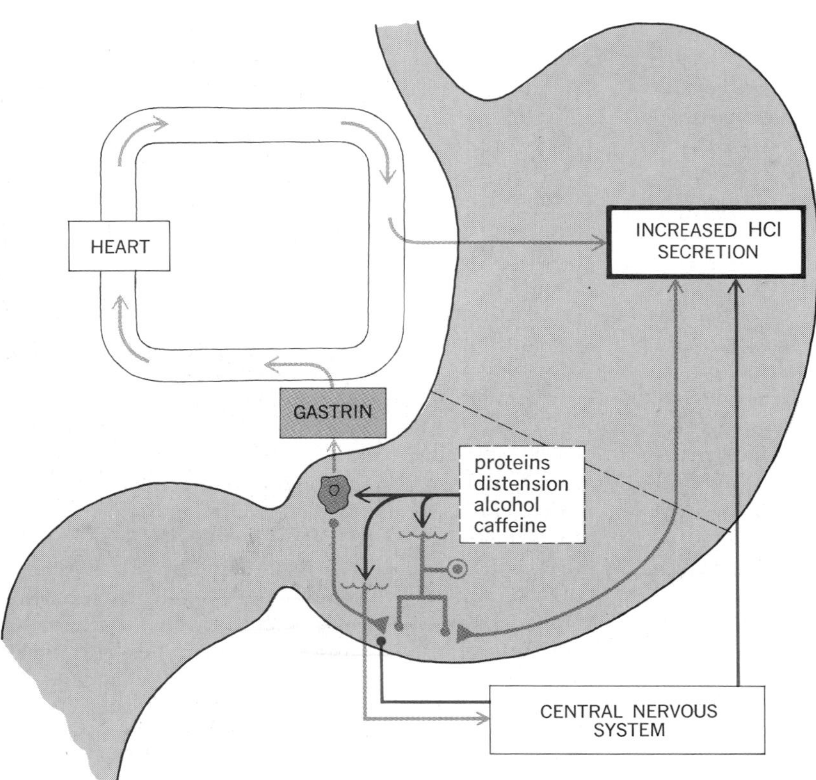

FIGURE 19-15 ***Pathways regulating increased acid secretion in the stomach.***

itself. Several factors protect the walls of the stomach and duodenum. The surface of the mucosa is lined with cells which secrete a slightly alkaline mucus, which in human beings forms a layer of 1.0 to 1.5 mm thick over the stomach surface. The protein content of mucus and its alkalinity tend to neutralize hydrogen ions in the immediate area of the epithelial cell layer, thus forming a chemical barrier between the highly acid contents of the lumen and the cell surface. In addition, the cell membranes lining the stomach have a very low permeability to hydrogen ions, preventing their entry into the underlying mucosa. Moreover, the lateral surfaces of the epithelial cells near the lumen are joined together by tight junctions so that there is no extracellular passage between the cells by which material could diffuse from the lumen into the mucosa. Finally, the epithelial mucous cells lining the walls of the stomach are continually being replaced over a period of 1 to 3 days through cell division. The mucous layer, the cell membranes, and cell replacement all contribute to maintaining a barrier between the contents of the lumen and the underlying tissues.

Yet, in some people these protective mechanisms are inadequate, and erosions (*ulcers*) of the gastric or duodenal mucosa occur. If severe enough, the ulcer may damage the underlying blood vessels and cause bleeding into the lumen. On occasion, the ulcer may penetrate the entire wall, with the leakage of luminal contents into the abdominal cavity. About 10 percent of the population of the United States are found at autopsy to have ulcers, which are about 10 times more frequent in the walls of the duodenum than in the stomach itself.

What causes the breakdown of this barrier and the formation of ulcers is unknown. In one unusual disease, huge quantities of acid are continually secreted by the stomach, and the patient has multiple ulcers in the esophagus and duodenum. The high acid secretion appears to be the result of a tumor in the pancreas which produces a gastrinlike hormone. However, many ulcer patients are found to have normal or below-normal

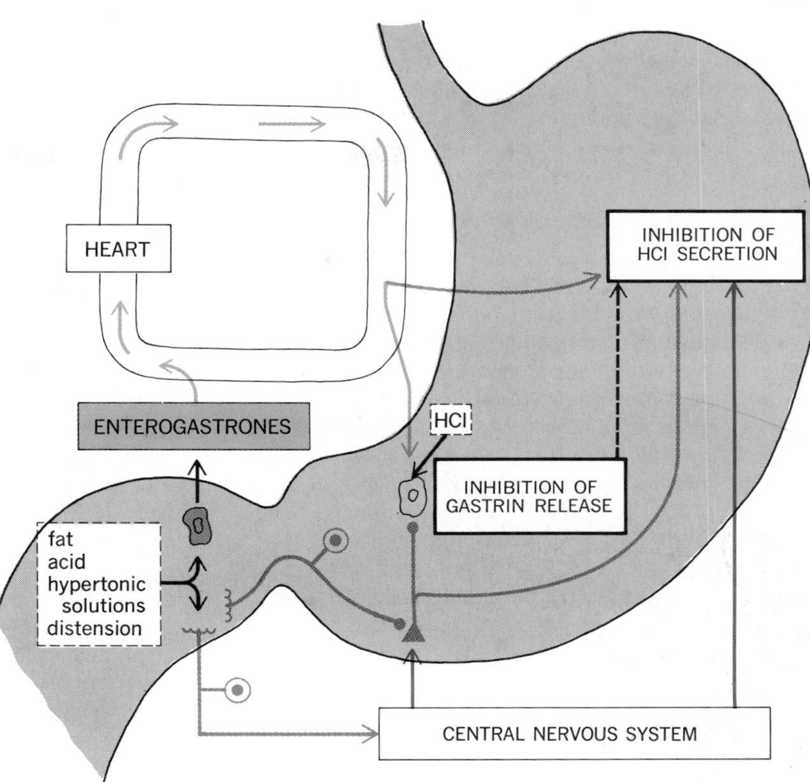

FIGURE 19-16 ***Pathways inhibiting HCl secretion. All except the local effect of HCl on gastrin release are involved in the intestinal phase of gastric secretion. Inhibition of gastrin release decreases HCl secretion.***

rates of acid secretion. A genetic factor appears to contribute to ulcer formation, since ulcers are often found in many members of the same family. Emotional stress and worry have often been implicated as contributing factors. Emotions are known to influence gastric motility and secretion by way of the autonomic nerves, and cutting the vagus nerves contributes to the relief of some, but not all, ulcer patients.

The pain associated with ulcers probably is the result of irritation of exposed nerve fibers and smooth muscle cells in the region of the ulcer wound as well as increased tension in the walls resulting from the contractile spasms of the smooth muscles irritated by acid. The pain from a duodenal ulcer often subsides following a meal, which buffers the acid secreted by the stomach and inhibits the emptying of acid from the stomach into the duodenum. The pain is most intense during the early hours of the morning, when unbuffered acid is entering the duodenum.

Absorption by the Stomach Very little food is absorbed from the stomach into the blood. There are no special transport systems for salts, amino acids, and sugars in the stomach walls like those found in the intestine, and most of the digestion products in the stomach are large, highly charged, and ionized so that they cannot diffuse across cell membranes. The lipids in the food reaching the stomach are not soluble in water and tend to separate into large droplets which do not mix with the acid contents of the stomach. Since little of this lipid comes into contact with the membranes lining the stomach, little is absorbed.

Several classes of molecules, however, can be absorbed directly by the stomach. A prime example is ethyl alcohol, CH_3CH_2OH. Alcohol is water-soluble, but since it is not ionized, it also has some degree of lipid solubility, allowing it to diffuse across lipid membranes and reach the bloodstream through the walls of the stomach. Although alcohol is absorbed from the stomach, it is absorbed more rapidly from the intestinal tract, which has a greater membrane surface area available for absorption. Accordingly, drinking a glass of milk before a cocktail party or eating cheese dip and high-fat hors d'oeuvres inhibits the rate of gastric emptying through the reflexes from the duodenum and slows down the rate of alcohol absorption but does not stop it.

Weak acids, most notably acetylsalicyclic acid (aspirin), are also absorbed directly across the walls of the stomach. A weak acid is fully ionized in solutions of low hydrogen-ion concentration and in its ionized form is unable to diffuse through cell membranes. In the highly acid environment of the stomach, however, a weak acid is almost totally converted to its un-ionized form, which is lipid-soluble and can cross the cell membrane. Once in the low-acid environment of the cell, the weak acid again ionizes, liberating a hydrogen ion, which tends to make the cell interior more acid. Therefore, if enough weak acid enters in a short time, the intracellular acidity may rise sufficiently to damage the cell. Although most people show few ill effects from the normal dosage of aspirin, small amounts of blood can be found in the stomachs of most individuals after taking aspirin, and in some individuals it can produce severe hemorrhaging. Since alcohol increases acid secretion, the combination of aspirin and alcohol increases the damage to the gastric mucosa.

THE PANCREAS AND PANCREATIC SECRETIONS

The *pancreas* is a soft, lobulated organ (Fig. 19-17) lying along the posterior abdominal wall behind the stomach. The pancreas is a mixed gland containing both endocrine and exocrine portions. The endocrine cells secrete the hormones insulin and glucagon into the blood (they will be discussed in Chap. 20). The exocrine portion of the pancreas secretes two solutions involved in the digestive process, one containing a high concentration of sodium bicarbonate and the other containing a large number of digestive enzymes. These solutions are secreted into ducts from the individual lobules, which converge into a single *main pancreatic duct* running from left to right through the pancreas (Fig. 19-17). Before emptying into the descending part of the duodenum, the main pancreatic duct joins the bile duct from the liver. The enzyme secretions of the pancreas are released from the *acinar cells* at the base of the exocrine glands. These cells have a high density of granular endoplasmic reticulum and many zymogen granules, indicating a high capacity for synthesizing and secreting proteins. The bicarbonate solution appears to be secreted by the cells lining the early portions of the ducts leading from the acinar cells.

Bicarbonate Secretion

The concentration of sodium bicarbonate in the pancreatic secretions rises with the rate of secretion and may approach values of 150 m*M* at maximal rates of secretion, 6 to 7 ml/min. Since the concentration of bicarbonate in the blood is about 27 m*M*, the secretion of bicarbonate by the pancreas is an active process. During the course of a day, 1.5 to 2.0 l of solution is secreted into the duodenum, most of which is reabsorbed.

The mechanism of bicarbonate secretion may be similar to the process of hydrochloric acid secretion by the stomach, the crucial difference being a reverse orientation of the transport systems, such that the bicarbonate ions are released into the lumen rather than the blood. The pancreas, like the stomach, contains a high concentration of carbonic anhydrase, and inhibition of this enzyme reduces bicarbonate secretion, as it does hydrogen-ion secretion in the stomach.

Secretion of an alkaline solution of bicarbonate ions necessities the formation of an equivalent amount of acid, and the blood leaving the pancreas is therefore more acid than the blood entering it. Accordingly, loss of large quantities of bicarbonate ions from the intestinal tract during periods of prolonged diarrhea leads to acidification of the blood, just as loss of acid from the stomach by vomiting leads to alkalinization of the blood. Normally there is no change in the acidity of the blood since the increase in blood bicarbonate by the stomach is balanced by the increase in blood acidity in the pancreas, and the acid secreted by the stomach is neutralized by the bicarbonate secreted by the pancreas (Fig. 19-18).

When a person takes a bicarbonate solution to relieve an upset stomach, the bicarbonate ions neutralize the acid in the stomach just as bicarbonate ions secreted by the pancreas neutralize acid in the duodenum. However, neutralization of the acid in the stomach actually leads to an *increase* in acid secretion by the stomach since it removes the inhibition of gastrin release. Provided adequate bicarbonate is ingested, the increased acid secreted is still neutralized by the bicarbonate. Since, as we shall see, acid in the duodenum is a major stimulant for the pancreatic secretion of bicarbonate, there is little pancreatic secretion and the bicarbonate formed during acid secretion in the stomach ac-

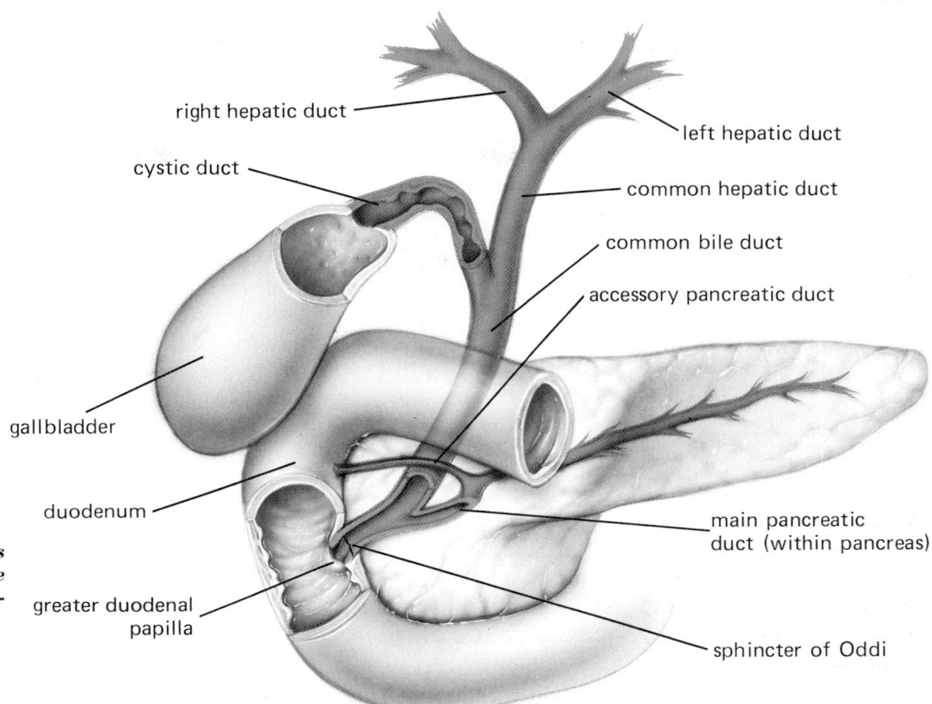

FIGURE 19-17 ***Pancreas and its connections to the common bile duct and intestine.***

cumulates in the blood. In this case the acid-base balance of the body is restored by the elimination of the excess bicarbonate ions through the kidneys.

Enzyme Secretion

Most of the enzymes which digest triglycerides, polysaccharides, and proteins to fatty acids, sugars, and amino acids are secreted by the pancreas. A partial list of these enzymes and their activity is given in Table 19-1. Many are secreted in an inactive form, similar to the secretion of pepsinogen by the stomach, and are then activated by ions and other enzymes in the duodenum. The secretion of enzymes in an inactive form is one mechanism for preventing these potent enzymes from digesting the cells in which they are formed. Over short periods of time, the relative proportions of the enzymes secreted by the pancreas do not vary with the composition of the diet. However, there is some evidence to suggest that if the diet contains a high proportion of protein for a period of weeks or months, the relative proportions of the proteolytic enzymes in the pancreatic secretion are increased. The mechanisms responsible for these changes are unknown.

TABLE 19-1
PANCREATIC ENZYMES

Enzyme	*Substrate*	*Action*
Trypsin, chymotrypsin	Proteins	Breaks amino acid bonds in the interior of proteins, forming peptide fragments.
Carboxypeptidase	Proteins	Splits off terminal amino acid from end of protein containing a free carboxyl group.
Lipase	Lipids (triglycerides)	Splits off fatty acids from positions 1 and 3 of triglycerides, forming free fatty acids and 2-monoglycerides.
Amylase	Polysaccharides	Similar to salivary amylase: splits polysaccharides into a mixture of glucose and maltose.
Ribonuclease, deoxyribonuclease	RNA, DNA	Splits nucleic acids into free mononucleotides.

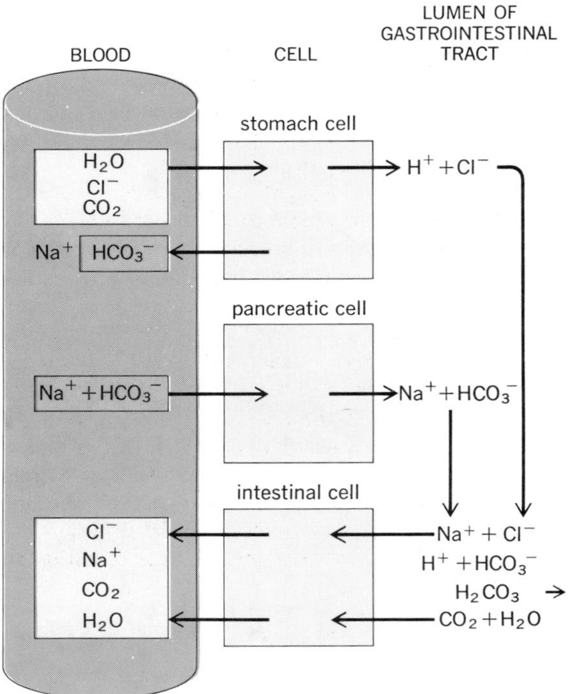

FIGURE 19-18 ***Overall pathway of HCl production in the stomach, neutralization by sodium bicarbonate secreted by the pancreas, and reabsorption in the small intestine of the products formed by the neutralization. Note that there is no net gain or loss of any substance.***

Control of Pancreatic Secretion

The exocrine secretions of the pancreas are controlled by the autonomic nerves to the pancreas (primarily the vagus nerve) and the three gastrointestinal hormones: secretin, CCK, and gastrin. As was true for gastric secretion, there is a cephalic phase to pancreatic secretion during which the sight and smell of food or its presence in the mouth produces increased pancreatic secretion mediated by the vagus nerve. This is primarily an enzymatic secretion containing little bicarbonate. The gastric phase of pancreatic secretion is mediated by the release of gastrin which stimulates primarily bicarbonate secretion. The major portion of pancreatic secretion occurs during the intestinal phase and is mediated by the release of secretin and CCK in response to chyme in the intestine.

Recall that the stimulus for secretin release is the presence of acid in the duodenum. Secretin[3] elicits a marked increase in the amount of bicarbonate and the volume of fluid secreted by the pancreas but only a slight stimulation of enzymatic secretion. Thus, the action of secretin is primarily on the duct cells of the pancreas which secrete bicarbonate. The bicarbonate ions neutralize the acid entering the intestine from the stomach; it is therefore appropriate that the most potent stimulus for bicarbonate secretion is secretin, which is released by acid in the duodenum, thus providing a negative-feedback control system to maintain the neutrality of the intestinal contents.

The second duodenal hormone, CCK,[4] produces a marked increase in pancreatic enzyme secretion but little increase in bicarbonate secretion. Recall that the stimulus for CCK release is the presence of the organic components in chyme—amino acids and fatty acids—rather than acid, and it is thus appropriate that CCK stimulates primarily enzyme secretion leading to the digestion of fat and protein.

In spite of the multiple number of hormones, nerves, and varieties of interactions between the stomach, duodenum, and pancreas, the overall adaptive significance of these interactions should be reemphasized. As chyme empties from the stomach into the duodenum, its content of acid, amino acids, and fatty acids stimulates the release of secretin and CCK from the wall of the duodenum; these hormones stimulate the secretion of bicarbonate and enzymes from the pancreas. The pancreatic enzymes act upon the large molecules in the chyme to produce amino acids, sugars, and fatty acids which can be absorbed and also maintain the chemical stimuli responsible for hormone release until such time as the nutrients are absorbed. The bicarbonate from the pancreas neutralizes the acid from the stomach, preventing ulcerative damage to the walls of the intestine, and also provides an environment in which the enzymes from the pancreas, which are inactive in a highly acid environment, can be ac-

[3] Secretin has the honor of being the first hormone ever to be discovered. In 1902 Bayliss and Starling, in England, noted that when food was placed in an isolated segment of the duodenum of a dog pancreatic secretion was increased and that extracts of the duodenum also increased secretion, indicating the release of a chemical substance from one tissue (the duodenum) which affected the activity of another tissue (the pancreas). Within a few years the term hormone was introduced into the biological literature to designate such chemical mediators.

[4] Prior to its isolation, the duodenal hormone stimulating the secretion of enzymes by the pancreas was called pancreozymin. However, upon purification it was found to be identical to the hormone cholecystokinin.

tive. Gastrin, which strongly stimulates acid secretion, also stimulates the pancreatic secretion of bicarbonate, assuring neutralization of the acid upon its arrival in the duodenum. The release of secretin and CCK also inhibits gastric secretion and motility, allowing time for the intestinal contents to be neutralized, digested, and absorbed. Figure 19-19 summarizes these pathways controlling pancreatic secretion.

THE LIVER AND SECRETION OF BILE

Bile, which is essential for the digestion of fat in the small intestine, is secreted by the liver. In addition to bile production, the cells of the liver (*hepatic cells*) play a role in iron metabolism, plasma-protein production, detoxication of substances circulating in the plasma, metabolism of breakdown products of hemoglobin, and numerous other biochemical pathways, some of which will be discussed in Chap. 20.

The liver is the largest gland in the body, weighing from 1 to 2.5 kg. It lies just under the diaphragm in the upper and upper right regions of the abdominal cavity. The liver comprises two lobes, the right lobe being much the larger and accounting for about five-sixths of the size of the total organ.

The blood supply to the liver is unusual in that it comes from two sources: the *hepatic artery* (a branch of the celiac trunk) and the *portal vein*, which receives blood from the stomach, intestines, pancreas, gallbladder, and spleen before passing to the liver (Fig. 19-20). Thus, blood returning to the heart from the abdominal portion of the digestive system first passes through the liver, giving the liver cells first crack at the foodstuffs absorbed into the bloodstream, a concept that will be important in the discussions in Chap. 20. (In contrast, fats absorbed into the lymph vessels in the intestinal wall do not pass directly to the liver but are released into the general circulation via the thoracic duct.)

The portal vein empties into sinusoids within the

FIGURE 19-19 ***Pathways regulating pancreatic secretion.***

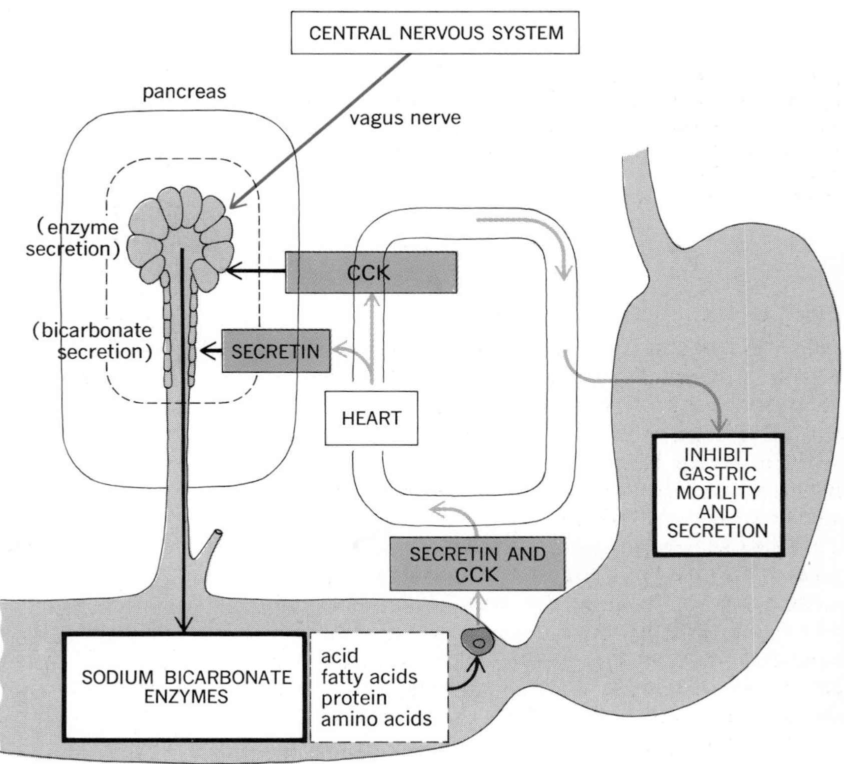

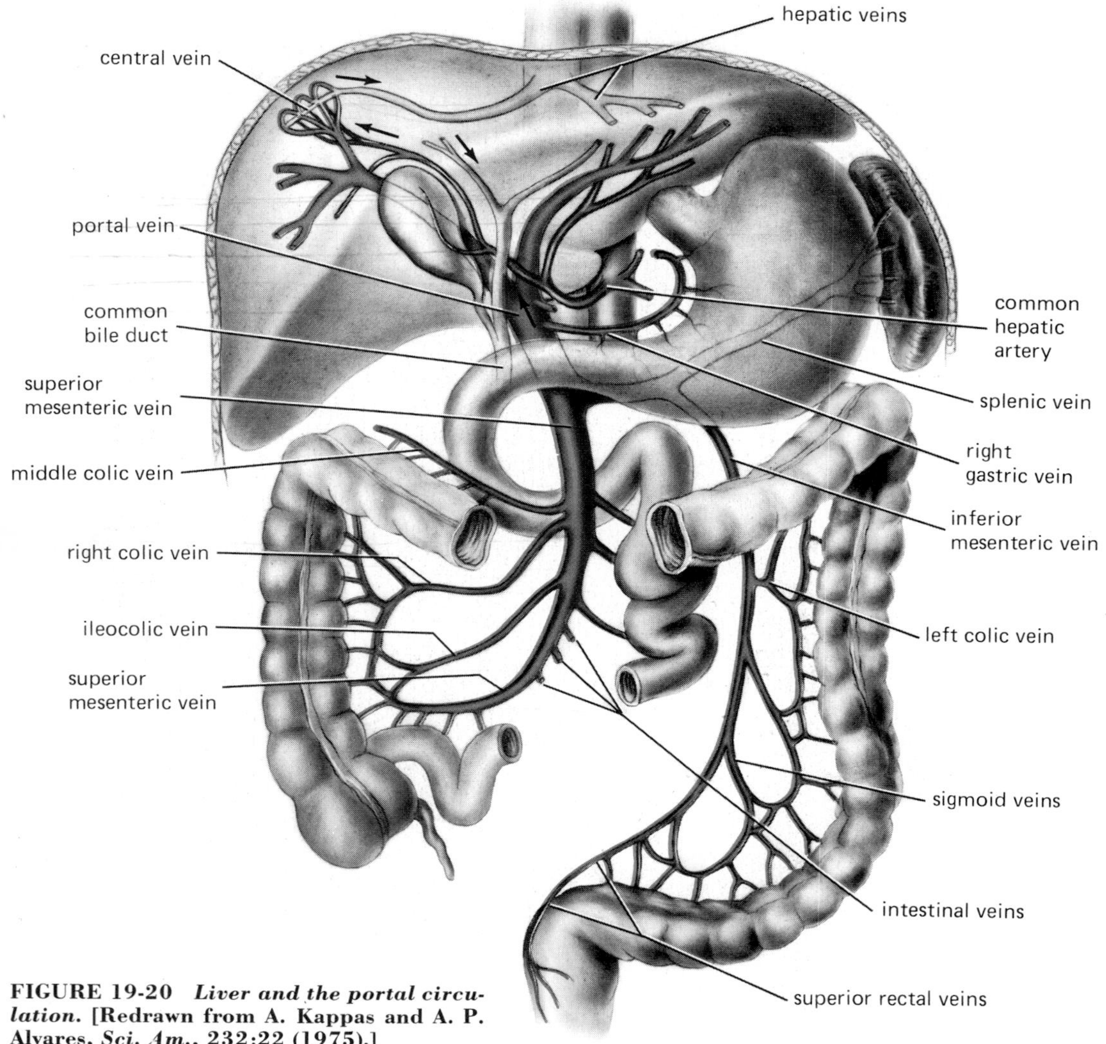

FIGURE 19-20 ***Liver and the portal circulation.*** **[Redrawn from A. Kappas and A. P. Alvares, *Sci. Am.*, 232:22 (1975).]**

liver (Fig. 19-21) which carry the blood from the portal vein to the *hepatic veins,* which in turn drain the blood from the liver into the inferior vena cava. The hepatic arteries also feed into the sinusoids, which therefore carry mixed venous and arterial blood (Fig. 19-21). It is this blood which nourishes the hepatic cells. The liver sinusoids are wider than ordinary capillaries, but they are capillarylike in that their endothelial lining allows the passage of substances from the blood to the interstitium. The sinusoid lining also contains phagocytic cells (the *hepatic macrophages,* or Kupffer cells).

The basic unit of the liver substance is the *hepatic lobule*. In the center of the lobule is the *central vein,* a small tributary of the hepatic vein (Fig. 19-21). The hepatic cells are arranged in sheets one cell thick that radiate out from the central vein to the periphery of the lobule to meet the *portal triad,* a cluster of three tubes consisting of a branch of the hepatic artery, a branch of the portal vein, and a small bile duct, all enclosed in a connective-tissue sheath. One surface of the liver cells faces the sinusoids; the other surfaces faces small *bile canaliculi,* which drain the bile produced by the hepatic

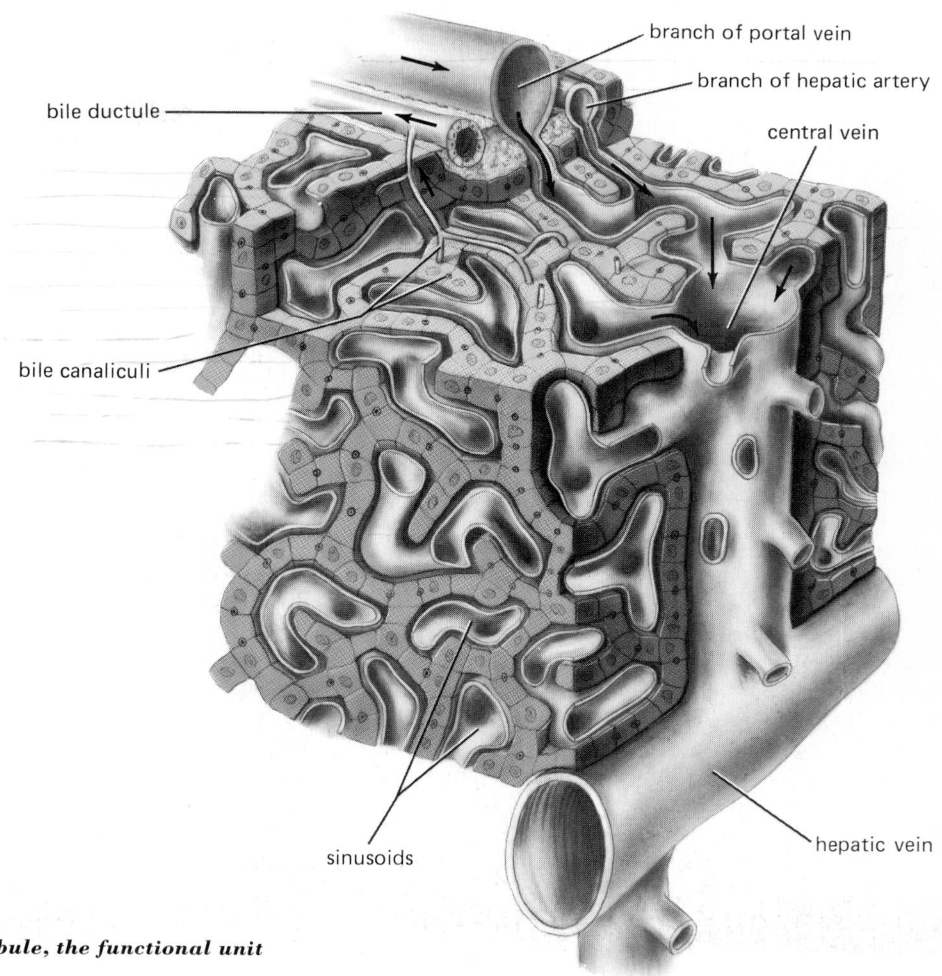

FIGURE 19-21 ***Liver lobule, the functional unit of liver tissue.***

cells into larger bile ducts (Fig. 19-21). Blood-borne materials have free access to the hepatic cells but are prevented from passing directly into the bile canaliculi by tight junctions between the cells forming their boundaries. Material does pass from the hepatic cells into the canaliculi.

The many bile ducts converge to form the common bile duct which finally empties into the duodenum (Figs. 19-17 and 19-20), often at the same site at which the pancreatic duct enters the duodenum. In human beings, a small sac branches off from the bile duct on the underside of the liver, forming the *gallbladder*. The gallbladder stores, concentrates, and releases bile into the intestinal tract. The gallbladder can be surgically removed without impairing the function of bile in the intestinal tract; in fact many animals which secrete bile do not even have a gallbladder; e.g., the rat and horse.

Bile is continually secreted by the liver, although the rate of secretion may vary, as described in the next section. Surrounding the bile duct at the point where it enters the duodenum is a ring of smooth muscle known as the *sphincter of Oddi*. When this sphincter is closed, the bile secreted by the liver is shunted into the gallbladder. The cells lining the gallbladder actively transport sodium from the bile back into the plasma. As solute is pumped out of the bile, water follows by osmosis. The net result is a five- to tenfold concentration of the organic constitutents of bile.

Bile is essential for the digestion of fat in the small intestine. Fat digestion presents special problems because, as fat is released from the breakdown of food particles in the stomach, the molecules of triglyceride aggregate together, forming large globules of fat that are immiscible with the chyme in the stomach. The function of bile is to break down these large globules into a suspension of very fine droplets about 1 μm in diameter, a process known as *emulsification*. This emulsion of fat can then be digested and absorbed. We shall discuss the details of this process in the section on the absorption of fat later in the chapter.

Bile consists of a salt solution containing four primary ingredients: (1) bile salts, (2) cholesterol, (3) lecithin (a phospholipid), and (4) bile pigments. The first three of these are involved in the emulsification of fat in the small intestine. The fourth, bile pigments, is one of the few substances that are normally excreted from the body through the intestinal tract. The major bile pigment is bilirubin, which is a breakdown product of hemoglobin (Chap. 13). The cells of the liver extract bilirubin from the circulation and secrete it into the bile by an active process. Bile pigments are yellow and give bile its golden color. The color of these pigments is modified in the intestinal tract by the digestive enzymes, and the mixture of these pigments gives feces their brown color. In the absence of bile secretion, feces are grayish white. Some of the pigments are reabsorbed during their passage through the intestinal tract and are eventually excreted in the urine, giving urine its yellow color.

Cholesterol, which is normally insoluble in water, is found in very high concentration in the bile. This is made possible by the presence of the bile salts and lecithin which solubilize cholesterol just as they solubilize fat in the intestine. If insufficient bile salts or lecithin is present in the bile, or if there is excessive cholesterol, the cholesterol precipitates out of solution, forming a *gallstone*. In human beings the concentration of bile salts, lecithin, and cholesterol in the bile is normally very near the point at which cholesterol precipitates out of solution. Slight changes in the concentrations of any of these may precipitate cholesterol, especially during the concentrating process in the gallbladder. If the gallstone is small, it may pass through the bile duct into the intestine with no complications. A larger stone may become lodged in the neck of the gallbladder, causing contractile spasms of the smooth muscle and pain. A more serious complication arises when the gallstone lodges in the bile duct, thereby preventing bile from entering the intestine and resulting in a failure to digest and absorb fat. In addition, a buildup of pressure in the blocked duct prevents further secretion of bile, and the bile pigments accumulate in the blood and tissues, giving them the deep yellow color known as *jaundice*. It is also possible for a gallstone to lodge at the junction of the bile and pancreatic duct in the duodenum, blocking the entry of both bile and pancreatic juices. Under these conditions, little digestion or absorption of any substance occurs in the intestinal tract.

From the standpoint of gastrointestinal function, the bile salts are the most important components of bile since they are involved in the digestion and absorption of fats. The total amount of bile salt in the body is about 3.6 g; yet during the digestion of a single fatty meal, as much as 4 to 8 g of bile salts may be emptied into the duodenum. This is possible because most of the bile salts entering the intestinal tract are reabsorbed in the lower part of the small intestine and returned to the liver via the portal vein, where they are again secreted into the bile. This circulatory pathway from the intestines to the liver and back to the intestines provides the route generally known as the *enterohepatic circulation* (Fig. 19-22), by which such substances as the bile salts can be recycled through the intestinal tract.

Control of Bile Secretion

Bile is secreted at the rate of 250 to 1,000 ml/day. Two components of bile secretion are under separate controls: the secretion of bile salts and the secretion of the isotonic fluid containing sodium, chloride, and bicarbonate in which the bile salts and other organic constituents are dissolved. The rate of bile-salt secretion by the liver is primarily determined by the concentration of the bile salts in the plasma; this concentration rises during a meal as a result of bile-salt reabsorption from the intestinal tract. Between meals, when there is little bile salt in the intestinal tract to be absorbed, the rate of secretion of bile salt by the liver is low.

In contrast, the secretion of fluid (with little increase in the rate of bile-salt secretion) is increased by secretin, cholecystokinin, and gastrin (as well as by the vagus nerves), secretin being the most active in this regard. Thus all the factors discussed previously which

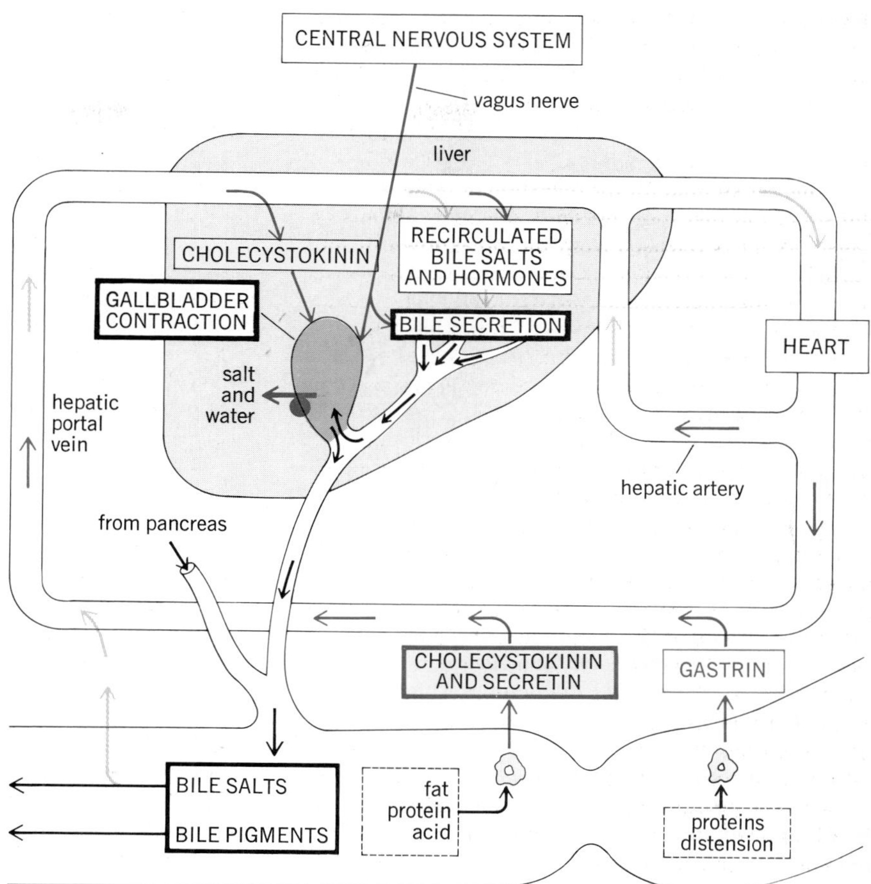

FIGURE 19-22 ***Pathways regulating bile secretion and release from the gallbladder.***

cause the release of these gastrointestinal hormones affect bile secretion.

Shortly after a meal, especially if it contains fat, the sphincter of Oddi relaxes and the gallbladder contracts, discharging concentrated bile into the duodenum. This response is mediated by both the vagus nerves and the three hormones, cholecystokinin being the most active. (It is from this ability to cause contraction of the gallbladder that cholecystokinin first received its name: *chole,* bile, *kystis,* bladder; *kinin,* to move.) The net result of these neural and hormonal influences is to assure the secretion and discharge of bile into the intestinal tract when food is present, the amount of bile discharge depending on the composition and amount of food entering the duodenum. Figure 19-22 summarizes the factors controlling the release of bile.

We have now discussed the major activities of the three gastrointestinal hormones, secretin, cholecystokinin, and gastrin. A summary of their activities is provided in Table 19-2.

THE SMALL INTESTINE

The small intestine consists of about 5 m of tubing, coiled in the central and lower parts of the abdominal cavity, leading from the stomach to the large intestine. It is encircled by the large intestine (Fig. 19-1) and separated from the anterior abdominal wall by the greater omentum. The first section of the small intestine is the *duodenum,* a short section which forms an incomplete circle around the head of the pancreas (Fig. 19-1). The *hepatopancreatic ampulla* (combined bile-pancreatic

TABLE 19-2
ACTIVITIES OF THE GASTROINTESTINAL HORMONES

	Secretin	*Cholecystokinin*	*Gastrin*
Secreted by:	Duodenum	Duodenum	Antrum of stomach
Primary stimulus for hormone release	Acid in duodenum	Amino acids and fatty acids in duodenum	Peptides in stomach Parasympathetic nerves to stomach
Effect on:			
Gastric motility	Inhibits	Inhibits	Stimulates
Gastric HCl secretion	Inhibits	Inhibits	STIMULATES †
Pancreatic secretion			
Bicarbonate	STIMULATES	Stimulates	Stimulates
Enzymes	Stimulates	STIMULATES	Stimulates
Bile secretion of bicarbonate	STIMULATES	Stimulates	Stimulates
Gallbladder contraction	Stimulates	STIMULATES	Stimulates

† STIMULATES denotes that this hormone is quantitatively more important than the other two.

duct) enters the duodenum at a small elevation, the *major duodenal papilla*. The remainder of the small intestine, comprising the *jejunum* and *ileum*, is longer and more highly coiled than the duodenum. Although there is a gradual change in morphology from the beginning of the jejunum to the end of the ileum, there is no sharp distinction between them and the division is somewhat arbitrary. The jejunum occupies approximately two-fifths of the length of the combined jejunal-ileal segment (i.e., about 2 m), and the ileum accounts for the remaining 3 m. It is in the small intestine that most digestion and absorption occur.

Motility

When the motion of the intestinal tract is observed by x-ray fluoroscopic examination, the contents of the lumen are seen to move back and forth with little apparent net movement toward the large intestine. The net flow of material toward the large intestine is normally quite slow; chyme begins to enter the large intestine at about the time that chyme from the next meal is entering the duodenum from the stomach. In contrast to the waves of peristaltic contraction that sweep over the surface of the stomach, the primary motion of the small intestine is an oscillating contraction and relaxation of the smooth muscle in the intestinal wall, known as *segmentation*. Rings of smooth muscle contract and relax at intervals along the intestine, dividing the chyme into segments (Fig. 19-23). The chyme in the lumen of a contracting segment is forced both up and down the intestine. This rhythmic contraction and relaxation of the intestinal segments produce a continuous division and subdivision of the intestinal contents which thoroughly mixes the chyme in the lumen and brings it into contact with the intestinal wall where absorption can occur.

These movements of the small intestine are initiated by electric activity generated by pacemaker cells located in the longitudinal smooth muscle. Just as in the stomach, these pacemaker cells produce a basic elctrical rhythm that is conducted through the gap junctions between smooth muscle cells. The frequency of segmentation follows the frequency of the basic electrical rhythm. The primary factor responsible for moving the chyme along the intestine is a gradient in the frequency of segmentation along the length of the intestine. Segmentation in the duodenum occurs at a frequency of about 12 contractions per minute whereas in the terminal portion of the ileum segmentation occurs at a rate of only 9 contractions per minute. Since the frequency of contraction is greater in the upper portion of the small intestine than in the lower, on the average, more chyme is forced downward than is forced upward. The gradient in segmentation frequency results from a sequence of pacemaker regions along the intestine. Each successive pacemaker has a slightly lower frequency than the one above.

The degree of contractile activity in the intestine

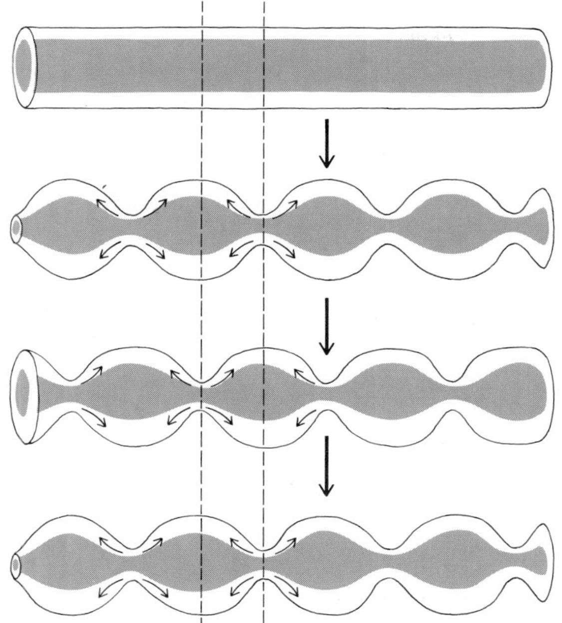

FIGURE 19-23 ***Segmentation movements of the small intestine. The small arrows indicate the movements of the luminal contents.***

can be altered by a number of factors which initiate reflexes in the internal nerve plexus or through external nerves and hormones. These reflexes produce changes in the intensity of the smooth muscle contraction and the rate of propagation of the basic electrical rhythm but do not change the natural frequency of the pacemaker regions. Local distension of the intestine produces a characteristic response. The distended portion contracts and the region just ahead (toward the large intestine) relaxes. The contracted segment then progresses several centimeters down the intestine, producing a short peristaltic wave. The coordination of this response appears to be mediated by the internal nerve plexus. These short peristaltic waves always proceed in the direction of the large intestine (a property known as the law of the intestine); this accounts for some of the net propulsion of chyme along the intestine. However, these peristaltic waves die out after traveling a very short distance; peristaltic waves which travel the entire length of the small intestine do not occur in human beings except under abnormal conditions.

Parasympathetic stimulation of the intestine increases contractile activity, and sympathetic stimulation decreases it. A person's emotional state can affect the contractile activity of the intestine and thus the rate of propulsion of chyme and the time available for digestion and absorption. Fear tends to decrease motility whereas hostility increases it, although these responses vary greatly in different individuals.

The contractile activity of an empty intestine is weak. Following a meal, distension increases the intensity of contractions but the actual rate of propulsion is decreased because the contracting segments narrow the lumen of the intestine, producing a greater resistance to flow. These changes in contractile activity are mediated by local reflexes, such as that produced by distending the lumen with chyme, and by external nerves to the intestine. In human beings, contractile activity in the ileum also increases during periods of gastric emptying. This is known as the *gastroileal reflex* (conversely, distension of the ileum produces decreased gastric motility, the *ileogastric reflex*). Large distensions of the intestine, injury to the intestinal wall, and various bacterial infections in the intestine lead to a complete cessation of motor activity, the *intestino-intestinal reflex*. All these later reflexes appear to be mediated primarily by external nerves.

Structure of the Intestinal Mucosa

The mucosa of the small intestine is highly folded (Fig. 19-24), and the surface of these folds is further convoluted by fingerlike projections known as *villi*. The folds (plicae circulares, or valves of Kerkring) are permanent, i.e., they are not flattened when the intestine is distended. The surface of each villus is covered with numerous epithelial cells, and the surface area of each cell is increased by small projections known as *microvilli* (Fig. 19-25). The combination of folded mucosa, villi, and microvilli increases the total surface area of the small intestine available for absorption about 600-fold over that of a flat-surfaced tube of the same length and diameter. The total surface area of the human small intestine has been estimated to be about 185 m^2, or equivalent to the area of a tennis court. The cells of the intestinal epithelium are joined together by tight junctions near their luminal border; thus the only route available for entry from lumen to the blood and lymph is by crossing a cell membrane.

The structure of a villus is illustrated in Fig. 19-24. The center of the villus is occupied by a capillary network which branches from an arteriole and drains into a venule. The blood flow to the small intestine of a

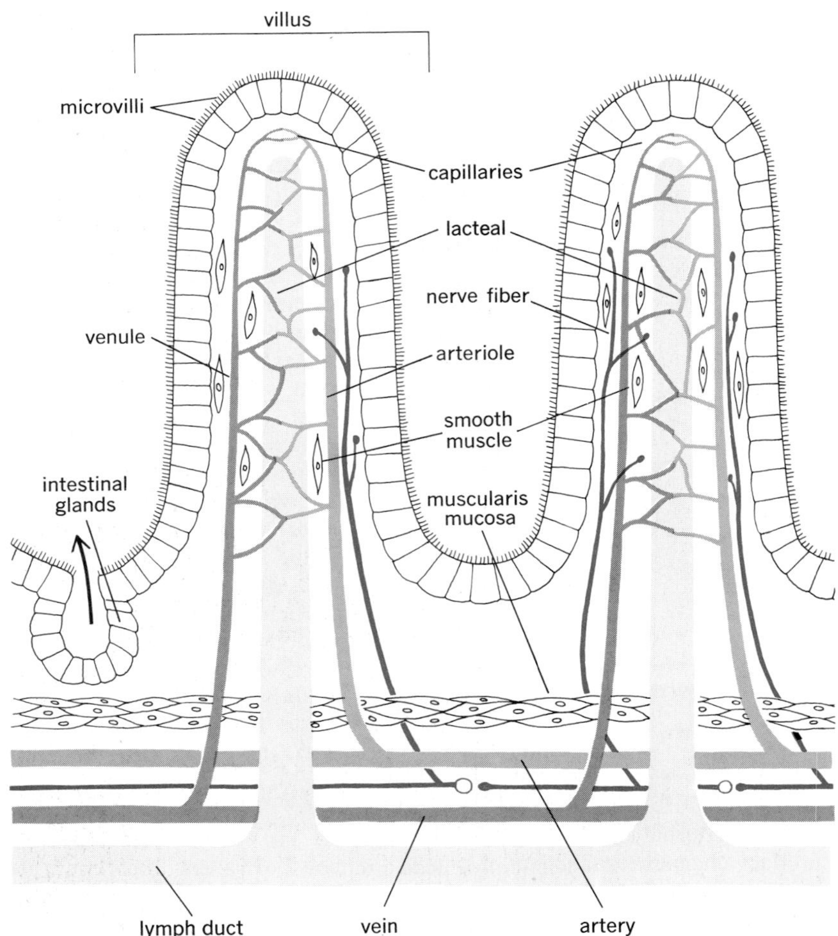

FIGURE 19-24 ***Structure of intestinal villi.***

normal man at rest averages about 1 l/min, or one-fifth the resting cardiac output. Intestinal blood flow increases during periods of digestive activity as a result of local autoregulation produced by the increased metabolic activity of the intestinal cells and by reflexes triggered by mechanical distension of the lumen by chyme. A single blind-ended lymph vessel, a *lacteal,* occupies the center of the villus. Nerve fibers and smooth muscle cells are also present. The epithelial cells at the tip of the villus have a much greater number of microvilli per cell than the cells at the base of the villus. These microvilli contain several enzymes important for digestion.

The villi of the intestine move back and forth independently of each other, presumably as a result of the contraction of the smooth muscle in them. The motion of the villi is increased following a meal, and associated with their increased motion is a greater flow of lymph through the lacteals. The folds of the intestinal mucosa are not permanent structures but can change their pattern with contraction of the underlying smooth muscle layer, the muscularis mucosae. Irritation of the mucosal wall by lumps of food or stimulation of the sympathetic nerves to the intestine causes contraction of the muscularis mucosae and greater mucosal folding.

The cells lining the intestinal tract are continually

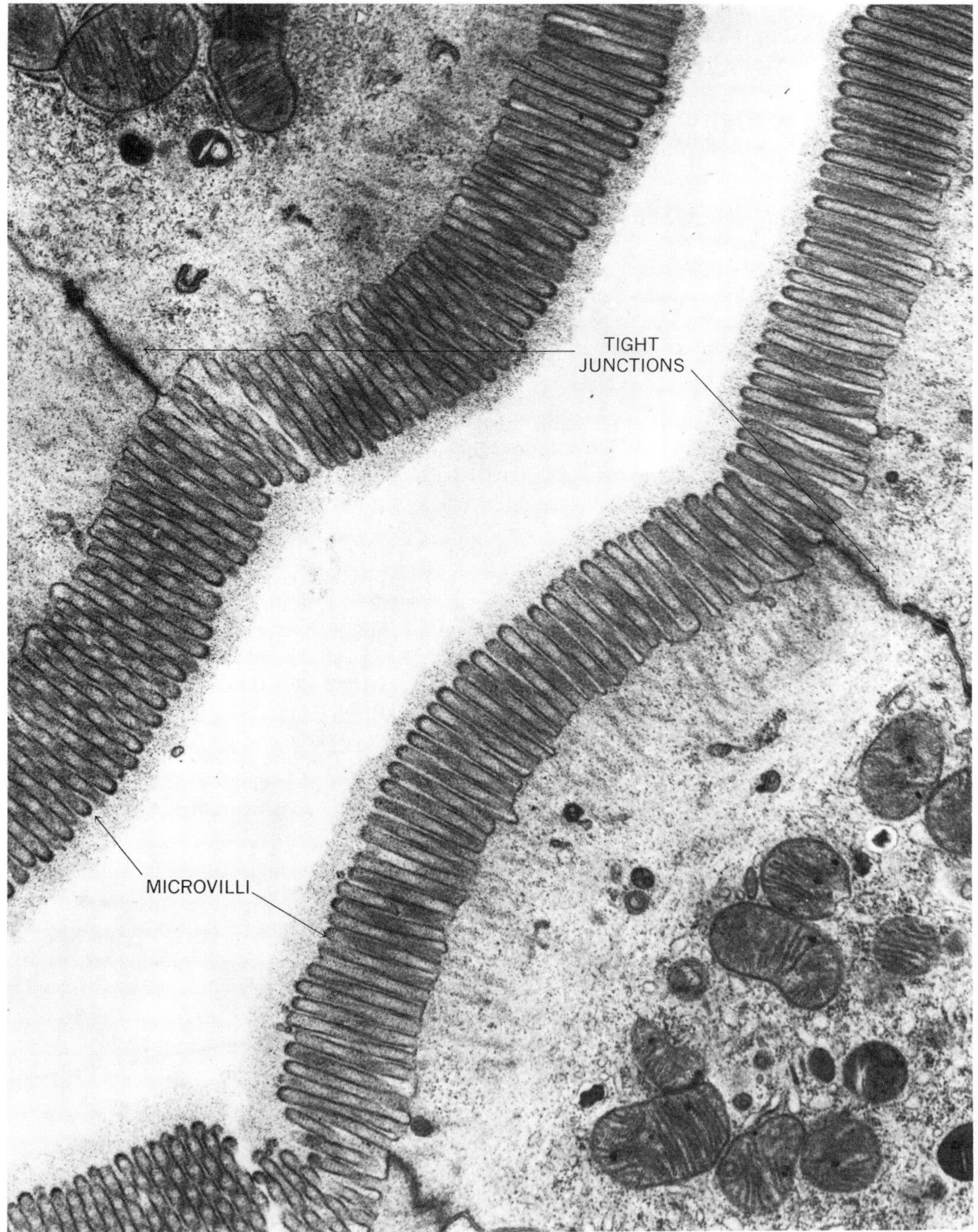

FIGURE 19-25 *Microvilli on the surface of intestinal epithelial cells.* [From D. W. Fawcett, *J. Histochem. Cytochem*, 13:75–91 (1965). Courtesy of Dr. Susumu Ito.]

being replaced as a result of the high mitotic activity of the cells at the base of the villi. The new cells migrate from the base of the villi to the top, replacing older cells, which disintegrate and are discharged into the lumen of the intestine. The entire epithelium of the intestine is replaced every 36 h. The continuous discharge of cells into the lumen amounts to about 250 g of cells every day.

The surface of the intestinal tract, along with the blood-forming regions of the bone marrow, because of their high rate of cell division, is very sensitive to damage by radiation during the period when DNA is being replicated. These areas are the first to be severely damaged by excessive x-ray radiation or atomic radiation.

Secretions

Glands located in the mucosa of the small intestine secrete about 2,000 ml of mucus and salt solutions into the lumen each day. These secretions were once believed to contain digestive enzymes, but it now appears that the enzymes found in the secretions of isolated intestinal segments are derived from the disintegration of the epithelial cells constantly being discharged from the tips of the villi. The true glandular secretions appear to be nearly enzyme-free.

Intestinal secretion rises following a meal, and mechanical stimulation of the intestinal wall enhances secretion. Secretion released in response to acid in the duodenum appears to be the primary stimulus for intestinal secretions. These secretions contain bicarbonate which may help to protect the surface of the intestine from the damaging effects of acid.

TABLE 19-3
TYPICAL FOOD AND WATER INTAKE PER DAY

	Average per day	*Percent solids*
Carbohydrate, g	500	62.5
Protein, g	200	25
Fat, g	80	10
Salt, g	20	2.5
Water, ml	1,200	

Digestion and Absorption

Almost all digestion and absorption of food and water occur in the small intestine. About 50 percent of the small intestine can be removed without interfering with the digestive and absorptive processes. Normally, most of the contents of the small intestine have been absorbed by the time the chyme has reached the middle of the jejunum.

During the course of a day, the average adult consumes about 2 lb of solid food and 2.5 lb of water (Table 19-3), but this is only a fraction of the total material entering the gastrointestinal tract each day. To the approximately 2,000 ml of ingested food and drink is added about 7,000 ml of fluid from the salivary glands, stomach, pancreas, liver, and intestinal tract (Fig. 19-26). Of this total volume of 9,000 ml, only about 500 ml passes into the large intestine each day, 94 percent being absorbed across the walls of the small intestine.

Carbohydrate The daily intake of carbohydrate varies considerably, ranging from 250 to 800 g/day in a typical American diet. Most of the carbohydrate is in

FIGURE 19-26 ***Average amounts of food and fluid ingested, secreted, absorbed, and excreted from the gastrointestinal tract daily.***

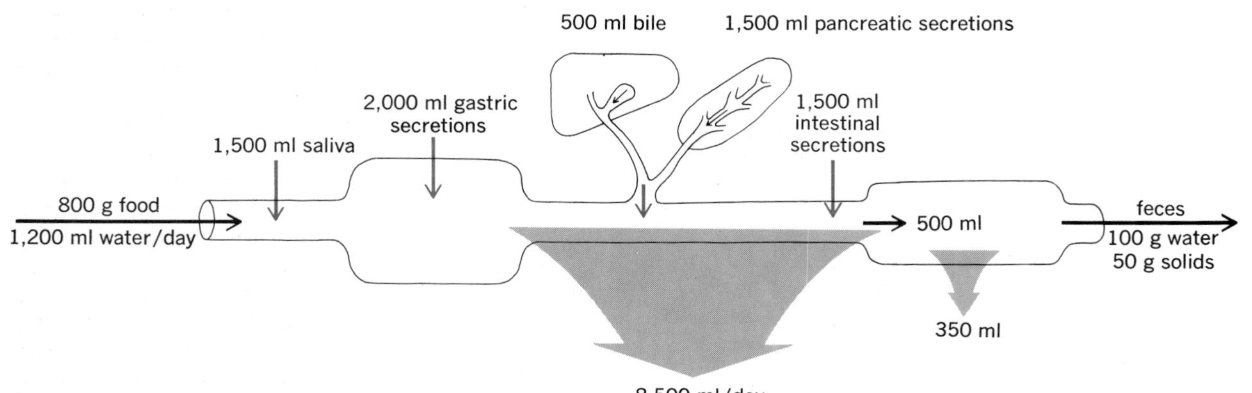

the form of the vegetable polysaccharide, starch. Small amounts of the disaccharides sucrose (glucose-fructose, table sugar) and lactose (glucose-galactose, milk sugar) may be present. In infants, lactose from milk makes up the majority of the carbohydrate in the diet. Only very small amounts of monosaccharides are normally present. The polysaccharide cellulose, which is present in plant cells, cannot be broken down into glucose units by the enzymes secreted by the gastrointestinal tract. Cellulose passes through the small intestine in undigested form and enters the large intestine, where it is partially digested by the bacteria which inhabit this region.

About 50 percent of the starch is digested by salivary amylase during passage through the stomach, and the remainder is digested in the duodenum by pancreatic amylase. Blockage of pancreatic secretions does not appreciably affect the digestion and absorption of carbohydrates because of the presence of salivary amylase. The primary product formed by these enzymes is the disaccharide maltose (glucose-glucose). Enzymes that split disaccharides into monosaccharides are located both in the microvilli of the cells lining the intestinal tract and in the luminal contents as a result of the continuous disintegration of intestinal epithelial cells. The monosaccharides glucose and galactose liberated from the breakdown of polysaccharides and disaccharides are actively transported across the intestinal epithelium into the blood. The presence of sodium in the lumen is required for monosaccharide transport. Apparently sodium and sugar combine with some carrier molecule in the epithelial cell membrane. The six-carbon sugar fructose, however, is not actively absorbed by the intestinal epithelium but probably crosses the membrane by a facilitated-diffusion process. The rapid digestion of polysaccharides and the transport of the resulting monosaccharides into the blood lead to the absorption of most of the dietary carbohydrate by the end of the jejunum.

Some children cannot hydrolyze or absorb lactose (milk sugar). Most of this sugar appears in the feces or is metabolized by bacteria in the large intestine. The production of acids by the fermenting bacteria stimulates motility resulting in diarrhea and dehydration. Because milk constitutes the major diet of infants, they may also become undernourished as a result of their inability to absorb lactose. Lactose intolerance has been shown to be a genetically determined error of metabolism in which the enzyme lactase is not synthesized by the intestinal epithelium.

Protein An intake of about 50 g of protein each day is required by an adult to supply essential amino acids and replace amino acid nitrogen loss in the urine. A typical American diet contains about 200 g of protein. In addition to the protein in the diet, 10 to 30 g of protein is secreted into the gastrointestinal tract by the various glands, and about 25 g of protein is derived from the disintegration of epithelial cells. The feces contain 10 to 20 g of protein, most of which comes from bacteria and disintegrated cells rather than from the diet.

Proteins are broken down to peptide fragments by proteolytic enzymes secreted by the stomach and pancreas. These peptide fragments are further digested to free amino acids by carboxypeptidase from the pancreas and aminopeptidase in the intestinal villi, which split off amino acids from the carboxyl and amino ends of the peptide chains, respectively. The free amino acids are actively transported across the walls of the intestine from lumen to blood.

Several different carrier systems are available for transporting different classes of amino acids. Some of these require sodium, just as in the case of carbohydrate transport. Several genetic diseases have been found in which there is a defective absorption of certain amino acids. These appear to result from the absence of specific carriers or their malfunction, and the same individual may show defective amino acid transport of the kidney tubules as well. In the absence of pancreatic secretion, protein digestion and absorption are decreased about 50 percent.

Fat The amount of fat in the diet varies from about 25 to 160 g/day. Most of it is in the form of triglycerides (neutral fat), the remainder being primarily phospholipids and cholesterol. There are very few free fatty acids in the average diet. Almost all the fat entering the digestive tract is absorbed, and the 3 to 4 g of fat in the feces is derived primarily from the bacteria in the large intestine.

Most fat digestion occurs in the small intestine under the influence of pancreatic lipase. To digest fat, the water-soluble lipase molecule must come into contact with a molecule of triglyceride, but neutral fats are not soluble in water and tend to separate from the water phase into droplets of lipid. In this form they enter the small intestine from the stomach. If the triglycerides remained aggregated in droplets, the only area for lipase action would be the surface of the droplet, which contains only a small fraction of all the molecules in the droplet. Digestion of lipid in this state is very slow.

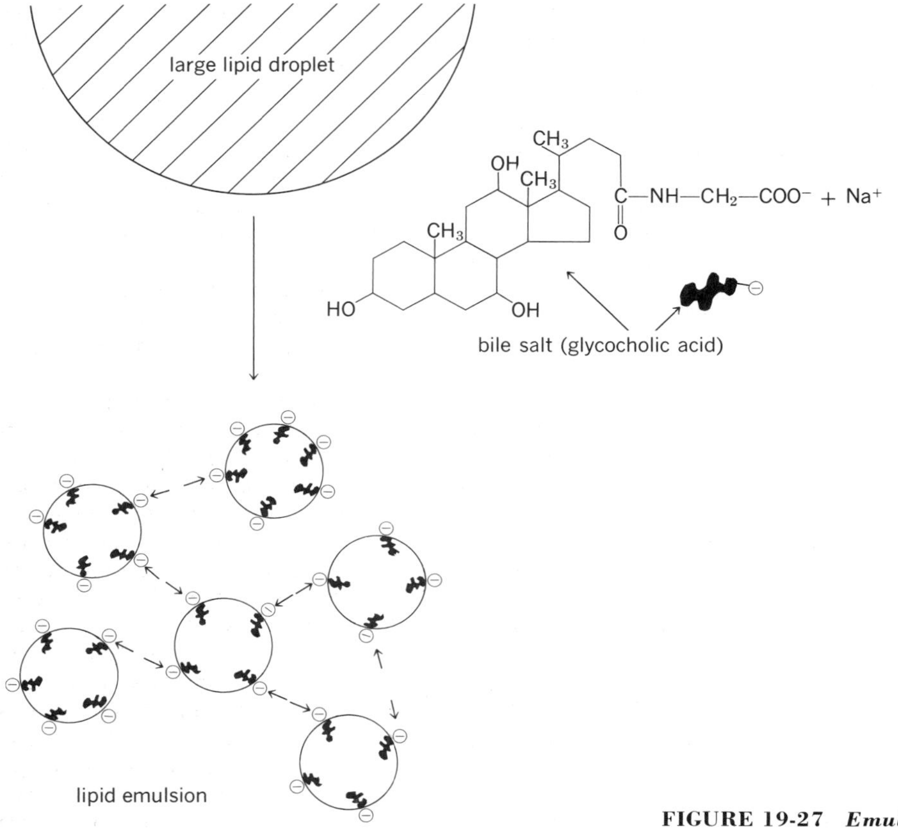

FIGURE 19-27 ***Emulsification of fat by bile salts.***

Bile salts along with the lecithin and cholesterol secreted by the liver speed up lipid digestion by emulsifying fat and increasing the surface area of the lipid droplets accessible to pancreatic lipase. Bile salts consist of a large steroid portion which is lipid-soluble and a small carbon chain containing an ionized carboxyl group (Fig. 19-27). The steroid portion dissolves in the large lipid droplets, leaving the ionized carboxyl groups on the surface, where they interact with polar water molecules. This produces a negative charge on the surface of the lipid droplet. Mechanical agitation in the intestine breaks up the large droplets into smaller ones, and the negatively charged groups on the droplet surface produce an electric repulsion between droplets, preventing them from recoalescing into large drops (Fig. 19-27).

Pancreatic lipase acts on the surface of the small droplets, forming free fatty acids and mixtures of mono- and diglycerides with only small amounts of free glycerol. The free fatty acids and monogylcerides, having polar and nonpolar segments like the bile salts, assist in the emulsification of the lipid droplets.

As digestion proceeds, the bile salts perform a second important function in promoting aggregation of the liberated free fatty acids and monoglycerides into water-soluble particles known as *micelles*. The structure of a micelle is similar to that of the lipid emulsion, only smaller. The nonpolar portions of the fatty acids, monoglycerides, and bile salts are oriented to the center of the micelle, with the polar portions at the surface. A single micelle is only about 30 to 100 Å in diameter and consists of only a few thousand molecules. Whereas a lipid emulsion appears cloudy because of the relatively large size of the emulsion droplets, a solution of micelles is perfectly clear.

Although free fatty acids and monoglycerides are

essentially insoluble in water, a few free molecules can exist in solution, and it is in this form that they are absorbed across the cell membrane by simple diffusion because of their high degree of lipid solubility in the lipid cell membrane. The small amounts of free fatty acids and monoglycerides that are present in solution are in equilibrium with the micelles. As the free components diffuse into the cells, some of the micelles liberate their fatty acids and monoglycerides into solution, thus maintaining a saturated solution with respect to the free fatty acids and monoglycerides. Since their concentration in free solution is very, very low and absorption takes place by diffusion down a concentration gradient, the rate of fat absorption is limited by diffusion. The ability of the micelles to maintain this solution in a saturated state allows fat to be absorbed relatively rapidly. In the absence of micelles, fat absorption is very slow.

If 90 percent of the pancreas is destroyed, the intestine loses little of its ability to digest and absorb fat; but if all the pancreas is destroyed so that there is no lipase, only about one-third of the dietary lipid is absorbed and two-thirds appears in the feces. The little digestion that does occur is probably the result of enzymes from the disintegrating epithelial cells. If, on the other hand, the bile duct is blocked so that bile salts do not enter the intestine to promote emulsification and micelle formation, about 50 percent of ingested fat appears in the feces.

Although fatty acids are the primary form of fat entering the epithelial cells, very little free fatty acid is released into the circulation. During their passage through the epithelial cells, fatty acids are resynthesized into triglycerides, and it is the triglycerides which are released into the circulation. Electron micrographs of the intestinal epithelium during fat absorption show the accumulation of fat droplets in the endoplasmic reticulum of these cells. The isolated membranes of the endoplasmic reticulum contain the enzymes involved in triglyceride synthesis. The droplets of lipid become larger as they proceed through the cell. Release of the lipid droplet is believed to be similar to the release of protein secretory granules: The membrane of the endoplasmic reticulum surrounding the lipid droplet fuses with the cell membrane, freeing the droplet into the extracellular space.

The triglycerides enter the lacteals (lymph vessels) rather than the capillaries, primarily in the form of small lipid droplets, 0.1 to 3.5 μm in diameter, known as *chylomicrons*. These small droplets contain about 90 percent triglyceride and small amounts of phospholipid, cholesterol, free fatty acids, and protein. The presence of chylomicrons in the blood after a fatty meal gives the plasma a milky appearance.

Entrance of the chylomicrons into the lacteals rather than into the capillaries appears to depend upon the relative permeabilities of the two vessels to the lipid droplets released from the intestinal epithelium. As in other capillaries of the body, a basement membrane composed of polysaccharides covers the outer surface of the capillary but not the lacteal and may be the barrier keeping the chylomicrons out of the capillaries. Figure 19-28 summarizes the pathway taken by fat in moving from the lumen into the lymphatic system.

Vitamins Most of the vitamins readily diffuse across the walls of the intestine into the blood. The fat-soluble vitamins are dissolved in micelles and their absorption is markedly decreased in the absence of bile. Vitamin B_{12}, a charged molecule with a high molecular weight, is unable to diffuse across cell membranes. Upon entering the stomach, the vitamin combines with a special protein secreted by the stomach, known as the *intrinsic factor*. In the absence of the intrinsic factor, very small amounts of B_{12} are absorbed. Patients with pernicious anemia lack the intrinsic factor and thus do not absorb adequate amounts of vitamin B_{12}, which is required for the formation of red blood cells (Chap. 13). A special system moves vitamin B_{12} across the wall of the ileum, where most of its absorption occurs. During the absorption process, vitamin B_{12} is released from the intrinsic factor and enters the blood. It has been suggested that vitamin B_{12} may cross the intestinal wall by pinocytosis. Vitamin B_{12} absorption is the only case known where a large molecule combines with a still larger molecule, the intrinsic factor, in order to cross a cell membrane. Vitamin B_{12} is probably the largest essential nutrient absorbed from the intestinal tract without first being digested into smaller molecules. The loss of intrinsic factor and the eventual development of pernicious anemia are consequences of the surgical removal of the stomach; in this sense the stomach is essential for the life of the individual, but not because of its digestive role. Normally, a large amount of B_{12} is stored in the liver; and the symptoms of pernicious anemia often do not develop for several years after the removal of the stomach.

Water and Salt The normal volume of 9,000 ml of fluid entering the intestinal tract each day is absorbed at the rate of about 200 to 400 ml/h. Chyme entering the

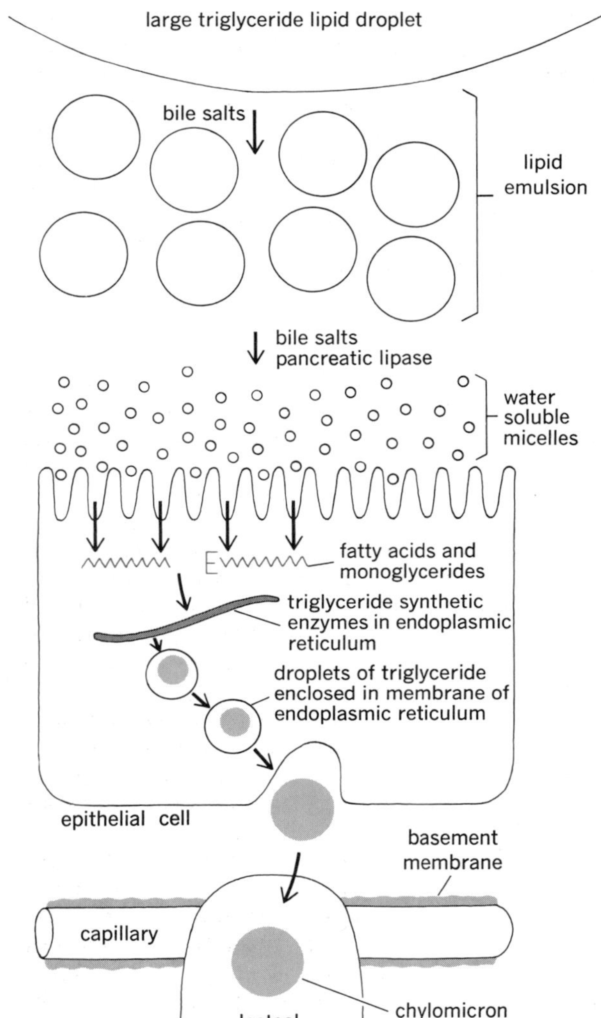

FIGURE 19-28 ***Summary of fat absorption across the walls of the small intestine.***

duodenum from the stomach is nearly isotonic. In the small intestine, sodium ions derived from the diet and from the gastrointestinal secretions are actively transported across the intestinal wall. Besides sodium, monosaccharides and amino acids are also actively transported from lumen to blood, often in combination with sodium, as we have seen. As a result of the active absorption of these solutes, the total concentration of solute in the lumen tends to drop below that of plasma and thus becomes relatively hypotonic. However, the intestinal wall has a very high permeability to water and thus cannot maintain a water concentration gradient between the lumen and blood. Water rapidly diffuses from the hypotonic solution in the lumen into the isotonic solution of the plasma. The movement of water is so rapid that it virtually accompanies the movement of solute, and thus the contents of the lumen remain essentially isotonic as they decrease in volume because of the net water and solute movement into the blood. This phenomenon is clearly similar to that for sodium and water reabsorption by the proximal renal tubules.

Water diffuses rapidly in both directions across the intestinal wall. Therefore, if a hypertonic solution is persent in the lumen, there is a net movement of water from blood to lumen, expanding the luminal contents until they become isotonic. Although the chyme entering the duodenum from the stomach is approximately isotonic, the duodenal contents may become hypertonic as a result of the rapid digestion of polysaccharides and proteins, forming a large number of osmotically active molecules from a few large molecules. If osmotically active molecules are formed by digestion faster than they are absorbed, the osmolarity of the solution increases and water flows from the blood into the lumen. In some cases, the movement of water into the intestine may be large enough to cause a significant decrease in blood volume, leading to cardiovascular complications. For example, when a small child consumes a lot of candy and soda pop, the high carbohydrate content has little inhibitory effect on gastric emptying, and the contents of the stomach soon enter the duodenum. Water rapidly diffuses from blood to lumen, diluting the high sugar concentration and increasing the volume of the lumen; distension of the intestinal wall and the decrease in blood volume lead to nausea, vomiting, pallor, sweating, and possibly fainting as a number of autonomic reflexes are triggered in the cardiovascular and gastrointestinal systems. The distension of the intestine triggers a reflexive increase in blood flow to the intestine which further aggravates the circulatory problems resulting from a decrease in blood volume.

A similar reaction, known as the *dumping syndrome,* is common in patients who have had large portions of their stomachs surgically removed because of disease. In these patients food may enter the duodenum directly from the esophagus without being retained by the stomach. The rapid increase in the osmolarity of the duodenum's contents, as a result of digestion, and the accompanying movement of water into the small intestine, produce the responses described above. Such patients must be fed small quantities of food many times a day.

Another example of fluid accumulation in the intestine occurs when the small intestine is obstructed. The rate of salt and water movement out of the intestine into the blood above the site of the obstruction is markedly decreased, with little changes in the rate at which salt and water enter the lumen from the blood. This may result from marked distension of the intestinal wall, which blocks the blood flow to the affected region, damaging the cells and altering membrane permeability. The result is a large net movement of fluid into the intestine and further distension of the intestinal walls. Sufficient quantities of fluid may be writhdrawn from the blood under these conditions to reduce blood pressure and cause severe complications, even death.

In addition to the active absorption of sodium, other inorganic ions such as Cl^-, K^+, Mg^{2+}, and Ca^{2+} are absorbed into the blood. Some are absorbed by active transport; others diffuse into the blood down their electrochemical gradients. The absorption of iron was described in Chap. 13.

THE LARGE INTESTINE

The colon (large intestine) forms the last 1.5 m of the gastrointestinal tract. It begins with a dilated portion, the *cecum*, which is a blind-ended pouch; the next three regions are the *colon, rectum*, and *anal canal* (Fig. 19-29A and B). The ileum joins the large intestine through the *ileocecal valve*, which is said to be a continuation of the muscle coats of the ileum. These form a sphincter that limits the rate of emptying of the ileal contents into the cecum and prevents their reflux into the small intestine. The sphincter is normally closed, but after a meal when the gastroileal reflex increases the contractile activity of the ielum, the sphincter relaxes each time the terminal portion of the ileum contracts, allowing chyme to enter the large intestine. Distension of the colon, on the other hand, produces a reflexive contraction of the sphincter, preventing further material from entering. Opening from the wall of the cecum below the ileocecal valve is the *vermiform appendix*, a worm-shaped tube about 9 cm long with no known function in human beings. The colon is not coiled but consists of three relatively straight segments, the ascending, transverse, and descending portions. The terminal portion of the descending colon is S-shaped, forming the sigmoid colon, which empties into a short section of tubing, the *rectum*. Although the large intestine has a greater diameter than the small intestine and is about half as long, its epithelial surface area is only about one-thirtieth that of the small intestine because the mucosa of the large intestine lacks villi and is not convoluted. The large intestine secretes no digestive enzymes and is responsible for the absorption of only about 4 percent of the total intestinal contents per day. Its primary function is to store and concentrate fecal material prior to defecation.

Secretion and Absorption

About 500 ml of chyme from the small intestine enters the colon each day. Most of this material is derived from the secretions of the small intestine, since most of the ingested food has been absorbed before reaching the large intestine. The secretions of the colon are very scanty and consist mostly of mucus.

The primary absorptive process in the large intestine is the active transport of sodium from the lumen to blood with the accompanying osmotic reabsorption of water. If fecal material remains in the large intestine for a long time, almost all the water is reabsorbed, leaving behind dry fecal pellets. The cells lining the large intestine are unable to actively transport either glucose or amino acids. There is a small net leakage of potassium into the colon, and severe depletion of total body potassium can occur as a result of repeated enemas and diarrhea.

The large intestine also absorbs some of the products synthesized by the bacteria in it. For example, small amounts of vitamins are synthesized by intestinal bacteria and absorbed into the body. Although this source of vitamins generally provides only a small part of the normal vitamin requirement per day, it may make a significant contribution when dietary intake of vitamins is low. The intestinal bacteria digest cellulose and utilize the glucose released for their own growth and reproduction.

Other bacterial products contribute to the production of intestinal gas (*flatus*). This gas is a mixture of nitrogen and carbon dioxide with small amounts of the inflammable gases hydrogen, methane, and hydrogen sulfide. Bacterial fermentation produces gas in the colon at the rate of about 400 to 700 ml/day.

Motility

The longitudinal smooth muscle in the human colon is incomplete; it is separated into three thick

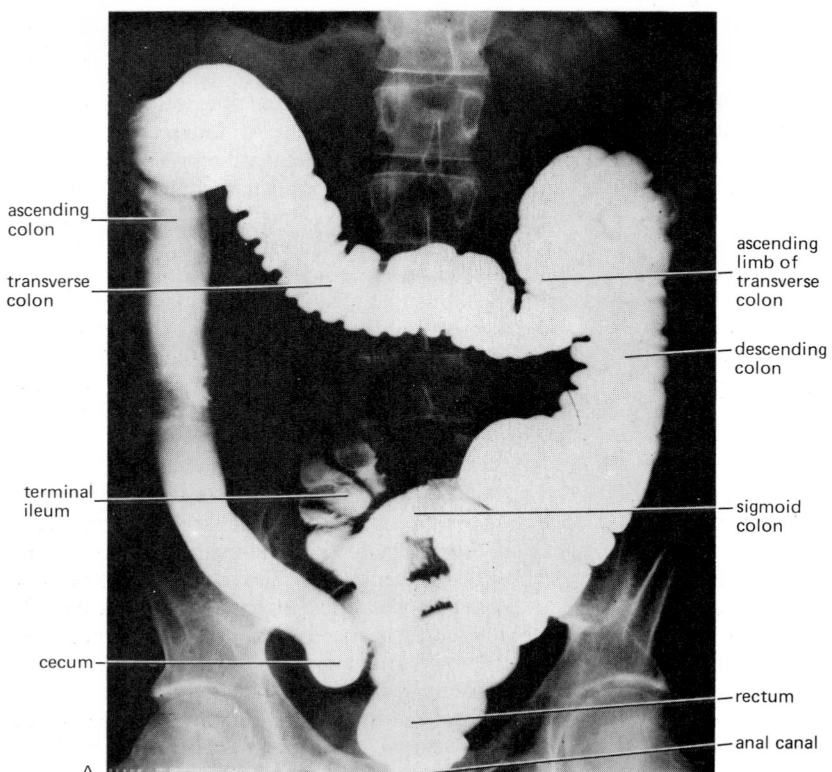

FIGURE 19-29 (A) ***X-ray of the large intestine. Haustra are apparent, especially in the transverse colon.*** (B) ***The cecum, ileocecal valve, and appendix. Part of the intestinal wall has been removed.***

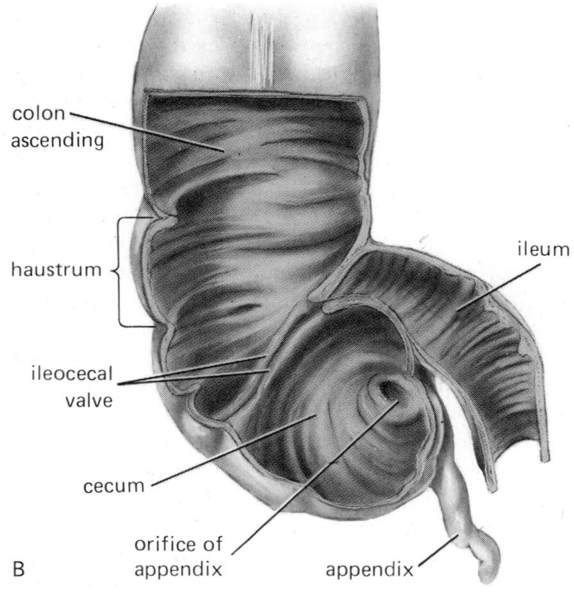

bands, the *teniae coli,* which pucker the walls of the large intestine into crosswise outpocketings called *haustra.* Contractions of the circular smooth muscle produce a segmentation motion which is not propulsive. This movement is considerably slower than in the small intestine, and a contraction may occur only once every 30 min. Because of this slow movement, material entering the colon from the small intestine remains for 18 to 24 h. Bacteria have time to grow and accumulate because of its slow movements; in the small intestine they do not have sufficient time to accumulate before being swept into the large intestine. During sleep and most of the day there is generally little or no movement in the large intestine, but three to four times a day, generally after meals, a marked increase in motility occurs. This usually coincides with the gastroileal reflex, described earlier, and probably has similar reflex mechanisms. This increased motility may lead to the phenomenon known as *mass movement,* in which large segments of the ascending and transverse colon contract simultaneously, propelling fecal material one-third to three-fourths of the length of the colon in a few seconds.

Defecation About 150 g of feces is eliminated from the body each day. This fecal material consists of about 100 g of water and 50 g of solid matter. The solid matter is made up mostly of bacteria, undigested cellulose, cell debris from the turnover of the intestinal epithelium, bile pigments, and small amounts of salt. The feces contain more potassium than the fluid entering the colon from the small intestine because some potassium is secreted by the cells lining the colon.

The rectum has the same diameter as the colon (about 4 cm when empty) but its lower portion enlarges to form the rectal ampulla. The mucous membrane forms longitudinal folds which flatten out when the rectum becomes distended with feces. The upper part of the rectum contains feces, but the lower part is normally empty except during defecation. The anal canal begins where the rectal ampulla becomes narrow; it is surrounded over its entire length by sphincter muscles which usually keep the canal closed. There is an abrupt transition here from simple columnar epithelium to stratified squamous epithelium.

The sudden distension of the walls of the rectum produced by the mass movement of fecal material is the normal stimulus for defecation. It initiates the defecation reflex, which is mediated primarily by the internal nerve plexuses but can be reinforced by external nerves to the terminal end of the large intestine. The reflex response consists of a contraction of the rectum, relaxation of the internal and external anal sphincters, and increased peristaltic activity in the sigmoid colon. This activity is sufficient to propel the feces through the anus. Defecation is normally assisted by a deep inspiration followed by closure of the glottis and contraction of the abdominal and chest muscles, causing a marked increase in intraabdominal pressure, which is transmitted to the contents of the large intestine and assists in the elimination of feces. This maneuver (*Valsalva's maneuver*) also causes a rise in intrathoracic pressure which leads to a sudden rise in blood pressure followed by a fall as the venous return to the heart is decreased (Chap. 15). In elderly people the cardiovascular stress resulting from the strain of defecation may precipitate a stroke or heart attack.

The *internal anal sphincter* is composed of smooth muscle, but the *external anal sphincter* is skeletal muscle under voluntary control. Higher brain centers in the central nervous system may, via descending pathways, override the afferent input from the rectum, therefore keeping the external sphincter closed and allowing a person to delay defecation.

The conscious urge to defecate accompanies the initial distension of the rectum. If defecation does not occur, the tension in the walls of the rectum slackens as the muscle relaxes, and the urge to defecate subsides until the next mass movement propels more feces into the rectum, increasing its volume and again initiating the defecation reflex.

Constipation and Diarrhea *Constipation* is the condition in which defecation is delayed for a variety of reasons. It may be due to consciously ignoring or preventing defecation or to decreased colonic motility, which most commonly is secondary to aging, emotion, or a low-bulk diet. Bulk refers to the content of fiber or other undigested materials in the diet, the volume of which is not decreased by absorption. The longer fecal material remains in the large intestine, the more water is reabsorbed, and the harder and drier the feces become, making defecation more difficult and sometimes painful. During this period additional material from the small intestine continues to enter the colon, progressively increasing the volume of its contents.

Many people have a mistaken belief that unless there is a bowel movement every day retention of fecal material and bacteria in the large intestine will somehow poison the body because of toxic products produced by the bacteria. Attempts to isolate such toxic agents from intestinal bacteria have been totally unsuccessful. In unusual cases where defecation has been prevented for a year or more by blockage of the rectum no ill effects from accumulated feces were noted except for the discomfort of carrying around the extra weight of 50 to 100 lb of feces retained in the large intestine. The symptoms of nausea, headache, loss of appetite, and general feeling of discomfort sometimes accompanying constipation appear to come from the distension of the rectum and large intestine. Experimentally inflating a balloon in the rectum of a normal individual produces similar sensations. Thus, there is no physiologic necessity for having bowel movements regulated by a clock; whatever maintains a person in a comfortable state is physiologically adequate, whether this means a bowel movement after every meal, or once a day, or once a week.

Cathartics, or laxatives, are sometimes necessary to relieve constipation. Several types are in common use. Cellulose in vegetable matter is a natural cathartic because of its ability to increase intestinal motility by providing bulk which stretches the smooth muscle of the intestinal wall, increasing its sensitivity to the basic electrical rhythm, and thus increasing its contractile activity. Castor oil acts by irritating the smooth muscle of the intestinal tract, increasing its motility. Some cathartics, such as mineral oil, act by lubricating hard, dry fecal material, thus easing defecation. Such agents as milk of magnesia are not absorbed or absorbed only slowly by the intestinal wall; the presence of nonabsorbable solute causes water to be retained in the intestinal tract and along with the increased motility resulting from the increased volume helps to flush out the large intestine.

Diarrhea, the opposite of constipation, is characterized by frequent defecation, usually of highly fluid fecal matter. A primary cause is greater intestinal motility with less time for absorption and thus the delivery of a large volume of fluid to the large intestine overloading its capacity to absorb salt and water. Disease-producing bacteria often irritate the intestinal wall, increase motility of the intestinal tract, and lead to diarrhea. Prolonged diarrhea can result in a serious loss of fluid and salt, especially potassium, from the body as well as upsetting the acid-base balance of the body due to loss of bicarbonate.

CHAPTER 20

Regulation of Organic Metabolism and Energy Balance

Section A. Control and Integration of Carbohydrate, Protein, and Fat Metabolism

In Chap. 1, we described the basic chemistry of living cells and their need for a continuous supply of nutrients. Although a certain fraction of these organic molecules is used in the synthesis of structural cell components, enzymes, coenzymes, hormones, antibodies, and other molecules serving specialized functions, most of the molecules in the food we eat are used by cells to provide the chemical energy required to maintain cell structure and function.

Essential for an understanding of organic metabolism is the remarkable ability of most cells, particularly those of the liver, to convert one type of molecule into another. These interconversions permit the human body to utilize the wide range of molecules found in different foods, but there are limits, and certain molecules must be present in the diet in adequate amounts. Enough protein must be ingested to provide the nitrogen needed for synthesis of protein and other nitrogenous substances in the body, and it must contain an adequate quantity of specific amino acids, called *essential* because they cannot be formed in the body by conversion from another molecule type. The other essential organic nutrients are a small group of fatty acids and the vitamins (Table 20-1).

The concept of a dynamic catabolic-anabolic steady state is also a critical component of organic metabolism. With few exceptions, e.g., DNA, virtually all organic molecules are being continuously broken down and rebuilt, usually at a rapid rate. The turnover rate of body protein is approximately 100 g/day; i.e., this quantity is broken down into amino acids and resynthesized each day. Few of the atoms present in a person's skeletal muscle a month ago are still there today.

With these basic concepts of *molecular interconvertibility* and *dynamic steady state* as foundation, we can discuss organic metabolism in terms of total body interactions. Figure 20-1 summarizes the major pathways of protein metabolism. The *amino acid pools,* which constitute the body's total free amino acids, are derived primarily from ingested protein (which is degraded to amino acids during digestion) and from the continuous breakdown of body protein. These pools are the source of amino acids for resynthesis of body protein and a host of specialized amino acid derivatives, such as nucleotides, epinephrine, etc. A very small quantity of amino acid and protein is lost from the body via the urine, skin, hair, and fingernails. The interactions between amino acids and the other nutrient types, carbohydrate and fat, are extremely important: Amino acids may be converted into carbohydrate (or fat) by removal of ammonia (deamination); one type of amino acid may participate in the formation of another by passing its nitrogen group to a carbohydrate. Both these processes were described in greater detail in Chap. 1 and are mentioned here to emphasize the interconvertibility of protein, carbohydrate, and fat. The ammonia, NH_3, formed during the first process is converted by the liver into urea, which is then excreted by the kidneys as the major end product of protein metabolism. Not all the events relating to amino acid metabolism occur in all cells; urea is formed in one organ (the liver) but excreted by another (the kidneys), but the concept of a pool is valid because all cells are interrelated by the vascular system and blood.

If any of the essential amino acids is missing from the diet, negative nitrogen balance always results. Apparently, the proteins for which that amino acid is essential cannot be synthesized and the other amino acids which would have been incorporated into the proteins are deaminated, their nitrogen being excreted as urea. It should be obvious, therefore, why a dietary requirement for protein cannot be specified without regard to the amino acid composition of that protein. Protein is graded in terms of how closely its ratio of essential amino acids approximates the ideal, which is their relative proportions in body protein. The highest-quality proteins are those found in animal products whereas the quality of most plant proteins is lower. Nevertheless, it is quite possible to obtain adequate quantities of all essential amino acids from plant protein alone although the total quantity of protein ingested must be larger.

Figure 20-2 summarizes the metabolic pathways for carbohydrate and fat, which are considered together because of their high potential rate of interconversion, particularly carbohydrate to fat (Chap. 1). The similarities between Figs. 20-1 and 20-2 are obvious, but there are several critical differences: (1) The major fate of both carbohydrate and fat is catabolism to yield energy, whereas amino acids can supply energy only after they are converted to carbohydrate or fat; and (2) excess carbohydrate and fat can be stored as such, whereas excess amino acids are not stored as protein but are converted to carbohydrate and fat.

In discussing the mechanisms which regulate the magnitude and direction of these molecular interconversions, we shall see that the *liver, adipose tissue* (the

TABLE 20-1
ESSENTIAL ORGANIC NUTRIENTS

Essential nutrient	RDA[6] for healthy adult male, mg	Dietary sources	Major body functions	Deficiency	Excess
Amino acids		From proteins: Good sources: Legume grains, dairy products, meat, fish; Adequate sources: Rice, corn, wheat; Poor sources: Cassava, sweet potato	Precursors of structural protein, enzymes and coenzymes, antibodies, hormones, metabolically active compounds, neurotransmitters, and porphyrins. Certain amino acids have specific functions: (a) Tyrosine is a precursor of epinephrine and other catecholamines and thyroxine (b) Arginine is a precursor of polyamines and urea (c) Methionine is required for methyl group metabolism (d) Tryptophan is a precursor of serotonin	Deficient protein intake leads to development of kwashiorkor and, coupled with low energy intake, to marasmus	Excess protein intake possibly aggravates or potentiates chronic disease states
Aromatic:					
Phenylalanine	1,100				
Tyrosine	1,100				
Basic:					
Arginine[1]	0				
Lysine	800				
Histidine[2]	0				
Branched chain:					
Isoleucine	700				
Leucine	1,000				
Valine	800				
Sulfur-containing:					
Methionine	1,100				
Cystine[3]	1,100				
Other:					
Tryptophan	250				
Threonine	500				
Fatty acids		Vegetable fats (corn, cottonseed, soy oils); wheat germ; vegetable shortenings	Involved in cell membrane structure and function; precursors of prostaglandins (regulation of gastric function, release of hormones, smooth-muscle activity)	Poor growth; skin lesions	Not known
Arachidonic	6,000				
Linoleic	6,000				
Linolenic	6,000				
Vitamins					
Water-soluble:					
Vitamin B-1 (thiamine)	1.5	Pork, organ meats, whole grains, legumes	Coenzyme (thiamine pyrophosphate) in reactions involving the removal of carbon dioxide	Beriberi (peripheral nerve changes, edema, heart failure)	None reported
Vitamin B-2 (riboflavin)	1.8	Widely distributed in foods	Constituent of two flavin nucleotide coenzymes involved in energy metabolism (FAD and FMN)	Reddened lips, cracks at corner of mouth (cheilosis), lesions of eye	None reported

TABLE 20-1
ESSENTIAL ORGANIC NUTRIENTS *(Continued)*

Essential nutrient	*RDA[6] for healthy adult male, mg*	*Dietary sources*	*Major body functions*	*Deficiency*	*Excess*
Niacin	20	Liver, lean meats, grains, legumes (can be formed from tryptophan)	Constituent of two coenzymes involved in oxidation-reduction reactions (NAD and NADP)	Pellagra (skin and gastrointestinal lesions; nervous, mental disorders)	Flushing, burning and tingling around neck, face, and hands
Vitamin B-6 (pyridoxine)	2	Meats, vegetables, whole-grain cereals	Coenzyme (pyridoxal phosphate) involved in amino acid metabolism	Irritability, convulsions, muscular twitching, dermatitis near eyes, kidney stones	None reported
Pantothenic acid	5–10	Widely distributed in foods	Constituent of coenzyme A, which plays a central role in energy metabolism	Fatigue, sleep disturbances, impaired coordination, nausea (rare in man)	None reported
Folacin	0.4	Legumes, green vegetables, whole-wheat products	Coenzyme (reduced form) involved in transfer of single-carbon units in nucleic acid and amino acid metabolism	Anemia, gastrointestinal disturbances, diarrhea, red tongue	None reported
Vitamin B-12	.003	Muscle meats, eggs, dairy products (not present in plant foods)	Coenzyme involved in transfer of single-carbon units in nucleic acid metabolism	Pernicious anemia, neurological disorders	None reported
Biotin	Not established; usual diet provides 0.15–0.3	Legumes, vegetables, meats	Coenzyme required for fat synthesis, amino acid metabolism, and glycogen formation	Fatigue, depression, nausea, dermatitis, muscular pains	None reported
Choline	Not established; usual diet provides 500–900	All foods containing phospholipids (egg yolk liver, grains, legumes)	Constituent of phospholipids; precursor of neurotransmitter acetylcholine	Not reported in man	None reported
Vitamin C (ascorbic acid)	45	Citrus fruits, tomatoes, green peppers, salad greens	Maintains intercellular matrix of cartilage, bone and dentine; important in collagen synthesis.	Scurvy (degeneration of skin, teeth, blood vessels; epithelial hemorrhages)	Relatively nontoxic; possibility of kidney stones

Vitamin	RDA (mg)	Sources	Functions	Deficiency	Excess
Fat-soluble:					
Vitamin A (retinol)	1	Provitamin A (beta-carotene) widely distributed in green vegetables; retinol present in milk, butter, cheese, fortified margarine	Constituent of rhodopsin (visual pigment); maintenance of epithelial tissues; role in mucopolysaccharide synthesis	Xerophthalmia (keratinization of ocular tissue), night blindness, permanent blindness	Headache, vomiting, peeling of skin, anorexia, swelling of long bones
Vitamin D[4]	0.001	Cod-liver oil, eggs, dairy products, fortified milk, and margarine	Promotes growth and mineralization of bones; increases absorption of calcium	Rickets (bone deformities) in children. Osteomalacia in adults	Vomiting, diarrhea, loss of weight, kidney damage
Vitamin E (tocopherol)	15	Seeds, green leafy vegetables, margarines, shortenings	Functions as an antioxidant to prevent cell-membrane damage	Possibly anemia	Relatively nontoxic
Vitamin K[5] (phylloquinone)	0.03	Green leafy vegetables; small amount in cereals, fruits and meats	Important in blood clotting (involved in formation of clotting factors)	Hemorrhage	Relatively nontoxic; synthetic forms at high doses may cause jaundice

SOURCE: **N. S. Scrimshaw and V. R. Young, "The Requirements of Human Nutrition,"** *Sci. Amer.*, 235(3):50, **Sept. 1976.**

[1] Nonessential if sufficient dietary phenylalanine.

[2] Unnecessary in adults in short-term studies but probably essential for normal growth of children.

[3] Nonessential if sufficient dietary methionine.

[4] If adequate sunlight is available, no dietary intake is required since skin synthesizes this "vitamin," which is more properly termed a "hormone" (see Chap. 18).

[5] Synthesized by intestinal bacteria; dietary intake normally unnecessary.

[6] RDA: Recommended Daily Allowance.

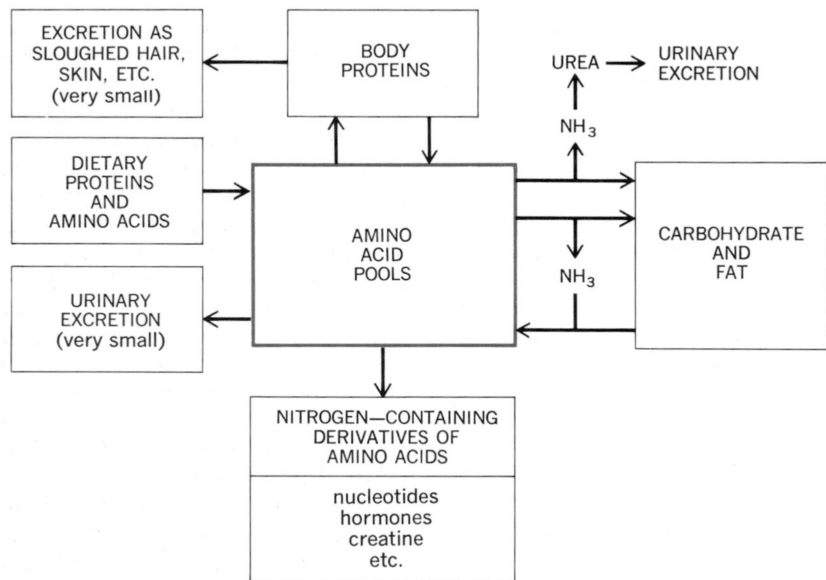

FIGURE 20-1 ***Amino acid pools and major pathways of protein metabolism.***

storage tissue for fat), and *muscle* are the dominant effectors and that the major controlling inputs to them are a group of hormones and the sympathetic nerves to adipose tissue and the liver. At this point, the reader should review the biochemical pathways described in Chap. 1, particularly those dealing with glucose and the interconversions of carbohydrate, protein, and fat.

EVENTS OF THE ABSORPTIVE AND POSTABSORPTIVE STATES

When food is readily available, human beings can get along by eating small amounts of food all day long if they wish; however, this is clearly not true for most other animals, for early human beings, or for most per-

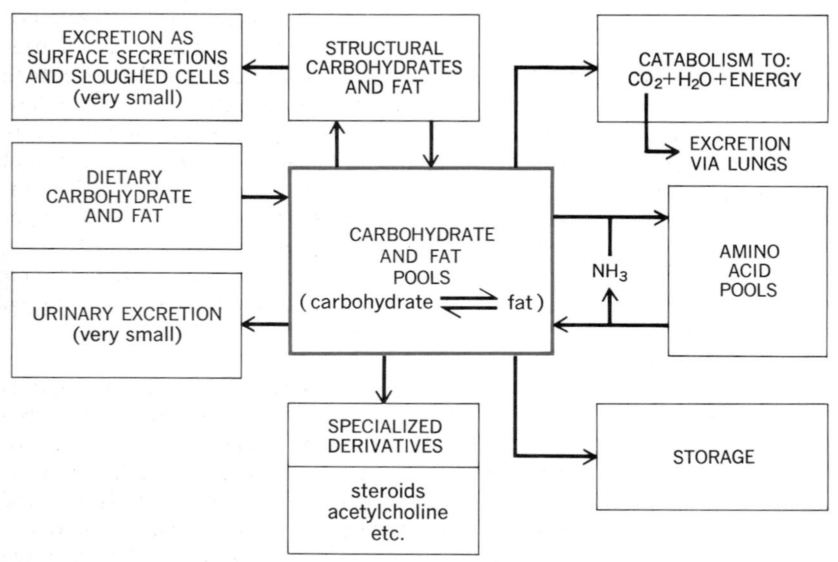

FIGURE 20-2 ***Major pathways of carbohydrate and fat metabolism. "Carbohydrate and fat pools" include the simple unspecialized carbohydrates and fats dissolved in the body fluids. Note that structural carbohydrate and fat are constantly being broken down and resynthesized.***

sons today, and animals have been forced to evolve mechanisms for survival during alternating periods of plenty and fasting. We speak of two functional states: the *absorptive state,* during which ingested nutrients are entering the blood from the gastrointestinal tract, and the *postabsorptive* (or *fasting*) *state,* during which the gastrointestinal tract is empty and energy must be supplied by the body's endogenous stores. Since an average meal requires approximately 4 h for complete absorption, our usual three-meal-a-day pattern places us in the postabsorptive state during the late morning and afternoon and almost the entire night. The average person can easily withstand a fast of many weeks (so long as water is provided), and extremely obese patients have been fasted for many months, being given only water and vitamins.

The absorptive state can be summarized as follows: During absorption of a normal meal, glucose provides the major energy source; only a small fraction of the absorbed amino acids and fat is ultimately utilized for energy; another fraction of amino acids and fat is used to resynthesize the continuously degraded body proteins and structural fat, respectively; most of the amino acids and fat as well as the large quantity of carbohydrate not oxidized for energy are transformed into adipose-tissue fat.

In the postabsorptive state, carbohydrate is synthesized in the body, but its utilization for energy is greatly reduced; the oxidation of endogenous fat provides most of the body's energy supply; fat and protein synthesis are curtailed and net breakdown occurs.

Figures 20-3 and 20-4 summarize the major pathways to be described. Although they may appear formidable at first glance, they should give little difficulty after we have described the component parts, and they should be referred to constantly during the following discussion. Mastery of this material is essential for an understanding of the hormonal mechanisms which control and integrate metabolism.

Absorptive State

We shall assume an average meal to contain approximately 65 percent carbohydrate, 25 percent protein, and 10 percent fat. Recall from Chap. 19 that these nutrients enter the blood and lymph from the gastrointestinal tract primarily as monosaccharides, amino acids, and triglycerides, respectively. The first two groups enter the blood, which leaves the gastrointestinal tract to go directly to the liver, allowing this remarkable biochemical factory to alter the composition of the blood before it is pumped to the rest of the body. In contrast, the fat droplets are absorbed into the lymph and not into the blood.

Glucose Some of the absorbed carbohydrate is galactose and fructose, but since the liver converts most of these carbohydrates immediately into glucose (and because fructose enters essentially the same metabolic pathways as does glucose), we shall simply refer to these sugars as glucose. As shown in Fig. 20-3, much of the absorbed carbohydrate enters the liver cells, but little of it is oxidized for energy, instead being built into the polysaccaride glycogen or transformed into fat. The importance of glucose as a precursor of fat cannot be overemphasized; note that glucose provides both the glycerol and the fatty acid moieties of triglycerides. Some of this fat synthesized in the liver may be stored there, but most is transported into the blood, from which it enters adipose-tissue cells. Much of the absorbed glucose which did not enter liver cells but remained in the blood enters adipose-tissue cells, where it is transformed into fat; another fraction is stored as glycogen in skeletal muscle and certain other tissues, and a very large fraction enters the various cells of the body and is oxidized to carbon dioxide and water, thereby providing the cells' energy requirements. Glucose is the body's major energy source during the absorptive state.

Triglycerides Almost all ingested fat is absorbed into the lymph as fat droplets (*chylomicrons*) containing primarily triglycerides, which enter adipose-tissue cells, where they are stored. Thus, there are three prominent sources of adipose-tissue triglyceride (TG): (1) ingested TG, (2) TG synthesized in adipose tissue from glucose, and (3) TG synthesized in the liver and transported via the blood to the adipose tissue. For simplicity, we have not shown in Fig. 20-3 that a fraction of fat is also oxidized during the absorptive state by various organs to provide energy. The actual amount utilized depends upon the content of the meal and the person's nutritional status.

Amino Acids Many of the absorbed amino acids enter liver cells and are entirely converted into carbohydrate (keto acids) by removal of the NH_3 portion of the molecule. The ammonia is converted by the liver into urea, which diffuses into the blood and is excreted by the kidneys. The keto acids can then enter the Krebs tricarboxylic acid cycle and be oxidized to provide

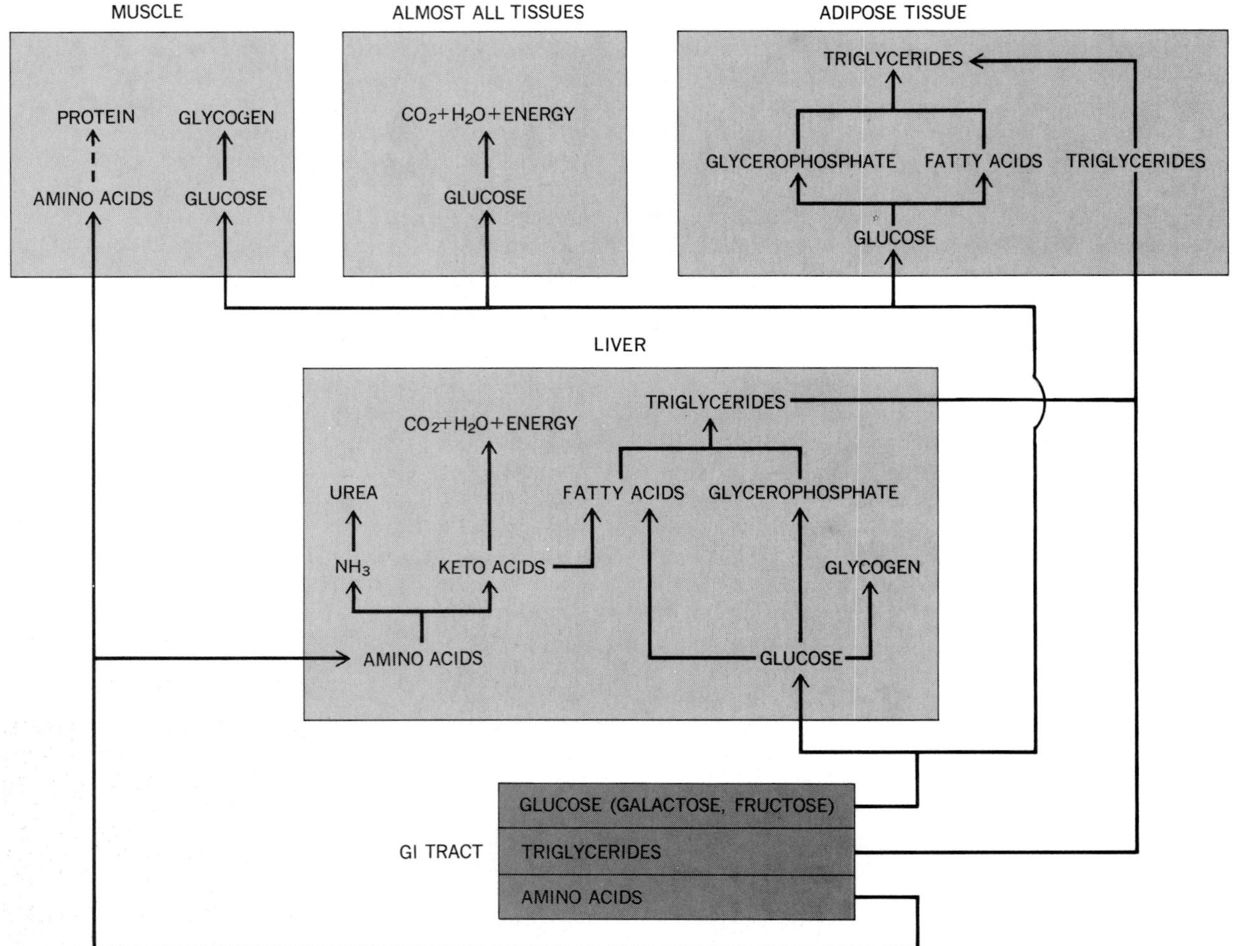

FIGURE 20-3 ***Major metabolic pathways of the absorptive phase.***

energy for the liver cells; indeed the liver is unusual in that by this interconversion amino acids provide much of its energy during the absorptive state. Finally, the keto acids can also be converted to fatty acids, thereby participating in fat synthesis by the liver. The ingested amino acids not taken up by the liver cells enter other cells of the body (Fig. 20-3). Although virtually all cells require a constant supply of amino acids for protein synthesis, we have simplified the diagram by showing only muscle because it constitutes the great preponderance of body mass and therefore contains the most important store, quantitatively, of body protein. Other organs of course participate, but to much lesser degree, in the amino acid exchanges occurring during the absorptive and postabsorptive states. After entering the cells, most of the amino acids are synthesized into protein. This process is represented by the dotted line in Fig. 20-3 to call attention to an important fact: Excess amino acids are *not* stored as protein, in the sense that glucose and fat are stored as fat and to a lesser degree as glycogen. Eating large amounts of protein does not significantly increase body protein; the excess amino acids are merely converted into carbohydrate or fat. On the other hand, a minimal supply of ingested amino acids is

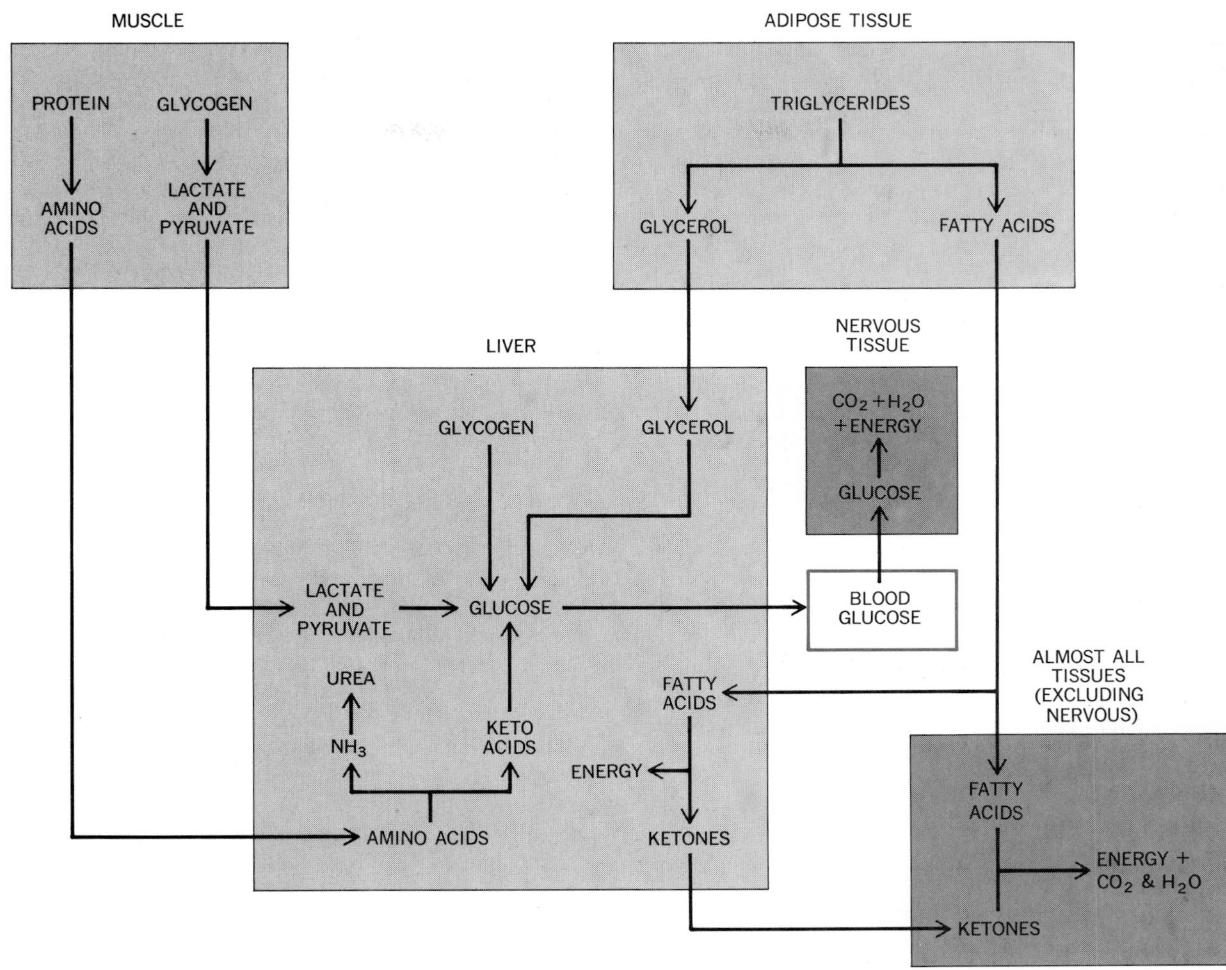

FIGURE 20-4 ***Major metabolic pathways of the postabsorptive (fasting) phase. The central focus is regulation of the blood glucose concentration.***

essential to maintain normal protein stores by preventing net protein breakdown in muscle and other tissues. During the usual alternating absorptive and postabsorptive states, the fluctuations in total body protein are relatively small.[1]

Summary During the absorptive period, energy is provided primarily by glucose, protein stores are maintained, and excess calories (regardless of source) are stored mostly as fat. Glycogen constitutes a quantitatively less important storage form for carbohydrate. The use of fat to store excess calories is an excellent adaptation for mobile animals, because 1 g of TG contains more than twice as many calories as 1 g of protein or glycogen and because there is very little water in adipose tissue.

[1] This discussion applies only to the adult; the growing child, of course, manifests a continuous increase of body protein.

Postabsorptive State

The essential problem during this period is that no glucose is being absorbed from the intestinal tract, yet the plasma glucose concentration must be maintained because the nervous system is an obligatory glucose utilizer; i.e., it is unable to oxidize any other nutrient for energy.[2] Lack of adequate glucose supply to the brain causes damage, coma, and death within minutes. Perhaps the most convenient way of viewing the events of the postabsorptive state is in terms of how the blood glucose concentration is maintained. These events fall into two categories: sources of glucose, and glucose sparing and fat utilization.

Sources of Blood Glucose The sources of blood glucose during fasting (Fig. 20-4) are as follows:

1 Glycogen stores in the liver are broken down to liberate glucose but are adequate only for a short time. After the absorptive period is completed, the normal liver contains less than 100 g of glycogen; at 4 kcal/g, this provides 400 kcal, enough to fulfill the body's total caloric need for only 4 h.

2 Glycogen in muscle (and to a lesser extent other tissues) provides approximately the same amount of glucose as the liver. A complication arises because muscle lacks the necessary enzyme to form free glucose from glycogen.[3] But glycolysis breaks the glycogen down into pyruvate and lactate, which are then liberated into the blood, circulate to the liver, and are synthesized into glucose. Thus, muscle glycogen contributes to the blood glucose indirectly via the liver.

3 As shown in Fig. 20-4, the catabolism of triglycerides yields glycerol and fatty acids. The former can be converted into glucose by the liver, but the latter cannot. Thus, a potential source of glucose is adipose-tissue TG breakdown, in which glycerol is liberated into the blood, circulates to the liver, and is converted into glucose.

4 The major source of blood glucose during prolonged periods of fasting comes from protein. Large quantities of protein in muscle and to a lesser extent other tissues are not absolutely essential for cell function; i.e., a sizable fraction of cell protein can be catabolized, as during prolonged fasting, without serious cellular malfunction. There are, of course, limits to this process, and continued protein loss ultimately means functional disintegration, sickness, and death. Before this point is reached, protein breakdown can supply large quantities of amino acids which are converted into glucose by the liver.

[2] There is an important exception to be described subsequently.

[3] Muscle glycogen is broken down in several steps to glucose 6-phosphate rather than free glucose, which is then catabolized via glycolysis and the Krebs cycle for energy.

To summarize, for survival of the brain, plasma glucose concentration must be maintained. Glycogen stores, particularly in the liver, form the first line of defense, are mobilized quickly, and can supply the body's needs for several hours, but they are inadequate for longer periods. Under such conditions, protein and fat supply amino acids and glycerol, respectively, for production of glucose by the liver. Hepatic synthesis of glucose from pyruvate, lactate, glycerol, and amino acids is known as *gluconeogenesis,* i.e., new formation of glucose. During a 24-h fast, it amounts to approximately 180 g of glucose. The kidneys are also capable of glucose synthesis, particularly in a prolonged fast (several weeks) at the end of which they may be contributing as much glucose as the liver.

Glucose Sparing and Fat Utilization A simple calculation reveals that even the 180 g of glucose per day produced by the liver during fasting cannot possibly supply all the body's energy needs: 180 g/day $\times$ 4 kcal/g $=$ 720 kcal/day, whereas normal total energy expenditure equals 1,500 to 3,000 kcal/day. The following essential adjustment must therefore take place during the transition from absorptive to postabsorptive state: The nervous system continues to utilize glucose normally, but virtually all other organs and tissues markedly reduce their oxidation of glucose and depend primarily on fat as their energy source, thus sparing the glucose produced by the liver to serve the obligatory needs of the nervous system. The essential step is the catabolism of adipose-tissue TG to liberate fatty acids into the blood. These fatty acids are picked up by virtually all tissues (excluding the nervous system), enter the Krebs cycle, and are oxidized to carbon dioxide and water, thereby providing energy. The liver, too, utilizes fatty acids for its energy source, thereby sparing amino acids (its usual energy source) for glucose synthesis. However, the liver's handling of fatty acids during fasting is unique; it oxidizes them to acetyl CoA, which instead of being oxidized further via the Krebs cycle is processed into a group of compounds called *ketone bodies*. One of these substances is acetone, some of which is expired and accounts for the distinctive breath odor of persons undergoing prolonged fasting or suffering from severe untreated diabetes mellitus. The significance of this process is that the ketone bodies are released into the blood and provide an important energy source for the many tissues capable of oxidizing them via the Krebs cycle.

The net result of adipose-tissue breakdown during

fasting (as much as 160 g/day) is provision of energy for the body and sparing of glucose for the brain. The combined effects of gluconeogenesis and the switch over to fat utilization are so efficient that, after several days of complete fasting, the plasma glucose concentration is reduced only by a few percent. After one month, it is decreased only 25 percent.

Recent studies have revealed an important change in brain metabolism with prolonged starvation. Apparently, after 2 to 4 days of fasting, the generalization that brain is an obligatory glucose utilizer is no longer valid, for the brain begins to utilize large quantities of ketone bodies, as well as glucose, for its energy source. The survival value of this phenomenon is very great; if the brain significantly reduces its glucose requirement (by utilizing ketones instead of glucose), much less protein need be broken down to supply the amino acids for gluconeogenesis. Accordingly, the protein stores will last longer, and the ability to withstand a long fast without serious tissue disruption is enhanced.

Thus far, our discussion has been purely descriptive; we now turn to the factors which so precisely control and integrate these metabolic pathways and transformations. Without question, the most important single factor is insulin. As before, the reader should constantly refer to Figs. 20-3 and 20-4. We shall focus primarily on the following questions (Fig. 20-5) raised by the previous discussion: (1) What controls the shift from the net anabolism of protein, glycogen, and TG to net catabolism? (2) What induces primarily glucose utilization during absorption and fat utilization during fasting; i.e., how do cells "know" they should start oxidizing fatty acids and ketones instead of glucose? (3) What drives net hepatic glucose uptake during absorption but net glucose synthesis (gluconeogenesis) and release during fasting?

INSULIN

Insulin is a protein hormone secreted by the islets of Langerhans, clusters of endocrine cells in the pancreas. Each of the 1 million or so islets in the human pancreas consists of polyhedral cells clustered together around a rich capillary network. Appropriate staining techniques reveal several types of islet cells, the most prominent being the β, or *beta*, and α, or *alpha, cells* (Fig. 20-6). The beta cells are the source of insulin and account for 80 percent of the islet cells. Insulin acts directly or indirectly on most tissues of the body, with the notable exception of brain. The effects of insulin are so important and widespread that an injection of this hormone into a fasting person duplicates the absorptive-state pattern of Fig. 20-3 (except, of course, for the absence of gastroin-

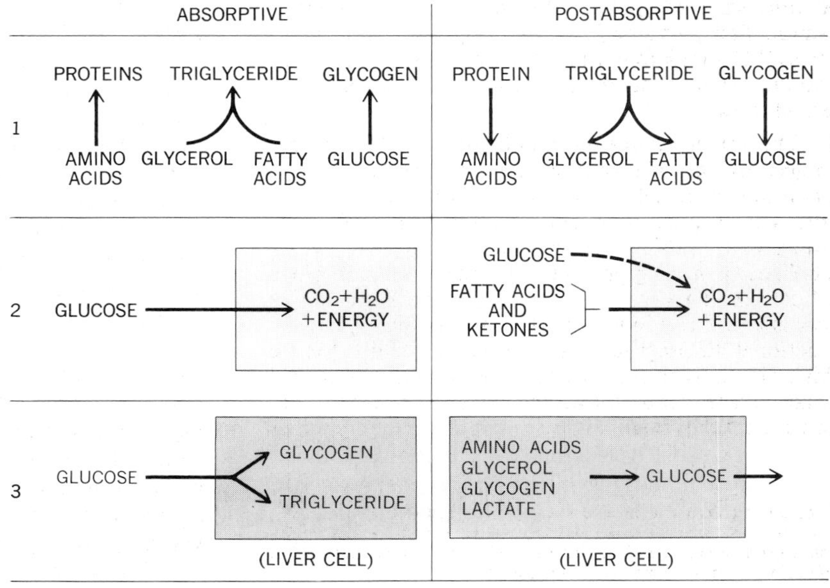

FIGURE 20-5 ***Summary of critical shifts in transition from absorptive to postabsorptive states.***

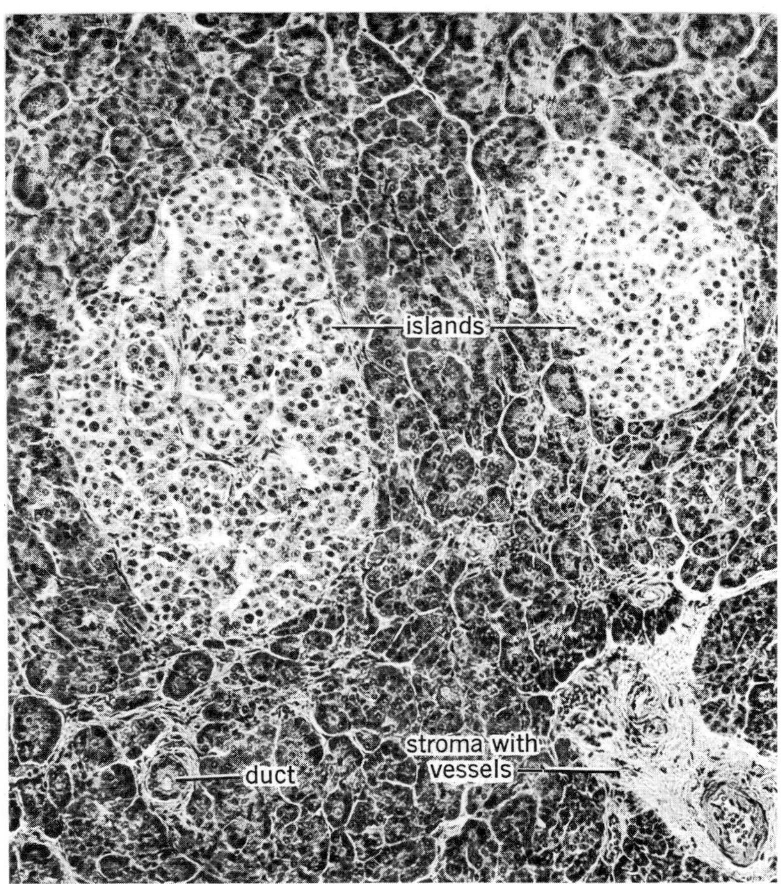

FIGURE 20-6 ***Section of the pancreas of a rhesus monkey, showing two islets of Langerhans and surrounding exocrine glandular tissue.*** **(From W. F. Windle, "Textbook of Histology," 5th ed., McGraw-Hill, New York, 1976.)**

testinal absorption); coversely, patients suffering from insulin deficiency (diabetes mellitus) manifest the postabsorptive pattern of Fig. 20-4. From these statements alone, it might appear (correctly) that secretion of insulin is stimulated by eating and inhibited by fasting.

Effects of Insulin

Glucose Uptake Glucose enters most cells by the carrier-mediated mechanism described as facilitated diffusion in Chap. 1. The most important single effect exerted by insulin is to stimulate this facilitated diffusion of glucose into certain cells, particularly muscle and adipose tissue (Fig. 20-7). The result should be apparent; greater glucose entry into cells increases the availability of glucose for all the reactions in which glucose participates. Thus, glucose oxidation, fat synthesis, and glycogen synthesis are all stimulated, in short, the major events of the absorptive state. It is important to note that insulin does *not* alter glucose uptake by the brain, nor does it influence the active transport of glucose across the renal tubule and gastrointestinal epithelium.

This single effect once seemed so allpowerful that many experts believed it to be the only one insulin exerted, but it has been demonstrated that insulin has other direct effects which reinforce its role as the dominant hormone of the absorptive period.

Stimulation of Glycogen Synthesis As shown above, increased glucose uptake per se stimulates glycogen synthesis. In addition, insulin also increases the activity of the enzyme which catalyzes the rate-limiting step in glycogen synthesis. Thus, insulin en-

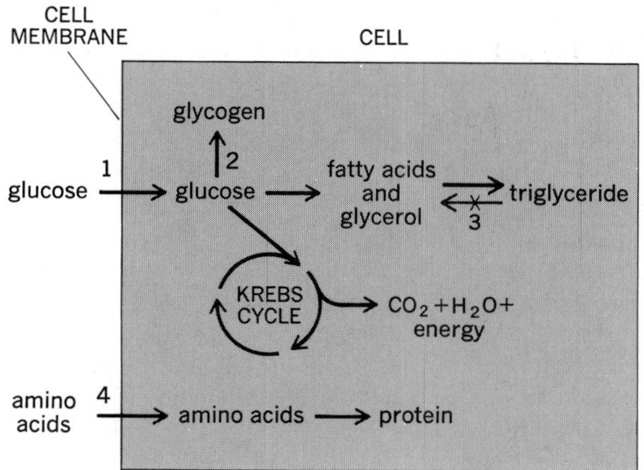

FIGURE 20-7 ***Major effects of insulin upon organic metabolism. The numbers denote distinct direct effects, the others being the indirect effects of increased glucose and amino acid entry. The X on arrow 3 denotes inhibition of TG breakdown.***

sures glucose transformation into glycogen by a double-barreled effect.

Inhibition of TG Breakdown Increased entry of glucose into adipose tissue facilitates fatty acid and glycerophosphate synthesis, which, by mass action, drives TG synthesis. In addition, insulin inhibits the enzyme which catalyzes TG breakdown. Insulin thus increases TG stroes by a double effect: driving TG synthesis by facilitating glucose entry and at the same time inhibiting TG breakdown via the enzyme.

Stimulation of Protein Synthesis Net protein synthesis is also increased by insulin, which stimulates the active membrane trasport of amino acids, particularly into muscle cells. Thus, in a manner analogous to that described for glucose, greater amino acid entry shifts the intracellular protein equilibrium toward net synthesis. Insulin also has important effects on the ribosomal protein-synthesizing machinery.

Effects on Other Liver Enzymes Insulin causes changes in the activities of concentrations of almost all the critical liver enzymes involved in both the utilization and synthesis of glucose. As might be predicted, the former are all stimulated whereas the latter are inhibited. The precise mechanisms by which insulin induces these changes are still poorly understood.

The Other Side of the Coin We have thus far dealt with the positive effects of insulin. Clearly insulin deficit will have just the opposite results. High rates of glucose entry and oxidation (except in brain) and the net anabolism of glycogen, protein, and TG all depend upon the presence of high blood concentrations of insulin; when the blood concentration decreases, the metabolic pattern is shifted toward decreased glucose entry and oxidation and a net catabolism of glycogen, protein, and TG. In other words, these metabolic pathways are in a dynamic state, capable of proceeding, in terms of net effect, in either direction. For this reason, energy metabolism can be shifted from the absorptive to the postabsorptive pattern merely by lowering the rate of insulin secretion: Glucose entry and oxidation decrease; glycogen breakdown increases; net protein catabolism liberates amino acids into the blood; net TG catabolism liberates glycerol and fatty acids into the blood, and the resulting higher fatty acid concentration is blood facilitates cellular uptake of fatty acids, which, in turn, stimulates fatty acid oxidation; and gluconeogenesis is stimulated not only by the increased availability of precursors (amino acids and glycerol) but by enzyme changes in the liver itself. The glucose can be utilized by the brain since its glucose uptake is not insulin-dependent. Despite its great importance, insulin is not the only hormone controlling these patterns. The role of the other hormones will be discussed below.

Control of Insulin Secretion

Insulin secretion is directly controlled by the glucose concentration of the blood flowing through the pancreas, a simple system requiring no participation of nerves or other hormones. An increase in blood glucose concentration stimulates insulin secretion; conversely, a reduction inhibits secretion. The feedback nature of this system is shown in Fig. 20-8. A rise in plasma glucose stimulates insulin secretion; insulin induces rapid entry of glucose into cells; this transfer of glucose out of the blood reduces the blood concentration of glucose, thereby removing the stimulus for insulin secretion, which returns to its previous level.

Figure 20-9 illustrates typical changes in plasma glucose and insulin concentrations following a normal carbohydrate-rich meal. Note the close association between the rising blood concentration (resulting from gastrointestinal absorption) and the plasma insulin increase induced by the glucose rise. The low postabsorptive values for plasma glucose and insulin concentrations are not the lowest attainable, and prolonged

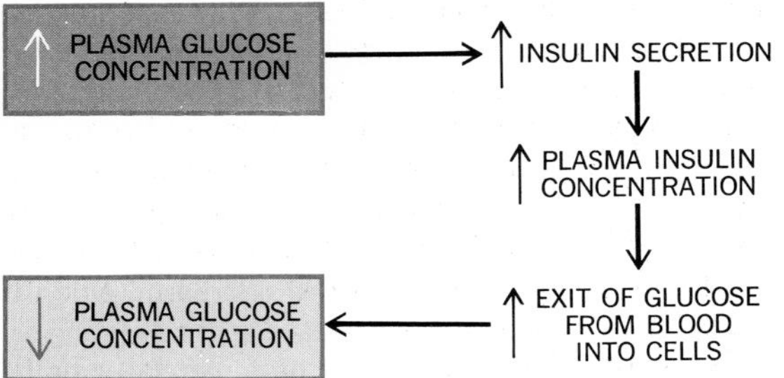

FIGURE 20-8 ***Negative-feedback nature of plasma glucose control over insulin secretion.***

fasting induces even further reductions of both variables until insulin is barely detectable in the blood.

Although it was once believed that plasma glucose constitutes the sole control over insulin secretion, such is not the case, for insulin secretion is sensitive to numerous other types of input. One of the most important is the plasma concentration of certain amino acids, an elevated amino acid concentration causing enhanced insulin secretion. This is easily understandable since amino acid concentrations increase after eating, particularly after a high-protein meal. The increased insulin stimulates cell uptake of these amino acids. There is also some direct neural and hormonal control over insulin secretion; e.g., the gastrointestinal hormones described in Chap. 19 may stimulate insulin secretion.

FIGURE 20-9 ***Blood concentrations of glucose and insulin following ingestion of 100 g of glucose. Study performed on normal human subjects.*** **(Adapted from Daughaday and Kipnis.)**

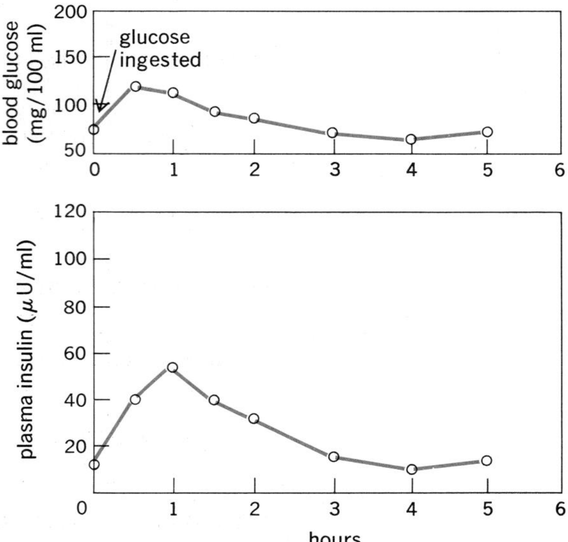

Diabetes Mellitus

The name diabetes, meaning syphon or running through, was used by the Greeks over 2,000 years ago to describe the striking urinary volume excreted by certain people. Mellitus, meaning sweet, distinguishes this urine from the large quantities of insipid urine produced by persons suffering from ADH deficiency (Chap. 18). This sweetness of the urine was first recorded in the seventeenth century, but in England the illness had long been called the pissing evil. Because of the marked weight loss despite huge food intake, the body's substance was believed to be dissolving and pouring out through the urinary tract, a view not far from the truth. In 1889, experimental diabetes was produced in dogs by surgical removal of the pancreas, and 32 years later, in 1921, Banting and Best discovered insulin.

A tendency toward diabetes can be inherited. We say tendency because diabetes often is not an all-or-none disease but may develop slowly, and overt signs may all but disappear with appropriate measures, e.g., weight reduction. The cause of diabetes is relative insulin deficiency. We have described how a lowered insulin concentration induces virtually all the metabolic changes characteristic of the fasting state (Fig. 20-4). The picture presented by an untreated diabetic is a gross caricature of this state (Fig. 20-10). The catabolism of triglyceride with resultant elevation of plasma fatty acids and ketones is an appropriate response because these substances must provide energy for the body's cells, which are prevented from taking up adequate glucose by the insulin deficiency. In contrast, glycogen and protein catabolism and the marked gluconeogenesis so important to maintain plasma glucose during fasting are completely inappropriate in diabetics since their plasma glucose is already high because it cannot enter into cells. These reactions serve only to raise the plasma glucose still higher, with disastrous consequences. This paradox is the essence of the diabetic situation: cell starvation in the presence of a markedly elevated plasma glucose. Only the brain is spared glucose deprivation since its uptake of glucose is not insulin-dependent. The obvious consequence of these catabolic processes is progressive loss of weight despite the increased food intake induced by constant

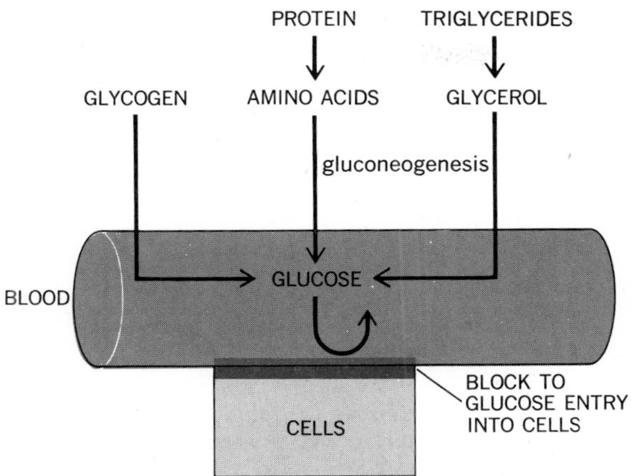

FIGURE 20-10 ***Factors which elevate blood glucose concentration in insulin deficiency.***

hunger. The ingested carbohydrate and protein simply are converted to glucose, which further increases the plasma glucose to no avail.

The elevated plasma glucose of diabetes induces changes in renal function of serious consequences. In Chap. 18, we pointed out that a normal person does not excrete glucose because all glucose filtered at the glomerulus is reabsorbed by the tubules. However, the elevated plasma glucose of diabetes may so increase the filtered load of glucose that the maximum tubular reabsorptive capacity is exceeded, and large amounts of glucose may be excreted. For the same reasons, large amounts of ketones may also appear in the urine. These urinary losses, of course, only aggravate the situation by further depleting the body of nutrients. Far worse, however, is the effect of these solutes on sodium and water excretion. In Chap. 18, we saw how tubular water reabsorption is a passive process induced by active solute reabsorption. In diabetes, the osmotic force exerted by unreabsorbed glucose and ketones holds water in the tubule, thereby preventing its reabsorption. For several reasons (the mechanisms are beyond the scope of this book) sodium reabsorption is also retarded. The net result is marked excretion of sodium and water, which leads, by the sequence of events shown in Fig. 20-11, to hypotension, brain damage, and death.

Another serious abnormality in diabetes is a markedly increased hydrogen-ion concentration, due primarily to the accumulation of ketone bodies, which, as moderately strong acids, generate large amounts of hydrogen ion by dissociation. Among other deleterious effects, the increased hydrogen ion concentration may induce brain malfunction and coma. The kidneys respond to this increase by excreting more hydrogen ion and are generally able to maintain balance fairly well, at least until the volume depletion described above interferes with renal function. What effect does the increased hydrogen-ion concentration have on respiration? A marked increase in ventilation occurs in response to stimulation of the medullary respiratory centers by hydrogen ion, and the resulting overexcretion of carbon dioxide further helps to keep the hydrogen-ion concentration below lethal limits.

Associated with diabetes are arteriosclerosis, small-vessel and nerve disease, susceptibility to infection, and a variety of other abnormalities. The precise causes of these long-term complications are not yet fully understood.

FIGURE 20-11 ***Effects of severe untreated insulin deficiency on renal function.***

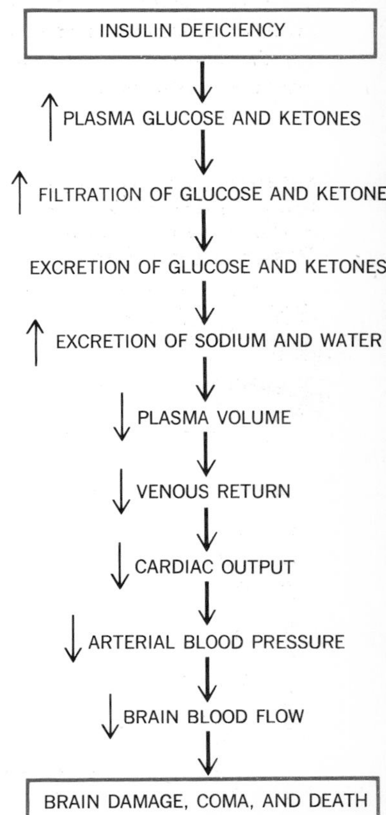

Presently being investigated is the observation that many diabetics actually manifest normal or elevated plasma insulin concentrations! Thus, in these cases, the diabetes is due not to an *absolute* insulin deficiency but to a *relative* deficiency. In other words, there must be some additional factor(s) present which antagonizes certain of the actions of insulin. Several hormones exert effects which oppose those of insulin, but in only a small number of patients can the disease be ascribed to excessive amounts of these hormones. An important clue may be the close relationship between obesity and development of diabetes in later life. Coupled with the fact that insulin exerts profound effects on adipose tissue, this association has stimulated much theorizing and research into the possible sequence of events. In any case, there is no question that simple weight reduction is frequently sufficient to eliminate the chemical manifestations of diabetes.

The treatment of diabetes is aimed at maintaining plasma glucose at a relatively normal value. The avoidance of concentrated sugars (candy, for example) helps, but the major therapy is administration of insulin, which must be given by injection since as a protein it is broken down by gastrointestinal enzymes. The dose must be determined carefully since an overdose abnormally lowers the plasma glucose concentration and causes brain damage, coma, and even death. Another type of therapy has proved useful; noninsulin drugs which can be taken by mouth act upon the islet cells to stimulate insulin secretion. Thus, therapy is actually accomplished with the patient's own insulin. Unfortunately, these drugs are effective in only a fraction of diabetics, the others having such severe islet-cell malfunction that drug stimulation of endogenous insulin secretion is not possible. Other types of oral medication are presently being introduced.

EPINEPHRINE, GLUCAGON, AND GROWTH HORMONE

Metabolic Effects

We have devoted so much space to the physiology of insulin because of its central role in regulating organic metabolism. Now we must turn to the three other hormones which play primary roles in controlling the metabolic adjustments required for feasting or fasting: epinephrine, the major hormone secreted by the adrenal medulla; glucagon, secreted by the α cells of the pancreatic islets; and growth hormone (GH), from the anterior pituitary. Most of the major effects of these hormones on organic metabolism are opposed to those of insulin and are listed in Table 20-2. The intensity of any given effect varies between hormones, and there still is controversy concerning the relative importance of some of these effects, but, for present purposes, we need not go into these details. Note that the overall

TABLE 20-2

MAJOR EFFECTS OF EPINEPHRINE, GLUCAGON, AND GROWTH HORMONE ON CARBOHYDRATE AND LIPID METABOLISM

	Increased glycogen breakdown (glycogenolysis)	***Increased liver gluconeogenesis***	***Increased breakdown of adipose-tissue triglyceride (fat mobilization)***	***Decreased glucose uptake by muscle and other tissues ("insulin antagonism")†***
Result:	↑Plasma glucose	↑Plasma glucose	↑Plasma fatty acids and glycerol	↑Plasma glucose
Epinephrine‡	Yes	Yes	Yes	No
Glucagon	Yes	Yes	Yes	No
Growth hormone	No	Yes	Yes	Yes

† This term has been used traditionally to denote the fact that GH interferes with insulin's stimulatory effect on the glucose transport system, but the mechanism of this effect is controversial.

‡ Activation of the sympathetic nerves to liver and adipose tissue produces effects virtually identical to those of epinephrine shown in the table.

results of all three hormones' effects are just the opposite of those of insulin: elevation of plasma glucose and fatty acid concentrations. The latter facilitates cellular utilization of fatty acids, glucose sparing, and maintenance of plasma glucose.

Control of Secretion

From a knowledge of these effects, one would logically suppose that the secretion of these hormones (and the activity of the sympathetic nerves to adipose tissue and hepatic cells) should be increased during the postabsorptive period and prolonged fasting, and such is the case. The stimulus is the same for all three, a decreased or decreasing plasma glucose concentration (although the receptors or pathways, or both, differ). The adaptive value of such reflexes is obvious; a decreasing plasma glucose stimulates increased release of the hormones which, by their effects on metabolism, serve to restore normal blood glucose levels and at the same time supply fatty acids for cell utilization. Conversely, an increased or increasing plasma glucose inhibits their secretion, thereby helping to return plasma glucose toward normal.

What are the receptors and pathways for these reflexes? That for glucagon is the simplest: The glucagon-secreting cells in the pancreas respond to changes in the glucose concentration of the blood perfusing the pancreas, no other nerves or hormones being involved. Thus the α and β cells of the pancreas constitute a push-pull system for regulating plasma glucose. As described in Chap. 8, epinephrine release is controlled entirely by the preganglionic sympathetic nerves to the adrenal medulla; the receptors initiating increased activity in these neurons (and the sympathetic pathways to the liver and adipose tissue) in response to changes in glucose are glucose receptors in the brain, probably in the hypothalamus (Fig. 20-12). As described in Chap. 12, growth-hormone secretion is directly controlled by a hypothalamic releasing factor. It is likely that the same brain glucose receptors described above for epinephrine communicate neurally with the hypothalamic neurons secreting GH-releasing factor so that a decreased glucose stimulates the release of GH-releasing factor, which then stimulates GH secretion by the anterior pituitary (Fig. 20-13); conversely, increased glucose inhibits the secretion of GH by decreasing the secretion of GH-releasing factor.

Thus far, the story is quite uncomplicated; the body produces two sets of hormones (insulin versus epinephrine, glucagon, and growth hormone) whose actions and controlling inputs are just the opposite of each other. However, we must now point out a complicating feature: A second major control of glucagon and growth-hormone secretion is the plasma amino acid concentration (acting on the α cells and brain amino acid receptors, respectively), and in this regard the effect is identical rather than opposite to that for insulin; glucagon and GH secretion, like that of insulin, is strongly stimulated by a rise in plasma amino acid concentration such as occurs following a protein-rich meal. Thus, during absorption of a carbohydrate-rich meal containing little protein, there occurs an increase in insulin secretion alone, caused by the rise in plasma glucose, but during absorption of a low-carbohydrate–high-protein meal, all three hormones increase, under the influence of the increased plasma amino acid concentration. The usual meal is somewhere between these extremes and is accompanied by a rise in insulin and relatively little change in glucagon and GH since the simultaneous increases in blood glucose and amino acids counteract each other so far as glucagon and GH secretion is concerned. Of course, regardless of the type of meal ingested, the postabsorptive period is always accompanied by a rise in glucagon and GH secretion.

What is the adaptive value of the amino acid–glucagon and GH relationship? Imagine what might occur were glucagon and GH secretion not part of the response to a high-protein meal: Insulin secretion would be increased by the amino acids but, since little carbohydrate was ingested and therefore available for absorption, the increase in plasma insulin could cause a marked and sudden drop in plasma glucose. In reality, the rise in glucagon and GH secretion caused by the amino acids permits the hyperglycemic effects of these hormones to counteract the hypoglycemic actions of insulin, and the net result is a stable plasma glucose. Thus, a high-protein meal virtually free of carbohydrate can be absorbed with little change in plasma glucose despite a marked increase in insulin secretion; moreover, the glucose derived from amino acids (under the influence of glucagon and GH) can be converted into fat for caloric storage (under the influence of insulin).

Finally, it should be noted that the secretion of growth hormone, epinephrine, and glucagon is stimulated by a variety of nonspecific "stresses," both physical and emotional. This constitutes a mechanism for the mobilization of energy stores for coping with a fight-or-flight situation. The GH-induced interference with glucose uptake might, at first thought, seem a maladaptive response since it might seem to reduce the uptake

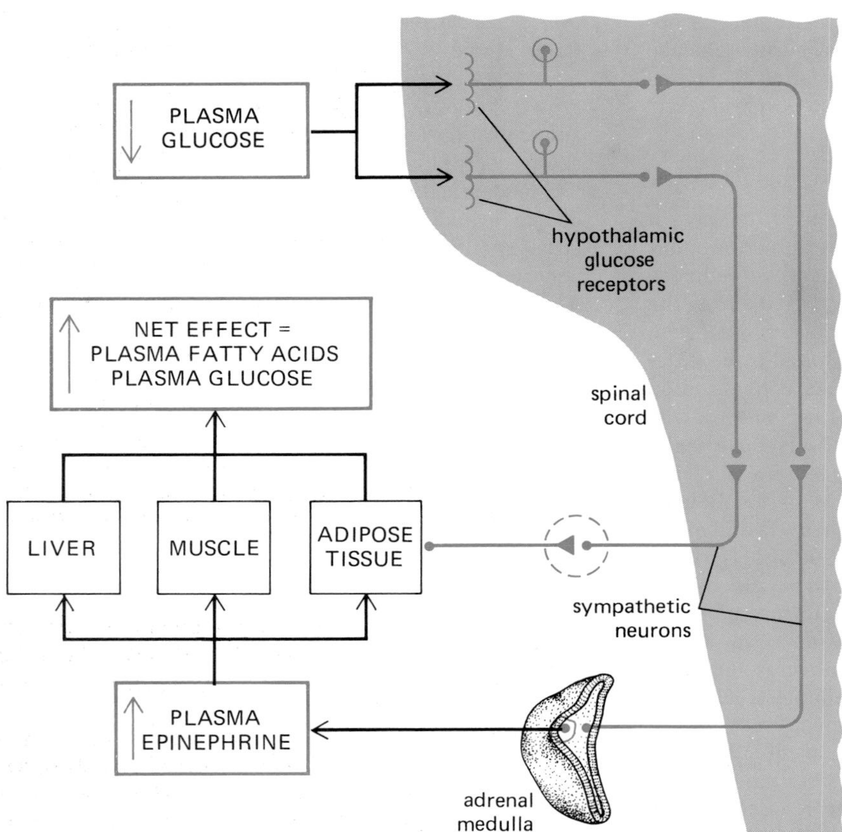

FIGURE 20-12 ***Control of epinephrine secretion and sympathetic nerves to adipose tissue by plasma glucose concentration. A decrease in plasma glucose stimulates the hypothalamic glucose receptors and, via the reflex chain shown, restores plasma glucose to normal while at the same time increasing plasma fatty acids.***

of glucose in exercising ("fighting" or "flighting") muscle; however, for reasons still poorly understood, glucose uptake by exercising muscle is very rapid and independent of circulating hormones. Moreover, mobilized fatty acids are also available as a major energy source.

Summary

The effects of epinephrine (and the sympathetic nerves to liver and adipose tissue), glucagon, and growth hormone are mainly opposed, in various ways, to the effects of insulin. To a great extent, insulin may be viewed as the "hormone of plenty" and the others as "hormones of fasting" (although, as described, this oversimplistic view must be qualified in several important ways). Insulin is increased during the absorptive period and decreased during fasting. In contrast, the other three hormones are increased during the postabsorptive period (and show varied responses to eating, depending upon the content of the meal). The influence of plasma glucose concentration is paramount in this regard, producing opposite effects on insulin secretion and on the secretion of the other three (Fig. 20-14).

In addition to these four hormones, there are others which have important effects on organic metabolism and the flow of nutrients. However, the secretion of these other hormones—cortisol, thyroxine, and the sex steroids—bears little or no relationship to the absorptive and postabsorptive states. This is not to say that they are secreted at constant rates but only that the rates are not determined by the state of glucose metabolism or by other indicators of the absorptive-postabsorptive state. Their physiology is described in detail in other chapters of this book.

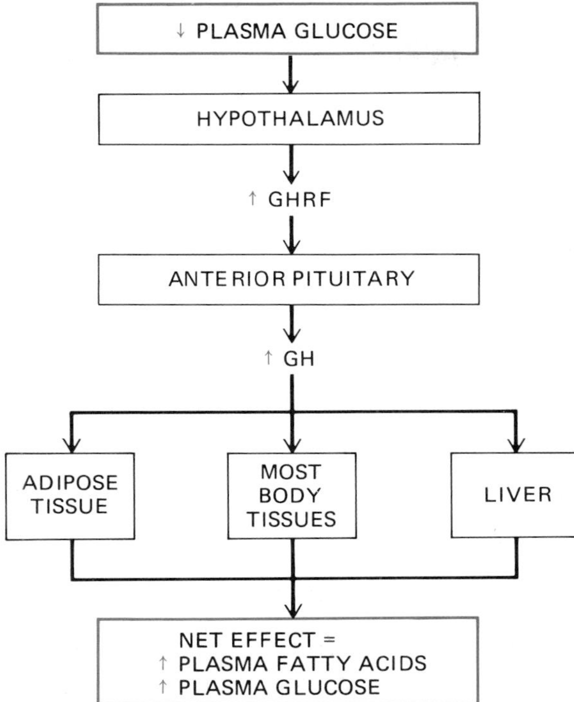

FIGURE 20-13 ***Control of growth-hormone secretion by plasma glucose concentration. A decrease in plasma glucose stimulates the hypothalamic glucose receptors, which in turn stimulate the increased release of GH-releasing factor into the hypothalamus–anterior pituitary portal capillaries. The resulting stimulation of growth-hormone secretion has the net effect of restoring normal plasma glucose while at the same time increasing plasma fatty acid concentration.***

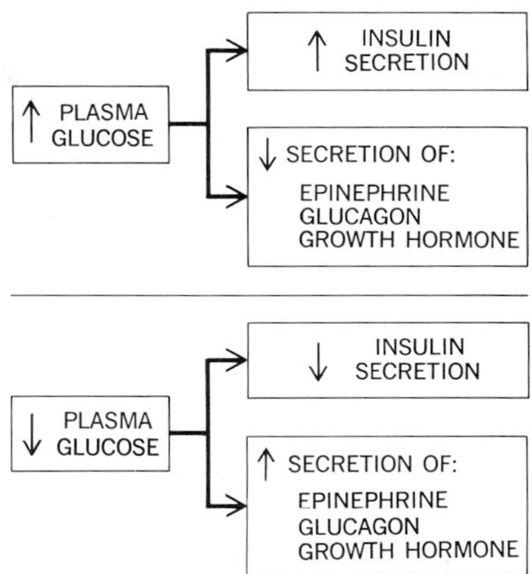

FIGURE 20-14 ***Summary of hormonal changes induced by changes in plasma glucose concentration.***

CONTROL OF GROWTH

A simple gain in body weight does not necessarily mean true growth since it may represent retention of either excess water or adipose tissue. In contrast, true growth usually involves lengthening of the long bones and increased cell division, but the real criteria are increased synthesis and accumulation of protein. Human beings manifest two periods of rapid growth (Fig. 20-15), one during the first 2 years of life, which is actually a continuation of rapid fetal growth, and the second during adolescence. Note, however, that total body growth may be a poor indicator of the rate of growth of specific organs (Fig. 20-15). We know relatively little about the control of many of these individual-organ growth patterns.

Another important implication of differential growth rates is that the so-called critical periods of development vary from organ to organ. Thus, a period of severe malnutrition during infancy when the brain is growing extremely rapidly may produce stunting of brain development which is irreversible, whereas reproductive organs would be little affected.

FIGURE 20-15 ***Rate of growth.***

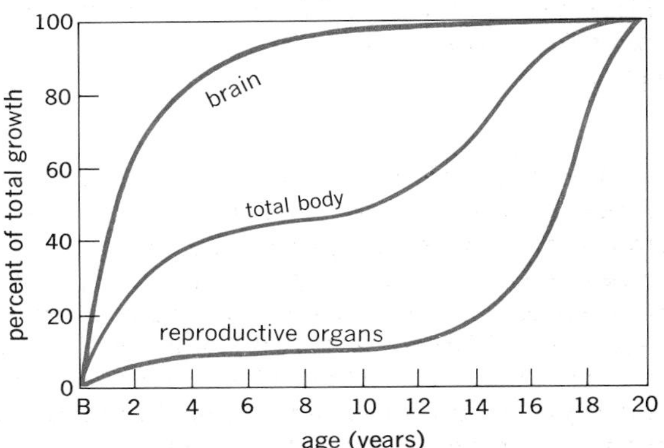

External Factors Influencing Growth

An individual's growth capacity is genetically determined, but there is no guarantee that the maximum capacity will be attained. Adequacy of food supply and freedom from disease are the primary external factors determining growth. Lack of sufficient amounts of any of the essential amino acids, essential fatty acids, vitamins, or minerals interferes with growth, and total protein and total calories must be adequate. No matter how much protein is ingested, growth cannot be normal if caloric intake is too low, since the protein is simply oxidized for energy.

Conversely, the profound stimulatory effects of good nutrition are well illustrated by the present teenage Japanese, who already seem to tower above their parents. On the other hand, it is important to realize that one cannot stimulate growth beyond the genetically determined maximum by eating more than adequate vitamins, protein, or total calories; this produces obesity, not growth.

Sickness can stunt growth, most likely because protein catabolism is enhanced by cortisol and other factors. If the illness is temporary, upon recovery the child manifests a remarkable growth spurt which rapidly brings him or her up to a normal growth curve. The mechanisms which control this important phenomenon are unknown, but the process illustrates the strength and precision of a genetically determined sequence.

Hormonal Influences on Growth

The hormones most important to human growth are growth hormone, thyroxine, insulin, androgens, and estrogen, which exert widespread effects. In addition, ACTH, TSH, prolactin, and FSH and LH selectively influence the growth and development of their target organs, the adrenal cortex, thyroid gland, breasts, and gonads, respectively.

Growth Hormone Removal of the pituitary in young animals arrests growth. Conversely, administration of large quantities of growth hormone to young animals causes excessive growth. When excess growth hormone is given to adult animals after the actively growing cartilaginous areas of the long bones have disappeared, it cannot lengthen the bones further, but it does produce the disfiguring bone thickening and overgrowth of other organs known as *acromegaly*. These experiments have spontaneously occurring human counterparts, as shown in Fig. 20-16. Thus growth hormone is essential for normal growth and in abnormally large amounts can cause excessive growth.

FIGURE 20-16 ***Progression of acromegaly.*** **(Top left)** ***Normal, age 9 years;*** **(top right)** ***age 16 years, with possible early coarsening of features;*** **(bottom left)** ***age 33 years, well-established acromegaly;*** **(bottom right)** ***age 52 years, end stage, acromegaly with gross disfigurement.*** **(From William H. Daughaday, The Adenohypophysis, in Robert H. Williams (ed.), "Textbook of Endocrinology," 4th ed., p. 74, fig. 2-28, Saunders, Philadelphia, 1968.)**

The effects of growth hormone on carbohydrate and fat metabolism have already been described; its growth-promoting effects are due primarily to its ability to stimulate protein synthesis. This it does by increasing membrane transport of amino acids into cells and by stimulating the synthesis of RNA, events essential for protein synthesis. Growth hormone also causes large increases of mitotic activity and cell division, the other major components of growth.

The effects on bone are dramatic. Growth hormone promotes bone lengthening by stimulating protein synthesis in both the cartilaginous center and bony edge of the epiphyseal plates as well as by increasing the rate of osteoblast mitosis (Chap. 7). It has now become clear that these effects on bone are mediated not by growth hormone, per se, but by a substance *somatomedin,* which growth hormone causes to be released by the liver.

Thyroxine Infants and children with deficient thyroid function manifest retarded growth, which can be restored to normal by administration of physiologic quantities of thyroxine. Administration of excess thyroxine, however, does not cause excessive growth (as was true of growth hormone) but marked catabolism of protein and other nutrients, as will be explained in the section on energy balance. The essential point is that normal amounts of thyroxine are necessary for normal growth, the most likely explanation apparently being that thyroxine somehow promotes the effects of growth hormone on protein synthesis; certainly the absence of thyroxine significantly reduces the ability of growth hormone to stimulate amino acid uptake and RNA synthesis. Thyroxine also plays a crucial role in the closely related area of organ development, particularly that of the central nervous system. Hypothyroid infants (*cretins*) are mentally retarded, a defect that can be completely repaired by adequate treatment with thyroid hormone although if the infant is untreated for long, the developmental failure is largely irreversible. The defect is probably due to failure of nerve myelination which occurs as a result of thyroid deficiency.

Insulin It should not be surprising that adequate amounts of insulin are necessary for normal growth since insulin is, in all respects, an anabolic hormone. Its effects on amino acid uptake and protein are particularly important in favoring growth.

Androgens and Estrogen In Chap. 21 we shall describe in detail the various functions of the sex hormones in directing the growth and development of the sexual organs and the obvious physical characteristics which distinguish male from female. Here we are concerned only with the effects of these hormones on general body growth.

Sex-hormone secretion begins in earnest at about the age of 8 to 10 and progressively increases to reach a plateau within 5 to 10 years. The testicular hormone, testosterone, is the major male sex hormone, but other androgens similar to it are also secreted in significant amounts by the adrenal cortex of both sexes. Females manifest a sizable increase of adrenal androgen secretion during adolescence. However, the adrenal androgens are not nearly so potent as testosterone. During adolescence the large increases in secretion of estrogen, the dominant female sex hormone, are virtually limited to the female. Thus the relative quantities of androgen and estrogen are very different between the sexes.

Androgens strongly stimulate protein synthesis in many organs of the body, not just the reproductive organs, and the adolescent growth spurt in both sexes is due, at least in part, to these anabolic effects. Similarly, the increased muscle mass of men compared with women may reflect their greater amount of more potent androgen. Androgens stimulate bone growth but also ultimately stop bone growth by inducing complete conversion of the epiphyseal plates. This accounts for the pattern seen in adolescence, i.e., rapid lengthening of the bones culminating in complete cessation of growth for life, and explains several clinical situations: (1) Unusually small children treated before puberty with large amounts of testosterone may grow several inches very rapidly but then stop completely and (2) eunuchs may be very tall because bone growth, although slower, continues much longer due to persistence of the epiphyseal plates.

Estrogen profoundly stimulates growth of the female sexual organs and sexual characteristics during adolescence (Chap. 21), but, unlike the androgens, it has relatively little direct anabolic effect on nonsexual organs and tissues. Present evidence suggests that the major function of estrogen in the adolescent growth spurt and ultimate closure of the epiphyses is to stimulate the adrenal secretion of androgens which then directly mediate these responses.

Compensatory Hypertrophy

We have dealt thus far only with growth during childhood. During adult life, maintenance of the status quo is achieved by the mechanisms described earlier in this chapter. In addition, a specific type of organ growth, known as *compensatory hypertrophy,* can occur in many human organs and is actually a type of regeneration. For example, within 24 h of the surgical removal of one kidney, the cells of the other begin to manifest increased mitotic activity, ultimately growing until the total mass approaches the initial mass of the two kidneys combined. What causes this compensatory growth? It certainly does not depend upon the nerves to the organ since it still occurs after their removal or destruction. Nor has it been possible to attribute it to any known hormone. Several types of experiments indicate, however, that some unidentified blood-borne agent is responsible. For example, if one removes 75 percent of the liver from one of two parabiotic rats, i.e., surgically created Siamese twins, the livers of both rats increase in size.

Section B. Regulation of Total Body Energy Balance

BASIC CONCEPTS OF ENERGY EXPENDITURE AND CALORIC BALANCE

The breakdown of organic molecules liberates the energy locked in their intramolecular bonds (Chap. 1). This is the source of energy utilized by cells in their performance of the various forms of biological work (muscle contraction, active transport, synthesis of molecules, etc.). As described in Chap. 1, the first law of thermodynamics states that energy can neither be created nor destroyed but can be converted from one form to another. Thus, internal energy liberated (ΔE) during breakdown of an organic molecule can either appear as heat (H) or be used for performing work (W).

$$\Delta E = H + W$$

In all animal cells, most of the energy appears immediately as heat, and only a small fraction is used for work.

(As described in Chap. 1, the energy used for work must first be incorporated into molecules of ATP, the subsequent breakdown of which serves as immediate energy source for the work.) It is essential to realize that the body is not a heat engine since it is totally incapable of converting heat into work. The heat is, of course, valuable for maintaining body temperature.

It is customary to divide biological work into two general cateogories: (1) *external work,* i.e., movement of external objects by contracting skeletal muscles, and (2) *internal work,* which comprises all other forms of biological work, including skeletal muscle activity not moving external objects. As we have seen, most of the energy liberated from the catabolism of nutrients appears immediately as heat, only a small fraction being used for performance of external or internal work. What may not be obvious is that all internal work is ultimately transformed into heat except during periods of growth (Fig. 20-17). Several examples will illustrate this essential point:

1 Internal work is performed during cardiac contraction, but this energy appears ultimately as heat generated by the resistance (friction) to flow offered by the blood vessels.

2 Internal work is performed during secretion of HCl by the stomach and $NaHCO_3$ by the pancreas, but this work appears as heat when the H^+ and HCO_3^- react in the small intestine.

3 The internal work performed during synthesis of a plasma protein is recovered as heat during the inevitable catabolism of the protein, since, with few exceptions, all bodily constituents are constantly being built up and broken down. However, during periods of net synthesis of protein, fat, etc., energy is stored in the bonds of these molecules and does not appear as heat.

FIGURE 20-17 ***General pattern of energy liberation in a biological system. Most of the energy released when nutrients, such as glucose, are broken down appears immediately as heat. A smaller fraction goes to form ATP, which can be subsequently broken down and the released energy coupled to biological work. Ultimately, the energy which performs this work is also completely converted into heat.***

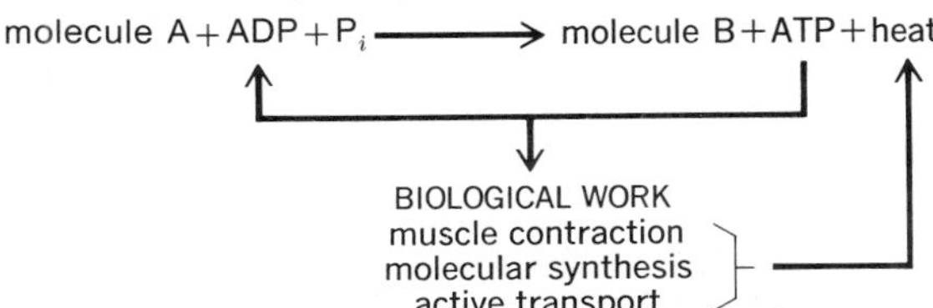

Thus, the total energy liberated when organic nutrients are catabolized by cells may be transformed into body heat, appear as external work, or be stored in the body in the form of organic molecules, the latter occurring only during periods of growth (or net fat deposition in obesity). The total energy expenditure of the body is therefore given by the equation

Total energy expenditure =
heat produced + external work + energy shortage

The units for energy are kilocalories (Chap. 1), and total energy expenditure per unit time is called the *metabolic rate*.[4]

In human beings the metabolic rate can be measured directly or indirectly. In either case the measurement is much simpler if the person is fasting and at rest; total energy expenditure then becomes equal to heat production since energy storage and external work are eliminated. The direct method is simple to understand but difficult to perform; the subject is placed in a *calorimeter,* an instrument large enough to accommodate people, and his or her heat production is measured by the temperature changes in the calorimeter. This is an excellent method in that it measures heat production directly, but calorimeters are found in only a few research laboratories; accordingly, a simple, indirect method has been developed for widespread use.

Using the indirect procedure, one simply measures the subject's oxygen uptake per unit time (by measuring total ventilation and Po_2 of inspired and expired air). From this value one calculates heat production based on the fundamental principle (Fig. 20-18) that the energy liberated by the catabolism of foods in the body must be the same as when they are catabolized outside the body. We know precisely how much heat is liberated when 1 l of oxygen is consumed in the oxidation of fat, protein, or carbohydrate outside the body; this same quantity of heat must be produced when 1 l of oxygen is consumed in the body. Fortunately, we do not need to know precisely which type of nutrient is being oxidized internally, because the quantities of heat produced per liter of oxygen consumed are reasonably similar for the oxidation of fat, carbohydrate, and protein, and average 4.8 kcal/l of oxygen. When more exact calculations are required, it is possible to estimate the relative quantity of each nutrient. Figure 20-18 presents values obtained by the indirect method for a

[4] In the field of nutrition, one Calorie implies, by convention, one *large* calorie, which is actually one kilocalorie.

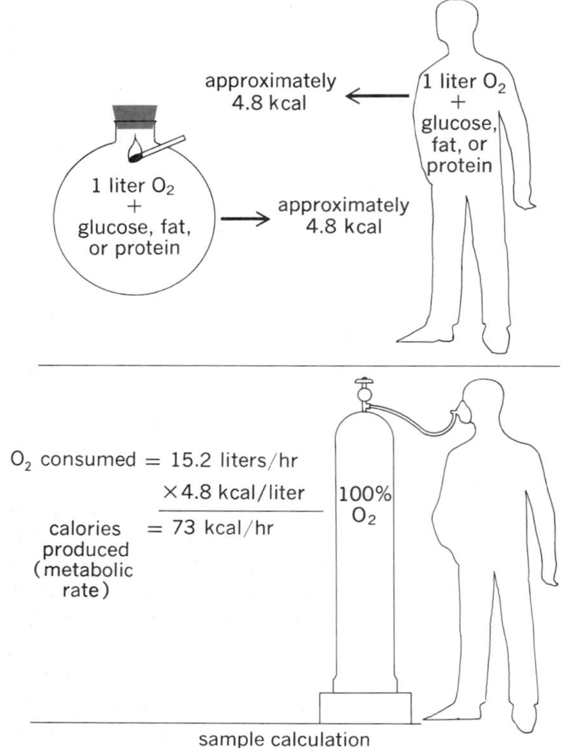

FIGURE 20-18 ***Indirect method for measuring metabolic rate. The calculation depends upon the basic principle that when 1 l of oxygen is utilized in the oxidation of organic nutrients approximately 4.8 kcal is liberated.***

normal fasted resting adult male. This man's rate of heat production is approximately equal to that of a single 100-W bulb.

Determinants of Metabolic Rate

Since many factors cause the metabolic rate to vary (Table 20-3), when one wishes to compare metabolic rates of different people, it is essential to control as many of the variables as possible. The test used clinically and experimentally to find the *basal metabolic rate* (BMR) tries to accomplish this by standardizing conditions: The subject is at mental and physical rest, in a room at comfortable temperatures, and has not eaten for at least 12 h. These conditions are arbitrarily designated basal, the metabolic rate during sleep being actually less than the BMR. The measured BMR is then compared with previously determined normal values for a person of the same weight, height, age, and sex.

TABLE 20-3
FACTORS AFFECTING THE METABOLIC RATE

Age
Sex
Height, weight, and surface area
Growth
Pregnancy, menstruation, lactation
Infection or other disease
Body temperature
Recent ingestion of food (SDA)
Prolonged fasting
Muscular activity
Emotional state
Sleep
Environmental temperature
Circulating levels of various hormones, especially epinephrine and thyroxine

BMR is often appropriately termed the metabolic *cost of living*. Under these conditions, most of the energy is expended, as might be imagined, by the heart, liver, kidneys, and brain. Its magnitude is related not only to physical size but to age and sex as well. Growing children's resting metabolic rate, relative to their size, is considerably higher than adults' because they expend a great deal of energy in net synthesis of new tissue. On the other end of the age scale, the metabolic cost of living gradually decreases with advancing age, for unknown reasons. The female's resting metabolic rate is generally less than that of the male (even taking into account size differences) but increases markedly, for obvious reasons, during pregnancy and lactation. The greater demands upon the body by infection or other disease generally increase total energy expenditure; moreover the presence of fever directly stimulates metabolic reaction rates and increases metabolic rate.

The ingestion of food also increases the metabolic rate, as shown by measuring the oxygen consumption or heat production of a resting man before and after eating; the metabolic rate is 10 to 20 percent higher after eating. This effect of food on metabolic rate is known as the *specific dynamic action* (SDA). Protein gives the greatest effect, carbohydrate and fat much less. The cause of SDA is not what one might expect, namely, the energy expended in the digestion and absorption of ingested food. These processes account for only a small fraction of the increased metabolic rate: Intravenous administration of amino acids produces almost the same SDA effect as oral ingestion of the same material. Most of the increased heat production appears to be secondary to the processing of exogenous nutrients by the liver, since it does not occur in an animal whose liver has been removed. On an average diet, SDA contributes to the total metabolic rate approximately one-tenth as much as the basal metabolism does. In contrast to eating, prolonged fasting causes a decrease in metabolic rate. This is due, in part, simply to reduction of body mass, but even when expressed on a per weight basis, metabolic rate is reduced. The mechanism is unclear, but the adaptive value of this change may be considerable since it decreases the amount of nutrient stores which must be catabolized each day.

All these influences on metabolic rate are small compared with the effects of *muscular activity* (Table 20-4). Even minimal increases in muscle tone significantly increase metabolic rate, and severe exercise may raise heat production more than fifteenfold. Changes in muscle activity also explain part of the effects on metabolic rate of sleep (decreased muscle tone), reduced environmental temperature (increased muscle tone and shivering), and emotional state (unconscious changes in muscle tone).

TABLE 20-4
ENERGY EXPENDITURE DURING DIFFERENT TYPES OF ACTIVITY FOR A 70-KG MAN

Form of activity	*kcal/h*
Awake, lying still	77
Sitting at rest	100
Typewriting rapidly	140
Dressing or undressing	150
Walking level at 2.6 mi/h	200
Sexual intercourse	280
Bicycling on level, 5.5 mi/h	304
Walking 3 percent grade at 2.6 mi/h	357
Sawing wood or shoveling snow	480
Jogging (5.3 mi/h)	570
Rowing 20 strokes/min	828
Maximal activity (untrained)	1,440

Metabolic rate is strongly influenced by the hormones *epinephrine* and *thyroxine*. The intravenous injection of epinephrine may promptly increase heat production by more than 30 percent. As we have seen, epinephrine has powerful effects on organic metabolism, and its calorigenic, i.e., heat-producing, effect is probably related to its stimulation of glycogen and triglyceride catabolism since ATP splitting and energy liberation occur in both the breakdown and the subsequent resynthesis of these molecules. Regardless of the mechanism, whenever epinephrine secretion is stimulated, the metabolic rate rises. This probably accounts for part of the greater heat production associated with emotional stress, although increased muscle tone is also contributory.

Thyroxine also increases the oxygen consumption and heat production of most body tissues, a notable exception being the brain. In contrast to epinephrine, this calorigenic effect does not begin for 6 to 12 h but lasts for many days, even after a single injection. So powerful is this effect that long-term excessive thyroxine, as in patients with hyperthroidism, induces a host of effects secondary to the hypermetabolism which well illustrate the interdependence of bodily functions. The increased metabolic demands markedly increase hunger and food intake; the greater intake frequently remains inadequate to meet the metabolic needs, and net catabolism of endogenous protein and fat stores leads to loss of body weight; excessive loss of skeletal muscle protein results in muscle weakness; catabolism of bone protein weakens the bones and liberates large quantities of calcium into the extracellular fluid, resulting in increased plasma and urinary calcium; the hypermetabolism increases the requirement for vitamins, and vitamin deficiency diseases may occur; respiration is increased to supply the required additional oxygen; cardiac output is also increased, and, if prolonged, the enhanced cardiac demands may cause heart failure; the greater heat production activates heat-dissipating mechanisms, and the patient suffers from marked intolerance to warm environments. These are only a few of the many results induced by thyroxine's calorigenic effect. The important effects of thyroxine relating to growth and development, described earlier, appear to be quite distinct from the calorigenic effect. The mechanisms by which thyroxine exerts this profound effect on oxygen consumption and heat production are presently unclear.

Determinants of Total Body Caloric Balance

Using the basic concepts of energy expenditure and metabolic rate as a foundation, we can consider total body caloric-fuel balance in much the same way as any other balance, i.e., in terms of input and output. The laws of thermodynamics dictate that, in the steady state, the total caloric expenditure of the body equals total body caloric-fuel input. We have already identified the ultimate forms of energy expenditure: internal heat production, external work, and net molecular synthesis (energy storage). The source of input, of course, is the energy contained in ingested food. Therefore the caloric-balance equation is

$$\text{Food energy intake} = \begin{matrix}\text{internal}\\ \text{heat}\\ \text{produced}\end{matrix} + \begin{matrix}\text{external}\\ \text{work}\end{matrix} + \begin{matrix}\text{energy}\\ \text{storage}\end{matrix}$$

Our equation includes no term for loss of fuel from the body via urinary excretion of nutrients. In a normal person, almost all the carbohydrate, amino acids, and lipids filtered at the glomerulus are reabsorbed by the tubules, so that the kidneys play no significant role in the regulation of caloric balance. In certain diseases, however, the most important being diabetes, urinary losses of organic molecules may be quite large and would have to be included in the equation. In all normal persons very small losses occur via the urine and feces and also as sloughed hair and skin, but we can ignore them as being negligible.

As predicted by this caloric-balance equation, three states are possible:

Food intake = internal heat production + external work (body weight constant)

Food intake > internal heat production + external work (body weight increases)

Food intake < internal heat production + external work (body weight decreases)

In most adults, body weight remains remarkably constant over long periods of time, implying that precise physiologic regulatory mechanisms operate to control (1) food intake or (2) internal heat production plus external work or (3) both. Actually, all these variables are subject to control in human beings, but the amount of food intake is the dominant factor. Control mechanisms for heat production are aimed primarily at regulating body temperature, rather than total caloric balance. For example, when someone is cold, her body produces additional heat by shivering even if she is starving; conversely, a fat person is not automatically impelled by his hypothalamus to run around the block—quite the reverse in most cases. It is essential to understand that as shown by the caloric-balance equation, an individual's degree of activity, i.e., heat production plus external work, *is* one of the essential determinants of total body energy balance, but its automatic physiologic control is not aimed primarily at achieving such a balance. Moreover, a man's total activity generally reflects the kind of work he does, his inclination toward sports, etc. The important generalization is that food intake is the major factor being automatically controlled so as to maintain caloric balance and constant body weight. To alter the two examples cited above: When exposure to cold or running around the block causes increased energy expenditure, the individual automatically increases his or her food intake by an amount sufficient to match the additional energy expended (this example, however, will be qualified later in the section on obesity).

CONTROL OF FOOD INTAKE

Hypothalamic Integration Centers and Satiety Signals

The structures primarily concerned in the control of food intake are several clusters of nerve cells (nuclei) in the hypothalamus. The lateral hypothalamus contains a *feeding center* which stimulates the efferent output controlling both the final motor acts of eating and such associated behavior as food seeking; the ventromedial hypothalamic neurons can, via synaptic input, inhibit the activity of this feeding center; thus, eating proceeds unless the midline centers are stimulated so as to inhibit the outer centers. The stimulatory inputs to the midline centers are appropriately known as *satiety signals,* and the centers themselves are termed the *satiety centers.*

The hypothalamic centers involved in control of food intake serve only as integrating centers processing afferent input and controlling efferent output. The afferent input must provide the critical information about the body's need for food. Although obviously food intake is often stimulated by the sight or smell of food, present thinking is that food intake is basically a tonic process which continues unless turned off by an input signaling satiety. In a sense, we start eating not because we become hungry but because we stop being satiated.

Early observations that hunger was generally associated with contractions of the empty stomach and that gastric distension could lessen hunger led to the hypothesis that hunger and satiety were signaled by afferent pathways from the stomach and other areas of the gastrointestinal tract. However, experiments have since proved that complete denervation of the upper gastrointestinal tract does not interfere with normal maintenance of energy balance, and thus gastrointestinal signals cannot constitute the major long-term regulators of food intake although they may play a modifying role. Nor does it seem possible, on purely theoretical grounds, that a bulk-detecting system could maintain energy balance, since the caloric content of food may bear no relationship to its bulk.

This leads us to the perplexing problem of what characteristics a receptor and environmental signal must have in order to detect total body energy content, which, after all, is the variable actually being regulated. Wide experimentation has led to three quite different but by no means mutually exclusive hypotheses, termed the glucostatic, lipostatic, and thermostatic theories. Most experts concede that none of them can explain all the observations and that probably all their postulated pathways contribute to the overall control of food intake.

Glucostatic Theory There is no doubt that some type of glucose receptor exists in the brain. Present evidence suggests that it is in the ventromedial hypothalamus and is perhaps the same group of neurons which constitute the satiety center. It is also possible that these hypothalamic glucose receptors are the same ones which initiate the reflexes leading to release of epinephrine and growth hormone when plasma glucose is reduced. For reasons already discussed, the plasma glucose concentration and, more specifically, the rate of cellular glucose utilization increase during or after eating and decrease during fasting. Detection of greater glucose utilization could signal satiety and inhibit eating (Fig. 20-19). Conversely, fasting would decrease glucose utilization, remove the input signaling satiety, and thereby promote eating.

Lipostatic Theory Basically this theory postulates that an ideal indicator of total body energy content would be a substance released from fat stores (adipose tissue) in direct proportion to their total mass. Thus, a positive energy balance would increase the amount of adipose tissue, which in turn would signal satiety. Such a mechanism would constitute an excellent long-term regulator, but it must be confessed that the concept is still highly theoretical with little experimental evidence either to support or disprove it.

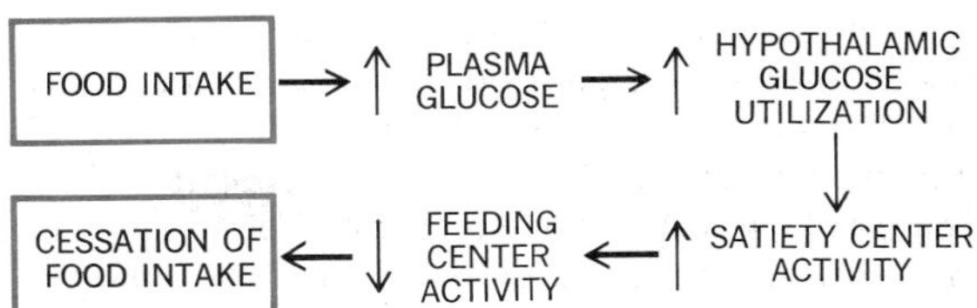

FIGURE 20-19 ***Glucostatic theory of food-intake control. Greater glucose utilization is the satiety signal.***

Thermostatic Theory The specific dynamic action of food, i.e., the increase in metabolic rate induced by eating, tends to raise body temperature, and it seems likely that temperature elevation may constitute a satiety signal. Such a mechanism would also be consistent with the fact that people eat more in colder climates than in warm ones.

Psychologic Components of Food-intake Control

Although total caloric balance unquestionably eflects, in large part, the reflex input from some combination of glucostats, lipostats, and thermostats, it also is strongly influenced by the reinforcement (both positive and negative) of such things as smell, taste, texture, psychologic associations, etc. Thus, the behavioral concepts of reinforcement, drive, and motivation described in Chap. 11 must be incorporated into any comprehensive theory of food-intake control. Obviously, these psychologic factors having little to do with energy balance are of very great importance in obese persons. It should be emphasized, however, that most people whose obesity is ascribed to psychologic factors do not continuously gain weight; their automatic homeostatic control mechanisms are operative but maintain total body energy content at supranormal levels. (See Table 20-5.)

TABLE 20-5
SUMMARY OF FACTORS WHICH INFLUENCE HYPOTHALAMIC CENTERS CONTROLLING FOOD INTAKE

Plasma glucose
Total body adipose tissue
Body temperature
State of gastrointestinal distension
Psychologic, social, and economic factors

Obesity

Obesity has been called the most common disease in America. The term disease is perfectly justified since obesity predisposes to illness and premature death from a multitude of causes. The seriousness of being overweight is underlined by statistics, which show a mortality rate more than 50 percent greater than normal in overweight persons in the same age groups. Our view of the etiology of obesity has undergone radical changes both in the past and recently. Most obesity was once ascribed to "glandular conditions." Later, the dictum became "You're fat because you eat too much." This is an unassailable fact but completely evades the issue of how much is "too much." The energy-balance equation clearly shows that too much simply means more than is needed to supply the energy needs of the body. Since beyond this point, additional food is stored as fat, the amount one eats must be viewed in relationship to one's activity pattern. For example, a study of obese high school girls revealed that they ate, on the average, *less* than a control group of normal-weight girls. The obese girls had much less physical activity than the control group and were eating "too much" only in relationship to their physical activity.

This example stresses the fact, so easily forgotten, that there are two sides to the energy-balance equation. Moreover, recent studies have revealed a startling physiologic relationship, namely, that low levels of physical activity may cause increased eating. The caloric intakes and body weights of large numbers of workers in the same factory in India were studied after grouping the men according to the physical exertion required by their jobs. Levels of activity below a certain arbitrary minimum were classified as sedentary. As shown in Fig. 20-20, men performing work loads above the sedentary range displayed the expected pattern; caloric intake was directly proportional to work level, and body weights for all groups of men were similar. The unexpected finding was that for men performing small work loads in the sedentary range, caloric intake varied inversely with work load, i.e., the less physical activity the men performed, the more they ate. Accordingly, these men were considerably fatter, on the average, than the other men. It is therefore apparent that very low levels of activity do not induce similar reductions of food intake but actually stimulate eating. The implications of these findings for energy balance in a society where so many of us fall into the sedentary category are obvious, and the elucidation of the factors responsible will be of considerable importance.

Having stressed the importance of physical activity (or rather its lack) in the etiology of obesity, we must admit that most cases of obesity can be explained only in part on this basis. *Ultimately, all obesity represents failure of normal food-intake control mechanisms.* The inappropriate effects of low activity levels on food intake merely represent one example of this failure.

Finally, it should be recognized that obesity is only one form of "overnutrition" endemic to westernized societies. There is no question that deficits of essential nutritients produce disease, but the question of whether excessive amounts of these same nutrients might be harmful has received far less attention.

REGULATION OF BODY TEMPERATURE

Animals capable of maintaining their body temperatures within very narrow limits are termed *homeothermic*. The adaptive significance of this ability stems primarily from the marked effects of temperature upon the rate of chemical reactions in general and enzyme activity in particular. Homeothermic animals are spared the slowdown of all bodily functions which occurs when the body temperature falls. However, the advantages obtained by a relatively high body temperature impose a great need for precise regulatory mechanisms since even moderate elevations of temperatures begin to cause nerve malfunction, protein denaturation, and death. Most people suffer convulsions at a body temperature of 106 to 107°F, and 110°F is the absolute limit for life. In contrast, most body tissues can withstand marked cooling (to less than 45°F), which has found an important place in surgery when the heart must be stopped, since the dormant cold tissues require little nourishment.

Figure 20-21 illustrates several important generalizations about normal body temperature in human beings: (1) Oral temperature averages about 1.0°F less than rectal; thus, all parts of the body do not have the same temperature. (2) Internal temperature is not absolutely constant but varies several degrees in perfectly normal persons in response to activity pattern and external temperature, and in addition, there is a characteristic diurnal fluctuation, so that temperature is lowest during sleep and slightly higher during the awake state even if the person remains relaxed in bed. An interesting variation in women is a higher temperature during the last half of the menstrual cycle (Chap. 21).

If temperature is viewed as a measure of heat "con-

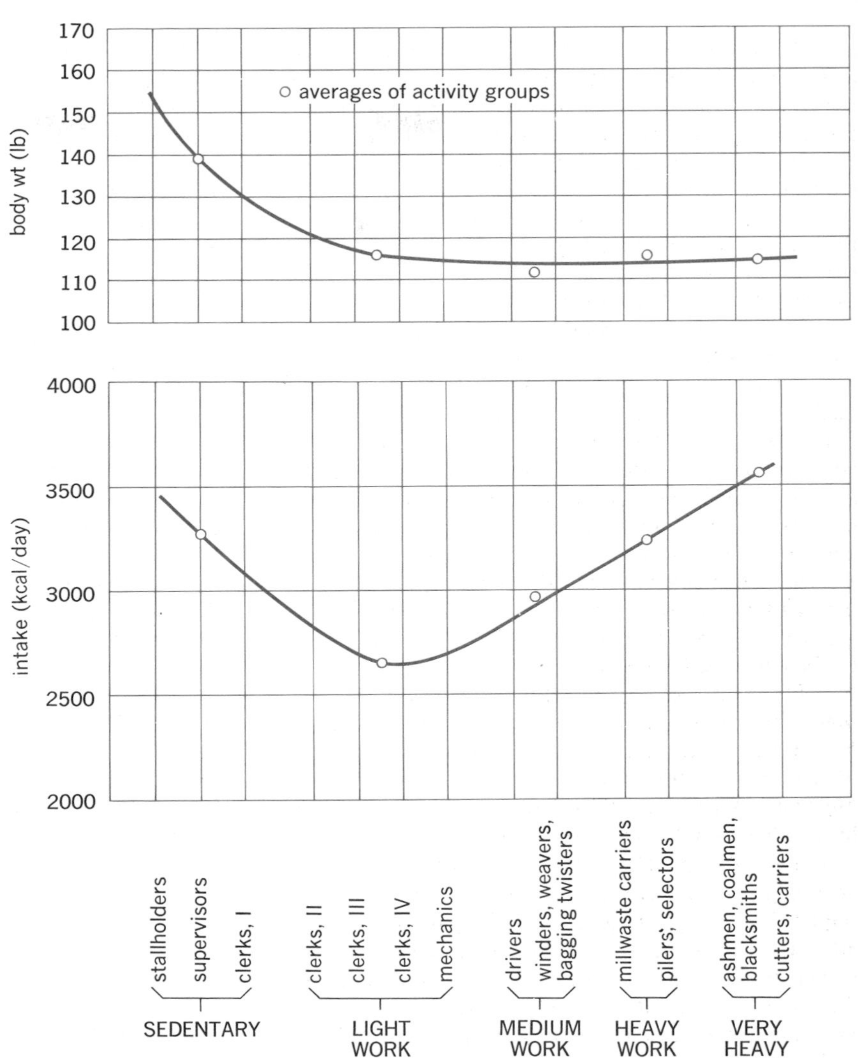

FIGURE 20-20 ***Body weight and caloric intake as functions of physical activity in workers at an Indian factory.*** **(Adapted from Mayer et al.)**

centration," temperature regulation can be studied by our usual balance methods. In this case, the total heat content of the body is determined by net difference between heat produced and heat lost from the body. Maintaining a constant body temperature implies that, overall, heat production must equal heat loss. Both these variables are subject to precise physiologic control.

Temperature regulation offers a classic example of a biological control system; its generalized components are shown in Fig. 20-22. The balance between heat production and heat loss is continuously being disturbed, either by changes in metabolic rate (exercise being the most powerful influence) or by changes in the external environment which alter heat loss. The resulting small changes in body temperature reflexly alter the

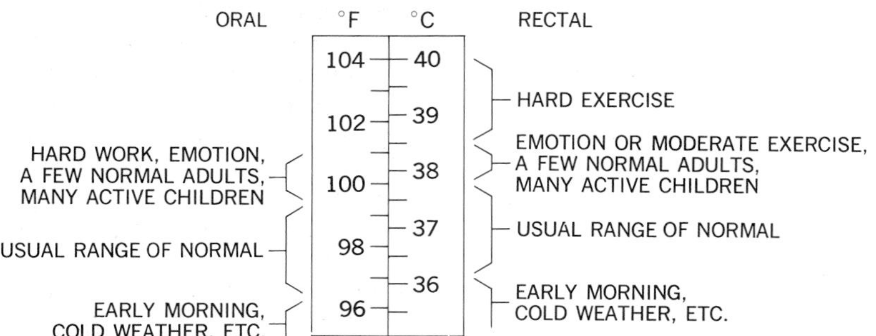

FIGURE 20-21 ***Ranges of body temperatures in normal persons.*** **(Adapted from Dubois.)**

output of the effector organs, which drive heat production or heat loss and restore normal body temperature.

Heat Production

The basic concepts of heat production have already been described. Recall that heat is produced by virtually all chemical reactions occurring in the body and that the cost-of-living metabolism by all organs sets the basal level of heat production, which can be increased as a result of skeletal muscular contraction or the action of several hormones.

Changes in Muscle Activity The first muscle changes in response to cold are a gradual and general increase in skeletal *muscle tone*. This soon leads to *shivering*, the characteristic muscle response to cold, which consists of oscillating rhythmic muscle tremors occurring at the rate of about 10 to 20 per second. So effective are these contractions that body heat production may be increased severalfold within seconds to minutes. Because no external work is performed, all the energy liberated by the metabolic machinery appears as internal heat. As always, the contractions are directly controlled by the efferent motor neurons to the muscles. During shivering these nerves are controlled by descending pathways under the primary control of the hypothalamus. Besides increased muscle tone and shivering, which are completely reflex in nature, human beings also use voluntary heat-production mechanisms such as foot stamping, hand clapping, etc.

FIGURE 20-22 ***Summary of temperature regulation. Heat loss from the body depends directly upon the external environment and upon changes controlled by temperature-regulating reflexes.***

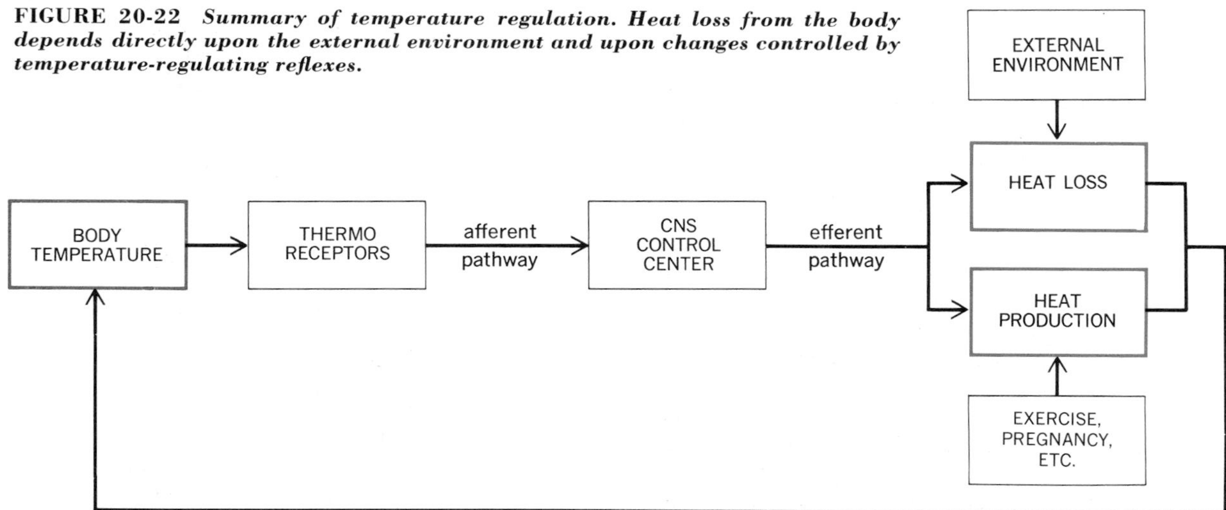

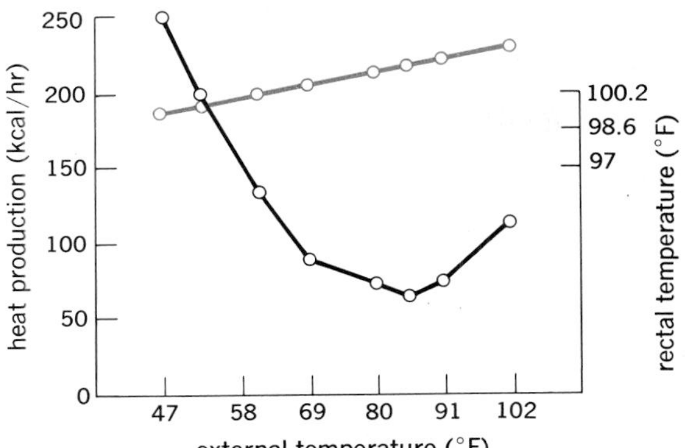

FIGURE 20-23 ***Effects of altering external temperature upon metabolic rate (heat production) and body (rectal) temperature. The subject was lightly clothed and seated quietly.***

Thus far, our discussion has focused primarily on the muscular response to cold; the opposite reactions occur in response to heat. Muscle tone is reflexly decreased and voluntary movement is also diminished ("It's too hot to move."). However, these attempts to reduce heat production are relatively limited in capacity both because muscle tone is already quite low normally and because of the *direct* effect of a body temperature increase on metabolic rate. As shown in Fig. 20-23, heat production rose as environmental temperature was cooled below normal, this response being due to increased muscle tone and shivering. As room temperature was heated above normal, there was a small drop in metabolic rate (decreased muscle tone), but the highest temperatures again elicited a higher metabolic rate, certainly an inappropriate response as far as temperature regulation is concerned. The explanation is that *body* temperature rose slightly at this point and directly increased the rate of all metabolic reactions; human beings' ability to regulate temperature by *decreasing* metabolic rate is seen to be quite limited.

Nonshivering ("Chemical") Thermogenesis In most experimental animals chronic cold exposure induces an increase in metabolic rate, which is not due to increased muscle activity. Indeed, as this so-called nonshivering thermogenesis increases over time, it is associated with a decrease in the degree of shivering. The cause of nonshivering thermogenesis has been the subject of considerable controversy; present evidence suggests that it is due mainly to an increased secretion of epinephrine and activity of the sympathetic nerves to adipose tissue. (Thyroxine may also be involved but this is much less clear.) Equally controversial is the question whether nonshivering thermogenesis is a significant phenomenon in human beings. Regardless of the outcomes of these debates, it seems clear that human hormonal changes and nonshivering thermogenesis are of secondary importance and that changes in muscle activity constitute the major control of heat production for temperature regulation, at least in the early response to temperature changes.

Heat-loss Mechanisms

The surface of the body exchanges heat with the external environment by radiation, conduction and convection (Fig. 20-24), and water evaporation.

FIGURE 20-24 ***Mechanisms of heat transfer. By radiation, heat is transferred by electromagnetic waves*** **(solid arrows);** ***in conduction and convection, heat moves by direct transfer of thermal energy from molecule to molecule*** **(dashed arrow).**

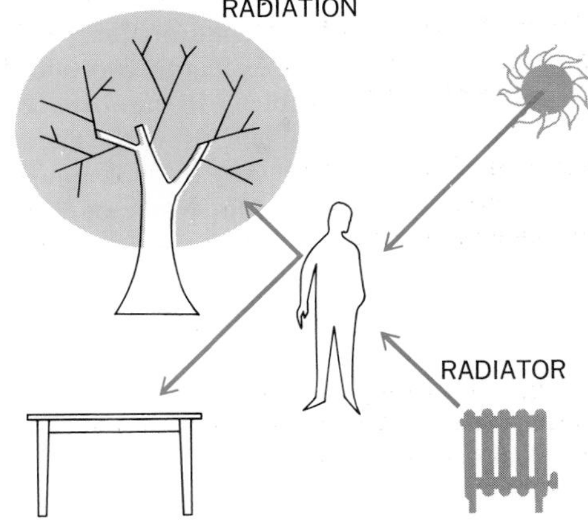

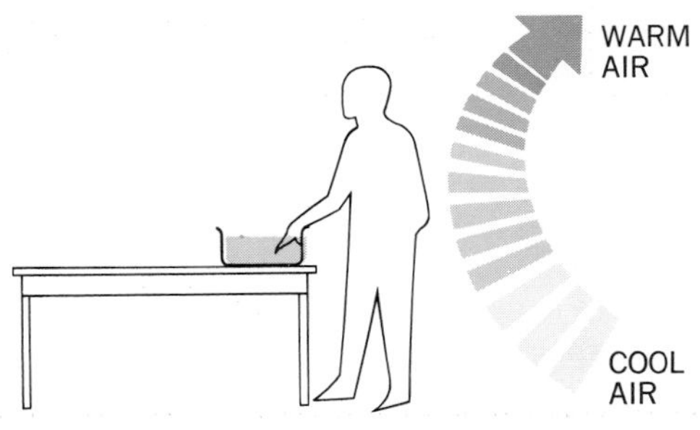

Radiation, Conduction, Convection The surface of the body constantly emits heat, in the form of electromagnetic waves. Simultaneously, all other dense objects are radiating heat. The rate of emissions is determined by the temperature of the radiating surface. Thus, if the surface of the body is warmer than the *average* of the various surfaces in the environment, net heat is lost, the rate being directly dependent upon the temperature difference. The sun, of course, is a powerful radiator, and direct exposure to it greatly decreases heat loss by radiation or may reverse it.

Conduction is the exchange of heat not by radiant energy but simply by transfer of thermal energy from atom to atom or molecule to molecule. Heat, like any other quantity, moves down a concentration gradient, and thus the body surface loses or gains heat by conduction only through direct contant with cooler or warmer substances, including, of course, the air.

Convection is the process whereby air (or water) next to the body is heated, moves away, and is replaced by cool air (or water), which in turn follows the same pattern. It is always occurring because warm air is less dense and therefore rises, but it can be greatly facilitated by external forces such as wind or fans. Thus, convection aids conductive heat exchange by continuously maintaining a supply of cool air. In the absence of convection, negligible heat would be lost to the air, and conduction would be important only in such unusual circumstances as immersion in cold water or sitting nude on a cold chair. (Because of the great importance of air movement in aiding heat loss, attempts have been made to quantitate the cooling effect of combinations of air speed and temperature; the most useful tool has been the wind-chill index.) Henceforth we shall also imply conduction when we use the term convection.

It should now be clear that heat loss by radiation and convection is largely determined by the temperature gradient between the body surface and the external environment. It is convenient to view the body as a central core surrounded by a shell consisting of skin and subcutaneous tissue (for convenience, we shall refer to the complex shell of tissues simply as skin) whose insulating capacity can be varied. It is the temperature of the central core which is being regulated at approximately 99°F; in contrast, as we shall see, the temperature of the outer surface of the skin changes markedly. If the skin were a perfect insulator, no heat would ever be lost from the core; the outer surface of the skin would equal the environmental temperature (except during direct exposure to the sun), and net convection or radiation would be zero. The skin, of course, is not a perfect insulator, so that the temperature of its outer surface generally lies somewhere between that of the external environment and the core. Of profound importance for temperature regulation of the core is that the skin's effectiveness as an insulator is subject to physiologic control by changing the blood flow to the skin. The more blood reaching the skin from the core, the more closely the skin's temperature approaches that of the core. In effect, the blood vessels diminish the insulating capacity of the skin by carrying heat to the surface (Fig. 20-25). These vessels are controlled primarily by vasoconstrictor sympathetic nerves. Vasoconstriction may be so powerful that the skin of the finger, for example, may undergo a 99 percent reduction in blood flow during exposure to cold.

The pattern of vasomotor response to cold and heat is diagramed in Fig. 20-26. Exposure to cold increases the gradient between core and environment; skin vasoconstriction increases skin insulation, reduces skin temperature, and lowers heat loss. Exposure to heat decreases (or may even reverse) the gradient between core and environment; in order to permit the required heat loss, skin vasodilatation occurs, the gradient between skin and environment increases, and heat loss increases. Although we have spoken of skin temperature as if it were uniform throughout the body, certain areas participate much more than others in the vasomotor responses; accordingly, skin temperatures vary with location.

What are the limits of this type of process? The lower limit is obviously the point at which maximal skin vasoconstriction has occurred; any further drop in environmental temperature increases the gradient and causes excessive heat loss. At this point, the body must increase its heat production to maintain temperature. The upper limit is set by the point of maximal vasodilatation, the environmental temperature, and the core temperature itself. As shown in Fig. 20-26, at high environmental temperatures, even maximal vasodilatation cannot establish a core-environment gradient large enough to eliminate heat as fast as it is produced. Another heat-loss mechanism, therefore, is brought strongly into play, sweating. Thus, the skin vasomotor contribution to temperature is highly effective in the midrange of environmental temperature (70 to 85°F), but the major burden is borne by increased heat production at lower temperatures and by increased heat loss via sweating at higher temperatures.

Two other important mechanisms for altering heat

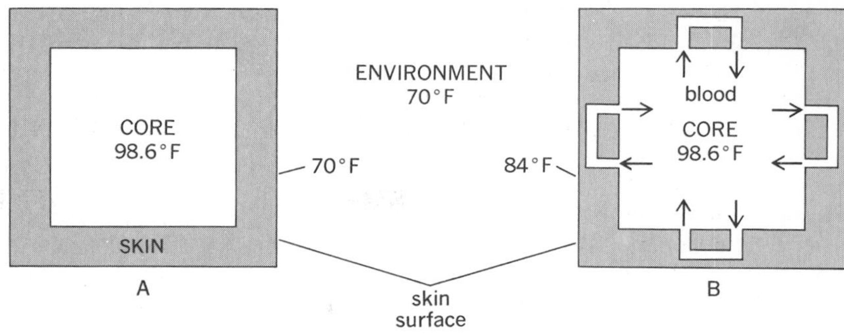

FIGURE 20-25 ***Relationship of skin's insulating capacity to its blood flow.*** **(A)** ***Skin as a perfect insulator, i.e., with zero blood flow, the temperature of the skin outer surface equaling that of the external environment. When the skin blood vessels dilate*** **(B)**, ***the increased flow carries heat to the body surface, i.e., reduces the insulating capacity of the skin, and the surface temperature becomes intermediate between that of the core and the external environment.***

loss by radiation and convection remain: changes in surface area and clothing. Curling up into a ball, hunching the shoulders, and similar maneuvers in response to cold reduce the surface area exposed to the environment, thereby decreasing radiation and convection. For human beings, clothing is also an important component of temperature regulation, substituting for the insulating effects of feathers in birds and fur in other mammals. The principle is similar in that the outer surface of the clothes now forms the true "exterior" of the

FIGURE 20-26 ***Effects of reflex vasoconstriction or vasodilatation in maintaining the heat-loss gradient for radiation and convection. Had the reflexes not occurred, core temperatures would have changed much more. However, the compensations are incomplete, and other mechanisms (changes in heat production and sweating) are required to stabilize body temperature completely during exposure to marked changes in environmental temperature.***

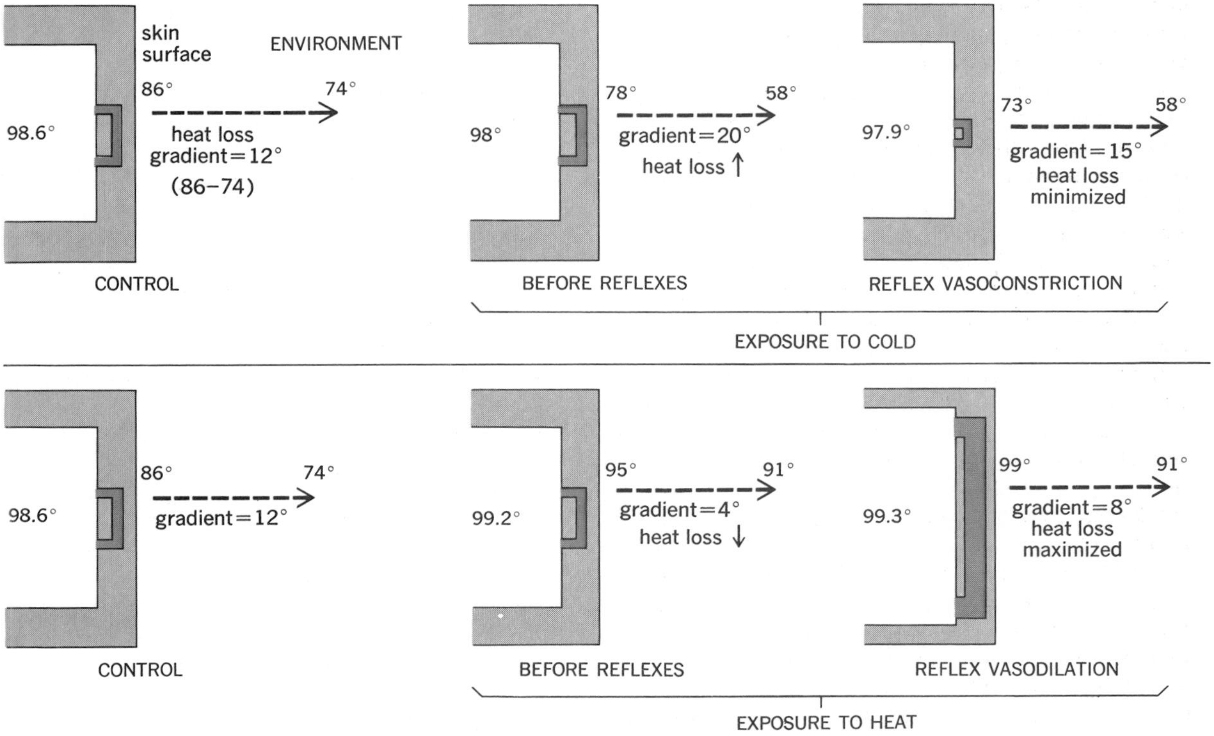

skin surface
ENVIRONMENT
98.6° 86° 74°
heat loss gradient=12° (86–74)
CONTROL
98° 78° 58°
gradient=20° heat loss ↑
BEFORE REFLEXES
97.9° 73° 58°
gradient=15° heat loss minimized
REFLEX VASOCONSTRICTION
EXPOSURE TO COLD

98.6° 86° 74°
gradient=12°
CONTROL
99.2° 95° 91°
gradient=4° heat loss ↓
BEFORE REFLEXES
99.3° 99° 91°
gradient=8° heat loss maximized
REFLEX VASODILATION
EXPOSURE TO HEAT

body surface. The skin loses heat directly to the air space trapped by the clothes; the clothes in turn pick up heat from the inner air layer and transfer it to the external environment. The insulating ability of clothing is determined primarily by its type and thickness as well as by the thickness of the trapped air layer. We have spoken thus far only of the ability of clothing to reduce heat loss; the converse is also desirable when the environmental temperature is greater than body temperature, since radiation and conduction then produce heat gain. Human beings therefore insulate themselves against temperatures which are greater than body temperature by wearing clothes. The clothing, however, must be loose so as to allow adequate movement of air to permit evaporation, the only source of heat loss under such conditions. White clothing is cooler since it reflects radiant energy, which dark colors absorb. Contrary to popular belief, loose-fitting light-colored clothes are far more cooling than going nude during direct exposure to the sun.

Evaporation Evaporation of water from the skin and lining membranes of the respiratory tract is the second major process for loss of body heat. Thermal energy must be supplied in order to transform water from the liquid to the gaseous state. Thus, whenever water vaporizes from the body's surface, the heat required to drive the process is abosrbed from the surface, thereby cooling it. Even in the absence of sweating, there is still loss of water by diffusion through the skin, which is not completely waterproof. A like amount is lost during expiration from the respiratory lining. This insensible water loss amounts to approximately 600 ml/day in human beings and accounts for a significant fraction of total heat loss. In contrast to this passive water diffusion, *sweating* requires the active secretion of fluid by sweat glands and its extrusion into ducts, which carry it to the skin surface. The sweat is pumped to the surface by periodic contraction of cells resembling smooth muscle in ducts. Production and delivery of sweat to the surface are stimulated by the sympathetic nerves. Sweat is a dilute solution containing primarily sodium chloride. The loss of this salt and water during severe sweating can cause diminution of plasma volume adequate to provoke hypotension, weakness, and fainting. It has been estimated that there are over 2.5 million sweat glands spread over the adult human body, and production rates of over 4 l/h have been reported. This is 9 lb of water, the evaporation of which would eliminate almost 1200 kcal from the body! It is essential to recognize that sweat must evaporate in order to exert its cooling effect. The most important factor determining evaporation is the water-vapor concentration of the air, i.e., the *humidity*. The discomfort suffered on humid days is due to the failure of evaporation; the sweat glands continue to secrete, but the sweat simply remains on the skin or drips off. Most other mammals differ from human beings in lacking sweat glands. They increase their evaporative losses primarily by panting, thereby increasing pulmonary air flow and increasing water losses from the lining of the respiratory tract, and they deposit water for evaporation on their fur or skin by licking.

Heat loss by evaporation of sweat gradually dominates as environmental temperature rises since radiation and convection decrease linearly as the body–environment temperature gradient diminishes. At environmental temperatures above that of the body, heat is actually gained by radiation and conduction, and evaporation is the sole mechanism for heat loss. The ability of human beings to survive such temperatures is determined by the humidity and by their maximal sweating rate. For example, when the air is completely dry, human beings can survive a temperature of 266°F for 20 min or longer, whereas very moist air at 115°F is not bearable for even a few minutes.

Changes in sweating determine human beings' chronic adaptation to high temperatures. Persons newly arrived in a hot environment have poor ability to do work initially, their body temperature rises, and severe weakness and illness may occur. After several days, there is great improvement in work tolerance with little increase in body temperature, and the persons are said to have *acclimatized* to the heat. Body temperature is kept low because there is an earlier onset of sweating and because of increased rates and lower sodium content of the sweat. The mechanisms remain unknown although heightened aldosterone secretion plays an important role in reducing the sodium.

Summary of Effector Mechanisms in Temperature Regulation

Table 20-6 summarizes the mechanisms regulating temperature, none of which is an all-or-none response but calls for a graded progressive increase or decrease in activity. As we have seen, heat production via skeletal muscle activity becomes extremely important at the cold end of the spectrum, whereas increased heat loss via sweating is critical at the hot end.

TABLE 20-6
SUMMARY OF EFFECTOR MECHANISMS IN TEMPERATURE REGULATION

Stimulated by cold		***Stimulated by heat***	
Decrease heat loss	Vasoconstriction of skin vessels; reduction of surface area (curling up, etc.)	Increase heat loss	Vasodilatation of skin vessels; sweating
Increase heat production	Shivering and increased voluntary activity; (?) increased secretion of thyroxine and epinephrine; increased appetite	Decrease heat production	Decreased muscle tone and voluntary activity; (?) decreased secretion of thyroxine and epinephrine; decreased appetite

Hypothalamic Centers Involved in Temperature Regulation

Removal of the cerebral cortex does not hinder an animal's ability to regulate temperature, but destruction of the hypothalamus severely disturbs temperature regulation. Neurons in this latter area, via descending pathways, control the output of somatic motor nerves to skeletal muscle (muscle tone and shivering) and of sympathetic nerves to skin arterioles (vasoconstriction and dilatation), sweat glands, and the adrenal medulla. In animals in which thyroxine is an important component of the response to cold, these centers also control the output of hypothalamic TSH-releasing factor (TRF).

Afferent Input to the Hypothalamic Centers

The final component of the human thermostat is the afferent input. Obviously, these temperature-regulating reflexes require receptors capable of detecting changes in the body temperature. There are two groups of receptors, one in the skin (*peripheral thermoreceptors*) and the other in deeper body structures (*central thermoreceptors*).

Peripheral Thermoreceptors In the skin (and certain mucous membranes) are nerve endings usually categorized as *cold* and *warm receptors*. In one sense, these are misleading terms since cold is not a separate entity but a lesser degree of warmth. Really there are two populations of temperature-sensitive skin receptors, one stimulated by a lower and the other by a higher range of temperatures (Fig. 20-27). Information from these receptors is transmitted via the afferent nerves and ascending pathways to the hypothalamus, which responds with appropriate efferent output; in this manner, the firing of cold receptors stimulates heat-producing and heat-conserving mechanisms, whereas enhanced firing of warmth receptors accomplishes just the opposite.

Central Thermoreceptors It should be clear that the skin thermoreceptors alone would be highly inefficient regulators of body temperature for the simple reason that it is the core temperature, not the skin temperature, which is actually being regulated. On theoretical grounds alone it was apparent that core, i.e., central, receptors had to exist somewhere in the body, and numerous experiments have localized them, to a large extent, to the hypothalamus itself (important ther-

FIGURE 20-27 ***Discharge rates of a typical skin cold receptor* (open circles) *and warm receptor* (closed circles) *in response to changes in temperature.*** **(Adapted from Dodt and Zotterman.)**

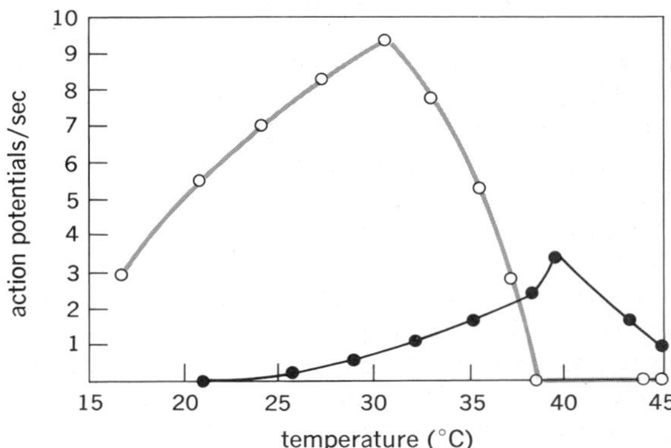

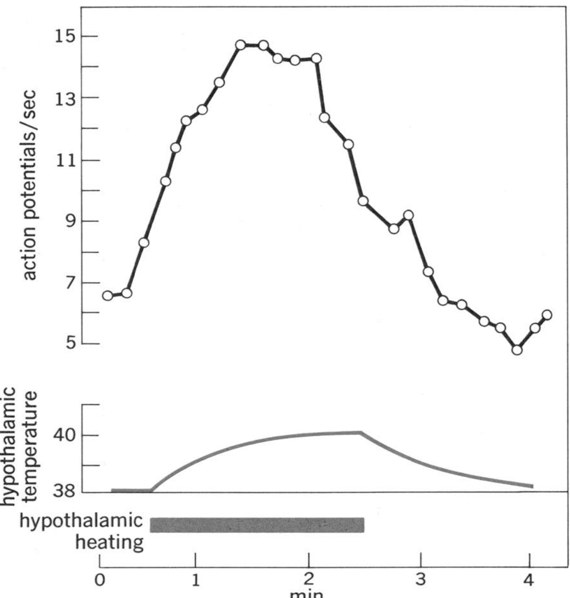

FIGURE 20-28 ***Effect of local heating of a discrete area of hypothalamus on the discharge rate of a single thermosensitive hypothalamic neuron.*** **(Adapted from Nakayama et al.)**

moreceptors also exist in other internal locations). In unanesthetized dogs, local warming of hypothalamic neurons through previously implanted thermodes causes them to fire rapidly (Fig. 20-28) and reproduces the entire picture of the dog's usual response to a warm environment: he becomes sleepy, stretches out to increase his surface area, pants heavily, salivates, and licks his fur. Conversely, local cooling induces vasoconstriction, intensive shivering, fluffing out of the fur, and curling up. These hypothalamic thermoreceptors have synaptic connections with hypothalamic integrating centers, which also receive input from other central thermoreceptors as well as from the skin thermoreceptors. The precise relative contributions of the various thermoreceptors remain the subject of considerable debate. Temperature-regulating reflexes are summarized in Fig. 20-29.

Fever

The elevation of body temperature so commonly induced by infection is due not to a breakdown of temperature-regulating mechanisms but to a resetting of the hypothalamic thermostat. Thus, a person with a fever regulates his or her temperature in response to heat or cold but always at a higher set point. The onset of fever is frequently gradual but it is most striking when it occurs rapidly in the form of a chill. It is as though the thermostat were suddenly raised; the person suddenly feels cold, and marked vasoconstriction and shivering occur. This association of heat conservation and increased heat production serves to drive body temperature up rapidly. All this time, the person may be putting on more blankets because of feeling cold. Later, the fever breaks as the thermostat is reset to normal; the person feels hot, throws off the covers, and manifests profound vasodilatation and sweating.

What is the basis for the resetting? As is described in the chapter on resistance to infection, certain endogenously produced chemicals known as *pyrogens* are released in the presence of infection or inflammation. These pyrogens act directly upon the thermoreceptors in the hypothalamus, altering their rate of firing and their input to the integrating centers. Recent evidence suggests that this effect of pyrogens is mediated via local release of prostaglandins which then directly alter thermoreceptor function. Consistent with this hypothesis is the fact that aspirin, which reduces fever by restoring thermoreceptor activity toward normal, inhibits the synthesis of prostaglandins. Aspirin does not lower the body temperature of a normal nonfebrile person, presumably because prostaglandin is not liberated in significant amounts, except when stimulated by pyrogen.

In addition to infection and inflammation, in which fever is induced by pyrogens, as described above, there are other situations in which fever is produced by quite different mechanisms. Excessive blood levels of epinephrine or thyroxine resulting from diseases of the adrenal medulla or thyroid gland elevate the body temperature by direct actions on heat-producing metabolic reactions rather than by altering the hypothalamic thermoreceptor setting. Certain lesions of the brain do not reset the hypothalamus but rather completely destroy its normal regulatory capacity; under such conditions lethal hyperthermia may occur very rapidly. *Heat stroke* is also characterized by a similar breakdown in function of the regulatory centers. It is frequently a positive-feedback state in which, because of inadequate balancing of heat loss and production, body temperature becomes so high that the hypothalamic regulatory centers are put out of commission and body temperature therefore rises even higher. Thus a patient suffering from heat stroke manifests a dry skin (absence of

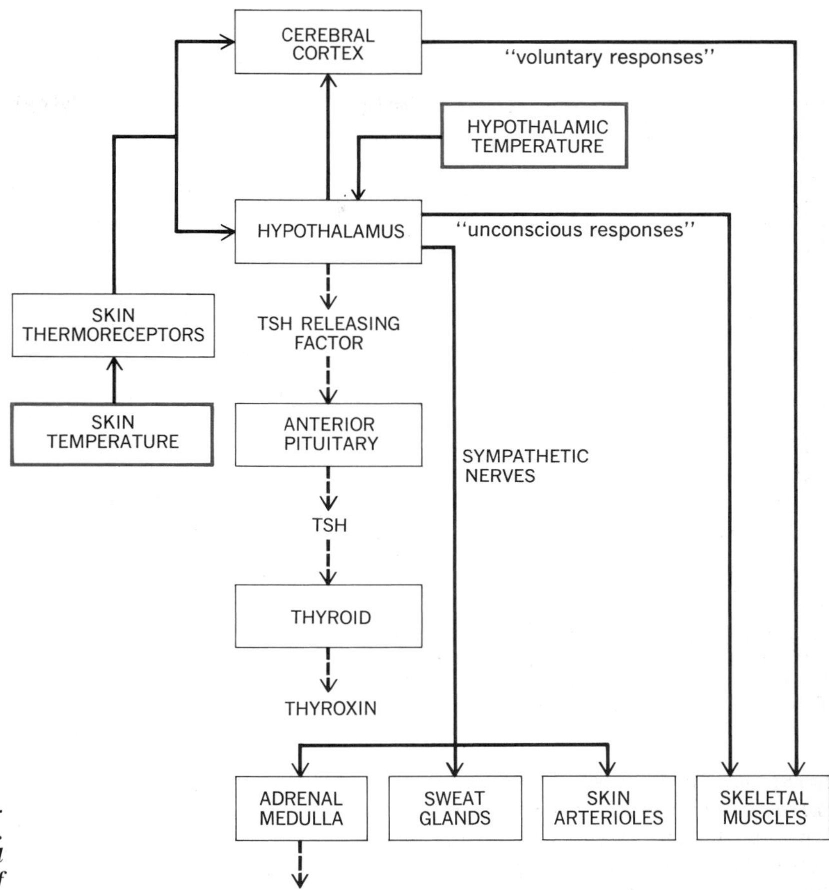

FIGURE 20-29 ***Summary of temperature-regulating mechanisms. The dashed lines are hormonal pathways, which are probably of minor importance in human beings.***

sweating) despite a markedly elevated body temperature.

Heat stroke, the attainment of a body temperature at which vital bodily functions are endangered, should be distinguished from *heat exhaustion*. The former is due to a breakdown in heat-regulating mechanisms, whereas the latter is not the result of failure of heat regulation but rather of the inability to meet the price of heat regulation. Heat exhaustion is a state of collapse due to hypotension brought on by depletion of plasma volume (secondary to sweating) and by extreme dilatation of skin blood vessels, i.e., by decreases in both cardiac output and peripheral resistance. Thus, heat exhaustion occurs as a direct consequence of the activity of heat-loss mechanisms; because these mechanisms have been so active, the body temperature is only modestly elevated. In a sense, heat exhaustion is a safety valve which, by forcing cessation of work when heat-loss mechanisms are overtaxed, prevents the larger rise in body temperature which would precipitate the far more serious condition of heat stroke.

CHAPTER 21

The Reproductive System

Before beginning detailed descriptions of male and female reproductive systems, it may be worthwhile to summarize some of the important terminology and patterns of classification. The primary reproductive organs are known as the *gonads,* the *testes* in the male and the *ovaries* in the female. In both sexes, the gonads serve dual functions: (1) production of the reproductive cells, *sperm* or *ova,* and (2) secretion of the *sex hormones.* The systems of ducts through which the sperm or ova are transported and the glands lining or emptying into the ducts are termed the *accessory reproductive organs* (in the female the breasts are also usually included in this category). Finally, the *secondary sexual characteristics* comprise the many external differences (hair, body contours, etc.) between male and female which are not directly involved in reproduction. The gonads and accessory reproductive organs are present at birth but remain relatively small and nonfunctional until the onset of *puberty,* at about 10 to 14 years of age. The secondary sexual characteristics are virtually absent until puberty. The term puberty actually means the attainment of sexual maturity in the sense that conception becomes possible; as commonly used, it usually refers to the entire period of sexual development culminating in the attainment of sexual maturity. The term *adolescence* has a much broader meaning and includes the total period of transition from childhood to adulthood in all respects, not just sexual.

Section A. Male Reproductive System

The essential male reproductive functions are the manufacture of sperm (*spermatogenesis*) and the deposition of the sperm in the female. The organs which carry out these functions are the testes, epididymides, vas deferens, ejaculatory ducts, and penis together with accessory glands: the seminal vesicles, prostate, and bulbourethral glands (Fig. 21-1). The *testes* are sur-

FIGURE 21-1 ***Anatomic organization of the male reproductive tract.***

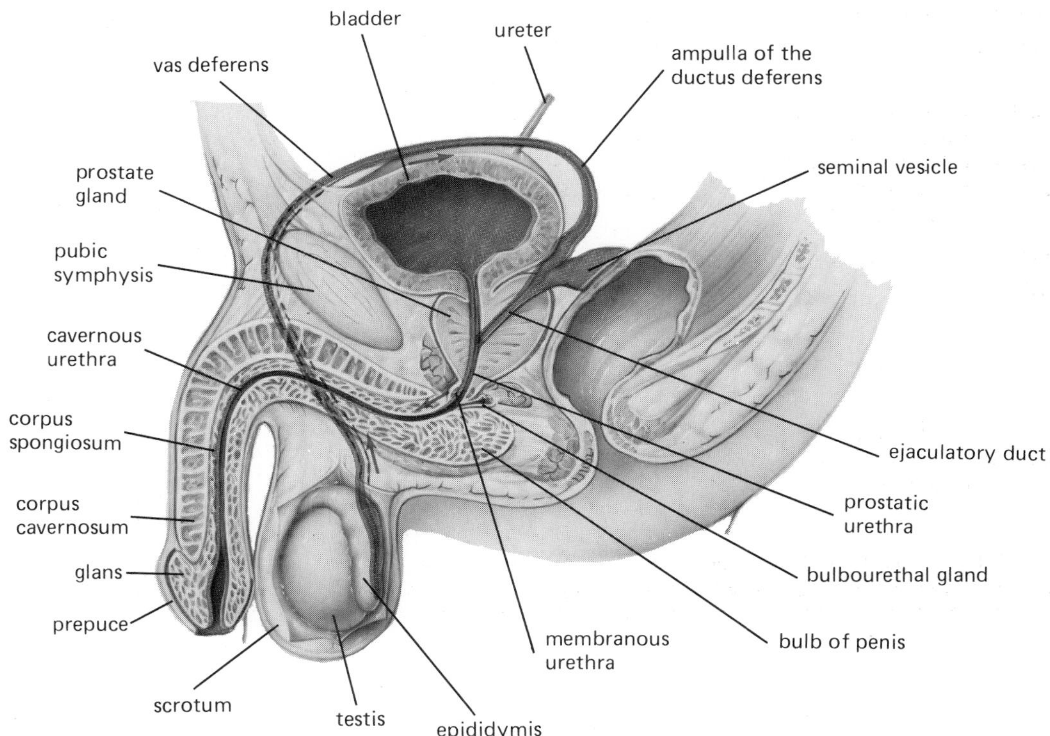

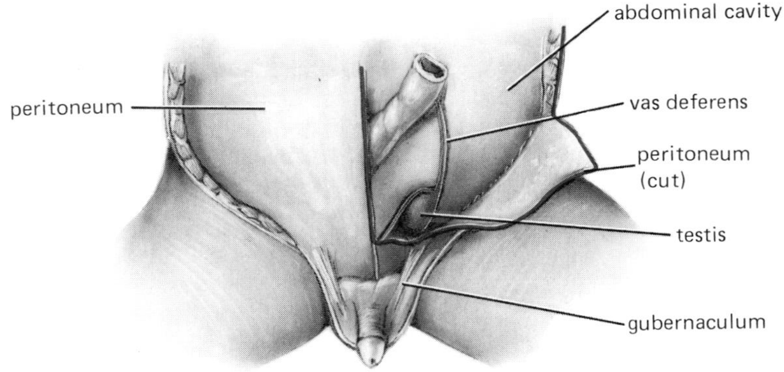

FIGURE 21-2 **(A)** ***Diagrams illustrating descent of the testis.*** **(B)** ***Lateral view of the testis. The enveloping membranes have been pulled aside.*** **(C)** ***Diagrammatic cross section of a human testis. The highly coiled seminiferous tubules are the sites of sperm production.***

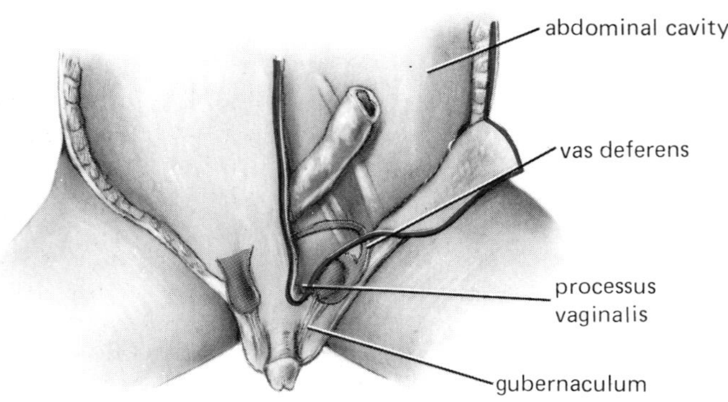

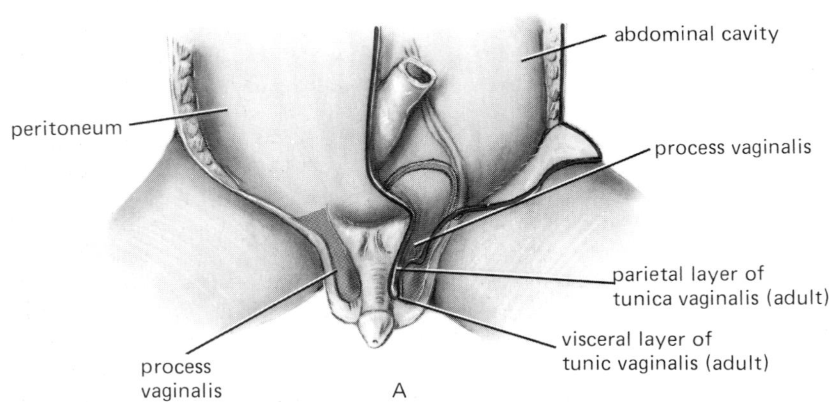

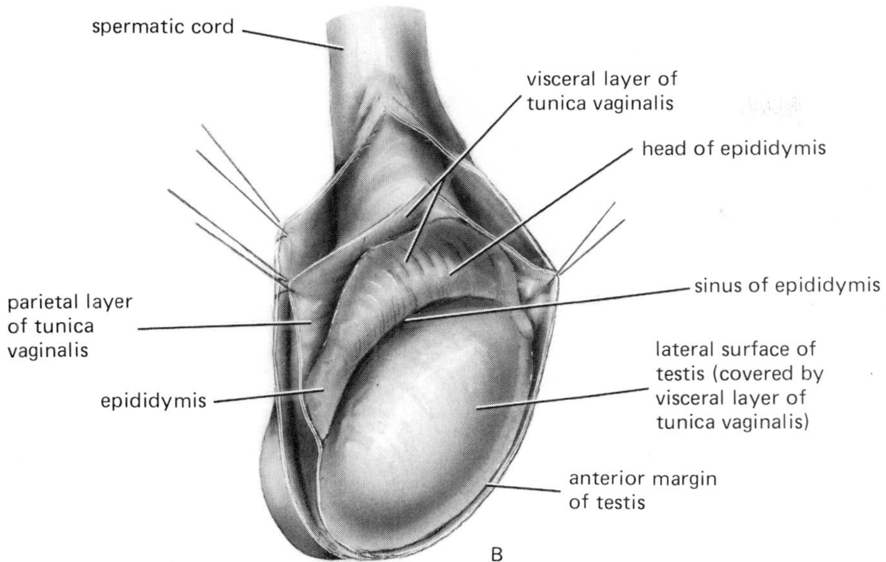

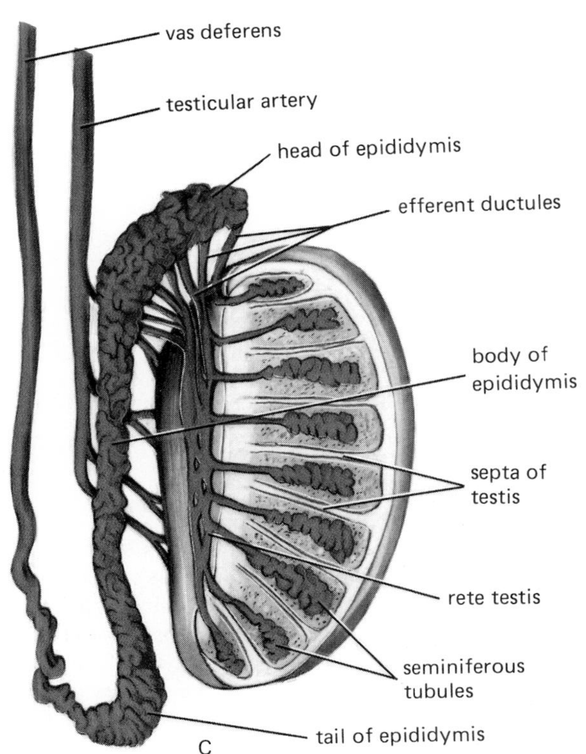

rounded by a tough, fibrous–connective-tissue capsule and suspended outside the body in a sac, the *scrotum*. During embryonic development, the testes lie in the posterior wall of the abdominal cavity, i.e., they are retroperitoneal. During the seventh month of intrauterine development, the testes descend into the scrotum, carrying with them the arteries, veins, and nerves which supply them and the ducts (vas deferens) connecting them with the urethra (Fig. 21-2*A*). These structures, surrounded by a connective-tissue sheath, form the *spermatic cord* (Fig. 21-2*B*). The actual mechanism of descent is unknown.

Connective-tissue projections from the testicular capsule pass into the substance of each testis and form *septa,* which incompletely divide the gland into some 200 to 300 *lobules* (Fig. 21-2*C*). Each lobule contains one or more tiny convoluted *seminiferous tubules,* which are lined with the sperm-producing *spermatogenic cells.* The testes also serve an endocrine function, the manufacture of the primary male sex hormone *testosterone.* The endocrine-secreting *interstitial cells* (Leydig cells) lie in the small connective-tissue spaces which surround the seminiferous tubules (Fig. 21-3). Of great importance is the relationship between the sperm-producing and endocrine functions; the process of spermatogenesis requires testosterone, but testosterone production does *not* depend upon spermatogenesis. In other words, testosterone deficiency produces sterility by interrupting spermatogenesis, but interference with

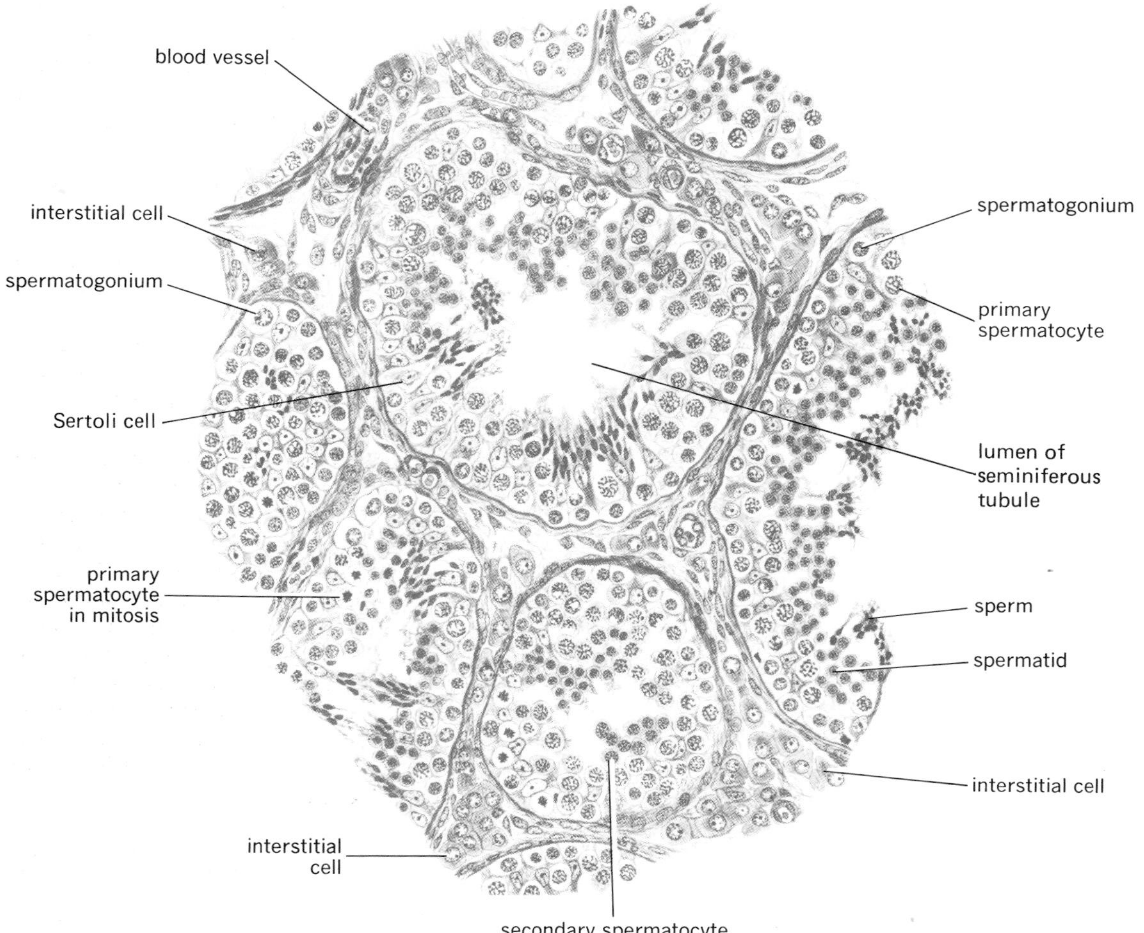

FIGURE 21-3 ***Section of human testis. The seminiferous tubules show various stages of spermatogenesis.*** **(Adapted from Bloom and Fawcett.)**

the function of the seminiferous tubules does not alter normal testosterone production by the interstitial cells. The great importance of this relationship is that a simple, effective method of sterilizing the male is vasectomy, surgical ligation and removal of a segment of the vas deferens, which carry sperm from the testes. This procedure prevents the delivery of sperm but does not appear to alter secretion of testosterone, which is the primary determinant of male sexual drive and maleness in general.

SPERMATOGENESIS

From Fig. 21-2*C*, a diagrammatic cross section of a human testis, it is evident that the testis is composed primarily of the many highly coiled seminiferous tubules, the combined length of which is approximately 375 m. In Fig. 21-3, a microscopic section of an adult human testis, it can be seen that the tubules contain many cells, the vast majority of which are in various stages of division; these are the spermatogenic cells.

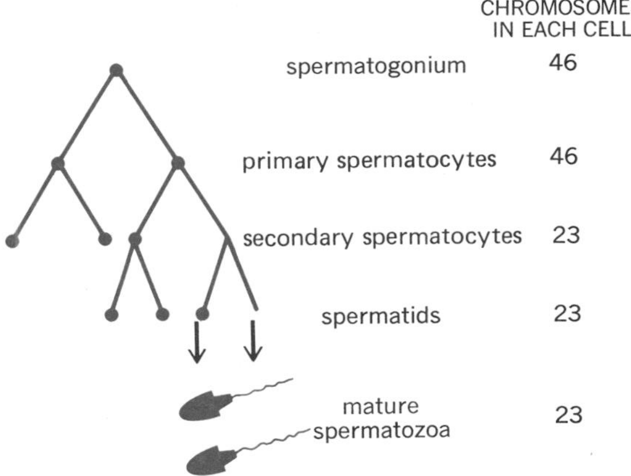

FIGURE 21-4 *Summary of spermatogenesis. Each spermatogonium yields eight mature sperm, each containing 23 chromosomes.*

Each seminiferous tubule is surrounded by a basement membrane. Only the outermost layer of spermatogenic cells is in contact with this membrane; these are undifferentiated germ cells termed *spermatogonia,* which, by dividing mitotically, provide a continuous source of new cells. Some spermatogonia move away from the basement membrane and increase markedly in size. Each large cell, now termed a *primary spermatocyte,* divides to form two *secondary spermatocytes,* each of which in turn divides into two *spermatids,* the latter ultimately being transformed into mature *spermatozoa* (Fig. 21-4). The division of the primary spermatocytes by *meiosis* differs from the ordinary mitotic division. During mitosis (Chap. 1) each daughter cell receives the full number of chromosomes; in meiosis, however, each daughter cell receives only half of the chromosomes present in the parent cell. The primary spermatocyte contains 46 chromosomes (the diploid number), 23 from each parent. Prior to the next division, these 46 chromosomes become duplicated; during the divison itself, half the chromosomes along with their duplicates pass to each secondary spermatocyte. When the latter cells divide, the chromosomes separate from their duplicates so that the resulting cells, the *spermatids,* have 23 nonduplicated chromosomes (the haploid number). Since development of the female germ cell follows a similar pattern, the eventual combination of sperm and ovum results in the reestablishment of the normal complement of 46 chromosomes.

The final phase of spermatogenesis, the transformation of the spermatids into mature spermatozoa, involves no further cell divisions. The head of a mature spermatozoon (Fig. 21-5) consists almost entirely of the nucleus, a dense mass of DNA bearing the sperm's genetic information. The tip of the nucleus is covered by a cap known as the *acrosome,* consisting of a protein-filled vesicle derived from the Golgi apparatus and believed to contain several lytic enzymes which enable the sperm to enter the ovum. The tail comprises a group of actomyosinlike contractile filaments, contraction of which produces a whiplike movement of the tail capable of propelling the sperm at a velocity of 1 to 4 mm/s. Finally, the mitochondria of the spermatid become rearranged to form the midpiece of the tail and probably provide its energy.

Throughout the entire process of maturation, the developing germ cells remain intimately associated with another type of cell, the *Sertoli cell.* Each Sertoli cell extends from the basement membrane of the seminiferous tubule all the way to the lumen. Intricate infolding of the luminal-border membranes forms recesses which contain spermatids and spermatozoa until they are mature enough for release. Apparently, the Sertoli cells serve as the route by which nutrients and chemical signals reach the developing germ cells as the latter move away from the basement membrane. They may also produce steroids which act locally to stimulate spermatogenesis.

FIGURE 21-5 *Mature human sperm.*

The mechanisms which guide the remarkable cellular transformation from spermatid to mature sperm remain uncertain. In any small segment of seminiferous tubules, the entire process of spermatogenesis proceeds in a regular sequence. For example, at any given time, virtually all the primary spermatocytes in one portion of the tubule are undergoing division, whereas in an adjacent segment, the secondary spermatocytes may be dividing. The entire process in a single area takes approximately 72 days. In mammals which breed seasonally, spermatogenesis is periodic, activity being followed by degeneration of the spermatogonia and shrinking of the seminiferous tubules. In contrast, nonseasonal breeders, such as human beings, manifest continuous activity, the cycles following each other without a break. Perhaps the most amazing characteristic of spermatogenesis is its sheer magnitude: the normal human male may manufacture several hundred million sperm per day (the ram produces billions).

DELIVERY OF SPERM

The seminiferous tubules unite as they leave the lobules of the testis to form larger ducts. As they join, they pack tightly together and through many cross branches, form a dense network, the *rete testis* (Fig. 21-2*C*). *Efferent ductules* leave the rete testis and pass to the epididymis where a single duct, the *duct of the epididymis,* is formed. The convoluted efferent ductules form the *head* of the epididymis; the equally tortuous convolutions of the duct of the epididymis form its *body* and *tail*. In the tail of the epididymis, the duct of the epididymis becomes the *vas deferens* (the *deferent duct*). Movement through these ducts is accomplished by two means: The pressure of the newly formed sperm-containing fluid pushes the spermatozoa forward; smooth muscle cells in the duct walls exert a peristaltic-like action; the sperm themselves are nonmotile at this time. Besides serving as a route for sperm exit, this system performs several important functions: (1) The epididymis and first portion of the vas deferens store sperm prior to ejaculation; (2) during passage through the epididymis or storage there, a final maturation makes the sperm both motile and fertile; this appears to be mediated by fluid secreted by the epithelium of the epididymis, but little is known about this process or the factors which influence it; (3) at ejaculation, the sperm are expelled from the epididymis and vas deferens by strong contractions of the smooth muscle lining the duct walls.

A pair of elongated glands, the *seminal vesicles,* drain into the vas deferens on each side – the duct now becomes the *ejaculatory duct* which enters the prostate and empties into the prostatic portion of the urethra. (Enlargement of the prostate later in life may obstruct the urethra and interfere with urination.) (Fig. 21-6.) The prostate and seminal vesicles, as well as the bulbourethral glands just below the prostate, secrete small quantities of fluid continuously and much larger quantities during sexual intercourse (*coitus*). The secretions constitute the bulk of the ejaculated fluid, the *semen,* which contains a large number of different chemical substances, the functions of which are presently being worked out. One function of the seminal fluid is that of sheer dilution of the sperm (in human beings, sperm constitutes only a few percent of the total ejaculated semen); without such dilution, motility is impaired. In addition, specific chemicals also contribute to the motility of the sperm; for example, the seminal vesicles secrete large quantities of the carbohydrate fructose utilized by the sperm contractile apparatus for energy. We shall describe later the possible contributions of the prostaglandins, a group of fatty acids found in very large concentrations in seminal fluid.

Erection

The primary components of the male sexual act are *erection* of the penis, which permits entry into the female vagina, and *ejaculation* of the sperm-containing semen into the vagina. Erection is a vascular phenomenon which can be understood from the structure of the penis (Fig. 21-7). This organ consists almost entirely of three cylindrical cords of *erectile tissue,* two *corpora cavernosa* and a single *corpus spongiosum* (surrounding the urethra) which are actually vascular spaces. Normally the arterioles supplying these chambers are constricted so that they contain little blood and the penis is flaccid; during sexual excitation, the arterioles dilate, the chambers become engorged with blood, and the penis becomes rigid. Moreover, as the erectile tissues expand, the veins emptying them are compressed, thus minimizing outflow and contributing to the engorgement. This entire process occurs rapidly, complete erection sometimes taking only 5 to 10 s. The vascular dilation is accomplished by stimulation of the parasympathetic nerves and inhibition of the sympathetic nerves to the arterioles of the penis (Fig. 21-8). This appears to be one of the few cases of direct parasympathetic control over high-resistance blood vessels. In addition to these vascular effects, the para-

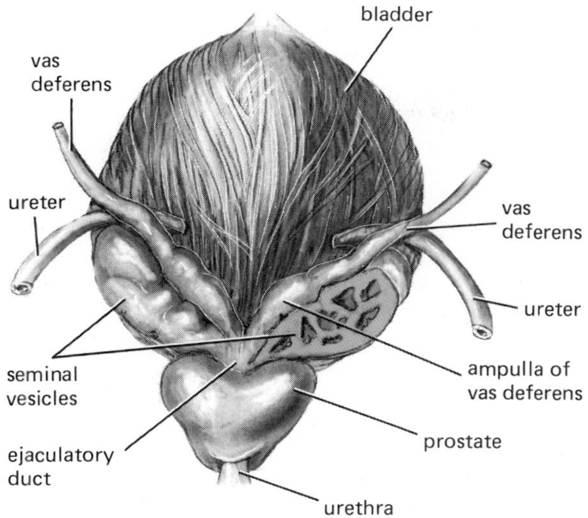

FIGURE 21-6 ***Posterior view of the male internal urogenital organs.***

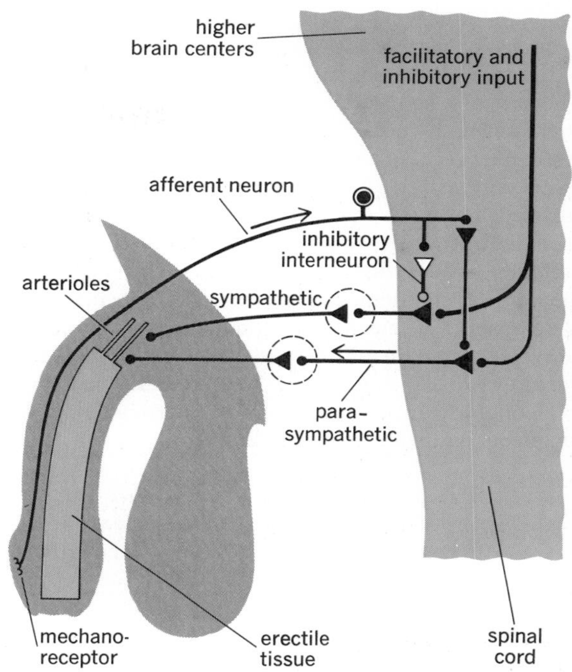

FIGURE 21-8 ***Reflex pathway for erection. The reflex is initiated by mechanoreceptors in the penis. Input from higher centers can facilitate or inhibit this reflex.***

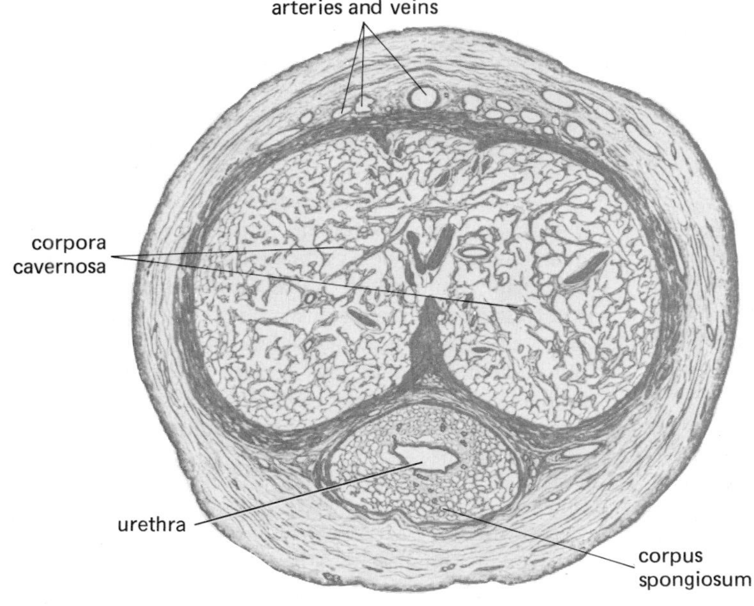

FIGURE 21-7 ***Cross section of a human penis. The large central area (the corpora cavernosa) and that surrounding the urethra (the corpus spongiosum) are vascular spaces which become filled with blood to cause erection.*** **(Adapted from Bloom and Fawcett.)**

sympathetic nerves stimulate urethral glands to secrete a mucuslike material which aids in lubrication. What receptors and afferent pathway initiate these reflexes? The primary input comes from highly sensitive mechanoreceptors located in the tip of the penis. The afferent fibers carrying the impulses synapse in the lower spinal cord and trigger the efferent outflow. It must be stressed, however, that higher brain centers, via descending pathways, may exert profound facilitative or inhibitory effects upon the efferent neurons. Thus, thoughts or emotions can cause erection in the complete absence of mechanical stimulation of the penis; conversely, failure of erection (*impotence*) may frequently be due to psychologic disturbances. The ability of alcohol to interfere with erection is probably due to its effects on higher brain centers.

Ejaculation

This process is basically a spinal reflex, the afferent pathways apparently being identical to those described for erection. When the level of stimulation reaches a critical peak, a patterned automatic sequence of efferent discharge is elicited to both the smooth muscle of the accessory ducts and the skeletal muscle at the base of the penis. The precise contribution of various pathways is complex, but the overall response can be divided into two phases: (1) The ducts contract, as a result of sympathetic stimulation to them, emptying their contents into the urethra (*emission*), and (2) the semen is then expelled from the penis by a series of rapid muscle contractions. During ejaculation the sphincter at the base of the bladder is closed so that sperm cannot enter the bladder nor can urine be expelled. The rhythmic contractions of the penis during ejaculation are associated with intense pleasure, the entire event being termed the *orgasm*. A simultaneous marked skeletal muscle contraction throughout the body is followed by the rapid onset of muscular and psychologic relaxation. At orgasm there is also a marked increase in heart rate and blood pressure. Once ejaculation has occurred, there is a so-called latent period during which a second erection is not possible; this period is quite variable but may be hours in perfectly normal men.

The average volume of fluid ejaculated is 3 ml, containing approximately 300 million sperm. However, the range of normal values is extremely large, and older ideas of the minimal concentration of sperm required for fertility are now being reevaluated. Although quantity is important, it is obvious that the quality of the sperm is another critical determinant of fertility.

HORMONAL CONTROL OF MALE REPRODUCTIVE FUNCTIONS

Virtually all aspects of male reproductive function are either directly controlled or indirectly influenced by testosterone or the anterior pituitary gonadotropins, *follicle-stimulating hormone* (FSH) and *luteinizing hormone* (LH). These pituitary hormones were named for their effects in the female, but their molecular structures are precisely the same in both sexes. (LH, in the male, is frequently called interstitial-cell-stimulating hormone, ICSH.) FSH and LH exert their effects only upon the testes, whereas testosterone manifests a broad spectrum of actions not only on the testes but on the accessory reproductive organs, the secondary sexual characteristics, sexual behavior, and organic metabolism in general.

Effects of Testosterone

Spermatogenesis Adequate amounts of the steroid hormone testosterone are essential for spermatogenesis, and sterility is an invariable result of severe testosterone deficiency. The cells which secrete testosterone are known as the *interstitial cells;* as shown in Fig. 21-3, they lie scattered between the seminiferous tubules. It seems likely that the stimulatory effects of testosterone on spermatogenesis are exerted locally by the hormone diffusing from the interstitial cells into the seminiferous tubules. Testosterone is not the only hormone required for spermatogenesis; the pituitary gonadotropins are also required, and their relationship with testosterone will be described subsequently.

Accessory Reproductive Organs The morphology and function of the entire male duct system, lining glands, and penis all depend upon testosterone. Following removal of the testes (*castration*) in the adult, all the accessory reproductive organs decrease in size, the glands markedly reduce their rates of secretion, and the smooth muscle activity of the ducts is inhibited. Erection and ejaculation are usually deficient. These defects disappear upon the administration of testosterone.

Behavior Most of our information comes from experiments on animals other than human beings, but even from our fragmentary information about human beings, there is little doubt that the development and maintenance of normal sexual drive and behavior in men are testosterone-dependent and may be seriously impaired by castration. However, it is a mistake to assume that deviant male sexual behavior must therefore be due to testosterone deficiency or excess. For example, most (but not all) male homosexuals have normal rates of testosterone secretion; although administration of exogenous testosterone may sometimes increase sexual activity in these men, it remains homosexual. To day, no clear-cut correlation has been established between homosexuality or hypersexuality and hormonal status in either men or women.

A question which has recently become the subject of enormous controversy is whether testosterone influences other human behavior in addition to sex, i.e., are there any inherent male-female differences or are the observed differences in behavior all socially conditioned. There is little doubt that behavioral differences based on sex do exist in other mammals; perhaps the best-studied is that of aggression which is clearly greater in males and is testosterone-dependent. Obviously, it will be difficult to answer such questions with respect to human beings but attempts are now being made to study them in a controlled scientific manner.

Secondary Sex Characteristics The masculine secondary characteristics are testosterone-dependent. For example, a male castrated before puberty does not develop a beard or axillary or public hair. A strange and unexplained finding is that baldness, although genetically determined in part, does not occur in castrated men. Other testosterone-dependent secondary sexual characteristics are the deepening of the voice (resulting from growth of the larynx), skin texture, thick secretion of the skin oil glands (predisposing to acne), and the masculine pattern of muscle and fat distribution. This leads us into an area of testosterone effects usually described as *general metabolic effects* but very difficult to separate from the secondary sex characteristics. It is obvious that the bodies of men and women (even excepting the breasts and external genitals) have very different appearances; a woman's curves are due in large part to the feminine distribution of fat, particularly in the region of the hips and lower abdomen but in the limbs as well. A castrated male gradually develops this pattern; conversely, a woman treated with testosterone loses it. A second very obvious difference is that of skeletal muscle mass; testosterone exerts a profound effect on skeletal muscle to increase its size. The overall relationship of testosterone to general body growth was described in Chap. 20.

Mechanism of Action of Testosterone The fact that testosterone exerts such extraordinarily widespread effects should not obscure what appears to be a common denominator, namely, that its major action is to promote growth. In cells capable of responding to testosterone, it, like other steroid hormones, acts upon the cell nucleus to promote transcription or translation, or both, of genetic information. Finally, how it affects behavior is completely unknown.

Testosteronelike substances, i.e., *androgens,* can also be found in the blood of normal women. In the female the site of production is primarily the adrenal gland, which also contributes some androgens in the male. The potency of the androgens secreted by the adrenal is normally low. These androgens play several important roles in the female, specifically stimulation of general body growth (Chap. 20) and maintenance of sexual drive. In several disease states, the female adrenal may secrete abnormally large quantities of androgen, which produce a remarkable virilism; the female fat distribution disappears, a beard appears along with the male body-hair distribution, the voice lowers in pitch, the skeletal muscle mass enlarges, the clitoris (homologue of the male penis) enlarges, and the breasts diminish in size. These changes illustrate the sex-hormone dependency of secondary characteristics. Sterily also results.

Anterior Pituitary and Hypothalamic Control of Testicular Function

FSH and LH are essential for normal spermatogenesis and testosterone secretion. Following removal of the anterior pituitary, the testes decrease greatly in weight, and spermatogenesis and testosterone secretion almost cease. FSH directly stimulates spermatogenesis, whereas LH stimulates testosterone secretion, but because testosterone is also required for spermatogenesis, it is evident that LH is indirectly involved in this process as well.

No discussion of the anterior pituitary is complete without inclusion of the hypothalamus. As described in Chap. 12, all anterior pituitary hormones are controlled by releasing factors secreted by discrete areas of the hypothalamus and reaching the pituitary via the hypothalamopituitary portal blood vessels. This input is essential for sexual function since destruction of the relevant hypothalamic areas stops spermatogenesis and markedly reduces testosterone secretion. Figure 21-9 summarizes the hypothalamic–anterior pituitary–testicular relationships.

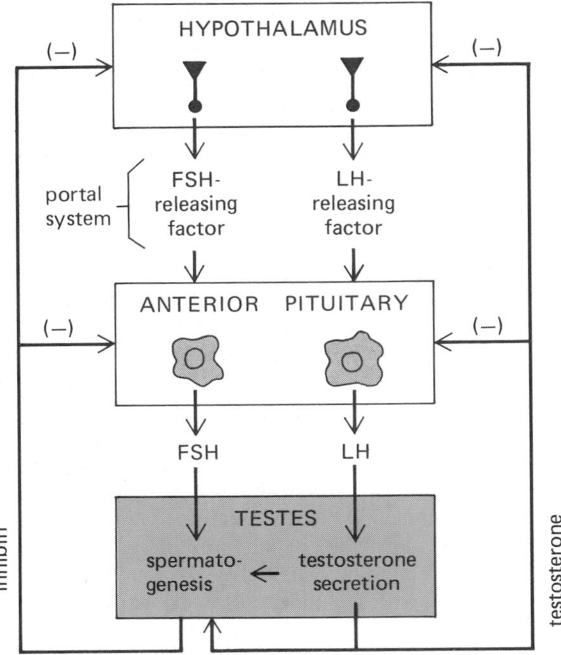

FIGURE 21-9 ***Summary of hormonal control of testicular function. The negative signs indicate that testosterone inhibits LH secretion in both the hypothalamus and the anterior pituitary. Testosterone reaches the seminiferous tubules to stimulate spermatogenesis both by local diffusion and by release into the blood and recirculation to the testes. See text for a discussion of FSH control.***

What controls the secretion of the hypothalamic releasing factors for FSH and LH? It is known that testosterone exerts a negative-feedback inhibition of LH secretion via both the anterior pituitary and the hypothalamus. Testosterone has less effect on the secretion of FSH, and it is presently unclear just how the testes exert their negative-feedback inhibition of FSH secretion. The inhibitory chemical signal, called *inhibin,* seems to arise from the seminiferous tubules themselves, most likely from the Sertoli cells, but its identity is not yet known. The Sertoli cell being the source of the negative-feedback inhibition of FSH secretion makes sense since present evidence also indicates that the stimulatory effect of FSH on spermatogenesis is exerted via the Sertoli cells. The solution to these questions is not merely of academic interest since an agent which inhibits FSH only and not LH would constitute an ideal male contraceptive; spermatogenesis, but not testosterone secretion, would be blocked. As emphasized in Chap. 12, such negative feedbacks are only modifiers and not the primary controllers of secretion. Recall that the hypothalamic cells which produce the releasing factors are nerve cells; we are left with the generalization that these nerve cells secrete their releasing factors either as a result of some inherent automaticity or secondary to stimuli reaching them via synaptic connections with other neurons. It is important to realize that, regardless of the triggering event, the secretion of releasing factors of LH and FSH in the male probably proceeds normally at a rather fixed, continuous rate during adult life; accordingly, FSH and LH release, spermatogenesis, and testosterone secretion also occur at relatively unchanging rates. This is unusual for hormonal systems and, more important, is completely different from the large cyclic swings of activity so characteristic of the female reproductive hormones. A word of caution may be in order here, however, for work in nonprimate mammals has indicated that testosterone levels can be made to vary in response to various sexual and social stimuli; the relevance of these studies for human beings is unknown.

Section B. Female Reproductive System

INTERNAL GENITALIA

The female reproductive system's primary organs—the ovaries and duct system—uterine tubes, uterus, and vagina—are located retroperitoneally in the lower pelvis and constitute the internal genitalia. Note that in the female, the urinary and reproductive duct systems are entirely separate.

Each ovary lies in an *ovarian fossa,* a depression on the lateral pelvic wall. The *uterine tubes* (*oviducts, Fallopian tubes*) are not directly connected to the ovaries but open into the peritoneal cavity close to them. This opening of each uterine tube is a trumpet-shaped expansion surrounded by long fingerlike projections (the *fimbriae*) which are lined with ciliated epithelium.

The other ends of the uterine tubes empty directly into the cavity of the *uterus* (womb), which is a hollow thick-walled muscular organ lying in the pelvis between the bladder and the rectum (Fig. 21-10A). The uterus can be divided approximately in its middle into the *corpus,* or *body,* above and the *cervix* below. The cervix terminates at the small opening into the vagina. The portion of the corpus above the entrance of the uterine tubes is the *fundus* (Fig. 21-10*B*). The wall of the uterus has three layers: an external peritoneal layer (*perimetrium*), a middle muscular layer (*myometrium*), and an inner mucous membrane (*endometrium*).

The vagina is a fibromuscular tube lined with stratified epithelium extending from the uterine cervix to the vestibule. Its walls are much thinner than those of the uterus and consist of an internal mucous membrane and a muscular coat.

All these internal genitalia are supported by ligaments covered by folds of peritoneum. The ovaries are supported by the *suspensory ligaments* (Fig. 21-10*A*), which contain the ovarian nerves and vessels, and a portion of the *broad ligaments of the uterus,* which are one of the numerous pairs of ligaments connecting the uterus to the bladder, rectum, and walls of the lower pelvis.

EXTERNAL GENITALIA

The external genitalia include the mons pubis, labia majora and minora, the clitoris, the vestibule of the vagina, and the greater vestibular glands. The term *vulva* includes all these parts. The *mons pubis* is a rounded elevation of adipose tissue in front of the symphysis pubis. The *labia majora,* the female analogue of the scrotum, are two prominent skin folds (Fig. 21-10*C*). The *labia minora* are small skin folds lying be-

FIGURE 21-10 **(A)** ***Sagittal section through a female pelvis.***

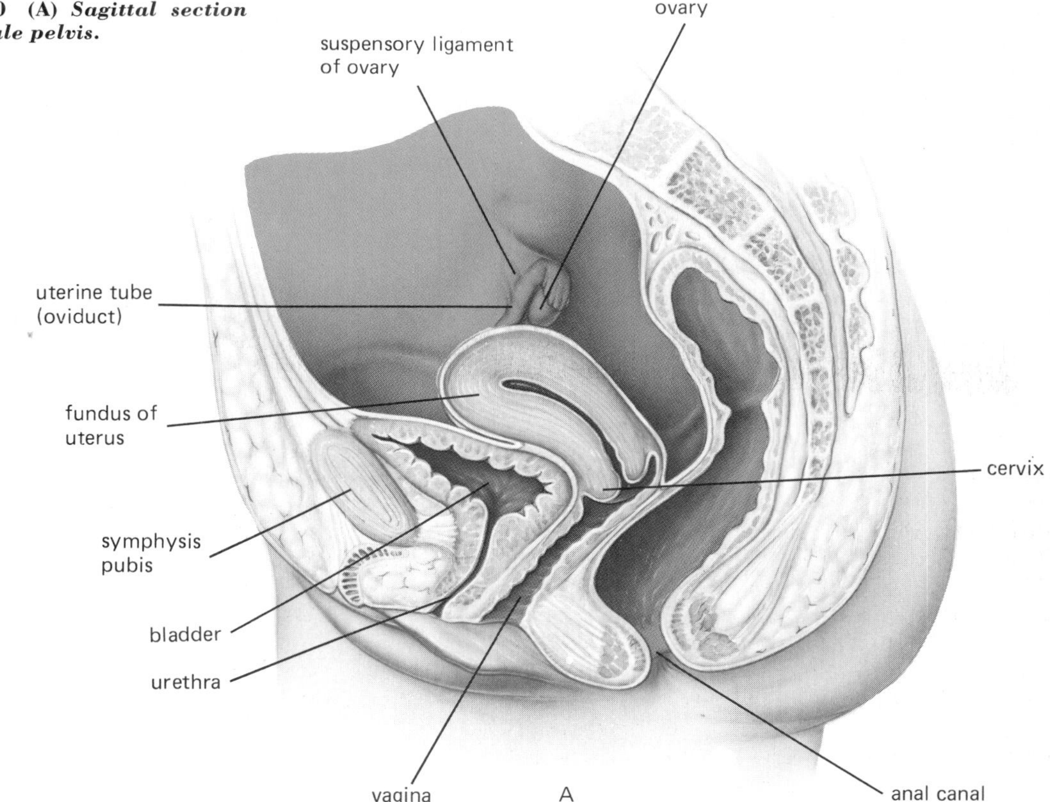

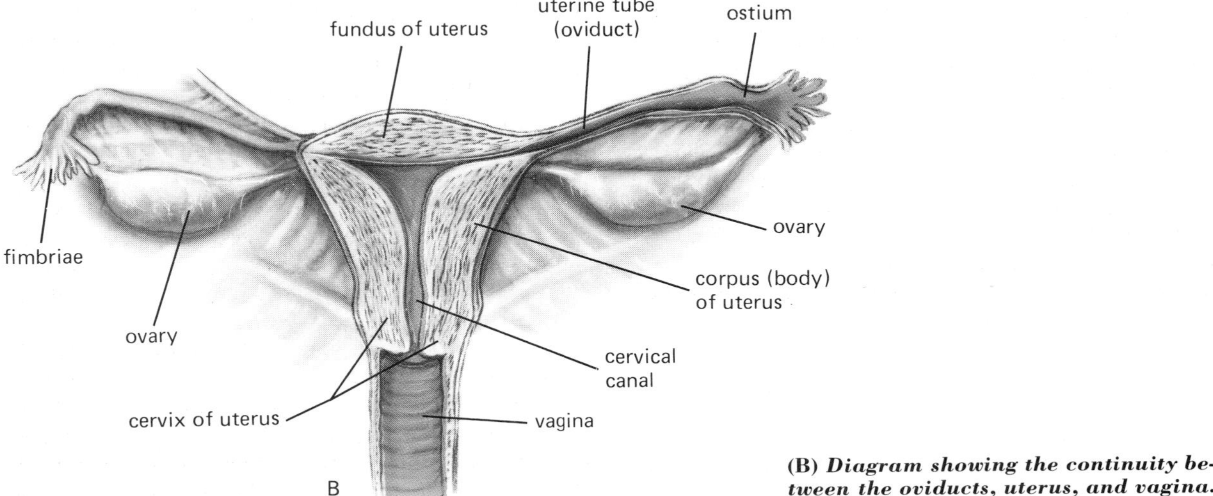

(B) ***Diagram showing the continuity between the oviducts, uterus, and vagina.***

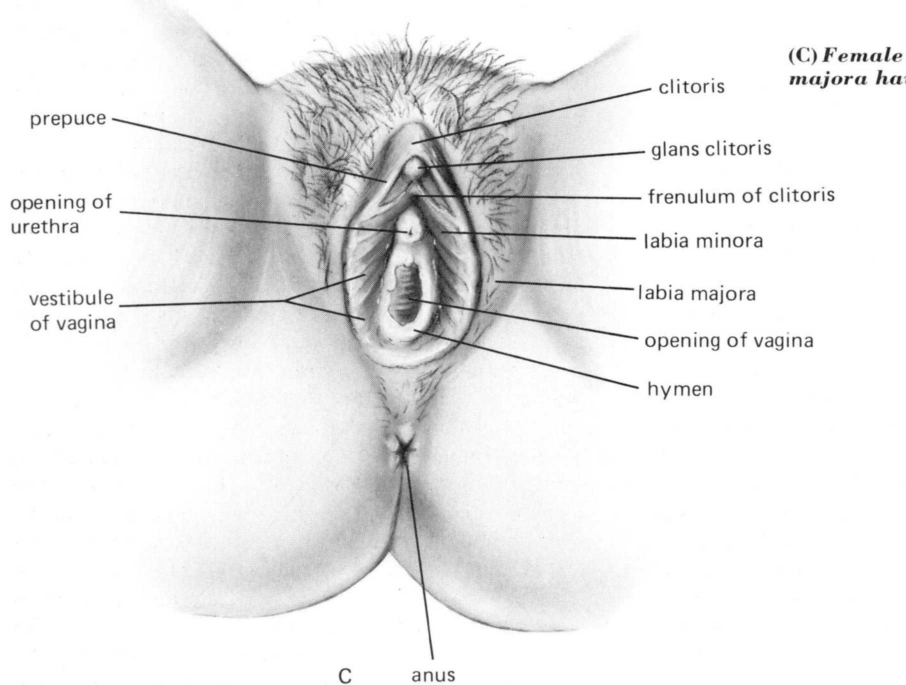

(C) ***Female external genitalia. The labia majora have been separated.***

tween the labia majora; they surround the urethral and vaginal openings, and the area thus enclosed is the *vestibule*. The vaginal opening (*introitus*) lies behind that of the urethra. Partially overlying the vaginal opening is a thin fold of mucous membrane, the *hymen*. Anteriorly, each labium minorus divides in two, the upper part passing above the clitoris, the lower part below. The *clitoris* is an erectile structure homologous to the penis. Finally, the floor underlying the pelvic girdle is known as the *perineum;* it is not part of the external genitalia but is stretched and may be torn or surgically incised (*episiotomy*) during childbirth.

OVARIAN FUNCTION

Unlike the continuous sperm production of the male, the maturation and release of the female germ cell, *the ovum,* are cyclic and intermittent. This pattern is true not only for ovum development but for the function and structure of virtually the entire female reproductive system. In human beings and other primates these cycles are called *menstrual cycles*.

The human ovary, like the testis, serves a dual purpose: (1) production of ova and (2) secretion of the female sex hormones, *estrogen* and *progesterone*.

Ovum and Follicle Growth

In Fig. 21-11, a cross section of an ovary, note the many discrete cell clusters known as *primordial follicles*. Each follicle is composed of one ovum surrounded by a single layer of flattened follicular cells. At birth, normal human ovaries each contain about 200,000 such follicles, and no new ones appear after birth. Thus, in marked contrast to the male, the newborn female already has all the germ cells she will ever have. Only a few, perhaps 400, are destined to reach full maturity during her active sexual life. All the others degenerate starting from birth on, so that few, if any, remain by the time she reaches menopause. One result of this is that the ova which are released (ovulated) near menopause are 30 to 35 years older than those ovulated just after puberty; it has been suggested that certain congenital defects much commoner among children of older women are the result of aging changes in the ovum.

In Fig. 21-11A there are still a large number of primordial (or primary) follicles, mainly on the periphery of the ovary, but there are also more immature follicles in different stages of development. The development of the follicle is characterized by an increase in size of the ovum and a proliferation of the surrounding follicular cells. The ovum becomes separated from the follicular, or *granulosa,* cells by a thick membrane, the *zona pellucida*. As the follicle grows, new cell layers are formed, not only from mitosis of the original follicle cells but from the growth of specialized ovarian connective-tissue cells. Thus, the follicle, originally composed only of the ovum and its surrounding layers of granulosa cells, becomes invested with additional outer layers of cells known as the *theca* (Fig. 21-11*B*). When the follicle reaches a certain diameter, a fluid-filled space, the *antrum,* begins to form in the midst of the granulosa cells as a result of fluid they secrete. We do not know what mechanisms stimulate development of certain primary follicles, leaving most unstimulated to degenerate without ever showing a growth phase.

By the time the antrum begins to form, the ovum has reached full size. From this point on, the follicle grows in part because of continued follicular-cell proliferation but largely because of the expanding antrum. Ultimately, the ovum, surrounded by the zona pellucida and several layers of granulosa cells, occupies a ridge projecting into the antrum. These granulosa layers immediately surrounding the ovum are termed the *cumulus ovaricus* whereas the outer granulosa cells are the *membrana granulosa*. Eventually, the antrum becomes so large that the completely mature follicle actually balloons out on the surface of the ovary. *Ovulation* occurs when the wall at the site of ballooning ruptures and the ovum, surrounded by its tightly adhering zona pellucida and cumulus, is carried out of the ovary by the antral fluid. Many women experience varying degrees of abdominal pain at approximately the midpoint of their menstrual cycles, which has generally been assumed to represent abdominal irritation induced by the entry of follicular contents at ovulation. However, recent evidence on the precise timing of ovulation has indicated that this time-honored concept may be wrong, and the cause of discomfort remains unclear.

In the human adult ovary, there are always several antrum-containing follicles of varying sizes, but during each cycle, normally only one follicle reaches the complete maturity just described, the process requiring approximately 2 weeks. All the other partially matured antral follicles undergo degeneration at some stage in their growth, the mechanism being unknown. On occasion (1 to 2 percent of all cycles), two or more follicles reach maturity, and more than one ovum may be ovulated. This is the commonest cause of multiple

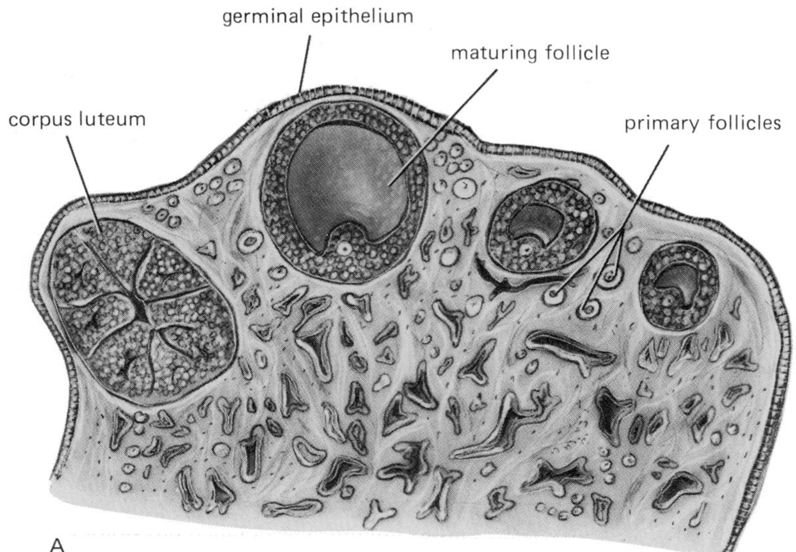

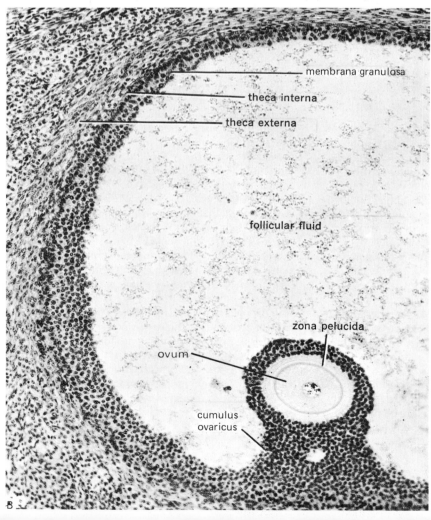

FIGURE 21-11 **(A)** ***Section of a human ovary.*** **(B)** ***Section through an ovarian follicle.*** **(Part A redrawn from R. Warwick and R. L. Williams, "Gray's Anatomy," 35 Brit. ed., Saunders, Philadelphia, 1973. Part B from W. F. Windle, "Textbook of Histology," 5th ed., McGraw-Hill, New York, 1976.)**

births; in such cases the siblings are fraternal, i.e., not identical.

Ovum Division An essential aspect of ovum development is its pattern of cell division. The ova present at birth are the result of numerous mitotic divisions which occurred during intrauterine life. After birth, no further division occurs until just before ovulation, at which time the mature ovum (a *primary oocyte*) about to be ovulated completes a division that was begun in intrauterine life. This division is meiotic and analogus to the division of the primary spermatocyte because each daughter cell receives only 23 chromosomes instead of the usual 46. However, in this division one cell, the *secondary oocyte,* retains virtually all the cytoplasm, the other, the *polar cell,* being only small and rudimentary. In this manner, the already full-size ovum loses half of its chromosomes but almost none of its nutrient-rich cytoplasm. A second cell division in the oviduct (fallopian tube) after ovulation (indeed, after penetration by a sperm) follows the usual mitotic pattern, and the daughter cells each retain 23 chromosomes. Once again, one daughter cell, now the *mature ovum,* retains nearly all the cytoplasm. The net result is that each primitive ovum is capable of producing only one mature fertilizable ovum (Fig. 21-12); in contrast, each primary spermatocyte produces four viable spermatozoa.

Formation of Corpus Luteum

After rupture of the follicle and discharge of the antral fluid and the ovum, a transformation occurs within the follicle, which collapses, and the antrum fills with partially clotted fluid. The follicular cells enlarge greatly and become filled with a yellowish pigment (lutein); the entire glandlike structure is known as the *corpus luteum.* If the discharged ovum is not fertilized, i.e., if pregnancy does not occur, the corpus luteum reaches its maximum development within approximately 10 days and then rapidly degenerates. If pregnancy does occur, the corpur luteum grows and persists until near the end of pregnancy.

Ovarian Hormones

The female sex hormones secreted by the ovary are the steroids estrogen[1] and progesterone. Estrogens are secreted to some extent by various ovarian cell types but primarily by the follicle cells (*not* the ovum) and corpus luteum. Progesterone may be secreted in very minute amounts by the follicle cells, but its major source is the corpus luteum. The detailed physiology of these hormones will be described subsequently.

Cyclic Nature of Ovarian Function

The length of a menstrual cycle varies considerably from woman to woman, averaging about 28 days. Day 1 is the first day of menstrual bleeding, and in a typical 28-day cycle, ovulation occurs around day 14. In terms of ovarian function, therefore, the menstrual cycle may be divided into two approximately equal phases: (1) the *follicular phase,* during which a single follicle and ovum develop to full maturity and (2) the *luteal phase,* during which the corpus luteum is the active ovarian structure. It must be stressed that the day of ovulation varies from woman to woman and frequently in the same woman from month to month.

[1] There are actually multiple estrogenic hormones—17β estradiol, estrone, and estriol—but we shall refer to them all as estrogen.

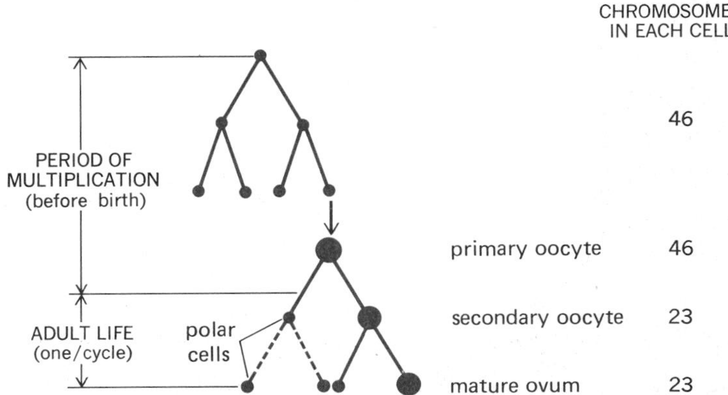

FIGURE 21-12 ***Summary of ovum development. Compare with the male pattern of Fig. 21-4. Each primitive ovum produces only one mature ovum containing 23 chromosomes.***

CONTROL OF OVARIAN FUNCTION

The basic pattern controlling ovum development, ovulation, and formation of the corpus luteum is analogous to the controls described for testicular function in that the anterior pituitary gonadotropins, FSH and LH, and the gonadal sex hormone, estrogen, play the primary roles. However, the overall schema is more complex in the female since it includes a second important gonadal hormone (progesterone) and a hormonal cycling, quite different from the more stable continuous rates of male hormone secretion.

For purposes of orientation, let us look first at the changes in the blood concentrations of all four participating hormones during a normal menstrual cycle (Fig. 21-13. Note that FSH is slightly higher in the early part of the follicular phase of the menstrual cycle and then steadily decreases throughout the remainder of the period except for a small transient midcycle peak. LH is quite constant during most of the follicular phase but then shows a very large midcycle surge (approximately 12 to 24 h before ovulation) followed by a progressive slow decline during the luteal phase. The estrogen pattern is more complex. After remaining fairly low and stable for the first week (as the follicle develops), it rises to reach a peak just before LH starts off on its surge. This peak is followed by a dip, a second rise (due to secretion by the corpus luteum), and, finally, a rapid decline during the last days of the cycle. The progesterone pattern is simplest of all; virtually no progesterone is secreted by the ovaries during the follicular phase, but very soon after ovaulation, the developing corpus luteum begins to secrete progesterone, and from this point the progesterone pattern is similar to that for estrogen. It is hoped that, after the following discussion, the reader will understand how these changes are all interrelated to yield a self-cycling pattern.

Control of Follicle and Ovum Development

Growth and development of the follicles depend upon follicle-stimulating hormone (FSH) and luteinizing hormone (LH). Accordingly, the ovary from an animal whose pituitary has been removed shows numerous early follicles but no late antral ones.

A second requirement for follicle development is estrogen, which may act, in large part, locally in the ovary. Estrogen is secreted largely by the follicle cells, so that its secretion rate progressively increases as the follicle enlarges. This secretion, however, also requires FSH and LH. Thus, FSH and LH are required both for normal follicle development and estrogen secretion, as summarized in Fig. 21-14.

Control of Ovulation

If one administers small quantities of FSH and LH each day to a woman whose pituitary has been removed (because of disease), she manifests normal follicle development, ovum maturation, and estrogen secretion, but she does not ovulate. If, on the other hand, after approximately 14 days of this therapy, she is given one or two larger injections of LH, ovulation occurs. This is precisely what happens in the normal woman; the ovum matures for 2 weeks under the influence of FSH, LH, and estrogen, and ovulation is triggered by a rapid brief outpouring from the pituitary of larger quantities of LH (Fig. 21-15). The specific mechanism by which LH then causes ovulation is unclear; best present evidence indicates that the hormone induces increased synthesis of enzymes which catalyze the chemical dissolution of the thin follicular and ovarian membranes at the bulge of the mature follicle.

Thus, the midcycle surge of LH emerges as perhaps the single most decisive event of the entire menstrual cycle; indeed, it is the presence of this surge which most strikingly distinguishes the female pattern of pituitary secretion from the relatively unchanging pattern of the male. What causes the LH surge?

It seems very likely that there exist two regulatory brain centers whose neurons synapse ultimately with the cells secreting LH-releasing factor. One of these centers – the "tonic" center – is located in the hypothalamus and exerts a continuous stimulatory effect on the secretion of LH-RF, the magnitude of which is dependent on the degree of negative-feedback inhibition exerted by estrogen on the center. It is this center which controls secretion of the gonadotropins during all the cycle except the period of the midcycle LH surge. The location of the other center – the "cyclic" center – is still uncertain but is probably the hypothalamus or the area just beyond it; its neurons are stimulated (rather than inhibited, as is the tonic center) by high levels of estrogen.[2] Moreover, this cyclic center is strongly inhibited by progesterone.

These facts nicely explain the LH surge and ovulation.[3] As estrogen secretion rises rapidly during the last half of the follicular phase, its blood concentration eventually becomes high enough to stimulate the cyclic

[2] This stimulatory effect of estrogen on LH release is often called a "positive-feedback" effect to distinguish it from the "negative feedback" usually exerted.

[3] Scattered bits of evidence suggest, however, that intercourse itself may influence the time of ovulation in human beings, particularly if there are relatively long intervals between intercourse. Thus, the rate of conception observed for wives of soldiers home on leave is much greater than that predicted if ovulation were not related to intercourse. This seems to be true for rape victims, also.

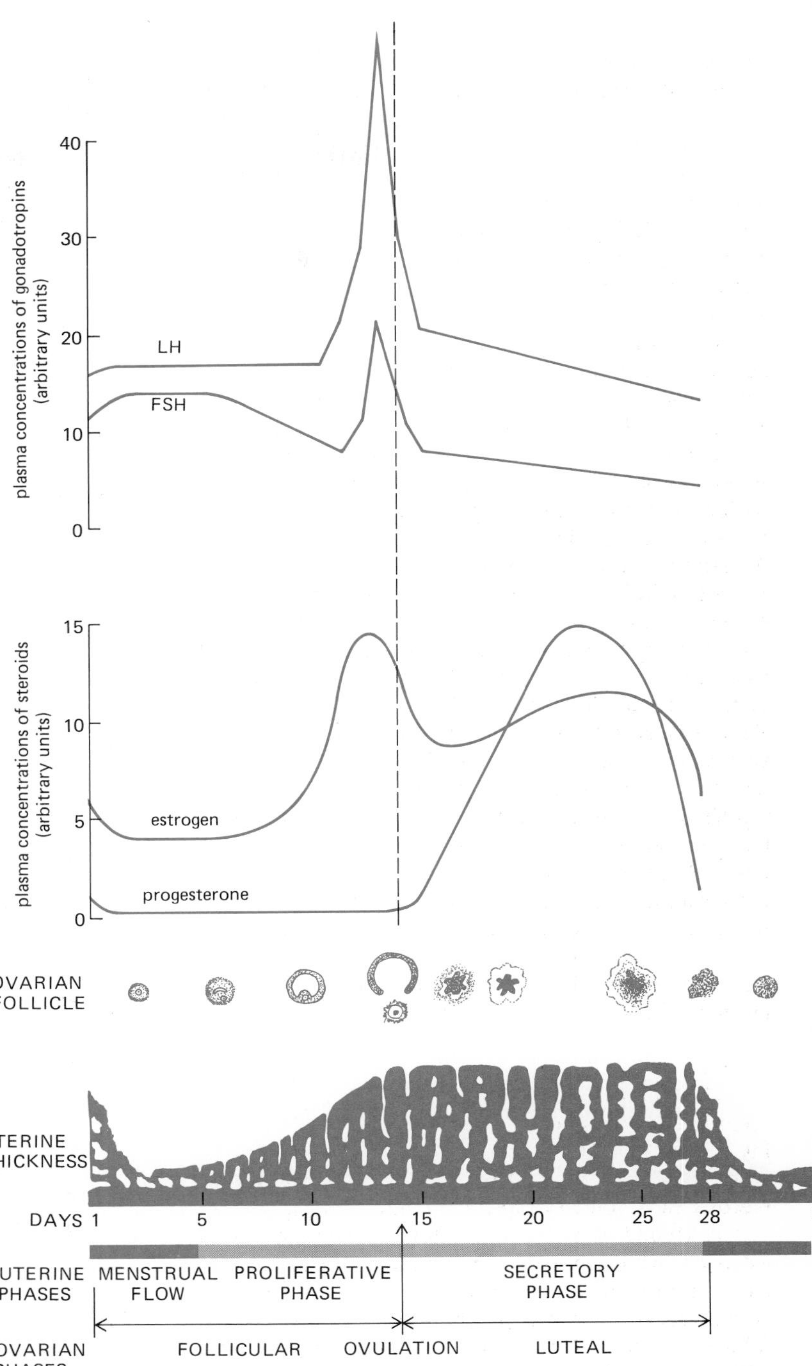

FIGURE 21-13 ***Summary of plasma hormone concentrations, ovarian events, and uterine changes during the menstrual cycle.***

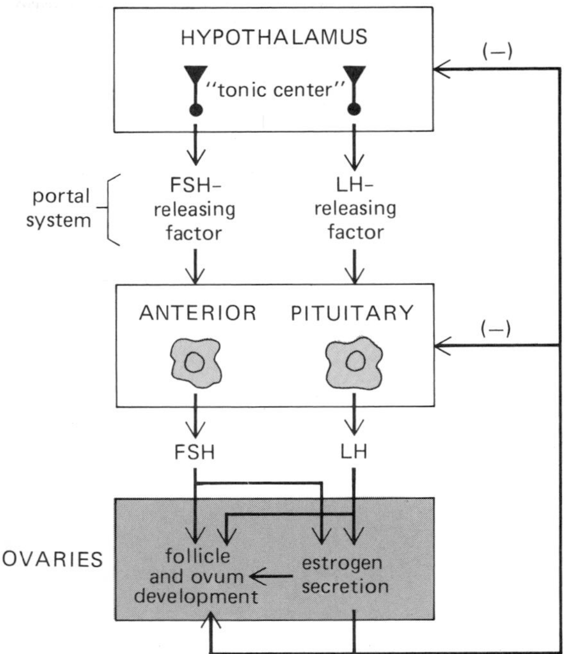

FIGURE 21-14 ***Summary of hormonal control of follicle and ovum development and estrogen secretion during the follicular phase of the menstrual cycle. Compare with the analogous pattern for the male (Fig. 21-9). The negative signs indicate that estrogen inhibits both the hypothalamus and the anterior pituitary. Estrogen reaches the developing ovum and follicle both by local diffusion and by release into the blood and recirculation to the ovaries.***

center, which, by way of its synaptic connections with the neurons secreting LH-RF, causes an outpouring of LH, which, in turn, induces ovulation. Why is there only one surge per cycle, i.e., why does not the elevated estrogen existing throughout most of the luteal phase keep inducing LH surges? The most likely answer is that, as a result of ovulation and succeeding corpus luteum formation, both induced by the LH surge, progesterone secretion begins and progesterone inhibits the cyclic center.

This discussion has so far dealt only with the effects of the sex hormones on the brain. There is, in addition, much evidence to support the view that estrogen exerts much of its inhibitory and stimulatory effects on LH secretion via a direct effect on the pituitary cells themselves. Thus, low levels of estrogen may inhibit not only the tonic brain center but the pituitary LH-secreting cells as well. Conversely, high levels of estrogen probably stimulate these pituitary cells directly (perhaps by enhancing their sensitivity to LH-RF) as well as indirectly through the cyclic center. The end result is, of course, the same.[4]

Control of the Corpus Luteum

Once formed, maintenance of the corpus luteum requires some stimulatory support from LH, but the amount of LH needed is quite small. What causes the corpus luteum to degenerate if no pregnancy results? The blood concentration of LH shows no sudden decrease during the late luteal phase so that it is difficult

[4] This question takes on added significance in light of evidence that there may be only a single releasing factor for both FSH and LH. If such turns out to be the case, then alteration by estrogen of the relative sensitivities of FSH- and LH-secreting cells to the single releasing factor would be critical. Answers to these questions should be forthcoming shortly.

FIGURE 21-15 ***Ovulation and corpus luteum formation are induced by the markedly increased LH, itself induced by the stimulatory effects of high levels of estrogen.***

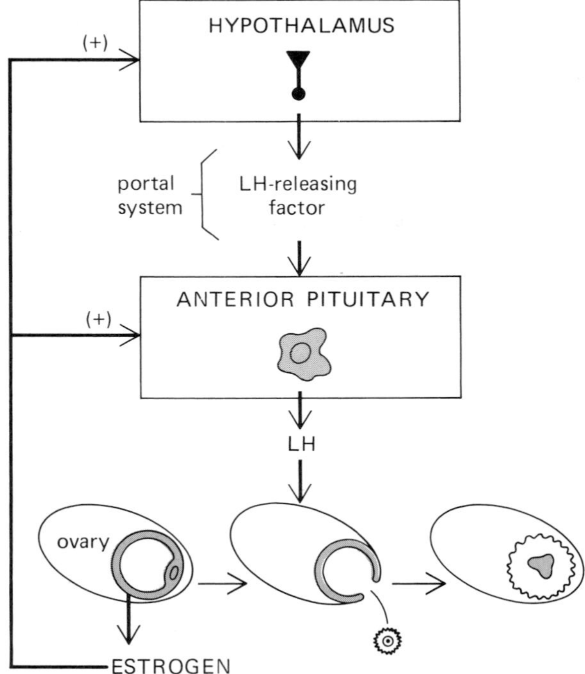

to blame regression on any sudden withdrawal of LH. In some mammalian species, it seems that regression is actively induced by some substance (perhaps prostaglandin) produced by the nonpregnant uterus, but such seems not to be the case in women. There are other hypotheses, including the idea that the corpus luteum has a "built-in" life-span of approximately 10 to 14 days and that it "self-destructs" unless prevented from doing so by the onset of pregnancy. How pregnancy does this will be described subsequently.

During its short life, the corpus luteum secretes large quantities of estrogen and progesterone. These hormones, particularly estrogen, exert a powerful negative-feedback inhibition (via the hypothalamus and anterior pituitary) of tonic FSH and LH secretion. Accordingly, during the luteal phase of the cycle, pituitary gonadotropin secretion is reduced, which explains the diminished rate of follicular maturation during this second half ot the cycle. With degeneration of the corpus luteum, blood estrogen and progresterone concentrations decrease, FSH and LH increase, and a new follicle is stimulated to mature.

Summary of Changes During the Menstrual Cycle

The events described thus far in this section are summarized in Fig. 21-13, showing the ovarian and hormonal changes during a normal menstrual cycle:

1 Under the influence of FSH, LH, and estrogen, a single follicle and ovum reach maturity at about 2 weeks.

2 During the second of these 2 weeks, under the influence of LH and FSH, estrogen secretion by the follicle cells and other ovarian cell types progressively increases.

3 For several days near midperiod production of LH (and FSH, to a lesser extent) increases sharply as a result of the stimulatory effects of high levels of estrogen on the cyclic center and the pituitary.

4 The high concentration of LH induces rupture of the follicular ovarian membranes, and ovulation occurs (it is not known what role if any the increased FSH plays).

5 Estrogen secretion decreases for several days after ovulation.

6 The ruptured follicle is rapidly transformed into the corpus luteum, which secretes large quantities of both estrogen and progesterone.

7 The high blood concentration of e[illegible] the release of LH and FSH, thereby [illegible] blood concentrations and preventing th[illegible] of a new follicle or ovum during the last [illegible] of the period; in addition, progesterone prevents any additional LH surge by inhibiting the cyclic center.

8 Failure of ovum fertilization leads, by unknown mechanisms, to the degeneration of the corpus luteum during the last days of the cycle.

9 The disintegrating corpus luteum is unable to maintain its secretion of estrogen and progesterone, and their blood concentrations drop rapidly.

10 The marked decrease of estrogen and progesterone removes the inhibition of FSH secretion.

11 The blood concentration of FSH begins to rise, follicle and ovum development are stimulated, and the cycle begins anew.

UTERINE CHANGES IN THE MENSTRUAL CYCLE

Profound changes in uterine morphology occur during the menstrual cycle and are completely attributable to the effects of estrogen and progesterone. Estrogen stimulates growth of the uterine smooth muscle (myometrium) and the glandular epithelium (endometrium) lining its inner surface. Progesterone acts upon this estrogen-primed endometrium to convert it to an actively secreting tissue: The glands become coiled and filled with secreted glycogen; the blood vessels become spiral and more numerous; various enzymes accumulate in the glands and connective tissue of the lining (Fig. 21-16). All these changes are ideally suited to provide a hospitable environment for implantation of a fertilized ovum.

Estrogen and progesterone also have important effects on the mucus secreted by the cervix. Under the influence of estrogen alone, this mucus is abundant, clear, and nonviscous; all these characteristics are most pronounced at the time of ovulation and facilitate penetration by sperm. In contrast, progesterone causes the cervical mucus to become thick and sticky, in essence a "plug" which may constitute an important blockade against the entry of bacteria from the vagina—a further protection for the fetus should conception occur.

The endometrial changes throughout the normal menstrual cycle should now be readily understandable (Fig. 21-13).

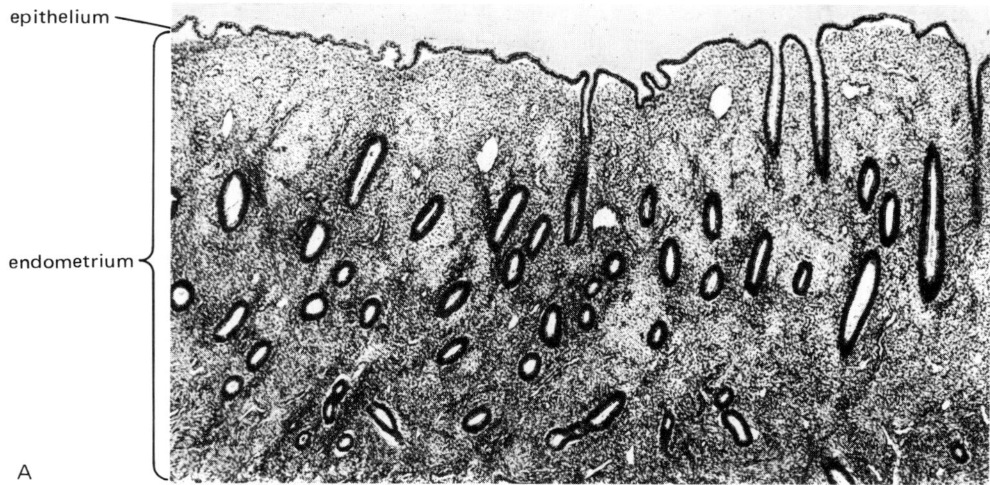

FIGURE 21-16 (A) ***Estrogen-primed endometrium (follicular phase). Parts A, B, and C*** **from W. F. Windle, "Textbook of Histology," 5th ed., McGraw-Hill, New York, 1976.**

1 The fall in blood progesterone and estrogen, which results from regression of the corpus luteum, deprives the highly developed endometrial lining of its hormonal support; the immediate result is profound constriction of the uterine blood vessels, which leads to diminished supply of oxygen and nutrients. Disintegration starts, the entire lining begins to slough, and the *menstrual flow* begins, marking the first day of the cycle.

2 After the initial period of vascular constriction, the endometrial arterioles dilate, resulting in hemorrhage through the weakened capillary walls; the menstrual flow consists of this blood mixed with endometrial debris. (Average blood loss per period equals 50 to 150 ml of blood.)

3 The *menstrual phase* continues for 3 to 5 days, during which time blood estrogen levels are low.

4 The menstrual flow ceases as the endometrium repairs itself and then grows under the influence of the rising blood estrogen concentration; this period, the *proliferative phase,* lasts for the 10 days between cessation of menstruation and ovulation.

5 Following ovulation and formation of the corpus luteum, progesterone, acting in concert with estrogen, induces the secretory type of endometrium described above.

6 This period, the *secretory phase,* is terminated by disintegration of the corpus luteum, and the cycle is completed.

It is evident that the phases of the menstrual cycle can be named either in terms of the ovarian or uterine events. Thus, the ovarian follicular phase includes the uterine menstrual and proliferative phases; the ovarian luteal phase is the same as the uterine secretory phase. The essential point is that the uterine changes simply reflect the effects of varying blood concentrations of estrogen and progesterone throughout the cycle. In turn, the secretory pattern of these hormones reflects the complex hypothalamic–anterior pituitary–ovarian interactions described previously.

NONUTERINE EFFECTS OF ESTROGEN AND PROGESTERONE

The uterine effects of the sex hormones described above represent only one set of a wide variety exerted by estrogen and progesterone (Table 21-1); all these effects are discussed in this chapter and are listed here for reference. The effects of estrogen in the female are analogous to those of testosterone in the male in that estrogen exerts dominant control over all the accessory sex organs and secondary sex characteristics. Estrogenic stimulation maintains the entire female genital tract—uterus, oviducts, vagina—the glands lining the tract, the external genitalia, and the breasts. It is

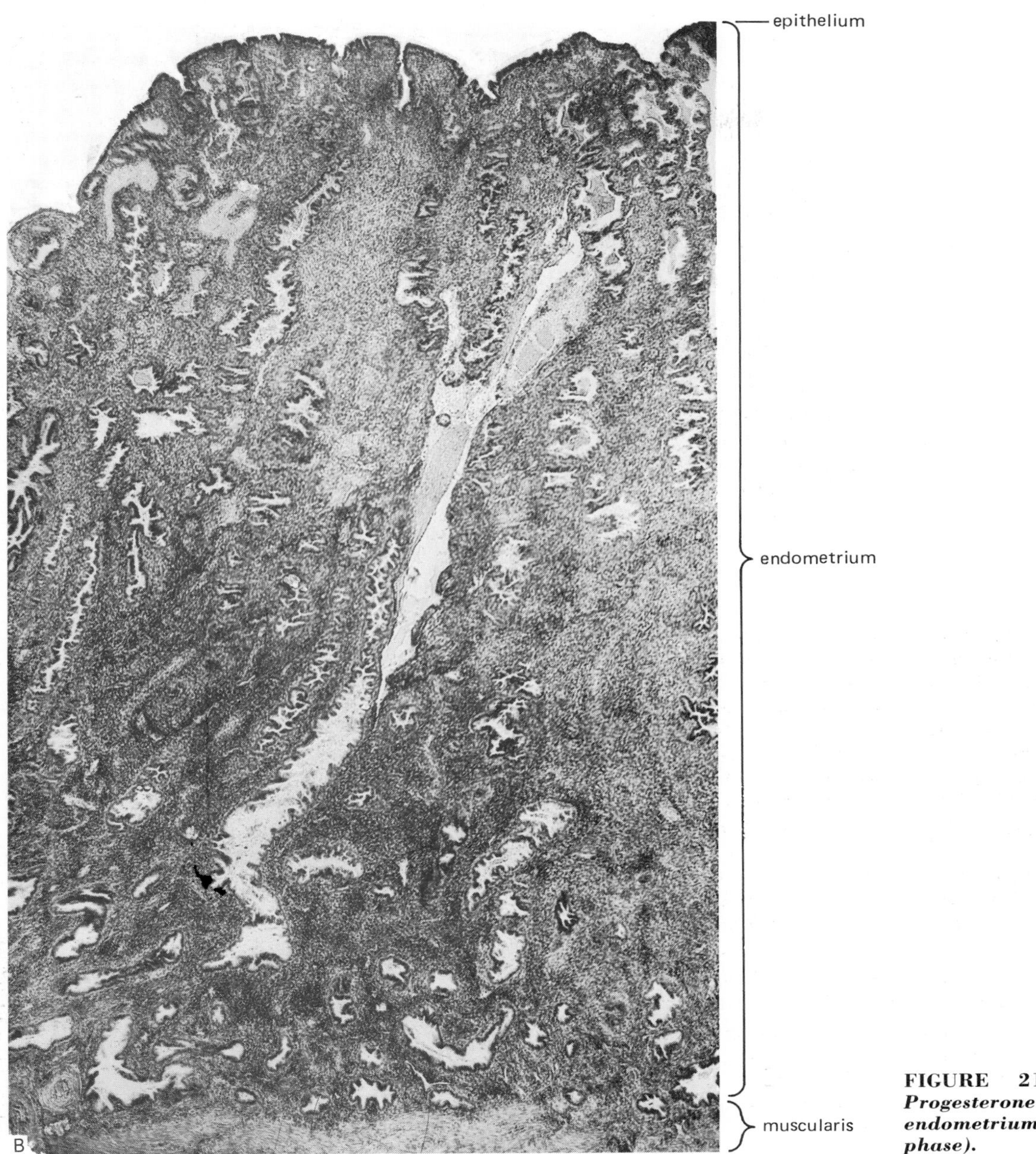

FIGURE 21-16 (B) ***Progesterone - primed endometrium (luteal phase).***

responsible for the female body-hair distribution and the general female body configuration: narrow shoulders, broad hips, and the characteristic female "curves," the result of fat deposition in the hips, abdomen, and other places. Estrogen has much less general anabolic effect on nonreproductive tissues than testosterone but probably contributes to the general body growth spurt at puberty. Finally, as described above, estrogen is required for follicle and ovum maturation; its increased secretion at puberty, in concert with that

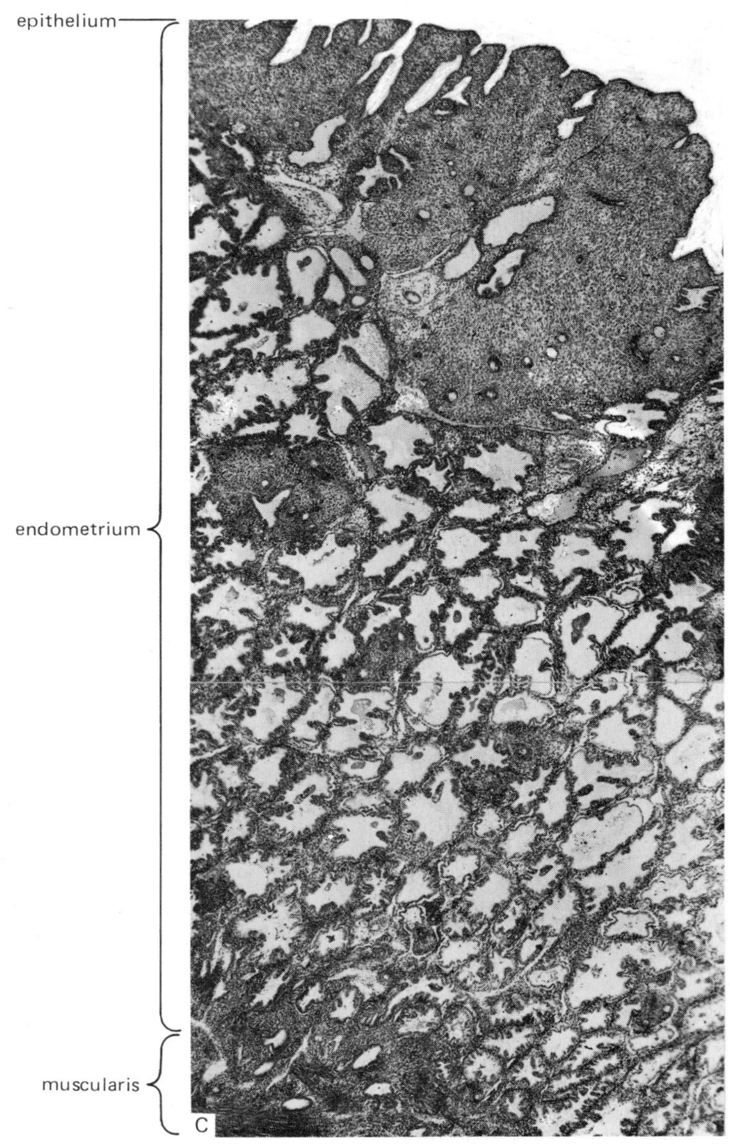

FIGURE 21-16 (C) ***Endometrium in early pregnancy. Magnification 1/3 less than Parts A and B.***

of the pituitary gonadotropins, permits ovulation and the onset of menstrual cycles. For the rest of the woman's reproductive life, estrogen continues to support the ovaries, accessory organs, and secondary sex characteristics. Because its blood concentration varies so markedly throughout the cycle, associated changes in all these dependent functions occur, the uterine manifestations being the most striking.

As was true for testosterone, estrogen acts on the cell nucleus, and its biochemical mechanism of action appears to be at the level of the genes themselves. It should be reemphasized that estrogen is not a uniquely female hormone. Small and usually insignificant quantities of estrogen are secreted by the male adrenal and the testicular interstitial cells (Fig. 21-17), the latter probably being responsible for the breast enlargement so

TABLE 21-1
EFFECTS OF FEMALE SEX STEROIDS

Effects of estrogens:
1 Growth of ovaries and follicles
2 Growth and maintenance of the smooth muscle and epithelial linings of the entire reproductive tract. Also:
 Oviducts: increased motility and ciliary activity
 Uterus: increased motility
  secretion of abundant, clear cervical mucus
 Vagina: increased "cornification" (layering of epithelial cells)
3 Growth of external genitalia
4 Growth of breasts (particularly ducts)
5 Development of female body configuration: narrow shoulders, broad hips, converging thighs, diverging arms
6 Stimulation of fluid sebaceous gland secretions ("antiacne")
7 Pattern of pubic hair (actual growth of pubic and axillary hair is androgen-stimulated)
8 Stimulation of protein anabolism and closure of the epiphyses (? due to stimulation of adrenal androgens)
9 Sex drive and behavior (? role of androgens)
10 Reduction of blood cholesterol
11 Vascular effects (deficiency → "hot flashes")
12 Feedback effects on hypothalamus and anterior pituitary
13 Fluid retention

Effects of progesterone:
1 Stimulation of secretion by endometrium; also induces thick, sticky cervical secretions
2 Stimulation of growth of myometrium (in pregnancy)
3 Decrease in motility of oviducts and uterus
4 Decrease in vaginal "cornification"
5 Stimulation of breast growth (particularly glandular tissue)
6 Inhibition of effects of prolactin on the breasts
7 Feedback effects on hypothalamus and anterior pituitary

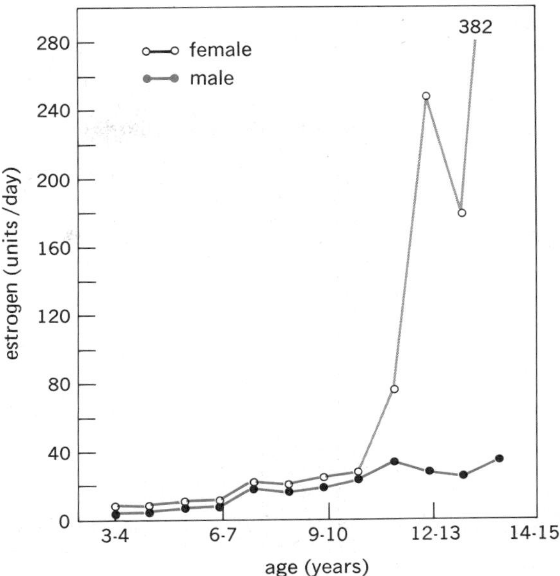

FIGURE 21-17 ***Excretion of estrogen in the urine of children, an indicator of the blood concentration of estrogen.*** **(Adapted from Nathanson et al.)**

commonly observed in pubescent boys; apparently, the rapidly developing interstitial cells release significant quantities of estrogen along with the much larger amounts of testosterone.

Progesterone is present in significant amounts only during the luteal phase of the menstrual cycle, and its effects are less widespread than those of estrogen, the endometrial changes being the most prominent. Progesterone also exerts important effects on the breasts, the oviducts, and the uterine smooth muscle, the significance of which will be described later. Progesterone also causes a transformation of the cells lining the vagina, and the microscopic examination of some of these cells provides an indicator that ovulation has or has not occurred. Another indicator is the frequently observed rise in body temperature which occurs shortly after ovulation and remains throughout the luteal phase; it has long been thought to be due to an effect of progesterone on the brain, but this may not be true.

FEMALE SEXUAL RESPONSE

The female response to coitus is very similar to that of the male in that it is characterized by marked vasocongestion and muscular contraction in many areas of the body. For example, mounting sexual excitement is associated with engorgement of the breasts and erection of the nipples, resulting from contraction of muscle filaments in them. During coitus, the vaginal epithelium also becomes highly congested and secretes a mucus-like lubricant.

The major sites of female sexual sensation are the vagina and the area of the external genitalia just above the vaginal opening, particularly the clitoris (Fig. 21-10*C*). This small shaft, homologue of the penis, is composed primarily of erectile tissue and endowed with a rich supply of sensory nerve endings. During sexual excitation, the clitoris becomes erect much as the penis does, and massaging it during coitus constitutes a primary source of sexual tension.

The final stage of female sexual excitement may be the process of orgasm, as in the male; if no orgasm occurs, there is a slow resolution of the physical changes and sexual excitement. The female has no counterpart to male ejaculation, but with this exception, the physical correlates of orgasm are very similar in the sexes: there is a sudden increase in skeletal muscle activity involving almost all parts of the body, followed by rapid relaxation; the heart rate and blood pressure increase; the female counterpart of male genital contraction is transient rhythmic contraction of the vagina and uterus.

A final question related to the female sexual response is sex drive. Incongruous as it may seem, sexual desire in adult women is more dependent upon androgens than estrogen. Thus libido is usually not altered by removal of the ovaries (or its physiologic analog, menopause). In contrast, sexual desire is greatly reduced by adrenalectomy, since these glands are the major source of androgens in women.

This completes our survey of normal reproductive physiology in the nonpregnant female. In weaving one's way through this maze, it is all too easy to forget the prime function subserved by this entire system, namely, reproduction. Accordingly, we must now return to the mature ovum we left free in the abdominal cavity, find it a mate, and carry it through pregnancy, delivery, and breast feeding.

PREGNANCY

Following ejaculation into the vagina, the sperm live approximately 48 h; after ovulation, the ovum remains fertile for 10 to 15 h. The net result is that for pregnancy to occur coitus must be performed no more than 48 h before or 15 h after ovulation (these are only average figures, and there is probably considerable variation in the survival time of both sperm and ovum). However, even these short time limits are probably too generous since, although fertile, the older ova manifest a variety of malfunctions after fertilization, frequently resulting in their rapid death. There seems little doubt that reflex ovulators like the rabbit are far more efficient than human beings are in ensuring a viable pregnancy.

Ovum Transport

At ovulation, the ovum is extruded onto the surface of the ovary and enters the uterine tube, which transmits the ovum from the ovary to the cavity of the uterus. At ovulation, smooth muscle of the fimbriae causes them to sweep over the ovary while the cilia beat in waves toward the interior of the duct; these motions immediately suck in the ovum as it emerges from the ovary and start it on its trip toward the uterus. The functional adaptability of this system has been strikingly demonstrated by the fact that women with a single ovary and a single oviduct on the opposite side, e.g., right ovary and a left oviduct, can become pregnant.

Once in the oviduct, the ovum moves rapidly for several minutes, propelled by cilia and by contractions of the duct's smooth muscle coating; the contractions soon diminish, and ovum movement becomes so slow that it takes several days to reach the uterus. Thus, fertilization must occur in the oviduct because of the short life-span of the unfertilized ovum. In this regard, the inhibitory effect of progesterone on the oviduct smooth muscle is probably of considerable importance in that ovum movement through the oviduct would otherwise be too rapid. Estrogen, in contrast, enhances oviduct motility so that, during the luteal phase, the actual degree of motility represents a subtle interplay between the opposing effects of the sex steroids.

Sperm Transport

Transport of the sperm to the site of fertilization within the oviduct is so rapid that the first sperm arrive within 30 min of ejaculation. This is far too rapid to be accounted for by the sperm's own motility; indeed, the movement produced by the sperm's tail is probably essential only for the final stages of approach and penetration of the ovum. The act of coitus itself provides some impetus for transport out of the vagina into the uterus because of the fluid pressure of the ejaculate and the pumping action of the penis during orgasm. After coitus, the primary transport mechanism is the contractions of the uterine and oviduct musculature. The factors controlling these muscular contractions remain obscure but may involve certain fatty acid derivatives (*prostaglandins*) in the semen which cause smooth muscle to contract. If the reader is puzzled by how the oviduct can move the ovum in one direction and simultaneously aid sperm movement in the other, so are physiologists. The mortality rate of sperm during the trip is huge; of the several hundred million deposited in the vagina, only a few thousand reach the oviduct. This is one of the major reasons that there must be so many sperm in the ejaculate to permit pregnancy; it may result in the selecting out of the most "fit" sperm.

In addition to aiding transport of sperm, the female reproductive tract exerts a second critical effect on them, namely, the conferring upon them of the capacity for fertilizing the egg. Although, as we have mentioned, sperm gain some degree of maturity during their stay in the epididymis, they are still not able to penetrate the zona pellucida surrounding the ovum until they have resided in the female tract for some period of time. The mechanism by which this process, known as *capacitation,* occurs is still very poorly understood.

Entry of the Sperm into the Ovum

The sperm makes initial contact with the ovum presumably by random motion, there being no good evidence for the existence of "attracting" chemicals. Having made contact, the sperm rapidly moves between adhering follicle cells and through the zona pellucida by releasing from its acrosomal cap, and perhaps other sites as well, enzymes which break down cell connections and intermolecular bonds.

Once through the zona pellucida, the sperm makes contact with the ovum cell membrane and by obscure mechanisms involving fusion of the membranes slowly passes through into the cytoplasm, frequently losing its tail in the process.[5] The nuclei of the sperm and ovum unite; the cell now contains 46 chromosomes, and *fertilization* is complete. However, viability depends upon stopping the entry of additional sperm, which inevitably prevents development of the fertilized egg. Although the mechanism is not known, it is clear that entry of a sperm causes a marked and rapid chemical transformation of the zona pellucida, making it impenetrable to other sperm.

The fertilized egg is now ready to begin its development as it continues its passage down the oviduct to the uterus. If fertilization had not occurred, the ovum would slowly disintegrate and usually be phagocytized by the lining of the uterus. Rarely a fertilized ovum remains in the oviduct, where implantation may take place. Such tubal pregnancies cannot succeed because of lack of space for the fetus to grow, and surgery may be necessary.

Early Development, Implantation, and Placentation

During the 3- to 4-day passage through the oviduct, the fertilized ovum undergoes a number of cell divisions[6] and after reaching the uterus, it floats free in the intrauterine fluid (from which it receives nutrients) for several more days, all the while undergoing cell division. This entire time span corresponds to days 14 to 21 of the typical menstrual cycle; thus while the ovum is undergoing fertilization and early development, the uterine lining is simultaneously being prepared by estrogen and progesterone to receive it. On approximately the twenty-first day of the cycle, i.e., 7 days after ovulation, *implantation* occurs. The fertilized ovum has by now developed into a ball of cells surrounding a recently formed central fluid-filled cavity and is known as a *blastocyst* (Chap. 2).

A section of the blastocyst is shown in Fig. 2-9. Note the disappearance of the zona pellucida, an event necessary for implantation. The inner cell mass is destined to develop into the fetus itself (as described in Chap. 2), whereas the outer lining of cells is already differentiating into the specialized cells, *trophoblasts,* which will form the nutrient membranes for the fetus, as described below. Once the zona pellucida has disintegrated, the trophoblastic layer rapidly enlarges and makes contact with the uterine wall. The trophoblast cells of blastocysts recovered from the uterus have been found to be quite sticky, particularly in the region overlying the innter cell mass; it is this portion which adheres to the endometrium upon contact and initiates implantation (Fig. 21-18). This initial contact somehow induces rapid development of the trophoblasts, which liberate lytic enzymes capable of breaking down the endometrial tissue. By this means, the embryo literally eats its way into the endometrium and is soon completely embedded within it (Fig. 21-19). The destructive powers of the trophoblastic layer serve a second important function: The breakdown of the nutrient-rich

[5] Upon penetration by the sperm, the ovum completes its last cell division, and the one daughter cell with practically no cytoplasm is extruded.

[6] Identical twins result when a single fertilized ovum at a very early stage of development becomes completely divided into two independently growing cell masses.

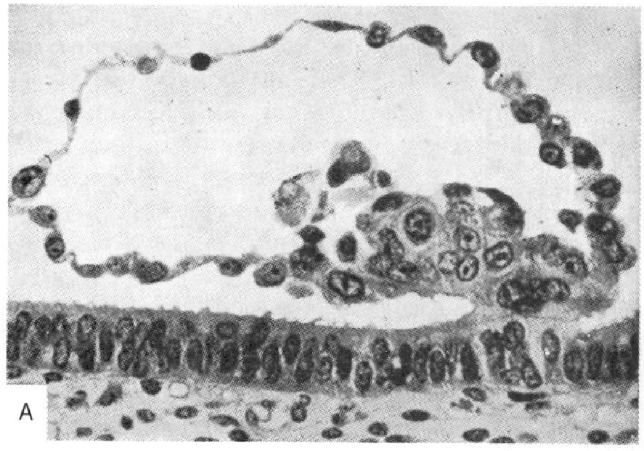

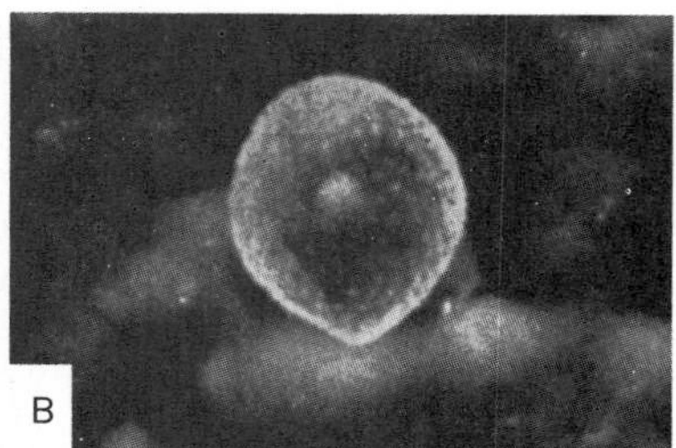

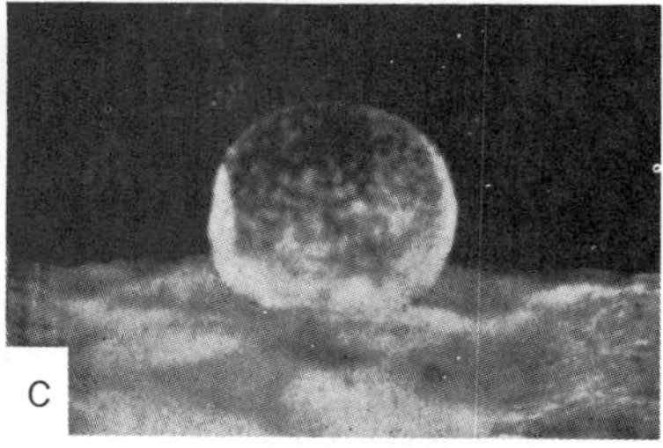

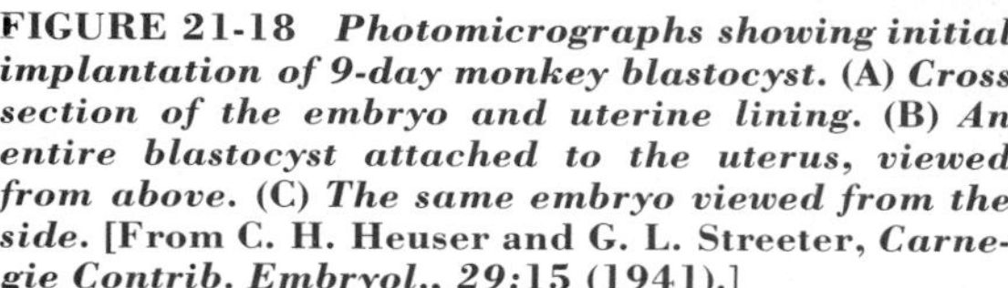

FIGURE 21-18 ***Photomicrographs showing initial implantation of 9-day monkey blastocyst.*** **(A)** ***Cross section of the embryo and uterine lining.*** **(B)** ***An entire blastocyst attached to the uterus, viewed from above.*** **(C)** ***The same embryo viewed from the side.*** **[From C. H. Heuser and G. L. Streeter, *Carnegie Contrib. Embryol., 29:*15 (1941).]**

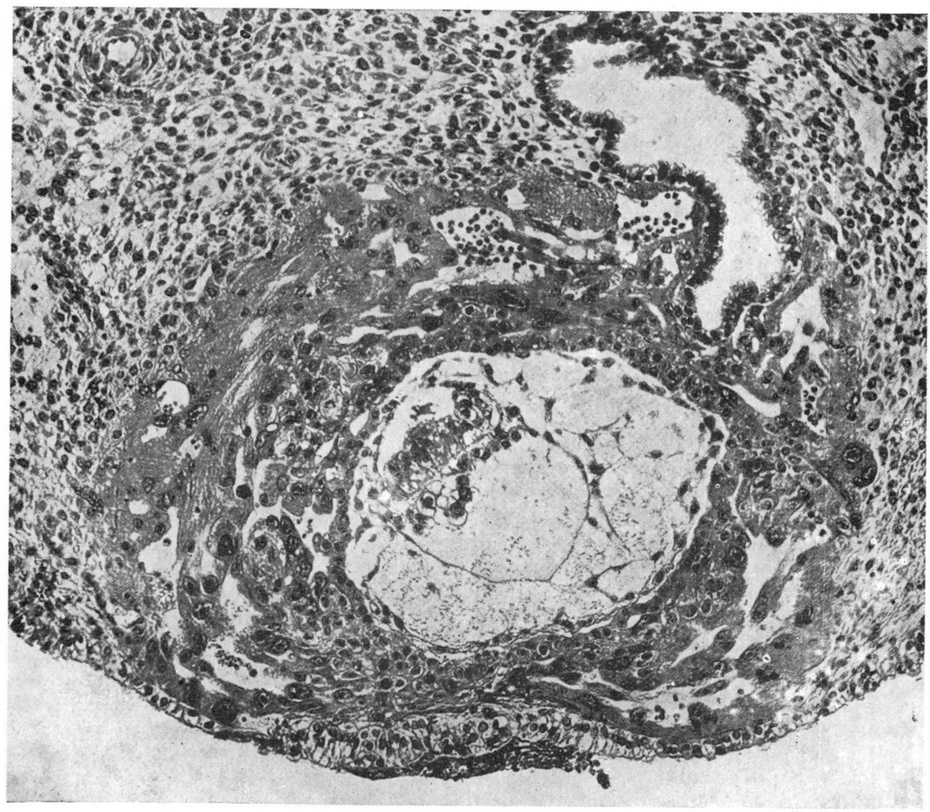

FIGURE 21-19 ***Eleven-day human embryo, completely embedded in the uterine lining.*** **[From A. T. Hertig and J. Rock, *Carnegie Contrib. Embryol.,* 29:127 (1941).]**

endometrial cells provides the metabolic fuel and raw materials for the developing embryo, and during this period, the embryo is completely dependent upon direct absorption of materials from this cell debris. This system, however, is adequate to provide for the embryo only during the first few weeks when it is very small. The system taking over after this is the fetal circulation and *placenta*.

The placenta is a combination of interlocking fetal and maternal tissues which serves as the organ of exchange between mother and fetus. The expanding trophoblastic layer breaks down endometrial capillaries, allowing maternal blood to ooze into the spaces surrounding it; clotting is prevented by the presence of some anticoagulant substance produced by the trophoblasts. Soon the trophoblastic layer completely surrounds and projects into these oozing areas (Fig. 21-19). By this time, the developing embryo has begun to send out into these trophoblastic projections blood vessels which are all branches of larger vessels, the *umbilical arteries* and *veins,* communicating with the main intraembryonic arteries and veins via the *umbilical cord* (Fig. 21-20). Five weeks after implantation, this system has become well established, the fetal heart has begun to pump blood, and the entire mechanism for nutrition of the fetus is in operation. Waste products move from the fetal blood across the placental membranes into the maternal blood; nutrients move in the opposite direction. Many substances, such as oxygen and carbon dioxide, move by simple passive diffusion, whereas other substances are carried by active-transport mechanisms in the placental membranes.

At first, as described above, the trophoblastic projections simply lie in endometrial spaces filled with blood, lymph, and some tissue debris. This basic pattern (Fig. 21-20) is retained throughout pregnancy, but many structural alterations have the net effect of making the system more efficient; e.g., the trophoblastic layer thins, the distance between maternal and fetal blood thereby being reduced. It must be emphasized that there is exchange of materials between the two blood streams but no actual mingling of the fetal and maternal blood; the maternal blood enters the placenta via the uterine artery, percolates through the spongelike endometrium, and then exits via the uterine veins; similarly, the fetal blood never leaves the fetal vessels. In several ways, the system is analogous to the artificial kidney described in Chap. 18, with the endometrial vascular spaces serving as the bath through which the fetal vessels course like the dialysis tubing. A major difference is that active transport is an important component of placental function. Moreover, the placenta must serve not only as the embryo's kidney but as its gastrointestinal tract and lungs as well. Finally, as will be

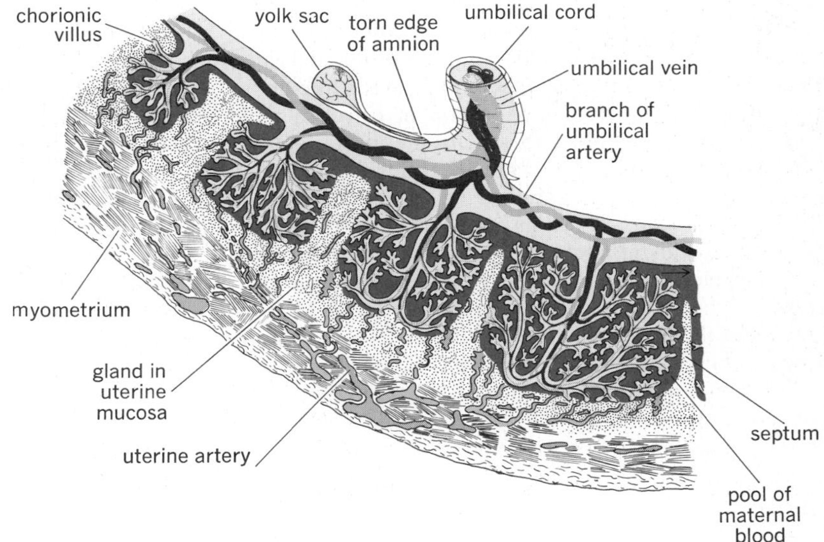

FIGURE 21-20 ***Schematic diagram: interrelations of fetal and maternal tissues in the formation of the placenta. The placenta becomes progressively more developed from left to right.*** **(Redrawn from B. M. Patten, "Human Embryology," 3d ed., McGraw-Hill, New York, 1968.)**

discussed below, the placenta (probably the trophoblasts) secretes several hormones of crucial importance for maintenance of pregnancy.

The fetus, floating in its completely fluid-filled cavity and attached by the umbilical cord to the placenta, develops into a viable infant during the next 9 months (babies born prematurely, often as early as 7 months, frequently survive). Intrauterine development was described in Chap. 2. A point of great importance is that during these 9 months the fetus is subject to considerable influence by a host of factors (noise, chemicals, etc.) affecting the mother. Via the placenta, drugs taken by the mother can reach the fetus and influence its growth and development. The thalidamide disaster was our major reminder of this fact in recent years, as is the growing number of babies who suffer heroin withdrawal symptoms after birth as a result of their mothers' drug use during pregnancy. Two other examples: Lead and DDT cross the placenta very easily. We do not know the potential effects, if any, on the fetus of many agents in the environment.

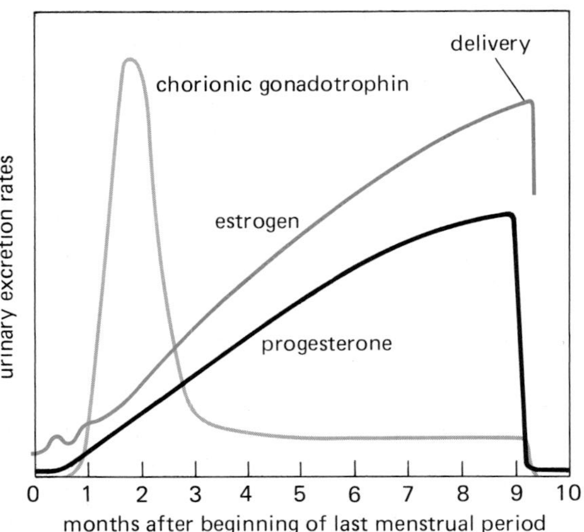

FIGURE 21-21 ***Urinary excretion of estrogen, progesterone, and chorionic gonadotropin during pregnancy. Urinary excretion rates are an indication of blood concentrations of these hormones.***

Hormonal Changes During Pregnancy

Throughout pregnancy, the specialized uterine structures and functions depend upon high concentrations of circulating estrogen and progesterone (Fig. 21-21). During approximately the first 3 months of pregnancy, almost all these steroid hormones are supplied by the extremely active corpus luteum formed after ovulation. Recall that if pregnancy had not occurred, this glandlike structure would have degenerated within 2 weeks after ovulation; in contrast, continued corpus luteum growth and steroid secretion are associated with a developing fetus. Persistence of the corpus luteum is essential since continued secretion of estrogen and progesterone is required to sustain the uterine lining and prevent menstruation (which does not occur during pregnancy). We have mentioned our lack of knowledge of the factors causing corpus luteum degeneration during a nonpregnant cycle; its persistence during pregnancy is due, at least in part, to a hormone from the placenta called *chorionic gonadotropin* (CG). Almost immediately after beginning their endometrial erosion, the trophoblastic cells start to secrete CG into the maternal blood. This protein hormone has properties very similar to those of LH although it is chemically different. What leads to its secretion is unknown, but CG strongly stimulates steroid secretion by the corpus luteum.

Recall that tonic secretion of both LH and FSH is powerfully inhibited by estrogen (it is also likely that CG itself inhibits the release of LH and FSH); therefore, the blood concentrations of these pituitary gonadotropins remain extremely low throughout pregnancy; by this means, further follicle development, ovulation, and menstrual cycles are eliminated for the duration of the pregnancy. In contrast, placental secretion of CG is inhibited to a much lesser degree by these steroids. The detection of CG in urine or plasma is the basis of most pregnancy tests.

The secretion of CG increases rapidly during early pregnancy, reaching a peak at 60 to 80 days after the end of the last menstrual period; it then falls just as rapidly, so that by the end of the third month, it has reached a low but definitely detectable level which remains relatively constant for the duration of the pregnancy. Associated with this falloff of CG secretion, the placenta itself begins to secrete large quantities of estrogen and progesterone. The very marked increases in blood steroids during the last 6 months of pregnancy are due almost entirely to the placental secretion. The corpus luteum remains, but its contribution is dwarfed by that of the placenta; indeed, removal of the ovaries during the last 6 months has no effect at all upon the pregnancy, whereas removal during the first 3 months causes immediate loss of the fetus (*abortion*).

An important and clinically useful aspect of placental steroid secretion is that the placenta is not capable of performing alone the entire sequence of reactions leading to the finished steroids. Nor are the fetal steroid-producing organs, but the steps that the one lacks are present in the other and vice versa. Therefore, the placenta and fetus complement each other, and it is possible to monitor the well-being of the fetus by measuring in the maternal blood the concentration of the major steroid, estriol, whose synthesis requires fetal participation.

Finally, this remarkable organ not only produces steroids and CG but several other hormones as well. Indeed, just as CG closely mimics the activity of LH, it may well be that the placenta produces a complete set of analogs of the anterior pituitary hormones. The best-documented one at present is a hormone which has effects very similar to those of growth hormone (and prolactin as well). This hormone, *human placental lactogen* (HPL), may play an important role in maintaining, in the mother, a positive protein balance, mobilizing fats for energy, and stabilizing plasma glucose at relatively high levels to meet the needs of the fetus.

Of the numerous other physiologic changes, hormonal and nonhormonal, in the mother during pregnancy, many, such as increased metabolic rate and appetite, are obvious results of the metabolic demands placed upon her by the growing fetus. Of great importance is salt and water metabolism because of its relationship to *toxemia of pregnancy*. During a normal pregnancy, body sodium and water increase considerably, the extracellular volume alone rising by approximately 1 l. At present, it appears that the major factor is greater secretion of renin, aldosterone, and ADH, but the mechanisms responsible remain obscure. Some women retain abnormally great amounts of fluid and manifest protein in the urine and hypertension, which if severe enough may cause convulsions. These are the symptoms of the disease known as toxemia of pregnancy. Since it can usually be well controlled by salt restriction, in the United States it is seen primarily in the lower socioeconomic segments of the population which do not obtain adequate medical care during pregnancy. Despite the obvious association with salt retention, all attempts to determine the factors responsible for the disease have failed; e.g., a causal role for aldosterone seems unlikely since full-blown toxemia can occur in adrenalectomized women.

Parturition (Delivery of the Infant)

A normal human pregnancy lasts approximately 40 weeks, although many babies are born 1 to 2 weeks earlier or later. Delivery of the infant, followed by the placenta, is produced by strong rhythmic contractions of the uterus. Actually, beginning at approximately 30 weeks, weak and infrequent uterine contractions occur, gradually increasing in strength and frequency. During the last month, the entire uterine contents shift downward so that the baby is brought into contact with the outlet of the uterus, the *cervix*. In over 90 percent of births, the baby's head is downward and acts as the wedge to dilate the cervical canal. By the onset of labor, the uterine contractions have become coordinated and quite strong (although usually painless at first) and occur at approximately 10- to 15-min intervals. Usually during this period or before, the membrane surrounding the fetus ruptures, and the intrauterine fluid escapes out the vagina. As the contractions, which begin in the upper portion and sweep down the uterus, increase in intensity and frequency, the cervical canal is forced open to a maximum diameter of approximately 10 cm. Until this point, the contractions have not moved the fetus out of the uterus but have served only to dilate the cervix. Now the contractions move the fetus through the cervix and vagina. At this time the mother, by bearing down to increase abdominal pressure, can help the uterine contractions to deliver the baby. The umbilical vessels and placenta are still functioning, so that the baby is not yet on its own, but within minutes of delivery both the infant's and mother's placental vessels completely contract, the entire placenta becomes separated from the underlying uterine wall, and a wave of uterine contractions delivers the placenta (the afterbirth) as well. Ordinarily, the entire process from beginning to end proceeds automatically and requires no real medical intervention, but in a small percentage of cases, the position of the baby or some maternal defect can interfere with normal delivery. The position is important for several reasons: (1) If the baby is not oriented head first, another portion of the body is in contact with the cervix and is generally a far less effective wedge; (2) because of its large diameter compared with the rest of the body, if the head went through the canal last, it might be obstructed by the cervical canal, leading to obvious problems when the baby attempts to breathe; and (3) if the umbilical cord becomes caught between the birth canal and the baby, mechanical compression of the umbilical vessels can result. Despite these potential difficulties, however, most babies who are not oriented head first are born normally and with little difficulty.

What mechanisms control the events of parturition? Let us consider a set of fairly well-established facts:

1 The uterus is composed of smooth muscle capable of autonomous contractions and having inherent rhythmicity, both of which are facilitated by stretching the muscle.

2 The efferent neurons to the uterus are of little importance in parturition since anesthetizing them in no way interferes with delivery.

3 Progesterone exerts a powerful inhibitory effect upon uterine contractility. Shortly before delivery, the secretion of progesterone (and estrogen) sometimes drops, perhaps due to "aging" changes in the placenta.

4 *Oxytocin,* one of the hormones released from the posterior pituitary, is an extremely potent uterine-muscle stimulant. Oxytocin is reflexly released as a result of input into the hypothalamus from receptors in the uterus, particularly the cervix.

5 The pregnant uterus contains several prostaglandins, one of which is a profound stimulator of uterine smooth muscle; an increase in the release of this substance has been demonstrated during labor.

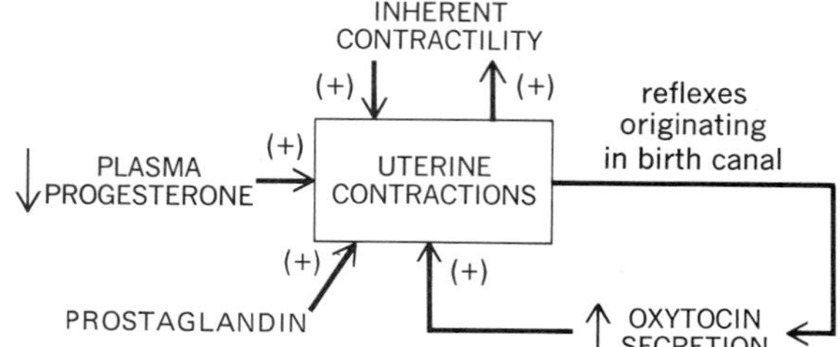

FIGURE 21-22 ***Factors stimulating uterine contractions during parturition. Note the positive-feedback nature of several of the inputs. What initiates parturition is not known.***

These facts can now be put together in a unified pattern, as shown in Fig. 21-22. The precise contributions of each of these factors are unclear; moreover, we cannot answer the crucial question: Which factor (if any) actually initiates the process? Once started, the uterine contractions exert a positive-feedback effect upon themselves via reflex stimulation of oxytocin and local facilitation of the muscle's inherent contractility; but what *starts* the contractions? The decrease in progesterone cannot be essential since it simply does not occur in most women. Nor is uterine distension or the presence of a fetus a requirement, as attested by the remarkable fact that typical "labor" begins at the expected time in animals from which the fetus has been removed weeks previously. A primary role for prostaglandin has recently been touted but the evidence is far from conclusive. Regardless of their relative contributions to normal parturition, both prostaglandins and oxytocin are useful clinically in artificially inducing labor.

Lactation

Perhaps no other process so clearly demonstrates the intricate interplay of various hormonal control mechanisms as milk production. The endocrine control has been established by numerous investigations and observations, none more striking than that in 1910 of Siamese twins; One twin became pregnant, and both women lactated after delivery.

The *breasts* are formed of glandular tissue (the *mammary glands*), fatty and fibrous connective tissue between the lobes of the glands, and blood vessels, nerves, and lymph vessels (Fig. 21-23). The epithelium-lined ducts, which form the glandular portion, converge at the *nipples*. These ducts, the *lactiferous ducts,* branch all through the breast tissue and terminate in saclike structures (*alveoli*) typical of exocrine glands. The alveoli, which secrete the milk, look like bunches of grapes with stems terminating in the ducts. Beneath the *areola* (the colored area of skin surrounding the nipple) the ducts are dilated and form lactiferous *sinuses,* which serve as reservoirs of milk. The alveoli and the ducts immediately adjacent to them are surrounded by specialized contractile cells called *myoepithelial cells.* Before puberty, the breasts are small and consist almost entirely of ducts; no alveoli are present. With the onset of puberty, estrogen and progesterone act in concert upon the gland to produce the basic architecture of the adult breast. However, in addition to these steroids, normal breast development at puberty requires *prolactin* and *growth hormone,* both secreted by the anterior pituitary.

During each menstrual cycle, breast morphology fluctuates in association with the changing blood concentrations of estrogen and progesterone, but these changes are small compared with the marked breast enlargement which occurs during pregnancy as a result of the prolonged stimulatory effects of estrogen and progesterone on both ducts and alveoli. Indeed, there also frequently occurs development *in utero* of the fetal breasts and production of so-called witches' milk; this infantile development is, of course, short-

FIGURE 21-23 (A) ***Dissection of the breast, showing increased secretory lobules during lactation.*** (B) ***Histology of an actively lactating human breast.*** (C) ***Alveolus of a lactating mammary gland of a rabbit. The cells contain fat droplets (stained black), which, together with the adjacent cytoplasm, are extruded into the lumen.*** **(Part C adapted from Bloom and Fawcett.)**

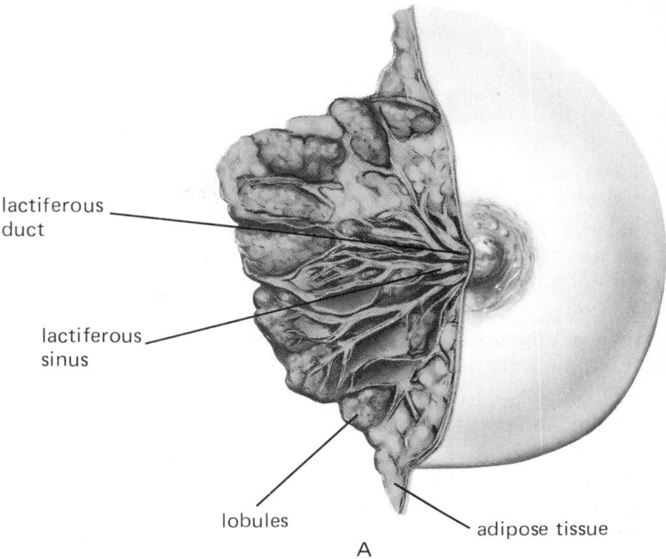

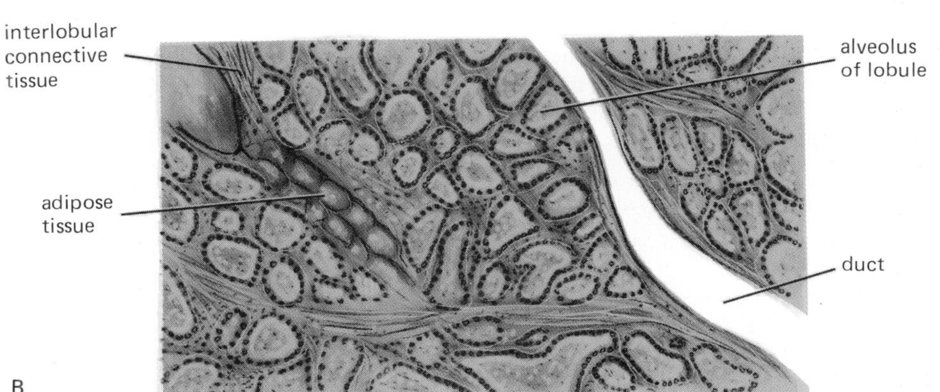

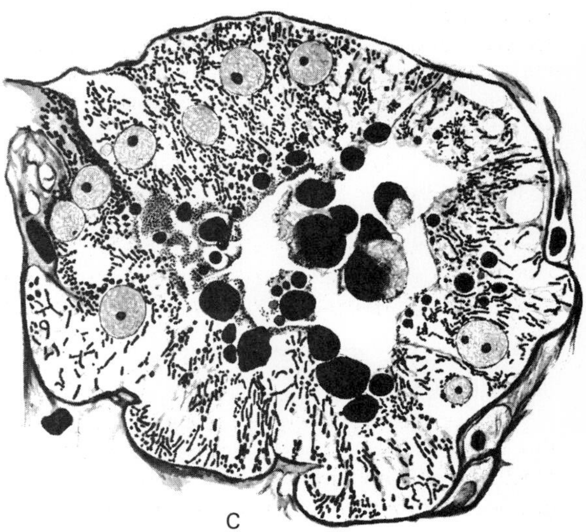

lived and quickly disappears when delivery removes the child from the effects of sex hormone stimulation.

The single most important hormone promoting milk production is prolactin from the anterior pituitary. Yet, despite the fact that prolactin is elevated and the breasts markedly enlarged as pregnancy progresses, there is no secretion of milk, probably because estrogen and progesterone in large concentration inhibit milk production by a direct action on the breasts. Delivery removes the sources of sex steroids (the placenta) and, thereby, the inhibition.

The stimulus to increased prolactin secretion during pregnancy is unclear (stimulation by estrogen is one factor), but the major factor maintaining the secretion during lactation is quite clear: reflex input to the hypothalamus from receptors in the nipples which are stimulated by suckling (Fig. 21-24). Thus, milk production ceases soon after the mother stops nursing her infant but continues uninterrupted for years if nursing is continued.

FIGURE 21-24 ***Nipple suckling reflex. The neural pathway is schematic; actually multiple interneurons are involved. The primary event is stimulation of the nipple mechano-receptors.***

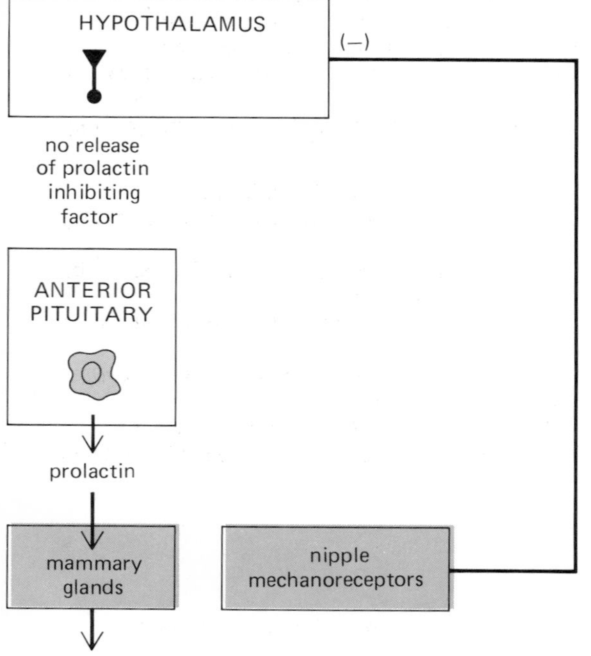

One final reflex process is essential for nursing. Milk is secreted into the lumen of the alveoli (Fig. 21-23*C*), but because of their structure the infant cannot suck the milk out. It must first be moved into the ducts and sinuses, from which it can be sucked. This process is called *milk let-down* and is accomplished by contraction of the myoepithelial cells surrounding the alveoli; the contraction is directly under the control of oxytocin, which is reflexly released by suckling (Fig. 21-25), just like prolactin.[7] Many women experience uterine contractions during nursing because of the uterine effects of oxytocin. Higher brain centers also exert important influence over oxytocin release: Many nursing mothers actually leak milk when the baby cries. In view of the central role of the nervous system in lactation reflexes, it is no wonder that psychologic factors can interfere with a woman's ability to nurse.

The end result of all these processes, the milk, contains four major constituents; water, protein, fat, and the carbohydrate lactose. The mammary alveolar cells must be capable of extracting the raw materials—amino acids, fatty acids, glycerol, glucose, etc.—from the blood and building them into the higher-molecular-weight substances. These synthetic processes require the participation of prolactin, insulin, growth hormone, cortisol, and probably other hormones.

Another important neuroendocrine reflex triggered by suckling is inhibition of FSH and LH release by the pituitary, with subsequent block of ovulation. This inhibition apparently is relatively short-lived in many women, and approximately 50 percent begin to ovulate despite continued nursing. Pregnancy is common in women lulled into false security by the mistaken belief that failure to ovulate is always associated with nursing.

Birth Control

Table 21-2 summarizes possible means of preventing fertility. We present it because it should serve to summarize a majority of the information presented in this chapter (there are no new facts in the table). For each possibility listed there are many possible preventive aspects; for example, implantation is extremely complex and requires a long sequence of events. Thus, for each possibility, scientific investigation revolves around the question: What are the normal physiologic

[7] The functions of prolactin and oxytocin in males are unknown.

TABLE 21-2
POSSIBLE MEANS OF PREVENTING FERTILITY

Possible means of preventing fertility in men†:

I. Interference with sperm survival
   A. Prevention of maturation process in epididymis
   B. Prevention of function of accessory glands
      1. Prevention of androgen action on accessory glands
      2. Prevention of formation of accessory-gland secretion
      3. Prevention of accessory-gland secretion from entering urethra
   C. Creation of hostile environment to sperm in vas deferens or urethra

II. Interference with testicular function at testicular level
   A. Prevention of androgen action on seminiferous tubules
   B. Prevention of action of FSH on seminiferous tubules
   C. Prevention of sperm division

III. Interference with pituitary function
   A. Prevention of FSH secretion at a pituitary level
   B. Prevention of action of FSH-releasing factor on pituitary
   C. Prevention of secretion of FSH-releasing factor
   D. Abolishment of extrahypothalamic central nervous factors required for normal reproductive function

Possible means of preventing fertility in women‡:

I. Interference with ovarian function at ovarian level
   A. Prevention of initiation of follicle growth (no response to early FSH rise)
   B. Prevention of response to ovulatory LH surge
   C. Prevention of maturation of follicle or maturation of ova (e.g., inhibition of meiotic division)
   D. Prevention of corpus luteum formation
   E. Prevention of ovarian estrogen secretion
   F. Prevention of ovarian progesterone secretion
   G. Prevention of the maintenance of the corpus luteum during early pregnancy

II. Interference with pituitary function
   A. Prevention of FSH or LH secretion, or both, at a pituitary level
   B. Prevention of action of releasing factors on pituitary
   C. Prevention of secretion of releasing factors
   D. Alteration of estrogen or progesterone action on tonic center (too much or too little)
   E. Prevention of estrogen action on cyclic center
   F. Abolishment of extrahypothalamic central nervous factors required for normal reproductive hypothalamic function

III. Interference with sperm action
   A. Prevention of sperm entrance to vagina
   B. Prevention of sperm entrance to uterus
   C. Prevention of sperm entrance to oviducts
   D. Creation of hostile environment to sperm in vagina, uterus, or oviducts
   E. Prevention of sperm capacitation
   F. Prevention of sperm penetration of ovum by action on sperm

IV. Interference with ovum action
   A. Prevention of ova release from ovary
   B. Prevention of ova entrance to oviducts
   C. Creation of hostile environment for ova in oviducts
   D. Prevention of sperm penetration of ovum by action on ova

V. Prevention of survival of fertilized ova
   A. Prevention of fertilized ova from undergoing mitosis
   B. Alteration of migration along fallopian tube (too fast or too slow)
   C. Prevention of implantation of blastocyst in uterine wall
   D. Destruction or expulsion of embryo after implantation in uterine wall
   E. Prevention of CG secretion by placenta
   F. Prevention of sex steroid secretion by placenta

† Not listed are surgical interventions removing the source of sperm, such as castration, or preventing egress of sperm, such as vasectomy.
‡ Not listed is surgical removal of any of the component organs (e.g., uterus, fallopian tubes, ovaries).

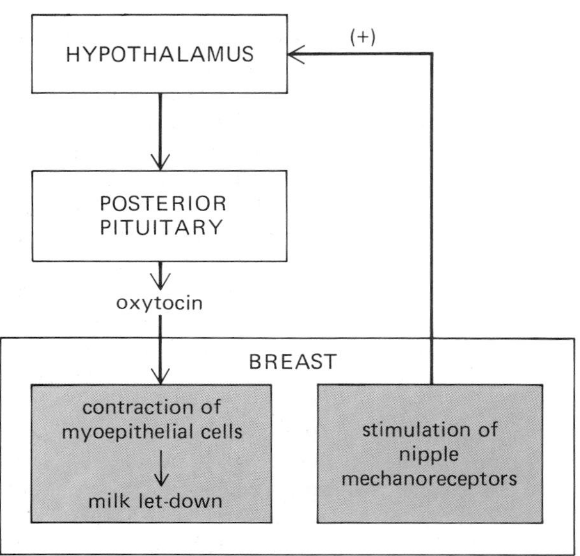

FIGURE 21-25 ***Suckling-reflex control of oxytocin secretion and milk let-down.***

events that make the process possible and how can we intervene to prevent them?

Until recently, techniques of birth control (*contraception*) were primarily those which prevent sperm from reaching the ovum: vaginal diaphragms, sperm-killing jellies, and male condoms. Each of these methods has aesthetic drawbacks, and, more important, they are frequently ineffective (Table 21-3). Another widely used method is the so-called rhythm method, in which couples merely abstain from coitus near the time of ovulation. Unfortunately, it is difficult to time ovulation precisely even with laboratory techniques; e.g., the rise in body temperature or change in cervical mucus and vaginal epithelium, all of which are indicators of ovulation, occur only *after* ovulation. This problem, combined with the marked variability of the time of ovulation in many women, explains why this technique is only partially effective.

TABLE 21-3
EFFECTIVENESS OF CONTRACEPTIVE METHODS

Method	*Pregnancies per 100 women per year*
None	115
Douche	31
Rhythm	24
Jelly alone	20
Withdrawal	18
Condom	14
Diaphragm	12
Intrauterine device	5
Oral contraceptive	1
correctly used	0

Since 1950, an intensive search has been made for a simple, effective contraceptive method, the first fruit of these studies being "the pill," the *oral contraceptive*. Its development was based on the knowledge that combinations of estrogen and progesterone inhibit pituitary gonadotropin release, thereby preventing ovulation. The most commonly used agents, at least at first, were combinations of an estrogen- and a progesterone-like substance (progestogens). Each month, the pill is taken for 20 days, then discontinued for 5 days; this steroid withdrawal produces menstruation, and the net result is a menstrual cycle without ovulation. The monthly withdrawal is required to avoid "breakthrough" bleeding which would occur if the steroids were administered continuously. Another type of regimen is the so-called sequential method in which estrogen is administered alone for 15 days followed by estrogen plus progestogen for 5 days, followed by withdrawal of both steroids. As with the combination pills, this regimen interferes with the orderly secretion of gonadotropins and prevents ovulation; no LH surge occurs, at least in part because of the constant estrogen levels, i.e., the absence of a rising estrogen level capable of exerting positive feedback.

Study of the mechanism of action of the oral contraceptives has revealed that they have multiple antifertility effects. In other words, the hormonal milieu required for normal pregnancy is such that these exogenous steroids interfere with many of the steps between coitus and implantation of the blastocyst. Taken correctly, they are almost 100 percent effective. Serious side effects, such as intravascular clotting, have been reported but only in a small number of women; however, only time can show whether undesirable effects will ultimately appear as a result of chronic alteration of normal hormonal balance.

Another type of contraceptive which is highly effective (although not 100 percent) and which illustrates the dependency of pregnancy upon the right conditions is the *intrauterine device* (Table 21-3). Placing one of these small objects in the uterus prevents pregnancy, apparently by somehow interfering with the endometrial preparation for acceptance of the blastocyst.

The search goes on for an effective method which will reduce even further the possibility of unwanted side effects and will still be easy to use. Almost every possible process shown in Table 21-2 is worth following up. Two recent developments are postcoital medications: prostaglandins and the estrogenlike diethylstilbestrol (DES). Both cause increased contractions of the female genital tract and may, therefore, cause expulsion of the fertilized ovum or failure to implant. Enthusiasm for DES has been tempered by the possibility that it may have cancer-producing properties.

Fertility prevention is of great importance, but the other side of the coin is the problem of unwanted infertility. Approximately 10 percent of the married couples in the United States are infertile. There are many reasons—some known, some unknown—for infertility; indeed, Table 21-2 also serves as a list of possible causes of infertility. Careful investigation of infertile couples frequently permits diagnosis and therapy of the basic problem.

Section C. The Chronology of Sex Development

SEX DETERMINATION

Sex is determined by the genetic inheritance of the individual, specifically by two chromosomes called the *sex chromosomes*. With appropriate tissue-culture techniques all the chromosomes in human cells can be made visible; such studies have demonstrated the presence of 46 chromosomes, 22 pairs of somatic (nonsex) chromosomes and 1 pair of sex chromosomes. The larger of the sex chromosomes is called the X chromosome and the smaller the Y chromosome. Genetic males possess one X and one Y, whereas females have two X chromosomes (Fig. 21-26). Thus the genetic difference between male and female is simply the difference in one chromosome. The reason for the approximately equal sex distribution of the population should be readily apparent (Fig. 21-27): The female can contribute only an X chromosome, whereas the male, during meiosis, produces sperm, half of which are X and half of which are Y. When the sperm and ovum join, 50 percent should have XX and 50 percent XY.

Interestingly, however, sex ratios at birth are not 1:1. Rather, there tends to be a slight preponderance of male births (in England, the ratio of male to female births is 1.06). Even more surprising, the ratio at the time of conception seems to be much higher. From various types of evidence, it has been estimated that there may be 30 percent more male conceptions than female. There are several implications of these facts. First, there must be a considerably larger *in utero* death rate for males. Second, the "male," that is, XY, sperm must have some advantage over the "female" sperm in reaching and fertilizing the egg. It has been suggested, for example, that since the Y chromosome is lighter than the X, the "male" sperm might be able to travel more rapidly. There are numerous other theories, but we are far from an answer. Moreover, it should be pointed out that conception and birth ratios show considerable variation in different parts of the world and, indeed, in rural and urban areas of the same country.

The methods of making human chromosomes visible are quite difficult. However, an easy method for distinguishing between the sex chromosomes was found quite by accident; the cells of female tissue (scrapings from the cheek mucosa are convenient) contain a readily detected nuclear mass believed to derive from the XX chromosomal combination. This has been called the *sex chromatin* and is not usually found in male cells. The method has proved valuable when genetic sex was in doubt. Its use and that of the more exacting tissue-culture visualization have revealed a group of genetic sex abnormalities characterized by such bizarre chromosomal combinations as XXX, XXY, X, and many others. Just how these combinations arise remains obscure, but the end result is usually the failure of normal anatomic and functional sexual development. For example, patients with only one sex chromosome, an X, show no gonadal development; apparently both X chromosomes are required for normal ovarian growth during embryologic development.

SEX DIFFERENTIATION

It is not surprising that persons with abnormal genetic endowment manifest abnormal sexual development, but careful study has also revealed patients with normal chromosomal combinations but abnormal sexual appearance and function. For example, a genetic male (XY) may have testes and female internal genitalia

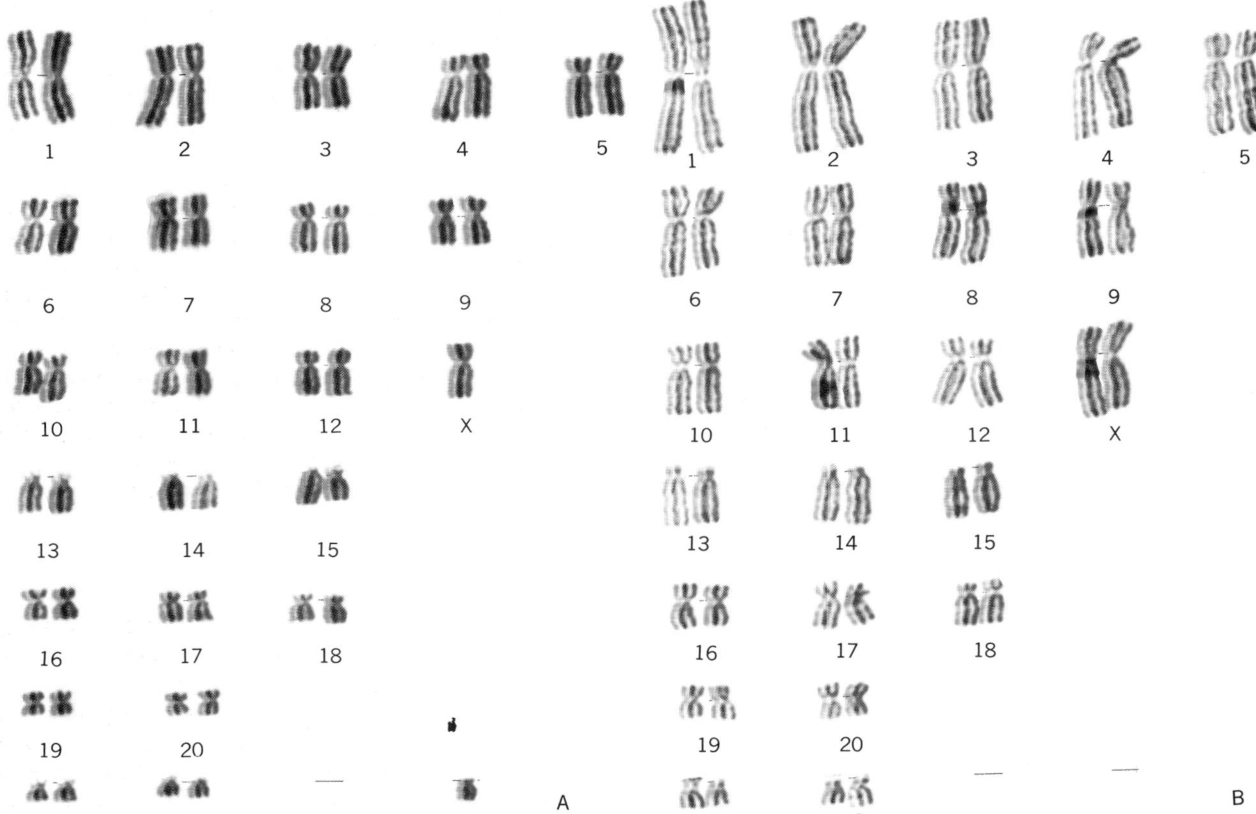

FIGURE 21-26 ***Normal chromosomes in male* (A) *and female* (B) *cells.* (Courtesy of J. Lejeune, Chaire de Genetique Fondamentale, Paris.)**

(vagina and uterus); such people are termed male pseudohermaphrodites. This kind of puzzle leads us into the realm of sex differentiation, i.e., the process by which the fetus develops the male or female characteristics directed by its genetic makeup. The genes *directly* determine only whether the individual will have testes or ovaries; virtually all the rest of sexual differentiation depends upon this genetically determined gonad.

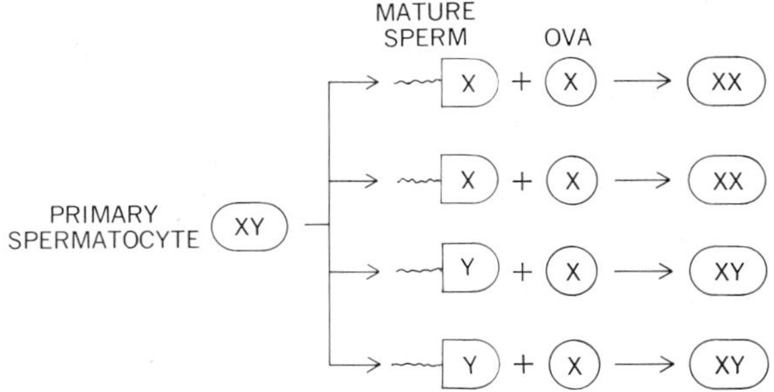

FIGURE 21-27 ***Basis of genetic sex determination.***

Differentiation of the Gonads

The male and female gonads derive embryologically from the same site in the body. Until the sixth week of life, there is no differentiation of this site. During the seventh week, in the genetic male, the testes begin to develop; in the genetic female, this does not take place, and several weeks later ovaries begin to develop instead. The embryonic gonad, testis or ovary, then regulates the remainder of the individual's sexual development.

Differentiation of Internal and External Genitalia

The very early fetus is sexually bipotential so far as its internal duct system and external genitalia are concerned. During differentiation, either the male or female duct system will fully develop, the other becoming vestigial. Similarly, externally, either a penis or clitoris will develop, and the tissue near it will either fuse, in the male, to become a scrotum or remain separate, in the female, as the labia (Fig. 21-28).

Until the sixth or seventh week of embryonic life, the primitive gonad is undifferentiated sexually and makes connections to two distinct sets of ducts, the *mesonephric ducts* (a part of the developing urinary system) and the *Mullerian ducts* (Fig. 21-29). As the gonad differentiates into a testis, the mesonephric duct remains and develops into the male duct system. In contrast, as the gonad differentiates into an ovary, the Mullerian system develops into the female duct system. In each case, the other set of ducts degenerates. In many species it is possible to remove or damage the area from which the gonads derive without interrupting the pregnancy. When this is done, regardless of genetic sex, the gonadless embryo develops *female* internal genitalia (uterus, vagina, etc.). The conclusion is that normally the presence of functioning testes represses the development of the female duct system and induces development of the male organs; in contrast, a female gonad need not be present for the female organs to develop. The testes exert these effects by secreting an unknown substance (probably not testosterone), which acts upon the developing genital tissue. This concept of inducers applies in virtually all areas of embryologic development. A similar analysis holds for the later development of the external genitalia except that here the inducer does appear to be testosterone secreted by the interstitial cells as a result of stimulation by chorionic gonadotropin from the placenta. The converse of this type of experiment was also done; a genetic female monkey whose mother had been given large doses of testosterone during the critical period of external genital development had a scrotum and markedly enlarged clitoris. To return to our previously described XY patient with the female duct system, it seems reasonable to suspect a failure of his gonadal function during the period of duct differentiation. Note that, depending upon the timing of gonadal failure, one could develop external and internal genitalia of opposite sexes.

Sexual Differentiation of the Central Nervous System

As we have seen, the male and female hypothalamus differ in that there is cyclic secretion of LH-releasing factor in the female but rather fixed continuous release in the male. In primates (as opposed to rats) it appears that this difference does not reflect any qualitative difference in hypothalamic "imprinting" between male and female; rather it is due simply to the fact, described previously, that large amounts of dominant female sex hormone, estrogen, can stimulate LH release whereas testosterone can only inhibit it. Thus, it has been demonstrated recently that the administration of estrogen to male castrated monkeys elicited LH surges indistinguishable from those shown by females.

The situation may be quite different for sexual behavior in that qualitative differences may be imprinted during development; genetic female monkeys given testosterone during late *in utero* life manifest not only masculinized external genitalia but pronounced masculine sex behavior as adults. Related to this is the question of whether exposure to androgens during *in utero* existence is necessary for development of other behavior patterns in addition to sex. Again, for other pri-

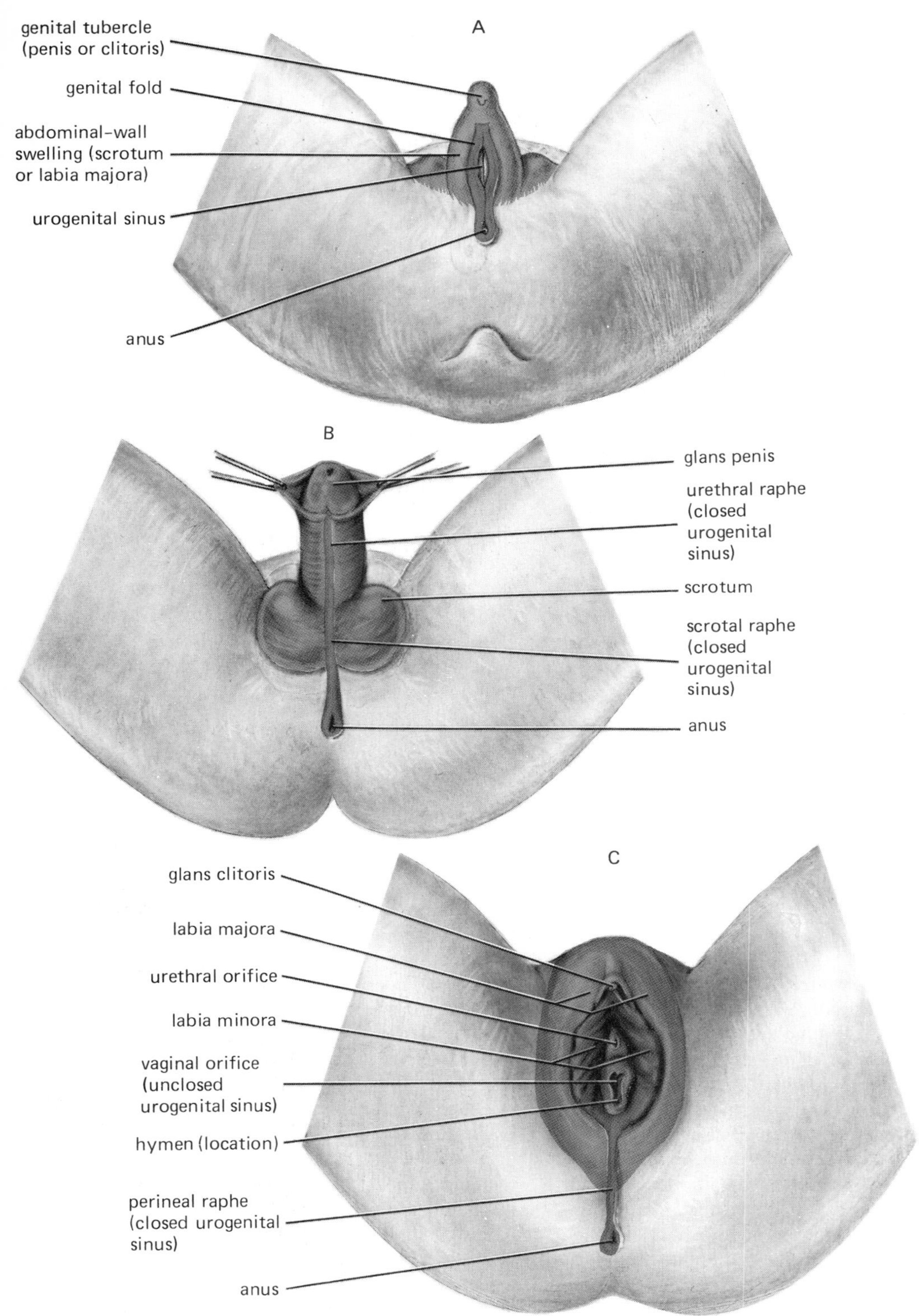

FIGURE 21-28 ***Stages in the development of the male* (B) *and female* (C) *external genitalia from the indifferent stage* (A). (Adapted from Patten.)**

FIGURE 21-29 ***Stages in the development of the internal genitalia.***

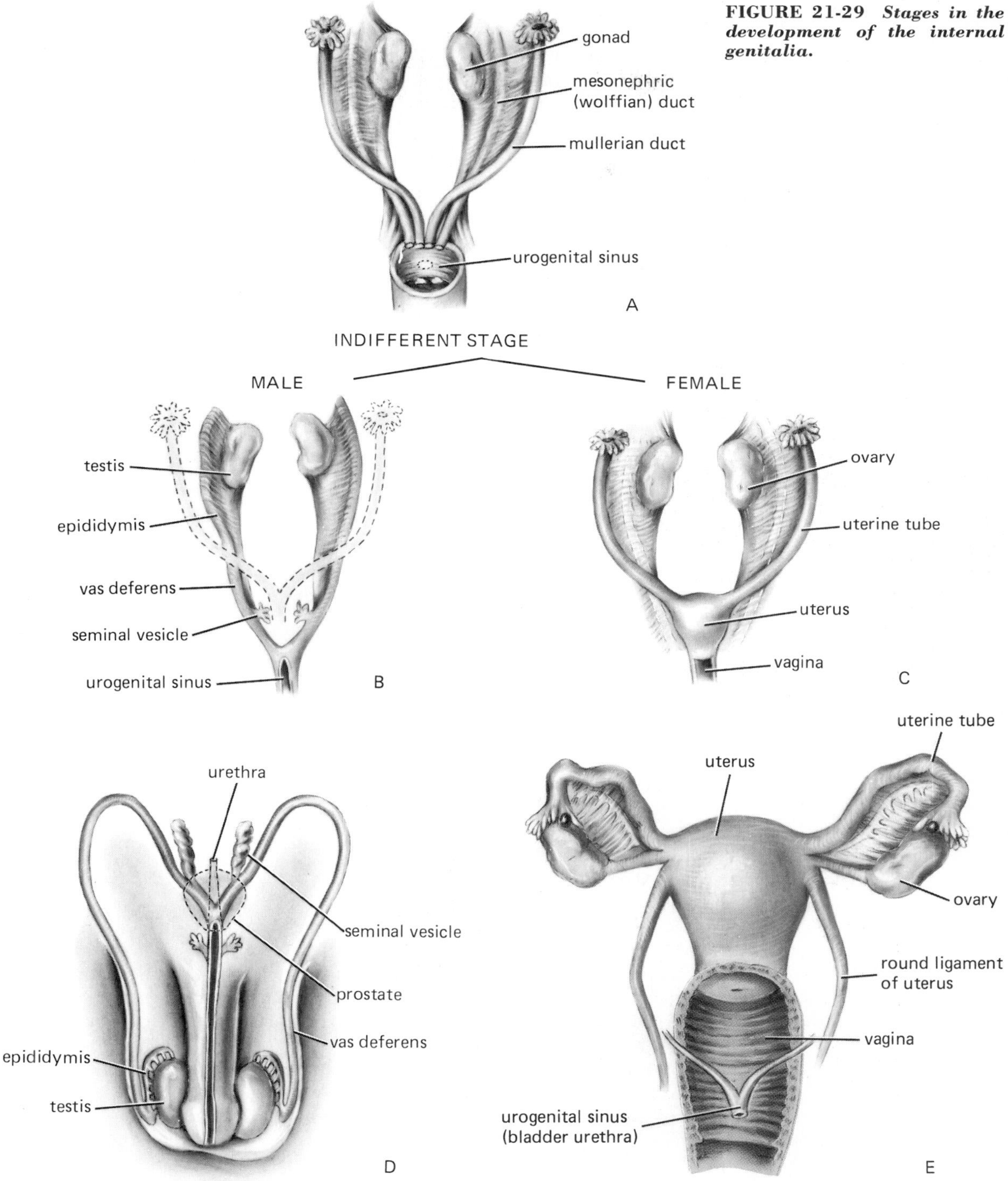

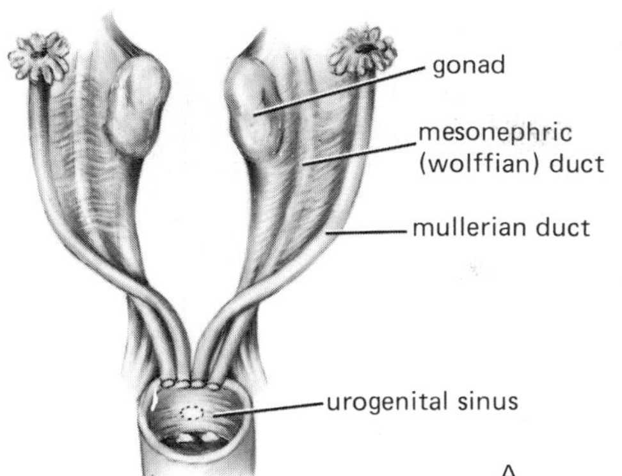

mates, the answer seems a clear-cut yes; the evidence for human beings is very scanty.

In summary (Fig. 21-30), it is evident that the male gonadal secretions play a crucial role in determining normal sexual differentiation *in utero* or just after birth. Shortly after delivery, when the gonad is deprived of the stimulant action of chorionic gonadotropin, testosterone secretion becomes very low until puberty, when it once again increases to stimulate the organs it had previously helped to differentiate. In contrast, estrogen probably takes little active part in *in utero* development, female differentiation requiring only the absence of testicular secretion. Estrogen is, of course, the stimulating agent for the female sex organs during puberty and adult life. Finally, the X and Y chromosomes appear to dictate directly only whether testes or ovaries develop.

PUBERTY

Puberty is the period, usually occurring sometime between the ages of 10 and 14, during which the reproductive organs mature and reproduction becomes possible. In the male, the seminiferous tubules begin to produce sperm; the genital ducts, glands, and penis enlarge and become functional; the secondary sex characteristics develop; and sexual drive is initiated. All these phenomena are effects of testosterone, and puberty in the male is the direct result of the onset of testosterone secretion by the testes (Fig. 21-31*A*). Although significant testosterone secretion occurs during late fetal life, within the first days of birth the interstitial cells disappear and testosterone secretion becomes very low until puberty.

The critical question is: What stimulates testosterone secretion at puberty? Or conversely: What inhibits it before puberty? Experiments with testis or pituitary transplantation in other mammals and studies of hormone injections in human beings have suggested the hypothalamus as the critical site of control. Before puberty, the hypothalamus fails to secrete significant quantities of releasing factors to stimulate secretion of FSH and LH (Fig. 21-31*B*); deprived of these gonadotropic hormones, the testes fail to produce sperm or large amounts of testosterone. Puberty is initiated by an unknown alteration of brain function which permits secretion of the hypothalamic releasing factors for LH and FSH; the mechanism of this change remains unknown. Children with brain tumors or other lesions of the hypothalamus or pineal may undergo precocious puberty, i.e., sexual maturation at an unusually early age, sometimes within the first 5 years of life, but the interpretation of these "experiments of nature" remains controversial. In particular, the possible role of that mysterious gland, the pineal, in control of reproduction is presently the subject of much investigation.

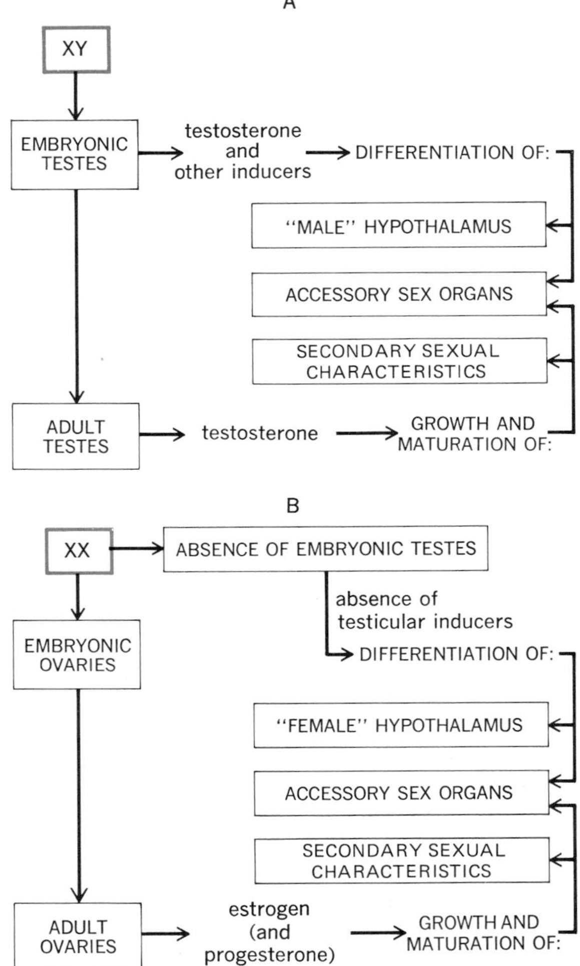

FIGURE 21-30 ***Summary of sex-organ differentiation in* (A) *male and* (B) *female.***

The picture for the female is completely analogous to that for the male. Throughout childhood, estrogen is secreted at very low levels (Fig. 21-17). Accordingly, the female accessory sex organs remain small and nonfunctional; there are minimal secondary sex character-

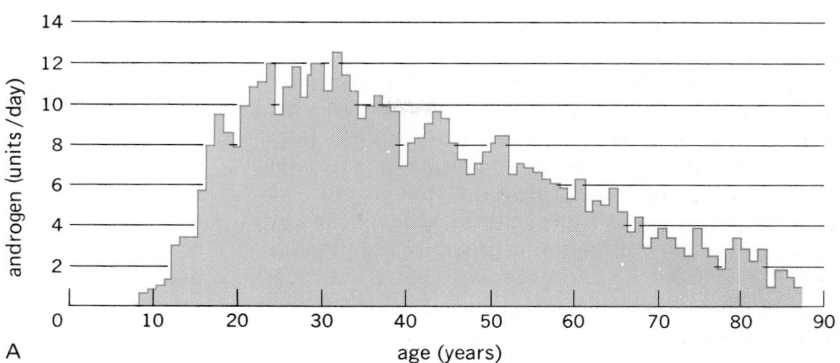

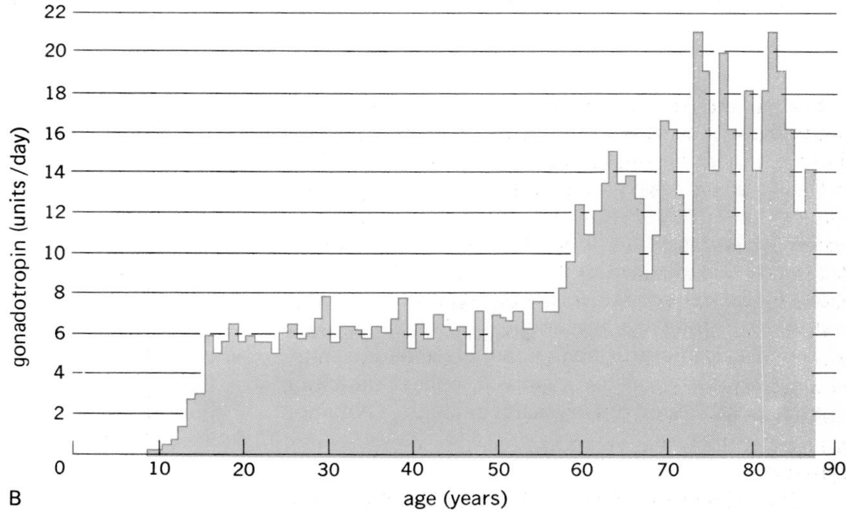

FIGURE 21-31 (A) ***Excretion of androgen in the urine of normal boys and men, an indicator of the blood concentration of androgen.*** (B) ***Excretion of anterior pituitary gonadotropins (FSH and LH) in the urine of normal boys and men, an indicator of the blood concentration of gonadotropins.*** **(Adapted from Pedersen-Bjergaard and Tonnesen)**

istics, and ovum maturation does not occur. As for the male, prepuberal dormancy is probably due mainly to deficient secretion of the hypothalamic releasing factors controlling FSH and LH secretion, and the onset of puberty is occasioned by an alteration in brain function which raises secretion of these releasing factors. Production of larger amounts of releasing factors at puberty raises secretion of pituitary gonadotropins and estrogen. The increased estrogen induces the striking changes associated with puberty and, through its stimulatory effect on the hypothalamic cyclic center, permits menstrual cycling to begin. Precocious puberty also occurs in females; the youngest mother on record gave birth to a full-term healthy infant by cesarean section at 5 years, 8 months.

It should be recognized that the maturational events of puberty usually proceed in an orderly sequence but that the ages at which they occur may vary among individuals. In boys, the first sign of puberty is acceleration of growth of testes and scrotum; pubic hair appears a trifle later, and axillary and facial hair still later. Acceleration of penis growth begins on the average at 13 (range 11 to 14.5 years) and is complete by 15 (13.5 to 17). But note that, because of the overlap in ranges, some boys may be completely mature whereas others at the same age (say 13.5) may be completely prepubescent. Obviously, this can lead to profound social and psychologic problems. In girls, appearance of "breast buds" is usually the first event (average age = 11) although pubic hair may, on occasion, appear first. Menarche, the first menstrual period, is a later event (average = 13) and occurs almost invariably after the peak of the total body growth spurt has passed. The early menstrual cycles are usually not accompanied by ovulation so that conception is generally not possible for 12 to 18 months after menarche. One of the most striking facts concerning menarche is the remarkable decrease over the past 150 years in the age at which it occurs in all industrialized countries. For example, the age of menarche in Norway has decreased from near 17.5 in 1830 to 13 at present. Improved nutrition may have played an important causal role in this phenomenon but other factors, as yet unknown, almost certainly contribute. Again, one can imagine the social and psychologic impact of such a change on young people.

MENOPAUSE

Ovarian function declines gradually from a peak usually reached before the age of 30, but significant problems, if they occur at all, do not usually arise until the forties. Figure 21-32*A* and *B* demonstrate that the cause of the decline is decreasing ability of the aging ovaries to respond to pituitary gonadotropins. Estrogen secretion drops despite the fact that the gonadotropins, partially released from the negative-feedback inhibition by estrogen, are secreted in greater amounts. Ovulation and the menstrual periods become irregular and ultimately cease completely. Some secretion of estrogen generally continues beyond these events but gradually diminishes until it is inadequate to maintain the estrogen-dependent tissues: the breasts and genital organs gradually atrophy; the decrease in protein anabolism causes thinning of the skin and bones; however, sexual drive is frequently not diminished and may even be increased. Severe emotional disturbances are not uncommon during menopause and are generally ascribed not to a direct effect of estrogen deficiency but to the disturbing nature of the entire period – the awareness that reproductive potential is ended, the hot flashes, etc. The hot flashes, so typical of menopause, result from dilatation of the skin arterioles, causing a feeling of warmth and marked sweating; why estrogen deficiency causes this is unknown.

Male changes with aging are much less drastic. Once testosterone and pituitary gonadotropin secretions are initiated at puberty, they continue throughout adult life (Fig. 21-31*A* and *B*). A steady decrease in testosterone secretion in later decades apparently reflects slow deterioration of testicular function. The mirror-image rise in gonadotropin secretion is due to diminishing negative-feedback inhibition from the decreasing plasma testosterone concentration. Despite the significant decrease, testosterone secretion remains high enough in most men to maintain sexual vigor throughout life, and fertility has been documented in men in their eighties. Thus, there is usually no complete cessation of reproductive function analogous to menopause.

FIGURE 21-32 (A) ***Excretion of estrogen in the urine of women from puberty to senescence, an indicator of blood concentration of estrogen:*** (1) ***before menopause and*** (2) ***menopause.*** (B) ***Excretion of gonadotropins in the urine of nonpregnant women, an indicator of blood concentrations of gonadotropins:*** (1) ***before menopause and*** (2) ***after menopause.*** **(Adapted from Pedersen-Bjergaard and Tonnesen.)**

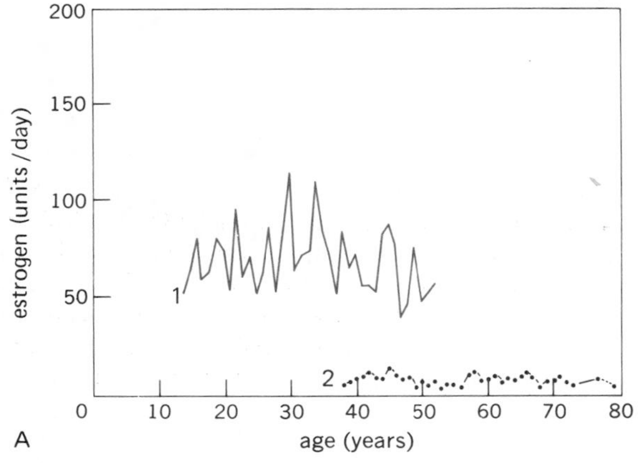

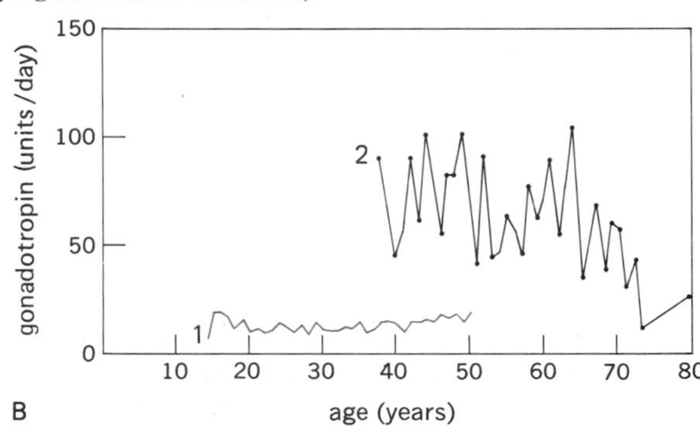

CHAPTER 22

Defense Systems: Immunology and Stress

Section A. Immunology: The Body's Defenses Against Foreign Material

Immunity constitutes all the physiologic mechanisms which allow the body to recognize materials as foreign to itself and to neutralize or eliminate them; in essence, these mechanisms maintain uniqueness of "self." Classically, immunity referred to the resistance of the body to microbes: viruses, bacteria, and other unicellular and multicellular organisms. It is now recognized, however, that the immune system has more diverse functions than this. It is involved both in the elimination of "worn-out" or damaged body cells (such as old erythrocytes) and in the destruction of abnormal or mutant cell types which arise within the body. This last function, known as *immune surveillance,* apparently constitutes one of the body's major defenses against cancer.

It has also become evident that immune responses are not always beneficial but may result in serious damage to the body. Such noxious effects may contribute to the development of a variety of diseases. In addition, the immune system seems to be involved in the process of aging. Finally, it constitutes the major obstacle to successful transplantation of organs.

Immune responses may be classified (Table 22-1) into two categories: (1) *nonspecific defense mechanisms* and (2) *specific immune mechanisms.* Specific immune responses depend upon prior exposure to the specific foreign substance, recognition of it upon subsequent exposure, and reaction to it. In contrast, the nonspecific or innate defense mechanisms do not depend upon previous exposure to a particular foreign substance; rather they nonselectively protect against foreign substances without having to recognize their specific identities. They are particularly important during the initial exposure to a foreign organism before the specific immune responses have been activated.

TABLE 22-1
CLASSIFICATION OF IMMUNE MECHANISMS

I. Nonspecific defense mechanisms
II. Specific immune mechanisms
    A. Humoral (antibodies)
    B. Cell-mediated ("sensitized" lymphocytes)

Specific immunity may be divided into two primary categories: *humoral* and *cell-mediated.* Both are the function of the *lymphoid cells,* and a critical difference between them, as we shall see, is that the former is mediated by circulating *antibodies* and the latter by *sensitized lymphocytes.*

Immune responses can be viewed in the same way as other homeostatic processes in the body, i.e., as stimulus-response sequences of events. In such an analysis, the groups of cells which mediate the final responses are effector cells, but before we begin our detailed description of immunology by introducing the various effector cells involved, let us first mention their major opponents: microbes (or microorganisms), i.e., bacteria, viruses, and fungi.

Bacteria are unicellular organisms belonging to the plant kingdom. As is typical of plant cells, they have not only a cell membrane but also an outer coating, the cell wall. Most bacteria are self-contained complete cells in that they have all the machinery required to sustain life and to reproduce themselves. In contrast, the *viruses* are essentially nucleic acid cores surrounded by a protein coat. They lack both the enzyme machinery for energy production and the ribosomes essential for protein synthesis. Thus, they cannot survive by themselves but must "live" inside other cells whose metabolic apparatus they make use of; i.e., they are obligatory parasites. Other types of microorganisms and multicellular parasites are potentially harmful to human beings, but we shall devote most of our attention to the body's defense mechanisms against bacteria and viruses.

How do microorganisms cause damage and endanger health? There are many answers to this question, depending upon the specific bacterium or virus involved. Some cause cellular destruction directly by locally releasing enzymes which break down cell membranes and organelles. Others give off toxins which may act throughout the body to disrupt the functions of the neuromuscular system and other organs and tissues. Moreover, the presence of microorganisms may constitute a continuous drain on the body's energy supplies. The viruses, in particular, often capture the metabolic and reproductive machinery of the cell which they inhabit; indeed, some forms of cancer in human beings may be caused by viruses. It must be admitted, however, that the damage-producing mechanisms in many infectious diseases are unknown.

EFFECTOR CELLS OF THE IMMUNE SYSTEM

The major effector-cell types (Table 22-2) are the white blood cells (described in Chap. 13) along with macrophages and plasma cells. Plasma cells are derived from lymphocytes and are responsible for specific immunity. They are the cells which secrete antibodies (humoral immunity). Macrophages have as their major function phagocytosis, and they are found scattered throughout the tissues of the body. In liver, spleen, lymph nodes, and bone marrow they form part of the lining of vascular and lymphatic channels. The structure of these cells varies from place to place, but their common distinctive features are numerous cytoplasmic granules and the ability to ingest almost any kind of foreign particle. Indeed, the method of identifying them is to inject a dye and observe which tissue cells display marked uptake

TABLE 22-2

MAJOR EFFECTOR-CELL TYPES OF THE IMMUNE SYSTEM

	White blood cells (leukocytes)						
	Polymorphonuclear granulocytes						
	Neutrophils	***Eosinophils*** 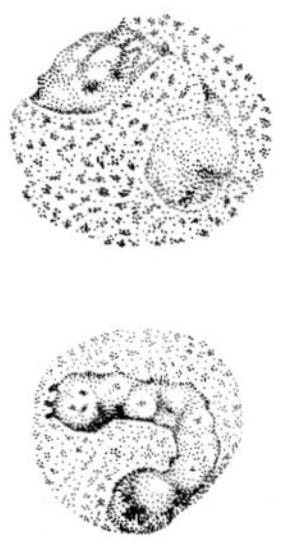	***Basophils*** 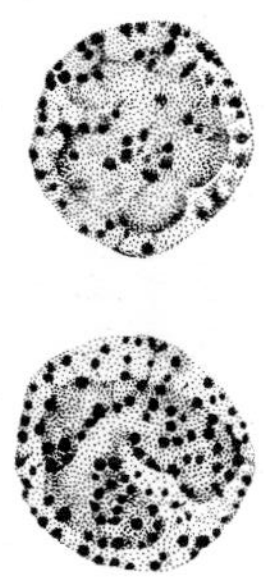	***Lymphocytes*** 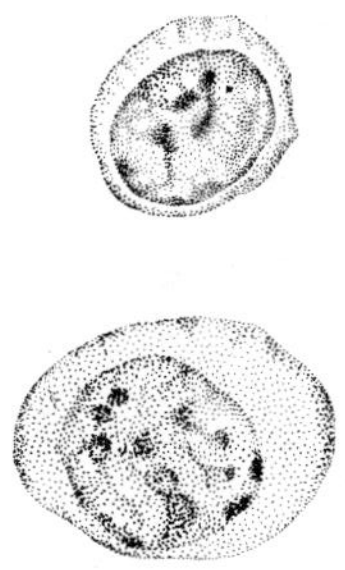	***Monocytes*** 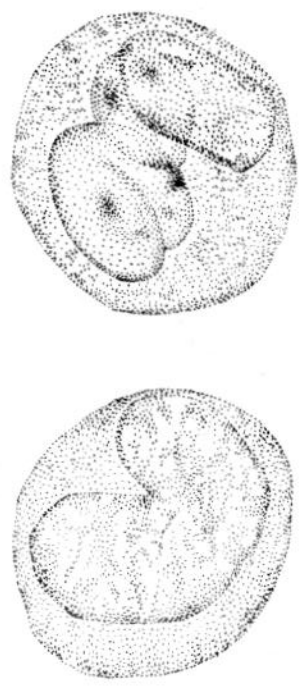	***Plasma cells*** 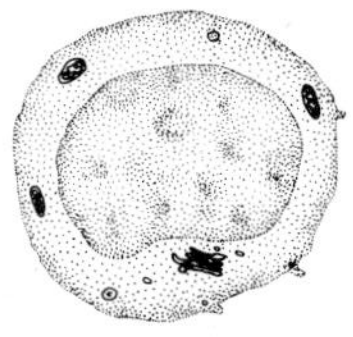	***Macrophages*** 
Percent of total leukocytes	50–70%	1–4%	0.1%	20–40%	2–8%		
Primary site of production	Bone marrow	Bone marrow	Bone marrow	Bone marrow and lymphoid tissue	Bone marrow	Derived from B lymphocytes in lymphoid tissue	Various sites; many formed from monocytes
Primary known function	Phagocytosis of bacteria, cell debris, and antibody-primed foreign matter	? ? ?	Release of histamine and other chemicals (similar to tissue mass cells)	B cells: production of antibodies (after transformation into plasma cells); T cells: responsible for cell-mediated immunity	Transformed into tissue macrophages with high phagocytic activity	Production of antibodies	Phagocytosis of bacteria, cell debris, and antibody-primed foreign matter

of it. Tissue macrophages are capable of mitotic activity, but division is not the only mechanism for increasing their number; in response to invasion by microorganisms or other foreign material, monocytes can be transformed into macrophages indistinguishable from the usual tissue macrophages.

The effector cells of the immune system are distributed throughout the organs and tissues of the body, but many are housed in the so-called lymphoid tissues: lymph nodes, spleen, thymus, tonsils, and aggregates of lymphoid follicles such as the Peyer's patches in the lining of the ileal region of the gastrointestinal tract.

Lymph nodes function as filters along the course of the lymph vessels (Fig. 15-30), lymph flowing through them before being returned to the general circulation. Lymph enters the node via afferent lymphatic vessels on the convex border of the node, trickles through the lymphatic sinuses of the node, and leaves via efferent lymphatic vessels on the other side (Fig. 22-1A). Each node is enclosed in a fibrous connective-tissue *capsule* from which partitions, or *trabeculae,* dip into the substance of the node. A reticular network extends from the trabeculae into all parts of the node. The lymphatic sinuses, which are relatively open channels in the reticular spaces, lie under the capsule and in the outer cortical region and dip down into the inner, medullary portion of the node. The spaces of the reticular network are packed with lymphocytes and plasma cells. Many of those in the outer, cortical region of the node are collected in *lymphatic follicles,* or *nodules,* containing *germinal centers* in which new lymphocytes are formed (Fig. 22-1B). The lymphocytes in the inner, medullary portion of the node form *medullary cords*. As the lymph flows through the lymphatic sinuses, some lymphocytes are removed from it to be stored temporarily in the node and others are added to the lymph. Some of the lymphocytes released into the lymph are those previously stored, and others are newly formed. The lymphatic sinuses are lined with macrophages, which phagocytize particulate matter, such as dust (inhaled into the lungs), cellular debris, and bacteria and other microorganisms. Some of the major nodes are indicated in Fig. 15-30.

The *spleen* is the largest of the lymphoid organs. It lies in the left part of the abdominal cavity between the stomach and diaphragm (Fig. 22-2). It is a highly vascular organ, purplish in color. Partitions, or *trabeculae,* from the connective-tissue capsule extend into the spleen and divide it into lobules; they also carry arteries, veins, and nerves into the spleen. The interior of the spleen is filled with a reticular meshwork, the *red pulp* and the *white pulp*. Blood, rather than lymph, percolates through the red pulp, and erythrocytes as well as lymphocytes and macrophages are collected in the spaces in the meshwork. The presence of the red blood cells is responsible for the name red pulp. This pulp accounts for much of the substance of the spleen. In the human fetus, the spleen is an important red-cell-forming organ, but in the adult only lymphocytes are formed there. They are produced in the *white pulp* of the spleen, areas of lymphoid tissue with germinal centers, which surround the arterioles. The macrophages of the spleen phagocytize many of the products of red cell degradation as well as the types of debris mentioned earlier.

The *thymus* lies in the mediastinum and consists of two unequally sized lobes connected by fatty tissue (Fig. 22-3). The size of the thymus varies with age; it is relatively large at birth and grows until puberty when it gradually atrophies and is replaced by fatty tissue. The thymus is pinkish gray and highly lobulated. The reticular framework of each lobule has an outer *cortex* containing densely packed lymphocytes and an inner *medulla* with lymphocytes in looser array. The thymus is an important lymphocyte-producing organ; it also secretes a group of hormones presently known collectively as *thymosin*. The role of the thymus in immunity will be described below.

Three groups of tonsils, the *palatine, pharyngeal,* and *lingual tonsils,* form a circular band of lymphoid tissue around the opening of the digestive and respiratory tubes. It is the palatine tonsils (Figs. 17-4 and 19-6) that are commonly called "the tonsils." A thin connective tissue capsule surrounds each tonsil, and a *crypt* indents its free surface. The crypt extends into the substance of the tonsil and is lined with lymphatic tissue containing lymphocyte-forming germinal centers and numerous leukocytes. The germinal centers are especially prominent in children and young adults.

NONSPECIFIC DEFENSE MECHANISMS

External Anatomic and Chemical "Barriers"

The body's first lines of defense against infection are the barriers offered by surfaces exposed to the external environment. Very few microorganisms can penetrate the intact skin, and the sweat, sebaceous, and lacrimal glands secrete chemical substances which are

FIGURE 22-1 (A) ***Diagram of a lymph node. Lymph enters the node via the afferent lymphatic vessels and leaves via the efferent lymphatic vessels.*** (B) ***Histologic cross section through a lymph node.*** **(From W. F. Windle, "Textbook of Histology," 5th ed., McGraw-Hill, New York, 1976, p. 211.)**

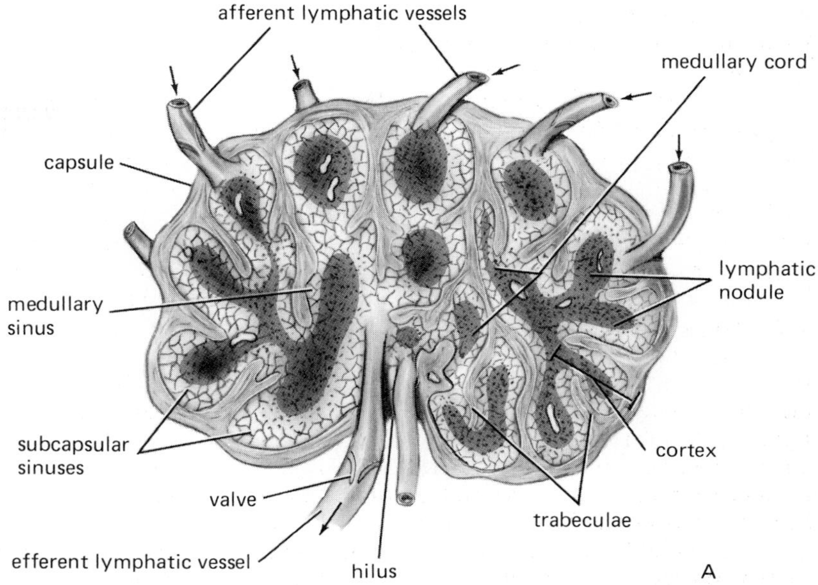

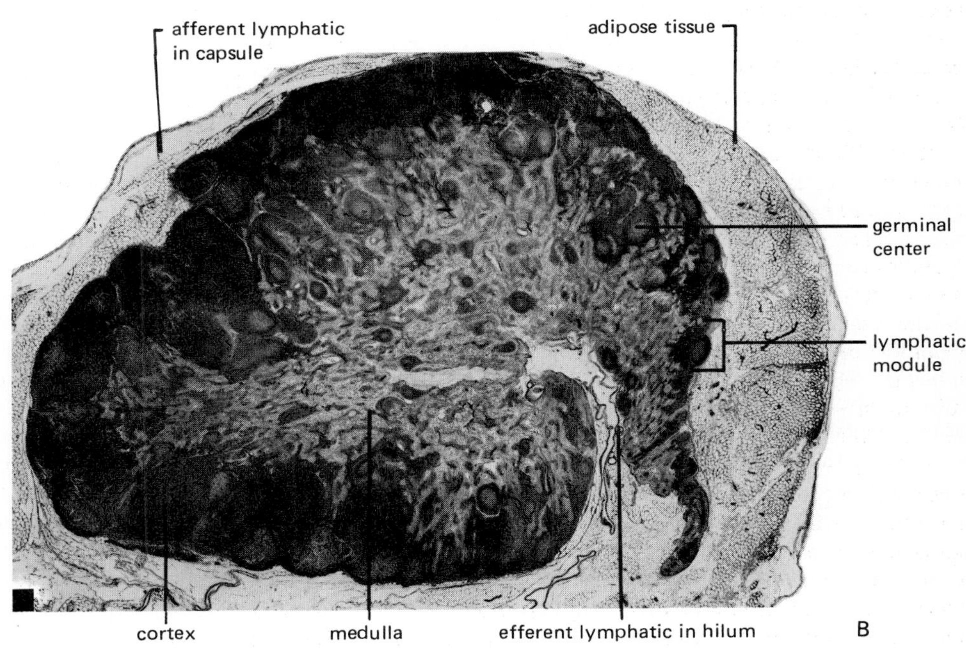

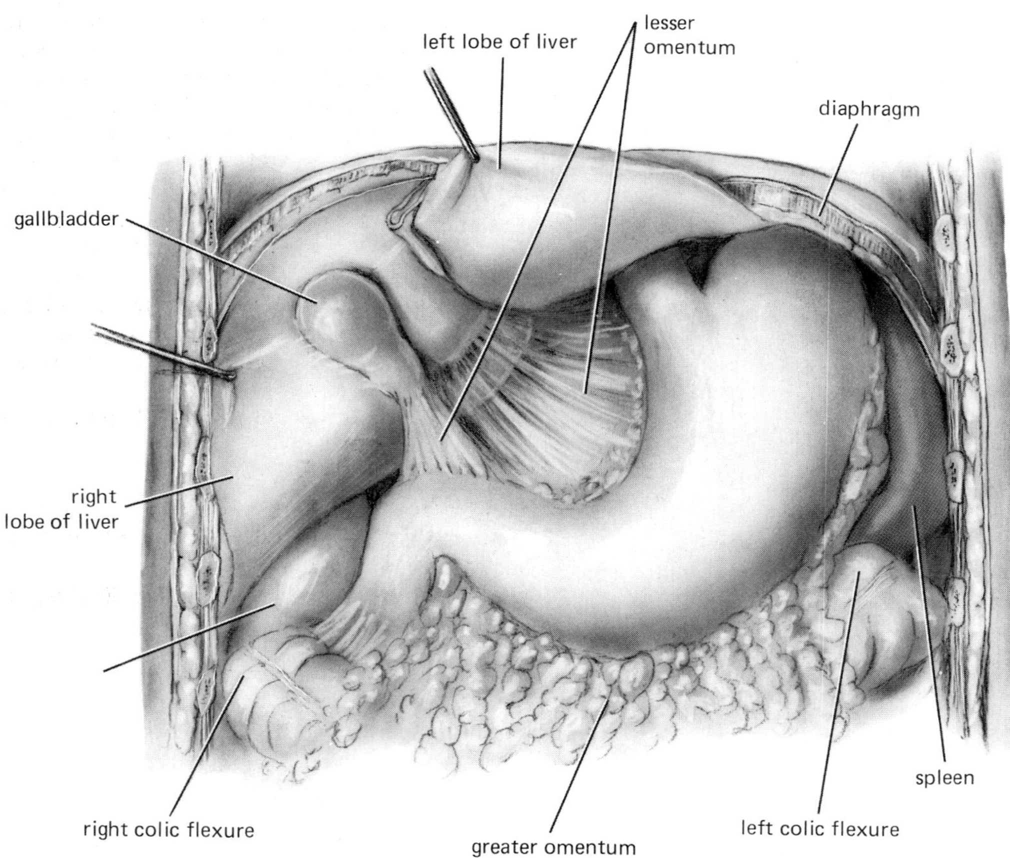

FIGURE 22-2 **(A)** ***The spleen.***

highly toxic to certain forms of bacteria. The mucous membranes also contain antimicrobial chemicals, but more important, mucus is sticky. When particles adhere to it, they can be swept away by ciliary action, as in the upper respiratory tract, or engulfed by phagocytic cells. Other specialized mechanisms related to the surface barriers are the hairs at the entrance to the nose, the cough reflexes, and the acid secretion of the stomach. Finally, a major "barrier" to infection is the normal microbial flora of the skin and other linings exposed to the external environment; these microbes suppress the growth of other potentially more virulent microorganisms.

Nonspecific Inflammatory Response

Despite the effectiveness of the external barriers, small numbers of microorganisms penetrate them every day. Once the invader has gained entry, it triggers off *inflammation,* the basic response to injury. The local manifestations of the inflammatory response are a complex sequence of highly interrelated events, the overall functions of which are to bring plasma proteins and phagocytes into the damaged area so that they can destroy the foreign invaders and set the stage for tissue repair. The nature of the sequence of events which constitute the inflammatory reaction varies, depending upon the injurious agent (bacteria, cold, heat, trauma,

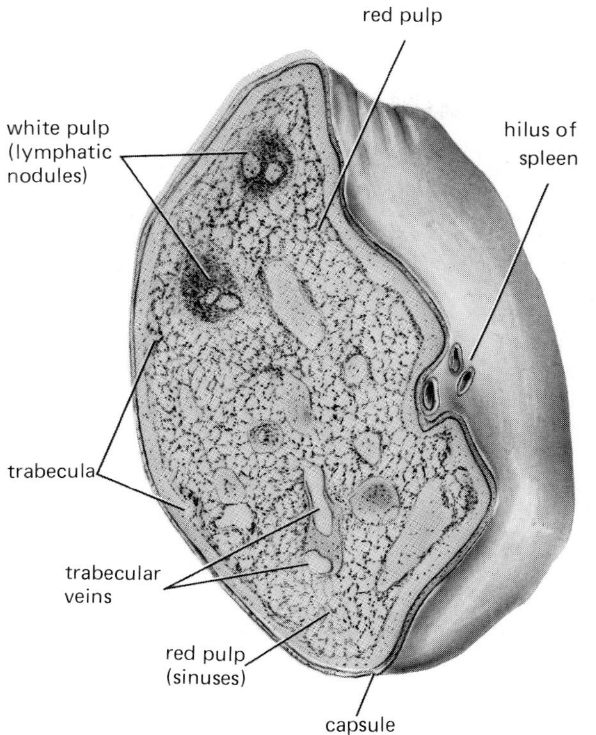

FIGURE 22-2 (B) ***Cross section of the spleen.***

etc.), the site of injury, and the state of the body. But since the similarities are in many respects more striking than the differences, inflammation can be viewed as a relatively stereotyped response to tissue damage, the precise manifestations of which differ according to the specific injurious agent and other important variables. It should be emphasized that, in this section, we describe inflammation in its most basic form, i.e., the nonspecific innate response to foreign material. As we shall see, inflammation remains the basic scenario for the acting out of specific immune responses as well, the difference being that the entire process is amplified and made more efficient by the participation of antibodies and sensitized lymphocytes.

The sequence of events in a local infection is briefly as follows:

1 Initial entry of microbes

2 Vasodilatation of the vessels of the microcirculation leading to increased blood flow

3 A marked increase in vascular permeability to protein

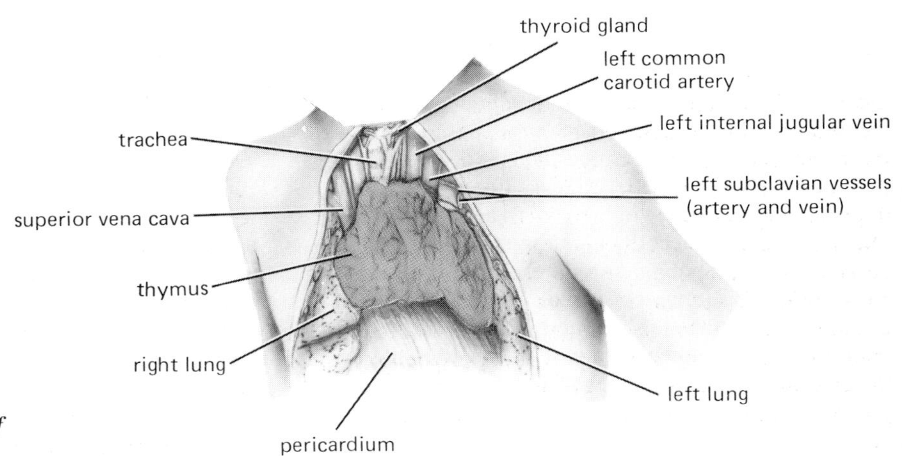

FIGURE 22-3 ***The thymus of a full-term infant.***

4 Filtration of fluid into the tissue with resultant swelling

5 Exit of neutrophils (and, later, monocytes) from the vessels into the tissues

6 Phagocytosis and destruction of the microbes

7 Tissue repair

The familiar gross manifestations of this process are redness, swelling, heat, and pain, the latter being the result both of distension and the direct effect of released substances on afferent nerve endings.

Vasodilatation and Increased Permeability to Protein The tissue swelling is directly related to an increased blood flow and vascular permeability to proteins, which occur immediately upon tissue damage. Chemical mediators dilate most of the vessels of the microcirculation and somehow alter the intercellular material so as to make it quite leaky to large molecules. The arteriolar dilatation associated with inflammation increases capillary hydrostatic pressure, which increases filtration of fluid from the capillaries, and the protein, which leaks out of the vessels as a result of increased permeability, builds up locally in the interstitium, thereby diminishing the protein difference between plasma and interstitium.

The adaptive value of all these vascular changes is twofold: (1) The increased blood flow ensures an adequate supply of phagocytic leukocytes and plasma proteins crucial for immune responses (to be described below) to the inflamed area, and (2) the increased capillary permeability to protein ensures that the relevant plasma proteins—all normally restrained by the capillary membranes—can gain entry to the inflamed area.

The major direct chemical mediators of these vascular changes in nonspecific inflammatory responses are *histamine* and a group of polypeptides known as *kinins*. Histamine is present in many tissues of the body but is particularly concentrated in mast cells, circulating basophils, and platelets. Release of histamine is induced by a wide-variety of factors. In an inflammatory response in which specific immune responses are absent, two of the major factors are simple mechanical disruption of histamine-containing cells and chemicals secreted by neutrophils attracted to the site. The released histamine, in addition to its vasodilating and permeability-increasing effects, has profound effects on nonvascular smooth muscle, perhaps the most significant being its constriction of the respiratory airways.

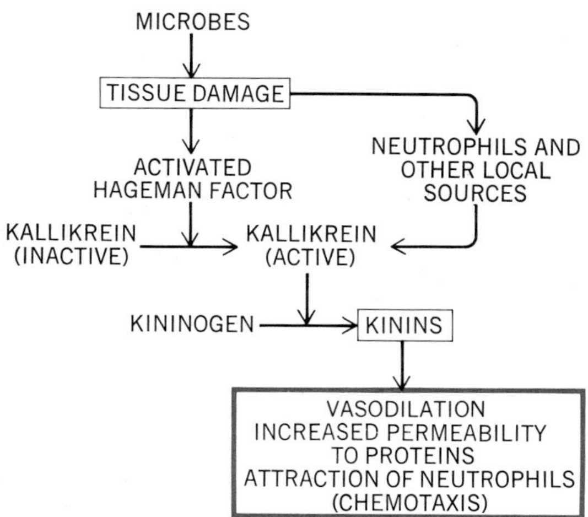

FIGURE 22-4 ***Pathways for kinin generation in nonspecific inflammatory reactions.***

The kinins are small polypeptides whose vascular effects are similar to those of histamine. The kinins are generated in plasma in the following ways (Fig. 22-4): There is present in plasma an enzyme, known as *kallikrein,* which exists normally in an inactive form but which, when activated, catalyzes the splitting off of the kinins from another normally occurring plasma protein known appropriately as *kininogen*. What is it that activates this enzyme? Again we find that multiple factors are capable of doing so but that, in the absence of specific immune responses, the most important is a chemical substance we already met in the discussion of hemostasis (Chap. 13): activated Hageman factor. Here is one of many connections between the hemostatic and immune systems. Recall that Hageman factor is, itself, inactive until activated locally by contact with altered vascular surfaces; once activated, it catalyzes the first steps in the cascade sequences leading to both clotting and plasmin formation. Now we see that, in addition, it leads to the activation of the key enzyme of the kinin-generating system, thereby contributing to the first phases of the inflammatory process. In addition to being generated from its inactive form in blood, active kallikrein is also found in many different tissues of the body as well as in neutrophils. Once inflammation has begun, the kallikrein present in these sources contributes to the generation of kinin. It should also be noted that, by their effects on afferent neuron terminals, the kinins account for much of the pain associated with inflammation.

Neutrophil Exudation Within 30 to 60 min after the onset of inflammation, a remarkable interaction occurs between the vascular endothelium and circulating neutrophils. First, the blood-borne neutrophils begin to stick to the inner surface of the endothelium. The process is quite specific since erythrocytes show no tendency to stick and other leukocytes do so only later, if at all. How the endothelium is made sticky by injury remains a mystery.

Following their surface attachment, the neutrophils begin to manifest considerable amebalike activity. Soon a narrow ameboid projection is inserted into the space between two endothelial cells (Fig. 22-5), and the entire neutrophil then squeezes through into the interstitium. The alterations of vessel structure induced by the released agents described above may facilitate this process by loosening the intercellular connections, or the neutrophil may simply pry the connection apart by the force of its ameboid movement.

By this process (neutrophil exudation), huge numbers of neutrophils move into the inflamed areas of tissue. This event is dependent upon *chemotaxis,* the attraction of neutrophils to an inflamed area by certain chemicals. Once again, we find a multitude of chemotactic chemicals from a variety of sources, including the bacteria, and neutrophils themselves. The kinins are also potent chemotactic agents.

Leukocyte exudation is usually not limited only to neutrophils. Monocytes follow, but much later (although their mode of entry is unclear), and once in the tissue are transformed into macrophages. Meanwhile the macrophages normally present in the tissue have begun to multiply by mitosis and to become motile. Thus, all the phagocytic types are present in the inflamed area. Usually the neutrophils predominate early in the infection but tend to die off more rapidly than the others, thereby yielding a predominantly mononuclear picture later. In certain types of allergic or parasitic inflammatory responses, the eosinophils are in striking preponderance. Indeed, one of the tests for allergically induced "runny nose" is to study a small amount of nasal discharge for the presence of eosinophils.

Phagocytosis Phagocytosis is the primary function of the inflammatory response, and the increased blood flow, vascular permeability, and leukocyte exudation serve only to ensure the presence of adequate numbers of phagocytes and to provide the milieu required for the performance of their function. The process of phagocytosis is demonstrated in Fig. 22-6. The phagocyte engulfs the organism by membrane invagination and pouch formation. Once inside, the microbe remains in its own pouch, a layer of phagocyte cell membrane separating it from the phagocyte cytoplasm.

The next step observed in vitro is known as *degranulation*. The membrane surrounding the pouch makes contact with the phagocyte granules, which are *lysosomes* filled with a variety of hydrolytic enzymes; the two membranes fuse, and the still intact granules enter the pouch; the lysosomal membranes separating the contents of the pouch and lysosome break down, and the powerful enzymes are discharged into the pouch. This process of degranulation, as observed microscopically, occurs explosively, and all these events subsequent to engulfment frequently require less than 10 min.

The substances released from the lysosomes kill the microorganism and catabolize it into low-molecular-weight products which can then be safely released from the phagocyte or actually utilized by the cell in its own metabolic processes. Nondegradable foreign particles (such as wood, tattoo dyes, or metal) may be retained indefinitely within macrophages. This entire process need not kill the phagocyte, which may repeat its function over and over before dying.

The neutrophils may also release entire lysosomal granules into the extracellular fluid; the enzymes released from these granules attack extracellular debris at the injury site, making it easier for the macrophages to phagocytize it at the battle's end, thus paving the way for repair of the damaged area. Figure 22-7 summarizes the basic events of the local inflammatory response to infection.

Tissue Repair The final stage of the inflammatory process is tissue repair. Depending upon the tissue involved, regeneration of organ-specific cells may or may not occur (for example, regeneration occurs in skin and liver but not in muscle or the central nervous system; the latter two tissues are incapable of cell division in adults). In addition, fibroblasts in the area divide rapidly and begin to secrete large quantities of collagen which endows the area with great tensile strength.

Systemic Manifestations of Inflammation We have thus far described the local inflammatory response. What are the systemic, overall body responses? Probably the single most common and strik-

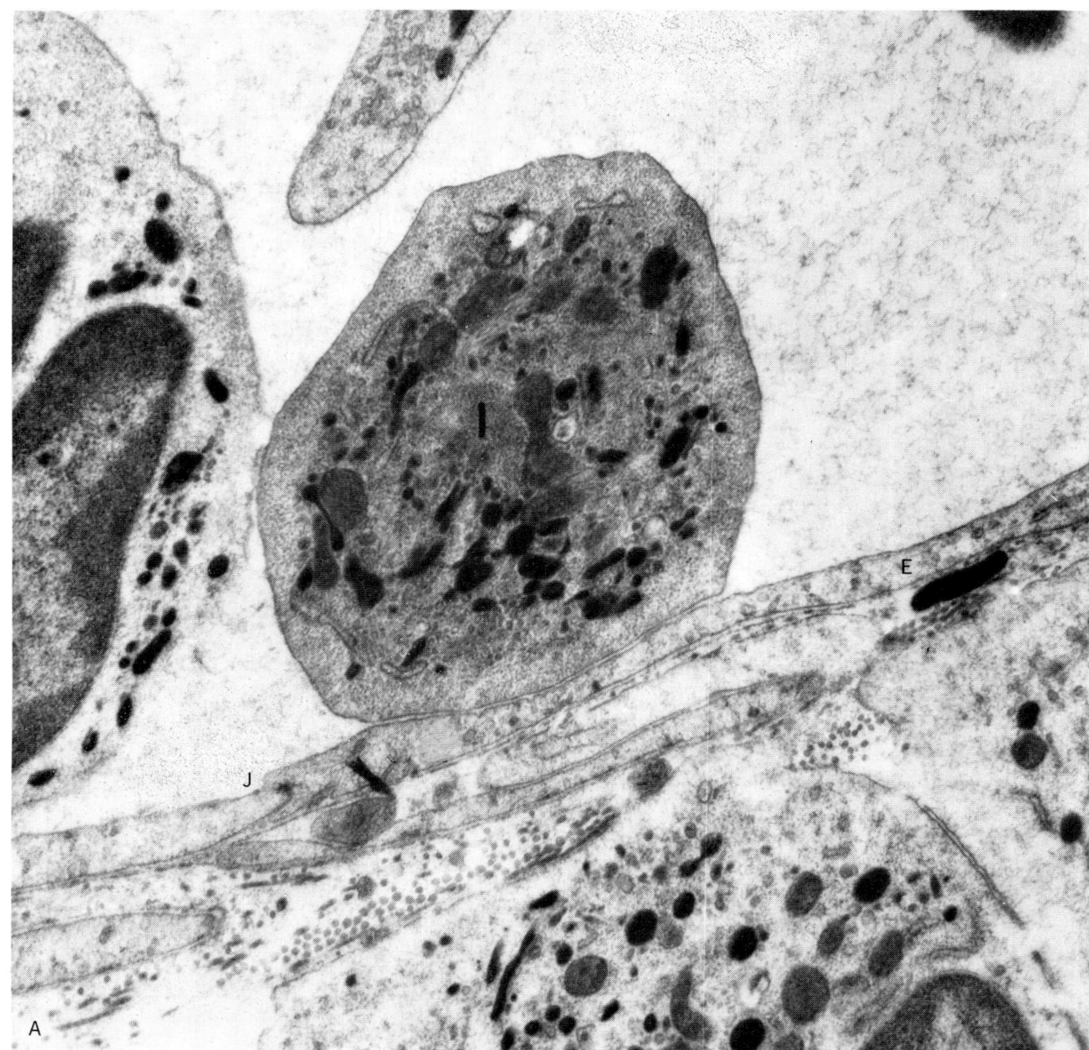

FIGURE 22-5 ***White blood cell emigration.*** **(A)** ***A polymorphonuclear granulocyte*** **(at center)** ***in the lumen of a capillary has just adhered to the endothelium (E); note the intact intercellular junction (J).*** **(B)** ***This cell is now protruding a pseudopod through the intercellular junction and out into the interstitial space. The entire cell will move through in this manner.*** **[From V. T. Marchesi, *Q. J. Exp. Physiol.,* 46:115 (1961).]**

ing systemic sign of injury is fever. The substance primarily responsible for the resetting of the hypothalamic thermostat, as described in Chap. 20, is a protein released by the neutrophils (and perhaps other cells) participating in the inflammatory response.

Another systemic manifestation of many bacterial diseases is *leukocytosis,* a marked increase in the synthesis and release of neutrophils by the bone marrow. In contrast, viral infections frequently are associated with decreased neutrophils. The factors controlling these phenomena are unknown.

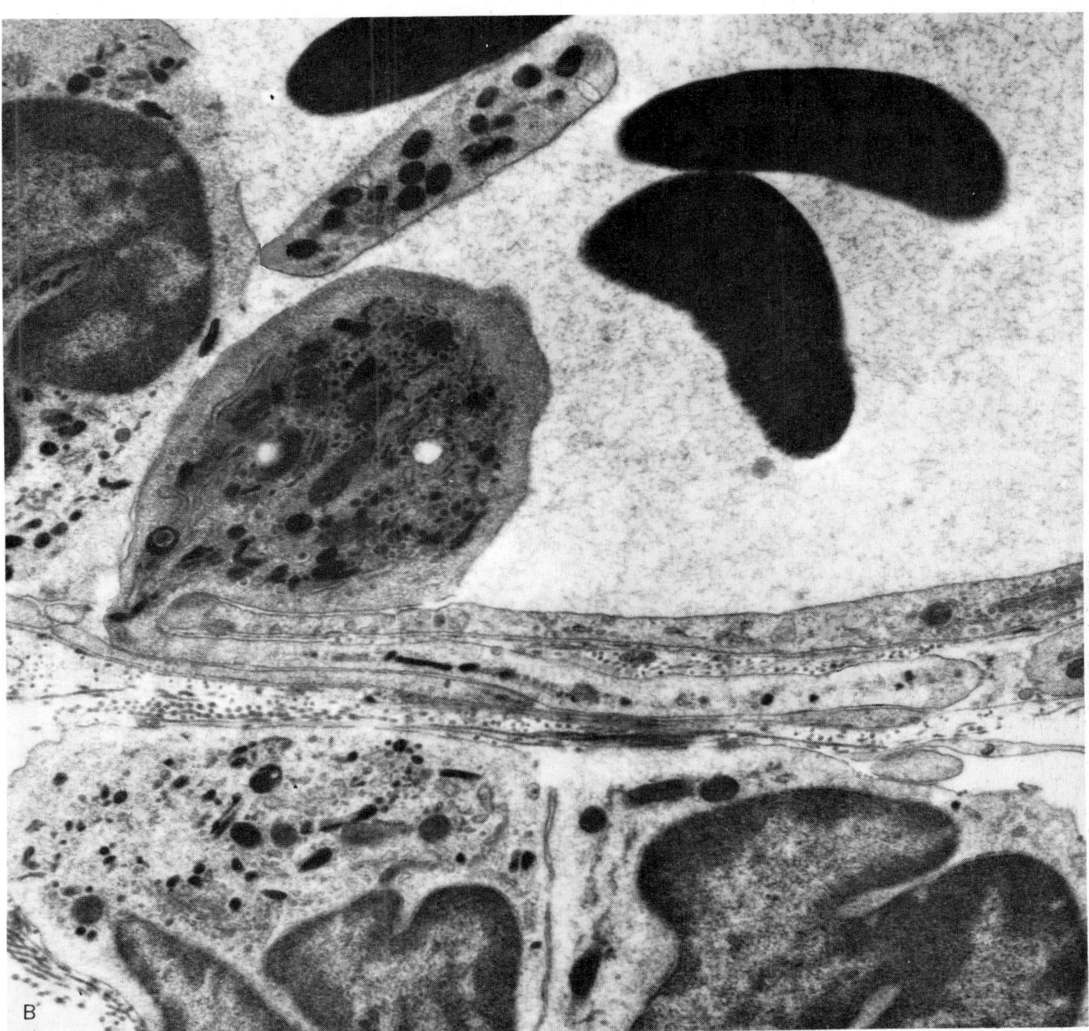

With few specific exceptions, we do not know what is responsible for the generalized malaise, aching, and weakness so frequently associated with infection. Our ignorance extends even to the most obvious symptoms, such as loss of appetite. It is hoped that this important and fertile general area will attract more researchers in the future.

Interferon

Interferon is a nonspecific defense mechanism against viral infection. It is a protein which inhibits viral growth and replication and can be produced by several different cell types of the body in response to a viral infection of the particular cell. Its production (or lack of production) can be understood from the basic principles of protein synthesis. Where there is no virus within the host cell, the potential for interferon synthesis exists but the actual synthesis is repressed. Entry of a virus into the cell in some manner eliminates the repressor, thereby inducing interferon synthesis by the usual DNA-RNA-ribosomal mechanisms. The inducer (or derepressor) may well be the viral nucleic acid. Once

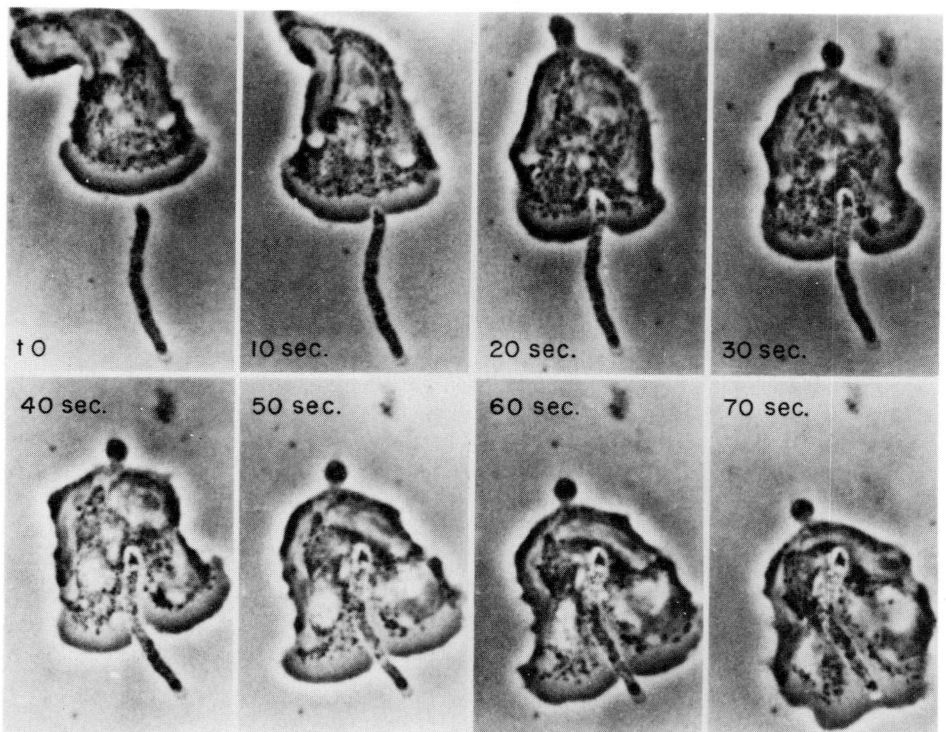

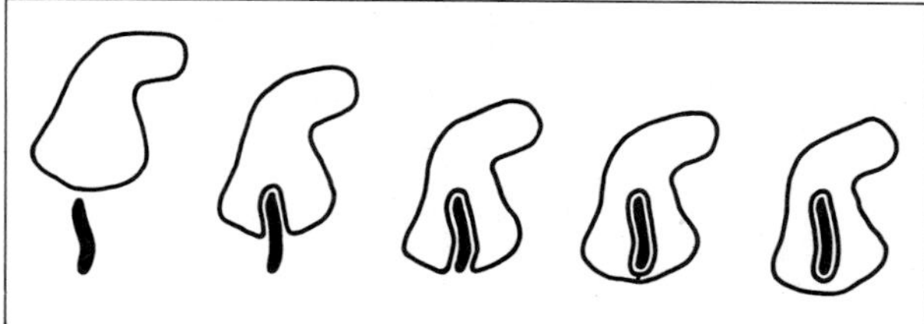

FIGURE 22-6 ***Prints from a motion picture of a human neutrophil engulfing a bacterium. The bottom shows the process in diagrammatic form.*** **(From R. J. Dubos and J. G. Hirsch, "Bacterial and Mycotic Infections of Man," 4th ed., Lippincott, Philadelphia, 1965.)**

synthesized, interferon may leave the cell, enter the circulation, and be picked up by another cell, despite the fact that it is a protein. Some interferon is always present in plasma (and in epithelial secretions) but a large increase occurs during viral infection. Interferon is not specific; all viruses induce the same kind of interferon synthesis, and interferon in turn can inhibit the multiplication of many different viruses.

SPECIFIC IMMUNE RESPONSES

There are two parts to the specific immune system, one mediated by antibodies secreted by B lymphocytes, and the other mediated directly by a second distinct population of lymphocytes, the T cells. Early in their development cells destined to be T cells travel to the thymus (thus the name T cell) which in some manner confers upon them the ability to differentiate and mature into cells competent to carry out the activities associated with cell-mediated immunity. The T cells then leave the thymus and take up residence in lymph nodes and other lymphoid tissues, but the thymus continues to influence them by means of a hormone which its epithelial cells secrete. Although most of this activity occurs prior to puberty when the thymus is largest and most active, most of the T cells are quite long-lived, so that removal of the thymus during adult life has little effect on a person's resistance.

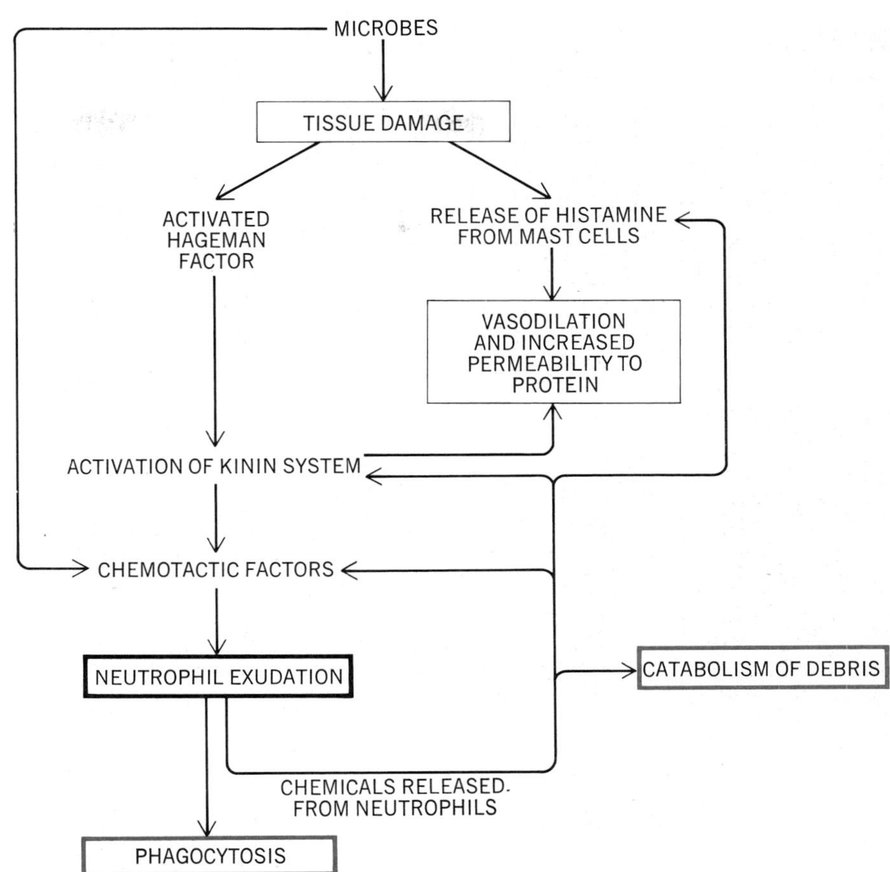

FIGURE 22-7 ***Summary of the nonspecific local inflammatory response to infection. This figure contains Fig. 22-4.***

In general, the B cells and T cells serve different functions. The B cells, through their secreted antibodies, confer specific immune resistance against most bacteria. The T cells are the major carriers of specific immunity against fungi, viruses, parasites, and the few bacteria which, to survive, must live inside cells. The T cells also mediate the destruction of cancer cells and the rejection of solid-tissue transplants. In the discussion to follow, we emphasize the separation of function of humoral immunity (B cells) and cell-mediated immunity (T cells). However, it is rapidly becoming apparent that a complete separation is not warranted. In ways still poorly understood, B cells and T cells influence the activity of each other in both synergistic and inhibitory ways. These interactions will certainly prove to be of great importance.

Antigens and Antibodies

An *antibody* is a specialized protein capable of combining chemically with the specific *antigen* which stimulated its production. Conversely, antigens are substances inducing the synthesis of antibodies, with which they can then specifically combine. The word specifically is essential in the definition since an antigenic substance reacts only with the type of antibodies elicited by its own kind or an extremely closely related kind of molecule. Specificity is thus related to the chemical structure of the antigen and its antibody. The same lock-and-key analogy used in discussing the enzyme-substrate combination probably applies.

Antibodies are all composed of polypeptide chains and are identical except for a relatively small number of amino acids occupying the first positions in the chains. These differences constitute the antibody's specificity. The three-dimensional structure conferred upon the tip of the chains by these amino acids permits the antibody to recognize the complementary structure in the antigen and combine with it.

Antibodies all belong to the family of proteins known as gamma globulins and are also known as immunoglobulins. They may be further subdivided into five classes according to differences in chemical structure and biological function. These are designated by the letters G, A, M, D, and E after the symbol Ig (for immunoglobulin). IgG and IgM antibodies provide the bulk of specific immunity against infectious microbes. The other class we shall be concerned with is IgE, for these antibodies mediate certain allergic responses. IgA antibodies are produced by lymphoid tissue lining the gastrointestinal, respiratory, and genitourinary tracts and exert their major activities in their secretions. The function of the IgD class is presently uncertain.

Since an essential determinant of antigenic capacity is size (most antigens have molecular weights greater than 10,000), many chemicals injected into the body do not induce antibody formation. However, many smaller molecules act as antigens after first attaching themselves to one of the host's proteins, thus forming a complex large enough to induce antibody formation. Still other low-molecular-weight substances incapable of inducing antibody synthesis because of their small size can combine with antibodies induced by another antigen; in such cases the structural unit of the large true antigen which was critical in the induction process must have been similar or identical to that of the small molecule. These last two phenomena explain why many small molecules can cause allergic attacks.

The antigens we shall be most concerned with occur on the surface of microbes or are microbial products, such as bacterial toxins. However, components of almost any foreign cell or molecule not normally present in the body, e.g., penicillin, may act as an antigen. Indeed, even normal body components can induce antibody formation under unusual circumstances. Once the antibodies have been formed (the mechanism will be discussed in a subsequent section), they are released into the blood, reach the site where the antigen is located, and combine with it.

Functions of Antibodies

Activation of Complement System In a previous section we described how a local inflammatory response is induced nonspecifically by any tissue damage. Now we shall see that the presence of antigen-antibody complexes triggers off events which profoundly amplify the inflammatory response. The most important mechanism for this enhancement involves the *complement* system. It is yet another example (the clotting, plasmin, and kinin systems are others) of a "system" consisting of a group of plasma proteins which normally circulate in the blood in an inactive state; upon activation of the first protein of the group, there occurs a sequential cascade in which active molecules are generated from the inactive precursors. The activators of the initial step in the complement sequence are antigen-antibody complexes, although how this occurs is presently unclear. The active complement components are generated in sequence, then act as the mediators (Fig. 22-8) for enhancing the various aspects of the inflammatory response (there are eleven proteins in the complement system mediating different effects, but we shall refer to the group collectively as complement).

It must be emphasized that the specificity in all these responses resides in the antigens and antibodies, not in complement. Complement is activated by almost any antigen-antibody complex (at least when the antibody is of the IgG or IgM class). In other words, there is only one set of complement molecules, and, once activated, they do essentially the same thing, regardless of the specific identity of the invader. In contrast, the formation of antibodies to antigens on the invader and their subsequent combination are highly specific. The function of the antibodies is to identify (i.e., "mark") the invading cells as foreign (by combining with antibody-specific antigens on the cells' surface) and to activate the complement system, which then mediates the actual attack. The "identification" function of the antibodies serves to "guide" those complement components which facilitate phagocytosis or kill the microbes outright; i.e., the antibodies must ensure that these complement components combine only with the invading cells and not randomly with the body's own cells (otherwise the latter might be phagocytized or destroyed). Somehow, the presence of the antibody combined with the antigen on the surface of the invading cell permits complement also to combine with surface sites and exert its effects. Moreover, in addition to enhanc-

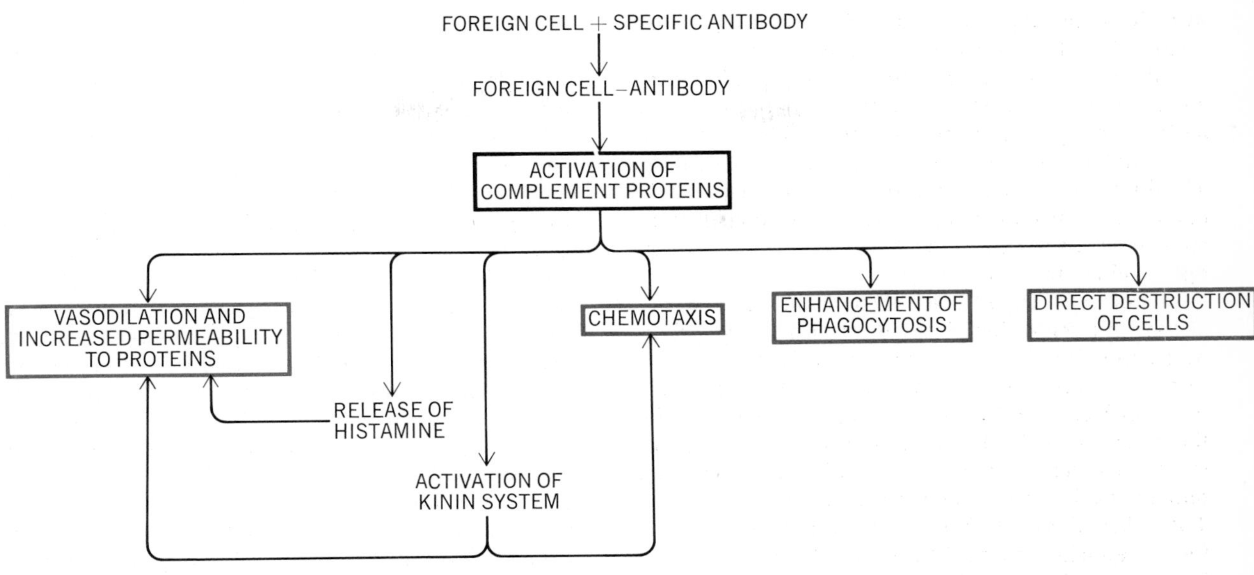

FIGURE 22-8 ***Functions of complement. Note how certain of these actions amplify the nonspecific inflammatory response illustrated in Fig. 22-7.***

ing cell destruction through phagocytosis, complement may directly destroy certain microbes apparently by enzymatically breaking down their cell membrane.

Direct Neutralization of Bacterial Toxins and Viruses Bacterial toxins and certain viral components act as antigens to induce specific antibody production. The antibodies then combine chemically with the toxins and viruses to neutralize them. Neutralization referred to a virus means that the combined antibody somehow prevents attachment of the virus to host cell membranes, thereby preventing virus entry into the cell. Since viruses are obligatory intracellular parasites, i.e., can live only in the host's cells, the antibody kills the virus indirectly by cutting off its access to possible host cells. Similarly, antibodies neutralize bacterial toxins by combining chemically with them, thus preventing the interaction of the toxin with susceptible cell-membrane sites. Antibodies generally have more than one potential site for combination with antigen so that an aggregate, or chain, of antigen-antibody complexes is formed (Fig. 22-9) which is then phagocytized.

Antibody Production

When a foreign antigen reaches lymph nodes, the spleen, or the lymphoid patches in the body's epithelial linings, it triggers off antibody synthesis. The antigen

FIGURE 22-9 ***Interlocking complex of antigens and antibodies. This type of chain, or clump, can be formed because antibodies have more than one potential site for combination with antigen.***

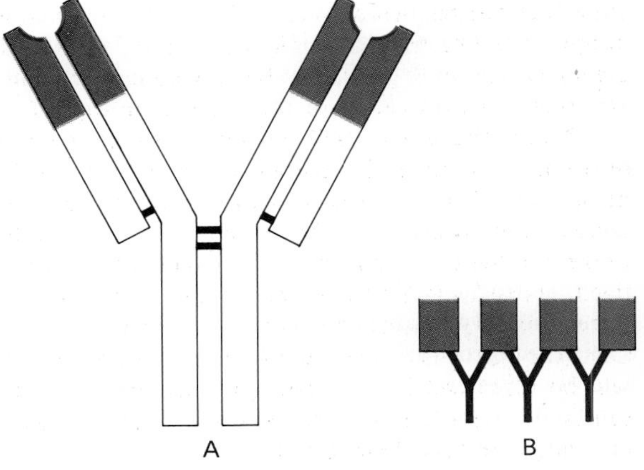

may reach the spleen via the blood, but much more commonly it is carried from its site of entry via the lymphatics to lymph nodes. There it stimulates a tiny fraction of the B lymphocytes to undergo rapid cell division, most of the progeny of which then differentiate rapidly into plasma cells, which are the active antibody producers. The most striking aspect of this transformation is a marked expansion of the cytoplasm, which consists almost entirely of the granular type of endoplasmic reticulum (Fig. 22-10) found in other cells which manufacture protein for export; after synthesis, the antibodies are released into the blood or lymph. Those B-cell progeny which do not fully differentiate into plasma cells constitute the "memory" of the occurrence, ready to respond more rapidly and forcefully should the antigen ever reappear.

Note that we stated that only a tiny fraction of the total B cells respond to any given antigen. It can also be demonstrated that different antigens stimulate entirely different populations (*clones*) of B cells. This is because the cells of any one lymphocyte clone (and the plasma cells it gives rise to) are capable of secreting only one kind of antibody. This limited synthetic capacity was probably determined by random mutations (during embryonic life) of the genes coding for the variable amino acids in the terminal portions of the antibody chains. According to this clonal theory, different antigens do not direct a single cell to produce different antibodies; rather each specific antigen triggers activity in the clone of cells already predetermined to secrete only antibody specific to that antigen. The antigen selects this particular clone and no other because the B cell displays on its surface the antibody molecules which it is capable of producing. These surface antibodies act as receptor sites with which the antigen can combine, thereby triggering off the entire process of division, differentiation, and antibody secretion just described. The staggering but statistically possible implication is that there must exist millions of different clones, one for each of the possible antigens an individual *might* encounter during his life.

Active and Passive Immunity

We have been discussing antibody formation without regard to the course of events in time. The response of the antibody-producing machinery to invasion by a foreign antigen varies enormously, depending upon whether it has previously been exposed to that antigen. Antibody response to the first contact with a microbial antigen occurs slowly over several days, with some circulating antibody remaining for long periods of time, but a subsequent infection elicits an immediate and marked outpouring of specific antibodies. It is evident that this type of "memory" confers a greatly enhanced resistance toward subsequent infection with a particular microorganism. This resistance, built up as a result of actual contact with microorganisms and their toxins or other antigenic components, is known as *active immunity*. Until modern times, the only way to develop active immunity was actually to suffer an infection, but now a variety of other medical techniques are used, i.e., the injection of vaccines or microbial derivatives. The actual material injected may be small quantities of living or weakened microbes, e.g., polio vaccine, small quantities of toxins, or harmless antigenic materials derived from the microorganism or its toxin. The general principle is always the same: Exposure of the body to the agent results in the induction of the antibody-synthesizing machinery required for rapid, effective response to possible future infection by that particular organism. However, it must be mentioned that not all microorganisms induce active immunity. For many microorganisms the memory component of the antibody response does not occur, and antibody formation follows the same course in time regardless of how often the body has been infected with the particular microorganism.

A second kind of immunity, known as *passive immunity,* is simply the direct transfer of actively formed antibodies from one person (or animal) to another, the recipient thereby receiving preformed antibodies. This exchange normally occurs between fetus and mother across the placenta and is an important source of protection for infants during the first months of life, when their own antibody-synthesizing capacity is relatively poor. The same principle is used clinically when specific antitoxin or pooled gamma globulin is given a person exposed to or actually suffering from certain infections, such as measles, hepatitis, or tetanus. The protection afforded by this passive transfer of antibodies is relatively short-lived, usually lasting only a few weeks. The procedure is not without danger since the injected antibodies (often of nonhuman origin) may themselves serve as antigens, eliciting active antibody production by the recipient and possibly severe allergic responses.

Cell-mediated Immunity

The T lymphocytes are responsible for cell-mediated immunity. Upon initial exposure to an appropriate antigen, a clone of T cells becomes "sensitized" to that particular antigen. The mechanism by which this occurs is unclear but it confers upon the lymphocyte the capacity to release locally a powerful battery of chemicals when the lymphocyte encounters that antigen again and combines with it (the antigen combines with receptor sites on the surface of the sensitized T cell). We must emphasize an important geographic difference between T-cell and B-cell function. Antibodies are secreted by B-cell progeny located in lymph nodes and other lymphoid organs far removed from the invasion site and reach the site via the blood; in contrast, the T cells travel to the invasion site where, upon combination with antigen, they release their chemicals.

The chemicals released when sensitized lymphocytes combine with specific antigen may either kill foreign cells directly or act as an amplification system for the facilitation of the inflammatory response and phagocytosis. Thus, cell-mediated immunity is analogous to humoral immunity in that it serves, in large part, to enhance and make more efficient the nonspecific defense mechanisms already elicited by the foreign material. The major difference is that humoral immunity utilizes a circulating group of plasma proteins (the complement system) as its major amplification system whereas the T cells literally produce and secrete their own chemical amplification system.

As might be predicted, some of these chemicals are

FIGURE 22-10 ***Electron micrograph of a guinea pig plasma cell. Note the extensive endoplasmic reticulum.*** **(From W. Bloom and D. W. Fawcett, "A Textbook of Histology," 9th ed., Saunders, Philadelphia, 1968.)**

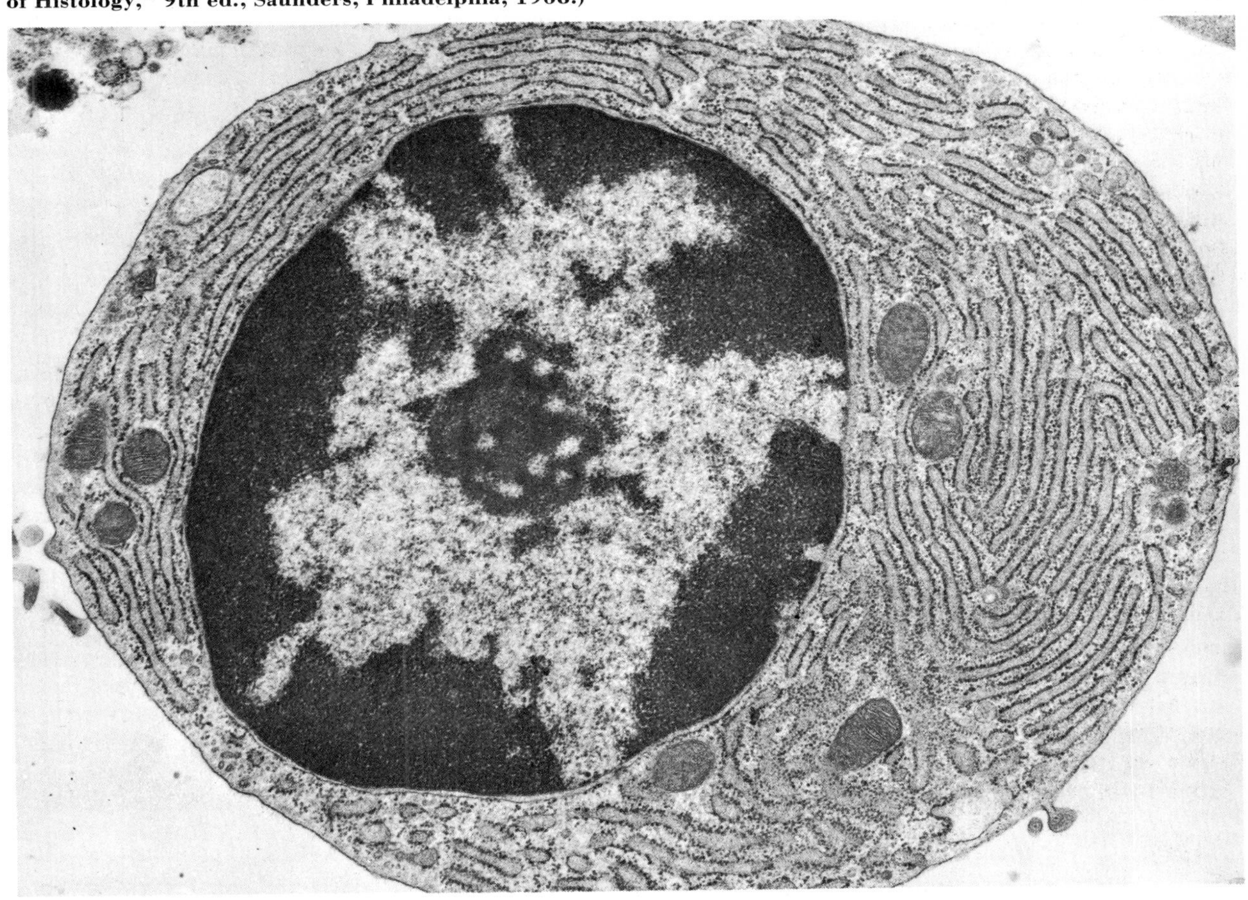

chemotactic factors. These serve to attract some neutrophils but many more monocytes to the area. The monocytes are converted to macrophages and begin their job of phagocytosis. But the lymphocytes go one step further: They not only secrete chemotactic factors to attract the macrophages-to-be; they secrete another substance which keeps the macrophages in the area and stimulates them to greater phagocytic activity (indeed, such "revved-up" macrophages are known as "angry" macrophages). In addition to facilitating the killing of target cells in this manner, the lymphocytes secrete so-called cytotoxins which are able to kill target cells directly, i.e., without phagocytosis. Here is another analogy to the complement system, with its ability to destroy cells directly as well as to facilitate phagocytosis.

As is true for the B system, some of the sensitized lymphocytes (this term specifically denotes T cells) do not actually participate in the immune response but serve as a "memory bank" which greatly speeds up and enhances the immune response if the person is ever exposed to the specific antigen again. Thus, active immunity exists for cell-mediated immune responses just as for antibody responses.

We may now summarize the interplay between nonspecific and specific immune mechanisms in resisting infection. When a microbe is encountered for the first time, nonspecific defense mechanisms resist its entry and, if entry is gained, attempt to eliminate it by phagocytosis. Simultaneously, the antigens on the foreign matter induce the final development of specific cell clones capable of antibody production or cell-mediated immune responses or both. If the nonspecific defenses are rapidly successful, these specific immune responses may never play an important role. If only partly successful, the infection may persist long enough for significant amounts of antibody or sensitized T cells, or both, to reach the scene; antibody activates its chemical amplification system—complement—which both enhances phagocytosis and directly destroys the foreign cells. Similar functions are served by the chemicals released from sensitized T cells. In either case, all subsequent encounters with that microbe will be associated with the same sequence of events, with the crucial difference that the specific immune responses are brought into play much sooner and with greater force; i.e., the person would enjoy active immunity against that microbe.

IMMUNE SURVEILLANCE: DEFENSE AGAINST CANCER

A major function of cell-mediated immunity is to recognize and destroy cancer cells. This is made possible by the fact that virtually all cancer cells have some surface antigens different from those of other body cells and can, therefore, be recognized as "foreign." It is likely that cancer arises as a result of genetic alteration (by viruses, chemicals, radiation, etc.) in previously normal body cells. One manifestation of the genetic change is the appearance of the new surface antigens. Circulating T cells encounter and become sensitized to these foreign cells, combine with the antigens on their surface, and release the effector chemicals which destroy the cells by the mechanisms described above. It is presently believed that such transformations occur very frequently, i.e., that we may "get cancer once a day" (one expert's estimate), but that the cells are destroyed as fast as they arise. According to this view, only when the cell-mediated system is ineffective in either recognizing or destroying the cells do they multiply and produce clinical cancer.

REJECTION OF TISSUE TRANSPLANTS

The cell-mediated immune system is also mainly responsible for the recognition and destruction, i.e., *rejection,* of tissue transplants. On the surfaces of all nucleated cells of an individual's body are protein molecules which are antigenic, known as *histocompatibility antigens.* The genes which code for these proteins are, of course, inherited from one's parents, so that the offspring's group of antigens are, in part, similar to his or her parents but not identical. Clearly, the more closely related two people are the more similar these antigens will be, but no two people (other than identical twins) have identical groups. Thus, the surface antigens constitute the basis for immune individuality. When tissue is transplanted from one individual to another, those surface antigens which differ from the recipient's are recognized as foreign and the cells are destroyed by circulating T cells which become sensitized to them.

Some of the most valuable tools aimed at reducing graft rejection are radiation and drugs which kill actively dividing lymphocytes and, thereby, decrease the T-cell population. Unfortunately, this also results in depletion of B cells as well so that antibody production is diminished and the patient becomes highly susceptible to infection. A more discriminating method presently being tried is to prepare and inject into the recipients antibodies against their T cells; by this means, their T cells would be destroyed but not the B cells.

Related to the general problem of graft rejection is one of the major unsolved questions of immunology: How does the body avoid producing antibodies or sensitized lymphocytes to its own cells; i.e., how does it distinguish self-antigens from nonself-antigens? In general, it appears that any antigens present during embryonic and very early neonatal life are recognized as self and no antibodies or sensitized lymphocytes are formed against them later in life, following maturation of specific immune mechanisms. This can be shown by fooling the embryo in the following manner: Foreign mouse cells are injected into an embryo mouse during intrauterine life; months later, when the mature mouse is given a graft from the same foreign species, the graft is not rejected.

Of course, the generalization stated above is empirical fact, not explanation. Present evidence warrants the generalization that the thymus, as might be predicted, is very much involved in this "imprinting" of self-recognition but just how is simply now known. The question is clearly of more than academic interest since, if we understood the mechanism by which tolerance for one's own tissues is established, then we might be able to confer tolerance for transplants.

Transfusion Reactions and Blood Types

Transfusion reactions are a special example of tissue rejection; moreover, they illustrate the fact that antibodies rather than sensitized T cells can sometimes be the major factor in leading to the destruction of nonmicrobial cells. It was very early recognized that the transfusion of blood into a person was, more often than not, rapidly followed by clumping and hemolysis of erythrocytes with the appearance of hemoglobin in the plasma. If severe, this was associated with jaundice, fever, and a variety of tissue damage because of liberation of erythrocyte contents. With the identification of erythrocyte surface antigens, it was ultimately demonstrated that the red cell damage was caused by antigen-antibody reaction.

Among the large numbers of erythrocyte membrane antigens, we still recognize those designated A, B, and O as most important. These antigens are inherited, A and B being dominant. Thus, an individual with the genes for either A and O or B and O will develop only the A or B antigen. Accordingly, the possible blood types are A, B, O, and AB. If the typical pattern of antibody induction were followed, one would expect that type A persons would develop antibodies against type B cells only if the B cells were introduced into their bodies. However, what is atypical of this system is that even without initial exposure type A persons always have a high plasma concentration of anti-B antibody. The sequence of events during early life which lead to the presence of the so-called natural antibodies in all type A persons is unknown. Similarly, type B persons have high levels of anti-A antibodies; type AB persons obviously have neither anti-A nor anti-B antibody; type O persons have both; anti-O antibodies are usually not present in anyone.

With this information as background, what will happen if a type A person is given type B blood? There are two incompatibilities: (1) The recipient's anti-B antibody causes the transfused cells to be attacked and (2) the anti-A antibody in the transfused plasma causes the recipient's cells to be attacked. The latter is generally of little consequence, however, because the transfused antibodies become so diluted in the recipient's plasma that they are ineffective. It is the destruction of the transfused cells which produces the transfusion reaction. The range of possibilities is shown in Table 22-3. It should be evident why type O people are frequently called universal donors whereas type AB people are universal recipients. These terms, however, are mis-

TABLE 22-3
SUMMARY OF ABO BLOOD-TYPE INCOMPATIBILITIES

Recipient	*Donor*	*Incompatible?*
A or AB	A	No
B or O	A	Yes
B or AB	B	No
A or O	B	Yes
A, B, or AB	AB	No
O	AB	Yes
A, B, AB, or O	O	No

leading and dangerous since there are a host of other incompatible erythrocyte antigens and plasma antibodies besides those of the ABO type. Therefore, except in dire emergency, the blood of donor and recipient must be matched carefully.

Another antigen of great medical importance is the so-called Rh factor (because it was first studied in rhesus monkeys) now known to be a group of erythrocyte membrane antigens. The Rh system follows the classic immunity pattern in that no one develops anti-Rh antibodies unless exposed to Rh-type cells (usually termed Rh-positive cells) from another person. Although this can be a problem in an Rh-negative person, i.e., one whose cells have no Rh antigen, subjected to multiple transfusions with Rh-positive blood, its major importance is in the mother-fetus relationship (Fig. 22-11). When an Rh-negative mother carries an Rh-positive fetus, apparently some of the fetal erythrocytes may cross the placental barriers into the maternal circulation, inducing her to synthesize anti-Rh antibodies. These, in turn, cross the placenta into the fetus against whose erythrocytes they cause an attack to be launched. The resulting hemolysis may be severe enough to kill the fetus or produce serious anemia. Moreover, irreversible brain damage may result in these infants. Fortunately, only about 5 percent of Rh-negative mothers actually produce anti-Rh antibodies while carrying an Rh-positive child. Moreover, the first baby is almost always safe, later pregnancies becoming more dangerous because of the memory component of immune mechanisms. In addition to the Rh problem, other maternal-fetal erythrocyte incompatilibities can produce a similar picture.

FIGURE 22-11 ***Sequence of events leading to hemolysis of fetal erythrocytes in Rh incompatibility between mother and fetus. The question mark indicates the unknown mechanism of erythrocyte movement across the placental membranes (1). The Rh antigen induces antibody formation by the mother's plasma cells (2 and 3), and these antibodies enter the mother's blood (4) and cross the placental membranes to enter the fetal blood (5). Then they react (6) with the antigens on the erythrocyte membrane, with resulting damage and hemolysis of the cell (7). The last step also involves complement, which is not shown.***

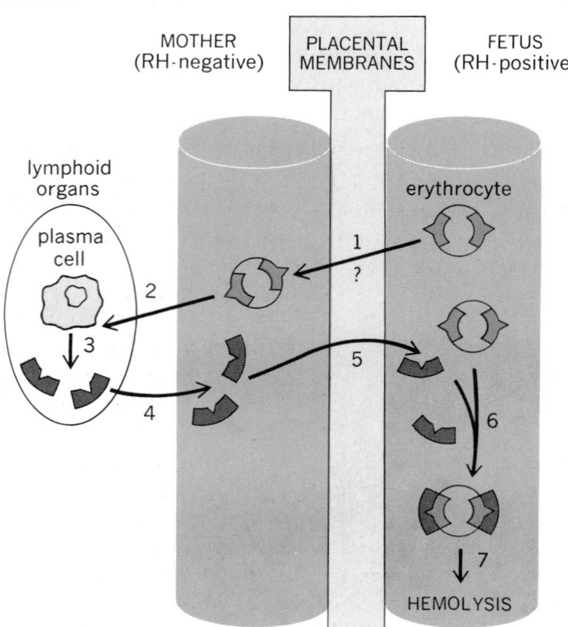

ALLERGY AND TISSUE DAMAGE

Immune responses obviously evolved to protect the body against invasion by foreign matter. Unfortunately, they frequently cause malfunction or damage to the body itself. The term *allergy* or *hypersensitivity* refers to an acquired reactivity to an antigen which can result in bodily damage upon subsequent exposure to that particular antigen. Allergic responses may be due to activation of either the humoral or cell-mediated system. There are a variety of types of allergic responses, and we shall describe only two categories: atopic allergy and autoimmune disease.

Atopic Allergy

This is the type of allergy popularly associated with the term allergy. A certain portion of the population is susceptible to sensitization by environmental antigens such as pollen, dusts, foods, etc. Initial exposure to the antigen leads to some antibody synthesis but, more important, to the memory storage which characterizes active immunity. Upon reexposure, the antigen elicits a more powerful antibody response. So far, none of this is unusual. What is it then that leads to body damage? The fact is that these particular antigens stimulate the production of the IgE class of antibodies which, upon their release from plasma cells, circulate to various parts of the body and attach themselves to mast cells (and basophils). When the antigen then combines with the IgE attached to the mast cell, the complex triggers, by unknown mechanisms (complement is not involved), a release of the mast cell's histamine and other vasoactive chemicals. These chemicals then initiate a local inflammatory response. Thus, the symptoms of atopic allergy are due to the various effects of these chemicals and the body site in which the antigen-IgE-mast cell combination occurs. For example, when a previously sensitized person inhales ragweed pollen, the antigen combines with IgE-mast cells in the respiratory passages. The histamine released causes increased mucus secretion, increased blood flow, leakage of protein, and constriction of the smooth muscle lining the airways. Thus, there follow the symptoms of congestion, running nose, sneezing, and difficulty in breathing which characterize hay fever.

In this manner, the symptoms of atopic allergy may be localized to the site of entry of the antigen. However, sometimes systemic symptoms may result if very large amounts of the vasoactive chemicals released enter the circulation and cause severe hypotension and bronchiolar constriction. This sequence of events, which can actually cause death, can be elicited in some sensitized people by the antigen in a single bee sting.

A major puzzle to biologists is the inappropriate nature of most atopic allergic responses, which are usually far more damaging to the body than the antigen triggering it. In other words, we clearly see the maladaptive nature of antigen-IgE reactions, but we do not know why such a system should have evolved, i.e., what normal physiologic function is subsumed by IgE antibodies. Similarly, it is not known why only a certain portion of the population is susceptible to atopic allergies.

Autoimmune Disease

We must qualify a generalization made previously by pointing out that the body does, all too often, produce antibodies or sensitized cells against its own tissues, the result being cell damage or destruction. A growing number of diseases in human beings are being recognized as *autoimmune* (a better word would be *autoallergic*) in origin.

There are multiple causes for the body's failure to recognize its own cells: (1) Normal antigens may be altered by combination with other chemicals (drugs or environmental pollutants, for example) from the external environment; (2) the cell may be infected by a virus whose DNA codes for a new protein (antigen); (3) genetic mutations may yield new antigens; (4) the body may encounter microbes whose antigens are so close in structure to certain self-antigens that the antibodies or sensitized lymphocytes produced against these antigens cross-react with the self-antigens; (5) components of certain tissues might never be exposed during embryonic life to whatever organs (? the thymus) must recognize and memorize the self-antigens; if they appear in the blood later in life, as the abnormal result of tissue disruption following injury or infection, they are treated as foreign. This list of possibilities is by no means complete, but whatever the cause, a breakdown in self-recognition results in turning the body's immune mechanisms against its own tissues.

The above description centers on the production of antibodies or sensitized lymphocytes against antigens on the body's own cells. However, autoimmune damage may also be brought about in several other quite different ways. An overzealous response (too much generation of complement or release of chemicals from platelets, neutrophils, or sensitized lymphocytes) may cause damage not only to invading foreign cells but to neighboring normal cells or membranes as well. For example, were a circulating antigen-antibody complex to be trapped within capillary membranes, the generation of complement or release of chemicals into the area might cause damage to the adjacent membranes. As might be predicted, the kidney glomeruli, with their larger filtering surface, are prime targets for such autoimmune destruction.

Section B. Resistance to Stress

Much of this book has been concerned with the body's response to stress in its broadest mening of an environmental change which must be adapted to if health and life are to be maintained. Thus, any change in external temperature, water intake, etc., sets into motion mechanisms designed to prevent a significant change in some physiologic variable. In this section, however, we describe the basic sterotyped response to stress in the more limited sense of noxious or potentially noxious stimuli. These comprise an immense number of situations, including physical trauma, prolonged heavy exercise, infection, shock, decreased oxygen supply, prolonged exposure to cold, pain, fright, and other emotional stresses. It is obvious that the overall response to cold exposure is very different from that to infection, but in one respect the response to all these situations is the same: Invariably secretion of cortisol is increased; indeed, the term stress has come to mean any event which elicits increased cortisol secretion. Also, sympathetic nervous activity is usually increased.

Historically, activation of the sympathetic nervous system was the first overall response to stress to be recognized and was labeled the fight-or-flight response. Only later did further work clearly establish the contribution of the adrenal cortical response. The increased cortisol secretion is mediated entirely by the hypothalamus–anterior pituitary system (Fig. 22-12) and does not occur in animals lacking a pictuitary or given a lesion in the hypothalamic area which secretes ACTH-releasing factor. Thus, afferent input to the hypothalamus induces secretion of ACTH-releasing factor, which is carried by the hypothalamopituitary portal vessels to the anterior pituitary and stimulates ACTH release. The ACTH, in turn, circulates to the adrenal and stimulates cortisol release. As described in Chap. 12, the hypothalamus receives input from virtually all areas of the brain and receptors of the body, and the pathway involved in any given situation depends upon the nature of the stress; e.g., ascending pathways from the arterial baroreceptors carry the input during hypotension, whereas pathways from other brain centers mediate the response to emotional stress. The destination is always the same, namely, synaptic connection with the hypothalamic neurons which secrete ACTH-releasing factor.

FIGURE 22-12 ***Pathway by which stressful stimuli elicit increased cortisol secretion.***

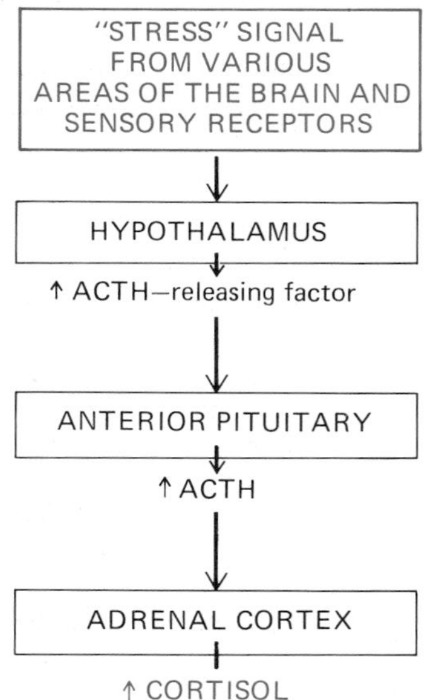

These same pathways also converge on the hypothalamic areas which control sympathetic nervous activity (including release of epinephrine from the adrenal medulla). However, it should be evident that certain types of stress can stimulate the sympathetic nervous system without the participation of the hypothalamus; e.g., the increased activity induced by hypotension requires the integrating activity only of the medullary cardiovascular centers (Chap. 16).

FUNCTIONS OF CORTISOL IN STRESS

Many of cortisol's most important effects are on organic metabolism. Cortisol (1) stimulates protein catabolism, (2) stimulates liver uptake of amino acids and their conversion to glucose (gluconeogenesis), (3) is permissive for stimulation of gluconeogenesis by other hormones (glucagon, growth hormone, etc.), and (4) inhibits glucose uptake and oxidation by many body cells ("insulin antagonism") but not by the brain. Indeed, so striking are these effects that cortisol is often called a glucocorticoid (to diminish it from the other major adrenal steroid, aldosterone, called a mineral-corticoid, because its major effects are on sodium and potassium metabolism).

These effects are ideally suited to meet a stressful situation. First, an animal faced with a potential threat is usually forced to forgo eating, and these metabolic changes are essential for survival during fasting—indeed, an adrenalectomized animal rapidly dies of hypoglycemia and brain dysfunction during fasting. Second, the amino acids liberated by catabolism of body protein stores not only provide energy, via gluconeogenesis, but also constitute a potential source of amino acids for tissue repair should injury occur.

A few of the many medically important implications of these cortisol-induced metabolic effects associated with stress are as follows: (1) Any patient ill or subjected to surgery catabolizes considerable quantities of body protein; (2) a diabetic who suffers an infection requires much more insulin than usual; (3) a child subjected to severe stress of any kind manifests retarded growth. The explanations for these phenomena should be evident.

Cortisol has important effects other than those on organic metabolism. One of the most important is that of enhancing vascular reactivity. A patient lacking cortisol faced with even a moderate stress may go into shock and die if untreated. This is due primarily to a marked decrease in total peripheral resistance. For unknown reasons, stress induces widespread arteriolar dilatation, despite massive sympathetic nervous system discharge, unless adequate amounts of cortisol are present. A large part of its counteracting effect is ascribable to the fact that cortisol permits norepinephrine to induce vasoconstriction.

Thus far we have presented the adaptive value of the stress-induced cortisol increase mainly in terms of its role in preparing the body physically for fight or flight, and there is no doubt that cortisol does function importantly in this way. However, in recent years, it has become apparent that cortisol may have other important functions. A large variety of psychosocial situations are associated with increased cortisol secretion; common denominators of many of them are novelty and challenge. Of great interest, therefore, are recent experiments which suggest that cortisol enhances learning in experimental animals, probably through direct actions on the brain. Thus, it may well be that the rise in cortisol secretion induced by psychosocial stress helps one to cope with the stress by facilitating the learning of appropriate responses.

Cortisol's Pharmacologic Effects and Disease

There are several situations in which adrenal corticosteroid levels in human beings become abnormally elevated. Patients with excessively hyperactive adrenals (there are several causes of this disease) represent one such situation, but the common occurrence is that of steroid administration for medical purposes. When cortisol is present in very high concentration, the previously described effects on organic metabolism are all magnified, but in addition there may appear one or more new effects, collectively known as the *pharmacologic effects* of cortisol. The most obvious is a profound reduction in the inflammatory response to injury or infection (indeed, reducing the inflammatory response in allergy, arthritis, or other diseases is the major reason for administering the cortisol to patients). Large amounts of cortisol inhibit almost every step of inflammation (vasodilatation, increased vascular permeability, phagocytosis) and may decrease antibody production as well. As might be expected, this decreases the ability of the person to resist infections. Large amounts of cortisol may also accelerate development of hypertension, atherosclerosis, and gastric ulcers, and may interfere with normal menstrual cycles.

As emphasized above, these pharmacologic effects are known to be elicited when cortisol levels are extremely elevated. Yet an unsettled question of great importance is whether long-standing lesser elevations of cortisol may do the same thing, albeit more slowly and less perceptibly. Put in a different way, do the psychosocial stresses, noise, etc., of everyday life contribute to disease production via increased cortisol?

FUNCTIONS OF THE SYMPATHETIC NERVOUS SYSTEM IN STRESS

A list of the major effects of increased general sympathetic activity almost constitutes a guide on how to meet emergencies. Since all these actions have been discussed in other sections of the book, they are listed here with little or no comment:

1 Increased hepatic and muscle glycogenolysis (provides a quick source of glucose)

2 Increased breakdown of adipose tissue triglyceride (provides a supply of glycerol for gluconeogenesis and of fatty acids for oxidation)

3 Increased central nervous system arousal and alertness

4 Increased skeletal muscle contractility and decreased fatigue

5 Increased cardiac output secondary to increased cardiac contractility and heart rate as well as increased venous return (venous constriction)

6 Shunting of blood from viscera to skeletal muscles by means of vasoconstriction in the former beds and vasodilatation in the latter

7 Increased ventilation

8 Increased coagulability of blood

The adaptive value of these responses in a fight-or-flight situation is obvious. But what purpose do they serve in the psychosocial stresses so common to modern life when neither fight nor flight is appropriate? A question yet to be answered is whether certain of these effects, like those of cortisol, might not, if prolonged, enhance the development of certain diseases, particularly atherosclerosis and hypertension. For example, one can easily imagine the increased blood fat concentration and cardiac work contributing to the former disease. Considerable work remains to be done to evaluate such possibilities.

OTHER HORMONES RELEASED DURING STRESS

Other hormones which are definitely released during many kinds of stress are aldosterone, antidiuretic hormone, and growth hormone. The increases in ADH and aldosterone ensure the retention of sodium and water within the body, an important adaptation in the face of potential losses by hemorrhage or sweating. Growth hormone reinforces the insulin antagonism effects of cortisol and the fat-mobilizing effects of epinephrine. Moreover, it probably stimulates the uptake of amino acids by an injured tissue and thereby facilitates tissue repair if needed; but since it cannot counteract the generalized protein catabolic effects of the increased cortisol, gluconeogenesis is not hampered.

Finally, recent evidence suggests that this list of hormones whose secretion rates are altered by stress is by no means complete. It is likely that the secretion of almost every known hormone may be influenced by stress. For example, prolactin, thyroxine, and glucagon are often increased whereas the pituitary gonadotropins (LH and FSH), insulin, and the sex steroids (testosterone or estrogen) are decreased. The adaptive significance of many of these changes is unclear but their possible contribution to stress-induced disease processes may be very important.

References

REFERENCES FOR FIGURE ADAPTATIONS

Anthony, C. P., and N. J. Kolthoff: "Textbook of Anatomy and Physiology," 8th ed., Mosby, St. Louis, 1971.

Avery, M. E., N. S. Wang, and H. W. Taeusch, Jr.: The Lung of the Newborn Infant, *Sci. Am.,* **228**:75, April, 1973.

Bekesy, G. von, and W. A. Rosenblith: In S. S. Stevens (ed.), "Handbook of Experimental Psychology," Wiley, New York, 1951.

Bloom, W., and D. W. Fawcett: "A Textbook of Histology," 9th ed., Saunders, Philadelphia, 1968.

Carlson, A. J., V. Johnson, and H. M. Cavert: "The Machinery of the Body," 5th ed., University of Chicago Press, Chicago, 1961.

Chapman, C. B., and J. H. Mitchell: *Sci. Am.,* May 1965.

Comroe, J. H.: "Physiology of Respiration," Year Book, Chicago, 1965.

Daughaday, W. H., and D. M. Kipnis: *Recent Prog. Horm. Res.,* **22**:49 (1966).

Dodt, E., and Y. Zotterman: *Acta Physiol. Scand.,* **26**:345 (1952).

Douglas, C. G., and J. S. Haldane: *J. Physiol.,* **38**:401 (1909).

Dubois, E. F.: "Fever and the Regulation of Body Temperature," Thomas, Springfield, Ill., 1948.

Ganong, W. F.: "Review of Medical Physiology," 4th ed., Lange, Los Altos, Calif., 1969.

Gordon, A. M., A. F. Huxley, and F. J. Julian: *J. Physiol.,* **184**:170 (1966).

Gregory, R. L.: "Eye and Brain: The Psychology of Seeing," McGraw-Hill, New York, 1966.

Grollman, S.: "The Human Body," 3d ed., Macmillan, 1974.

Guillemin, R., and R. Burgus: The Hormones of the Hypothalamus, *Sci. Am.,* October 1972.

Guyton, A. C.: "Functions of the Human Body," 3d ed., Saunders, Philadelphia, 1969.

Hensel, H., and K. K. A. Bowman: *J. Neurophysiol.,* **20**:564 (1960).

Hoffman, B. F., and P. E. Cranefield: "Electrophysiology of the Heart," McGraw-Hill, New York, 1960.

Hubel, D. H., and T. N. Wiesel: *J. Physiol.,* **154**:572 (1960); **160**:106 (1962).

Ito, S.: *J. Cell Biol.,* **16**:541 (1963).

Lambertsen, C. J.: In P. Bard (ed.), "Medical Physiology," 11th ed., Mosby, St. Louis, 1961.

Landau, Barbara R.: "Essential Human Anatomy and Physiology," Scott, Foresman, Glenview, Ill., 1976.

Langley, L. L., et al.: "Dynamic Anatomy and Physiology," 4th ed., McGraw-Hill, New York, 1974.

Langman, Jan: "Medical Embryology," Williams & Wilkins, Baltimore, 1975.

Lentz, Thomas L.: "Cell Fine Structure," Saunders, Philadelphia, 1971.

Livingston, R. B.: in G. C. Quarton, T. Melnechuk, and F. W. Schmitt (eds.), "The Neurosciences: A Study Program," Rockefeller University Press, New York, 1967.

Loewenstein, W. R.: *Sci. Am.,* August 1960.

Mayer, J., P. Roy, and K. P. Mitra: *Am. J. Nutr.,* **4**:169 (1956).

McNaught, A. B., and R. Callender: "Illustrated Physiology," Williams & Wilkins, Baltimore, 1963.

Nakayama, T., H. T. Hammel, J. D. Hardy, and J. S. Eisenman: *Am. J. Physiol.,* **204**:1122 (1963).

Nathanson, I. T., L. E. Towne, and J. C. Aub: *Endocrinology,* **28**:851 (1941).

Pedersen-Biergaard, K., and M. Ionnesen: *Acta Endocrinol.,* **1**:38 (1948); *Acta Med. Scand.,* **131** (suppl. 213):284 (1948).

Pitts, R. F.: "Physiology of the Kidney and Body Fluids," 2d ed., Year Book, Chicago, 1968.

Rasmussen, A. T.: "Outlines of Neuro-anatomy," 2d ed., Brown, Dubuque, Iowa, 1943.

Rushmer, R. F.: "Cardiovascular Dynamics," 2d ed., Saunders, Philadelphia, 1961.

Warwick, Roger, and Peter L. Williams: "Gray's Anatomy," 35th British ed., Saunders, Philadelphia, 1973.

Wersall, J., L. Gleisner, and P. G. Lundquist: in A. V. S. de Reuck and J. Knight (eds.), "Myotatic, Kinesthetic, and Vestibular Mechanisms," Ciba Foundation Symposium, Little, Brown, Boston, 1967.

Whitby, L. E. H. III, and C. J. C. Britton: "Disorders of the Blood," 10th ed., Churchill, London, 1969.

Woodburne, R. T.: "Essentials of Human Anatomy," 3d ed., Oxford, New York, 1965.

Zweifach, B. W.: *Sci. Am.*, January 1959.

SUGGESTED READING

The books and articles listed below constitute only a tiny fraction of the readings available, but they should provide a source of additional information on the topics covered in this book. Their bibliographies will serve as a further entry into the scientific literature, particularly for original research reports.

We have suggested the relative difficulty of entries by asterisks. No asterisk indicates that, after finishing the relevant chapter in this book, the reader should be able to understand the entry with little difficulty. One asterisk indicates that the work is more difficult but should be comprehensible with some effort. Two asterisks denote a work which is highly detailed or technical and may require a strong background in mathematics, chemistry, or biology for a full understanding; this type of reading has been included for students who wish to pursue a subject in depth and are willing to expend the effort required.

We list first a group of books, mainly textbooks, which cover large areas of physiology and anatomy. Since each can serve as additional reading for many chapters, they are grouped here according to field and are not mentioned again for individual chapters.

Books Covering Wide Areas of Physiology and Anatomy

Cell Physiology and Biochemistry

Baker, Jeffrey J. W., and Garland E. Allen: "Matter, Energy, and Life," 3d ed., Addison-Wesley, Reading, Mass., 1975 (paperback).

** Beck, Felix, and John B. Lloyd (eds.): "The Cell in Medical Science," 4 vols., Academic, New York, 1974–1975.

** Davson, H.: "A Textbook of General Physiology," 4th ed., vols. I and II, Little, Brown, Boston, 1970.

* DeRobertis, E. D. P., Francisco A. Saez, and E. M. F. DeRobertis: "Cell Biology," 6th ed., Saunders, Philadelphia, 1975.

* Giese, Arthur C.: "Cell Physiology," 4th ed., Saunders, Philadelphia, 1973.

* Lehninger, Albert L.: "Biochemistry: The Molecular Basis of Cell Structure and Function," 3d ed., Worth, New York, 1975.

McElroy, William D., and Carl P. Swanson: "Modern Cell Biology," 2d ed., Prentice-Hall, Englewood Cliffs, N.J., 1976 (paperback).

Swanson, Carl P., and Peter L. Webster: "The Cell," 4th ed., Prentice-Hall, Englewood Cliffs, N.J., 1977 (paperback).

* Wolfe, Stephen L.: "Biology of the Cell," Wadsworth, Belmont, Calif., 1972.

Wooldridge, Dean E.: "The Machinery of Life," McGraw-Hill, New York, 1966 (paperback).

Organ System Physiology

Astrand, P., and K. Rodahl: "Textbook of Work Physiology," McGraw-Hill, New York, 1970.

* Ganong, W. F.: "Review of Medical Physiology," 8th ed., Lange, Los Altos, Calif., 1977.

* Guyton, A. C.: "Textbook of Medical Physiology," 5th ed., Saunders, Philadelphia, 1976.

** "Handbook of Physiology," Williams & Wilkins, Baltimore, 1959–(continual publication of new volumes).

** Quarton, G. C., T. Melnechuk, and F. O. Schmitt (eds.): "The Neurosciences: A Study Program," Rockefeller University Press, New York, 1967.

* Ruch, T. C., and H. D. Patton: "Medical Physiology and Biophysics," 20th ed., Saunders, Philadelphia, 1973.

** Schmitt, F. O. (ed.): "The Neurosciences: Second Study Program," Rockefeller University Press, New York, 1970.

** Schmitt, F. O., and F. G. Worden (eds.): "The Neurosciences: Third Study Program," M.I.T., Cambridge, Mass., 1974.

Anatomy

Basmajian, J. V.: "Grant's Method of Anatomy," 9th ed., Williams & Wilkins, Baltimore, 1975.

Goss, C. M.: "Gray's Anatomy of the Human Body," 29th American ed., Lea & Febiger, Philadelphia, 1973.

Hamilton, W. J., and N. W. Mossman: "Human Embryology," 4th ed., Williams & Wilkins, Baltimore, 1972.
Hollinshead, W. H.: "Textbook of Anatomy," 3d ed., Harper & Row, New York, 1974.
Leeson, T. S., and C. R. Leeson: "Human Structure," Saunders, Philadelphia, 1972.
* Woodburne, R. T.: "Essentials of Human Anatomy," 5th ed., Oxford, New York, 1973.

Pathophysiology

** Frohlich, Edward D. (ed.): "Pathophysiology," 2d ed., Lippincott, Philadelphia, 1976.
* Roddie, Ian C., and William F. M. Wallace: "The Physiology of Disease," Year Book, Chicago, 1975.
Snively, W. D., Jr., and Donna R. Beshear: "Textbook of Pathophysiology," Lippincott, Philadelphia, 1972.
* Sodeman, William A., and William A. Sodeman, Jr.: "Pathologic Physiology: Mechanisms of Disease," 5th ed., Saunders, Philadelphia, 1974.

SUGGESTIONS FOR INDIVIDUAL CHAPTERS

Chapter 1 The Cell: Structure and Function

Allison, Anthony: Lysosomes and Disease, *Sci. Am.,* November 1967.
* Anfinsen, Christian B.: "The Molecular Basis of Evolution," Wiley, New York, 1963 (paperback).
Baltimore, David: The Molecular Biology of Poliovirus, *Sci. Am.,* May 1975.
* Bernhard, Sidney A.: "The Structure and Function of Enzymes," Benjamin, New York, 1968 (paperback).
Bolin, Bert: The Carbon Cycle, *Sci. Am.,* September 1970.
** Bretscher, Mark S.: Membrane Structure: Some General Principles, *Science,* **181:**662 (1973).
Campbell, Allan M.: How Viruses Insert their DNA into the DNA of the Host Cell, *Sci. Am.,* December 1976.
Capaldi, Roderick A.: A Dynamic Model of Cell Membranes, *Sci. Am.,* March 1974.
Changeux, J.: The Control of Biochemical Reactions, *Sci. Am.,* April 1965.
Clark, B. F. C., and K. A. Marcker: How Proteins Start, *Sci. Am.,* January 1968.
Cohen, Stanley N.: The Manipulation of Genes, *Sci. Am.,* July 1975.
* Conn, E. E., and P. K. Stumpf: "Outlines of Biochemistry," 4th ed., Wiley, New York, 1976.
Crick, F. H. C.: The Genetic Code, *Sci. Am.,* October 1962; October, 1966.
** Darnell, James E., Warren R. Jelinek, and George R. Molloy: Biogenesis of mRNA: Genetic Regulation in Mammalian Cells, *Science,* **181:**1215 (1973).
** Dick, D. A. T.: "Cell Water," Butterworth, Washington, D.C., 1966.
Doty, P.: Proteins, *Sci. Am.,* September, 1967.
Frieden, Earl: The Chemical Elements of Life, *Sci. Am.,* July 1972.
Frieden, Earl: The Enzyme-Substrate Complex, *Sci. Am.,* August 1969.
German, James: Studying Human Chromosomes Today, *Am. Sci.,* **58:**182 (1970).
Goodenough, Ursula W., and R. P. Levine: The Genetic Activity of Mitochondria and Chloroplasts, *Sci. Am.,* November 1970.
Green, D. E.: The Mitochondrion, *Sci. Am.,* January 1964.
Green, D. E.: The Synthesis of Fat, *Sci. Am.,* August, 1960.
Grobstein, Clifford: The Recombinant-DNA Debate, *Sci. Am.,* July 1977.
* Hendler, Richard W.: "Protein Biosynthesis and Membrane Biochemistry," Wiley, New York, 1968.
Hurwitz, J., and J. J. Furth: Messenger RNA, *Sci. Am.,* February 1962.
* Ingram, V. M.: "Biosynthesis of Macromolecules," 2d ed., Benjamin, New York, 1972 (paperback).
Kendrew, J. C.: Three-dimensional Study of a Protein, *Sci. Am.,* December 1961.
Kornberg, A.: The Synthesis of DNA, *Sci. Am.,* September 1961.
Koshland, Daniel E., Jr.: Protein Shape and Biological Control, *Sci. Am.,* October 1973.
** Kotyk, Arnost, and Karel Janacek: "Cell Membrane Transport," 2d ed., Plenum, New York, 1975.
Lambert, Joseph B.: The Shapes of Organic Molecules, *Sci. Am.,* January 1970.
* Lehninger, Albert L.: "Bioenergetics: The Molecular Basis of Biological Energy Transformations: 2d ed., Benjamin, New York, 1971 (paperback).
Maniatis, T., and M. Ptashne: A DNA Operator-Repressor System, *Sci. Am.,* **234:**64 (January 1976).
Mazia, Daniel: The Cell Cycle, *Sci. Am.,* January 1974.
Miller, O. L., Jr.: The Visualization of Genes in Action, *Sci. Am.,* March 1973.
Mirsky, A. E.: The Discovery of DNA, *Sci. Am.,* June 1968.
Mott-Smith, Morton: "The Concept of Energy Simply Explained," Dover, New York, 1964 (paperback).
Neutra, M., and C. P. Leblond: The Golgi Apparatus, *Sci. Am.,* February 1969.
** Newsholme, E. A., and C. Start: "Regulation in Metabolism," Wiley, New York, 1975.
Nirenberg, M. W.: The Genetic Code, II, *Sci. Am.,* March 1963.
* Racker, Efraim: The Inner Mitochondrial Membrane: Basic and Applied Aspects, *Hosp. Pract.,* February 1974.
Rich, A.: Polyribosomes, *Sci. Am.,* December 1963.
Rustad, R. C.: Pinocytosis, *Sci. Am.,* April 1961.
* Saier, Milton H., Jr., and Charles D. Stiles: "Molecular Dynamics in Biological Membranes," Springer-Verlag, New York, 1975 (paperback).

Satir, B.: The Final Steps in Secretion, *Sci. Am.*, **233**:29 (October 1975).

Schmitt, F. O.: Giant Molecules in Cells and Tissues, *Sci. Am.*, September 1957.

** Segal, Harold L.: Enzymatic Interconversion of Active and Inactive Forms of Enzymes, *Science*, **180**:25 (1973).

** Singer, S. J., and Garth L. Nicolson: The Fluid Mosaic Model of the Structure of Cell Membranes, *Science*, **175**:720 (1972).

Solomon, Arthur K.: The State of Water in Red Cells, *Sci. Am.*, February 1971.

* Speakman, J. C.: "Molecules," McGraw-Hill, New York, 1966 (paperback).

** Stein, W. D.: "The Movement of Molecules Across Cell Membranes," Academic, New York, 1967.

* Watson, James D.: "Molecular Biology of the Gene," 3d ed., Benjamin, New York, 1976 (paperback).

Watson, James D.: "The Double Helix," Atheneum, New York, 1968. (Autobiographical account of Watson's role in discovering the DNA double helix.)

Weissman, Gerald, and Robert Claiborne (eds.): "Cell Membranes: Biochemistry, Cell Biology, and Pathology," Hospital Practice Publishing, New York, 1976.

Yanofsky, C.: Gene Structure and Protein Structure, *Sci. Am.*, May 1967.

Chapter 2 Organization of the Body

Cairns, John: The Cancer Problem, *Sci. Am.*, November 1975.

Edwards, R., and R. Fowler: Human Embryos in the Laboratory, *Sci. Am.*, December 1970.

* Elkinton, J. B., and T. S. Danowsky: "The Body Fluids: Basic Physiology and Practical Therapeutics," Williams & Wilkins, Baltimore, 1955.

Langman, Jan: "Medical Embryology," 3d ed., Williams & Wilkins, Baltimore, 1975.

** Mitchison, J. M.: "Biology of the Cell Cycle," Cambridge, New York, 1972.

** Stephens, R. E., and K. T. Edds: Microtubules: Structure, Chemistry and Function, *Physiol. Rev.* **56**:709, 1976.

Trinkaus, J. P.: "Cells into Organs," Prentice-Hall, Englewood Cliffs, N.J., 1969.

Wessells, N. K., and W. J. Rutter: Phases in Cell Differentiation, *Sci. Am.*, March 1969.

Wolf, A. V.: Body Water, *Sci. Am.*, November 1958.

Chapter 3 Epithelium, Connective Tissue, and Skin

Bickers, D. R., and A. Kappas: Metabolic and Pharmacologic Properties of the Skin, *Hosp. Pract.*, May 1974.

* Bloom, W., and D. W. Fawcett: "A Textbook of Histology," 10th ed., Saunders, Philadelphia, 1975.

** Mier, Paul D., and Dennis W. K. Cotton: "The Molecular Biology of Skin," Blackwell Scientific Publications, Oxford, 1976.

Nicolaides, N.: Skin Lipids: Their Biochemical Uniqueness, *Science*, **186**:19 (1974).

* Rhodin, Johannes A. G.: "Histology: A Text and Atlas," Oxford, New York, 1974.

Ross, R.: Wound Healing, *Sci. Am.*, June 1969.

Ross, R., and P. Bornstein: Elastic Fibers in the Body, *Sci. Am.*, June 1971.

Satir, Birgit: The Final Steps in Secretion, *Sci. Am.*, October 1975.

Satir, P.: How Cilia Move, *Sci. Am.*, October 1974.

** Vaughan, Janet: "The Physiology of Bone," 2d ed., Clarendon, Oxford, 1975.

Windle, William F.: "Textbook of Histology," 5th ed., McGraw-Hill, New York, 1976.

Chapter 4 Neural Tissue

Baker, P. F.: The Nerve Axon, *Sci. Am.*, March 1966.

* Bodian, D.: The Generalized Vertebrate Neuron, *Science*, **137**:323 (1962).

* Bullock, T. H.: Neuron Doctrine and Electrophysiology, *Science*, **129**:997 (1959).

* DeRobertis, E.: Ultrastructure and Cytochemistry of the Synaptic Region, *Science*, **156**:907 (1967).

Eccles, J. C.: The Synapse, *Sci. Am.*, January 1965.

** Eccles, J. C.: "The Physiology of Nerve Cells," Johns Hopkins, Baltimore, 1957 (paperback).

* Eccles, J. C.: "The Understanding of the Brain," McGraw-Hill, New York, 1973 (paperback), chaps. 1–3.

* Eyzaguirre, C., and S. J. Fidone: "Physiology of the Nervous System," 2d ed., Year Book, Chicago, 1975.

* Goodman, L. S., and A. Gilman: Neurohumoral Transmission and the Autonomic Nervous System, in "The Pharmacological Basis of Therapeutics," 5th ed., Macmillan, New York, 1975.

* Hodgkin, A. L.: "The Conduction of the Nervous Impulse," Thomas, Springfield, Ill., 1964.

* Katz, B.: Quantal Mechanism of Neural Transmitter Release, Nobel Prize Lecture, 1970, *Science*, **173**:123 (1971).

* Katz, B.: "Nerve, Muscle, and Synapse," McGraw-Hill, New York, 1966 (paperback).

Katz, B.: How Cells Communicate, *Sci. Am.*, September 1961.

Keynes, R. D.: The Nerve Impulse in the Squid, *Sci. Am.*, December 1958.

** Krnjevic, K.: Chemical Nature of Synaptic Transmission in Vertebrates, *Physiol. Rev.*, **54**:418 (1974).

** Kuffler, S. W., and J. G. Nicholls: "From Neuron to Brain," Sinauer, Sunderland, Mass., 1976 (paperback).

Loewenstein, W. R.: Biological Transducers, *Sci. Am.*, August 1960.

* Ochs, S.: "Elements of Neurophysiology," Wiley, New York, 1965 pp. 302–341.

** Ruch, T. C., H. D. Patton, W. Woodbury, and A. L.

Towe: "Neurophysiology," Saunders, Philadelphia, 1965, chaps. 1 and 2.

** Schmitt, Francis O., Parvati Dev, and Barry H. Smith: Electronic Processing of Information by Brain Cells, *Science,* **193:**114 (1976).

* Shepherd, J. M.: "The Synaptic Organization of the Brain: An Introduction," Oxford, London, 1974 (paperback), Part I.

Whittaker, V. P.: Membranes in Synaptic Function, *Hosp. Pract.,* **9:**111 (April 1974).

Chapter 5 Muscle Tissue

* Bendall, J. R.: "Muscles, Molecules and Movement," American Elsevier, New York, 1969.

** Bourne, G. H. (ed.): "The Structure and Function of Muscle," 2d ed., Academic, New York, vol. 1 (1972), vols. II and III (1973), vol. IV (1974).

** Carlson, Francis D., and Douglas R. Wilkie: "Muscle Physiology," Prentice-Hall, Englewood Cliffs, N.J., 1974.

** Close, R. I.: Dynamic Properties of Mammalian Skeletal Muscles, *Physiol. Rev.,* **52:**129 (1972).

Cohen, Carolyn: The Protein Switch of Muscle Contraction, *Sci. Am.,* November 1975.

** Ebashi, Setsuro: Excitation-Contraction Coupling, *Ann. Rev. Physiol.,* **38:**293 (1976).

** Endo, Makoto: Calcium Release from the Sarcoplasmic Reticulum, *Physiol. Rev.,* **57:**71 (1977).

** Guth, Lloyd: Trophic Influences of Nerve on Muscle, *Physiol. Rev.,* **44:**645 (1968).

** Holloszy, J. O., Booth, F. W.: Biochemical Adaptations to Endurance Exercise in Muscle, *Ann. Rev. Physiol.,* **38:**273 (1976).

Hoyle, Graham: How is Muscle Turned On and Off?, *Sci. Am.,* April 1970.

** Huddart, Henry, and Stephen Hunt: "Visceral Muscle," Halstead, New York, 1975.

** Huxley, H. E.: The Mechanism of Muscular Contraction, *Science,* **164:**1356 (1969).

Huxley, H. E.: The Contraction of Muscle, *Sci. Am.,* November 1968.

Lester, Henry A.: The Response to Acetylcholine, *Sci. Am.,* February 1977.

Margaria, R.: The Sources of Muscular Energy, *Sci. Am.,* March 1972.

Merton, P. A.: How We Control the Contraction of Our Muscles, *Sci. Am.,* May 1972.

** Mommaerts, W. F. H. M.: Energetics of Muscle Contraction, *Physiol. Rev.,* **49:**427 (1969).

Murray, John M., and Annemarie Weber: The Cooperative Action of Muscle Proteins, *Sci. Am.,* February 1974.

* Needham, Dorothy M.: "Machina Carnis: The Biochemistry of Muscular Contraction in Its Historical Development," Cambridge, New York, 1971.

Porter, K. R., and C. Franzini-Armstrong: The Sarcoplasmic Reticulum, *Sci. Am.,* March 1965.

** Weber, Annemarie, and John M. Murray: Molecular Control Mechanisms in Muscle Contraction, *Physiol. Rev.,* **53:**612 (1973).

Chapter 6 Homeostasis and Control Systems

** Adolph, E. F.: Early Concepts of Physiological Regulations, *Physiol. Rev.,* **41:**737 (1961).

Bernard, C.: "An Introduction to the Study of Experimental Medicine," Dover, New York, 1957 (paperback).

* Bullough, W. S.: "The Evolution of Differentiation," Academic, New York, 1967.

Cannon, W. B.: "The Wisdom of the Body," Norton, New York, 1939 (paperback).

* Fox, S. W., and K. Dose: "Molecular Evolution and the Origin of Life," Freeman, San Francisco, 1972 (paperback).

Langley, L. L. (ed.): "Homeostasis: Origin of the Concept," Dowden, Hutchinson, & Ross, Inc., Stroudsburg, Pa., 1973.

** Milhorn, H. T., Jr.: "The Application of Control Theory to Physiological Systems," Saunders, Philadelphia, 1966.

** Oparin, A. I.: "The Origin of Life," Dover, New York, 1953 (paperback).

Tustin, A.: Feedback, *Sci. Am.,* September 1952.

Wald, G.: The Origin of Life, *Sci. Am.,* August 1954.

* Weiner, N.: Concept of Homeostasis in Medicine, *Trans. Coll. Physicians Phila,* **20:**87 (1948).

Chapter 7 The Musculoskeletal System

Basmajian, J. V.: "Grant's Method of Anatomy," 9th ed., Williams & Wilkins, Baltimore, 1975.

Goss, C. M.: "Gray's Anatomy of the Human Body," 29th American ed., Lea & Febiger, Philadelphia, 1973.

Hollinshead, W. H.: "Textbook of Anatomy," 3d ed., Harper & Row, New York, 1974.

Chapter 8 The Nervous System: I. Structure

Barr, M. L.: "The Human Nervous System," 2d ed., Harper & Row, New York, 1974.

Cannon, W. B.: "Bodily Changes in Pain, Hunger, Fear, and Rage," Harper & Row, New York, 1963.

Dicara, L. V.: Learning in the Autonomic Nervous System, *Sci. Am.,* January 1970.

Llinás, R. R.: The Cortex of the Cerebellum, *Sci. Am.,* **232:**56 (January 1975).

* Miller, N. E.: Learning of Visceral and Glandular Responses, *Science,* **163:**434 (1969).

* Shepherd, G. M.: "The Synaptic Organization of the Brain: An Introduction," Oxford, London, 1974 (paperback), Part II.

Chapter 9 The Nervous System: II. Sensory Systems

* Alpern, M., M. Lawrence, and D. Wolsk: "Sensory Processes," Brooks/Cole, Belmont, Calif., 1967 (paperback).

Attneave, F.: Multistability in Perception, *Sci. Am.,* December 1971.

Bekesy, G. von: The Ear, *Sci. Am.,* August 1957.

Casey, K. L.: Pain: A Current View of Neural Mechanisms, *Am. Sci.,* **61:**194 (1973).

** Daw, N. W.: Neurophysiology of Color Vision, *Physiol. Rev.,* **53:**571 (1973).

Day, R. H.: Visual Spatial Illusions, A General Explanation, *Science,* **175:**1335 (1972).

Deregowski, J. B.: Pictorial Perception and Culture, *Sci. Am.,* November 1972.

* De Valois, R. L., and G. H. Jacobs: Primate Color Vision, *Science,* **162:**533 (1968).

Favreau, Olga Eizner, and Michael C. Corballis: Negative Aftereffects in Visual Perception, *Sci. Am.,* December 1976.

Glickstein, Mitchell, and Alan R. Gibson: Visual Cells in the Pons of the Brain, *Sci. Am.,* November 1976.

Gombrich, E. H.: The Visual Image, *Sci. Am.,* September 1972.

Gordon, B.: The Superior Colliculus of the Brain, *Sci. Am.,* December 1972.

* Granit, R.: The Development of Retinal Neurophysiology, Nobel Prize Lecture, 1967, *Science,* **160:**1192 (1968).

Gregory, R. L.: "Eye and Brain: The Psychology of Seeing," 2d ed., McGraw-Hill, New York, 1973.

* Hartline, H. K.: Visual Receptors and Retinal Interactions, Nobel Prize Lecture, 1967, *Science,* **164:**270 (1969).

Heyningen, Ruth van: What Happens to the Human Lens in Cataract, *Sci. Am.,* December 1975.

Hubel, D. H.: The Visual Cortex of the Brain, *Sci. Am.,* November 1963.

Johansson, Gunnar: Visual Motion Perception, *Sci. Am.,* June 1975.

MacNichol, E. F.: Three-Pigment Color Vision, *Sci. Am.,* December 1964.

Michael, C. R.: Retinal Processing of Visual Images, *Sci. Am.,* May 1969.

Miller, W. H., F. Ratliff, and H. K. Hartline: How Cells Receive Stimuli, *Sci. Am.,* September 1961.

** Moulton, David G.: Spatial Patterning of Response to Odors in the Peripheral Olfactory System, *Physiol. Rev.,* **56:**578 (1976).

** Moulton, D. G., and L. M. Beidler: Structure and Function in the Peripheral Olfactory System, *Physiol. Rev.,* **47:**1 (1967).

Noton, D., and L. Stark: Eye Movements and Visual Perception, *Sci. Am.,* June 1971.

** Oakley, B., and R. M. Benjamin: Neural Mechanisms of Taste, *Physiol. Rev.,* **46:**173 (1966).

Pettigrew, J. D.: The Neurophysiology of Binocular Vision, *Sci. Am.,* August 1972.

Ratliff, F.: Contour and Contrast, *Sci. Am.,* June 1972.

Robinson, D. A.: Eye Movement Control in Primates, *Science,* **161:**1219 (1968).

Ruston, W. A. H.: Visual Pigments and Color Blindness, *Sci. Am.,* March 1975.

** Singer, Wolf: Control of Thalamic Transmission by Corticofugal and Ascending Reticular Pathways in the Visual System, *Physiol. Rev.,* **57:**386 (1977).

Stent, G. S.: Cellular Communication, *Sci. Am.,* September 1972.

* Stevens, S. S.: Neural Events and the Psychophysical Law, *Science,* **170:**1043 (1970).

** Symposium: Structure and Function of the Visual System, *Fed. Proc.,* **35:**36 (1976).

Toates, F. M.: Accommodation Function of the Human Eye, *Physiol. Rev.,* **52:**828 (1972).

* Wald, G.: Molecular Basis of Visual Excitation, Nobel Prize Lecture, 1967, *Science,* **162:**230 (1968).

Warren, R. M., and R. P. Warren: Auditory Illusions and Confusions, *Sci. Am.,* December 1970.

Warshofsky, F., and S. S. Stevens: "Sound and Hearing," Time-Life, New York, 1969.

Werblin, F. S.: The Control of Sensitivity in the Retina, *Sci. Am.,* September 1972.

Young, R. W.: Visual Cells, *Sci. Am.,* October 1970.

Chapter 10 The Nervous System: III. Motor Control

** Asanuma, Hiroshi: Recent Developments in the Study of the Columnar Arrangement of Neurons within the Motor Cortex, *Physiol. Rev.,* **55:**143 (1975).

** Duvoisin, Roger: Parkinsonism, CIBA Found. Clinical Symp. 28, No. 1 (1976).

* Eccles, J. C.: "The Understanding of the Brain," McGraw-Hill, New York, 1973 (paperback), chap. 4.

Evarts, E. V.: Brain Mechanisms in Movement, *Sci. Am.,* July 1973.

* Eyzaguirre, C., and S. J. Fidone: "Physiology of the Nervous System," 2d ed., Year Book, Chicago, 1975.

** Granit, R.: "The Basis of Motor Control," Academic, New York, 1970.

Lippold, O.: Physiological Tremor, *Sci. Am.,* May 1971.

** Matthews, P. B. C.: Muscle Spindles and Their Motor Control, *Physiol. Rev.,* **44:**219 (1964).

Merton, P. A.: How We Control the Contraction of Our Muscles, *Sci. Am.,* May 1972.

* Ochs, S.: "Elements of Neurophysiology," Wiley, New York, 1965, pp. 280–301, 342–363, and 493–526.

Pearson, Kei: The Control of Walking, *Sci. Am.,* December 1976.

** Ruch, T. C., H. D. Patton, J. W. Woodbury, and A. L. Towe: "Neurophysiology," 2d ed., Saunders, Philadelphia, 1965, pp. 153–225, 252–300.

* Sherrington, Sir Charles: "The Integrative Action of the Nervous System," Yale, New Haven, Conn., 1961 (originally published in 1906; paperback).

** Shik, M. L., and G. N. Orlovsky: Neurophysiology of Locomotor Automatism, *Physiol. Rev.*, **56**:465 (1976).

Wilson, V. J.: Inhibition in the Central Nervous System, *Sci. Am.*, May 1966.

* Wilson, V. J.: The Labyrinth, the Brain and Posture, *Am. Sci.*, **63**:325 (1975)

Chapter 11 The Nervous System: IV. Consciousness and Behavior

Agranoff, B. W.: Memory and Protein Synthesis, *Sci. Am.*, June 1967.

"Altered States of Awareness," (readings from *Scientific American*), Freeman, San Francisco, 1972 (paperback).

Atkinson, R. C., and R. M. Schiffrin: The Control of Short-term Memory, *Sci. Am.*, August 1971.

Broadbent, D. E.: Attention and the Perception of Speech, *Sci. Am.*, April 1962.

Butter, C. M.: "Neuropsychology: The Study of Brain and Behavior," Brooks/Cole, Belmont, Calif., 1968 (paperback).

Ceraso, J.: The Interference Theory of Forgetting, *Sci. Am.*, October 1967.

** Cooper, Jack R., Floyd E. Bloom, and Robert H. Roth: "The Biochemical Basis of Neuropharmacology," 2d ed., Oxford, New York, 1974 (paperback).

Corballis, M. C., and I. L. Beale: On Telling Right from Left, *Sci. Am.*, March 1971.

Deutsch, D.: Musical Illusion, *Sci. Am.*, **233**:92 (October, 1975).

de Wied, D.: Hormonal Influences on Motivation, Learning, and Memory, *Hosp. Pract.*, **11** (January, 1976).

"Eccles, J. C. (ed.): "Brain and Conscious Experience," Springer-Verlag, New York, 1966.

* J. C. Eccles: "The Understanding of the Brain," McGraw-Hill, New York, 1973 (paperback), chaps. 5 and 6.

Fernstron, J. D., and R. L. Wurtman: Nutrition and the Brain, *Sci. Am.*, February 1974.

Fromkin, V. A.: Slips of the Tongue, *Sci. Am.*, December 1973.

Gazzaniga, M. S.: The Split Brain in Man, *Sci. Am.*, August 1967.

Geschwind, N.: Language and the Brain, *Sci. Am.*, April 1972.

** Goldstein, Avram: Opioid Peptides (Endorphins) in Pituitary and Brain, *Science*, **193**:1081 (1976).

Haber, R. N.: How We Remember What We See, *Sci. Am.*, May 1970.

Hess, W. R.: Causality, Consciousness, and Cerebral Organization, *Science*, **158**:1279 (1967).

Hollister, Leo E., Kenneth L. Davis, and Bonnie Morrison Davis: Hormones in the Treatment of Psychiatric Disorders, *Hosp. Pract.*, November 1975.

Horn, G., S. P. R. Rose, and P. P. G. Bateson: Experience and Plasticity in the Central Nervous System, *Science*, **181**:506 (1973).

Jacobson, M., and R. K. Hunt: The Origins of Nerve-Cell Specificity, *Sci. Am.*, February 1973.

Jerison, Harry J.: Paleoneurology and the Evolution of Mind, *Sci. Am.*, January 1976.

Kimura, D.: The Asymmetry of the Human Brain, *Sci. Am.*, March 1973.

Lenneberg, E. H.: On Explaining Language, *Science*, **164**:635 (1969).

Luria, A. R.: The Functional Organization of the Brain, *Sci. Am.*, March 1970.

** Meyer, David E., and Roger W. Schvaneveldt: Meaning, Memory Structure and Mental Processes, *Science*, **192**:27 (1976).

** Miller, Neal E., and Barry R. Dworkin: Effects of Learning on Visceral Functions—Biofeedback, *N. Engl. J. Med.*, **296**:1274 (1977).

** Moskowitz, Michael A., and Richard J. Wurtman: Catecholamines and Neurologic Diseases, *N. Engl. J. Med.*, **293**:274, 332 (1975).

* Nathanson, James A., and Paul Greengard: "Second Messenger" in the Brain, *Sci. Am.*, August 1977.

Oatley, K.: "Brain Mechanisms and Mind," Dutton, New York, 1972 (paperback).

Olton, David S.: Spatial Memory, *Sci. Am.*, June 1977.

Pappenheimer, John R.: The Sleep Factor, *Sci. Am.*, August 1976.

* Penfield, W., and L. Roberts: "Speech and Brain-mechanisms," Atheneum, New York, 1966 (paperback).

Premack, A. J., and D. Premack: Teaching Language to an Ape, *Sci. Am.*, October 1972.

Pribram, K. H.: The Neurophysiology of Remembering, *Sci. Am.*, January 1969.

"Psychobiology: The Biological Basis of Behavior," (readings from *Scientific American*), Freeman, San Francisco, 1972 (paperback).

Rosenzweig, M. R., E. L. Bennett, and M. C. Diamond: Brain Changes in Response to Experience, *Sci. Am.*, February 1972.

Sachar, Edward J.: Hormonal Changes in Stress and Mental Illness, *Hosp. Pract.*, July 1975.

Schwartz, Gary E.: Biofeedback, Self-Regulation and the Patterning of Physiological Processes, *Am. Sci.*, **63**:314 (1975).

Scott, J. P.: Critical Periods in Behavioral Development, *Science*, **138**:949 (1962).

Siegel, R. K.: Hallucinations, *Sci. Am.*, **237**:132 (October, 1977).

Snyder, Solomon H.: Opiate Receptors and Internal Opiates, *Sci. Am.*, March 1977.

** Snyder, Solomon H.: Opiate Receptors in the Brain, *N. Engl. J. Med.,* **296:**266 (1977).
* Sperry, R. W.: A Modified Concept of Consciousness, *Psychol. Rev.,* **76:**532 (1969).
Stent, Gunther S.: Limits to the Scientific Understanding of Man, *Science,* **187:**1052 (1975).
Tart, C. T. (ed.): "Altered States of Consciousness," Anchor Books, Doubleday, Garden City, N.Y., 1972 (paperback).
Wallace, R. K., and H. Benson: The Physiology of Meditation, *Sci. Am.,* February 1972.
Wooldridge, D. E.: "Mechanical Man: The Physical Basis of Intelligent Life," McGraw-Hill, New York, 1968 (paperback).
Wooldridge, D. E.: "The Machinery of the Brain," McGraw-Hill, New York, 1963 (paperback).

Chapter 12 The Endocrine System

** Catt, K. J., and M. L. Dufau: Peptide Hormone Receptors, *Ann. Rev. Physiol.,* **39:**529 (1977).
* Davidson, J. M., and S. Levine: Endocrine Regulation of Behavior, *Ann. Rev. Physiol.,* **35:**375 (1972).
Frohman, L. A.: Neurotransmitters as Regulators of Endocrine Function, *Hosp. Pract.,* **10:**54 (April 1975).
Ganong, W. F., L. C. Alpert, and T. C. Lee: ACTH and the Regulation of Adrenocortical Secretion, *N. Engl. J. Med.,* **290:**1006 (1974).
Gillie, R. B.: Endemic Goiter, *Sci. Am.,* June 1971.
Goldberg, N. D.: Cyclic Nucleotides and Cell Function, *Hosp. Pract.,* **9:**127 (May 1974).
* Lefkowitz, R. J.: Isolated Hormone Receptors, *N. Engl. J. Med.,* **288:**1061 (1973).
McEwen, B. S.: Interactions between Hormones and Nerve Tissue, *Sci. Am.,* **235:**48 (July 1976).
McEwen, B. S.: The Brain as a Target Organ of Endocrine Hormones, *Hosp. Pract.,* **10:**95 (May 1975).
Pastan, I.: Cyclic AMP, *Sci. Am.,* August 1972.
Pike, J. E.: Prostaglandins, *Sci. Am.,* November 1971.
Rasmussen, H.: Ions as "Second Messengers," *Hosp. Pract.,* **9:**99 (June 1974).
** Reichlin, S., et al.: Hypothalamic Hormones, *Ann. Rev. Physiol.,* **38:**389 (1976).
** Segal, H. L.: Enzymatic Interconversion of Active and Inactive Forms of Enzumes, *Science,* **180:**25 (1973).
Stent, G. S.: Cellular Communication, *Sci. Am.,* September 1972.
* Sutherland, E. W.: Studies on the Mechanism of Hormone Action, *Science,* **177:**401 (1972).
* Tepperman, J.: "Metabolic and Endocrine Physiology," 3d ed., Year Book, Chicago, 1973.
* Turner, C. D.: "General Endocrinology," 6th ed., Saunders, Philadelphia, 1976.
** Vale, W., et al.: Regulatory Peptides Hypothalamus, *Ann. Rev. Physiol.,* **39:**473 (1977).
Zuckerman, S.: Hormones, *Sci. Am.,* March 1957.

Chapter 13 The Circulatory System: I. Blood

* Davie, E. W., and O. D. Ratnoff: Waterfall Sequence for Intrinsic Blood Clotting, *Science,* **145:**1310 (1964).
* Weiss, H. J.: Platelet Physiology and Abnormalities of Platelet Function, *N. Engl. J. Med.,* **293:**531, 580 (1975).
Zucker, M. B.: Blood Platelets, *Sci. Am.,* February 1961.

Chapter 14 The Circulatory System: II. The Heart

Adolph, E. F.: The Heart's Pacemaker, *Sci. Am.,* March 1967.
* Berne, R. M., and M. N. Levy: "Cardiovascular Physiology," 3d ed., Mosby, St. Louis, 1977.
* Braunwald, E.: "The Myocardium: Failure and Infarction," HP Publishing, New York, 1974.
* Cranefield, P. F.: Ventricular Fibrillation, *N. Engl. J. Med.,* **289:**732 (1973).
** Fozzard, H. A.: Heart: Excitation-Contraction Coupling, *Ann. Rev. Physiol.,* **39:**201 (1977).
* Katz, A. M.: Congestive Heart Failure, *N. Engl. J. Med.,* **293:**1184 (1975).
Scher, A. M.: The Electrocardiogram, *Sci. Am.,* November 1961.
Wiggers, C. J.: The Heart, *Sci. Am.,* May 1957.

Chapter 15 The Circulatory System: III. The Vascular System

Baez, S.: Microcirculation. *Ann. Rev. Physiol.,* **39:**391 (1977).
* Berne, R. M., and M. N. Levy: "Cardiovascular Physiology," 3d ed., Mosby, St. Louis, 1977.
Carrier, O., Jr.: The Local Control of Blood Flow: An Illustration of Homeostasis, *Bioscience,* **15:**665 (1965).
* Folkow, B., and E. Neil: "Circulation," Oxford, New York, 1971.
Mayerson, H.: The Lymphatic System, *Sci. Am.,* June 1963.
* Rushmer, R. F.: "Structure and Function of the Cardiovascular System," 2d ed., Saunders, Philadelphia, 1976.
Spain, D. M.: Atherosclerosis, *Sci. Am.,* August 1966.
Wood, J. E.: The Venous System, *Sci. Am.,* January 1968.
Zweifach, B. J.: The Microcirculation of the Blood, *Sci. Am.,* January 1959.

Chapter 16 The Circulatory System: IV. Integration of Cardiovascular Function in Health and Disease

Benditt, E. P.: The Origin of Atherosclerosis, *Sci. Am.,* **236:**74 (February 1977).
* Berne, R. M., and M. N. Levy: "Cardiovascular Physiology," 3d ed., Mosby, St. Louis, 1977.
** Bevegard, B. S., and J. T. Shepherd: Regulation of the

Circulation During Exercise in Man, *Physiol. Rev.,* **47**:178 (1967).

Braunwald, E.: Regulation of the Circulation, *N. Engl. J. Med.,* **290**:1124, 1420 (1974).

Chapman, C. B., and J. H. Mitchell: The Physiology of Exercise, *Sci. Am.,* May 1965.

** Chien, S.: Role of the Sympathetic Nervous System in Hemorrhage, *Physiol. Rev.,* **47**:214 (1967).

* Folkow, B., and E. Neil: "Circulation," Oxford, New York, 1971.

Öberg, B.: Overall Cardiovascular Regulation, *Ann. Rev. Physiol.,* **38**:537 (1976).

* Rushmer, R. F.: "Structure and Function of the Cardiovascular System," 2d ed., Saunders, Philadelphia, 1976.

Scheuer, J., and C. M. Tipton: Cardiovascular Adaptations to Physical Training, *Ann. Rev. Physiol.* **39**:221 (1977).

Vatner, S. F., E. Braunwald: Cardiovascular Control Mechanisms in the Conscious State, *N. Engl. J. Med.,* **293**:970 (1975).

Chapter 17 The Respiratory System

Avery, M. E., N. Wang, and H. W. Taeusch, Jr.: The Lung of the Newborn Infant, *Sci. Am.,* April 1973.

Baker, P. T.: Human Adaptation to High Altitude, *Science,* **163**:1149 (1969).

* Brewer, G. J., and T. W. Eaton: Erythrocyte Metabolism: Interaction with Oxygen Transport, *Science,* **26**:1205 (1971).

Clements, J. A.: Surface Tension in the Lungs, *Sci. Am.,* December 1962.

Cohen, A. B., and W. M. Gold: Defense Mechanisms of the Lungs, *Ann. Rev. Physiol.,* **37**:325 (1975).

Comroe, J. H., Jr.: The Lung, *Sci. Am.,* February 1966.

Crandall, E. D.: Pulmonary Gas Exchange, *Ann. Rev. Physiol.,* **38**:69 (1976).

* Davenport, H. W.: "The ABC of Acid-Base Chemistry," 6th ed., University of Chicago Press, Chicago, 1973.

Fenn, W. O.: The Mechanism of Breathing, *Sci. Am.,* January 1960.

Finch, C. A., and C. Lenfant: Oxygen Transport in Man, *N. Engl. J. Med.,* **286**:407 (1972).

Guz, A.: Regulation of Respiration in Man, *Ann. Rev. Physiol.,* **37**:303 (1975).

* Lenfant, C., and K. Sullivan: Adaptation to High Altitude, *N. Engl. J. Med.,* **284**:1298 (1971).

* Mitchell, T. H., and G. Blomqvist: Maximal Oxygen Uptake, *N. Engl. J. Med.,* **284**:1018 (1971).

** Morgan, T. E.: Pulmonary Surfactant, *N. Engl. J. Med.,* **284**:1185 (1971).

Newhouse, M., et al.: Lung Defense Mechanisms, *N. Engl. J. Med.,* **295**:990, 1045 (1976).

Perutz, M. F.: The Hemoglobin Molecule, *Sci. Am.,* November 1964.

Ponder, E.: The Red Blood Cell, *Sci. Am.,* January 1957.

Smith, C. A.: The First Breath, *Sci. Am.,* October 1963.

Wyman, R. J.: Neural Generation of the Breathing Rhythm, *Ann. Rev. Physiol.,* **39**:417 (1977).

Chapter 18 The Urinary System

** Anderson, B.: Regulation of Body Fluids, *Ann. Rev. Physiol.,* **39**:185 (1977).

* Davenport, H. W.: "The ABC of Acid-Base Chemistry," 6th ed., University of Chicago Press, Chicago, 1973.

* Elkinton, J. R., and T. S. Danowsky: "The Body Fluids: Basic Physiology and Practical Therapeutics," Williams & Wilkins, Baltimore, 1955.

** Gauer, O. H., and J. P. Henry: Circulatory Basis of Fluid Volume Control, *Physiol. Rev.,* **43**:423 (1963).

* Hays, R. M.: Antidiuretic Hormone, *N. Engl. J. Med.,* **295**:659 (1976).

* Jamison, R. L., and R. H. Maffly: The Urinary Concentrating Mechanism, *N. Engl. J. Med.,* **295**:1059 (1976).

** Kuru, M.: Nervous Control of Micturition, *Physiol. Rev.,* **45**:425 (1965).

** Lassiter, W. E.: Kidney, *Ann. Rev. Physiol.,* **37**:371 (1975).

Merrill, J. P.: The Artificial Kidney, *Sci. Am.,* July 1961.

Merrill, J. P., and C. L. Hampers: Uremia, *N. Engl. J. Med.,* **282**:953 (1970).

** Peart, W. S.: Renin-Angiotensin System, *N. Engl. J. Med.,* **292**:302 (1975).

* Ramsay, D. J., and W. F. Ganong: CNS Regulation of Salt and Water Intake, *Hosp. Pract.,* **12** (March 1977).

Rasmussen, H.: The Parathyroid Hormone, *Sci. Am.,* April 1961.

Rasmussen, H., and M. M. Pechet: Calcitonin, *Sci. Am.,* October 1970.

** Schrier, R. W., and H. E. DeWardener: Tubular Reabsorption of Sodium Ion, *N. Engl. J. Med.,* **285**:1731 (1971).

** Schwartz, I. L., and W. B. Schwartz (eds.): Symposium on on Antidiuretic Hormones, *Am. J. Med.,* May 1967.

Smith, H. W.: "From Fish to Philosopher," Anchor Books, Doubleday, Garden City, N.Y., 1961 (paperback).

Smith, H. W.: The Kidney, *Sci. Am.,* January 1953.

Solomon, A. K.: Pumps in the Living Cell, *Sci. Am.,* August 1962.

* Valtin, H.: "Renal Function," Little, Brown, Boston, 1973.

Vander, A. J.: "Renal Physiology," McGraw-Hill, New York, 1975.

Chapter 19 The Gastrointestinal System

** Bortoff, Alexander: Myogenic Control of Intestinal Motility, *Physiol. Rev.,* **56**:418 (1976).

** Brooks, Frank P. (ed.): "Gastrointestinal Pathophysiology," Oxford, New York, 1974 (paperback).

Davenport, Horace W.: Why the Stomach Does Not Digest Itself, *Sci. Am.,* January 1972.

* Davenport, Horace W.: "A Digest of Digestion," Year Book, Chicago, 1975.

* Davenport, Horace W.: "Physiology of the Digestive Tract," 4th ed., Year Book, Chicago, 1977.

** Javitt, Norman B.: Hepatic Bile Formation, *N. Engl. J. Med.,* **295:**1464, 1511 (1976).

* Javitt, Norman B., and Charles K. McSherry: Pathogenesis of Cholesterol Gallstones, *Hosp. Pract.,* July 1973.

Kappas, Attallah, and Alvito P. Alvares: How the Liver Metabolizes Foreign Substances, *Sci. Am.,* June 1975.

Kretchmer, Norman: Lactose and Lactase, *Sci. Am.,* October 1972.

Neurath, H.: Protein Digesting Enzymes, *Sci. Am.,* December 1974.

* Phillips, Sidney F.: Fluid and Electrolyte Fluxes in the Gut, *Hosp. Pract.,* March 1973.

** Rayford, Phillip L., Thomas A. Miller, and James C. Thompson: Secretin, Cholecystokinin and Newer Gastrointestinal Hormones, *N. Engl. J. Med.,* **294:**1093, 1157 (1976).

* Stahlgren, Leroy H.: The Dumping Syndrome: A Study of Its Hemodynamics, *Hosp. Pract.,* December 1970.

** Symposium: Gastrointestinal Hormones: Physiological Implications, *Fed. Proc.,* **36:**1929 (1977).

** Wilson, T. H.: "Intestinal Absorption," Saunders, Philadelphia, 1962.

** Wood, J. D.: Neurophysiology of Auerbach's Plexus and Control of Intestinal Motility, *Physiol. Rev.,* **55:**307 (1975).

Chapter 20 Regulation of Organic Metabolism and Energy Balance

* Adolph, E. F.: "Physiology of Man in the Desert," Interscience, New York, 1947.

Benzinger, T. H.: The Human Thermostat, *Sci. Am.,* January 1961.

** Cabanac, M.: Temperature Regulation, *Ann. Rev. Physiol.,* **37:**415 (1975).

Cannon, W. B.: "The Wisdom of the Body," Norton, New York, 1939 (paperback).

** Daughaday, W. H. et al.: The Regulation of Growth by Endocrines, *Ann. Rev. Physiol.,* **37:**211 (1977).

Dole, V. P.: Body Fat, *Sci. Am.,* December 1959.

* Edelman, I. S.: Thyroid Thermogenesis, *N. Engl. J. Med.,* **290:**1303 (1974).

Eichenwald, H. F., and P. C. Fry: Nutrition and Learning, *Science,* **163:**644 (1969).

* Felig, P., and J. Wahren: Fuel Homeostasis in Exercise, *N. Engl. J. Med.,* **293:**1078 (1975).

** Gerich, J. E., et al.: Regulation of Pancreatic Insulin and and Glucagon, *Ann. Rev. Physiol.,* **38:**353 (1976).

Gray, G. W.: Human Growth, *Sci. Am.,* October 1953.

Irving, L.: Adaptations to Cold, *Sci. Am.,* January 1966.

Mayer, J.: "Overweight: Causes, Cost, and Control," Prentice-Hall, Englewood Cliffs, N.J., 1968.

Mayer, J., and D. W. Thomas: Regulation of Food Intake and Obesity, *Science,* **156:**327 (1967).

* Randle, P. J.: The Interrelationships of Hormones, Fatty Acid and Glucose in the Provision of Energy, *Postgrad. Med. J.,* **40:**457 (1964).

* Tepperman, J.: "Metabolic and Endocrine Physiology," 3d ed., Year Book, Chicago, 1973.

* Unger, R. H.: Glucagon: Physiology and Pathophysiology, *N. Engl. J. Med.,* **285:**443 (1971).

** Unger, R. H., and L. Orci: Physiology and Pathophysiology of Glucagon, *Physiol. Rev.,* **56:**778 (1976).

* Van Wyk, J. J., et al.: The Somatomedins, *Am. J. Dis. Child,* **126** (November 1973).

Wilkins, L.: The Thyroid Gland, *Sci. Am.,* March 1960.

Young, V. C., and N. S. Scrimshaw: The Physiology of Starvation, *Sci. Am.,* October 1971.

Chapter 21 The Reproductive System

Allen, R. D.: The Moment of Fertilization, *Sci. Am.,* July 1959.

* Bremner, W. J., and D. M. deKretser: The Prospects for New, Reversible Male Contraceptives, *N. Engl. J. Med.,* **295:**1111 (1976).

** Brenner, R. M., and N. B. West: Hormonal Regulation of the Reproductive Tract in Female Mammals, *Ann. Rev. Physiol.,* **37:**273 (1975).

* Chan, L., and B. W. O'Malley: Mechanism of Action of the Sex Steroid Hormones, *N. Engl. J. Med.,* **294:**1322 (1976).

Csapo, A.: Progesterone, *Sci. Am.,* April 1958.

Davidson, J. M.: Hormones and Sexual Behavior in the Male, *Hosp. Pract.,* **10:**126 (September 1975).

Edwards, R. G., and R. E. Fowler: Human Eggs in the Laboratory, *Sci. Am.,* December 1970.

Epel, D.: The Program of Fertilization, *Sci. Am.,* **237:**129 (November 1977).

Klopfer, P. H.: Mother Love: What Turns It On?, *Am. Sci.,* **59:**404 (1971).

* Lloyd, C. W.: "Human Reproduction and Sexual Behavior," Lea & Febiger, Philadelphia, 1964.

* Macleod, J.: The Parameters of Male Fertility, *Hosp. Pract.,* December 1973.

Masters, W. H., and V. E. Johnson: "Human Sexual Response," Little, Brown, Boston, 1966.

* McCann, S. M.: Luteinizing-hormone-releasing Hormone, *N. Engl. J. Med.,* **296:**797 (1977).

* Michael, R. P.: Hormones and Sexual Behavior in the Female, *Hosp. Pract.,* December 1975.

Mittwoch, U.: Sex Differences in Cells, *Sci. Am.,* July 1963.

Money, J., and A. E. Ehrhardt: "Man and Woman, Boy and Girl. The Differentiation and Dimorphism of Gender Identity from Conception to Maturity," Johns Hopkins, Baltimore, 1973 (paperback).

* Odell, W. D., and D. L. Moyer: "Physiology of Reproduction," Mosby, St. Louis, 1971.

* Patten, B. M., and B. M. Carlson: "Foundations of Embryology," 3d ed., McGraw-Hill, New York, 1974.
Segal, S. J.: The Physiology of Human Reproduction, *Sci. Am.,* **231**:53 (September, 1974).
* Sherwood, L. M.: Human Prolactin, *N. Engl. J. Med.,* **284**:774 (1971).
* Tepperman, J.: "Metabolic and Endocrine Physiology," 3d ed., Year Book, Chicago, 1973.

Chapter 22 Defense Systems: Immunology and Stress

Abramoff, P., and M. La Via: "Biology of the Immune Response," McGraw-Hill, New York, 1970.
Beer, A. E., and R. E. Billingham: The Embryo as a Transplant, *Sci. Am.,* April 1974.
Burke, D. C.: The Status of Interferon, *Sci. Am.,* **236**:42 (April 1977).
Capra, J. D., and A. B. Edmundson: The Antibody Combining Site, *Sci. Am.,* **236**:50 (January 1977).
Clowes, R. C.: The Molecule of Infectious Drug Resistance, *Sci. Am.,* April 1973.
Constandinides, P. E., and N. Carey: The Alarm Reaction, *Sci. Am.,* March 1949.
Cummingham, B. A.: The Structure and Function of Histocompatibility Antigens, *Sci. Am.,* **237**:96 (October 1977).
* Dannenberg, A. M., Jr.: Macrophages in Inflammation and Infection, *N. Engl. J. Med.,* **293**:489 (1975).
Dubos, R.: "Man Adapting," Yale, New Haven, Conn., 1965 (paperback).
Good, R. A., and D. W. Fisher: "Immunobiology," Sinauer, Stamford, 1973.
Holland, J. J.: Slow, Inapparent and Recurrent Viruses, *Sci. Am.,* February 1974.
Isaacs, A.: Interferon, *Sci. Am.,* May 1961.
Laki, K.: The Clotting of Fibrinogen, *Sci. Am.,* March 1962.
Lerner, R. A., and F. J. Dixon, The Human Lymphocyte as Experimental Animal, *Sci. Am.,* June 1973.
** Leung, K., A. Munck.: Peripheral Actions of Glucocorticoids, *Ann. Rev. Physiol.,* **37**:245 (1975).
Levey, R. H.: The Thymus Hormone, *Sci. Am.,* July 1964.
Levine, S.: Stress and Behavior, *Sci. Am.,* January 1971.
Mason, J. W.: Organization of Psychoendocrine Mechanisms, *Psychosom. Med.,* **30** (II):(1968).
Mayer, M. M.: The Complement System, *Sci. Am.,* November 1973.
Merigan, T. C.: Host Defenses Against Viral Disease, *N. Engl. J. Med.,* **290**:323 (1974).
* Müller-Eberhard, Hans J.: Chemistry and Function of the Complement System, *Hosp. Pract.,* **12**:33 (August 1977).
Nossal, G. J. V.: How Cells Make Antibodies, *Sci. Am.,* December 1964.
* Notkins, A. L.: Viral Infections: Mechanisms of Immunologic Defense and Injury, *Hosp. Pract.,* September 1974.
Notkins, A. L., and H. Koprowski: How the Immune Response to a Virus Can Cause Disease, *Sci. Am.,* January 1973.
* Parker, C. W.: Control of Lymphocyte Function, *N. Engl. J. Med.,* **295**:1180 (1976).
Porter, R. R.: The Structure of Antibodies, *Sci. Am.,* October 1967.
Raff, M. C.: Cell-Surface Immunology, *Sci. Am.,* May 1976.
Ratnoff, O. D.: The Interrelationship of Clotting and Immunologic Mechanisms, *Hosp. Pract.,* April 1971.
Rensfeld, R. A., and B. D. Kahan: Markers of Biological Individuality, *Sci. Am.,* June 1972.
Ross, R.: Wound Healing, *Sci. Am.,* June 1969.
Terne, N. K.: The Immune System, *Sci. Am.,* January 1973.
Walker, W. A., and K. J. Isselbacher: Intestinal Antibodies, *N. Engl. J. Med.,* **297**:767 (1977).
Weiss J. M.: Psychological Factors in Stress and Disease, *Sci. Am.,* June 1972.

Glossary

prepared by
John P. Harley
Eastern Kentucky University

The task of learning—and remembering—many of the terms in anatomy and physiology can be made easier if you know some of the key parts of words that are found in these terms. Some terms consist entirely of such parts (e.g., apnea, bronchitis, cardiovascular). The prefixes, suffixes, and combining forms listed below are most commonly used in anatomy and physiology. The effort required to learn them will be repaid many times over in the ease with which one can learn and remember new vocabulary.

PREFIXES

a-, ab-	away from, outside of, deviating from
a-, an-	without, lacking, not
ad-	to, toward
ante-	before
anti-	against
auto-	self
bi-	two
bi-, bio-	life (biology)
brady-	slow
chrom-, chromo-	color, pigment
circum-	around
contra-	opposed, against
de-	remove, decrease
di-	two
dia-	through
dys-	bad, difficult
ect-, ecto-	on, without, on outside
end-, endo-	in, within
ep-, epi-	upon, above, among
erythr-, erythro-	red
eu-	good, well (euphoria)
ex-	out, away from
hemi-	half
hetero-	varied, unlike, different
hist-, histo-	tissue
homoeo-, homoio-, homeo-	similar, like
hyper-	above, beyond, excess
hyp, hypo-	below, under, deficient
in-	into, not
infra-	below, under, within
inter-	between, among, mutual
intra-	within, inside
ipse-	same
is-, iso-	equal, like
juxta-	next to
leuc-, leuco-, leuk-, leuko-	white
macr-, macro-	large, long
mal-	bad
meg-, mega- megal-, megalo-	large, great enlarged
mes-, meso-	medium, middle
micr-, micro-	small, minute, undersized
mon-, mono-	one, single
morph-, morpho-	shape, form
necr-, necro-	dead

noci-	injurious, pain
olig-, oligo	scant, sparse, deficiency
par-, para-	beside, near, closely resembling, beyond
peri-	around
phag-, phago-	eat
poly-	many
post-	behind, after
pre-	before, in front of
pro-	before, giving rise to
retro-	backward, behind
scler-, sclero-	hard
semi-	half
sub-	below
super-, supra-	above, excess
syn, sym-	with, binding together
tachy-	swift, accelerated
trans-	across
ultra-	beyond, excessive
uni-	one

SUFFIXES

-algia	pain
-ase	enzyme
-ectomy	cut out, surgically remove
-emia, -aemia,	blood
-gen, -gene	producing
-graph	writing
-itis	inflammation
-lysis	dissolving, destruction, separation
-meter	measure
-oid	like, similar to
-ole	small
-opia, -opy	vision
-osis	a condition, a process
-plegia -plexy	paralysis, stroke
-pnea, pnoea	respiration
-rrhage	to burst forth, excessive discharge
-rrhea, rrhoea	flow
-soma, -some	body
-stomy	make an opening
-tome, -tomy	cutting
-trope	to turn
-trophe	nourishment
-ule, -ulus	small
-uria	urine

COMBINING FORMS

aden-, adeno-	gland
amyl-, amylo-	starch
angi-, angio-	vessel
ano-	anus
arthr-, arthro-	joint
bili-	bile
brachi-, brachio-	arm
branchi-, bronchio-	bronchus, trachea
cardi-, cardia-, cardio-	heart
cephal-, cephalo-	head
cerebr-, cerebri-, cerebro-	brain
chol-, chole-, cholo-	bile
chondr-, chondri-, chondro-	cartilage
cor-, core-, coro-	eye
corpus	body
cost-, costi-, costo-	rib
cyt-, cyto-	cell
derm-, derma-, dermo-	skin
enter-, entero-	intestine, intestinal
gastr-, gastro-	stomach
gloss-, glosso-	tongue
gluc-, gluco- glyc-, glyco-	glucose, sugar
hem-, hema-, hemo-, haem-, haema-, haemo-	blood
hepat-, hepato-	liver
hydr-, hydro-	water
hyster-, hystero-	uterus
ili-, ilio-	ilium
my-, myo-	muscle
myel-, myelo-	marrow
nephr-, nephros-	kidney
neur-, neuro-	nerve
ophthalm-, ophthalmo-	eye
ost-, oste-, osteo-	bone
ot-, oto-	ear
ovi-, ovo-	egg
path-, patho-	disease
phleb-, phlebo-	vein
pneum-, pneumo-	air, lungs
psych-, psycho-	mind
pulmo-	lung
ren-, reni-, reno-	kidney
sacr-, sarco-	flesh
therm-, thermo-	heat
thromb-, thrombo-	clot
ur-, uro	urine; tail
vas-, vasi-, vaso-	vessel

a-, an-. Prefixes indicating without, away from, or not.

A-band. The middle band of a muscle fiber sarcomere composed of overlapping thick and thin filaments.

abdomen (ab-do′mun). The portion of the body that lies between the diaphragm and pelvis.

abducens (ab-dew′sunz). Sixth cranial nerve which innervates the lateral rectus muscle of the eyeball.

abduct (ab-dukt′). To draw away from the midline.

abductor (ab-duck′tur). A muscle performing the function of abduction.

abortion (uh-bor′shun). Premature expulsion of the embryo or fetus from the uterus.

abscess (ab′sess). Localized collection of pus.

absorption (ub-sorp′shun). Passage of materials across an epithelial layer of cells from a body cavity or compartment toward the blood.

acclimatization (a-klye′muh-ti-zay′shun). The physiologic adjustment of the body to changing environmental conditions.

accommodation (a-kom′uh-da′shun). Adjustment of the eye by changing the shape of the lens for viewing various distances.

acetabulum (as′e-tab′yoo-lum). The large cup-shaped cavity on the innominate bone (os coxae) in which the femur articulates.

acetate (as′e-tate). The ionized form of acetic acid (CH_3COOH).

acetylcholine (as′e-til-ko′leen). A chemical transmitter agent released from many peripheral nerve endings and from some neurons in the central nervous system, e.g., from postganglionic parasympathetic fibers and the neuromuscular junction.

acetyl coenzyme A (a-see′til ko-en′zime ay). A two-carbon fragment attached to coenzyme A; a metabolic intermediate that supplies carbon atoms to the Krebs cycle and to various synthetic pathways such as those for fatty acid synthesis.

achalasia (ack′uh-lay′zhuh). Failure of the gastroesophageal sphincter to relax during swallowing.

Achilles tendon (ah-kil′eez ten′dun). The tendon connecting the muscles of the calf of the leg (gastrocnemius and soleus) to the calcaneus bone of the heel.

acid (as′id). A substance that releases hydrogen ions (H^+) to a solution; the higher the hydrogen ion concentration of a solution, the greater its acidity.

acidosis (as′i-do′sis). An abnormal condition characterized by an increased acidity of the body's extracellular fluid.

acinus (as′i-nus). A small saclike dilatation as in glands.

acne (ack′nee). An inflammatory disease of the sebaceous glands.

acoustic (uh-koos′-tick). Pertaining either to the sense of hearing or to sound.

acromegaly (ack′ro-meg′uh-lee). A disease caused by excessive secretion of growth hormone from the anterior pituitary; characterized by an enlargement of skeletal extremities.

acromion (a-kro′mee-on). The process of the scapula that forms the highest point of the shoulder.

acrosome (ack′ro-sohm). A vesicle filled with digestive enzymes located at the tip of the sperm cell.

ACTH. Abbreviation for adrenocorticotropic hormone, which is released from the anterior pituitary and stimulates the adrenal cortex.

actin (ack′tin). A muscle contractile protein located in the thin myofilaments.

active immunity (ack′tiv i-mew′ni-tee). Immunity resulting from the stimulation of antibody formation in an organism.

action potential (ack′shun po-ten′chal). The rapid change in the membrane potential of excitable nerve and muscle cells associated with the conduction of impulses along the cell membrane.

active transport (ack′tiv trans′port). The movement of materials across a cell membrane from low to high concentration in combination with membrane-bound carrier molecules and requiring a source of energy.

acuity (a-kew′i-tee). Acuteness or clearness of vision.

Adam's apple. The prominence in the front of the neck formed by the thyroid cartilage of the larynx.

Addison's disease (ad′i-sunz di-zeez′). A disease resulting from a deficiency in the secretion of hormones from the adrenal cortex.

adduct (a-dukt′). To move toward the midline of the body.

adduction (a-duck′shun). Movement toward the midline of the body.

adductor (a-duck′-tur). A muscle that adducts.

adenine (ad′i-neen). One of the four bases found in the nucleotides which constitute the subunits of DNA and RNA.

adenohypophysis (ad′e-no-high-pof′-i-sis). The anterior portion of the pituitary gland (hypophysis) which secretes the hormones FHS, LH, ACTH, TSH, GH, and prolactin.

adenoid (ad′e-noyd). Resembling a gland; hypertrophy of the pharyngeal tonsils.

adenosine triphosphate (ah-den′-o-seen trye-fos′fate). A nucleotide containing adenine, ribose, and three phosphate groups; used by all living cells to transfer chemical energy from one reaction to another.

adenyl cyclase (ad′e-nil sigh′klace). An enzyme which converts ATP to 3′,5′-cyclic 3′,5′-AMP.

adhesion (ad-hee′zhun). The abnormal attachment of adjacent surfaces by fibrous connective tissue.

ADH. Abbreviation for antidiuretic hormone, which is released from the posterior pituitary and acts on the kidney tubules.

adipose tissue (ad′i-poce tish′oo). Fatty in nature; composed of specialized fat-storing cells.

adolescence (ad′uh-les′unce). The period of transition from childhood to adulthood.

adrenal glands (a-dree′nul). The suprarenal glands located just above each kidney.

adrenalin (a-dren′uh-lin). The major hormone released by the adrenal medulla; also known as epinephrine.

adrenergic (ad′re-nur′jick). A term applied to those nerve fibers which liberate norepinephrine from their terminals; i.e., most postganglionic sympathetic fibers.

adrenocorticotropic hormone (a-dree′no-kor′ti-ko-tro′pick hor′mone). A hormone secreted by the anterior pituitary which stimulates some of the cells in the adrenal cortex.

adventitia (ad′ven-tish′ee-uh). The outermost connective tissue covering of a structure.

aerobic (ay′ur-o′bick). Requiring air or free oxygen.

afferent (af′ur-unt). Conveying to or toward a central point. When applied to the nervous system, it refers to those nerve fibers that conduct action potentials from receptors in the periphery to the central nervous system.

afterbirth (af′tur-burth). The placenta and membranes which are extruded from the placenta after birth.

agglutination (a-gloo′ti-na′shun). The clumping together of cells distributed in a fluid; generally a result of antigen-antibody reactions.

alba (al′buh). White substance.

albumin (al-bew′min). A class of water soluble proteins found in almost all animal serum; albumin makes up the largest fraction of the proteins in blood plasma.

aldehyde (al′de-hide). A chemical group found in organic molecules; contains the —CHO group.

aldosterone (al′do-ste-rone′). The principal mineralocorticoid hormone; secreted by the adrenal cortex; regulates electrolytic balance.

alkaline (al′kuh-lyne). Having a lower hydrogen-ion concentration (acidity) than pure water.

alkalosis (al-kuh-lo′sis). An abnormal condition characterized by a decrease in acidity of the body's extracellular fluid below normal levels.

alveolus (al-vee′o-lus). A small cavity; usually refers to the smallest air cavities in the lungs or to clusters of cells in the mammary glands. Plural, alveoli; adjective, alveolar.

all-or-none principle. A single stimulus applied to an excitable tissue (nerve or muscle) which produces either a complete response (all) or no response (none); there are no intermediate levels of response.

allantois (a-lan′to-is). A tubular sac in the posterior part of the hind gut of the embryo; in lower forms it serves as a reservoir for excretory products; in humans it is rudimentary and plays no role in development.

amenorrhea (a-men′o-ree′-uh). The absence of menstruation.

amine (a-meen′). A number of organic compounds that contain nitrogen, usually in the form of amino groups, $—NH_2$.

amino acid (a-mee′no as′id). An organic compound containing both an amino ($—NH_2$) and a carboxyl (—COOH) group attached to the same carbon atom; twenty different amino acids form the structural subunits of proteins.

AMP. Abbreviation for adenosine monophosphate.

amphiarthrosis (am′fee-arh-thro′sis). A slightly movable joint.

ampulla (am-pul′luh). A flasklike dilatation of a tubular structure.

amnesia (am-nee′zhuh). The inability to recall past experiences.

amylase (am′i-lace). A digestive enzyme that breaks down starch into smaller polysaccharides.

anabolism (a-nab′uh-lizm). The synthetic process of metabolism whereby simple substances are assembled into more complex structures.

anaerobic (an′air-o-bick). Not requiring molecular oxygen.

analgesia (an′al-jee′-zee-uh). Insensitivity to pain without loss of consciousness.

anaphylaxis (an′uh-fi-lack′sis). An exaggerated immunological reaction of an organism to a foreign protein (antigen).

anastomosis (a-nas′tuh-mo′sis). A connection between two structures, e.g., between two blood vessels.

anatomy (uh-nat′uh-mee). The science of the structure or form of plants and animals.

androgen (an′dro-jin). Any chemical that produces effects in the body similar to those of testosterone, the male sex hormone produced by the testes.

anemia (uh-nee′mee-uh). A reduction below normal levels in the number of erythrocytes in the blood or in the amount of hemoglobin or both.

anesthesia (an′es-theezh′uh). The absence of sensation in some body area.

aneurysm (an′yoo-rism). An abnormal, blood-filled, saclike dilation of the wall of an artery.

angina (an′ji-nuh). Any pathological condition characterized by spasmodic choking or suffocative pain; especially *angina pectoris,* a chest pain associated with inadequate blood flow to the heart muscle.

angiotensin (an-jee-o-ten′sin). A hormone present in the blood formed by the reaction of renin and a plasma globulin; causes the release of aldosterone from the adrenal cortex, and vasoconstriction.

angstrom (ang′strum). The unit of length equal to 0.1 μm or $^1/_{100}$-millionth of a centimeter.

anion (an′eye-on). An ion or particle carrying a net negative charge.

annulus (an′yoo-lus). A ringlike or circular structure.

anorexia (an′o-reck′see-uh). The loss of appetite.

anoxia (an-ock′see-uh). The absence of molecular oxygen.

antagonistic muscles (an-tag-uh-nist′ik mus′uls). Two muscles which exert opposite forces on a bone and produce opposite movements around a joint.

anterior (an-teer′ee-ur). The front of the erect human body or toward the front of the body.

anterior root (an-teer′ee-ur root). The bundle of axons of efferent (motor) fibers that emerge from the anterior (ventral) sides of the spinal cord.

antibody (an′ti-bod′ee). A protein that is produced by lymphoid tissue in response to a specific foreign protein; it is able to bind to the foreign protein.

anticodon (an′tee-ko′don). The trinucleotide sequence complementary to a codon in mRNA.

antidiuretic hormone (ADH) (an′tee-dye-yoo-ret′ick). A hormone released by the posterior pituitary; acts on the kidney tubules to increase the reabsorption of water and thus decrease the volume of urine secreted.

antigen (an′ti-jin). A protein or protein-polysaccharide complex which causes formation of antibodies against itself when introduced into the body.

antihistamine (an′tee-hiss′tuh-meen). A chemical that blocks the action of histamine.

antrum (an′trum). A cavity or chamber, e.g., the gastric antrum.

anus (a′nus). The distal opening of the rectum through which the end of the gastrointestinal tract opens to the exterior.

aorta (ay-or′tuh). The main systemic blood vessel arising from the left ventricle of the heart.

aortic valve (ay-or′tick valv). The valve located between the left ventricle of the heart and the aorta.

aperture (ap′ur-chur). An opening or orifice.

apex (ay′pecks). The top of an organ or part.

aphasia (ah-fay′zhuh). The impairment or loss of either verbal expression or comprehension or both; caused by lesion in cerebral cortex.

apnea (ap′nee-uh). A temporary cessation of breathing.

apocrine (ap′o-krin). A type of secretion in which the secretory products accumulate at the apical end of the cell; the entire free end of the cell is then pinched off leaving the remainder of the cell to repeat the process.

apoenzyme (ap′o-en′zime). The protein portion of enzymes that contains additional nonprotein components.

appendicular skeleton (ap′en-dick′yoo-lur skel′e-tun). That portion of the skeleton consisting of the limbs (appendages) and their girdles.

aponeurosis (ap′o-new-ro′sis). The flat sheet of fibrous connective tissue that serves as an attachment for muscles.

appendix (appen′-dicks). A supplementary part attached to a main structure; e.g., the small wormlike sac attached to the cecum of the large intestine.

aqueduct (ack′we-dukt). A passage or channel in a body structure.

aqueous humor (ay′kwee-us hew′mur). The watery fluid in the anterior part of the eye.

arachnoid (uh-rack′noyd). Resembling a spider web; the middle of the three meningeal membranes covering the surface of the brain and spinal cord.

arborvitae (ahr′bur-vye′tee). The white matter in the cerebellum.

areola (a-ree′o-luh). A small space; the ringlike pigmented area around the nipple.

arrhythmia (ah-rith′mee-uh). Any variation from the normal heartbeat rhythm.

arteriole (ahr-teer′ee-ole). A small arterial blood vessel just proximal to the capillaries containing a large proportion of smooth muscle relative to its size.

arteriosclerosis (ahr-teer′ee-o-skle-ro′-sis). The hardening of arterial blood vessel walls due to the accumulation of connective tissue, smooth muscle cells, and fatty deposits.

arthritis (ahr-thrigh′tis). The inflammation of a joint.

articular cartilage. (ahr-tick′yoo-lur kahr′ti-lidge). The cushioning layer of cartilage covering the ends of bones that form synovial joints.

articulation (ahr-tick′yoo-lay′shun). A joint; the union between two or more bones.

arytenoid cartilage (ar′i-tee′noyd kahr′ti-lidj). Two small cartilages of the larynx.

ascites (a-sigh′teez). The accumulation of fluid in the abdominal cavity.

association area (a-so′see-ay′shun a′ree-uh). A region of the cerebral cortex that integrates sensory, motor, and other information from other parts of the brain.

asphyxia (as-fick′see-uh). A decrease in the amount of oxygen in the blood accompanied by an increase in carbon dioxide, leading to suffocation.

asthma (az′muh). A disease characterized by recurrent attacks of dyspnea as a result of constriction of the bronchi.

astigmatism (a-stig′muh-tizm). A defective curvature of the cornea of the eye which causes an improper focusing of light on the retina.

astrocytes (as′tro-sights). Star-shaped neuroglial cells in the brain and spinal cord.

ataxia (a-tack′see-uh). The loss, weakness, or incoordination of postural muscle coordination.

atelectasis (at′e-leck′tuh-sis). A collapsed, airless portion of lung.

atherosclerosis (ath′ur-o′skle-ro′sis). The hardening and narrowing of vessels due to the buildup of lipid deposits in the arterial wall.

atlas (at′lus). The first cervical vertebra.

atom (at′um). The smallest particle of matter that has distinct chemical parts; there are 105 different types of atoms corresponding to the 105 elements.

atrium (ay′tree-um). A chamber or cavity, as the right and left atrium of the heart.

ATP. Abbreviation of adenosine triphosphate.

atrophy (at′ruh-fee). A wasting away or decrease in the size of a part due to poor nutrition or lack of use.

auditory (aw′di-tor′ee). Pertaining to the sense of hearing.

auricle (aw′ri-kul). A portion of the external ear; an ear-shaped structure or appendage; also a projection from the atria of the heart.

autoimmunity (aw′to-i-mew′ni-tee). The production by an organism of antibodies against its own proteins.

autonomic nervous system (aw′tuh-nom′-ick nur′vus sis′tum). That portion of the peripheral nervous system which innervates heart and smooth muscle and glands; comprises sympathetic and parasympathetic divisions.

axial (ack′see-ul). Pertaining to the axis or central portion of the body; opposite of appendicular.

axilla (ack-sil′uh). The armpit.

axon (acks′on). The cylindrical process extending from a nerve cell; conducts action potential away from the cell body toward the axon terminals.

azygos (az′i-gos). An unpaired process, muscle, vein or artery.

baroreceptor (bar′o-re-sep′tur). A neural receptor sensitive to a change in pressure; e.g., the carotid sinus and aortic arch baroreceptors.

Bartholin's glands (bahr′to-linz glands). Two mucous-secreting glands located on either side of the vaginal opening in the female.

basal ganglia (bay′sul gang′glee-uh). Several nuclei (clusters of cell bodies and gray matter) located in the cerebral hemispheres that code and relay information associated with the fine control of muscle movements.

basal metabolism (bay′sul me-tab′o-lism). The energy expenditure of the body under resting (not sleeping) conditions at a normal room temperature. Abbreviated *BMR* (basal metabolic rate).

base (bace). A substance that will bond hydrogen ions; the opposite of an acid.

basilar membrane (bas′i-lur mem′brane). The membrane that divides the cochlea of the inner ear into two passageways and supports the spiral organ of corti.

basophil (bay′suh-fil). A white blood cell (leukocyte) that stains with basic dyes; characterized by coarse granules in the cytoplasm. Also, a type of cell found in the anterior pituitary.

benign (be-nine′). Not malignant.

bicarbonate (bi-kahr′buh-nate). The anion resulting from the ionization of carbonic acid (HCO_3^{-1}).

biceps (bye′seps). A muscle that has two heads.

biconcave (bye′kon′kave). Having two concave surfaces.

bicuspid (bye-kus′pid). Having two cusps; e.g., the bicuspid valve or a premolar tooth.

bile (bile). The yellow-green fluid secreted by the liver into the small intestine.

bilirubin (bil′i-roo′bin). A product of hemoglobin breakdown which is excreted in the bile.

bipolar (bye-po′lur). Having two poles or processes.

blastocele (blas′to-seel). In embryology, the cavity of the blastula.

blastocyst (blas′to-sist). A stage in embryology following the morula.

blastoderm (blas′to-durm). The mass of cells forming the embryonic blastocyst.

BMR. Abbreviation for basal metabolic rate.

bolus (bo′lus). A small mass or clump of partially chewed or digested food.

Bowman's capsule (bo′manz cap′sool). A sac at the beginning of a nephron that encloses the glomerulus in the kidney.

Boyle's law. The chemical principle that, at a constant temperature, the volume of a gas is inversely proportional to its pressure.

brachial (bray′kee-ul). Pertaining to the arm.

brachial plexus (bray-kee-ul pleck′sus). The network of nerve fibers formed from the ventral branch of nerves C5-C8 and T1.

bradycardia (brad′ee-kahr′dee-uh). The slowing of the heart rate to 60 beats per minute or less.

bradykinin (brad′ee-kigh′nin). A vasoconstrictor composed of nine amino acids found in the blood plasma.

bronchi (bronk′ee); singular **bronchus.** The passageways that convey air to the lungs, located between the trachea and bronchioles.

bronchiole (bronk′ee-ole). A small airway connecting the bronchi to the alveolar ducts.

buccal (buck′ul). Referring to the cheek or mouth.

buffer (buf′ur). A solution of weak acids or bases and their salts that minimizes changes in hydrogen ion concentration.

bulbourethral glands (bul′bo-yoo-ree′-thrul glandz). A pair of glands, also known as Cowper's glands, located below the prostate gland. They secrete an alkaline fluid into the urethra.

bundle of His. A bundle of specialized cardiac muscle fibers that conduct action potentials from the AV node to the Purkinje fibers in the heart.

bursa (bur′sa). A fluid-filled cavity situated in tissues where friction would otherwise develop.

bursitis (bur-sigh′tis). Inflammation of the bursa.

buttocks (but′ucks). The prominence formed by the gluteal muscles.

calcaneus (kal-kay′nee-us). The heel bone of the foot.

calcification (kal′si-fi-kay′shun). The process of depositing calcium salts within tissues.

calcitonin (kal′si-to′nin). A peptide hormone secreted by the thyroid gland; regulates calcium homeostasis.

callus (kal′us). New bone tissue.

calorie (kal′o-ree). The amount of heat required to raise the temperature of 1 gram of water 1°C. (One large calorie, used to represent the calorie-equivalent of various nutrients, is equivalent to 1,000 calories.)

calyx (kay′licks). A cup-shaped division of the renal pelvis.

canaliculus (kan′uh-lick′yoo-lus). A small passage or canal.

cancellous (kan′se-lus). Spongy or trabecular bone tissue.

cancer (kan′sur). An uncontrolled malignant growth of cells in a multicellular organism; neoplasm.

canine (kay′nine). A fanglike or pointed tooth.

capacitation (ka-pas′i-ta′shun). The process whereby spermatozoa acquire the ability to fertilize ova.

capillary (kap′i-lar′e). The smallest blood vessels that connect the arterioles to the venules across whose walls materials move into and out of the interstitial fluid.

capsule. (kap′sool). A fibrous or membranous connective tissue envelope.

carbaminohemoglobin (kahr′buh-mee′-no-hee′muh-glo′bin). Hemoglobin combined with carbon dioxide.

carbohydrate (kahr′bo-high′drate). Organic molecules having the general formula $(CH_2O)_n$, where n is any whole number, e.g., glucose, $C_6H_{12}O_6$.

carbonic acid (kahr-bon′ick as′id). H_2CO_3; formed from H_2O and CO_2.

carbonic anhydrase (kahr-bon′ick an-high′drace). The enzyme that catalyzes the reversible reaction between CO_2 and H_2O to form carbonic acid (H_2CO_3).

carboxyhemoglobin (kahr-bock′see-hee′-mo-glo′bin). The relatively stable compound formed when carbon monoxide (CO) binds to hemoglobin.

carcinogen (kahr′si-no-jen). A chemical substance that produces cancerous growth.

cardiovascular (kahr′dee-o-vas′kew-lur). Referring to the heart and blood vessels.

caries (kair′eez). Tooth decay.

carotene (kar′oteen). A pigment found in vegetables that is converted to vitamin A in the body.

carotid arteries (ka-rot′id). The two major arteries in the neck, delivering blood to the head.

carpal (kahr′pul). Pertaining to the wrist.

cartilage (kahr′ti-lidj). A semisolid type of connective tissue found in the external ear, nose, trachea and joints in adults; forms the major skeletal tissue of the embryo prior to its replacement by bone tissue.

casein (ka′se-in). A phosphorus-containing protein found in milk.

castration (kas-tray′shun). Removal of gonads, i.e., testes or ovaries.

catabolism (ka′tab′uh-lizm). That portion of metabolism associated with the breakdown of organic molecules, in contrast to anabolism, which is associated with chemical synthesis.

catalyst (kat′uh-list). Any substance that increases the rate of a chemical reaction without itself undergoing any net chemical change.

cataract (kat′uh-rakt). An opacity of the lens of the eye or its capsule.

cathartic (ka-thahr′tick). An agent causing increased intestinal motility.

catecholamine (kat′e-kol-uh-meen). A term for the three neurotransmitter agents (norepinephrine, epinephrine, and dopamine) which are formed from the amino acid tyrosine and are structurally related to the substance catechol.

cation (kat′eye-on). A positively charged ion.

caudal (kaw′dul). Denoting a position toward the tail.

cecum (see′kum). The blind pouch at the beginning of the large intestine.

celiac (see′lee-ack). Pertaining to the abdomen.

cell (sel). The smallest functional unit of living matter.

cellulose (sel′yoo-loce). A polysaccharide found in plants.

centriole (sen′tree-ole). A small organelle composed of microtubules which forms the poles of the spindle during cell division.

centromere (sen′tro-meer). The site on a chromosome to which the spindle fibers are attached during cell division.

cephalic (se-fal′ick). Pertaining to the head.

cerebellum (serr′e-bel′um). The division of the brain posterior to the cerebrum and above the pons and fourth ventricle; primarily involved in coordinating skeletal muscle activity in posture and movement.

cerebrum (serr′e-brum). The two hemispheres forming the upper and larger portions of the brain, containing the regions associated with the higher cognitive functions of the mind.

cervix (sur′vicks). The neck of an organ, e.g., the cervix of the uterus.

CG. Abbreviation for chorionic gonadotropin.

chemotaxis (kem′o-tack′sis). The orientation and movement of cells in response to a chemical stimulus.

chiasm (kigh′azm). An x-shaped crossing as of the optic nerves.

chief cells. Cells in the lining of the stomach that secrete digestive enzymes.

cholecystokinin (kol′e-sis′to-kigh′nin). A peptide hormone secreted by the intestinal mucosa; regulates several gastrointestinal activities including motility and secretion of the stomach, gallbladder, and pancreas.

cholesterol (ko-les′tur-ol). A lipid belonging to the steroid class; present in plasma membranes; a precursor of steroid hormones.

cholinergic (ko′lin-ur′jick). A term applied to nerves that liberate acetylcholine at their synapses.

cholinesterase (ko′li-nes′tur-ace). An enzyme which splits acetylcholine into acetate and choline.

chondrocyte (kon′dro-sight). A cartilage-forming cell.

chorion (ko′ree-on). A membranous layer composed of trophoblast and mesoderm layers; forms the extra-embryonic membrane that makes contact with the uterus and forms the placenta.

choroid (kor′oyd). Skinlike; resembling the chorion.

choroid plexus (kor′oyd pleck′sus). Vascular structure lining portions of the brain ventricles; secretes cerebrospinal fluid.

chromatid (kro′muh-tid). The duplicated filament constituting half a chromosome. During cell division, each chromatid of the chromosome goes to a different pole of the dividing cell.

chromatin (kro′muh-tin). DNA plus protein; forms the chromosomes.

chromosome (kro′muh-sohm). One of the dark-staining bodies composed of condensed chromatin that appears in the nucleus during cell division. Human cells have 46 chromosomes.

chyle (kile). The fatty material taken up by the central lacteals in the villi of the duodenum after a fatty meal.

chylomicron (ki′lomi′kron). An emulsified fat particle that enters the blood during the process of fat absorption from the intestines.

chyme (kime). A semifluid, homogeneous material in the lumen of the gastrointestinal tract; produced by the digestion of food.

chymotrypsin (kigh′mo-trip′sin). An enzyme secreted by the pancreas into the intestine; breaks polypeptides into amino acids.

cilia (sil′ee-a). Hairlike projections found on epithelial cell surfaces, capable of whiplike beating motions, propelling the extracellular fluid past the cell surface.

circadian (surkay′dee-un). A cycle of approximately 24 hours duration.

cirrhosis (sir-ro′sis). A degeneration of the liver tissue characterized by progressive destruction of liver cells and an increase in connective tissue.

cisterna (sistur′nuh). A fluid-filled space or reservoir.

citric acid (sit′rick as′id). An organic intermediate found in the Krebs cycle.

climacteric (klye-mack′tur′ick). That period accompanying the normal diminution of the reproductive cycle in the female and sexual activity in the male.

clitoris (klit′o-ris). A small erectile body of the female external genitalia; homologous with the penis in the male.

cochlea (kock′lee-uh). The spirally coiled labyrinth of the inner ear; contains the organ of corti.

codon (ko′don). The sequence of three nucleotides in messenger RNA that specify one amino acid. The sequence of codons in RNA determines the sequence of amino acids in the synthesized protein.

coenzyme (ko-en′zime). A nonprotein substance that must be present in order for some enzymes to function; many of the vitamins function as coenzymes.

cofactor (ko′fack-tur). A general term similar to coenzyme that includes both organic and inorganic substances required for enzyme activity.

coitus (ko′i-tus). Sexual intercourse.

collagen (kol′uh-jin). The major fibrous protein found in connective tissue; tendons are composed primarily of collagen.

colloid (kol′oyd). A state of matter in which small multimolecular particles are suspended in a liquid medium.

colon (ko′lun). The large intestine.

colostrum (ko-los′trum). The thin milky fluid secreted by the mammary glands for a few days after birth.

commissure (kom′i-shur). A joining together.

concha (kong′kuh). Resembling a shell in shape.

condyle (kon′dile). A rounded projection on a bone.

cones (kones). The photoreceptor cells in the retina that are responsible for color vision.

congenital (kun-jen′i-tul). Existing at or before birth.

conjunctiva (kon′junk-tye′vuh). The membrane that covers the surface of the eyeball and eyelids.

constipation (kon′sti-pay′shun). A condition associated with a decreased frequency of defecation.

coracoid (kor′uh-koyd). Shaped like a raven's beak; applied to the processes of certain bones, e.g., the coracoid process of the scapula.

cornea (kor′nee′uh). The transparent anterior part of the fibrous tunic of the eye.

corpus callosum (kor′pus ka-lo′sum). A large band of nerve fibers connecting the two cerebral hemispheres below the longitudinal fissure.

corpus luteum (kor′pus lew′tee-um). The glandular mass in the ovary formed by the remnant of the graffian follicle following ovulation.

corpus striatum (kor′pus strye-ay′tum). An area within each cerebral hemisphere composed of the caudate and lentiform nuclei (basal ganglia) and arranged in a striated manner.

cortex (kor′tecks). The outermost layer of an organ.

cortisol (kor′ti-sol). A steroid hormone secreted by the cortex of the adrenal gland; regulates various aspects of organic metabolism, especially carbohydrate metabolism.

cranium (kray′nee-um). The bones of the skull that enclose and protect the brain.

creatine (kree′uh-teen). An organic compound that combines with phosphate to form creatinphosphate, which serves as a source of energy for anaerobic muscle contraction.

cretinism (kree′tin-izm). A condition caused by a congenital lack of thyroid secretion.

cribriform (krib′ri-form). Perforated with small holes like a sieve.

cricoid (krye′koyd). Ring-shaped cartilage in the larynx.

crystalloid (kris′tul-oid). Resembling a crystal.

cubital (ku′bi-tal). Pertaining to the space anterior to the elbow joint.

curare (koo-rah′re). A drug that blocks neuromuscular transmission and produces muscular paralysis.

cusp (kusp). A tapering, pointed projection.

cutaneous (kew-tay′nee-us). Pertaining to the skin.

cyanosis (sigh′uh-no′sis). A condition in which the skin appears blue due to an excessive accumulation of reduced hemoglobin in the blood.

cyst (sist). Any normal or abnormal sac.

cytochromes (sigh′to-kromes). Enzymes located in the inner membrane of mitochondria associated with oxygen utilization and ATP formation.

cytokinesis (sigh′to-ki-nee′sis). The division of the cytoplasm during cell division.

cytology (sigh-tol′uh-jee). The science that concerns itself with the structure of cells.

cytoplasm (sigh′to-plazm). The region in a cell located outside the nucleus.

cytosine (sigh′to-seen). A pyramidine base found in nucleic acids.

deamination (dee-am′i-nay′shun). The removal of an amide ($—NH_2$) group from an amino acid.

decussation (dee′kuh-say′shun). A crossing over, especially of symmetrical parts.

defecation (def′e-kay′shun). The discharge of feces from the rectum.

deferent (def′er-ent). Carrying away from.

deglutition (dee-gloo-tish′un). The act of swallowing.

dendrite (den′drite). A branched or treelike process of a nerve cell.

dentin (den′tin). The hard substance of a tooth, which surrounds the pulp and is covered by enamel.

deoxyribonucleic acid (DNA) (dee-ock′-see-rye′bo-new-klee′ick as′id). The nucleic acid that transmits the genetic code; structurally it consists of a double strand (double helix) composed of units of deoxyribose sugar, phosphate, and a nitrogen base.

depolarization (dee-po′lur-i′zay′shun). A decrease in the magnitude of the electric potential across a cell membrane; the neutralization of polarity.

dermis (dur′mis). The deep, thick, inner layer of skin underlying the epidermis.

descending colon (de-sen′ding ko′lon). The part of the large intestine on the left side of the abdomen.

desmosome (dez′mo-sohm). A type of junction between two cells.

detoxification (de-tock′si-fi-kay′shun). The reduction of the toxic properties of a substance.

dextrin (decks′trin). A product of starch hydrolyses.

dextrose (deks′troce). A monosaccharide similar to glucose.

diabetes insipidus (dye′uh-bee′teez in-sip′i-dus). A disease resulting from an inadequate secretion of antidiuretic hormone marked by great thirst and the passage of a large volume of urine.

diabetes mellitus (dye′uh-bee′teez mel-eye′tus). A disease caused by an insufficient amount of insulin marked by the passage of excessive amounts of urine containing glucose.

diagnosis (dye′ug-no′sis). The process of determining what disease is present.

dialysis (dye-al′i-sis). The process of separating crystalloids from colloids by the difference in their rates of diffusion through a semipermeable membrane.

diaphysis (dye-af′-i-sis). The shaft of a long bone.

diarrhea (dye′uh-ree′uh). The passage of watery stools.

diarthrosis (dye′ahr-thro′sis). A freely movable joint.

diastole (dye-as′tuh-lee). The period of the cardiac cycle when the ventricles of the heart are not contracting.

diencephalon (dye′en-sef′uh-lon). The posterior part of the brain, consisting of the hypothalamus, thalamus, epithalamus, and metathalamus.

diethylstilbesterol (dye-eth′il-stil-bes′-trol). A compound possessing estrogenic activity. Abbreviation DES.

diffusion (di-few′zhun). The movement of molecules from an area of high concentration to an area of low concentration due to random thermal motion.

digestion (di-jes′chun). The process whereby ingested food is broken down to form that can be absorbed across the intestinal wall.

digitalis (didj′i-tal′is). The drug obtained from the purple foxglove plant; used to increase the force of cardiac contractions in a weakened heart.

dipeptide (dye-pep′tide). A peptide which, upon hydrolysis, yields two amino acids.

diploid (dip′loyd). Having two sets of genes, one of maternal and one of paternal origin.

disaccharide (dye-sack′uh-ride). A carbohydrate composed of two monosaccharides linked together, e.g., sucrose.

diuretic (dye′yoo-ret′ick). Any substance that causes an increase in urine excretion.

diurnal (dye-ur′nul). Occurring in a 24-hour cycle.

diverticulum (dye′vur-tick′yoo-lum). A pouch or sac formed from a main cavity or tube.

DNA. Abbreviation for deoxyribonucleic acid.

dopamine (do′puh-meen). A catecholamine neurotransmitter; also the precursor of epinephrine and norepinephrine.

dorsal (dor′sul). Pertaining to or toward the back.

DPG. The abbreviation for diphosphoglycerate.

duct (dukt). A passage or tube.

duodenum (dew′o-dee′num). The first portion of the small intestine extending from the stomach to the jejunum.

dura mater (dew′ruh may′tur). The outermost fibrous membrane covering the brain.

dysmenorrhea (dis′men′o-ree′uh). Painful menstruation.

dyspnea (disp-nee′uh). The sensation of being unable to breathe adequately.

dystrophy (dis′truh-fee). The degeneration of an organ or tissue often caused by faulty nutrition.

ECG. The abbreviation for electrocardiogram.

ectoderm (eck′to-durm). The outermost of the three germ layers of the embryo.

ectopic (eck-top′ick). Located at a site other than the normal location.

edema (e-dee′muh). The presence of excess fluid in the interstitial spaces of the body.

EEG. The abbreviation for electroencephalogram.

effector (ef-feck′tur). The end organ in a reflex; usually muscle or gland.

efferent (ef′ur-unt). To convey away from a central point.

effluent (ef′lew-ùnt). Something that flows out.

effusion (e-few′zhun). Escape of fluid into a part or tissue.

ejaculation (e-jack′yoo-lay′shun). The release of sperm cells accompanying orgasm.

elasticity (elas′ti′si-tee). The property of a body that enables it to return to its original shape after being deformed.

electrocardiogram (e-leck′tro-kahr′dee-o-gram). A recording of the electric activities of the heart, recorded from the surface of the skin; abbreviated ECG or EKG.

electroencephalogram (e-leck′tro-en-sef′uh-lo-gram). A recording of the electric activity of the brain, recorded from the surface of the scalp; abbreviated EEG.

electron (e-leck′tron). A negatively charged subatomic particle that revolves around the nucleus of atoms.

electrophoresis (e-leck′tro-fo-ree′sis). The movement of charged particles between the negative and positive poles of an electric field.

embolus (em′buh-lus). A clump of matter (blood clot) or air bubble that is carried by the blood until it becomes lodged in a vessel and obstructs blood flow.

embryo (em′bree-o). The earliest stages of development; the first 8 weeks following conception.

embryology (em′bree-ol′uh-jee). The study of the development of the embryo.

emphysema (em′fi-see′muh). A pathological condition of the lungs associated with enlargement of the alveoli.

emulsion (e-mul′shun). A suspension of very small droplets of one immiscible liquid in another.

enamel (e-nam′ul). The white-hard substance that covers the dentin and crown of a tooth.

endocarditis (en′do-kahr-dye′tis). An inflammation of the inner wall of the heart.

endocardium (en′do-kahr′dee-um). The endothelial lining of the inside of the heart.

endocrine gland (en′do-krin). A gland that secretes hormones into the interstitial fluid, from which they enter the blood and are distributed throughout the body; also known as ductless glands.

endocytosis (en′do-sigh-to-sis). The invagination of small portions of the plasma membrane which breaks off, carrying a portion of the extracellular medium into the cell.

endometrium (en′do-mee′tree-um). The epithelial lining of the uterus.

endomysium (en′do-mis′ee-um). The connective tissue surrounding muscle fibers.

endothelium (en′do-theel-ee-um). A layer of epithelial cells lining the cavities of the heart and the blood vessels.

enterokinase (en′tur-o-kigh′nace). A proteolytic enzyme that activates the inactive pancreatic enzyme, trypsinogen, to active trypsin following its secretion into the intestinal tract.

entoderm (en′to-durm). The innermost germ layer of the embryo.

enzyme (en′zime). A protein catalyst that accelerates specific chemical reactions in the body.

eosinophil (ee′o-sin′uh-fil). A white blood cell readily stained by eosin; contains cytoplasmic granules.

epicardium (ep′i-kahr′dee-um). The outer layer of the heart wall.

epidermis (ep′i-dur′mis). The outer epithelial layer of the skin which overlies the dermis.

epididymis (ep′i-did′i-mis). That portion of the genital duct preceding the vas deferens.

epiglottis (ep′i-glot′is). The cartilaginous appendage overlying the opening of the larynx.

epilepsy (ep′i-lep′see). A disease of the central nervous system indicated by sudden attacks of uncontrolled muscular activity, sometimes associated with unconsciousness.

epimysium (ep′i-miz′ee-um). The fibrous sheath around an entire muscle.

epinephrine (ep′i-nef′rin). A hormone secreted by the adrenal medulla; stimulates cardiac contraction and regulates fat and glucose metabolism; also called adrenalin.

epineurium (ep′i-new′ree-um). The sheath of connective tissue surrounding a peripheral nerve.

epithelium (ep′i-theel′ee-um). Tissue which covers all body surfaces and which forms the glands of the body.

EPSP. Abbreviations for excitatory postsynaptic potential.

ergosterol (ur-gos′tur-ol). A sterol which, upon irradiation with ultraviolet light, becomes vitamin D_2.

erythema (err′i-theem-uh). A redness of the skin produced by congestion of the capillaries.

erythroblastosis fetalis (e-rith′ro-blas-to′sis fee-tay′lis). A hemolytic anemia of the fetus caused by the transplacental transfer of maternally formed Rh antibodies against an Rh positive fetus.

erythrocyte (e-rith′ro-sight). A red blood cell.

erythropoiesis (e-rith′ro-poy-ee′sis). The formation of red blood cells.

erythropoietin (e-rith′ro-poy′e-tin). A glycoprotein hormone that stimulates the production of red blood cells.

esophagus (e-sof′uh-gus). The tubular passage extending from the pharynx to the stomach.

estriol (es′tree-ole). The predominant estrogenic hormone of pregnancy.

estrogen (es′tro-jin). A term for several female sex hormones that promote the development of the female reproductive tract and secondary sex characteristics.

estrus (es′trus). The recurrent period of sexual activity in nonprimate mammals; the mating period in such animals.

eunuch (yoo′nuck). An adult male whose testes have never developed or have been removed.

eupnea (yoop-nee′uh). Normal respiration.

eustachian tube (yoo-stay′kee-un, -stay′shun). The auditory tube; a narrow channel connecting the middle ear with the naso-pharynx.

exocrine gland (eck′so-krin). A gland that secretes its products into a duct or tube.

expiration (eck′spi-ray′shun). Exhalation or breathing out.

extension (eks-ten′shun). A movement which brings two parts farther apart.

external (ecks-tur′nul). Situated on the outside.

external auditory meatus (ecks-tur′nul aw′di-tor′ee mee-ay′tus). The canal in the temporal bone that leads into the middle ear.

extracellular (ecks′truh-sel′yoo-lur). Outside the cell.

extrasystole (eks′truh-sis′tuh-lee). A heartbeat which occurs before its normal time in the cardiac cycle.

extrapyramidal (ecks′truh-pi-ram′i-dul). The descending nerve tracts that lie outside the medulla oblongata pyramids.

extrinsic (eck-strin′sick). Originating from the outside.

exudate (ecks′yoo-date). The accumulation of material in a body cavity or on a body surface.

facet (fas′it). A smooth plane or surface for articulation.

facilitated diffusion (fa-sil′i-tay-tid dif-fyoo′shun). Movement of molecules through a membrane in combination with a carrier molecule; not linked to metabolic use of energy. Net movement always occurs down a concentration gradient from high to low concentration.

FAD. The abbreviation for flavine adenine dinucleotide, a coenzyme that functions as a hydrogen carrier.

fallopian tube (fa-lo′pee-un tewb). The oviduct; the tube connecting the ovary to the uterus.

fascia (fash′ee-uh). The thin layers of connective tissue that form sheets of supporting tissue to which other tissues are attached.

fascicle (fas′i-kul). A small bundle of nerve or muscle fibers.

fats (fats). A class of organic molecules characterized by their insolubility in water; composed primarily of fatty acids and glycerol.

fatty acid (fat′ee as′id). A long hydrocarbon chain (C_{14} to C_{20}) with a carboxyl (COOH) group at one end.

feces (fee′seez). The material eliminated from the large intestine during defecation; composed primarily of water, bacteria, and indigestible matter.

feedback (feed′bak). Stimulation of the receptor input to a control system by the response produced by the effector output of the same control system.

fermentation (fur′men-tay′shun). The metabolism of carbohydrate in the absence of oxygen; the end products are alcohol and energy.

fertilization (fur′ti-li-zay′shun). The union of male and female gametes.

ferritin (ferr′i-tin). One of the forms in which iron is stored in the body; an iron-protein complex.

fetus (fee′tus). An unborn human after the second month of gestation.

fever (fee′ver). An abnormally high body temperature.

fiber (figh′bur). A threadlike structure.

fibrillation (figh′bri-lay′shun). The extremely rapid, uncoordinated contraction of segments of the heart wall; prevents the development of any net pressure within those chambers of the heart.

fibril (figh′bril). A minute fiber.

fibrin (figh′brin). The main fibrous protein in a blood clot.

fibrinogen (figh-brin′o-jin). The plasma protein that reacts with thrombin to form fibrin, the essential component of a blood clot.

fibroblast (figh′bro-blast). A connective tissue cell that synthesizes and secretes the protein that forms extracellular collagen.

filtration (fil-tray′shun). The process whereby pressure produces a flow of fluid through a semipermeable membrane.

fimbria (fim′bre-ah). A fingerlike projection surrounding an open area.

fissure (fish′ur). Any fold, cleft, or groove; may be normal or abnormal.

fistula (fis′tu′lah). An abnormal opening between two internal structures or between an internal structure and the surface of the body.

flagellum (fla-jel′um). A long hairlike motile process on the surface of a cell; usually only one per cell, in contrast to cilia, which are short, motile, hairlike projections, with hundreds present on a single cell.

flatus (flay′tus). Gas in the stomach or intestine.

flexion (fleck′shun). The act of bending; movement that closes the angle between two bones at a joint.

follicle (fol′i-kul). A very small sac or gland.

fontanelle (fon′tuh-nel′). An unossified area of the cranium of an infant (soft spot).

foramen (fo-ray′men). A small opening.

fornix (for′nicks). A general term designating an archlike structure.

fossa (fos′uh). A pit or hollow area.

fovea (fo′vee-uh). A pit or depression.

frenulum (fren′yoo-lum). A membranous fold.

fructose (frook′toce). A six-carbon monosaccharide; one of several sugars in the body.

FSH. The abbreviation for follicle-stimulating hormone; produced by the anterior pituitary.

fundus (fun′dus). The base of a hollow organ.

fusiform few′zi-form). Spindle-shaped; tapering at both ends.

fusion (few′zhun). The act or process of uniting.

galactose (ga-lack′toce). A six-carbon monosaccharide; one of several types of sugars.

gamete (gam′eet). The sex cells; male sperm of female ovum.

gametogenesis (gam′e-to-jen′e-sis). The process of forming the male (sperm) and female (ovum) sex cells.

gamma globulin (gam′uh glob′yoo-lin). The fraction of blood proteins containing antibodies.

ganglion (gang′glee-un). A group of nerve cell bodies located outside the central nervous system.

gastric (gas′trick). Pertaining to the stomach.

gastrin (gas′trin). A peptide secreted by the pyloric mucosa; stimulates gastric HCl secretion.

gene (jeen). The biologic unit of hereditary; corresponds to the sequence of nucleotides in DNA which specifies the sequence of amino acids in a protein.

genetics (je-net′iks). The science of heredity.

genotype (jen′o-type). The complete hereditary constitution of an individual.

germ cells (jurm). Sperm and ova.

gestation (jes-tay′shun). The duration of pregnancy or period of intrauterine development.

GFR. The abbreviation for glomerular filtration rate.

giantism (jye′un-tizm). An abnormal enlargement of the body or its parts.

gingivae (jin′ji-vuh). The gums.

gland (gland). A collection of cells specialized for secretion.

glenoid (glee′noid). A pit or socket.

glial cells (gli′al). Nonneural cells of the central nervous system.

globulin (glob′u-lin). A class of proteins which are insoluble in pure water but soluble in neutral salt solutions.

glomerular filtration rate (glom-err′yoo-lur fil-tray′shun rayt). The volume of fluid filtered from the blood plasma into the kidney tubules by the glomeruli per unit time. Abbreviation GFR.

glomerulonephritis (glo-merr′yoo-lo-ne-frye′tis). An inflammation of the capillary loops in the glomeruli of the kidney.

glomerulus (glom-err′yoo-lus). A tuft or cluster of blood vessels or nerves.

glossal (glos′ul). Pertaining to the tongue.

glottis (glot′is). The opening between the vocal cords of the larynx and the walls surrounding the opening.

glucagon (gloo′kuh-gon). Pancreatic peptide hormone secreted in response to hypoglycemia; stimulates liver to produce glucose via glycogenolysis.

glucocorticoid (gloo′ko-kor′ti-koid). Several steriod hormones secreted by the adrenal cortex which increase gluconeogenesis, causing an increase in blood sugar.

glucokinase (gloo′ko-kigh′nace). An enzyme that catalyzes the phosphorylation of glucose.

gluconeogenesis (gloo′ko-nee′o-jen′e-sis). The process by which glucose is synthesized from materials such as amino acids.

glucose (gloo′koce). A six-carbon monosaccharide sugar; blood sugar.

glycerol (glis′ur-ole). A three-carbon carbohydrate which combines with fatty acids to form triglycerides (neutral fat.)

glycine (glye′seen). One of the twenty amino acids found in proteins.

glycogen (glye′kuh-jin). A polysaccharide composed of glucose units; the major storage form of carbohydrate in the body.

glycogenesis (glye′ko-jen′e-sis). The formation of glycogen from carbohydrates.

glycogenolysis (glye-ko-je-nol′i-sis). The breakdown of glycogen into glucose.

glycolysis (glye-kol′i-sis). The metabolic pathway that converts glucose to lactic acid; forms small amounts of ATP in the absence of oxygen.

glycoprotein (glye-ko-pro′teen). A class of proteins in which carbohydrate is bound to some of the amino acids.

glycoside (glye′ko-sid). Any compound which contains a carbohydrate as part of its structure.

goblet cell (gob′lit sel). A type of epithelial cell which secretes mucus.

goiter (goi′ter). An enlargement of the thyroid gland.

Golgi apparatus or complex (gol′je ap′uh-ray′tus). An organelle composed of a series of cup-shaped membranes; functions in the secretion of proteins from cells.

gonads (go′nads). The organs of the body which produce the gametes; the ovaries and testes.

gonadotropins (gon-nad′o-tro′pins). Two hormones (FSH and LH) produced by the anterior pituitary; stimulate the gonads.

grand mal (grahn mal). A major epileptic seizure.

granulocyte (gran′yoo-lo-sight). Any of several types of white blood cells that contain granules in their cytoplasm—eosinophils, neutrophils, and basophils.

granuloma (gran′yoo-lo′muh). A localized accumulation of granular inflammatory tissue.

groin (groin). The junction of the abdomen and thighs.

growth hormone (groath hor′mone). The anterior pituitary protein hormone which stimulates growth. Abbreviated GH.

guanine (gwah′neen). One of the nitrogenous bases making up the structure of nucleic acids.

gustatory (gus′tuh-to′ree). Pertaining to the sense of taste.

gyrus (ji′rus). A convolution of the brain's surface.

hamate (hay′mate). The hooked carpal bone in the wrist.

haploid (hap′loyd). Having a chromosome number equivalent to that of the gametes, i.e., half the diploid number.

haptene (hap′teen). The portion of an antigenic molecule that determines its immunologic specificity.

haustra (haw′stra). The sacculations of the wall of the large intestine.

Haversian canal (ha-vur′zhun kuhnal). A canal in bone tissue containing nerves and blood vessels; surrounded by concentric layers of calcified bone tissue.

heart attack (hahrt uh-tack′). The general term for a myocardial infarct.

hematocrit (hem′uh-to-krit). The percentage of the blood volume occupied by blood cells, primary erythrocytes.

hematoma (hee′muh-to′muh). A localized mass of blood that has entered the interstitial spaces of a tissue.

hematopoiesis (hem′uh-to-poy-ee′sis). The formation of blood cells; also called hemopoiesis.

hemoglobin (hee′muh-glo′bin). The oxygen-carrying protein found within erythrocytes.

hemoglobinuria (hee′muh-glo′bi-new′ree-uh). The presence of hemoglobin in the urine.

hemolysis (he-mol′i-sis). The release of hemoglobin from red blood cells.

hemolyze (hee′mo-liz). To cause hemolysis.

hemophilia (hee′mo-fil′ee-uh). Sex-linked hereditary disorder characterized by a defect in the clotting mechanism of the blood.

hemorrhage (hem′uh-ridj). The escape of blood from a vessel.

hemorrhoid (hem′uh-royd). The varicose dilation of veins in the hemorrhoidal plexus of the rectum.

heparin (hep′uh-rin). A mucopolysaccharide acid that prevents blood from clotting.

hepatic (he-pat′ick). Pertaining to the liver.

hepatitis (hep′uh-tye′-tis). An inflammation of the liver due to either a virus or toxic substance(s).

heredity (he-red′i-tee). The genetic transmission of characteristics from a parent to a child.

hernia (hur′nee-uh). The abnormal protrusion of an organ or tissue through an opening.

heterozygous (het′ur-o-zye′gus). Having two different allelic genes specifying the same genetically determined component.

hexose (heck′soce). A six-carbon sugar.

hiatus (high-ay′tus). A gap or opening in an organ.

hilus (high′lus). A depression or pit, especially where vessels exit or enter an organ.

hippocampus (hip′o-kam′pus). That region of the brain which is closely associated with eating, smelling, and emotions; part of the rhinencephalon.

histidine (his′ti-deen). A basic amino acid.

histamine (his′tuh-meen). An amine derived from histadine; stimulates smooth muscle and glands when released locally in a tissue.

histiocyte (his′te-o′sit). A phagocytic cell formed by the reticuloendothelial system; corresponds to a macrophage.

histology (his-tol′o-jee). The study of the structure of tissues.

hives (hive′z). An allergic skin condition characterized by the appearance of intensely itchy whelts with raised white centers and a surrounding area of erythema; also known as urticaria.

holocrine gland (hol′o-krin). A gland whose secretion results from the disintegration of the entire gland cell.

hemeostasis (ho′mee-os′tuh-sis). The state of approximate consistency in the physical and chemical composition of an organism's internal environment in the face of a varying external environment.

homogeneous (ho′modj′e-nus). Composed of similar elements or ingredients.

homotherm (ho′mo′thurm). Animals which maintain a relatively constant internal temperature.

homozygous (ho′mo-zye′gus). Possessing an identical pair of allelic genes specifying the same genetically determined component.

hormone (hor′mone). A chemical secreted by an endocrine gland into the blood which transports it to its target tissue where it alters the functional activity of that tissue.

hyaline (high′uh-lin). Glassy and transparent.

hyaluronidase (high′uh-lew-ron′i-dace). An enzyme that catalyzes the hydrolysis hyaluronic acid in connective tissue.

hydration (high-dray′shun). Combination with water.

hydrocephalus (high-dro-sef′uh-lus). A condition characterized by an abnormal accumulation of fluid in the cranial cavity.

hydrolysis (high-drol′i-sis). The breaking of a chemical bond with the addition of the elements of water to the products formed.

hydrostatic pressure (high′dro-stat′ick presh′ur). The pressure exerted by the weight of a column of water.

hyperemia (high′pur-ee′mee-uh). An increase in the flow of blood through a region of the body.

hyperglycemia (high′pur-glye-see′mee-uh). An elevated concentration of glucose in the blood.

hyperopic (high′pur-o′pick). Farsighted vision.

hyperplasia (high′pur-play′zhuh). An increase in the number of cells in a tissue or organ.

hypertension (high′pur-ten′shun). A resting blood pressure that is maintained above normal levels.

hypertonic (high′pur-ton′ick). Having a solute concentrate greater than the normal solute concentration of body fluids; solutions causing cells to shrink through loss of water.

hypertrophy (high-pur′truh-fee). An increase in the size of individual cells within a tissue or organ.

hyperventilation (high-pur-ven′ti-lay′shun). The state in which a greater volume of air enters the lungs than is required for normal oxygen–carbon dioxide exchange; reduces the partial pressure or carbon dioxide in the blood.

hypoglycemia (high′po-glye-see′mee-uh). A low concentration of glucose in the blood.

hypophysis (high-pof′i-sis). The pituitary gland.

hypotension (high′po-ten′shun). Below normal blood pressure.

hypothalamus (high′po-thal′uh-mus). An area of the brain concerned primarily with autonomic nervous activities; that portion of the diencephalon that forms the floor and walls of the third ventricle.

hypotonic (high′po-ton′ick). Having a total solute concentration less than that of normal body fluids; solutions causing cells to swell through gain of water.

hypoxia (high-pock′see-uh). Low oxygen content.

hysterectomy (his′tur-eck′tuh-mee). The surgical removal of part or all of the uterus.

ileum (il′ee-um). The distal part of the small intestine.

iliac (il′ee-ack). One of the bones of each half of the pelvis.

immune (i-mewn′). Protected against the damaging effects of a foreign substance.

immunity (i-mew′ni-tee). The condition of being immune.

immunization (im′yoo-ni-zay′shun). The process of rendering a person immune.

immunology (im′yoo-nol′uh-jee). The science of immunity.

impotence (im′puh-tunce). Inability in the male to perform the sexual act.

impulse (im′pulce). An action potential transmitted along a nerve or muscle fiber.

incisure (in-sigh′zhur). A cut, notch, or incision.

inclusion (in-kloo′zhun). Any thing or substance that is enclosed.

incus (ing′kus). The middle anvil-shaped bone of the middle ear.

infarction (in-fahrk′shun). The presence of a localized area of necrotic tissue produced as a result of inadequate oxygenation.

infundibulum (in′fun-dib′yoo-lum). The stalk that attaches the pituitary gland to the hypothalamus of the brain.

ingestion (in-jes′chun). The act of taking material into the body via the mouth.

inguinal region (ing′gwi-nul). Pertaining to the groin area.

inhalant (in-hay′lunt). A substance taken into the body by way of the nose and trachea through the respiratory system.

innervate (in′ur-vate). To supply with nerves.

inosine (in′o-seen). A nucleotide base formed by the deamination of adenosine.

in situ (in sigh′too). In place.

insulin (in′suh-lin). A proteinaceous pancreatic hormone that lowers blood sugar levels.

integument (in-teg′yoo-ment). The skin.

intercalated discs (in-tur′kuh-lay′ted). The area along the surface of heart cells where they are joined together by desmosomes.

interferon (in′tur-feer′on). A protein that is formed nonspecifically which stimulates the body's defenses against viruses.

internuncial (in′tur-nun′see-ul). Connecting two nerves or nerve centers.

interphase (in′tur-faze). That period of a cell's life cycle in which it is not dividing.

interpolar (in′ter-po′lar). Situated between two poles.

interstitial (in′tur-stish′ul). Lying between; the extracellular space of a tissue.

intima (in′ti-muh). The innermost layer of a blood vessel.

intracellular (in′truh-sel′yoo-lar). Situated within a cell.

intramembranous ossification (in′truh-mem′bruh-nus os′i-fi′kay′shun). The physiologic process of bone formation whereby bone is formed directly in membranous tissue.

intravenous (in′truh-vee′nus). Within a vein or veins.

intrinsic (in-trin′sick). Situated within a part.

inulin (in′yoo-lin). A polysaccharide composed of fructose; used in the determination of glomerular filtration rate.

in utero (in yoo′te-ro). Within the uterus.

in vivo (L) (in vee′vo). Within the body.

IPSP. Abbreviation for inhibitory postsynaptic potential.

iris (eye′ris). The circular pigmented membrane behind the cornea of the eye.

irritability (irr′i-tuh-bil′it-ee). The quality of response to a stimulus.

ischemia (is-kee′mee-uh). A deficiency in the blood flow to a body part.

ischium (is′kee-um). The bone of the hip.

islets of Langerhans (eye′lit ov lang′erhans). The clusters of endocrine cells in the pancreas that produce insulin.

isometric (eye′so-met′rick). The generation of force by a muscle under conditions in which the muscle length remains constant.

isoosmotic (eye′so-oz-mot′ick). Having the same osmotic pressure as the body fluids.

isotonic (eye′so-ton′ick). Having the same solute concentration as the normal fluids in the body; the contraction of muscles with no change in tension.

isotope (eye′suh-tope). A chemical element having the same atomic number as another but possessing a different atomic mass.

isotropic (eye′so-trop′ick). Having similar properties in all directions.

isthmus (is′mus). A narrow connection between two larger bodies.

jaundice (jawn′dis). Yellow coloration of the skin due to the abnormal accumulation of bile pigment.

jejunum (je-joo′num). That portion of the small intestine which extends from the duodenun to the ileum.

joint (joint). The union between two or more bones of the skeleton.

jugular (jug′yoo-lur). Pertaining to the neck.

keratin (kerr′uh-tin). A scleroprotein found in hair and nails and on the surface of the epidermis.

keto (kee′to). A prefix which denotes the possession of a carbonyl (CO) group.

ketone (kee′tone). A compound containing the carbonyl (CO) group.

keytone bodies (kee′tone bod-ee). Acetoacetic acid and acetone; found in increased amounts in untreated diabetes.

ketonuria (kee′to-new′ree-uh). The presence of ketone bodies in the urine.

kilocalorie (kil′o-kal′o-ree). 100 calories; equal to one large calorie.

kinesiology (ki-nee′see-ol′uh-jee). The science concerned with human motion.

labia majora (lay′bee-uh ma-jo′ruh). The two large folds of skin that form the outer lips of the vulva.

labia minora (lay′bee-uh mi-no′ruh). The inner highly vascular folds of the vulva.

labyrinth (lab′i-rinth). A system of connecting cavities or canals; as the bony labyrinth in the temporal bone of the skull.

lacrimal (lack′ri-mul). Pertaining to the tears.

lactase (lack′tace). An enzyme that splits milk sugar, lactose, into glucose and galactose.

lactation (lack-tay′shun). The secretion of milk.

lacteal (lack′tee-ul). A small lymph vessel in the middle of intestinal villae.

lactic acid (lack′tick as′id). A three-carbon sugar; the end product of anaerobic glycolysis.

lactose (lack′toce). A disaccharide; known as milk sugar.

lacuna (lah-kew′nuh). A small, hollow cavity.

lamella (lah-mel′uh). A thin leaf or plate, e.g., a narrow concentric ring surrounding the Haversian canal in compact bone.

larynx (lar′inks). The voice box.

latent (lay′tunt). A state of inactivity.

lecithin (les′i-thin). One type of phospholipid.

lemniscus (lem-nis′kus). A band of longitudinal nerve fibers running between the medulla and pons; carries impulses to the thalamus.

lens (lenz). A transparent organ lying posterior to the pupil and iris of the eyeball and anterior to the vitreous body.

lesion (lee′zhun). Pathological or damaged tissue.

leucocyte (lew′ko-sight). Leukocyte; a general term for the several different types of white blood cells. A white blood cell.

leukemia (lew-kee′mee-uh). A disease of the bone marrow, spleen, and lymph nodes characterized by a marked increase in the number of white blood cells and their precursors.

leukocytosis (lew′ko-sigh-to′sis). An increase in the number of leukocytes in the blood (above 10,000 per mm^3).

leukopenia (lew′ko-pee′nee-uh). A reduction in the number of leukocytes in the blood (below 5,000 per mm^3).

LH. Abbreviation for luteinizing hormone.

libido (li-bee′do). A state of sexual desire.

ligament (lig′uh-munt). A band of connective tissue that connects bones or supports the viscera.

limbic system (lim′bick). That portion of the forebrain concerned with various aspects of emotion and behavior.

lingual (ling′gwul). Pertaining to the tongue.

lipase (lye′pace). An enzyme that catalyzes the hydrolysis of fatty acids from triglycerides and phospholipids.

lipemia (li-pee′mee-uh). The presence of abnormally high concentrations of lipids (fats) in the blood.

lipid (lip′id). A group of organic compounds which are insoluble in water but soluble in alcohol and other fat solvents.

lipoprotein (lip′o-pro′tee′in). A molecule composed of both lipid and protein.

lobule (lob′yool). A small lobe.

locus (lo′kus). A specific site; in genetics, the location of a gene on a chromosome.

loin (loin). That portion of the back between the thorax and pelvis.

lumen (lew′min). The cavity or channel within a tube or organ.

luteinizing hormone (lew′tee-in-iz′ing). An anterior pituitary protein hormone that stimulates ovulation and formation of the corpus luteum in the ovary and the secretion of testosterone by the interstitial cells in the testes.

lymph (limf). The transparent, slightly yellow fluid filling lymphatic vessels.

lymphocyte (lim′fo-sight). A type of white blood cell responsible for specific immunity.

lysine (lye′seen). One of twenty different amino acids found in proteins.

lysis (lye′sis). The destruction of cells resulting from damage to the plasma membrane.

lysosome (lye′so-sohm). An organelle in the cytoplasm that contains a number of hydrolytic enzymes; associated with the digestion of ingested material.

lysozyme (lye′so-zyme). A protein found in tears and saliva; functions as an antibacterial enzyme.

macrophage (mack′ro-faidj). A large phagocytic cell.

macula (mack′u-luh). A small stain or spot differing from the surrounding tissue.

malabsorption (mal′ub-sorp′shun). Abnormal absorption of food from the intestine.

malformation (mal′for-may′shun). A defective or abnormal formation.

malignancy (muh-lig′nun-se). Cancer.

malleolus (mal-lee′o-lus). A rounded process on either side of the ankle joint.

malleus (mal′ee-us). A hammerlike bone of the middle ear.

maltase (mawl′tace). An enzyme which catalyzes the hydrolysis of the disaccharide maltose into two molecules of glucose.

maltose (mawl′toce). A disaccharide formed during the digestion of starch.

mast cells (mast sel). Connective tissue cells which release histamine and other compounds; act locally within a tissue.

mastication (mas′tick-ay′shun). Chewing.

mastoid (mas′toid). Nipple-shaped.

matrix (may′tricks). A term often used to refer to the extracellular materials that are formed of connective tissue; the semidense medium surrounding these cells.

meatus (mee-ay′tus). An opening or passageway.

mediastinum (mee′dee-as-tye′num). The extrapulmonary space between the two lungs that extends from the sternum to the backbone.

medulla (me-dul′uh). The central portion of an organ, as opposed to the periphery or cortex.

megakaryocyte (meg′uh-kar′ee-o-sight). The large bone marrow cell that gives rise to blood platelets.

meiosis (migh-o′sis). The process by which the chromosome number is halved during the formation of gametes (sex cells).

melanin (mel′uh-nin). The dark pigment found in the skin and hair.

melanocyte (mel′uh-no-sight). The cell that produces melanin.

membrane (mem′brane). Either the lipoprotein bilayer that surrounds cells and organelles or a multicellular layer of epithelial cells that separate two compartments.

menarche (me-nahr′kee). The time of life when the first menstrual cycle begins.

meninges (me-nin′jeez). The three membranes covering the brain and spinal cord: the dura mater, pia mater, and arachnoid.

meningitis (men′in-jye′tis). An inflammation of the meninges.

menopause (men′o-pawz). The cessation of menstruation in the female, usually occurring between the ages of 45 and 55.

menstruation (men′stroo-ay′shun). The cyclic uterine bleeding which recurs at approximately four-week intervals.

merocrine (merr′o-krin). A type of gland in which the secreting cells remain intact throughout the process of formation and discharge of the secretory products; in contrast to apocrine and holocrine secretion.

mesencephalon (mez′en-sef′uh-lon). The midbrain.

mesenchyme (mes′in-kime). The network of embryonic connective tissue in the mesoderm; forms the connective tissue, blood vessels, and lymphatic vessels.

mesentery (mes′un-terr′ee). The fold of peritoneum attaching the intestinal tract to the posterior abdominal wall.

metabolism (me-tab′o-lizm). All the chemical processes that occur in an organism.

metaphase (met′uh-faze). The stage of mitosis in which the chromosomes line up on the equitorial plate.

metastasis (me-tas′tuh-sis). The spread of diseased tissue from one part of the body to another part, e.g., the spread of cancer cells from their site of origin throughout the body.

micelle (mi-sel′). A small aggregate of lipid molecules which forms a clear solution even though the lipids remain in a separate phase in the water.

microglia (migh-krog′lee-uh). A type of neuroglial cell in the nervous system.

microvillus (migh′kro-vil′us). The small fingerlike extensions on the surface of a cell.

micturition (mick′tew-rish′un). The expulsion of urine from the bladder; urination.

midbrain (mid′brane). The part of the brain between the pons and cerebrum.

mineralocorticoid (min′ur-uh-lo′kor′ti-koid). The type of adrenal cortical hormone that regulates sodium and potassium balance in the body.

mitochondrion (migh′to-kon′dree-un). The organelle in which ATP is formed by oxidative phosphorylation. The primary site of oxygen utilization and carbon dioxide production in cells.

mitosis (migh-to′sis). The separation of individual sets of chromosomes during the process of somatic cell division.

mitral (migh′trul). Pertaining to the mitral or bicuspid valve of the heart; the left atrioventricular valve.

mole (mole). The weight of a substance in grams divided by its molecular weight gives the number of moles of a substance.

molarity (mo-lar′i-tee). A unit of concentration; the number of moles of solute per liter of solution.

molecule (mol′e-kyool). A combination of two or more atoms which form a specific chemical substance.

monocyte (mon′o-sight). A large leukocyte that has an oval nucleus with no granules in the cytoplasm.

monosaccharide (mon′o-sack′uh-ride). A simple sugar, such as glucose.

mons pubis (monz pew′bis). The rounded fatty prominence over the symphysis pubis, covered with pubic hair.

morula (mor′yoo-luh). A solid globular mass of cells.

motor end plate (mo′tor end playt). The specialized region of the skeletal muscle membrane underlying the motor nerve endings.

motor nerve cells (mo′tor nerve sels). Nerve cells which stimulate skeletal muscles.

motor unit (mo′tor yoo′nit). A single motor neuron plus all the skeletal muscle fibers innervated by it.

mucin (mew′sin). A mixture of mucopolysaccharides which forms a thick, viscous solution known as mucus.

mucopolysaccharide (mew′ko-pol′ee-sack′uh-ride). A polysaccharide that contains an amino sugar; found in mucous secretions and various types of connective tissue secretions.

mucosa (mew-ko′suh). The epithelial and underlying glandular layers of the gastrointestinal tract, as distinct from the muscle layers.

mucous (mew′kus). The adjectival form of the word mucus.

mucus (mew′kus). A viscous liquid consisting of inorganic salts, water, epithelial cells, and other substances held in suspension by glycoproteins and mucopolysaccharides.

mutation (mew-tay′shun). To change or delete the coded neucleotide sequence of a gene.

myelin (migh′e-lin). The membranous layers that form a sheath around certain nerve fibers.

myeloid (migh′e-loyd). Pertaining to bone marrow.

myocardium (migh′o-kahr′dee-um). The muscular layer of the heart wall.

myofibril (migh′o-figh′bril). The slender thread composed of thick and thin filaments found in striated muscle tissue; runs parallel to the long axis of the muscle.

myoglobin (migh′o-glo′bin). The oxygen-binding hemoglobinlike protein found in muscle cells.

myometrium (migh-o-mee′tree-um). The smooth muscle layer of the uterus.

myoneural junction (migh′o-new′rul junk′shun). The neuromuscular junction; area where a nerve ends on a muscle.

myopia (migh-o′pee-uh). A defect in vision resulting in nearsightedness; objects can be seen only when very close to the eyes.

myosin (migh′o-sin). One of the contractile proteins in muscle; located in the thick filaments of myofibrils; contains the enzymatic site for splitting ATP.

NAD. The abbreviation for nicotinamide adenine dinucleotide; a coenzyme that functions as a hydrogen carrier in cells.

nares (nair′eez). The openings of the nasal cavity.

nasal (nay′zul). Pertaining to the nose.

navicular (nah-vick′yoo-lur). Boat-shaped.

necrosis (ne-kro′sis). The death of a tissue or portion of an organ.

neoplasm (nee′o-plazm). A tumor (benign or malignant).

nephron (nef′ron). The anatomic and physiologic unit of the kidney made up of a glomerulus, proximal convoluted tubule, loop of Henle, distal convoluted tubule, and collecting duct; each kidney is composed of about one million nephrons.

nerve (nerv). The anatomic structure composed of many nerve axons; conveys nerve impulses between the central nervous and some other part of the body.

nervous (nur′vus). Pertaining to the nervous system; excitable.

neural (new′rul). Pertaining to the nervous system.

neuralgia (new-ral′juh). A pain extending along the course of one or more nerves.

neurofibril (new′ro-figh′bril). Thin fibers found in the axons of nerve cells.

neuroglia (new-rog′lee-uh). The supportive, nonneural cells in the nervous system.

neurohypophysis (new′ro-high-pof′i-sis). The posterior lobe of the pituitary gland.

neuromuscular junction (new′ro-mus′kew-lur junk′shun). The junction between a nerve and skeletal muscle cell.

neuron (new′ron). A nerve cell.

neurotransmitter (new′ro-tranz-mit′ur). A chemical that diffuses from a nerve cell ending to an adjacent cell where it affects the membrane activity of that cell.

neutral fat (new′trul fat). Another term for triglyceride.

neutrophil (new′truh-fil). A white blood cell that stains readily with neutral stains.

Nissl bodies (nis-sul′ bod′eez). The rough endoplasmic reticulum of neurons.

node of Ranvier (node ov ran′vee′ur). The regions between the myelein segments; located at intervals along the sheath of myelinated nerve fibers where the axon membrane is in contact with the extracellular fluids.

nodule (nod′yool). A small, solid node which can be detected by touch.

norepinephrine (nor-ep′i-nef′reen). The neurotransmitter released by most postganglionic neurons in the sympathetic nervous system.

nucleolus (new-klee′uh-lus). The densely staining organelle in the nucleus of a cell; a major site of RNA synthesis.

nucleotide (new′klee-o-tide). The compound composed of a base, a monosaccharide, and phosphate; the subunit of nucleic acids.

nucleus (new′klee-us). The membrane-bound area of a cell exclusive of the cytoplasm; contains the cell DNA.

nutrition (nu-trish′un). The value of different foods.

nystagmus (nis-tag′mus). The rapid involuntary oscillations of the eyeball.

obesity (o-bee′si-tee). An excessive accumulation of fat in the body.

obstruction (ub-struk′shun). Something that blocks or clogs.

occlusion (uh-klew′zhun). Something that has closed or is closing.

olfaction (ol-fack′shun). The sense of smelling.

oligodendrocyte (ol′i-go-den′dro-sight). Neuroglial cells that have a few delicate processes; probably form myelin in the central nervous system.

oliguria (ol′ig-yoo′ree-uh). A decrease in the normal amount of urine production.

omentum (o-men′tum). A fold of peritoneum (mesentery) attached to the stomach.

oncotic (ong-kot′ik). Synonomous with colloid; i.e., colloid osmotic pressure is the same as oncotic pressure.

ontogeny (on-todj′e-nee). The developmental life history of an organism.

oocyte (o′o-site). The ovum or egg cell.

operon (op′ur-on). A region of the DNA molecule that includes both the structural genes and the operator genes controlling the transcription of the structural genes.

ophthalmic (off-thal′mick). Pertaining to the eye.

oral (or′ul). Pertaining to the mouth.

orbit (or′bit). The bony cavity of the skull that contains the eyeball.

organ (or′gun). A body component of two or more tissues that has a specific function.

organelle (or′guh-nel′). A structure located in cells which performs a specialized function.

organic (or′gan′ick). Pertaining to compounds containing the element carbon.

organism (or′guh-nizm). The total animal or plant.

orifice (or′i-fis). The entrance or outlet of any body cavity.

os (os). Latin for bone.

osmolarity (oz′mo-lar′i-tee). The sum of the concentrations of all solutes in a solution; total solute concentration.

osmol (oz′mol). The molecular weight of a substance divided by the number of particles it releases in solution.

osmoreceptor (oz′mo-re-sep′tur). A cell which responds to changes in the osmolarity of the surrounding fluid.

osmosis (os-mo′sis). The net diffusion of water molecules through a semipermeable membrane from an area of higher water concentration to an area of lower water concentration.

osmotic pressure (oz-mot′ick presh′ur). The pressure that must be applied to a solution to prevent the net flow of water (osmosis) into the solution.

ossicle (os′si-kul). Any small bone such as those found in the middle ear.

osteoarthritis (oss′tee-o-ahr-thrigh′tis). A chronic degenerative joint disease.

osteoblast (os′tee-o-blast). A cell that produces the extracellular matrix of bone tissue.

osteoclast (os′tee-o-klast). A cell that degrades the extracellular matrix of bone and cartilagenous tissues; osteoblasts and osteoclasts may be different functional stages of the same cell type.

osteocyte (os′te-o-site). An osteoblast that has become embedded within bone matrix.

osteogenesis (os′tee-o-jen′e-sis). The formation of bony tissue.

osteomalacia (os′tee-o-muh-lay′shee-uh). A progressive softening of bone.

osteomyelitis (os′tee-o-migh′e-lye′tis). Inflammation of bone caused by microorganisms.

osteoporosis (os′tee-o-po-ro′sis). The abnormal porosity of bone due to the failure of osteoblasts to lay down bone matrix.

otolith (o′to-lith). Calcium carbonate particles in the labyrinth of the ear.

ova (o′vah). Latin plural of ovum.

ovary (o′vur-ee). The organ in the female that produces hormones and ova.

oviduct (o′vi-dukt). The tube leading from the ovaries to the uterus; also fallopian tube.

ovulation (o′vyoo-lay′shun). The release of an ovum from the surface of the ovary.

ovum (o′vum). The female reproductive cell; mature egg cell.

oxidation (ock′si-day′shun). The chemical process in which electrons are removed from a substance; the combining of a substance with oxygen.

oxidative phosphorylation (ock′si-day′tiv fos′for-i-lay′shun). The metabolic pathway in the mitochondria where the synthesis of ATP is coupled to the energy released from the combination of hydrogen and oxygen to form water.

oxyhemoglobin (ock′si-hee′muh-glo′bin). The compound formed when hemoglobin combines with molecular oxygen.

oxytocin (ock′si-to′sin). A peptide hormone secreted by the posterior lobe of the pituitary gland; stimulates the uterus to contract and the mammary glands to release milk.

pacemaker (pace′may′kur). The sinoatrial node located in the right atrium; initiates the electric activity that determines the heart rate (pace).

pacinian corpuscle (pa-sin′ee-un kor′pusul). Specialized sensory receptor endings that respond to changes in pressure.

palate (pal′ut). The roof of the mouth.

pancreas (pan′kree-us). A gland located in the abdomen near the stomach and connected by a duct to the small intestine. Contains both endocrine gland cells, which secrete the hormones insulin and glucogen into the blood, and exocrine gland cells, which secrete digestive enzymes and bicarbonate into the intestine.

pancreozymin (pan′kree-o-zye′min). Cholecystokinin; the name pancreozymin is no longer used.

pantothenic acid (pan′to-then′ic as′id). A member of the vitamin-B complex.

papilla (pa-pil′uh). A small nipple-shaped elevation.

paralysis (puh-ral′i-sis). Failure of muscle contraction due to a lesion of the neural or muscular system.

paraplegia (par′uh-plee′jee-uh). Paralysis of the legs and lower part of the body.

parasympathetic (par′uh-sim′puh-thet′ick). The craniosacral portion of the autonomic nervous system; releases acetylcholine from its postganglionic nerve endings.

parasympathomimetic (par′uh-sim′puh-tho-mi-met′ick). A chemical that produces effects similar to those of a parasympathetic nerve.

parathyroid gland (par′uh-thigh′royd). Four small endocrine glands situated beside the thyroid gland; their hormonal secretion (parathormone) regulates the homeostasis of calcium and phosphorus.

parietal cells (puh-rye′e-tul). The cells in the stomach that secrete hydrochloric acid.

pars (pahrs). A portion of a larger area, organ or structure.

parturition (pahr′tew-rish′un). The process of giving birth.

patella (pa-tel′uh). The kneecap.

patent (pay′tunt). Open; not blocked or obstructed.

pathology (pa-thol′uh-jee). The study of disease and disease processes.

pedicle (ped′i-kul). The process connecting the lamina of a vertebra with the centrum.

peduncle (pe-dunk′ul). A stemlike part.

pentose (pen′toce). A monosaccharide consisting of five carbon atoms.

pepsin (pep′sin). An enzyme secreted by the stomach; digests protein.

pepsinogen (pep-sin′o-jen). The inactive protein that is secreted by the stomach; is changed into pepsin by the action of hydrocholoric acid.

peptide (pep′tide). A series of amino acids linked by peptide bonds; usually refers to small proteins or protein fragments.

perfusion (per-few′zhun). The passage of fluid through a vessel or organ.

pericardium (perr′i-kahr′dee-um). The fibroserous sac surrounding the heart.

perichondrium (perr′i-kon′dree-um). The connective tissue membrane which covers the surface of cartilage.

perimysium (perr′i-mis′ee-um). The connective tissue layers binding a number of muscle fibers into bundles.

perineum (perr′i-nee′um). The floor of the pelvis; the anatomical space between the anus and scrotum or urethral opening.

periodicity (peer′ee-uh-dis′i-tee). Recurrence at regular intervals of time.

periosteum (perr′ee-os′tee-um). The connective tissue membrane covering the outer surface of bones.

peristalsis (perr′i-stal′sis). A wave of contraction passing along a muscular tube.

peritoneum (perr′i-to-nee′um). The epithelial membrane that lines the abdominal cavity and covers the viscera.

peritonitis (perr′i-to-nigh′tis). An inflammation of the peritoneum.

permeability (perr′mee-uh-bil′i-tee). The degree to which molecules are able to move through a barrier such as a cell membrane or epithelial layer of cells.

pernicious anemia (pur-nish′us). A decreased synthesis of hemoglobin resulting from a deficiency of vitamin B_{12}.

petit mal (puh-tee′ mal′). A particular form of epileptic seizure.

pH. A measure of the hydrogen-ion concentration of a solution; defined as the negative logarithm, to the base 10, of the hydrogen-ion concentration. The smaller the pH, the larger the hydrogen-ion concentration. A pH above 7 represents alkalinity in an aqueous medum; below 7, acidity.

phagocyte (fag′o-sight). A cell that ingests foreign material by endocytosis.

phagocytosis (fag′o-sigh-to′sis). The engulfing of foreign material by phagocytes.

phalanx (fay′lanks). Any bone of a finger or toe. Plural, phalanges.

phallus (fal′us). The penis or clitoris.

pharynx (far′inks). The tube connecting the back of the nose and mouth to the larynx and esophagus, serving as a common passageway for both food and air.

phenotype (fee′no-tipe). The visible expression of the hereditary constitution of an organism.

phenylalanine (fen′il-al′uh-neen). One of the twenty amino acids found in proteins.

phenylketonuria (fen′il-kee′to-new′ree-uh). A genetic disease involving a defect in the metabolism of phenylalanine leading to excessive excretion in the urine; associated with impaired mental development.

phlebitis (fle-bye′tis). Inflammation of a vein.

phlegm (flem). Excessive mucus secreted by the glands lining the airways to the lungs.

phosphatase (fos′fuh-taze). An enzyme which removes phosphate from organic molecules.

phosphocreatine (fos′fo-kree′uh-teen). A compound used to store small amounts of metabolic energy that can be rapidly transfered to ATP; found in large amounts in muscle tissue.

phospholipid (fos′fo-lip′id). A lipid which contains phosphorus; the major type of lipid in cell membranes.

photon (fo′ton). A quantum of electromagnetic radiation.

phrenic (fren′ick). The nerve that innervates the diaphragm muscle.

physiology (fiz′ee-ol′uh-jee). The study of the mechanism of normal body functions.

pia mater (pye′uh may′tur). The meningeal covering closest to the surface of the brain or spinal cord.

pineal body (pin′ee-ul bod′ee). A gland which is an outgrowth from the roof of the diencephalon.

pinna (pin′uh). The outer ear.

pinocytosis (pin′o-sigh-to′sis). The endocytosis of liquid material by cells.

pituitary (pi-tew′i-ter′ee). The hypophysis; also known as the "master" endocrine gland of the body. Located above the roof of the mouth and beneath the hypothalamus of the brain.

placenta (pluh-sen′tuh). The organ connecting the developing fetus to the uterus across which nutrients and wastes are exchanged between the fetus and mother.

plaque (plack). A patch or flat area.

plasma (plaz′mah). The liquid portion of blood in which the blood cells and platelets are suspended.

platelet (plait′lit). Small cell fragments found in the blood which play an essential role in blood clotting.

pleura (ploor′uh). A membrane lining the thoracic cavity which covers each lung.

pleurisy (ploor′i-see). Inflammation of the pleura.

plexus (pleck′sus). A network of interlacing nerve fibers or blood vessels.

pneumonia (new-mo′nyuh). A microbial disease of the lungs.

pneumothorax (new′mo-tho′racks). An accumulation of air in the pleural cavity causing the collapse of a lung.

polarity (po-lar′i-tee). The condition of having two electric poles, one positive and one negative.

poliomyelitis (po′lee-o-migh′e-lye′tis). A viral disease of motor neurons.

polycythemia (pol′ee-sigh′theem′ee-uh). An excess number of red blood cells.

polymer (pol′i-mer). A large molecule

formed by the sequential linkage of many similar smaller molecules.

polymerize (pol′im′ur-ize). To form a polymer by linking together many small subunits.

polymorphonuclear (pol′ee-mor′fo-new′klee-ur). Having nuclei of many shapes.

polyp (pol′ip). A protruding growth from a mucous membrane.

polysaccharide (pol′ee-sack′uh-ride). A large carbohydrate polymer formed by linking many monosaccharides in sequence.

polyunsaturated fat (pol′ee-un-satch′uh-ray′tid). A fatty acid that contains more than one double bond.

polyuria (pol′ee-yoo′ree-uh). The excretion of large volumes of urine.

pons (ponz). Any bridge of tissue connecting two parts; a region of the brain located between the medulla and the midbrain.

porphyrin (por′fi-rin). A ring-shaped molecule that is usually complexed with a metal such as iron, e.g., the heme group in hemoglobin.

portal (por′tul). Pertaining to a gateway; a vascular connection between two capillary beds in series.

postganglionic fiber (pohst′gang′glee-on′ick). The axons of neurons whose cell bodies are located in the ganglia of the autonomic nervous system.

potential (po-ten′chul). A separation of electric charge whose magnitude is measured in volts.

precipitate (pre-sip′i-tate). To cause materials in a solution to settle out in an insoluble aggregate.

precocious (pre-ko′shus). To develop earlier than usual.

prepuce (pree′pewce). The covering of skin over the glans penis.

presbyopia (prez′bee-o′pee-uh). Impaired vision due to loss of elasticity of the lens; common during advancing years or old age.

pressure (presh′ur). The force acting on a unit area of surface.

process (pro′sess). A slender outgrowth of a structure; e.g., the axonal process of a nerve cell.

progesterone (pro-jes′tur-ohn). The steroid hormone produced by the corpus luteum of the ovaries; stimulates the growth and vascular development of the uterus in preparation for the implantation of the fertilized ovum.

prolactin (pro-lack′tin). A peptide hormone from the anterior pituitary; stimulates milk secretion by the mammary glands.

prophylaxis (pro′fi-lack′sis). A procedure designed to prevent the onset of disease.

prostaglandins (pros′tuh-glan′dinz). A class of lipids which stimulates a variety of glands and muscles when released locally in a tissue.

prostate gland (pros′tate gland). A glandular organ in the male surrounding the upper portion of the urethra; secretes fluids and nutrients which form the bulk of the seminal fluid.

protein (pro′tee-in). A polymer made up of amino acids.

proteolytic (pro′tee-olit′ick). Causing the breakage of the peptide bonds linking amino acids in proteins.

prothrombin (pro-throm′bin). The plasma protein that is converted to thrombin during the process of blood coagulation.

proton (pro′ton). The positively charged subatomic particle located in the nucleus.

psychosis (sigh-ko′sis). A disorder of the mind characterized by a distorted interpretation and response to reality.

ptyalin (tye′uh-lin). An enzyme in saliva that converts starch to maltose and dextrose.

puberty (pew′bur-tee). The age at which the reproductive organs become functional as a result of gonadotropins released by the pituitary.

pulse (pulse). The cyclical rise and fall in blood pressure with each beat of the heart.

pupil (pew′pil). The hole in the center of the iris through which light enters the posterior part of the eyeball.

purines (pew′reens). Bases found in nucleic acid; composed of two nitrogenous rings, e.g., adenine and quanine.

Purkinje fibers (poor-kin′-ye). The muscle fibers in the ventricles of the heart that rapidly conduct action potentials throughout the ventricles.

pus (pus). A fluid made up of microbes, leukocytes, and inflammatory products as a result of cellular destruction.

pyrimidine (pye-rim′i-deen). Bases found in nucleic acids; composed of a single nitrogen ring, e.g., cytosine, thymine, and uracil.

pyrogen (pye′ro-jen). A fever-producing substance.

pyruvate (pye-roo′vate). The anion formed by dissociation of pyruvic acid.

quadriceps (kwah′dri-seps). A muscle having four heads.

radiation (ray-dee-ay′shun). The propagation of energy through space in the form of electromagnetic waves.

radical (rad′i-kul). A specific grouping of atoms that is found in many types of molecules; e.g., the carboxyl and amino radicals.

rale (rahl). An abnormal respiratory sound; indicates a pathological condition.

ramus (ray′mus). Designating a branch.

reagin (ree-ay′-jin). A substance behaving like an antibody in a complement fixation reaction.

receptor (re-sep′tur). An afferent nerve terminal which responds to a stimulus from the environment; also used to refer to a specific binding site on a cell membrane to which hormones and neurotransmitters bind.

rectum (reck′tum). The terminal portion of the intestine lying between the colon and anus.

recumbent (re-kum′bunt). Lying down.

reduction (re-duck′shun). The chemical process whereby one or more electrons are added to a substance; also, restoring a broken bone to normal relationship.

reflex (ree′flecks). The response to a stimulus that is built into a system by the nature of the anatomic links between receptor and effector organs.

reflex arc (ree′flecks ark). The pathway linking a receptor to an effector.

refractory period (re-frack′tuh-ree peer′ee-ud). That period following an action potential when a second stimulus will not elicit a second action potential.

regeneration (ree-jen er-ay′shun). The renewal of a tissue or part.

regurgitation (re-gur′ji-tay′shun). A backward flow of fluid or food.

relapse (re-laps′). The return of a disease after its cessation.

renal (ree′nul). Pertaining to the kidney.

renin (ree′nin). An enzyme liberated by the kidney into the blood; converts angiotensinogen to angiotensin.

rennin (ren′in). A milk-curdling enzyme; also known as chymosin.

repressor (re-pres′ur). The compound that interacts with operator genes to inhibit the transcription of structural genes and thus repress specific protein synthesis.

resonance (rez′o-nunce). The prolongation and intensification of sound by transmission through a cavity.

respiration (res′pi-ray′shun). The exchange of gases between the atmosphere and the internal environment. Also the metabolic utilization of oxygen and production of carbon dioxide by cells.

resuscitation (re-sus′i-tay′shun). Bringing back to life.

rete (ree′tee). A new or meshwork.

reticulum (re-tick′yoo-lum). A network.

retina (ret′i-nuh). The cellular layer lining the inner surface of the eye; contains the cells that respond to light.

rheumatism (roo′muh-tizm). An inflammatory disease of connective tissue structures, especially joints.

rhinencephalon (rye′nen-sef′uh-lon). The portion of the brain processing olfactory information.

rhinitis (rye-nigh′tis). An inflammation of the mucous membrane of the nose.

rhodopsin (ro-dop′sin). The reddish-purple pigment in rod cells of the retina which are photosensitive; also known as visual purple.

rhombencephalon (rom′ben-sef′uh-lon). The hindbrain.

rhomboid (rom′boyd). Shaped like a kite.

riboflavin (rye′bo-flay′vin). A heat-stable B vitamin that promotes growth; vitamin B².

ribonucleic acid (rye′bo-new-klee-ic as′id). A nucleic acid that functions in the translocation of genetic information in protein synthesis.

ribose (rye′boce). A monosaccharide found in RNA.

ribosome (rye′bo-sohm). A particle composed of protein and RNA found either free in the cytoplasm or attached to the endoplasmic reticulum; the site of protein synthesis.

rickets (rick′its). A disease of the bone caused by a deficiency of vitamin D.

ruga (roo′guh). The ridge or folds that appear in the wall of a hollow organ.

sac (sak). Any baglike organ or structure.

sacrum (say′krum). The last fused vertebra of the spinal column.

saliva (suh-lye′vuh). The clear, alkaline viscid fluid secreted from the parotid, submandibular, and sublingual glands of the mouth.

sarcoma (sahr-ko′muh). A tumor derived from nonepithelial tissues.

sarcomere (sahr′ko-meer). The repeating contractile unit in a myofibril extending from one Z line to the next.

sarcoplasm (sahr′ko-plazm). The cytoplasm of a skeletal muscle cell.

sarcoplasmic reticulum (sahr′ko-plaz′mik retik′ulum). The endoplasmic reticulum of a muscle fiber; stores and releases calcium.

saturated fatty acid (satch′uh-ray′tid fat′ee as′id). A fatty acid that does not contain any double bonds.

sclera (skleer′uh). The white fibrous tissue that forms the outer protective covering of the eyeball.

scleroid (skleer′oid). Having a hard or dense texture.

scoliosis (sko′lee-o′sis). The abnormal lateral curvature of the spine.

scrotum (skro′tum). The pouch which contains the testes.

scurvy (skur′vee). A pathological condition caused by a deficiency of vitamin C.

sebum (see′bum). The secretion of the sebaceous glands.

secrete (se-kreet′). To release substances from a cell into the extracellular spaces surrounding the cell.

secretin (se-kree′tin). A peptide hormone produced in the wall of the small intestine; stimulates the secretion of pancreatic juice and bile.

semen (see′mun). The fluid ejaculated during a male orgasm composed of sperm and secretions from the prostate gland, seminal vesicles, and various other glands lining the reproductive tract.

semilunar (sem′ee-lew-nur). A crescent-shaped or half-moon structure, e.g., the semilunar valves of the heart.

semimembranous (sem′ee-mem′bruh-nus). Partly made of membrane or fascia.

sepsis (sep′sis). The reaction of the body to bacterial infection.

septicemia (sep′ti-see′mee-uh). The presence of bacteria and their toxins in the blood.

septum (sep′tum). A dividing partition.

serology (se-rol′uh-jee). The in vitro study of antigen-antibody reactions.

serosa (se-ro′suh). A serous membrane, lining the pleural and peritoneal cavities.

serotonin (seer′o-to′nin). An amine found in platelets that induces vasoconstriction; functions as a neurotransmitter at some neural endings.

serum (see′rum). The clear portion of the blood remaining after coagulation has removed the cellular elements and some of the proteins.

sex chromosomes (secks kro′muh-sohmz). The X and Y chromosomes which contain sex-determining genes; XX-female; XY-male.

shunt (shunt). To divert from one location to another.

sickle cell (sik′el sel). A crescent-shaped erythrocyte containing an abnormal form of hemoglobin.

sigmoid (sig′moid). Shaped like the letter S.

sinoatrial (sigh′no-ay′tree-ul). A region of the right atrium near the junction with the venae cava composed of specialized cardiac cells; site of the cardiac pacemaker.

sinus (sigh′nus). A hollow cavity or channel.

sinusoid (si′nuh-soid). Resembling a sinus.

solar plexus (so′lur pleck′sus). The celiac ganglion located behind the stomach consisting of sympathetic nerve fibers.

solute (sol′yoot). Any substance dissolved in a solvent.

solution (suh′lew′shun). The mixture of solute and solvent.

solvent (sol′vunt). The fluid in which solutes are dissolved.

soma (so′muh). The tissues of the body excluding the reproductive tissues.

somatotropic (so′muh-to-trop′ick). Having a stimulating effect on body growth.

somatotropin (so′muh-to-tro′pin). A protein hormone secreted by the anterior pituitary; also called growth hormone.

somesthetic (so′mes-thet′ick). Pertaining to general body feelings.

somite (so′mite). An embryonic segment of mesoderm along the neural tube of the embryo.

spasm (spaz′um). An involuntary muscle contraction.

sperm (spurm). The reproductive gamete of the male.

spermatid (sper′muh-tid). The cell derived from the secondary spermatocyte by fission; develops into a mature spermatozoon.

spermatogenesis (spur′muh-to-jen′e-sis). The formation of spermatozoa.

spermatozoon (spur′muh-to-zo′on). A mature male germ cell; sperm.

sphincter (sfink′tur). A ring-shaped muscle that contracts and closes an opening.

spine (spyne). A thornlike process or projection.

splanchnic (splank′nik). Pertaining to the viscera.

spleen (spleen). The large oblong organ lying under the rib cage in the upper left quadrant of the body; composed of lymphoid tissue.

squamous (skway′mus). Scaly or platelike.

starch (stahrch). A polysaccharide found in plant tissue; resembles glycogen in animal tissues.

stasis (stay′sis). The stoppage of fluid flow in a body part.

stenosis (ste-no′sis). The narrowing of an opening.

steroid (steer′oid). A term applied to a class of lipids composed of multiple rings of carbon atoms including cholesterol and various hormones secreted by adrenal cortex and gonads.

stimulus (stim′yoo-lus). Any physical or chemical change in the environment that can be detected by an organism which leads to a response.

strabismus (stra-biz′mus). An abnormal condition of the eye.

stroke (strowk). The local disruption of blood flow in the brain as a result of a blood clot or ruptured blood vessel.

styloid (stye′loid). Resembling a long, pointed pillar.

sub-. Latin prefix meaning under.

sucrase (su′kras). Invertin; an enzyme that splits the disaccharide sucrose into glucose and fructose.

sucrose (sue′kroce). A disaccharide composed of glucose and fructose; table sugar.

sudoriferous (sue′dur-if′ur-us). Producing sweat.

sugar (shoog′ur). Any carbohydrate having a sweet taste.

sulcus (sul′kus). A groove, trench, or depression such as found on the surface of the brain, separating the gyri.

summation (sumay′shun). The addition of multiple synaptic potentials or mechanical responses in the muscle.

super- (sue′pur). A Latin prefix signifying above or in excess.

supine (suh′pine). Lying on one's back.

surfactant (sur-fack′tunt). A mixture of lipids and proteins produced by lung tissue that reduces the surface tension of the aveolar wall.

suspension (sus-pen′shun). A cessation of any vital process; also, a state in which solute particles are mixed in a liquid but are not dissolved.

suture (sue′chur). A joint between bones of the cranium that is immovable; also, a stitch or series of stitches that are made to secure opposite edges of a wound.

sympathetic (sim′puh-thet′ick). Pertaining to the thoracolumbar portion of the autonomic nervous system whose postganglionic neurons release norepinephrine as a neurotransmitter.

sympathomimetic (sim′puh-tho-mi-met′ick). Producing effects similar to the sympathetic nervous system.

symphysis (sim′fi-sis). A growing together; especially, the junction of the two pubic bones.

synapse (sin′aps). The anatomical area that serves as the functional junction between two neurons.

syncytium (sin-sish′ee-um). A fusion of multiple cells to form one continuous interconnected unit.

syndrome (sin′drome). A set of symptoms that occur together.

synergist (sin′ur-jist). An agent which cooperates functionally with another agent; e.g., one muscle aiding the actions of another.

synovial (si′no-vee′ul). A fluid usually found in diarthrotic joints.

systemic (sis-tem′ick). Pertaining to or affecting the body as a whole, especially the circulation from the left ventricle to the right atrium.

systerole (sis′tu-lee). The contraction phase of the cardiac cycle; when used alone, refers to contraction of the ventricles.

tachycardia (tack′e-kahr′de-uh). A rapid heart rate.

tactile (tack′til). Pertaining to touch.

taenia coli (tee′nee-uh ko′lye). One of the bands of smooth muscle on the large intestine.

talus (tay′lus). A bone of the ankle.

tamponade (tam′puh-nade′). A cardiac condition resulting from an accumulation of excessive fluid in the pericardium.

taste buds (taist budz). The regions on the tongue containing receptors that respond to various chemical agents producing the sensation of taste.

taxis (tack′sis). The movement of a cell or organism in response to a stimulus.

telencephalon (tel′en-sef′uh-lon). The anterior portion of the brain.

tendon (ten′dun). A fibrous band of collagen fibers which attaches muscle to muscle or muscle to bone.

testis (tes′tis). The male gonad.

testosterone (tes-tos′ter-ohn). The major male steroid sex hormone produced by the testes.

tetany (tet′uh-nee). A condition marked by muscle cramps, twitching, and convulsions as a result of low extracellular calcium.

thalamus (thal′uh-mus). The bilobed structure in the diencephalon which is the major sensory relay center of the brain.

thorax (tho′racks). Pertaining to the chest.

threshold (thresh′oald). The minimum magnitude of a stimulus strength which will elicit a response.

thrombin (throm′bin). The enzyme which converts fibrinogen to fibrin in the blood.

thrombocyte (throm′bo-sight). A blood platelet.

thromboplastin (throm′bo-plas′tin). A substance produced in the blood; plays a role in the coagulation process.

thrombosis (throm-bo′sis). The formation or presence of a blood clot in a vessel.

thrombus (throm′bus). The plug or clot in a blood vessel.

thymine (thigh′meen). One of the pyrimidine bases found in DNA.

thymus (thy′mus). A bilobed gland located in the mediastinum of the thorax; plays a role in the immune mechanisms of the body.

thyroxine (thigh-rock′seen). A thyroid hormone.

tidal volume (tye′dul vol′yoom). The volume of air inspired or expired during inspiration or expiration.

tissue (tish′oo). A collection of similar cells specialized to perform a specific function.

tone. The degree of tension in a muscle.

tonic (ton′ick). A continuous activity, as opposed to a phasic one.

tonicity (to-nis′i-tee). Referring to the osmolarity of a solution.

tonsil (ton′sil). A small, rounded mass of lymphoid tissue in the mucosa of the pharynx.

toxic (tok′sick). Pertaining to the nature of a poison.

trabecula (trah-beck′yoo-luh). A septum passing into the substance of a structure from its margin or surface.

trachea (tray′kee-uh). The windpipe, connecting the larynx to the lungs.

tracheotomy (tray′kee-ot′uh-mee). An incision that opens the trachea through the neck to allow breathing when the upper respiratory tract is blocked.

transamination (trans-am′i-nay′shun). The transfer of an amino group from an amino acid to an alpha-keto acid forming a new amino acid and keto acid.

transferrin (trans-ferr′in). A serum beta-globulin that binds and transports iron.

transplantation (trans′plan-tay′shun). The grafting of a tissue or organ from one part of the body to another or from one individual to another.

trauma (traw′muh). A wound or injury.

tremor (trem′ur). An involuntary quivering or trembling of muscle.

triglyceride (trye-glis′ur-ide). A neutral fat made up of three fatty acids and glycerol.

trigone (trye′gohn). A smooth triangular area on the inner surface of the urinary bladder.

trochanter (tro-kan′tur). Either of the two processes of the superior femoral shaft.

trochlea (trock′lee-uh). A pulley-shaped structure.

tropomyosin (tro′po-migh′o-sin). A muscle protein located in the filaments; regulates the activity of the contractible proteins.

trypsin (trip′sin). One of the proteolytic enzymes of pancreatic juice.

TSH. Abbreviation for thyroid stimulating hormone.

t-tubule. A transverse invagination in the plasma membrane of the muscle fiber, responsible for the conduction of action potentials into the muscle fiber.

tubercle (tew′bur-kul). A nodule.

tuberosity (tew′bur-os′i-tee). An elevation or protuberance.

tumor (tew′mur). An abnormal enlargement; may be benign or malignant.

tunica (tew′ni-kuh). A covering.

turbinate (tur′bin-ut). One of the bony processes on the lateral wall of the nasal cavity.

twitch (twich). The contraction of a muscle in response to a single action potential.

tympanic (tim-pan′ick). Pertaining to the ear drum (the tympanic membrane).

tyrosine (tye′ro-seen). One of the twenty amino acids found in proteins.

ulcer (ul′sur). A localized erosion of a tissue; an open sore.

umbilicus (um′bil-i-kus). The site of the former attachment of the umbilical cord in the fetus; the navel.

urea (yoo-ree′uh). The waste product of nitrogen metabolism; produced in the liver and excreted in the urine.

uremia (yoo-ree′me-uh). The accumulation of compounds in the blood that are normally excreted in the urine.

ureter (yoo-ree′tur). The tube that connects a kidney to the urinary bladder.

urethra (yoo-ree′thruh). The tube that connects the bladder to the exterior of the body.

uric acid (u′rick). The end product of purine metabolism.

urine (yoor′in). The fluid produced by the kidneys.

uterine (yoo′tur-in). Pertaining to the uterus.

uterus (yoo′tur-us). The hollow pear-shaped female structure which provides the protective nutrient environment for the developing fetus and embryo. The womb.

utricle (yoo′tri-kul). A little sac.

uvula (yoo′vew-luh). The small, fleshy mass hanging from the soft palate in the pharynx.

vaccine (vack′seen). The dead or attenuated microorganisms administered medically for the prevention of infectious disease.

vacuole (vack′yoo-ole). Any small space or cavity found in a cell.

vagina (va-jye′nuh). The tubular muscular organ that leads from the uterus to the exterior in the female.

vaginitis (vadj′i-nigh′tis). An inflammation of the vagina.

vagus (vay′gus). The tenth cranial nerve; the major parasympathetic nerve in the body, whose path is traced through most thoracic and abdominal organs.

valve (valv). A structure in a tube or passage which prevents the backflow of its contents.

varicose (var'i-koce). A distended vein resulting from the failure of the valves to prevent the backflow of blood.

vas (vas). A vessel or canal that carries a fluid.

vascular (vas'kew-lur). Pertaining to vessels.

vasopressin (vay'zo-pres'in). A peptide hormone also known as antidiuretic formed by neurons in the hypothalamus and released by nerve endings in the posterior pituitary; increases the reabsorption of water by the renal tubules and causes the contraction of blood vessels.

vein (vain). A vessel carrying blood from an organ to the heart.

ventilation (ven'ti-lay'shun). The process whereby atmospheric gases are moved into the lungs and waste gases are expelled.

ventricle (ven'tri-kul). Any small cavity, e.g., the ventricles of the heart and brain.

venule (ven'yool). A small vein, located between the capillaries and larger veins, linking them.

vermiform (vur'mi-form). Shaped like a worm.

vertebra (vur'te-bruh). One of the bones of the spinal column.

vertigo (vur'ti-go). Dizziness.

vesicle (ves'i-kul). A small fluid-containing sac.

vessel (ves'ul). Any channel used for carrying fluid.

villus (vil'us). A minute projection. Plural, villi.

virus (vye'rus). A very small infectious particle composed primarily of nucleic acid and proteins; lacks the capacity for metabolism and can reproduce only in living cells.

viscera (vis'er-a). Referring to the internal organs of the body.

viscosity (vis-kos'i-tee). The physical property of a liquid that is characterized by the degree of friction between its component molecules; determines the ease with which the liquid will flow.

viscous (vis'kus). Characterized by a high degree of viscosity.

vital capacity (vye'tul kuh-pas'i-tee). The maximum volume of air that can be moved in or out of the lungs.

vitamin (vye'tuh-min). An organic molecule that occurs in foods in small amounts and is necessary for the normal metabolic functioning of the body.

vitreous (vit'ree-us). The vitreous body of the eyeball; lies between the lens and the retina.

vomer (vo'mur). The flat bone that forms the inferior and posterior part of the nasal septum.

vomiting (vom'it-ing). The evacuation of matter from the stomach through the mouth.

vulva (vul'vuh). The external genitalia of the female.

wart (wort). A benign epidermal tumor of viral origin.

wheal (wheel). A red raised area on the skin.

wormian (wur'mee-un). Pertaining to one of the sutural bones of the skull.

xiphoid (zif'oid). Shaped like a sword; the last cartilagenous segment of the sternum.

xylose (zye'loce). A five-carbon monosaccharide.

yolk (yoke). The part of the egg which provides nutrients for the developing embryo.

Z line. The structure to which the thin filaments of the myofibrils are attached at the end of the sarcomere.

zona pellucida (zo'nuh pe-lew'si-duh). A clear zone surrounding a fertilized egg cell.

zygote (zye'gote). The fertilized ovum.

zymogen (zye'mo-jen). The inactive form of a protein in a cell; converted to an active enzyme after being secreted by the cell.

Index

Page numbers in **boldface** refer to illustrations or tables.